CONCISE INORGANIC CHEMISTRY

CONCISE INORGANIC CHEMISTRY

FOURTH EDITION

J.D. Lee

Senior Lecturer in Inorganic Chemistry
Loughborough University of Technology

CHAPMAN & HALL
University and Professional Division
London · New York · Tokyo · Melbourne · Madras

UK	Chapman & Hall, 2–6 Boundary Row, London SE1 8HN
USA	Chapman & Hall, 29 West 35th Street, New York NY10001
JAPAN	Chapman & Hall Japan, Thomson Publishing Japan, Hirakawacho Nemoto Building, 7F, 1-7-11 Hirakawa-cho, Chiyoda-ku, Tokyo 102
AUSTRALIA	Chapman & Hall Australia, Thomas Nelson Australia, 102 Dodds Street, South Melbourne, Victoria 3205
INDIA	Chapman & Hall India, R. Seshadri, 32 Second Main Road, CIT East, Madras 600 035

First published 1964.
Fourth edition 1991

Typeset in 10/12 Times by Best-set Typesetter Ltd.
Printed in Singapore by Fong & Sons Printers Pte. Ltd.

ISBN 0 412 40290 4

British Library Cataloguing in Publication Data
Lee, J.D. (John David) *1931–*
Concise inorganic chemistry. – 4th ed.
1. Inorganic chemistry
I. Title
546

ISBN 0–412–40290–4

Library of Congress Cataloging-in-Publication Data
Lee, J.D. (John David), 1931–
Concise inorganic chemistry / J.D. Lee. – 4th ed.
p. cm.
Rev ed. of: A new concise inorganic chemistry. 3rd ed. 1977.
Includes bibliographical references and index.
ISBN 0–412–40290–4 (pbk.)
1. Chemistry, Physical and theoretical. 2. Chemical bonds.
I. Lee, J.D. (John David), 1931– New concise inorganic chemistry.
II. Title.
QD453.2.L45 1991
546—dc20

91-9816
CIP

Contents

Preface to the fourth edition

It is 25 years since the first edition of *Concise Inorganic Chemistry* was published. This is a remarkable life for any textbook, and it seemed appropriate to mark the Silver Jubilee with a new edition. This, the fourth edition, has taken three years to write, and was made possible by the authorities at Loughborough University who granted me a year's study leave, and by my colleagues who shouldered my teaching duties during this time. I am greatly indebted to them. The new edition is inevitably larger than its predecessors, though the publishers were reluctant to change the title to *A Less Concise Inorganic Chemistry*! Einstein said 'all things are relative', and the book is still concise compared with other single volumes and with multi-volume series on the subject.

The aim of the fourth edition remains exactly the same as that for the first edition of the book. That is to provide a modern textbook of inorganic chemistry that is long enough to cover the essentials, yet short enough to be interesting. It provides a simple and logical framework into which the reader should be able to fit factual knowledge, and extrapolate from this to predict unknown facts. The book is intended to fill the gap between school books and final year honours degree chemistry texts. The need for an appropriate and sympathetically written text has increased significantly now that the first cohorts of GCSE students are applying to read chemistry at degree and diploma level. It is aimed primarily at first or second year degree students in chemistry, but will also be useful for those doing chemistry as ancillary subjects at university, and also for BTEC courses and Part I Grad RIC in polytechnics and technical colleges. Some parts will be usable by good sixth form students. Above all it is intended to be easy to read and understand.

The structure of the book is largely unchanged, and is based on descriptive chemistry combined with some of the reasons why elements and compounds behave in the way they do. For convenience the book is divided into six 'parts' covering theoretical concepts and hydrogen, the *s*-block, the *p*-block, the *d*-block, the *f*-block and other topics. Every chapter has been completely rewritten, updated and enlarged. The section on theoretical concepts and hydrogen contains introductory chapters on atomic structure, ionic, covalent and metallic bonding and general properties, which make up about one fifth of the book. The original chapter on coordination compounds has been moved into this section since it is mainly about the coordinate bond and crystal field theory. These are followed by a systematic coverage of hydrogen, the main group elements, the transition elements, the lanthanides and the actinides in turn. There are separate chapters on the nucleus and spectroscopy. To make it easier to find the appropriate section, the text has been divided into a larger number of chapters. Thus, the original chapter on bonding has been split into an introduction to bonding and chapters on ionic, covalent and metallic bonding. The original chapter on the *s*-block has been split into chapters on Groups I and II. That on the *p*-block has been split into chapters on Groups III, IV, V, VI, VII and 0. The original chapter on the *d*-block has been split into an introduction to the transition elements followed by ten smaller chapters on the triads of elements. I have retained a very large and comprehensive index, and a large table of contents as previously. The descriptive material necessarily has a large place, but the book attempts to show the reasons for the structure, properties and reactions of compounds, wherever this is possible with elementary methods.

At the end of most chapters is a section on further reading, and almost 600 references are given to other work. The references may be used at several different levels. In increasing order of complexity these are:

1. Easy to understand articles in journals such as the *Journal of Chemical Education, Chemistry in Britain* and *Education in Chemistry*.
2. References to specialized textbooks.
3. Review articles such as Quarterly Reviews, Coordination Chemistry Reviews, and the proceedings of specialist conferences and symposia.
4. A small number of references are made to original articles in the primary literature. In general such references are beyond the scope of this text, but those given have special (often historical) significance. Examples include the use of Ellingham diagrams, the Sidgwick–Powell theory of molecular shape, and the discovery of ferrocene and of warm superconductors.

Chemistry is still a practical subject. In the chemical industry, as with many others, the adage 'where there's muck there's money' holds particularly true. Unless chemicals were needed and used in large amounts there would be no chemical industry, hence no students in chemistry, no teachers of chemistry, and no need for textbooks. An American professor told me he divided inorganic chemistry books into two types: *theoretical* and *practical*. In deciding how to classify any particular book he first looked to see if the extraction of the two most produced metals (Fe and Al) was adequately covered, what impurities were likely to be present, and how the processing was adapted to remove them. Second, he looked to see if the treatment of the bonding in xenon compounds and ferrocene was longer than that on the production of ammonia. Third, he looked to see if the production and uses of phosphates were covered adequately. For some years there has been a trend for chemistry teaching to become more theoretical. There is always theoretical interest in another interesting oxidation state or another unusual complex, but the balance of this book is tilted to ensure that it does not exclude the commonplace, the mundane and the commercially important. This book is intentionally what my American friend would call the 'practical' type.

It is distressing to find both teachers and students who show little idea of which chemicals are commercially important and produced in very large tonnages. What are the products used for? What processes are used now as opposed to processes used 30 or more years ago? Where do the raw materials come from, and in what ways are the processes actually used related to likely impurities in the raw materials? Many books give scant coverage to these details. Though this is not intended to be an industrial chemistry book, it relates to chemistry in the real world, and this edition contains rather more on large tonnage chemicals. I have contacted about 250 firms to find what processes are currently in use. Production figures are quoted to illustrate which chemicals are made in large amounts and where the minerals come from. The figures quoted are mainly from *World Mineral Statistics*, published by the British Geological Survey in 1988, and from the *Industrial Statistics Yearbook 1985* Vol. II, published by the United Nations, 1987, New York. Both are mines of information. Inevitably these figures will vary slightly from year to year, but they illustrate the general scale of use, and the main sources of raw materials. Thus, the production of major chemicals such as H_2SO_4, NH_3, NaOH, Cl_2, O_2 and N_2 are adequately covered. Other important materials such as cement and steel, polymers such as polythene, silicones and Teflon, soap and detergents are also covered. In addition, many smaller scale but fascinating applications are described and explained. These include baking powder, photography, superconductors, transistors, photocopiers, carbon dating, the atomic bomb and uses of radioisotopes.

There is currently a greater awareness of environmental issues. These are discussed in more detail than in previous editions. Problems such as freons and the ozone layer, the greenhouse effect, acid rain, lead pollution, the toxic effects of tin and mercury, asbestos, excessive use of phosphates and nitrates and the toxic effects of various materials in drinking water are discussed. The section on the development of the atomic bomb and the peaceful uses of atomic energy is also enlarged.

While much inorganic chemistry remains the same, it is a living subject and the approach to our current thinking and the direction of future work have altered. In particular our ideas on bonding have changed. Until 1950 inorganic chemistry was largely descriptive. The research and development which led to the production of the atomic bomb in 1946 is probably the greatest chemical achievement of the century. The impetus from this led to the discovery of many new elements in the actinide and lanthanide series. This was followed by a period of great interest in physical inorganic chemistry, where instead of just observing what happened we looked for the reasons why. Thermodynamics and kinetics were applied to chemical reactions, and magnetism and UV-visible spectroscopy were explored.

There was a flurry of activity when it was found that the noble gases really did form compounds. This was followed by a concentrated phase of preparing organometallic compounds and attempting to explain the bonding in these compounds, many of which defied rational explanation by existing theories. Future developments seem likely to fall in two main areas – bioinorganic chemistry and new materials. Much bioinorganic work is in progress: how enzymes and catalysts function; how haemoglobin and chlorophyll really work; and how bacteria incorporate atmospheric nitrogen so easily when we find it so difficult. Work on new materials includes the production of polymers, alloys, superconductors and semiconductors.

This book is mainly about the chemistry of the elements, which is properly regarded as *inorganic* chemistry. I consider it unhelpful for students to put information into rigid compartments, since the ideas in one subject may well relate to other subjects and the boundaries between subjects are partly artificial. The book incorporates information on the chemistry of the elements regardless of the source of that chemistry. Thus, in places the book crosses boundaries into analytical chemistry, biochemistry, materials science, nuclear chemistry, organic chemistry, physics and polymer chemistry. It is worth remembering that in 1987 the Nobel Prize for Chemistry was given for work on complexes using crowns and crypts which have biological overtones, and the Nobel Prize for Physics was for discoveries in the field of warm superconductors. Both involve chemistry.

I am extremely grateful to Dr A.G. Briggs for help and constructive criticism in the early stages of writing the book. In addition I am greatly indebted to Dr A.G. Fogg for his help and encouragement in correcting and improving the manuscript, and to Professor F. Wilkinson for valuable advice. In a book of this size and complexity it is inevitable that an occasional mistake remains. These are mine alone, and where they are shown to be errors they will be corrected in future editions. I hope that the new edition will provide some interest and understanding of the subject for future generations of students, and that having passed their examinations they may find it useful in their subsequent careers. The final paragraph from the Preface to the First Edition is printed unchanged:

> A large amount of chemistry is quite easy, but some is enormously difficult. I can find no better way to conclude than that by the late Professor Silvanus P. Thompson in his book *Calculus Made Easy*, 'I beg to present my fellow fools with the parts that are not hard. Master these thoroughly, and the rest will follow. What one fool can do, another can'.

J.D. Lee
Loughborough, 1991

SI UNITS

SI units for energy are used throughout the fourth edition, thus making a comparison of thermodynamic properties easier. Ionization energies are quoted in $kJ\,mol^{-1}$, rather than ionization potentials in eV. Older data from other sources use eV and may be converted into SI units ($1\,kcal = 4.184\,kJ$, and $1\,eV = 23.06 \times 4.184\,kJ\,mol^{-1}$).

Metres are strictly the SI units for distance, and bondlengths are sometimes quoted in nanometres ($1\,nm = 10^{-9}\,m$). However ångström units Å ($10^{-10}\,m$) are a permitted unit of length, and are widely used by crystallographers because they give a convenient range of numbers for bondlengths. Most bonds are between 1 and 2 Å (0.1 to 0.2 nm). Ångström units are used throughout for bondlengths.

The positions of absorption peaks in spectra are quoted in wave numbers cm^{-1}, because instruments are calibrated in these units. It must be remembered that these are not SI units, and should be multiplied by 100 to give SI units of m^{-1}, or multipled by 11.96 to give $J\,mol^{-1}$.

The SI units of density are $kg\,m^{-3}$, making the density of water 1000 $kg\,m^{-3}$. This convention is not widely accepted, so the older units of $g\,cm^{-3}$ are retained so water has a density of $1\,g\,cm^{-3}$.

In the section on magnetism both SI units and Debye units are given, and the relation between the two is explained. For inorganic chemists who simply want to find the number of unpaired electron spins in a transition metal ion then Debye units are much more convenient.

NOMENCLATURE IN THE PERIODIC TABLE

For a long time chemists have arranged the elements in groups within the periodic table in order to relate the electronic structures of the elements to their properties, and to simplify learning. There is however no uniform and universally accepted method of naming the groups. A number of well known books, including Cotton and Wilkinson and Greenwood and Earnshaw, name the main groups and the transition elements as A and B subgroups. Though generally accepted in North America until 1984 and fairly widely accepted up till the present time in most of the world, the use of A and B subgroups dates back to the older Mendeleef periodic table of half a century ago. Its disadvantages are that it may over emphasize slight similarities between the A and B subgroups, and there are a large number of elements in Group VIII. IUPAC have suggested that the main groups and the transition metals should be numbered from 1 to 18. The IUPAC system has gained some acceptance in the USA, but has encountered strong opposition elsewhere, particularly in Europe. It seems inconsistent that the groups of elements in the *s*-block, *p*-block and *d*-block are numbered, but the elements in the *f*-block are not. As in earlier editions of this book, these arguments are avoided, and the main group elements, that is the *s*-block and the *p*-block, are numbered as groups I to VII and 0, depending on the number of electrons in the outer shell of the atoms, and the transition elements are dealt with as triads of elements and named as the top element in each group of three.

Names of the various groups

I	II											III	IV	V	VI	VII	0
IA	IIA	IIIA	IVA	VA	VIA	VIIA	⟨---- VIII ----⟩			IB	IIB	IIIB	IVB	VB	VIB	VIIB	0
H																	He
Li	Be											B	C	N	O	F	Ne
Na	Mg											Al	Si	P	S	Cl	Ar
K	Ca	Sc	Ti	V	Cr	Mn	Fe	Co	Ni	Cu	Zn	Ga	Ge	As	Se	Br	Kr
Rb	Sr	Y	Zr	Nb	Mo	Tc	Ru	Rh	Pd	Ag	Cd	In	Sn	Sb	Te	I	Xe
Cs	Ba	La	Hf	Ta	W	Re	Os	Ir	Pt	Au	Hg	Tl	Pb	Bi	Po	At	Rn
1	2	3	4	5	6	7	8	9	10	11	12	13	14	15	16	17	18

Theoretical Concepts and Hydrogen

Part One

Atomic structure and the periodic table

1

THE ATOM AS A NUCLEUS WITH ORBITAL ELECTRONS

All atoms consist of a central nucleus surrounded by one or more orbital electrons. The nucleus always contains protons and all nuclei heavier than hydrogen contain neutrons too. The protons and neutrons together make up most of the mass of the atom. Both protons and neutrons are particles of unit mass, but a proton has one positive charge and a neutron is electrically neutral (i.e. carries no charge). Thus the nucleus is always positively charged. The number of positive charges on the nucleus is exactly balanced by an equal number of orbital electrons, each of which carries one negative charge. Electrons are relatively light – about 1/1836 the mass of a proton. The 103 or so elements at present known are all built up from these three fundamental particles in a simple way.

Hydrogen is the first and most simple element. It consists of a nucleus containing one proton and therefore has one positive charge, which is balanced by one negatively charged orbital electron. The second element is helium. The nucleus contains two protons, and so has a charge of +2. The nuclear charge of +2 is balanced by two negatively charged orbital electrons. The nucleus also contains two neutrons, which minimize the repulsion between the protons in the nucleus, and increase the mass of the atom. All nuclei heavier than hydrogen contain neutrons, but the number present cannot be predicted reliably.

This pattern is repeated for the rest of the elements. Element 3, lithium, has three protons in the nucleus (plus some neutrons). The nuclear charge is +3 and is balanced by three orbital electrons. Element 103, lawrencium, has 103 protons in the nucleus (plus some neutrons). The nuclear charge is +103 and is balanced by 103 orbital electrons. The number of positive charges on the nucleus of an atom always equals the number of orbital electrons, and is called the atomic number of the element.

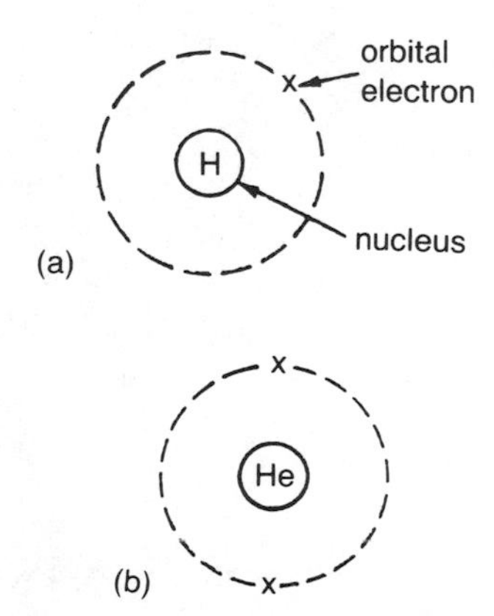

Figure 1.1 Structures of (a) hydrogen, symbol H, atomic number 1; and (b) helium, symbol H, atomic number 2.

In the simple planetary theory of the atom, we imagine that these electrons move round the nucleus in circular orbits, in much the same way as the planets orbit round the sun. Thus hydrogen and helium (Figure 1.1) have one and two electrons respectively in their first orbit. The first orbit is then full. The next eight atoms are lithium, beryllium, boron, carbon,

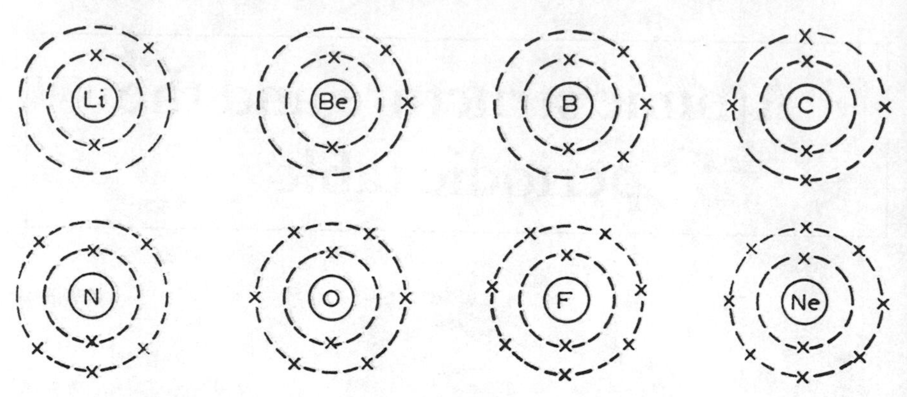

Figure 1.2 Structures of the elements lithium to neon.

nitrogen, oxygen, fluorine and neon. Each has one more proton in the nucleus than the preceding element, and the extra electrons go into a second orbit (Figure 1.2). This orbit is then full. In the next eight elements (with atomic numbers 11 to 18), the additional electrons enter a third shell.

The negatively charged electrons are attracted to the positive nucleus by electrostatic attraction. An electron near the nucleus is strongly attracted by the nucleus and has a low potential energy. An electron distant from the nucleus is less firmly held and has a high potential energy.

ATOMIC SPECTRA OF HYDROGEN AND THE BOHR THEORY

When atoms are heated or subjected to an electric discharge, they absorb energy, which is subsequently emitted as radiation. For example, if sodium chloride is heated in the flame of a Bunsen burner, sodium atoms are produced which give rise to the characteristic yellow flame coloration. (There are two lines in the emission spectrum of sodium corresponding to wavelengths of 589.0 nm and 589.6 nm.) Spectroscopy is a study of either the radiation absorbed or the radiation emitted. Atomic spectroscopy is an important technique for studying the energy and the arrangement of electrons in atoms.

If a discharge is passed through hydrogen gas (H_2) at a low pressure, some hydrogen atoms (H) are formed, which emit light in the visible region. This light can be studied with a spectrometer, and is found to comprise a series of lines of different wavelengths. Four lines can be seen by eye, but many more are observed photographically in the ultraviolet region. The lines become increasingly close together as the wavelength (λ) decreases, until the continuum is reached (Figure 1.3). Wavelengths, in metres, are related to the frequency, ν, in Hertz (cycles/second) by the equation:

$$\nu = \frac{c}{\lambda}$$

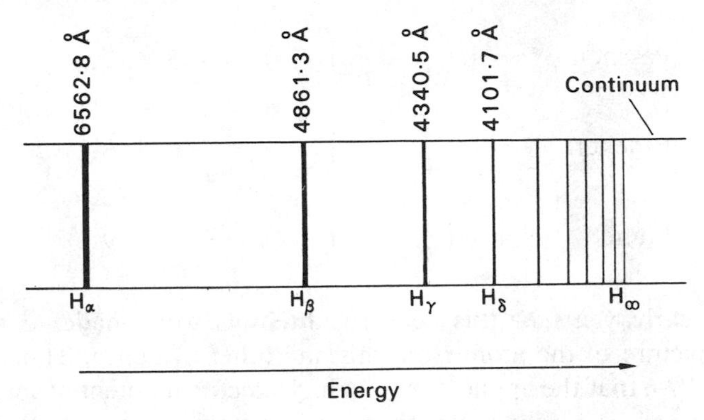

Figure 1.3 Spectrum of hydrogen in the visible region (Balmer series.)

where c is the velocity of light ($2.9979 \times 10^8\,\mathrm{m\,s^{-1}}$). In spectroscopy, frequencies are generally expressed as wave numbers $\bar{\nu}$, where $\bar{\nu} = 1/\lambda\,\mathrm{m}^{-1}$.

In 1885 Balmer showed that the wave number $\bar{\nu}$ of any line in the visible spectrum of atomic hydrogen could be given by the simple empirical formula:

$$\bar{\nu} = R\left(\frac{1}{2^2} - \frac{1}{n^2}\right)$$

where R is the Rydberg constant and n has the values 3, 4, 5. . . , thus giving a series of lines.

The lines observed in the visible region are called the Balmer series, but several other series of lines may be observed in different regions of the spectrum (Table 1.1).

Similar equations were found to hold for the lines in the other series in the hydrogen spectrum.

Lyman $\quad \bar{\nu} = R\left(\frac{1}{1^2} - \frac{1}{n^2}\right) \quad n = 2, 3, 4, 5\ldots$

Balmer $\quad \bar{\nu} = R\left(\frac{1}{2^2} - \frac{1}{n^2}\right) \quad n = 3, 4, 5, 6\ldots$

Table 1.1 Spectral series found in atomic hydrogen

	Region of spectrum
Lyman series	ultraviolet
Balmer series	visible/ultraviolet
Paschen series	infrared
Brackett series	infrared
Pfund series	infrared
Humphries series	infrared

$$\text{Paschen} \quad \bar{\nu} = R\left(\frac{1}{3^2} - \frac{1}{n^2}\right) \qquad n = 4, 5, 6, 7 \ldots$$

$$\text{Brackett} \quad \bar{\nu} = R\left(\frac{1}{4^2} - \frac{1}{n^2}\right) \qquad n = 5, 6, 7, 8 \ldots$$

$$\text{Pfund} \quad \bar{\nu} = R\left(\frac{1}{5^2} - \frac{1}{n^2}\right) \qquad n = 6, 7, 8, 9 \ldots$$

In the early years of this century, attempts were made to obtain a physical picture of the atom from this and other evidence. Thomson had shown in 1896 that the application of a high electrical potential across a gas gave electrons, suggesting that these were present in atoms. Rutherford suggested from alpha particle scattering experiments that an atom consisted of a heavy positively charged nucleus with a sufficient number of electrons round it to make the atom electrically neutral. In 1913, Niels Bohr combined these ideas and suggested that the atomic nucleus was surrounded by electrons moving in orbits like planets round the sun. He was awarded the Nobel Prize for Physics in 1922 for his work on the structure of the atom. Several problems arise with this concept:

1. The electrons might be expected to slow down gradually.
2. Why should electrons move in an orbit round the nucleus?
3. Since the nucleus and electrons have opposite charges, they should attract each other. Thus one would expect the electrons to spiral inwards until eventually they collide with the nucleus.

To explain these problems Bohr postulated:

1. An electron did not radiate energy if it stayed in one orbit, and therefore did not slow down.
2. When an electron moved from one orbit to another it either radiated or absorbed energy. If it moved towards the nucleus energy was radiated and if it moved away from the nucleus energy was absorbed.
3. For an electron to remain in its orbit the electrostatic attraction between the electron and the nucleus which tends to pull the electron towards the nucleus must be equal to the centrifugal force which tends to throw the electron out of its orbit. For an electron of mass m, moving with a velocity v in an orbit of radius r

$$\text{centrifugal force} = \frac{mv^2}{r}$$

If the charge on the electron is e, the number of charges on the nucleus Z, and the permittivity of a vacuum ε_0

$$\text{Coulombic attractive force} = \frac{Ze^2}{4\pi\varepsilon_0 r^2}$$

so

$$\frac{mv^2}{r} = \frac{Ze^2}{4\pi\varepsilon_0 r^2} \tag{1.1}$$

hence

$$v^2 = \frac{Ze^2}{4\pi\varepsilon_0 mr} \tag{1.2}$$

According to Planck's quantum theory, energy is not continuous but is discrete. This means that energy occurs in 'packets' called quanta, of magnitude $h/2\pi$, where h is Planck's constant. The energy of an electron in an orbit, that is its angular momentum mvr, must be equal to a whole number n of quanta.

$$mvr = \frac{nh}{2\pi}$$

$$v = \frac{nh}{2\pi mr}$$

$$v^2 = \frac{n^2h^2}{4\pi^2m^2r^2}$$

Combining this with equation (1.2)

$$\frac{Ze^2}{4\pi\varepsilon_0 mr} = \frac{n^2h^2}{4\pi^2m^2r^2}$$

hence

$$r = \frac{\varepsilon_0 n^2h^2}{\pi me^2Z} \tag{1.3}$$

For hydrogen the charge on the nucleus $Z = 1$, and if

$n = 1$ this gives a value $r = 1^2 \times 0.0529\,\text{nm}$
$n = 2$ $\quad r = 2^2 \times 0.0529\,\text{nm}$
$n = 3$ $\quad r = 3^2 \times 0.0529\,\text{nm}$

This gives a picture of the hydrogen atom where an electron moves in circular orbits of radius proportional to 1^2, 2^2, 3^2... The atom will only radiate energy when the electron jumps from one orbit to another. The kinetic energy of an electron is $-\frac{1}{2}mv^2$. Rearranging equation (1.1)

$$E = -\tfrac{1}{2}mv^2 = -\frac{Ze^2}{8\pi\varepsilon_0 r}$$

Substituting for r using equation (1.3)

$$E = -\frac{Z^2e^4m}{8\varepsilon_0^2n^2h^2}$$

If an electron jumps from an initial orbit i to a final orbit f, the change in energy ΔE is

$$\Delta E = \left(-\frac{Z^2 e^4 m}{8\varepsilon_0^2 n_i^2 h^2}\right) - \left(-\frac{Z^2 e^4 m}{8\varepsilon_0^2 n_f^2 h^2}\right)$$

$$= \frac{Z^2 e^2 m}{8\varepsilon_0^2 h^2}\left(\frac{1}{n_f^2} - \frac{1}{n_i^2}\right)$$

Energy is related to wavelength ($E = hc\bar{\nu}$ so this equation is of the same form as the Rydberg equation:

$$\bar{\nu} = \frac{Z^2 e^4 m}{8\varepsilon_0^2 h^3 c}\left(\frac{1}{n_f^2} - \frac{1}{n_i^2}\right) \quad (1.4)$$

$$\bar{\nu} = R\left(\frac{1}{n_1^2} - \frac{1}{n_2^2}\right) \quad \text{(Rydberg equation)}$$

Thus the Rydberg constant

$$R = \frac{Z^2 e^4 m}{8\varepsilon_0^2 h^3 c}$$

The experimental value of R is $1.097373 \times 10^7\,m^{-1}$, in good agreement with the theoretical value of $1.096776 \times 10^7\,m^{-1}$, The Bohr theory provides an explanation of the atomic spectra of hydrogen. The different series of spectral lines can be obtained by varying the values of n_i and n_f in equation (1.4). Thus with $n_f = 1$ and $n_i = 2, 3, 4\ldots$ we obtain the Lyman series of lines in the UV region. With $n_f = 2$ and $n_i = 3, 4, 5\ldots$ we get the Balmer series of lines in the visible spectrum. Similarly, $n_f = 3$ and $n_i = 4, 5, 6\ldots$ gives the Paschen series, $n_f = 4$ and $n_i = 5, 6, 7\ldots$ gives the Brackett series, and $n_f = 6$ and $n_i = 7, 8, 9\ldots$ gives the Pfund series. The various transitions which are possible between orbits are shown in Figure 1.4.

REFINEMENTS TO THE BOHR THEORY

It has been assumed that the nucleus remains stationary except for rotating on its own axis. This would be true if the mass of the nucleus were infinite, but the ratio of the mass of an electron to the mass of the hydrogen nucleus is 1/1836. The nucleus actually oscillates slightly about the centre of gravity, and to allow for this the mass of the electron m is replaced by the reduced mass μ in equation (1.4):

$$\mu = \frac{mM}{m + M}$$

where M is the mass of the nucleus. The inclusion of the mass of the nucleus explains why different isotopes of an element produce lines in the spectrum at slightly different wavenumbers.

The orbits are sometimes denoted by the letters K, L, M, N... counting outwards from the nucleus, and they are also numbered 1, 2, 3, 4... This number is called the principal quantum number, which is given the symbol

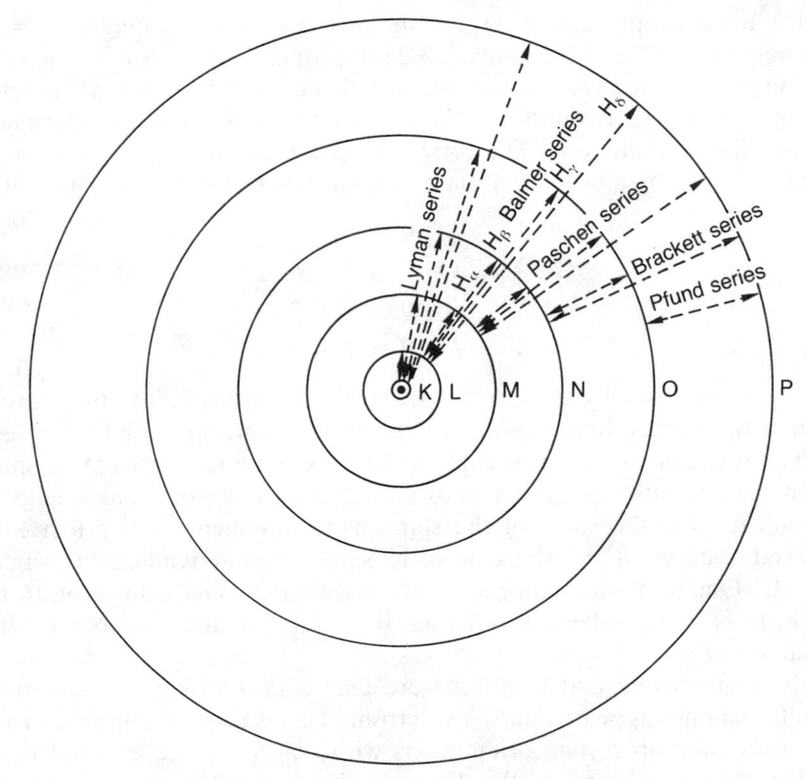

Figure 1.4 Bohr orbits of hydrogen and the various series of spectral lines.

n. It is therefore possible to define which circular orbit is under consideration by specifying the principal quantum number.

When an electron moves from one orbit to another it should give a single sharp line in the spectrum, corresponding precisely to the energy difference between the initial and final orbits. If the hydrogen spectrum is observed with a high resolution spectrometer it is found that some of the lines reveal 'fine structure'. This means that a line is really composed of several lines close together. Sommerfeld explained this splitting of lines by assuming that some of the orbits were elliptical, and that they precessed in space round the nucleus. For the orbit closest to the nucleus, the principal quantum number $n = 1$, and there is a circular orbit. For the next orbit, the principal quantum number $n = 2$, and both circular and elliptical orbits are possible. To define an elliptical orbit, a second quantum number k is needed. The shape of the ellipse is defined by the ratio of the lengths of the major and minor axes. Thus

$$\frac{\text{major axis}}{\text{minor axis}} = \frac{n}{k}$$

k is called the azimuthal or subsidiary quantum number, and may have values from $1, 2 \ldots n$. Thus for $n = 2$, n/k may have the values 2/2 (circular

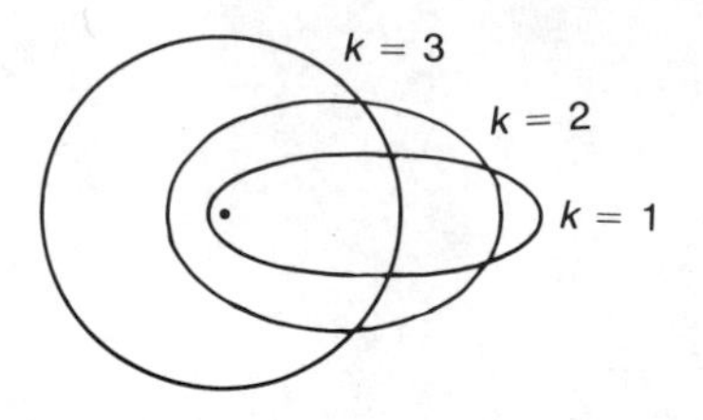

Figure 1.5 Bohr–Sommerfield orbits when $n = 3$.

orbit) and 2/1 (elliptical orbit). For the principal quantum number $n = 3$, n/k may have values 3/3 (circular), 3/2 (ellipse) and 3/1 (narrower ellipse).

The presence of these extra orbits, which have slightly different energies from each other, accounts for the extra lines in the spectrum revealed under high resolution. The original quantum number k has now been replaced by a new quantum number l, where $l = k - 1$. Thus for

$$
\begin{array}{ll}
n = 1 & l = 0 \\
n = 2 & l = 0 \text{ or } 1 \\
n = 3 & l = 0 \text{ or } 1 \text{ or } 2 \\
n = 4 & l = 0 \text{ or } 1 \text{ or } 2 \text{ or } 3
\end{array}
$$

This explained why some of the spectral lines are split into two, three, four or more lines. In addition some spectral lines are split still further into two lines (a doublet). This is explained by assuming that an electron spins on its axis in either a clockwise or an anticlockwise direction. Energy is quantized, and the value of the spin angular momentum was first considered to be $m_s \cdot h/2\pi$, where m_s is the spin quantum number with values of $\pm\frac{1}{2}$. (Quantum mechanics has since shown the exact expression to be $\sqrt{s(s+1)} \cdot h/2\pi$, where s is either the spin quantum number or the resultant of several spins.)

Zeeman showed that if atoms were placed in a strong magnetic field additional lines appeared on the spectrum. This is because elliptical orbits can only take up certain orientations with respect to the external field, rather than precessing freely. Each of these orientations is associated with a fourth quantum number m which can have values of l, $(l - 1), \ldots 0 \ldots (-l + 1), -l$.

Thus a single line in the normal spectrum will appear as $(2l + 1)$ lines if a magnetic field is applied.

Thus in order to explain the spectrum of the hydrogen atom, four quantum numbers are needed, as shown in Table 1.2. The spectra of other atoms may be explained in a similar manner.

THE DUAL NATURE OF ELECTRONS – PARTICLES OR WAVES

The planetary theory of atomic structure put forward by Rutherford and Bohr describes the atom as a central nucleus surrounded by electrons in

Table 1.2 The four main quantum numbers

	Symbol	Values
Principal quantum number	n	$1, 2, 3 \ldots$
Azimuthal or subsidiary quantum number	l	$0, 1, \ldots (n-1)$
Magnetic quantum number	m	$-l, \ldots 0, \ldots +l$
Spin quantum number	m_s	$\pm\frac{1}{2}$

certain orbits. The electron is thus considered as a particle. In the 1920s it was shown that moving particles such as electrons behaved in some ways as waves. This is an important concept in explaining the electronic structure of atoms.

For some time light has been considered as either particles or waves. Certain materials such as potassium emit electrons when irradiated with visible light, or in some cases with ultraviolet light. This is called the photoelectric effect. It is explained by light travelling as particles called photons. If a photon collides with an electron, it can transfer its energy to the electron. If the energy of the photon is sufficiently large it can remove the electron from the surface of the metal. However, the phenomena of diffraction and interference of light can only be explained by assuming that light behaves as waves. In 1924, de Brogie postulated that the same dual character existed with electrons – sometimes they are considered as particles, and at other times it is more convenient to consider them as waves. Experimental evidence for the wave nature of electrons was obtained when diffraction rings were observed photographically when a stream of electrons was passed through a thin metal foil. Electron diffraction has now become a useful tool in determining molecular structure, particularly of gases. Wave mechanics is a means of studying the build-up of electron shells in atoms, and the shape of orbitals occupied by the electrons.

THE HEISENBERG UNCERTAINTY PRINCIPLE

Calculations on the Bohr model of an atom require precise information about the position of an electron and its velocity. It is difficult to measure both quantities accurately at the same time. An electron is too small to see and may only be observed if perturbed. For example, we could hit the electron with another particle such as a photon or an electron, or we could apply an electric or magnetic force to the electron. This will inevitably change the position of the electron, or its velocity and direction. Heisenberg stated that the more precisely we can define the position of an electron, the less certainly we are able to define its velocity, and vice versa. If Δx is the uncertainty in defining the position and Δv the uncertainty in the velocity, the uncertainty principle may be expressed mathematically as:

$$\Delta x \,.\, \Delta v \geqslant \frac{\mathrm{h}}{4\pi}$$

where h = Planck's constant = 6.6262×10^{-34} J s. This implies that it is impossible to know both the position and the velocity exactly.

The concept of an electron following a definite orbit, where its position and velocity are known exactly, must therefore be replaced by the probability of finding an electron in a particular position, or in a particular volume of space. The Schrödinger wave equation provides a satisfactory description of an atom in these terms. Solutions to the wave equation are called wave functions and given the symbol ψ. The probability of finding an electron at a point in space whose coordinates are x, y and z is $\psi^2(x, y, z)$.

THE SCHRÖDINGER WAVE EQUATION

For a standing wave (such as a vibrating string) of wavelength λ, whose amplitude at any point along x may be described by a function $f(x)$, it can be shown that:

$$\frac{d^2f(x)}{dx^2} = -\frac{4\pi^2}{\lambda_2} f(x)$$

If an electron is considered as a wave which moves in only one dimension then:

$$\frac{d^2\psi}{dx^2} = -\frac{4\pi^2}{\lambda^2} \psi$$

An electron may move in three directions x, y and z so this becomes

$$\frac{\partial^2\psi}{\partial x^2} + \frac{\partial^2\psi}{\partial y^2} + \frac{\partial^2\psi}{\partial z^2} = -\frac{4\pi^2}{\lambda^2} \psi$$

Using the symbol ∇ instead of the three partial differentials, this is shortened to

$$\nabla^2\psi = -\frac{4\pi^2}{\lambda^2} \psi$$

The de Broglie relationship states that

$$\lambda = \frac{h}{mv}$$

(where h is Planck's constant, m is the mass of an electron and v its velocity); hence:

$$\nabla^2\psi = -\frac{4\pi^2 m^2 v^2}{h^2} \psi$$

or

$$\nabla^2\psi + \frac{4\pi^2 m^2 v^2}{h^2} \psi = 0 \qquad (1.5)$$

However, the total energy of the system E is made up of the kinetic energy K plus the potential energy V

$$E = K + V$$

so

$$K = E - V$$

But the kinetic energy $= \frac{1}{2}mv^2$ so

$$\tfrac{1}{2}mv^2 = E - V$$

and

$$v^2 = \frac{2}{m}(E - V)$$

Substituting for v^2 in equation (1.5) gives the well-known form of the Schrödinger equation

$$\nabla^2\psi + \frac{8\pi^2 m}{h^2}(E - V)\psi = 0$$

Acceptable solutions to the wave equation, that is solutions which are physically possible, must have certain properties:

1. ψ must be continuous.
2. ψ must be finite.
3. ψ must be single valued.
4. The probability of finding the electron over all the space from plusinfinity to minus infinity must be equal to one.

The probability of finding an electron at a point x, y, z is ψ^2, so

$$\int_{-\infty}^{+\infty} \psi^2 \, dx \, dy \, dz = 1$$

Several wave functions called ψ_1, ψ_2, ψ_3... will satisfy these conditions to the wave equation, and each of these has a corresponding energy E_1, E_2, E_3.... Each of these wave functions ψ_1, ψ_2, etc. is called an *orbital*, by analogy with the *orbits* in the Bohr theory. In a hydrogen atom, the single electron normally occupies the lowest of the energy levels E_1. This is called the ground state. The corresponding wave function ψ_1 describes the orbital, that is the volume in space where there is a high probability of finding the electron.

For a given type of atom, there are a number of solutions to the wave equation which are acceptable, and each orbital may be described uniquely by a set of three quantum numbers, n, l and m. (These are the same quantum numbers – principal, subsidiary and magnetic – as were used in the Bohr theory).

The subsidiary quantum number l describes the shape of the orbital occupied by the electron. l may have values 0, 1, 2 or 3. When $l = 0$, the orbital is spherical and is called an s orbital; when $l = 1$, the orbital is dumb-bell shaped and is called a p orbital; when $l = 2$, the orbital is double dumb-bell shaped and is called a d orbital; and when $l = 3$ a more complicated f orbital is formed (see Figure 1.6). The letters s, p, d and f come from the spectroscopic terms *s*harp, *p*rincipal, *d*iffuse and *f*undamental, which were used to describe the lines in the atomic spectra.

Examination of a list of all the allowed solutions to the wave equation shows that the orbitals fall into groups.

In the first group of solutions the value of the wave function ψ, and

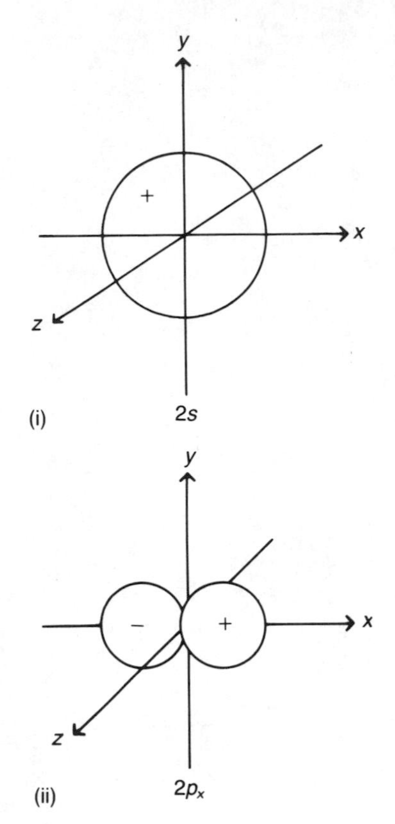

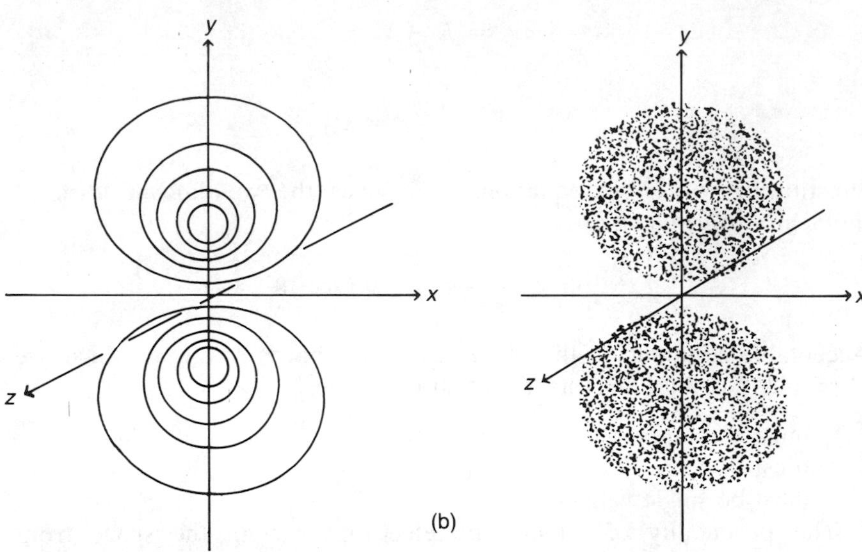

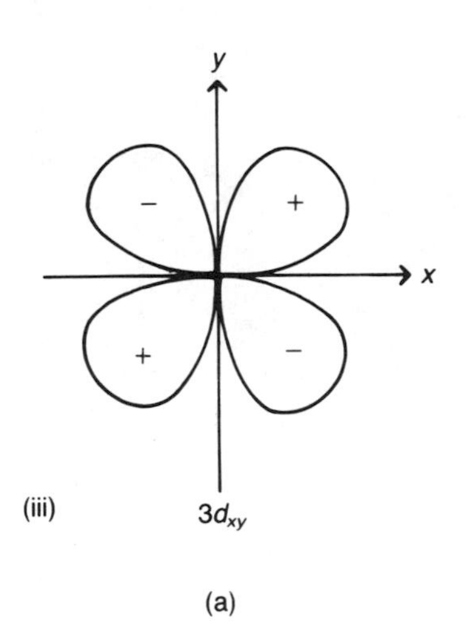

Figure 1.6 (a) Wave functions ψ for *s*, *p* and *d* atomic orbitals: (i) *s* orbital, 2*s*; (ii) *p* orbital, $2p_x$; (iii) *d* orbital, $3d_{xy}$. (Note that the + and − signs refer to symmetry, not charge.) (b) Different ways of representing ψ^2 for a 2*p* orbital (as a contour diagram or as a 90% boundary surface).

hence the probability of finding the electron ψ^2, depends only on the distance *r* from the nucleus, and is the same in all directions.

$$\psi = f(r)$$

This leads to a spherical orbital, and occurs when the subsidiary quantum number *l* is zero. These are called *s* orbitals. When $l = 0$, the magnetic quantum number $m = 0$, so there is only one such orbital for each value of *n*.

In the second group of solutions to the wave equation, ψ depends both on the distance from the nucleus, and on the direction in space (*x*, *y* or *z*). Orbitals of this kind occur when the subsidiary quantum number $l = 1$. These are called *p* orbitals and there are three possible values of the magnetic quantum number ($m = -1$, 0 and +1). There are therefore three orbitals which are identical in energy, shape and size, which differ only in their direction in space. These three solutions to the wave equation may be written

$$\psi_x = f(r) . f(x)$$
$$\psi_y = f(r) . f(y)$$
$$\psi_z = f(r) . f(z)$$

Orbitals that are identical in energy are termed degenerate, and thus three degenerate *p* orbitals occur for each of the values of $n = 2, 3, 4 \ldots$

The third group of solutions to the wave equation depend on the

Table 1.3 Atomic orbitals

Principal quantum number n	Subsidiary quantum number l	Magnetic quantum numbers m	Symbol
1	0	0	1*s* (one orbital)
2	0	0	2*s* (one orbital)
2	1	−1, 0, +1	2*p* (three orbitals)
3	0	0	3*s* (one orbital)
3	1	−1, 0, +1	3*p* (three orbitals)
3	2	−2, −1, 0, +1, +2	3*d* (five orbitals)
4	0	0	4*s* (one orbital)
4	1	−1, 0, +1	4*p* (three orbitals)
4	2	−2, −1, 0, +1, +2	4*d* (five orbitals)
4	3	−3, −2, −1, 0, +1, +2, −3	4*f* (seven orbitals)

distance from the nucleus r and also on two directions in space, for example

$$\psi = \mathrm{f}(r) \, . \, \mathrm{f}(x) \, . \, \mathrm{f}(y)$$

This group of orbitals has $l = 2$, and these are called *d* orbitals. There are five solutions corresponding to $m = -2, -1, 0, +1$ and $+2$, and these are all equal in energy. Thus five degenerate *d* orbitals occur for each of the values of $n = 3, 4, 5 \ldots$.

A further set of solutions occurs when $l = 3$, and these are called *f* orbitals. There are seven values of m: $-3, -2, -1, 0, +1, +2$ and $+3$, and seven degenerate *f* orbitals are formed when $n = 4, 5, 6 \ldots$.

RADIAL AND ANGULAR FUNCTIONS

The Schrödinger equation can be solved completely for the hydrogen atom, and for related ions which have only one electron such as He^+ and Li^{2+}. For other atoms only approximate solutions can be obtained. For most calculations, it is simpler to solve the wave equation if the cartesian coordinates x, y and z are converted into polar coordinates r, θ and ϕ. The coordinates of the point A measured from the origin are x, y, and z in cartesian coordinates, and r, θ and ϕ in polar coordinates. It can be seen that the two sets of coordinates are related by the following expressions:

$$z = r \cos \theta$$
$$y = r \sin \theta \sin \phi$$
$$x = r \sin \theta \cos \phi$$

The Schrödinger equation is usually written:

$$\nabla^2\psi + \frac{8\pi^2 m}{h^2}(E - V)\psi = 0$$

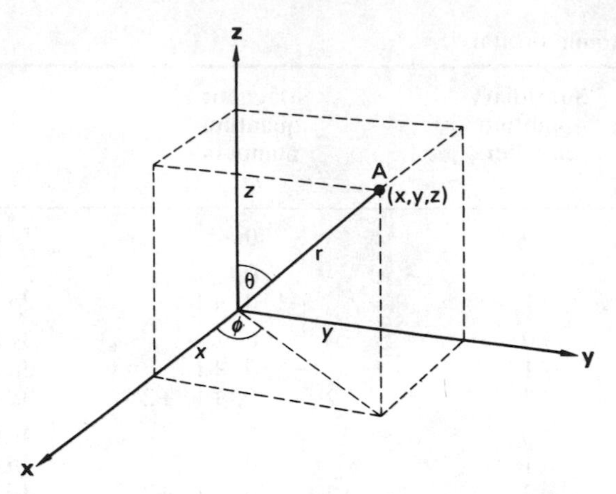

Figure 1.7 The relationship between cartesian and polar coordinates.

where

$$\nabla^2\psi = \frac{\partial^2\psi}{\partial x^2} + \frac{\partial^2\psi}{\partial y^2} + \frac{\partial^2\psi}{\partial z^2}$$

Changing to polar coordinates, $\nabla^2\psi$ becomes

$$\frac{1}{r^2}\frac{\partial}{\partial r}\left(r^2\frac{\partial\psi}{\partial r}\right) + \frac{1}{r^2\sin^2\theta}\cdot\frac{\partial^2\psi}{\partial\phi^2} + \frac{1}{r^2\sin\theta}\cdot\frac{\partial}{\partial\theta}\left(\sin\theta\frac{\partial\psi}{\partial\theta}\right)$$

The solution of this is of the form

$$\psi = R(r)\,.\,\Theta(\theta)\,.\,\Phi(\phi) \qquad (1.6)$$

$R(r)$ is a function that depends on the distance from the nucleus, which in turn depends on the quantum numbers n and l
$\Theta(\theta)$ is a function of θ, which depends on the quantum numbers l and m
$\Phi(\phi)$ is a function of ϕ, which depends only on the quantum number m

Equation (1.6) may be rewritten

$$\psi = R(r)_{nl}\,.\,A_{ml}$$

This splits the wave function into two parts which can be solved separately:

1. $R(r)$ the radial function, which depends on the quantum numbers n and l.
2. A_{ml} the total angular wave function, which depends on the quantum numbers m and l.

The radial function R has no physical meaning, but R^2 gives the probability of finding the electron in a small volume dv near the point at which R is measured. For a given value of r the number of small volumes is $4\pi r^2$, so the probability of the electron being at a distance r from the nucleus is $4\pi r^2R^2$. This is called the radial distribution function. Graphs of the

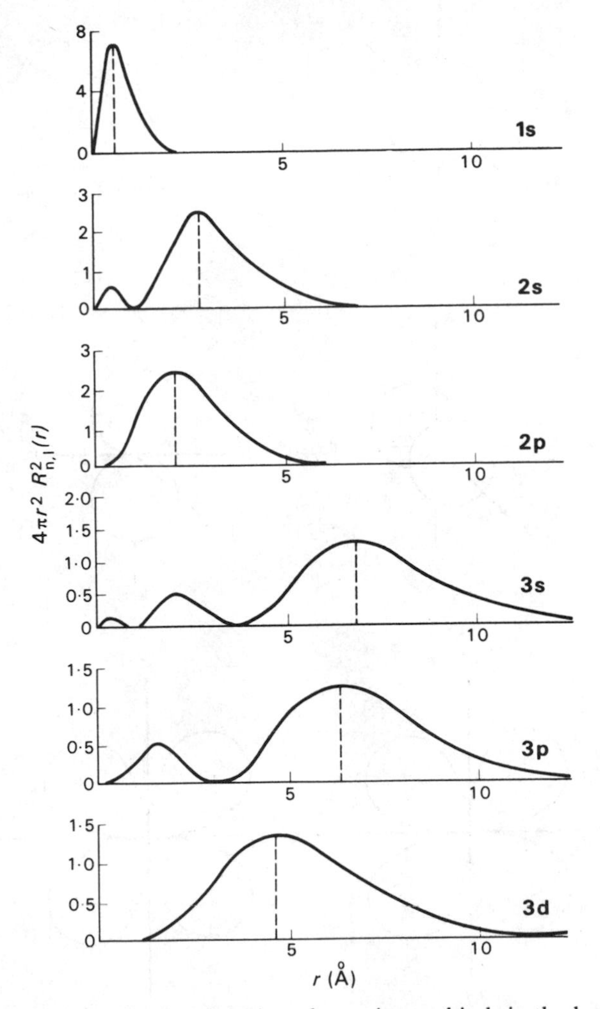

Figure 1.8 Radial distribution functions for various orbitals in the hydrogen atom.

radial distribution function for hydrogen plotted against r are shown in Figure 1.8.

These diagrams show that the probability is zero at the nucleus (as $r = 0$), and by examining the plots for 1*s*, 2*s* and 3*s* that the most probable distance increases markedly as the principal quantum number increases. Furthermore, by comparing the plots for 2*s* and 2*p*, or 3*s*, 3*p* and 3*d* it can be seen that the most probable radius decreases slightly as the subsidiary quantum number increases. All the *s* orbitals except the first one (1*s*) have a shell-like structure, rather like an onion or a hailstone, consisting of concentric layers of electron density. Similarly, all but the first *p* orbitals (2*p*) and the first *d* orbitals (3*d*) have a shell structure.

The angular function A depends only on the direction, and is independent of the distance from the nucleus (r). Thus A^2 is the probability of

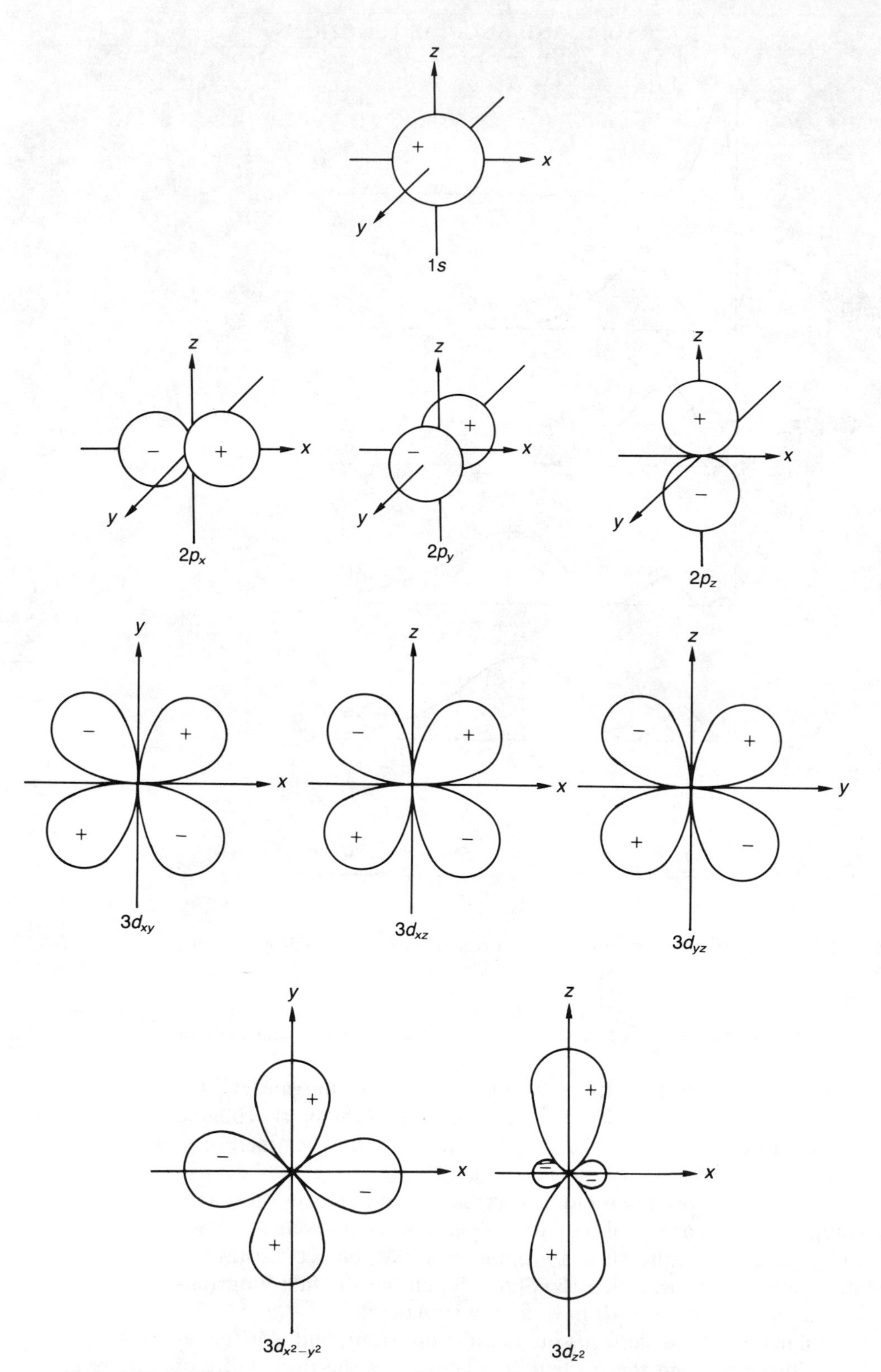

Figure 1.9 Boundary surface for the angular part of the wave function $A(\theta, \phi)$ for the $2s$, $2p$ and $3d$ orbitals for a hydrogen atom shown as polar diagrams.

finding an electron at a given direction θ, ϕ at any distance from the nucleus to infinity. The angular functions A are plotted as polar diagrams in Figure 1.9. It must be emphasized that *these polar diagrams do* **not** *represent the total wave function* υ, but only the angular part of the wave function. (The total wave function is made up from contributions from both the radial and the angular functions.)

$$\psi = R(r) \,.\, A$$

Thus the probability of finding an electron simultaneously at a distance r and in a given direction θ, ϕ is $\psi^2_{r,\theta,\phi}$.

$$\psi^2_{r,\theta,\phi} = R^2(r) \,.\, A^2(\theta,\phi)$$

Polar diagrams, that is drawings of the the angular part of the wave function, are commonly used to illustrate the overlap of orbitals giving bonding between atoms. Polar diagrams are quite good for this purpose, as they show the signs + and − relating to the symmetry of the angular function. For bonding like signs must overlap. These shapes are slightly different from the shapes of the total wave function. There are several points about such diagrams:

1. It is difficult to picture an angular wave function as a mathematical equation. It is much easier to visualize a boundary surface, that is a solid shape, which for example contains 90% of the electron density. To emphasize that ψ is a continuous function, the boundary surfaces have been extended up to the nucleus in Figure 1.9. For p orbitals the electron density is zero at the nucleus, and some texts show a p orbital as two spheres which do not touch.
2. These drawings show the symmetry for the $1s$, $2p$, $3d$ orbitals. However, in the others, $2s$, $3s$, $4s$ . . . , $3p$, $4p$, $5p$ . . . , $4d$, $5d$ . . . the sign (symmetry) changes inside the boundary surface of the orbital. This is readily seen as nodes in the graphs of the radial functions (Figure 1.8).

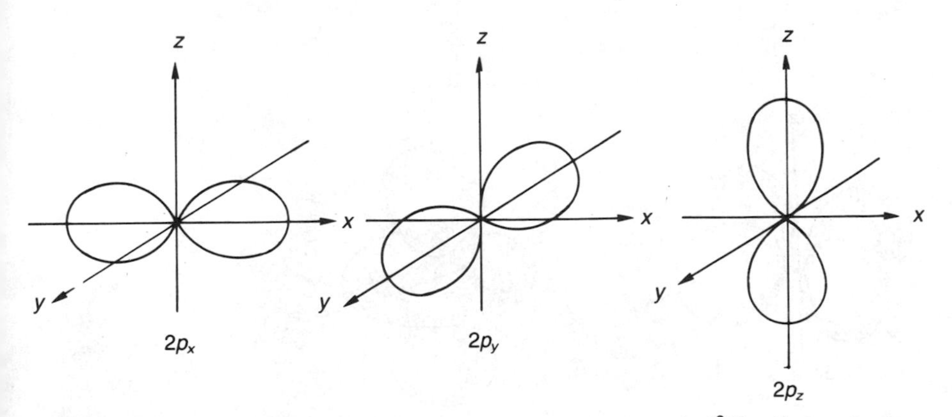

Figure 1.10 The angular part of the wave function squared $A^2(\theta, \phi)$ for the $2p$ orbitals for a hydrogen atom.

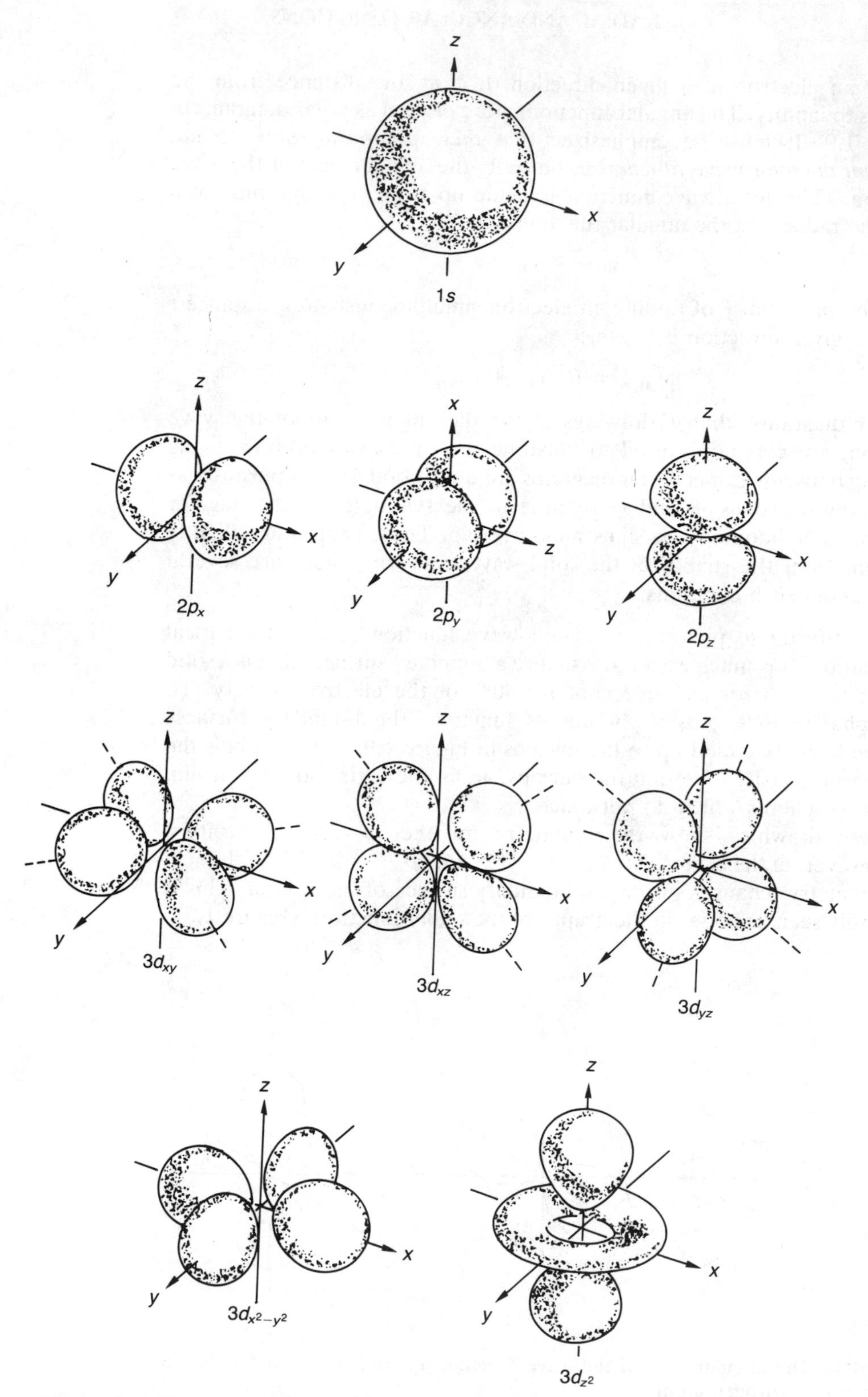

Figure 1.11 Total wave function (orbitals) for hydrogen.

3. The probability of finding an electron at a direction θ, ϕ is the wave function squared, A^2, or more precisely $\psi_\theta^2\psi_\phi^2$. The diagrams in Figure 1.9 are of the angular part of the wave function A, not A^2. Squaring does not change the shape of an *s* orbital, but it elongates the lobes of *p* orbitals (Figure 1.10). Some books use elongated *p* orbitals, but strictly these should **not** have signs, as squaring removes any sign from the symmetry. Despite this, many authors draw shapes approximating to the probabilities, i.e. squared wave functions, and put the signs of the wave function on the lobes, and refer to both the shapes and the wave functions as orbitals.
4. A full representation of the probability of finding an electron requires the total wave function squared and includes both the radial and angular probabilities squared. It really needs a three-dimensional model to display this probability, and show the shapes of the orbitals. It is difficult to do this adequately on a two-dimensional piece of paper, but a representation is shown in Figure 1.11. The orbitals are not drawn to scale. Note that the *p* orbitals are not simply two spheres, but are ellipsoids of revolution. Thus the $2p_x$ orbital is spherically symmetrical about the *x* axis, but is not spherical in the other direction. Similarly the p_y orbital is spherically symmetrical about the *y* axis, and both the p_z and the $3d_{z^2}$ are spherically symmetrical about the *z* axis.

PAULI EXCLUSION PRINCIPLE

Three quantum numbers *n*, *l* and *m* are needed to define an orbital. Each orbital may hold up to two electrons, provided they have opposite spins. An extra quantum number is required to define the spin of an electron in an orbital. Thus four quantum numbers are needed to define the energy of an electron in an atom. The Pauli exclusion principle states that no two electrons in one atom can have all four quantum numbers the same. By permutating the quantum numbers, the maximum number of electrons which can be contained in each main energy level can be calculated (see Figure 1.12).

BUILD-UP OF THE ELEMENTS, HUND'S RULE

When atoms are in their ground state, the electrons occupy the lowest possible energy levels.

The simplest element, hydrogen, has one electron, which occupies the 1*s* level; this level has the principal quantum number $n = 1$, and the subsidiary quantum number $l = 0$.

Helium has two electrons. The second electron also occupies the 1*s*level. This is possible because the two electrons have opposite spins. This level is now full.

The next atom lithium has three electrons. The third electron occupies the next lowest level. This is the 2*s* level, which has the principal quantum number $n = 2$ and subsidiary quantum number $l = 0$.

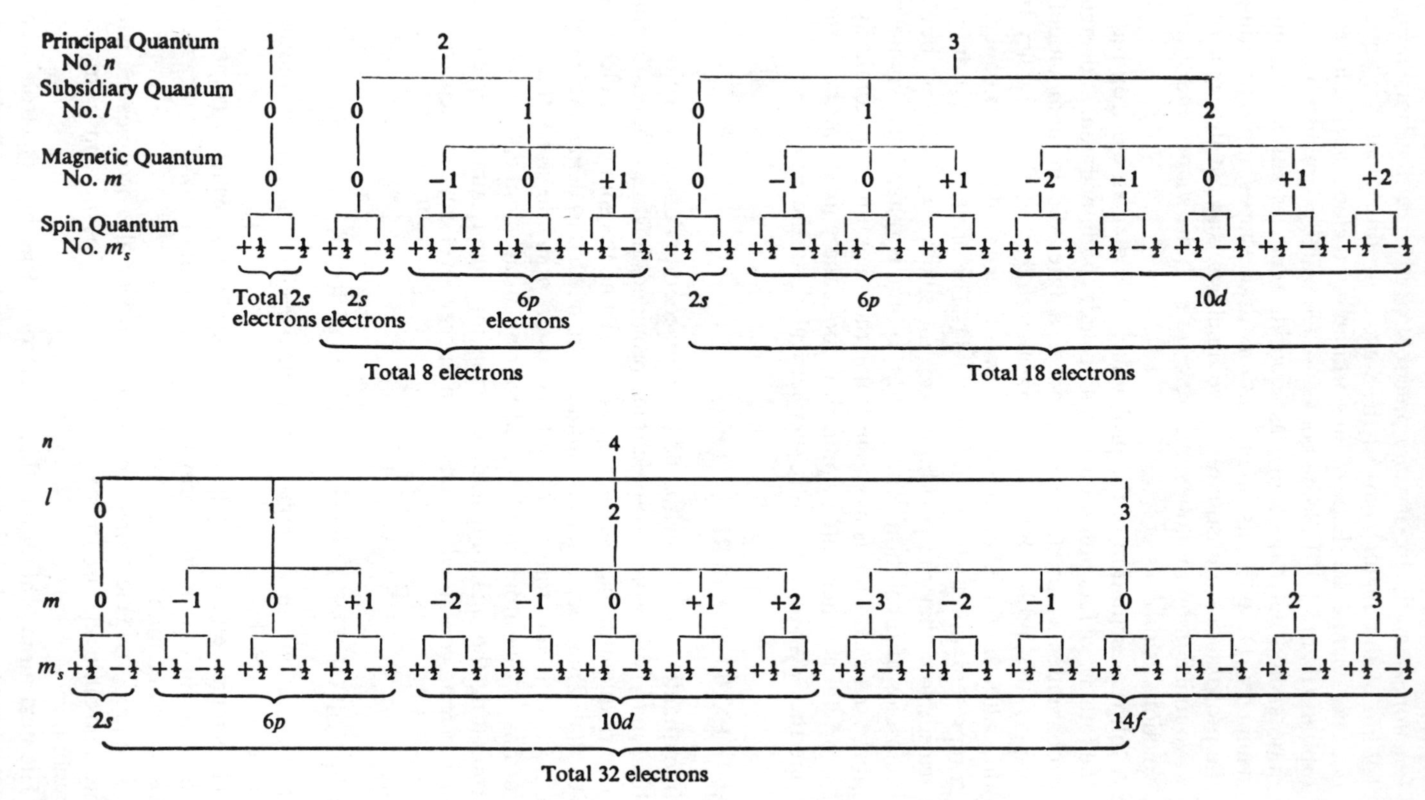

Figure 1.12 Quantum numbers, the permissible number of electrons and the shape of the periodic table.

The fourth electron in beryllium also occupies the $2s$ level. Boron must have its fifth electron in the $2p$ level as the $2s$ level is full. The sixth electron in carbon is also in the $2p$ level. Hund's rule states that the number of unpaired electrons in a given energy level is a maximum. Thus in the ground state the two p electrons in carbon are unpaired. They occupy separate p orbitals and have parallel spins. Similarly in nitrogen the three p electrons are unpaired and have parallel spins.

To show the positions of the electrons in an atom, the symbols $1s$, $2s$, $2p$, etc. are used to denote the main energy level and sub-level. A superscript indicates the number of electrons in each set of orbitals. Thus for hydrogen, the $1s$ orbital contains one electron, and this is shown as $1s^1$. For helium the $1s$ orbital contains two electrons, denoted $1s^2$. The electronic structures of the first few atoms in the periodic table may be written:

H	$1s^1$
He	$1s^2$
Li	$1s^2\ 2s^1$
Be	$1s^2\ 2s^2$
B	$1s^2\ 2s^2\ 2p^1$
C	$1s^2\ 2s^2\ 2p^2$
N	$1s^2\ 2s^2\ 2p^3$
O	$1s^2\ 2s^2\ 2p^4$
F	$1s^2\ 2s^2\ 2p^5$
Ne	$1s^2\ 2s^2\ 2p^6$
Na	$1s^2\ 2s^2\ 2p^6\ 3s^1$

An alternative way of showing the electronic structure of an atom is to draw boxes for orbitals, and arrows for the electrons.

	1s	2s	2p		
Electronic structure of H atom in the ground state	↑				
Electronic structure of He atom in the ground state	↑↓				
Electronic structure of Li atom in the ground state	↑↓	↑			
Electronic structure of Be atom in the ground state	↑↓	↑↓			
Electronic structure of B atom in the ground state	↑↓	↑↓	↑		

	1s	2s	2p		
Electronic structure of C atom in the ground state	↑↓	↑↓	↑	↑	

	1s	2s	2p		
Electronic structure of N atom in the ground state	↑↓	↑↓	↑	↑	↑

	1s	2s	2p		
Electronic structure of O atom in the ground state	↑↓	↑↓	↑↓	↑	↑

	1s	2s	2p		
Electronic structure of F atom in the ground state	↑↓	↑↓	↑↓	↑↓	↑

	1s	2s	2p			3s	3p		
Electronic structure of Ne atom in the ground state	↑↓	↑↓	↑↓	↑↓	↑↓				

	1s	2s	2p			3s	3p		
Electronic structure of Na atom in the ground state	↑↓	↑↓	↑↓	↑↓	↑↓	↑			

The process continues in a similar way.

SEQUENCE OF ENERGY LEVELS

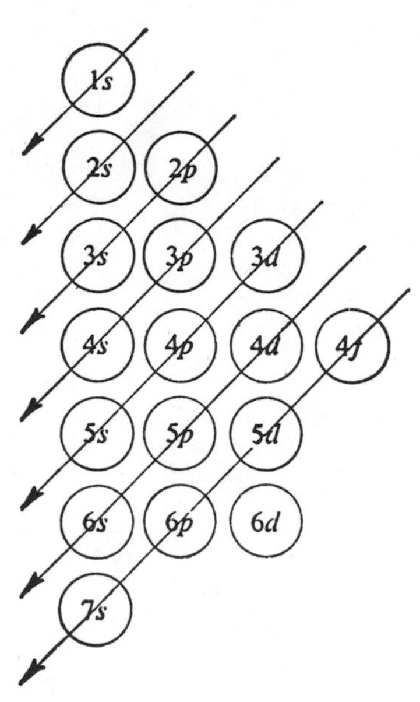

Figure 1.13 Sequence of filling energy levels.

It is important to know the sequence in which the energy levels are filled. Figure 1.13 is a useful aid. From this it can be seen that the order of filling of energy levels is: 1*s*, 2*s*, 2*p*, 3*s*, 3*p*, 4*s*, 3*d*, 4*p*, 5*s*, 4*d*, 5*p*, 6*s*, 4*f*, 5*d*, 6*p*, 7*s*, etc.

After the 1*s*, 2*s*, 2*p*, 3*s* and 3*p* levels have been filled at argon, the next two electrons go into the 4*s* level. This gives the elements potassium and calcium. Once the 4*s* level is full the 3*d* level is the next lowest in energy, not the 3*p* level. Thus the 3*d* starts to fill at scandium. The elements from scandium to copper have two electrons in the 4*s* level and an incomplete 3*d* level, and all behave in a similar manner chemically. Such a series of atoms is known as a transition series.

A second transition series starts after the 5*s* orbital has been filled, at strontium, because in the next element, yttrium, the 4*d* level begins to fill up. A third transition series starts at lanthanum where the electrons start to fill the 5*d* level after the 6*d* level has been filled with two electrons.

A further complication arises here because after lanthanum, which has one electron in the 5*d* level, the 4*f* level begins to fill, giving the elements from cerium to lutetium with from one to 14*f* electrons. These are sometimes called the inner transition elements, but are usually known as the lanthanides or rare earth metals.

ARRANGEMENT OF THE ELEMENTS IN GROUPS IN THE PERIODIC TABLE

The chemical properties of an element are largely governed by the number of electrons in the outer shell, and their arrangement. If the elements are arranged in groups which have the same outer electronic arrangement, then elements within a group should show similarities in chemical and physical properties. One great advantage of this is that initially it is only necessary to learn the properties of each group rather than the properties of each individual element.

Elements with one *s* electron in their outer shell are called Group I (the alkali metals) and elements with two *s* electrons in their outer shell are called Group II (the alkaline earth metals). These two groups are known as the *s*-block elements, because their properties result from the presence of *s* electrons.

Elements with three electrons in their outer shell (two *s* electrons and one *p* electron) are called Group III, and similarly Group IV elements have four outer electrons, Group V elements have five outer electrons, Group VI elements have six outer electrons and Group VII elements have seven outer electrons. Group 0 elements have a full outer shell of electrons so that the next shell is empty; hence the group name. Groups III, IV, V, VI, VII and 0 all have *p* orbitals filled and because their properties are dependent on the presence of *p* electrons, they are called jointly the *p*-block elements.

In a similar way, elements where *d* orbitals are being filled are called the *d*-block, or transition elements. In these, *d* electrons are being added to the penultimate shell.

Finally, elements where *f* orbitals are filling are called the *f*-block, and here the *f* electrons are entering the antepenultimate (or second from the outside) shell.

In the periodic table (Table 1.4), the elements are arranged in order of increasing atomic number, that is in order of increased nuclear charge, or increased number of orbital electrons. Thus each element contains one more orbital electron than the preceding element. Instead of listing the 103 elements as one long list, the periodic table arranges them into several horizontal rows or periods, in such a way that each row begins with an alkali metal and ends with a noble gas. The sequence in which the various energy levels are filled determines the number of elements in each period, and the periodic table can be divided into four main regions according to whether the *s*, *p*, *d* or *f* levels are being filled.

1st period	1*s*				elements in this period 2
2nd period	2*s*			2*p*	elements in this period 8
3rd period	3*s*			3*p*	elements in this period 8
4th period	4*s*		3*d*	4*p*	elements in this period 18
5th period	5*s*		4*d*	5*p*	elements in this period 18
6th period	6*s*	4*f*	5*d*	6*p*	elements in this period 32

Table 1.4 The periodic table

	s-block												*p*-block					
Group / Period	I	II											III	IV	V	VI	VII	0
1	^{1}H																^{1}H	^{2}He
2	^{3}Li	^{4}Be											^{5}B	^{6}C	^{7}N	^{8}O	^{9}F	^{10}Ne
3	^{11}Na	^{12}Mg					*d*-block						^{13}Al	^{14}Si	^{15}P	^{16}S	^{17}Cl	^{18}Ar
4	^{19}K	^{20}Ca	^{21}Sc	^{22}Ti	^{23}V	^{24}Cr	^{25}Mn	^{26}Fe	^{27}Co	^{28}Ni	^{29}Cu	^{30}Zn	^{31}Ga	^{32}Ge	^{33}As	^{34}Se	^{35}Br	^{36}Kr
5	^{37}Rb	^{38}Sr	^{39}Y	^{40}Zr	^{41}Nb	^{42}Mo	^{43}Tc	^{44}Ru	^{45}Rh	^{46}Pd	^{47}Ag	^{48}Cd	^{49}In	^{50}Sn	^{51}Sb	^{52}Te	^{53}I	^{54}Xe
6	^{55}Cs	^{56}Ba	^{57}La	^{72}Hf	^{73}Ta	^{74}W	^{75}Re	^{76}Os	^{77}Ir	^{78}Pt	^{79}Au	^{80}Hg	^{81}Tl	^{82}Pb	^{83}Bi	^{84}Po	^{85}At	^{86}Rn
7	^{87}Fr	^{88}Ra	^{89}Ac															
									f-block									
Lanthanides				^{58}Ce	^{59}Pr	^{60}Nd	^{61}Pm	^{62}Sm	^{63}Eu	^{64}Gd	^{65}Tb	^{66}Dy	^{67}Ho	^{68}Er	^{69}Tm	^{70}Yb	^{71}Lu	
Actinides				^{90}Th	^{91}Pa	^{92}U	^{93}Np	^{94}Pu	^{95}Am	^{96}Cm	^{97}Bk	^{98}Cf	^{99}Es	^{100}Fm	^{101}Md	^{102}No	^{103}Lr	

The alkali metals appear in a vertical column labelled Group I, in which all elements have one *s* electron in their outer shell, and hence have similar properties. Thus when one element in a group reacts with a reagent, the other elements in the group will probably react similarly, forming compounds which have similar formulae. Thus reactions of new compounds and their formulae may be predicted by analogy with known compounds. Similarly the noble gases all appear in a vertical column labelled Group 0, and all have a complete outer shell of electrons. This is called the long form of the periodic table. It has many advantages, the most important being that it emphasizes the similarity of properties within a group and the relation between the group and the electron structure. The *d*-block elements are referred to as the transition elements as they are situated between the *s*- and *p*-blocks.

Hydrogen and helium differ from the rest of the elements because there are no *p* orbitals in the first shell. Helium obviously belongs to Group 0, the noble gases, which are chemically inactive because their outer shell of electrons is full. Hydrogen is more difficult to place in a group. It could be included in Group I because it has one *s* electron in its outer shell, is univalent and commonly forms univalent positive ions. However, hydrogen is not a metal and is a gas whilst Li, Na, K, Rb and Cs are metals and are solids. Similarly, hydrogen could be included in Group VII because it is one electron short of a complete shell, or in Group IV because its outer shell is half full. Hydrogen does not resemble the alkali metals, the halogens or Group IV very closely. Hydrogen atoms are extremely small, and have many unique properties. Thus there is a case for placing hydrogen in a group on its own.

FURTHER READING

Karplus, M. and Porter, R.N. (1971) *Atoms and Molecules*, Benjamin, New York.
Greenwood, N.N. (1980) *Principles of Atomic Orbitals*, Royal Institute of Chemistry Monographs for Teachers No. 8, 3rd ed., London.

PROBLEMS

1. Name the first five series of lines that occur in the atomic spectrum of hydrogen. Indicate the region in the electromagnetic spectrum where these series occur, and give a general equation for the wavenumber applicable to all the series.
2. What are the assumptions on which the Bohr theory of the structure of the hydrogen atom is based?
3. Give the equation which explains the different series of lines in the atomic spectrum of hydrogen. Who is the equation named after? Explain the various terms involved.
4. (a) Calculate the radii of the first three Bohr orbits for hydrogen. (Planck's constant $h = 6.6262 \times 10^{-34}$ J s; mass of electron

$m = 9.1091 \times 10^{-31}$ kg; charge on electron $e = 1.60210 \times 10^{-19}$ C; permittivity of vacuum $\varepsilon_0 = 8.854185 \times 10^{-12}\,kg^{-1}\,m^{-3}\,A^2$.)
(Answers: 0.529×10^{-10} m; 2.12×10^{-10} m; 4.76×10^{-10} m; that is 0.529 Å 2.12 Å and 4.76 Å.)

(b) Use these radii to calculate the velocity of an electron in each of these three orbits.
(Answers: $2.19 \times 10^6\,m\,s^{-1}$; $1.09 \times 10^6\,m\,s^{-1}$; $7.29 \times 10^5\,m\,s^{-1}$.)

5. The Balmer series of spectral lines for hydrogen appear in the visible region. What is the lower energy level that these electronic transitions start from, and what transitions correspond to the spectral lines at 379.0 nm and 430.0 nm respectively?

6. What is the wavenumber and wavelength of the first transition in the Lyman, Balmer and Paschen series in the atomic spectra of hydrogen?

7. Which of the following species does the Bohr theory apply to? (a) H, (b) H^+, (c) He, (d) He^+, (e) Li, (f) Li^+, (g) Li^{+2}, (h) Be, (g) Be^+, (h) Be^{2+}, (i) Be^{3+}.

8. How does the Bohr theory of the hydrogen atom differ from that of Schrödinger?

9. (a) Write down the general form of the Schrödinger equation and define each of the terms in it.
(b) Solutions to the wave equation that are physically possible must have four special properties. What are they?

10. What is a radial distribution function? Draw this function for the 1*s*, 2*s*, 3*s*, 2*p*, 3*p* and 4*p* orbitals in a hydrogen atom.

11. Explain (a) the Pauli exclusion principle, and (b) Hund's rule. Show how these are used to specify the electronic arrangements of the first 20 elements in the periodic table.

12. What is an orbital? Draw the shapes of the 1*s*, 2*s*, $2p_x$, $2p_y$, $2p_z$, $3d_{xy}$, $3d_{xz}$, $3d_{yz}$, $3d_{x^2-y^2}$ and $3d_{z^2}$ orbitals.

13. Give the names and symbols of the four quantum numbers required to define the energy of electrons in atoms. What do these quantum numbers relate to, and what numerical values are possible for each? Show how the shape of the periodic table is related to these quantum numbers.

14. The first shell may contain up to 2 electrons, the second shell up to 8, the third shell up to 18, and the fourth shell up to 32. Explain this arrangement in terms of quantum numbers.

15. Give the values of the four quantum numbers for each electron in the ground state for (a) the oxygen atom, and (b) the scandium atom. (Use positive values for m_l and m_s first.)

16. Give the sequence in which the energy levels in an atom are filled with electrons. Write the electronic configurations for the elements of atomic number 6, 11, 17 and 25, and from this decide to which group in the periodic table each element belongs.

17. Give the name and symbol for each of the atoms which have the ground state electronic configurations in their outer shells: (a) $2s^2$, (b) $3s^2 3p^5$, (c) $3s^2 3p^6 4s^2$, (d) $3s^2 3p^6 3d^6 4s^2$, (e) $5s^2 5p^2$, (f) $5s^2 5p^6$.

2 Introduction to bonding

ATTAINMENT OF A STABLE CONFIGURATION

How do atoms combine to form molecules and why do atoms form bonds? A molecule will only be formed if it is more stable, and has a lower energy, than the individual atoms.

To understand what is happening in terms of electronic structure, consider first the Group 0 elements. These comprise the noble gases, helium, neon, argon, krypton, xenon and radon, which are noteworthy for their chemical inertness. Atoms of the noble gases do not normally react with any other atoms, and their molecules are monatomic, i.e. contain only one atom. The lack of reactivity is because the atoms already have a low energy, and it cannot be lowered further by forming compounds. The low energy of the noble gases is associated with their having a complete outer shell of electrons. This is often called a *noble gas structure*, and it is an exceptionally stable arrangement of electrons.

Normally only electrons in the outermost shell of an atom are involved in forming bonds, and by forming bonds each atom acquires a stable electron configuration. The most stable electronic arrangement is a noble gas structure, and many molecules have this arrangement. However, less stable arrangements than this are commonly attained by transition elements.

TYPES OF BONDS

Atoms may attain a stable electronic configuration in three different ways: by losing electrons, by gaining electrons, or by sharing electrons.

Elements may be divided into:

1. Electropositive elements, whose atoms give up one or more electrons fairly readily.
2. Electronegative elements, which will accept electrons.
3. Elements which have little tendency to lose or gain electrons.

Three different types of bond may be formed, depending on the electropositive or electronegative character of the atoms involved.

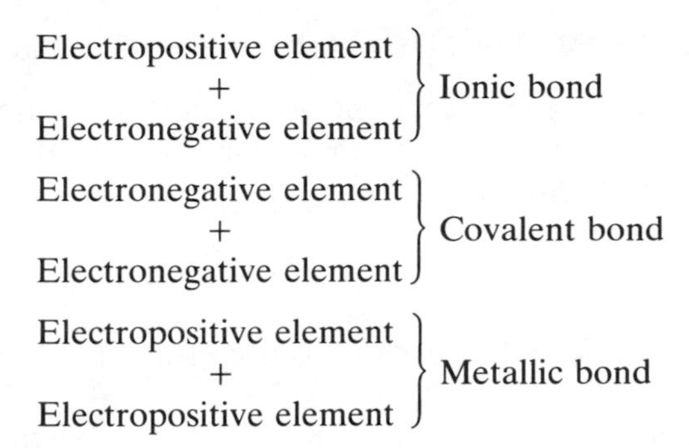

Ionic bonding involves the complete transfer of one or more electrons from one atom to another. Covalent bonding involves the sharing of a pair of electrons between two atoms, and in metallic bonding the valency electrons are free to move throughout the whole crystal.

These types of bonds are idealized or extreme representations, and though one type generally predominates, in most substances the bond type is somewhere between these extreme forms. For example, lithium chloride is considered to be an ionic compound, but it is soluble in alcohol, which suggests that it also possesses a small amount of covalent character. If the three extreme bond types are placed at the corners of a triangle, then compounds with bonds predominantly of one type will be represented as points near the corners. Compounds with bonds intermediate between two types will occur along an edge of the triangle, whilst compounds with bonds showing some characteristics of all three types are shown as points inside the triangle.

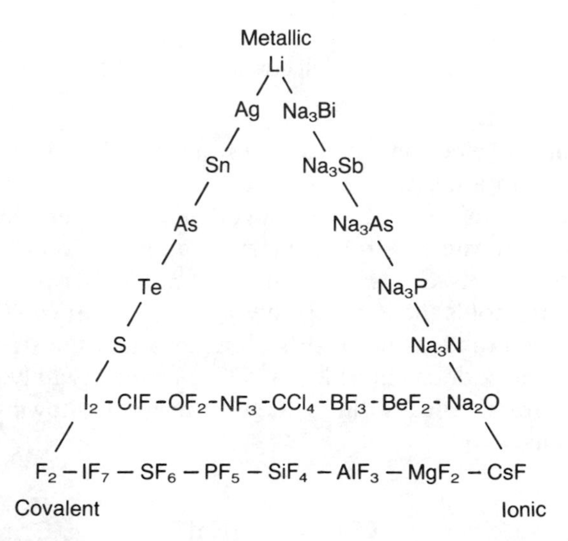

Figure 2.1 Triangle illustrating the transitions between ionic, covalent and metallic bonding. (Reproduced from *Chemical Constitution*, by J.A.A. Ketelaar, Elsevier.)

TRANSITIONS BETWEEN THE MAIN TYPES OF BONDING

Few bonds are purely ionic, covalent or metallic. Most are intermediate between the three main types, and show some properties of at least two, and sometimes of all three types.

Ionic bonds

Ionic bonds are formed when electropositive elements react with electronegative elements.

Consider the ionic compound sodium chloride. A sodium atom has the electronic configuration $1s^2\ 2s^2\ 2p^6\ 3s^1$. The first and second shells of electrons are full, but the third shell contains only one electron. When this atom reacts it will do so in such a way that it attains a stable electron configuration. The noble gases have a stable electron arrangement and the nearest noble gas to sodium is neon, whose configuration is $1s^2\ 2s^2\ 2p^6$. If the sodium atom can lose one electron from its outer shell, it will attain this configuration and in doing so the sodium acquires a net charge of +1 and is called a sodium ion Na^+. The positive charge arises because the nucleus contains 11 protons, each with a positive charge, but there are now only 10 electrons. Sodium atoms tend to lose an electron in this way when they are supplied with energy, and so sodium is an electropositive element:

$$\underset{\text{sodium atom}}{\mathrm{Na}} \rightarrow \underset{\text{sodium ion}}{\mathrm{Na}^+} + \text{electron}$$

Chlorine atoms have the electronic configuration $1s^2\ 2s^2\ 2p^6\ 3s^2\ 3p^5$. They are only one electron short of the stable noble gas configuration of argon $1s^2\ 2s^2\ 2p^6\ 3s^2\ 3p^6$, and when chlorine atoms react, they gain an electron. Thus chlorine is an electronegative element.

$$\underset{\text{chlorine atom}}{\mathrm{Cl}} + \text{electron} \rightarrow \underset{\text{chloride ion}}{\mathrm{Cl}^-}$$

Through gaining an electron, an electrically neutral chlorine atom becomes a chloride ion with a net charge of −1.

When sodium and chlorine react together, the outer electron of the sodium atoms is transferred to the chlorine atoms to produce sodium ions Na^+ and chloride ions Cl^-. Electrostatic attraction between the positive and negative ions holds the ions together in a crystal lattice. The process is energetically favourable as both sorts of atoms attain the stable noble gas configuration, and sodium chloride Na^+Cl^- is formed readily. This may be illustrated diagrammatically in a Lewis diagram showing the outer electrons as dots:

$$\underset{\text{sodium atom}}{\mathrm{Na}\,\cdot} + \underset{\text{chlorine atom}}{\cdot\ddot{\underset{\cdot\cdot}{\mathrm{Cl}}}:} \rightarrow \underset{\text{sodium ion}}{[\mathrm{Na}]^+} + \underset{\text{chloride ion}}{\left[:\ddot{\underset{\cdot\cdot}{\mathrm{Cl}}}:\right]^-}$$

The formation of calcium chloride $CaCl_2$ may be considered in a similar way. Ca atoms have two electrons in their outer shell. Ca is an electropositive element, so each Ca atom loses two electrons to two Cl atoms, forming a calcium ion Ca^{2+} and two chloride ions Cl^-. Showing the outer electrons only, this may be represented as follows:

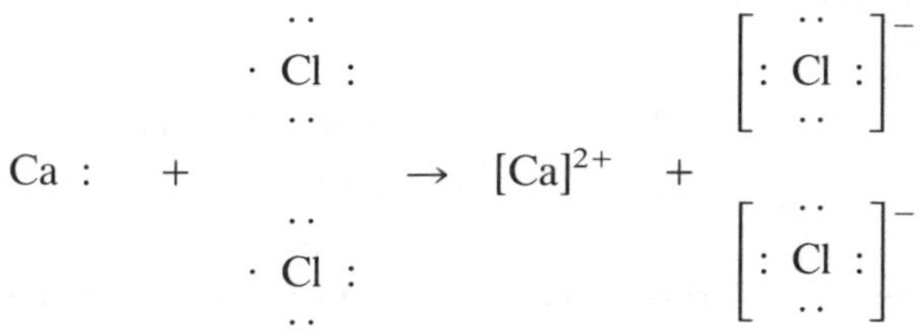

$$\underset{\text{calcium atom}}{\mathrm{Ca}\,:} \quad + \quad \underset{\text{chlorine atoms}}{\begin{matrix} \cdot\,\underset{\cdot\cdot}{\overset{\cdot\cdot}{\mathrm{Cl}}}\,: \\ \\ \cdot\,\underset{\cdot\cdot}{\overset{\cdot\cdot}{\mathrm{Cl}}}\,: \end{matrix}} \quad \rightarrow \quad \underset{\text{calcium ion}}{[\mathrm{Ca}]^{2+}} \quad + \quad \underset{\text{chloride ions}}{\begin{matrix} \left[:\underset{\cdot\cdot}{\overset{\cdot\cdot}{\mathrm{Cl}}}:\right]^- \\ \\ \left[:\underset{\cdot\cdot}{\overset{\cdot\cdot}{\mathrm{Cl}}}:\right]^- \end{matrix}}$$

Covalent bonds

When two electronegative atoms react together, both atoms have a tendency to gain electrons, but neither atom has any tendency to lose electrons. In such cases the atoms share electrons so as to attain a noble gas configuration.

First consider diagrammatically how two chlorine atoms Cl react to form a chlorine molecule Cl_2 (only the outer electrons are shown in the following diagrams):

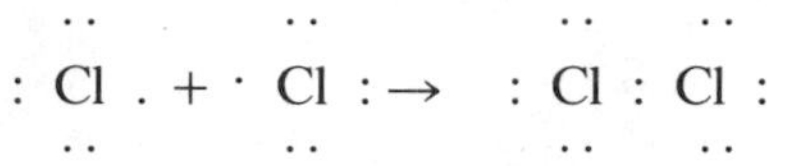

$$\underset{\text{chlorine atoms}}{:\underset{\cdot\cdot}{\overset{\cdot\cdot}{\mathrm{Cl}}}\,. + \cdot\,\underset{\cdot\cdot}{\overset{\cdot\cdot}{\mathrm{Cl}}}\,:} \rightarrow \underset{\text{chlorine molecule}}{:\underset{\cdot\cdot}{\overset{\cdot\cdot}{\mathrm{Cl}}}:\underset{\cdot\cdot}{\overset{\cdot\cdot}{\mathrm{Cl}}}:}$$

Each chlorine atom gives a share of one of its electrons to the other atom. A pair of electrons is shared equally between both atoms, and each atom now has eight electrons in its outer shell (a stable octet) – the noble gas structure of argon. In this electron dot picture (Lewis structure), the shared electron pair is shown as two dots between the atoms Cl : Cl. In the valence bond representation, these dots are replaced by a line, which represents a bond Cl—Cl.

In a similar way a molecule of tetrachloromethane CCl_4 is made up of one carbon and four chlorine atoms:

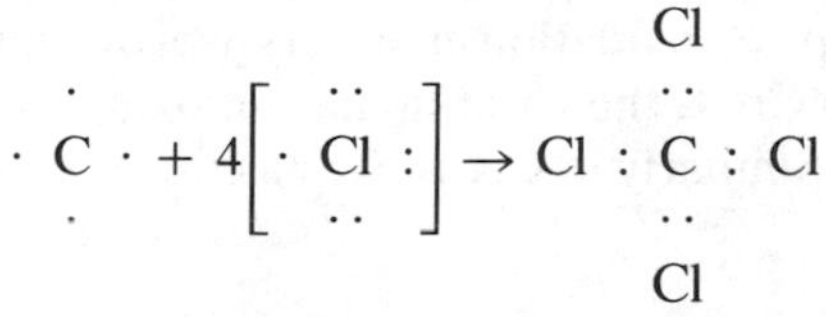

$$\cdot\,\underset{\cdot}{\overset{\cdot}{\mathrm{C}}}\,\cdot + 4\left[\cdot\,\underset{\cdot\cdot}{\overset{\cdot\cdot}{\mathrm{Cl}}}\,:\right] \rightarrow \mathrm{Cl} : \underset{\underset{\mathrm{Cl}}{\cdot\cdot}}{\overset{\overset{\mathrm{Cl}}{\cdot\cdot}}{\mathrm{C}}} : \mathrm{Cl}$$

The carbon atom is four electrons short of the noble gas structure, so it forms four bonds, and the chlorine atoms are one electron short, so they each form one bond. By sharing electrons in this way, both the carbon and all four chlorine atoms attain a noble gas structure. It must be emphasized that although it is possible to build up molecules in this way in order to understand their electronic structures, it does not follow that the atoms will react together directly. In this case, carbon and chlorine do not react directly, and tetrachloromethane is made by indirect reactions.

A molecule of ammonia NH_3 is made up of one nitrogen and three hydrogen atoms:

$$\cdot\overset{\cdot\cdot}{\underset{\cdot}{\mathrm{N}}}\cdot + 3[\mathrm{H}\cdot] \rightarrow \mathrm{H} : \overset{\cdot\cdot}{\underset{\cdot\cdot}{\mathrm{N}}} : \mathrm{H}$$
$$\mathrm{H}$$

The nitrogen atom is three electrons short of a noble gas structure, and the hydrogen atoms are one electron short of a noble gas structure. Nitrogen forms three bonds, and the hydrogen atoms one bond each, so all four atoms attain a stable configuration. One pair of electrons on the N atom is not involved in bond formation, and this is called a *lone pair* of electrons.

Other examples of covalent bonds include water (with two covalent bonds and two lone pairs of electrons), and hydrogen fluoride (one covalent bond and three lone pairs):

$$\mathrm{H} : \overset{\cdot\cdot}{\underset{\cdot\cdot}{\mathrm{O}}} : \qquad \mathrm{H} : \overset{\cdot\cdot}{\underset{\cdot\cdot}{\mathrm{F}}} :$$
$$\mathrm{H}$$

Oxidation numbers

The oxidation number of an element in a covalent compound is calculated by assigning shared electrons to the more electronegative element, and then counting the theoretical charge left on each atom. (Electronegativity is described in Chapter 6.) An alternative approach is to break up (theoretically) the molecule by removing all the atoms as ions, and counting the charge left on the central atom. It must be emphasized that molecules are not really broken, nor electrons really moved. For example, in H_2O, removal of two H^+ leaves a charge of -2 on the oxygen atom, so the oxidation state of O in H_2O is $(-II)$. Similarly in H_2S the oxidation state of S is $(-II)$; in F_2O the oxidation state of O is $(+II)$; in SF_4 the oxidation state of S is $(+IV)$; whilst in SF_6 the oxidation state of S is $(+VI)$. The concept of oxidation numbers works equally well with ionic compounds, and in $CrCl_3$ the Cr atom has an oxidation state of $(+III)$ and it forms Cr^{3+} ions. Similarly in $CrCl_2$, Cr has the oxidation state $(+II)$, and exists as Cr^{2+} ions.

Coordinate bonds

A covalent bond results from the sharing of a pair of electrons between two atoms, where each atom contributes one electron to the bond. It is also possible to have an electron pair bond where both electrons originate from one atom and none from the other. Such bonds are called coordinate bonds or dative bonds. Since, in coordinate bonds, two electrons are shared by two atoms, they differ from normal covalent bonds only in the way they are formed, and once formed they are identical to normal covalent bonds.

Even though the ammonia molecule has a stable electron configuration, it can react with a hydrogen ion H^+ by donating a share in the lone pair of electrons, forming the ammonium ion NH_4^+:

$$\begin{array}{c} \mathrm{H} \\ \cdot\cdot \\ \mathrm{H : N :} \\ \cdot\cdot \\ \mathrm{H} \end{array} + [\mathrm{H}]^+ \rightarrow \left[\begin{array}{c} \mathrm{H} \\ \cdot\cdot \\ \mathrm{H : N : H} \\ \cdot\cdot \\ \mathrm{H} \end{array}\right]^+ \text{ or } \left[\begin{array}{c} \mathrm{H} \\ | \\ \mathrm{H—N{\rightarrow}H} \\ | \\ \mathrm{H} \end{array}\right]^+$$

Covalent bonds are usually shown as straight lines joining the two atoms, and coordinate bonds as arrows indicating which atom is donating the electrons. Similarly ammonia may donate its lone pair to boron trifluoride, and by this means the boron atom attains a share in eight electrons:

$$\begin{array}{c} \mathrm{H} \\ \cdot\cdot \\ \mathrm{H : N :} \\ \cdot\cdot \\ \mathrm{H} \end{array} + \begin{array}{c} \mathrm{F} \\ \cdot\cdot \\ \mathrm{B : F} \\ \cdot\cdot \\ \mathrm{F} \end{array} \rightarrow \begin{array}{c} \mathrm{H\quad F} \\ |\quad\ | \\ \mathrm{H—N{\rightarrow}B—F} \\ |\quad\ | \\ \mathrm{H\quad F} \end{array}$$

In a similar way, a molecule of BF_3 can form a coordinate bond by accepting a share in a lone pair from a F^- ion.

$$\left[\begin{array}{c} \cdot\cdot \\ \mathrm{: F :} \\ \cdot\cdot \end{array}\right]^- + \begin{array}{c} \mathrm{F} \\ \cdot\cdot \\ \mathrm{B : F} \\ \cdot\cdot \\ \mathrm{F} \end{array} \rightarrow \left[\begin{array}{c} \mathrm{F} \\ | \\ \mathrm{F{\rightarrow}B—F} \\ | \\ \mathrm{F} \end{array}\right]^-$$

There are many other examples, including:

$$PCl_5 + Cl^- \rightarrow [PCl_6]^-$$
$$SbF_5 + F^- \rightarrow [SbF_6]^-$$

Double and triple bonds

Sometimes more than two electrons are shared between a pair of atoms. If four electrons are shared, then there are two bonds, and this arrangement is called a double bond. If six electrons are shared then there are three bonds, and this is called a triple bond:

```
H    H          H     H
 ..  ..          \   /
  C :: C          C=C        Ethene molecule
 ..  ..          /   \       (double bond)
H    H          H     H

H : C ::: C : H    H—C≡C—H   Ethyne molecule
                             (triple bond)
```

Metallic bonds and metallic structures

Metals are made up of positive ions packed together, usually in one of the three following arrangements:

1. Cubic close-packed (also called face-centred cubic).
2. Hexagonal close-packed.
3. Body-centred cubic.

Negatively charged electrons hold the ions together. The number of positive and negative charges are exactly balanced, as the electrons originated from the neutral metal atoms. The outstanding feature of metals is their extremely high electrical conductivity and thermal conductivity, both of which are because of the mobility of these electrons through the lattice.

The arrangements of atoms in the three common metallic structures are shown in Figure 2.2. Two of these arrangements (cubic close-packed and hexagonal close-packed) are based on the closest packing of spheres. The metal ions are assumed to be spherical, and are packed together to fill the space most effectively, as shown in Figure 2.3a. Each sphere touches six other spheres within this one layer.

A second layer of spheres is arranged on top of the first layer, the protruding parts of the second layer fitting into the hollows in the first layer as shown in Figure 2.4a. A sphere in the first layer touches three spheres in the layer above it, and similarly touches three spheres in the layer below it, plus six spheres in its own layer, making a total of 12. The coordination number, or number of atoms or ions in contact with a given atom, is therefore 12 for a close-packed arrangement. With a close-packed arrangement, the spheres occupy 74% of the total space.

When adding a third layer of spheres, two different arrangements are possible, each preserving the close-packed arrangement.

If the first sphere of the third layer is placed in the depression X shown in Figure 2.4a, then this sphere is exactly above a sphere in the first layer. It follows that every sphere in the third layer is exactly above a sphere in the first layer as shown in Figure 2.2a. If the first layer is represented by A, and the second layer by B, the repeating pattern of close-packed sheets is ABABAB. . . . This structure has hexagonal symmetry, and it is therefore said to be hexagonal close-packed.

Alternatively, the first sphere of the third layer may be placed in a

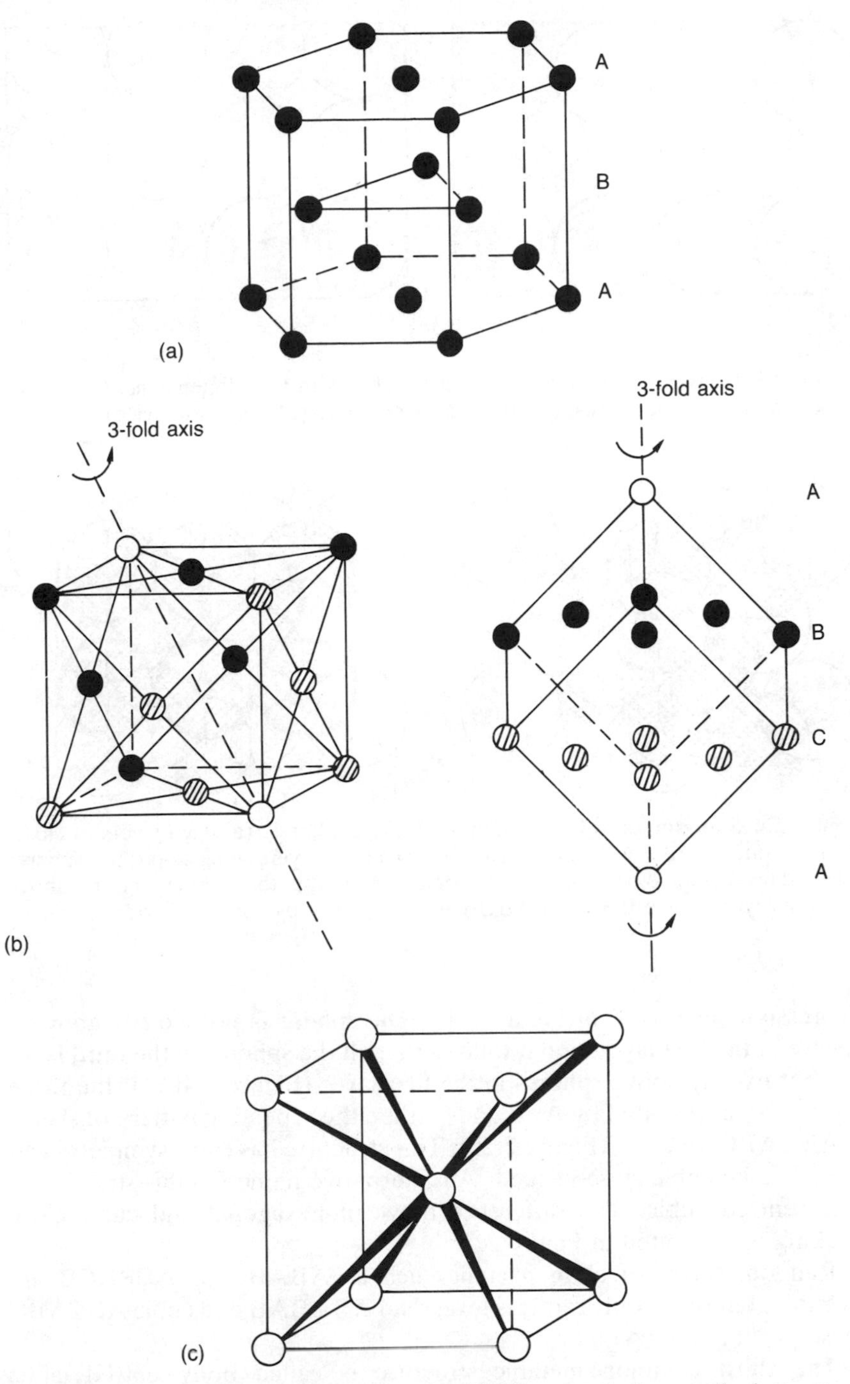

Figure 2.2 The three metallic structures. (a) Hexagonal close-packed structure showing the repeat pattern of layers ABABAB...and the 12 neighbours surrounding each sphere. (b) Cubic close-packed structure (coordination number is also 12) showing repeat pattern of layers ABCABC. (c) Body-centred cubic structure showing the 8 neighbours surrounding each sphere.

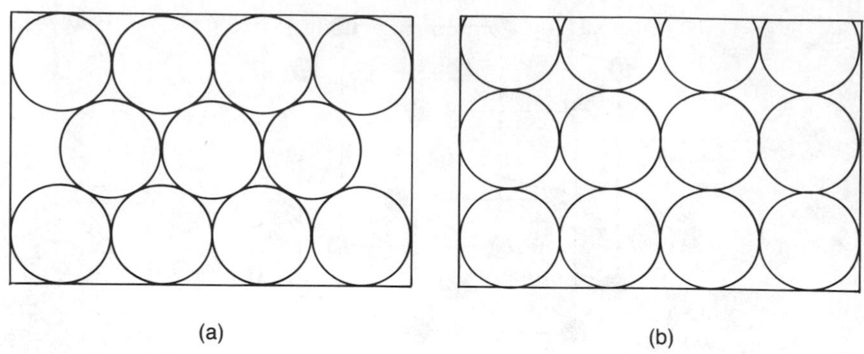

Figure 2.3 Possible ways of packing equal spheres in two dimensions. (a) Close-packed (fills 74% of space). (b) Body-centred cubic (fills 68% of space).

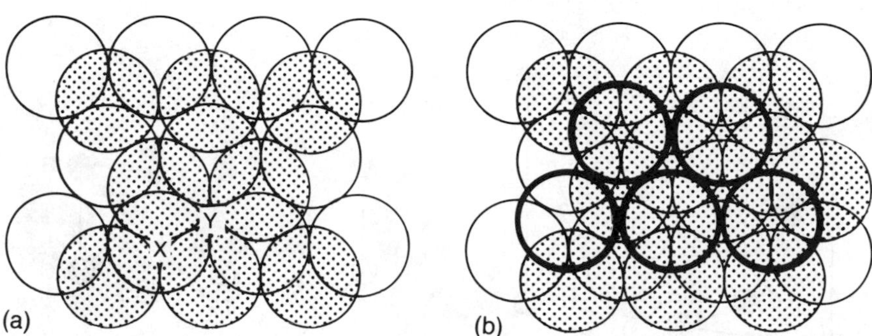

Figure 2.4 Superimposed layers of close-packed spheres. (a) Two layers of close-packed spheres (second layer is shaded). (b) Three layers of close-packed spheres (second layer shaded, third layer bold circles). Note that the third layer is not above the first layer, hence this is an ABCABC. . . (cubic close-packed) arrangement.

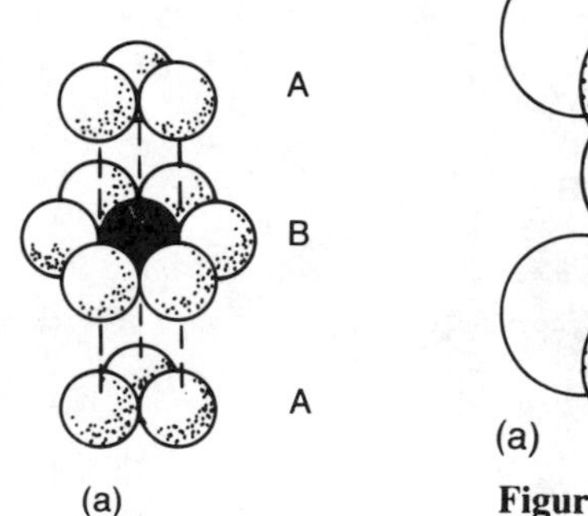

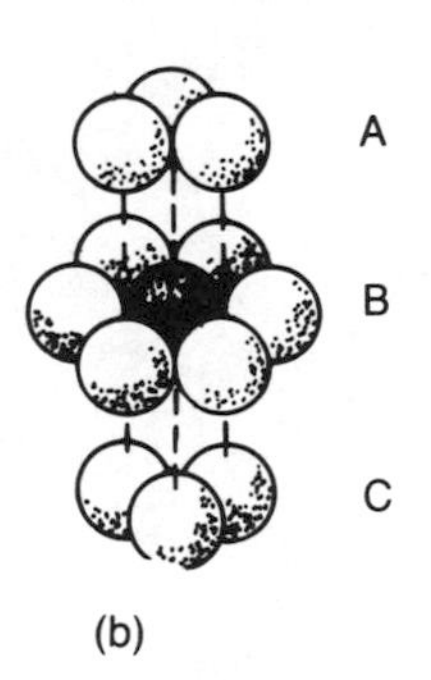

Figure 2.5 Arrangement of 12 nearest neighbours in hexagonal and cubic close-packed arrangements. (Note that the top and middle layers are the same, but in the cubic close-packed structure the bottom layer is rotated 60° relative to the hexagonal close-packed. (a) Hexagonal close-packed. (b) Cubic close-packed.

depression such as Y in Figure 2.4a. The sphere is not exactly above a sphere in the first layer, and it follows that all the spheres in the third layer are not exactly above spheres in the first layer (Figure 2.4b). If the three layers are represented by A, B and C, then the repeating pattern of sheets is ABCABCABC. . . (Figure 2.2b). This structure has cubic symmetry and is said to be cubic close-packed. An alternative name for this structure is face-centred cubic. The difference between hexagonal and cubic close packing is illustrated in Figure 2.5.

Random forms of close packing such as ABABC or ACBACB are possible, but occur only rarely. Hexagonal ABABAB and cubic ABCABC close packing are common.

The third common metallic structure is called body-centred cubic (Figure 2.2c). The spheres are packed in sheets as shown in Figure 2.3b. The second layer occupies the hollows in this first sheet. The third layer occupies hollows in the second layer, and the third layer is immediately above the first layer. This form of packing is less efficient at filling the space than closest packing (compare Figures 2.3a and b). In a body-centred cubic structure the spheres occupy 68% of the total space and have a coordination number of 8, compared with close-packed structures where 74% of the space is occupied and the coordination number is 12. Metallic structures always have high coordination numbers.

The theories of bonding in metals and alloys are described in Chapter 5.

Metallic bonding is found not only in metals and alloys, but also in several other types of compound:

1. Interstitial borides, carbides, nitrides and hydrides formed by the transition elements (and by some of the lanthanides too). Some low oxidation states of transition metal halides also belong to this group, where the compounds show electrical conductivity, and are thought to contain free electrons in conduction bands.
2. Metal cluster compounds of the transition metals, and cluster compounds of boron, where the covalent bonding is delocalized over several atoms, and is equivalent to a restricted form of metallic bonding.
3. A group of compounds including the metal carbonyls which contain a metal–metal bond. The cluster compounds, and the compounds with metal–metal bonds, may help to explain the role of metals as catalysts.

Melting points

Ionic compounds are typically solids and usually have high melting and boiling points. In contrast covalent compounds are typically gases, liquids or low melting solids. These differences occur because of differences in bonding and structure.

Ionic compounds are made up of positive and negative ions arranged in a regular way in a lattice. The attraction between ions is electrostatic, and is non-directional, extending equally in all directions. Melting the compound involves breaking the lattice. This requires considerable energy, and so the melting point and boiling point are usually high, and the compounds are very hard.

Compounds with covalent bonds are usually made up of discrete molecules. The bonds are directional, and strong covalent bonding forces hold the atoms together to make a molecule. In the solid, molecules are held together by weak van der Waals forces. To melt or boil the compound we only need supply the small amount of energy needed to break the van der Waals forces. Hence covalently bonded compounds are often gases, liquids or soft solids with low melting points.

In a few cases such as diamond, or silica SiO_2, the structures are covalent giant lattices instead of discrete molecules. In these cases there is a three-dimensional lattice, with strong covalent bonds in all directions. It requires a large amount of energy to break this lattice, and so diamond, silica and other materials with giant three-dimensional lattices are very hard and have high melting points.

Conductivity

Ionic compounds conduct electricity *when the compound is melted, or in solution*. Conduction is achieved by the ions migrating towards the electrodes under the influence of an electric potential. If an electric current is passed through a solution of sodium chloride, Na^+ ions are attracted to the negatively charged electrode (cathode), where they gain an electron and form sodium atoms. The Cl^- ions are attracted to the positive electrode (anode), where they lose an electron and become chlorine atoms. This process is called electrolysis. The changes amount to the transfer of electrons from cathode to anode, but *conduction occurs by an ionic mechanism involving the migration of both positive and negative ions in opposite directions*.

In the solid state, the ions are trapped in fixed places in the crystal lattice, and as they cannot migrate, they cannot conduct electricity in this way. It is, however, wrong to say that ionic solids do not conduct electricity without qualifying the statement. The crystal may conduct electricity *to a very small extent* by semiconduction if the crystal contains some defects. Suppose that a lattice site is unoccupied, and there is a 'hole' where an ion is missing. An ion may migrate from its lattice site to the vacant site, and in so doing it makes a 'hole' somewhere else. The new 'hole' is filled by another ion, and so on, so eventually the hole migrates across the crystal, and a charge is carried in the other direction. Plainly the amount of current carried by this mechanism is extremely small, but semiconductors are of great importance in modern electronic devices.

Metals conduct electricity better than any other material, but the mechanism is by the movement of electrons instead of ions.

Covalent compounds contain neither ions (as in ionic compounds) nor mobile electrons (as in metals), so they are unable to conduct electricity in either the solid, liquid or gaseous state. Covalent compounds are therefore insulators.

Solubility

If they dissolve at all, ionic compounds are usually soluble in polar solvents. These are solvents of high dielectric constant such as water, or the mineral acids. Covalent compounds are not normally soluble in these solvents but if they dissolve at all they are soluble in non-polar (organic) solvents of low dielectric constant, such as benzene and tetrachloromethane. The general rule is sometimes stated that 'like dissolves like', and so ionic compounds usually dissolve in ionic solvents, and covalent compounds usually dissolve in covalent solvents.

Speed of reactions

Ionic compounds usually react very rapidly, whilst covalent compounds usually react slowly. For ionic reactions to occur, the reacting species are ions, and as these already exist, they have only to collide with the other type of ion. For example, when testing a solution for chloride ions (by adding silver nitrate solution), precipitation of AgCl is very rapid.

$$Ag^+ + Cl^- \rightarrow AgCl$$

Reactions of covalent compounds usually involve breaking a bond and then substituting or adding another group. Energy is required to break the bond. This is called the activation energy, and it often makes reactions slow. Collisions between the reactant molecules will only cause reaction if they have enough energy. For example, reduction of preparative amounts of nitrobenzene to aniline takes several hours. Similarly the reaction of H_2 and Cl_2 is typically slow except in direct sunlight when the mixture may explode!

$$C_6H_5NO_2 + 6[H] \rightarrow C_6H_5NH_2 + 2H_2O$$

$$\left\{\begin{aligned} H_2 &\rightarrow 2H \\ Cl_2 &\rightarrow 2Cl \\ H + Cl &\rightarrow HCl \end{aligned}\right.$$

It is important to realize that bonds are not necessarily 100% covalent or 100% ionic, and that bonds of intermediate character exist. If a molecule is made up of two identical atoms, both atoms have the same electronegativity, and so have an equal tendency to gain electrons. (See Chapter 6.) In such a molecule the electron pair forming the bond is equally shared by both atoms. This constitutes a 100% covalent bond, and is sometimes called a *non-polar covalent bond*.

More commonly molecules are formed between different types of atoms, and the electronegativity of the two atoms differs. Consider for example the molecules ClF and HF. Fluorine is the most electronegative atom, and it attracts electrons more strongly than any other element when covalently bonded. The bonding electrons spend more time round the F than round the other atom, so the F atom has a very small negative charge δ− and the atom (Cl or H) has a small positive charge δ+.

$$\overset{\delta+}{Cl}\text{——}\overset{\delta-}{F} \qquad \overset{\delta+}{H}\text{——}\overset{\delta-}{F}$$

Though these bonds are largely covalent, they possess a small amount of ionic character, and are sometimes called *polar covalent bonds*. In such molecules, a positive charge, and an equal negative charge, are separated by a distance. This produces a permanent dipole moment in the molecule.

The dipole moment measures the tendency of the molecule to turn and line up its charges when placed in an electric field. Polar molecules have a high dielectric constant, and non-polar molecules have a low dielectric constant. The dielectric constant is the ratio of the capacitance of a condenser with the material between the plates, to the capacitance of the same condenser with a vacuum between them. By measuring the capacitance with the substance between the plates and then with a vacuum, we can obtain the dielectric constant. Its size indicates whether the material is polar or non-polar.

Ionic, covalent and metallic bonds are considered in more detail in the following chapters.

The ionic bond

3

STRUCTURES OF IONIC SOLIDS

Ionic compounds include salts, oxides, hydroxides, sulphides, and the majority of inorganic compounds. Ionic solids are held together by the electrostatic attraction between the positive and negative ions. Plainly there will be repulsion if ions of the same charge are adjacent, and attraction will occur when positive ions are surrounded by negative ions, and vice versa. The attractive force will be a maximum when each ion is surrounded by the greatest possible number of oppositely charged ions. The number of ions surrounding any particular ion is called the coordination number. Positive and negative ions will both have the same coordination number when there are equal numbers of both types of ions, as in NaCl, but the coordination numbers for positive and negative ions are different when there are different numbers of the ions, as in $CaCl_2$.

RADIUS RATIO RULES

The structures of many ionic solids can be accounted for by considering the relative sizes of the positive and negative ions, and their relative numbers. Simple geometric calculations allow us to work out how many ions of a given size can be in contact with a smaller ion. Thus we can predict the coordination number from the relative sizes of the ions.

When the coordination number is three in an ionic compound AX, three X^- ions are in contact with one A^+ ion (Figure 3.1a). A limiting case arises (Figure 3.1b) when the X^- ions are also in contact with one another. By simple geometry this gives the ratio (radius A^+/radius X^-) = 0.155. This is the lower limit for a coordination number of 3. If the radius ratio is less than 0.155 then the positive ion is not in contact with the negative ions, and it 'rattles' in the hole, and the structure is unstable (Figure 3.1c). If the radius ratio is greater than 0.155 then it is possible to fit three X^- ions round each A^+ ion. As the difference in the size of the two ions increases, the radius ratio also increases, and at some point (when the ratio exceeds 0.225), it becomes possible to fit four ions round one, and so on for six ions round one, and eight ions round one. Coordination numbers of 3, 4, 6 and

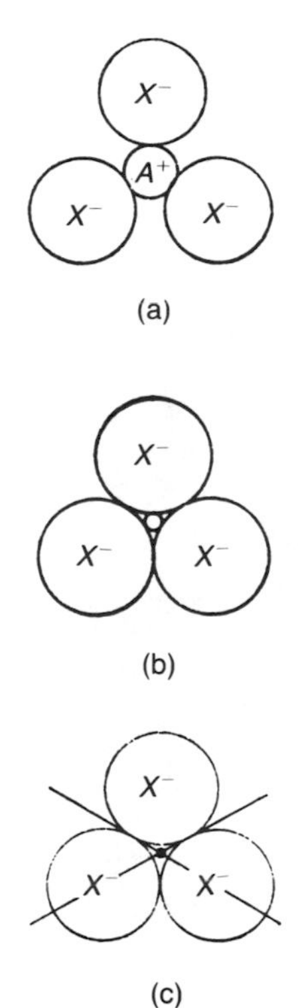

Figure 3.1 Sizes of ions for coordination number 3.

Table 3.1 Limiting radius ratios and structures

Limiting radius ratio r^+/r^-	Coordination number	Shape
<0.155	2	Linear
0.155 → 0.225	3	Planar triangle
0.225 → 0.414	4	Tetrahedral
0.414 → 0.732	4	Square planar
0.414 → 0.732	6	Octahedral
0.732 → 0.999	8	Body-centred cubic

8 are common, and the appropriate limiting radius ratios can be worked out by simple geometry, and are shown in Table 3.1.

If the ionic radii are known, the radius ratio can be calculated and hence the coordination number and shape may be predicted. This simple concept predicts the correct structure in many cases.

CALCULATION OF SOME LIMITING RADIUS RATIO VALUES

This section may be skipped except by those interested in the origin of the limiting radius ratio values.

Coordination number 3 (planar triangle)

Figure 3.2a shows the smaller positive ion of radius r^+ in contact with three larger negative ions of radius r^-. Plainly AB = BC = AC = $2r^-$, BE = r^-, BD = $r^+ + r^-$. Further, the angle A–B–C is 60°, and the angle D–B–E is 30°. By trigonometry

$$\cos 30° = BE/BD$$

$$BD = BE/\cos 30°$$

$$r^+ + r^- = r^-/\cos 30° = r^-/0.866 = r^- \times 1.155$$

$$r^+ = (1.155r^-) - r^- = 0.155r^-$$

hence $$r^+/r^- = 0.155$$

Coordination number 4 (tetrahedral)

Figure 3.2b shows a tetrahedral arrangement inscribed in a cube with sides of length d. The diagonal on the bottom face XX is $2 \times r^-$. By Pythagoras, on the triangle VXY,

$$XY^2 = VX^2 + VY^2$$

$$XY^2 = d^2 + d^2 = 2d^2$$

hence $$XY = d\sqrt{2}$$

and $$2r^- = d\sqrt{2}$$

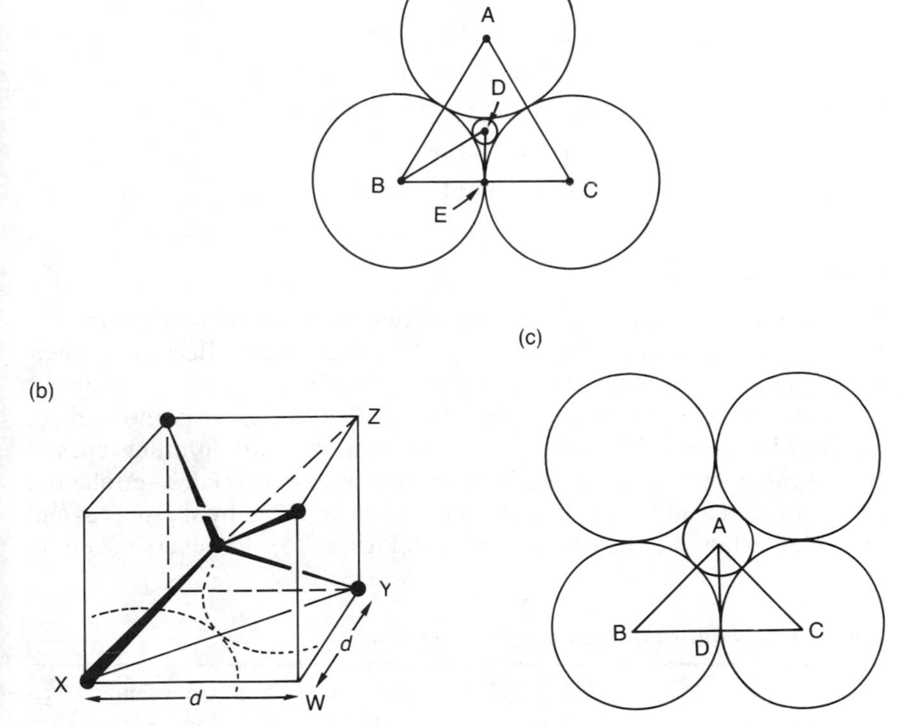

Figure 3.2 Limiting radius ratios for coordination numbers 3, 4 and 6. (a) Cross-section through a planar triangle site; (b) tetrahedron inscribed in a cube; and (c) cross-section through an octahedral site.

In the triangle XZY, by Pythagoras

$$XZ^2 = XY^2 + YZ^2$$

$$= (d\sqrt{2})^2 + d^2 = 3d^2$$

so $$XZ = d\sqrt{3}$$

However $$XZ = 2r^+ + 2r^-$$

so $$2r^+ + 2r^- = d\sqrt{3}$$

and $$2r^+ = d\sqrt{3} - 2r^-$$

$$= d\sqrt{3} - d\sqrt{2}$$

$$r^+/r^- = \tfrac{1}{2}(d\sqrt{3} - d\sqrt{2})/\tfrac{1}{2}d\sqrt{2} = \sqrt{(3/2)} - 1 = 1.225 - 1 = 0.225$$

Coordination number 6 (octahedral)

A cross-section through an octahedral site is shown in Figure 3.2c, and the smaller positive ion (of radius r^+) touches the six larger negative ions (of radius r^-). (Note that only four negative ions are shown in this section, and

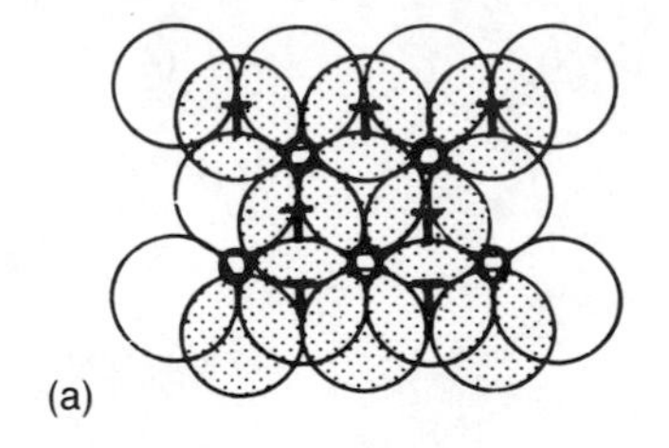

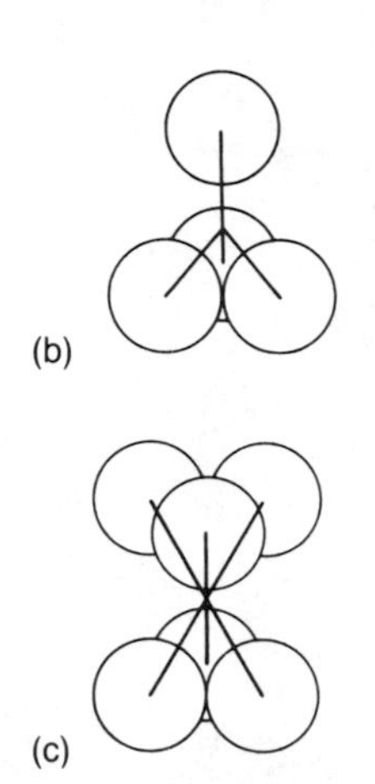

Figure 3.3 Tetrahedral and octahedral holes: (a) tetrahedral and octahedral sites in a close-packed lattice; (b) tetrahedral site; and (c) octahedral site.

one is above and another one below the plane of the paper.) It is obvious that $AB = r^+ + r^-$, and that $BD = AD = r^-$. By Pythagoras

$$AB^2 = AD^2 + BD^2$$

i.e.

$$(r^+ + r^-)^2 = (r^-)^2 + (r^-)^2 = 2(r^-)^2$$

hence

$$r^+ + r^- = \surd[2(r^-)^2] = 1.414r^-$$

$$r^+ = 0.414r^-$$

hence

$$r^+/r^- = 0.414$$

CLOSE PACKING

Many common crystal structures are related to, and may be described in terms of, hexagonal or cubic close-packed arrangements. Because of their shape, spheres cannot fill space completely. In a close-packed arrangement of spheres, 74% of the space is filled. Thus 26% of the space is unoccupied, and may be regarded as holes in the crystal lattice. Two different types of hole occur. Some are bounded by four spheres and are called tetrahedral holes (marked T in Figure 3.3), and others are bounded by six spheres and are called octahedral holes (marked O in Figure 3.3). For every sphere in

Table 3.2 Some structures based on close packing

Formula		Type of cp	Tetrahedral	Octahedral	Coordination No. A	:	X
AX	NaCl	ccp	none	all	6	:	6
	NiAs	hcp	none	all	6	:	6
	ZnS zinc blende	ccp	$\frac{1}{2}$	none	4	:	4
	ZnS wurtzite	hcp	$\frac{1}{2}$	none	4	:	4
AX_2	F_2Ca* fluorite	ccp*	all	none	8	:	4
	CdI_2	hcp	none	$\frac{1}{2}$	6	:	3
	$CdCl_2$	ccp	none	$\frac{1}{2}$	6	:	3
	β-$ZnCl_2$	hcp	$\frac{1}{4}$	none	4	:	2
	HgI_2	ccp	$\frac{1}{4}$	none	4	:	2
MX_3	BiI_3	hcp	none	$\frac{1}{3}$	6	:	2
	$CrCl_3$	ccp	none	$\frac{1}{3}$	6	:	2
MX_4	SnI_4	hcp	$\frac{1}{8}$	none	4	:	1
MX_6	α-WCl_6 and UCl_6	ccp	none	$\frac{1}{6}$	6	:	1
M_2X_3	α-Al_2O_3 corundum	hcp	none	$\frac{2}{3}$	6	:	4

* The metal ions adopt a face-centred cubic arrangement, which is exactly like cubic close packing except that the ions do not touch. (Note it is the M^+ ions that are almost close packed, not the negative ions as with the other examples.)

the close-packed arrangement there is one octahedral hole and two tetrahedral holes. The octahedral holes are larger than the tetrahedral holes.

An ionic structure is composed of oppositely charged ions. If the larger ions are close packed, then the smaller ions may occupy either the octahedral holes or the tetrahedral holes depending on their size. Normally the type of hole occupied can be determined from the radius ratio. An ion occupying a tetrahedral hole has a coordination number of 4, whilst one occupying an octahedral hole has a coordination number of 6. In some compounds the relative sizes of the ions are such that the smaller ions are too large to fit in the holes, and they force the larger ions out of contact with each other so that they are no longer close packed. Despite this, the relative positions of the ions remain unchanged, and it is convenient to retain the description in terms of close packing.

CLASSIFICATION OF IONIC STRUCTURES

It is convenient to divide ionic compounds into groups AX, AX_2, AX_3 depending on the relative numbers of positive and negative ions.

IONIC COMPOUNDS OF THE TYPE AX (ZnS, NaCl, CsCl)

Three structural arrangements commonly found are the zinc sulphide, sodium chloride and caesium chloride structures.

Structures of zinc sulphide

In zinc sulphide, ZnS, the radius ratio of 0.40 suggests a tetrahedral arrangement. Each Zn^{2+} ion is tetrahedrally surrounded by four S^{2-} ions and each S^{2-} ion is tetrahedrally surrounded by four Zn^{2+} ions. The coordination number of both ions is 4, so this is called a 4:4 arrangement. Two different forms of zinc sulphide exist, zinc blende and wurtzite (Figure 3.4). Both are 4:4 structures.

These two structures may be considered as close-packed arrangements of S^{2-} ions. Zinc blende is related to a cubic close-packed structure whilst wurtzite is related to a hexagonal close-packed structure. In both structures the Zn^{2+} ions occupy tetrahedral holes in the lattice.Since there are twice as many tetrahedral holes as there are S^{2-} ions, it follows that to obtain a formula ZnS only half of the tetrahedral holes are occupied by Zn^{2+} ions (that is every alternate tetrahedral site is unoccupied).

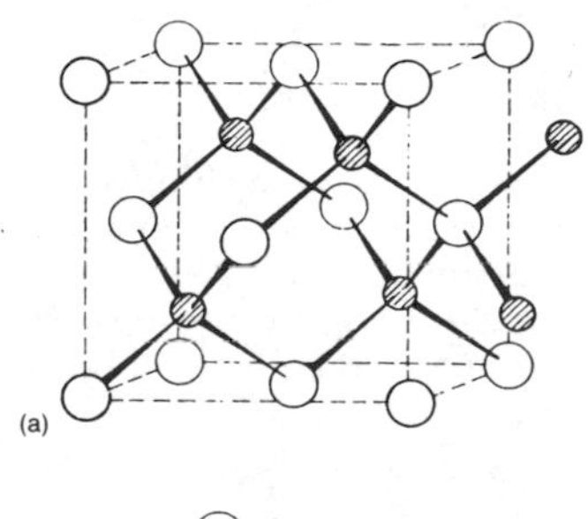

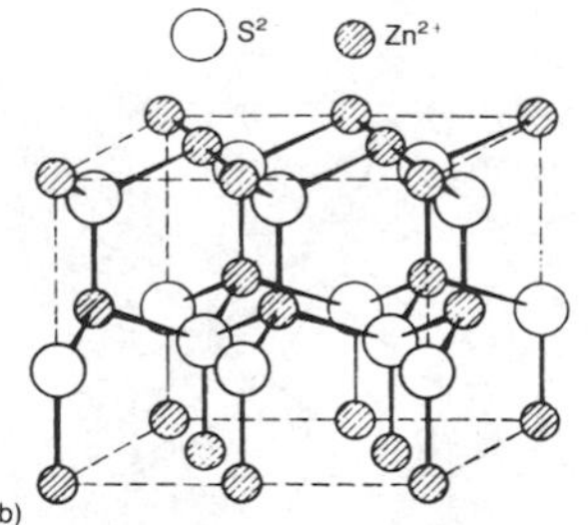

Figure 3.4 Structures of ZnS: (a) zinc blende and (b) wurtzite. (Reproduced with permission from Wells, A.F., *Structural Inorganic Chemistry*, 5th ed., Oxford University Press, Oxford, 1984.)

Sodium chloride structure

For sodium chloride, NaCl, the radius ratio is 0.52 and this suggests an octahedral arrangement. Each Na^+ ion is surrounded by six Cl^- ions at the corners of a regular octahedron and similarly each Cl^- ion is surrounded by six Na^+ ions (Figure 3.5). The coordination is thus 6:6. This structure

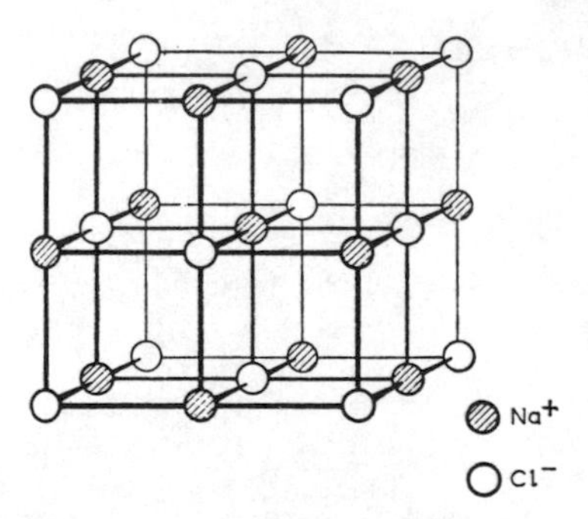

Figure 3.5 Rock salt (NaCl) structure. (Reproduced by permission of Wells, A.F., *Structural Inorganic Chemistry*, 5th ed., Oxford University Press, Oxford, 1984.)

may be regarded as a cubic close-packed array of Cl^- ions, with Na^+ ions occupying all the octahedral holes.

Caesium chloride structure

In caesium chloride, CsCl, the radius ratio is 0.93. This indicates a body-centred cubic type of arrangement, where each Cs^+ ion is surrounded by eight Cl^- ions, and vice versa (Figure 3.6). The coordination is thus 8 : 8. Note that this structure is not close packed, and is not strictly body-centred cubic.

In a body-centred cubic arrangement, the atom at the centre of the cube is identical to those at the corners. This structure is found in metals, but in CsCl if the ions at the corners are Cl^- then there will be a Cs^+ ion at the body-centred position, so it is not strictly body-centred cubic. The caesium chloride structure should be described as a *body-centred cubic type of arrangement* and not *body-centred cubic*.

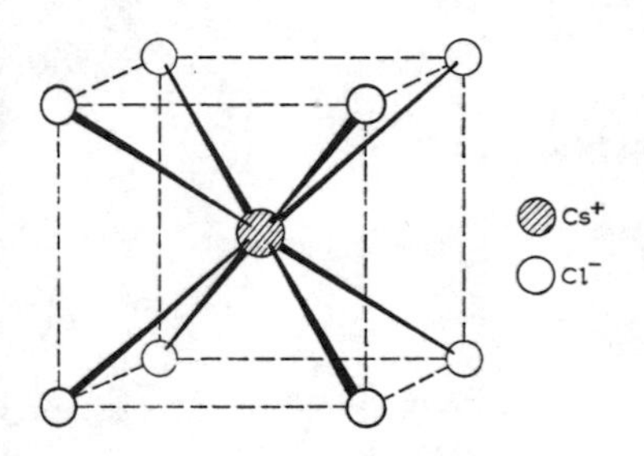

Figure 3.6 Caesium chloride (CsCl) structure. (Reproduced by permission of Wells, A.F., *Structural Inorganic Chemistry*, 5th ed., Oxford University Press, 1984.)

IONIC COMPOUNDS OF THE TYPE AX_2 (CaF_2, TiO_2, SiO_2)

The two most common structures are fluorite, CaF_2 (Figure 3.7), and rutile, TiO_2 (Figure 3.8), and many difluorides and dioxides have one of these structures. Another fairly common structure is one form of SiO_2 called β-cristobalite (Figure 3.9). These are true ionic structures. Layer structures are formed instead if the bonding becomes appreciably covalent.

Calcium fluoride (fluorite) structure

In fluorite, each Ca^{2+} ion is surrounded by eight F^- ions, giving a body-centred cubic arrangement of F^- round Ca^{2+}. Since there are twice as many F^- ions as Ca^{2+} ions, the coordination number of both ions is different, and four Ca^{2+} ions are tetrahedrally arranged around each F^- ion. The coordination numbers are therefore 8 and 4, so this is called an 8 : 4 arrangement. The fluorite structure is found when the radius ratio is 0.73 or above.

An alternative description of the structure is that the Ca^{2+} ions form a face-centred cubic arrangement. The Ca^{2+} ions are too small to touch each other, so the structure is not close packed. However, the structure is related to a close-packed arrangement, since the Ca^{2+} occupy the same relative positions as for a cubic close-packed structure, and the F^- ions occupy all the tetrahedral holes.

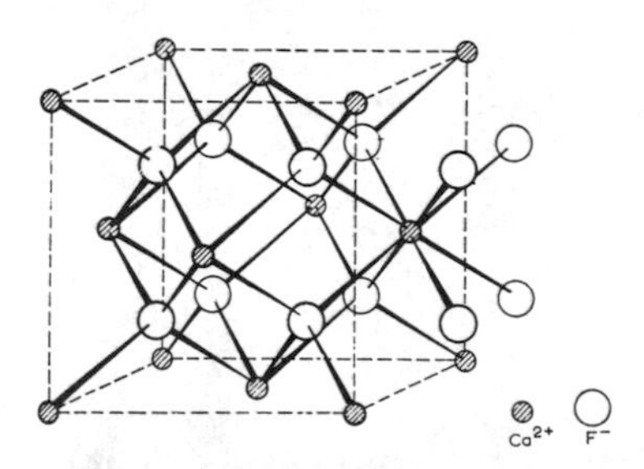

Figure 3.7 Fluorite (CaF_2) structure. (Reproduced by permission of Wells, A.F., *Structural Inorganic Chemistry*, 5th ed., Oxford University Press, Oxford, 1984.)

Rutile structure

TiO_2 exists in three forms called anatase, brookite and rutile. The rutile structure is found in many crystals where the radius ratio is between 0.41 and 0.73. This suggests a coordination number of 6 for one ion, and from the formula it follows that the coordination number of the other ion must be 3. This is a 6 : 3 structure. Each Ti^{4+} is octahedrally surrounded by six O^{2-} ions and each O^{2-} ion has three Ti^{4+} ions round it in a plane triangular arrangement.

The rutile structure is not close packed. The unit cell, i.e. the repeating unit of this structure, is not a cube, since one of the axes is 30% shorter than the other two. It is convenient to describe it as a considerably distorted cube (though the distortion is rather large). The structure may then be described as a considerably distorted body-centred cubic lattice of Ti^{4+} ions. Each Ti^{4+} ion is surrounded octahedrally by six O^{2-} ions, and the O^{2-} are in positions of threefold coordination, that is each O^{2-} is surrounded by three Ti^{4+} ions at the corners of an equilateral triangle. Three-coordination is not common in solids. There are no examples of three-coordination in compounds of the type AX, but there is another example in the compounds of type AX_2, that is CdI_2, though in this case the shape is not an equilateral triangle. The structure of $CaCl_2$ is also a 6 : 3 structure, and is similar to CdI_2. These are described later.

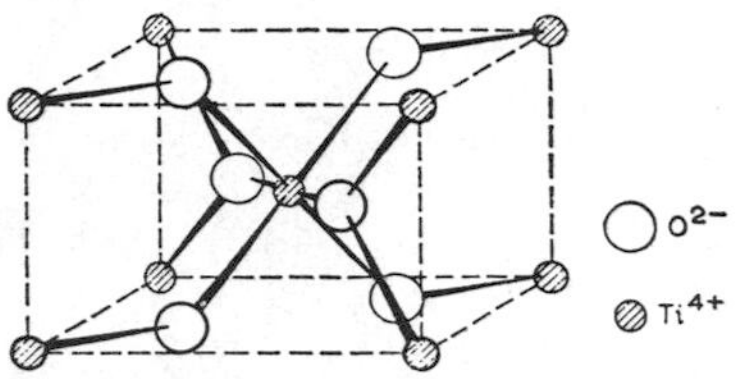

Figure 3.8 Rutile (TiO_2) structure.

There are only a few cases where the radius ratio is below 0.41. Examples include silica SiO_2 and beryllium fluoride BeF_2. These have coordination numbers of 4 and 2, but radius ratio predictions are uncertain since they are appreciably covalent.

β-cristobalite (silica) structure

Silica SiO_2 exists in six different crystalline forms as quartz, cristobalite and tridymite, each with an α and β form. β-cristobalite is related to zinc blende, with two interpenetrating close-packed lattices, one lattice arising from Si occupying the S^{2-} positions, and the other lattice from Si occupying the Zn^{2+} positions (i.e. the tetrahedral holes in the first lattice). The oxygen atoms lie midway between the Si atoms, but are shifted slightly off the line joining the Si atoms, so the bond angle Si—O—Si is not 180°. The radius ratio predicts a coordination number of 4, and this is a 4 : 2 structure.

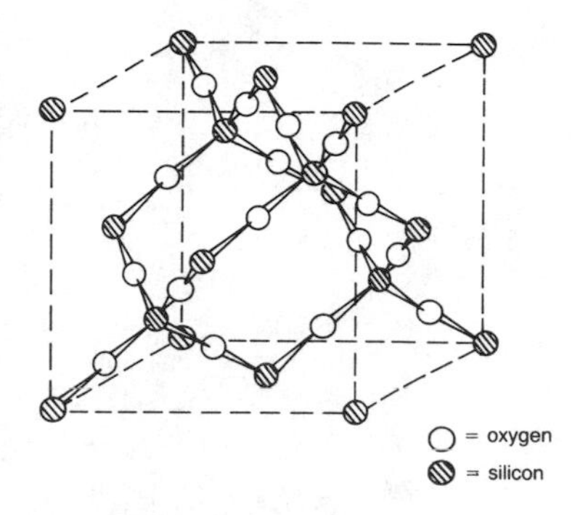

Figure 3.9 β-cristobalite structure.

LAYER STRUCTURES (CdI_2, $CdCl_2$, [NiAs])

Cadmium iodide structure

Many AX_2 compounds are not sufficiently ionic to form the perfectly regular ionic structures described. Many chlorides, bromides, iodides and sulphides crystallize into structures which are very different from those described. Cadmium fluoride CdF_2 forms an ionic lattice with the CaF_2 structure, but in marked contrast cadmium iodide CdI_2 is much less ionic, and does not form the fluorite structure. The radius ratio for CdI_2 is 0.45, and this indicates a coordination number of 6 for cadmium. The structure is made up of electrically neutral layers of Cd^{2+} ions with layers of I^- ions on either side – rather like a sandwich where a layer of Cd^{2+} corresponds to the meat in the middle, and layers of F^- correspond to the bread on either side. This is called a layer structure, and it is not a completely regular ionic structure. With a sandwich, bread is separated from bread by the meat, but in a pile of sandwiches, bread from one sandwich touches bread from the next sandwich. Similarly, in CdI_2 two sheets of I^- ions are separated by Cd^{2+} within a 'sandwich', but between one 'sandwich' and the next, two I^- layers are in contact. Whilst there is strong electrostatic bonding between Cd^{2+} and I^- layers, there are only weak van der Waals forces holding the adjacent layers of I^- together. The packing of layers in the crystal structure is not completely regular, and the solid is flaky, and it cleaves into two parallel sheets quite easily. This structure is adopted by many transition metal diiodides (Ti, V, Mn, Fe, Co, Zn, Cd) and by some main group diiodides and dibromides (Mg, Ca, Ge and Pb). Many hydroxides have similar layer structures ($Mg(OH)_2$, $Ca(OH)_2$, $Fe(OH)_2$, $Co(OH)_2$, $Ni(OH)_2$, and $Cd(OH)_2$.

In cadmium iodide, the third layer of I^- ions is directly above the first layer, so the repeating pattern is ABABAB... The I^- ions may be regarded as an approximately hexagonal close-packed arrangement. The Cd^{2+} ions occupy half of the octahedral sites. Rather than half filling the octahedral sites in a regular way throughout the whole structure, all of the octahedral sites are filled between two I^- layers, and none of the octahedral sites is filled between the next two layers of I^- ions. All of the octahedral holes are filled between the next two layers of I^- ions, none between the next pair, and so on.

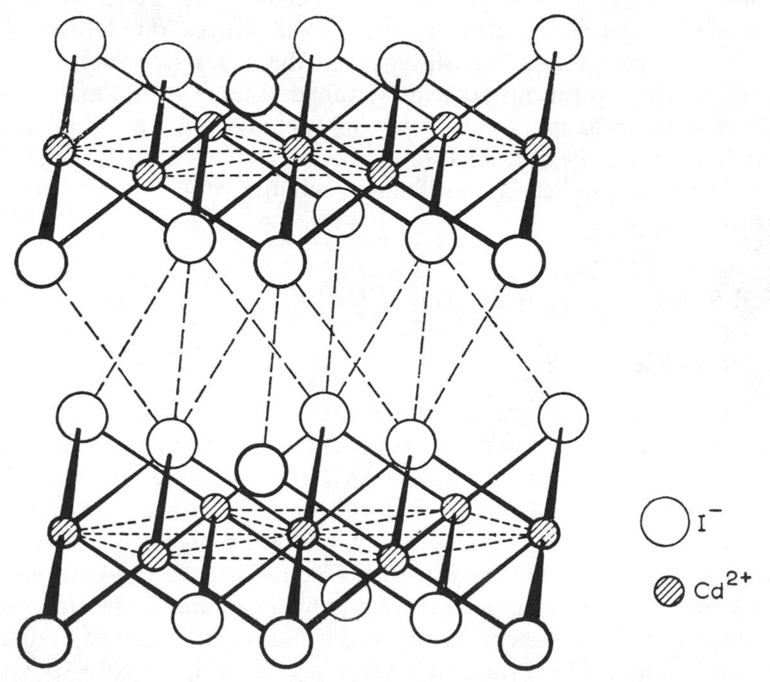

Figure 3.10 Part of two layers of cadmium iodide (CdI_2) structure.

Cadmium chloride structure

Cadmium chloride forms a closely related layer structure, but in this the chloride ions occur approximately in a cubic close-packed arrangement (ABCABC...).

Layer structures are intermediate in type between the extreme cases of:

1. A totally ionic crystal with a regular arrangement of ions and strong electrostatic forces in all directions.
2. A crystal in which small discrete molecules are held together by weak residual forces such as van der Waals forces and hydrogen bonds.

Nickel arsenide structure

The structure of nickel arsenide NiAs is related to the structure of CdI_2. In NiAs (Figure 3.11), the arsenic atoms form a hexagonal close-packed type of lattice with nickel atoms occupying all of the octahedral sites between all of the layers of arsenic atoms. (In CdI_2 all of the octahedral sites between half of the layers are filled, whilst with NiAs all of the octahedral sites between all of the layers are filled.)

In the nickel arsenide structure each atom has six nearest neighbours of the other type of atom. Each arsenic atom is surrounded by six nickel atoms at the corners of a trigonal prism. Each nickel atom is surrounded octahedrally by six arsenic atoms, but with two more nickel atoms sufficiently close to be bonded to the original nickel atom. This structure is adopted by many transition elements combined with one of the heavier elements from the *p*-block (Sn, As, Sb, Bi, S, Se, Te) in various alloys. These are better regarded as intermetallic phases rather than true compounds. They are opaque, have metallic lustre, and sometimes have a variable composition.

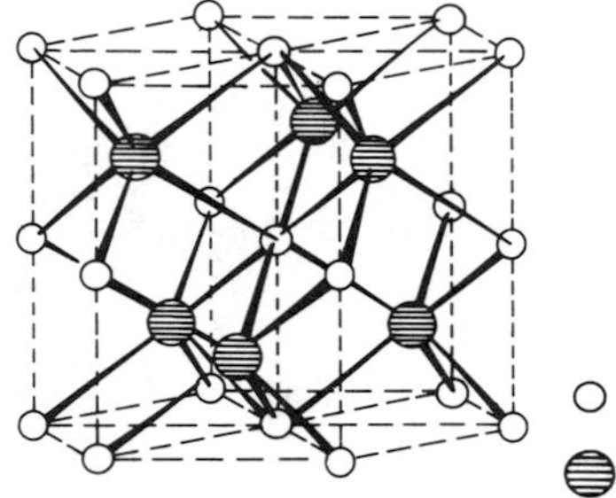

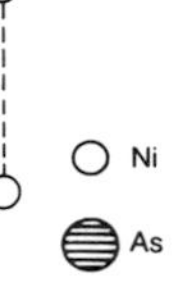

Figure 3.11 Nickel arsenide structure.

For details of other ionic structures, such as perovskite and spinels, see Chapter 20 and the Further Reading (Adams, Addison, Douglas McDaniel and Alexander, Greenwood, Wells) at the end of this chapter.

Structures containing polyatomic ions

There are many ionic compounds of types AX and AX_2 where A, or X, or both ions are replaced by complex ions. When the complex ion is roughly spherical, the ions often adopt one of the more symmetrical structures described above. Ions such as SO_4^{2-}, ClO_4^- and NH_4^+ are almost spherical. In addition, the transition metal complex $[Co(NH_3)_6]I_2$ adopts the CaF_2 (fluorite) structure. $K_2[PtCl_6]$ adopts an anti-fluorite structure, which is the same as a fluorite structure except that the sites occupied by positive and negative ions are interchanged. Both ions may be complex: $[Ni(H_2O)_6]$ $[SnCl_6]$, for example, forms a slightly distorted CsCl structure. Other ions (CN^- and SH^-) sometimes attain effective spherical symmetry by free rotation, or by random orientation. Examples include CsCN, TlCN and CsSH.

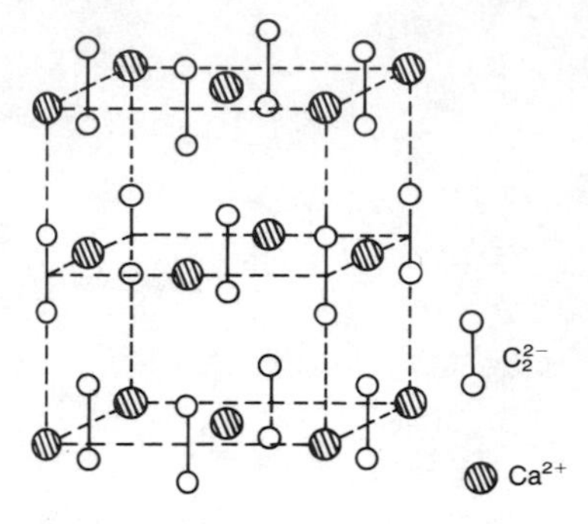

Figure 3.12 Calcium carbide structure.

Sometimes the presence of non-spherical ions simply distorts the lattice. Calcium carbide has a face-centred structure like NaCl, except that the linear C_2^{2-} ions are all oriented in the same direction along one of the unit cell axes. This elongates the unit cell in that direction (Figure 3.12). Similarly calcite, $CaCO_3$, has a structure related to NaCl, but the planar triangular CO_3^{2-} ion distorts the unit cell along a threefold axis of symmetry, rather than along one of the cell axes. Several divalent metal carbonates, a number of nitrates, $LiNO_3$ and $NaNO_3$, and some borates, $ScBO_3$, YBO_3 and $InBO_3$, also have the calcite structure.

A MORE CRITICAL LOOK AT RADIUS RATIOS

To a first approximation, the relative numbers and sizes of the ions will determine the structure of the crystal. The radius ratios of the alkali metal halides and the alkaline earth metal oxides, sulphides, selenides and tellurides are shown in Table 3.3.

All of the crystals with a radius ratio between 0.41 and 0.73 (enclosed by full line in Table 3.3) would be expected to have the sodium chloride structure. In fact all but four of the compounds listed have a sodium chloride structure at normal temperatures. A lot more compounds adopt the NaCl structure than would be predicted. The exceptions are CsCl, CsBr and CsI, which have a caesium chloride structure, and MgTe, which has a zinc sulphide structure. RbCl and RbBr are unusual since they both form a NaCl structure with a coordination number of 6 when crystallized at normal room temperatures and pressures, but they adopt a CsCl structure with a coordination number of 8 if crystallized at high pressures or temperatures. The fact that they can form both structures indicates that the difference in lattice energy between the two structures is small, and hence there is only a small difference in stability between them.

A CAUTIONARY WORD ON RADIUS RATIOS

Radius ratios provide a useful guide to what is possible on geometric grounds, and also a first guess at the likely structure, but there are other factors involved. Radius ratios do not necessarily provide a completely reliable method for predicting which structure is actually adopted.

Table 3.3 Radius ratios of Group I halides and Group II oxides

	F^-	Cl^-	Br^-	I^-		O^{2-}	S^{2-}	Se^{2-}	Te^{2-}
Li^+	0.57	0.41	0.39	0.35	Be				
Na^+	0.77	0.55	0.52	0.46	Mg^{2+}	0.51	0.39	0.36	0.33
K^+	0.96*	0.75	0.70	0.63	Ca^{2+}	0.71	0.54	0.51	0.45
Rb^+	0.88*	0.83	0.78	0.69	Sr^{2+}	0.84	0.64	0.60	0.53
Cs^+	0.80*	0.91	0.85	0.76	Ba^{2+}	0.96	0.73	0.68	0.61

* Indicates reciprocal value of r^-/r^+ since the normal ratio is greater than unity.

Though radius ratios indicate the correct structure in many cases, there are a significant number of exceptions where they predict the wrong structure. It is therefore worth examining the assumptions behind the radius ratio concept, to see if they are valid. The assumptions are:

1. That accurate ionic radii are known.
2. That ions behave as hard inelastic spheres.
3. That stable arrangements are only possible if the positive and negative ions touch.
4. That ions are spherical in shape.
5. That ions always adopt the highest possible coordination number.
6. That bonding is 100% ionic.

Values for ionic radii cannot be measured absolutely, but are estimated. They are not completely accurate or reliable. Though it is possible to measure the interatomic distance between two different ions very accurately by X-ray crystallography, it is much less certain how to divide the distance between the two ions to obtain ionic radii. Furthermore the radius of an ion is not constant but changes depending on its environment. In particular the radius changes when the coordination number changes. The radii usually quoted are for a coordination number of 6, but the radius effectively increases 3% when the coordination number is changed from 6 to 8, and decreases 6% when the coordination number changes from 6 to 4.

Ions are not hard inelastic spheres. They are sometimes fitted into 'holes' that are slightly too small, that is the ions are compressed, and the lattice may be distorted.

The assumption that the ions touch is necessary to calculate the critical lower limit for radius ratios. In principle positive and negative ions should touch, so as to get the ions close together, and get the maximum electrostatic attraction. (Electrostatic attraction depends on the product of the charges on the ions divided by the distance between them.) Theoretically structures where the smaller metal ion 'rattles' in its hole (that is, it does not touch the neighbouring negative ions) should be unstable. A more favourable electrostatic attraction should be obtained by adopting a different geometric arrangement with a smaller coordination number, so that the ions can get closer. It has already been shown that in the alkali halides and alkaline earth oxides the NaCl structure with coordination numbers of 6:6 is sometimes adopted when other structures are predicted by radius ratios. It follows that, since the smaller ion no longer fits the site it occupies, it must either 'rattle', or be compressed.

Are ions spherical? It is reasonable to consider ions with a noble gas structure as spherical. This includes the majority of the ions formed by elements in the main groups. There are a small number of exceptions where the ions have an inert pair (Ga^+, In^+ Tl^+, Sn^{2+}, Pb^{2+}, I^+, I^{3+}). These ions do not have a centre of symmetry, and the structures they form usually show some distortion, with the metal ion slightly displaced off-centre from its expected position. Transition metal ions with partially filled *d* orbitals are not spherical, though in contrast to inert pair distortion they usually have a centre of symmetry. The arrangement of electrons in these *d* orbitals gives rise to Jahn–Teller distortion. (See Chapter 28.) A partially filled *d* orbital pointing towards a coordinated ion will repel it. A completely filled *d* orbital will repel the ion even more. This can give rise to a structure with some long and some short bonds, depending on both the electronic structure of the metal ion, and the crystal structure adopted, i.e. the positions of the coordinating ions.

It is most unlikely that bonding is ever 100% ionic. The retention of a NaCl structure by a number of compounds which might be expected to adopt a CsCl structure is largely because there is a small covalent contribution to the bonding. The three *p* orbitals are at 90° to each other, and in a NaCl structure they point towards the six nearest neighbours, so covalent overlap of orbitals is possible. The geometric arrangement of the NaCl structure is ideally suited to allow some covalent contribution to bonding. This is not so for the CsCl structure.

Thus radius ratios provide a rough guide to what structures are geometrically possible. Radius ratios often predict the correct structure, but they do not always predict the correct structure. Ultimately the reason why any particular crystal structure is formed is that it gives the most favourable lattice energy.

LATTICE ENERGY

The lattice energy (U) of a crystal is the energy evolved when one gram molecule of the crystal is formed from gaseous ions:

$$Na^{+}_{(g)} + Cl^{-}_{(g)} \rightarrow NaCl_{(crystal)} \qquad U = -782\,\text{kJ mol}^{-1}$$

Lattice energies cannot be measured directly, but experimental values are obtained from thermodynamic data using the Born–Haber cycle (see Chapter 6).

Theoretical values for lattice energy may be calculated. The ions are treated as point charges, and the electrostatic (coulombic) energy E between two ions of opposite charge is calculated:

$$E = -\frac{z^{+}z^{-}e^{2}}{r}$$

where

z^{+} and z^{-} are the charges on the positive and negative ions
e is the charge on an electron
r is the inter-ionic distance

For more than two ions, the electrostatic energy depends on the number of ions, and also on A their arrangement in space. For one mole, the attractive energy is:

$$E = -\frac{N_{o}Az^{+}z^{-}e^{2}}{r}$$

Table 3.4 Madelung constants

Type of structure	A	M
zinc blende ZnS	1.63806	1.63806
wurtzite ZnS	1.64132	1.64132
sodium chloride NaCl	1.74756	1.74756
caesium chloride CsCl	1.76267	1.76267
rutile TiO_2	2.408	4.816
fluorite CaF_2	2.51939	5.03878
corundum Al_2O_3	4.17186	25.03116

where

N_o is the Avogadro constant – the number of molecules in a mole – which has the value $6.023 \times 10^{23}\,mol^{-1}$

A is the Madelung constant, which depends on the geometry of the crystal

Values for the Madelung constant have been calculated for all common crystal structures, by summing the contributions of all the ions in the crystal lattice. Some values are given in Table 3.4. (It should be noted that different values from these are sometimes given where the term z^+z^- is replaced by z^2, where z is the highest common factor in the charges on the ions. The Madelung constant is rewritten $M = Az^+z^-/z^2$. This practice is not recommended.)

The equation for the attractive forces between the ions gives a negative value for energy, that is energy is given out when a crystal is formed. The inter-ionic distance r occurs in the denominator of the equation. Thus the smaller the value of r, the greater the amount of energy evolved when the crystal lattice is formed, and hence the more stable the crystal will be. Mathematically, the equation suggests that an infinite amount of energy should be evolved if the distance r is zero. Plainly this is not so. When the inter-ionic distance becomes small enough for the ions to touch, they begin to repel each other. This repulsion originates from the mutual repulsion of the electron clouds on the two atoms or ions. The repulsive forces increase rapidly as r decreases. The repulsive force is given by B/r^n, where B is a

Table 3.5 Average values for the Born exponent

Electronic structure of ion	n	Examples
He	5	Li^+, Be^{2+}
Ne	7	Na^+, Mg^{2+}, O^{2-}, F^-
Ar	9	K^+, Ca^{2+}, S^{2-}, Cl^-, Cu^+
Kr	10	Rb^+, Br^-, Ag^+
Xe	12	Cs^+, I^-, Au^+

Average values are used, e.g. in LiCl, $Li^+ = 5$, $Cl^- = 9$, hence for LiCl, $n = (5 + 9)/2 = 7$

constant that depends on the structure, and n is a constant called the Born exponent. For one gram molecule the total repulsive force is $(N_oB)/r^n$. The Born exponent may be determined from compressibility measurements. Often chemists use a value of 9, but it is better to use values for the particular ions in the crystal.

The total energy holding the crystal together is U the lattice energy. This is the sum of the attractive and the repulsive forces.

$$U = \underbrace{-\frac{N_oAz^+z^-e^2}{r}}_{\text{attractive force}} + \underbrace{\frac{N_oB}{r^n}}_{\text{repulsive force}} \tag{3.1}$$

(A is the Madelung constant and B is a repulsion coefficient, which is a constant which is approximately proportional to the number of nearest neighbours.)

The equilibrium distance between ions is determined by the balance between the attractive and repulsion terms. At equilibrium, $dU/dr = 0$, and the equilibrium distance $r = r_o$

$$\frac{dU}{dr} = \frac{N_oAz^+z^-e^2}{r_o^2} - \frac{nN_oB}{r_o^{n+1}} = 0 \tag{3.2}$$

Rearranging this gives an equation for the repulsion coefficient B.

$$B = \frac{Az^+z^-e^2r_o^{n-1}}{n}$$

Substituting equation (3.3) into (3.1)

$$U = -\frac{N_oAz^+z^-e^2}{r_o}\left(1 - \frac{1}{n}\right)$$

This equation is called the *Born–Landé equation*. It allows the lattice energy to be calculated from a knowledge of the geometry of the crystal, and hence the Madelung constant, the charges z^+ and z^-, and the interionic distance. When using SI units, the equation takes the form:

$$U = -\frac{N_oAz^+z^-e^2}{4\pi\varepsilon_or_o}\left(1 - \frac{1}{n}\right) \tag{3.4}$$

where ε_o is the permittivity of free space $= 8.854 \times 10^{-12}\,\mathrm{F\,m^{-1}}$.

This equation gives a calculated value of $U = -778\,\mathrm{kJ\,mol^{-1}}$ for the lattice energy for sodium chloride, which is close to the experimental value of $-77\,\mathrm{kJ\,mol^{-1}}$ at 25 °C (obtained using the Born–Haber cycle). The experimental and theoretical values for the alkali metal halides and the oxides and halides of the alkaline earths (excluding Be), all agree within 3%.

Other expressions, for example the Born–Mayer and Kapustinskii equations, are similar, but calculate the repulsive contribution in a slightly different way. Agreement is even better if allowances are made for van der Waals forces and zero point energy.

Several important points arise from the Born–Landé equation:

1. The lattice becomes stronger (i.e. the lattice energy U becomes more negative), as r the inter-ionic distance decreases. U is proportional to $1/r$.

	r (Å)	U (kJ mol^{-1})
LiF	2.01	−1004
CsI	3.95	−527

2. The lattice energy depends on the product of the ionic charges, and U is proportional to $(z^+ . z^-)$.

	r (Å)	$(z^+ . z^-)$	U (kJ mol^{-1})
LiF	2.01	1	−1004
MgO	2.10	4	−3933

3. The close agreement between the experimental lattice energies and those calculated by the Born–Landé equation for the alkali metal halides does not of itself prove that the equation itself, or the assumptions on which it is based, are correct. The equation is remarkably self-compensating, and tends to hide errors. There are two opposing factors in the equation. Increasing the inter-ionic distance r reduces the lattice energy. It is almost impossible to change r without changing the structure, and therefore changing the Madelung constant A. Increasing A *increases the lattice energy: hence the effects of changing* r and A may largely cancel each other.

   This may be illustrated by choosing a constant value for n in the Born–Landé equation. Then changes in inter-ionic distance can be calculated for either changes in the coordination number, or in crystal structure. Taking a constant value of $n = 9$, we may compare the inter-ionic distances with those for six-coordination:

Coordination number	12	8	6	4
Ratio of inter-ionic distance	1.091	1.037	1.000	0.951

   For a change of coordination number from 6 (NaCl structure) to 8 (CsCl structure) the inter-ionic distance increases by 3.7%, and the Madelung constants (NaCl $A = 1.74756$, and CsCl $A = 1.76267$) change by only 0.9%. Thus a change in coordination number from 6 to 8 would result in a reduction in lattice energy, and in theory the NaCl structure should always be more stable than the CsCl structure. In a similar way reducing the coordination number from 6 to 4 decreases r by 4.9%. The decrease in A is 6.1% or 6.3% (depending on whether a zinc blende or wurtzite structure is formed), but in either case it more than compensates for the change in r, and in theory coordination number 6 is more stable than 4.

   This suggests that neither four- nor eight-coordinate structures should exist, since the six-coordinate NaCl structure is more stable. Since ZnS is known (coordination number 4), and CsCl, CsBr and CsI have a coordination number of 8, this suggestion is plainly incorrect. We must

Table 3.6 Inter-ionic distances and ionic charges related to m.p. and hardness

	r(Å)	($z^+ . z^-$)	m.p. (°C)	Hardness (Mohs' scale)
NaF	2.310	1	990	3.2
BeO	1.65	4	2530	9.0
MgO	2.106	4	2800	6.5
CaO	2.405	4	2580	4.5
SrO	2.580	4	2430	3.5
BaO	2.762	4	1923	3.3
TiC	2.159	16	3140	8–9

therefore look for a mistake in the theoretical assumptions made. First the value of n was assumed to be 9, when it may vary from 5 to 12. Second, the calculation of electrostatic attraction assumes that the ions are point charges. Third, the assumption is made that there is no reduction in charge because of the interaction (i.e. the bonds are 100% ionic).

4. Crystals with a high lattice energy usually melt at high temperatures, and are very hard. Hardness is measured on Mohs' scale. (See Appendix N.) High lattice energy is favoured by a small inter-ionic distance, and a high charge on the ions.

It has been seen that a number of salts which might be expected from radius ratio considerations to have a CsCl structure in fact adopt a NaCl structure. The Madelung constant for CsCl is larger than for NaCl, and would give an increased lattice energy. However, the inter-ionic distance r will be larger in a CsCl type of structure than in a NaCl type of structure, and this would decrease the lattice energy. These two factors work in opposite directions and partly cancel each other. This makes the lattice energy more favourable for a NaCl type of lattice in some cases where a CsCl structure is geometrically possible. Consider a case such as RbBr, where the radius ratio is close to borderline between six-coordination (NaCl structure) and eight-coordination (CsCl structure). If the CsCl structure is adopted, the Madelung constant is larger than for NaCl, and *this increases the lattice energy by 0.86%*. At the same time the inter-ionic distance in a CsCl structure increases by 3%, and *this decreases the lattice energy by 3%*. Clearly the NaCl structure is preferred.

FEATURES OF SOLIDS

The essential feature of crystalline solids is that the constituent molecules, atoms or ions are arranged in a completely regular three-dimensional pattern. Models built to show the detailed structure of crystalline materials are usually grossly misleading, for they imply a perfect static pattern. Since the atoms or ions have a considerable degree of thermal vibration, the crystalline state is far from static, and the pattern is seldom perfect. Many of the most useful properties of solids are related to the thermal vibrations of atoms, the presence of impurities and the existence of defects.

STOICHIOMETRIC DEFECTS

Stoichiometric compounds are those where the numbers of the different types of atoms or ions present are exactly in the ratios indicated by their chemical formulae. They obey the law of constant composition that '*the same chemical compound always contains the same elements in the same composition by weight*'. At one time these were called Daltonide compounds, in contrast to Berthollide or nonstoichiometric compounds where the chemical composition of a compound was variable, not constant.

Two types of defects may be observed in stoichiometric compounds, called Schottky and Frenkel defects respectively. At absolute zero, crystals tend to have a perfectly ordered arrangement. As the temperature increases, the amount of thermal vibration of ions in their lattice sites increases, and if the vibration of a particular ion becomes large enough, it may jump out of its lattice site. This constitutes a point defect. The higher the temperature, the greater the chance that lattice sites may be unoccupied. Since the number of defects depends on the temperature, they are sometimes called thermodynamic defects.

Schottky defects

A Schottky defect consists of a pair of 'holes' in the crystal lattice. One positive ion and one negative ion are absent (see Figure 3.13). This sort of defect occurs mainly in highly ionic compounds where the positive and negative ions are of a similar size, and hence the coordination number is high (usually 8 or 6), for example NaCl, CsCl, KCl and KBr.

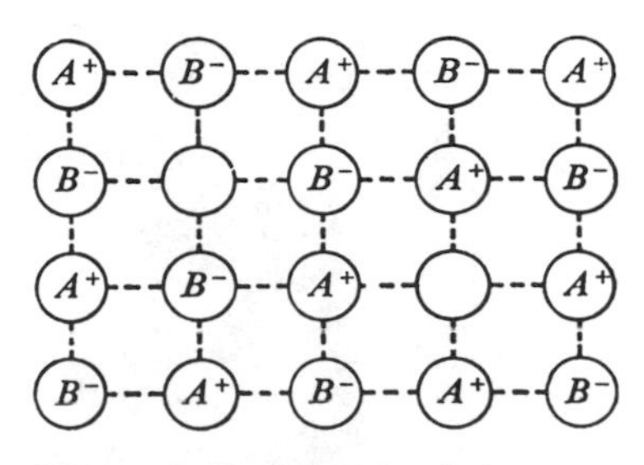

Figure 3.13 Schottky defect.

The number of Schottky defects formed per cm^3 (n_s) is given by

$$n_s = N\exp\left(-\frac{W_s}{2kT}\right)$$

where N is the number of sites per cm^3 that could be left vacant, W_s is the work necessary to form a Schottky defect, k is the gas constant and T the absolute temperature.

Frenkel defects

A Frenkel defect consists of a vacant lattice site (a 'hole'), and the ion which ideally should have occupied the site now occupies an interstitial position (see Figure 3.14).

Metal ions are generally smaller than the anions. Thus it is easier to squeeze A^+ into alternative interstitial positions, and consequently it is more common to find the positive ions occupying interstitial positions. This type of defect is favoured by a large difference in size between the positive

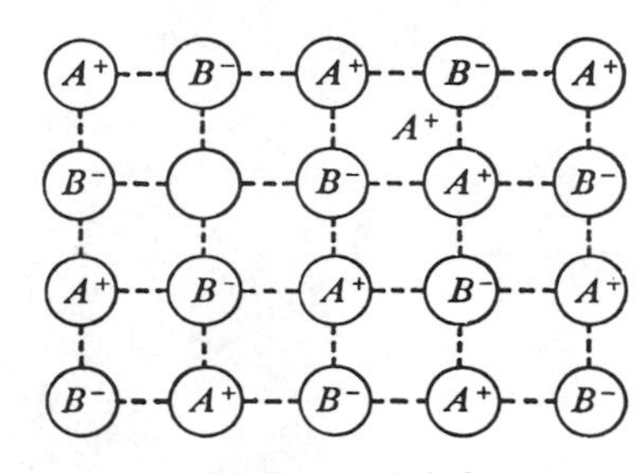

Figure 3.14 Frenkel defect.

and negative ions, and consequently the coordination number is usually low (4 or 6). Since small positive ions are highly polarizing and large negative ions are readily polarized, these compounds have some covalent character. This distortion of ions, and the proximity of like charges, leads to a high dielectric constant. Examples of this type of defect are ZnS, AgCl, AgBr and AgI.

The number of Frenkel defects formed per cm^3 (n_f) is given by

$$n_f = \sqrt{NN'} \exp\left(-\frac{W_f}{2kT}\right)$$

where N is the number of sites per cm^3 that could be left vacant, N' is the number of alternative interstitial positions per cm^3, W_f is the work necessary to form a Frenkel defect, k is the gas constant and T the absolute temperature.

The energy needed to form either a Schottky defect or a Frenkel defect depends on the work needed to form the defect, and on the temperature. In a given compound one type generally predominates.

In NaCl, the energy to form a Schottky defect is about 2003 kJ mol^{-1} compared with a lattice energy of approximately 750 kJ mol^{-1}. It is therefore much easier to form a defect than to break the lattice.

The number of defects formed is relatively small, and at room temperature NaCl has only one defect in 10^{15} lattice sites, this value rising to one in 10^6 sites at 500 °C and one in 10^4 sites at 800 °C.

A consequence of these defects is that a crystalline solid that has defects may conduct electricity to a small extent. Electrical conductivity in a chemically pure, stoichiometric semiconductor is called '*intrinsic semiconduction*'. In the above cases, intrinsic semiconduction occurs by an ionic mechanism. If an ion moves from its lattice site to occupy a 'hole', it creates a new 'hole'. If the process is repeated many times, a 'hole' may migrate across a crystal, which is equivalent to moving a charge in the opposite direction. (This type of semiconduction is responsible for the unwanted background noise produced by transistors.)

Crystals with Frenkel defects have only one type of hole, but crystals containing Schottky defects have holes from both positive and negative ions, and conduction may arise by using either one type of hole or both types. Migration of the smaller ion (usually the positive ion) into the appropriate holes is favoured at low temperatures, since moving a small

Table 3.7 Percentage of conduction by cations and anions

Temp. (°C)	NaF		NaCl		NaBr	
	cation %	anion %	cation %	anion %	cation %	anion %
400	100	0	100	0	98	2
500	100	0	98	2	94	6
600	92	8	91	9	89	11

ion requires less energy. However, migration of both types of ions in opposite directions (using both types of holes) occurs at high temperatures. For example, at temperatures below 500 °C the alkali halides conduct by migration of the cations, but at higher temperatures both anions and cations migrate. Further, the amount of anionic conduction increases with temperature, as shown in Table 3.7.

The density of a defect lattice should be different from that of a perfect lattice. The presence of 'holes' should lower the density, but if there are too many 'holes' there may be a partial collapse or distortion of the lattice – in which case the change in density is unpredictable. The presence of ions in interstitial positions may distort (expand) the lattice and increase the unit cell dimensions.

NONSTOICHIOMETRIC DEFECTS

Nonstoichiometric or Berthollide compounds exist over a range of chemical composition. The ratio of the number of atoms of one kind to the number of atoms of the other kind does not correspond exactly to the ideal whole number ratio implied by the formula. Such compounds do not obey the law of constant composition. There are many examples of these compounds, particularly in the oxides and sulphides of the transition elements. Thus in FeO, FeS or CuS the ratio of Fe : O, Fe : S or Cu : S differs from that indicated by the ideal chemical formula. If the ratio of atoms is not exactly 1 : 1 in the above cases, there must be either an excess of metal ions, or a deficiency of metal ions (e.g. $Fe_{0.84}O$–$Fe_{0.94}O$, $Fe_{0.9}S$). Electrical neutrality is maintained either by having extra electrons in the structure, or changing the charge on some of the metal ions. This makes the structure irregular in some way, i.e. it contains defects, which are in addition to the normal thermodynamic defects already discussed.

Metal excess

This may occur in two different ways.

F-centres

A negative ion may be absent from its lattice site, leaving a 'hole' which is occupied by an electron, thereby maintaining the electrical balance (see Figure 3.15). This is rather similar to a Schottky defect in that there are 'holes' and not interstitial ions, but only one 'hole' is formed rather than a pair. This type of defect is formed by crystals which would be expected to form Schottky defects. When compounds such as NaCl, KCl, LiH or δ-TiO are heated with excess of their constituent metal vapours, or treated with high energy radiation, they become deficient in the negative ions, and their formulae may be represented by $AX_{1-\delta}$, where δ is a small fraction. The nonstoichiometric form of NaCl is yellow, and the nonstoichiometric

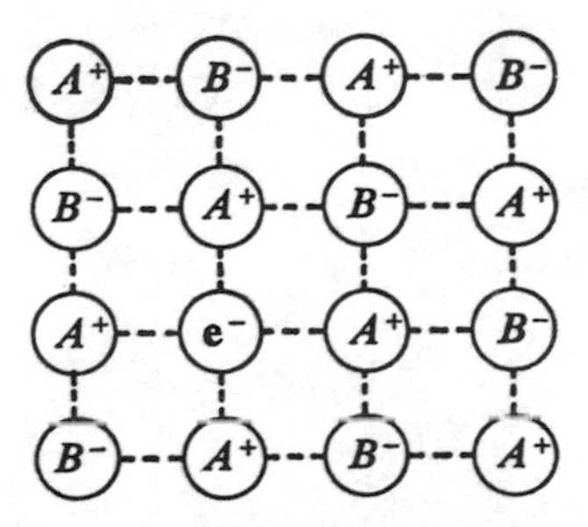

Figure 3.15 Metal excess defect because of absent anion.

form of KCl is blue–lilac in colour. Note the similarity with the flame colorations for Na and K.

The crystal lattice has vacant anion sites, which are occupied by electrons. Anion sites occupied by electrons in this way are called *F*-centres. (*F* is an abbreviation for *Farbe*, the German word for colour.) These *F*-centres are associated with the colour of the compound and the more *F*-centres present, the greater the intensity of the coloration. Solids containing *F*-centres are paramagnetic, because the electrons occupying the vacant sites are unpaired. When materials with *F*-centres are irradiated with light they become photoconductors. When electrons in the *F*-centres absorb sufficient light (or heat) energy, the electron is promoted into a conduction band, rather similar to the conduction bands present in metals. Since conduction is by electrons it is *n-type semiconduction*.

Interstitial ions and electrons

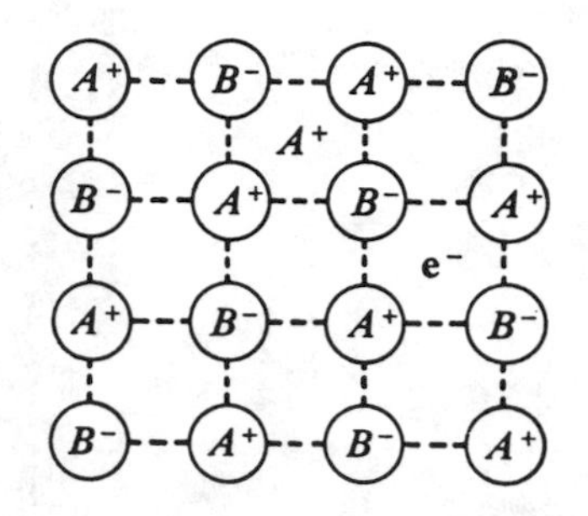

Figure 3.16 Metal excess defects caused by interstitial cations.

Metal excess defects also occur when an extra positive ion occupies an interstitial position in the lattice, and electrical neutrality is maintained by the inclusion of an interstitial electron (see Figure 3.16). Their composition may be represented by the general formula $A_{1+\delta}X$.

This type of defect is rather like a Frenkel defect in that ions occupy interstitial positions, but there are no 'holes', and there are also interstitial electrons. This kind of metal excess defect is much more common than the first, and is formed in crystals which would be expected to form Frenkel defects (i.e. the ions are appreciably different in size, have a low co-ordination number, and have some covalent character). Examples include ZnO, CdO, Fe_2O_3 and Cr_2O_3.

If this type of defect oxide is heated in oxygen, then cooled to room temperature, its conductivity decreases. This is because the oxygen oxidizes some of the interstitial ions, and these subsequently remove interstitial electrons, which reduces the conductivity.

Crystals with either type of metal excess defect contain free electrons, and if these migrate they conduct an electric current. Since there are only a small number of defects, there are only a few free electrons that can conduct electricity. Thus the amount of current carried is very small compared with that in metals, fused salts or salts in aqueous solutions, and these defect materials are called *semiconductors*. Since the mechanism is normal electron conduction, these are called *n-type semiconductors*. These free electrons may be excited to higher energy levels giving absorption spectra, and in consequence their compounds are often coloured, e.g. nonstoichiometric NaCl is yellow, nonstoichiometric KCl is lilac, and ZnO is white when cold but yellow when hot.

Metal deficiency

Metal-deficient compounds may be represented by the general formula $A_{1-\delta}X$. In principle metal deficiency can occur in two ways. Both require variable valency of the metal, and might therefore be expected with the transition metals.

Positive ions absent

If a positive ion is absent from its lattice site, the charges can be balanced by an adjacent metal ion having an extra positive charge (see Figure 3.17). Examples of this are FeO, NiO, δ-TiO, FeS and CuI. (If an Fe^{2+} is missing from its lattice site in FeO, then there must be two Fe^{3+} ions somewhere in the lattice to balance the electrical charges. Similarly if a Ni^{2+} is missing from its lattice site in NiO, there must be two Ni^{3+} present in the lattice.)

Crystals with metal deficiency defects are semiconductors. Suppose the lattice contains A^+ and A^{2+} metal ions. If an electron 'hops' from an A^+ ion to the positive centre (an A^{2+} ion), the original A^+ becomes a new positive centre. There has been an apparent movement of A^{2+}. With a series of similar 'hops', an electron may be transferred in one direction across the structure, and at the same time the positive hole migrates in the opposite direction across the structure. This is called positive hole, or *p-type semiconduction*.

If a defect oxide of this type is heated in oxygen, its room temperature conductivity increases, because the oxygen oxidizes some of the metal ions, and this increases the number of positive centres.

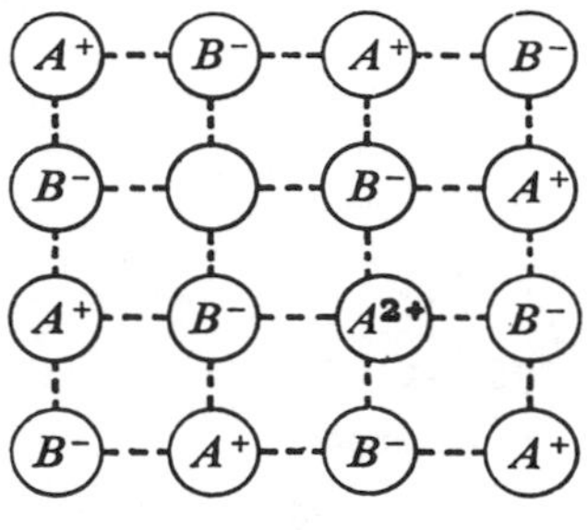

Figure 3.17 Metal deficiency caused by missing positive ion.

Extra interstitial negative ions

In principle it might be possible to have an extra negative ion in an interstitial position and to balance the charges by means of an extra charge on an adjacent metal ion (see Figure 3.18). However, since negative ions are usually large, it would be difficult to fit them into interstitial positions. No examples of crystals containing such negative interstitial ions are known at present.

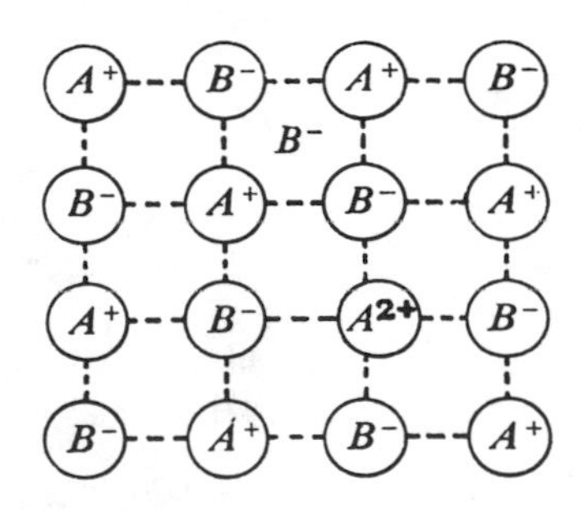

Figure 3.18 Metal deficiency caused by interstitial negative ions.

SEMICONDUCTORS AND TRANSISTORS

Semiconductors are solids where there is only a small difference in energy, called a *band gap*, between the filled valency band of electrons and a conduction band. If cooled to absolute zero, the electrons occupy their lowest possible energy levels. The conduction band is empty, and the material is a perfect insulator. At normal temperatures, some electrons are thermally excited from the valency band to the conduction band, and hence they can conduct electricity by the passage of electrons at normal temperatures. The conductivity is in between that of a metal and an insulator and depends on the number of electrons in the conduction band.

Germanium and, to an even greater extent, silicon are the most important commercial examples of semiconductors. The crystal structures of both are like diamond. Atoms of Si and Ge both have four electrons in their outer shell, which form four covalent bonds to other atoms. In both

Table 3.8 Band gaps of some semiconductors at absolute zero

Compound	Energy gap ($kJ\,mol^{-1}$)	Compound	Energy gap ($kJ\,mol^{-1}$)
α-Sn	0	GaAs	145
PbTe	19	Cu_2O	212
Te	29	CdS	251
PbS	29	GaP	278
Ge	68	ZnO	328
Si	106	ZnS	376
InP	125	Diamond	579

Si and Ge at very low temperatures, the valence band is filled and the conduction band is empty. Under these conditions, Si and Ge are both insulators, and cannot carry any electric current.

The band gaps are only $68\,kJ\,mol^{-1}$ for Ge, and $106\,kJ\,mol^{-1}$ for Si, and at room temperature a few valence electrons gain sufficient energy from the thermal vibration of the atoms to be promoted into the conduction band. If the crystal is connected in an electric circuit, these thermally excited electrons carry a small current, and make the Si or Ge crystal slightly conducting. This is termed *intrinsic semiconduction*. Expressed in another way, some bonds are broken, and these valence electrons can migrate, and conduct electricity.

As the temperature is increased, the conductivity increases, that is the electrical resistance decreases. (This is the opposite of the situation with metals.) Above 100°C, so many valence electrons are promoted to the conduction band in Ge that the crystal lattice disintegrates. With Si the maximum working temperature is 150°C. This intrinsic semiconduction is undesirable, and precautions must be taken to limit the working temperature of transistors.

Pure Si and Ge can be made semiconducting in a controlled way by adding impurities which act as charge carriers. Si or Ge are first obtained extremely pure by zone refining. Some atoms with five outer electrons, such as arsenic As, are deliberately added to the silicon crystal. This process is called 'doping' the crystal. A minute proportion of Si atoms are randomly replaced by As atoms with five electrons in their outer shell. Only four of the outer electrons on each As atom are required to form bonds in the lattice. At absolute zero or low temperatures, the fifth electron is localized on the As atom. However, at normal temperatures, some of these fifth electrons on As are excited into the conduction band, where they can carry current quite readily. This is *extrinsic conduction*, and it increases the amount of semiconduction far above that possible by intrinsic conduction. Since the current is carried by excess electrons, it is *n-type semiconduction*.

Alternatively a crystal of pure Si may be doped with some atoms with only three outer electrons, such as indium In. Each indium atom uses its three outer electrons to form three bonds in the lattice, but they are unable to form four bonds to complete the covalent structure. One bond is incomplete, and the site normally occupied by the missing electron is called a 'positive hole'. At absolute zero or low temperatures, the positive holes are localized around the indium atoms. However, at normal temperatures a valence electron on an adjacent Si atom may gain sufficient energy to move into the hole. This forms a new positive hole on the Ge atom. The positive hole seems to have moved in the opposite direction to the electron. By a series of 'hops', the 'positive hole' can migrate across the crystal. This is equivalent to moving an electron in the opposite direction, and thus current is carried. Since current is carried by the migration of positive centres, this is *p-type semiconduction*.

Silicon must be ultra-highly purified before it can be used in semiconductors. First impure silicon (98% pure) is obtained by reducing SiO_2 with carbon in an electric furnace at about 1900 °C. This may be purified by reacting with HCl, forming trichlorosilane $SiHCl_3$, which may be distilled to purify it, then decomposed by heating to give pure silicon.

$$SiO_2 + C \rightarrow Si + CO_2$$

$$Si + 3HCl \xrightarrow{350\,^\circ C} H_2 + SiHCl_3 \xrightarrow{\text{strong heat}} Si$$

The final purification is by zone refining, where a rod of silicon is melted near one end by an electric furnace. As the furnace is slowly moved along the rod, the narrow molten zone gradually moves to the other end of the rod. The impurities are more soluble in the liquid melt than in the solid, so they concentrate in the molten zone, and eventually move to the end of the rod. The impure end is removed, leaving an ultra-purified rod, with a purity of at least 1 part in 10^{10}. Purified silicon (or germanium) crystals can be converted to *p*-type or *n*-type semiconductors by high temperature diffusion of the appropriate dopant element, up to a concentration of 1 part in 10^8. In principle any of the Group III elements boron, aluminium, gallium or indium can be used to make *p*-type semiconductors, though indium is the most used because of its low melting point. Similarly Group V elements such as phosphorus or arsenic can be used to make *n*-type semiconductors, but because of its low melting point arsenic is most used.

If a single crystal is doped with indium at one end, and with arsenic at the other end, then one part is a *p*-type semiconductor and the other an *n*-type semiconductor. In the middle there will be a boundary region where the two sides meet, which is a *p-n* junction. Such junctions are the important part of modern semiconductor devices.

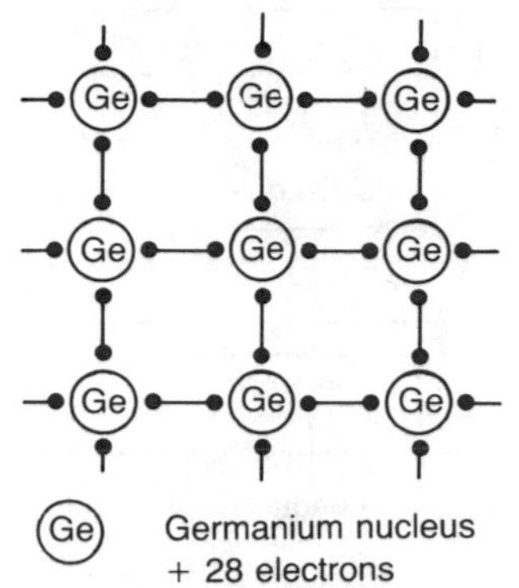

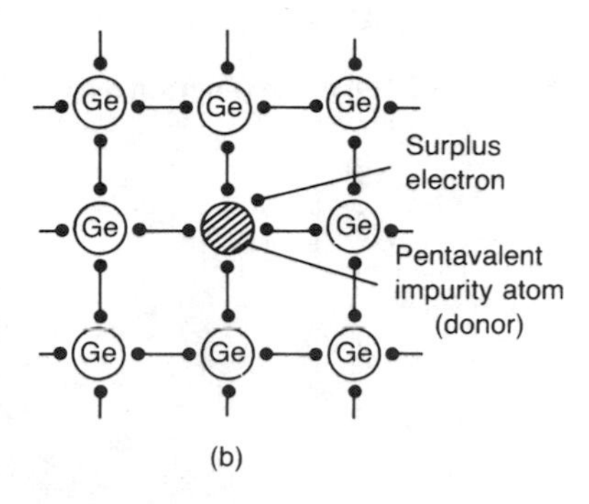

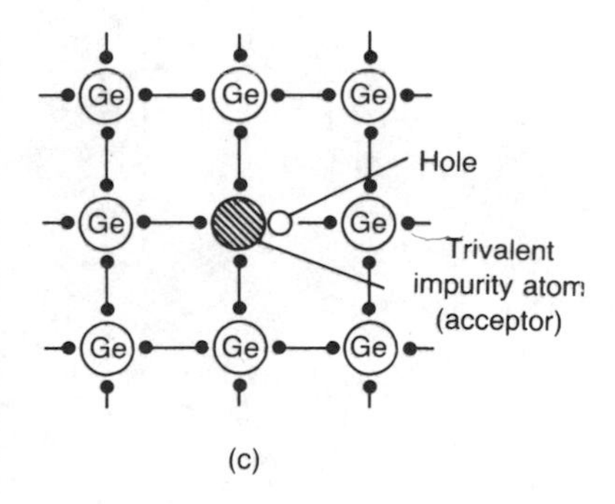

Figure 3.19 (a) Pure germanium. (b) *n*-type germanium. (c) *p*-type germanium.

RECTIFIERS

A rectifier will only allow current from an outside source to flow through it in one direction. This is invaluable in converting alternating current AC into direct current DC, and it is common to use a square of four diodes in a

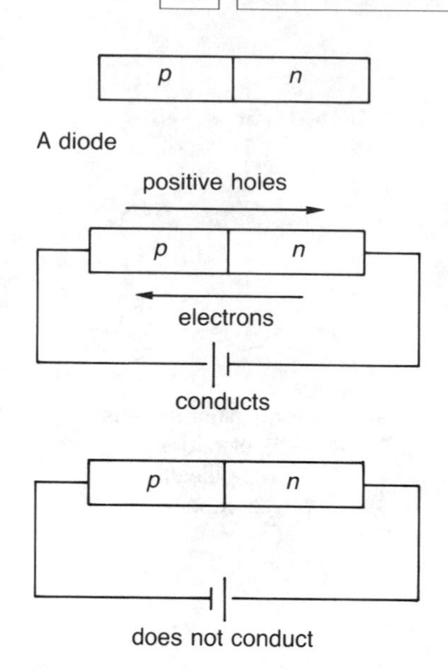

Figure 3.20 An *n-p* junction as a rectifier.

circuit to do this. A diode is simply a transistor with two zones, one *p*-type, and the other *n*-type, with a *p-n* junction in between.

Suppose that a positive voltage is applied to the *p*-type region, and a negative (or more negative) voltage applied to the *n*-type region. In the *p*-type region, positive holes will migrate towards the *p-n* junction. In the *n*-type region, electrons will migrate towards the junction. At the junction, the two destroy each other. Expressed in another way, at the junction the migrating electrons from the *n*-type region move into the vacant holes in the valence band of the *p*-type region. The migration of electrons and holes can continue indefinitely, and a current will flow for as long as the external voltage is applied.

Consider what will happen if the voltages are reversed, so the *p*-type region is negative, and the *n*-type region positive. In the *p*-type region, positive holes migrate away from the junction, and in the *n*-type region electrons migrate away from the junction. At the junction there are neither positive holes nor electrons, so no current can flow.

PHOTOVOLTAIC CELL

If a *p-n* junction is irradiated with light, provided that the energy of the light photons exceeds the band gap, then some bonds will break, giving electrons and positive holes, and these electrons are promoted from the valence band to the conduction band. The extra electrons in the conduction band make the *n*-type region more negative, whilst in the *p*-type region the electrons are trapped by some positive holes. If the two regions are connected in an external circuit, then electrons can flow from the *n*-type region to the *p*-type region, that is current flows from the *p*-type to the *n*-type region. Such a device acts as a battery that can generate electricity from light. Efforts are being made to make efficient cells of this type to harness solar energy.

TRANSISTORS

Transistors are typically single crystals of silicon which have been doped to give three zones. In Britain *p-n*-p transistors are mainly used, whilst in the USA *n-p-n* transistors are most widely used. Both types have many uses, for example as amplifiers and oscillators in radio, TV and hi-fi circuits and in computers. They are also used as phototransistors, tunnel diodes, solar cells, thermistors, and in the detection of ionizing radiation.

Different voltages must be applied to the three regions of a transistor to make it work. Typical bias potentials for a *p-n*-p transistor are shown in Figure 3.21. The base is typically −0.2 volts and the collector is typically −2.0 volts with respect to the emitter. The charge carriers in the emitter are positive holes, and these migrate from the emitter at 0 volts to the base at −0.2 volts. The positive holes cross the emitter/base *p-n* junction, and in the *n*-type base region some positive holes combine with electrons and are destroyed. There is a flow of electrons in the reverse direction, from the

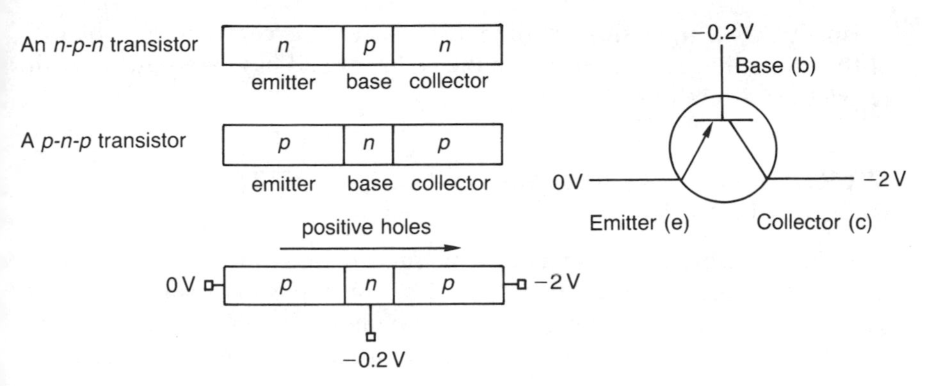

Figure 3.21 *n-p-n* and *p-n-p* transistors.

base to the emitter. There is thus a small base current. However, the base is very thin, and the collector has a much greater negative voltage, so most of the positive holes pass through the base to the collector, where they combine with electrons from the circuit. At the emitter, electrons leave the *p*-type semiconductor and enter the circuit, and in doing so they produce more positive holes. Typically, if the emitter current is 1 mA, the base current is 0.02 mA, and the collector current 0.98 mA.

The most common method of using a transistor as an amplifier is the common or grounded emitter circuit (Figure 3.22a). The emitter is common to both the base and collector circuits, and is sometimes grounded (earthed). The base current is the input signal, and the collector current is the output signal. If the base current is reduced, for example by increasing R_1, the base becomes positively charged, and this reduces the movement of positive holes to the collector. In a typical transistor, a change in the base current can produce a change 50 times as great in the collector current, giving a current amplification factor of 50. A small change in input current to the base produces a much larger change in the collector current, so the original signal is amplified.

In practice the bias for both the base and the collector are often obtained from one battery by having the resistance of R_1 much greater than that of R_2 (Figure 3.22b).

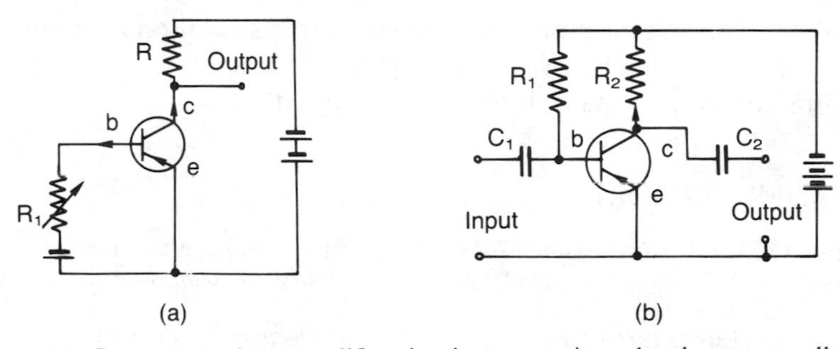

Figure 3.22 Common emitter amplifier circuits. e = emitter, b = base, c = collector.

Finally, *n-p-n* transistors work in a similar way, except that the polarity of the bias voltages is reversed, so the collector and base are positive with respect to the emitter.

MICRO-MINATURIZED SEMICONDUCTOR DEVICES INTEGRATED CIRCUITS

It is now possible to manufacture computer chips with the equivalent of many thousands of single crystal transistor junctions on a small wafer of silicon, only a few millimetres square. (Memory chips for computers are readily available which store 64K, 256K, 1 megabyte and even 4 megabytes of data on a single chip.)

The steps in the manufacture of such chips is:

1. A fairly large single crystal of Si is doped to make it an *n*-type semiconductor, and then it is carefully cut into thin slices.
2. A slice is heated in air to form a thin surface layer of SiO_2.
3. The oxide layer is then coated with a photosensitive film, sometimes called a photoresist.
4. A mask is placed over the photoresist, and the slice is exposed to UV light. Those parts of the photoresist exposed to light are changed, and are removed by treatment with acid, but the unexposed parts remain protected by the photoresist.
5. The slice is then treated with HF, which etches (removes) the exposed areas of SiO_2. After this, the unchanged photoresist is removed.
6. The surface is exposed to the vapour of a Group III element. Some of the surface is covered by a film of SiO_2, and some has exposed silicon. The parts covered by a SiO_2 film are unaffected, but in the parts where the silicon itself is exposed, some Si atoms are randomly replaced, forming a layer of *p*-type semiconductor.
7. The steps (2) to (5) are repeated using a different mask, and the exposed areas of Si exposed to the vapour of a Group V element, to produce another layer of *n*-type semiconductor.
8. Steps (2) to (5) are repeated using a mask to produce the openings into which metal can be deposited to 'wire together' the various semiconductors so produced into an integrated circuit.
9. Finally the chip is packaged in plastic or ceramic, connecting pins are soldered on so that it may be plugged in to a socket on a circuit board, and the chip is tested. A significant number turn out to be faulty. Faulty chips cannot be repaired, and are discarded.

FURTHER READING

Adams, D.M. (1974) *Inorganic Solids*, Wiley-Interscience, New York.

Addison, W.E. (1961) *Structural Principles in Inorganic Compounds*, Longmans, London.

Bamfield, P. (ed.) (1986) *Fine Chemicals for the Electronic Industry*, Royal Society of Chemistry, Special Publication No. 40, London.

Burdett, J.K. (1982) New ways to look at solids, *Acc. Chem. Res.*, **15**, 34.
Cartmell, E. and Fowles, G.W.A. (1977) *Valency and Molecular Structure*, 4th ed., Butterworths, London.
Cox, P.A. (1987) *The Electronic Structure and Chemistry of Solids*, Oxford University Press.
Dasent, W.E. (1982) *Inorganic Energetics: An Introduction* (Cambridge Texts in Chemistry and Biochemistry Series), Cambridge University Press.
Douglas, B., McDaniel, D.H. and Alexander J.J. (1983) *Concepts and Models in Inorganic Chemistry*, 2nd ed., Wiley, New York.
Ebsworth, E.A.V., Rankin, D.W.H. and Cradock, S. (1987) *Structural Methods in Inorganic Chemistry*, Blackwell Scientific, Oxford.
Galasso, F.S. (1970) *Structure and Properties of Inorganic Solids*, Pergamon, Oxford. (Contains extensive tables.)
Galwey, A.K. (1967) *Chemistry of Solids*, Chapman & Hall, London.
Greenwood, N.N. (1968) *Ionic Crystals, Lattice Defects and Non-Stoichiometry*, Butterworths, London. (Still the best single volume book on the subject.)
Ho, S.M. and Douglas, B.E. (1972) Structures of the elements and the PTOT system, *J. Chem. Ed.*, **49**, 74.
Hyde, B.G. and Andersson, S. (1989) *Inorganic Crystal Structures*, Wiley, New York.
Jenkins, H.D.B. (1979) The calculation of lattice energy: some problems and some solutions, *Revue de Chimie Minerale*, **16**, 134–150.
Ladd, M.F.C. (1974) *Structure and Bonding in Solid State Chemistry*, Wiley, London.
Moss, S.J and Ledwith, A. (eds) (1987) *The Chemistry of the Semiconductor Industry*, Blackie.
Parish, R.V. (1976) *The Metallic Elements*, Longmans, London.
Rao, C.N.R. (ed.) (1974) *Solid State Chemistry*, Dekker, New York.
Rao, C.N.R. and Gopalakrishnan, J. (1986) *New Directions in Solid State Chemistry*, Cambridge University Press, Cambridge.
Shannon R.D. (1976) Revised effective ionic radii. *Acta Cryst.*, **A32**, 751–767. (The most up to date and generally accepted values for ionic radii.)
Walton, A. (1978) *Molecular and Crystal Structure Models*, Ellis Horwood. Chichester.
Wells, A.F. (1984) *Structural Inorganic Chemistry*, 5th ed., Oxford University Press, Oxford. (The standard text, with excellent diagrams.)
West, A.R. (1984) *Solid State Chemistry and its Applications*, Wiley, New York.

PROBLEMS

1. Relate the tendency of atoms to gain or lose electrons to the types of bonds they form.
2. Indicate to what extent the following will conduct electricity, and give the mechanism of conduction in each case:
   (a) NaCl (fused)
   (b) NaCl (aqueous solution)
   (c) NaCl (solid)
   (d) Cu (solid)
   (e) CCl_4 (liquid).
3. Why are ionic compounds usually high melting, whilst most simple covalent compounds have low melting points? Explain the high melting point of diamond.

4. How are the minimum values of radius ratio arrived at for various coordination numbers, and what are these limits? Give examples of the types of crystal structure associated with each coordination number.

5. Show by means of a diagram, and a simple calculation, the minimum value of the radius ratio r^+/r^- which permits a salt to adopt a caesium chloride type of structure.

6. Give the coordination numbers of the ions and describe the crystal structures of zinc blende, wurtzite and sodium chloride in terms of close packing and the occupancy of tetrahedral and octahedral holes.

7. CsCl, CsI, TlCl and TlI all adopt a caesium chloride structure. The inter-ionic distances are: Cs–Cl 3.06 Å, Cs–I 3.41 Å. Tl–Cl 2.55 Å and Tl–I 2.90 Å. Assuming that the ions behave as hard spheres and that the radius ratio in TlI has the limiting value, calculate the ionic radii for Cs^+, Th^+, Cl^-, I^- in eight-coordination.

8. Write down the Born–Landé equation and define the terms in it. Use the equation to show why some crystals, which according to the radius ratio concept should adopt a coordination number of 8, in fact have a coordination number of 6.

9. Outline a Born–Haber cycle for the formation of an ionic compound MCl. Define the terms used and state how these might be measured or calculated. How do these enthalpy terms vary throughout the periodic table? Use these variations to suggest how the properties of NaCl might differ from those of CuCl.

10. Explain the term lattice energy as applied to an ionic solid. Calculate the lattice energy of caesium chloride using the following data:

$$Cs(s) \rightarrow Cs(g) \qquad \Delta H = +79.9\,\text{kJ mol}^{-1}$$
$$Cs(g) \rightarrow Cs^+(g) \qquad \Delta H = +374.05\,\text{kJ mol}^{-1}$$
$$Cl_2(g) \rightarrow 2Cl(g) \qquad \Delta H = +241.84\,\text{kJ mol}^{-1}$$
$$Cl(g) + e \rightarrow Cl^-(g) \qquad \Delta H = -397.90\,\text{kJ mol}^{-1}$$
$$Cs(s) + \tfrac{1}{2}Cl_2 \rightarrow CsCl(s) \qquad \Delta H = -623.00\,\text{kJ mol}^{-1}$$

11. (a) Draw the structures of CsCl and TiO_2, showing clearly the coordination of the cations and anions. (b) Show how the Born–Haber cycle may be used to estimate the enthalpy of the hypothetical reaction:

$$Ca(s) + \tfrac{1}{2}Cl_2(g) \rightarrow CaCl(s)$$

Explain why CaCl(s) has never been made even though the enthalpy for this reaction is negative.

12. The standard enthalpy changes ΔH° at 298 K for the reaction:

$$MCl_2(s) + \tfrac{1}{2}Cl_2(g) \rightarrow MCl_3(s)$$

are given for the first row transition metals:

	Sc	Ti	V	Cr	Mn	Fe	Co	Ni	Cu
ΔH°/kJ mol^{-1}	−339	−209	−138	−160	+22	−59	+131	+280	+357

Use a Born–Haber cycle to account for the change in ΔH° as the atomic number of the metal increases. Comment on the relative stabilities of the +II and +III oxidation states of the 3*d* metals.

13. List the types of defect that occur in the solid state and give an example of each. Explain in each case if any electrical conduction is possible and by what mechanism.

4 The covalent bond

INTRODUCTION

There are several different theories which explain the electronic structures and shapes of known molecules, and attempt to predict the shape of molecules whose structures are so far unknown. Each theory has its own virtues and shortcomings. None is rigorous. Theories change in the light of new knowledge and fashion. If we knew or could prove what a bond was, we would not need theories, which by definition cannot be proved. The value of a theory lies more in its usefulness than in its truth. Being able to predict the shape of a molecule is important. In many cases all the theories give the correct answer.

THE LEWIS THEORY

The octet rule

The Lewis theory was the first explanation of a covalent bond in terms of electrons that was generally accepted. If two electrons are shared between two atoms, this constitutes a bond and binds the atoms together. For many light atoms a stable arrangement is attained when the atom is surrounded by eight electrons. This octet can be made up from some electrons which are 'totally owned' and some electrons which are 'shared'. Thus atoms continue to form bonds until they have made up an octet of electrons. This is called the 'octet rule'. The octet rule explains the observed valencies in a large number of cases. There are exceptions to the octet rule; for example, hydrogen is stable with only two electrons. Other exceptions are discussed later. A chlorine atom has seven electrons in its outer shell, so by sharing one electron with another chlorine atom both atoms attain an octet and form a chlorine molecule Cl_2.

$$:\overset{..}{\underset{..}{Cl}}\,.\quad\cdot\,\overset{..}{\underset{..}{Cl}}: \;\rightarrow\; :\overset{..}{\underset{..}{Cl}}:\overset{..}{\underset{..}{Cl}}:$$

A carbon atom has four electrons in its outer shell, and by sharing all four electrons and forming four bonds it attains octet status in CCl_4.

$$\cdot\dot{\underset{\cdot}{\mathrm{C}}}\cdot + 4\left[\cdot\ddot{\underset{\cdot\cdot}{\mathrm{Cl}}}:\right] \rightarrow \begin{array}{c} \mathrm{Cl} \\ \cdot\cdot \\ \mathrm{Cl} : \mathrm{C} : \mathrm{Cl} \\ \cdot\cdot \\ \mathrm{Cl} \end{array}$$

In a similar way, a nitrogen atom has five outer electrons, and in NH_3 it shares three of these, forming three bonds and thus attaining an octet. Hydrogen has only one electron, and by sharing it attains a stable arrangement of two electrons.

$$\cdot\ddot{\underset{\cdot}{\mathrm{N}}}\cdot + 3\left[\mathrm{H}\cdot\right] \rightarrow \begin{array}{c} \cdot\cdot \\ \mathrm{H} : \mathrm{N} : \mathrm{H} \\ \cdot\cdot \\ \mathrm{H} \end{array}$$

In a similar way an atom of oxygen attains an octet by sharing two electrons in H_2O and an atom of fluorine attains an octet by sharing one electron in HF.

$$\begin{array}{c} \cdot\cdot \\ \mathrm{H} : \mathrm{O} : \\ \cdot\cdot \\ \mathrm{H} \end{array} \qquad \mathrm{H} : \ddot{\underset{\cdot\cdot}{\mathrm{F}}} :$$

Double bonds are explained by sharing four electrons between two atoms, and triple bonds by sharing six electrons.

$$\cdot\dot{\underset{\cdot}{\mathrm{C}}}\cdot + 2\left[.\ddot{\underset{\cdot}{\mathrm{O}}}:\right] \rightarrow :\underset{\cdot\cdot}{\mathrm{O}} \vdots \mathrm{C} \vdots \underset{\cdot\cdot}{\mathrm{O}} :$$

Exceptions to the octet rule

The octet rule is broken in a significant number of cases:

1. For example, for atoms such as Be and B which have less than four outer electrons. Even if all the outer electrons are used to form bonds an octet cannot be attained.

$$\cdot\,\mathrm{Be}\,\cdot + 2\left[\cdot\ddot{\underset{\cdot\cdot}{\mathrm{F}}}:\right] \rightarrow :\ddot{\underset{\cdot\cdot}{\mathrm{F}}} : \mathrm{Be} : \ddot{\underset{\cdot\cdot}{\mathrm{F}}} :$$

$$\cdot\dot{\mathrm{B}}\cdot + 3\left[\cdot\ddot{\underset{\cdot\cdot}{\mathrm{F}}}:\right] \rightarrow \begin{array}{c} \cdot\cdot \\ :\mathrm{F}: \\ \cdot\cdot \\ :\ddot{\underset{\cdot\cdot}{\mathrm{F}}} : \mathrm{B} : \ddot{\underset{\cdot\cdot}{\mathrm{F}}} : \end{array}$$

2. The octet rule is also broken where atoms have an extra energy level which is close in energy to the p level, and may accept electrons and be

used for bonding. PF_3 obeys the octet rule, but PF_5 does not. PF_5 has ten outer electrons, and uses one 3*s*, three 3*p* and one 3*d* orbitals. Any compound with more than four covalent bonds must break the octet rule, and these violations become increasingly common in elements after the first two periods of eight elements in the periodic table.
3. The octet rule does not work in molecules which have an odd number of electrons, such as NO and ClO_2, nor does it explain why O_2 is paramagnetic and has two unpaired electrons.

Despite these exceptions, the octet rule is surprisingly reliable and did a great deal to explain the number of bonds formed in simple cases. However, it gives no indication of the shape adopted by the molecule.

SIDGWICK–POWELL THEORY

In 1940 Sidgwick and Powell (see Further Reading) reviewed the structures of molecules then known. They suggested that for molecules and ions that only contain single bonds, the approximate shape can be predicted from the number of electron pairs in the outer or valence shell of the central atom. The outer shell contains one or more bond pairs of electrons, but it may also contain unshared pairs of electrons (lone pairs). Bond pairs and lone pairs were taken as equivalent, since all electron pairs take up some space, and since all electron pairs repel each other. Repulsion is minimized if the electron pairs are orientated in space as far apart as possible.

1. If there are two pairs of electrons in the valence shell of the central atom, the orbitals containing them will be oriented at 180° to each other. It follows that if these orbitals overlap with orbitals from other atoms to form bonds, then the molecule formed will be linear.
2. If there are three electron pairs on the central atom, they will be at 120° to each other, giving a plane triangular structure.
3. For four electron pairs the angle is 109°28′, and the shape is tetrahedral.
4. For five pairs, the shape is a trigonal bipyramid.
5. For six pairs the angles are 90° and the shape is octahedral.

VALENCE SHELL ELECTRON PAIR REPULSION (VSEPR) THEORY

In 1957 Gillespie and Nyholm (see Further Reading) improved the Sidgwick–Powell theory to predict and explain molecular shapes and bond angles more exactly. The theory was developed extensively by Gillespie as the Valence Shell Electron Pair Repulsion (VSEPR) theory. This may be summarized:

1. The shape of the molecule is determined by repulsions between all of the electron pairs present in the valence shell. (This is the same as the Sidgwick–Powell theory.)

Table 4.1 Molecular shapes predicted by Sidgwick–Powell theory

Number of electron pairs in outer shell	Shape of molecule	Bond angles
2	linear	180°
3	plane triangle	120°
4	tetrahedron	109°28′
5	trigonal bipyramid	120° and 90°
6	octahedron	90°
7	pentagonal bipyramid	72° and 90°

2. A lone pair of electrons takes up more space round the central atom than a bond pair, since the lone pair is attracted to one nucleus whilst the bond pair is shared by two nuclei. It follows that repulsion between two lone pairs is greater than repulsion between a lone pair and a bond pair, which in turn is greater than the repulsion between two bond pairs. Thus the presence of lone pairs on the central atom causes slight distortion of the bond angles from the ideal shape. If the angle between a lone pair, the central atom and a bond pair is increased, it follows that the actual bond angles between the atoms must be decreased.
3. The magnitude of repulsions between bonding pairs of electrons depends on the electronegativity difference between the central atom and the other atoms.
4. Double bonds cause more repulsion than single bonds, and triple bonds cause more repulsion than a double bond.

Effect of lone pairs

Molecules with four electron pairs in their outer shell are based on a tetrahedron. In CH_4 there are four bonding pairs of electrons in the outer shell of the C atom, and the structure is a regular tetrahedron with bond angles H—C—H of 109°28′. In NH_3 the N atom has four electron pairs in the outer shell, made up of three bond pairs and one lone pair. Because of the lone pair, the bond angle H—N—H is reduced from the theoretical tetrahedral angle of 109°28′ to 107°48′. In H_2O the O atom has four electron pairs in the outer shell. The shape of the H_2O molecule is based on a tetrahedron with two corners occupied by bond pairs and the other two corners occupied by lone pairs. The presence of two lone pairs reduces the bond angle further to 104°27′.

In a similar way SF_6 has six bond pairs in the outer shell and is a regular octahedron with bond angles of exactly 90°. In BrF_5 the Br also has six outer pairs of electrons, made up of five bond pairs and one lone pair. The lone pair reduces the bond angles to 84°30′. Whilst it might be expected that two lone pairs would distort the bond angles in an octahedron still further, in XeF_4 the angles are 90°. This is because the lone pairs are *trans* to each other in the octahedron, and hence the atoms have a regular square planar arrangement.

Molecules with five pairs of electrons are all based on a trigonal bipyramid. Lone pairs distort the structures as before. The lone pairs always occupy the equatorial positions (in the triangle), rather than the apical positions (up and down). Thus in the I_3^- ion the central I atom has five electron pairs in the outer shell, made of two bond pairs and three lone pairs. The lone pairs occupy all three equatorial positions and the three atoms occupy the top, middle, and bottom positions in the trigonal bipyramid, thus giving a linear arrangement with a bond angle of exactly 180° (Table 4.2).

Effect of electronegativity

NF_3 and NH_3 both have structures based on a tetrahedron with one corner occupied by a lone pair. The high electronegativity of F pulls the bonding electrons further away from N than in NH_3. Thus repulsion between bond pairs is less in NF_3 than in NH_3. Hence the lone pair in NF_3 causes a greater distortion from tetrahedral and gives a F—N—F bond angle of 102°30′, compared with 107°48′ in NH_3. The same effect is found in H_2O (bond angle 104°27′) and F_2O (bond angle 102°).

Table 4.2 The effects of bonding and lone pairs on bond angles

	Orbitals on central atom	Shape	Number of bond pairs	Number of lone pairs	Bond angle
$BeCl_2$	2	Linear	2	0	180°
BF_3	3	Plane triangle	3	0	120°
CH_4	4	Tetrahedral	4	0	109°28′
NH_3	4	Tetrahedral	3	1	107°48′
NF_3	4	Tetrahedral	3	1	102°30′
H_2O	4	Tetrahedral	2	2	104°27′
F_2O	4	Tetrahedral	2	2	102°
PCl_5	5	Trigonal bipyramid	5	0	120° and 90°
SF_4	5	Trigonal bipyramid	4	1	101°36′ and 86°33′
ClF_3	5	Trigonal bipyramid	3	2	87°40′
I_3^-	5	Trigonal bipyramid	2	3	180°
SF_6	6	Octahedral	6	0	90°
BrF_5	6	Octahedral	5	1	84°30′
XeF_4	6	Octahedral	4	2	90°

Isoelectronic principle

Isoelectronic species usually have the same structure. This may be extended to species with the same number of valence electrons. Thus BF_4^-, CH_4 and NH_4^+ are all tetrahedral, CO_3^{2-}, NO_3^- and SO_3 are all planar triangles, and CO_2, N_3^- and NO_2^+ are all linear.

SOME EXAMPLES USING THE VSEPR THEORY

BF_3 and the $[BF_4]^-$ ion

Consider BF_3 first. The VSEPR theory only requires the number of electron pairs in the outer shell of the central atom. Since B is in Group III it has three electrons in the outer shell. (Alternatively the electronic structure of B (the central atom), is $1s^2\,2s^2\,2p^1$, so there are three electrons in the outer valence shell.) If all three outer electrons are used to form bonds to three F atoms, the outer shell then has a share in six electrons, that is three electron pairs. Thus the structure is a planar triangle.

The $[BF_4]^-$ ion may be regarded as being formed by adding a F^- ion to a BF_3 molecule by means of a coordinate bond. Thus the B atom now has three electron pairs from the BF_3 plus one electron pair from the F^-. There are therefore four electron pairs in the outer shell: hence the BF_4^- ion has a tetrahedral structure.

Ammonia NH_3

N is the central atom. It is in Group V and has five electrons in the outer valence shell. (The electronic structure of N is $1s^2\,2s^2\,2p^3$.) Three of these electrons are used to form bonds to three H atoms, and two electrons take no part in bonding and constitute a 'lone pair'. The outer shell then has a share in eight electrons, that is three bond pairs of electrons and one lone pair. Four electron pairs give rise to a tetrahedral structure and in this case three positions are occupied by H atoms and the fourth position is occupied by the lone pair (Figure 4.1). The shape of NH_3 may either be described as tetrahedral with one corner occupied by a lone pair, or alternatively as pyramidal. The presence of the lone pair causes slight distortion from 109°28′ to 107°48′.

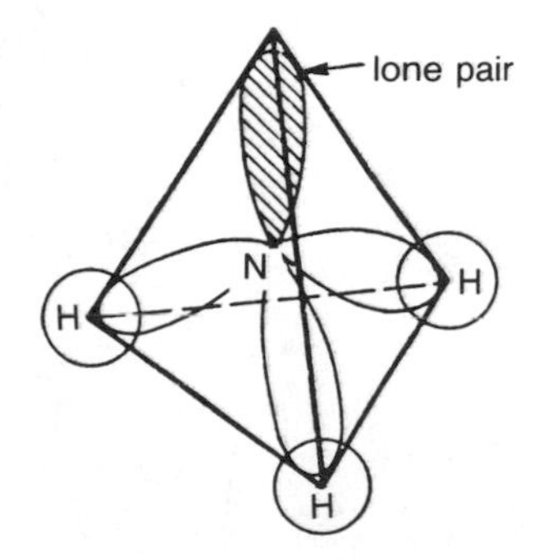

Figure 4.1 Structure of NH_3.

Water H_2O

O is the central atom. It is in Group VI and hence has six outer electrons. (The electronic structure of O is $1s^2\,2s^2\,2p^4$.) Two of these electrons form bonds with two H atoms, thus completing the octet. The other four outer electrons on O are non-bonding. Thus in H_2O the O atom has eight outer electrons (four electron pairs) so the structure is based on a tetrahedron. There are two bond pairs and two lone pairs. The structure is described as tetrahedral with two positions occupied by lone pairs. The two lone pairs distort the bond angle from 109°28′ to 104°27′ (Figure 4.2).

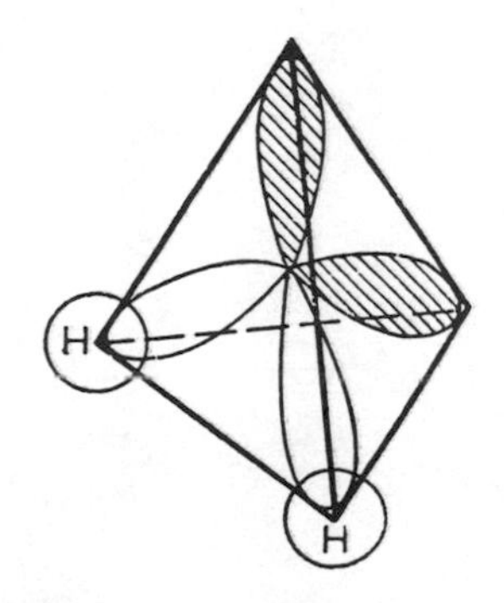

Figure 4.2 Structure of H_2O.

Any triatomic molecule must be either linear with a bond angle of 180°, or else angular, that is bent. In H_2O the molecule is based on a tetrahedron, and is therefore bent.

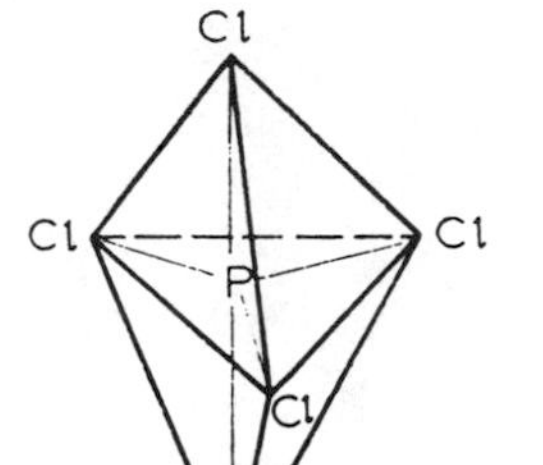

Figure 4.3 Structure of PCl_5 molecule.

Phosphorus pentachloride PCl_5

Gaseous PCl_5 is covalent. P (the central atom) is in Group V and so has five electrons in the outer shell. (The electronic structure of P is $1s^2\ 2s^2\ 2p^6\ 3s^2\ 3p^3$.) All five outer electrons are used to form bonds to the five Cl atoms. In the PCl_5 molecule the valence shell of the P atom contains five electron pairs: hence the structure is a trigonal bipyramid. There are no lone pairs, so the structure is not distorted. However, a trigonal bipyramid is not a completely regular structure, since some bond angles are 90° and others 120°. Symmetrical structures are usually more stable than asymmetrical ones. Thus PCl_5 is highly reactive, and in the solid state it splits into $[PCl_4]^+$ and $[PCl_6]^-$ ions, which have tetrahedral and octahedral structures respectively.

Chlorine trifluoride ClF_3

The chlorine atom is at the centre of the molecule and determines its shape. Cl is in Group VII and so has seven outer electrons. (The electronic structure of Cl is $1s^2\ 2s^2\ 2p^6\ 3s^2\ 3p^5$.) Three electrons form bonds to F, and four electrons do not take part in bonding. Thus in ClF_3 the Cl atom has five electron pairs in the outer shell: hence the structure is a trigonal bipyramid. There are three bond pairs and two lone pairs.

It was noted previously that a trigonal bipyramid is not a regular shape since the bond angles are not all the same. It therefore follows that the corners are not equivalent. Lone pairs occupy two of the corners, and F atoms occupy the other three corners. Three different arrangements are theoretically possible, as shown in Figure 4.4.

The most stable structure will be the one of lowest energy, that is the one with the minimum repulsion between the five orbitals. The greatest repulsion occurs between two lone pairs. Lone pair–bond pair repulsions are next strongest, and bond pair–bond pair repulsions the weakest.

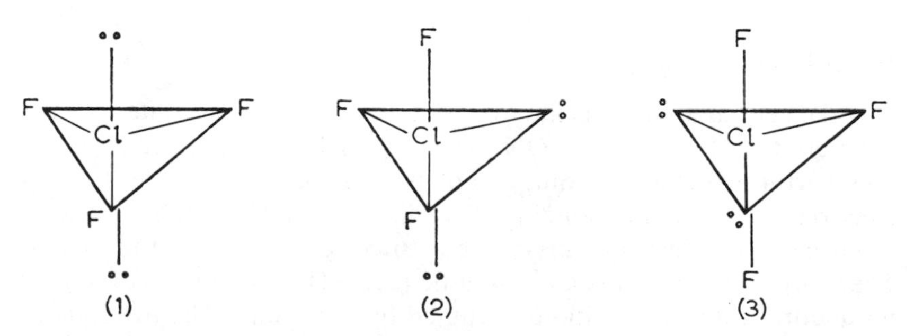

Figure 4.4 Chlorine trifluoride molecule.

Groups at 90° to each other repel each other strongly, whilst groups 120° apart repel each other much less.

Structure 1 is the most symmetrical, but has six 90° repulsions between lone pairs and and atoms. Structure 2 has one 90° repulsion between two lone pairs, plus three 90° repulsions between lone pairs and atoms. Structure 3 has four 90° repulsions between lone pairs and atoms. These factors indicate that structure 3 is the most probable. The observed bond angles are 87°40′, which is close to the theoretical 90°. This confirms that the correct structure is (3), and the slight distortion from 90° is caused by the presence of the two lone pairs.

As a general rule, if lone pairs occur in a trigonal bipyramid they will be located in the equatorial positions (round the middle) rather than the apical positions (top and bottom), since this arrangement minimizes repulsive forces.

Sulphur tetrafluoride SF_4

S is in Group VI and thus has six outer electrons. (The electronic configuration of S is $1s^2\,2s^2\,2p^6\,3s^2\,3p^4$.) Four outer electrons are used to form bonds with the F atoms, and two electrons are non-bonding. Thus in SF_4 the S has five electron pairs in the outer shell: hence the structure is based on a trigonal bipyramid. There are four bond pairs and one lone pair. To minimize the repulsive forces the lone pair occupies an equatorial position, and F atoms are located at the other four corners, as shown in Figure 4.5.

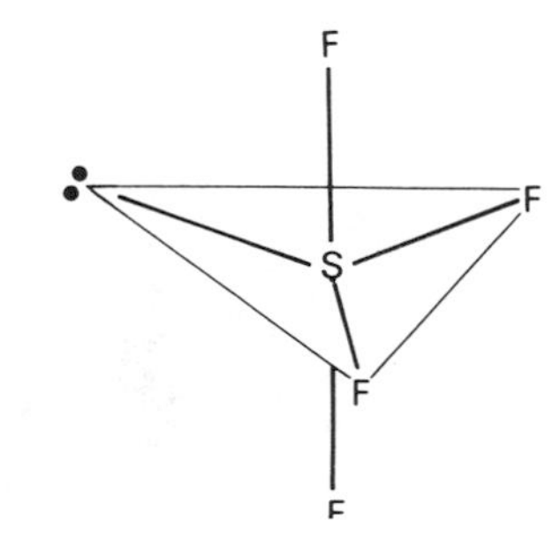

Figure 4.5 Sulphur tetrafluoride molecule.

The triiodide ion I_3^-

If iodine is dissolved in aqueous potassium iodide, the triiodide ion I_3^- is formed. This is an example of a polyhalide ion, which is similar in structure to $BrICl^-$ (see Chapter 15). The I_3^- ion (Figure 4.6) has three atoms, and must be either linear or angular in shape. It is convenient to consider the structure in a series of stages – first an I atom, then an I_2 molecule, and then the I_3^- ion made up of an I_2 molecule with an I^- bonded to it by means of a coordinate bond.

$$I_2 + I^- \rightarrow [I{-}I{\leftarrow}I]^-$$

Iodine is in Group VII and so has seven outer electrons. (The electronic configuration of I is $1s^2\,2s^2\,2p^6\,3s^2\,3p^6\,3d^{10}\,4s^2\,4p^6\,4d^{10}\,5s^2\,5p^5$.) One of the outer electrons is used to bond with another I atom, thus forming an I_2 molecule. The I atoms now have a share in eight electrons. One of the I atoms in the I_2 molecule accepts a lone pair from an I^- ion, thus forming an I_3^- ion. The outer shell of the central I atom now contains ten electrons, that is five electron pairs. Thus the shape is based on a trigonal bipyramid. There are two bond pairs and three lone pairs. To minimize the repulsive forces the three lone pairs occupy the equatorial positions, and I atoms are located at the centre and in the two apical positions. The ion is therefore linear in shape, with a bond angle of exactly 180°.

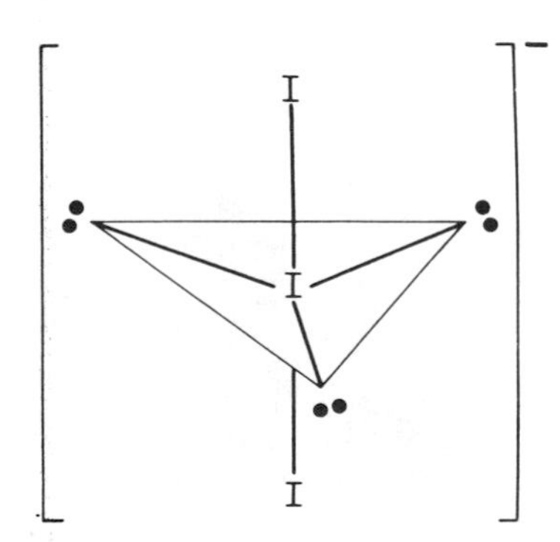

Figure 4.6 The triiodide ion.

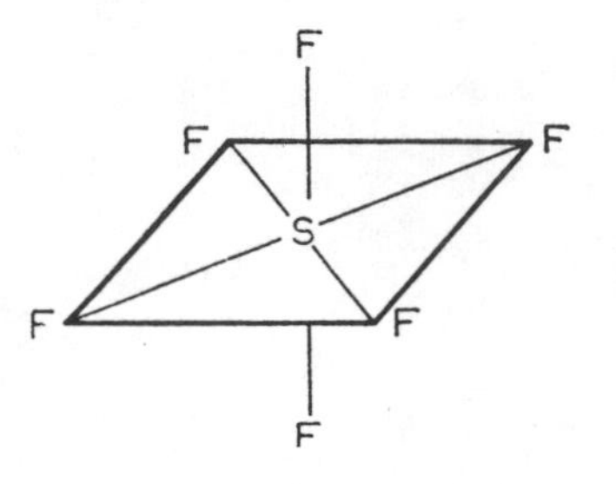

Figure 4.7 Sulphur hexafluoride molecule.

Sulphur hexafluoride SF_6

Sulphur is in Group VI and thus has six outer electrons. (The electronic structure of S is $1s^2\ 2s^2\ 2p^6\ 3s^2\ 3p^4$.) All six of the outer electrons are used to form bonds with the F atoms. Thus in SF_6 the S has six electron pairs in the outer shell: hence the structure is octahedral. There are no lone pairs so the structure is completely regular with bond angles of 90°.

Iodine heptafluoride IF_7

This is the only common example of a non-transition element using seven orbitals for bonding giving a pentagonal bipyramid. (See Chapter 15).

The total numbers of outer orbitals, bonding orbitals and lone pairs are related to the commonly occurring shapes of molecules in Table 4.4.

VALENCE BOND THEORY

This theory was proposed by Linus Pauling, who was awarded the Nobel Prize for Chemistry in 1954. The theory was very widely used in the period 1940–1960. Since then it has to some extent fallen out of fashion. However, it is still much used by organic chemists, and it provides a basis for simple description of small inorganic molecules.

Atoms with unpaired electrons tend to combine with other atoms which also have unpaired electrons. In this way the unpaired electrons are paired up, and the atoms involved all attain a stable electronic arrangement. This is usually a full shell of electrons (i.e. a noble gas configuration). Two electrons shared between two atoms constitute a bond. The number of bonds formed by an atom is usually the same as the number of unpaired electrons in the ground state, i.e. the lowest energy state. However, in some cases the atom may form more bonds than this. This occurs by excitation of the atom (i.e. providing it with energy) when electrons which were paired in the ground state are unpaired and promoted into suitable empty orbitals. This increases the number of unpaired electrons, and hence the number of bonds which can be formed.

The shape of the molecule is determined primarily by the directions in which the orbitals point. Electrons in the valence shell of the original atom which are paired are called lone pairs.

A covalent bond results from the pairing of electrons (one from each atom). The spins of the two electrons must be opposite (antiparallel) because of the Pauli exclusion principle that no two electrons in one atom can have all four quantum numbers the same.

Consider the formation of a few simple molecules.

1. In HF, H has a singly occupied *s* orbital that overlaps with a singly filled $2p$ orbital on F.
2. In H_2O, the O atom has two singly filled $2p$ orbitals, each of which overlaps with a singly occupied *s* orbital from two H atoms.

3. In NH_3, there are three singly occupied p orbitals on N which overlap with s orbitals from three H atoms.
4. In CH_4, the C atom in its ground state has the electronic configuration $1s^2$, $2s^2$, $2p_x^1$, $2p_y^1$ and only has two unpaired electrons, and so can form only two bonds. If the C atom is excited, then the $2s$ electrons may be unpaired giving $1s^2$, $2s^1$, $2p_x^1$, $2p_y^1$, $2p_z^1$. There are now four unpaired electrons which overlap with singly occupied s orbitals on four H atoms.

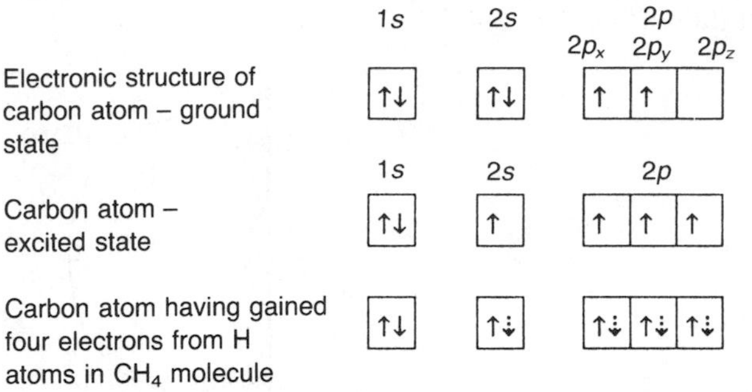

	1s	2s	2p ($2p_x$ $2p_y$ $2p_z$)
Electronic structure of carbon atom – ground state	↑↓	↑↓	↑ | ↑ |
Carbon atom – excited state	↑↓	↑	↑ | ↑ | ↑
Carbon atom having gained four electrons from H atoms in CH_4 molecule	↑↓	↑⇣	↑⇣ | ↑⇣ | ↑⇣

The shape of the CH_4 molecule is not immediately apparent. The three p orbitals p_x, p_y and p_z are mutually at right angles to each other, and the s orbital is spherically symmetrical. If the p orbitals were used for bonding then the bond angle in water should be 90°, and the bond angles in NH_3 should also be 90°. The bond angles actually found differ appreciably from these:

CH_4	H—C—H	= 109°28′
NH_3	H—N—H	= 107°48′
H_2O	H—O—H	= 104°27′

Hybridization

The chemical and physical evidence indicates that in methane CH_4 there are four equivalent bonds. If they are equivalent, then repulsion between electron pairs will be a minimum if the four orbitals point to the corners of a tetrahedron, which would give the observed bond angle of 109°28′.

Each electron can be described by its wave function ψ. If the wave functions of the four outer atomic orbitals of C are ψ_{2s}, ψ_{2p_x}, ψ_{2p_y}, and ψ_{2p_z}, then the tetrahedrally distributed orbitals will have wave functions ψ_{sp^3} made up from a linear combination of these four atomic wave functions.

$$\psi_{sp^3} = c_1\psi_{2s} + c_2\psi_{2p_x} + c_3\psi_{2p_y} + c_4\psi_{2p_z}$$

There are four different combinations with different weighting constants c_1, c_2, c_3 and c_4.

$$\psi_{sp^3(1)} = \tfrac{1}{2}\psi_{2s} + \tfrac{1}{2}\psi_{2p_x} + \tfrac{1}{2}\psi_{2p_y} + \tfrac{1}{2}\psi_{2p_z}$$
$$\psi_{sp^3(2)} = \tfrac{1}{2}\psi_{2s} + \tfrac{1}{2}\psi_{2p_x} - \tfrac{1}{2}\psi_{2p_y} - \tfrac{1}{2}\psi_{2p_z}$$

$$\psi_{sp^3(3)} = \tfrac{1}{2}\psi_{2s} - \tfrac{1}{2}\psi_{2p_x} + \tfrac{1}{2}\psi_{2p_y} - \tfrac{1}{2}\psi_{2p_z}$$
$$\psi_{sp^3(4)} = \tfrac{1}{2}\psi_{2s} - \tfrac{1}{2}\psi_{2p_x} - \tfrac{1}{2}\psi_{2p_y} + \tfrac{1}{2}\psi_{2p_z}$$

Combining or mixing the wave functions for the atomic orbitals in this way is called hybridization. Mixing one *s* and three *p* orbitals in this way gives four sp^3 hybrid orbitals. The shape of an sp^3 orbital is shown in Figure 4.8. Since one lobe is enlarged, it can overlap more effectively than an *s* orbital or a *p* orbital on its own. Thus sp^3 hybrid orbitals form stronger bonds than the original atomic orbitals. (See Table 4.3.)

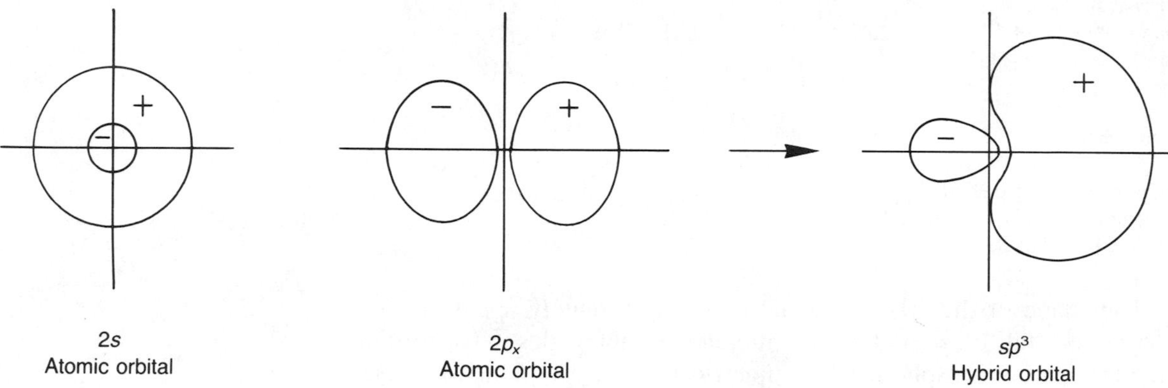

Figure 4.8 Combination of *s* and *p* atomic orbitals to give an sp^3 hybrid orbital: (a) 2*s* atomic orbital, (b) $2p_x$ atomic orbital and (c) sp^3 hybrid orbital.

Table 4.3 Approximate strengths of bonds formed by various orbitals

Orbital	Relative bond strength
s	1.0
p	1.73
sp	1.93
sp^2	1.99
sp^3	2.00

It is possible to mix other combinations of atomic orbitals in a similar way. The structure of a boron trifluoride BF_3 molecule is a planar triangle with bond angles of 120°. The B atom is the central atom in the molecule, and it must be excited to give three unpaired electrons so that it can form three covalent bonds.

	1s	2s	2p		
Boron atom – ground state	↑↓	↑↓	↑		
Boron atom – excited state	↑↓	↑	↑	↑	

BF_3 molecule having gained a share in three electrons by bonding to three F atoms

↑↓ ↑⇣ ↑⇣ ↑⇣

sp^2 hybridization of the three orbitals in outer shell, hence structure is a planar triangle

Combining the wave functions of the $2s$, $2p_x$ and $2p_y$ atomic orbitals gives three hybrid sp^2 orbitals.

$$\psi_{sp^2(1)} = \frac{1}{\sqrt{3}}\psi_{2s} + \frac{2}{\sqrt{6}}\psi_{2p_x}$$

$$\psi_{sp^2(2)} = \frac{1}{\sqrt{3}}\psi_{2s} + \frac{1}{\sqrt{6}}\psi_{2p_x} + \frac{1}{\sqrt{2}}\psi_{2p_y}$$

$$\psi_{sp^2(3)} = \frac{1}{\sqrt{3}}\psi_{2s} + \frac{1}{\sqrt{6}}\psi_{2p_x} - \frac{1}{\sqrt{2}}\psi_{2p_y}$$

These three orbitals are equivalent, and repulsion between them is minimized if they are distributed at 120° to each other giving a planar triangle. In the hybrid orbitals one lobe is bigger than the other, so it can overlap more effectively and hence form a stronger bond than the original atomic orbitals. (See Table 4.3.) Overlap of the sp^2 orbitals with p orbitals from F atoms gives the planar triangular molecule BF_3 with bond angles of 120°.

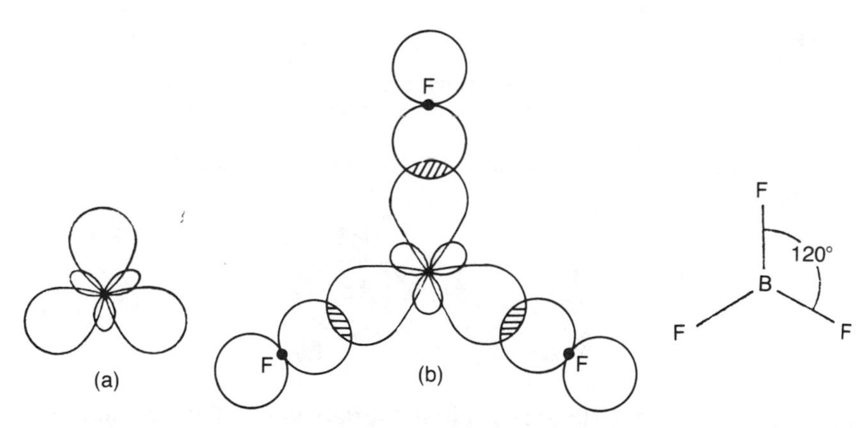

Figure 4.9 (a) sp^2 hybrid orbitals and (b) the BF_3 molecule.

The structure of a gaseous molecule of beryllium fluoride BeF_2 is linear F—Be—F. Be is the central atom in this molecule and determines the shape of the molecule formed. The ground state electronic configuration of Be is $1s^2\ 2s^2$. This has no unpaired electrons, and so can form no bonds. If energy is supplied, an excited state will be formed by unpairing and promoting a $2s$ electron to an empty $2p$ level, giving $1s^2\ 2s^1\ 2p_x^1$. There are now two unpaired electrons, so the atom can form the required two bonds.

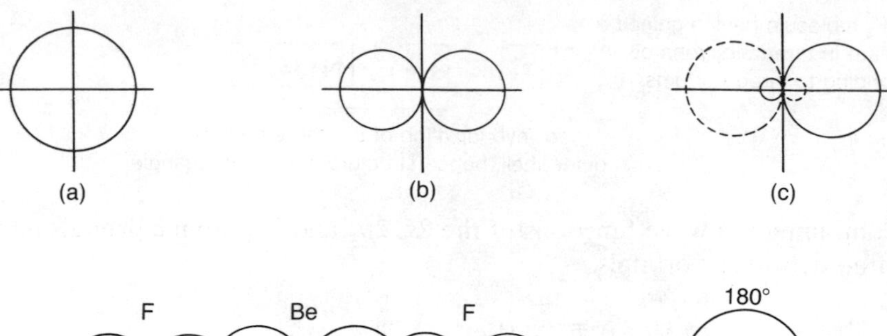

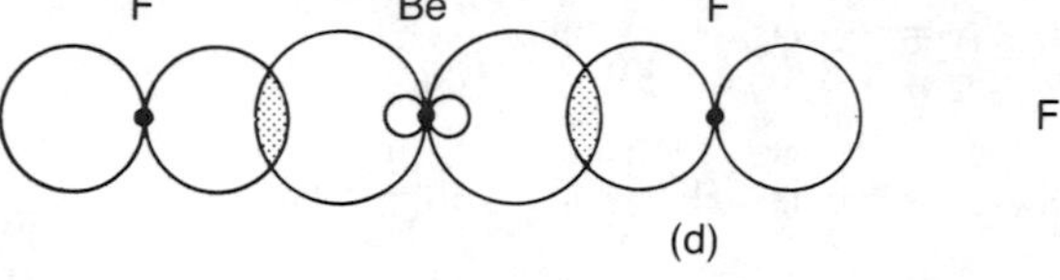

Figure 4.10 (a) *s* orbital, (b) *p* orbital, (c) formation of two *sp* hybrid orbitals and (d) their use in forming beryllium difluoride.

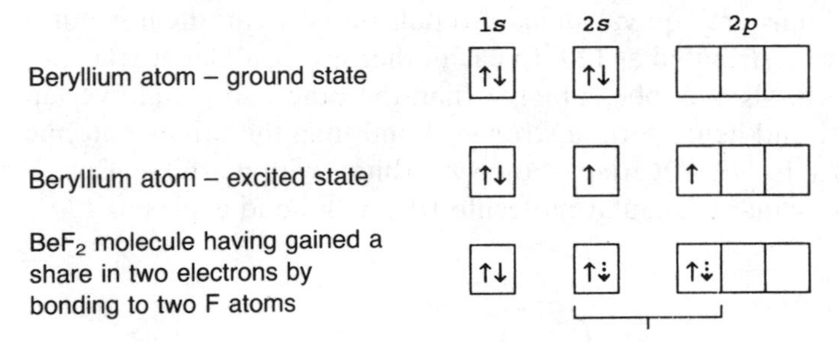

Hybridizing the 2*s* and $2p_x$ atomic orbitals gives two equivalent *sp* hybrid orbitals.

$$\psi_{sp(1)} = \frac{1}{\sqrt{2}}\psi_{2s} + \frac{1}{\sqrt{2}}\psi_{2p_x}$$

$$\psi_{sp(2)} = \frac{1}{\sqrt{2}}\psi_{2s} - \frac{1}{\sqrt{2}}\psi_{2p_x}$$

Because of their shape these *sp* orbitals overlap more effectively and result in stronger bonds than the original atomic orbitals. Repulsion is minimized if these two hybrid orbitals are oriented at 180° to each other. If these orbitals overlap with *p* orbitals on F atoms, a linear BeF_2 molecule is obtained.

It should in principle be possible to calculate the relative strength of bonds formed using *s*, *p* or various hybrid orbitals. However, the wave equation can only be solved exactly for atoms containing one electron, that is hydrogen-like species H, He^+, Li^{2+}, Be^{3+} etc. Attempts to work out the relative bond strengths involve approximations, which may or may not be valid. On this basis it has been suggested that the relative strengths of

Table 4.4 Number of orbitals and type of hybridization

Number of outer orbitals	Type of hybridization	Distribution in space of hybrid orbitals
2	sp	Linear
3	sp^2	Plane triangle
4	sp^3	Tetrahedron
5	sp^3d	Trigonal bipyramid
6	sp^3d^2	Octahedron
7	sp^3d^3	Pentagonal bipyramid
(4	dsp^2)	Square planar

bonds using *s*, *p* and various hybrid atomic orbitals may be as shown in Table 4.3.

Hybridization and the mixing of orbitals is a most useful concept. Mixing of *s* and *p* orbitals is well accepted, but the involvement of *d* orbitals is controversial. For effective mixing, the energy of the orbitals must be nearly the same.

It is a common misconception that hybridization is the cause of a particular molecular shape. This is not so. The reason why any particular shape is adopted is its energy. It is also important to remember that the hybridized state is a theoretical step in going from an atom to a molecule, and the hybridized state never actually exists. It cannot be detected even spectroscopically, so the energy of hybrid orbitals cannot be measured and can only be estimated theoretically.

THE EXTENT OF *d* ORBITAL PARTICIPATION IN MOLECULAR BONDING

The bonding in PCl_5 may be described using hybrids of the 3*s*, 3*p* and 3*d* atomic orbitals for P – see below. However, there are doubts as to whether *d* orbitals can take part and this has led to the decline of this theory.

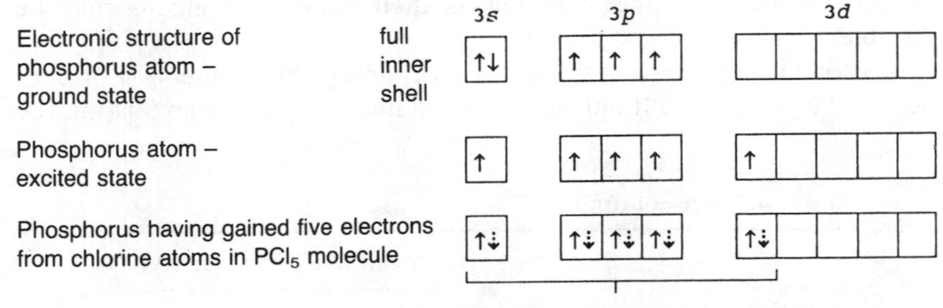

sp^3d hybridization, trigonal bipyramid

However, *d* orbitals are in general too large and too high in energy to mix completely with *s* and *p* orbitals. The difference in size is illustrated by the mean values for the radial distance for different phosphorus orbitals: 3*s* = 0.47 Å, 3*p* = 0.55 Å and 3*d* = 2.4 Å. The energy of an orbital is proportional to its mean radial distance, and since the 3*d* orbital is much larger it is much higher in energy than the 3*s* and 3*p* orbitals. It would at first seem unlikely that hybridization involving *s*, *p* and *d* orbitals could possibly occur.

Several factors affect the size of orbitals. The most important is the charge on the atom. If the atom carries a formal positive charge then all the electrons will be pulled in towards the nucleus. The effect is greatest for the outer electrons. If the central P atom is bonded to a highly electronegative element such as F, O or Cl, then the electronegative element attracts more than its share of the bonding electrons and the F or Cl atom attains a $\delta-$ charge. This leaves a $\delta+$ charge on P, which makes the orbitals contract. Since the 3*d* orbital contracts in size very much more than the 3*s* and 3*p* orbitals, the energies of the 3*s*, 3*p* and 3*d* orbitals may become close enough to allow hybridization to occur in PCl_5. Hydrogen does not cause this large contraction, so PH_5 does not exist.

In a similar way the structure of SF_6 can be described by mixing the 3*s*, three 3*p* and two 3*d* orbitals, that is sp^3d^2 hybridization.

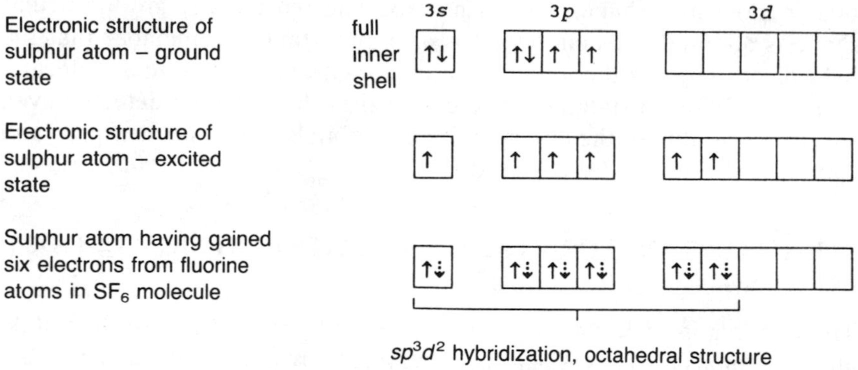

		3s	3p	3d
Electronic structure of sulphur atom – ground state	full inner shell	↑↓	↑↓ ↑ ↑	☐ ☐ ☐ ☐ ☐
Electronic structure of sulphur atom – excited state		↑	↑ ↑ ↑	↑ ↑ ☐ ☐ ☐
Sulphur atom having gained six electrons from fluorine atoms in SF_6 molecule		↑↓	↑↓ ↑↓ ↑↓	↑↓ ↑↓ ☐ ☐ ☐

sp^3d^2 hybridization, octahedral structure

The presence of six highly electronegative F atoms causes a large contraction of the *d* orbitals, and lowers their energy, so mixing may be possible.

A second factor affecting the size of *d* orbitals is the number of *d* orbitals occupied by electrons. If only one 3*d* orbital is occupied on an S atom, the

Table 4.5 Sizes of orbitals

	Mean radial distance (Å)		
(sp^3d^2 configuration)	3*s*	3*p*	3*d*
S atom (neutral, no charge)	0.88	0.94	1.60
S atom (charge +0.6)	0.87	0.93	1.40

average radial distance is 2.46 Å, but when two $3d$ orbitals are occupied the distance drops to 1.60 Å. The effect of changing the charge can be seen in Table 4.5.

A further small contraction of d orbitals may arise by coupling of the spins of electrons occupying different orbitals.

It seems probable that d orbitals do participate in bonding in cases where d orbital contraction occurs.

SIGMA AND PI BONDS

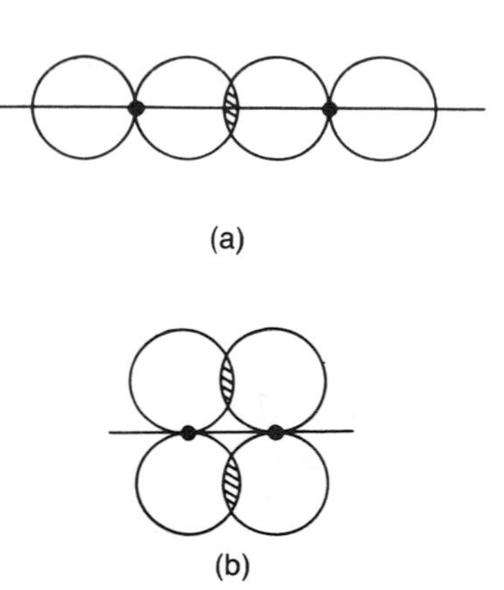

Figure 4.11 Sigma and pi overlap: (a) sigma overlap (lobes point along the nuclei); (b) pi overlap (lobes are at right angles to the line joining the nuclei).

All the bonds formed in these examples result from end to end overlap of orbitals and are called sigma σ bonds. In σ bonds the electron density is concentrated in between the two atoms, and on a line joining the two atoms. Double or triple bonds occur by the sideways overlap of orbitals, giving pi π bonds. In π bonds the electron density also concentrates between the atoms, but on either side of the line joining the atoms. The shape of the molecule is determined by the σ bonds (and lone pairs) but **not** by the π bonds. Pi bonds merely shorten the bond lengths.

Consider the structure of the carbon dioxide molecule. Since C is typically four-valent and O is typically two-valent, the bonding can be simply represented

$$O{=}C{=}O$$

Triatomic molecules must be either linear or angular. In CO_2, the C atom must be excited to provide four unpaired electrons to form the four bonds required.

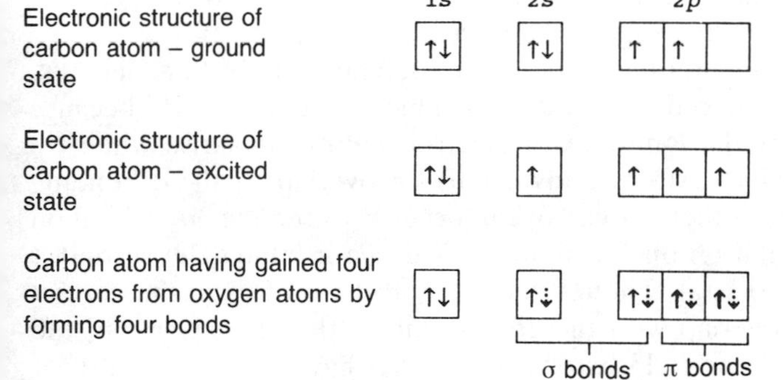

There are two σ bonds and two π bonds in the molecule. Pi orbitals are ignored in determining the shape of the molecule. The remaining s and p orbitals are used to form the σ bonds. (These could be hybridized and the two sp^2 orbitals will point in opposite directions. Alternatively VSEPR theory suggests that these two orbitals will be oriented as far apart as possible.) These two orbitals overlap with p orbitals from two O atoms, forming a linear molecule with a bond angle of 180°. The $2p_y$ and $2p_z$ orbitals on C used for π bonding are at right angles to the bond, and overlap sideways with p orbitals on the O atoms at either side. This π overlap shortens the C—O distances, but does not affect the shape.

The sulphur dioxide molecule SO_2 may be considered in a similar way. S shows oxidation states of (+II), (+IV) and (+VI), whilst O is two-valent. The structure may be represented:

$$O{=}S{=}O$$

Triatomic molecules are either linear or bent. The S atom must be excited to provide four unpaired electrons.

		3s	3p	3d
Electronic structure of sulphur atom – ground state	full inner shell	↑↓	↑↓ | ↑ | ↑	| | | |
Electronic structure of sulphur atom – excited state		↑↓	↑ | ↑ | ↑	↑ | | | |
Sulphur atom having gained four electrons from four bonds to oxygen atoms in SO_2 molecule		↑↓	↑↓ | ↑↓ | ↑↓	↑↓ | | | |

two σ bonds and one lone pair   two π bonds

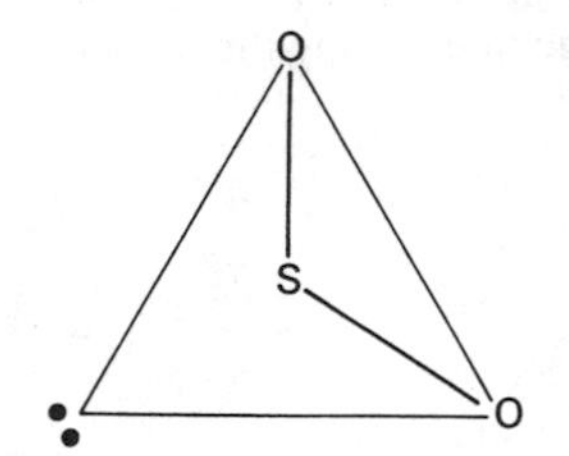

Figure 4.12 Sulphur dioxide molecule.

The two electron pairs which form the π bonds do not affect the shape of the molecule. The remaining three orbitals point to the corners of a triangle, and result in a planar triangular structure for the molecule with two corners occupied by O atoms and one corner occupied by a lone pair. The SO_2 molecule is thus angular or V shaped (Figure 4.12).

The π bonds do not alter the shape, but merely shorten the bond lengths. The bond angle is reduced from the ideal value of 120° to 119°30′ because of the repulsion by the lone pair of electrons. Problems arise when we examine exactly which AOs are involved in π overlap. If the σ bonding occurs in the xy plane then π overlap can occur between the $3p_z$ orbital on S and the $2p_z$ orbital on one O atom to give one π bond. The second π bond involves a d orbital. Though the $3d_{z^2}$ orbital on S is in the correct orientation for π overlap with the $2p_z$ orbital on the other O atom, the symmetry of the $3d_{z^2}$ orbital is wrong (both lobes have a + sign) whilst for a p orbital one lobe is + and the other −. Thus overlap of these orbitals does not result in bonding. The $3d_{xz}$ orbital on S is in the correct orientation, and has the correct symmetry to overlap with the $2p_z$ orbital on the second O atom, and could give the second π bond. It is surprising that π bonds involving p and d orbitals both have the same energy (and bond length). This calls into question whether it is correct to treat molecules with two π bonds as containing two discrete π bonds. A better approach is to treat the π bonds as being delocalized over several atoms. Examples of this treatment are given near the end of this chapter.

In the sulphur trioxide molecule SO_3 valency requirements suggest the structure

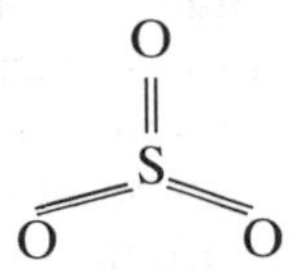

The central S atom must be excited to provide six unpaired electrons to form six bonds.

	3s	3p	3d
Electronic structure of sulphur atom – excited state	↑	↑ ↑ ↑	↑ ↑
Sulphur atom having gained six electrons from six bonds to oxygen atoms in SO_3 molecule	↑↓	↑↓ ↑↓ ↑↓	↑↓ ↑↓

three σ bonds (3s, 3p) — three π bonds (3d)

The three π bonds are ignored in determining the shape of the molecule. The three σ orbitals are directed towards the corners of an equilateral triangle, and the SO_3 molecule is a completely regular plane triangle (Figure 4.13). The π bonds shorten the bond lengths, but do not affect the shape. This approach explains the σ bonding and shape of the molecule, but the explanation of π bonding is unsatisfactory. It presumes:

1. That one $3p$ and two $3d$ orbitals on S are in the correct orientation to overlap sideways with the $2p_y$ or $2p_z$ orbitals on three different O atoms, and
2. That the π bonds formed are all of equal strength.

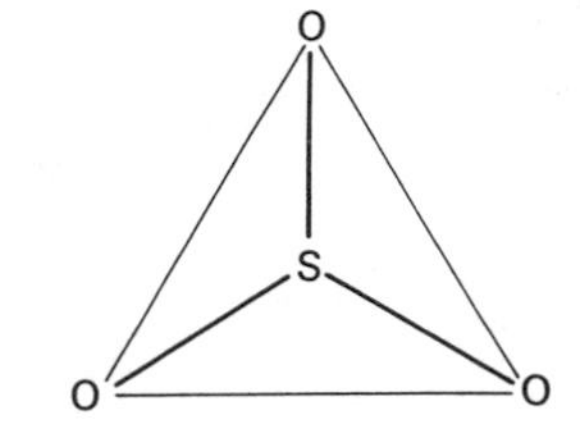

Figure 4.13 Sulphur trioxide molecule.

This calls into question the treatment of π bonds. In molecules with more than one π bond, or molecules where the π bond could equally well exist in more than one position, it is better to treat the π bonding as being delocalized over several atoms rather than localized between two atoms. This approach is developed near the end of this chapter.

MOLECULAR ORBITAL METHOD

In the valence bond (electron pair) theory, a molecule is considered to be made up of atoms. Electrons in atoms occupy atomic orbitals. These may or may not be hybridized. If they are hybridized, atomic orbitals *from the same atom* combine to produce hybrid orbitals which can overlap more effectively with orbitals from other atoms, thus producing stronger bonds. Thus the atomic orbitals (or the hybrid orbitals) are thought to remain even when the atom is chemically bonded in a molecule.

In the molecular orbital theory, the valency electrons are considered to be associated with all the nuclei in the molecule. Thus the atomic orbitals *from different atoms* must be combined to produce molecular orbitals.

Electrons may be considered either as particles or waves. An electron in an atom may therefore be described as occupying an atomic orbital, or by a wave function ψ, which is a solution to the Schrödinger wave equation. Electrons in a molecule are said to occupy molecular orbitals. The wave function describing a molecular orbital may be obtained by one of two procedures:

1. Linear combination of atomic orbitals (LCAO).
2. United atom method.

LCAO METHOD

Consider two atoms A and B which have atomic orbitals described by the wave functions $\psi_{(A)}$ and $\psi_{(B)}$. If the electron clouds of these two atoms overlap when the atoms approach, then the wave function for the molecule (molecular orbital $\psi_{(AB)}$) can be obtained by a linear combination of the atomic orbitals $\psi_{(A)}$ and $\psi_{(B)}$:

$$\psi_{(AB)} = N(c_1\psi_{(A)} + c_2\psi_{(B)})$$

where N is a normalizing constant chosen to ensure that the probability of finding an electron in the whole of the space is unity, and c_1 and c_2 are constants chosen to give a minimum energy for $\psi_{(AB)}$. If atoms A and B are similar, then c_1 and c_2 will have similar values. If atoms A and B are the same, then c_1 and c_2 are equal.

The probability of finding an electron in a volume of space dv is ψ^2dv, so the probability density for the combination of two atoms as above is related to the wave function squared:

$$\psi^2_{(AB)} = (c_1^2\psi^2_{(A)} + 2c_1c_2\psi_{(A)}\psi_{(B)} + c_2^2\psi^2_{(B)})$$

If we examine the three terms on the right of the equation, the first term $c_1^2\psi^2_{(A)}$ is related to the probability of finding an electron on atom A if A is an isolated atom. The third term $c_2^2\psi^2_{(B)}$ is related to the probability of finding an electron on atom B if B is an isolated atom. The middle term becomes increasingly important as the overlap between the two atomic orbitals increases, and this term is called the overlap integral. This term represents the main difference between the electron clouds in individual atoms and in the molecule. The larger this term the stronger the bond.

s – s combinations of orbitals

Suppose the atoms A and B are hydrogen atoms; then the wave functions $\psi_{(A)}$ and $\psi_{(B)}$ describe the 1s atomic orbitals on the two atoms. Two combinations of the wave functions $\psi_{(A)}$ and $\psi_{(B)}$ are possible:

1. Where the signs of the two wave functions are the same
2. Where the signs of the two wave functions are different.

(If one of the wave functions $\psi_{(A)}$ is arbitrarily assigned a +ve sign, the other may be either +ve or −ve.) Wave functions which have the same sign may be regarded as waves that are in phase, which when combined add up to give a larger resultant wave. Similarly wave functions of different signs correspond to waves that are completely out of phase and which cancel each other by destructive interference. (The signs + and − refer to signs of the wave functions, which determine their symmetry, and have nothing to do with electrical charges.) The two combinations are:

$$\psi_{(g)} = N\{\psi_{(A)} + \psi_{(B)}\}$$

and

$$\psi_{(u)} = N\{\psi_{(A)} + [-\psi_{(B)}]\} \equiv N\{\psi_{(A)} - \psi_{(B)}\}$$

The latter equation should be regarded as the summation of the wave functions and *not* as the mathematical difference between them.

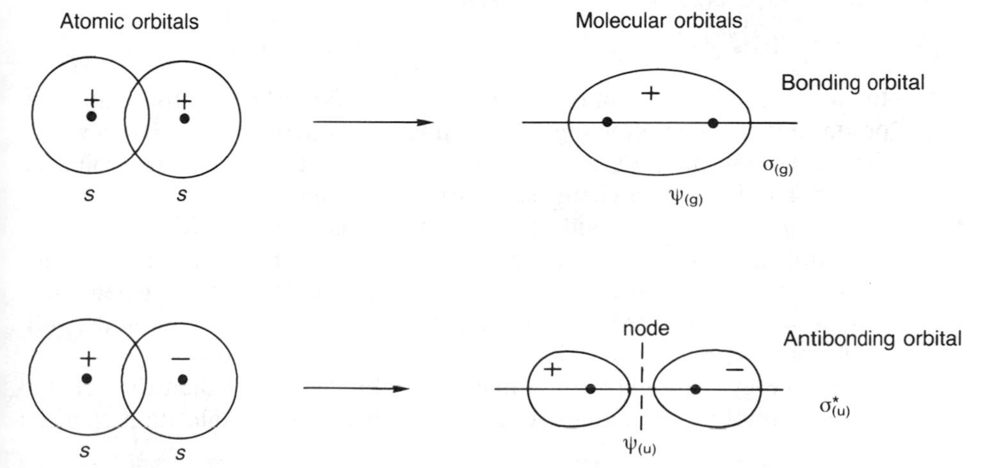

Figure 4.14 *s–s* combinations of atomic orbitals.

When a pair of atomic orbitals $\psi_{(A)}$ and $\psi_{(B)}$ combine, they give rise to a pair of molecular orbitals $\psi_{(g)}$ and $\psi_{(u)}$. The number of molecular orbitals produced must always be equal to the number of atomic orbitals involved. The function $\psi_{(g)}$ leads to increased electron density in between the nuclei, and is therefore a bonding molecular orbital. It is lower in energy than the original atomic orbitals. Conversely $\psi_{(u)}$ results in two lobes of opposite sign cancelling and hence giving zero electron density in between the nuclei. This is an antibonding molecular orbital which is higher in energy (Figure 4.15).

The molecular orbital wave functions are designated $\psi_{(g)}$ and $\psi_{(u)}$; g stands for *gerade* (even) and u for *ungerade* (odd). g and u refer to the symmetry of the orbital about its centre. If the sign of the wave function is unchanged when the orbital is reflected about its centre (i.e. x, y and z are replaced by $-x$, $-y$ and $-z$) the orbital is *gerade*. An alternative method

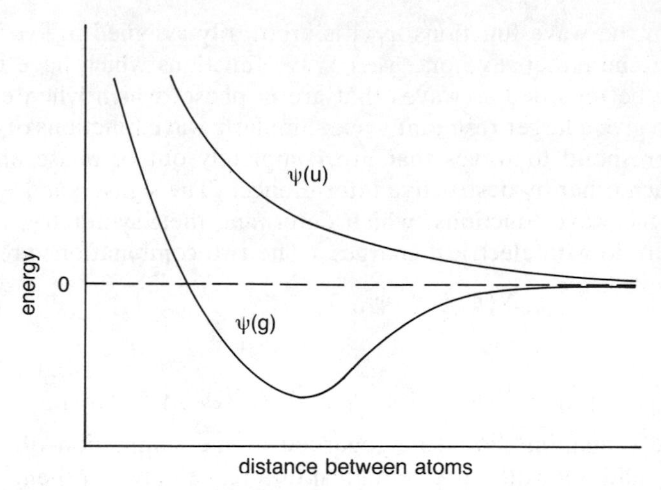

Figure 4.15 Energy of $\psi_{(g)}$ and $\psi_{(u)}$ molecular orbitals.

for determining the symmetry of the molecular orbital is to rotate the orbital about the line joining the two nuclei and then about a line perpendicular to this. If the sign of the lobes remains the same, the orbital is *gerade*, and if the sign changes, the orbital is *ungerade*.

The energy of the bonding molecular orbital $\psi_{(g)}$ passes through a minimum (Figure 4.15), and the distance between the atoms at this point corresponds to the internuclear distance between the atoms when they form a bond. Consider the energy levels of the two 1*s* atomic orbitals, and of the bonding $\psi_{(g)}$ and antibonding $\psi_{(u)}$ orbitals (Figure 4.16).

The energy of the bonding molecular orbital is lower than that of the atomic orbital by an amount Δ. This is known as the stabilization energy.

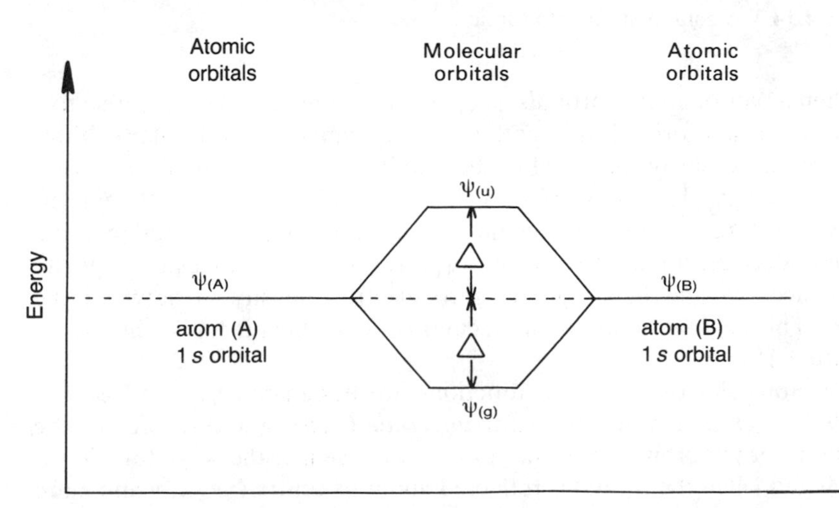

Figure 4.16 Energy levels of *s*–*s* atomic and molecular orbitals.

Similarly the energy of the antibonding molecular orbital is increased by Δ. Atomic orbitals may hold up to two electrons (provided that they have opposite spins) and the same applies to molecular orbitals. In the case of two hydrogen atoms combining, there are two electrons to be considered: one from the 1*s* orbital of atom A and one from the 1*s* orbital of atom B. When combined, these two electrons both occupy the bonding molecular orbital $\psi_{(g)}$. This results in a saving of energy of 2Δ, which corresponds to the bond energy. It is only because the system is stabilized in this way that a bond is formed.

Consider the hypothetical case of two He atoms combining. The 1*s* orbitals on each He contain two electrons, making a total of four electrons to put into molecular orbitals. Two of the electrons occupy the bonding MO, and two occupy the antibonding MO. The stabilization energy 2Δ derived from filling the bonding MO is offset by the 2Δ destabilization energy from using the antibonding MO. Since overall there is no saving of energy, He_2 does not exist, and this situation corresponds to non-bonding.

Some further symbols are necessary to describe the way in which the atomic orbitals overlap. Overlap of the orbitals along the axis joining the nuclei produces σ molecular orbitals, whilst lateral overlap of atomic orbitals forms π molecular orbitals.

s – p combinations of orbitals

An *s* orbital may combine with a *p* orbital provided that the lobes of the *p* orbital are pointing along the axis joining the nuclei. When the lobes which overlap have the same sign this results in a bonding MO with an increased electron density between the nuclei. When the overlapping lobes have

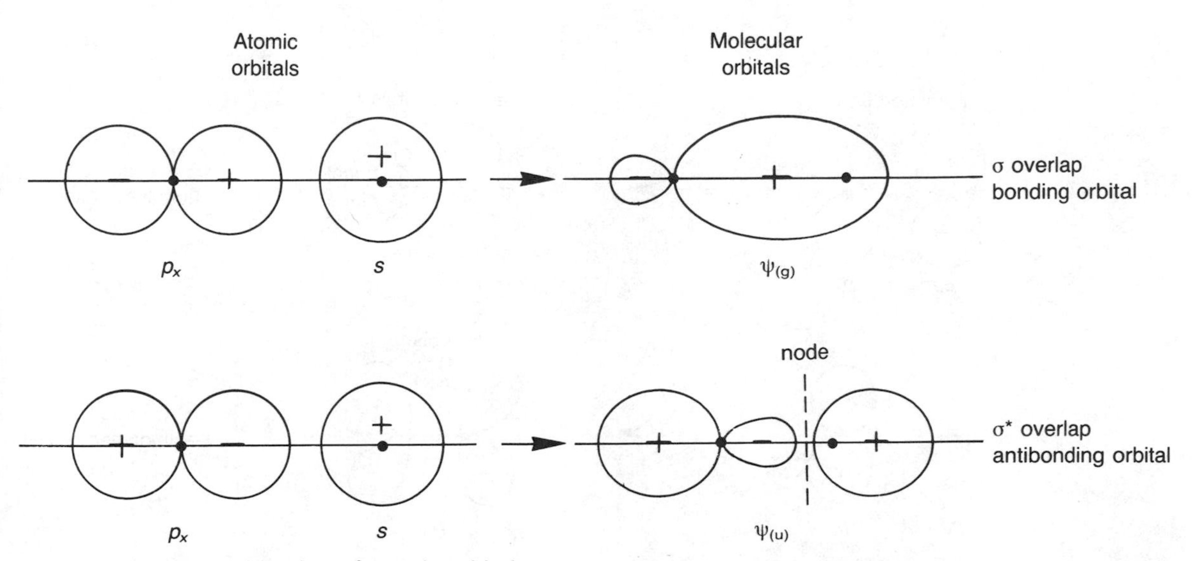

Figure 4.17 *s–p* combination of atomic orbitals.

opposite signs this gives an antibonding MO with a reduced electron density in between the nuclei (Figure 4.17).

p – p combinations of orbitals

Consider first the combination of two *p* orbitals which both have lobes pointing along the axis joining the nuclei. Both a bonding MO and an antibonding MO are produced (Figure 4.18).

Next consider the combination of two *p* orbitals which both have lobes perpendicular to the axis joining the nuclei. Lateral overlap of orbitals will

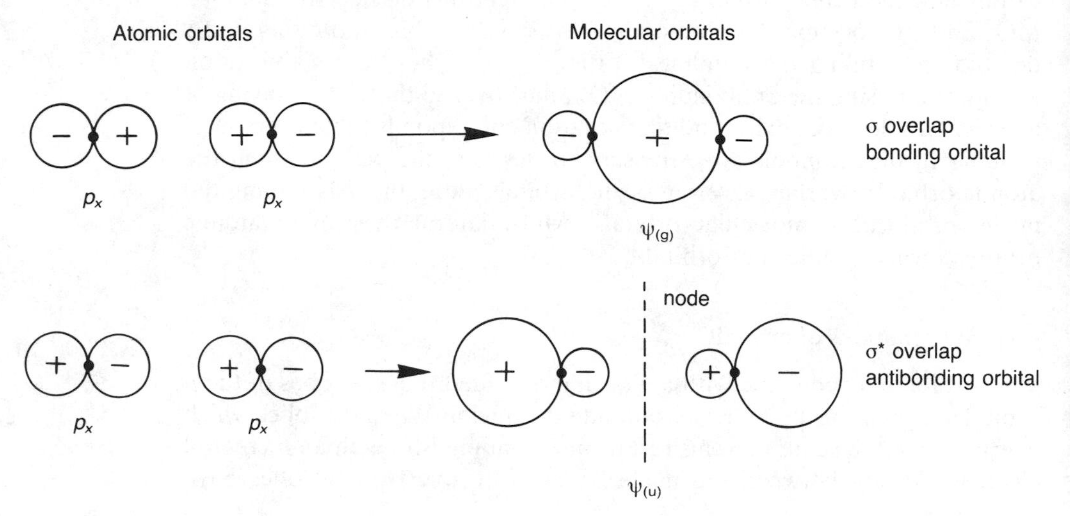

Figure 4.18 *p–p* combination of atomic orbitals.

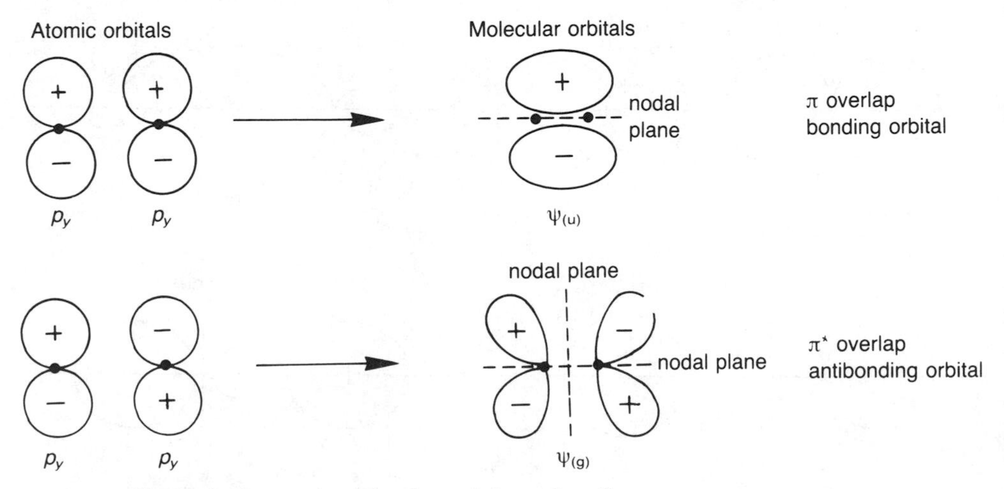

Figure 4.19 *p–p* combinations giving π bonding.

occur, resulting in π bonding and π^* antibonding MOs being produced (Figure 4.19).

There are three points of difference between these molecular orbitals and the σ orbitals described previously:

1. For π overlap the lobes of the atomic orbitals are perpendicular to the line joining the nuclei, whilst for σ overlap the lobes point along the line joining the two nuclei.
2. For π molecular orbitals, ψ is zero along the internuclear line and consequently the electron density ψ^2 is also zero. This is in contrast to σ orbitals.
3. The symmetry of π molecular orbitals is different from that shown by σ orbitals. If the bonding π MO is rotated about the internuclear line a change in the sign of the lobe occurs. The π bonding orbitals are therefore *ungerade*, whereas all σ bonding MOs are *gerade*. Conversely the antibonding π MO is *gerade* whilst all σ antibonding MOs are *ungerade*.

Pi bonding is important in many organic compounds such as ethene (where there is one σ bond and one π bond between the two carbon atoms), ethyne (one σ and two π), and benzene, and also in a number of inorganic compounds such as CO_2 and CN^-.

Ethene contains a localized double bond, which involves only the two carbon atoms. Experimental measurements show that the two C atoms and the four H atoms are coplanar, and the bond angles are close to 120°. Each C atom uses its $2s$ and two $2p$ orbitals to form three sp^2 hybrid orbitals that form σ bonds to the other C atom and two H atoms. The remaining p orbital on each C atom is at right angles to the σ bonds so far formed. In the valence bond theory these two p orbitals overlap sideways to give a π bond. This sideways overlap is not as great as the end to end overlap in σ bonds so a C=C, though stronger than a C—C bond, is not twice as strong (C—C in ethane 598 kJ mol^{-1}, C=C in ethene 346 kJ mol^{-1}). The molecule can be twisted about the C—C bond in ethane, but it cannot be twisted in ethene since this would reduce the amount of π overlap. In the molecular orbital theory the explanation of the π bonding is slightly different. The two p orbitals involved in π bonding combine to form two π molecular orbitals, one bonding and one antibonding. Since there are only two electrons involved, these occupy the π bonding MO since this has the lower energy. The molecular orbital explanation becomes more important in cases where there is non-localized π bonding, that is where π bonding covers several atoms as in benzene, NO_3^- and CO_3^{2-}.

In ethyne each C atom uses sp hybrid orbitals to form σ bonds to the other C atom and a H atom. These four atoms form a linear molecule. Each C atom has two p orbitals at right angles to one another, and these overlap sideways with the equivalent p orbitals on the other C atom, thus forming two π bonds. Thus a C≡C triple bond is formed, which is stronger than a C=C double bond (C≡C in ethyne 813 kJ mol^{-1}).

The majority of strong π bonds occur between elements of the first short period in the periodic table, for example C≡C, C≡N, C≡O, C=C and C=O. This is mainly because the atoms are smaller and hence the orbitals involved are reasonably compact, so it is possible to get substantial overlap of orbitals. There are a smaller number of cases where π bonding occurs between different types of orbitals, for example the $2p$ and $3d$ orbitals. Though these orbitals are much larger, the presence of nodes may concentrate electron density in certain parts of the orbitals.

p–d combinations of orbitals

A p orbital on one atom may overlap with a d orbital on another atom as shown, giving bonding and antibonding combinations. Since the orbitals do not point along the line joining the two nuclei, overlap must be of the π type (Figure 4.20). This type of bonding is responsible for the short bonds found in the oxides and oxoacids of phosphorus and sulphur. It also occurs in transition metal complexes such as the carbonyls and cyanides.

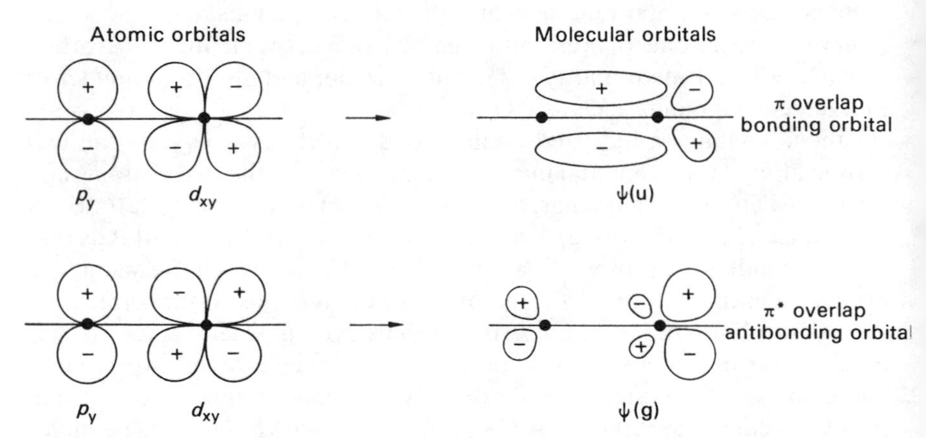

Figure 4.20 p–d combinations of atomic orbitals.

d–d combinations of orbitals

It is possible to combine two d atomic orbitals, producing bonding and antibonding MOs which are called δ and δ* respectively. On rotating these orbitals about the internuclear axis, the sign of the lobes changes four times compared with two changes with π overlap and no change for σ overlap.

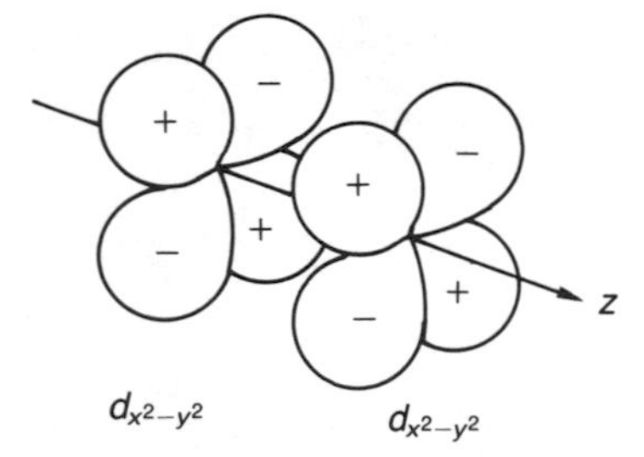

Figure 4.21 δ bonding by d orbitals (sideways overlap of two $d_{x^2-y^2}$ orbitals.

Non-bonding combinations of orbitals

All the cases of overlap of atomic orbitals considered so far have resulted in a bonding MO of lower energy, and an antibonding MO of higher energy. To obtain a bonding MO with a concentration of electron density in between the nuclei, the signs (symmetry) of the lobes which overlap

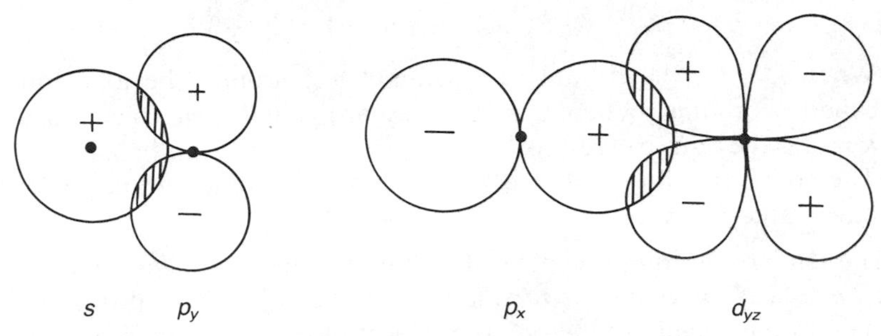

Figure 4.22 Some non-bonding combinations of atomic orbitals.

must be the same. Similarly for antibonding MOs the signs of the overlapping lobes must be different. In the combinations shown in Figure 4.22 any stabilization which occurs from overlapping + with + is destabilized by an equal amount of overlap of + with −. There is no overall change in energy, and this situation is termed non-bonding. It should be noted that in all of these non-bonding cases the symmetry of the two atomic orbitals is different, i.e. rotation about the axis changes the sign of one.

RULES FOR LINEAR COMBINATION OF ATOMIC ORBITALS

In deciding which atomic orbitals may be combined to form molecular orbitals, three rules must be considered:

1. The atomic orbitals must be roughly of the same energy. This is important when considering overlap between two different types of atoms.
2. The orbitals must overlap one another as much as possible. This implies that the atoms must be close enough for effective overlap and that the radial distribution functions of the two atoms must be similar at this distance.
3. In order to produce bonding and antibonding MOs, either the symmetry of the two atomic orbitals must remain unchanged when rotated about the internuclear line, or both atomic orbitals must change symmetry in an identical manner.

In the same way that each atomic orbital has a particular energy, and may be defined by four quantum numbers, each molecular orbital has a definite energy, and is also defined by four quantum numbers.

1. The principal quantum number n has the same significance as in atomic orbitals.
2. The subsidiary quantum number l also has the same significance as in atomic orbitals.
3. The magnetic quantum number of atomic orbitals is replaced by a new nuantum number λ. In a diatomic molecule, the line joining the nuclei is taken as a reference direction and λ represents the quantization of angular momentum in $h/2\pi$ units with respect to this axis. λ takes the same values as m takes for atoms, i.e.

$$\lambda = -l, \ldots, -3, -2, -1, 0, +1, +2, +3, \ldots, +l$$

When $\lambda = 0$, the orbitals are symmetrical around the axis and are called σ orbitals. When $\lambda = \pm 1$ they are called π orbitals and when $\lambda = \pm 2$ they are called δ orbitals.

4. The spin quantum number is the same as for atomic orbitals and may have values of $\pm\frac{1}{2}$.

The Pauli exclusion principle states that *in a given atom no two electrons can have all four quantum numbers the same*. The Pauli principle also applies to molecular orbitals: *No two electrons in the same molecule can have all four quantum numbers the same.*

The order of energy of molecular orbitals has been determined mainly from spectroscopic data. In simple homonuclear diatomic molecules, the order is:

$$\sigma 1s^2, \sigma^* 1s^2, \sigma 2s^2, \sigma^* 2s^2, \sigma 2p_x^2, \begin{cases} \pi 2p_y^2, \\ \pi 2p_z^2, \end{cases} \begin{cases} \pi^* 2p_y^2, \\ \pi^* 2p_z^2 \end{cases} \sigma^* 2p_x^2$$

$$\xrightarrow{\text{increasing energy}}$$

Note that the $2p_y$ atomic orbital gives π bonding and π^* antibonding MOs and the $2p_z$ atomic orbital gives π bonding and π^* antibonding MOs. The bonding $\pi 2p_y$ and $\pi 2p_z$ MOs have exactly the same energy and are said to be double degenerate. In a similar way the antibonding $\pi^* 2p_y$ and $\pi^* 2p_z$ MOs have the same energy and are also doubly degenerate.

A similar arrangement of MOs exists from $\sigma 3s$ to $\sigma^* 3p_x$, but the energies are known with less certainty.

The energies of the $\sigma 2p$ and $\pi 2p$ MOs are very close together. The order of MOs shown above is correct for nitrogen and heavier elements, but for the lighter elements boron and carbon the $\pi 2p_y$ and $\pi 2p_z$ are probably lower than $\sigma 2p_x$. For these atoms the order is:

$$\sigma 1s^2, \sigma^* 1s^2, \sigma 2s^2, \sigma^* 2s^2, \begin{cases} \pi 2p_y^2, \\ \pi 2p_z^2, \end{cases} \sigma 2p_x^2, \sigma^* 2p_x^2, \begin{cases} \pi^* 2p_y^2 \\ \pi^* 2p_z^2 \end{cases}$$

$$\xrightarrow{\text{increasing energy}}$$

EXAMPLES OF MOLECULAR ORBITAL TREATMENT FOR HOMONUCLEAR DIATOMIC MOLECULES

In the build-up of atoms, electrons are fed into atomic orbitals. The *Aufbau* principle is used:

1. Orbitals of lowest energy are filled first.
2. Each orbital may hold up to two electrons, provided that they have opposite spins.

Hund's rule states that when several orbitals have the same energy (that is they are degenerate), electrons will be arranged so as to give the maximum number of unpaired spins.

In the molecular orbital method, we consider the whole molecule rather than the constituent atoms, and use molecular orbitals rather than atomic orbitals. In the build-up of the molecule, the total number of electrons from all the atoms in the molecule is fed into molecular orbitals. The *Aufbau* principle and Hund's rule are used as before.

For simplicity homonuclear diatomic molecules will be examined first. Homonuclear means that there is only one type of nucleus, that is one element present, and diatomic means that the molecule is composed of two atoms.

H_2^+ molecule ion

This may be considered as a combination of a H atom with a H^+ ion. This gives one electron in the molecular ion which occupies the lowest energy MO:

$$\sigma 1s^1$$

The electron occupies the $\sigma 1s$ bonding MO. The energy of this ion is thus lower than that of the constituent atom and ion, by an amount Δ, so there is some stabilization. This species exists but it is not common since H_2 is much more stable. However, H_2^+ can be detected spectroscopically when H_2 gas under reduced pressure is subjected to an electric discharge.

H_2 molecule

There is one electron from each atom, and hence there are two electrons in the molecule. These occupy the lowest energy MO:

$$\sigma 1s^2$$

This is shown in Figure 4.23. The bonding $\sigma 1s$ MO is full, so the stabilization energy is 2Δ. A σ bond is formed, and the H_2 molecule exists and is well known.

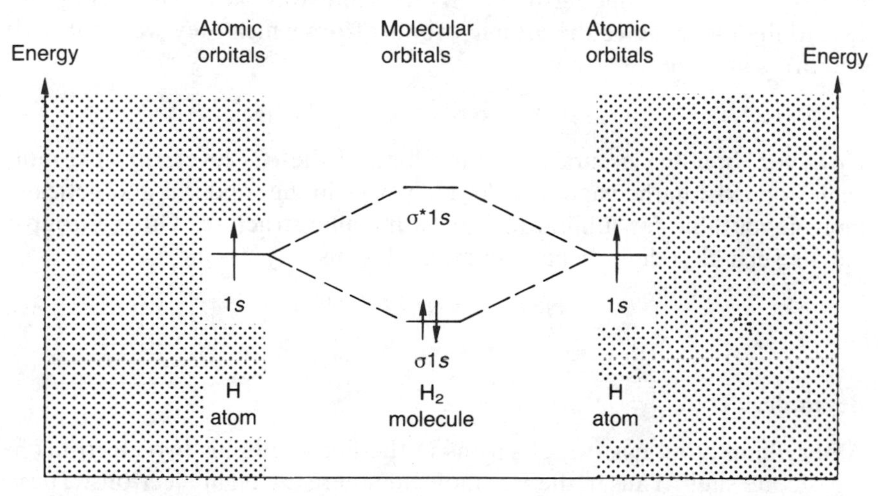

Figure 4.23 Electronic configuration, atomic and molecular orbitals for hydrogen.

He_2^+ molecule ion

This may be considered as a combination of a He atom and a He^+ ion. There are three electrons in the molecular ion, which are arranged in MOs:

$$\sigma 1s^2,\ \sigma^* 1s^1$$

The filled $\sigma 1s$ bonding MO gives 2Δ stabilization, whilst the half filled $\sigma 1s^*$ gives Δ destabilization. Overall there is Δ stabilization. Thus the helium molecule ion can exist. It is not very stable, but it has been observed spectroscopically.

He_2 molecule

There are two electrons from each atom, and the four electrons are arranged in MOs:

$$\sigma 1s^2,\ \sigma^* 1s^2$$

The 2Δ stabilization energy from filling the $\sigma 2s$ MO is cancelled by the 2Δ destabilization energy from filling the $\sigma^* 1s$ MO. Thus a bond is not formed, and the molecule does not exist.

Li_2 molecule

Each Li atom has two electrons in its inner shell, and one in its outer shell, giving three electrons. Thus there is a total of six electrons in the molecule, and these are arranged in MOs:

$$\sigma 1s^2,\ \sigma^* 1s^2,\ \sigma 2s^2$$

This is shown in Figure 4.24. The inner shell of filled $\sigma 1s$ MOs does not contribute to the bonding in much the same way as in He_2. They are essentially the same as the atomic orbitals from which they were formed, and are sometimes written:

$$KK,\ \sigma 2s^2$$

However, bonding occurs from the filling of the $\sigma 2s$ level, and Li_2 molecules do exist in the vapour state. However, in the solid it is energetically more favourable for lithium to form a metallic structure. Other Group I metals such as sodium behave in an analogous way:

$$Na_2 \quad KK,\ LL,\ \sigma 3s^2$$

Be_2 molecule

A beryllium atom has two electrons in the first shell plus two electrons in the second shell. Thus in the Be_2 molecule there are eight electrons. These are arranged in MOs:

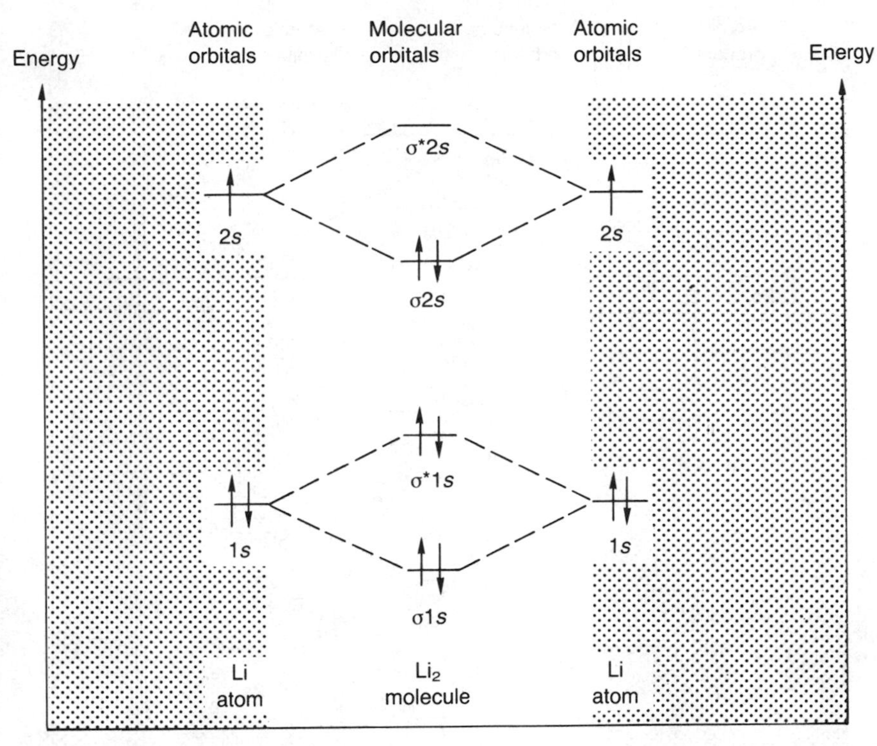

Figure 4.24 Electronic configuration, atomic and molecular orbitals for lithium.

$$\sigma 1s^2,\ \sigma^*1s^2,\ \sigma 2s^2,\ \sigma^*2s^2$$

or

$$\text{KK},\ \sigma 2s^2,\ \sigma^*2s^2$$

Ignoring the inner shell as before, it is apparent that the effects of the bonding $\sigma 2s$ and antibonding σ^*2s levels cancel, so there is no stabilization and the molecule would not be expected to exist.

B_2 molecule

Each boron atom has 2 + 3 electrons. The B_2 molecule thus contains a total of ten electrons, which are arranged in MOs:

$$\sigma 1s^2,\ \sigma^*1s^2,\ \sigma 2s^2,\ \sigma^*2s^2,\ \begin{cases} \pi 2p_y^1 \\ \pi 2p_z^1 \end{cases}$$

This may be shown diagrammatically (Figure 4.25). Note that B is a light atom and the order of energies of MOs is different from the 'usual' arrangement. Thus the $\pi 2p$ orbitals are lower in energy than the $\sigma 2p_x$. Since the $\pi 2p_y$ and $\pi 2p_z$ orbitals are degenerate (identical in energy), Hund's rule applies, and each is singly occupied. The inner shell does not

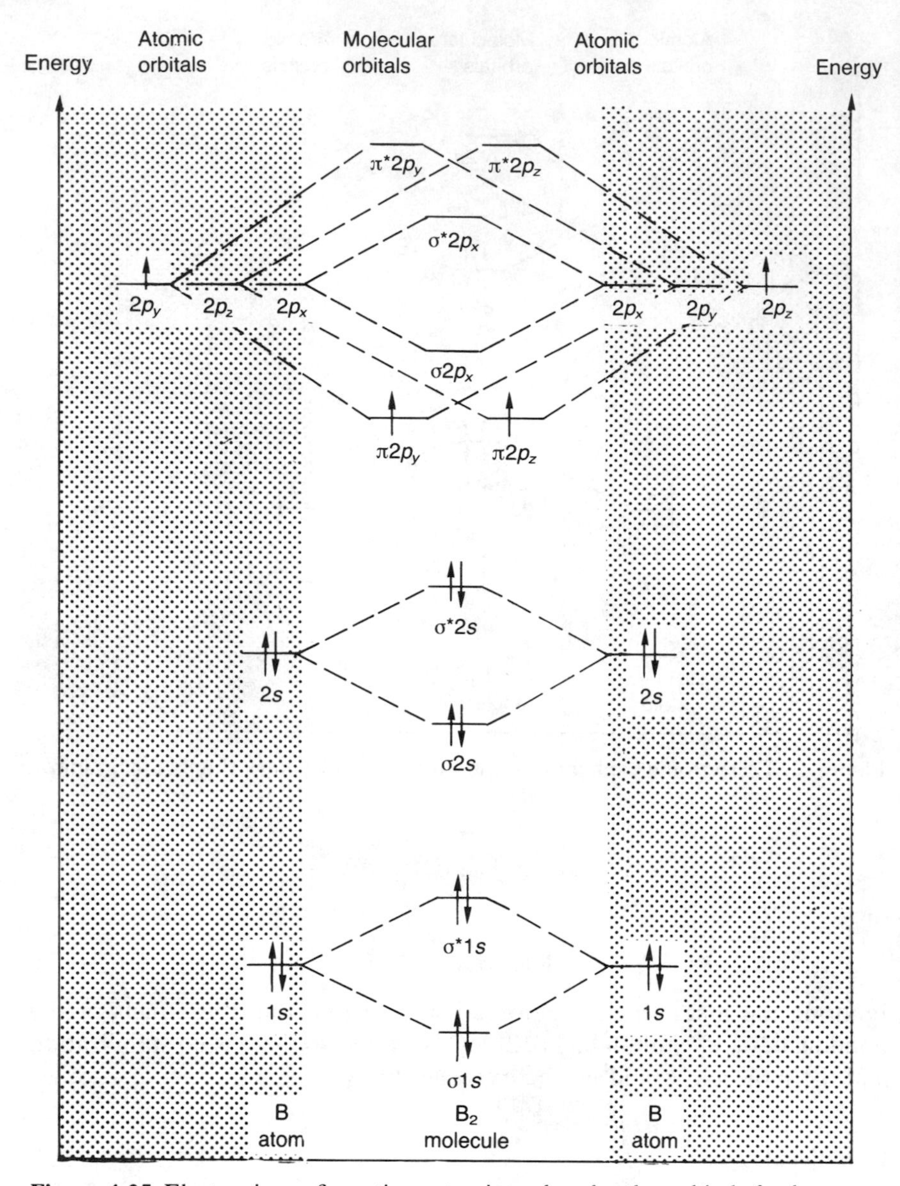

Figure 4.25 Electronic configuration, atomic and molecular orbitals for boron.

participate in bonding. The effects of bonding and antibonding $\sigma 2s$ orbitals cancel but stabilization occurs from the filling of the $\pi 2p$ orbitals, and hence a bond is formed and B_2 exists.

C_2 molecule

A carbon atom has 2 + 4 electrons. A C_2 molecule would contain a total of 12 electrons, and these would be arranged in MOs:

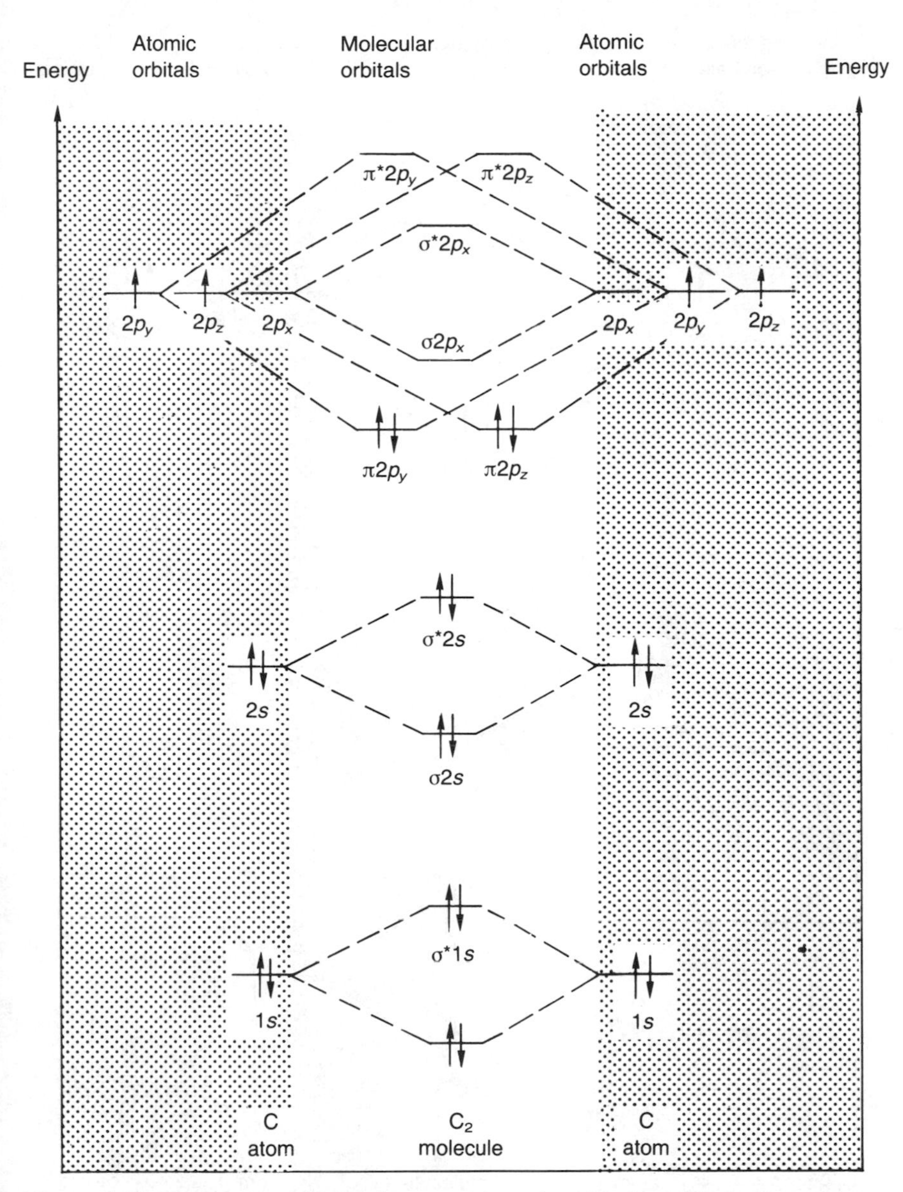

Figure 4.26 Electronic configuration, atomic and molecular orbitals for carbon.

$$\sigma 1s^2,\ \sigma^* 1s^2,\ \sigma 2s^2,\ \sigma^* 2s^2,\ \left\{ \begin{array}{l} \pi 2p_y^2 \\ \pi 2p_z^2 \end{array} \right.$$

This is shown diagrammatically in Figure 4.26.

The molecule should be stable, since the two π2*p* bonding orbitals provide 4Δ of stabilization energy, giving two bonds. In fact carbon exists as a macromolecule in graphite and diamond, since these are an even more stable arrangement (where each carbon forms four bonds): hence diamond and graphite are formed in preference to C_2.

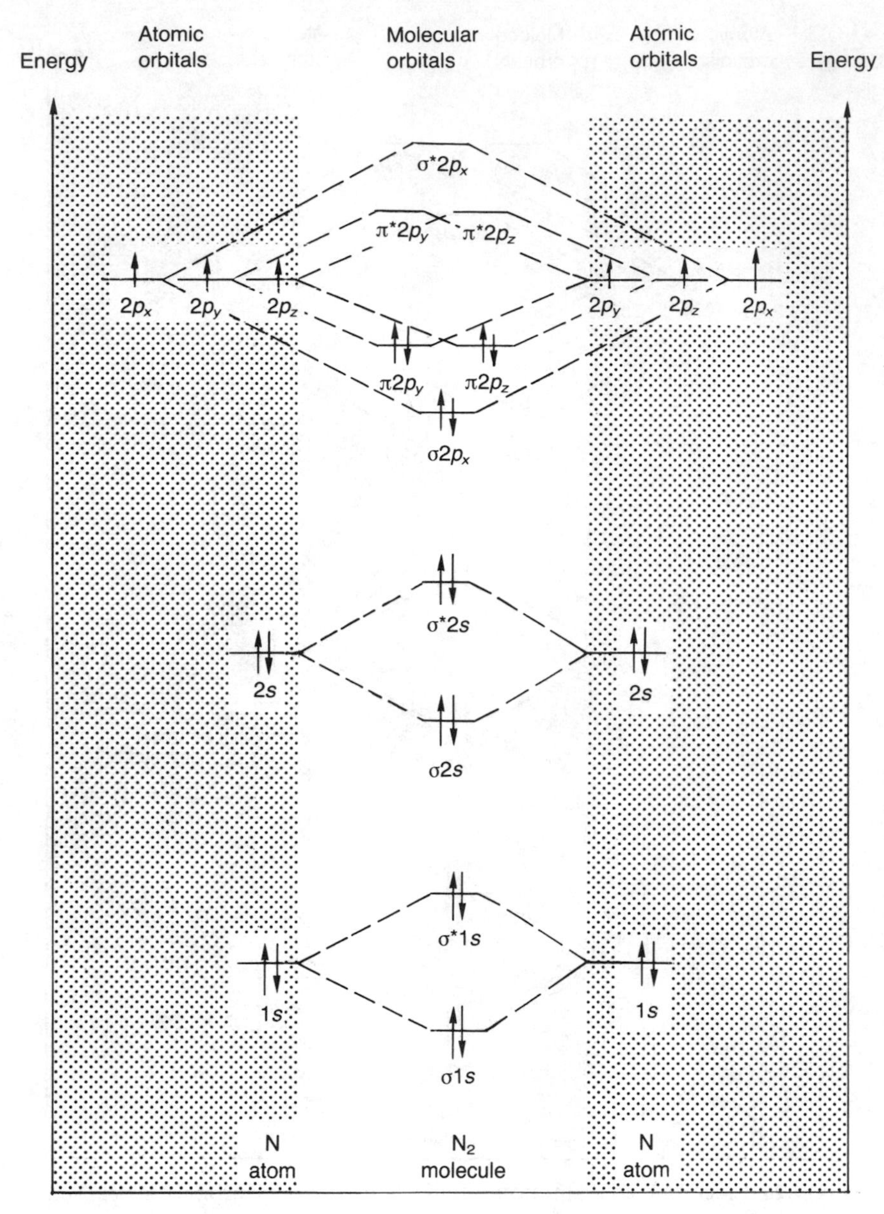

Figure 4.27 Electronic configuration, atomic and molecular orbitals for nitrogen.

N_2 molecule

A nitrogen atom has $2 + 5 = 7$ electrons. Thus the N_2 molecule contains 14 electrons. These are arranged in MOs:

$$\sigma 1s^2,\ \sigma^* 1s^2,\ \sigma 2s^2,\ \sigma^* 2s^2,\ \sigma 2p_x^2,\ \begin{cases} \pi 2p_y^2 \\ \pi 2p_z^2 \end{cases}$$

This is shown diagrammatically (Figure 4.27).

Assuming that the inner shell does not participate in bonding, and that the bonding and antibonding $2s$ levels cancel, one σ and two π bonding pairs remain, giving a total of three bonds. This is in agreement with the valence bond formulation as N≡N.

O_2 molecule

Each oxygen atom has $2 + 6 = 8$ electrons. Thus the O_2 molecule contains a total of 16 electrons. These are arranged in MOs:

$$\sigma 1s^2,\ \sigma^* 1s^2,\ \sigma 2s^2,\ \sigma^* 2s^2,\ \sigma 2p_x^2,\ \begin{cases} \pi 2p_y^2, \\ \pi 2p_z^2, \end{cases} \begin{cases} \pi^* 2p_y^1 \\ \pi^* 2p_z^1 \end{cases}$$

This is shown diagrammatically in Figure 4.28.

The antibonding π^*2p_y and π^*2p_z orbitals are singly occupied in accordance with Hund's rule. Unpaired electrons give rise to paramagnetism. Since there are two unpaired electrons with parallel spins, this explains why oxygen is paramagnetic. If this treatment is compared with the Lewis electron pair theory or the valence bond theory, these do not predict unpaired electrons or paramagnetism.

$$:\ddot{\underset{\cdot}{O}}\cdot + \cdot\ddot{\underset{\cdot}{O}}: \rightarrow :\underset{\cdot\cdot}{O} \vdots \underset{\cdot\cdot}{O}:$$

This was the first success of the Molecular orbital theory in successfully predicting the paramagnetism of O_2, a fact not even thought of with a valence bond representation of O=O.

As in the previous examples, the inner shell does not participate in bonding and the bonding and antibonding $2s$ orbitals cancel each other. A σ bond results from the filling of $\sigma 2p_x^2$. Since $\pi^*2p_y^1$ is half filled and therefore cancels half the effect of the completely filled $\pi 2p_y^2$ orbital, half of a π bond results. Similarly another half of a π bond arises from $\pi 2p_z^2$ and $\pi^*2p_z^1$, giving a total of $1 + \frac{1}{2} + \frac{1}{2} = 2$ bonds. The bond order is thus two.

Instead of working out the bond order by cancelling the effects of filled bonding and antibonding MOs, the bond order may be calculated as half the difference between the number of bonding and antibonding electrons:

$$\text{Bond order} = \frac{\begin{pmatrix}\text{number of electrons} \\ \text{occupying bonding orbitals}\end{pmatrix} - \begin{pmatrix}\text{number of electrons} \\ \text{in antibonding orbitals}\end{pmatrix}}{2}$$

In the case of O_2 the bond order calculates as $(10 - 6)/2 = 2$, which corresponds to a double bond.

O_2^- ion

The compound potassium superoxide KO_2 contains the superoxide ion O_2^-. The O_2^- ion has 17 electrons, and has one more electron than the O_2

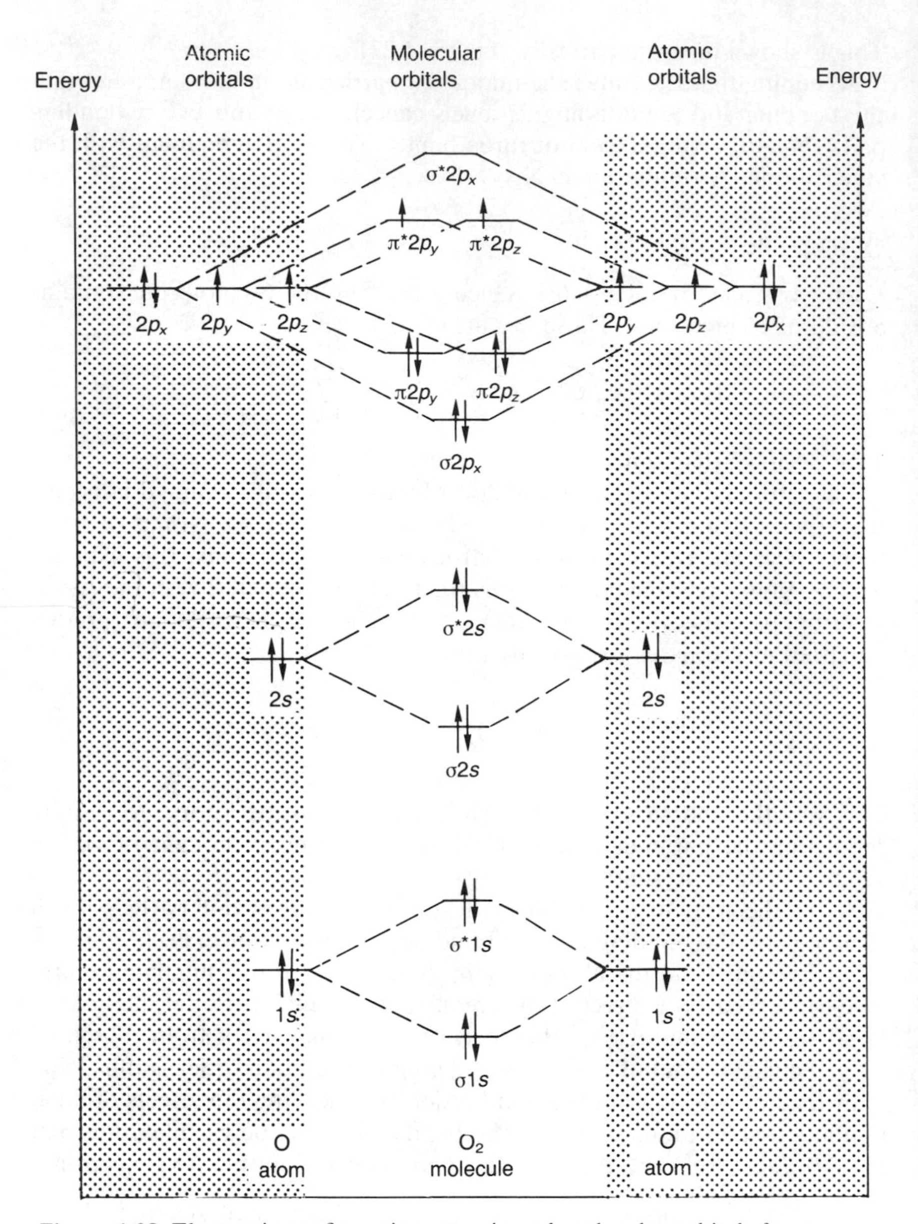

Figure 4.28 Electronic configuration, atomic and molecular orbitals for oxygen.

molecule. This extra electron occupies either the π^*2p_y or π^*2p_z orbital. It does not matter which it occupies since they are the same energy.

$$\sigma 1s^2,\ \sigma^*1s^2,\ \sigma 2s^2,\ \sigma^*2s^2,\ \sigma 2p_x^2,\ \begin{cases} \pi 2p_y^2, \\ \pi 2p_z^2, \end{cases} \begin{cases} \pi^*2p_y^2 \\ \pi^*2p_z^1 \end{cases}$$

The inner shell of electrons does not take part in bonding. The bonding $\sigma 2s^2$ and antibonding σ^*2s^2 cancel. The $\sigma 2p_x^2$ orbital is filled and forms a

σ bond. The effects of the bonding $\pi 2p_y^2$ and antibonding $\pi 2p_y^2$ orbitals cancel, and the completely filled bonding $\pi 2p_z^2$ is half cancelled by the half filled antibonding $\pi 2p_z^1$, thus giving half a π bond. The bond order is thus $1 + \frac{1}{2} = 1\frac{1}{2}$. Alternatively the bond order may be calculated like this: (bonding − antibonding)/2, that is $(10 - 7)/2 = 1\frac{1}{2}$. This corresponds to a bond that is intermediate in length between a single and a double bond. The superoxide ion has an unpaired electron and is therefore paramagnetic. (A bond order of $1\frac{1}{2}$ is well accepted in benzene.)

O_2^{2-} ion

In a similar way sodium peroxide Na_2O_2 contains the peroxide ion O_2^{2-}. This ion has 18 electrons, arranged:

$$\sigma 1s^2, \sigma^* 1s^2, \sigma 2s^2, \sigma^* 2s^2, \sigma 2p_x^2, \begin{cases} \pi 2p_y^2, \\ \pi 2p_z^2, \end{cases} \begin{cases} \pi^* 2p_y^2 \\ \pi^* 2p_z^2 \end{cases}$$

Once again the inner shell takes no part in bonding. The bonding and antibonding $2s$ orbitals completely cancel each other. One σ bond forms from the filled $2p_x$ orbital. Both the bonding $2p_y$ and $2p_z$ orbitals are cancelled out by their corresponding antibonding orbitals. Thus the bond order is one, that is a single bond. Alternatively the bond order may be calculated as (bonding − antibonding)/2, that is $(10 - 8)/2 = 1$.

F_2 molecule

Fluorine atoms have 2 + 7 electrons, so an F_2 molecule contains 18 electrons. These are arranged:

$$\sigma 1s^2, \sigma^* 1s^2, \sigma 2s^2, \sigma^* 2s^2, \sigma 2p_x^2, \begin{cases} \pi 2p_y^2, \\ \pi 2p_z^2, \end{cases} \begin{cases} \pi^* 2p_y^2 \\ \pi^* 2p_z^2 \end{cases}$$

This is shown diagrammatically in Figure 4.29.

The inner shell is non-bonding, and the filled bonding $2s$, $2p_y$ and $2p_z$ are cancelled by the equivalent antibonding orbitals. This leaves a σ bond from the filled $\sigma 2p_x^2$ orbital, and thus a bond order of one. Alternatively the bond order may be calculated as (bonding − antibonding)/2, that is $(10 - 8)/2 = 1$.

It should be noted that Cl_2 and Br_2 have structures analogous to F_2, except that additional inner shells of electrons are full.

The F—F bond is rather weak (see Chapter 15) and this is attributed to the small size of fluorine and repulsion between lone pairs of electrons on adjacent atoms.

EXAMPLES OF MOLECULAR ORBITAL TREATMENT FOR HETERONUCLEAR DIATOMIC MOLECULES

The same principles apply when combining atomic orbitals from two different atoms as applied when the atoms were identical, that is:

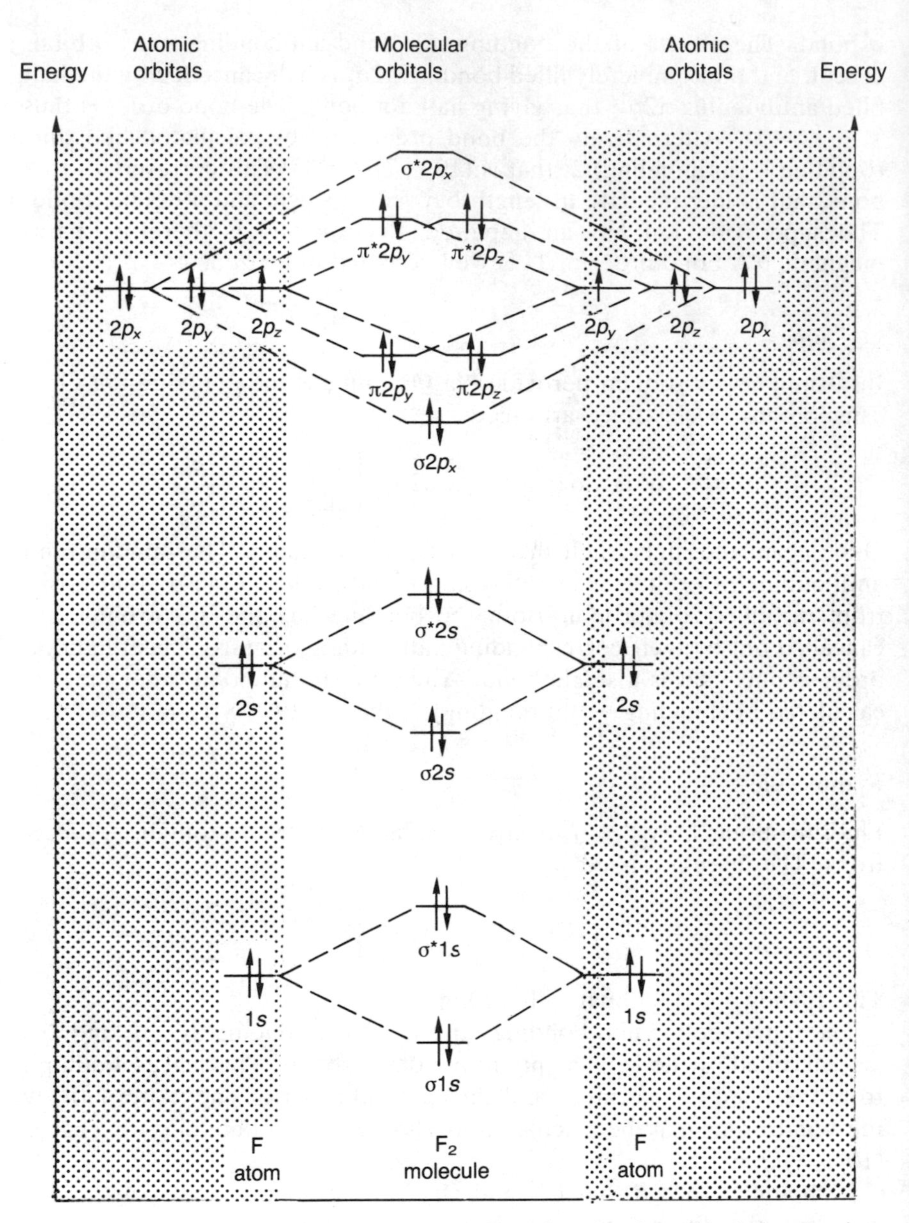

Figure 4.29 Electronic configuration, atomic and molecular orbitals for fluorine.

1. Only atomic orbitals of about the same energy can combine effectively.
2. They should have the maximum overlap.
3. They must have the same symmetry.

Since the two atoms are different, the energies of their atomic orbitals are slightly different. A diagram showing how they combine to form molecular orbitals is given in Figure 4.30.

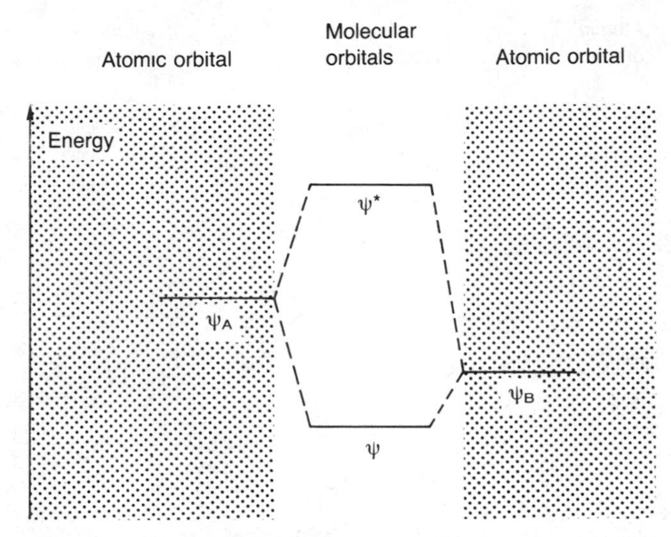

Figure 4.30 The relative energy levels of atomic orbitals and molecular orbitals for a heteronuclear diatomic molecule AB.

The problem is that in many cases the order of MO energy levels is not known with certainty. Thus we will consider first some examples where the two different atoms are close to each other in the periodic table, and consequently it is reasonable to assume that the order of energies for the MOs are the same as for homonuclear molecules.

NO molecule

The nitrogen atom has 2 + 5 = 7 electrons, and the oxygen atom has 2 + 6 = 8 electrons, making 15 electrons in the molecule. The order of energy levels of the various MOs are the same as for homonuclear diatomic molecules heavier than C_2, so the arrangement is:

$$\sigma 1s^2,\ \sigma^* 1s^2,\ \sigma 2s^2,\ \sigma^* 2s^2,\ \sigma 2p_x^2,\ \begin{cases} \pi 2p_y^2, \\ \pi 2p_z^2, \end{cases} \begin{cases} \pi^* 2p_y^1 \\ \pi^* 2p_z^0 \end{cases}$$

This is shown in Figure 4.31.

The inner shell is non-bonding. The bonding and antibonding $2s$ orbitals cancel, and a σ bond is formed by the filled $\sigma 2p_x^2$ orbital. A π bond is formed by the filled $\pi 2p_z^2$ orbital. The half filled $\pi^* 2p_y^1$ half cancels the filled $\pi 2p_y^2$ orbital, thus giving half a bond. The bond order is thus $2\frac{1}{2}$, that is in between a double and a triple bond. Alternatively the bond order may be worked out as (bonding − antibonding)/2, that is $(10 - 5)/ = 2\frac{1}{2}$. The molecule is paramagnetic since it contains an unpaired electron. In NO there is a significant difference of about 250 kJ mol^{-1} in the energy of the AOs involved, so that combination of AOs to give MOs is less effective than in O_2 or N_2. The bonds are therefore weaker than might be expected. Apart from this the molecular orbital pattern (Figure 4.31)

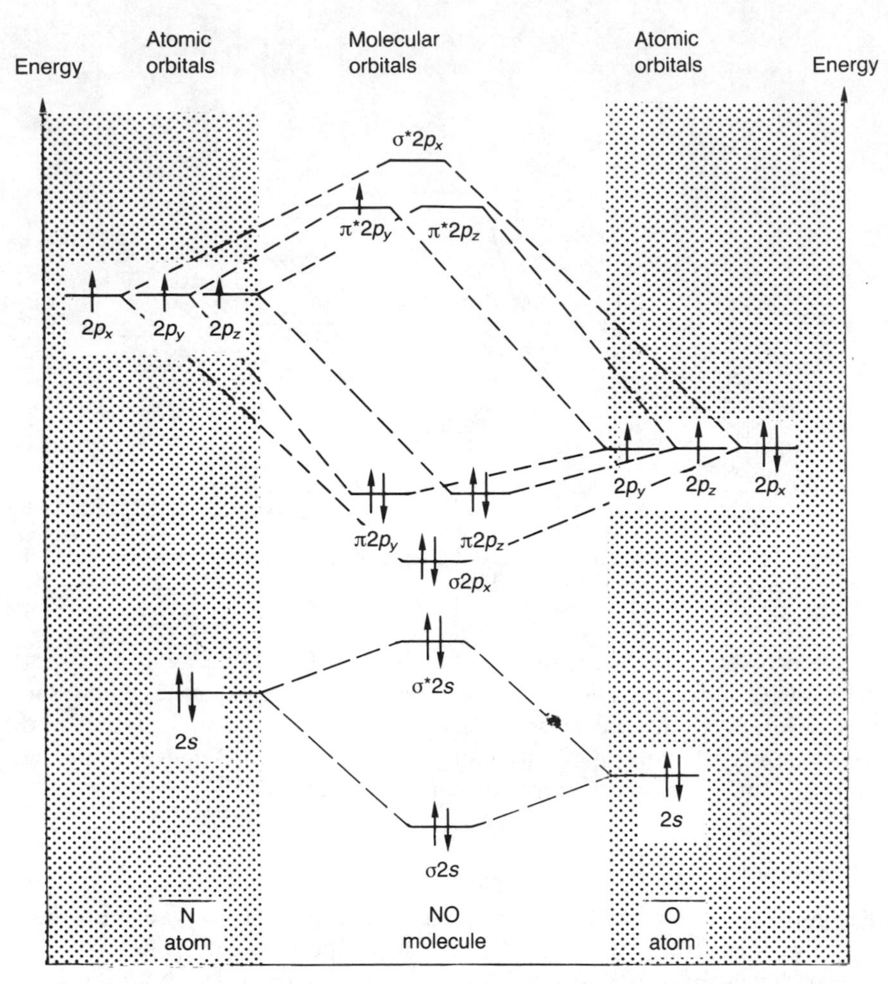

Figure 4.31 Electronic configuration, atomic orbitals and molecular orbitals for nitric oxide. (This diagram is essentially the same as that for homonuclear diatomic molecules such as N_2, O_2 or F_2. The difference is that the atomic energy levels of N and O are not the same.)

is similar to that for homonuclear diatomic molecules. Removal of one electron to make NO^+ results in a shorter and stronger bond because the electron is removed from an antibonding orbital, thus increasing the bond order to 3.

CO molecule

The carbon atom has 2 + 4 = 6 electrons, and the O atom has 2 + 6 = 8 electrons, so the CO molecule contains 14 electrons. In this case we are rather less certain of the order of energies of the MOs, since they are different for C and O. Assume the order is the same as for light atoms like C:

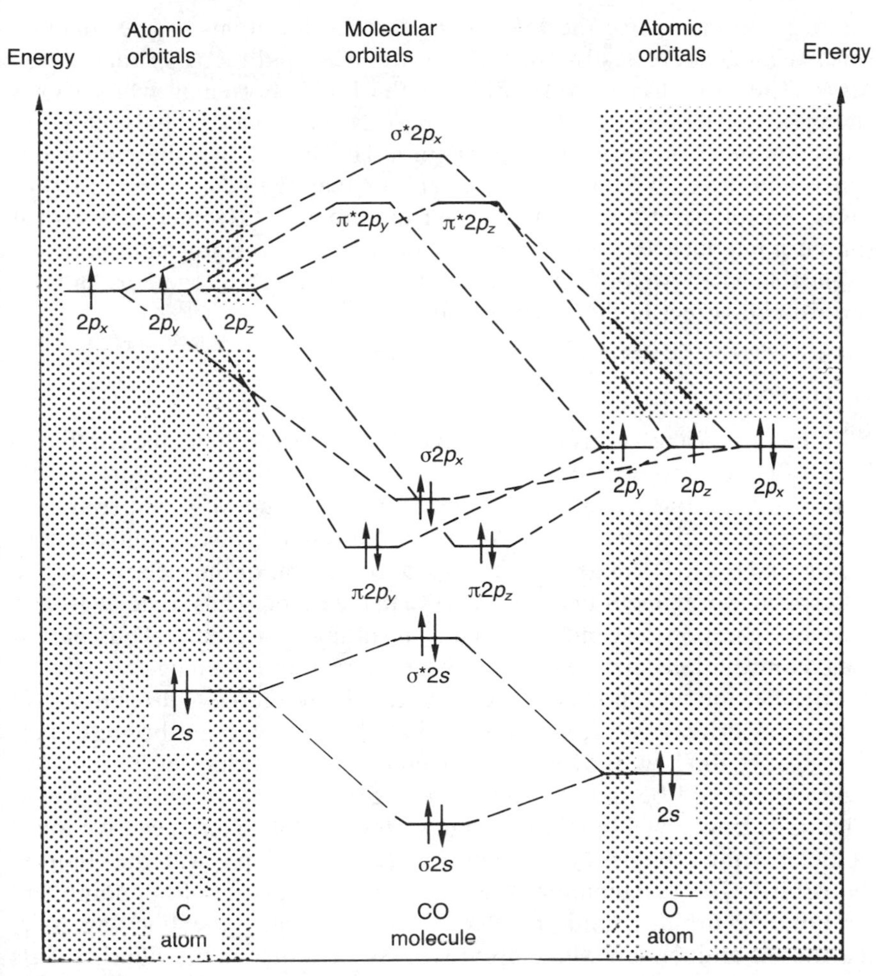

Figure 4.32 Electronic configuration, atomic orbitals and molecular orbitals for carbon monoxide.

$$\sigma 1s^2,\ \sigma^*1s^2,\ \sigma 2s^2,\ \sigma^*2s^2,\ \begin{Bmatrix} \pi 2p_y^2 \\ \pi 2p_z^2 \end{Bmatrix}\ \sigma 2p_x^2$$

This is shown in Figure 4.32.

The inner shell is non-bonding, and the bonding and antibonding $2s$ orbitals cancel, leaving one σ and two π bonds – and thus a bond order of 3. Alternatively the bond order may be calculated using the formula (bonding − antibonding)/2, that is (10 − 4)/2 = 3. This simple picture is not adequate, since if CO is ionized to give CO^+ by removal of one electron from the $\sigma 2p_x$ orbital then the bond order should be reduced to $2\frac{1}{2}$ and the bond length increased. In fact the bond length in CO is 1.128 Å and in CO^+ it is 1.115 Å. Thus the bond length decreases when we expected it to increase, and it indicates that the electron must have been removed from an antibonding orbital. The problem remains if we assume the order of energy for the MOs is the same as for atoms heavier than C, since this only reverses the position of the $\sigma 2p_x$ and the ($\pi 2p_y$ and $\pi 2p_z$) MOs. The most likely explanation of the bond shortening when CO is changed to CO^+ is that the $\sigma 2s$ and $\sigma^* 2s$ molecular orbitals differ in energy more than is shown in the figure. This means that they are wider apart, and the $\sigma^* 2s$ MO is higher in energy than the $\sigma 2p_x$, $\pi 2p_y$ and $\pi 2p_z$ MOs. This illustrates very plainly that the order of MO energy levels for simple homonuclear diatomic molecules used above is not automatically applicable when two different types of atoms are bonded together, and it is certainly incorrect in this particular heteronuclear case.

HCl molecule

With heteronuclear atoms it is not obvious which AOs should be combined by the LCAO method to form MOs. In addition because the energy levels of the AOs on the two atoms are not identical, some MOs will contain a bigger contribution from one AO than the other. This is equivalent to saying that the MO 'bulges' more towards one atom, or the electrons in the MO spend more time round one atom than the other. Thus some degree of charge separation $\delta+$ and $\delta-$ occurs, resulting in a dipole. Thus partial ionic contributions may play a significant part in the bonding.

Consider the HCl molecule. Combination between the hydrogen $1s$ AO and the chlorine $1s$, $2s$, $2p$ and $3s$ orbitals can be ruled out because their energies are too low. If overlap occurred between the chlorine $3p_y$ and $3p_z$ orbitals it would be non-bonding (see Figure 4.22) because the positive lobe of hydrogen will overlap equally with the positive and negative lobes of the chlorine orbitals. Thus the only effective overlap is with the chlorine $3p_x$ orbital. The combination of H $1s^1$ and Cl $3p_x^1$ gives both bonding and antibonding orbitals, and the two electrons occupy the bonding MO, leaving the antibonding MO empty. It is assumed that all the chlorine AOs except $3p_x$ are localized on the chlorine atom and retain their original AO status, and the $3s$, $3p_y$ and $3p_z$ orbitals are regarded as non-bonding lone pairs.

This over-simplification ignores any ionic contribution such as can be shown with the valence bond resonance structures

$$H^+Cl^- \text{ and } H^-Cl^+$$

The former would be expected to contribute significantly, resulting in a stronger bond.

EXAMPLES OF MOLECULAR ORBITAL TREATMENT INVOLVING DELOCALIZED π BONDING

Carbonate ion CO_3^{2-}

The structure of the carbonate ion is a planar triangle, with bond angles of 120°. The C atom at the centre uses sp^2 orbitals. All three oxygen atoms are equivalent, and the C—O bonds are shorter than single bonds. A single valence bond structure such as that shown would have different bond lengths, and so fails to describe the structure adequately.

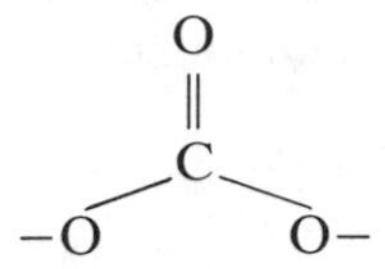

The problem is simply that an electron cannot be represented as a dot, or a pair of electrons as a line (bond). The fourth electron pair that makes up the double bond is not localized in one of the three positions, but is somehow spread out over all three bonds, so that each bond has a bond order of $1\frac{1}{3}$.

Pauling adapted the valence bond notation to cover structures where electrons are delocalized. Three contributing structures can be drawn for the carbonate ion:

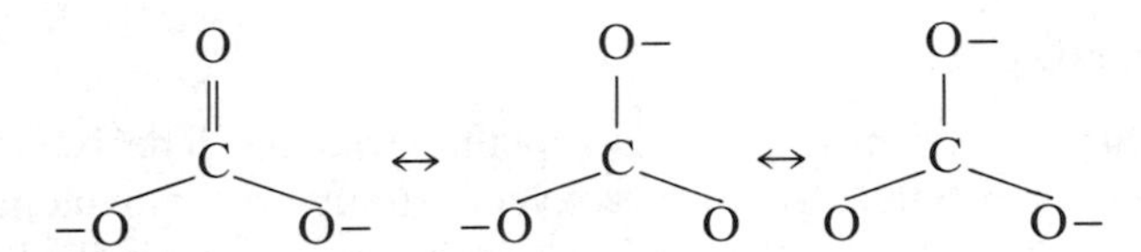

These contributing structures do not actually exist. The CO_3^{2-} does not consist of a mixture of these structures, nor is there an equilibrium between them. The true structure is somewhere in between, and is called a *resonance hybrid*. Resonance was widely accepted in the 1950s but is now regarded at best as clumsy and inadequate, and at worst as misleading or wrong!

Delocalized π bonding is best described by multi-centre bonds, which involve π molecular orbitals. The steps in working this out are:

1. Find the basic shape of the molecule or ion, either experimentally, or from the VSEPR theory using the number of σ bonds and lone pairs on the central atom.
2. Add up the total number of electrons in the outer (valence) shell of all the atoms involved, and add or subtract electrons as appropriate to form ions.
3. Calculate the number of electrons used in σ bonds and lone pairs, and by subtracting this from the total determine the number of electrons which can participate in π bonding.
4. Count the number of atomic orbitals which can take part in π bonding. Combine these to give the same number of molecular orbitals which are delocalized over all of the atoms. Decide whether MOs are bonding, non-bonding or antibonding, and feed the appropriate number of π electrons into the MOs (two electrons per MO). The orbitals with lowest energy are filled first. The number of π bonds formed can easily be determined from the MOs which have been filled.

The structure of the CO_3^{2-} will be examined in this way. There are 24 electrons in the valence shell (four from C, six from each of the three O atoms and two from the charges on the ion).

Of these, six are used to form the σ bonds between C and the three O atoms. Each O has four non-bonding electrons. This leaves six electrons available for π bonding.

The atomic orbitals available for π bonding are the $2p_z$ orbital on C and the $2p_z$ orbitals from the three O atoms. Combining these four atomic orbitals gives four four-centre π molecular orbitals. Each of these covers all four atoms in the ion. The lowest energy MO is bonding, the highest is antibonding, and the remaining two are non-bonding (and are also degenerate, i.e. the same in energy). The six π electrons occupy the MOs of lowest energy. Two electrons fill the bonding MO and four electrons fill both of the non-bonding MOs and thus contribute one π bond to the molecule. Each of the C—O bonds has a bond order of $1\frac{1}{3}$, 1 from the σ bond and $\frac{1}{3}$ from the π bond.

Nitrate ion NO_3^-

The structure of the nitrate ion is a planar triangle. The N atom at the centre uses sp^2 orbitals. All three oxygen atoms are equivalent, and the bond lengths N—O are all a little shorter than for a single bond. This cannot be explained by a valence bond structure:

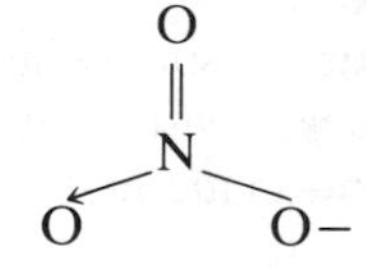

There 24 electrons in the valence shell (five from N, six from each of the three O atoms and one from the charge on the ion).

Of these, six are used to form the σ bonds between N and the three O atoms. Each O has four non-bonding electrons. This leaves six electrons available for π bonding.

The atomic orbitals used for π bonding are the $2p_z$ orbitals on N and the three O atoms. Combining these four atomic orbitals gives four four-centre π molecular orbitals. The lowest in energy is bonding, the highest is antibonding, and the remaining two are degenerate (the same in energy) and are non-bonding. The six π electrons fill the bonding MO and both of the non-bonding MOs and thus contribute one π bond to the molecule. Each of the N—O bonds has a bond order of $1\frac{1}{3}$, 1 from the σ bond and $\frac{1}{3}$ from the π bond.

Sulphur trioxide SO_3

The structure of SO_3 is a planar triangle. The S atom at the centre uses sp^2 orbitals. All three oxygen atoms are equivalent, and the S—O bonds are much shorter than single bonds. The valence bond structure is:

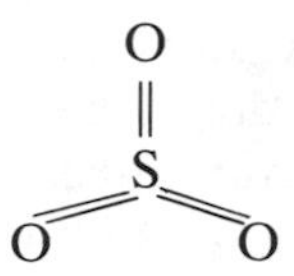

The multi-centre π MO explanation is as follows. There are 24 electrons in the valence shell (six from S and six from each of the three O atoms).

Of these, six are used to form the σ bonds between S and the three O atoms. Each O has four non-bonding electrons. This leaves six electrons available for π bonding.

SO_3 has 24 outer electrons like the NO_3^- ion. If SO_3 followed the same pattern as the NO_3^- ion and used the $3p_z$ AO on S and the $2p_z$ AOs on the three O atoms, four MOs would be formed, one bonding, two non-bonding and one antibonding, and the six π electrons would occupy the bonding and non-bonding MOs, thus contributing one π bond to the molecule and giving a S—O bond order of $1\frac{1}{3}$. The bonds are much shorter than this would imply. Though SO_3 has the same number of outer electrons as NO_3^-, the two are **not** isoelectronic. The S atom has three shells of electrons, so there is the possibility of using *d* orbitals in the bonding scheme.

The six atomic orbitals available for π bonding are the $2p_z$ orbitals on the three O atoms and the $3p_z$, $3d_{xz}$ and $3d_{yz}$ orbitals on S. Combining one $2p_z$ AO with the $3p_z$ AO gives two MOs, one bonding and the other antibonding. Similarly, combining the second $2p_z$ AO with the $3d_{xz}$ AO gives one bonding MO and one antibonding MO, and combining the third $2p_z$ AO with the $3d_{yz}$ AO gives one bonding MO and one antibonding MO. Thus we obtain three bonding MOs and three antibonding MOs. The six electrons available for π bonding occupy the three bonding MOs, and thus contribute three π bonds to the molecule. Each of the S—O bonds has a bond order of approximately 2, 1 from the σ bond and approximately 1 from the π bond. The reason why the bond order is approximate is that the extent of *d* orbital participation depends on the number of electrons and the size and energy of the orbitals involved. This involves detailed calculation.

Ozone O_3

Ozone O_3 forms a V-shaped molecule. Both bond lengths are 1.278 Å, and the bond angle is 116°48′. We assume that the central O atom uses roughly sp^2 orbitals for σ bonding. The valence bond representation of the structure is inadequate since it suggests that the bonds are of different lengths, though it could be explained in terms of resonance hybrids.

```
      O
     // ↘        single valence bond structure
    O    O

    O                O
   // ↘     ↔      ↙ \\       resonance hybrids
  O    O          O    O
```

The double bonding in the structure is best explained by means of delocalized three-centre π bonding. There is a total of 18 electrons in the valence shell, made up of six from each of the three O atoms.

The central O atom forms a σ bond with each of the other O atoms, which accounts for four electrons. The central O atom uses sp^2 orbitals, one of which is a lone pair. If the 'end' O atoms also use sp^2 atomic orbitals, each O contains two non-bonding pairs of electrons. Thus lone pairs account for 10 electrons. Sigma bonds and lone pairs together account for 14 electrons, thus leaving four electrons for π bonding.

The atomic orbitals involved in π bonding are the $2p_z$ orbitals on each of the three O atoms. These give rise to three molecular orbitals. These are three-centre π molecular orbitals. The lowest energy MO is bonding, the highest energy MO is antibonding, and the middle one is non-bonding. There are four π electrons and two fill the bonding MO and two fill the non-bonding MO, thus contributing one π bond over the molecule. This gives a bond order of 1.5 for the O—O bonds. The π system is thus a four-electron three-centre bond.

Nitrite ion NO_2^-

The nitrite ion NO_2^- is V-shaped. This is based on a plane triangular structure, with N at the centre, two corners occupied by O atoms, and the third corner occupied by a lone pair. Thus the N atom is roughly sp^2 hybridised.

In the NO_2^- ion there are 18 electrons in the valence shell. These are made up of five from N, six from each of the two O atoms, and one from the charge on the ion.

The N atom forms σ bonds to each of the O atoms, which accounts for four electrons, and the N atom has a lone pair accounting for two electrons. If the O atoms also use sp^2 atomic orbitals (one for bonding and two for lone pairs), the lone pairs on the O atoms account for eight more electrons. A total of 14 electrons has been accounted for, leaving four electrons for π bonding.

Three atomic orbitals are involved in π bonding: the $2p_z$ orbitals on the N atom and on both of the O atoms. These three atomic orbitals form three molecular orbitals. These are three-centre π molecular orbitals. The lowest in energy is bonding, the highest is antibonding, and the middle one is non-bonding. Two of the four π electrons fill the bonding MO and two fill the non-bonding MO, thus contributing one π bond over the molecule. The bond order of the N—O bonds is thus 1.5, and the N—O distances are in between those for a single and double bond.

Carbon dioxide CO_2

The structure of CO_2 is linear O—C—O, and the C atom uses sp hybrid orbitals for σ bonds. Both C—O bonds are the same length, but are much shorter than a single bond. This is best explained by delocalized π bonding, and involves multi-centre π molecular orbitals. The molecule contains 16 outer shell electrons, made up from six electrons from each of the two O atoms and four electrons from the C atom.

The C atom forms σ bonds to both the O atoms, thus accounting for four electrons. There are no lone pairs of electrons on the C atom. If the O atoms also use *sp* hybrid orbitals then there is one lone pair of electrons on each O atom, accounting for a further four electrons. This accounts for eight electrons altogether, leaving eight electrons available for π bonding.

If the σ bonding and lone pairs of electrons occupy the $2s$ and $2p_x$ atomic orbitals on each O atom, then the $2p_y$ and $2p_z$ atomic orbitals can be used for π bonding. Thus there are six atomic orbitals available for π bonding. The three $2p_y$ atomic orbitals (one from C and one from each of the O atoms) form three three-centre π molecular orbitals which cover all three atoms. The MO with the lowest energy is called a bonding molecular orbital. The MO with the highest energy is called an antibonding MO, and the remaining MO is non-bonding. In a similar way, the three $2p_z$ atomic orbitals also form bonding, non-bonding and antibonding three-centre π molecular orbitals. Each of these MOs covers all three atoms in the molecule. The eight π electrons occupy the MOs of lowest energy, in this case two electrons in the bonding $2p_y$ MO, two electrons in the bonding $2p_z$ MO, then two electrons in the non-bonding $2p_y$ MO and two electrons in the non-bonding $2p_z$ MO. This gives a net contribution of two π bonds to the molecule, in addition to the two σ bonds. Thus the bond order C—O is thus two.

Azide ion N_3^-

The N_3^- ion has 16 outer electrons (five from each N and one from the charge on the ion). It is isoelectronic with CO_2, and is linear N—N—N like CO_2. We assume the central N uses *sp* hybrid orbitals for σ bonding.

Four electrons are used for the two σ bonds. Each of the end N atoms has one non-bonding pair of electrons, accounting for four more electrons. This leaves eight electrons for π bonding.

If the bonding and non-bonding electrons are assumed to use the $2s$ and $2p_x$ orbitals, this leaves six atomic orbitals for π bonding. These are three $2p_y$ AOs and three $2p_z$ AOs. The three $2p_y$ orbitals form three three-centre π molecular orbitals. The lowest in energy is bonding, the highest is antibonding, and the remaining MO is non-bonding. In a similar way the three $2p_z$ atomic orbitals give bonding, non-bonding and antibonding MOs. The eight π electrons fill both of the bonding MOs, and both of the non-bonding MOs. Thus there are two σ and two π bonds, giving a bond order of 2. Thus both N—N bonds are the same length, 1.16 Å.

SUMMARY OF MULTI-CENTRE π BONDED STRUCTURES

Isoelectronic species have the same shape and the same bond order (Table 4.6).

Table 4.6 Multi-centre bonded structures

Species	Number of outer electrons	Shape	Bond order
CO_2	16	Linear	2
N_3^-	16	Linear	2
O_3	18	V-shaped	1.5
NO_2^-	18	V-shaped	1.5
CO_3^{2-}	24	Plane triangle	1.33
NO_3^-	24	Plane triangle	1.33

UNITED ATOM METHOD

The LCAO method described above is tantamount to bringing the atoms from infinity to their equilibrium positions in the molecule. The united atom method is an alternative approach. It starts with a hypothetical 'united atom' where the nuclei are superimposed, and then moved to their equilibrium distance apart. The united atom has the same number of orbitals as a normal atom, but it contains the electrons from two atoms. Thus some electrons must be promoted to higher energy levels in the

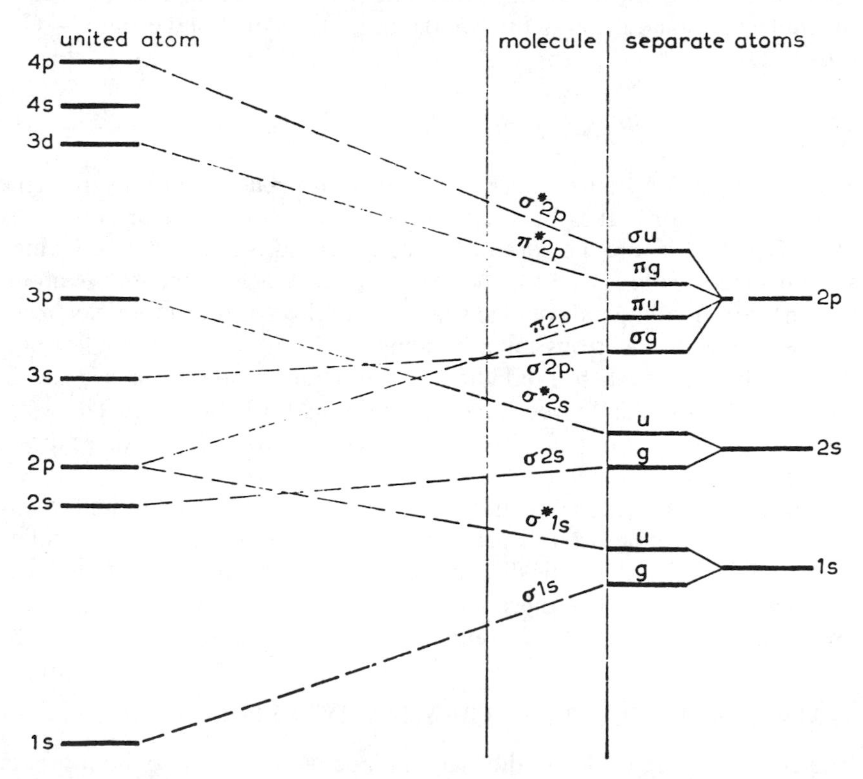

Figure 4.33 Mulliken correlation for like atoms forming a diatomic molecule.

united atom. Further, the energy of the united atom orbitals differs from that of the atomic orbitals because of the greater nuclear charge. Thus the molecular orbitals are in an intermediate position between the orbitals in the united atom and those in the separate atom. If lines are drawn between the energies of the electrons in the separate atoms and in the united atom (that is a graph of internal energy against the distance between the nuclei from $r = 0$ to $r = \infty$), a correlation diagram is obtained (Figure 4.33).

FURTHER READING

Atkins, P.W. (1983) *Molecular Quantum Mechanics*, Oxford University Press, Oxford.

Ballhausen, C.J. and Gray, H.B. (1964) *Molecular Orbital Theory*, Benjamin, Menlo Park, California.

Ballhausen, C.J. and Gray, H.B. (1980) *Molecular Electronic Structures*, Benjamin-Cummings, Menlo Park, California.

Brown, I.D. (1978) A simple structural model for inorganic chemistry, *Chem. Soc. Rev.*, **7**, 359.

Burdett, J.K. (1980) *Molecular Shapes: Theoretical Models for Inorganic Stereochemistry*, Wiley-Interscience, New York.

Cartmell, E. and Fowles, G.W.A. (1977) *Valency and Molecular Structure*, 4th ed., Butterworths, London.

Coulson, C.A. (1982) revised by McWeeny, R., *The Shape and Structure of Molecules*, 2nd ed., Clarendon Press, Oxford.

Coulson, C.A. (1979) revised by McWeeny, R., *Valence*, 3rd ed., Oxford University Press, Oxford. (An updated version of Coulson's 1969 book.)

DeKock, R.L. and Bosma, W.B. (1988) The three-center, two-electron chemical bond, *J. of Chem. Ed.*, **65**, 194–197.

DeKock, R.L. and Gray, H.B. (1980) *Chemical Structure and Bonding*, Benjamin/Cummings, Menlo Park, California.

Douglas, B., McDaniel, D.H. and Alexander J.J. (1983) *Concepts and Models in Inorganic Chemistry*, 2nd ed., Wiley, New York.

Ebsworth, E.A.V., Rankin, D.W.H. and Cradock, S. (1987) *Structural Methods in Inorganic Chemistry*, Blackwell Scientific, Oxford.

Ferguson, J.E. (1974) *Stereochemistry and Bonding in Inorganic Chemistry*, Prentice Hall. Englewood Cliffs, N.J.

Gillespie, R.J. (1972) *Molecular Geometry*, Van Nostrand Reinhold, London. (The latest on the VSEPR theory.)

Gillespie, R.J. and Nyholm, R.S. (1957) *Q. Rev. Chem. Soc.*, **11**, 339. (Develops the Sidgwick–Powell theory into the modern VSEPR theory.)

Karplus, M. and Porter, R.N. (1970) *Atoms and Molecules*, Benjamin, New York.

Kettle, S.F.A. (1985) *Symmetry and Structure*, Wiley, London.

Kutzelnigg, W. (1984) Chemical bonding in higher main group elements, *Angew. Chemie* (International edition in English), **23**, 272.

Murrell, J.N., Kettle, S.F.A. and Tedder, J.M. (1985) *The Chemical Bond*, 2nd ed., Wiley, London.

O'Dwyer, M.F., Kent, J.E. and Brown, R.D. (1978) *Valency*, 2nd ed., Springer (reprinted 1986).

Pauling, L. (1961) *The Nature of the Chemical Bond*, 3rd ed., Oxford University Press, Oxford. (A classical text on bonding.)

Pauling, L. (1967) *The Chemical Bond*, Oxford University Press, Oxford. (A shorter and updated book on bonding.)

Sidgwick, N.V. and Powell H.M. (1940) *Proc. R. Soc.*, **176A**, 153. (The original paper on electron pair repulsion theory.)

Speakman, J.C. (1977) *Molecular Structure: Its Study by Crystal Diffraction*, Royal Society for Chemistry, Monographs for Teachers 30.
Urch, D.S. (1970) *Orbitals and Symmetry*, Penguin.
Wade, K. (1971) *Electron Deficient Compounds*, Nelson, London.
Worral, J. and Worral I.J. (1969) *Introduction to Valence Theory*, American Elsevier Publishing Co., New York.

Bond lengths and bond angles of molecular structures in the crystalline and gaseous states are given in The Chemical Society's Special Publication 11 (*Interatomic Distances*) and Special Publication 18 (*Interatomic Distances Supplement*).

PROBLEMS

1. Show by drawings how an *s* orbital, a *p* orbital or a *d* orbital on one atom may overlap with *s*, *p* or *d* orbitals of an adjacent atom.
2. List three rules for the linear combination of atomic orbitals.
3. Show how the LCAO approximation gives rise to bonding and anti-bonding orbitals. Illustrate your answer by reference to three different diatomic molecules.
4. Use the molecular orbital theory to explain why the bond strength in a N_2 molecule is greater than that in a F_2 molecule.
5. Use the MO theory to predict the bond order and the number of unpaired electrons in O_2^{2-}, O_2^-, O_2, O_2^+, NO and CO.
6. Draw MO energy level diagrams for C_2, O_2 and CO. Show which orbitals are occupied, and work out the bond orders and magnetic properties of these molecules.
7. Name the three types of hybrid orbital that may be formed by an atom with only *s* and *p* orbitals in its valence shell. Draw the shapes and stereochemistry of the hybrid orbitals so produced.
8. What are the geometric arrangements of sp^3d^2, sp^3d and dsp^2 hybrid orbitals?
9. Predict the structure of each of the following, and indicate whether the bond angles are likely to be distorted from the theoretical values: (a) $BeCl_2$; (b) BCl_3; (c) $SiCl_4$; (d) PCl_5 (vapour); (e) PF_3; (f) F_2O; (g) SF_4; (h) IF_5; (i) SO_2; (j) SF_6.
10. How and why does the cohesive force in metals change on descending a group, or on moving from one group to another? What physical properties follow these changes in cohesive force?
11. Use energy level diagrams and the band theory to explain the difference between conductors, insulators and semiconductors.

The metallic bond

5

GENERAL PROPERTIES OF METALS

All metals have characteristic physical properties:

1. They are exceptionally good conductors of electricity and heat.
2. They have a characteristic metallic lustre – they are bright, shiny and highly reflective.
3. They are malleable and ductile.
4. Their crystal structures are almost always cubic close-packed, hexagonal close-packed, or body-centred cubic.
5. They form alloys readily.

Conductivity

All metals are exceptionally good conductors of heat and electricity. Electrical conduction arises by the movement of electrons. This is in contrast to the movement of ions which is responsible for conduction in aqueous solution or fused melts of ionic compounds like sodium chloride, where sodium ions migrate to the cathode, and chloride ions migrate to the anode. In the solid state, ionic compounds may conduct to a very small extent (semiconduction) if defects are present in the crystal. There is an enormous difference in the conductivity between metals and any other type of solid (Table 5.1).

Table 5.1 Electrical conductivity of various solids

Substance	Type of bonding	Conductivity (ohm cm^{-1})
Silver	Metallic	6.3×10^5
Copper	Metallic	6.0×10^5
Sodium	Metallic	2.4×10^5
Zinc	Metallic	1.7×10^5
Sodium chloride	Ionic	10^{-7}
Diamond	Covalent giant molecule	10^{-14}
Quartz	Covalent giant molecule	10^{-14}

Most of the elements to the left of carbon in the periodic table are metals. A carbon atom has four outer electrons. If these are all used to form four bonds, the outer shell is complete and there are no electrons free to conduct electricity.

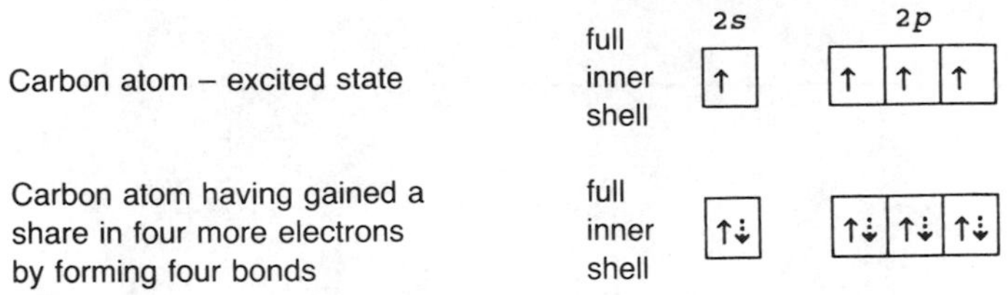

		2s	2p		
Carbon atom – excited state	full inner shell	↑	↑	↑	↑
Carbon atom having gained a share in four more electrons by forming four bonds	full inner shell	↑↓	↑↓	↑↓	↑↓

Elements to the left of carbon have fewer electrons, and so they must have vacant orbitals. Both the number of electrons present in the outer shell, and the presence of vacant orbitals in the valence shell, are important features in explaining the conductivity and bonding of metals.

The conductivity of metals decreases with increasing temperature. Metals show some degree of paramagnetism, which indicates that they possess unpaired electrons.

Lustre

Smooth surfaces of metals typically have a lustrous shiny appearance. All metals except copper and gold are silvery in colour. (Note that when finely divided most metals appear dull grey or black.) The shininess is rather special, and is observed at all viewing angles, in contrast to the shininess of a few non-metallic elements such as sulphur and iodine which appear shiny when viewed at low angles. Metals are used as mirrors because they reflect light at all angles. This is because of the 'free' electrons in the metal, which absorb energy from light and re-emit it when the electron drops back from its excited state to its original energy level. Since light of all wavelengths (colours) is absorbed, and is immediately re-emitted, practically all the light is reflected back – hence the lustre. The reddish and golden colours of copper and gold occur because they absorb some colours more readily than others.

Many metals emit electrons when exposed to light – the photoelectric effect. Some emit electrons when irradiated with short-wave radiation, and others emit electrons on heating (thermionic emission).

Malleability and cohesive force

The mechanical properties of metals are that they are typically malleable and ductile. This shows that there is not much resistance to deformation of the structure, but that a large cohesive force holds the structure together.

$$M_{crystal} \xrightarrow{\Delta H} M_{gas}$$

Table 5.2 Enthalpies of atomization $\Delta H°$ (kJ mol^{-1}) (Measured at 25 °C except for Hg)

Metal	$\Delta H°$	Melting point (°C)	Boiling point (°C)
Li	162	181	1331
Na	108	98	890
K	90	64	766
Rb	82	39	701
Cs	78	29	685
Be	324	1277	2477
Mg	146	650	1120
Ca	178	838	1492
Sr	163	768	1370
Ba	178	714	1638
B	565	2030	3927
Al	326	660	2447
Ga	272	30	2237
Sc	376	1539	2480
Ti	469	1668	3280
V	562	1900	3380
Cr	397	1875	2642
Mn	285	1245	2041
Fe	415	1537	2887
Co	428	1495	2887
Ni	430	1453	2837
Cu	339	1083	2582
Zn	130	420	908

Enthalpies of atomization from Brewer, L., *Science*, 1968, **161**, 115, with some additions.

The cohesive force may be measured as the heat of atomization. Some numerical values of $\Delta H°$, the heats of atomization at 25 °C, are given in Table 5.2. The heats of atomization (cohesive energy) decrease on descending a group in the periodic table Li–Na–K–Rb–Cs, showing that they are inversely proportional to the internuclear distance.

The cohesion energy increases across the periodic table from Group I to Group II to Group III. This suggests that the strength of metallic bonding is related to the number of valency electrons. The cohesive energy increases at first on crossing the transition series Sc–Ti–V as the number of unpaired *d* electrons increases. Continuing across the transition series the number of electrons per atom involved in metallic bonding eventually falls, as the *d* electrons become paired, reaching a minimum at Zn.

The melting points and to an even greater extent the boiling points of the metals follow the trends in the cohesive energies. The cohesive energies vary over an appreciable range, and they approach the magnitude of the lattice energy which holds ionic crystals together. The cohesive energies are much larger than the weak van der Waals forces which hold discrete covalent molecules together in the solid state.

There are two rules about the cohesive energy and structure of metals (or alloys), and these are examined below:

Rule 1. The bonding energy of a metal depends on the average number of unpaired electrons available for bonding on each atom.
Rule 2. The crystal structure adopted depends on the number of s and p orbitals on each atom that are involved with bonding.

Consider the first rule – Group I metals have the outer electronic configuration ns^1, and so have one electron for bonding. In the ground state (lowest energy), Group II elements have the electronic configuration ns^2, but if the atom is excited, an outer electron is promoted, giving the configuration ns^1, np^1, with two unpaired electrons, which can form two bonds. Similarly Group III elements in the ground state have the configuration ns^2, np^1, but when excited to ns^1, np^2, they can use three electrons for metallic bonding.

The second rule attempts to relate the number of *s* and *p* electrons available for bonding to the crystal structure adopted (Table 5.3). Apart from Group I metals, the atoms need to be excited, and the structures adopted are shown in Table 5.4.

Table 5.3 Prediction of metal structures from the number of *s* and *p* electrons involved in metallic bonding

Number of *s* and *p* electrons per atom involved in bonding	Structure
Less than 1.5	Body-centred cubic
1.7–2.1	Hexagonal close-packed
2.5–3.2	Cubic close-packed
Approaching 4	Diamond structure – not metallic

Group I elements have a body-centred cubic structure, and follow the rule. In Group II, only Be and Mg have a hexagonal close-packed structure and strictly follow the rule. In Group III, Al has a cubic close-packed structure as expected. However, not all the predictions are correct. There is no obvious reason why Ca and Sr form cubic close-packed structures. However, the high temperature forms of Ca and Sr, and the room temperature form of Ba form body-centred cubic structures (like Group I), instead of the expected hexagonal close-packed structure. The explanation is probably that the paired *s* electron is excited to a *d* level instead of a *p* level, and hence there is only one *s* or *p* electron per atom participating in metallic bonding. This also explains why the first half of the transition metals also form body-centred cubic structures. In the second half of the transition series, the extra electrons may be put in the *p* level, to avoid

Table 5.4 Type of structure adopted by metals in the periodic table (The room temperature structure is shown at the bottom. Other structures which occur at higher temperatures are listed above this in order of temperature stability)

Li bcc	Be hcp											B	C	N	
Na bcc	Mg hcp											Al ccp	Si d	P	S
K bcc	Ca bcc ccp	Sc bcc hcp	Ti bcc hcp	V bcc	Cr bcc	Mn bcc ccp β χ	Fe bcc ccp bcc	Co ccp hcp	Ni ccp	Cu ccp	Zn hcp	Ga ●	Ge d	As ● α	Se
Rb bcc	Sr bcc hcp ccp	Y bcc hcp	Zr bcc hcp	Nb bcc	Mo bcc	Tc hcp	Ru hcp	Rh ccp	Pd ccp	Ag ccp	Cd hcp	In ccp*	Sn d	Sb ● α	Te
Cs bcc	Ba bcc	La bcc ccp hcp	Hf bcc hcp	Ta bcc	W bcc	Re hcp	Os hcp	Ir ccp	Pt ccp	Au ccp	Hg	Tl bcc hcp	Pb ccp	Bi ● α	Po

bcc = body-centred cubic
ccp = cubic close-packed
ccp* = distorted cubic close-packed
hcp = hexagonal close-packed

d = diamond structure
α = rhombohedral – puckered sheets
χ = other structure
● = special case (see individual group)

pairing d electrons, and so allow the maximum participation of d orbitals in metallic bonding. This increases the number of s and p electrons involved in metallic bonding, and for example in Cu, Ag and Au the excited electronic state involved in bonding is probably d^8, s^1, p^2, giving a cubic close-packed structure and five bonds per atom (two d, one s and two p electrons). At Zn the d orbitals are full, and the excited state used for bonding is $3d^{10}$, $4s^1$, $4p^1$, giving two bonds per atom and a body-centred cubic structure. The enthalpies of atomization are in general agreement with these ideas on bonding.

Crystal structures of metals

Metallic elements usually have a close-packed structure with a coordination number of 12. There are two types of close packing depending on the arrangement of adjacent layers in the structure: cubic close packing ABCABC and hexagonal close packing ABAB (see Metallic bonds and metallic structures in Chapter 2). However, some metals have a body-centred cubic type of structure (which fills the space slightly less efficiently) where there are eight nearest neighbours, with another six next-nearest neighbours about 15% further away. If this small difference in distance between nearest and next-nearest neighbours is disregarded, the coordination number for a body-centred cubic structure may be regarded loosely as 14. The mechanical properties of malleability and ductility depend on the ease with which adjacent planes of atoms can glide over each other, to give an equivalent arrangement of spheres. These properties are also affected by physical imperfections such as grain boundaries and dislocations, by point defects in the crystal lattice and by the presence of traces of impurity in the lattice. The possibility of planes gliding is greatest in cubic close-packed structures, which are highly symmetrical and have possible slip planes of close-packed layers in four directions (along the body diagonals), compared with only one direction in the hexagonal close-packed structure. This explains why cubic close-packed structures are generally softer and more easily deformed than hexagonal or body-centred cubic structures. Impurities may cause dislocations in the normal metal lattice, and the localized bonding increases the hardness. Some soft metals like Cu become work hardened – it is harder to bend the metal a second time. This is because dislocations are caused by the first bending, and these disrupt the slip planes. Other metals such as Sb and Bi are brittle. This is because they have directional bonds, which pucker layers, preventing one layer from slipping over another.

The type of packing varies with the position of the element in the periodic table (Table 5.4), which is related to the number of s and p electrons on each atom that can take part in metallic bonding. This has been described earlier.

Metallic elements commonly react with other metallic elements, often over a wide range of composition, forming a variety of alloys which look like metals, and have the properties of metals.

Table 5.5 Interatomic distances in M_2 molecules and metal crystals

	Distance in metal (Å)	Distance in M_2 molecule (Å)
Li	3.04	2.67
Na	3.72	3.08
K	4.62	3.92
Rb	4.86	4.22
Cs	5.24	4.50

Bond lengths

If the valence electrons in a metal are spread over a large number of bonds, each bond should be weaker and hence longer. The alkali metals exist as diatomic molecules in the vapour state, and the interatomic distances in the metal crystal are longer than in the diatomic molecule (Table 5.5).

Though the bonds in the metal are longer and weaker, there are many more of them than in the M_2 molecule, so the total bonding energy is greater in the metal crystal. This can be seen by comparing the enthalpy of sublimation of the metal crystal with the enthalpy of dissociation of the M_2 molecules (Table 5.6).

THEORIES OF BONDING IN METALS

The bonding and structures adopted by metals and alloys are less fully understood than those with ionic and covalent compounds. Any successful theory of metallic bonding must explain both the bonding between a large number of identical atoms in a pure metal, and the bonding between widely different metal atoms in alloys. The theory cannot involve directional bonds, since most metallic properties remain even when the metal is in the liquid state (for example mercury), or when dissolved in a suitable

Table 5.6 Comparison of enthalpies of sublimation and dissociation

	Enthalpy of sublimation of metal ($kJ\,mol^{-1}$)	$\frac{1}{2}$ enthalpy of dissociation of M_2 molecule ($kJ\,mol^{-1}$)
Li	161	54
Na	108	38
K	90	26
Rb	82	24
Cs	78	21

solvent (for example solutions of sodium in liquid ammonia). Further, the theory should explain the great mobility of electrons.

Free electron theory

As early as 1900, Drude regarded a metal as a lattice with electrons moving through it in much the same way as molecules of a gas are free to move. The idea was refined by Lorentz in 1923, who suggested that metals comprised a lattice of rigid spheres (positive ions), embedded in a gas of free valency electrons which could move in the interstices. This model explains the free movement of electrons, and cohesion results from electrostatic attraction between the positive ions and the electron cloud. Whilst it does explain in a rough qualitative way why an increased number of valency electrons results in an increased cohesive energy, quantitative calculations are much less successful than similar calculations for the lattice energies of ionic compounds.

Valence bond theory

Consider a simple metal such as lithium, which has a body-centred cubic structure, with eight nearest neighbours and six next-nearest neighbours at a slightly greater distance. A lithium atom has one electron in its outer shell, which may be shared with one of its neighbours, forming a normal two-electron bond. The atom could equally well be bonded to any of its other eight neighbours, so many different arrangements are possible, and Figures 5.1a and b are two examples.

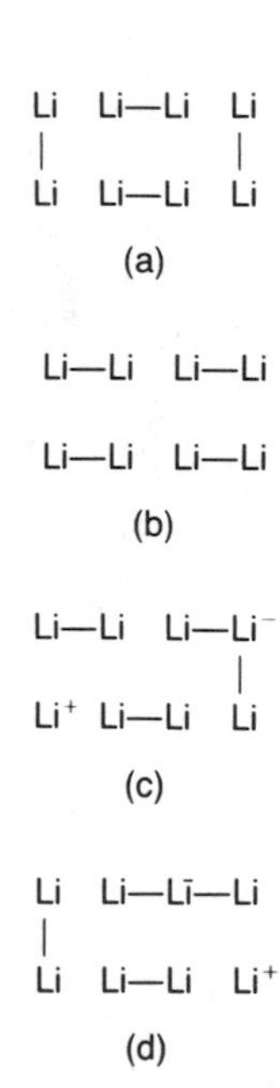

Figure 5.1 Representations of some bonding possibilities in lithium.

A lithium atom may form two bonds if it ionizes, and it can then form many structures similar to those in Figures 5.1c and d. Pauling suggested that the true structure is a mixture of all the many possible bonding forms. The more possible structures there are, the lower the energy. This means that the cohesive force which holds the structure together is large, and in metallic lithium the cohesive energy is three times greater than in a Li_2 molecule. The cohesive energy increases from Group I to II to III, and this is explained by the atoms being able to form an increased number of bonds, and give an even larger number of possible structures. The presence of ions could explain the electrical conduction, but the theory does not explain the conduction of heat in solids, or the lustre, or the retention of metallic properties in the liquid state or in solution.

Molecular orbital or band theory

The electronic structure of a lithium atom is

1s	2s	2p		
↑↓	↑			

The Li_2 molecule exists in the vapour state, and bonding occurs using the 2*s* atomic orbital. There are three empty 2*p* orbitals in the valence shell, and the presence of empty AOs is a prerequisite for metallic properties. (Carbon in its excited state, nitrogen, oxygen, fluorine, and neon all lack empty AOs in the valence shell and are all non-metals.)

The valence shell has more AOs than electrons, so even if the electrons are all used to form normal two-electron bonds, the atom cannot attain a noble gas structure. Compounds of this type are termed 'electron deficient'.

Empty AOs may be utilized to form additional bonds in two different ways:

1. Empty AOs may accept lone pairs of electrons from other atoms or ligands, forming coordinate bonds.
2. Cluster compounds may be formed, where each atom shares its few electrons with several of its neighbours, and obtains a share in their electrons. Clustering occurs in the boron hydrides and carboranes, and is a major feature of metals.

The molecular orbital description of an Li_2 molecule has been discussed earlier in Chapter 4, in the examples of MO treatment. There are six electrons arranged in molecular orbitals:

$$\sigma 1s^2,\ \sigma^* 1s^2,\ \sigma 2s^2$$

Bonding occurs because the $\sigma 2s$ bonding MO is full and the corresponding antibonding orbital is empty. Ignoring any inner electrons, the 2*s* AOs on each of the two Li atoms combine to give two MOs – one bonding and one antibonding. The valency electrons occupy the bonding MO (Figure 5.2a).

Suppose three Li atoms joined to form Li_3. Three 2*s* AOs would combine to form three MOs – one bonding, one non-bonding and one antibonding. The energy of the non-bonding MO is between that for the bonding and antibonding orbitals. The three valency electrons from the three atoms would occupy the bonding MO (two electrons) and the non-bonding MO (one electron) (Figure 5.2b).

In Li_4, the four AOs would form four MOs – two bonding, and two antibonding. The presence of two non-bonding MOs between the bonding and antibonding orbitals reduces the energy gap between the orbitals. The four valency electrons would occupy the two lowest energy MOs, which are both bonding orbitals, as shown in Figure 5.2c.

As the number of electrons in the cluster increases, the spacing between the energy levels of the various orbitals decreases further, and when there are a large number of atoms, the energy levels of the orbitals are so close together that they almost form a continuum (Figure 5.2d).

The number of MOs must by definition be equal to the number of constituent AOs. Since there is only one valence electron per atom in lithium, and a MO can hold two electrons, it follows that only half the MOs in the 2*s* valence band are filled – i.e. the bonding MOs. It requires

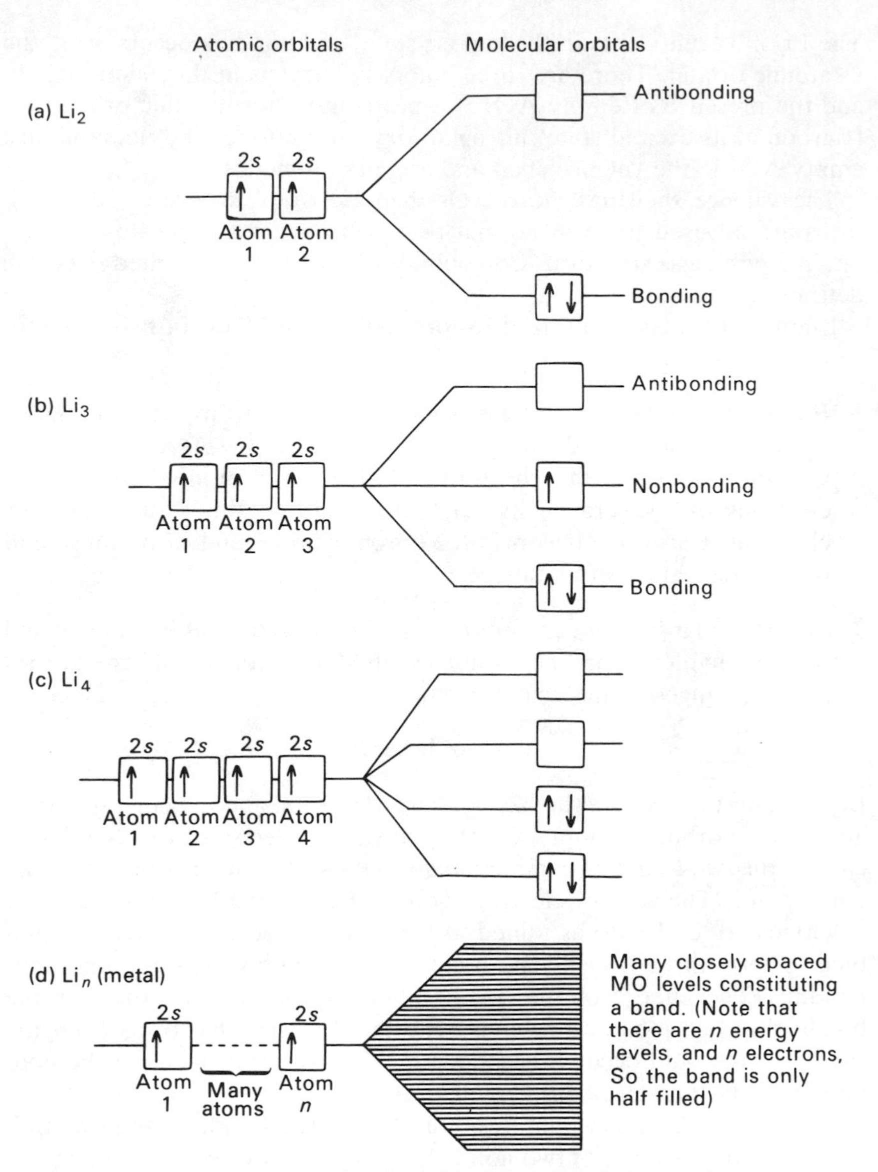

Figure 5.2 Development of molecular orbitals into bands in metals.

only a minute amount of energy to perturb an electron to an unoccupied MO.

The MOs extend in three dimensions over all the atoms in the crystal, so electrons have a high degree of mobility. The mobile electrons account for the high thermal and electrical conduction of metals.

If one end of a piece of metal is heated, electrons at that end gain energy and move to an unoccupied MO where they can travel rapidly to any other part of the metal, which in turn becomes hot. In an analogous manner, electrical conduction takes place through a minor perturbation in energy

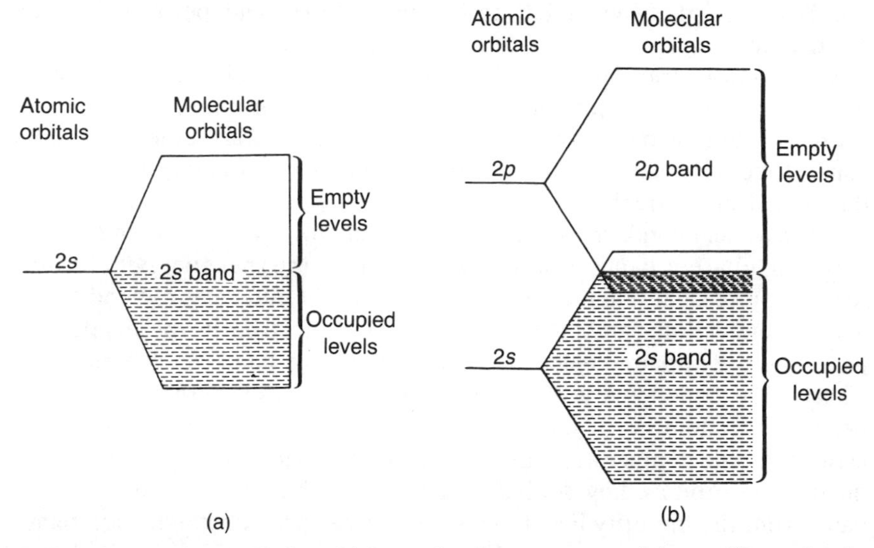

Figure 5.3 Two methods by which conduction can occur: (a) metallic molecular orbitals for lithium showing half filled band; (b) metallic molecular orbitals for beryllium showing overlapping bands.

promoting an electron to an unfilled level, where it can move readily. In the absence of an electric field, equal numbers of electrons will move in all directions. If a positive electrode is placed at one end, and a negative electrode at the other, then electrons will move towards the anode much more readily than in the opposite direction; hence an electric current flows.

Conduction occurs because the MOs extend over the whole crystal, and because there is effectively no energy gap between the filled and unfilled MOs. The absence of an energy gap in lithium is because only half the MOs in the valence band are filled with electrons (Figure 5.3a).

In beryllium there are two valence electrons, so the valence electrons would just fill the 2*s* valence band of MOs. In an isolated beryllium atom, the 2*s* and 2*p* atomic orbitals differ in energy by $160\,kJ\,mol^{-1}$. In much the same way as the 2*s* AOs form a band of MOs, the 2*p* AOs form a 2*p* band of MOs. The upper part of the 2*s* band overlaps with the lower part of the 2*p* band (Figure 5.3b). Because of this overlap of the bands some of the 2*p* band is occupied and some of the 2*s* band is empty. It is both possible and easy to perturb electrons to an unoccupied level in the conduction band, where they can move throughout the crystal. Beryllium therefore behaves as a metal. It is only because the bands overlap that there is no energy gap, so perturbation from the filled valence band to the empty conduction band can occur.

CONDUCTORS, INSULATORS AND SEMICONDUCTORS

In electrical conductors (metals), either the valence band is only partly full, or the valence and conduction bands overlap. There is therefore no significant gap between filled and unfilled MOs, and perturbation can occur readily.

In insulators (non-metals), the valence band is full, so perturbation within the band is impossible, and there is an appreciable difference in energy (called the band gap) between the valence band and the next empty band. Electrons cannot therefore be promoted to an empty level where they could move freely.

Intrinsic semiconductors are basically insulators, where the energy gap between adjacent bands is sufficiently small for thermal energy to be able to promote a small number of electrons from the full valence band to the empty conduction band. Both the promoted electron in the conduction band and the unpaired electron left in the valence band can conduct electricity. The conductivity of semiconductors increases with temperature, because the number of electrons promoted to the conduction band increases as the temperature increases. Both *n*-type and *p*-type semiconductors are produced by doping an insulator with a suitable impurity. The band from the impurity lies in between the valence and conduction bands in the insulator, and acts as a bridge, so that electrons may be excited from the insulator bands to the impurity bands, or vice versa (Figure 5.4). (Defects and semiconductors are discussed at the end of Chapter 3.)

ALLOYS

When two metals are heated together, or a metal is mixed with a non-metallic element, then one of the following will occur:

1. An ionic compound may be formed.
2. An interstitial alloy may be formed.
3. A substitutional alloy may be formed.
4. A simple mixture may result.

Which of these occurs depends on the chemical nature of the two elements concerned, and on the relative sizes of the metal atoms and added atoms.

Ionic compounds

Consider first the chemical nature of the two elements. If an element of high electronegativity (e.g. F 4.0, Cl 3.0 or O 3.5) is added to a metal of low electronegativity (e.g. Li 1.0, Na 0.9), the product will be ionic, not metallic.

Interstitial alloys and related compounds

Next consider the relative sizes of the atoms. The structure of many metals is a close-packed lattice of spherical atoms or ions. There are therefore many tetrahedral and octahedral holes. If the element added has small atoms, they can be accommodated in these holes without altering the

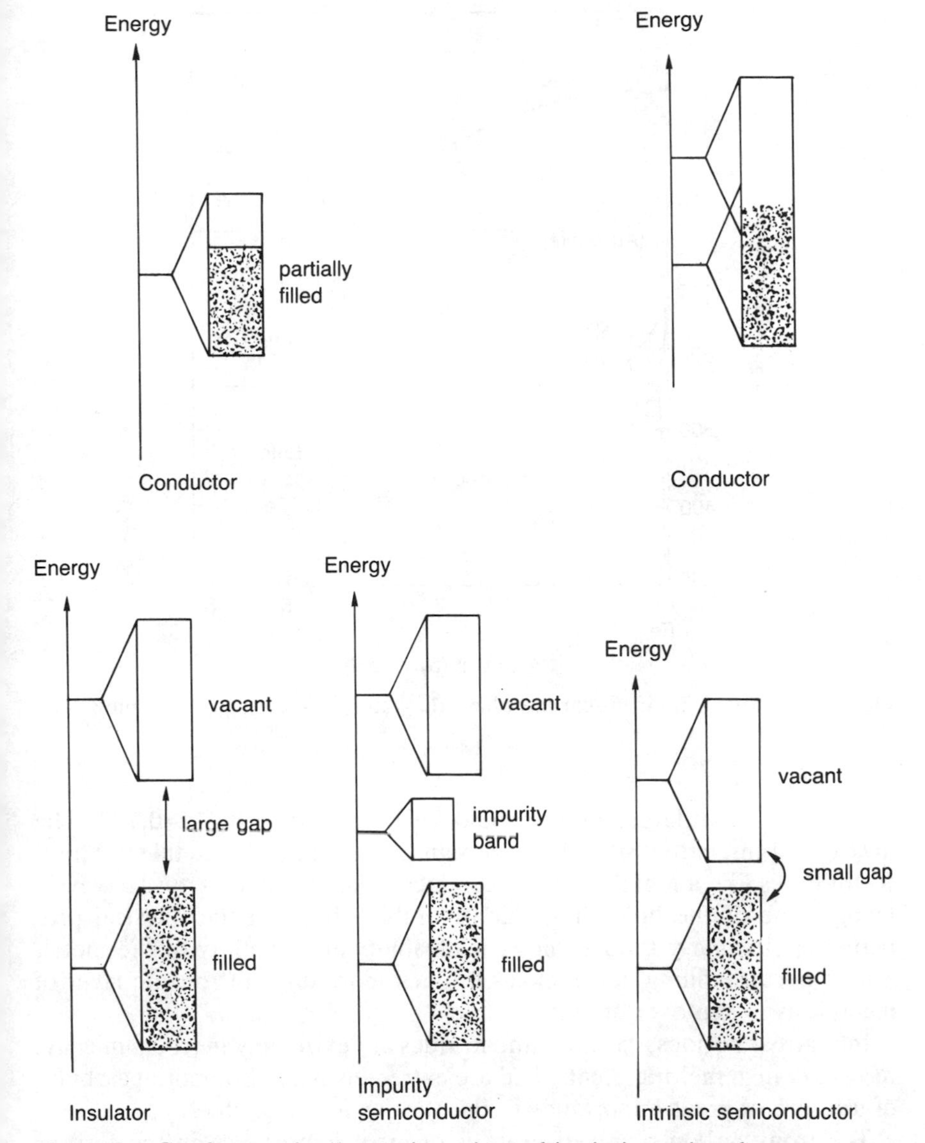

Figure 5.4 Conductors, insulators, impurity and intrinsic semiconductors.

structure of the metal. Hydrogen is small enough to occupy tetrahedral holes, but most other elements occupy the larger octahedral holes.

The invading atoms occupy interstitial positions in the metal lattice, instead of replacing the metal atoms. The chemical composition of compounds of this type may vary over a wide range depending on how many holes are occupied. Such alloys are called interstitial solid solutions, and are formed by a wide range of metals with hydrogen, boron, carbon, nitrogen and other elements. The most important factor is the size of the invading atoms. For octahedral holes to be occupied, the radius ratio of

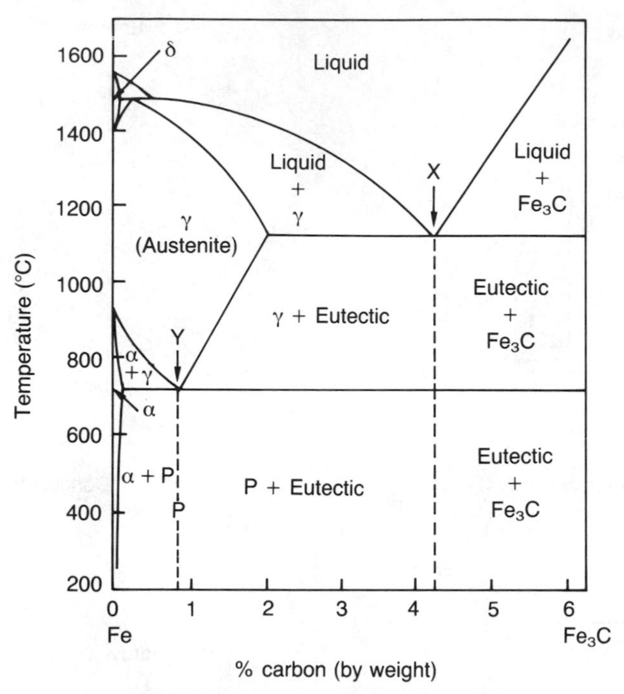

Figure 5.5 Part of the iron–carbon phase diagram (X = eutectic, Y = eutectoid, P = pearlite).

the smaller atom/larger atom should be in the range 0.414–0.732. The invasion of interstitial sites does not significantly alter the metal structure. It still looks like a metal, and still conducts heat and electricity. However, filling some of the holes has a considerable effect on the physical properties, particularly the hardness, malleability and ductility of the metal. This is because filling holes makes it much more difficult for one layer of metal ions to slip over another.

Interstitial borides, carbides and nitrides are extremely inert chemically, have very high melting points, and are extremely hard. Interstitial carbides of iron are of great importance in the various forms of steel.

The iron–carbon phase diagram is of great importance in the ferrous metal industry, and part of this is shown in Figure 5.5. The most important part is from pure Fe to the compound iron carbide or cementite, Fe_3C. Pure Fe exists as two allotropic forms: one is α-ferrite or austenite, with a body-centred cubic structure, which is stable up to 910 °C; above this temperature it changes to γ-ferrite with a face centred–cubic structure. Above 1401 °C γ-ferrite changes back to a body-centred cubic structure, but is now called δ-ferrite.

The upper part of the curve is typical of two solids which are only partly miscible, and a eutectic point occurs at X, between γ-ferrite, iron carbide and liquid. A similar triple point occurs at Y, but since it occurs in a completely solid region it is called a eutectoid point. A solid with the eutectoid composition (a mixture of γ-ferrite and iron carbide) is called pearlite. This is a mixture, not a compound, and is marked P in the diagram. The name pearlite refers to the mother-of-pearl-like appearance when examined under a microscope. The various solid regions α, γ, δ are the different allotropic forms of iron and all contain varying amounts of carbon in interstitial positions.

Steel contains up to 2% carbon. The more carbon present, the harder and more brittle the alloy. When steel is heated, the solid forms austenite, which can be hot rolled, bent or pressed into any required shape. On cooling, the phases separate, and the way in which the cooling is carried out affects the grain size and the mechanical properties. The properties of steel can be changed by heat treatment such as annealing and tempering.

Cast iron contains more than 2% carbon. Iron carbide is extremely hard, and brittle. Heating cast iron does not produce a homogeneous solid solution (similar to austenite for steel), so cast iron cannot be worked mechanically, and the liquid must be cast into the required shape.

Substitutional alloys

If two metals are completely miscible with each other they can form a continuous range of solid solutions. Examples include Cu/Ni, Cu/Au, K/Rb, K/Cs and Rb/Cs. In cases like these, one atom may replace another at random in the lattice.

In the Cu/Au case at temperatures above 450 °C a disordered structure exists (Figure 5.7c), but on slow cooling the more ordered superlattice may be formed (Figure 5.7d). Only a few metals form this type of continuous solid solution, and Hume-Rothery has shown that for complete miscibility the following three rules should apply.

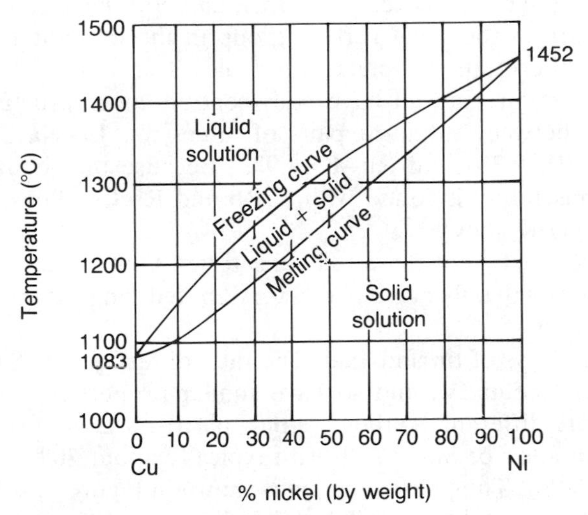

Figure 5.6 Cu/Ni – a continuous series of solid solutions. (After W.J. Moore, *Physical Chemistry*.)

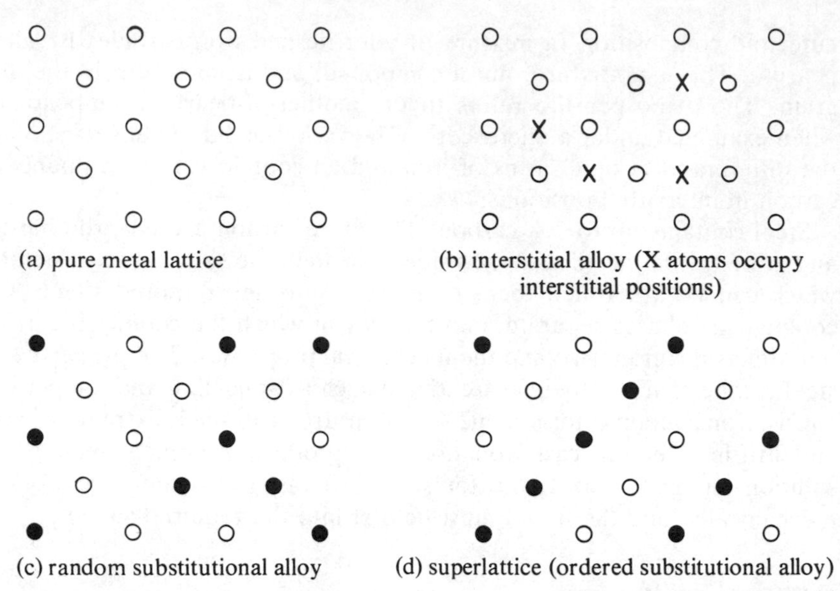

Figure 5.7 Metal and alloy structures: (a) pure metal lattice; (b) interstitial alloy (X atoms occupy interstitial positions); (c) random substitutional alloy and (d) superlattice (ordered substitutional alloy).

1. The two metals must be similar in size – their metallic radii must not differ by more than 14–15%.
2. Both metals must have the same crystal structure.
3. The chemical properties of the metals must be similar – in particular the number of valency electrons should be the same.

Consider an alloy of Cu and Au. The metallic radii differ by only 12.5%, both have cubic close-packed structures, and both have similar properties since they are in the same vertical group in the periodic table. The two metals are therefore completely miscible. The Group I elements are chemically similar, and all have body-centred cubic structures. The size differences between adjacent pairs of atoms are Li–Na 22.4%, Na–K 22.0%, K–Rb 9.3% and Rb–Cs 6.9%. Because of the size difference, complete miscibility is found with K/Rb and Rb/Cs alloys, but not with Li/Na and Na/K alloys.

If only one *or* two of these rules is satisfied then random substitutional solid solutions will only occur over a very limited range at the two extremes of composition.

Consider alloys of tin and lead. The radii differ by only 8.0%, and they are both in Group IV, and so have similar properties. However, their structures are different, so they are only partly miscible. (See Figure 5.8.) Solder is an alloy of Sn and Pb with typically about 30% Sn, but it may have 2–63% Sn. The phase diagram is shown in Figure 5.8. There are two small areas of complete miscibility, labelled α and β, at the extremes of composition at the extreme left and right of the diagram. With plumbers'

Table 5.7 Metallic radii of the elements (Å) (for 12-coordination)

Li 1.52	Be 1.12											B 0.89	C 0.91	N 0.92	
Na 1.86	Mg 1.60											Al 1.43	Si 1.32	P 1.28	S 1.27
K 2.27	Ca 1.97	Sc 1.64	Ti 1.47	V 1.35	Cr 1.29	Mn 1.37	Fe 1.26	Co 1.25	Ni 1.25	Cu 1.28	Zn 1.37	Ga 1.23	Ge 1.37	As 1.39	Se 1.40
Rb 2.48	Sr 2.15	Y 1.82	Zr 1.60	Nb 1.47	Mo 1.40	Tc 1.35	Ru 1.34	Rh 1.34	Pd 1.37	Ag 1.44	Cd 1.52	In 1.67	Sn 1.62	Sb 1.59	Te 1.60
Cs 2.65	Ba 2.22	La 1.87	Hf 1.59	Ta 1.47	W 1.41	Re 1.37	Os 1.35	Ir 1.36	Pt 1.39	Au 1.44	Hg 1.57	Tl 1.70	Pb 1.75	Bi 1.70	Po 1.76

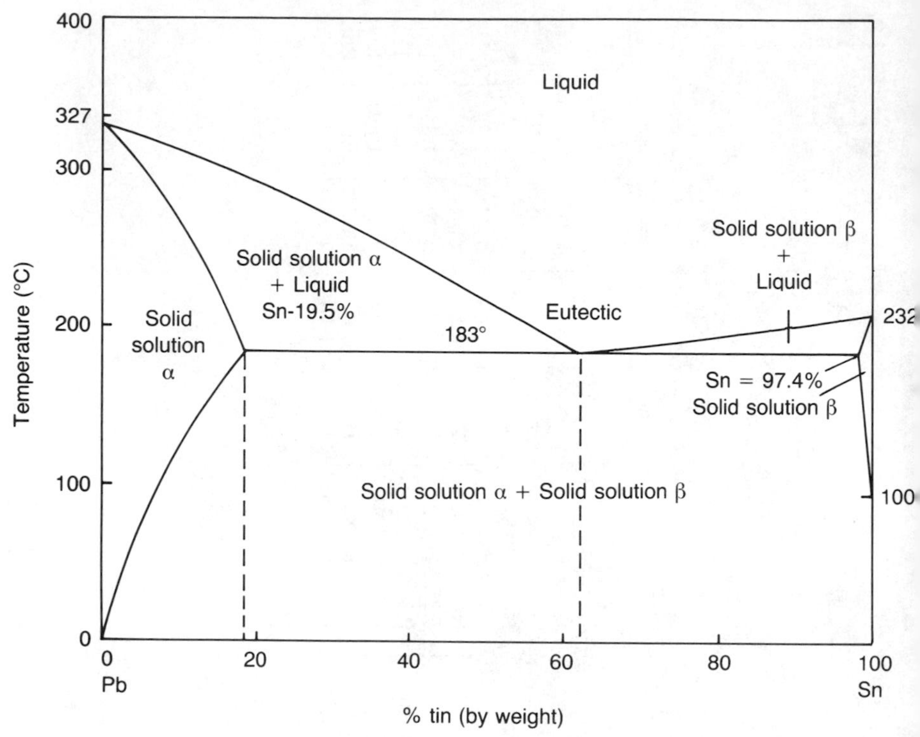

Figure 5.8 Phase diagram for Sn/Pb showing partial miscibility, and only a limited range of solid solutions. (The eutectic occurs at 62% Sn, and eutectoid points occur at 19.5% Sn and 97.4% Sn.)

solder (30% Sn, 70% Pb), the liquid and solid curves are far apart, so that there is a temperature interval of nearly 100 °C over which the solder is pasty, with solid solution suspended in liquid. When in this part-solid part-liquid state, a solder joint can be 'wiped' smooth.

Similar behaviour is found with the Na/K alloy, and the Al/Cu alloy. The metallic radii of Na and K differ by 22.0%, so despite their structural and chemical similarities they only form solid solutions over a limited range of composition.

In other cases where only a limited range of solid solutions are formed, the tendency of the different metals to form compounds instead of solutions is important. One or more intermetallic phases may exist, each of which behaves as a compound of the constituent metals, though the exact stoichiometry may vary over a limited range. For example, in the Cu/Zn system the metallic radii differ by only 7.0%, but they have different structures (Cu is cubic close-packed and Zn is hexagonal close-packed), and they have a different number of valence electrons. Only a limited range of solid solutions is expected, but the atoms have a strong tendency to form compounds, and five different structures may be distinguished, as shown in Table 5.8.

Table 5.8 Table of intermetallic phases

Phase	Zn Composition	Structure
α	0–35%	Random substitutional solid solution of Zn in Cu
β	45–50%	Intermetallic compound of approximate stoichiometry CuZn. Structure body-centred cubic
γ	60–65%	Intermetallic compound of approximate stoichiometry Cu_5Zn_8. Structure complex cubic
ε	82–88%	Intermetallic compound of approximate stoichiometry $CuZn_3$. Structure hexagonal close-packed
η	97–100%	Random substitutional solid solution of Cu in Zn

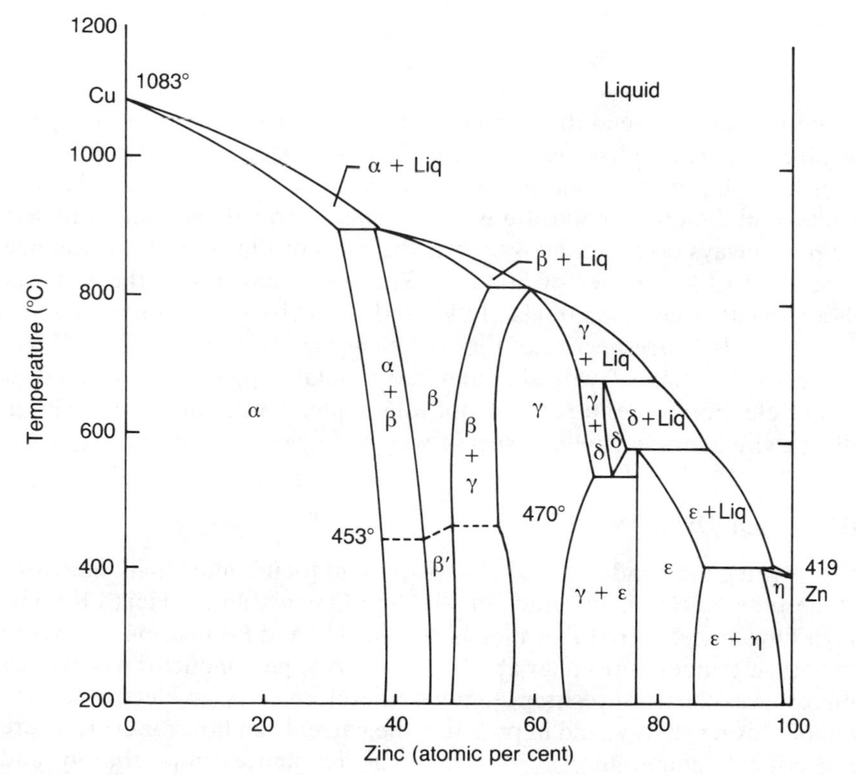

Figure 5.9 Phase diagram for Cu/Zn alloy systems. (Copyright Bohm and Klemm, *Z. Anorg. Chem.*, **243**, 69, 1939.)

Table 5.9 Some intermetallic compounds with various ratios of valency electrons to number of atoms

Ideal formula	No. of valency electrons / No. of atoms	
CuZn	3/2	β phases
Cu_3Al	6/4 = 3/2	
Cu_5Sn	9/6 = 3/2	
AgZn	3/2	
Cu_5Si	9/6 = 3/2	
Ag_3Al	6/4 = 3/2	
$CoZn_3$	3/2*	
Cu_5Zn_8	21/13	γ phases
Cu_9Al_4	21/13	
$Na_{31}Pb_8$	21/13	
Co_5Zn_{21}	21/13*	
$CuZn_3$	7/4	ε phases
Cu_3Si	7/4	
Ag_5Al_3	14/8 = 7/4	
Au_5Al_3	14/8 = 7/4	

* Metals of the Fe, Co and Ni groups are assumed to have zero valence electrons for metallic bonding.

The relation between the various phases is shown in the phase diagram (Figure 5.9). Each phase can be represented by a typical composition or ideal formula, even though it exists over a range of composition. Hume-Rothery studied the compositions of the phases formed and found that the β phase always occurs in alloys when the ratio of the sum of the valency electrons to the number of atoms is 3:2. In a similar way the γ phase always occurs when the ratio is 21 : 13, and the η phase always occurs when the ratio is 7 : 4, irrespective of the particular metals involved (Table 5.9).

The explanation of why similar binary metallic phases are formed at similar electron to atom ratios is not fully understood, but seems to lie in filling the electronic bands in such a way as to give the minimum energy.

SUPERCONDUCTIVITY

Metals are good conductors of electricity, and their conductivity increases as the temperature is lowered. In 1911 the Dutch scientist Heike Kamerlingh Onnes discovered that metals such as Hg and Pb became superconductors at temperatures near absolute zero. A superconductor has zero or almost zero electrical resistance. It can therefore carry an electric current without losing energy, and in principle the current can flow for ever. There is a critical temperature T_c at which the resistance drops sharply and superconduction occurs. Later, Meissner and Ochsenfeld found that some superconducting materials will not permit a magnetic field to penetrate their bulk. This is now called the Meissner effect, and gives rise to 'levitation'. Levitation occurs when objects float on air. This can be achieved by the mutual repulsion between a permanent magnet and a superconductor. A superconductor also expels all internal magnetic fields (arising from unpaired electrons), so superconductors are diamagnetic. In many cases the change in magnetic properties is easier to detect than the increased electrical conductivity, since the passage of high currents or strong magnetic fields may destroy the superconductive state. Thus there is also a critical current and critical magnetization which are linked to T_c.

A superconducting alloy of niobium and titanium, which has a T_c of about 4 K and requires liquid helium to cool it, has been known since the 1950s. Considerable effort has been put into finding alloys which are superconductors at higher temperatures. Alloys of Nb_3Sn, Nb_3Ge, Nb_3Al and V_3Si all show superconductivity and have T_c values of about 20 K. It is interesting that these alloys all have the same β-tungsten structure. The Nb_3Sn and and Nb_3Ge alloys have T_c values of 22 K and 24 K respectively. These alloys are used to make the wire for extremely powerful electromagnets. These magnets have a variety of uses:

1. In linear accelerators used as atom smashers for high energy particle physics research
2. In nuclear fusion research to make powerful magnetic fields which act as a magnetic bottle for a plasma
3. For military purposes
4. For nuclear magnetic resonance imaging (which is used in diagnostic medicine).

An extremely high current can be passed through a very fine wire made of a superconductor. Thus small coils with a large number of turns can be used to make extremely powerful high field electromagnets. Because the superconductor has effectively zero resistance, the wire does not get hot. Since there is no current loss, once the current is flowing in the coil it continues indefinitely. For example, in large superconducting magnets used in plasma research, the current used by a Nb/Ta superconducting alloy at 4 K was only 0.3% of the current used in an electromagnet of similar power using copper wire for the metal turns. A major obstacle to the widespread use of these *low temperature superconductors* has been the very low value of the transition temperature T_c. The only way of attaining these low temperatures was to use liquid helium, which is very expensive.

The first non-metallic superconductor was found in 1964. This was a metal oxide with a perovskite crystal structure and is a different type of superconductor from the alloys. It was of no practical use since the T_c is only 0.01 K.

The perovskite structure is formed by compounds of formula ABO_3, where the oxidation states of A and B add up to 6. Examples include $BaTiO_3$, $CaTiO_3$ and $NaNb^VO_3$. The perovskite crystal structure is cubic. A Ca^{2+} ion is located at the body-centred position (at the centre of the cube), the smaller Ti^{4+} ions are located at each corner, and the O^{2-} are located half-way along each of the edges of the cube. Thus the Ca^{2+} has a coordination number of 12 since it is surrounded by 12 O atoms, and the Ti^{4+} are surrounded octahedrally by 6 O atoms. This structure is illustrated in Figure 19.2.

Superconductivity has also been observed in certain organic materials with flat molecules stacked on top of each other, and in certain sulphides called Chevrel compounds.

In 1986 Georg Bednorz and Alex Müller (who were working for IBM in Zurich, Switzerland) reported a new type of superconductor with a T_c value of 35 K. This temperature was appreciably higher than that for the alloys. This compound is a mixed oxide in the Ba–La–Cu–O system. Though originally given a different formula, it has now been reformulated as $La_{(2-x)}Ba_xCuO_{(4-y)}$ where x is between 0.15 and 0.20 and y is small. This compound has a perovskite structure based on La_2CuO_4. Though La_2CuO_4 itself is non-conducting, superconductors can be made by replacing 7.5–10% of the La^{3+} ions by Ba^{2+}. There is a small deficiency of O^{2-}. It seems reasonable that the oxygen loss from the lattice is balanced by the reduction of an easily reducible metal cation, in this case Cu^{3+}.

$$O^{2-}_{(\text{lattice})} \rightarrow \tfrac{1}{2}O_2 + 2e$$
$$2Cu^{3+} + 2e \rightarrow 2Cu^{2+}$$

The publication of this paper stimulated enormous interest in 'ceramic' superconductors and a flood of papers was published in 1987. Different laboratories prepared similar compounds, replacing Ba^{2+} with Ca^{2+} or Sr^{2+}, substituting different lanthanides, and varying the preparative conditions to control the amount of oxygen. In the main syntheses stoichiometric quantities of the appropriate metal oxides or carbonates are heated in air, cooled, ground, heated in oxygen and annealed. Compounds were made with T_c values of about 50 K. Bednorz and Müller were awarded the Nobel Prize for Physics in 1987.

Another very significant superconducting system based on the Y–Ba–Cu–O system was reported in March 1987 by Wu, Chu and coworkers. This was important because it was the first report of a superconductor which worked at 93 K. This temperature was significant for practical reasons. It allows liquid nitrogen (boiling point 77 K) to be used as coolant rather than the more expensive liquid helium. The compound is formulated as $YBa_2Cu_3O_{7-x}$. This is called the 1–2–3 system because of the ratio of the metals present. Like the previous La_2CuO_4 system, the 1–2–3 structure contains Cu and is based on a perovskite structure. This comprises three cubic perovskite units stacked one on top of the other, giving an elongated (tetragonal) unit cell.

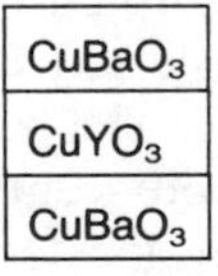

The upper and lower cubes have a Ba^{2+} ion at the body-centred position and the smaller Cu^{2+} ions at each corner. The middle cube is similar but has a Y^{3+} ion at the body centre. A perovskite structure has the formula ABO_3, and the stoichiometry of this compound would be $YBa_2Cu_3O_9$. Since the formula actually found is $YBa_2Cu_3O_{7-x}$, there is a massive oxygen deficiency, and about one quarter of the oxygen sites in the crystal are vacant. In a perovskite cube, O^{2-} are located half-way along each of the 12 edges of the cube. Neutron diffraction shows that the O vacancies are ordered. All the O which should be present at the same height up the z axis as the Y atom are absent: half of the O atoms around Cu and between the Ba planes are also missing.

Several lanthanides, including Sm, Eu, Nd, Dy and Yb, have been substituted for Y in 1–2–3 structures. Values of T_c up to 93 K are well established. These are called *warm superconductors*.

In 1988 new systems were reported using Bi or Tl instead of the lanthanides. For example, in the system $Bi_2Sr_2Ca_{(n-1)}Cu_nO_{(2n+4)}$ compounds are known where n is 1, 2, 3 and 4. These all have a perovskite structure and have T_c values of 12 K, 80 K, 110 K and 90 K respectively. A similar range of compounds $Tl_2Ba_2Ca_{(n-1)}Cu_nO_{(2n+4)}$ are known with T_c values of 90 K, 110 K, 122 K and 119 K respectively. There are claims that the compound $Bi_{1.7}Pb_{0.2}Sb_{0.1}Sr_2Ca_2Cu_{2.8}O_y$ has a T_c value of 164 K.

$BaBiO_3$ has a perovskite structure, but is not a superconductor. However, replacing some of the Ba sites with K, or replacing some of the Bi sites with Pb, gives other superconducting phases such as $K_xBa_{(1-x)}Bio_3$ and $BaPb_{(1-x)}Bi_xO_3$. These compounds have relatively low T_c values, but are of theoretical interest because they do not contain Cu or a lanthanide element.

The race to discover superconductors which work at higher temperatures continues. The prospect of making superconductors which work at room temperature will continue to attract attention, since its technical applications have great financial benefits. What are these potential uses?

1. The possibility of power transmission using a superconductor is highly attractive. There are obvious difficulties about making long cables from a ceramic material. However, low loss transmission of DC through resistanceless cables from electricity generating power stations rather than AC through normal wire is economically attractive.
2. Use in computers. One of the biggest difficulties in further miniaturization of computer chips is how to get rid of unwanted heat. If superconductors were used, the heat problems would be dramatically reduced. The greater speed of chips is hindered by the time it takes to charge a capacitor, due to the resistance of the interconnecting metal film. Superconductors could lead to faster chips.
3. Powerful electromagnets using superconducting windings are already used. It would be much easier to do this at higher temperatures.
4. Levitation – much pioneering work was done by Eric Laithwaite at

Imperial College on linear motors, and a prototype of a train which floats on a magnetic field has been built in Japan.

Superconductivity of metals and alloys is thought to involve two electrons at a time (Bardeen *et al.*, 1957; Ogg, 1946). There is no one accepted explanation of how high temperature superconduction occurs in these mixed oxide (ceramic) systems. However, it seems appropriate to draw together the apparent facts at this time:

1. Many, but not all, warm superconductors *contain Cu*. Two features of Cu chemistry are that it exists in three oxidation states, (+I), (+II) and (+III), and that Cu(II) forms many tetragonally distorted octahedral complexes. Both of these factors may be important. In the La_2CuO_4 compounds some Ba^{2+} ions are substituted for La^{3+}. To balance the charges some Cu(II) atoms change into Cu(III). Superconductivity in this system is thought to involve the transfer of electrons from Cu(II) to Cu(III), but if the process involves two electrons as in the metal superconductors it could involve electron transfer from Cu(I) to Cu(III).
2. It is also significant that *these superconductors are all related to the perovskite structure*.
3. Another common feature is that *the oxygen deficiency seems to be critical*. There is strong evidence from neutron diffraction that the vacancies left by missing O are ordered. It seems reasonable to suppose that, since Cu is normally octahedrally surrounded by six O atoms, when an O vacancy occurs (that is when an O is omitted), then two Cu atoms may interact directly with each other. Interactions such as $Cu^{II}-Cu^{III}$ or $Cu^{I}-Cu^{III}$ could occur by transferring an electron between the two Cu atoms. Similarly superconductivity in the $YBa_2Cu_3O_{7-x}$ is thought to be associated with the ready transfer of electrons between Cu(I), Cu(II) and Cu(III).

FURTHER READING

Adams, D.M. (1974) *Inorganic Solids*, Wiley, New York.
Addison, C.C. (1974) The chemistry of liquid metals, *Chemistry in Britain*, **10**, 331.
Brewer, L. (1968) *Science*, **161**, 115. (Enthalpies of atomisation .)
Burdett, J.K. (1982) New ways to look at solids, *Acc. Chem. Res.*, **15**, 34.
Chemistry in Britain, May 1969 – The whole issue is devoted to metals and alloys.
Cox, P.A. (1987) *The Electronic Structure and Chemistry of Solids*, Oxford University Press, Oxford.
Duffy, J.A. (1983) Band theory of conductors, semiconductors and insulators, *Education in Chemistry*, **20**, 14–18.
Galwey, A.K. (1967) *Chemistry of Solids*, Chapman Hall, London.
Ho, S.M. and Douglas, B.E. (1972) Structures of the elements and the PTOT system, *J. Chem. Ed.*, **49**, 74.
Hume-Rothery, W. (1964) Review of bonding in metals, *Metallurgist*, **3**, 11.
Hume-Rothery, W. (1964) A note on the intermetallic chemistry of the later transition elements, *J. Less-Common Metals*, **7**, 152.
Hume-Rothery, W.J. and Raynor, G.V. (1962) *The Structure of Metals and Alloys*, 4th ed., Institute of Metals, London.

Hume-Rothery, W.J., Christian, J.W. and Pearson, W.B. (1952) *Metallurgical Equilibrium Diagrams*, Institute of Physics, London.
Jolly, W.L. (1976) *The Principles of Inorganic Chemistry*, (Chapter 11: Metals; Chapter 12: Semiconductors), McGraw Hill, New York.
Metal Structures Conference (Brisbane 1983), (ISBN 0-85825-183-3), Gower Publishing Company. Parish, R.V. (1976) *The Metallic Elements*, Longmans, London.
Parish, R.V. (1976) The Metallic Elements, Longmans, London.

Superconductivity

Bardeen, J., Cooper, L.N. and Schreiffer, J.R. (1957) *Phys. Rev.*, **106**, 162. (Development of the BCS theory of superconductivity in metals arising from the movement of electron pairs.)
Bednorz, J.G. and Müller, A. (1986) Possible high T_c superconductivity in the Ba–La–Cu–O system, *Z. Phys.*, B., **64**, 189. (The paper which started interest in metal oxide superconductors.)
Edwards, P.P., Harrison, M.R. and Jones, R. (1987) Superconductivity returns to chemistry, *Chemistry in Britain*, **23**, 962–966.
Ellis, A.B. (1987) Superconductors, *J. Chem. Ed.*, **64**, 836–841.
Khurana, A. (1989) *Physics Today*, April, 17–19.
Murray Gibson, J. (1987) Superconducting ceramics, *Nature*, **329**, 763.
Ogg, R.A. (1946) *Phys. Rev.*, **69**, 243. (First suggestion that superconduction in alloys is by electron pairs.)
Sharp, J.H. (1990) A review of the crystal chemistry of mixed oxide superconductors, *Br. Ceram. Trans. J.*, **89**, 1–7. (An understandable review of warm superconductors which attempts to relate properties and structure – the best so far.)
Tilley, D.R. and Tilley, J. (1986) *Superfluidity and Superconductivity*, 2nd ed., Hilger, Bristol.
Wu, M.K. *et al.*, (1987) Superconductivity at 93 K in a new mixed phase Y–Ba–Cu–O compound system at ambient pressure, *Phys. Rev. Lett.*, **58**, 908–910.

PROBLEMS

1. List the physical and chemical properties associated with metals.
2. Name and draw the three common crystal structures adopted by metals.
3. Aluminium has a face-centred cubic structure. The unit cell length is 4.05 Å. Calculate the radius of Al in the metal. (Answer: 1.43 Å.)
4. Explain why the electrical conductivity of a metal decreases as the temperature is raised, but the opposite occurs with semiconductors.
5. Describe the structures of interstitial and substitutional alloys and outline the factors determining which is formed.
6. What is superconductivity? What uses and potential uses are there for superconductors? What types of materials are superconductors?

6 General properties of the elements

SIZE OF ATOMS AND IONS

Size of atoms

The size of atoms decreases from left to right across a period in the periodic table. For example, on moving from lithium to beryllium one extra positive charge is added to the nucleus, and an extra orbital electron is also added. Increasing the nuclear charge results in all of the orbital electrons being pulled closer to the nucleus. In a given period, the alkali metal is the largest atom and the halogen the smallest. When a horizontal period contains ten transition elements the contraction in size is larger, and when in addition there are 14 inner transition elements in a horizontal period, the contraction in size is even more marked.

On descending a group in the periodic table such as that containing lithium, sodium, potassium, rubidium and caesium, the sizes of the atoms increase due to the effect of extra shells of electrons being added: this outweighs the effect of increased nuclear charge.

Size of ions

Metals usually form positive ions. These are formed by removing one or more electrons from the metal atom. Metal ions are smaller than the atoms from which they were formed for two reasons:

1. The whole of the outer shell of electrons is usually ionized, i.e. removed. This is one reason why cations are much smaller than the original metal atom.
2. A second factor is the effective nuclear charge. In an atom, the number of positive charges on the nucleus is exactly the same as the number of orbital electrons. When a positive ion is formed, the number of positive charges on the nucleus exceeds the number of orbital electrons, and the effective nuclear charge (which is the ratio of the number of charges on the nucleus to the number of electrons) is increased. This results in the remaining electrons being more strongly attracted by the nucleus. Thus the electrons are pulled in – further reducing the size.

Table 6.1 Covalent radii of the elements

Period \ Group	I	II											III	IV	V	VI	VII	0
1	H ~0.30																H ~0.30	He 1.20*
2	Li 1.23	Be 0.89											B 0.80	C 0.77	N 0.74	O 0.74	F 0.72	Ne 1.60*
3	Na 1.57	Mg 1.36											Al 1.25	Si 1.17	P 1.10	S 1.04	Cl 0.99	Ar 1.91*
4	K 2.03	Ca 1.74	Sc 1.44	Ti 1.32	V 1.22	Cr 1.17	Mn 1.17	Fe 1.17	Co 1.16	Ni 1.15	Cu 1.17	Zn 1.25	Ga 1.25	Ge 1.22	As 1.21	Se 1.14	Br 1.14	Kr 2.00*
5	Rb 2.16	Sr 1.91	Y 1.62	Zr 1.45	Nb 1.34	Mo 1.29	Tc –	Ru 1.24	Rh 1.25	Pd 1.28	Ag 1.34	Cd 1.41	In 1.50	Sn 1.40	Sb 1.41	Te 1.37	I 1.33	Xe 2.20*
6	Cs 2.35	Ba 1.98	La 1.69	Hf 1.44	Ta 1.34	W 1.30	Re 1.28	Os 1.26	Ir 1.26	Pt 1.29	Au 1.34	Hg 1.44	Tl 1.55	Pb 1.46	Bi 1.52	Po	At	Rn
7	Fr	Ra	Ac															

Lanthanides	Ce	Pr	Nd	Pm	Sm	Eu	Gd	Tb	Dy	Ho	Er	Tm	Yb	Lu
	1.65	1.64	1.64	–	1.66	1.85	1.61	1.59	1.59	1.58	1.57	1.56	1.70	1.56

COVALENT RADII OF THE ELEMENTS
(Numerical values are given in Ångström units. * The values for the noble gases are atomic radii, i.e. non-bonded radii, and should be compared with van der Waals radii rather than with covalent bonded radii. Large circles indicate large radii and small circles small radii.)
After Moeller, T., *Inorganic Chemistry*, Wiley 1952

A positive ion is always smaller than the corresponding atom, and the more electrons which are removed (that is, the greater the charge on the ion), the smaller the ion becomes.

Metallic radius	Na	1.86 Å	Atomic radius	Fe	1.17 Å
Ionic radius	Na^+	1.02 Å	Ionic radius	Fe^{2+}	0.780 Å (high spin)
			Ionic radius	Fe^{3+}	0.645 Å (high spin)

When a negative ion is formed, one or more electrons are added to an atom, the effective nuclear charge is reduced and hence the electron cloud expands. Negative ions are bigger than the corresponding atom.

Covalent radius	Cl	0.99 Å
Ionic radius	Cl^-	1.84 Å

Problems with ionic radii

There are several problems in obtaining an accurate set of ionic radii.

1. Though it is possible to measure the internuclear distances in a crystal very accurately by X-ray diffraction, for example the distance between Na^+ and F^- in NaF, there is no universally accepted formula for apportioning this to the two ions. Historically several different sets of ionic radii have been estimated. The main ones are by Goldschmidt, Pauling and Ahrens. These are all calculated from observed internuclear distances, but differ in the method used to split the distance between the ions. The most recent values, which are probably the most accurate, are by Shannon (1976).
2. Corrections to these radii are necessary if the charge on the ion is changed.
3. Corrections must also be made for the coordination number, and the geometry.
4. The assumption that ions are spherical is probably true for ions from the *s*- and *p*-blocks with a noble gas configuration, but is probably untrue for transition metal ions with an incomplete *d* shell.
5. In some cases there is extensive delocalization of *d* electrons, for example in TiO where they give rise to metallic conduction, or in cluster compounds. This also changes the radii.

Thus ionic radii are not absolute constants, and are best seen as a working approximation.

Trends in ionic radii

Irrespective of which set of ionic radii are used, the following trends are observed:

1. In the main groups, radii increase on descending the group, e.g. Li^+ = 0.76 Å, Na^+ = 1.02 Å, K^+ = 1.38 Å, because extra shells of electrons are added.

2. The ionic radii decrease moving from left to right across any period in the periodic table, e.g. Na^+ = 1.02 Å, Mg^{2+} = 0.720 Å and Al^{3+} = 0.535 Å. This is partly due to the increased number of charges on the nucleus, and also to the increasing charge on the ions.
3. The ionic radius decreases as more electrons are ionized off, that is as the valency increases, e.g. Cr^{2+} = 0.80 Å (high spin), Cr^{3+} = 0.615 Å, Cr^{4+} = 0.55 Å, Cr^{5+} = 0.49 Å and Cr^{6+} = 0.44 Å.
4. The *d* and *f* orbitals do not shield the nuclear charge very effectively. Thus there is a significant reduction in the size of ions just after 10*d* or 14*f* electrons have been filled in. The latter is called the lanthanide contraction, and results in the sizes of the second and third row transition elements being almost the same. This is discussed in Chapter 30.

IONIZATION ENERGIES

If a small amount of energy is supplied to an atom, then an electron may be promoted to a higher energy level, but if the amount of energy supplied is sufficiently large the electron may be completely removed. The energy required to remove the most loosely bound electron from an isolated gaseous atom is called the ionization energy.

Ionization energies are determined from spectra and are measured in $kJ\,mol^{-1}$. It is possible to remove more than one electron from most atoms. The first ionization energy is the energy required to remove the first electron and convert M to M^+; the second ionization energy is the energy required to remove the second electron and convert M^+ to M^{2+}; the third ionization energy converts M^{2+} to M^{3+}, and so on.

The factors that influence the ionization energy are:

1. The size of the atom.
2. The charge on the nucleus.
3. How effectively the inner electron shells screen the nuclear charge.
4. The type of electron involved (*s*, *p*, *d* or *f*).

These factors are usually interrelated. In a small atom the electrons are tightly held, whilst in a larger atom the electrons are less strongly held. Thus the ionization energy decreases as the size of the atoms increases.

Table 6.2 Ionization energies for Group I and II elements ($kJ\,mol^{-1}$)

	1st	2nd		1st	2nd	3rd
Li	520	7296	Be	899	1757	14847
Na	496	4563	Mg	737	1450	7731
K	419	3069	Ca	590	1145	4910
Rb	403	2650	Sr	549	1064	4207
Cs	376	2420	Ba	503	965	
Fr			Ra	509	979	3281*

* Estimated value.

This trend is shown, for example, by Group I and Group II elements (See Table 6.2), and also by the other main groups.

Comparison of the first and second ionization energies for the Group I elements shows that removal of a second electron involves a great deal more energy, between 7 and 14 times more than the first ionization energy. Because the second ionization energy is so high, a second electron is not removed. The large difference between the first and second ionization energies is related to the structure of the Group I atoms. These atoms have just one electron in their outer shell. Whilst it is relatively easy to remove the single outer electron, it requires much more energy to remove a second electron, since this involves breaking into a filled shell of electrons.

The ionization energies for the Group II elements show that the first ionization energy is almost double the value for the corresponding Group I element. This is because the increased nuclear charge results in a smaller size for the Group II element. Once the first electron has been removed, the ratio of charges on the nucleus to the number of orbital electrons (the effective nuclear charge) is increased, and this reduces the size. For example, Mg^+ is smaller than the Mg atom. Thus the remaining electrons in Mg^+ are even more tightly held, and consequently the second ionization energy is greater than the first. Removal of a third electron from a Group II element is very much harder for two reasons:

1. The effective nuclear charge has increased, and hence the remaining electrons are more tightly held.
2. Removing another electron would involve breaking a completed shell of electrons.

The ionization energy also depends on the type of electron which is removed. *s*, *p*, *d* and *f* electrons have orbitals with different shapes. An *s* electron penetrates nearer to the nucleus, and is therefore more tightly held than a *p* electron. For similar reasons a *p* electron is more tightly held than a *d* electron, and a *d* electron is more tightly held than an *f* electron. Other factors being equal, the ionization energies are in the order $s > p > d > f$. Thus the increase in ionization energy is not quite smooth on moving from left to right in the periodic table. For example, the first ionization energy for a Group III element (where a *p* electron is being removed) is actually less than that for the adjacent Group II element (where an *s* electron is being removed).

In general, the ionization energy decreases on descending a group and increases on crossing a period. Removal of successive electrons becomes

Table 6.3 Comparison of some first ionization energies ($kJ\,mol^{-1}$)

Li 520	Be 899	B 801	C 1086	N 1403	O 1410	F 1681	Ne 2080
Na 496	Mg 737	Al 577	Si 786	P 1012	S 999	Cl 1255	Ar 1521

Table 6.4 First ionization energies of the elements

Period \ Group	I	II											III	IV	V	VI	VII	0
1	H ● 1311																	He ● 2372
2	Li • 520	Be • 899											B • 801	C ● 1086	N ● 1403	O ● 1410	F ● 1681	Ne ● 2080
3	Na • 496	Mg • 737											Al • 577	Si • 786	P ● 1012	S ● 999	Cl ● 1255	Ar ● 1521
4	K • 419	Ca • 590	Sc • 631	Ti • 656	V • 650	Cr • 652	Mn • 717	Fe • 762	Co • 758	Ni • 736	Cu • 745	Zn • 906	Ga • 579	Ge • 760	As ● 947	Se • 941	Br ● 1142	Kr ● 1351
5	Rb • 403	Sr • 549	Y • 616	Zr • 674	Nb • 664	Mo • 685	Tc • 703	Ru • 711	Rh • 720	Pd • 804	Ag • 731	Cd • 876	In • 558	Sn • 708	Sb • 834	Te • 869	I ● 1191	Xe ● 1170
6	Cs • 376	Ba • 503	La • 541	Hf • 760	Ta • 760	W • 770	Re • 759	Os • 840	Ir • 900	Pt • 870	Au • 889	Hg ● 1007	Tl • 589	Pb • 715	Bi • 703	Po • 813	At ● 912	Rn ● 1037
7	Fr	Ra	Ac															

FIRST IONIZATION ENERGIES OF THE ELEMENTS
(Numerical values are given in kJ mol^{-1}.)
(Large circles indicate high values and small circles low values.)
After Sanderson, R.T., *Chemical Periodicity*, Reinhold, New York.

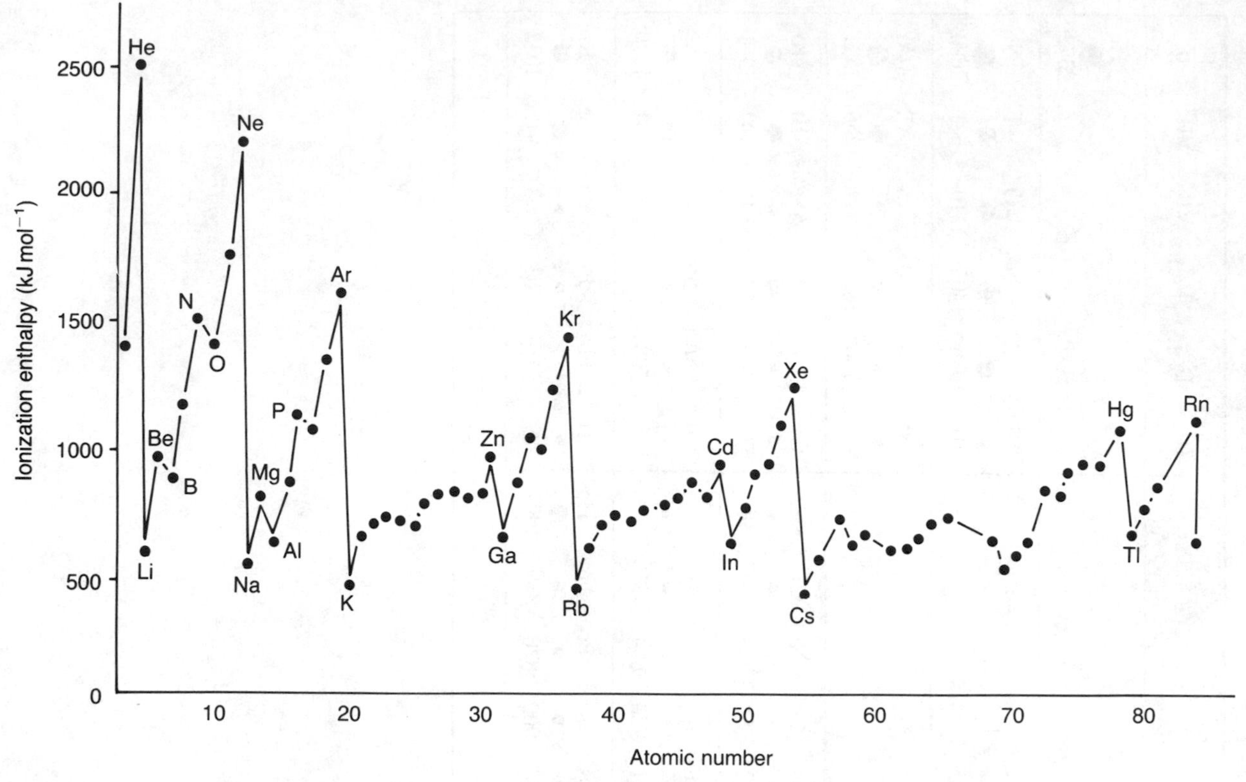

Figure 6.1 First ionization energies of the elements.

more difficult and first ionization energy < second ionization energy < third ionization energy. There are a number of deviations from these generalizations.

The variation in the first ionization energies of the elements are shown in Figure 6.1. The graph shows three features:

1. The noble gases He, Ne, Ar, Kr, Xe and Rn have the highest ionization energies in their respective periods.
2. The Group I metals Li, Na, K and Rb have the lowest ionization energies in their respective periods.
3. There is a general upward trend in ionization energy within a horizontal period, for example from Li to Ne or from Na to Ar.

The values for Ne and Ar are the highest in their periods because a great deal of energy is required to remove an electron from a stable filled shell of electrons.

The graph does not increase smoothly. The values for Be and Mg are high, and this is attributed to the stability of a filled *s* level. The values for N and P are also high, and this indicates that a half filled *p* level is also particularly stable. The values for B and Al are lower because removal of one electron leaves a stable filled *s* shell, and similarly with O and S a stable half filled *p* shell is left.

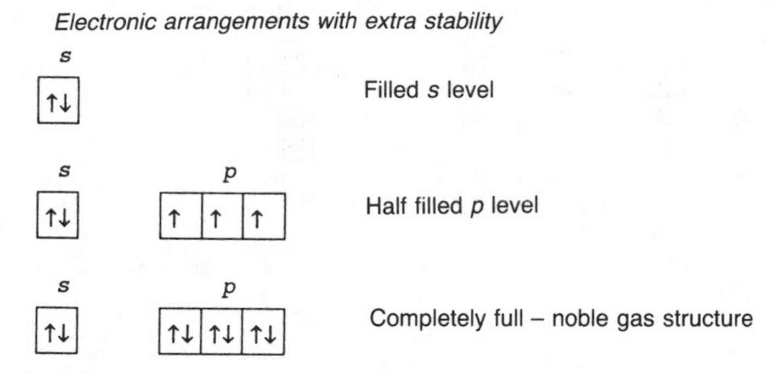

In general the first ionization energy decreases in a regular way on descending the main groups. A departure from this trend occurs in Group III, where the expected decrease occurs between B and Al, but the values for the remaining elements Ga, In and Tl do not continue the trend, and are irregular. The reason for the change at Ga is that it is preceded by ten elements of the first transition series (where the 3*d* shell is being filled). This makes Ga smaller than it would otherwise be. A similar effect is observed with the second and third transition series, and the presence of the three transition series not only has a marked effect on the values for Ga, In and Tl, but the effect still shows in Groups IV and V.

Table 6.5 Ionization energies for Group III elements ($kJ\,mol^{-1}$)

	1st	2nd	3rd
B	801	2427	3659
Al	577	1816	2744
Ga	579	1979	2962
In	558	1820	2704
Tl	589	1971	2877

The ionization energies of the transition elements are slightly irregular, but the third row elements starting at Hf have lower values than would be expected due to the interpolation of the 14 lanthanide elements between La and Hf.

ELECTRON AFFINITY

The energy released when an extra electron is added to a neutral gaseous atom is termed the electron affinity. Usually only one electron is added, forming a uninegative ion. This repels further electrons and energy is needed to add on a second electron: hence the negative affinity of O^{2-}. Electron affinities depend on the size and effective nuclear charge. They

Table 6.6 Some electron affinity values ($kJ\,mol^{-1}$)

	$H \rightarrow H^-$ −72		
	$He \rightarrow He^-$ 54		
$Li \rightarrow Li^-$ −57	$Na \rightarrow Na^-$ −21		
$Be \rightarrow Be^-$ 66	$Mg \rightarrow Mg^-$ 67		
$B \rightarrow B^-$ −15	$Al \rightarrow Al^-$ −26		
$C \rightarrow C^-$ −121	$Si \rightarrow Si^-$ −135		
$N \rightarrow N^-$ 31	$P \rightarrow P^-$ −60		
$O \rightarrow O^-$ −142	$S \rightarrow S^-$ −200		
$O \rightarrow O^{2-}$ 702	$S \rightarrow S^{2-}$ 332		
$F \rightarrow F^-$ −333	$Cl \rightarrow Cl^-$ −348	$Br \rightarrow Br^-$ −324	$I \rightarrow I^-$ −295
$Ne \rightarrow Ne^-$ 99			

cannot be determined directly, but are obtained indirectly from the Born–Haber cycle.

Negative electron affinity values indicate that energy is given out when the atom accepts an electron. The above values show that the halogens all evolve a large amount of energy on forming negative halide ions, and it is not surprising that these ions occur in a large number of compounds.

Energy is evolved when one electron is added to an O or S atom, forming the species O^- and S^-, but a substantial amount of energy is absorbed when two electrons are added to form O^{2-} and S^{2-} ions. Even though it requires energy to form these divalent ions, compounds containing these ions are known. It follows that the energy required to form the ions must come from some other process, such as the lattice energy when the ions are packed together in a regular way to form a crystalline solid, or from solvation energy in solution. It is always dangerous to consider one energy term in isolation, and a complete energy cycle should be considered whenever possible.

BORN–HABER CYCLE

This cycle devised by Born and Haber in 1919 relates the lattice energy of a crystal to other thermochemical data. The energy terms involved in building a crystal lattice such as sodium chloride may be taken in steps. The elements in their standard state are first converted to gaseous atoms, and then to ions, and finally packed into the crystal lattice.

The enthalpies of sublimation and dissociation and the ionization energy are positive since energy is supplied to the system. The electron affinity and lattice energy are negative since energy is evolved in these processes.

According to Hess's law, the overall energy change in a process depends only on the energy of the initial and final states and not on the route taken. Thus the enthalpy of formation ΔH_f is equal to the sum of the terms going the other way round the cycle.

$$-\Delta H_f = \Delta H_s + I + \tfrac{1}{2}\Delta H_d - E - U$$

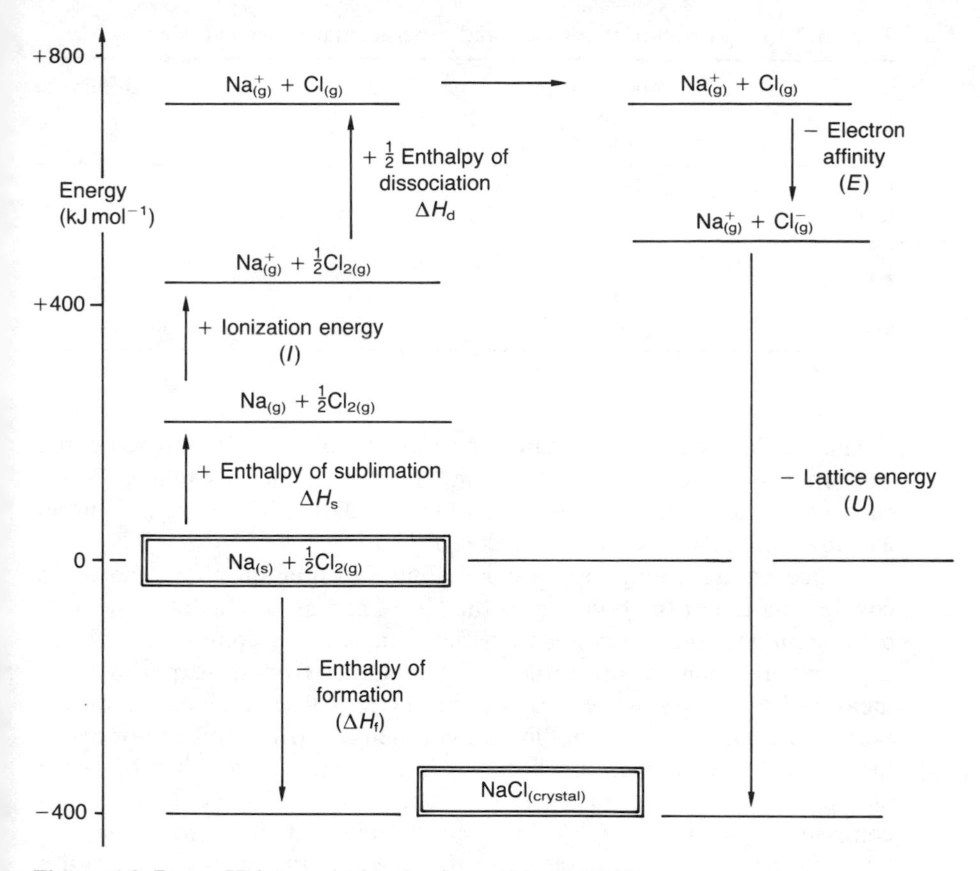

Figure 6.2 Born–Haber cycle for the formation of NaCl.

All the terms except the lattice energy and electron affinity can be measured. Originally the cycle was used to calculate electron affinities. By using known crystal structures, it was possible to calculate the lattice energy, and hence values were obtained for the electron affinity.

$$-\Delta H_f = +\Delta H_s + I \quad + \tfrac{1}{2}H_d \quad -E - U$$

For NaCl $\quad -381.2 = +108.4 + 495.4 + 120.9 - E - 757.3$

hence $\quad E = -348.6\,\text{kJ mol}^{-1}$

Now that some electron affinity values are known, the cycle is used to calculate the lattice energy for unknown crystal structures.

It is useful to know the lattice energy, as a guide to the solubility of the crystal. When a solid dissolves, the crystal lattice must be broken up (which requires that energy is put in). The ions so formed are solvated (with the evolution of energy). When the lattice energy is high a large amount of energy is required to break the lattice. It is unlikely that the enthalpy of solvation will be big enough (and evolve sufficient energy to offset this), so the substance will probably be insoluble.

Table 6.7 Comparison of theoretical and experimental lattice energies

	Theoretical lattice energy ($kJ\,mol^{-1}$)	Born–Haber lattice energy ($kJ\,mol^{-1}$)	% difference
LiCl	−825	−817	0.8
NaCl	−764	−764	0.0
KCl	−686	−679	1.0
KI	−617	−606	1.8
CaF_2	−2584	−2611	1.0
CdI_2	−1966	−2410	22.6

The 'noble behaviour' of many transition metals, that is their resistance to chemical attack, is related to a similar series of energy changes. Noble character is favoured by a high heat of sublimation, high ionization energy and low enthalpy of solvation of the ions.

Lattice energies may also provide some information about the ionic/covalent nature of the bonding. If the lattice energy is calculated theoretically assuming ionic bonding then the value can be compared with the experimental value for the lattice energy obtained from the experimentally measured quantities in the Born–Haber cycle. Close agreement indicates that the assumption that bonding is ionic is in fact true, whilst poor agreement may indicate that the bonding is not ionic. A number of lattice energies are compared in Table 6.7. The agreement is good for all the compounds listed except for CdI_2, confirming that these are ionic. The large discrepancy for CdI_2 indicates that the structure is not ionic, and in fact it forms a layer structure which is appreciably covalent.

POLARIZING POWER AND POLARIZABILITY – FAJANS' RULES

Consider making a bond theoretically by bringing two ions A^+ and B^- together to their equilibrium distance. Will the bond remain ionic, or will it become covalent? Ionic and covalent bonding are two extreme types of bonding, and almost always the bonds formed are intermediate in type, and this is explained in terms of polarizing (that is deforming) the shape of the ions.

The type of bond between A^+ and B^- depends on the effect one ion has on the other. The positive ion attracts the electrons on the negative ion and at the same time it repels the nucleus, thus distorting or polarizing the negative ion. The negative ion will also polarize the positive ion, but since anions are usually large, and cations small, the effect of a large ion on a small one will be much less pronounced. If the degree of polarization is quite small, then the bond remains largely ionic. If the degree of polarization is large, electrons are drawn from the negative ion towards the positive ion, resulting in a high concentration of electrons between the two nuclei, and a large degree of covalent character results.

The extent to which ion distortion occurs depends on the power of an ion to distort the other ion (that is on its polarizing power) and also on how susceptible the ion is to distortion (that is on its polarizability). Generally the polarizing power increases as ions become smaller and more highly charged. The polarizability of a negative ion is greater than that of a positive ion since the electrons are less firmly bound because of the differences in effective nuclear charge. Large negative ions are more polarizable than small ones.

Fajans put forward four rules which summarize the factors favouring polarization and hence covalency.

1. *A small positive ion favours covalency.*
   In small ions the positive charge is concentrated over a small area. This makes the ion highly polarizing, and very good at distorting the negative ion.
2. *A large negative ion favours covalency.*
   Large ions are highly polarizable, that is easily distorted by the positive ion, because the outermost electrons are shielded from the charge on the nucleus by filled shells of electrons.
3. *Large charges on either ion, or on both ions, favour covalency.*
   This is because a high charge increases the amount of polarization.
4. *Polarization, and hence covalency, is favoured if the positive ion does* **not** *have a noble gas configuration.*
   Examples of ions which do not have a noble gas configuration include a few main group elements such as Tl^{+}, Pb^{2+} and Bi^{3+}, many transition metal ions such as Ti^{3+}, V^{3+}, Cr^{2+}, Mn^{2+} and Cu^{+}, and some lanthanide metal ions such as Ce^{3+} and Eu^{2+}. A noble gas configuration is the most effective at shielding the nuclear charge, so ions without the noble gas configuration will have high charges at their surfaces, and thus be highly polarizing.

ELECTRONEGATIVITY

In 1931, Pauling defined the electronegativity of an atom as the tendency of the atom to attract electrons to itself *when combined in a compound.*

The implication of this is that when a covalent bond is formed, the electrons used for bonding need not be shared equally by both atoms. If the bonding electrons spend more time round one atom, that atom will have a δ^{-} charge, and consequently the other atom will have a δ^{+} charge. In the extreme case where the bonding electrons are round one atom all of the time, the bond is ionic. Pauling and others have attempted to relate the electronegativity difference between two atoms to the amount of ionic character in the bond between them.

Generally, small atoms attract electrons more strongly than large ones, and hence small atoms are more electronegative. Atoms with nearly filled shells of electrons tend to have higher electronegativities than those with sparsely occupied ones. Electronegativity values are very difficult to measure. Even worse, a particular type of atom in different molecules may well be in a different environment. It is unlikely that the electronegativity of an atom remains constant regardless of its environment, though it is invariably assumed that it is constant. Some of the more important approaches to obtaining electronegativity values are outlined below.

Pauling

Pauling pointed out that since reactions of the type:

$$A_2 + B_2 \rightarrow 2AB$$

are almost always exothermic, the bond formed between the two atoms A and B must be stronger than the average of the single bond energies of A—A and B—B molecules. For example:

$$H_{2(gas)} + F_{2(gas)} \rightarrow 2HF_{(gas)} \qquad \Delta H = -5393\,\text{kJ mol}^{-1}$$
$$H_{2(gas)} + Cl_{2(gas)} \rightarrow 2HCl_{(gas)} \qquad \Delta H = -1852\,\text{kJ mol}^{-1}$$
$$H_{2(gas)} + Br_{2(gas)} \rightarrow 2HBr_{(gas)} \qquad \Delta H = -727\,\text{kJ mol}^{-1}$$

The bonding molecular orbital for AB (ϕ_{AB}) is made up from contributions from the wave functions for the appropriate atomic orbitals (ψ_A and ψ_B).

$$\phi_{AB} = (\psi_A) + \text{constant}\ (\psi_B)$$

If the constant is greater than 1, the molecular orbital is concentrated on the B atom, which therefore acquires a partial negative charge, and the bond is partly polar.

$$\overset{\delta^+}{A}\text{——}\overset{\delta^-}{B}$$

Conversely, if the constant is less than 1, atom A gains a partial negative charge. Because of this partial ionic character, the A—B bond is stronger than would be expected for a pure covalent bond. Theextra bond energy is called delta Δ.

$$\Delta = (\text{actual bond energy}) - (\text{energy for 100\% covalent bond})$$

The bond energy can be measured, but the energy of a 100% covalent bond must be calculated. Pauling suggested the 100% covalent bond energy be calculated as the the geometric mean of the covalent energies of A—A and B—B molecules.

$$E_{100\%\ \text{covalent A—B}} = \sqrt{(E_{A—A} \cdot E_{B—B})}$$

The bond energy in A—A and B—B molecules can be measured and so:

$$\Delta = (\text{actual bond energy}) - \sqrt{(E_{A—A} \cdot E_{B—B})}$$

Pauling states that the electronegativity difference between two atoms is equal to $0.208\sqrt{\Delta}$, where Δ is the extra bond energy in kcal mol^{-1}.

(Converting the equation to SI units gives $0.1017\sqrt{\Delta}$, where Δ is measured in kJ mol^{-1}.)

Pauling evaluated $0.208\sqrt{\Delta}$ for a number of bonds and called this the electronegativity difference between A and B. Repeating Pauling's calculation with SI units for energy, we can evaluate $0.1017\sqrt{\Delta}$:

Bond	Δ(kJ mol^{-1})	$0.1017\sqrt{\Delta}$	
C—H	24.3	0.50	i.e. χC $-$ χH = 0.50
H—Cl	102.3	1.02	i.e. χCl $-$ χH = 1.02
N—H	105.9	1.04	i.e. χN $-$ χH = 1.04

(χ (chi) = electronegativity of atom)

If χH = 0 then the electronegativity values for C, Cl and N would be 0.50, 1.02 and 1.04 respectively. Pauling changed the origin of the scale from χH = 0 to χH = 2.05 to avoid having any negative values in the table of values, and this made the value for C become 2.5 and the value for F become 4.0. At the same time the values for a number of other elements approximated to whole numbers: Li = 1.0, B = 2.0, N = 3.0. Thus by adding 2.05 to the values calculated in this way we can obtain the usually accepted electronegativity values (Table 6.8).

If two atoms have similar electronegativities, that is a similar tendency to attract electrons, the bond between them will be predominantly covalent. Conversely a large difference in electronegativity leads to a bond with a high degree of polar character, that is a bond that is predominantly ionic.

Rather than have two extreme forms of bond (ionic and covalent), Pauling introduced the idea that the ionic character of a bond varies with

Table 6.8 Pauling's electronegativity coefficients (for the most common oxidation states of the elements)

						H 2.1
Li 1.0	Be 1.5	B 2.0	C 2.5	N 3.0	O 3.5	F 4.0
Na 0.9						Cl 3.0
K 0.8						Br 2.8
Rb 0.8						I 2.5
Cs 0.7						

Table 6.9 Pauling's electronegativity values

Period \ Group	I	II											III	IV	V	VI	VII	0
1	H 2.1																H 2.1	He
2	Li 1.0	Be 1.5											B 2.0	C 2.5	N 3.0	O 3.5	F 4.0	Ne
3	Na 0.9	Mg 1.2											Al 1.5	Si 1.8	P 2.1	S 2.5	Cl 3.0	Ar
4	K 0.8	Ca 1.0	Sc 1.3	Ti 1.5	V 1.6	Cr 1.6	Mn 1.5	Fe 1.8	Co 1.8	Ni 1.8	Cu 1.9	Zn 1.6	Ga 1.6	Ge 1.8	As 2.0	Se 2.4	Br 2.8	Kr
5	Rb 0.8	Sr 1.0	Y 1.2	Zr 1.4	Nb 1.6	Mo 1.8	Tc 1.9	Ru 2.2	Rh 2.2	Pd 2.2	Ag 1.9	Cd 1.7	In 1.7	Sn 1.8	Sb 1.9	Te 2.1	I 2.5	Xe
6	Cs 0.7	Ba 0.9	La 1.1	Hf 1.3	Ta 1.5	W 1.7	Re 1.9	Os 2.2	Ir 2.2	Pt 2.2	Au 2.4	Hg 1.9	Tl 1.8	Pb 1.8	Bi 1.9	Po 2.0	At 2.2	Rn
7	Fr 0.7	Ra 0.9	Ac 1.1															

PAULING'S ELECTRONEGATIVITY VALUES
Electronegativity varies with the oxidation state of the element. The values gives are for the most common oxidation states.
(Large circles indicate high values and small circles small values.)

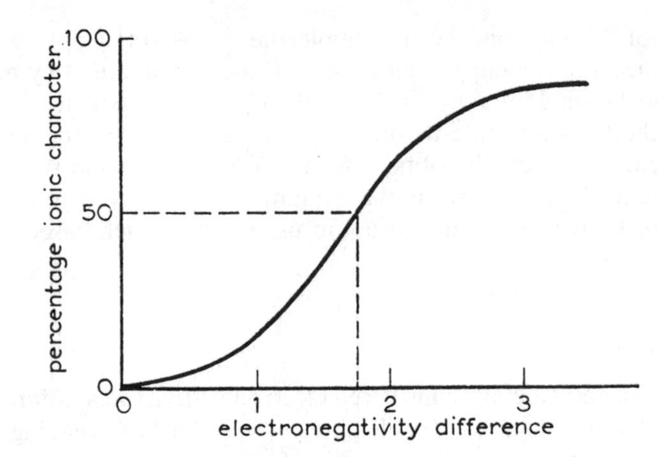

Figure 6.3 Electronegativity difference.

the difference in electronegativity as shown in Figure 6.3. This graph is based on the ionic characters HI 4% ionic, HBr 11%, HCl 19% and HF 45%, which are known from dipole measurements. Fifty per cent ionic character occurs when the electronegativity difference between the atoms is about 1.7, so for a larger difference than this a bond is more ionic than covalent. Similarly, if the electronegativity difference is less than 1.7, the bond is more covalent than ionic. It is better to describe a bond such as one of those in BF_3 as 63% ionic, rather than just ionic.

Mulliken

In 1934, Mulliken suggested an alternative approach to electronegativity based on the ionization energy and electron affinity of an atom. Consider two atoms A and B. If an electron is transferred from A to B, forming ions A^+ and B^-, then the energy change is the ionization energy of atom A (I_A) minus the electron affinity of atom B (E_B), that is $I_A - E_B$. Alternatively, if the electron was transferred the other way to give B^+ and A^- ions, then the energy change would be $I_B - E_A$. If A^+ and B^- are actually formed, then this process requires less energy, and

$$(I_A - E_B) < (I_B - E_A)$$

Rearranging

$$(I_A + E_A) < (I_B + E_B)$$

Thus Mulliken suggested that electronegativity could be regarded as the average of the ionization energy and the electron affinity of an atom.

$$\text{Electronegativity} = \frac{(I + E)}{2}$$

Mulliken used I and E values measured in electron volts, and the values were about 2.8 times larger than the Pauling values. We now measure I and

E in $kJ\,mol^{-1}$. The energy 1 eV/molecule = 96.48 $kJ\,mol^{-1}$, so the commonly accepted Pauling values are more nearly obtained by performing this calculation $(I + E)/(2 \times 2.8 \times 96.48)$ or $(I + E)/540$.

This method has a simple theoretical basis, and also has the advantage that different values can be obtained for different oxidation states of the same element. It suffers from the limitation that only a few electron affinities are known. It is more usual to use the approach based on bond energies.

Allred and Rochow

In 1958 Allred and Rochow considered electronegativity in a different way, and worked out values for 69 elements. (See Further Reading.) They defined electronegativity as the attractive force between a nucleus and an electron at a distance equal to the covalent radius. This force F is electrostatic, and is given by:

$$F = \frac{e^2 \,.\, Z_{\text{effective}}}{r^2}$$

where e is the charge on an electron, r is the covalent radius, and $Z_{\text{effective}}$ is the effective nuclear charge. The latter is the nuclear charge modified by screening factors for the orbital electrons. The screening factors vary depending on the principal quantum number (the shell that the electron occupies), and the type of electron, s, p, d or f. Screening factors have been worked out by Slater, so this provides a convenient method of calculating electronegativity values. These F values may be converted to electronegativity values on the Pauling scale of values using an empirical relationship:

$$\chi = 0.744 + \frac{0.359 Z_{\text{effective}}}{r^2}$$

The electronegativity values so obtained agree quite closely with those obtained by Pauling and Mulliken.

As the oxidation number of an atom increases, the attraction for the electrons increases, so the electronegativity should also increase. Allred and Rochow's method gives slightly different values:

Mo(II)	2.18	Fe(II)	1.83	Tl(I)	1.62	Sn(II)	1.80
Mo(III)	2.19	Fe(III)	1.96	Tl(III)	2.04	Sn(IV)	1.96
Mo(IV)	2.24						
Mo(V)	2.27						
Mo(VI)	2.35						

Allred and Rochow's method depends on measuring covalent radii (and these are obtained with great accuracy by X-ray crystallography) so it might be expected to yield very accurate electronegativity values. This is not so, because although the interatomic distances can be measured very precisely, covalent radii are much less well known because the multiplicity of the bond is not known for certain, that is the bond may possess some double bond character.

The electronegativity values given in this book are those due to Pauling, but others have been calculated from different theoretical assumptions by Mulliken, Allred and Rochow and Sanderson. For details of these and several modern reviews of electronegativity values see Further Reading. *It is now considered that attempts to measure very accurate values for electronegativity are unjustified, and it is better to retain a loose definition of electronegativity, and use it for a more qualitative description of bonds.* For this purpose, it is worth remembering a few electronegativity values (see Table 6.8). From these it is possible to make a reasonable guess at the values for other elements, and hence predict the nature of the bonds formed. Bonds between atoms with similar electronegativity values will be largely non-polar (covalent), and bonds between atoms with a large electronegativity difference will be largely polar (ionic). Predictions using electronegativity in general agree with those made using Fajans' rules.

The basic properties of elements are inversely related to the electronegativity. Thus on descending one of the main groups, the electronegativity decreases, and basic properties increase. Similarly, on going across a period the elements become more electronegative, and less basic.

METALLIC CHARACTER

Metals are electropositive and have a tendency to lose electrons, if supplied with energy:

$$M \rightarrow M^+ + e^-$$

The stronger this tendency, the more electropositive and more metallic an element is. The tendency to lose electrons depends on the ionization energy. It is easier to remove an electron from a large atom than from a small one, so metallic character increases as we descend the groups in the periodic table. Thus in Group IV, carbon is a non-metal, germanium shows some metallic properties, and tin and lead are metals. Similarly, metallic character decreases from left to right across the periodic table because the size of the atoms decreases and the ionization energy increases. Thus sodium and magnesium are more metallic than silicon, which, in turn, is more metallic than chlorine. The most electropositive elements are found in the lower left of the periodic table and the most non-metallic in the top right.

Electropositivity is really the converse of electronegativity, but it is convenient to use the concept of electropositivity when describing metals. Strongly electropositive elements give ionic compounds. Metallic oxides and hydroxides are basic since they ionize, and give hydroxyl ions:

$$NaOH \rightarrow Na^+ + OH^-$$
$$CaO + H_2O \rightarrow Ca^{2+} + 2OH^-$$

Oxides which are insoluble in water cannot produce OH^- in this way, and these are regarded as basic if they react with acids to form salts. Thus in the main groups of the periodic table, basic properties increase on descending a group because the elements become more electropositive and more ionic. However, this generalization does not hold for the *d*-block, and particularly for the central groups of transition elements (Cr, Mn, Fe, Co, Ni) where basicity and the ability to form simple ions decreases on descending the group.

The degree of electropositivity is shown in a variety of ways. Strongly electropositive elements react with water and acids. They form strongly basic oxides and hydroxides, and they react with oxoacids to give stable salts such as carbonates, nitrates and sulphates. Weakly electropositive elements are unaffected by water and are much less readily attacked by acids. Their oxides are frequently amphoteric, and react with both acids and alkalis. They are not basic enough to form stable carbonates.

The electropositive nature of a metal is also shown in the degree of hydration of the ions. In the change M^+ to $[(H_2O)_n \rightarrow M]^+$ the positive charge becomes spread over the whole complex ion. Since the charge is no longer localized on the metal, this is almost the same as the change $M^+ \rightarrow M$. Strongly electropositive metals have a great tendency to the opposite change, $M \rightarrow M^+$, so that they are not readily hydrated. The less electropositive the metal, the weaker the tendency $M \rightarrow M^+$ and the stronger the degree of hydration. Thus the elements in Group II are less electropositive than those of Group I, and Group II ions are more heavily hydrated than those in Group I. The degree of hydration also decreases down a group, e.g. $MgCl_2 \cdot 6H_2O$ and $BaCl_2 \cdot 2H_2O$.

Salts of strongly electropositive metals have little tendency to hydrolyse and form oxosalts. Since the metal ion is large, it has little tendency to form complexes. On the other hand, salts of weakly electropositive elements hydrolyse and may form oxosalts. Because they are smaller, the metal ions have a greater tendency to form complexes.

VARIABLE VALENCY AND OXIDATION STATES

In the *s*-block the oxidation state is always the same as the group number. For *p*-block elements, the oxidation state is normally the group number or eight minus the group number. Variable valency does occur to a limited extent in the *p*-block. In these cases the oxidation state always changes by two, e.g. $TlCl_3$ and TlCl, $SnCl_4$ and $SnCl_2$, PCl_5 and PCl_3, and is due to a pair of electrons remaining paired and not taking part in bonding (the inert pair effect). The term oxidation state is preferred to valency. The oxidation state may be defined as the charge left on the central atom when all the other atoms of the compound have been removed in their usual oxidation states. Thus Tl shows oxidation states of (+III) and (+I), Sn of (+IV) and (+II), and P of (+V) and (+III). The oxidation number can be calculated equally well for ionic or covalent compounds, and without knowing the types of bonds. The oxidation number of S in H_2SO_4 can be worked out as follows. O usually has an oxidation state of (−II) (except in O_2 and O_2^{2-}). H usually has an oxidation state of (+I) (except in H_2 and H^-). The sum of the oxidation numbers of all the atoms in H_2SO_4 is zero, so:

$$(2 \times 1) + (S^x) + (4 \times -2) = 0$$

Thus x, the oxidation state of S, is (+VI). In the case of the oxidation state of Mn in $KMnO_4$, the compound ionizes into K^+ and MnO_4^- ions. In MnO_4^- the sum of the oxidation states is equal to the charge on the ion, so:

$$Mn^x + (4 \times -2) = -1$$

Thus x, the oxidation state of Mn, is 7, i.e. (+VII).

One of the most striking features of the transition elements is that the elements usually exist in several different oxidation states. Furthermore, the oxidation states change in units of 1, e.g. Fe^{3+} and Fe^{2+}, Cu^{2+} and Cu^+. This is in contrast to the *s*-block and *p*-block elements. The reason why this occurs is that a different number of *d* electrons may take part in bonding.

Though the oxidation number is the same as the charge on the ion for ions such as Tl^+ and Tl^{3+}, the two are not necessarily the same. Thus Mn exists in the oxidation state (+VII) but Mn^{7+} does not exist, as $KMnO_4$ ionizes into K^+ and MnO_4^-.

STANDARD ELECTRODE POTENTIALS AND ELECTROCHEMICAL SERIES

When a metal is immersed in water, or a solution containing its own ions, the metal tends to lose positive metal ions into the solution. Thus the metal acquires a negative charge.

$$M^{n+} + ne \rightleftharpoons M$$

The size of the electric potential E set up between the two depends on the particular metal, the number of electrons involved, the activity of the ions in solution, and the temperature. $E°$ is the standard electrode potential, which is a constant for any particular metal and is in fact the electrode potential measured under standard conditions of temperature and with unit activity.These terms are related by the equation:

$$E = E° + \frac{RT}{nF} \ln (a)$$

(where R is the gas constant, T the absolute temperature, a the activity of the ions in solution, n the valency of the ion and F the Faraday). For most purposes, the activity, a, may be replaced by the concentration of ions in solution.

The potential of a single electrode cannot be measured, but if a second electrode of known potential is placed in the solution, the potential difference between the two electrodes can be measured. The standard against which all electrode potentials are compared is the hydrogen electrode.

Table 6.10 Standard electrode potentials (volts at 25 °C)

Li^+	Li	−3.05
K^+	K	−2.93
Ca^{2+}	Ca	−2.84
Al^{3+}	Al	−1.66
Mn^{2+}	Mn	−1.08
Zn^{2+}	Zn	−0.76
Fe^{2+}	Fe	−0.44
Cd^{2+}	Cd	−0.40
Co^{2+}	Co	−0.27
Ni^{2+}	Ni	−0.23
Sn^{2+}	Sn	−0.14
Pb^{2+}	Pb	−0.13
H^+	H_2	0.00
Cu^{2+}	Cu	+0.35
Ag^+	Ag	+0.80
Au^{3+}	Au	+1.38

Table 6.11 Standard electrode potentials (V)

$O_2 \mid OH^-$	+0.40
$I_2 \mid I^-$	+0.57
$Br_2 \mid Br^-$	+1.07
$Cl_2 \mid Cl^-$	+1.36
$F_2 \mid F^-$	+2.85

(This comprises a platinized platinum electrode, which is saturated with hydrogen at one atmosphere pressure and immersed in a solution of H_3O^+ at unit activity. The potential developed by this electrode is arbitrarily fixed as zero.)

If the elements are arranged in order of increasing standard electrode potentials, the resulting Table 6.10 is called the electrochemical series.

Electrode potentials can also be measured for elements such as oxygen and the halogens which form negative ions (Table 6.11).

In the electrochemical series the most electropositive elements are at the top and the least electropositive at the bottom. The greater the negative value of the potential, the greater is the tendency for a metal to ionize. Thus a metal high in the electrochemical series will displace another metal lower down the series from solution. For example, iron is above copper in the electrochemical series, and scrap iron is sacrificed to displace Cu^{2+} ions from solution of $CuSO_4$ in the recovery of metallic copper.

$$Fe + Cu^{2+} \rightarrow Cu + Fe^{2+}$$

In the Daniell cell zinc displaces copper from copper salts in solution. This causes the potential difference between the plates.

Table 6.12 Some standard reduction potentials in acid solution at 25 °C (volts)

Group I	E°	*Group V*	E°	*Group VIII*	E°
$Li^+ + e \rightarrow Li$	−3.05	$As + 3e \rightarrow AsH_3$	−0.60	$I_3^- + 2e \rightarrow 3I^-$	+0.54
$K^+ + e \rightarrow K$	−2.93	$Sb + 3e \rightarrow SbH_3$	−0.51	$Br_3^- + 2e \rightarrow 3Br^-$	+1.05
$Rb^+ + e \rightarrow Rb$	−2.93	$H_3PO_2 + e \rightarrow P$	−0.51	$2ICl_2^- + 2e \rightarrow I_2$	+1.06
$Cs^+ + e \rightarrow Cs$	−2.92	$H_3PO_3 + 2e \rightarrow H_3PO_2$	−0.50	$Br_2 + 2e \rightarrow 2Br^-$	+1.07
$Na^+ + e \rightarrow Na$	−2.71	$H_3PO_4 + 2e \rightarrow H_3PO_3$	−0.28	$2IO_3^- + 10e \rightarrow I_2$	+1.20
		$\frac{1}{2}N_2 + 3e \rightarrow NH_4^+$	−0.27	$Cl_2 + 2e \rightarrow 2Cl^-$	+1.36
Group II		$\frac{1}{2}N_2 + 2e \rightarrow \frac{1}{2}N_2H_5^+$	−0.23	$2HOI + 2e \rightarrow I_2$	+1.45
$Ba^{2+} + 2e \rightarrow Ba$	−2.90	$P + 3e \rightarrow PH_3$	+0.06	$H_5IO_6 + 2e \rightarrow IO_3^-$	+1.60
$Sr^{2+} + 2e \rightarrow Sr$	−2.89	$\frac{1}{2}Sb_2O_3 + 3e \rightarrow Sb$	+0.15	$2HOCl + 2e \rightarrow Cl_2$	+1.63
$Ca^{2+} + 2e \rightarrow Ca$	−2.87	$HAsO_2 + 3e \rightarrow As$	+0.25	$F_2 + 2e \rightarrow 2F^-$	+2.65
$Mg^{2+} + 2e \rightarrow Mg$	−2.37	$H_3AsO_4 + 2e \rightarrow HAsO_2$	+0.56		
$Be^{2+} + 2e \rightarrow Be$	−1.85	$HN_3 + 8e \rightarrow 3NH_4^+$	+0.69	*Transition Metals*	
		$NO_3^- + 3e \rightarrow NO$	+0.96	$La^{3+} + 3e \rightarrow La$	−2.52
Group III		$HNO_2 + e \rightarrow NO$	+1.00	$Sc^{3+} + 3e \rightarrow Sc$	−2.08
$Al^{3+} + 3e \rightarrow Al$	−1.66	$\frac{1}{2}N_2O_4 + 2e \rightarrow NO$	+1.03	$Mn^{2+} + 2e \rightarrow Mn$	−1.18
$Ga^{3+} + 3e \rightarrow Ga$	−0.53	$\frac{1}{2}N_2H_5^+ + 2e \rightarrow NH_4^+$	+1.28	$Zn^{2+} + 2e \rightarrow Zn$	−0.76
$In^{3+} + 3e \rightarrow In$	−0.34	$NH_3OH + 2e \rightarrow NH_4^+$	+1.35	$Cr^{3+} + 3e \rightarrow Cr$	−0.74
$Tl^+ + e \rightarrow Tl$	−0.34			$Fe^{2+} + 2e \rightarrow Fe$	−0.44
$Tl^{3+} + 2e \rightarrow Tl^+$	+1.25	*Group VI*		$Cr^{3+} + e \rightarrow Cr^{2+}$	−0.41
		$Te + 2e \rightarrow H_2Te$	−0.72	$Cd^{2+} + 2e \rightarrow Cd$	−0.40
Group IV		$Se + 2e \rightarrow H_2Se$	−0.40	$Ni^{2+} + 2e \rightarrow Ni$	−0.25
$SiO_2 + 4e \rightarrow Si$	−0.86	$S_4O_6^{2-} + 2e \rightarrow 2S_2O_3^{2-}$	+0.08	$Cu^{2+} + e \rightarrow Cu^+$	+0.15
$PbSO_4 + 2e \rightarrow Pb$	−0.36	$S + 2e \rightarrow H_2S$	+0.14	$Hg_2Cl_2 + 2e \rightarrow 2Hg$	+0.27
$CO_2 + 4e \rightarrow C$	−0.20	$HSO_4^- + 2e \rightarrow H_2SO_3$	+0.17	$Cu^{2+} + 2e \rightarrow Cu$	+0.35
$GeO_2 + 4e \rightarrow Ge$	−0.15	$H_2SO_3 + 2e \rightarrow \frac{1}{2}S_2O_3^{2-}$	+0.40	$[Fe(CN)_6]^{3-} + e \rightarrow [Fe(CN)_6]^{4-}$	+0.36
$Sn^{2+} + 2e \rightarrow Sn$	−0.14	$H_2SO_3 + 4e \rightarrow S$	+0.45	$Cu^+ + e \rightarrow Cu$	+0.50
$Pb^{2+} + 2e \rightarrow Pb$	−0.13	$4H_2SO_3 + 6e \rightarrow S_4O_6^{2-}$	+0.51	$Cu^{2+} + e \rightarrow CuCl$	+0.54
$Si + 4e \rightarrow SiH_4$	+0.10	$S_2O_6^{2-} + 2e \rightarrow 2H_2SO_4$	+0.57	$MnO_4^- + e \rightarrow MnO_4^{2-}$	+0.56
$C + 4e \rightarrow CH_4$	+0.13	$O_2 + 2e \rightarrow H_2O_2$	+0.68	$Fe^{3+} + e \rightarrow Fe^{2+}$	+0.77
$Sn^{4+} + 2e \rightarrow Sn^{2+}$	+0.15	$H_2SeO_3 + 4e \rightarrow Se$	+0.74	$Hg_2^{2+} + 2e \rightarrow 2Hg$	+0.79
$PbO_2 + 2e \rightarrow PbSO_4$	+1.69	$SeO_4^{2-} + 2e \rightarrow H_2SeO_3$	+1.15	$2Hg^{2+} + 2e \rightarrow Hg_2^{2+}$	+0.92
		$\frac{1}{2}O_2 + 2e \rightarrow H_2O$	+1.23	$MnO_2 + 2e \rightarrow Mn^{2+}$	+1.23
		$H_2O_2 + 2e \rightarrow 2H_2O$	+1.77	$\frac{1}{2}Cr_2O_7^{2-} + 3e \rightarrow Cr^{3+}$	+1.33
		$S_2O_8^{2-} + 2e \rightarrow 2SO_4^{2-}$	+2.01	$MnO_4^- + 5e \rightarrow Mn^{2+}$	+1.54
		$O_3 + 2e \rightarrow O_2$	+2.07	$NiO_2 + 2e \rightarrow Ni^{2+}$	+1.68
				$MnO_4^- + 3e \rightarrow MnO_2$	+1.70

Table 6.12 is a table of standard reduction potentials. From this table we can see that the standard reduction potential for Cu^{2+}/Cu is 0.35 V. What does this mean?

Cu^{2+}/Cu is referred to as a redox couple and as written it refers to the half reaction (or electrode reaction)

$$Cu^{2+} + 2e^- \rightarrow Cu$$

In general, redox couples are written *ox/red* where *ox* is the oxidized form and is written on the left and *red* is the reduced form and is written on the right.

Standard reduction potential values are determined relative to a hydrogen electrode, that is the redox couple H^+/H_2 at 25 °C for 1 M concentrations (or one atmosphere pressure) of all chemical species in the equations. (The concentration of water is included in the constant.)

Thus, Cu^{2+}/Cu $E^\circ = +0.35$ V really means that the standard reduction potential of the reaction is 0.35 V.

$$Cu^{2+} + H_2 \rightarrow 2H^+ + Cu \qquad E^\circ = +0.35\text{ V} \qquad (6.1)$$

Similarly the standard reduction potential of the couple Zn^{2+}/Zn is -0.76 V.

$$Zn^{2+} + H_2 \rightarrow 2H^+ + Zn \qquad E^\circ = -0.76\text{ V} \qquad (6.2)$$

Subtracting equation (6.2) from (6.1) gives

$$Cu^{2+} + Zn \rightarrow Cu + Zn^{2+} \qquad E^\circ = +0.35 - (-0.76) = +1.10\text{ V}$$

Both of the standard potentials are relative to the H^+/H_2 couple and therefore H^+ and H_2 disappear when the Cu^{2+}/Cu couple is combined with the Zn^{2+}/Zn couple.

From experience the oxidized forms of couples of high positive potential, for example $MnO_4 + 5e \rightarrow Mn^{2+}$ $E^\circ = +1.54$ V, are termed strong oxidizing agents. Conversely the reduced forms of couples of high negative potential, for example $Li^+ + e \rightarrow Li$ $E^\circ = -3.05$ V, are termed strong reducing agents. It follows that at some intermediate potential the oxidizing power of the oxidized form and the reducing power of the reduced form are similar. What is the value of this potential at which there is a change-over from oxidizing to reducing properties? The first point to note is that it is **not** at 0 V, the value assigned arbitrarily to the H^+/H couple: hydrogen is known to be a reducing agent. A group of chemical species which are used in classical (analytical) chemistry as weak reducing agents (e.g. sulphite and tin(II)) are the reduced forms of couples with potentials between 0 and about +0.6 V. On the other hand VO^{2+} is the stable form of vanadium and VO_2^+ is a weak oxidizing agent: the potential VO_2^+/VO^{2+} is +1.00 V. Thus from experience, as a general rule of thumb we can say that if $E^\circ \approx 0.8$ V, then the oxidized and reduced forms are of about equal stability in redox processes.

It is not very discriminating to term a metal a reducing agent: most metals may be called reducing agents. It is useful to divide metals into four groups in regard to the ease of reduction of their metal ions.

1. The noble metals (with E° more positive than 0 V).
2. Metals which are easily reduced (e.g. with coke) (E° 0 − (−0.5) V).
3. Typically reactive transition metals (E° (−0.5) − (−1.5) V) which are often prepared by reduction with electropositive metals.
4. The electropositive metals (E° more negative than −1.5 V) which can be prepared by electrochemical reduction.

When a solution is electrolysed the externally applied potential must overcome the electrode potential. The minimum voltage necessary to cause deposition is equal and opposite in sign to the potential between the solution and the electrode. Elements low down in the series discharge first; thus Cu^{2+} discharges before H^+, so copper may be electrolysed in aqueous solution. However, hydrogen and other gases often require a considerably higher voltage than the theoretical potential before they discharge. For hydrogen, this extra or over-voltage may be 0.8 volts, and thus it is possible to electrolyse zinc salts in aqueous solution.

Several factors affect the value of the standard potential. The conversion of M to M^+ in aqueous solution may be considered in a series of steps:

1. sublimation of a solid metal
2. ionization of a gaseous metal atom
3. hydration of a gaseous ion

These are best considered in a Born–Haber type of cycle (Figure 6.4).

The enthalpy of sublimation and the ionization energy are positive since energy must be put into the system, and the enthalpy of hydration is negative since energy is evolved. Thus

$$E = +\Delta H_s + I - \Delta H_h$$

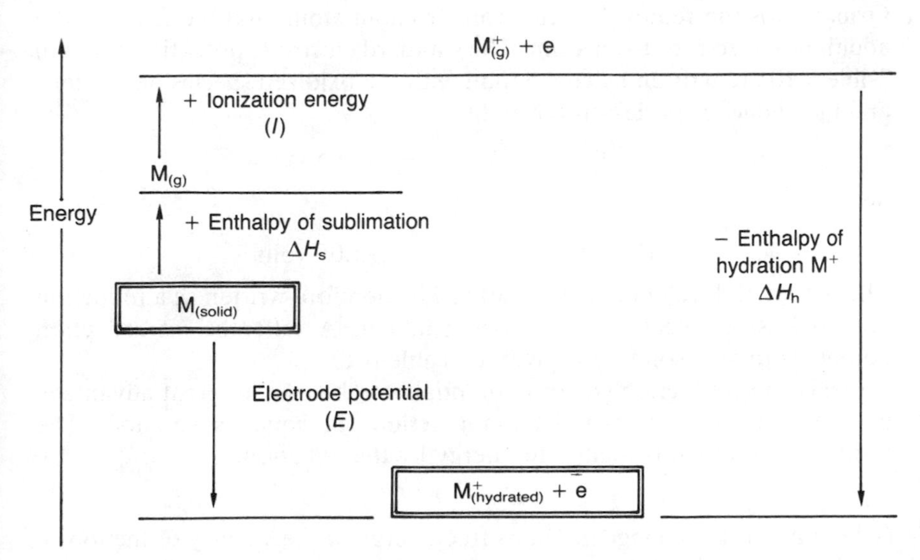

Figure 6.4 Energy cycle for electrode potentials.

Consider first a transition metal. Most transition metals have high melting points: hence the enthalpy of sublimation is high. Similarly they are fairly small atoms and have high ionization energies. Thus the value for the electrode potential E is low, and the metal has little tendency to form ions: hence it is unreactive or noble.

In contrast the *s*-block metals (Groups I and II) have low melting points (hence low enthalpies of sublimation), and the atoms are large and therefore have low ionization energies. Thus the electrode potential E is high and the metals are reactive.

Electrons are lost when a substance is oxidized and electrons are gained when it is reduced. A reducing agent must therefore supply electrons, and elements having large negative electrode potentials are strong reducing agents. The strengths of oxidizing and reducing agents may be measured by the size of the potential between a solution and an inert electrode. Standard reduction potentials are obtained when the concentrations of oxidized and reduced forms are 1 M, and the potential developed is measured against a standard hydrogen electrode. The most powerful oxidizing agents have a large positive oxidation potential and strong reducing agents have a large negative potential. Standard oxidation potentials allow us to predict which ions should oxidize or reduce other ions. The potentials indicate if the energy changes for the process are favourable or unfavourable. It is important to realize that though the potentials may suggest that a reaction is possible, they do not give any kinetic information concerning the rate of the reaction. The rate of the reaction may be very fast or slow, and in some cases a catalyst may be required for it to occur at all – for example in the oxidation of sodium arsenite by ceric sulphate.

OXIDATION–REDUCTION REACTIONS

Oxidation is the removal of electrons from an atom, and reduction is the addition of electrons to an atom. The standard electrode potentials given in Table 6.10 are written by convention with the oxidized species on the left, and the reduced species on the right.

$$Li^+|Li \qquad E^\circ = -3.05 \text{ volts}$$

or

$$Li^+ + e \rightarrow Li \qquad E^\circ = -3.05 \text{ volts}$$

The potential developed by the half cell is therefore written as a reduction potential, since electron(s) are being added. A fuller list of reduction potentials in acid solution is given in Table 6.12.

Oxidation–reduction (redox) potentials can be used to great advantage in explaining oxidation–reduction reactions in aqueous solution. The reduction potential is related to energy by the equation:

$$\Delta G = -nFE^\circ$$

(where ΔG is the change in Gibbs free energy, n the valency of the ion, F the Faraday and E° the standard electrode potential). This is really an application of thermodynamics. Ultimately whether a reaction occurs or not depends on energy. A reaction will not proceed if the free energy change ΔG is positive, and thus thermodynamics saves us the trouble of trying the reaction. If ΔG is negative, then the reaction is thermodynamically possible. It does not follow that because a reaction is thermodynamically possible, it will necessarily occur. Thermodynamics does not give any information on the rate of a reaction, which may be fast, slow, or infinitely slow, nor does it indicate if another reaction is even more favourable.

Consider the corrosion that may occur when a sheet of galvanized iron is scratched. (Galvanized iron is iron which has been coated with zinc to prevent rusting.) Half reactions and the corresponding reduction potentials are shown below.

$$Fe^{2+} + 2e \rightarrow Fe \qquad E^\circ = -0.44 \text{ volts}$$

$$Zn^{2+} + 2e \rightarrow Zn \qquad E^\circ = -0.76 \text{ volts}$$

When in contact with water, either metal might be oxidized and lose metal ions, so we require the reverse reactions, and the potentials for these are called oxidation potentials, and have the same magnitude but the opposite sign to the reduction potentials.

$$Fe \rightarrow Fe^{2+} + 2e \qquad E^\circ = +0.44 \text{ volts}$$

$$Zn \rightarrow Zn^{2+} + 2e \qquad E^\circ = +0.76 \text{ volts}$$

Plainly, since $Zn \rightarrow Zn^{2+}$ produces the largest positive E° value, and since $\Delta G = nFE^\circ$, it will produce the largest negative ΔG value. Thus it is energetically more favourable for the Zn to dissolve, and hence the Zn will corrode away in preference to the Fe.

It is possible that when the galvanized steel is scratched, the air may oxidize some iron. The Fe^{2+} so produced is immediately reduced to iron by the zinc, and rusting does not occur.

$$Zn + Fe^{2+} \rightarrow Fe + Zn^{2+}$$

Similar applications in which one metal is sacrificed to protect another are the attaching of sacrificial blocks of magnesium to underground steel pipelines and the hulls of ships to prevent the rusting of iron.

Thus the coating of zinc serves two purposes – first it covers the iron and prevents its oxidation (rather like a coat of paint) and second it provides anodic protection.

A table of standard reduction potentials (Table 6.12) may be used to predict if a reaction is possible, and what the equilibrium constant will be. Consider for example if the triiodide ion I_3^- will oxidize As(III) in arsenious acid $HAsO_2$ into As(V).

$$HAsO_2 + I_3^- + 2H_2O \rightarrow H_3AsO_4 + 3I^- + 2H^+$$

Since the table lists reduction potentials, we must find the half reactions for $H_3AsO_4 + 2e \rightarrow$ products, and $I_3^- + 2e \rightarrow$ products.

$$H_3AsO_4 + 2e + 2H^+ \rightarrow HAsO_2 + 2H_2O \qquad E^\circ = +0.56 \text{ volts}$$
$$I_3^- + 2e \rightarrow 3I^- \qquad E^\circ = +0.54 \text{ volts}$$

The reaction we are investigating requires the first half reaction in the reverse direction, added to the second half reaction. E° values for half reactions must not be added together, since they do not take account of the number of electrons involved. However, E° values may be converted to the corresponding ΔG values, which may be added to give ΔG for the overall reaction.

$$HAsO_2 + 2H_2O \rightarrow H_3AsO_4 + 2e + 2H^+ \qquad E^\circ = -0.56\text{ V} \qquad \Delta G = +(2 \times F \times 0.56)$$
$$I_3^- + 2e \rightarrow 3I^- \qquad E^\circ = +0.54\text{ V} \qquad \Delta G = -(2 \times F \times 0.54)$$
$$HAsO_2 + I_3^- + 2H_2O \rightarrow H_3AsO_4 + 3I^- + 2H^+ \qquad \Delta G = +0.04F$$

The ΔG free energy change so calculated is positive, which indicates that the reaction will not proceed spontaneously in the forward direction, and suggests that it is energetically feasible for the reaction to proceed in the reverse direction. It should be noted that the value of ΔG is very small, and thus it is unwise to draw very firm conclusions. The E° values relate to standard conditions, and since ΔG is small, a small change in conditions, such as varying the concentration, or the pH, or the temperature, could change the potentials and hence change ΔG sufficiently to make the reaction proceed in either direction. There are volumetric methods of analysis for reducing arsenic acid with iodide ions in 5 M acid, and for oxidizing arsenious acid by triiodide ion at pH 7.

THE USE OF REDUCTION POTENTIALS

Enormous use may be made of reduction potentials for summarizing what species will oxidize or reduce something else, what the products of the reaction will be, and what oxidation states are stable with respect to the solvent, and also with respect to disproportionation. This topic is often insufficiently understood, so a number of examples are given.

A great deal of useful information about an element can be shown by the appropriate half reactions and reduction potentials. Consider some half reactions involving iron:

$$Fe^{2+} + 2e \rightarrow Fe \qquad E^\circ = -0.47 \text{ volts}$$
$$Fe^{3+} + 3e \rightarrow Fe \qquad E^\circ = -0.057$$
$$Fe^{3+} + e \rightarrow Fe^{2+} \qquad E^\circ = +0.77$$
$$FeO_4^{2-} + 3e + 8H^+ \rightarrow Fe^{3+} + 4H_2O \qquad E^\circ = +2.20$$

Where an element exists in several different oxidation states (in this case Fe(VI), Fe(III), Fe(II), and Fe(0)), it is convenient to display all of the reduction potentials for the half reactions in a single reduction potential diagram. In this the highest oxidation state is written at the left, and the lowest state at the right, and species such as electrons, H^+ and H_2O are omitted.

oxidation state	VI		III		II		0
$E°$ (V)	FeO_4^{2-}	$\xrightarrow{2.20}$	Fe^{3+}	$\xrightarrow{0.77}$	Fe^{2+}	$\xrightarrow{-0.47}$	Fe

$Fe^{3+} \xrightarrow{-0.057} Fe$

The potential for the reduction of FeO_4^{2-} to Fe^{3+} is +2.20 volts. Since $\Delta G = -nFE°$, it follows that ΔG for this change will be large and negative. This means that the reaction is thermodynamically possible since it releases a large amount of energy, and FeO_4^{2-} is a strong oxidizing agent.

Standard electrode potentials are measured on a scale with

$$H^+ + e \rightarrow H \qquad E° = 0.00 \text{ volts}$$

Since hydrogen is normally regarded as a reducing agent, reactions with negative value for $E°$ are more strongly reducing than hydrogen, that is they are strongly reducing. Materials which are generally accepted as oxidizing agents have $E°$ values above +0.8 volts, those such as $Fe^{3+} \rightarrow Fe^{2+}$ of about 0.8 volts are stable (equally oxidizing and reducing), and those below +0.8 volts become increasingly reducing.

For the change Fe^{3+}/Fe^{2+}, $E°$ is +0.77 V. This is close to the value of 0.8 V, and therefore Fe^{3+} and Fe^{2+} are of almost equal stability with respect to oxidation and reduction.The $E°$ values for the changes $Fe^{3+} \rightarrow Fe$ and for $Fe^{2+} \rightarrow Fe$ are both negative: hence ΔG is positive, so neither Fe^{3+} nor Fe^{2+} have any tendency to reduce to Fe.

One of the most important facts which can be obtained from a reduction potential diagram is whether any of the oxidation states are unstable with regard to disproportionation. Disproportionation is where one oxidation state decomposes, forming some ions in a higher oxidation state, and some in a lower oxidation state. This happens when a given oxidation state is a stronger oxidizing agent than the next highest oxidation state, and this situation occurs when a reduction potential on the right is more positive than one on the left. In the diagram of iron reduction potentials, the values become progressively more negative on moving from left to right, and hence Fe^{3+} and Fe^{2+} are stable with respect to disproportionation.

At first sight the potential of −0.057 V for $Fe^{3+} \rightarrow Fe$ seems wrong since the potentials for $Fe^{3+} \rightarrow Fe^{2+}$ and $Fe^{2+} \rightarrow Fe$ are 0.77 V and −0.47 V respectively, and adding 0.77 and −0.47 does *not* give −0.057. Potentials for complete reactions may be added since there are no electrons left over in the process. Potentials may not be added for half reactions since the electrons may not balance. However, potentials can always be converted into free energies using the equation $\Delta G = -nFE°$ where n is the number of electrons involved and F is the Faraday. Since the Gibbs free energy G is a thermodynamic function, free energies may be added, and the final total free energy converted back to an $E°$ value:

$$e + Fe^{3+} \rightarrow Fe^{2+} \quad E^\circ = +0.77\,V \quad \Delta G = -1(+0.77)F = -0.77F$$
$$2e + Fe^{2+} \rightarrow Fe \quad E^\circ = -0.47\,V \quad \Delta G = -2(-0.47)F = +0.94F$$

adding $\quad \Delta G = +0.17F$

$3e + Fe^{3+} \rightarrow Fe$

Hence E° can be calculated for the reaction $Fe^{3+} \rightarrow Fe$

$$E^\circ = \frac{\Delta G}{-nF} = \frac{0.17F}{-3F} = -0.057\,V$$

The reduction potential diagram for copper in acid solution is

oxidation state	II		I		0
E°(V)	Cu^{2+}	+0.15	Cu^{+} *	+0.50	Cu
			+0.35 (Cu^{2+} → Cu)		

* Disproportionates

The potential, and hence the energy released when Cu^{2+} is reduced to Cu^{+}, are both very small, and so Cu^{2+} is not an oxidizing agent but is stable. On moving from left to right the potentials $Cu^{2+}-Cu^{+}-Cu$ become more positive. Whenever this is found, the species in the middle (Cu^{+} in this case) disproportionates, that is it behaves as both a self-oxidizing and self-reducing agent because it is energetically favourable for the following two changes to occur together

$$Cu^{+} \rightarrow Cu^{2+} + e \quad E_{oxidation} = -0.15 \quad \Delta G = +0.15F$$
$$Cu^{+} + e \rightarrow Cu \quad E_{reduction} = +0.50 \quad \Delta G = -0.50F$$

overall $2Cu^{+} \rightarrow Cu^{2+} + Cu \quad \Delta G = -0.35F$

Thus in solution Cu^{+} disproportionates into Cu^{2+} and Cu, and hence Cu^{+} is only found in the solid state.

The reduction potential diagram for oxygen is shown.

oxidation state	0		−I		−II
E°(V)	O_2	+0.682	H_2O_2 *	+1.776	H_2O
			+1.229 (O_2 → H_2O)		

* Disproportionates

On moving from left to right, the reduction potentials increase, and hence H_2O_2 is unstable with respect to disproportionation.

$$\overset{-I}{2H_2O_2} \rightarrow \overset{0}{O_2} + \overset{-II}{H_2O}$$

It must be remembered that the solvent may impose a limitation on what species are stable, or exist at all. Very strong oxidizing reagents will oxidize water to O_2, whilst strong reducing agents will reduce it to H_2. Thus very strong oxidizing or reducing agents can not exist in aqueous solution. The following half reactions are of special importance:

Reduction of water

neutral solution	$H_2O + e^- \rightarrow OH^- + \frac{1}{2}H_2$	$E^\circ = -0.414\,V$
1.0 M acid solution	$H_3O^+ + e^- \rightarrow H_2O + \frac{1}{2}H_2$	$E^\circ = 0.000\,V$
1.0 M base solution	$H_2O + e^- \rightarrow OH^- + \frac{1}{2}H_2$	$E^\circ = -0.828\,V$

Oxidation of water

neutral solution	$\frac{1}{2}O_2 + 2H^+ + 2e^- \rightarrow H_2O$	$E^\circ = +0.185\,V$
1.0 M acid solution	$\frac{1}{2}O_2 + 2H^+ + 2e^- \rightarrow H_2O$	$E^\circ = +1.229\,V$
1.0 M base solution	$\frac{1}{2}O_2 + H_2O + 2e^- \rightarrow 2OH^-$	$E^\circ = +0.401\,V$

These reactions limit the *thermodynamic stability* of any species in aqueous solution.

Thus the minimum reduction potentials required to oxidize water to oxygen is $E^\circ > +0.185\,V$ in neutral solution, $E^\circ > +1.229\,V$ in 1.0 M acid solution and $E^\circ > +0.401\,V$ in 1.0 M basic solution.

In the same way half reactions with E° potentials less than zero (that is negative values) should reduce water to H_2 in 1.0 M acid solution, whilst an $E^\circ < -0.414\,V$ is required in neutral solution, and $E^\circ < -0.828\,V$ in 1.0 M basic solution.

Often when the E° values are just large enough to suggest that a reaction is thermodynamically possible, we find that it does not appear to happen. It must be remembered that a substance may be thermodynamically unstable, but kinetically stable, since the activation energy for the reaction is high. This means that the rates of these reactions are very slow. If the potentials are appreciably more positive or negative than these limits then reaction with the solvent is usually observed.

The reduction potentials for americium show that Am^{4+} is unstable with regard to disproportionation.

$$\begin{array}{ccccccccc} +VI & & +V & & +IV & & +III & & 0 \\ AmO_2^{2+} & \xrightarrow{+1.70} & \overset{*}{AmO_2^+} & \xrightarrow{+0.86} & \overset{*}{Am^{4+}} & \xrightarrow{+2.62} & Am^{3+} & \xrightarrow{-2.07} & Am \end{array}$$

* Disproportionates

The potential for the couple $AmO_2^+ \rightarrow Am^{3+}$ can be calculated by converting the values of 0.86 and 2.62 volts into free energies, adding them, then converting back to give a potential of 1.74 volts. When this step is added to the diagram it becomes apparent that the potentials do not decrease from AmO_2^{2+} to AmO_2^- to Am^{3+}, and hence AmO_2^+ is unstable with regard to disproportionation to AmO_2^{2+} and Am^{3+}. Finally, the potential for the couple $AmO_2^{2+} \rightarrow Am^{3+}$ can be worked out to be +1.726 volts. Thus considering $AmO_2^{2+} \rightarrow Am^{3+} \rightarrow Am$, Am^{3+} is stable.

$$\begin{array}{ccccc} +VI & +V & +IV & +III & 0 \end{array}$$

$$AmO_2^{2+} \xrightarrow{+1.70} AmO_2^{+\,*} \xrightarrow{+0.86} Am^{4+\,*} \xrightarrow{+2.62} Am^{3+} \xrightarrow{-2.07} Am$$

$AmO_2^{+} \rightarrow Am^{3+}$: +1.74; $AmO_2^{2+} \rightarrow Am^{3+}$: +1.726

* Disproportionates

It is important to include all the possible half reactions in a reduction potential diagram, or incorrect conclusions may be drawn. Examination of the incomplete diagram for chlorine in basic solution would indicate that ClO_2 should disproportionate into ClO_3^- and OCl^-, and that Cl_2 should disproportionate into OCl^- and Cl^-. Both of these deductions are correct.

$$\begin{array}{cccccc} +VII & +V & +III & +I & 0 & -I \end{array}$$

$$ClO_4^- \xrightarrow{+0.36} ClO_3^- \xrightarrow{+0.33} ClO_2^{\,*} \xrightarrow{+0.66} OCl^- \xrightarrow{+0.40} \tfrac{1}{2}Cl_2^{\,*} \xrightarrow{+1.36} Cl^-$$

* Disproportionates

The incomplete data also suggest that OCl^- should be stable with regard to disproportionation, but this is not true. The species which disproportionate are 'ignored', and a single potential calculated for the change $ClO_3^- \rightarrow OCl^-$ to replace the values +0.33 V and +0.66 V. Similarly a single potential is calculated for $OCl^- \rightarrow Cl^-$.

$$\begin{array}{cccccc} +VII & +V & +III & +I & 0 & -I \end{array}$$

$$ClO_4^- \xrightarrow{+0.36} ClO_3^{-\,*} \xrightarrow{+0.33} ClO_2^{\,*} \xrightarrow{+0.66} OCl^{-\,*} \xrightarrow{+0.40} \tfrac{1}{2}Cl_2^{\,*} \xrightarrow{+1.36} Cl^-$$

$ClO_3^- \rightarrow OCl^-$: +0.50; $OCl^- \rightarrow Cl^-$: +0.88

* Disproportionates

When the complete diagram is examined, it is apparent that the potentials around OCl^- do not decrease from left to right, and hence OCl^- is unstable with respect to disproportionation into ClO_3^- and Cl^-.

$$ClO_3^- \xrightarrow{+0.50} OCl^- \xrightarrow{+0.88} Cl^-$$

In the same way, the potentials round ClO_3^- do not decrease from left to right

$$ClO_4^- \xrightarrow{+0.36} ClO_3^- \xrightarrow{+0.50} OCl^-$$

Similarly ClO_3^- should disproportionate into ClO_4^- and OCl^-, and OCl^- should disproportionate to give Cl^- and more ClO_3^-.

Reduction potential diagrams may also be used to predict the products of reactions in which the elements have several oxidation states. Consider for example the reaction between an acidified solution of $KMnO_4$ and KI. The reduction potential diagrams are:

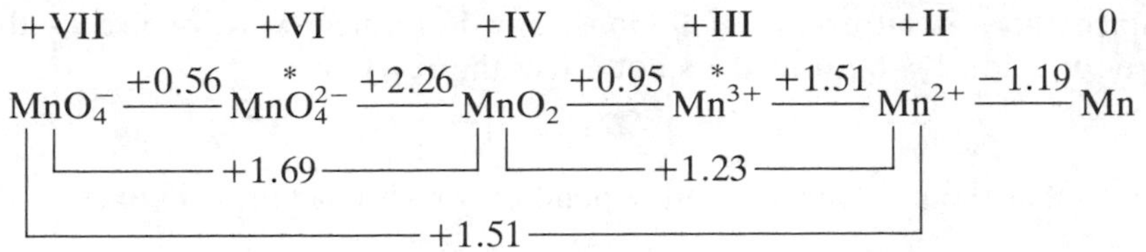

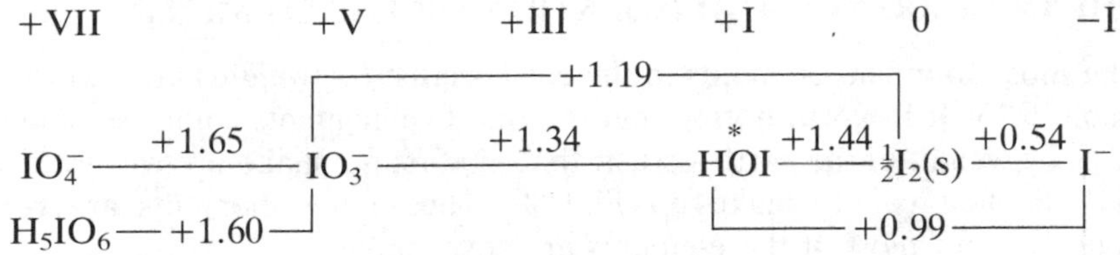

+VII +VI +IV +III +II 0

$MnO_4^- \xrightarrow{+0.56} MnO_4^{2-}$* $\xrightarrow{+2.26} MnO_2 \xrightarrow{+0.95} Mn^{3+}$* $\xrightarrow{+1.51} Mn^{2+} \xrightarrow{-1.19} Mn$

$MnO_4^- \rightarrow MnO_2$: +1.69; $MnO_2 \rightarrow Mn^{2+}$: +1.23; $MnO_4^- \rightarrow Mn^{2+}$: +1.51

+VII +V +III +I 0 −I

$IO_4^- \xrightarrow{+1.65} IO_3^- \xrightarrow{+1.34} HOI$* $\xrightarrow{+1.44} \frac{1}{2}I_2(s) \xrightarrow{+0.54} I^-$

H_5IO_6 — +1.60 — IO_3^-; $IO_3^- \rightarrow \frac{1}{2}I_2(s)$: +1.19; $HOI \rightarrow I^-$: +0.99

* Disproportionates

If we assume that the reactions are thermodynamically controlled, that is equilibrium is reached fairly quickly, then since MnO_4^{2-}, Mn^{3+} and HOI disproportionate, they need not be considered. The half reaction $Mn^{2+} \rightarrow Mn$ has a large negative $E°$ value, and hence ΔG will have a large positive value, so this will not occur, and can be ignored. Thus the reduction potential diagrams may be simplified:

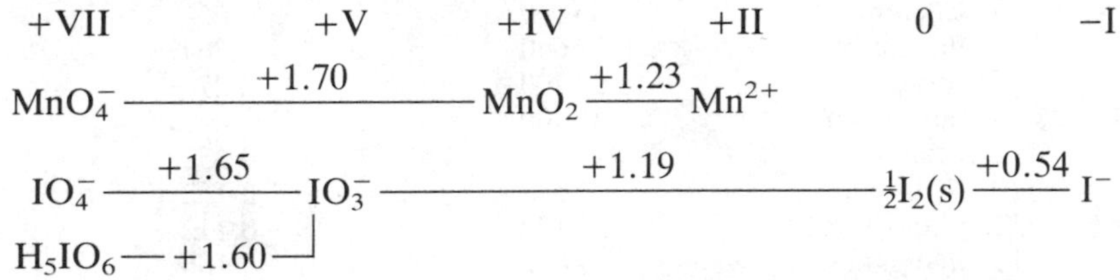

+VII +V +IV +II 0 −I

$MnO_4^- \xrightarrow{+1.70} MnO_2 \xrightarrow{+1.23} Mn^{2+}$

$IO_4^- \xrightarrow{+1.65} IO_3^- \xrightarrow{+1.19} \frac{1}{2}I_2(s) \xrightarrow{+0.54} I^-$

H_5IO_6 — +1.60 — IO_3^-

If the reaction is carried out by adding KI solution dropwise to an acidified solution of $KMnO_4$, the products of the reaction must be stable in the presence of $KMnO_4$. Thus Mn^{2+} cannot be formed, since $KMnO_4$ would oxidize it to MnO_2. In a similar way, I_2 cannot be formed, since $KMnO_4$ would oxidize it. The fact that the half reaction potentials for $IO_4^- \rightarrow IO_3^-$ and $H_5IO_6 \rightarrow IO_3^-$ are close to the $MnO_4^- \rightarrow MnO_2$ potential is a complication, and it is not obvious whether IO_3^-, IO_4^- or H_5IO_6 will be the product. In fact I^- is oxidized to a mixture of IO_3^- and IO_4^-.

$$2MnO_4^- + I^- + 2H^+ \rightarrow 2MnO_2 + IO_3^- + H_2O$$
$$8MnO_4^- + 3I^- + 8H^+ \rightarrow 8MnO_2 + 3IO_4^- + 4H_2O$$

If the reaction is carried out in a different way, by adding the $KMnO_4$ dropwise to the KI solution, then the products formed must be stable in the presence of I^-. Thus MnO_2 cannot be formed, since it would oxidize I^- to I_2. Similarly, IO_3^- cannot be formed since it would oxidize any excess I^- to I_2. The reaction which takes place is

$$2MnO_4 + 10I^- + 16H^+ \rightarrow 2Mn^{2+} + 5I_2 + 8H_2O$$

Since there is an excess of I^- ions, any I^2 formed will dissolve as the triiodide ion I_3^-, but this does not affect the reaction

$$I_2 + I^- \rightarrow I_3^-$$

Note that the products formed depend on which reactant is in excess.

THE OCCURRENCE AND ISOLATION OF THE ELEMENTS

The most abundant elements in the earth's crust (by weight) are shown in Table 6.13. It is worth noting that the first five elements comprise almost 92% by weight of the earth's crust, that the first ten make up over 99.5%, and the first twenty make up 99.97%. Thus a few elements are very abundant but most of the elements are very scarce.

Table 6.13 The most abundant elements

	Parts per million of earth's crust	% of earth's crust
1. oxygen	455 000	45.5
2. silicon	272 000	27.2
3. aluminium	83 000	8.3
4. iron	62 000	6.2
5. calcium	46 000	4.66
6. magnesium	27 640	2.764
7. sodium	22 700	2.27
8. potassium	18 400	1.84
9. titanium	6 320	0.632
10. hydrogen	1 520	0.152
11. phosphorus	1 120	0.112
12. manganese	1 060	0.106

A full table of abundances is given in Appendix A.

Other very abundant elements are nitrogen (78% of the atmosphere) and hydrogen, which occurs as water in the oceans. The chemistry of these abundant elements is well known, but some elements which are rare are also well known, because they occur in concentrated deposits – for example, lead as PbS (galena) and boron as $Na_2B_4O_7 . 10H_2O$ (borax).

The different methods for separating and extracting elements may be divided into five classes (see Ives, D.J.G. in Further Reading).

Mechanical separation of elements that exist in the native form

A surprisingly large number of elements occur in the free elemental state. They have remained in the native form because they are unreactive. Only the least reactive of the metals, those of the copper/silver/gold group and the platinum metals, occur in significant amounts as native elements.

1. Gold is found in the native form, as grains in quartz, as nuggets and in the silt of river beds. Gold has a density of $19.3\,g\,cm^{-2}$, which is very

much higher than that of the rocks or silt it is mixed with, and gold can be separated by 'panning'. (In recent times it has been more commonly extracted by amalgamating with mercury.) Silver and copper are some times found in the native form as 'nuggets'. All three metals are noble or unreactive, and this is associated with their position in the electro chemical series below hydrogen, and with the non-metals.

2. Palladium and platinum are also found as native metals. In addition natural alloys of the Pt group are found.

   The platinum metals are Ru Rh Pd
   Os Ir Pt

   The names of these natural alloys indicate their composition: osmididium, iridosmine.
3. Liquid droplets of mercury are found associated with cinnabar HgS. Non-metals which occur as native elements in the earth's crust are from the carbon and sulphur groups, but the atmosphere comprises N_2, O_2 and the noble gases.
4. Diamonds are found in the earth, and are obtained by mechanical separation of large amounts of earth and rock. The largest deposits are in Australia, Zaire, Botswana, the USSR and South Africa. Diamonds are mostly used for making cutting tools, and some for jewellery. Graphite is mined mainly in China, South Korea, the USSR, Brazil and Mexico. It is used for making electrodes, in steel making, as a lubricant, and in pencils, brake linings and brushes for electric motors. It is also used as the moderator in the cores of gas cooled nuclear reactors.
5. Deposits of sulphur are also found deep underground in Louisiana (USA), Poland, Mexico and the USSR. These are extracted by the *Frasch process*. Small amounts of selenium and tellurium are often present in sulphur.
6. The atmosphere is made up of about 78% nitrogen, 22% oxygen and traces of the noble gases argon, helium and neon. These may be separated by fractional distillation of liquid air. Helium is also obtained from some natural gas deposits.

Thermal decomposition methods

A few compounds will decompose into their constituent elements simply by heating.

1. A number of hydrides will decompose in this way, but since hydrides are usually made from the metal itself, the process is of no commercial significance. The hydrides arsine AsH_3 and stibine SbH_3 are produced in Marsh's test, where an arsenic or antimony compound is converted to the hydride with Zn/H_2SO_4 and the gaseous hydrides are decomposed to give a silvery mirror of metal by passing the hydride through a heated tube.
2. Sodium azide NaN_3 decomposes to give sodium and pure nitrogen on

gentle heating. Considerable care is needed as azides are often explosive. This method is not used commercially, but it is useful for making small quantities of very pure nitrogen in the laboratory.

$$2NaN_3 \rightarrow 2Na + 3N_2$$

3. Nickel carbonyl $Ni(CO)_4$ is gaseous and may be produced by warming Ni with CO at 50 °C. Any impurities in the Ni sample remain solid and the gas is heated to 230 °C, when it decomposes to give pure metal and CO which is recycled. This was the basis of the *Mond process* for purifying nickel which was used in South Wales from 1899 until the 1960s. A new plant in Canada uses the same principle but uses 150 °C and 20 atmospheres pressure to form $Ni(CO)_4$.

$$Ni + 4CO \xrightarrow{50\,°C} Ni(CO)_4 \xrightarrow{230\,°C} Ni + 4CO$$

4. The iodides are the least stable of the halides, and the *van Arkel-de-Boer process* has been used to purify small quantities of zirconium and boron. The impure element is heated with iodine, producing a volatile iodide ZrI_4 or BI_3. These are decomposed by passing the gas over an electrically heated filament of tungsten or tantalum which is white hot. The element is deposited on the filament and the iodine is recycled. The filament grows fatter, and is eventually removed. The tungsten core is drilled out of the centre, and a small amount of high purity Zr or B is obtained.
5. Most oxides are thermally stable at temperatures up to 1000 °C but the metals below hydrogen in the electrochemical series decompose fairly easily. Thus HgO and Ag_2O decompose on heating. The mineral cinnabar HgS is roasted in air to give the oxide, which then decomposes on heating. Silver residues from the laboratory and photographic processing are collected as AgCl and treated with Na_2CO_3, giving Ag_2CO_3, which decomposes on heating, first to Ag_2O and then to Ag.

$$2HgO \rightarrow 2Hg + O_2$$
$$Ag_2CO_3 \rightarrow CO_2 + Ag_2O \rightarrow 2Ag + \tfrac{1}{2}O_2$$

6. Oxygen may be produced by heating hydrogen peroxide H_2O_2, barium peroxide BaO_2, silver oxide Ag_2O or potassium chlorate $KClO_3$.

$$2H_2O_2 \rightarrow 2H_2O + O_2$$
$$2BaO_2 \rightleftharpoons 2BaO + O_2$$
$$2Ag_2O \rightarrow 2Ag + O_2$$
$$2KClO_3 \rightarrow 2KCl + 3O_2$$

Displacement of one element by another

In principle any element may be displaced from solution by another element which is higher in the electrochemical series. The method is inapplicable to elements which react with water, and to be economic must involve sacrificing a cheap element to obtain a more expensive element.

1. Copper ores which are too lean in CuS for the Cu to be extracted by roasting in air are left to be weathered by air and rain to form a solution of $CuSO_4$. The Cu^{2+} ions are displaced as Cu metal by sacrificing scrap iron which turns into Fe^{2+} because iron is above copper in the electrochemical series.

$$Fe + Cu^{2+} \rightarrow Fe^{2+} + Cu$$

2. Cadmium occurs in small amounts with zinc ores. The Zn is recovered by electrolysing a solution of $ZnSO_4$ which contains traces of $CdSO_4$. After a time the amount of Cd^{2+} has concentrated, and since Zn is above Cd in the electrochemical series some Zn metal is sacrificed to displace the Cd^{2+} from solution as Cd metal. The Zn which was sacrificed is subsequently recovered by electrolysis.

$$Zn + Cd^{2+} \rightarrow Zn^{2+} + Cd$$

3. Sea water contains Br^- ions. Chlorine is above bromine in the electrochemical series, and bromine is obtained by passing chlorine into sea water.

$$Cl_2 + 2Br^- \rightarrow 2Cl^- + Br_2$$

High temperature chemical reduction methods

A large number of commercial processes comc into this group. Carbon can be used to reduce a number of oxides and other compounds, and because of the low cost and availability of coke this method is widely used. The disadvantages are that a high temperature is needed, which is expensive and necessitates the use of a blast furnace, and many metals combine with carbon, forming carbides. Some examples are:

Reduction by carbon

$$Fe_2O_3 + C \xrightarrow{\text{blast furnace}} Fe$$

$$ZnO + C \xrightarrow{1200\,^\circ C} Zn$$

$$Ca_3(PO_4)_2 + C \xrightarrow{\text{electric furnace}} P$$

$$MgO + C \xrightarrow[\text{electric furnace}]{2000\,^\circ C} Mg \text{ (process now obsolete)}$$

$$PbO + C \longrightarrow Pb$$

Reduction by another metal

If the temperature needed for carbon to reduce an oxide is too high for economic or practical purposes, the reduction may be effected by another highly electropositive metal such as aluminium, which liberates a large amount of energy ($1675\,kJ\,mol^{-1}$) on oxidation to Al_2O_3. This is the basis of the *Thermite process*:

$$3Mn_3O_4 + 8Al \rightarrow 9Mn + 4Al_2O_3$$
$$B_2O_3 + Al \rightarrow 2B + Al_2O_3$$
$$Cr_2O_3 + Al \rightarrow 2Cr + Al_2O_3$$

Magnesium is used in a similar way to reduce oxides. In certain cases where the oxide is too stable to reduce, electropositive metals are used to reduce halides.

$$TiCl_4 + 2Mg \xrightarrow[1000-1150\,^{\circ}C]{\text{Kroll process}} Ti + 2MgCl_2$$

$$TiCl_4 + 4Na \xrightarrow{\text{IMI process}} Ti + 4NaCl$$

Self-reduction

A number of metals occur as sulphide ores (for example PbS, CuS and Sb_2S_3) which may be roasted first in air to partially convert them to the oxide, and then further roasted in the absence of air, causing self-reduction:

$$CuS \xrightarrow{\text{roast in air}} \left\{ \begin{array}{l} CuO \\ \;+ \\ CuS \end{array} \right. \xrightarrow[\text{without air}]{\text{roast}} Cu + SO_2$$

Reduction of oxides with hydrogen

$$Co_3O_4 + 4H_2 \rightarrow 3Co + 4H_2O$$
$$GeO_2 + 2H_2 \rightarrow Ge + 2H_2O$$
$$NH_4[MoO_4] + 2H_2 \rightarrow Mo + 4H_2O + NH_3$$
$$NH_4[WO_4] + 2H_2 \rightarrow W + 4H_2O + NH_3$$

This method is not widely used, because many metals react with hydrogen at elevated temperatures, forming hydrides. There is also a risk of explosion from hydrogen and oxygen in the air.

Electrolytic reduction

The strongest possible reducing agent is an electron. Any ionic material may be electrolysed, and reduction occurs at the cathode. This is an excellent method, and gives very pure products, but electricity is expensive. Electrolysis may be performed:

In aqueous solution

Provided that the products do not react with water, electrolysis can be carried out conveniently and cheaply in aqueous solution. Copper and zinc are obtained by electrolysis of aqueous solutions of their sulphates.

In other solvents

Electrolysis can be carried out in solvents other than water. Fluorine reacts violently with water, and it is produced by electrolysis of KHF_2 dissolved in anhydrous HF. (The reaction has many technical difficulties in that HF is corrosive, the hydrogen produced at the cathode must be kept separate from the fluorine produced at the anode or an explosion will occur, water must be rigorously excluded, and the fluorine produced attacks the anode and the reaction vessel.)

In fused melts

Elements that react with water are often extracted from fused melts of their ionic salts. These melts are frequently corrosive, and involve large fuel bills to maintain the high temperatures required. Aluminium is obtained by electrolysis of a fused mixture of Al_2O_3 and cryolite $Na_3[AlF_6]$. Both sodium and chlorine are obtained from the electrolysis of fused NaCl: in this case up to two thirds by weight of $CaCl_2$ is added as an impurity to lower the melting point from 803 °C to 505 °C.

Factors influencing the choice of extraction process

The type of process used commercially for any particular element depends on a number of factors.

1. Is the element unreactive enough to exist in the free state?
2. Are any of its compounds unstable to heat?
3. Does the element exist as an ionic compound, and is the element stable in water? If both are true, is there a cheap element above it in the electrochemical series which can be sacrificed to displace it from solution?
4. Does the element occur as sulphide ores which can be roasted, or oxide ores which can be reduced – using carbon is the cheapest whilst the use of Mg, Al and Na as reducing agents is more expensive.
5. If all other methods fail, electrolysis usually works for ionic materials, but is expensive. If the element is stable in water, electrolysing aqueous solutions is cheaper than using fused melts.

Thermodynamics of reduction processes

The extraction of metals from their oxides using carbon or other metals, and by thermal decomposition, involves a number of points which merit detailed discussion.

Table 6.14 Reduction potentials and extraction methods

Element	$E°$ (V)		Materials	Extraction method
Lithium	$Li^+ \mid Li$	−3.05	LiCl	Electrolysis of fused salts, usually chlorides
Potassium	$K^+ \mid K$	−2.93	KCl, $[KCl \cdot MgCl_2 \cdot 6H_2O]$	
Calcium	$Ca^{2+} \mid Ca$	−2.84	$CaCl_2$	
Sodium	$Na^+ \mid Na$	−2.71	NaCl	
Magnesium	$Mg^{2+} \mid Mg$	−2.37	$MgCl_2$, MgO	Electrolysis of $MgCl_2$ High temperature reduction with C
Aluminium	$Al^{3+} \mid Al$	−1.66	Al_2O_3	Electrolysis of Al_2O_3 dissolved in molten $Na_3[AlF_6]$
Manganese	$Mn^{2+} \mid Mn$	−1.08	Mn_3O_4, MnO_2	Reduction with Al Thermite process
Chromium	$Cr^{3+} \mid Cr$	−0.74	$FeCr_2O_4$	
Zinc	$Zn^{2+} \mid Zn$	−0.76	ZnS	Chemical reduction of oxides by C Sulphides are converted to oxides then reduced by C, or sometimes H_2
Iron	$Fe^{2+} \mid Fe$	−0.44	Fe_2O_3, Fe_3O_4	
Cobalt	$Co^{2+} \mid Co$	−0.27	CoS	
Nickel	$Ni^{2+} \mid Ni$	−0.23	NiS, $NiAs_2$	
Tin	$Sn^{2+} \mid Sn$	−0.14	SnO_2	
Lead	$Pb^{2+} \mid Pb$	−0.13	PbS	
Copper	$Cu^{2+} \mid Cu$	+0.35	Cu(metal), CuS	Found as native metal, or compounds easily decomposed by heat. (Also cyanide extraction)
Silver	$Ag^+ \mid Ag$	+0.80	Ag(metal), Ag_2S, AgCl	
Mercury	$Hg^{2+} \mid Hg$	+0.85	HgS	
Gold	$Au^{3+} \mid Au$	+1.38	Au(metal)	

For a spontaneous reaction, the free energy change ΔG must be negative.

$$\Delta G = \Delta H - T\Delta S$$

ΔH is the enthalpy change during the reaction, T is the absolute temperature, and ΔS is the change in entropy during the reaction. Consider a reaction such as the formation of an oxide:

$$M + O_2 \rightarrow MO$$

Oxygen is used up in the course of this reaction. Gases have a more random structure (less ordered) than liquids or solids. Consequently gases have a higher entropy than liquids or solids. In this reaction S the entropy or randomness decreases, and hence ΔS is negative. Thus if the temperature is raised then $T\Delta S$ becomes more negative. Since $T\Delta S$ is subtracted in the equation, then ΔG becomes less negative. *Thus the free energy change decreases with an increase of temperature.*

The free energy changes that occur when one gram molecule of a

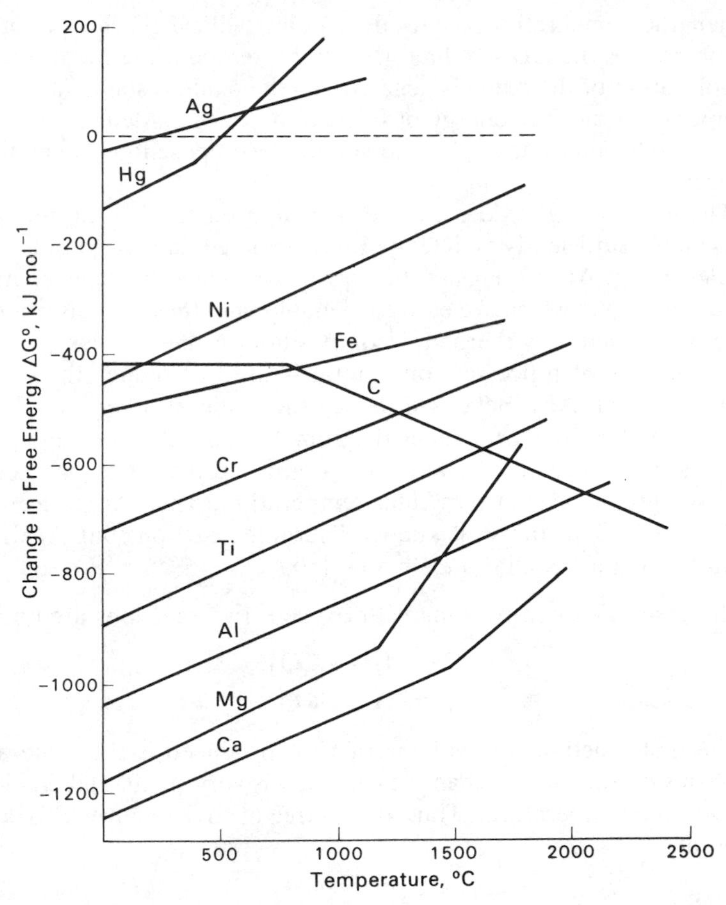

Figure 6.5 Ellingham diagram showing the change in free energy ΔG with temperature for oxides (based on 1 g mol of oxygen in each case).

common reactant (in this case oxygen) is used may be plotted graphically against temperature for a number of reactions of metals to their oxides. This graph is shown in Figure 6.5 and is called an Ellingham diagram (for oxides). Similar diagrams can be produced for one gram molecule of sulphur, giving an Ellingham diagram for sulphides, and similarly for halides.

The Ellingham diagram for oxides shows several important features:

1. The graphs for metal to metal oxide all slope upwards, because the free energy change decreases with an increase of temperature as discussed above.
2. The free energy changes all follow a straight line unless the materials melt or vaporize, when there is a large change in entropy associated with the change of state, which changes the slope of the line (for example the Hg–HgO line changes slope at 356 °C when Hg boils, and similarly Mg–MgO changes at 1120 °C).

3. When the temperature is raised, a point will be reached where the graph crosses the $\Delta G = 0$ line. Below this temperature the free energy of formation of the oxide is negative, so the oxide is stable. Above this temperature the free energy of formation of the oxide is positive, and the oxide becomes unstable, and should decompose into the metal and oxygen.

   Theoretically all oxides can be decomposed to give the metal and oxygen if a sufficiently high temperature can be attained. In practice the oxides of Ag, Au and Hg are the only oxides which can be decomposed at temperatures which are easily attainable, and these metals can therefore be extracted by thermal decomposition of their oxides.
4. In a number of processes, one metal is used to reduce the oxide of another metal. Any metal will reduce the oxide of other metals which lie above it in the Ellingham diagram because the free energy will become more negative by an amount equal to the difference between the two graphs at that particular temperature. Thus Al reduces FeO, CrO and NiO in the well known Thermite reaction, but Al will not reduce MgO at temperatures below 1500 °C.

In the case of carbon reacting with oxygen, two reactions are possible:

$$C + O_2 \rightarrow CO_2$$
$$C + \tfrac{1}{2}O_2 \rightarrow CO$$

In the first reaction, the volume of CO_2 produced is the same as the volume of O_2 used, so the change in entropy is very small, and ΔG hardly changes with temperature. Thus the graph of ΔG against T is almost horizontal.

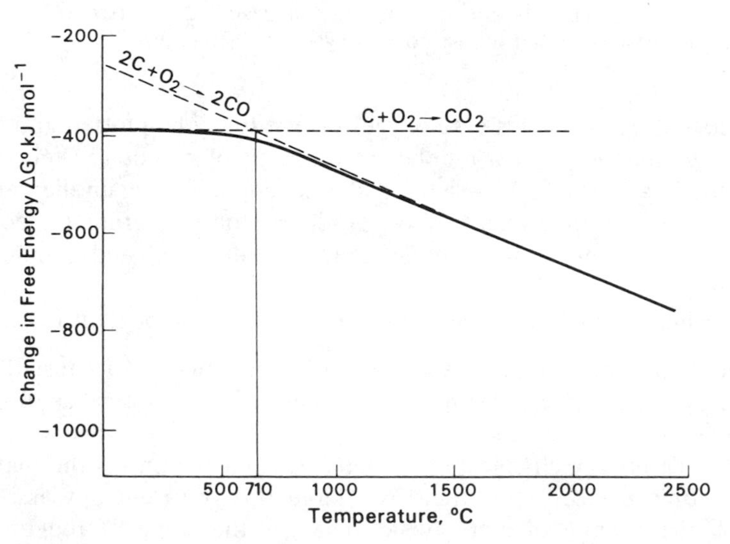

Figure 6.6 Ellingham diagram for carbon. (The composite curve is the solid line.)

The second reaction produces two volumes of CO for every one volume of oxygen used. Thus ΔS is positive, and hence ΔG becomes increasingly negative as T increases. Consequently the line on the Ellingham diagram slopes downwards (Figure 6.6). The two lines for $C \rightarrow CO_2$ and $C \rightarrow CO$ cross at about 710 °C. Below this temperature the reaction to form CO_2 is energetically more favourable, but above 710 °C the formation of CO is preferred.

Carbon is extensively used to reduce iron oxide in the extraction of iron, but it may also be used to reduce any other of the oxides above it on the Ellingham diagram. Since the ΔG line slopes downwards it will eventually cross and lie below all the other graphs for metal/metal oxide. Thus in principle carbon could be used to reduce any metal oxide if a sufficiently high temperature were used. At one time MgO was reduced by C at 2000 °C, followed by shock (i.e. rapid) cooling, though this process is now obsolete. Similarly the reduction of very stable oxides like TiO_2, Al_2O_3 and MgO is theoretically possible, but is not attempted because of the high cost and practical difficulties of using extremely high temperatures. A further limitation on the use of carbon for extracting metals is that at high temperatures many metals react with carbon, forming carbides.

Many metals occur as sulphide ores. Though carbon is a good reducing agent for oxides, it is a poor reducing agent for sulphides. The reason why carbon reduces so many oxides at elevated temperatures is that the $\Delta G°/T$ line for CO has a negative slope. There is no compound CS analogous to CO with a steep negative $\Delta G°/T$ line. Thus sulphides are normally roasted in air to form oxides before reducing with carbon.

$$2MS + 3O_2 \rightarrow 2MO + 2SO_2$$

In a similar way hydrogen is of limited use as a reducing agent for extracting metals from their oxides since the $\Delta G°/T$ line has a positive slope, and runs parallel to many metal oxide lines.

$$2H_2 + [O_2] \rightarrow 2H_2O$$

Thus only those metals with metal → metal oxide lines above the hydrogen line will be reduced, and this does not change with temperature. A further problem with H_2 is that many metals react with hydrogen, forming hydrides, and if hydrogen remains dissolved in the metal (interstitial hydrides) it significantly affects the properties of the metal.

Thermodynamic arguments about what will reduce a given compound have two limitations. They assume that the reactants and products are in equilibrium, which is often untrue, and they indicate whether a reaction is possible but do not predict the rate of reaction, or if some alternative reaction is even more favourable.

Further details of extraction processes and Ellingham diagrams for halides and sulphides are given in Further Reading see Ives D.J.G., and Ellingham, H.J.T.

Table 6.15 Extraction methods and the periodic table

	s-block												p-block					
Period \ Group	I	II											III	IV	V	VI	VII	0
1	^{1}H																	^{2}He
2	^{3}Li	^{4}Be											^{5}B	^{6}C	^{7}N	^{8}O	^{9}F	^{10}Ne
3	^{11}Na	^{12}Mg					d-block						^{13}Al	^{14}Si	^{15}P	^{16}S	^{17}Cl	^{18}Ar
4	^{19}K	^{20}Ca	^{21}Sc	^{22}Ti	^{23}V	^{24}Cr	^{25}Mn	^{26}Fe	^{27}Co	^{28}Ni	^{29}Cu	^{30}Zn	^{31}Ga	^{32}Ge	^{33}As	^{34}Se	^{35}Br	^{36}Kr
5	^{37}Rb	^{38}Sr	^{39}Y	^{40}Zr	^{41}Nb	^{42}Mo	^{43}Tc	^{44}Ru	^{45}Rh	^{46}Pd	^{47}Ag	^{48}Cd	^{49}In	^{50}Sn	^{51}Sb	^{52}Te	^{53}I	^{54}Xe
6	^{55}Cs	^{56}Ba	^{57}La	^{72}Hf	^{73}Ta	^{74}W	^{75}Re	^{76}Os	^{77}Ir	^{78}Pt	^{79}Au	^{80}Hg	^{81}Tl	^{82}Pb	^{83}Bi			

Fractional distillation of liquid air (N, O; group 0)

Electrolysis of fused salts (often chlorides) (groups I, II and Sc, Y, La)

Electrolysis or chemical reduction (Ti, V, Zr, Nb, Hf, Ta)

Found free in nature or compounds easily decomposed by heat (Ru, Rh, Pd, Ag, Os, Ir, Pt, Au, Hg, part of Cu)

Oxides reduced by carbon or sulphides converted to oxides then reduced by carbon (Cr, Mn, Mo, W, Re, Fe to Zn, and p-block groups III to VI)

Notes
1. Al, F and Cl are obtained by electrolysis of solutions
2. Br is obtained by displacement
3. I is obtained by reduction
4. Tc does not occur in nature

HORIZONTAL, VERTICAL AND DIAGONAL RELATIONSHIPS IN THE PERIODIC TABLE

On moving across a period in the periodic table, the number of electrons in the outer shell increases from one to eight. Thus Group I elements all have one electron in their outer shell. When they react they are univalent, because the loss of one electron leaves a noble gas structure. Similarly Group II elements have two electrons in their outer shell and are divalent. The valency of an element in one of the main groups is either the group number, which is the same as the number of outer electrons, or eight minus the group number. Group V elements (e.g. nitrogen) have five outer electrons. If three of these are shared in covalent bonds with other atoms, the nitrogen atom has a share in eight electrons and has a stable configuration. Thus nitrogen is trivalent, for example in ammonia NH_3. The halogens are in Group VII and have seven outer electrons. The valency should be $8 - 7 = 1$. A stable structure is attained by gaining one electron either by forming an ionic or a covalent bond. The number of outer electrons thus determines the valency of the element.

On moving from left to right across a period, the size of the atoms decreases because of the additional nuclear charge. Thus the orbital electrons are more tightly held, and the ionization energy increases. The metallic character of the element also decreases, and the oxides of the elements become less basic. Thus Na_2O is strongly basic; Al_2O_3 is amphoteric and reacts with both acids and bases; SO_2 is an acidic oxide since it dissolves in water to form sulphurous acid (H_2SO_3) and reacts with bases to form sulphites. Generally, metallic oxides are basic, whilst non-metallic oxides are acidic.

On descending a group in the periodic table, the elements all have the same number of outer electrons and the same valency, but the size increases. Thus the ionization energy decreases and the metallic character increases. This is particularly apparent in Groups IV and V which begin with the non-metals carbon and nitrogen and end with the metals lead and bismuth. The oxides become increasingly basic on descending the group.

On moving diagonally across the periodic table the elements show certain similarities. These are usually weaker than the similarities within a group, but are quite pronounced in the following pairs of elements:

```
Li   Be   B   C
   \    \   \
Na   Mg   Al   Si
```

On moving across a period, the charge on the ions increases and the size decreases, causing the polarizing power to increase. On moving down a group, the size increases and the polarizing power decreases. On moving diagonally these two effects partly cancel each other, so that there is no marked change in properties. The type and strength of bond formed and the properties of the compounds are often similar, although the valency is different. Thus lithium is similar to magnesium in many of its properties and beryllium is similar to aluminium. These similarities are examined in more detail in the chapters on Groups I, II and III. Diagonal similarities are most important among the lighter elements, but the line separating the metals from the non-metals also runs diagonally.

FURTHER READING

Size, ionization energy, electron affinity, energetics and Born–Haber cycle, electronegativity

Allred, A.L. and Rochow, E.G. (1958) *J. Inorg. Nucl. Chem.*, **5**, 264. (Original paper on Allred and Rochow scale of electronegativity values.)
Allred, A.L. (1961) *J. Inorg. Nucl. Chem.*, **17**, 215. (More on electronegativity values.)
Ashcroft, S.J. and Beech, G. (1973) *Inorganic Thermodynamics*, Van Nostrand.
Bratsch, S.G. (1988) Revised Mulliken electronegativities, *J. Chem. Ed.*, Part I: **65**, 34–41; Part II: **65**, 223–226.
Blustin, P.H. and Raynes, W.T. (1981) An electronegativity scale based on geometry changes on ionization, *J. Chem. Soc. (Dalton)*, 1237.
Emeléus, H.J. and Sharpe A.G. (1973) *Modern Aspects of Inorganic Chemistry*, 4th ed., (Chapter 5: Structures and energetics of inorganic molecules; Chapter 6: Inorganic chemistry in aqueous media), Routledge and Kegan Paul, London.
Huheey, J.E. (1972) *Inorganic Chemistry*, Harper and Row, New York. (Discussion on electronegativity.)
Lieberman, J.F. (1973) Ionization enthalpies and electron attachment enthalpies, *J. Chem. Ed.*, **50**, 831.
Mulliken, R.S. (1934, 1935) *J. Chem. Phys.*, **2**, 782; **3**, 573. (Mulluken electronegativity scale.)
Pauling, L. (1960) *The Nature of the Chemical Bond*, 3rd ed., Oxford University Press, London. (An old but classic text.)
Sanderson, R.T. (1945) A scale of electronegativity, *J. Chem. Ed.*, **31**, 2. (Original paper on Sanderson electronegativity scale.)
Sanderson, R.T. (1986) The Inert Pair Effect and Electronegativity, *Inorganic Chemistry*, **25**, 1856–1858.
Sanderson, R.T. (1988) Principles of electronegativity, *J. Chem. Ed.*, Part I: **65**, 112–118; Part II: **65**, 227–231.
Shannon, R.D. (1976) Revised effective ionic radii, *Acta Cryst.*, **A32**, 751–767. (The most recent and widely accepted values for ionic radii.)
Sharpe, A.G. (1981) *Inorganic Chemistry* (Chapter 3: Electronic configurations and some physical properties of atoms.), Longmans, London.
Zhang, Y. (1982) Electronegativities of elements in valence states, *Inorganic Chemistry*, **21**, 3886–3889.

Standard electrode potentials, redox reactions

Baes, C.F. and Mesmer, R.E. (1976) *The Hydrolysis of Cations*, Wiley-Interscience, London, 1976. (Comprehensive but understandable.)
Bard, A.J., Parsons, R. and Jordan, J. (1985) *Standard Potentials in Aqueous Solution* (Monographs in Electroanalytical Chemistry and Electrochemistry Series, Vol. 6), Marcel Dekker, New York. (Commissioned by IUPAC to replace the earlier values in Latimer's book.)
Burgess, J. (1988) *Ions in Solution*, Ellis Horwood, Chichester.
Fromhold, A.T., Jr (1980) *Theory of Metal Oxidation*, North Holland Publishing Co., Amsterdam and Oxford.

Jolly, W.L. (1976) *Inorganic Chemistry*, McGraw Hill, New York. (Redox reactions, and aqueous solutions.)
Johnson, D.A. (1968) *Some Thermodynamic Aspects of Inorganic Chemistry*, Cambridge University Press, Cambridge. (Lattice energies etc.)
Latimer, W.M. (1952) *The Oxidation States of the Elements and Their Potentials in Aqueous Solution*, 2nd ed., Prentice Hall, New York. (Old, but until very recently the standard source of oxidation potential data.)
Rosotti, H. (1978) *The Study of Ionic Equilibria in Aqueous Solution*, Longmans, London. (Redox reactions, solubility.)
Sanderson, R.T. (1966) The significance of electrode potentials, *J. Chem. Ed.*, **43**, 584–586.
Sharpe, A.G. (1969) *Principles of Oxidation and Reduction* (Royal Institute of Chemistry Monographs for Teachers No. 2), London.
Sharpe A.G. (1981) *Inorganic Chemistry*, (Chapter 7: Inorganic chemistry in aqueous media), Longmans, London.
Vincent, A. (1985) *Oxidation and Reduction in Inorganic and Analytical Chemistry: A Programmed Introduction*, John Wiley, Chichester.

Abundance and extraction of the elements

Cox, P.A. (1989) *The Elements: Their origins, Abundance and Distribution*, Oxford University Press, Oxford.
Ellingham, H.J.T. (1944, 1948) *J. Soc. Chem. Ind. Lond.*, **63**, 125; *Disc. Faraday Soc.*, **4**, 126, 161. (Original paper on Ellingham diagrams.)
Fergusson, J.E. (1982) *Inorganic Chemistry and the Earth: Chemical Resources, Their Extraction, Use and Environmental Impact* (Pergamon Series on Environmental Science, Vol. 6), Pergamon Press, Oxford.
Ives, D.J.G. (1969) *Principles of the Extraction of Metals* (Royal Institute of Chemistry Monographs for Teachers No. 3), London.
Jeffes, J.H.E. (1969) Extraction Metallurgy, *Chemistry in Britain*, **5**, 189–192.

PROBLEMS

1. (a) How does the size of atoms vary from left to right in a period, and on descending a group in the periodic table? What are the reasons for these changes?
   (b) Can you explain the large atomic radii of the noble gases?
   (c) Why is the decrease in size between Li and Be much greater than that between Na and Mg or K and Ca?

2. Explain what is meant by the ionization energy of an element. How does this vary between hydrogen and neon in the periodic table? Discuss how the variation can be related to the electronic structure of the atoms.

3. (a) What is the correlation between atomic size and ionization energy?
   (b) Account for the fact that there is a decrease in first ionization energy from Be to B, and Mg to Al.
   (c) Suggest the reason for the decrease in first ionization energy from N to O, and P to S.
   (d) Explain why the substantial decrease in first ionization energy

observed between Na and K, and Mg and Ca, is not observed between A and Ga.
(e) What is the significance of the large increase in the third ionization energy of Ca and the fifth ionization energy of Si?
(f) Why is the first ionization energy of the transition elements reasonably constant?

4. (a) What is electronegativity, and how is it related to the type of bond formed?
(b) What are Fajans' rules?
(c) Predict the type of bonds formed in HCl, CsCl, NH_3, CS_2 and $GeBr_4$.

5. (a) List the different scales of electronegativity and briefly describe the theoretical basis behind each.
(b) Give four examples to show how electronegativity values may be used to predict the type of bond formed in a compound.

6. Use a modified Born–Haber cycle suitable for the estimation of electrode potentials to explain:
(a) Why Li is as strong a reducing agent as Cs
(b) Why Ag is a noble metal and K a highly reactive metal.

7. (a) What are the standard electrode potentials, and how are they related to the electrochemical series?
(b) Explain the recovery of copper from solution using scrap iron.
(c) How is it possible to preferentially deposit metals electrolytically, e.g. Cu, Ni, and Zn from a solution containing all three?
(d) Why is it possible to obtain zinc by electrolysis of an aqueous solution even though the electrode potentials would suggest that the water should decompose first?

8. (a) Explain why Cu^+ disproportionates in solution.
(b) Explain why the standard reduction potentials for $Cu^2 \rightarrow Cu^+$ and $Cu^+ \rightarrow Cu$ are +0.15 and +0.50 volt, respectively, yet that for $Cu^{2+} \rightarrow Cu$ is + 0.34 volt.

9. Name the eight most abundant elements in the earth's crust and place them in the correct order.

10. Describe the following named metallurgical processes: (a) Bessemer, (b) BOP, (c) Kroll, (d) Van Arkel, (e) Hall-Héroult, (f) Parkes.

11. Which elements occur in the native state?

12. List five ores which are smelted, and give equations to show what occurs during smelting.

13. Describe the extraction of three different elements using carbon as the reducing agent.

14. Draw an Ellingham diagram for metal oxides and explain what information can be obtained from it. In addition explain why most of

the lines slope upwards from left to right, why the lines change in slope, and what happens when a line crosses the $\Delta G = 0$ axis.

15. Use the Ellingham diagram for oxides to find:
    (a) if Al will reduce chromium oxide
    (b) at what temperature C will reduce magnesium oxide, and
    (c) at what temperature mercuric oxide will decompose into its elements.
16. Explain in detail the processes involved in the production of pig iron and steel.
17. Describe the extraction of two metals and two non-metals by electrolysis.
18. Describe the extraction of magnesium and bromine from sea water.

7 Coordination compounds

DOUBLE SALTS AND COORDINATION COMPOUNDS

Addition compounds are formed when stoichiometric amounts of two or more stable compounds join together. For example:

$$KCl + MgCl_2 + 6H_2O \rightarrow KCl \cdot MgCl_2 \cdot 6H_2O$$
(carnallite)

$$K_2SO_4 + Al_2(SO_4)_3 + 24H_2O \rightarrow K_2SO_4 \cdot Al_2(SO_4)_3 \cdot 24H_2O$$
(potassium alum)

$$CuSO_4 + 4NH_3 + H_2O \rightarrow CuSO_4 \cdot 4NH_3 \cdot H_2O$$
(tetrammine copper(II) sulphate monhydrate)

$$Fe(CN)_2 + 4KCN \rightarrow Fe(CN)_2 \cdot 4KCN$$
(potassium ferrocyanide)

Addition compounds are of two types:

1. Those which lose their identity in solution (double salts)
2. Those which retain their identity in solution (complexes)

When crystals of carnallite are dissolved in water, the solution shows the properties of K^+, Mg^{2+} and Cl^-, ions. In a similar way, a solution of potassium alum shows the properties of L^+, Al^{3+} and SO_4^{2-} ions. These are both examples of double salts which exist only in the crystalline state.

When the other two examples of coordination compounds dissolve they do not form simple ions – Cu^{2+}, or Fe^{2+} and CN^- – but instead their complex ions remain intact. Thus the cuproammonium ion $[Cu(H_2O)_2(NH_3)_4]^{2+}$ and the ferrocyanide ion $[Fe(CN)_6]^{4-}$ exist as distinct entities both in the solid and in solution. Complex ions are shown by the use of square brackets. Compounds containing these ions are called coordination compounds. *The chemistry of metal ions in solution is essentially the chemistry of their complexes. Transition metal ions, in particular, form many stable complexes.* In solution 'free' metal ions are coordinated either to water or to other ligands. Thus Cu^{2+} exists as the pale blue complex ion $[Cu(H_2O)_6]^{2+}$ in aqueous solution (and also in hydrated crystalline salts). If aqueous ammonia is added to this solution, the familiar deep blue cuproammonium ion is formed:

$$[Cu(H_2O)_6]^{2+} + 4NH_3 \rightleftharpoons [Cu(H_2O)_2(NH_3)_4]^{2+} + 4H_2O$$

Note that this reaction is a substitution reaction, and the NH_3 replaces water in the complex ion.

WERNER'S WORK

Werner's coordination theory in 1893 was the first attempt to explain the bonding in coordination complexes. It must be remembered that this imaginative theory was put forward before the electron had been discovered by J.J. Thompson in 1896, and before the electronic theory of valency. This theory and his painstaking work over the next 20 years won Alfred Werner the Nobel Prize for Chemistry in 1913.

Complexes must have been a complete mystery without any knowledge of bonding or structure. For example, why does a stable salt like $CoCl_3$ react with a varying number of stable molecules of a compound such as NH_3 to give several new compounds: $CoCl_3 \cdot 6NH_3$, $CoCl_3 \cdot 5NH_3$ and $CoCl_3 \cdot 4NH_3$? What are their structures? At that time X-ray diffraction, which is the most powerful method of determining the structures of crystals, had yet to be discovered. Werner did not have at his disposal any of the modern instrumental techniques, and all his studies were made using simple reaction chemistry. *Werner was able to explain the nature of bonding in complexes, and he concluded that in complexes the metal shows two different sorts of valency:*

1. *Primary valencies.* These are non-directional. The modern explanation would be as follows. The complex commonly exists as a positive ion. The primary valency is the number of charges on the complex ion. In compounds, this charge is matched by the same number of charges from negative ions. Primary valency applies equally well to simple salts and to complexes. Thus in $CoCl_2$ ($Co^{2+} + 2Cl^-$) there are two primary valencies, i.e. two ionic bonds. The complex $[Co(NH_3)_6]Cl_3$ actually exists as $[Co(NH_3)_6]^{3+}$ and $3Cl^-$. Thus the primary valency is 3, as there are three ionic bonds.
2. *Secondary valencies.* These are directional. In modern terms the number of secondary valencies equals the number of ligand atoms coordinated to the metal. This is now called the coordination number. Ligands are commonly negative ions such as Cl^-, or neutral molecules such as NH_3. Less commonly, ligands may be positive ions such as NO^+. Each metal has a characteristic number of secondary valencies. Thus in $[Co(NH_3)_6]Cl_3$ the three Cl^- are held by primary valencies. The six NH_3 groups are held by secondary valencies.

Secondary valencies are directional, and so a complex ion has a particular shape, e.g. the complex ion $[Co(NH_3)_6]^{3+}$ is octahedral. Werner deduced the shapes of many complexes. He did this by preparing as many different isomeric complexes of a system as was possible. He noted the number of isomers formed and related this number to the number of isomers predicted for different geometric shapes. The most common coordination number in transition metal complexes is 6, and the shape is usually octahedral. The coordination number 4 is also common, and this gives rise to either tetrahedral or square planar complexes.

Werner treated cold solutions of a series of coordination complexes with an excess of silver nitrate, and weighed the silver chloride precipitated. The stoichiometries of complex–AgCl formed were as follows:

$$CoCl_3 \cdot 6NH_3 \rightarrow 3AgCl$$
$$CoCl_3 \cdot 5NH_3 \rightarrow 2AgCl$$
$$CoCl_3 \cdot 4NH_3 \rightarrow 1AgCl$$

Werner deduced that in $CoCl_3 \cdot 6NH_3$ the three chlorines acted as primary valencies, and the six ammonias as secondary valencies. Inmodern terms the complex is written $[Co(NH_3)_6]Cl_3$. The three Cl^- are ionic and hence are precipitated as AgCl by $AgNO_3$. The six NH_3 ligands form coordinate bonds to Co^{3+}, forming a complex ion $[Co(NH_3)]^{3+}$ (Figure 7.1a).

Werner deduced that loss of one NH_3 from $CoCl_3 \cdot 6NH_3$ should give $CoCl_3 \cdot 5NH_3$, and at the same time one Cl changed from being a primary valency to a secondary valency. Thus this complex had two primary valencies and six secondary valencies. In modern terms the complex $[Co(NH_3)_5Cl]Cl_2$ ionizes to give $[Co(NH_3)_5Cl]^{2+}$ and two Cl^- ions. Thus only two of the three chlorine atoms are ionic and thus only two are precipitated as AgCl with $AgNO_3$. Five NH_3 and one Cl form coordinate bonds to Co^{3+}, forming a complex ion (Figure 7.1b).

Similarly in $CoCl_3 \cdot 4NH_3$ Werner deduced that one Cl formed a primary valency, and that there were six secondary valencies (two Cl and four NH_3). In modern terms the complex $[Co(NH_3)_4Cl_2]Cl$ ionizes to give $[Co(NH_3)_4Cl_2]^+$ and Cl^- and so only one Cl^- can be precipitated as AgCl. The coordination number of Co^{3+} is 6; in this case four NH_3 and two Cl^- form coordinate bonds to Co^{3+}. The old and modern ways of writing the formulae of these complexes are shown in Table 7.1.

Thus Werner established that the number of secondary valencies (that is the coordination number) was 6 in these complexes. He then attempted to

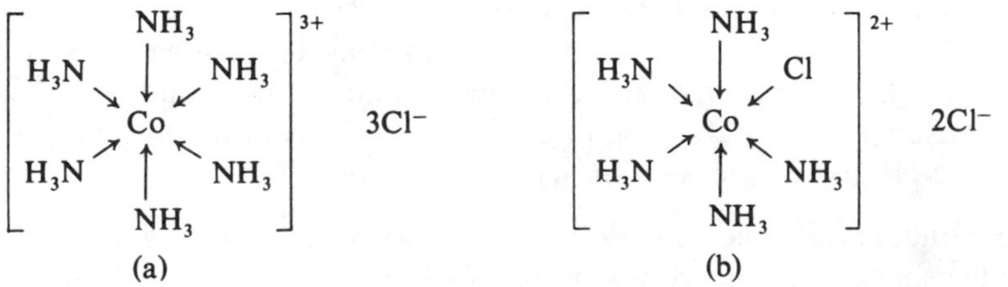

Figure 7.1 Structures of (a) $[Co(NH_3)_6]Cl_3$ and (b) $[Co(NH_3)_5Cl]Cl_2$.

Table 7.1 Formulae of some cobalt complexes

Old	New
$CoCl_3 \cdot 6NH_3$	$[Co(NH_3)_6]^{3+}$ $3Cl^-$
$CoCl_3 \cdot 5NH_3$	$[Co(NH_3)_5Cl]^{2+}$ $2Cl^-$
$CoCl_3 \cdot 4NH_3$	$[Co(NH_3)_4Cl_2]^+$ Cl^-

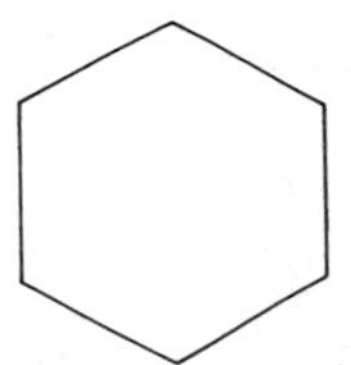

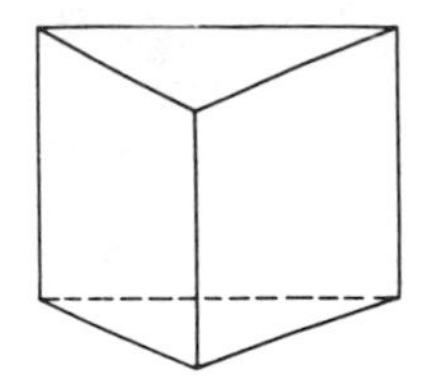

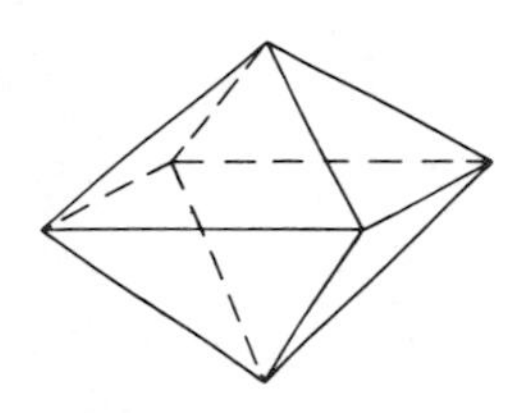

Figure 7.2 Possible geometric shapes for six-coordination.

find the shapes of the complexes. The possible arrangements of six groups round one atom are a planar hexagon, a trigonal prism, and an octahedron (Figure 7.2). Werner then compared the number of isomeric forms he had obtained with the theoretical number for each of the possible shapes (Table 7.2).

Table 7.2 Number of isomers predicted and actually found

Complex	Observed	Predicted		
		Octahedral	Planar hexagon	Trigonal prism
$[MX_6]$	1	1	1	1
$[MX_5Y]$	1	1	1	1
$[MX_4Y_2]$	2	2	3	3
$[MX_3Y_3]$	2	2	3	3

These results strongly suggested that these complexes have an octahedral shape. This proof was not absolute proof, as it was just possible that the correct experimental conditions had not been found for preparing all the isomers. More recently the X-ray structures have been determined, and these establish that the shape is octahedral (Figure 7.3).

More recently, with a bidentate ligand such as ethylenediamine (1,2-diaminoethane), two optically active isomers have been found (Figure 7.4).

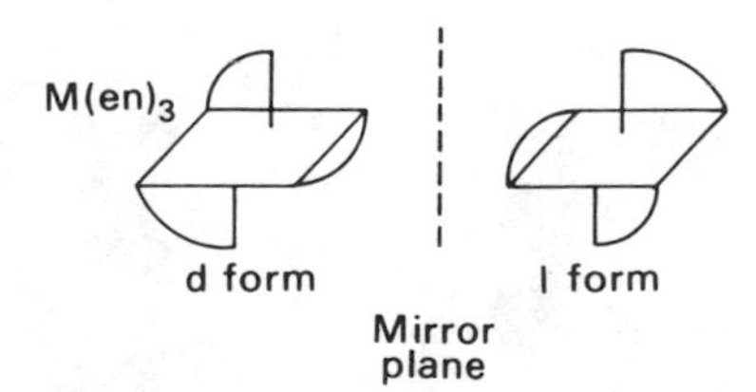

Figure 7.4 Optical isomerism in octahedral complexes.

In a similar way, Werner studied a range of complexes which included $[Pt^{II}(NH_3)_2Cl_2]$ and $[Pd^{II}(NH_3)_2Cl_2]$. The coordination number is 4, and the shape could be either tetrahedral or square planar. Werner was able to prepare two different isomers for these complexes. A tetrahedral complex can only exist in one form, but a square planar complex can exist in two isomeric forms. This proved these complexes are square planar rather than tetrahedral (Figure 7.5).

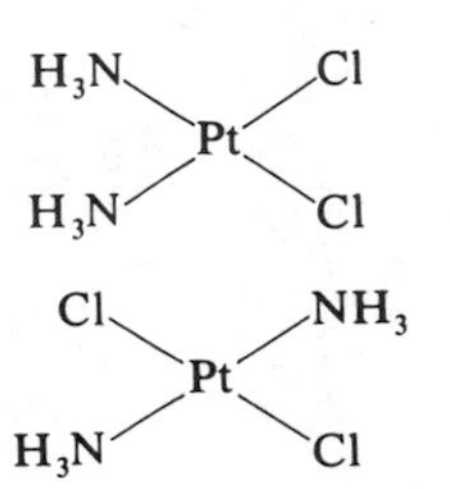

Figure 7.5 Isomerism in square planar complexes.

MORE RECENT METHODS OF STUDYING COMPLEXES

The electrical conductivity of a solution of an ionic material depends on:

1. The concentration of solute.
2. The number of charges on the species which are formed on dissolution.

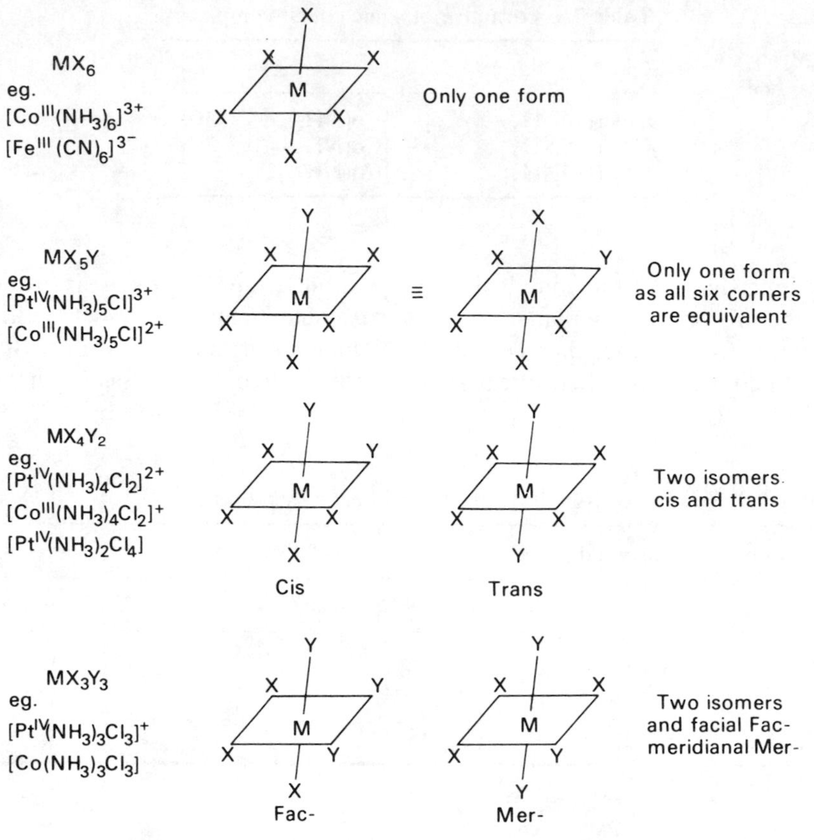

Figure 7.3 Isomers in octahedral complexes.

Molar conductivities relate to a 1 M solution and thus the concentration factor is removed. The total number of charges on the species formed when the complex dissolves can be deduced by comparison of its molar conductivity with that of known simple ionic materials (Table 7.3). These conductivities suggest the same structures for the cobalt/ammonia/chlorine

Table 7.3 Conductivities of salts and complexes (Molar conductivities measured at 0.001 M concentration)

			$ohm^{-1}\,cm^2\,mol^{-1}$
$LiCl$	$\rightarrow Li^+\ Cl^-$	(total of 2 charges)	112.0
$CaCl_2$	$\rightarrow Ca^{2+}\ 2Cl^-$	(total of 4 charges)	260.8
$CoCl_3 \cdot 5NH_3$			261.3
$CoBr_3 \cdot 5NH_3$			257.6
$LaCl_3$	$\rightarrow La^{3+}\ 3Cl^-$	(total of 6 charges)	393.5
$CoCl_3 \cdot 6NH_3$			431.6
$CoBr_3 \cdot 6NH_3$			426.9

Table 7.4 Number of charges related to modern and Werner structures

	Charges	Primary valency ionizable chlorines	Secondary valency
$[Co(NH_3)_6]^{3+}$ $3Cl^-$	6	3	$6NH_3 = 6$
$[Co(NH_3)_5Cl]^{2+}$ $2Cl^-$	4	2	$5NH_3 + 1Cl^- = 6$
$[Co(NH_3)_4Cl_2]^+$ Cl^-	2	1	$4NH_3 + 2Cl^- = 6$

complexes mentioned earlier, as do the results from Werner's AgCl experiments, shown in Table 7.4.

The freezing point of a liquid is lowered when a chemical substance is dissolved in it. Cryoscopic measurements involve measuring how much the freezing point is lowered. The depression of freezing point obtained depends on the number of particles present. Cryoscopic measurements can be used to find if a molecule dissociates, and how many ions are formed. If a molecule dissociates into two ions it will give twice the expected depression for a single particle. If three ions are formed this will give three times the expected depression. Thus:

$$
\begin{array}{llll}
LiCl \rightarrow Li^+ + Cl^- & \text{(2 particles)} & \text{(2 charges)} \\
MgCl_2 \rightarrow Mg^{2+} + 2Cl^- & \text{(3 particles)} & \text{(4 charges)} \\
LaCl_3 \rightarrow La^{3+} + 3Cl^- & \text{(4 particles)} & \text{(6 charges)}
\end{array}
$$

The number of particles formed from a complex molecule determines the size of the depression of freezing point. Note that the number of particles formed may be different from the total number of charges which can be obtained from conductivity measurements. The two types of information can be used together to establish the structure (Table 7.5).

The magnetic moment can be measured (see Chapter 18 – Magnetic properties). This provides information about the number of unpaired electron spins present in a complex. From this it is possible to decide how the electrons are arranged and which orbitals are occupied. Sometimes the structure of the complex can be deduced from this. For example, the compound $Ni^{II}(NH_3)_4(NO_3)_2 \cdot 2H_2O$ might contain four ammonia mole-

Table 7.5 Establishing the structure of complexes

Formula	Cryoscopic measurement	Molar conductivity	Structure
$CoCl_3 \cdot 6NH_3$	4 particles	6 charges	$[Co(NH_3)_6]^{3+}$ $3Cl^-$
$CoCl_3 \cdot 5NH_3$	3 particles	4 charges	$[Co(NH_3)_5Cl]^{2+}$ $2Cl^-$
$CoCl_3 \cdot 4NH_3$	2 particles	2 charges	$[Co(NH_3)_4Cl_2]^+$ Cl^-
$CoCl_3 \cdot 3NH_3$	1 particle	0 charge	$[Co(NH_3)_3Cl_3]$
$Co(NO_2)_3 \cdot KNO_2 \cdot 2NH_3$	2 particles	2 charges	K^+ $[Co(NH_3)_2(NO_2)_4]^-$
$Co(NO_2)_3 \cdot 2KNO_2 \cdot NH_3$	3 particles	4 charges	$2K^+$ $[Co(NH_3)(NO_2)_5]^{2-}$
$Co(NO_2)_3 \cdot 3KNO_2$	4 particles	6 charges	$3K^+$ $[Co(NO_2)_6]^{3-}$

cules coordinated to Ni in a square planar $[Ni(NH_3)_4]^{2+}$ ion and two molecules of water of crystallization and have no unpaired electrons. Alternatively the water might be coordinated to the metal, giving an octahedral $[Ni(H_2O)_2(NH_3)_4]^{2+}$ complex with two unpaired electrons. Both these complex ions exist and their structures can be deduced from magnetic measurements.

Dipole moments may also yield structural information but only fornon-ionic complexes. For example, the complex $[Pt(NH_3)_2Cl_2]$ is square planar, and can exist as *cis or trans* forms. The dipole moments from the various metal–ligand bonds cancel out in the *trans* configuration. However, a finite dipole moment is given by the *cis* arrangement.

Electronic spectra (UV and visible) also provide valuable information on the energy of the orbitals, and on the shape of thecomplex. By this means it is possible to distinguish betweentetrahedral and octahedral complexes, and whether the shape isdistorted or regular.

The most powerful method, however, is the X-ray determination of the crystal structure. This provides details of the exact shape and the bond lengths and angles of the atoms in the structure.

EFFECTIVE ATOMIC NUMBERS

The number of secondary valencies in the Werner theory is now called the coordination number of the central metal in the complex. This is the number of ligand atoms bonded to the central metal ion. Each ligand donates an electron pair to the metal ion, thus forming a coordinate bond. Transition metals form coordination compounds very readily because they have vacant *d* orbitals which can accommodate these electron pairs. The electronic arrangement of the noble gases is known to be very stable. Sidgwick, with his effective atomic number rule, suggested that electron pairs from ligands were added until the central metal was surrounded by the same number of electrons as the next noble gas. Consider potassium hexacyanoferrate(II) $K_4[Fe(CN)_6]$ (formerly called potassium ferrocyanide). An iron atom has 26 electrons, and so the central metal ion Fe^{2+} has 24 electrons. The next noble gas Kr has 36 electrons. Thus the addition of six electron pairs from six CN^- ligands adds 12 electrons, thus raising the effective atomic number (EAN) of Fe^{2+} in the complex $[Fe(CN)_6]^{4-}$ to 36.

$$[24 + (6 \times 2) = 36]$$

Further examples are given in Table 7.6.

The EAN rule correctly predicts the number of ligands in many complexes. There are, however, a significant number of exceptions where the EAN is not quite that of a noble gas. If the original metal ion has an odd number of electrons, for example, the adding of electron pairs cannot result in a noble gas structure. The tendency to attain a noble gas configuration is a significant factor but not a necessary condition for complex formation. It is also necessary to produce a symmetrical structure (tetra-

Table 7.6 Effective atomic numbers of some metals in complexes

Atom	Atomic number	Complex	Electrons lost in ion formation	Electrons gained by coordination	EAN	
Cr	24	$[Cr(CO)_6]$	0	12	36	(Kr)
Fe	26	$[Fe(CN)_6]^{4-}$	2	12	36	(Kr)
Fe	26	$[Fe(CO)_5]$	0	10	36	(Kr)
Co	27	$[Co(NH_3)_6]^{3+}$	3	12	36	(Kr)
Ni	28	$[Ni(CO)_4]$	0	8	36	(Kr)
Cu	29	$[Cu(CN)_4]^{3-}$	1	8	36	(Kr)
Pd	46	$[Pd(NH_3)_6]^{4+}$	4	12	54	(Xe)
Pt	78	$[PtCl_6]^{2-}$	4	12	86	(Rn)
Fe	26	$[Fe(CN)_6]^{3-}$	3	12	35	
Ni	28	$[Ni(NH_3)_6]^{2+}$	2	12	38	
Pd	46	$[PdCl_4]^{2-}$	2	8	52	
Pt	78	$[Pt(NH_3)_4]^{2+}$	2	8	84	

hedral, square planar, octahedral) irrespective of the number of electrons involved.

SHAPES OF *d* ORBITALS

Since *d* orbitals are often used in coordination complexes it is important to study their shapes and distribution in space. The five *d* orbitals are not identical and the orbitals may be divided into two sets. The three t_{2g} orbitals have identical shape and point between the axes, *x*, *y* and *z*. The

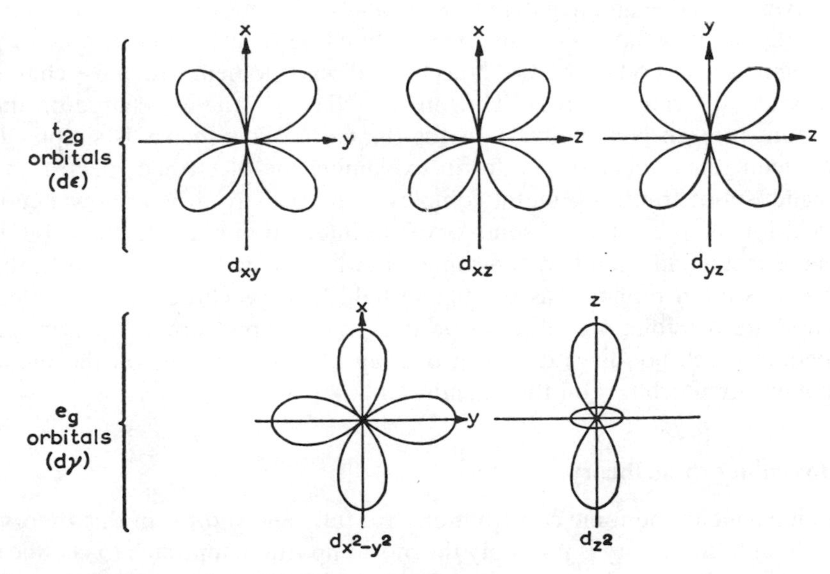

Figure 7.6 Shapes of *d* orbitals.

two e_g orbitals have different shapes and point along the axes (Figure 7.6). Alternative names for t_{2g} and e_g are $d\varepsilon$ and $d\gamma$ respectively.

BONDING IN TRANSITION METAL COMPLEXES

There are three theories of metal to ligand bonding in complexes, all dating back to the 1930s.

Valence bond theory

This theory was developed by Pauling. Coordination compounds contain complex ions, in which ligands form coordinate bonds to the metal. Thus the ligand must have a lone pair of electrons, and the metal must have an empty orbital of suitable energy available for bonding. The theory considers which atomic orbitals on the metal are used for bonding. From this the shape and stability of the complex are predicted. The theory has two main limitations. Most transition metal complexes are coloured, but the theory provides no explanation for their electronic spectra. Further, the theory does not explain why the magnetic properties vary with temperature. For these reasons it has largely been superseded by the crystal field theory. However, it is of interest for study as it shows the continuity of the development of modern ideas from Werner's theory.

Crystal field theory

This theory was proposed by Bethe and van Vleck. The attraction between the central metal and ligands in the complex is considered to be purely electrostatic. Thus bonding in the complex may be ion–ion attraction (between positive and negative ions such as Co^{3+} and Cl^-). Alternatively, ion–dipole attractions may give rise to bonding (if the ligand is a neutral molecule such as NH_3 or CO). NH_3 has a dipole moment with a $\delta-$ charge on N and $\delta+$ charges on H. Thus in $[Co(NH_3)_6]^{3+}$ the $\delta-$ charge on the N atom of each NH_3 points towards the Co^{3+}. This theory is simple. It has been remarkably successful in explaining the electronic spectra and magnetism of transition metal complexes, particularly when allowance is made for the possibility of some covalent interaction between the orbitals on the metal and ligand. When some allowance is made for covalency, the theory is often renamed as the ligand field theory. Three types of interaction are possible: σ overlap of orbitals, π overlap of orbitals, or $d\pi-p\pi$ bonding (back bonding) due to π overlap of full d orbitals on the metal with empty p orbitals on the ligands.

Molecular orbital theory

Both covalent and ionic contributions are fully allowed for in this theory.

Though this theory is probably the most important approach to chemical bonding, it has not displaced the other theories. This is because the quantitative calculations involved are difficult and lengthy, involving the use of extensive computer time. Much of the qualitative description can be obtained by other approaches using symmetry and group theory.

VALENCE BOND THEORY

The formation of a complex may be considered as a series of hypothetical steps. First the appropriate metal ion is taken, e.g. Co^{3+}. A Co atom has the outer electronic structure $3d^7 4s^2$. Thus a Co^{3+} ion will have the structure $3d^6$, and the electrons will be arranged:

full inner shell	3d	4s	4p	4d
	↑↓ ↑ ↑ ↑ ↑	☐	☐☐☐	☐☐☐☐☐

If this ion forms a complex with six ligands, then six empty atomic orbitals are required on the metal ion to receive the coordinated lone pairs of electrons. The orbitals used are the 4*s*, three 4*p* and two 4*d*. These are hybridized to give a set of six equivalent sp^3d^2 hybrid orbitals. A ligand orbital containing a lone pair of electrons forms a coordinate bond by overlapping with an empty hybrid orbital on the metal ion. In this way a σ bond is formed with each ligand. The *d* orbitals used are the $4d_{x^2-y^2}$ and $4d_{z^2}$. In the diagrams below, electron pairs from the ligands are shown as ↑↓.

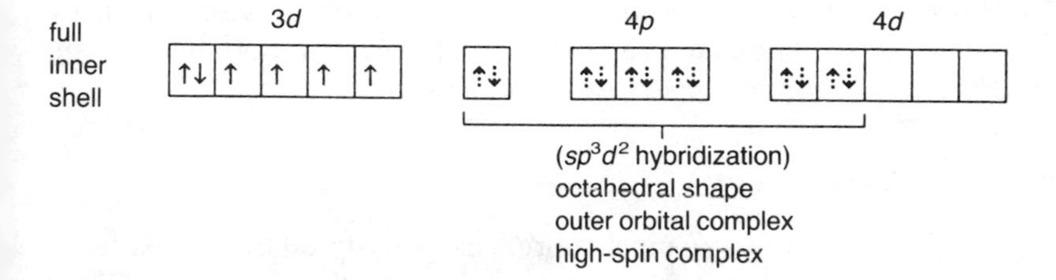

Since the outer 4*d* orbitals are used for bonding this is called an *outer orbital complex*. The energy of these orbitals is quite high, so that the complex will be reactive or labile. The magnetic moment depends on the number of unpaired electrons. The 3*d* level contains the maximum number of unpaired electrons for a d^6 arrangement, so this is sometimes called a *high-spin* or a *spin-free complex*. An alternative octahedral arrangement is possible when the electrons on the metal ion are rearranged as shown below. As before, lone pairs from the ligands are shown as ↑↓.

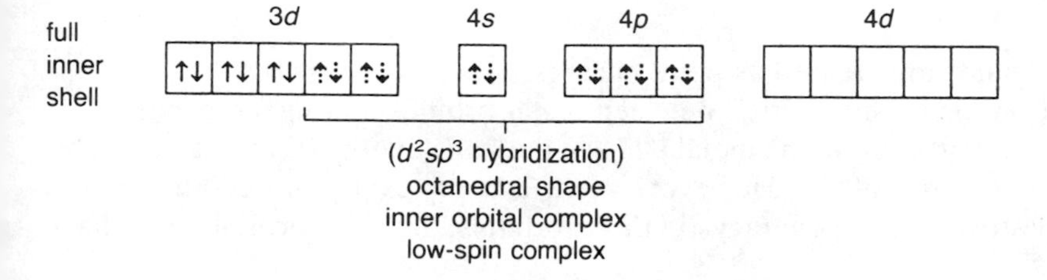

Since low energy inner *d* orbitals are used this is called an *inner orbital complex*. Such complexes are more stable than the outer orbital complexes. The unpaired electrons in the metal ion have been forced to pair up, and so this is now a low-spin complex. In this particular case all the electrons are paired, so the complex will be diamagnetic.

The metal ion could also form four-coordinate complexes, and two different arrangements are possible. *It must be remembered that hybrid orbitals do not actually exist*. Hybridization is a mathematical manipulation of the wave equations for the atomic orbitals involved.

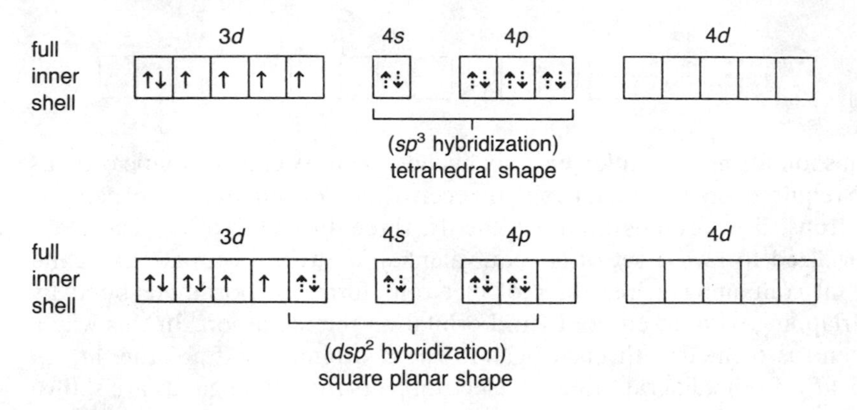

The theory does not explain the colour and spectra of complexes. The theory shows the number of unpaired electrons. From this the magnetic moment can be calculated (see Chapter 18). However, it does not explain why the magnetic moment varies with temperature.

CRYSTAL FIELD THEORY

The crystal field theory is now much more widely accepted than the valence bond theory. It assumes that the attraction between the central metal and the ligands in a complex is purely electrostatic. The transition metal which forms the central atom in the complex is regarded as a positive ion of charge equal to the oxidation state. This is surrounded by negative ligands or neutral molecules which have a lone pair of electrons. If the ligand is a neutral molecule such as NH_3, the negative end of the dipole in the molecule is directed towards the metal ion. The electrons on the central metal are under repulsive forces from those on the ligands. Thus the electrons occupy the *d* orbitals furthest away from the direction of approach of ligands. In the crystal field theory the following assumptions are made.

1. Ligands are treated as point charges.
2. There is no interaction between metal orbitals and ligand orbitals.
3. The *d* orbitals on the metal all have the same energy (that is degenerate) in the free atom. However, when a complex is formed the ligands destroy the degeneracy of these orbitals, i.e. the orbitals now have

different energies. In an isolated gaseous metal ion, the five d orbitals do all have the same energy, and are termed degenerate. If a spherically symmetrical field of negative charges surrounds the metal ion, the d orbitals remain degenerate. However, the energy of the orbitals is raised because of repulsion between the field and the electrons on the metal. In most transition metal complexes, either six or four ligands surround the metal, giving octahedral or tetrahedral structures. In both of these cases the field produced by the ligands is not spherically symmetrical. Thus the d orbitals are not all affected equally by the ligand field.

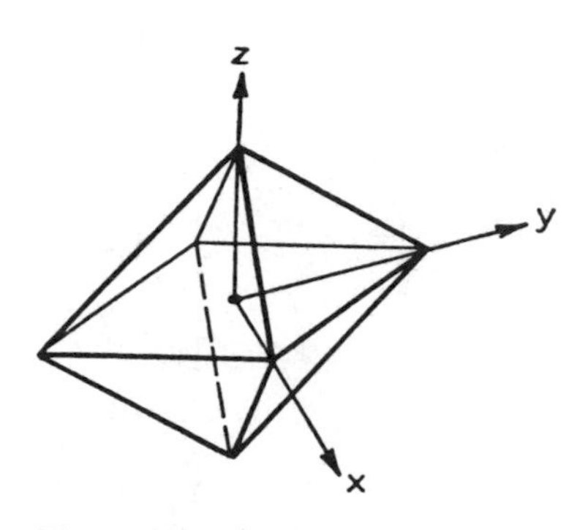

Figure 7.7 The directions in an octahedral complex.

Octahedral complexes

In an octahedral complex, the metal is at the centre of the octahedron, and the ligands are at the six corners. The directions x, y and z point to three adjacent corners of the octahedron as shown in Figure 7.7.

The lobes of the e_g orbitals ($d_{x^2-y^2}$ and d_{z^2}) point along the axes x, y and z. The lobes of the t_{2g} orbitals (d_{xy}, d_{xz} and d_{yz}) point in between the axes. It follows that the approach of six ligands along the x, y, z, $-x$, $-y$ and $-z$ directions will increase the energy of the $d_{x^2-y^2}$ and d_{z^2} orbitals (which point along the axes) much more than it increases the energy of the d_{xy}, d_{xz} and d_{yz} orbitals (which point between the axes). Thus under the influence of an octahedral ligand field the d orbitals split into two groups of different energies (Figure 7.8).

Rather than referring to the energy level of an isolated metal atom, the weighted mean of these two sets of perturbed orbitals is taken as the zero: this is sometimes called the Bari centre. The difference in energy between the two d levels is given either of the symbols Δ_o or 10 Dq. It follows that

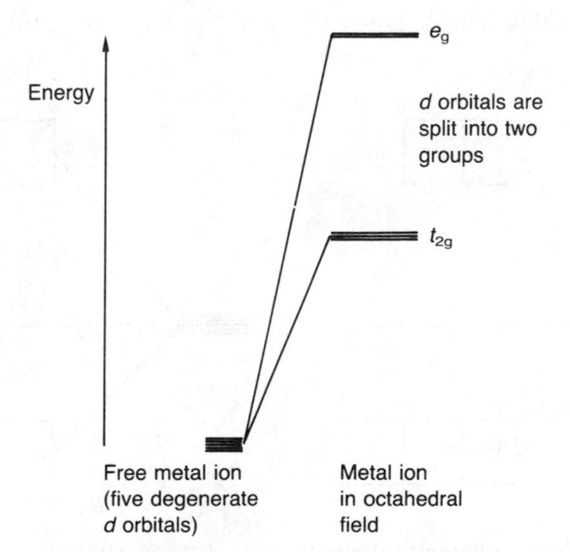

Figure 7.8 Crystal field splitting of energy levels in an octahedral field.

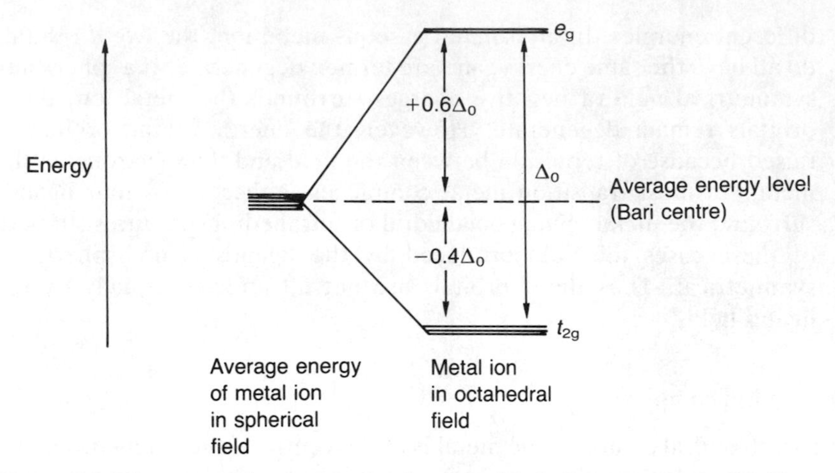

Figure 7.9 Diagram of the energy levels of *d* orbitals in an octahedral field.

the e_g orbitals are $+0.6\Delta_o$ above the average level, and the t_{2g} orbitals are $-0.4\Delta_o$ below the average (Figure 7.9).

The size of the energy gap Δ_o between the t_{2g} and e_g levels can be measured easily by recording the UV–visible spectrum of the complex. Consider a complex like $[Ti(H_2O)_6]^{3+}$. The Ti^{3+} ion has one *d* electron. In the complex this will occupy the orbital with the lowest energy, that is one of the t_{2g} orbitals (Figure 7.10a). The complex absorbs light of the correct wavelength (energy) to promote the electron from the t_{2g} level to the e_g level (Figure 7.10b).

The electronic spectrum for $[Ti(H_2O)]^{3+}$ is given in Figure 7.11. The steep part of the curve from 27 000 to 30 000 cm^{-1} (in the UV region) is due to charge transfer. The *d*–*d* transition is the single broad peak with a maximum at 20 300 cm^{-1}. Since 1 kJ mol^{-1} = 83.7 cm^{-1}, the value of Δ_o

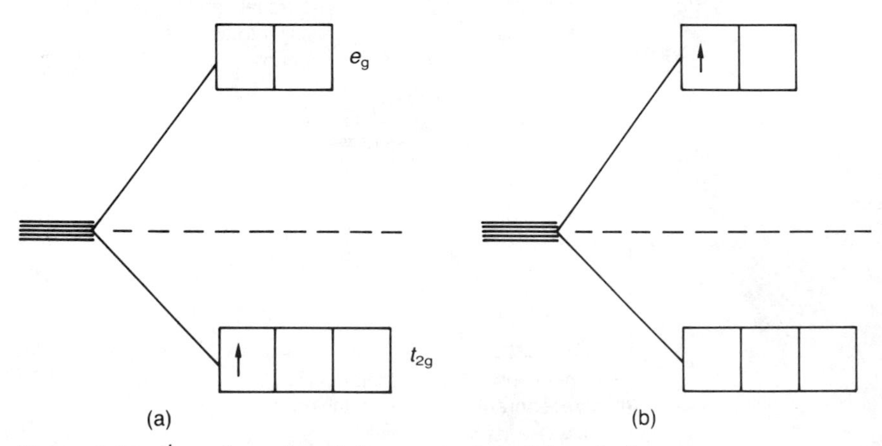

Figure 7.10 d^1 configuration: (a) ground state, (b) excited state.

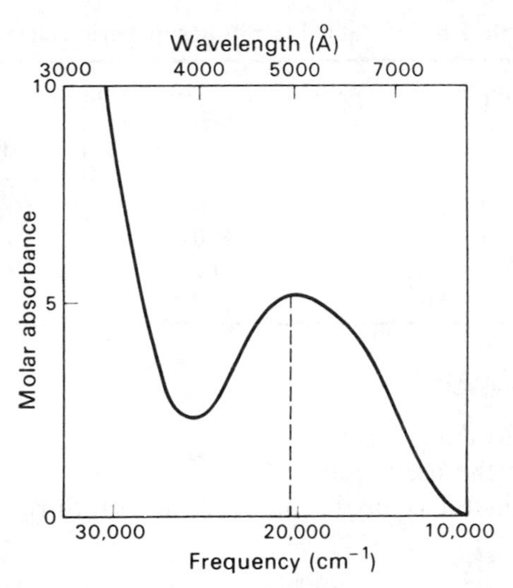

Figure 7.11 Ultraviolet and visible absorption spectrum of $[Ti(H_2O)_6]^{3+}$.

for $[Ti(H_2O)]^{3+}$ is 20 300/83.7 = 243 kJ mol^{-1}. This is much the same as the energy of many normal single bonds (see Appendix F).

The above method is the most convenient way of measuring Δ_o values. However, Δ_o values can also be obtained from values of observed lattice energies and those calculated using the Born–Landé equation (see Chapter 3).

Solutions containing the hydrated Ti^{3+} ion are reddish violet coloured. This is because yellow and green light are absorbed to excite the electron. Thus the transmitted light is the complementary colour red–violet (Table 7.7).

Because of the crystal field splitting of *d* orbitals, the single *d* electron in $[Ti(H_2O)]^{3+}$ occupies an energy level $2/5\Delta_o$ below the average energy of the *d* orbitals. As a result the complex is more stable. The crystal field stabilization energy (CFSE) is in this case $2/5 \times 243 = 97$ kJ mol^{-1}.

The magnitude of Δ_o depends on three factors:

Table 7.7 Colours absorbed and colours observed

Colour absorbed	Colour observed	Wavenumber observed (cm^{-1})
yellow–green	red–violet	24 000–26 000
yellow	indigo	23 000–24 000
orange	blue	21 000–23 000
red	blue–green	20 000–21 000
purple	green	18 000–20 000
red–violet	yellow–green	17 300–18 000
indigo	yellow	16 400–17 300
blue	orange	15 300–16 400
blue–green	red	12 800–15 300

Table 7.8 Crystal field splittings by various ligands

Complex	Absorption peak	
	(cm^{-1})	($kJ\,mol^{-1}$)
$[Cr^{III}Cl_6]^{3-}$	13 640	163
$[Cr^{III}(H_2O)_6]^{3+}$	17 830	213
$[Cr^{III}(NH_3)_6]^{3+}$	21 680	259
$[Cr^{III}(CN)_6]^{3-}$	26 280	314

1. The nature of the ligands.
2. The charge on the metal ion.
3. Whether the metal is in the first, second or third row of transition elements.

Examination of the spectra of a series of complexes of the same metal with different ligands shows that the position of the absorption band (and hence the value of Δ_o) varies depending on the ligands which are attached (Table 7.8).

Ligands which cause only a small degree of crystal field splitting are termed weak field ligands. Ligands which cause a large splitting are called strong field ligands. Most Δ values are in the range $7000\,cm^{-1}$ to $30\,000\,cm^{-1}$. The common ligands can be arranged in ascending order of crystal field splitting Δ. The order remains practically constant for different metals, and this series is called the spectrochemical series (see Further Reading Tsuchida, 1938; Jørgensen, 1962).

Spectrochemical series

weak field ligands

$I^- < Br^- < S^{2-} < Cl^- < NO_3^- < F^- < OH^- <$ EtOH $<$ oxalate $< H_2O$
$<$ EDTA $<$ (NH_3 and pyridine) $<$ ethylenediamine $<$ dipyridyl
$<$ o-phenanthroline $< NO_2^- < CN^- <$ CO

strong field ligands

The spectrochemical series is an experimentally determined series. It is difficult to explain the order as it incorporates both the effects of σ and π bonding. The halides are in the order expected from electrostatic effects. In other cases we must consider covalent bonding to explain the order. A pattern of increasing σ donation is followed:

halide donors $<$ O donors $<$ N donors $<$ C donors

The crystal field splitting produced by the strong field CN^- ligand is about double that for weak field ligands like the halide ions. This is attributed to π bonding in which the metal donates electrons from a filled t_{2g} orbital into a vacant orbital on the ligand. In a similar way, many unsaturated N donors and C donors may also act as π acceptors.

The magnitude of Δ_o increases as the charge on the metal ion increases.

Table 7.9 Crystal field splittings for hexa-aqua complexes of M^{2+} and M^{3+}

Oxidation state		Ti	V	Cr	Mn	Fe	Co	Ni	Cu
(+II)	Electronic configuration	d^2	d^3	d^4	d^5	d^6	d^7	d^8	d^9
	Δ_o in cm^{-1}	–	12 600	13 900	7 800	10 400	9 300	8500	12 600
	Δ_o in kJ mol^{-1}	–	151	(166)	93	124	111	102	(151)
(+III)	Electronic configuration	d^1	d^2	d^3	d^4	d^5	d^6	d^7	d^8
	Δ_o in cm^{-1}	20 300	18 900	17 830	21 000	13 700	18 600	–	–
	Δ_o in kJ mol^{-1}	243	226	213	(251)	164	222	–	–

Values for d^4 and d^9 are approximate because of tetragonal distortion.

Table 7.10 Δ_o crystal field splittings in one group

	cm^{-1}	kJ mol^{-1}
$[Co(NH_3)_6]^{3+}$	24 800	296
$[Rh(NH_3)_6]^{3+}$	34 000	406
$[Ir(NH_3)_6]^{3+}$	41 000	490

For first row transition metal ions, the values of Δ_o for M^{3+} complexes are roughly 50% larger than the values for M^{2+} complexes (Table 7.9).

The value of Δ_o also increases by about 30% between adjacent members down a group of transition elements (Table 7.10). The crystal field stabilization energy in $[Ti(H_2O)_6]^{3+}$, which has a d^1 configuration, has previously been shown to be $-0.4\Delta_o$. In a similar way, complexes containing a metal ion with a d^2 configuration will have a CFSE of $2 \times -0.4\Delta_o = -0.8\Delta_o$ by singly filling two of the t_{2g} orbitals. (This is in agreement with Hund's rule that the arrangement with the maximum number of unpaired electrons is the most stable.) Complexes of d^3 metal ions have a CFSE of $3 \times -0.4\Delta_o = -1.2\Delta_o$.

Complexes with a metal ion with a d^4 configuration would be expected to have an electronic arrangement in accordance with Hund's rule (Figure 7.12a) with four unpaired electrons, and the CFSE will be $(3 \times -0.4\Delta_o) + (0.6\Delta_o) = -0.6\Delta_o$. An alternative arrangement of electrons which does not comply with Hund's rule is shown in Figure 7.12b. This arrangement has two unpaired electrons, and the CFSE is $(4 \times -0.4\Delta_o) = -1.6\Delta_o$. The CFSE is larger than in the previous case. However, the energy P used to pair the electrons must be allowed for, so the total stabilization energy is $-1.6\Delta_o + P$. These two arrangements differ in the number of unpaired electrons. The one with the most unpaired electrons is called 'high-spin' or 'spin-free', and the other one the 'low-spin' or 'spin-paired' arrangement. Both arrangements have been found to exist. Which arrangement occurs for any particular complex depends on whether the energy to promote an

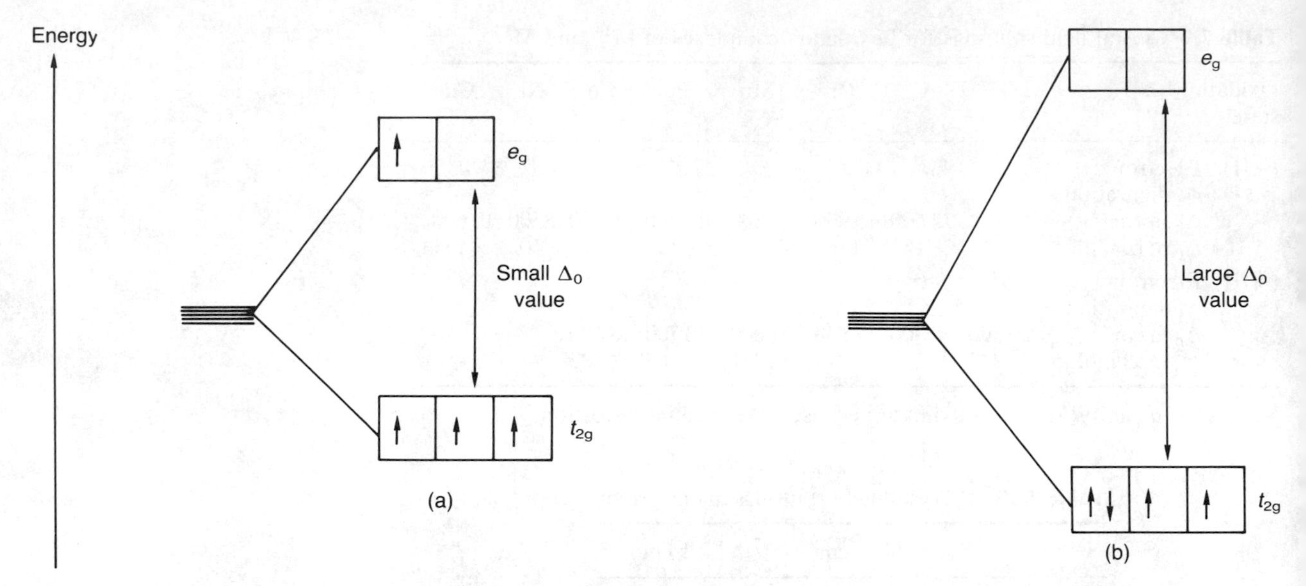

Figure 7.12 High- and low-spin complexes: (a) d^4 high-spin arrangement (weak ligand field); (b) d^4 low-spin arrangement (strong ligand field).

Table 7.11 CFSE and pairing energy for some complexes

Complex	Configuration	Δ_o (cm^{-1})	P (cm^{-1})	Predicted	Found
$[Fe^{II}(H_2O)_6]^{2+}$	d^6	10 400	17 600	high spin	high spin
$[Fe^{II}(CN)_6]^{4-}$	d^6	32 850	17 600	low spin	low spin
$[Co^{III}F_6]^{3-}$	d^7	13 000	21 000	high spin	high spin
$[Co^{III}(NH_3)_6]^{3+}$	d^7	23 000	21 000	low spin	low spin

electron to the upper e_g level (that is the crystal field splitting Δ_o) is greater than the energy to pair electrons (that is P) in the lower t_{2t} level. For a given metal ion P is constant. Thus the amount of crystal field splitting is determined by the strength of the ligand field. A weak field ligand such as Cl^- will only cause a small splitting of energy levels Δ_o. Thus it will be more favourable energetically for electrons to occupy the upper e_g level and have a high-spin complex, rather than to pair electrons. In a similar way, strong field ligands such as CN^- cause a large splitting Δ_o. In this case it requires less energy to pair the electrons and form a low-spin complex.

Similar arguments apply to high- and low-spin complexes of metal ions with d^5, d^6 and d^7 configurations. These are summarized in Table 7.12.

EFFECTS OF CRYSTAL FIELD SPLITTING

In octahedral complexes, the filling of t_{2g} orbitals decreases the energy of a complex, that is makes it more stable by $-0.4\Delta_o$ per electron. Filling e_g

Table 7.12 CFSE and electronic arrangements in octahedral complexes

Number of d electrons	Arrangement in weak ligand field				Arrangement in strong ligand field			
	t_{2g}	e_g	CFSE Δ_o	Spin only magnetic moment μ_s(D)	t_{2g}	e_g	CFSE Δ_o	Spin only magnetic moment μ_s(D)
d^1	[↑][ ][ ]	[ ][ ]	−0.4	1.73	[↑][ ][ ]	[ ][ ]	−0.4	1.73
d^2	[↑][↑][ ]	[ ][ ]	−0.8	2.83	[↑][↑][ ]	[ ][ ]	−0.8	2.83
d^3	[↑][↑][↑]	[ ][ ]	−1.2	3.87	[↑][↑][↑]	[ ][ ]	−1.2	3.87
d^4	[↑][↑][↑]	[↑][ ]	−1.2 +0.6 = −0.6	4.90	[↑↓][↑][↑]	[ ][ ]	−1.6	2.83
d^5	[↑][↑][↑]	[↑][↑]	−1.2 +1.2 = −0.0	5.92	[↑↓][↑↓][↑]	[ ][ ]	−2.0	1.73
d^6	[↑↓][↑][↑]	[↑][↑]	−1.6 +1.2 = −0.4	4.90	[↑↓][↑↓][↑↓]	[ ][ ]	−2.4	0.00
d^7	[↑↓][↑↓][↑]	[↑][↑]	−2.0 +1.2 = −0.8	3.87	[↑↓][↑↓][↑↓]	[↑][ ]	−2.4 +0.6 = −1.8	1.73
d^8	[↑↓][↑↓][↑↓]	[↑][↑]	−2.4 +1.2 = −1.2	2.83	[↑↓][↑↓][↑↓]	[↑][↑]	−2.4 +1.2 = −1.2	2.83
d^9	[↑↓][↑↓][↑↓]	[↑↓][↑]	−2.4 +1.8 = −0.6	1.73	[↑↓][↑↓][↑↓]	[↑↓][↑]	−2.4 +1.8 = −0.6	1.73
d^{10}	[↑↓][↑↓][↑↓]	[↑↓][↑↓]	−2.4 +2.4 = 0.0	0.00	[↑↓][↑↓][↑↓]	[↑↓][↑↓]	−2.4 +2.4 = 0.0	0.00

orbitals increases the energy by $+0.6\Delta_o$ per electron. The total crystal field stabilization energy is given by

$$\text{CFSE}_{(\text{octahedral})} = -0.4n_{(t_{2g})} + 0.6n_{(e_g)}$$

where $n_{(t_{2g})}$ and $n_{(e_g)}$ are the number of electrons occupying the t_{2g} and e_g orbitals respectively. The CFSE is zero for ions with d^0 and d^{10} configurations in both strong and weak ligand fields. The CFSE is also zero for d^5 configurations in a weak field. All the other arrangements have some

Table 7.13 Measured and calculated lattice energies

Compound	Structure	Measured lattice energy ($kJ\,mol^{-1}$)	Calculated lattice energy ($kJ\,mol^{-1}$)	Difference (measured − calculated) ($kJ\,mol^{-1}$)
NaCl	Sodium chloride	−764	−764	0
AgCl	Sodium chloride	−916	−784	−132
AgBr	Sodium chloride	−908	−759	−149
MgF_2	Rutile	−2908	−2915	+7
MnF_2	Rutile	−2770	−2746	−24
FeF_2	Rutile	−2912	−2752	−160
NiF_2	Rutile	−3046	−2917	−129
CuF_2	Rutile	−3042	−2885	−157

CFSE, which increases the thermodynamic stability of the complexes. Thus many transition metal compounds have a higher measured lattice energy (obtained by calculations using the terms in the Born–Haber cycle) than is calculated using the Born–Landé, Born–Meyer or Kapustinskii equations. In contrast, the measured (Born–Haber) and calculated values for compounds of the main groups (which have no CFSE) are in close agreement (Table 7.13). There is also close agreement in MnF_2 which has a d^5 configuration and a weak field ligand: hence there is no CFSE.

A plot of the lattice energies of the halides of the first row transition elements in the divalent state is given in Figure 7.13. In the solid, the coordination number of these metals is 6, and so the structures are analogous to octahedral complexes. The graphs for each halide show a minimum at Mn^{2+}, which has a d^5 configuration. In a weak field this has a high-spin arrangement with zero CFSE. The configurations d^0 and d^{10} also have zero CFSE. The broken line through Ca^{2+}, Mn^{2+} and Zn^{2+} represents zero stabilization. The heights of other points above this line are the crystal field stabilization energies.

The hydration energies of the M^{2+} ions of the first row transition elements are plotted in Figure 7.14a.

$$M^{2+}_{(g)} + \text{excess } H_2O \rightarrow [M(H_2O)_6]^{2+}$$

The ions Ca^{2+}, Mn^{2+} and Zn^{2+} have d^0, d^5 and d^{10} configurations, and have zero CFSE. An almost straight line can be drawn through these points. The distance of the other points above this line corresponds to the CFSE. Values obtained in this way agree with those obtained spectroscopically. A similar graph of the M^{3+} ions is shown in Figure 7.14b: here the d^0, d^5 and d^{10} species are Sc^{3+}, Fe^{3+} and Ga^{3+}.

The ionic radii for M^{2+} ions might be expected to decrease smoothly from Ca^{2+} to Zn^{2+} because of the increasing nuclear charge, and the poor shielding by d electrons. A plot of these radii is given in Figure 7.15. The change in size is not regular.

A smooth (broken) line is drawn through Ca^{2+}, Mn^{2+} and Zn^{2+}. These

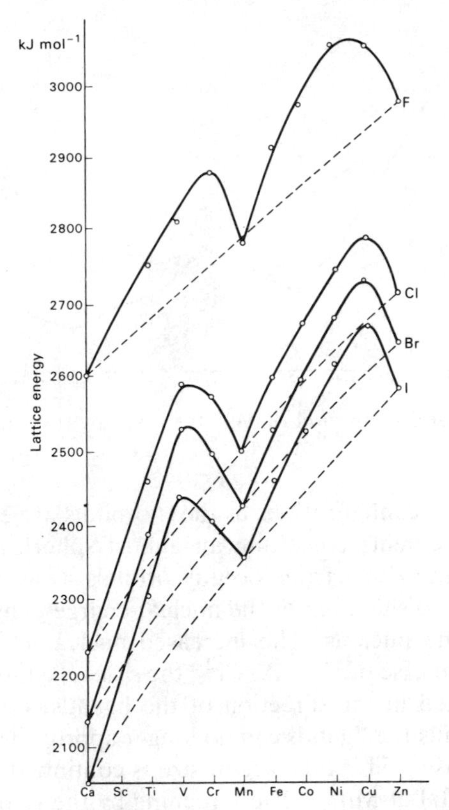

Figure 7.13 CFSE of dihalides of the first transition series. (After T.C. Waddington, Lattice energies and their significance in inorganic chemistry, *Advances in Inorganic Chemistry and Radiochemistry*, **1**, Academic Press, New York, 1959.)

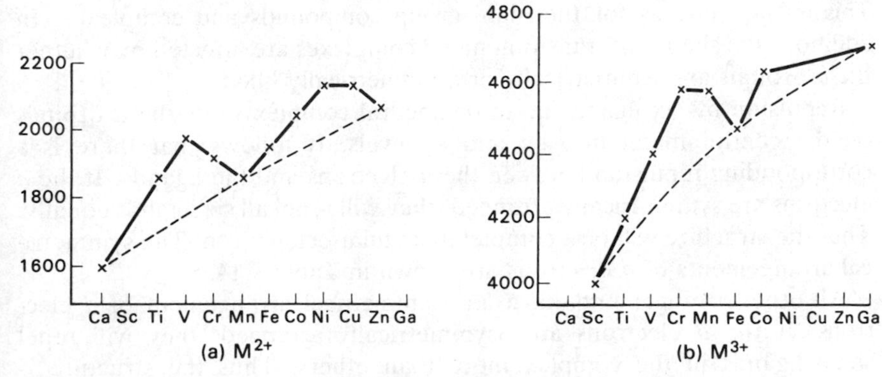

Figure 7.14 Enthalpies of hydration for M^{2+} and M^{3+}, in kJ mol^{-1}.

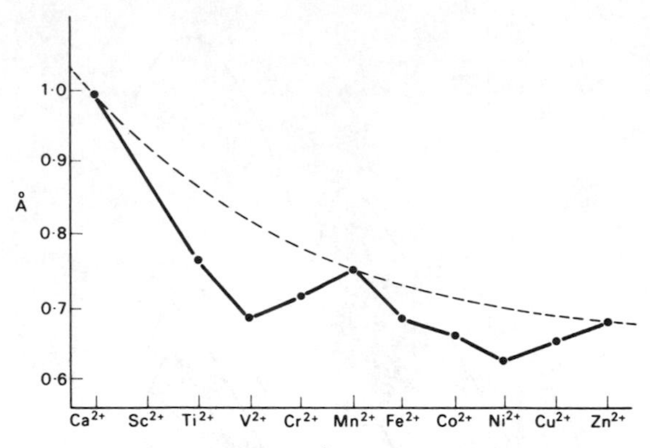

Figure 7.15 Octahedral ionic radii of M^{2+} for first row transition elements.

have d^0, d^5 and d^{10} configurations as the d orbitals are empty, half full or full. These arrangements constitute an almost spherical field round the nucleus. In Ti^{2+} the d electrons occupy orbitals away from the ligands, providing little or no shielding of the nuclear charge. Thus the ligands are drawn closer to the nucleus. The increased nuclear charge has an even greater effect in the case of V^{2+}. At Cr^{2+} the e_g level contains one electron. This is concentrated in the direction of the ligands, thus providing very good shielding. Thus the ligands can no longer approach so closely and the ionic radius increases. This increase in size is continued with the filling of the second e_g orbital at Mn^{2+}. The screening by the e_g orbitals is so good that the radius of Mn^{2+} is slightly smaller than it would be if it were in a truly spherical field. The same sequence of size changes is repeated in the second half of the series.

TETRAGONAL DISTORTION OF OCTAHEDRAL COMPLEXES (JAHN–TELLER DISTORTION)

The shape of transition metal complexes is determined by the tendency of electron pairs to occupy positions as far away from each other as possible. This is the same as for the main group compounds and complexes. In addition, the shapes of transition metal complexes are affected by whether the d orbitals are symmetrically or asymmetrically filled.

Repulsion by six ligands in an octahedral complex splits the d orbitals on the central metal into t_{2g} and e_g levels. It follows that there is a corresponding repulsion between the d electrons and the ligands. If the d electrons are symmetrically arranged, they will repel all six ligands equally. Thus the structure will be a completely regular octahedron. The symmetrical arrangements of d electrons are shown in Table 7.14.

All other arrangements have an asymmetrical arrangement of d electrons. If the d electrons are asymmetrically arranged, they will repel some ligands in the complex more than others. Thus the structure is distorted because some ligands are prevented from approaching the metal

Table 7.14 Symmetrical electronic arrangements

Electronic configuration	t_{2g}	e_g	Nature of ligand field	Examples
d^0	[ ][ ][ ]	[ ][ ]	Strong or weak	$Ti^{IV}O_2$, $[Ti^{IV}F_6]^{2-}$ $[Ti^{IV}Cl_6]^{2-}$
d^3	[↑][↑][↑]	[ ][ ]	Strong or weak	$[Cr^{III}(oxalate)_3]^{3-}$ $[Cr^{III}(H_2O)_6]^{3+}$
d^5	[↑][↑][↑]	[↑][↑]	Weak	$[Mn^{II}F_6]^{4-}$ $[Fe^{III}F_6]^{3-}$
d^6	[↑↓][↑↓][↑↓]	[ ][ ]	Strong	$[Fe^{II}(CN)_6]^{4-}$ $[Co^{III}(NH_3)_6]^{3+}$
d^8	[↑↓][↑↓][↑↓]	[↑][↑]	Weak	$[Ni^{II}F_6]^{4-}$ $[Ni^{II}(H_2O)_6]^{2+}$
d^{10}	[↑↓][↑↓][↑↓]	[↑↓][↑↓]	Strong or weak	$[Zn^{II}(NH_6)_6]^{2+}$ $[Zn^{II}(H_2O)_6]^{2+}$

as closely as others. The e_g orbitals point directly at the ligands. Thus asymmetric filling of the e_g orbitals results in some ligands being repelled more than others. This causes a significant distortion of the octahedral shape. In contrast the t_{2g} orbitals do not point directly at the ligands, but point in between the ligand directions. Thus asymmetric filling of the t_{2g} orbitals has only a very small effect on the stereochemistry. Distortion caused by asymmetric filling of the t_{2g} orbitals is usually too small to measure. The electronic arrangements which will produce a large distortion are shown in Table 7.15.

The two e_g orbitals $d_{x^2-y^2}$ and d_{z^2} are normally degenerate. However, if they are asymmetrically filled then this degeneracy is destroyed, and the two orbitals are no longer equal in energy. If the d_{z^2} orbital contains one

Table 7.15 Asymmetrical electronic arrangements

Electronic configuration	t_{2g}	e_g	Nature of ligand field	Examples
d^4	[↑][↑][↑]	[↑][ ]	Weak field (high-spin complex)	Cr(+II), Mn(+III)
d^7	[↑↓][↑↓][↑↓]	[↑][ ]	Strong field (low-spin complex)	Co(+II), Ni(+III)
d^9	[↑↓][↑↓][↑↓]	[↑↓][↑]	Either strong or weak	Cu(+II)

more electron than the $d_{x^2-y^2}$ orbital then the ligands approaching along $+z$ and $-z$ will encounter greater repulsion than the other four ligands. The repulsion and distortion result in elongation of the octahedron along the z axis. This is called tetragonal distortion. Strictly it should be called tetragonal elongation. This form of distortion is commonly observed.

If the $d_{x^2-y^2}$ orbital contains the extra electron, then elongation will occur along the x and y axes. This means that the ligands approach more closely along the z axis. Thus there will be four long bonds and two short bonds. This is equivalent to compressing the octahedron along the z axis, and is called tetragonal compression. Tetragonal elongation is much more common than tetragonal compression, and it is not possible to predict which will occur.

For example, the crystal structure of CrF_2 is a distorted rutile (TiO_2) structure. Cr^{2+} is octahedrally surrounded by six F^-, and there are four Cr—F bonds of length 1.98–2.01 Å, and two longer bonds of length 2.43 Å. The octahedron is said to be tetragonally distorted. The electronic arrangement in Cr^{2+} is d^4. F^- is a weak field ligand, and so the t_{2g} level contains three electrons and the e_g level contains one electron. The $d_{x^2-y^2}$ orbital has four lobes whilst the d_{z^2} orbital has only two lobes pointing at the ligands. To minimize repulsion with the ligands, the single e_g electron will occupy the d_{z^2} orbital. This is equivalent to splitting the degeneracy of the e_g level so that d_{z^2} is of lower energy, i.e. more stable, and $d_{x^2-y^2}$ is of higher energy, i.e. less stable. Thus the two ligands approaching along the $+z$ and $-z$ directions are subjected to greater repulsion than the four ligands along $+x$, $-x$, $+y$ and $-y$. This causes tetragonal distortion with four short bonds and two long bonds. In the same way MnF_3 contains Mn^{3+} with a d^4 configuration, and forms a tetragonally distorted octahedral structure.

Many Cu(+II) salts and complexes also show tetragonally distorted octahedral structures. Cu^{2+} has a d^9 configuration:

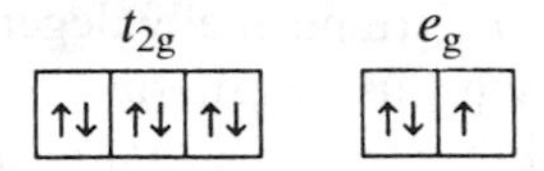

To minimize repulsion with the ligands, two electrons occupy the d_{z^2} orbital and one electron occupies the $d_{x^2-y^2}$ orbital. Thus the two ligands along $-z$ and $-z$ are repelled more strongly than are the other four ligands (see Chapter 27, under +II state for copper).

The examples above show that whenever the d_{z^2} and $d_{x^2-y^2}$ orbitals are unequally occupied, distortion occurs. This is known as Jahn–Teller distortion. The Jahn–Teller theorem states that 'Any non-linear molecular system in a degenerate electronic state will be unstable, and will undergo some sort of distortion to lower its symmetry and remove the degeneracy'. More simply, molecules or complexes (of any shape except linear), which have an unequally filled set of orbitals (either t_{2g} or e_g), will be distorted. In octahedral complexes distortions from the t_{2g} level are too small to be detected. However, distortions resulting from uneven filling of the e_g orbitals are very important.

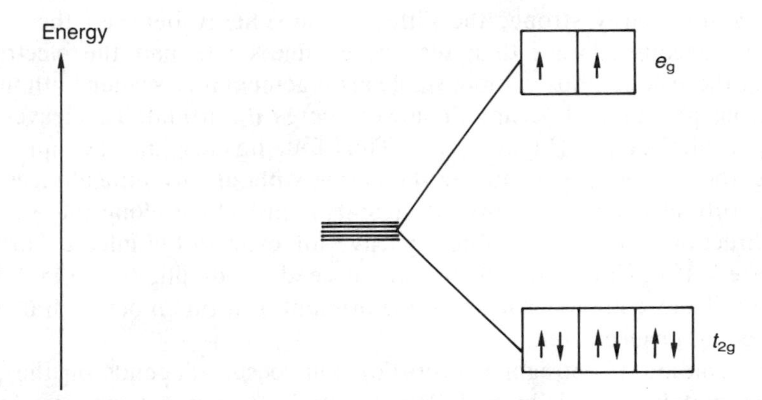

Figure 7.16 d^8 arrangement in weak octahedral field.

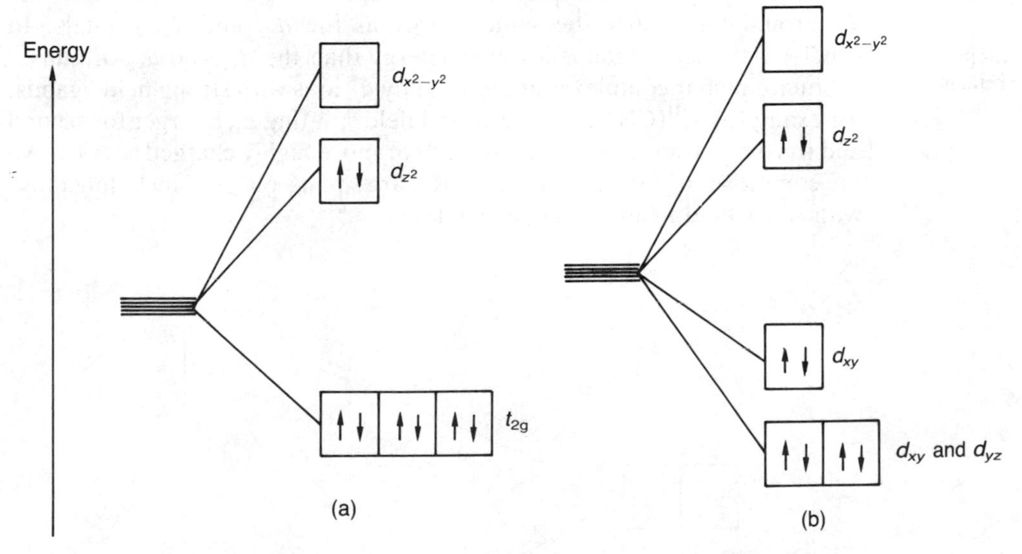

Figure 7.17 d^8 arrangement in very strong octahedral field. Tetragonal distortion splits (a) the e_g level; and (b) also splits the t_{2g} level. The d_{xy} orbital is higher in energy than the d_{xz} or d_{yz}. (For simplicity this is sometimes ignored.)

SQUARE PLANAR ARRANGEMENTS

If the central metal ion in a complex has a d^8 configuration, six electrons will occupy the t_{2g} orbitals and two electrons will occupy the e_g orbitals. The arrangement is the same in a complex with weak field ligands. The electrons are arranged as shown in Figure 7.16. The orbitals are symmetrically filled, and a regular octahedral complex is formed, for example by $[Ni^{II}(H_2O)_6]^{2+}$ and $[Ni^{II}(NH_3)_6]^{2+}$.

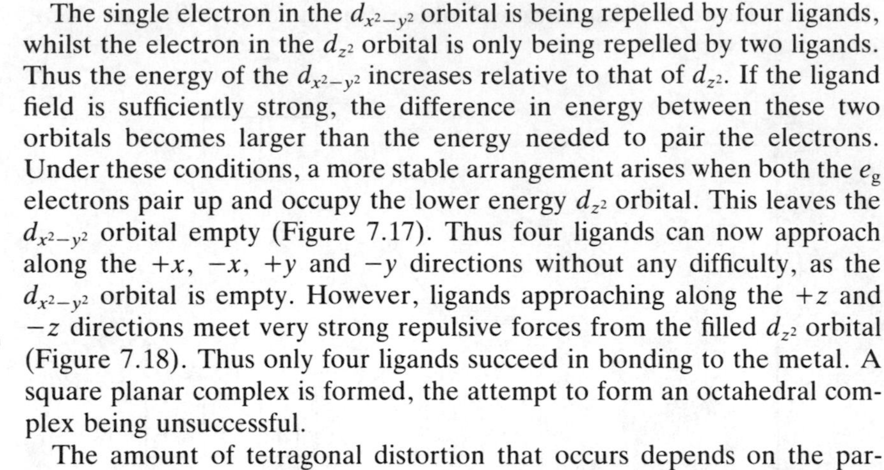

The single electron in the $d_{x^2-y^2}$ orbital is being repelled by four ligands, whilst the electron in the d_{z^2} orbital is only being repelled by two ligands. Thus the energy of the $d_{x^2-y^2}$ increases relative to that of d_{z^2}. If the ligand field is sufficiently strong, the difference in energy between these two orbitals becomes larger than the energy needed to pair the electrons. Under these conditions, a more stable arrangement arises when both the e_g electrons pair up and occupy the lower energy d_{z^2} orbital. This leaves the $d_{x^2-y^2}$ orbital empty (Figure 7.17). Thus four ligands can now approach along the $+x$, $-x$, $+y$ and $-y$ directions without any difficulty, as the $d_{x^2-y^2}$ orbital is empty. However, ligands approaching along the $+z$ and $-z$ directions meet very strong repulsive forces from the filled d_{z^2} orbital (Figure 7.18). Thus only four ligands succeed in bonding to the metal. A square planar complex is formed, the attempt to form an octahedral complex being unsuccessful.

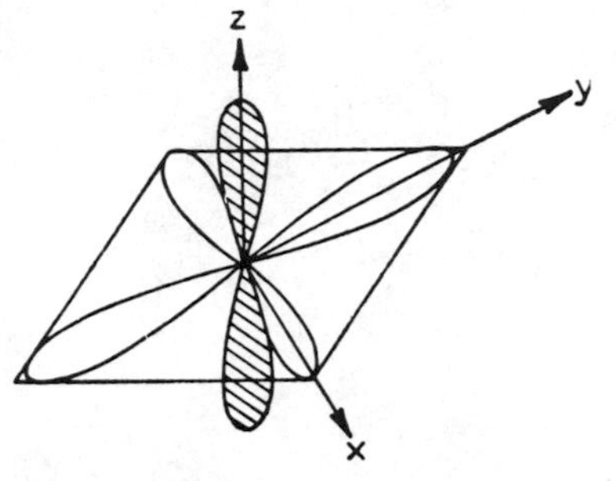

Figure 7.18 d^8 arrangement, strong field. (The d_{z^2} orbital is full, the $d_{x^2-y^2}$ empty.)

The amount of tetragonal distortion that occurs depends on the particular metal ion and ligands. Sometimes the tetragonal distortion may become so large that the d_{z^2} orbital is lower in energy than the d_{xy} orbital as shown in Figure 7.19. In square planar complexes of Co^{II}, Ni^{II} and Cu^{II} the d_{z^2} orbital has nearly the same energy as the d_{xz} and d_{yz} orbitals. In $[PtCl_4]^{2-}$ the d_{z^2} orbital is lower in energy than the d_{xz} and d_{yz} orbitals.

Square planar complexes are formed by d^8 ions with strong field ligands, for example $[Ni^{II}(CN)_4]^{2-}$. The crystal field splitting Δ_o is larger for second and third row transition elements, and for more highly charged species. All the complexes of Pt(+II) and Au(+III) are square planar – including those with weak field ligands such as halide ions.

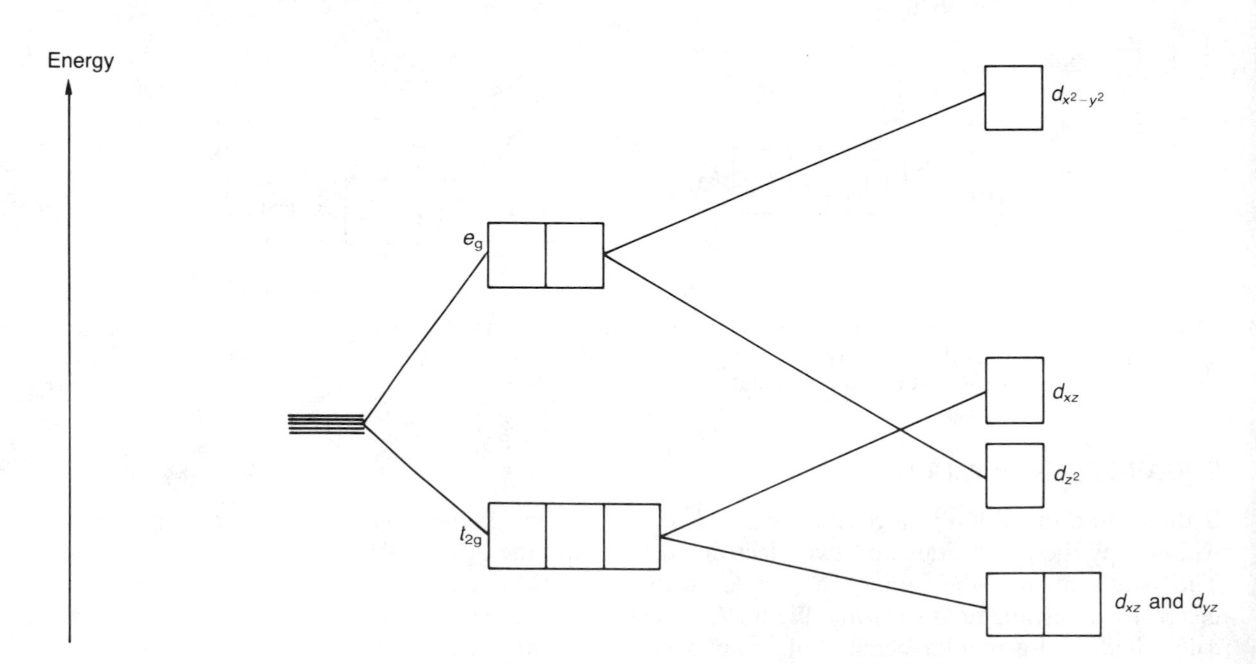

Figure 7.19 Tetragonal distortion.

Table 7.16 Ions that form square planar complexes

Electronic configuration	Ions	Type of field	Number of unpaired electrons
d^4	Cr(+II)	Weak	4
d^6	Fe(+II)	(Haem)	2
d^7	Co(+II)	Strong	1
d^8	Ni(+II), Rh(+I), Ir(+I)	Strong	0
	Pd(+II), Pt(+II), Au(+III)	Strong and weak	0
d^9	Cu(+II), Ag(+II)	Strong and weak	1

Square planar structures can also arise from d^4 ions in a weak ligand field. In this case the d_{z^2} orbital only contains one electron.

TETRAHEDRAL COMPLEXES

A regular tetrahedron is related to a cube. One atom is at the centre of the cube, and four of the eight corners of the cube are occupied by ligands as shown in Figure 7.20.

The directions x, y and z point to the centres of the faces of the cube. The e_g orbitals point along x, y and z (that is to the centres of the faces). The t_{2g} orbitals point between x, y and z (that is towards the centres of the edges of the cube) (Figure 7.21).

The direction of approach of the ligands does not coincide exactly with

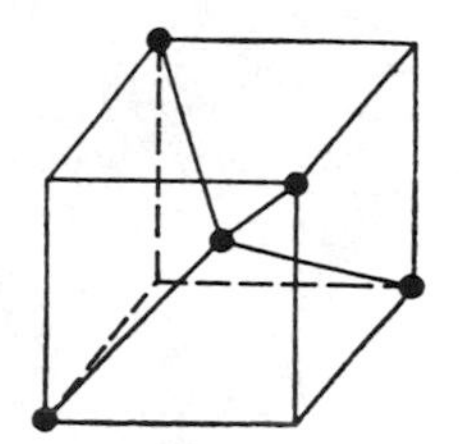

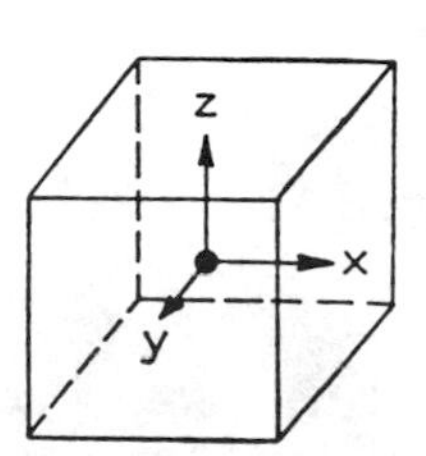

Figure 7.20 Relation of a tetrahedron to a cube.

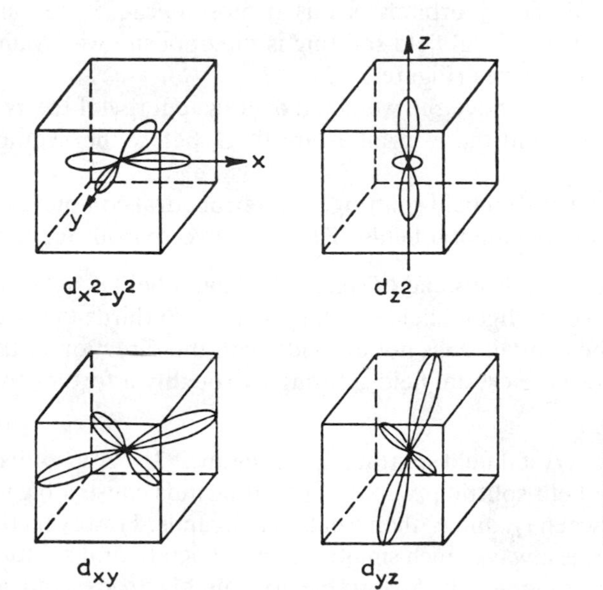

Figure 7.21 Orientation of d orbitals relative to a cube.

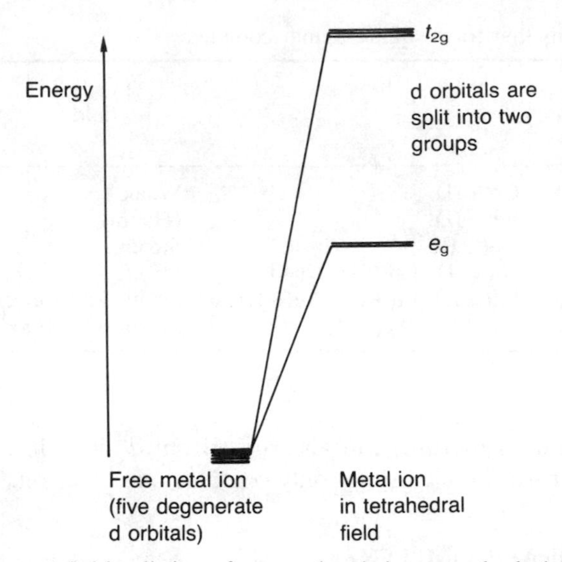

Figure 7.22 Crystal field splitting of energy levels in a tetrahedral field.

either the e_g or the t_{2g} orbitals. The angle between an e_g orbital, the central metal and the ligand is half the tetrahedral angle = 109°28′/2 = 54°44′. The angle between a t_{2g} orbital, the central metal and the ligand is 35°16′. Thus the t_{2g} orbitals are nearer to the direction of the ligands than the e_g orbitals. (Alternatively the t_{2g} orbitals are half the side of the cube away from the approach of the ligands, whilst the e_g orbitals are half the diagonal of the cube away.) The approach of the ligands raises the energy of both sets of orbitals. The energy of the t_{2g} orbitals is raised most because they are closest to the ligands. This crystal field splitting is the opposite way round to that in octahedral complexes (Figure 7.22).

The t_{2g} orbitals are $0.4\Delta_t$ above the weighted average energy of the two groups (the Bari centre) and the e_g orbitals are $0.6\Delta_t$ below the average (Figure 7.23).

The magnitude of the crystal field splitting Δ_t in tetrahedral complexes is considerably less than in octahedral fields. There are two reasons for this:

1. There are only four ligands instead of six, so the ligand field is only two thirds the size: hence the ligand field splitting is also two thirds the size.
2. The direction of the orbitals does not coincide with the direction of the ligands. This reduces the crystal field splitting by roughly a further two thirds.

Thus the tetrahedral crystal field splitting Δ_t is roughly $2/3 \times 2/3 = 4/9$ of the octahedral crystal field splitting Δ_o. Strong field ligands cause a bigger energy difference between t_{2g} and e_g than weak field ligands. However, the tetrahedral splitting Δ_t is always much smaller than the octahedral splitting Δ_o. Thus it is never energetically favourable to pair electrons, and all tetrahedral complexes are high-spin.

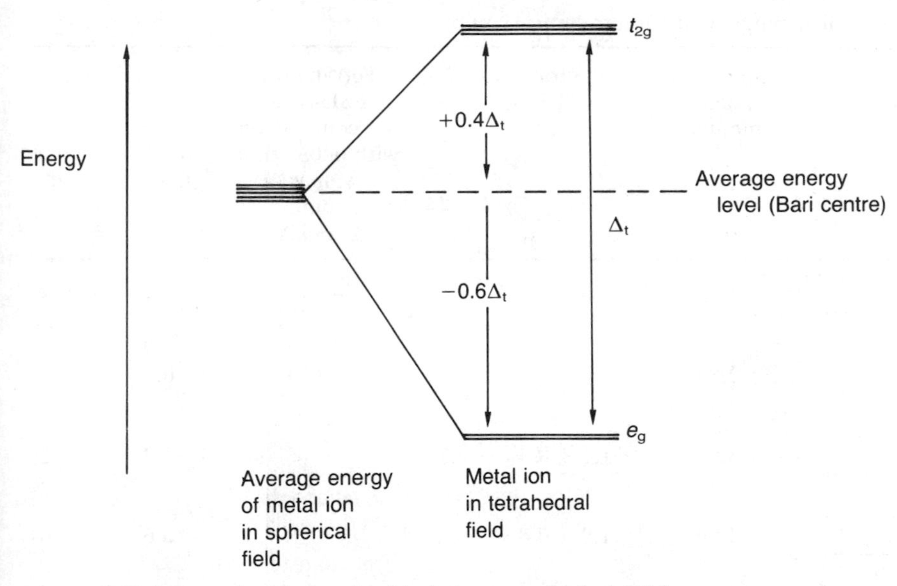

Figure 7.23 Energy levels for *d* orbitals in a tetrahedral field.

The CFSE in both octahedral and tetrahedral environments is given in Table 7.17. This shows that for d^0, d^5 and d^{10} arrangements the CFSE is zero in both octahedral and tetrahedral complexes. For all other electronic arrangements there is some CFSE, and the octahedral CFSE is greater than the tetrahedral CFSE. It follows that octahedral complexes are generally more stable and more common than tetrahedral complexes. This is partly because there are six bond energy terms rather than four, and partly because there is a larger CFSE term. Despite this some tetrahedral complexes are formed, and are stable. Tetrahedral complexes are favoured:

1. Where the ligands are large and bulky and could cause crowding in an octahedral complex.
2. Where attainment of a regular shape is important. For tetrahedral structures d^0, d^2, d^5, d^7 and d^{10} configurations are regular. Some tetrahedral complexes which are regular are: $Ti^{IV}Cl_4$ (e_g^0, t_{2g}^0), $[Mn^{VII}O_4]^-$ (e_g^0, t_{2g}^0), $[Fe^{VI}O_4]^{2-}$ (e_g^2, t_{2g}^0), $[Fe^{III}Cl_4]^-$ (e_g^2, t_{2g}^3), $[Co^{II}Cl_4]^{2-}$ (e_g^4, t_{2g}^3) and $[Zn^{II}Cl_4]^{2-}$ (e_g^4, t_{2g}^{10}).
3. When the ligands are weak field, and the loss in CFSE is thus less important.
4. Where the central metal has a low oxidation state. This reduces the magnitude of Δ.
5. Where the electronic configuration of the central metal is d^0, d^5 or d^{10} as there is no CFSE.
6. Where the loss of CFSE is small, e.g. d^1 and d^6 where the loss in CFSE is $0.13\Delta_o$ or d^2 and d^7 where the loss is $0.27\Delta_o$.

Many transition metal chlorides, bromides and iodides form tetrahedral structures.

Table 7.17 CFSE and electronic arrangements in tetrahedral complexes

Number of d electrons	Arrangement of electrons		Spin only magnetic moment	Tetrahedral CFSE	Tetrahedral CFSE scaled for comparison with octahedral values, assuming	Octahedral CFSE Δ_o	
	e_g	t_{2g}	μ(D)	Δ_t	$\Delta_t = \frac{4}{9}\Delta_o$	Weak field	Strong field
d^1	[↑][ ]	[ ][ ][ ]	1.73	−0.6	−0.27	−0.4	−0.4
d^2	[↑][↑]	[ ][ ][ ]	2.83	−1.2	−0.53	−0.8	−0.8
d^3	[↑][↑]	[↑][ ][ ]	3.87	−1.2 + 0.4 = −0.8	−0.36	−1.2	−1.2
d^4	[↑][↑]	[↑][↑][ ]	4.90	−1.2 + 0.8 = −0.4	−0.18	−0.6	−1.6
d_5	[↑][↑]	[↑][↑][↑]	5.92	−1.2 + 1.2 = 0.0	0.00	0.0	−2.0
d^6	[↑↓][↑]	[↑][↑][↑]	4.90	−1.8 + 1.2 = −0.6	−0.27	−0.4	−2.4
d_7	[↑↓][↑↓]	[↑][↑][↑]	3.87	−2.4 + 1.2 = −1.2	−0.53	−0.8	−1.8
d^8	[↑↓][↑↓]	[↑↓][↑][↑]	2.83	−2.4 + 1.6 = −0.8	−0.36	−1.2	−1.2
d^9	[↑↓][↑↓]	[↑↓][↑↓][↑]	1.73	−2.4 + 2.0 = −0.4	−0.18	−0.6	−0.6
d^{10}	[↑↓][↑↓]	[↑↓][↑↓][↑↓]	0.00	−2.4 + 2.4 = 0.0	0.00	0.0	0.0

CHELATES

Some of the factors that favour complex formation have already been mentioned:

1. Small highly charged ions with suitable vacant orbitals of the right energy.
2. The attainment of a noble gas structure (effective atomic number rule).
3. The attainment of a symmetrical shape and a high CFSE.

In some complexes a ligand occupies more than one coordination position. Thus more than one atom in the ligand is bonded to the central metal. For example, ethylenediamine forms a complex with copper ions:

$$\mathrm{Cu^{2+}} + 2\left|\begin{array}{l}\mathrm{CH_2 \cdot NH_2}\\ \\ \mathrm{CH_2 \cdot NH_2}\end{array}\right. \rightarrow \left[\begin{array}{ccccc}\mathrm{CH_2{-}NH_2} & & & & \mathrm{NH_2{-}CH_2}\\ | & \searrow & & \swarrow & |\\ | & & \mathrm{Cu} & & |\\ | & \nearrow & & \nwarrow & |\\ \mathrm{CH_2{-}NH_2} & & & & \mathrm{NH_2{-}CH_2}\end{array}\right]^{2+}$$

In this complex the copper is surrounded by four $-NH_2$ groups. Thus each ethylenediamine molecule is bonded to the copper in two places. For this reason ethylenediamine is called a bidentate group or ligand. (Bidentate means literally two teeth!) A ring structure is thus formed (in this case a pair of five-membered rings) and such ring structures are called chelates. (Chelos is the Greek word for crab.) Chelated complexes are more stable than similar complexes with unidentate ligands, as dissociation of the complex involves breaking two bonds rather than one. Some common polydentate ligands are listed in Figure 7.24.

The more rings that are formed, the more stable the complex is. Chelating agents with three, four and six donor atoms are known and are termed tridentate, tetradentate and hexadentate ligands. An important example of the latter is ethylenediaminetetraacetic acid. This bonds through two N and four O atoms to the metal, and so forms five rings. Due to this

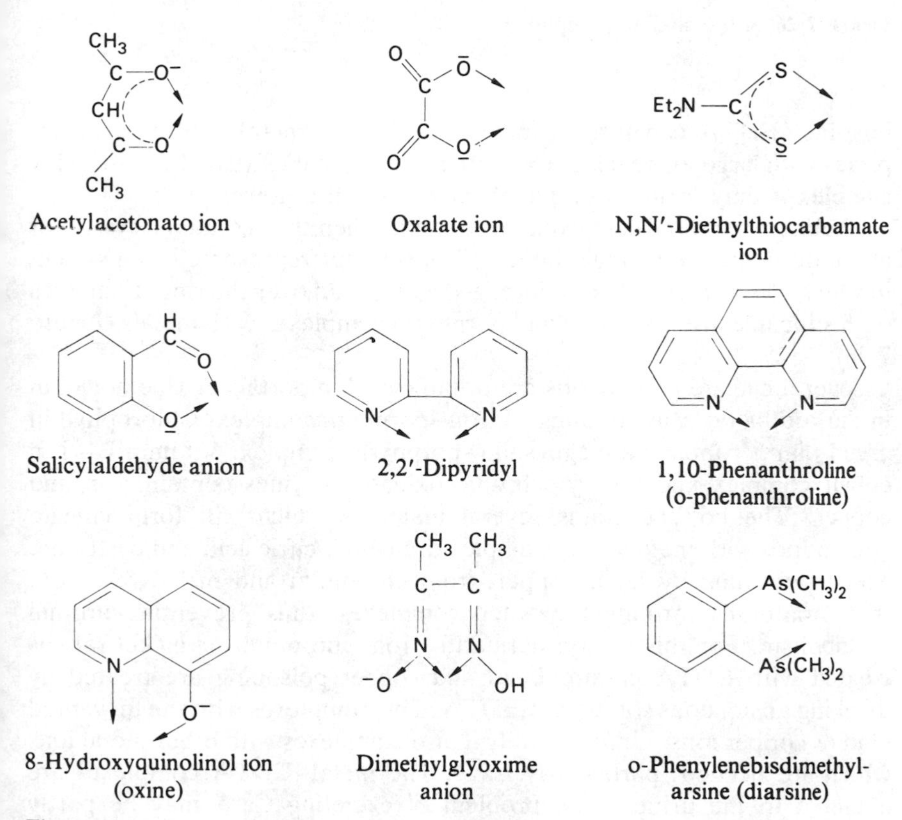

Figure 7.24 Some common polydentate ligands.

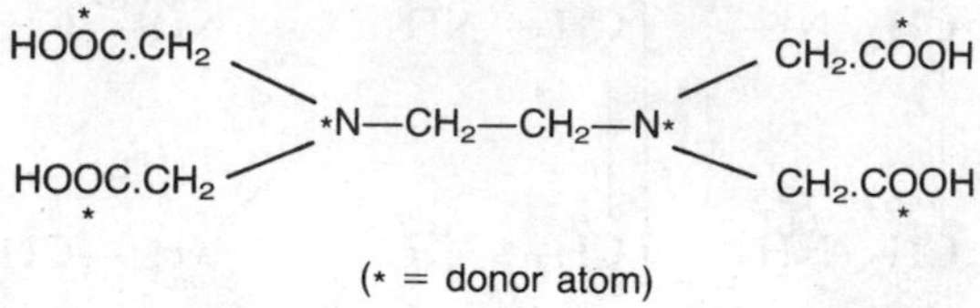

HOOC.CH₂ and CH₂.COOH groups on each N: $(HOOC.CH_2)_2N—CH_2—CH_2—N(CH_2.COOH)_2$

(* = donor atom)

Figure 7.25 EDTA.

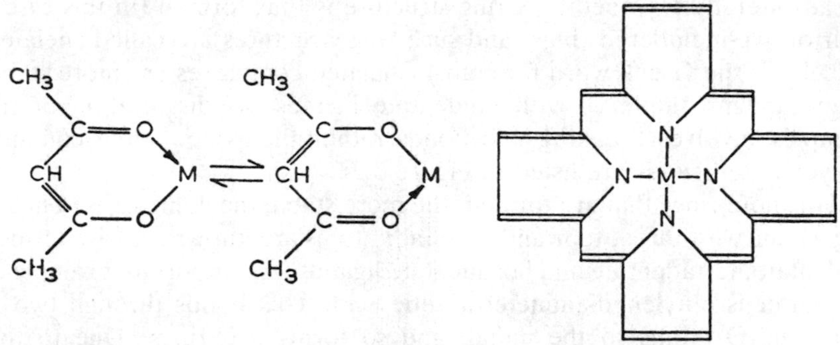

Resonance in acetylacetone chelate

Porphyrin complex

Figure 7.26 Some chelate complexes.

bonding, EDTA can form complexes with most metal ions. Even complexes with large ions such as Ca^{2+} are relatively stable. (The Ca^{2+}–EDTA complex is only formed completely at pH 8, not at lower pH.)

Chelate compounds are even more stable when they contain a system of alternate double and single bonds. This is better represented as a system in which electron density is delocalized and spread over the ring. Examples of this include acetylacetone and porphyrin complexes with metals (Figure 7.26).

Several chelate compounds are of biological importance. Haemoglobin in the red blood cells contains an iron–porphyrincomplex. Chlorophyll in green plants contains a magnesium–porphyrin complex. Vitamin B_{12} is a cobalt complex and the cytochrome oxidase enzymes contain iron and copper. The body contains several materials which will form chelate compounds with metals, for example adrenaline, citric acid and cortisone. Metal poisoning by lead, copper, iron, chromium and nickel results in these materials forming unwanted complexes, thus preventing normal metabolism. For this reason dermatitis from chromium or nickel salts is treated with EDTA cream. Lead and copper poisoning are treated by drinking an aqueous solution of EDTA. This complexes with the unwanted lead or copper ions. Unfortunately it also complexes with other metal ions which are needed, particularly Ca^{2+}. The metal–EDTA complexes are excreted in the urine. (The problem of excreting Ca^{2+} may be partly overcome by using the Ca–EDTA complex rather than EDTA itself.)

MAGNETISM

The magnetic moment can be measured using a Gouy balance (see Chapter 18). If we assume that the magnetic moment arises entirely from unpaired electron spins then the 'spin only' formula can be used to estimate n, the number of unpaired electrons. This gives reasonable agreement for complexes of the first row of transition metals.

$$\mu_S = \sqrt{n(n+2)}$$

Once the number of unpaired electrons is known, either the valence bond or the crystal field theory can be used to work out the shape of the complex, the oxidation state of the metal, and, for octahedral complexes, whether inner or outer d orbitals are used. For example, Co(+III) forms many complexes, all of which are octahedral. Most of them are diamagnetic, but $[CoF_6]^{3-}$ is paramagnetic with an observed magnetic moment of 5.3 BM. Crystal field theory explains this (Figure 7.27).

Co(+II) forms both tetrahedral and square planar four-coordinate complexes. These can be distinguished by magnetic measurements (Figure 7.28).

However, orbital angular momentum also contributes to a greater or lesser degree to the magnetic moment. For the second and third row transition elements not only is this contribution significant, but spin orbit coupling may occur. Because of this, the 'spin only' approximation is no longer valid, and there is extensive temperature dependent paramagnetism. Thus the simple interpretation of magnetic moments in terms of the number of unpaired electrons cannot be extended from the first row of

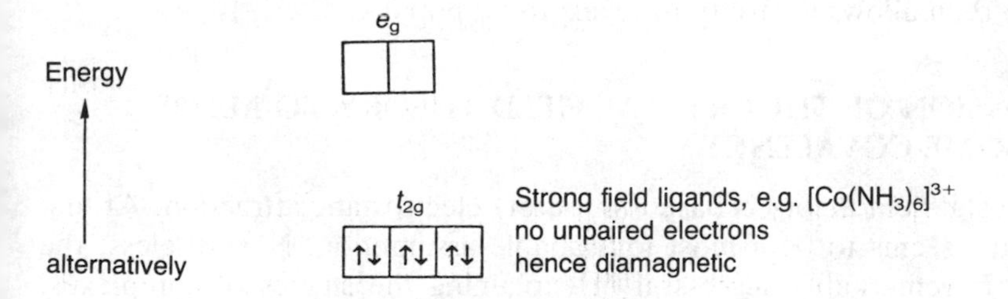

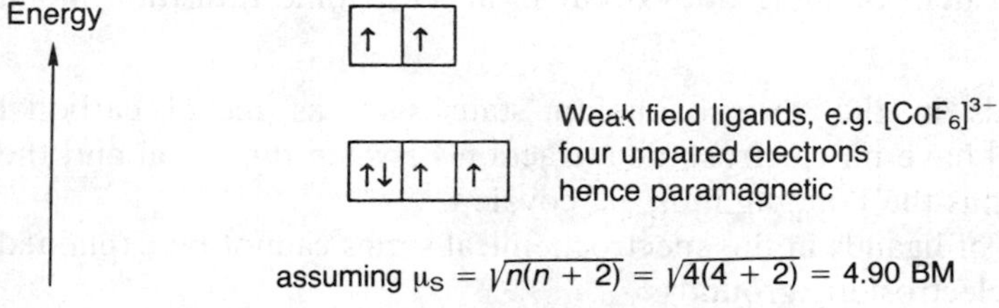

Figure 7.27 Co^{3+} in high-spin and low-spin complexes.

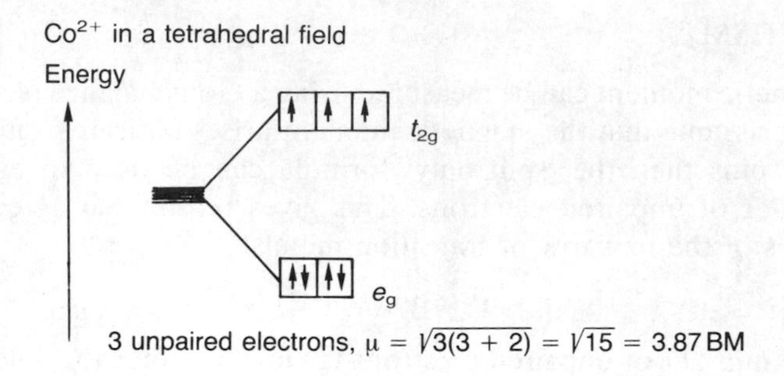

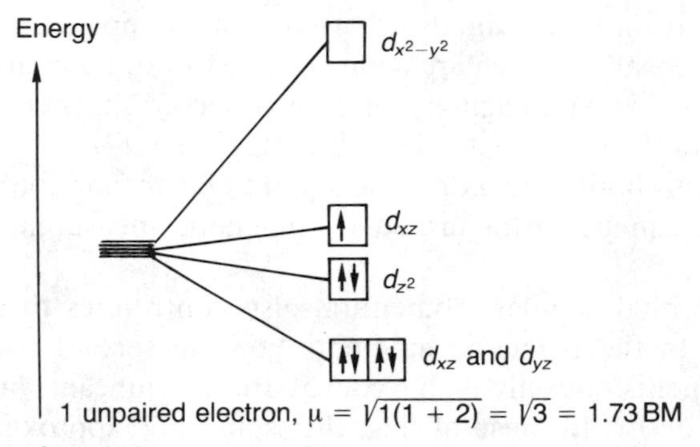

Figure 7.28 Co^{2+} in tetrahedral and square planar complexes.

transition elements to the second and third rows. The temperature dependence is explained by the spin orbit coupling. This removes the degeneracy from the lowest energy level in the ground state. Thermal energy then allows a variety of levels to be populated.

EXTENSION OF THE CRYSTAL FIELD THEORY TO ALLOW FOR SOME COVALENCY

The crystal field theory is based on purely electrostatic attraction. At first sight this seems to be a most improbable assumption. Nevertheless, the theory is remarkably successful in explaining the shapes of complexes, their spectra and their magnetic properties. Calculations can be carried out quite simply. The disadvantage of the theory is that it ignores evidence that some covalent bonding does occur in at least some transition metal complexes:

1. Compounds in the zero oxidation state such as nickel carbonyl $[Ni^0(CO)_4]$ have no electrostatic attraction between the metal and the ligands. Thus the bonding must be covalent.
2. The order of ligands in the spectrochemical series cannot be explained solely on electrostatic grounds.
3. There is some evidence from nuclear magnetic resonance and electron

spin resonance that there is some unpaired electron density on the ligands. This suggests the sharing of electrons, and hence some covalency.

The Racah interelectron repulsion parameter B is introduced into the interpretation of spectra. This makes some allowance for covalency arising from the delocalization of d electrons from the metal onto the ligand. If B is reduced below the value for a free metal ion, the d electrons are delocalized onto the ligand. The more B is reduced the greater the delocalization and the greater the amount of covalency. In a similar way an electron delocalization factor k can be used in interpreting magnetic measurements.

MOLECULAR ORBITAL THEORY

The molecular orbital theory incorporates covalent bonding. Consider a first row transition element forming an octahedral complex, for example

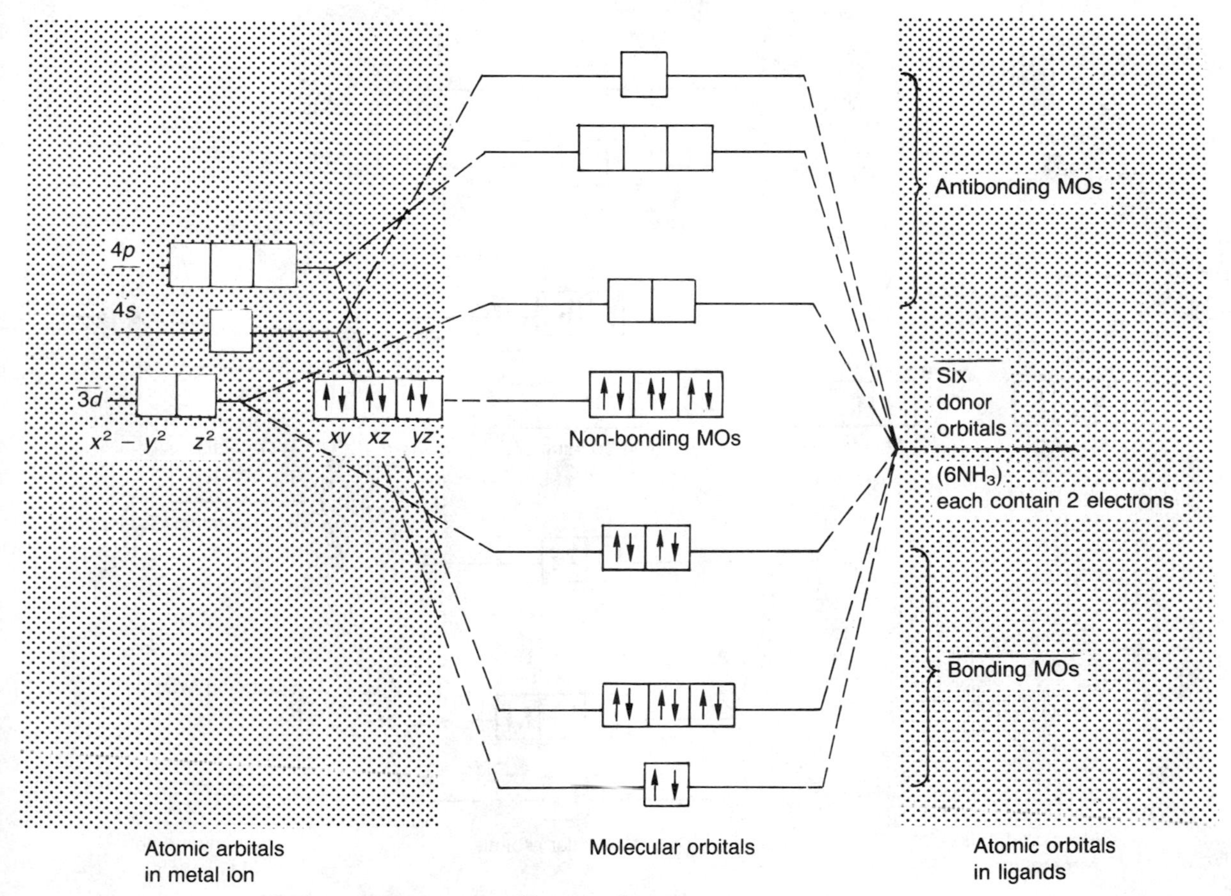

Figure 7.29 Molecular orbital diagram for $[Co^{III}(NH_3)_6]^{3+}$.

$[Co^{III}(NH_3)_6]^{3+}$. The atomic orbitals on Co^{3+} which are used to make molecular orbitals are $3d_{x^2-y^2}$, $3d_{z^2}$, $4s$, $4p_x$, $4p_y$ and $4p_z$. A $2p$ atomic orbital from each NH_3 containing a lone pair is also used to make molecular orbitals. Thus there are 12 atomic orbitals, which combine to give 12 molecular orbitals (six bonding MOs and six antibonding MOs). The 12 electrons from the six ligand lone pairs are placed in the six bonding MOs. This accounts for the six bonds. The transition metal Co^{3+} has other d orbitals, which have so far been ignored. These are the $3d_{xy}$, $3d_{xz}$ and $3d_{yz}$ orbitals. These form non-bonding MOs, and in Co^{3+} they contain six electrons, but contribute nothing to the bonding. The antibonding MOs are all empty. The arrangement is shown in Figure 7.29. We would predict that the complex should be diamagnetic as all the electrons are paired. The complex should be coloured since promotion of electrons from the non-bonding MOs to the antibonding e_g^* MOs is feasible. The energy jump Δ_o is $23\,000\,cm^{-1}$. The six non-bonding d electrons are paired in this complex because Δ_o is larger than the pairing energy of $19\,000\,cm^{-1}$.

A similar MO diagram can be drawn for the complex $[Co^{III}F_6]^{3-}$.

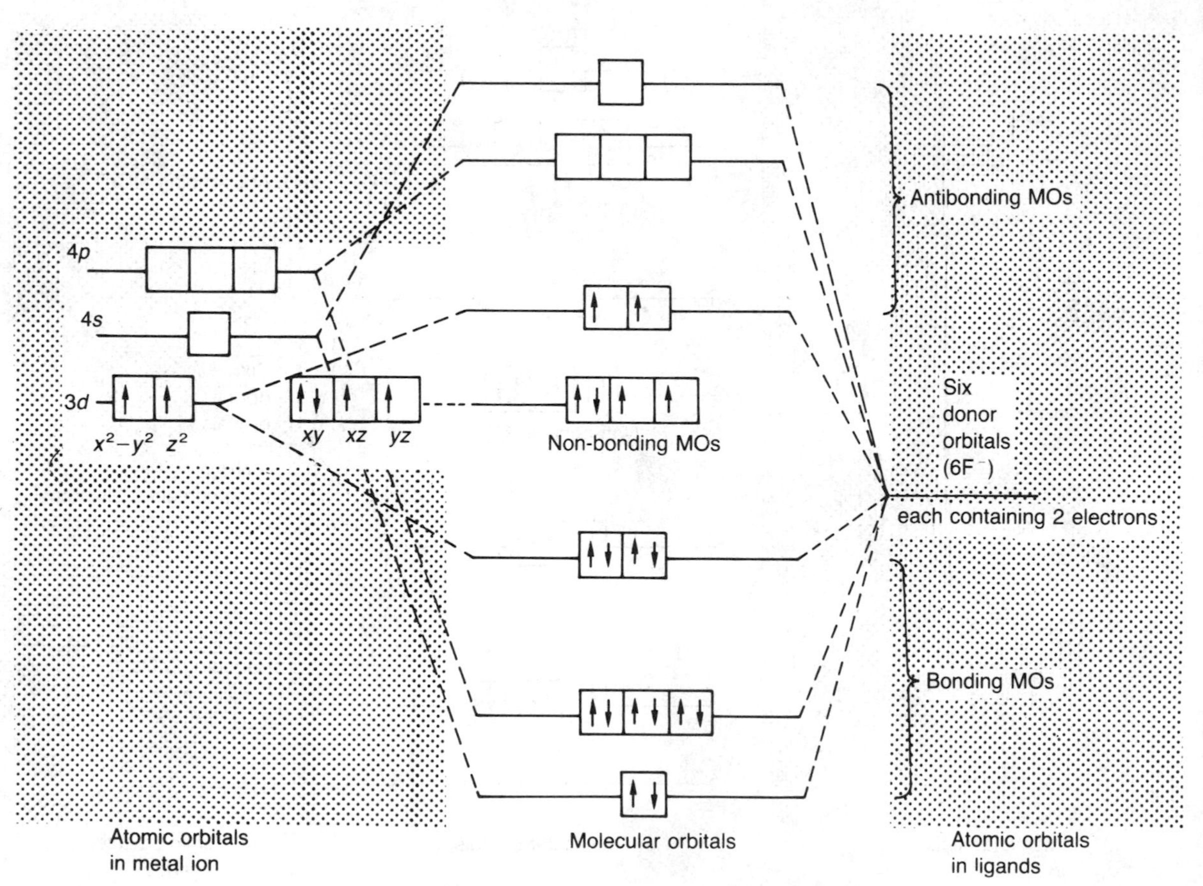

Figure 7.30 Molecular orbital diagram for $[CoF_6]^{3-}$.

However, the energies of the $2p$ orbitals on F^- are much lower than the energy of the corresponding orbital on N in NH_3. This alters the spacing of the MO energy levels (Figure 7.30). Spectroscopic measurements show that Δ_o is 13 000 Cm^{-1}. Thus the gap between the non-bonding MOs and the antibonding e_g^* MOs is less than the pairing energy of 19 000 cm^{-1}. Thus the non-bonding d electrons do not pair up as in the $[Co(NH_3)_6]^{3+}$ complex because there is a net gain in energy if electrons are left unpaired. Thus $[CoF_6]^{3-}$ has four unpaired electrons and is a high-spin complex, whilst $[Co(NH_3)_6]^{3+}$ has no unpaired electrons and is a low-spin complex.

Thus the MO theory explains the magnetic properties and spectra of complexes equally as well as the crystal field theory. Both theories rely on spectra to measure the energy of Δ_o. Either theory may be used depending on which is the most convenient.

The MO theory is based on wave mechanics and so has the disadvantage that enthalpies of formation and bond energies cannot be calculated directly. So far we have considered σ bonding between ligands and the central metal. The MO theory has the great advantage that it is easily extended to cover π bonding. Pi bonding helps to explain how metals in low oxidation states (e.g. $[Ni^0(CO)_4]$) can form complexes. It is impossible to explain any attractive force in such a complex using the crystal field theory because of the lack of charge on the metal. Pi bonding also helps to explain the position of some ligands in the spectrochemical series. There are two cases:

1. Where the ligands act as π acceptors, by accepting electrons from the central metal. Examples include CO, CN^-, NO^+ and phosphines.
2. Where the ligands act as π donors and transfer charge from ligand to metal in π interactions as well as σ interactions. Pi bonding of this kind commonly occurs in oxoions of metals in high oxidation states, e.g. $[Mn^{VII}O_4]^-$ and $[Cr^{VI}O_4]^{2-}$.

π acceptors

Ligands such as CO, CN^- and NO^+ have empty π orbitals with the correct symmetry to overlap with the metal t_{2g} orbitals, forming π bonds. This is often described as back bonding. Normally the π orbitals on the ligands are of higher energy than the metal t_{2g} orbitals. No more electrons are added to the scheme as the ligand π orbitals are empty, but the π interaction increases the value of Δ_o. This accounts for the position of these ligands as 'strong field ligands' at the right of the spectrochemical series.

π donors

The ligand has filled π orbitals which overlap with the metal t_{2g} orbitals, giving a π bond. Thus electron density is transferred from the ligand to the metal. The σ bonding also transfers charge to the metal. This type of complex is favoured when the central metal has a high oxidation state, and 'is short of electrons'. The ligand π orbitals are lower in energy than the metal t_{2g} orbitals. Delocalizing π electrons from the ligand to the metal in this way reduces the value of Δ. It is not always clear if π donor bonding has occurred, but it is most likely with ligands at the left of the spectrochemical series.

NOMENCLATURE OF COORDINATION COMPOUNDS

The International Union of Pure and Applied Chemistry (IUPAC) publication *Nomenclature of Inorganic Chemistry* (1989), Blackwell Scientific Publishers, contains the rules for the systematic naming of coordination compounds. The basic rules are summarized here.

1. The positive ion is named first followed by the negative ion.
2. When writing the name of a complex, the ligands are quoted in alphabetical order, regardless of their charge (followed by the metal).
3. When writing the formula of complexes, ligands are named before the metal. The coordinated groups are listed in the order: negative ligands, neutral ligands, positive ligands (and alphabetically according to the first symbol within each group).
   (a) The names of negative ligands end in -o, for example:

F^-	fluoro	H^-	hydrido	HS^-	mercapto
Cl^-	chloro	OH^-	hydroxo	S^{2-}	thio
Br^-	bromo	O^{2-}	oxo	CN^-	cyano
I^-	iodo	O_2^{2-}	peroxo	NO_2^-	nitro

   (b) Neutral groups have no special endings. Examples include NH_3 ammine, H_2O aqua, CO carbonyl and NO nitrosyl. The ligands N_2 and O_2 are called dinitrogen and dioxygen. Organic ligands are usually given their common names, for example phenyl, methyl, ethylenediamine, pyridine, triphenylphosphine.
   (c) Positive groups end in -ium, e.g. NH_2—NH_2 hydrazinium.
4. Where there are several ligands of the same kind, we normally use the prefixes di, tri, tetra, penta and hexa to show the number of ligands of that type. An exception occurs when the name of the ligand includes a number, e.g. dipyridyl or ethylenediamine. To avoid confusion in such cases, bis, tris and tetrakis are used instead of di, tri and tetra and the name of the ligand is placed in brackets.
5. The oxidation state of the central metal is shown by a Roman numeral in brackets immediately following its name (i.e. no space, e.g. titanium(III)).
6. Complex positive ions and neutral molecules have no special ending but complex negative ions end in -ate.
7. If the complex contains two or more metal atoms, it is termed polynuclear. The bridging ligands which link the two metal atoms together are indicated by the prefix μ-. If there are two or more bridging groups of the same kind, this is indicated by di-μ-, tri-μ- etc. Bridging

groups are listed alphabetically with the other groups unless the symmetry of the molecule allows a simpler name. If a bridging group bridges more than two metal atoms it is shown as μ_3, μ_4, μ_5 or μ_6 to indicate how many atoms it is bonded to.

8. Sometimes a ligand may be attached through different atoms. Thus M—NO_2 is called nitro and M—ONO is called nitrito. Similarly the SCN group may bond M—SCN thiocyanato or M—NCS isothiocyanato. These may be named systematically thiocyanato-S or thiocyanato-N to indicate which atom is bonded to the metal. This convention may be extended to other cases where the mode of linkage is ambiguous.
9. If any lattice components such as water or solvent of crystallization are present, these follow the name, and are preceded by the number of these groups in Arabic numerals.

These rules are illustrated by the following examples:

Complex anions	
$[Co(NH_3)_6]Cl_3$	Hexaamminecobalt(III) chloride
$[CoCl(NH_3)_5]^{2+}$	Pentaamminechlorocobalt(III) ion
$[CoSO_4(NH_3)_4]NO_3$	Tetraamminesulphatocobalt(III) nitrate
$[Co(NO_2)_3(NH_3)_3]$	Triamminetrinitrocobalt(III)
$[CoCl \cdot CN \cdot NO_2 \cdot (NH_3)_3]$	Triamminechlorocyanonitrocobalt(III)
$[Zn(NCS)_4]^{2+}$	Tetrathiocyanato-N-zinc(II)
$[Cd(SCN)_4]^{2+}$	Tetrathiocyanato-S-cadmium(II)
Complex cations	
$Li[AlH_4]$	Lithium tetrahydridoaluminate(III) (*lithium aluminium hydride*)
$Na_2[ZnCl_4]$	Sodium tetrachlorozincate(II)
$K_4[Fe(CN)_6]$	Potassium hexacyanoferrate(II)
$K_3[Fe(CN)_5NO]$	Potassium pentacyanonitrosylferrate(II)
$K_2[OsCl_5N]$	Potassium pentachloronitridoosmate(VI)
$Na_3[Ag(S_2O_3)_2]$	Sodium bis(thiosulphato)argentate(I)
$K_2[Cr(CN)_2O_2(O_2)NH_3]$	Potassium amminedicyanodioxoperoxo chromate(VI)
Organic groups	
$[Pt(py)_4][PtCl_4]$	Tetrapyridineplatinum(II) · tetrachloroplatinate(II)
$[Cr(en)_3]Cl_3$	*d* or *l* Tris(ethylenediamine)chromium(III) chloride
$[CuCl_2(CH_3NH_2)_2]$	Dichlorobis(dimethylamine)copper(II)
$Fe(C_5H_5)_2$	Bis(cyclopentadienyl)iron(II)
$[Cr(C_6H_6)_2]$	Bis(benzene)chromium(0)

Bridging groups

$[(NH_3)_5Co \cdot NH_2 \cdot Co(NH_3)_5](NO_3)_5$	μ-amidobis[pentaamminecobalt(III)] nitrate
$[(CO)_3Fe(CO)_3Fe(CO)_3]$	Tri-μ-carbonyl-bis(tricarbonyliron(0)) (*di iron enneacarbonyl*)
$[Be_4O(CH_3COO)_6]$	Hexa-μ-acetato(O,O′)-μ_4-oxo-tetraberyllium(II) (*basic beryllium acetate*)

Hydrates

$AlK(SO_4)_2 \cdot 12H_2O$	Aluminium potassium sulphate 12-water

ISOMERISM

Compounds that have the same chemical formula but different structural arrangements are called isomers. Because of the complicated formulae of many coordination compounds, the variety of bond types and the number of shapes possible, many different types of isomerism occur. Werner's classification into polymerization, ionization, hydrate linkage, coordination, coordination position, and geometric and optical isomerism is still generally accepted.

Polymerization isomerism

This is not true isomerism because it occurs between compounds having the same empirical formula, but different molecular weights. Thus $[Pt(NH_3)_2Cl_2]$, $[Pt(NH_3)_4][PtCl_4]$, $[Pt(NH_3)_4][Pt(NH_3)Cl_3)_2$ and $[Pt(NH_3)_3Cl]_2[PtCl_4]$ all have the same empirical formula. Polymerization isomerism may be due to a different number of nuclei in the complex, as shown in Figure 7.31.

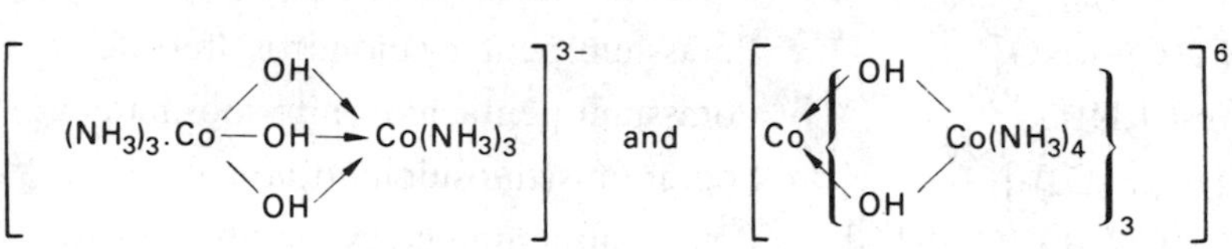

$$\left[(NH_3)_3Co(OH)_3Co(NH_3)_3\right]^{3-} \quad \text{and} \quad \left[Co\left\{(OH)_2Co(NH_3)_4\right\}_3\right]^{6-}$$

Figure 7.31 Polymerization isomers.

Ionization isomerism

This type of isomerism is due to the exchange of groups between the complex ion and the ions outside it. $[Co(NH_3)_5Br]SO_4$ is red–violet. An aqueous solution gives a white precipitate of $BaSO_4$ with $BaCl_2$ solution, thus confirming the presence of free SO_4^{2-} ions. In contrast $[Co(NH_3)_5SO_4]Br$ is red. A solution of this complex does not give a positive sulphate test with $BaCl_2$. It does give a cream-coloured precipitate of AgBr with $AgNO_3$, thus confirming the presence of free Br^- ions. Note that the sulphate ion occupies only one coordination position even though it has two negative charges. Other examples of ionization isomerism are $[Pt(NH_3)_4Cl_2]Br_2$ and $[Pt(NH_3)_4Br_2]Cl_2$, and $[Co(en)_2NO_2 \cdot Cl]SCN$, $[Co(en)_2NO_2 \cdot SCN]Cl$ and $[Co(en)_2Cl \cdot SCN]NO_2$.

Hydrate isomerism

Three isomers of $CrCl_3 \cdot 6H_2O$ are known. From conductivity measurements and quantitative precipitation of the ionized chlorine, they have been given the following formulae:

$[Cr(H_2O)_6]Cl_3$	violet	(three ionic chlorines)
$[Cr(H_2O)_5Cl]Cl_2 \cdot H_2O$	green	(two ionic chlorines)
$[Cr(H_2O)_4Cl_2] \cdot Cl \cdot 2H_2O$	dark green	(one ionic chlorine)

Linkage isomerism

Certain ligands contain more than one atom which could donate an electron pair. In the NO_2^- ion, either N or O atoms could act as the electron pair donor. Thus there is the possibility of isomerism. Two different complexes $[Co(NH_3)_5NO_2]Cl_2$ have been prepared, each containing the NO_2^- easily decomposed by acids to give nitrous acid. It contains Co—ONO and is a nitrito complex. The other complex is yellow and is stable to acids. It contains the Co—NO_2 group and is a nitro compound. The two materials are represented in Figure 7.32. This type of isomerism also occurs with other ligands such as SCN^-.

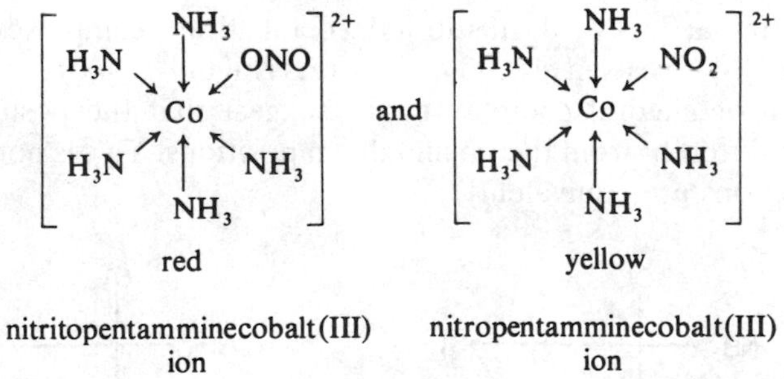

$[Co(NH_3)_5(ONO)]^{2+}$ (ligands NH_3, H_3N, ONO, H_3N, NH_3, NH_3 around Co) and $[Co(NH_3)_5(NO_2)]^{2+}$ (ligands NH_3, H_3N, NO_2, H_3N, NH_3, NH_3 around Co)

red — nitritopentamminecobalt(III) ion

yellow — nitropentamminecobalt(III) ion

Figure 7.32 Nitrito and nitro complexes.

Coordination isomerism

When both the positive and negative ions are complex ions, isomerism may be caused by the interchange of ligands between the anion and cation, for example $[Co(NH_3)_6][Cr(CN)_6]$ and $[Cr(NH_3)_6][Co(CN)_6]$. Intermediate types between these extremes are also possible.

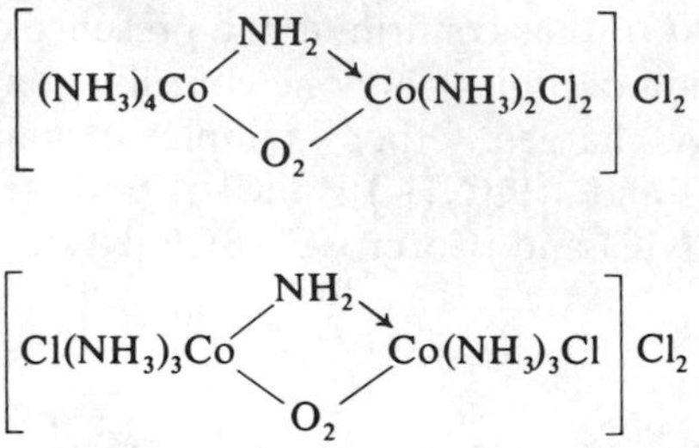

$$\left[(NH_3)_4Co\genfrac{}{}{0pt}{}{\overset{NH_2}{\diagup\quad\searrow}}{\underset{O_2}{\diagdown\quad\diagup}}Co(NH_3)_2Cl_2\right]Cl_2$$

$$\left[Cl(NH_3)_3Co\genfrac{}{}{0pt}{}{\overset{NH_2}{\diagup\quad\searrow}}{\underset{O_2}{\diagdown\quad\diagup}}Co(NH_3)_3Cl\right]Cl_2$$

Figure 7.33 Coordination position isomers.

Coordination position isomerism

In polynuclear complexes an interchange of ligands between the different metal nuclei gives rise to positional isomerism. An example is given in Figure 7.33.

Geometric isomerism or stereoisomerism

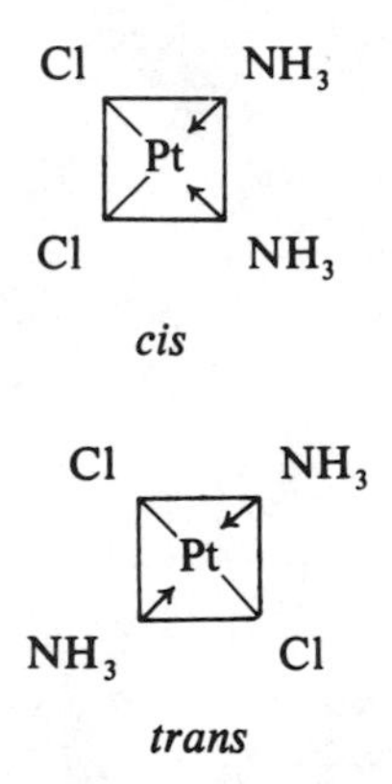

Figure 7.34 *Cis* and *trans* isomers.

In disubstituted complexes, the substituted groups may be adjacent or opposite to each other. This gives rise to geometric isomerism. Thus square planar complexes such as $[Pt(NH_3)_2Cl_2]$ can be prepared in two forms, *cis* and *trans*. If the complex is prepared by adding NH_4OH to a solution of $[PtCl_4]^{2-}$ ions, the complex has a finite dipole moment and must therefore be *cis*. The complex prepared by treating $[Pt(NH_3)_4]^{2+}$ with HCl has no dipole, and must therefore be *trans*. The two complexes are shown in Figure 7.34. The same sort of isomerism can also occur in square planar chelate complexes if the chelating group is not symmetrical. An example of *cis–trans* isomerism is found in the complex between glycine and platinum (Figure 7.35).

In a similar way disubstituted octahedral complexes such as $[Co(NH_3)_4Cl_2]^+$ exists in *cis* and *trans* forms (Figure 7.36). (This method of drawing an octahedral complex might suggest that the positions in the square are different from the up and down positions. This is not the case as all six positions are equivalent.)

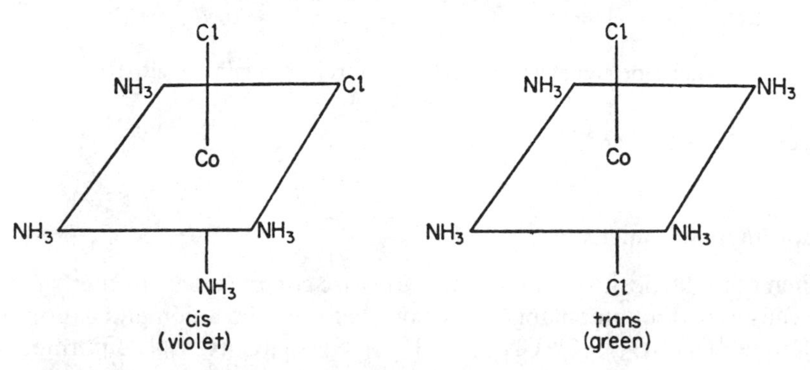

Figure 7.36 *Cis* and *trans* octahedral complexes.

Optical isomerism

At one time it was thought that optical isomerism was associated only with carbon compounds. It exists in inorganic molecules as well. If a molecule is

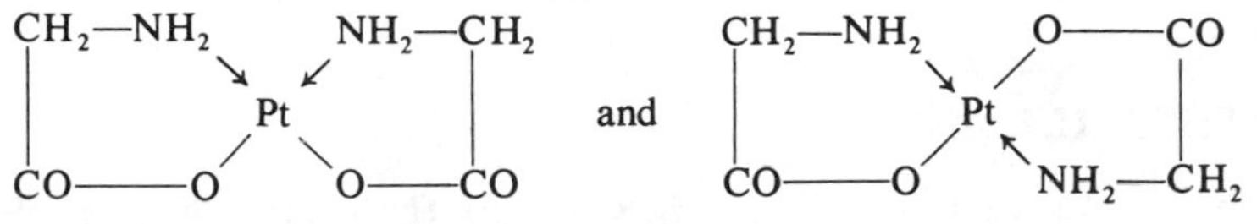

CH_2—NH_2 → Pt ← NH_2—CH_2 ; CO—O—Pt—O—CO   and CH_2—NH_2 → Pt — O—CO ; CO—O—Pt ← NH_2—CH_2

Figure 7.35 *Cis* and *trans* glycine complexes.

asymmetric, it cannot be superimposed on its mirror image. The two forms have the type of symmetry shown by the left and right hands and are called an enantiomorphic pair. The two forms are optical isomers. They are called either *dextro* or *laevo* (often shortened to *d* or *l*). This depends on the direction they rotate the plane of polarized light in a polarimeter. (*d* rotates to the right, *l* to the left.) Optical isomerism is common in octahedral complexes involving bidentate groups. For example, $[Co(en)_2Cl_2]^+$

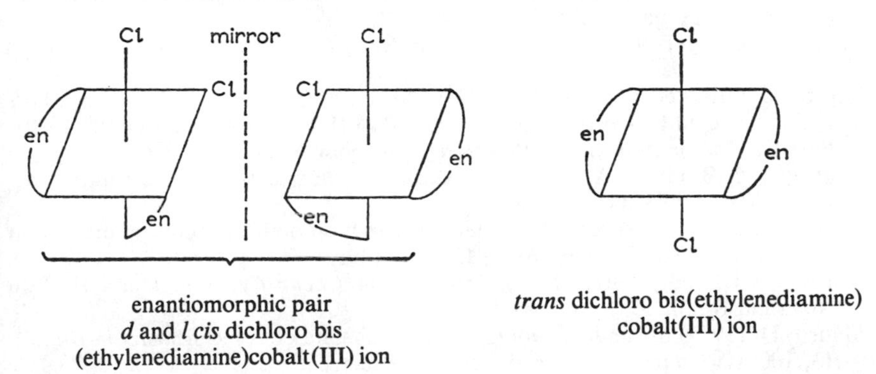

Figure 7.37 Isomers of $[Co(en)_2Cl_2]^+$.

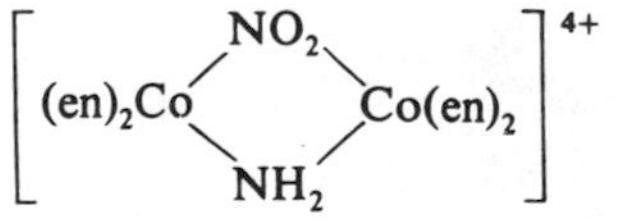

$$\left[(en)_2Co\overset{NO_2}{\underset{NH_2}{\diamond}}Co(en)_2\right]^{4+}$$

Figure 7.38

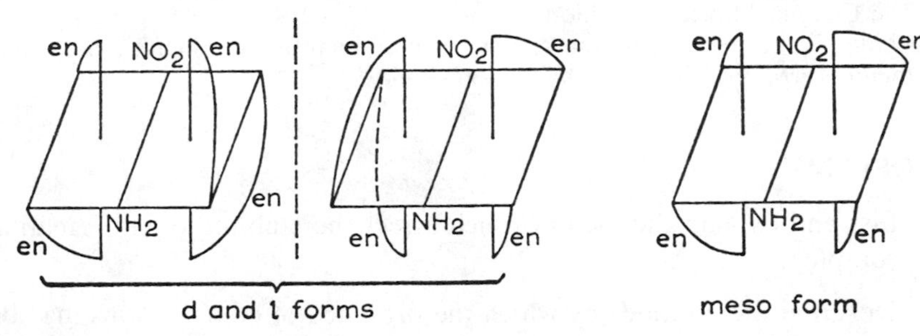

Figure 7.39 *d*, *l* and *meso* forms.

shows *cis* and *trans* forms (geometric isomerism). In addition the *cis* form is optically active and exists in *d* and *l* forms, making a total of three isomers (Figure 7.37). Optical activity occurs also in polynuclear complexes, such as that shown in Figure 7.38. This has been resolved into two optically active forms (*d* and *l*) and an optically inactive form which is internally compensated and is called the *meso* form (Figure 7.39).

FURTHER READING

Ahrland, S., Chatt, J. and Davies, N.R. (1958) The relative affinities of ligand atoms for acceptor molecules and ions, *Q. Rev. Chem. Soc.*, **12**, 265–276.

Bell, C.F. (1977) *Principles and Applications of Metal Chelation*, Oxford University Press, Oxford.

Emeléus H.J. and Sharpe, A.G. (1973) *Modern Aspects of Inorganic Chemistry*, 4th ed. (Complexes of Transition Metals: Chapter 14, Structure; Chapter 15, Bonding; Chapter 16, Magnetic Properties; Chapter 17, Electronic Spectra), Routledge and Kegan Paul, London.

Gerloch, M. (1981) The sense of Jahn–Teller distortions in octahedral copper(II) and other transition metal complexes, *Inorg. Chem.*, **20**, 638–640.

Hogfeldt, E. (ed.) (1982) *Stability Constants of Metal-ion Complexes*, Pergamon, Oxford. (Inorganic ligands.)

Johnson, B.F.G. (1973) *Comprehensive Inorganic Chemistry*, Vol. IV (Chapter 52: Transition metal chemistry), Pergamon Press, Oxford.

Jørgensen, C.K. (1962) *Absorption Spectra and Chemical Bonding in Complexes* (Chapter 7) Pergamon Press, Oxford.

Kauffman, G.B. (1966) *Alfred Werner Founder of Coordination Theory*, Springer, Berlin.

Kauffman, G.B. (ed.) (1968, 1976, 1978) *Classics in Coordination Chemistry*, Part I, The Selected Papers of Alfred Werner; Part II, Selected Papers (1798–1899); Part III, Twentieth Century Papers. Dover, New York.

Kauffman, G.B. (1973) Alfred Werner's research on structural isomerism, *Coord. Chem. Rev.*, 1973, **11**, 161–188.

Kauffman, G.B. (1974) Alfred Werner's research on optically active coordination compounds, *Coord. Chem. Rev.*, **12**, 105–149.

Martell, A.E. (ed.) (1971, 1978) *Coordination Chemistry*, Vol. I and II, Van Nostrand Reinhold, New York.

Munro, D. (1977) Mjsunderstandings over the chelate effect, *Chemistry in Britain*, **13**, 100. (A simple article on the chelate effect.)

Perrin, D. (ed.) (1979) *Stability Constants of Metal-ion Complexes*, Chemical Society, Pergamon. (Organic ligands.)

Sillen, L.G. and Martell, A.E. (1964, 1971) *Stability Constants of Metal-ion Complexes* (Special Publications of the Chemical Society, no. 17 and no. 25), The Chemical Society, London.

Tsuchida, R. (1938) Absorption spectra of coordination compounds, *Bull. Soc. Japan*, 1938, 388–400, 434–450 and 471–480.

PROBLEMS

1. List and explain the factors which affect the stability of coordination complexes.

2. Describe the methods by which the presence of complex ions may be detected in solution.

3. Draw all of the isomers of an octahedral complex which has six unidentate ligands, two of type A and four of type B.
4. Draw all of the isomers of an octahedral complex which has three unidentate ligands of type A and three unidentate ligands of type B.
5. Draw all of the isomers of an octahedral complex which has three identical bidentate ligands.
6. Draw all of the isomers of both tetrahedral and square planar complexes which have two unidentate ligands of type A and two unidentate ligands of type B.
7. Draw each of the possible stereoisomers of the octahedral complexes listed: (a) Ma_3bcd, (b) Ma_2bcde and (c) M(AA)(AA)cd. The lower case letters a, b, c, d, and e represent monodentate ligands, and upper case letters (AA) represent the donor atoms of a bidentate ligand. Indicate which isomers are optically active (chiral).
8. Draw the shapes of the various *d* orbitals, and explain why they are split into two groups t_{2g} and e_g in an octahedral ligand field.
9. Draw a diagram to show how the *d* orbitals are split into groups with different energy in an octahedral ligand field. Some electronic configurations may exist in both high-spin and low-spin arrangements in an octahedral field. Draw all of these cases, and suggest which metal ions and which ligands might give rise to each.
10. Draw an energy level diagram to show the lifting of the degeneracy of the 3*d* orbitals in a tetrahedral ligand field.
11. Draw energy level diagrams and indicate the occupancy of the orbitals in the following complexes:
    (a) d^6, octahedral, low-spin
    (b) d^9, octahedral with tetragonal elongation
    (c) d^8, square planar
    (d) d^6, tetrahedral.
    Calculate in units of Δ_o the difference in crystal field stabilization energy between complexes (a) and (d) assuming that the ligands are strong field ligands.
    (Answer: octahedral $-2.4\Delta_o$, tetrahedral $-0.27\Delta_o$, difference $-2.13\Delta_o$.)
12. Calculate the crystal field stabilization energy for a d^8 ion such as Ni^{2+} in octahedral and tetrahedral complexes. Use units of Δ_o in both cases. Which is the most stable? State any assumptions made.
13. Calculate the spin only magnetic moment for a d^8 ion in octahedral, square planar and tetrahedral ligand fields.
14. Show by means of a diagram how the pattern of *d* orbital splitting changes as an octahedral complex undergoes tetragonal distortion and eventually becomes a square planar complex.

15. Why are *d–d* electronic transitions forbidden? Why are they weakly absorbing and why do they occur at all?

16. Why are compounds of Ti^{4+} and Zn^{2+} typically white? Why are Mn^{2+} compounds very pale in colour? What *d–d* transitions are spin allowed for a d^5 ion?

17. What is the spectrochemical series, and what is its importance?

18. Given that the maximum absorption in the *d–d* peak for $[Ti(H_2O)_6]^{3+}$ occurs at 20300 cm^{-1}, predict where the peaks will occur for $[Ti(CN)_6]^{3-}$ and $[Ti(Cl)_6]^{3-}$.

19. Describe how Δ_o changes as the charge on the central metal changes from M^{2+} to M^{3+}, and how it changes in a vertical group or triad between a first row, second row or third row transition element.

20. What would you expect the crystal field stabilization energy to be, and what value of magnetic moment would you expect, for the following complexes: (a) $[CoF_6]^{3-}$, (b) $[Co(NH_3)_6]^{3+}$, (c) $[Fe(H_3O]_6^{2+}$, (d) $[Fe(CN)_6]^{4-}$ and (e) $[Fe(CN)_6]^{3-}$.

21. In the crystal structure of CuF_2, the Cu^{2+} is six-coordinate with four F^- at a distance of 1.93 Å and two F^- at 2.27 Å. Explain the reason for this.

22. Describe and explain the Jahn–Teller effect in octahedral complexes of Cr^{2+} and Cu^{2+}.

23. The complex $[Ni(CN)_4]^{2-}$ is diamagnetic, but $[NiCl_4]^{2-}$ is paramagnetic and has two unpaired electrons. Explain these observations and deduce the structures of the two complexes.

24. What methods could be used to distinguish between *cis* and *trans* isomers of a complex?

25. Name the individual isomers of each of the following:
    (a) $[Pt(NH_3)_2Cl_2]$
    (b) $CrCl_36H_2O$
    (c) $[Co(NH_3)_5NO_2](NO_3)_2$
    (d) $Co(NH_3)_5(SO_4)(Cl)$
    (e) $\left[(en)_2Co \overset{NH_2}{\underset{NO_2}{\ }} Co(en)_2 \right] Br_4$
    (f) $Co(en)_2NH_3BrSO_4$
    (g) $[Pt(NH_3)(H_2O)(C_5H_5N)(NO_2)]Cl$.

26. Account for the following:
    (a) $Ni(CO)_4$ is tetrahedral
    (b) $[Ni(CN)_4]^{2-}$ is square planar
    (c) $[Ni(NH_3)_6]^{2+}$ is octahedral.

27. What is the oxidation number of the metal in each of the following complexes:
    (a) $[Co(NH_3)_6]Cl_3$
    (b) $[CoSO_4(NH_3)_4]NO_3$
    (c) $[Cd(SCN)_4]^{2+}$
    (d) $[Cr(en)_3]Cl_3$
    (e) $[CuCl_2(CH_3NH_2)_2]$
    (f) $[AlH_4]^-$
    (g) $[Fe(CN)_6]^{4-}$
    (h) $[OsCl_5N]^{2-}$
    (i) $[Ag(S_2O_3)_2]^{3-}$

28. Write the formula for each of the following complexes:
    (a) hexamminecobalt(III) chloride
    (b) potassium iron(III) hexacyanoferrate(II)
    (c) diamminedichloroplatinum(II)
    (d) tetracarbonylnickel(0)
    (e) triamminechlorocyanonitrocobalt(III)
    (f) lithium tetrahydridoaluminate(III)
    (g) sodium bis(thiosulphato)argentate(I)
    (h) nickel hexachloroplatinate(IV)
    (i) tetraammineplatinum(II) amminetrichloroplatinate(II)

29. Write the formula for each of the following complexes:
    (a) tetraamminecopper(II) sulphate
    (b) potassium tetracyanonickelate(0)
    (c) bis(cyclopentadienyl)iron(II)
    (d) tetrathiocyanato-N-zinc(II)
    (e) diamminebis(ethylenediamine)cobalt(III) chloride
    (f) tetraamminedithiocyanatochromium(III)
    (g) potassium tetraoxomanganate(VII)
    (h) potassium trioxalatoaluminate(III)
    (i) tetrapyridineplatinum(II) tetrachloroplatinate(II)

8 Hydrogen and the hydrides

ELECTRONIC STRUCTURE

Hydrogen has the simplest atomic structure of all the elements, and consists of a nucleus containing one proton with a charge +1 and one orbital electron. The electronic structure may be written as $1s^1$. Atoms of hydrogen may attain stability in three different ways:

1. *By forming an electron pair (covalent) bond with another atom*
   Non-metals typically form this type of bond with hydrogen, for example H_2, H_2O, $HCl_{(gas)}$ or CH_4, and many metals do so too.
2. *By losing an electron to form H^+*
   A proton is extremely small (radius approximately 1.5×10^{-5} Å, compared with 0.7414 Å for hydrogen, and 1–2 Å for most atoms). Because H^+ is so small, it has a very high polarizing power, and therefore distorts the electron cloud on other atoms. Thus protons are always associated with other atoms or molecules. For example, in water or aqueous solutions of HCl and H_2SO_4, protons exist as H_3O^+, $H_9O_4^+$ or $H(H_2O)_n^+$ ions. Free protons do not exist under 'normal conditions', though they are found in low pressure gaseous beams, for example in a mass spectrometer.
3. *By gaining an electron to form H^-*
   Crystalline solids such as LiH contain the H^- ion and are formed by highly electropositive metals (all of Group I, and some of Group II). However, H^- ions are uncommon.

Since hydrogen has an electronegativity of 2.1, it may use any of the three methods, but the most common way is forming covalent bonds.

POSITION IN THE PERIODIC TABLE

Hydrogen is the first element in the periodic table, and is unique. There are only two elements in the first period, hydrogen and helium. Hydrogen is quite reactive, but helium is inert. There is no difficulty relating the structure and properties of helium to those of the other noble gases in

Group 0, but the properties of hydrogen cannot be correlated with any of the main groups in the periodic table, and hydrogen is best considered on its own.

The structure of hydrogen atoms is in some ways like that of the alkali metals. The alkali metals (Group I) also have just one electron in their outer shell, but they tend to lose this electron in reactions and form positive ions M^+. Though H^+ are known, hydrogen has a much greater tendency to pair the electron and form a covalent bond.

The structure of hydrogen atoms is in some ways like that of the halogens (Group VII), since both are one electron short of a noble gas structure. In many reactions the halogens gain an electron and so form negative ions X^-. Hydrogen does not typically form a negative ion, although it does form ionic hydrides M^+H^- (e.g. LiH and CaH_2) with a few highly electropositive metals .

In some ways the structure of hydrogen resembles that of the Group IV elements, since both have a half filled shell of electrons. There are a number of similarities between hydrides and organometallic compounds since the groups CH_3— and H— both have one remaining valency. Thus the hydride is often considered as part of a series of organometallic compounds, for example LiH, LiMe, LiEt; NH_3, NMe_3, NEt_3; or SiH_4, CH_3SiH_3, $(CH_3)_2SiCl_2$, $(CH_3)_3SiCl$, $(CH_3)_4Si$. However, hydrogen is best treated as a group on its own.

ABUNDANCE OF HYDROGEN

Hydrogen is the most abundant element in the universe. Some estimates are that 92% of the universe is made up of hydrogen, and 7% helium, leaving only 1% for all of the other elements. However, the abundance of H_2 in the earth's atmosphere is very small. This is because the earth's gravitational field is too small to hold so light an element, though some H_2 is found in volcano gases. In contrast, hydrogen is the tenth most abundant element in the earth's crust (1520 ppm or 0.152% by weight). It also occurs in vast quantities as water in the oceans. Compounds containing hydrogen are very abundant, particularly water, living matter (carbohydrates and proteins), organic compounds, fossil fuels (coal, petroleum, and natural gas), ammonia and acids. In fact hydrogen is present in more compounds than any other element.

PREPARATION OF HYDROGEN

Hydrogen is manufactured on a large scale by a variety of methods:

1. Hydrogen is made cheaply, and in large amounts, by passing steam over red hot coke. The product is water gas, which is a mixture of CO and H_2. This is an important industrial fuel since it is easy to make and it burns, evolving a lot of heat.

$$\mathrm{C} + \mathrm{H_2O} \xrightarrow{1000\,^\circ\mathrm{C}} \underbrace{\mathrm{CO} + \mathrm{H_2}}_{\text{water gas}}$$

$$\mathrm{CO} + \mathrm{H_2} + \mathrm{O_2} \rightarrow \mathrm{CO_2} + \mathrm{H_2O} + \text{heat}$$

It is difficult to obtain pure H_2 from water gas, since CO is difficult to remove. The CO may be liquified at a low temperature under pressure, thus separating it from H_2. Alternatively the gas mixture can be mixed with steam, cooled to 400 °C and passed over iron oxide in a shift converter, giving H_2 and CO_2. The CO_2 so formed is easily removed either by dissolving in water under pressure, or reacting with K_2CO_3 solution, giving $KHCO_3$, and thus giving H_2 gas.

$$\underbrace{\mathrm{CO} + \mathrm{H_2}}_{\text{water gas}} \xrightarrow[450\,^\circ\mathrm{C}\ \mathrm{Fe_2O_3}]{+\mathrm{H_2O}} 2\mathrm{H_2} + \mathrm{CO_2}$$

2. Hydrogen is also made in large amounts by the steam reformer process. The hydrogen produced in this way is used in the Haber process to make NH_3, and for hardening oils. Light hydrocarbons such as methane are mixed with steam and passed over a nickel catalyst at 800–900 °C. These hydrocarbons are present in natural gas, and are also produced at oil refineries when 'cracking' hydrocarbons.

$$\mathrm{CH_4} + \mathrm{H_2O} \rightarrow \mathrm{CO} + 3\mathrm{H_2}$$
$$\mathrm{CH_4} + 2\mathrm{H_2O} \rightarrow \mathrm{CO_2} + 4\mathrm{H_2}$$

The gas emerging from the reformer contains CO, CO_2, H_2 and excess steam. The gas mixture is mixed with more steam, cooled to 400 °C and passed into a shift converter. This contains an iron/copper catalyst and CO is converted into CO_2.

$$\mathrm{CO} + \mathrm{H_2O} \rightarrow \mathrm{CO_2} + \mathrm{H_2}$$

Finally the CO_2 is absorbed in a solution of K_2CO_3 or ethanolamine $HOCH_2CH_2NH_2$. The K_2CO_3 or ethanolamine are regenerated by heating.

$$\mathrm{K_2CO_3} + \mathrm{CO_2} + \mathrm{H_2O} \rightarrow 2\mathrm{KHCO_3}$$
$$2\mathrm{HOCH_2CH_2NH_2} + \mathrm{CO_2} + \mathrm{H_2O} \rightarrow (\mathrm{HOCH_2CH_2NH_3})_2\mathrm{CO_3}$$

3. In oil refineries, natural hydrocarbon mixtures of high molecular weight such as naphtha and fuel oil are 'cracked' to produce lower molecular weight hydrocarbons which can be used as petrol. Hydrogen is a valuable by-product.
4. Very pure hydrogen (99.9% pure) is made by electrolysis of water or solutions of NaOH or KOH. This is the most expensive method. Water does not conduct electricity very well, so it is usual to electrolyse aqueous solutions of NaOH or KOH in a cell with nickel anodes and

iron cathodes. The gases produced in the anode and cathode compartments must be kept separate.

$$\text{Anode} \qquad 2OH^- \rightarrow H_2O + \tfrac{1}{2}O_2 + 2e^-$$
$$\text{Cathode} \qquad 2H_2O + 2e^- \rightarrow 2OH^- + H_2$$
$$\text{Overall} \qquad H_2O \rightarrow H_2 + \tfrac{1}{2}O_2$$

5. A large amount of pure hydrogen is also formed as a by-product from the chlor-alkali industry, in which aqueous NaCl is electrolysed to produce NaOH, Cl_2 and H_2.
6. The usual laboratory preparation is the reaction of dilute acids with metals, or of an alkali with aluminium.

$$Zn + H_2SO_4 \rightarrow ZnSO_4 + H_2$$
$$2Al + 2NaOH + 6H_2O \rightarrow 2Na[Al(OH)_4] + 3H_2$$

7. Hydrogen can be prepared by the reaction of salt-like hydrides with water.

$$LiH + H_2O \rightarrow LiOH + H_2$$

PROPERTIES OF MOLECULAR HYDROGEN

Hydrogen is the lightest gas known, and because of its low density, it is used instead of helium to fill balloons for meteorology. It is colourless, odourless and almost insoluble in water. Hydrogen forms diatomic molecules H_2, and the two atoms are joined by a very strong covalent bond (bond energy 435.9 kJ mol^{-1}).

Hydrogen is not very reactive under normal conditions. The lack of reactivity is due to kinetics rather than thermodynamics, and relates to the strength of the H—H bond. An essential step in H_2 reacting with another element is the breaking of the H—H bond to produce atoms of hydrogen. This requires 435.9 kJ mol^{-1}: hence there is a high activation energy to such reactions. Consequently many reactions are slow, or require high temperatures, or catalysts (often transition metals). Many important reactions of hydrogen involve heterogeneous catalysis, where the catalyst first reacts with H_2 and either breaks or weakens the H—H bond, and thus lowers the activation energy. Examples include:

1. The Haber process for the manufacture of NH_3 from N_2 and H_2 using a catalyst of activated Fe at 380–450 °C and 200 atmospheres pressure.
2. The hydrogenation of a variety of unsaturated organic compounds, (including the hardening of oils), using finely divided Ni, Pd or Pt as catalysts.
3. The production of methanol by reducing CO with H_2 over a Cu/Zn catalyst at 300 °C.

Thus hydrogen will react directly with most elements *under the appropriate conditions*.

Hydrogen burns in air or oxygen, forming water, and liberates a large amount of energy. This is used in the oxy-hydrogen flame for welding and cutting metals. Temperatures of almost 3000 °C can be attained. Care should be taken with these gases since mixtures of H_2 and O_2 close to a 2 : 1 ratio are often explosive.

$$2H_2 + O_2 \rightarrow 2H_2O \qquad \Delta H = -485\,\text{kJ mol}^{-1}$$

Hydrogen reacts with the halogens. The reaction with fluorine is violent, even at low temperatures. The reaction with chlorine is slow in the dark, but the reaction is catalysed by light (photocatalysis), and becomes faster in daylight, and explosive in sunlight. Direct combination of the elements is used to produce HCl.

$$H_2 + F_2 \rightarrow 2HF$$
$$H_2 + Cl_2 \rightarrow 2HCl$$

A number of metals react with H_2, forming hydrides. The reactions are not violent, and usually require a high temperature. These are described in a later section.

Large quantities of H_2 are used in the industrial production of ammonia by the Haber process. The reaction is reversible, and the formation of NH_3 is favoured by high pressure, the presence of a catalyst (Fe), and a low temperature. In practice a high temperature of 380–450 °C and a pressure of 200 atmospheres are used to get a reasonable conversion in a reasonable time.

$$N_2 + 3H_2 \rightleftharpoons 2NH_3 \qquad \Delta G_{298°C} = -33.4\,\text{kJ mol}^{-1}$$

Large amounts of H_2 are used for hydrogenation reactions, in which hydrogen is added to a double bond in an organic compound. An important example is the hardening of fats and oils. Unsaturated fatty acids are hydrogenated with H_2 and a palladium catalyst, forming saturated fatty acids which have higher melting points. By removing double bonds in the carbon chain in this way, edible oils which are liquid at room temperature may be converted into fats which are solid at room temperature. The reason for doing this is that solid fats are more useful than oils, for example in the manufacture of margarine.

$$CH_3 \cdot (CH_2)_n \cdot CH{=}CH \cdot COOH + H_2 \rightarrow CH_3 \cdot (CH_2)_n \cdot CH_2 \cdot CH_2 \cdot COOH$$

Hydrogen is also used to reduce nitrobenzene to aniline (dyestuffs industry), and in the catalytic reduction of benzene (the first step in the production of nylon-66). It also reacts with CO to form methyl alcohol.

$$CO + 2H_2 \xrightarrow{\text{catalyst}} CH_3OH$$

The hydrogen molecule is very stable, and has little tendency to dissociate at normal temperatures, since the dissociation reaction is highly endothermic.

$$H_2 \rightarrow 2H \qquad \Delta H = 435.9\,\text{kJ mol}^{-1}$$

However, at high temperatures, in an electric arc, or under ultraviolet light, H_2 does dissociate. The atomic hydrogen produced exists for less than half a second, after which it recombines to give molecular hydrogen and a large amount of heat. This reaction has been used in welding metals. Atomic hydrogen is a strong reducing agent, and is commonly prepared in solution by means of a zinc–copper couple or a mercury–aluminium couple.

There has been much talk of *the hydrogen economy*. (See Further Reading.) The idea is that hydrogen could replace coal and oil as the major source of energy. Burning hydrogen in air or oxygen forms water and liberates a great deal of energy. In contrast to burning coal or oil in power stations, or petrol or diesel fuel in motor engines, burning hydrogen produces no pollutants like SO_2 and oxides of nitrogen that are responsible for acid rain, nor CO_2 that is responsible for the greenhouse effect, nor carcinogenic hydrocarbons, nor lead compounds. Hydrogen can be produced readily by electrolysis, and chemical methods. Hydrogen can be stored and transported as gas in cylinders, as liquid in very large cryogenic vacuum flasks, or 'dissolved' in various metals. (For example, the alloy $LaNi_5$ can absorb seven moles of hydrogen per mole of alloy at 2.5 atmospheres pressure and room temperature.) Liquid hydrogen is used as a fuel in space rockets for the Saturn series and the space shuttle in the US space programme. Car engines have been modified to run on hydrogen. Note that the use of hydrogen involves the risk of an explosion, but so does the use of petrol.

ISOTOPES OF HYDROGEN

If atoms of the same element have different mass numbers they are called isotopes. The difference in mass number arises because the nucleus contains a different number of neutrons. Naturally occurring hydrogen contains three isotopes: protium ^1_1H or H, deuterium ^2_1H or D, and tritium ^3_1H or T. Each of the three isotopes contains one proton and 0, 1 or 2 neutrons respectively in the nucleus. Protium is by far the most abundant.

Naturally occurring hydrogen contains 99.986% of the ^1_1H isotope, 0.014% of ^2_1D and $7 \times 10^{-16}\%$ ^3_1T, so the properties of hydrogen are essentially those of the lightest isotope.

These isotopes have the same electronic configuration and have essentially the same chemical properties. The only differences in chemical properties are the rates of reactions, and equilibrium constants. For example:

1. H_2 is more rapidly adsorbed on to surfaces than D_2.
2. H_2 reacts over 13 times faster with Cl_2 than D_2, because H_2 has a lower energy of activation.

Differences in properties which arise from differences in mass are called *isotope effects*. Because hydrogen is so light, the percentage difference in mass between protium ^1_1H, deuterium ^2_1H and tritium ^3_1H is greater than

Table 8.1 Physical constants for hydrogen, deuterium and tritium

Physical constant	H_2	D_2	T_2
Mass of atom (amu)	1.0078	2.0141	3.0160
Freezing point (°C)	−259.0	−254.3	−252.4
Boiling point (°C)	−252.6	−249.3	−248.0
Bond length (Å)	0.7414	0.7414	(0.7414)
Heat of dissociation† ($kJ\,mol^{-1}$)	435.9	443.4	446.9
Latent heat of fusion ($kJ\,mol^{-1}$)	0.117	0.197	0.250
Latent heat of vaporisation ($kJ\,mol^{-1}$)	0.904	1.226	1.393
Vapour pressure* (mm Hg)	54	5.8	–

* Measured at −259.1 °C.
† Measured at 25 °C.

between the isotopes of any other element. Thus the isotopes of hydrogen show much greater differences in physical properties than are found between the isotopes of other elements. Some physical constants for H_2, D_2 and T_2 are given in Table 8.1.

Protium water H_2O dissociates to about three times the extent that heavy water D_2O does. The equilibrium constant for the dissociation of H_2O is 1.0×10^{-14} whilst for D_2O it is 3.0×10^{-15}.

$$H_2O \rightleftharpoons H^+ + OH^-$$

$$D_2O \rightleftharpoons D^+ + OD^-$$

Protium bonds are broken more readily than deuterium bonds (up to 18 times more readily in some cases). Thus when water is electrolysed, H_2 is liberated much faster than D_2, and the remaining water thus becomes enriched in heavy water D_2O. If the process is continued until only a small volume remains, then almost pure D_2O is obtained. About 29 000 litres of water must be electrolysed to give 1 litre of D_2O that is 99% pure. This is the normal way of separating deuterium. Heavy water D_2O undergoes all of the reactions of ordinary water, and is useful in the preparation of other deuterium compounds. Because D_2O has a lower dielectric constant, ionic

Table 8.2 Physical constants for water and heavy water

Physical constant	H_2O	D_2O
Freezing point (°C)	0	3.82
Boiling point (°C)	100	101.42
Density at 20 °C ($g\,cm^{-3}$)	0.917	1.017
Temperature of maximum density (°C)	4	11.6
Ionic product K_w at 25 °C	1.0×10^{-14}	3.0×10^{-15}
Dielectric constant at 20 °C	82	80.5
Solubility g NaCl/100 g water at 25 °C	35.9	30.5
Solubility g $BaCl_2$/100 g water at 25 °C	35.7	28.9

compounds are less soluble in it than in water. Some physical properties of H_2O and D_2O are compared in Table 8.2.

Deuterium compounds are commonly prepared by 'exchange' reactions where under suitable conditions deuterium is exchanged for hydrogen in compounds. Thus D_2 reacts with H_2 at high temperatures, forming HD, and it also exchanges with NH_3 and CH_4 to give NH_2D, NHD_2, ND_3 and $CH_3D—CD_4$. It is usually easier to prepare deuterated compounds using D_2O rather than D_2. The D_2O may be used directly in the preparation instead of H_2O, or exchange reactions may be carried out using D_2O.

Exchange reactions

$$NaOH + D_2O \rightarrow NaOD + HDO$$

$$NH_4Cl + D_2O \rightarrow NH_3DCl + HDO$$

$$Mg_3N_2 + 3D_2O \rightarrow 2ND_3 + 3MgO$$

Direct reactions

$$SO_3 + D_2O \rightarrow D_2SO_4$$

$$P_4O_{10} + 6D_2O \rightarrow 4D_3PO_4$$

Tritium is radioactive and decays by β emission.

$$^3_1T \rightarrow {}^3_2He + {}^{\ 0}_{-1}e$$

It has a relatively short half life time of 12.26 years. Thus any T present when the earth was formed has decayed already, and the small amount now present has been formed recently by reactions induced by cosmic rays in the upper atmosphere.

$$^{14}_7N + {}^1_0n \rightarrow {}^{12}_6C + {}^3_1T$$

$$^{14}_7N + {}^1_1H \rightarrow {}^3_1T + \text{other fragments}$$

$$^2_1D + {}^2_1D \rightarrow {}^3_1T + {}^1_1H$$

Tritium only occurs to the extent of one part T_2 to 7×10^{17} parts H_2. It was first made by bombarding D_3PO_4 and $(ND_4)_2SO_4$ with deuterons D^+.

$$^2_1D + {}^2_1D \rightarrow {}^3_1T + {}^1_1H$$

It is now produced on a large scale by irradiating lithium with slow neutrons in a nuclear reactor.

$$^6_3Li + {}^1_0n \rightarrow {}^4_2He + {}^3_1T$$

Tritium is used to make thermonuclear devices, and for research into fusion reactions as a means of producing energy. The gas is usually stored by making UT_3, which on heating to 400 °C releases T_2. Tritium is widely used as a radioactive tracer, since it is relatively cheap, and it is easy to work with. It only emits low energy β radiation, with no γ radiation. The β radiation is stopped by 0.6 cm of air, so no shielding is required. It is non-toxic, except if labelled compounds are swallowed.

Tritiated compounds are made from T_2 gas. T_2O is made as follows:

$$T_2 + CuO \rightarrow T_2O + Cu$$

or

$$2T_2 + O_2 \xrightarrow{\text{Pd catalyst}} 2T_2O$$

Many tritiated organic compounds can be made by storing the compound under T_2 gas for a few weeks, when exchange of H and T occurs. Many compounds can be made by catalytic exchange in solution using either T_2 gas dissolved in the water, or T_2O.

$$NH_4Cl + T_2O \text{ (or HTO)} \rightleftharpoons NH_3TCl$$

ORTHO AND *PARA* HYDROGEN

The hydrogen molecule H_2 exists in two different forms known as *ortho* and *para* hydrogen. The nucleus of an atom has nuclear spin, in a similar way to electrons having a spin. In the H_2 molecule, the two nuclei may be spinning in either the same direction, or in opposite directions. This gives rise to spin isomerism, that is two different forms of H_2 may exist. These are called *ortho* and *para* hydrogen. Spin isomerism is also found in other symmetrical molecules whose nuclei have spin momenta, e.g. D_2, N_2, F_2, Cl_2. There are considerable differences between the physical properties (e.g. boiling points, specific heats and thermal conductivities) of the *ortho* and *para* forms, because of differences in their internal energy. There are also differences in the band spectra of the *ortho* and *para* forms of H_2.

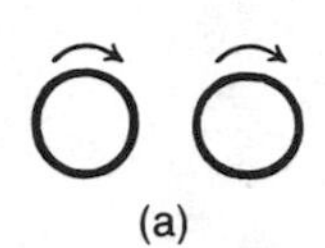

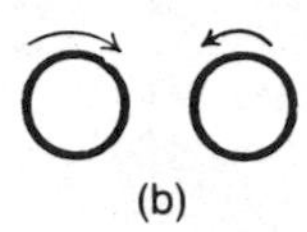

Figure 8.1 *Ortho* and *para* hydrogen: (a) *ortho*, parallel spins; (b) *para*, opposite.

The *para* form has the lower energy, and at absolute zero the gas contains 100% of the *para* form. As the temperature is raised, some of the *para* form changes into the *ortho* form. At high temperatures the gas contains about 75% *ortho* hydrogen.

Para hydrogen is usually prepared by passing a mixture of the two forms of hydrogen through a tube packed with charcoal cooled to liquid air temperature. *Para* hydrogen prepared in this way can be kept for weeks at room temperature in a glass vessel, because the *ortho–para* conversion is slow in the absence of catalysts. Suitable catalysts include activated charcoal, atomic hydrogen, metals such as Fe, Ni, Pt and W and paramagnetic substances or ions (which contain unpaired electrons) such as O_2, NO, NO_2, Co^{2+} and Cr_2O_3.

HYDRIDES

Binary compounds of the elements with hydrogen are called hydrides. The type of hydride which an element forms depends on its electronegativity, and hence on the type of bond formed. Whilst there is not a sharp division between ionic, covalent and metallic bonding, it is convenient to consider hydrides in three classes (Figure 8.2):

1. ionic or salt-like hydrides
2. covalent or molecular hydrides
3. metallic or interstitial hydrides

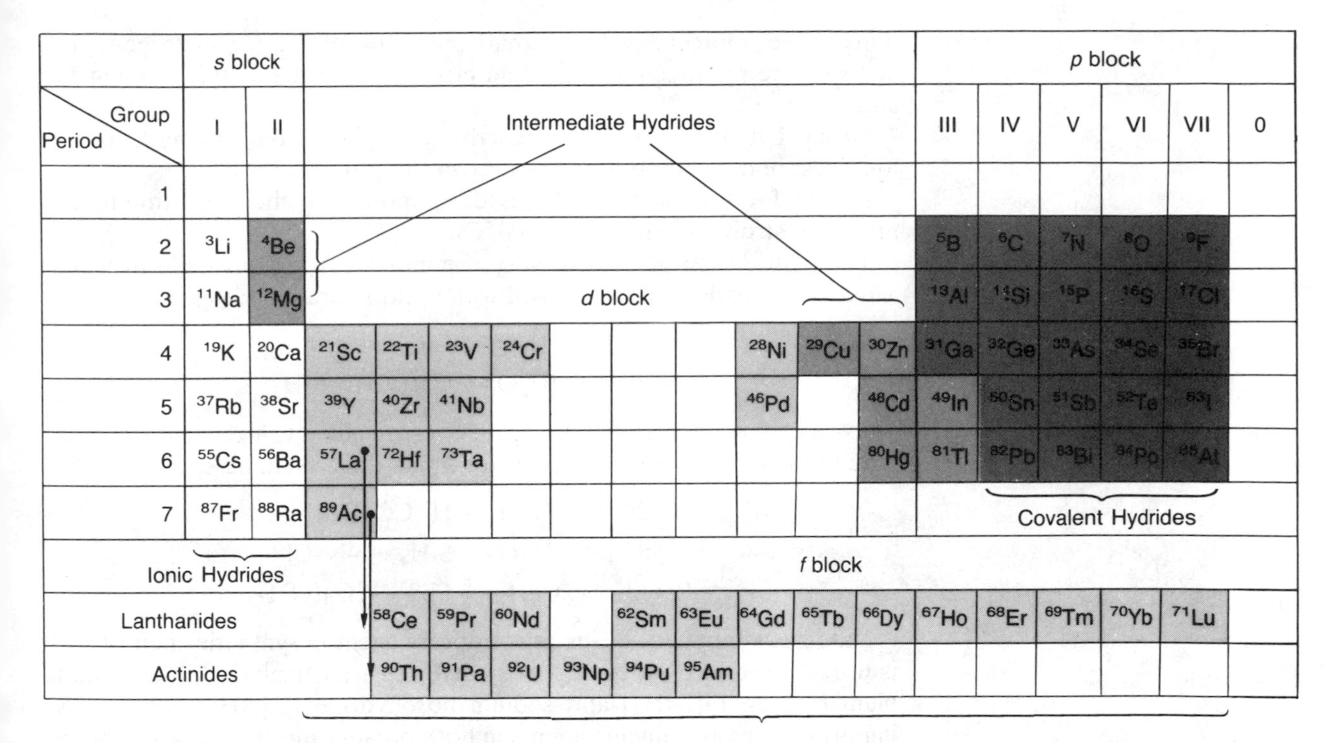

Figure 8.2 Types of hydride and the periodic table.

Ionic or salt-like hydrides

At high temperatures the metals of Group I (alkali metals) and the heavier Group II metals (alkaline earth metals) Ca, Sr and Ba form ionic hydrides such as NaH and CaH_2. These compounds are solids with high melting points, and are classified as ionic (salt-like) hydrides. The evidence that they are ionic is:

1. Molten LiH (m.p. 691 °C) conducts electricity, and H_2 is liberated at the anode, thus confirming the presence of the hydride ion H^-.
2. The other ionic hydrides decompose before melting, but they may be dissolved in melts of alkali halides (e.g. CaH_2 dissolves in a eutectic mixture of LiCl/KCl), and when the melt is electrolysed then H_2 is evolved at the anode.
3. The crystal structures of these hydrides are known, and they show no evidence of directional bonding.

Lithium is more polarizing and hence more likely to form covalent compounds than the other metals. Thus if LiH is largely ionic, the others must be ionic, and thus contain the hydride ion H^-.

The density of these hydrides is greater than that of the metal from which they were formed. This is explained by H^- ions occupying holes in the lattice of the metal, without distorting the metal lattice. Ionic hydrides have high heats of formation, and are always stoichiometric.

This type of hydride is only formed by elements with an electronegativity value appreciably lower than the value of 2.1 for hydrogen, thus allowing the hydrogen to attract an electron from the metal, forming M^+ and H^-.

Group I hydrides are more reactive than the corresponding Group II hydrides, and reactivity increases down the group.

Except for LiH, ionic hydrides decompose into their constituent elements on strong heating (400–500 °C).

The hydride ion H^- is not very common, and it is unstable in water. Thus ionic hydrides all react with water and liberate hydrogen.

$$LiH + H_2O \rightarrow LiOH + H_2$$
$$CaH_2 + 2H_2O \rightarrow Ca(OH)_2 + H_2$$

They are powerful reducing agents, especially at high temperatures, though their reactivity towards water limits their usefulness.

$$2CO + NaH \rightarrow H\cdot COONa + C$$
$$SiCl_4 + 4NaH \rightarrow SiH_4 + 4NaCl$$
$$PbSO_4 + 2CaH_2 \rightarrow PbS + 2Ca(OH)_2$$

NaH has a number of uses as a reducing agent in synthetic chemistry. It is used to produce other important hydrides, particularly lithium aluminium hydride $Li[AlH_4]$ and sodium borohydride $Na[BH_4]$, which have important uses as reducing agents in both organic and inorganic syntheses.

$$4LiH + AlCl_3 \rightarrow Li[AlH_4] + 3LiCl$$
$$4NaH + B(OCH_3)_3 \rightarrow Na[BH_4] + 3NaOCH_3$$

Covalent hydrides

Hydrides of the *p*-block elements are covalent. This would be expected since there is only a small difference in electronegativity between these atoms and hydrogen. The compounds usually consist of discrete covalent molecules, with only weak van der Waals forces holding the molecules together, and so they are usually volatile, and have low melting and boiling points. They do not conduct electricity. The formula of these hydrides is XH_n or $XH_{(8-n)}$ where n is the group in the periodic table to which X belongs. These hydrides are produced by a variety of synthetic methods:

1. A few may be made by direct action.

Group	III	IV	V	VI	VII
	B	C	N	O	F
	Al	Si	P	S	Cl
	Ga	Ge	As	Se	Br
	In	Sn	Sb	Te	I
		Pb	Bi	Po	

Figure 8.3 Covalent hydrides.

$$3H_2 + N_2 \rightarrow 2NH_3 \text{ (high temperature and pressure + catalyst, Haber process)}$$

$$2H_2 + O_2 \rightarrow 2H_2O \text{ (spark – explosive)}$$

$$H_2 + Cl_2 \rightarrow 2HCl \text{ (burn – preparation of pure HCl)}$$

2. Reaction of a halide with $Li[AlH_4]$ in a dry solvent such as ether.

$$4BCl_3 + 3Li[AlH_4] \rightarrow 2B_2H_6 + 3AlCl_3 + 3LiCl$$

$$SiCl_4 + Li[AlH_4] \rightarrow SiH_4 + AlCl_3 + LiCl$$

3. Treating the appropriate binary compound with acid.

$$2Mg_3B_2 + 4H_3PO_4 \rightarrow B_4H_{10} + 2Mg_3(PO_4)_2 + H_2$$

$$Al_4C_3 + 12HCl \rightarrow 3CH_4 + 4AlCl_3$$

$$FeS + H_2SO_4 \rightarrow H_2S + FeSO_4$$

$$Ca_3P_2 + 3H_2SO_4 \rightarrow 2PH_3 + 3CaSO_4$$

4. Reaction of an oxoacid with $Na[BH_4]$ in aqueous solution.

$$4H_3AsO_3 + 3Na[BH_4] \rightarrow 4AsH_3 + 3H_3BO_3 + 3NaOH$$

5. Converting one hydride into another by pyrolysis (heating).

$$B_4H_{10} \rightarrow B_2H_6 + \text{other products}$$

6. A silent electric discharge or microwave discharge may produce long chains from simple hydrides.

$$GeH_4 \rightarrow Ge_2H_6 \rightarrow Ge_3H_8 \rightarrow \text{up to } Ge_9H_{20}$$

Table 8.3 Melting and boiling points of some covalent hydrides

Compound	m.p. (°C)	b.p. (°C)
B_2H_6	−165	−90
CH_4	−183	−162
SiH_4	−185	−111
GeH_4	−166	−88
SnH_4	−150	−52
NH_3	−78	−33
PH_3	−134	−88
AsH_3	−117	−62
SbH_3	−88	−18
H_2O	0	+100
H_2S	−86	−60
HF	−83	+20
HCl	−115	−84
HBr	−89	−67
HI	−51	−35

The Group III hydrides are unusual in that they are electron deficient and polymeric, although they do not contain direct bonds between the Group III elements. The simplest boron hydride is called diborane B_2H_6, though more complicated structures such as B_4H_{10}, B_5H_9, B_5H_{11}, B_6H_{10} and $B_{10}H_{14}$ are known. Aluminium hydride is polymeric $(AlH_3)_n$. In these structures, hydrogen appears to be bonded to two or more atoms, and this is explained in terms of multi-centre bonding. This is discussed in Chapter 12.

In addition to the simple hydrides, the rest of the lighter elements except the halogens form polynuclear hydrides. The tendency to do this is strongest with the elements C, N and O, and two or more of the non-metal atoms are directly bonded to each other. The tendency is greatest with C which catenates (forms chains) of several hundreds of atoms. These are grouped into three homologous series of aliphatic hydrocarbons, and aromatic hydrocarbons based on benzene.

CH_4, C_2H_6, C_3H_8, C_4H_{10} ... C_nH_{2n+2} (alkanes)
C_2H_4, C_3H_6, C_4H_8 .... C_nH_{2n} (alkenes)
C_2H_2, C_3H_4, C_4H_6 .... C_nH_{2n-2} (alkynes)
C_6H_6 (aromatic)

The alkanes are saturated, but alkenes have double bonds, and alkynes have triple bonds. Si and Ge only form saturated compounds, and the maximum chain length is $Si_{10}H_{22}$. The longest hydride chains formed by other elements are Sn_2H_6, N_2H_4 and HN_3, P_3H_5, As_3H_5, H_2O_2 and H_2O_3, and H_2S_2, H_2S_3, H_2S_4, H_2S_5 and H_2S_6.

The melting point and boiling point of water stand out in Table 8.3 as being much higher than the others, but on closer examination the values for NH_3 and HF also seem higher than would be expected in their respective groups. This is due to hydrogen bonding, which is discussed later in this chapter.

Metallic (or interstitial) hydrides

Many of the elements in the *d*-block, and the lanthanide and actinide elements in the *f*-block, react with H_2 and form metallic hydrides. However, the elements in the middle of the *d*-block do not form hydrides. The absence of hydrides in this part of the periodic table is sometimes called *the hydrogen gap*. (See Figure 8.2.)

Metallic hydrides are usually prepared by heating the metal with hydrogen under high pressure. (If heated to higher temperatures the hydrides decompose, and this may be used as a convenient method of making very pure hydrogen.)

These hydrides generally have properties similar to those of the parent metals: they are hard, have a metallic lustre, conduct electricity, and have magnetic properties. The hydrides are less dense than the parent metal, because the crystal lattice has expanded through the inclusion of hydrogen.

This distortion of the crystal lattice may make the hydride brittle. Thus when the hydride is formed a solid piece of metal turns into finely powdered hydride. If the finely powdered hydrides are heated they decompose, giving hydrogen and very finely divided metal. These finely divided metals may be used as catalysts. They are also used in metallurgy in powder fabrication, and zirconium hydride has been used as a moderator in nuclear reactors.

In many cases the compounds are nonstoichiometric, for example LaH_n, TiH_n and PdH_n, where the chemical composition is variable. Typical formulae are $LaH_{2.87}$, $YbH_{2.55}$, $TiH_{1.8}$, $ZrH_{1.9}$, $VH_{1.6}$, $NbH_{0.7}$ and $PdH_{0.7}$. Such compounds were originally called *interstitial hydrides*, and it was thought that a varying number of interstitial positions in the metal lattice could be filled by hydrogen.

The nonstoichiometric compounds may be regarded as solid solutions. Metals which can 'dissolve' varying amounts of hydrogen in this way can act as catalysts for hydrogenation reactions. The catalysts are thought to be effective through providing H atoms rather than H_2 molecules. It is not certain whether the hydrogen is present in the interstitial sites as atoms of hydrogen, or alternatively as H^+ ions with delocalized electrons, but they have strongly reducing properties.

Even small amounts of hydrogen dissolved in a metal adversely affect its strength and make it brittle. Titanium is extracted by reducing $TiCl_4$ with Mg or Na in an inert atmosphere. If an atmosphere of H_2 is used, the Ti dissolves H_2, and is brittle. Titanium is used to make supersonic aircraft, and since strength is important, it is produced in an atmosphere of argon.

The bonding is more complicated than was originally thought, and is still the subject of controversy.

1. Many of the hydrides have structures where hydrogen atoms occupy tetrahedral holes in a cubic close-packed array of metal atoms. If all of the tetrahedral sites are occupied then the formula is MH_2, and a fluorite structure is formed. Generally some sites are unoccupied, and hence the compounds contain less hydrogen. This accounts for the compounds of formula $MH_{1.5-2}$ formed by the scandium and titanium groups, and most of the lanthanides and actinides.
2. Two of the lanthanide elements, europium and ytterbium, are unusual in that they form ionic hydrides EuH_2 and YbH_2, which are stoichiometric and resemble CaH_2. The lanthanides are typically trivalent, but Eu and Yb form divalent ions (associated with stable electronic structures Eu(+II) $4f^7$ (half filled *f* shell), and Yb(+II) $4f^{14}$ (filled *f* shell)).
3. The compounds YH_2 and LaH_2, as well as many of the lanthanide and actinide hydrides MH_2, can absorb more hydrogen, forming compounds of limiting composition MH_3. Compositions such as $LaH_{2.76}$ and $CeH_{2.69}$ are found. The structures of these are complex, sometimes cubic and sometimes hexagonal. The third hydrogen atom is more loosely held than the others, and rather surprisingly it may occupy an octahedral hole.

4. Uranium is unusual and forms two different crystalline forms of UH_3 that are stoichiometric.
5. Some elements (V, Nb, Ta, Cr, Ni and Pd) form hydrides approximating to MH. Formulae such as $NbH_{0.7}$ and $PdH_{0.6}$ are typical. These are less stable than the other hydrides, are nonstoichiometric and exist over a wide range of composition.

The Pd/H_2 system is both extraordinary and interesting. When red hot Pd is cooled in H_2 it may absorb or occlude up to 935 times its own volume of H_2 gas. This may be used to separate H_2 or deuterium D_2 from He or other gases. The hydrogen is given off when the metal is heated, and this provides an easy method of weighing H_2. The limiting formula is $PdH_{0.7}$, but neither the structure nor the nature of the interaction between Pd and H are understood. As hydrogen is absorbed, the metallic conductivity decreases, and the material eventually becomes a semiconductor. The hydrogen is mobile and diffuses throughout the metal. It is possible that the erroneous reports of producing energy by 'cold-fusion' by electrolysis of D_2O at room temperature between Pd electrodes was really energy from the reaction between Pd and D_2 rather than nuclear fusion of hydrogen or deuterium to give helium. (See Chapter 31.)

Intermediate hydrides

A few hydrides do not fit easily into the above classification. Thus $(BeH_2)_n$ is polymeric, and is thought to be a chain polymer with hydrogen bridges. MgH_2 has properties in between those of ionic and covalent hydrides.

CuH, ZnH_2, CdH_2 and HgH_2 have properties intermediate between metallic and covalent hydrides. They are probably electron deficient like $(AlH_3)_n$. CuH is endothermic, that is energy must be put in to make the compound, and is formed by reducing Cu^{2+} with hypophosphorous acid. The hydrides of Zn, Cd and Hg are made by reducing the chlorides with $Li[AlH_4]$.

THE HYDROGEN ION

The energy required to remove the electron from a hydrogen atom (i.e. the ionization energy of hydrogen) is $1311\,kJ\,mol^{-1}$. This is a very large amount of energy, and consequently the bonds formed by hydrogen in the gas phase are typically covalent. Hydrogen fluoride is the compound most likely to contain ionic hydrogen (i.e. H^+), since it has the greatest difference in electronegativity, but even here the bond is only 45% ionic.

Thus compounds containing H^+ will only be formed if the ionization energy can be provided by some other process. Thus if the compound is dissolved, for example in water, then the hydration energy may offset the very high ionization energy. In water H^+ are solvated, forming H_3O^+, and the energy evolved is $1091\,kJ\,mol^{-1}$. The remainder of the $1311\,kJ\,mol^{-1}$ ionization energy comes from the electron affinity (the energy evolved in forming the negative ion), and also the solvation energy of the negative ion.

Compounds which form solvated hydrogen ions in a suitable solvent are called acids. Even though the ions present are H_3O^+ (or even $H_9O_4^+$), it is customary to write the ion as H^+, indicating a hydrated proton.

HYDROGEN BONDING

In some compounds a hydrogen atom is attracted by rather strong forces to two atoms, for example in $[F—H—F]^-$. (Sometimes hydrogen is attracted to more than two atoms.) It was at first thought that hydrogen formed two covalent bonds, but it is now recognized that, since hydrogen has the electronic structure $1s^1$, it can only form one covalent bond. The hydrogen bond is most simply regarded as a weak electrostatic attraction between a lone pair of electrons on one atom, and a covalently bonded hydrogen atom that carries a fractional charge $\delta+$.

Hydrogen bonds are formed only with the most electronegative atoms. (Of these, F, O, N and Cl are the four most important elements.) These bonds are very weak, and are typically about $10\,kJ\,mol^{-1}$, though hydrogen bonds may have a bond energy from 4 to $45\,kJ\,mol^{-1}$. This must be compared with a C—C covalent bond of $347\,kJ\,mol^{-1}$. Despite their low bond energy, hydrogen bonds are of great significance both in biochemical systems and in normal chemistry. They are extremely important because they are responsible for linking polypeptide chains in proteins, and for linking pairs of bases in large nucleic acid-containing molecules. The hydrogen bonds maintain these large molecules in specific molecular configurations, which is important in the operation of genes and enzymes. Hydrogen bonds are responsible for water being liquid at room temperature, and but for this, life as we know it would not exist. Since hydrogen bonds have a low bond energy, they also have a low activation energy, and this results in their playing an important part in many reactions at normal temperatures.

Hydrogen bonding was first used to explain the weakness of trimethylammonium hydroxide as a base compared with tetramethylammonium hydroxide. In the trimethyl compound the OH group is hydrogen bonded to the Me_3NH group (shown by a dotted line in Figure 8.4), and this makes it more difficult for the OH group to ionize, and hence it is a weak base. In the tetramethyl compound, hydrogen bonding cannot occur, so the OH group ionizes and the tetramethyl compound is thus a much stronger base.

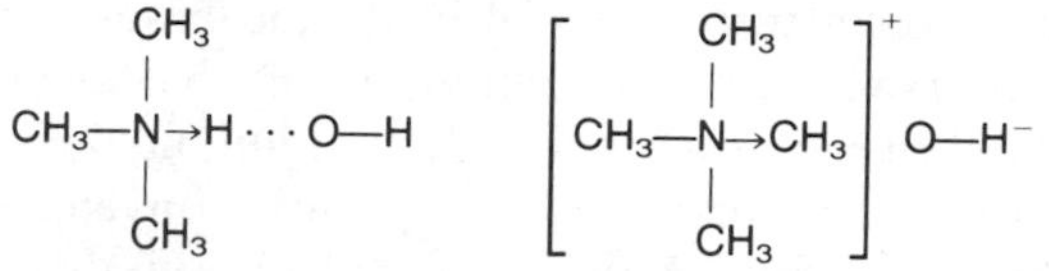

```
     CH3                     ┌      CH3      ┐+
      |                      │       |       │
CH3—N→H···O—H                │ CH3—N→CH3     │ O—H-
      |                      │       |       │
     CH3                     └      CH3      ┘
```

Figure 8.4 Structures of trimethyl and tetramethyl ammonium hydroxide.

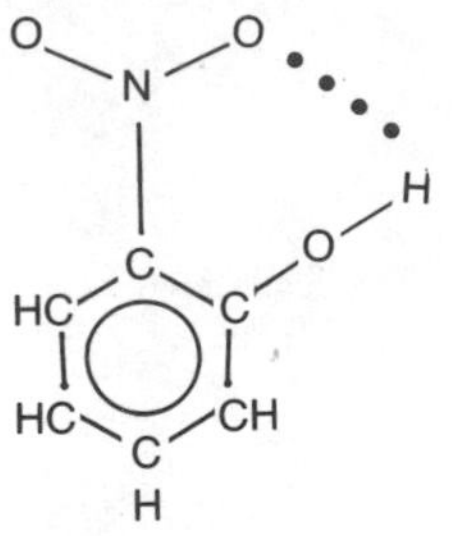

In a similar way the formation of an intramolecular hydrogen bond in *o*-nitrophenol reduces its acidity compared with *m*-nitrophenol and *p*-nitrophenol where the formation of a hydrogen bond is not possible (Figure 8.5).

Intermolecular hydrogen bonding may also take place, and it has a striking effect on the physical properties such as melting points, boiling points, and the enthalpies of vaporization and sublimation (Figure 8.6). In general the melting and boiling points for a related series of compounds increase as the atoms get larger, owing to the increase in dispersive force. Thus by extrapolating the oiling points of H_2Te, H_2Se and H_2S one would predict that the boiling point of H_2O should be about −100 °C, whilst it is actually +100 °C. Thus water boils about 200 °C higher than it would in the absence of hydrogen bonding.

In much the same way the boiling point of NH_3 is much higher than would be expected by comparison with PH_3, AsH_3 and SbH_3, and similarly HF boils much higher than HCl, HBr and HI. The reason for the higher than expected boiling points is hydrogen bonding. Note that the boiling points of the Group IV hydrides CH_4, SiH_4, GeH_4 and SnH_4, and also those of the noble gases, change smoothly, as they do not involve hydrogen bonding.

The hydrogen bonds in HF link the F atom of one molecule with the H atom of another molecule, thus forming a zig-zag chain $(HF)_n$ in both the liquid and also in the solid. Some hydrogen bonding also occurs in the gas, which consists of a mixture of cyclic $(HF)_6$ polymers, dimeric $(HF)_2$, and monomeric HF. (The hydrogen bond in F—H. . .F is 29 kJ mol^{-1} in $HF_{(gas)}$.)

A similar pattern can be seen in the melting points and the enthalpies of vaporization of the hydrides, indicating hydrogen bonding in NH_3, H_2O and HF, but not in CH_4 or Ne.

Strong evidence for hydrogen bonding comes from structural studies. Examples include ice, which has been determined both by X-ray and neutron diffraction, the dimeric structure of formic acid (determined in the gas phase by electron diffraction), X-ray structures of the solids for sodium hydrogencarbonate and boric acid (Figure 8.8), and many others.

Another technique for studying hydrogen bonds is infra-red absorption spectra in CCl_4 solution, which allows the O—H and N—H stretching frequencies to be studied.

ortho

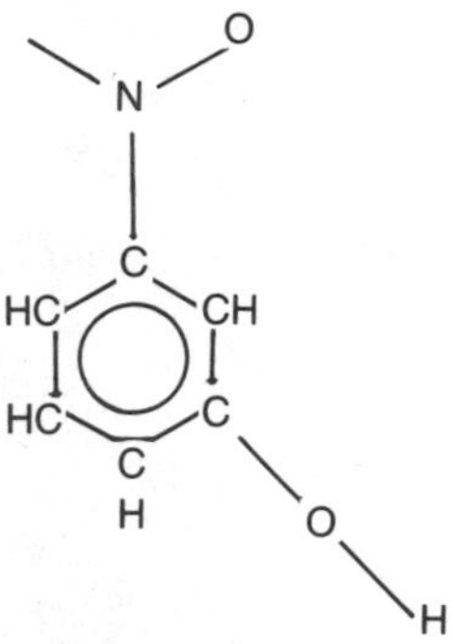

meta

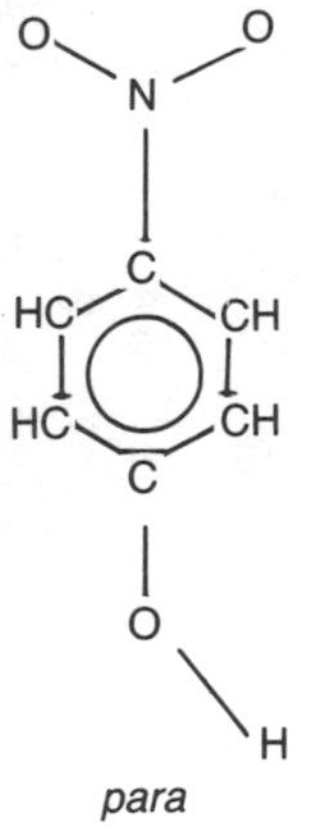

para

Figure 8.5 Structures of *ortho*, *meta* and *para* nitrophenol.

ACIDS AND BASES

There are several so-called *theories* of acids and bases, but they are not really theories but merely different definitions of what we choose to call an acid or a base. Since it is only a matter of definition, no theory is more right or wrong than any other, and we use the most convenient theory for a particular chemical situation. Which is the most useful theory or definition of acids and bases? There is no simple answer to this. The answer depends on whether we are considering ionic reactions in aqueous solution, in non-aqueous solution, or in a fused melt, and whether we require a measure of the strengths of acids and bases. For this reason we need to know several theories.

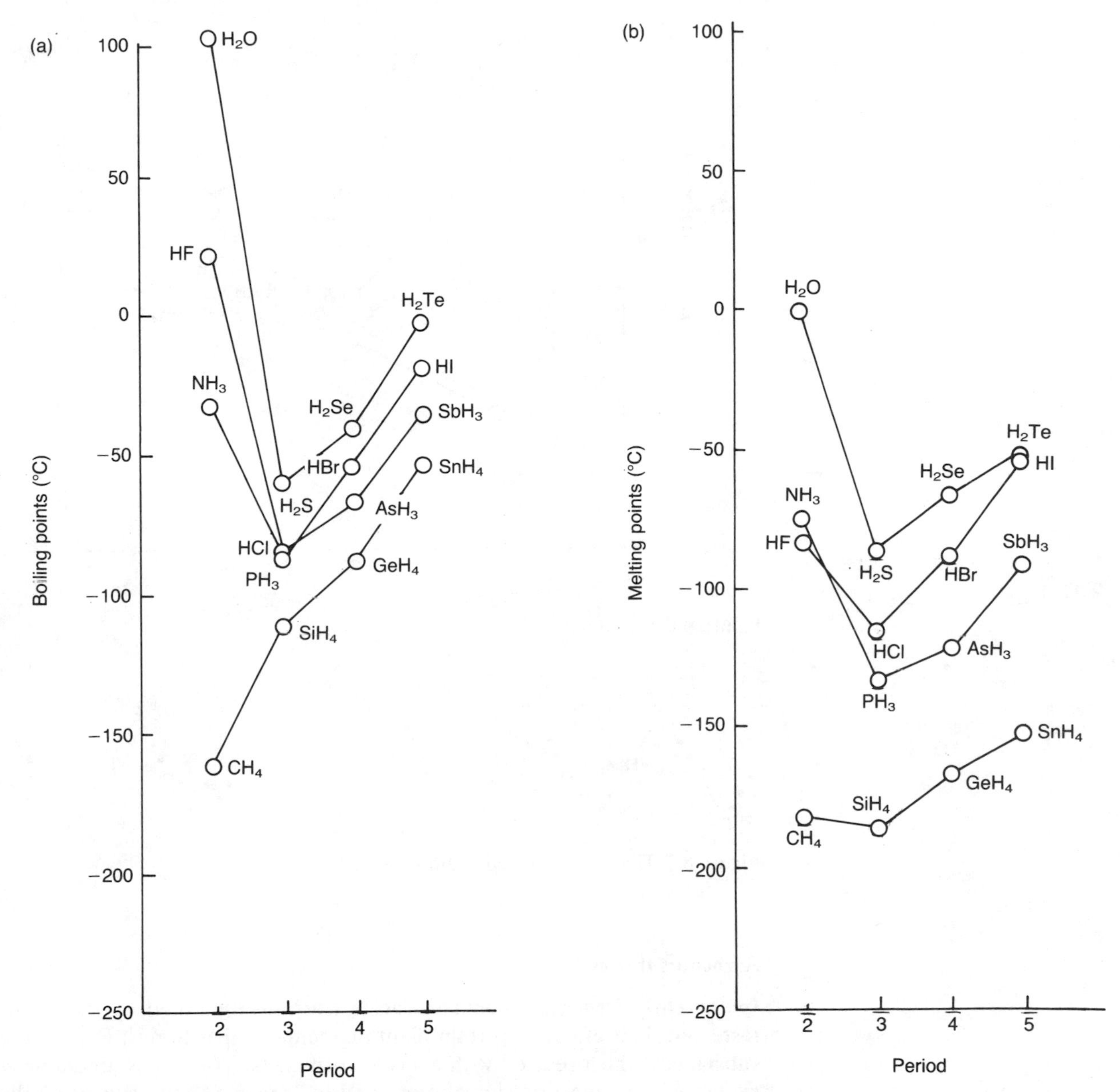

Figure 8.6 (a) Boiling points of hydrides. (b) Melting points of hydrides. (c) Enthalpies of vaporization of hydrides. (Adapted from Lagowski, J.J., *Modern Inorganic Chemistry*, Marcel Dekker, New York, p. 174.)

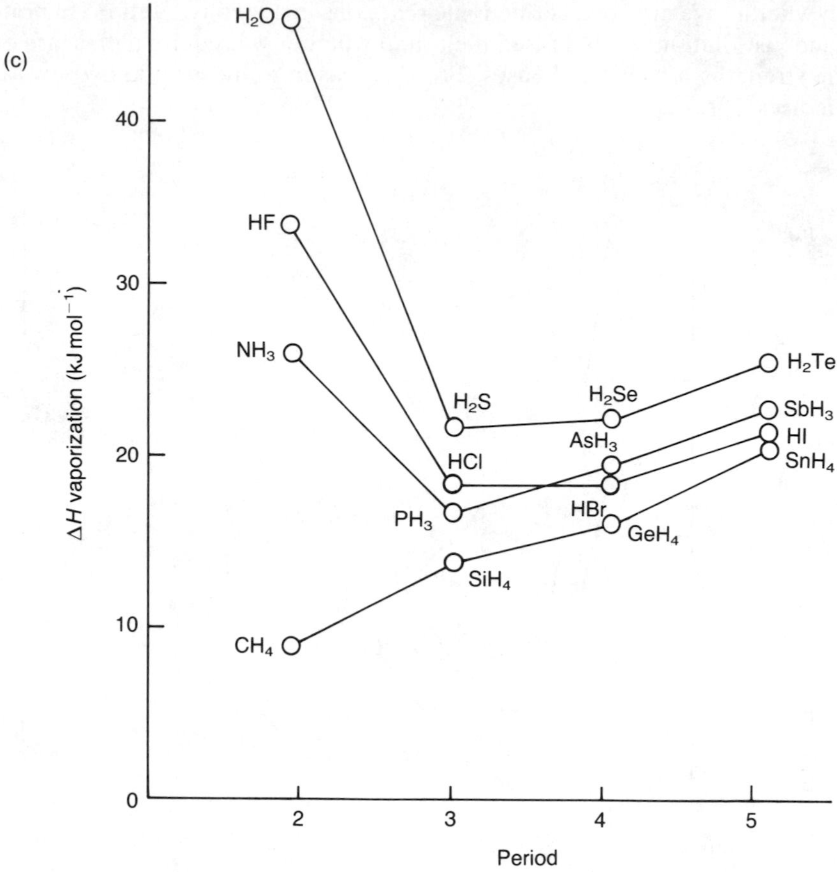

Figure 8.6 continued.

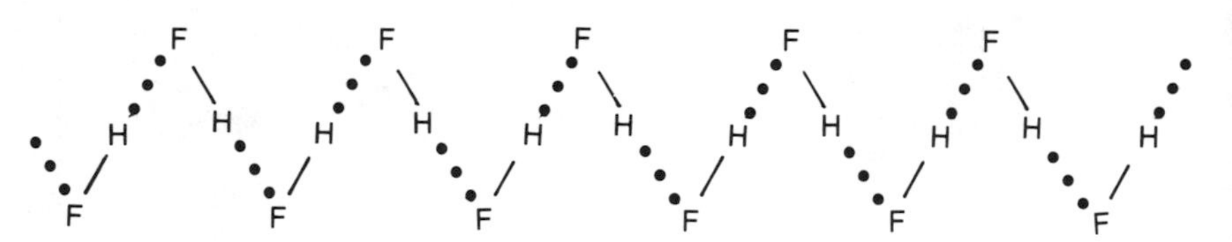

Figure 8.7 Hydrogen bonded chain in solid HF.

Arrhenius theory

In the early stages of chemistry, acids were distinguished by their sour taste and their effect on certain plant pigments such as litmus. Bases were substances which reacted with acids to form salts. Water was used almost exclusively for reactions in solution, and in 1884 Arrhenius suggested the theory of electrolytic dissociation and proposed the self–ionization of water:

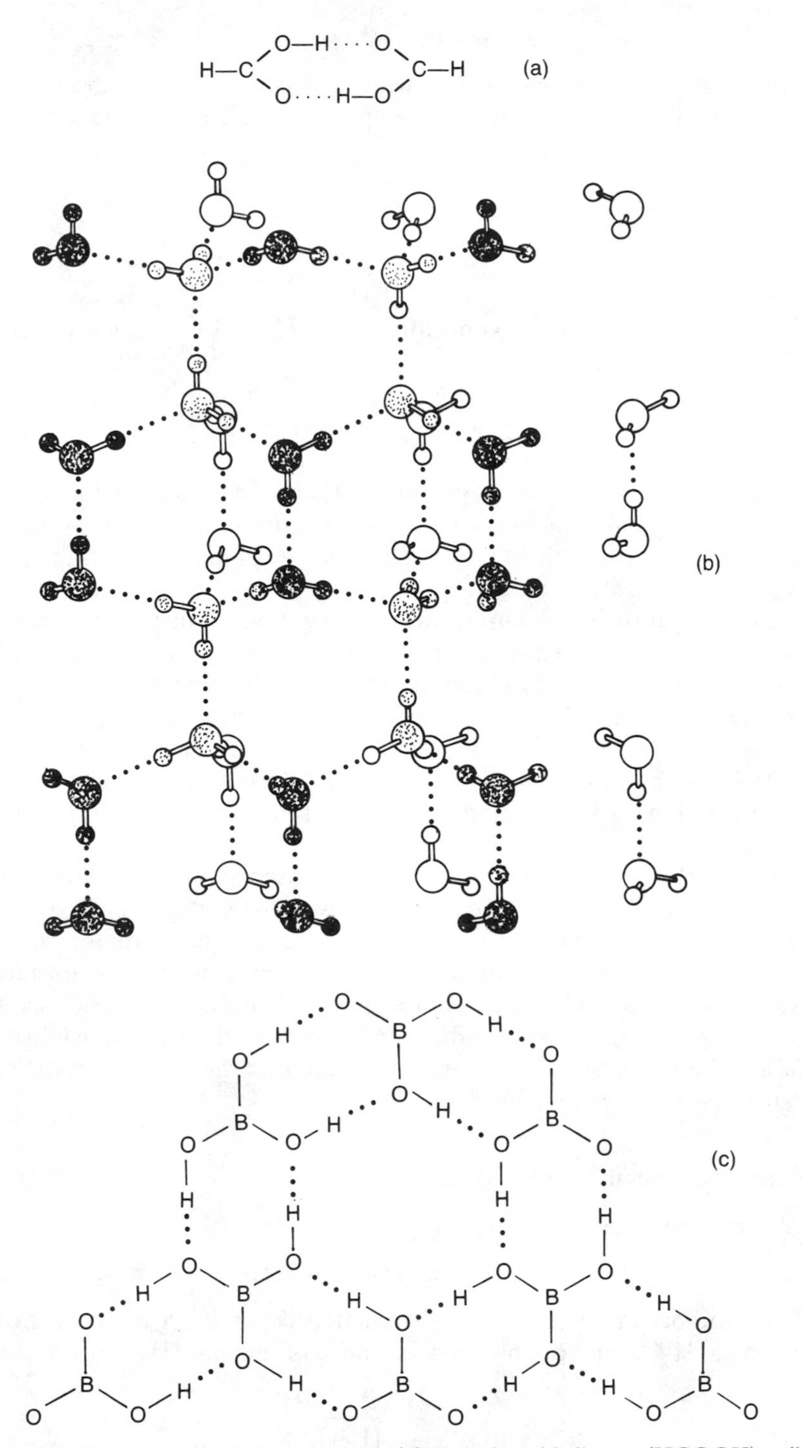

Figure 8.8 Hydrogen bonded structures. (a) Formic acid dimer, $(HCOOH)_2$. (b) Ice. (From L. Pauling, *The Nature of the Chemical Bond*, 3rd ed., pp. 449–504, Cornell University Press, Ithaca, 1960.) (c) A layer of crystalline H_3BO_3.

$$H_2O \rightleftharpoons H^+ + OH^-$$

Thus substances producing H^+ were called acids, and substances producing OH^- were called bases. A typical neutralization reaction is

$$\underset{\text{acid}}{HCl} + \underset{\text{base}}{NaOH} \rightarrow \underset{\text{salt}}{NaCl} + \underset{\text{water}}{H_2O}$$

or simply

$$H^+ + OH^- \rightarrow H_2O$$

In aqueous solutions the concentration of H^+ is often given in terms of pH, where:

$$pH = \log_{10}\frac{1}{[H^+]} = -\log_{10}[H^+]$$

where $[H^+]$ is the hydrogen ion concentration. More strictly the activity of the hydrogen ions should be used. This logarithmic scale is very useful for expressing concentrations over several orders of magnitude (e.g. 1 M H^+ is pH 0, 10^{-14} M H^+ is pH 14).

Until the turn of the nineteenth century it was thought that water was the only solvent in which ionic reactions could occur. Studies made by Cady in 1897 and by Franklin and Kraus in 1898 on reactions in liquid ammonia, and by Walden in 1899 on reactions in liquid sulphur dioxide, revealed many analogies with reactions in water. These analogies suggested that the three media were ionizing solvents and could be useful for ionic reactions, and that acids, bases and salts were common to all three systems.

Although water is still the most widely used solvent, its exclusive use limited chemistry to those compounds which are stable in its presence. Non-aqueous solvents are now used increasingly in inorganic chemistry because many new compounds can be prepared which are unstable in water, and some anhydrous compounds can be prepared, such as anhydrous copper nitrate, which differ markedly from the well known hydrated form. The concepts of acids and bases based on the aqueous system need extending to cover non-aqueous solvents.

Acids and bases in proton solvents

Water self-ionizes:

$$2H_2O \rightleftharpoons H_3O^+ + OH^-$$

The equilibrium constant for this reaction depends on the concentration of water $[H_2O]$, and on the concentrations of the ions $[H_3O^+]$ and $[OH^-]$.

$$K_1 = \frac{[H_3O^+][OH^-]}{[H_2O]^2}$$

Since water is in large excess, its concentration is effectively constant, so the ionic product of water may be written:

Table 8.4 Ionic product of water at various temperatures

Temperature (°C)	K_W(mol^2 l^{-2})
0	0.12×10^{-14}
10	0.29×10^{-14}
20	0.68×10^{-14}
25	1.00×10^{-14}
30	1.47×10^{-14}
40	2.92×10^{-14}
100	47.6×10^{-14}

$$K_w = [H_3O^+][OH^-] = 10^{-14}\,\text{mol}^2\,\text{l}^{-2}$$

The value of K_w is $1.00 \times 10^{-14}\,\text{mol}^2\,\text{l}^{-2}$ at 25 °C, but it varies with temperature. Thus at 25 °C there will be $10^{-7}\,\text{mol}^{-1}$ of H_3O^+ and $10^{-7}\,\text{mol}^{-1}$ of OH^- in pure water.

Acids such as HA increase the concentration of H_3O^+:

$$HA + H_2O \rightleftharpoons H_3O^+ + A^-$$

$$K_w = \frac{[H_3O^+][A^-]}{[HA][H_2O]}$$

In dilute solution water is in such a large excess that the concentration of water is effectively constant (approximately 55 M), and this constant can be incorporated in the constant at the left hand side. Thus:

$$K_a = \frac{[H_3O^+][A^-]}{[HA]}$$

Table 8.5 Relation between pH, [H$^+$] and [OH$^-$]

pH	[H$^+$] (mol l^{-1})	[OH$^-$] (mol l^{-1})	
0	10^0	10^{-14}	Acidic
1	10^{-1}	10^{-13}	
2	10^{-2}	10^{-12}	
3	10^{-3}	10^{-11}	
4	10^{-4}	10^{-10}	
5	10^{-5}	10^{-9}	
6	10^{-6}	10^{-8}	
7	10^{-7}	10^{-7}	←Neutral
8	10^{-8}	10^{-6}	Basic
9	10^{-9}	10^{-5}	
10	10^{-10}	10^{-4}	
11	10^{-11}	10^{-3}	
12	10^{-12}	10^{-2}	
13	10^{-13}	10^{-1}	
14	10^{-14}	10^0	

The pH scale is used to measure the activity of hydrogen ions ($pH = -\log[H^+]$), and it refers to the number of powers of ten used to express the concentration of hydrogen ions. In a similar way the acid dissociation constant K_a may be expressed as a pK_a value:

$$pK_a = \log \frac{1}{K_a} = -\log K_a$$

Thus pK_a is a measure of the strength of an acid. If the acid ionizes almost completely (high acid strength) then K_a will be large, and thus pK_a will be small. The pK_a values given below show that acid strength increases on moving from left to right in the periodic table:

	CH_4	NH_3	H_2O	HF
pK_a	46	35	16	3

Acid strength also increases on moving down a group:

	HF	HCl	HBr	HI
pK_a	3	−7	−9	−10

With oxoacids containing more than one hydrogen atom, successive dissociation constants rapidly become more positive, i.e. the phosphate species formed on successive removal of H^+ become less acidic:

$$H_3PO_4 \rightleftharpoons H^+ + H_2PO_4^- \qquad pK_1 = 2.15$$
$$H_2PO_4^- \rightleftharpoons H^+ + HPO_4^{2-} \qquad pK_2 = 7.20$$
$$HPO_4^{2-} \rightleftharpoons H^+ + PO_4^{3-} \qquad pK_3 = 12.37$$

If an element forms a series of oxoacids, then the more oxygen atoms present, the more acidic it will be. The reason for this is that the electrostatic attraction for the proton decreases as the negative charge is spread over more atoms, thus facilitating ionization.

very weak acid	*weak acid*	*strong acid*	*very strong acid*
	HNO_2 $pK_a = 3.3$	HNO_3 $pK_a = -1.4$	
	H_2SO_3 $pK_a = 1.9$	H_2SO_4 $pK_a = (-1)$	
HOCl $pK_a = 7.2$	$HClO_2$ $pK_a = 2.0$	$HClO_3$ $pK_a = -1$	$HClO_4$ $pK_a = (-10)$

Bronsted–Lowry theory

In 1923, Bronsted and Lowry independently defined acids as proton donors, and bases as proton acceptors.

$$\underset{\text{solvent}}{2H_2O} \rightleftharpoons \underset{\text{acid}}{H_3O^+} + \underset{\text{base}}{OH^-}$$

For aqueous solutions, this definition does not differ appreciably from the Arrhenius theory. Water self-ionizes as shown above. Substances that increase the concentration of $[H_3O]^+$ in an aqueous solution above the value of $10^{-7}\,mol^2\,l^{-2}$ from the self-ionization are acids, and those that decrease it are bases.

The Bronsted–Lowry theory is useful in that it extends the scope of acid–base systems to cover solvents such as liquid ammonia, glacial acetic acid, anhydrous sulphuric acid, and all hydrogen-containing solvents. It should be emphasized that bases accept protons, and there is no need for them to contain OH^-.

In liquid ammonia:

$$\underset{\text{acid}}{NH_4Cl} + \underset{\text{base}}{NaNH_2} \rightarrow \underset{\text{salt}}{Na^+Cl^-} + \underset{\text{solvent}}{2NH_3}$$

or simply:

$$\underset{\substack{\text{acid}\\\text{(donates}\\\text{a proton)}}}{NH_4^+} + \underset{\substack{\text{base}\\\text{(accepts}\\\text{a proton)}}}{NH_2^-} \rightarrow \underset{\text{solvent}}{2NH_3}$$

Similarly in sulphuric acid:

$$\underset{\text{acid}}{H_3SO_4^+} + \underset{\text{base}}{HSO_4^-} \rightarrow \underset{\text{solvent}}{2H_2SO_4}$$

Chemical species that differ in composition only by a proton are called 'a conjugate pair'. Thus every acid has a conjugate base, which is formed when the acid donates a proton. Similarly every base has a conjugate acid.

$$\underset{\text{acid}}{A} \rightleftharpoons \underset{\substack{\text{conjugate}\\\text{base}}}{B^-} + H^+$$

$$\underset{\text{base}}{B} + H^+ \rightleftharpoons \underset{\substack{\text{conjugate}\\\text{acid}}}{A^+}$$

In water

$$\underset{\text{acid}}{HCl} + \underset{\text{base}}{H_2O} \rightleftharpoons \underset{\substack{\text{conjugate}\\\text{acid}}}{H_3O^+} + \underset{\substack{\text{conjugate}\\\text{base}}}{Cl^-}$$

In the above reaction, HCl is an acid since it donates protons, and in doing so forms Cl^-, its conjugate base. Since H_2O accepts protons it is a base, and forms H_3O^+, its conjugate acid. A strong acid has a weak conjugate base and vice versa.

In liquid ammonia:

$$\underset{\text{acid}}{NH_4^+} + \underset{\text{base}}{S^{2-}} \rightleftharpoons \underset{\substack{\text{conjugate}\\\text{acid}}}{HS^-} + \underset{\substack{\text{conjugate}\\\text{base}}}{NH_3}$$

In liquid ammonia all ammonium salts act as acids since they can donate protons, and the sulphide ion acts as a base since it accepts protons. The reaction is reversible, and it will proceed in the direction that produces the weaker species, in this case HS^- and NH_3.

A limitation of the Bronsted–Lowry theory is that the extent to which a dissolved substance can act as an acid or a base depends largely on the solvent. The solute only shows acidic properties if its proton-donating properties exceed those of the solvent. This sometimes upsets our traditional ideas on what are acids, which are based on our experience of what happens in water. Thus $HClO_4$ is an extremely strong proton donor. If liquid $HClO_4$ is used as a solvent, then HF dissolved in this solvent is forced to accept protons, and thus act as a base.

$$HClO_4 + HF \rightleftharpoons H_2F^+ + ClO_4^-$$

In a similar way HNO_3 is forced to accept protons and thus act as a base in both $HClO_4$ and liquid HF as solvent.

Water has only a weak tendency to donate protons. The mineral acids (HCl, HNO_3, H_2SO_4 etc.) all have a much stronger tendency to donate protons. Thus in aqueous solutions the mineral acids all donate protons to the water, thus behaving as acids, and in the process the mineral acids ionize completely.

In liquid ammonia as solvent, the acids which were strong acids in water all react completely with the ammonia, forming NH_4^+.

$$HClO_4 + NH_3 \rightarrow NH_4^+ + ClO_4^-$$

$$HNO_3 + NH_3 \rightarrow NH_4^+ + NO_3^-$$

Acids which were slightly less strong in water also react completely with NH_3, forming NH_4^+.

$$H_2SO_4 + 2NH_3 \rightarrow 2NH_4^+ + SO_4^-$$

Weak acids in water, such as oxalic acid, also react completely with NH_3.

$$(COOH)_2 + 2NH_3 \rightarrow 2NH_4^+ + (COO)_2^{2-}$$

The acid strengths have all been levelled by the solvent liquid ammonia: hence liquid ammonia is called a levelling solvent. It even makes some molecules, such as urea, which show no acidic properties in water, behave as weak acids.

$$NH_2CONH_2 + NH_3 \rightarrow NH_4^+ + NH_2CONH^-$$

Differentiating solvents such as glacial acetic acid emphasize the difference in acid strength, and several mineral acids are only partially ionized in this solvent. This is because acetic acid itself is a proton donor, and if a substance dissolved in acetic acid is to behave as an acid, it must donate protons more strongly than acetic acid. Thus the dissolved material must force the acetic acid to accept protons (i.e. the acetic acid behaves as a base). Thus the solvent acetic acid makes it more difficult for the usual acids to ionize, and conversely it will encourage the usual bases to ionize completely. It follows that a differentiating solvent for acids will act as a levelling solvent for bases, and vice versa.

Lewis theory

Lewis developed a definition of acids and bases that did not depend on the presence of protons, nor involve reactions with the solvent. He defined acids as materials which accept electron pairs, and bases as substances which donate electron pairs. Thus a proton is a Lewis acid and ammonia is a Lewis base since the lone pair of electrons on the nitrogen atom can be donated to a proton:

$$H^+ + :NH_3 \rightarrow [H \leftarrow :NH_3]^+$$

Similarly hydrogen chloride is a Lewis acid because it can accept a lone pair from a base such as water though this is followed by ionization:

$$H_2O + HCl \rightarrow [H_2O: \rightarrow HCl] \rightarrow H_3O^+ + Cl^-$$

Though this is a more general approach than that involving protons, it has several drawbacks:

1. Many substances, such as BF_3 or metal ions, that are not normally regarded as acids, behave as Lewis acids. This theory also includes reactions where no ions are formed, and neither hydrogen ions nor any other ions are transferred (e.g. $Ni(CO)_4$).

acid	base	
BF_3	$+ NH_3$	$\rightarrow [H_3N: \rightarrow BF_3]$
Ag^+	$+ 2NH_3$	$\rightarrow [H_3N: \rightarrow Ag \leftarrow :NH_3]^+$
Co^{3+}	$+ 6Cl^-$	$\rightarrow [CoCl_6]^{3-}$
Ni	$+ 4CO$	$\rightarrow Ni(CO)_4$
O	$+ C_6H_5$	$\rightarrow C_6H_5N: \rightarrow O$ (pyridine oxide)

2. There is no scale of acid or basic strength, since the strength of an acid or a base compound is not constant, and varies from one solvent to another, and also from one reaction to another.
3. Almost all reactions become acid–base reactions under this system.

The solvent system

Perhaps the most convenient general definition of acids and bases is due to Cady and Elsey, and can be applied in all cases where the solvent undergoes self-ionization, regardless of whether it contains protons or not.

Many solvents undergo self-ionization, and form positive and negative ions in a similar way to water:

$$2H_2O \rightleftharpoons H_3O^+ + OH^-$$
$$2NH_3 \rightleftharpoons NH_4^+ + NH_2^-$$
$$2H_2SO_4 \rightleftharpoons H_3SO_4^+ + HSO_4^-$$
$$2POCl_3 \rightleftharpoons POCl_2^+ + POCl_4^-$$
$$2BrF_3 \rightleftharpoons BrF_2^+ + BrF_4^-$$
$$N_2O_4 \rightleftharpoons NO^+ + NO_3^-$$

Acids are defined as substances that increase the concentration of the positive ions characteristic of the solvent (H_3O^+ in the case of water, NH_4^+ in liquid ammonia, and NO^+ in N_2O_4). Bases are substances that increase the concentration of the negative ions characteristic of the solvent (OH^- in water, NH_2^- in ammonia, NO_3^- in N_2O_4).

There are two advantages to this approach. First, most of our traditional ideas on what are acids and bases in water remain unchanged, as do neutralization reactions. Second, it allows us to consider non-aqueous solvents by analogy with water.

Thus water ionizes, giving H_3O^+ and OH^- ions. Substances providing H_3O^+ (e.g. HCl, KNO_3 and H_2SO_4) are acids, and substances providing OH^- (e.g. NaOH and NH_4OH) are bases. Neutralization reactions are of the type *acid + base → salt + water*.

$$\underset{\text{acid}}{HCl} + \underset{\text{base}}{NaOH} \rightarrow \underset{\text{salt}}{NaCl} + \underset{\text{water}}{H_2O}$$

Similarly liquid ammonia ionizes, giving NH_4^+ and NH_2^- ions. Thus ammonium salts are acids since they provide NH_4^+ ions, and sodamide $NaNH_2$ is a base since it provides NH_2^- ions. Neutralization reactions are of the type acid + base → salt + *solvent*.

$$\underset{\text{acid}}{NH_4Cl} + \underset{\text{base}}{NaNH_2} \rightarrow \underset{\text{salt}}{NaCl} + \underset{\text{solvent}}{2NH_3}$$

N_2O_4 self-ionizes into NO^+ and NO_3^- ions. Thus in N_2O_4 as solvent, NOCl is an acid since it produces NO^+, and $NaNO_3$ is a base since it produces NO_3^-.

$$\underset{\text{acid}}{NOCl} + \underset{\text{base}}{NaNO_3} \rightarrow \underset{\text{salt}}{NaCl} + \underset{\text{solvent}}{N_2O_4}$$

Clearly this definition applies equally well to proton and non-proton systems. This broader definition also has advantages when considering protonic solvents, since it explains why the acidic or basic properties of a solute are not absolute, and depend in part on the solvent. We normally regard acetic acid as an acid, because in water it produces H_3O^+.

$$CH_3COOH + H_2O \rightarrow H_3O^+ + CH_3COO^-$$

However, acetic acid behaves as a base when sulphuric acid is the solvent since H_2SO_4 is a stronger proton donor than CH_3COOH. In a similar way HNO_3 is forced to behave as a base in H_2SO_4 as solvent, and this is important in producing nitronium ions NO_2^+ in the nitration of organic compounds by a mixture of concentrated H_2SO_4 and HNO_3.

$$H_2SO_4 + CH_3COOH \rightarrow CH_3COOH_2^+ + HSO_4^-$$

$$H_2SO_4 + HNO_3 \rightarrow [H_2NO_3]^+ + HSO_4^-$$

$$[H_2NO_3]^+ \rightarrow H_2O + NO_2^+$$

The Lux–Flood definition

Lux originally proposed a different definition of acids and bases which was extended by Flood. Instead of using protons, or ions characteristic of the solvent, they defined acids as oxides which accept oxygen, and bases as oxides which donate oxygen. Thus:

$$\begin{array}{lll} CaO & + CO_2 & \rightarrow Ca^{2+}[CO_3]^{2-} \\ SiO_2 & + CaO & \rightarrow Ca^{2+}[SiO_3]^{2-} \\ \underset{\text{acid}}{6Na_2O} & + \underset{\text{base}}{P_4O_{10}} & \rightarrow 4Na_3^+[PO_4]^{3-} \end{array}$$

This system is very useful in dealing with anhydrous reactions in fused melts of oxides, and other high temperature reactions such as are found in metallurgy and ceramics.

This theory has an inverse relationship to aqueous chemistry, since Lux–Flood acids are oxides which react with water, giving bases in water, and Lux–Flood bases react with water, giving acids.

$$Na_2O + H_2O \rightarrow 2NaOH$$
$$P_4O_{10} + 6H_2O \rightarrow 4H_3PO_4$$

The Usanovich definition

This defines an acid as any chemical species which reacts with bases, gives up cations, or accepts anions or electrons. Conversely a base is any chemical species which reacts with acids, gives up anions or electrons, or combines with cations. This is a very wide definition and includes all the Lewis acid–base type of reactions, and in addition it includes redox reactions involving the transfer of electrons.

Hard and soft acids and bases

Metal ions may be divided into two types depending on the strength of their complexes with certain ligands.

Type (a) metals include the smaller ions from Groups I and II, and the left hand side of the transition metals, particularly when in high oxidation states, and these form the most stable complexes with nitrogen and oxygen donors (ammonia, amines, water, ketones, alcohols), and also with F^- and Cl^-.

Type (b) metals include ions from the right hand side of the transition series, and also transition metal complexes with low oxidation states, such as the carbonyls. These form the most stable complexes with ligands such as I^-, SCN^- and CN^-.

This empirical classification was useful in predicting the relative stabilities of complexes. Pearson extended the concept into a broad range of acid–base interactions. Type (a) metals are small and not very polariz-

Table 8.6 Some hard and soft acids and bases

Hard acids	*Soft acids*
H^+	Pd^{2+}, Pt^{2+}, Cu^+, Ag^+, Au^+, Hg^{2+},
Li^+, Na^+, K^+,	$(Hg_2)^{2+}$, Tl^+
Be^{2+}, Mg^{2+}, Ca^{2+}, Sr^{2+},	$B(CH_3)_3$, B_2H_6, $Ga(CH_3)_3$, $GaCl_3$,
Al^{3+}, BF_3, $Al(CH_3)_3$, $AlCl_3$,	$GaBr_3$, GaI_3
Sc^{3+}, Ti^{4+}, Zr^{4+}, VO^{2+}, Cr^{3+},	$[Fe(CO)_5]$, $[Co(CN)_5]^{3-}$
MoO^{3+}, WO^{4+},	
Ce^{3+}, Lu^{3+},	
CO_2, SO_3	
Hard bases	*Soft bases*
NH_3, RNH_2, N_2H_4	H^-, CN^-, SCN^-, $S_2O_3^{2-}$, I^-, RS^-,
H_2O, ROH, R_2O	R_2S, CO, B_2H_6, C_2H_4, R_3P, $P(OR)_3$
OH^-, NO_3^-, ClO_4^-, CO_3^{2-}, SO_4^{2-},	
PO_4^{3-}, CH_3COO^-, F^-, Cl^-	

able, and these prefer ligands that are also small and not very polarizable. Pearson called these metals hard acids, and the ligands hard bases. In a similar way, type (b) metals and the ligands they prefer are larger and more polarizable, and he called these soft acids and soft bases. He stated the relationship *hard acids prefer to react with hard bases, and soft acids react with soft bases*. This definition takes in the usually accepted acid–base reactions (H^+ strong acid, OH^- and NH_3 strong bases), and in addition a great number of reactions involving the formation of simple complexes, and complexes with π bonding ligands.

FURTHER READING

Hydrogen

Brown, H.C. (1979) Hydride reductions: A 40 year revolution in organic chemistry, *Chem. Eng. News*, March 5, 24–29.

Emeléus, H.J. and Sharpe A.G. (1973) *Modern Aspects of Inorganic Chemistry*, 4th ed. (Chapter 8: Hydrogen and the Hydrides), Routledge and Kegan Paul, London.

Evans, E.A. (1974) *Tritium and its Compounds*, 2nd ed., Butterworths, London, (Contains over 4000 references.)

Grant, W.J. and Redfearn, S.L. (1977) Industrial gases, in *The Modern Inorganic Chemicals Industry* (ed. Thompson, R.), The Chemical Society, London, Special Publication no. 31.

Jolly, W.L. (1976) *The Principles of Inorganic Chemistry*, (Chapters 4 and 5), McGraw Hill, New York.

Mackay, K.M. (1966) *Hydrogen Compounds of the Metallic Elements*, Spon, London.

Mackay, K.M. (1973) *Comprehensive Inorganic Chemistry*, Vol. 1 (Chapter 1: The element hydrogen; Chapter 2 Hydrides), Pergamon Press, Oxford.

Mackay, K.M. and Dove, M.F.A. (1973) *Comprehensive Inorganic Chemistry*, Vol. 1 (Chapter 3: Deuterium and Tritium), Pergamon Press, Oxford.
Moore, D.S. and Robinson, S.D. (1983) Hydrido complexes of the transition metals, *Chem. Soc. Rev.*, **12**, 415–452.
Muetterties, E.L. (1971) *Transition Metal Hydrides*, Marcel Dekker, New York.
Sharpe, A.G. (1981) *Inorganic Chemistry* (Chapter 9), Longmans, London.
Stinson, S.C. (1980) Hydride reducing agents, use expanding, *Chem. Eng. News*, Nov 3, 18–20.
Wiberg, E. and Amberger, E. (1971) *Hydrides*, Elsevier.

The hydrogen economy

McAuliffe, C.A. (1973) The hydrogen economy, *Chemistry in Britain*, **9**, 559–563.
Marchetti, C. (1977) The hydrogen economy and the chemist, *Chemistry in Britain*, **13**, 219–222.
Williams, L.O. (1980) *Hydrogen Power: An Introduction to Hydrogen Energy and Its Applications*, Pergamon Press, Oxford.

Hydrogen bonding

Coulson, C. A. (1979) *Valence*, Oxford University Press, Oxford. 3rd ed. by McWeeny, R. (This is an updated version of Coulson's 1952 book.)
DeKock, R.L. and Gray, H.B. (1980) *Chemical Structure and Bonding*, Benjamin/Cummins, Menlo Park, California.
Douglas, B., McDaniel, D.H. and Alexander, J.J. (1982) *Concepts and Models of Inorganic Chemistry*, 2nd ed (Chapter V: The Hydrogen Bond), John Wiley, New York.
Emsley, J. (1980) Very strong hydrogen bonds, *Chem. Soc. Rev.*, **9**, 91–124.
Joesten, M.D. and Schaad, L.J. (1974) *Hydrogen Bonding*, Marcel Dekker, New York.
Pauling, L. (1960) *The Nature of the Chemical Bond* (Chapter 12), 3rd ed., Oxford University Press, London.
Pimentel, G.C. and McClellan, A.L. (1960) *The Hydrogen Bond*, W.H. Freeman, San Francisco. (A well written monograph, with over 2200 references. Dated but thorough.)
Vinogradov, S.N. and Linnell, R.H. (1971) *Hydrogen Bonding*, Van Nostrand Reinhold, New York. (Good general treatment.)
Wells, A.F. (1984) *Structural Inorganic Chemistry*, 5th ed., Oxford University Press, Oxford.

Acids and bases

Bell, R.P. (1973) *The Proton in Chemistry*, 2nd ed., Chapman and Hall, London.
Bronsted, J.N. (1923) *Rec. Trav. Chim.*, **42**, 718. (Original paper on the Bronsted theory.)
Cady. H.P and Elsey, H.M. (1928) A general definition of acids, bases and salts, *J. Chem. Ed.*, **5**, 1425. (Original paper on the solvent system.)
Drago, R.S. (1974) A Modern approach to acid–base chemistry, *J. Chem. Ed.*, **51**, 300.
Finston, H.L and Rychtman, A.C. (1982) *A New View of Current Acid–Base Theories*, John Wiley, Chichester.
Fogg, P.G.T. and Gerrard, W. (eds) (1990) *Hydrogen Halides in Non-Aqueous Solvents*, Pergamon, New York.
Gillespie, R.J. (1973) The chemistry of the superacid system, *Endeavour*, **32**, 541.

Gillespie, R.J. (1975) Proton acids, Lewis acids, hard acids, soft acids and superacids, Chapter 1 in *Proton Transfer Reactions* (ed. Caldin, E. and Gold, V.), Chapman and Hall, London.

Hand, C.W. and Blewitt, H.L. (1986) *Acid–Base Chemistry*, Macmillan, New York; Collier Macmillan, London.

Huheey, J.E. (1978) *Inorganic Chemistry*, 2nd ed. (Chapter 7), Harper and Row, New York.

Jensen, W.B. (1980) *The Lewis Acid–Base Concepts*, Wiley, New York and Chichester.

Koltoff, I.M. and Elving, P.J. (1986) (eds), *Treatise on Analytical Chemistry*, 2nd ed., Vol. 2, Part 1, Wiley, Chichester, 157–440.

Olah, G.A., Surya Prakask, G.K., and Sommer, J. (1985) *Superacids*, Wiley, Chichester and Wiley-Interscience, New York.

Pearson, R.G. (1987) Recent advances in the concept of hard and soft acids and bases, *J. Chem. Ed.* **64**, 561–567.

Smith, D.W. (1987) An acidity scale for binary oxides, *J. Chem. Ed.*, **64**, 480–481.

Vogel, A.I., Jeffery, G.H., Bassett, J., Mendham, J. and Denney, R.C. (1990) *Vogel's Textbook of Quantitative Chemical Analysis*, 5th ed., Halstead Press, (Indicators, acid–base titrations, weak acids and bases, buffers etc.)

Water and solutions

Burgess, J. (1988) *Ions in Solution*, Ellis Horwood, Chichester.

Franks, F. (1984) *Water*, 1st revised ed., Royal Society for Chemistry, London.

Hunt, J.P. and Friedman, H.L. (1983) Aquo complexes of metal ions, *Progr. in Inorg. Chem.*, **30**, 359–387.

Murrell, J.N. and Boucher, E.A. (1982) *Properties of Liquids and Solutions*, John Wiley, Chichester.

Nielson, G.W. and Enderby, J.E. (eds) (1986) *Water and Aqueous Solutions*, Colston Research Society 37th Symposium (University of Bristol), Adam Hilger, Bristol.

Symons, M.C.R. (1989) Liquid water – the story unfolds, *Chemistry in Britain*, **25**, 491–494.

Non-aqueous solvents

Addison, C.C. (1980) Dinitrogen tetroxide, nitric acid and their mixtures as media for inorganic reactions, *Chem. Rev.*, **80**, 21–39.

Addison, C.C. (1984) *The Chemistry of the Liquid Alkali Metals*, John Wiley, Chichester.

Burger, K. (1983) *Ionic Solvation and Complex Formation Reactions in Non-Aqueous Solvents*, Elsevier, New York.

Emeléus, H.J. and Sharpe, A.G. (1973) *Modern Aspects of Inorganic Chemistry*, 4th ed. (Chapter 7: Reactions in Non–Aqueous Solvents; Chapter 8: Hydrogen and the hydrides), Routledge and Kegan Paul, London.

Gillespie, R.J. and Robinson, E.A. (1959) The sulphuric acid solvent system, *Adv. Inorg. Chem. Radiochem.*, **1**, 385.

Lagowski, J. (ed.) (1978) *The Chemistry of Non-aqueous Solvents*, Academic Press, New York.

Nicholls, D. (1979) *Inorganic Chemistry in Liquid Ammonia* (Topics in Inorganic and General Chemistry, Monograph 17), Elsevier, Oxford.

Popovych, O. and Tomkins, R.P.T. (1981) *Non Aqueous Solution Chemistry*, Wiley, Chichester.

Waddington, T.C. (ed.) (1969) *Non Aqueous Solvent Systems*, Nelson.

PROBLEMS

1. Suggest reasons for and against the inclusion of hydrogen in the main groups of the periodic table.
2. Describe four ways in which hydrogen is produced on an industrial scale. Give one convenient method of preparing hydrogen in the laboratory.
3. Give an account of the main uses of hydrogen.
4. Give equations to show the reaction of hydrogen with: (a) Na, (b) Ca, (c) CO, (d) N, (e) S, (f) Cl, (g) CuO.
5. Describe the different types of hydrides which are formed.
6. Give examples of six proton solvents other than water, and show how they self-ionize.
7. What species are characteristic of acids and of bases in the following solvents: (a) liquid ammonia, anhydrous acetic acid, (b) anhydrous nitric acid, (c) anhydrous HF, (d) anhydrous perchloric acid, (e) anhydrous sulphuric acid, (f) dinitrogen tetroxide.
8. Describe how the various physical properties of a liquid affect its usefulness as a solvent.
9. How are the properties of H_2O, NH_3 and HF affected by hydrogen bonding?
10. Explain the variation in boiling points of the hydrogen halides (HF 20 °C, HCl −85 °C, HBr −67 °C and HI −36 °C.
11. Discuss the theoretical background, practical uses and theoretical limitations of liquid hydrogen fluoride as a non-aqueous solvent. List materials which behave as acids and bases in this solvent, and explain what happens when SbF_5 is dissolved in HF.
12. Discuss the theoretical background, practical uses and limitations of liquid ammonia non-aqueous solvent. Explain what happens when $^{15}NH_4Cl$ is dissolved in unlabelled liquid ammonia and the solvent evaporated.

PROBLEMS

[illegible] in the main [illegible]

[illegible]

[illegible]

The *s*-Block Elements

Part Two

Group I – the alkali metals

9

Table 9.1 Electronic structures

Element	Symbol	Electronic structure	
Lithium	Li	$1s^2 2s^1$	or [He] $2s^1$
Sodium	Na	$1s^2 2s^2 2p^6 3s^1$	or [Ne] $3s^1$
Potassium	K	$1s^2 2s^2 2p^6 3s^2 3p^6 4s^1$	or [Ar] $4s^1$
Rubidium	Rb	$1s^2 2s^2 2p^6 3s^2 3p^6 3d^{10} 4s^2 4p^6 5s^1$	or [Kr] $5s^1$
Caesium	Cs	$1s^2 2s^2 2p^6 3s^2 3p^6 3d^{10} 4s^2 4p^6 4d^{10} 5s^2 5p^6 6s^1$	or [Xe] $6s^1$
Francium	Fr		[Rn] $7s^1$

INTRODUCTION

The elements of Group I illustrate, more clearly than any other group of elements, the effects of increasing the size of atoms or ions on the physical and chemical properties. They form a closely related group, and probably have the least complicated chemistry of any group in the periodic table. The physical and chemical properties of the elements are closely related to their electronic structures and sizes. The elements are all metals, excellent conductors of electricity, and are typically soft and highly reactive. They have one loosely held valence electron in their outer shell, and typically form univalent, ionic and colourless compounds. The hydroxides and oxides are very strong bases, and the oxosalts are very stable.

Lithium, the first element in the group, shows considerable differences from the rest of the group. In all of the main groups the first element shows a number of differences from the later elements in the group.

Sodium and potassium together make up over 4% by weight of the earth's crust. Their compounds are very common, and have been known and used from very early times. Some of their compounds are used in very large amounts. World production of NaCl was 179.6 million tonnes in 1988. (Most was used to make NaOH and Cl_2.) Thirty-four million tonnes of NaOH were produced in 1985. About 22 million tonnes/year of Na_2CO_3 is used, and $NaHCO_3$, Na_2SO_4 and NaOCl are also of industrial importance. World production of potassium salts (referred to as 'potash' and measured as K_2O) was 32.1 million tonnes in 1988. Much of it was used as fertilizers, but KOH, KNO_3 and K_2O are also important. In addition sodium and potassium are essential elements for animal life. The metals were first isolated by Humphrey Davy in 1807 by the electrolysis of KOH and NaOH.

OCCURRENCE AND ABUNDANCE

Despite their close chemical similarity, the elements do not occur together, mainly because their ions are of different size.

Lithium is the thirty-fifth most abundant element by weight and is

Table 9.2 Abundance of the elements in the earth's crust, by weight

	Abundance in earth's crust		Relative abundance
	(ppm)	(%)	
Li	18	0.0018	35
Na	22 700	2.27	7
K	18 400	1.84	8
Rb	78	0.0078	23
Cs	2.6	0.00026	46

mainly obtained as the silicate minerals, spodumene $LiAl(SiO_3)_2$, and lepidolite $Li_2Al_2(SiO_3)_3(FOH)_2$. World production of lithium minerals was 7800 tonnes in 1988. The main sources are the USSR 42%, and Zimbabwe, China, Canada and Portugal 11% each.

Sodium and potassium are the seventh and eighth most abundant elements by weight in the earth's crust. NaCl and KCl occur in large amounts in sea water. The largest source of sodium is rock salt (NaCl). Various salts including NaCl, $Na_2B_4O_7.10H_2O$ (borax), (Na_2CO_3. $NaHCO_3.2H_2O$) (trona), $NaNO_3$ (saltpetre) and Na_2SO_4 (mirabilite) are obtained from deposits formed by the evaporation of ancient seas such as the Dead Sea and the Great Salt Lake at Utah USA. Sodium chloride is extremely important, and is used in larger tonnages than any other chemical. World production was 179.6 million tonnes in 1988. The main sources are the USA 19%, China 10%, the USSR 9%, India 7%, West Germany 8%, Canada 6%, the UK and Australia 5% each, and France and Mexico 4% each. In most places it is mined as rock salt. In the UK (the Cheshire salt field) about 75% is extracted in solution as brine, and similarly in Germany over 70% is extracted as brine. 'Solar' salt is obtained by evaporating sea water in some hot countries. Ninety-two per cent of the salt produced in India is by evaporation, and 26% of that from Spain and France. This method is also used in Australia.

Potassium occurs mainly as deposits of KCl (sylvite), a mixture of KCl and NaCl (sylvinite), and the double salt $KCl \cdot MgCl_2 \cdot 6H_2O$ (carnallite). Soluble potassium salts are collectively called 'potash'. World production of potash was 32.1 million tonnes in 1988, measured as K_2O. The main sources are from mined deposits (the USSR 35%, Canada 25%, East Germany 11%, West Germany 7%, France and the USA 5% each and Israel 4%). Large amounts are recovered from brines such as the Dead Sea (Jordan) and the Great Salt Lake (Utah USA), where the concentration may be 20–25 times higher than in sea water, but it is not economic to recover potassium salts from 'normal' sea water.

There is no convenient source of rubidium and only one of caesium andthese elements are obtained as by-products from lithium processing.

All of the elements heavier than bismuth (atomic number 83) $_{83}Bi$ are radioactive. Thus francium (atomic number 89) is radioactive, and as it has a short half life period of 21 minutes it does not occur appreciably in nature. Any that existed when the earth was formed will have disappeared, and any formed now from actinium will have a transitory existence.

$$^{227}_{89}Ac \xrightarrow{99\%} {}^{0}_{-1}e + {}^{227}_{90}Th \quad \text{(beta decay)}$$

$$^{227}_{89}Ac \xrightarrow{1\%} {}^{4}_{2}He + {}^{223}_{87}Fr \quad \text{(alpha decay)}$$

$$^{223}_{87}Fr \xrightarrow{\text{half life 21 min}} {}^{0}_{-1}e + {}^{223}_{88}Ra \quad \text{(beta decay)}$$

EXTRACTION OF THE METALS

The metals of this group are too reactive to be found in the free state. Their compounds are amongst the most stable to heat, so thermal decomposition is impractical. Since the metals are at the top of the electrochemical series they react with water, so displacement of one element from solution by another higher in the electrochemical series will be unsuccessful. The metals are the strongest chemical reducing agents known, and so cannot be prepared by reducing the oxides. Electrolysis of aqueous solutions in order to obtain the metal is also unsuccessful unless a mercury cathode is used, when it is possible to obtain amalgams, but recovery of the pure metal from the amalgam is difficult.

The metals may all be isolated by electrolysis of a fused salt, usually the fused halide, often with impurity added to lower the melting point.

Sodium is made by the electrolysis of a molten mixture of about 40% NaCl and 60% $CaCl_2$ in a Downs cell (Figure 9.1). This mixture melts at about 600 °C compared with 803 °C for pure NaCl. The small amount of calcium formed during the electrolysis is insoluble in the liquid sodium, and dissolves in the eutectic mixture. There are three advantages to electrolysing a mixture.

1. It lowers the melting point and so reduces the fuel bill.
2. The lower temperature results in a lower vapour pressure for sodium, which is important as sodium vapour ignites in air.
3. At the lower temperature the liberated sodium metal does not dissolve in the melt, and this is important because if it dissolved it would short-circuitthe electrodes and thus prevent further electrolysis.

continued overleaf

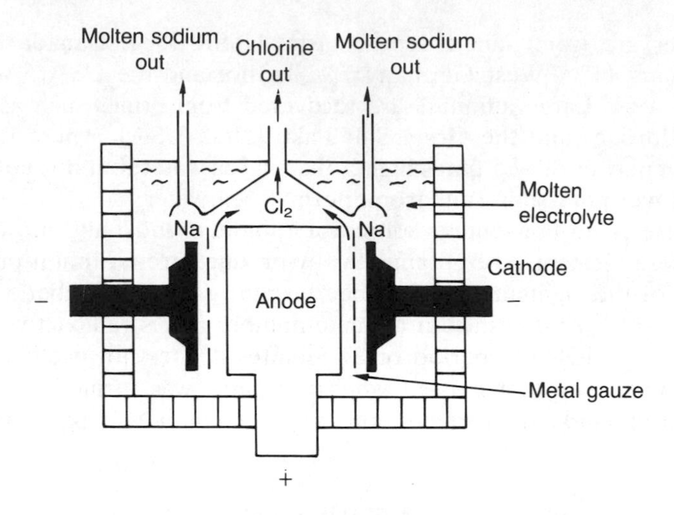

Figure 9.1 Downs cell for the production of sodium.

A Downs cell comprises a cylindrical steel vessel lined with firebrick, measuring about 2.5 m in height and 1.5 m in diameter. The anode is a graphite rod in the middle, and is surrounded by a cast steel cathode. A metal gauze screen separates the two electrodes, and prevents the Na formed at the cathode from recombining with Cl_2 produced at the anode. The molten sodium rises, as it is less dense than the electrolyte, and it is collected in an inverted trough and removed, and packed into steel drums.

A similar cell can be used to obtain potassium by electrolysing fused KCl. However, the cell must be operated at a higher temperature because the melting point of KCl is higher, and this results in the vaporization of the liberated potassium. Since sodium is a more powerful reducing agent than potassium and is readily available, the modern method is to reduce molten KCl with sodium vapour at 850 °C in a large fractionating tower. This gives K of 99.5% purity.

$$Na + KCl \rightarrow NaCl + K$$

Rb and Cs are produced in a similar way by reducing the chlorides with Ca at 750 °C under reduced pressure.

USES OF GROUP I METALS AND THEIR COMPOUNDS

Lithium stearate $C_{17}H_{35}COOLi$ is used in making automobile grease. Li_2CO_3 is added to bauxite in the electrolytic production of aluminium, as it lowers the melting point. Li_2CO_3 is also used to toughen glass. It also has uses in medicine, as it affects the balance between Na^+ and K^+ and Mg^{2+} and Ca^{2+} in the body. Lithium metal is used to make alloys, for example with lead to make 'white metal' bearings for motor engines, with aluminium to make aircraft parts which are light and strong, and with magnesium to make armour plate. There is great interest in lithium for thermonuclear purposes, since when bombarded with neutrons it produces tritium (see the section on Nuclear Fusion in Chapter 31). Lithium is also used to make electrochemical cells (both primary and secondary batteries). Primary batteries produce electricity by a chemical change, and are discarded when they 'run down'. These have Li anodes, carbon cathodes and $SOCl_2$ as the electrolyte. There is interest in Li/S batteries which could power battery cars in the future, and in secondary cells, which may provide a practical way of storing off-peak electricity. LiH is used to generate hydrogen, and LiOH to absorb CO_2.

Caustic soda NaOH is the most important alkali used in industry and is used for a wide variety of purposes including making many inorganic and organic compounds, paper making, neutralizations, and making alumina, soap and rayon. Soda ash Na_2CO_3 may be used interchangeably with NaOH in many applications such as making paper, soap and detergents. Large amounts are used in making glass, phosphates, silicates, and cleaning preparations and removing SO_2 pollution from the flue gases at coal-fired electricity generating stations. Large amounts of Na_2SO_4 are used to make paper, detergents and glass. NaOCl is used as a bleach and a disinfectant, and production is about 180 000 tonnes/year. $NaHCO_3$ is used in baking powder. Sodium metal is used in large quantities. About 200 000 tonnes/year is produced in the USA alone. Globally about 60% of the Na produced is used to make a Na/Pb alloy which is used to make $PbEt_4$ and $PbMe_4$ which are used as anti-knock additives to petrol, but this will decrease with the increasing use of lead-free petrol. About 20% is used to reduce $TiCl_4$ and $ZrCl_4$ to the metals, and the remainder is used to make compounds such as Na_2O_2 and NaH. Liquid sodium metal is used as a coolant in one type of nuclear reactor. It is used to transfer heat from the reactor to turbines where it produces steam which is used to generate electricity. Fast breeder nuclear reactors, such as those at Dounreay (Scotland) and Grenoble (France), operate at a temperature of about 600 °C; being a metal, sodium conducts heat very well, and as its boiling point is 881 °C it is ideal for this purpose. Small amounts of the metal are used in organic synthesis, and for drying organic solvents.

Potassium is an essential element for life. Roughly 95% of potassium compounds are used as fertilizers for plants – KCl 90%, K_2SO_4 9%, KNO_3 1%. Potassium salts are always more expensive than sodium salts, usually by a factor of 10 or more. KOH (which is prepared by electrolysing aqueous KCl) is used to make potassium phosphates and also soft soap, e.g. potassium stearate, both of which are used in liquid detergents. KNO_3 is used in explosives. $KMnO_4$ is used in the manufacture of saccharin, as an oxidizing agent and for titrations. K_2CO_3 is used in ceramics, colour TV tubes and fluorescent light tubes. Potassium superoxide KO_2 is used in breathing apparatus and in submarines, and KBr is used in photography. Not much potassium metal is produced, and most of it is used to make KO_2.

ELECTRONIC STRUCTURE

Group I elements all have one valency electron in their outer orbital – an *s* electron which occupies a spherical orbital. Ignoring the filled inner shells the electronic structures may be written: $2s^1$, $3s^1$, $4s^1$, $5s^1$, $6s^1$ and $7s^1$. The single valence electron is a long distance from the nucleus, is only weakly held and is readily removed. In contrast the remaining electrons are closer to the nucleus, more tightly held, and are removed only with great difficulty. Because of similarities in the electronic structures of these elements, many similarities in chemical behaviour would be expected.

SIZE OF ATOMS AND IONS

Group I atoms are the largest in their horizontal periods in the periodic table. When the outer electron is removed to give a positive ion, the size decreases considerably. There are two reasons for this.

1. The outermost shell of electrons has been completely removed.
2. Having removed an electron, the positive charge on the nucleus is now greater than the charge on the remaining electrons, so that each of the remaining electrons is attracted more strongly towards the nucleus. This reduces the size further.

Positive ions are always smaller than the parent atom. Even so, the ions are very large, and they increase in size from Li^+ to Fr^+ as extra shells of electrons are added.

The Li^+ is much smaller than the other ions. For this reason, Li only mixes with Na above 380 °C, and it is immiscible with the metals K, Rb and Cs, even when molten; nor will Li form substitutional alloys with them. In contrast the other metals Na, K, Rb and Cs are miscible with each other in all proportions.

DENSITY

The atoms are large, so Group I elements have remarkably low densities. Lithium metal is only about half as dense as water, whilst sodium and potassium are slightly less dense than water (see Table 9.3). It is unusual

Table 9.3 Size and density

	Metallic radius (Å)	Ionic radius M^+ six-coordinate (Å)	Density $(g\,cm^{-3})$
Li	1.52	0.76	0.54
Na	1.86	1.02	0.97
K	2.27	1.38	0.86
Rb	2.48	1.52	1.53
Cs	2.65	1.67	1.90

for metals to have low densities, and in contrast most of the transition metals have densities greater than $5\,g\,cm^{-3}$, for example iron $7.9\,g\,cm^{-3}$, mercury $13.6\,g\,cm^{-3}$, and osmium and iridium (the two most dense elements) 22.57 and $22.61\,g\,cm^{-3}$ respectively.

IONIZATION ENERGY

The first ionization energies for the atoms in this group are appreciably lower than those for any other group in the periodic table. The atoms are very large so the outer electrons are only held weakly by the nucleus: hence the amount of energy needed to remove the outer electron is not very large. On descending the group from Li to Na to K to Rb to Cs, the size of the atoms increases: the outermost electrons become less strongly held, so the ionizationenergy decreases.

The second ionization energy – that is the energy to remove a second electron from the atoms – is extremely high. The second ionization energy is always larger than the first, often by a factor of two, because it involves removing an electron from a smaller positive ion, rather than from a larger neutral atom. The difference between first and second ionization energies is much larger in this case since in addition it corresponds to removing an electron from a closed shell. A second electron is never removed under normal conditions, as the energy required is greater than that needed to ionize the noble gases. The elements commonly form M^+ ions.

ELECTRONEGATIVITY AND BOND TYPE

The electronegativity values for the elements in this group are very small – in fact the smallest values of any element. Thus when these elements react with other elements to form compounds, a large electronegativity difference between the two atoms is probable, and ionic bonds are formed.

Na electronegativity	0.9
Cl electronegativity	3.0
Electronegativity difference	2.1

Table 9.4 Ionization energies

	First ionization energy ($kJ\,mol^{-1}$)	Second ionization energy ($kJ\,mol^{-1}$)
Li	520.1	7296
Na	495.7	4563
K	418.6	3069
Rb	402.9	2650
Cs	375.6	2420

Table 9.5 Electronegativity values

	Pauling's electronegativity
Li	1.0
Na	0.9
K	0.8
Rb	0.8
Cs	0.7

An electronegativity difference of approximately 1.7–1.8 corresponds to 50% ionic character. The value 2.1 exceeds this, so the bonding in NaCl is predominantly ionic. Similar arguments apply to other compounds: for example, the electronegativity difference in LiF is 3.0, and in KBr is 2.0, and both compounds are ionic.

The chemistry of the alkali metals is largely that of their ions.

BORN–HABER CYCLE: ENERGY CHANGES IN THE FORMATION OF IONIC COMPOUNDS

When elements react to form compounds, ΔG (the free energy of formation) is negative. For a reaction to proceed spontaneously, the free energy of the products must be lower than that of the reactants.

Usually the energy changes are measured as enthalpy values ΔH, and ΔG is related to ΔH by the equation:

$$\Delta G = \Delta H - T\Delta S$$

In many cases enthalpy values are used instead of free energy values, and the two are almost the same if the term $T\Delta S$ is small. At room temperature T is almost 300 K, so ΔG and ΔH are only similar when the change in entropy ΔS is very small. Entropy changes are large when there is a change in physical state, e.g. solid to liquid, or liquid to gas, but otherwise entropy changes are usually small.

A whole series of energy changes is involved when one starts from the elements and finishes with an ionic crystal. These changes are shown in the Born–Haber cycle (Figure 9.2). The cycle serves two purposes. First it explains how these various energy changes are related, and second, if all but one of the terms can be measured, then the remaining value can be calculated. There is no direct way of obtaining electron affinity values, and these have been calculated from this type of energy cycle.

Hess's law states that the energy change occurring during a reaction depends only on the energy of the initial reactants and the energy of the final products, and not on the reaction mechanism, or the route taken. Thus, by Hess's law, the energy change for the reaction of solid sodium and chlorine gas to form a sodium chloride crystal by the direct route (measured as the enthalpy of formation) must be the same as the sum of

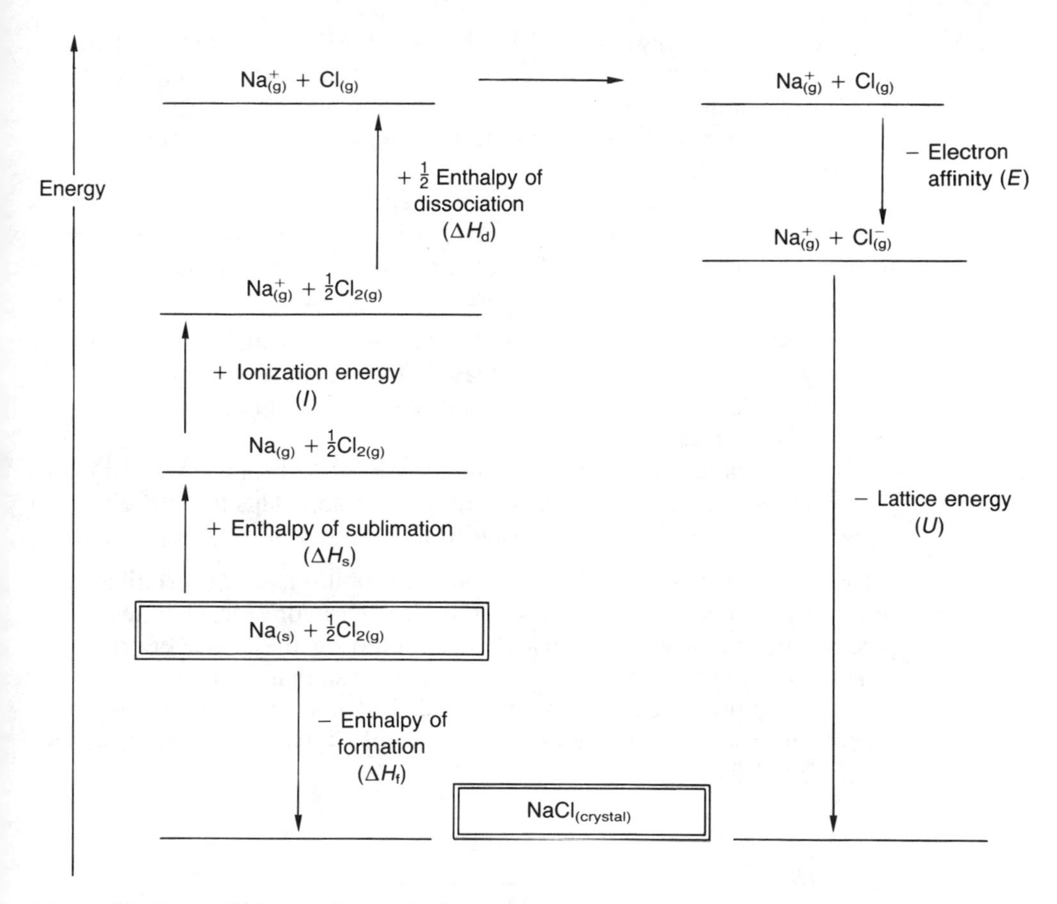

Figure 9.2 Born–Haber cycle for the formation of NaCl.

all the energy changes going round the cycle by the long route, i.e. by producing first gaseous atoms of the elements, then gaseous ions, and finally packing these to give the crystalline solid. This may be expressed as:

$$-\Delta H_f = +\Delta H_s + I + \tfrac{1}{2}\Delta H_d - E - U$$

Details of these energy terms are shown in Table 9.6. A considerable amount of energy (the enthalpies of sublimation and dissociation, and

Table 9.6 Enthalpy (ΔH) values for MCl (all values in $kJ\,mol^{-1}$)

	Sublimation energy $M_{(s)}-M_{(g)}$	$\frac{1}{2}$ enthalpy of dissociation $\frac{1}{2}Cl_2-Cl$	Ionization energy $M-M^+$	Electron affinity $Cl-Cl^-$	Lattice energy	Total = enthalpy of formation
Li	161	121.5	520	−355	−845	−397.5
Na	108	121.5	496	−355	−770	−399.5
K	90	121.5	419	−355	−703	−427.5
Rb	82	121.5	403	−355	−674	−422.5
Cs	78	121.5	376	−355	−644	−423.5

the ionization energy) is used to produce the ions, so these terms are positive. Ionic solids are formed because an even larger amount of energy is evolved, mainly coming from the lattice energy and to a smaller extent from the electron affinity, resulting in a negative value for the enthalpy of formation ΔH_f.

All the halides MCl have negative enthalpies of formation, which indicates that thermodynamically (that is in terms of energy) it is feasible to form the compounds MCl from the elements. The values are shown in Table 9.7. Several trends are apparent in these values:

1. The most negative enthalpies of formation occur with the fluorides. For any given metal, the values decrease in the sequence fluoride > chloride > bromide > iodide. Thus the fluorides are the most stable, and the iodides the least stable.
2. The enthalpies of formation for the chlorides, bromides and iodides become more negative on descending the group. This trend is observed with most salts, but the opposite trend is found in the fluorides.

Ionic compounds may also be formed in solution, when a similar cycle of energy changes must be considered, but the hydration energies of the positive and negative ions must be substituted for the lattice energy.

The energy cycle shown in Figure 9.3 is very similar to the Born–Haber cycle. The enthalpy of formation of hydrated ions from the elements in their natural state must be equal to the sum of all the other energy changes going round the cycle.

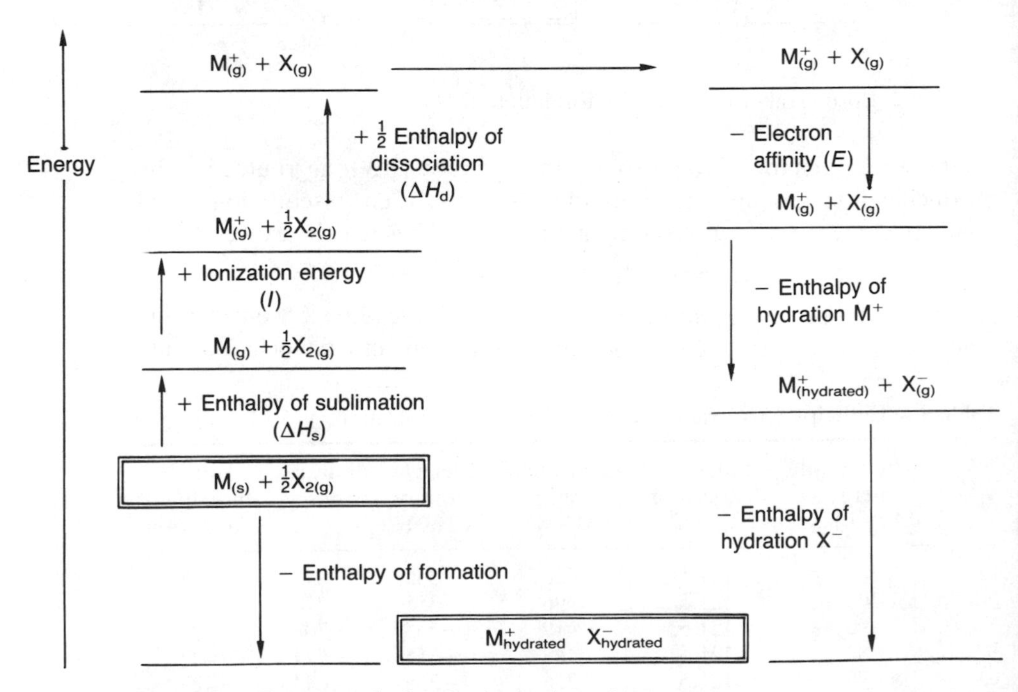

Figure 9.3 Energy cycle for the hydration of ions.

Table 9.7 Standard enthalpies of formation for Group I halides (all values in $kJ\,mol^{-1}$)

	MF	MCl	MBr	MI
Li	−612	−398	−350	−271
Na	−569	−400	−360	−288
K	−563	−428	−392	−328
Rb	−549	−423	−389	−329
Cs	−531	−424	−395	−337

STRUCTURES OF THE METALS, HARDNESS, AND COHESIVE ENERGY

At normal temperatures all the Group I metals adopt a body-centred cubic type of lattice with a coordination number of 8. However, at very low temperatures lithium forms a hexagonal close-packed structure with a coordination number of 12.

The metals are all very soft, and can be cut quite easily with a knife. Lithium is harder than the others, but is softer than lead. Bonding in metals is discussed in Chapters 2 and 5 in terms of delocalized molecular orbitals or bands, extending over the whole crystal.

The cohesive energy is the force holding the atoms or ions together in the solid. (This is the same in magnitude, but the opposite in sign, to the enthalpy of atomization, which is the energy required to break the solid up into gaseous atoms.) The cohesive energies of Group I metals are about half of those for Group II, and one third of those for Group III elements. The magnitude of the cohesive energy determines the hardness, and it depends on the number of electrons that can participate in bonding and on the strength of the bonds formed. The softness, low cohesive energy and weak bonding in Group I elements are consequences of these metals having only one valency electron which can participate in bonding (compared with two or more electrons in most other metals), and of the large size and diffuse nature of the outer bonding electron. The atoms become larger on descending the group from lithium to caesium, so the bonds are weaker, the cohesive energy decreases and the softness of the metals increases.

Table 9.8 Cohesive energy

	Cohesive energy (Enthalpy of atomization) ($kJ\,mol^{-1}$)
Li	161
Na	108
K	90
Rb	82
Cs	78

MELTING AND BOILING POINTS

The generally low values for cohesive energy are reflected in the very low values of melting and boiling points in the group. The cohesive energy decreases down the group, and the melting points decrease correspondingly.

The melting points range from lithium 181 °C to caesium 28.5 °C. These are extremely low values for metals, and contrast with the melting points of the transition metals, most of which are above 1000 °C.

The melting point of lithium is nearly twice as high (in °C) as that for sodium, though the others are close together. With many properties it is found that the first element in each group differs appreciably from the rest of the group. (Differences between lithium and the other Group I elements are discussed near the end of this chapter.)

Table 9.9 Melting and boiling points

	Melting point (°C)	Boiling point (°C)
Li	181	1347
Na	98	881
K	63	766
Rb	39	688
Cs	28.5	705

FLAME COLOURS AND SPECTRA

A result of the low ionization energies is that when these elements are irradiated with light, the light energy absorbed may be sufficient to make an atom lose an electron. Electrons emitted in this way are called photoelectrons, and this explains the use of caesium and potassium as cathodes in photoelectric cells.

Electrons may also be quite readily excited to a higher energy level, for example in the flame test. To perform this test, a sample of the metal chloride, or any salt of the metal moistened with concentrated HCl, is heated on a platinum or nichrome wire in a Bunsen burner flame. The heat from the burner excites one of the orbital electrons to a higher energy

Table 9.10 Flame colours and wavelengths

	Colour	Wavelength (nm)	Wavenumber (cm^{-1})
Li	crimson	670.8	14 908
Na	yellow	589.2	16 972
K	lilac	766.5	13 046
Rb	red–violet	780.0	12 821
Cs	blue	455.5	21 954

level. When the excited electron drops back to its original energy level it gives out the extra energy it obtained. The energy E is related to the wave number ν by the Einstein relationship:

$$E = h\nu \text{ (where h is Planck's constant)}$$

For Group I metals, the energy emitted appears as visible light, thus giving the characteristic flame colorations.

The colour actually arises from electronic transitions in short-lived species which are formed momentarily in the flame. The flame is rich in electrons, and in the case of sodium the ions are temporarily reduced to atoms.

$$Na^+ + e \rightarrow Na$$

The sodium D-line (which is actually a doublet at 589.0 nm and 589.6 nm) arises from the electronic transition $3s^1 \rightarrow 3p^1$ **in sodium atoms** formed in the flame. The colours from different elements do not all arise from the same transition, or from the same species. Thus the red line for lithium arises from a short-lived LiOH species formed in the flame.

These characteristic flame colorations of the *emission spectra* are used for the analytical determination of these elements by flame photometry. A solution of a Group I salt is aspirated into an oxygen–gas flame in a flame photometer. The energy from the flame excites an electron to a higher energy level, and when it falls back to the lower energy level the extra energy is given out as light. The intensity of the flame coloration is measured with a photoelectric cell. The intensitydepends on the concentration of metal present. A calibration graph is produced by measuring intensities with known standard solutions, and the exact concentration of the unknown solution can be found by comparison with the standard graph.

Alternatively *atomic absorption spectroscopy* may be used to estimate Group I metals.Here a lamp that emits a wavelength appropriate for a particular electronic transition is used to irradiate the sample in the flame. Thus a sodium lamp is used to detect sodium in the sample: other lamps are used to detect other elements. The amount of light absorbed, this time by the ground state atoms, is measured, and is proportional to the amount of the particular element being tested for.

COLOUR OF COMPOUNDS

Colour arises because the energy absorbed or emitted in electronic transitions corresponds to a wavelength in the visible region. The Group I metal ions all have noble gas configurations in which all the electrons are paired. Thus promoting an electron requires some energy to unpair an electron, some to break a full shell of electrons and some to promote the electron to a higher level. The total energy is large: hence there are no suitable transitions and the compounds are typically white. Any transitions which do occur will be of high energy, will appear in the ultraviolet region rather than in the visible region, and will be invisible to the human eye. Compounds of Group I metals are typically white, except those where the anion is coloured, for example sodium chromate $Na_2[CrO_4]$ (yellow), potassium dichromate $K_2[Cr_2O_7]$ (orange), and potassium permanganate $K[MnO_4]$ (deep purple). In these cases the colour comes from the anions $[CrO_4]^-$, $[Cr_2O_7]^{2-}$ or $[MnO_4]^-$ and not from the Group I metal ion.

When Group I elements form compounds (usually ionic, but there are a few covalent compounds), all the electrons are paired. Because of this Group I compounds are diamagnetic. There is one notable exception – the superoxides, which are discussed later.

CHEMICAL PROPERTIES

Table 9.11 Some reactions of Group I metals

Reaction	Comment
$M + H_2O \rightarrow MOH + H_2$	The hydroxides are the strongest bases known
with excess oxygen	
$Li + O_2 \rightarrow Li_2O$	Monoxide is formed by Li and to a small extent by Na
$Na + O_2 \rightarrow Na_2O_2$	Peroxide formed by Na and to a small extent by Li
$K + O_2 \rightarrow KO_2$	Superoxide formed by K, Rb, Cs
$M + H_2 \rightarrow MH$	Ionic 'salt-like' hydrides
$Li + N_2 \rightarrow Li_3N$	Nitride formed only by Li
$M + P \rightarrow M_3P$	All the metals form phosphides
$M + As \rightarrow M_3As$	All the metals form phosphides
$M + Sb \rightarrow M_3Sb$	All the metals form phosphides
$M + S \rightarrow M_2S$	All the metals form sulphides
$M + Se \rightarrow M_2Se$	All the metals form selenides
$M + Te \rightarrow M_2Te$	All the metals form tellurides
$M + F_2 \rightarrow MF$	All the metals form fluorides
$M + Cl_2 \rightarrow MCl$	All the metals form chlorides
$M + Br_2 \rightarrow MBr$	All the metals form bromides
$M + I_2 \rightarrow MI$	All the metals form iodides
$M + NH_3 \rightarrow MNH_2$	All the metals form amides

Reaction with water

Group I metals all react with water, liberating hydrogen and forming the hydroxides. The reaction becomes increasingly violent on descending the group. Thus lithium reacts gently, sodium melts on the surface of the water and the molten metal skates about vigorously and may catch fire (especially if localized), and potassium melts and always catches fire.

$$2Li + 2H_2O \rightarrow 2LiOH + H_2$$
$$2Na + 2H_2O \rightarrow 2NaOH + H_2$$
$$2K + 2H_2O \rightarrow 2KOH + H_2$$

The standard electrode potentials $E°$ are $Li^+|Li = -3.05$ volts, $Na^+|Na = -2.71$, $K^+|K = -2.93$, $Rb^+|Rb = -2.92$, $Cs^+|Cs = -2.92$. Lithium has the most negative standard electrode potential of any element in the periodic table, largely because of its high hydration energy. Standard electrode potentials $E°$ and Gibbs free energy ΔG are related by the equation:

$$\Delta G = -nFE°$$

where n is the number of electrons removed from the metal to produce the ion, and F is the Faraday constant.

The reaction $Li^+ + e \rightarrow Li$ has the largest negative $E°$ value, and hence the largest positive ΔG value. Thus the reaction does **not** occur. However, the reverse reaction $Li \rightarrow Li^+ + e$ has a large negative value of ΔG, so lithium liberates more energy than the other metals when it reacts with water. In view of this it is at first sight rather surprising that lithium reacts gently with water, whereas potassium, which liberates less energy, reacts violently and catches fire. The explanation lies in the kinetics (that is the rate at which the reaction proceeds), rather than in the thermodynamics (that is the total amount of energy liberated). Potassium has a low melting point, and the heat of reaction is sufficient to make it melt, or even vaporize. The molten metal spreads out, and exposes a larger surface to the water, so it reacts even faster, gets even hotter and catches fire.

Reaction with air

Chemically Group I elements are very reactive, and tarnish rapidly in dry air. Sodium, potassium, rubidium and caesium form oxides of various types, but lithium forms a mixture of the oxide and the nitride, Li_3N.

Reaction with nitrogen

Lithium is the only element in the group that reacts with nitrogen to form a nitride. Lithium nitride, Li_3N, is ionic ($3Li^+$ and N^{3-}), and is ruby red. Two reactions of the nitride are of interest. First, on heating to a high temperature it decomposes to the elements, and second, it reacts with water, giving ammonia.

$$2Li_3N \xrightarrow{\text{heat}} 6Li + N_2$$
$$Li_3N + 3H_2O \rightarrow 3LiOH + NH_3$$

OXIDES, HYDROXIDES, PEROXIDES AND SUPEROXIDES

Reaction with air

The metals all burn in air to form oxides, though the the product varies depending on the metal. Lithium forms the monoxide Li_2O (and some peroxide Li_2O_2), sodium forms the peroxide Na_2O_2 (and some monoxide Na_2O), and the others form superoxides of the type MO_2.

All five metals can be induced to form the normal oxide, peroxide or superoxide by dissolving the metal in liquid ammonia and bubbling in the appropriate amount of oxygen.

Normal oxides – monoxides

The monoxides are ionic, for example $2Li^+$ and O^{2-}. Li_2O and Na_2O are pure white solids as expected, but surprisingly K_2O is pale yellow, Rb_2O is bright yellow and Cs_2O is orange. Metallic oxides are usually basic. The typical oxides M_2O are strongly basic oxides, and they react with water, forming strong bases.

$$Li_2O + H_2O \rightarrow 2LiOH$$
$$Na_2O + H_2O \rightarrow 2NaOH$$
$$K_2O + H_2O \rightarrow 2KOH$$

The crystal structures of Li_2O, Na_2O, K_2O and Rb_2O are anti-fluorite structures. The anti-fluorite structure is like that for fluorite CaF_2, except that the positions of the positive and negative ions are interchanged. Thus Li^+ fill the sites occupied by F^-, and O^{2-} fill sites occupied by Ca^{2+}. Cs_2O has an anti-$CdCl_2$ layer structure.

Hydroxides

Sodium hydroxide NaOH is often called caustic soda, and potassium hydroxide is called caustic potash, because of their corrosive properties (for example on glass or on skin). These caustic alkalis are the strongest bases known in aqueous solution. The hydroxides of Na, K, Rb and Cs are very soluble in water, but LiOH is much less soluble (see Table 9.12). At

Table 9.12 Solubility of Group I hydroxides

Element	Solubility ($g/100\,g\ H_2O$)
Li	13.0 (25 °C)
Na	108.3 (25 °C)
K	112.8 (25 °C)
Rb	197.6 (30 °C)
Cs	385.6 (15 °C)

25 °C a saturated solution of NaOH is about 27 molar, whilst saturated LiOH is only about 5 molar.

The bases react with acids to form salts and water, and are used for many neutralizations.

$$NaOH + HCl \rightarrow NaCl + H_2O$$

The bases also react with CO_2, even traces in the air, forming the carbonate. LiOH is used to absorb carbon dioxide inclosed environments such as space capsules (where its light weight is an advantage in reducing the launching weight).

$$2NaOH + CO_2 \rightarrow Na_2CO_3 + H_2O$$

They also react with the amphoteric oxides, Al_2O_3, forming aluminates, SiO_2 (or glass), forming silicates, SnO_2, forming stannates, PbO_2, forming plumbates and ZnO, forming zincates.

The bases liberate ammonia from both ammonium salts and coordination complexes where ammonia is attached to a transition metal ion (ammine complexes).

$$NaOH + NH_4Cl \rightarrow NH_3 + NaCl + H_2O$$

$$6NaOH + \underset{\text{hexammine cobalt(III) chloride}}{2[Co(NH_3)_6]Cl_3} \rightarrow 12NH_3 + Co_2O_3 + 3NaCl + 3H_2O$$

NaOH reacts with H_2S to form sulphides S^{2-}, and hydrogen sulphides SH^-, and it is used to remove mercaptans from petroleum products.

$$NaOH + H_2S \rightarrow NaSH \rightarrow Na_2S$$

The hydroxides react with alcohols, forming alkoxides.

$$NaOH + EtOH \rightarrow \underset{\text{sodium ethoxide}}{NaOEt} + H_2O$$

KOH resembles NaOH in all its reactions, but as KOH is much more expensive it is seldom used. However, KOH is much more soluble in alcohol, thus producing $OC_2H_5^-$ ions by the equilibrium

$$C_2H_5OH + OH^- \rightleftharpoons OC_2H_5^- + H_2O$$

This accounts for the use of alcoholic KOH in organic chemistry. Group I hydroxides are thermally stable, illustrating the strong electropositive nature of the metals. On heating, many hydroxides decompose, losing water and forming the oxide.

Peroxides and superoxides

The peroxides all contain the $[—O—O—]^{2-}$ ion. They are diamagnetic (all the electrons are paired), and are oxidizing agents. They may be regarded as salts of the dibasic acid H_2O_2, and they react with water and acid, giving hydrogen peroxide H_2O_2.

$$Na_2O_2 + 2H_2O \rightarrow 2NaOH + H_2O_2$$

Na_2O_2 is pale yellow in colour. It is used industrially for bleaching wood pulp, paper and fabrics such as cotton and linen. It is a powerful oxidant, and many of its reactions are dangerously violent, particularly with materials that are reducing agents such as aluminium powder, charcoal, sulphur and many organic liquids. Because it reacts with CO_2 in the air it has been used to purify the air in submarines and confined spaces, as it both removes CO_2 and produces O_2. Potassium superoxide KO_2 is even better for this purpose. Some typical reactions are:

$$Na_2O_2 + Al \rightarrow Al_2O_3$$
$$Na_2O_2 + Cr^{3+} \rightarrow CrO_4^{2-}$$
$$Na_2O_2 + CO \rightarrow Na_2CO_3$$
$$2Na_2O_2 + 2CO_2 \rightarrow Na_2CO_3 + O_2$$

The industrial process for forming sodium peroxide is a two-stage reaction in the presence of excess air:

$$2Na + \tfrac{1}{2}O_2 \rightarrow Na_2O$$
$$Na_2O + \tfrac{1}{2}O_2 \rightarrow Na_2O_2$$

The superoxides contain the ion $[O_2]^-$, which has an unpaired electron, and hence they are paramagnetic and are all coloured (LiO_2 and NaO_2 yellow, KO_2 orange, RbO_2 brown and CsO_2 orange).

NaO_2 has three different crystal structures, the marcasite structure at liquid air temperatures, the pyrites structure FeS_2 between $-77\,°C$ and $-50\,°C$, and a calcium carbide CaC_2 structure at room temperature. Both the pyrites and calcium carbide structures are related to the NaCl structure in that the metal ions occupy the Na^+ sites, and O_2^-, S_2^{2-} and C_2^{2-} ions are centred on the Cl^- sites. Since the negative ions contain two atoms, their shape is an elongated rod rather than a sphere. In the CaC_2 structure, the C_2^{2-} ions are all oriented along one of the cubic axes, and thus the unit cell is elongated in that direction: hence the unit cell is cubic in NaCl but tetragonal in CaC_2. The pyrites structure is similar, but the C_2^{2-} ions are not all in alignment, and the cubic structure is retained.

Superoxides are even stronger oxidizing agents than peroxides, and give both H_2O_2 and O_2 with either water or acids.

$$KO_2 + 2H_2O \rightarrow KOH + H_2O_2 + \tfrac{1}{2}O_2$$

KO_2 is used in space capsules, submarines, and breathing masks, because it both produces oxygen and removes carbon dioxide.Both functions are important in life support systems.

$$4KO_2 + 2CO_2 \longrightarrow 2K_2CO_3 + 3O_2$$

$$4KO_2 + 4CO_2 + 2H_2O \xrightarrow{\text{more } CO_2} 4KHCO_3 + 3O_2$$

Sodium superoxide cannot be prepared by burning the metal in oxygen at atmospheric pressure, but it is made commercially and in good yields by reacting sodium peroxide with oxygen at a high temperature and pressure (450 °C and 300 atmospheres) in a stainless steel bomb.

$$Na_2O_2 + O_2 \rightarrow 2NaO_2$$

The bonding in peroxides and superoxides is described in the examples of molecular orbital treatment in Chapter 4. The peroxide ion $[—O—O—]^{2-}$ has 18 electrons, which occupy the molecular orbitals as shown:

$$\sigma 1s^2, \sigma^* 1s^2, \sigma 2s^2, \sigma^* 2s^2, \sigma 2p_x^2, \begin{cases} \pi 2p_y^2, \\ \pi 2p_z^2, \end{cases} \begin{cases} \pi^* 2p_y^2 \\ \pi^* 2p_z^2 \end{cases}$$

increasing energy →

Thus the bond order is one, corresponding to a single bond.

The superoxide ion $[O_2]^-$ has only 17 electrons, which give a bond order of 1.5.

$$\sigma 1s^2, \sigma^* 1s^2, \sigma 2s^2, \sigma^* 2s^2, \sigma 2p_x^2, \begin{cases} \pi 2p_y^2, \\ \pi 2p_z^2, \end{cases} \begin{cases} \pi^* 2p_y^2 \\ \pi^* 2p_z^1 \end{cases}$$

Generally, large atoms or ions form weaker bonds than small ones. The peroxide and superoxide ions are large, and it is noteworthy that the stability of the peroxides and superoxides increases as the metal ions become larger. This shows that large cations can be stabilized by large anions, since if both ions are similar in size the coordination number will be high, and this gives a high lattice energy.

SULPHIDES

The metals all react with sulphur, forming sulphides such as Na_2S, and polysulphides Na_2S_n where $n = 2, 3, 4, 5$ or 6. The polysulphide ions are made from zig-zag chains of sulphur atoms.

```
    S          S   S−      S   S          S   S   S−
   / \        / \ /       / \ / \        / \ / \ /
−S    S−  −S    S      −S    S    S−  −S    S    S
```

Sodium sulphide can also be made by heating sodium sulphate with carbon, or by passing H_2S into NaOH solution.

$$Na_2SO_4 + 4C \rightarrow Na_2S + 4CO$$
$$NaOH + H_2S \rightarrow NaHS + H_2O$$
$$NaOH + NaHS \rightarrow Na_2S + H_2O$$

Group I sulphides hydrolyse appreciably in water, giving strongly alkaline solutions:

$$Na_2S + H_2O \rightarrow NaSH + NaOH$$

Na_2S is used to make organic sulphur dyestuffs, and in the leather industry to remove hair from hides. Na_2S is readily oxidized by air to form sodium thiosulphate, which is used in photography to dissolve silver halides, and as a laboratory reagent for iodine titrations.

$$2Na_2S + 2O_2 + H_2O \rightarrow Na_2S_2O_3 + 2NaOH$$
$$2Na_2S_2O_3 + I_2 \rightarrow Na_2S_4O_6 + 2NaI$$

SODIUM HYDROXIDE

Sodium hydroxide is the most important alkali used in industry. It is produced on a large scale (34 million tonnes in 1985) by the electrolysis of an aqueous solution of NaCl (brine) using either a diaphragm cell or a mercury cathode cell. At one time it was also made from Na_2CO^3 by the lime–caustic soda process, but this is only used a little nowadays as other methods are cheaper. Details of the industrial methods, uses, and tonnages are given in Chapter 10.

SODIUM HYDROGENCARBONATE (SODIUM BICARBONATE)

About 200 000 tonnes of $NaHCO_3$ are produced annually in the USA, of which 40% is used for baking powder, 15% to make other chemicals, 12% in pharmaceutical products including anti-acid preparations for indigestion, and 10% in fire extinguishers.

$NaHCO_3$ can be used on its own to make cakes or bread 'rise' since it decomposes between 50 °C and 100 °C, giving bubbles of CO_2.

$$2NaHCO_3 \xrightarrow{\text{gentle heat}} Na_2CO_3 + H_2O + CO_2$$

Baking powder is more commonly used, and contains $NaHCO_3$, $Ca(H_2PO_4)_2$ and starch. The $Ca(H_2PO_4)_2$ is acidic and when moistened it reacts with $NaHCO_3$, giving CO_2. The starch is a filler. An improved 'combination baking powder' contains about 40% starch, 30% $NaHCO_3$, 20% $NaAl(SO_4)_2$ and 10% $Ca(H_2PO_4)_2$. The $NaAl(SO_4)_2$ slows the reaction down so the CO_2 is given off more slowly.

SODIUM SULPHATE

About 4.2 million tonnes of Na_2SO_4 are used annually. About 55% of this is made synthetically, as a by-product from the manufacture of HCl, and also from many neutralization processes that use H_2SO_4. About 45%, mainly Glauber's salt $Na_2SO_4 \cdot 10H_2O$, is mined.

The major use of Na_2SO_4 – some 70% – is in the paper industry, and about 10% is used in detergents, and 10% in glass manufacture. In the Kraft paper making process, a strong alkaline solution of Na_2SO_4 is used to dissolve the lignin that holds the cellulose fibres together in wood chips.

The cellulose fibres are then turned into corrugated cardboard and brown paper.

OXOSALTS – CARBONATES, BICARBONATES, NITRATES AND NITRITES

Group I metals are highly electropositive and thus form very strong bases, and have quite stable oxosalts.

The carbonates are remarkably stable, and will melt before they eventually decompose into oxides at temperatures above 1000°C. Li_2CO_3 is considerably less stable and decomposes more readily.

Because Group I metals are so strongly basic, they also form solid bicarbonates (also called hydrogencarbonates). No other metals form solid bicarbonates, though NH_4HCO_3 also exists as a solid. Bicarbonates evolve carbon dioxide and turn into carbonates on gentle warming. This is one test for bicarbonates in qualitative analysis. The crystal structures of $NaHCO_3$ and $KHCO_3$ both show hydrogen bonding, but are different. In $NaHCO_3$ the HCO_3^- ions are linked into an infinite chain, whilst in $KHCO_3$ a dimeric anion is formed.

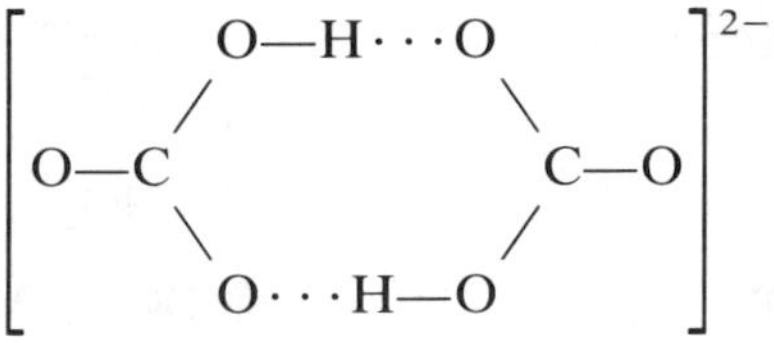

```
[        O—H···O          ]2−
        /       \
   O—C           C—O
        \       /
         O···H—O
]
```

Lithium is exceptional in that it does not form a solid bicarbonate, though $LiHCO_3$ can exist in solution. All the carbonates and bicarbonates are soluble in water.

Over 50 000 tonnes of Li_2CO_3 are produced annually. Most of it is added as an impurity to Al_2O_3 to lower its melting point in the extraction of aluminium by electrolysis. Some is used to toughen glass (sodium in the glass is replaced by lithium). Na_2CO_3 is used as washing soda to soften water in hard water areas, and $NaHCO_3$ is used as baking powder.

The nitrates can all be prepared by the action of HNO_3 on the corresponding carbonate or hydroxide, and they are all very soluble in water. $LiNO_3$ is used for fireworks and red-coloured distress flares. Large deposits of $NaNO_3$ are found in Chile, and are used as a nitrogenous fertilizer. Solid $LiNO_3$ and $NaNO_3$ are deliquescent, and because of this KNO_3 is used in preference to $NaNO_3$ in gunpowder (gunpowder is a mixture of KNO_3, sulphur and charcoal). KNO_3 is usually obtained from synthetic nitric acid and K_2CO_3, but at one time it was made from $NaNO_3$:

$$2HNO_3 + K_2CO_3 \rightarrow 2KNO_3 + CO_2 + H_2O$$
$$NaNO_3 + KCl \rightarrow KNO_3 + NaCl$$

Group I nitrates are fairly low melting solids, and are amongst the most stable nitrates known. However, on strong heating they decompose into nitrites, and at higher temperatures to the oxide. $LiNO_3$ decomposes more readily than the others, forming the oxide.

$$2NaNO_3 \xrightleftharpoons{500\,°C} 2NaNO_2 + O_2$$
$$4NaNO_3 \xrightleftharpoons{800\,°C} 2Na_2O + 5O_2 + 2N_2$$

Alkali metal nitrates are widely used as molten salts as a solvent in which to carry out high temperature oxidations, and also as a heat transfer medium. They are used up to around 600 °C, but molten salt baths are often used at much lower temperatures. For example, a 1:1 mixture of $LiNO_3/KNO_3$ melts at the surprisingly low temperature of 125 °C.

Nitrites are important in the manufacture of organonitrogen compounds, the most important being the azo dyes. Small amounts of $NaNO_2$ are used in molten salt baths with $NaNO_3$, and some is used as a food preservative. Nitrites are easily recognized in the laboratory, because on treatment with dilute acids they produce brown fumes of NO_2.

$$2NaNO_2 + 2HCl \rightarrow 2NaCl + H_2O + NO_2 + NO$$
$$2NO + O_2 \rightarrow 2NO_2$$

$NaNO_2$ is manufactured by absorbing oxides of nitrogen in Na_2CO_3 solution.

$$Na_2CO_3 + NO_2 + NO \rightarrow 2NaNO_2 + CO_2$$

They can also be made by thermal decomposition of nitrates and the chemical reduction of nitrates:

$$2NaNO_3 + C \rightarrow 2NaNO_2 + CO_2$$
$$KNO_3 + Zn \rightarrow KNO_2 + ZnO$$

or by reacting NO with a hydroxide.

$$2KOH + 4NO \rightarrow 2KNO_2 + N_2O + H_2O$$
$$4KOH + 6NO \rightarrow 4KNO_2 + N_2 + 2H_2O$$

HALIDES AND POLYHALIDES

Since Li^+ is the smallest ion in the group, it would be expected to form hydrated salts more readily than the other metals. LiCl, LiBr and LiI form trihydrates $LiX.3H_2O$, but the other alkali metal halides form anhydrous crystals.

All the halides adopt a NaCl type of structure with a coordination number of 6 except for CsCl, CsBr and CsI. The latter have a CsCl type of structure with a coordination number of 8. Rather more compounds adopt the NaCl type of structure than would be expected from the radius ratios of the ions r^+/r^-, and the reason for this structure being adopted is that it gives the highest lattice energy (see the sections on Ionic compounds of the type AX, and Lattice energy in Chapter 3).

The alkali metal halides react with the halogens and interhalogen compounds forming ionic polyhalide compounds:

$$KI + I_2 \rightarrow K[I_3]$$
$$KBr + ICl \rightarrow K[BrICl]$$
$$KF + BrF_3 \rightarrow K[BrF_4]$$

HYDRIDES

Group I metals all react with hydrogen, forming ionic or salt-like hydrides M^+H^-. However, the ease with which they do so decreases from lithium to caesium. These hydrides contain the H^- ion (which is not commonly found, since hydrogen usually forms H^+ ions). It can be proved that H^- ions exist because on electrolysis hydrogen is liberated at the anode.

The hydrides react with water, liberating hydrogen, and lithium hydride is used as a source of hydrogen for military purposes and for filling meteorological balloons.

$$LiH + H_2O \rightarrow LiOH + H_2$$

Lithium also forms a complex hydride $Li[AlH_4]$, calledlithium aluminium hydride, which is a useful reducing agent. It is made from lithium hydride in dry ether solution.

$$4LiH + AlCl_3 \rightarrow Li[AlH_4] + 3LiCl$$

Lithium aluminium hydride is ionic, and the $[AlH_4]^-$ ion is tetrahedral. $Li[AlH_4]$ is a powerful reducing agent and is widely used in organic chemistry, as it reduces carbonyl compounds to alcohols. It reacts violently with water, so it is necessary to use absolutely dry organic solvents, for example ether which has been dried over sodium. $Li[AlH_4]$ will also reduce a number of inorganic compounds.

$$BCl_3 + Li[AlH_4] \rightarrow B_2H_6 \text{ diborane}$$
$$PCl_3 + Li[AlH_4] \rightarrow PH_3 \text{ phosphine}$$
$$SiCl_4 + Li[AlH_4] \rightarrow SiH_4 \text{ silane}$$

Sodium tetrahydridoborate (sodium borohydride) $Na[BH_4]$ is another hydride complex. It is ionic, comprising tetrahedral $[BH_4]^-$ ions. It is best obtained by heating sodium hydride with trimethyl borate:

$$4NaH + B(OCH_3)_3 \xrightarrow{230-270\,^\circ C} Na[BH_4] + 3NaOCH_3$$

Other tetrahydridoborates for Group I and II metals, aluminium and some transition metals can be made from the sodium salt. These tetrahydridoborates are used as reducing agents, and the alkali metal compounds (particularly those of Na and K) are becoming increasingly used as they are much less sensitive to water than $Li[AlH_4]$. Thus $Na[BH_4]$ can be crystallized from cold water, and $K[BH_4]$ from hot water, so they have the advantage that they can be used in aqueous solutions. The others react with water. (See Group III.)

$$[BH_4]^- + 2H_2O \rightarrow BO_2^- + 4H_2$$

SOLUBILITY AND HYDRATION

All the simple salts dissolve in water, producing ions, and consequently the solutions conduct electricity. Since Li^+ ions are small, it might be expected that solutions of lithium salts would conduct electricity better than solutions of the same concentration of sodium, potassium, rubidium or caesium salts. The small ions should migrate more easily towards the cathode, and thus conduct more than the larger ions. However, ionic mobility or conductivity measurements in aqueous solution (Table 9.13) give results in the opposite order $Cs^+ > Rb^+ > K^+ > Na^+ > Li^+$. The reason for this apparent anomaly is that the ions are hydrated in solution. Since Li^+ is very small, it is heavily hydrated. This makes the radius of the hydrated ion large, and hence it moves only slowly. In contrast, Cs^+ is the least hydrated, and the radius of the hydrated Cs^+ ion is smaller than the radius of hydrated Li^+, and hence hydrated Cs^+ moves faster, and conducts electricity more readily.

Table 9.13 Ionic mobilities and hydration

	Ionic radius (Å)	Ionic mobility at infinite dilution	Approx. radius hydrated ion (Å)	Approx. hydration number	Hydration terms		
					$\Delta H°$	$\Delta S°$ (kJ mol^{-1})	$\Delta G°$
Li^+	0.76	33.5	3.40	25.3	−544	−134	−506
Na^+	1.02	43.5	2.76	16.6	−435	−100	−406
K^+	1.38	64.5	2.32	10.5	−352	−67	−330
Rb^+	1.52	67.5	2.28	10.0	−326	−54	−310
Cs^+	1.67	68.0	2.28	9.9	−293	−50	−276

The hydration number is the average number of water molecules associated with the metal ion. The values need not be whole numbers, and are obtained by measuring the transference of water in a conductivity cell.

Some water molecules touch the metal ion and bond to it, forming a complex. These water molecules constitute the *primary shell* of water. Thus Li^+ is tetrahedrally surrounded by four water molecules. This may be explained by the oxygen atoms of the four water molecules using a lone pair to form a coordinate bond to the metal ion. With four electron pairs in the valence shell the VSEPR theory predicts a tetrahedral structure. Alternatively, using valence bond theory, the $2s$ orbital and the three $2p$ orbitals form four sp^3 hybrid orbitals which are filled by the lone pairs from the oxygen atoms.

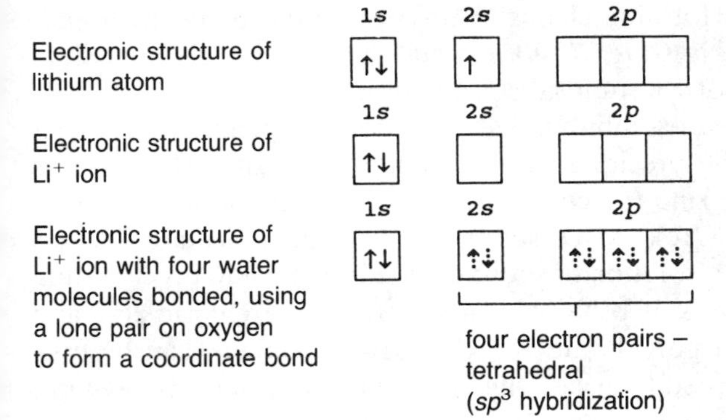

With the heavier ions, particularly Rb^+ and Cs^+, the number of water molecules increases to six. VSEPR theory predicts an octahedral structure. Valence bond theory also indicates an octahedral arrangement using one *s* orbital, three *p* orbitals and two *d* orbitals for bonding.

A *secondary layer* of water molecules further hydrates the ions, though these are only held by weak ion–dipole attractive forces. The strength of such forces is inversely proportional to the distance, that is to the size of the metal ion. Thus the secondary hydration decreases from lithium to caesium, and accounts for Li^+ being the most heavily hydrated.

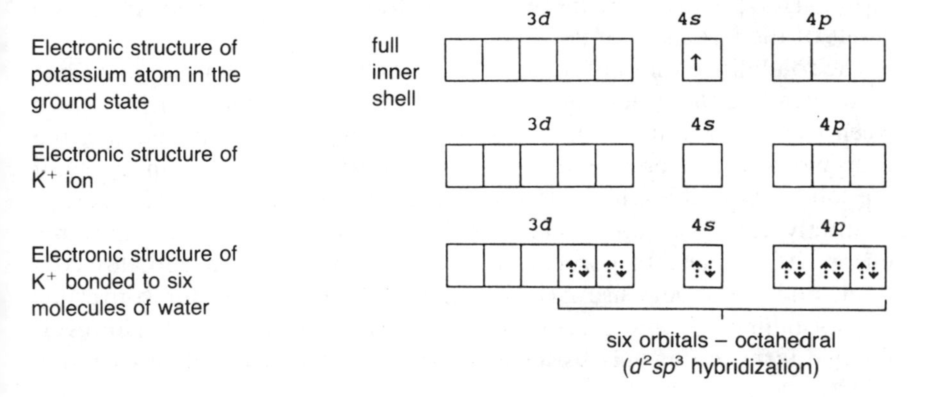

Note that the *d* orbitals comprise a group of three (called t_{2g} orbitals), and a group of two (called e_g orbitals). Only the group of two is used for bonding.

The size of the hydrated ions is an important factor affecting the passage of these ions through cell walls. It also explains their behaviour on cation-exchange columns, where hydrated Li^+ ions are attached less strongly, and hence eluted first.

The decrease in hydration from Li^+ to Cs^+ is also shown in the crystalline salts, for nearly all lithium salts are hydrated, commonly as trihydrates. In these hydrated Li salts Li^+ is coordinated to $6H_2O$, and the octahedra share faces, forming chains. Many sodium salts are hydrated, e.g. $Na_2CO_3 \cdot 10H_2O$, $Na_2CO_3 \cdot 7H_2O$ and $Na_2CO_3 \cdot H_2O$. Few potassium salts and no rubidium or caesium salts are hydrated.

The simple salts are all soluble in water, and so in qualitative analysis these metals need to be precipitated as less common salts. Thus Na^+ is precipitated by adding zinc (or copper) uranyl acetate solution and precipitating $NaZn(UO_2)(Ac)_9 \cdot H_2O$ sodium zinc uranyl acetate. K^+ is precipitated by adding a solution of sodium cobaltinitrite and precipitating potassium cobaltinitrite $K_3[Co(NO_2)_6]$ or by adding perchloric acid and precipitating potassium perchlorate $KClO_4$. Group I metals can be estimated gravimetrically, sodium as the uranylacetate, and potassium, rubidium and caesium as tetraphenylborates. However, modern instrumental methods such as flame photometry and atomic absorption spectrometry are much quicker and easier to use and are now used in preference to gravimetric analysis.

$$K^+ + Na_3[Co(NO_2)_6] \rightarrow Na^+ + K_3[Co(NO_2)_6] \text{ potassium cobaltinitrite}$$

$$K^+ + NaClO_4 \rightarrow Na^+ + KClO_4 \text{ potassium perchlorate}$$

$$K^+ + Na[B(C_6H_5)_4] \rightarrow Na^+ + K[B(C_6H_5)_4] \text{ potassium tetraphenylborate quantitative precipitate}$$

If a salt is insoluble its lattice energy is greater than the hydration energy. $K[B(C_6H_5)_4]$ is insoluble because the hydration energy is very small as a result of the large size of its ions.

The solubility of most of the salts of Group I elements in water decreases on descending the group. For a substance to dissolve the energy evolved when the ions are hydrated (hydration energy) must be larger than the energy required to break the crystal lattice (lattice energy). Conversely, if the solid is insoluble, the hydration energy is less than the lattice energy.

Strictly in the two cycles shown in Figure 9.4 we should use Gibbs free energy ΔG values. In particular, the lattice energy is an enthalpy $\Delta H°$ term, and we should use $\Delta G°$ the standard free energy for converting the crystalline salt into gaseous ions an infinite distance apart. However, the two terms differ only by a small term for the entropy of vaporization

Table 9.14 Hydration and lattice energy values for Group I halides at 25 °C.

	Free energy of hydration $\Delta G°$ ($kJ\,mol^{-1}$)	Lattice energy ($kJ\,mol^{-1}$)			
		MF	MCl	MBr	MI
Li^+	−506	−1035	−845	−800	−740
Na^+	−406	−908	−770	−736	−690
K^+	−330	−803	−703	−674	−636
Rb^+	−310	−770	−674	−653	−515
Cs^+	−276	−720	−644	−623	−590

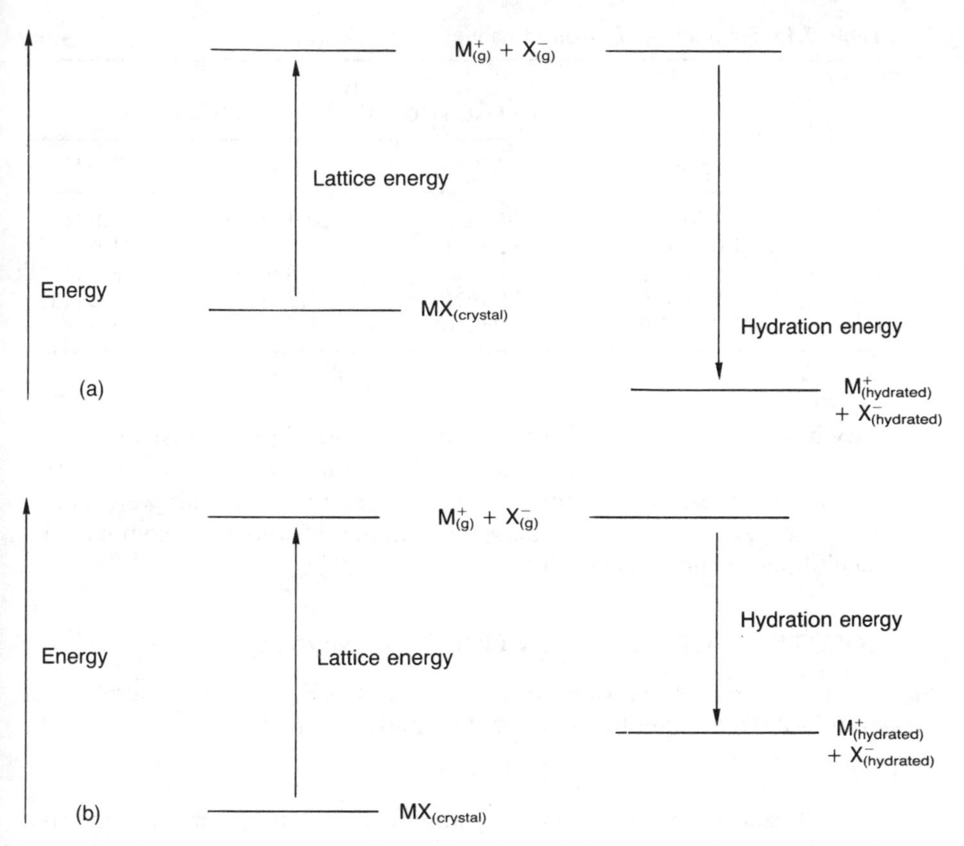

Figure 9.4 Solubility related to lattice energy and hydration energy. (a) The solid dissolves, (b) the solid is insoluble.

of the ions. It should in principle be possible to predict solubilities from lattice energies and hydration energies. In practice there are difficulties in predicting solubilities because the values for the data are not known very accurately, and the result depends on a small difference between two large values.

The reason why the solubility of most Group I metals decreases on descending the group is that the lattice energy only changes slightly, but the free energy of hydration changes rather more. For example, the difference in lattice energy between NaCl and KCl is $67\,kJ\,mol^{-1}$, and yet the difference in $\Delta G_{(hydration)}$ for Na^+ and K^+ is $76\,kJ\,mol^{-1}$. Thus KCl is less soluble than NaCl.

The Group I fluorides and carbonates are exceptional in that their solubilities increase rapidly on descending the group. The reason for this is that their lattice energies change more than the hydration energies on descending the group. The lattice energy depends on electrostatic attraction between ions, and is proportional to the distance between the ions, that is proportional to $1/(r^+ + r^-)$. It follows that the lattice energy will

Table 9.15 Solubilities of Group I halides

	Solubility (Molar value first, g/100 g H_2O given in brackets)			
	MF	MCl	MBr	MI
Li	0.1 (0.27)	19.6 (830)	20.4 (177)	8.8 (165)
Na	1.0 (4.22)	6.2 (36)	8.8 (91)	11.9 (179)
K	15.9 (92.3)	4.8 (34.7)	7.6 (67)	8.7 (144)
Rb	12.5 (130.6)	7.5 (91)	6.7 (110)	7.2 (152)
Cs	24.2 (367.0)	11.0 (186)	5.1 (108)	3.0 (79)

vary most when r^- is small, that is with F^-, and will vary least when r^+ is large (with I^-). The weight of solute dissolving does not provide a very useful comparison of the solubilities, because the molecular weights differ. The easiest way to compare the number of ions is to compare the solubilities as molar quantities.

SOLUTIONS OF METALS IN LIQUID AMMONIA

In the presence of impurities or catalysts such as Fe, the alkali metals react with liquid ammonia to form a metal amide and hydrogen.

$$M + NH_3 \rightarrow MNH_2 + \tfrac{1}{2}H_2$$

If all catalysts and impurities are absent, then Group I metals, and to a lesser extent the elements Ca, Sr and Ba in Group II and the lanthanide elements Eu and Yb, dissolve directly in very high concentration in liquid ammonia. The metal can be recovered simply by allowing the ammonia to boil off.

Dilute solutions of alkali metals in liquid ammonia are dark blue in colour, and the main species present are solvated metal ions and solvated electrons. If the blue solution is allowed to stand, the colour slowly fades until it disappears owing to the formation of a metal amide. At concen-

Table 9.16 Solubilities in liquid ammonia

Element	Solubility (g metal/100 g NH_3)	
	−33.4 °C	0 °C
Li	10.9	11.3
Na	25.1	23.0
K	47.1	48.5

Note that −33.4 °C is the boiling point of liquid ammonia at one atmosphere pressure. The 0 °C data were measured under pressure.

trations above 3 M, solutions are copper–bronze coloured and have a metallic lustre because metal ion clusters are formed.

These solutions of metals in liquid ammonia conduct electricity better than any salt in any liquid, and the conductivity is similar to that of pure metals (specific conductivity of Hg = 10^4 ohm^{-1}; Na/NH_3 = 0.5 × 10^4 ohm^{-1}; K/NH_3 = 0.45 × 10^4 ohm^{-1}). Conduction is due mainly to the presence of solvated electrons.

The metals are also soluble in other amines and these solutions are used in organic and inorganic syntheses.These solutions of metals in liquid ammonia act as powerful reducing agents for the elements of Groups IV, V and VI, for many compounds and coordination complexes, and they will even reduce an aromatic ring.

These reductions can be carried out in liquid ammonia, but not in water, because the alkali metals are stronger reducing agents than is hydrogen, and so will react with water and liberate hydrogen. The metals can exist for some time in liquid ammonia.

$$Bi + Na/NH_3 \rightarrow Na_3Bi \text{ (Bi reduced from oxidation state 0 to } -\text{III)}$$

$$S + Na/NH_3 \rightarrow Na_2S \text{ (S reduced from oxidation state 0 to } -\text{II)}$$

$$[Ni(CN)_4]^{2-} + 2e \rightarrow [Ni(CN)_4]^{4-} \text{ (Ni reduced from +II to 0)}$$

COMPOUNDS WITH CARBON

If lithium is heated with carbon, an ionic carbide Li_2C_2 is formed. The other metals do not react with carbon directly, but do form similar carbides when heated with ethyne (formerly called acetylene), or when ethyne is passed through a solution of the metal in liquid ammonia.

$$2Li + 2C \rightarrow Li_2C_2$$

$$Na + C_2H_2 \rightarrow NaHC_2 \rightarrow Na_2C_2$$

These compounds contain the carbide ion $[C{\equiv}C]^{2-}$ or hydridocarbide ion $[C{\equiv}C{-}H]^-$. The most important reaction of carbides is with water, when they give ethyne (acetylene). Thus they are termed acetylides.

$$Na_2C_2 + 2H_2O \rightarrow 2NaOH + C_2H_2$$

LiC_2H is used in the industrial manufacture of vitamin A.

The metals potassium, rubidium and caesium react with graphite by invading the space between the layers of carbon in the graphite lattice. They form highly coloured interstitial carbides that are nonstoichiometric, (that is of variable composition), ranging from $C_{60}K$ (grey), to $C_{36}K$ (blue), to a maximum invasion corresponding to C_8K (bronze). (See Chapter 12.)

ORGANIC AND ORGANOMETALLIC COMPOUNDS

The alkali metals replace hydrogen in organic acids, forming salts such as sodium acetate (sodium ethanoate) CH_3COONa and potassium benzoate

C_6H_5COOK. Soap is a mixture of the sodium salts of palmitic, oleic and stearic acids. (Palmitic acid $C_{15}H_{31} \cdot COOH$ occurs in palm oil, oleic acid $C_{17}H_{33} \cdot COOH$ occurs in olive oil and stearic acid $C_{17}H_{35} \cdot COOH$ occurs in beef and mutton fat and tallow.) Soap is made by the saponification (hydrolysis) of naturally occurring fats and oils. These fats and oils are esters of glycerol, and their hydrolysis with NaOH first breaks the ester to glycerol and fatty acids, neutralizing the fatty acid to give the sodium salts, i.e. the soap. World production of soap was 7.8 million tonnes in 1985.

$$\begin{array}{lcll} CH_2 \cdot O \cdot OC \cdot C_{15}H_{31} & & CH_2 \cdot OH & \\ | & & | & \\ CH \cdot O \cdot OC \cdot C_{15}H_{31} + 3NaOH & \rightarrow & CH \cdot OH & + 3C_{15}H_{31} \cdot COOH \\ | & & | & \\ CH_2 \cdot O \cdot OC \cdot C_{15}H_{31} & & CH_2 \cdot OH & \\ \text{glyceryl tripalmitate} & & \text{glycerol} & \text{palmitic acid} \\ \text{(palm oil)} & & & \end{array}$$

$$C_{15}H_{31} \cdot COOH + NaOH \rightarrow C_{15}H_{31} \cdot COONa + H_2O$$

Lithium stearate is also a 'soap', and is made from LiOH and some natural fat such as tallow. It is widely used to thicken hydrocarbon oils used as lubricants (the so-called detergent oils), and it is also used to make greases for motor vehicles.

Lithium shows a stronger tendency to covalency than the other alkali metals. Lithium also shows a diagonal relationship with magnesium. Magnesium forms a number of alkyl and aryl compounds called Grignard compounds which are very important in making organometallic compounds. It is not surprising that lithium also forms a number of covalent alkyls and aryls which are of great importance in the preparation of organometallic compounds. For example, $(LiCH_3)_4$ is typical of a range of compounds: it is covalent, soluble in organic solvents, and can be sublimed or distilled. These compounds are frequently tetrameric or hexameric. They are made from the alkyl or aryl halide, usually the chloride, in a solvent such as light petroleum, cyclohexane, toluene or ether.

$$RCl + Li \rightarrow LiR + LiCl$$

The structure of the $(LiCH_3)_4$ cluster is unusual. The four Li atoms occupy the corners of a tetrahedron. Each methyl C atom is above a face of the tetrahedron, and forms a triple bridge to the three Li atoms that make up the face of the tetrahedron. The intramolecular Li–C distance is 2.31 Å. The C is bonded to the three H atoms in the methyl group. The C is also bonded to a Li atom in another tetrahedron (with an intermolecular Li–C distance of 2.36 Å). The coordination number for the C atom is therefore 7. This cannot be explained by classical bonding theories as the C atom has only one *s* and three *p* orbitals available for bonding. The simplest explanation involves a four-centre two-electron bond covering the three Li atoms at the corners of a face, and the C atom above it. In a similar way the coordination number of Li is also 7, made up by three Li in the tetrahedron, three C at the centres of faces of the tetrahedron, and one Li in another tetrahedron.

Lithium ethyl is tetrameric in the solid $(LiEt)_4$, but is hexameric $(LiEt)_6$ when dissolved in hydrocarbons. The solid is similar in structure to $(LiCH_3)_4$, and the hexamer is thought to comprise an octahedron of Li atoms with Et groups above six of the eight faces, involving multi-centre bonding.

n-Butyl lithium is also tetrameric in the solid $(LiBu)_4$. It is commercially available. Production is about 1000 tonnes/year. The main uses are as a polymerization catalyst and for alkylation. It is a very versatile reagent in the laboratory for the synthesis of aromatic derivatives and unsaturated derivatives such as vinyl and allyl lithium. Many of these reactions are similar to those using Grignard reagents.

$$\text{LiBu} + \text{ArI} \rightarrow \text{LiAr} + \text{BuI} \quad (\text{Bu} = \text{butyl, Ar} = \text{aryl})$$

$$4\text{LiAr} + \text{Sn(CH=CH)}_4 \rightarrow 4\text{LiCH=CH}_2 + \text{Sn(Ar)}_4$$

From these an extremely wide range of organometallic and organic compounds can be prepared.

(R = alkyl or aryl)

$3LiR + BCl_3$	$\rightarrow BR_3$	$+ 3LiCl$	(organoboron compounds)
$4LiR + SnCl_4$	$\rightarrow SnR_4$	$+ 4LiCl$	(organotin compounds)
$3LiR + P(OEt)_3$	$\rightarrow PR_3$	$+ 3LiOEt$	(organophosphorus compounds)
$2LiR + HgI_2$	$\rightarrow HgR_2$	$+ 2LiI$	(organomercury compounds)
$LiR + R'I$	$\rightarrow R\text{-}R'$	$+ LiI$	(hydrocarbon)
$LiR + H^+$	$\rightarrow R\text{-}H$	$+ Li^+$	(hydrocarbon)
$LiR + Cl_2$	$\rightarrow R\text{-}Cl$	$+ LiCl$	(alkyl/aryl halide)
$LiR + HCONMe_2$	$\rightarrow R \cdot CHO$	$+ LiNMe_2$	(aldehydes)
$LiR + 3CO$	$\rightarrow R_2CO$	$+ 2LiCO$	(ketones)
$LiR + CO_2$	$\rightarrow R \cdot COOH$	$+ LiOH$	(carboxylic acids)

Alkyls of Na, K, Rb and Cs are usually prepared from the corresponding organomercury compound.

$$2\text{K} + \text{HgR}_2 \rightarrow \text{Hg} + 2\text{KR}$$

These compounds are ionic M^+R^-, and are extremely reactive. They catch fire in air, react violently with most compounds except nitrogen and saturated hydrocarbons, and are consequently difficult to handle.

COMPLEXES, CROWNS AND CRYPTS

Group I metals stand out from the other groups in their weak tendency to form complexes. This is predictable because the factors favouring complex formation are small size, high charge, and empty orbitals of low energy for forming the bonds, and Group I metal ions are very large and have a low charge (+1).

A number of aqua complexes are known such as $[Li(H_2O)_4]^+$ and a primary hydration shell of four H_2O molecules arranged tetrahedrally is found in various crystalline salts. Na^+ and K^+ also have the same primary hydration shell, but Rb^+ and Cs^+ coordinate six H_2O molecules. Stable complexes are formed with phosphine oxides; for example, complexes of formula $[LiX \cdot 4Ph_3PO]$, $[LiX \cdot 5Ph_3PO]$ and $[NaX \cdot 5Ph_3PO]$ are known where X is a large anion such as ClO_4^-, I^-, NO_3^- or SbF_6^-. There is a slight tendency to form ammine complexes such as $[Li(NH_3)_4]I$. Weak complexes ofsulphates, peroxosulphates and thiosulphates, and also hexacyanoferrates, are known in solution.

However, some organic chelating agents (particularly salicaldehyde and β-diketones) are extremely strong complexing agents, and Group I ions form complexes with these. These ligands are very strong complexing agents because they are multidentate, that is they have more than one donor group so they form more than one bond to the metal, and also because they form a ring or chelate compound by bonding to the metal. Examples include salicaldehyde, acetylacetone, benzoylacetone, methyl salicylate, o-nitrophenol, and o-nitrocresol. The metal usually attains a coordination number of 4 or 6 (see Figure 9.5).

An important development in the chemistry of the alkali metals is the discovery of complexes with polyethers, and 'cryptate complexes' with macrocyclic molecules with nitrogen and oxygen.

The crown ethers are an interesting class of complexing agents first synthesized by Pedersen in 1967. An example is dibenzo-18-crown-6, and the name indicates that there are two benzene rings in the compound, 18 atoms make up a crown-shaped ring, and six of the ring atoms are oxygen. These six oxygen atoms may complex with a metal ion, even with large ions like Group I ions that are not very good at forming complexes. The organic part of the molecule is puckered to give the crown arrangement, and the oxygen atoms with their lone pairs are nearly planar about the metal ion at the centre of the ring. The bonding of the metal ion to the polyether is largely electrostatic, and a close fit between the size of the metal ion and the size of the hole in the centre of the polyether is essential. Cyclic polyethers can have varying sizes of ring; for example, benzo-12-crown-4 has a ring of 12 atoms, four of which are oxygen. The polyethers form complexes selectively with the alkali metal ions. The size of the ring opening in the crown determines the size of the metal ion which may be accommodated. Thus a crown-4 (a cyclic polyether with four oxygens) is selective for Li^+, Na^+ prefers crown-5, and K^+ prefers crown-6. It is possible to get complexes with the unusual coordination number of 10, for example K^+(dibenzo-30-crown-10). Crown ethers form a number of crystalline complexes, but more importantly they are sometimes added to organic solvents to make them dissolve inorganic salts which, being ionic, would not normally dissolve. Polyethers of this type act as ion carriers

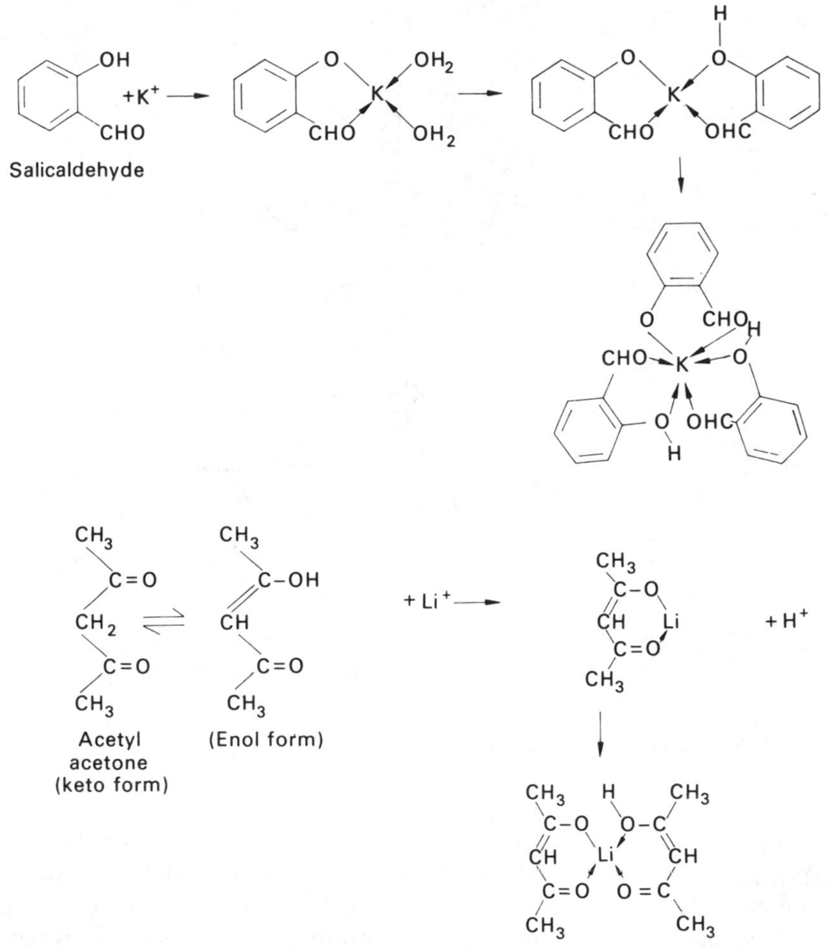

Salicaldehyde

Acetyl acetone (keto form)

(Enol form)

Figure 9.5 Salicaldehyde and acetylacetone complexes.

inside living cells to transport ions across cell membranes, and thus maintain the balance between Na^+ and K^+ inside and outside cells.

The crown ethers also form some unusual complexes called electrides. These are black and paramagnetic, and have formulae such as Cs^+[(crown ether) · e^-]. The structure consists of a Cs^+ ion, and the crown ether with an electron in the central hole instead of a metal ion.

The cryptates are even more selective and evenstronger complexing agents than are the crown ethers. They differ from the crown ethers by using nitrogen atoms as well as oxygen atoms to bond to the metal ion, and as they are polycyclic they can surround the metal ion completely. A typical crypt is the molecule $N[CH_2CH_2OCH_2CH_2OCH_2CH_2]_3N$. This is called (2, 2, 2-crypt) and forms a complex [Rb(crypt)]CNS · H_2O in which six oxygen atoms and two nitrogen atoms in the crypt molecule bond to the metal ion, giving the metal ion a coordination number of 8. The

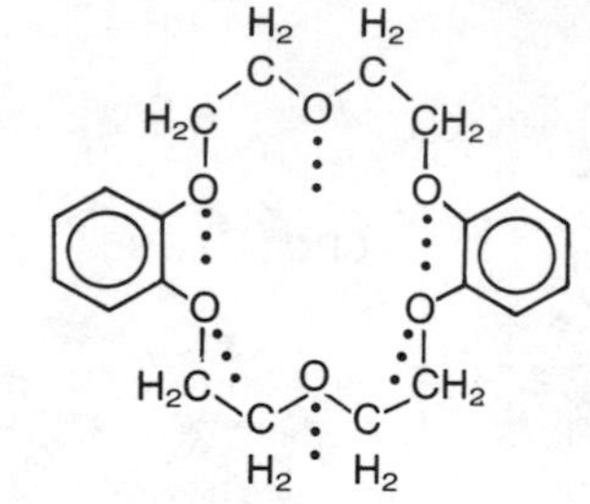

dibenzo-18-crown-6

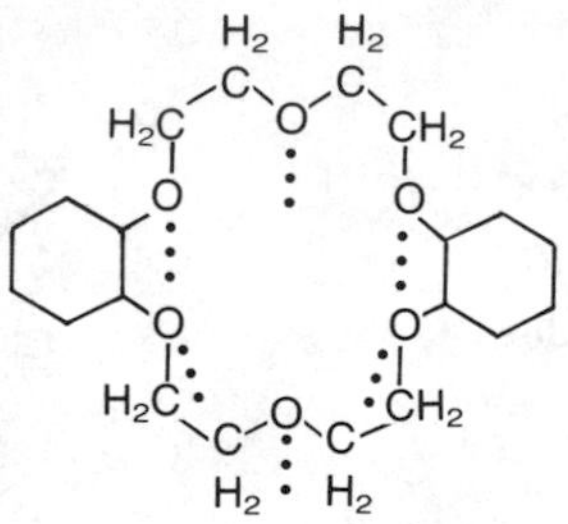

dicyclohexyl-18-crown-6

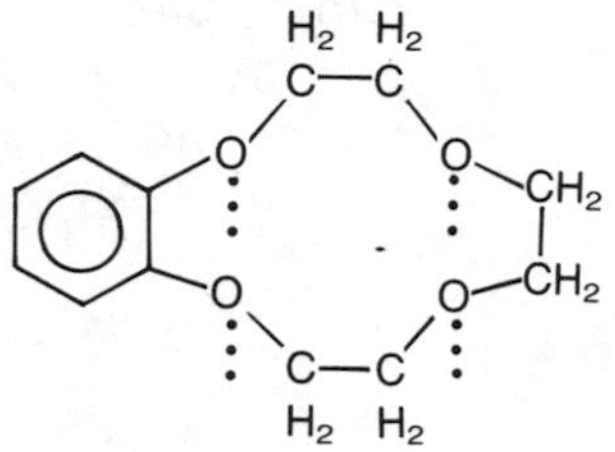

benzo-12-crown-4

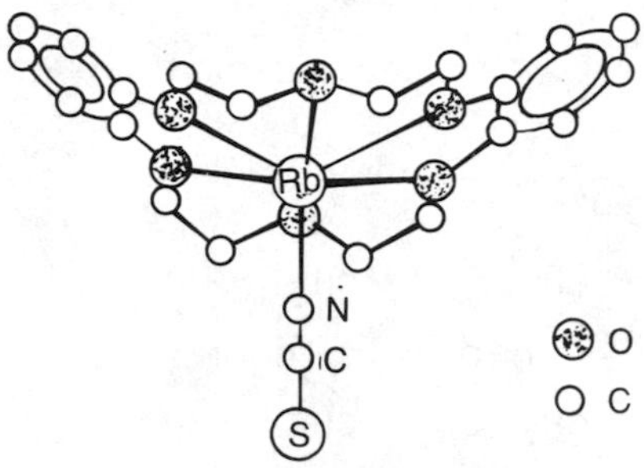

RbCNS (dibenzo-18-crown-6) complex

Figure 9.6 Structures of some crown ethers.

ligand completely wraps round the metal ion, hiding it: hence the name crypt. The complex presents a hydrocarbon exterior, and so is soluble in organic solvents. Such complexes are used for solvent extraction, stabilizing uncommon oxidation states, and promoting otherwise improbable reactions.

An unusual compound $[Na(2,2,2\text{-crypt})]^+Na^-$ can be formed by cooling a solution of Na in ethylamine with 2,2,2-crypt. The compound is crystalline and is endothermic. Presumably it is only formed because of the complexing power of the crypt, and the lattice energy of the crystal. It is stable below −10 °C. The interesting feature is the formation of the Na^- sodide ion. The K^- potasside ion has been made in a similar way, but is less stable. These alkalide compounds are yellow–brown in colour, and are diamagnetic.

BIOLOGICAL IMPORTANCE

Living organisms require at least 27 elements, of which 15 are metals. Metals required in major quantities are K, Mg, Na and Ca. Minor quantities of Mn, Fe, Co, Cu, Zn and Mo, and trace amounts of V, Cr, Sn, Ni and Al, are required by at least some organisms.

Bulk quantities of Group I and II metals are required, mainly to balance the electrical charges associated with negatively charged organic macromolecules in the cell, and also to maintain the osmotic pressure inside the cell, to keep it turgid and prevent its collapse.

In view of the close similarity of chemical properties between Na and K, it is surprising that their biological functions are very different. Na^+ are actively expelled from cells, whereas K^+ are not. This ion transport is sometimes called a sodium pump, and it involves both the active expulsion of Na^+ and the active take-up of K^+. Analysis of the fluids inside and outside animal cells shows that ion transport really does occur. In animal cells the concentration of K^+ is about 0.15 M and the concentration of Na^+ is about 0.01 M. In body fluids (lymph and blood) the concentrations of K^+ and Na^+ are about 0.003 M and 0.15 M respectively. The transport of ions requires energy, and this is obtained by the hydrolysis of ATP. It is estimated that hydrolysis of one ATP molecule to ADP provides enough energy to move three Na^+ ions out of the cell, and two K^+ and one H^+ ions back in to the cell. The mechanism for ion transport involves polyethers natural to the organism.

The different ratio of Na^+ to K^+ inside and outside cells produces an electrical potential across the cell membrane, which is essential for the functioning of nerve and muscle cells. The movement of glucose into cells is associated with Na^+ ions; they enter the cell together. This is favoured by a high concentration gradient. The Na^+ ions entering the cell in this way must then be expelled.The movement of amino acids is similar. K^+ ions inside the cell are essential for the metabolism of glucose, the synthesis of proteins, and the activation of some enzymes.

The 1987 Nobel Prize for Chemistry was awarded to C.J. Pedersen, J.M. Lehn and D. Cram for their work on the discovery and applications of crown ethers and cryptates.

DIFFERENCES BETWEEN LITHIUM AND THE OTHER GROUP I ELEMENTS

The properties of lithium and its compounds differ far more from those of the other Group I elements than the other Group I elements and compounds differ among themselves. Apart from having the same oxidation number as the rest of Group I, lithium compounds may show closer similarities with Group II elements (particularly magnesium) than they show towards their own group. Some of the differences are set out below:

1. The melting and boiling points of lithium metal are much higher than those for the other Group I elements.
2. Lithium is much harder than the other Group I metals.
3. Lithium reacts the least readily with oxygen, forming the normal oxide. It forms a peroxide only with great difficulty, and the higher oxides are unstable.
4. Lithium hydroxide is less basic than the other hydroxides in the group, and therefore many of its salts are less stable. Li_2CO_3, $LiNO_3$ and

LiOH all form the oxide on gentle heating, though the analogous compounds of the rest of the group are stable. Another example of the less basic nature is that though lithium forms a bicarbonate in solution, it does not form a solid bicarbonate, whereas the others all form stable solid bicarbonates.

5. Lithium forms a nitride Li_3N. None of the other Group I elements forms a nitride, but Group II elements form nitrides.
6. Lithium reacts directly with carbon to form an ionic carbide. None of the other Group I elements do this, but Group II elements all react similarly with carbon.
7. Lithium has a greater tendency to form complexes than have the heavier elements, and ammoniated salts such as $[Li(NH_3)_4]I$ exist as solids.
8. Li_2CO_3, Li_3PO_4 and LiF are all insoluble in water, and LiOH is only sparingly soluble. The rest of Group I form soluble compounds, but the corresponding magnesium salts are insoluble or sparingly soluble.
9. The halides and alkyls of lithium are far more covalent than the corresponding sodium compounds, and because of this covalency they are soluble in organic solvents. Similarlylithium perchlorate and to a lesser extent sodium perchlorate resemble magnesium perchlorate in their high solubility inacetone (propanone).
10. The lithium ion itself, and also its compounds, are more heavily hydrated than those of the rest of the group.

Several generalizations may be drawn from this apparently anomalous behaviour of lithium.

The first element in each of the main groups (Li, Be, B, C, N, O and F) differs from the rest of the group. This is partly because the first element is much smaller than the subsequent elements, and consequently it is more likely to form covalent compounds (Fajans' rules) and complexes.

The first element in a group can form a maximum of four conventional electron pair bonds. This is because the outer shell of electrons contains only one *s* orbital and three *p* orbitals. The subsequent elements can use *d* orbitals for bonding: they can attain a coordination number of 6, by using one *s*, three *p* and two *d* orbitals. For this reason the coordination number attained by a complex or a covalent compound of the first element in a group is commonly 4, and for the subsequent elements the coordination number is commonly 6. This simple concept is based on a bond consisting of two electrons shared between two atoms. Exceptions occur when multi-centre bonds are formed (as in $Li_4(CH_3)_4$).

The similarity between lithium (the first member of Group I) and magnesium (the second element in Group II) is called a diagonal relationship.Diagonal relationships also exist between other pairs of elements Be and Al, B and Si as shown:

```
Li   Be   B   C
   \    \   \
Na   Mg   Al  Si
```

The diagonal relationship arises because of the effects of both size and charge. On descending a group, the atoms and ions increase in size. On moving from left to right in the periodic table, the size decreases. Thus on moving diagonally, the size remains nearly the same. For example, lithium is smaller than sodium, and magnesium is also smaller than sodium, and hence lithium and magnesium are similar in size. The sizes of $Li^+ = 0.76$ Å and $Mg^{2+} = 0.72$ Å are close, and so in situations where size is important their behaviour should be similar.

Beryllium and aluminium also show a diagonal relationship. In this case the sizes are not so close ($Be^{2+} = 0.45$ Å and $Al^{3+} = 0.535$ Å), but the charge per unit area is similar (Be^{2+} 2.36 and Al^{3+} 2.50) because the charges are 2+ and 3+ respectively.

$$\text{Charge per unit area} = \frac{\text{(ionic charge)}}{\frac{4}{3}\cdot\pi\cdot\text{(ionic radius)}^2}$$

It is sometimes suggested that the diagonal relationship arises because of a diagonal similarity in electronegativity values.

Li	Be	B	C
1.0	1.5	2.0	2.5
\	\	\	
Na	Mg	Al	Si
0.9	1.2	1.5	1.8

Since ionic size and electronegativity are closely related, this is part of the same picture.

FURTHER READING

Addison, C.C. (1984) *The Chemistry of the Liquid Alkali Metals*, John Wiley, Chichester.

Bach, R.O. (ed.) (1985) *Lithium: Current Applications in Science, Medicine and Technology*, John Wiley, Chichester and New York, (Conference proceedings.)

Dietrich, B. (1985) Coordination chemistry of alkali and alkaline earth cations with macrocyclic ligands, *J. Chem. Ed.*, **62**, 954–964. (Crowns and crypts.)

Gockel, G.W. (1990) *Crown Ethers and Cryptands* (one of a series on Supramolecular Chemistry, ed. Stoddart, J.F.), Royal Society for Chemistry, London.

Hanusa, T.P. (1987) Re-examining the diagonal relationships, *J. Chem. Ed.*, **64**, 686–687.

Hart, W.A. and Beumel, O.F. (1973) *Comprehensive Inorganic Chemistry*, Vol. I (Chapter 7: Lithium and its compounds), Pergamon Press, Oxford.

Hughes, M.N. and Birch, N.J. (1982) IA and IIA cations in biology, *Chemistry in Britain*, **18**, 196–198.

Jolly, W.L. (1972) *Metal Ammonia Solutions*, Dowden, Hutchinson and Row, Stroudburg, PA.

Lagowski, J. (ed.) (1967) *The Chemistry of Non-aqueous Solvents* (Chapter 6), Academic Press, New York. (Solutions of metals in liquid ammonia.)

Lehn, J.M. (1973) Design of organic complexing agents. *Structure and Bonding*, **16**, 1–69.

Lippard, S. (ed.) (1984) *Progress in Inorganic Chemistry*, Vol. 32 by Dye, J.L.,

Electrides, Negatively Charged Metal Ions, and Related Phenomena, Wiley-Interscience, New York.
March, N.N. (1990) *Liquid Metals*, Cambridge University Press.
Parker, D. (1983) Alkali and alkaline earth cryptates, *Adv. Inorg. and Radiochem.*, **27**, 1–26.
Pedersen, C.J. (1967) *J. Am. Chem. Soc.*, **89**, 2495, 7017–7036. (Cyclic polyethers and their complexes with metals.)
Pedersen, C.J. and Frensdorf, H.K. (1972) *Angew. Chem.*, **11**, 16–25. (Cyclic polyethers and their complexes with metals.)
Sargeson, A.M. (1979) Caged metal ions, *Chemistry in Britain*, **15**, 23–27. (A straightforward account of crown ethers, crypts etc.)
The Chemical Society (1967) *The Alkali Metals*, (Special Publication No. 22), London.
Waddington, T.C. (ed.) (1969) *Non Aqueous Solvent Systems* (Chapter 1 by Jolly, W.L. and Hallada, C.J.), Nelson. (Solutions of metals in liquid ammonia.)
Wakefield, B.J. (1976) *The Chemistry of Organolithium Compounds*, Pergamon Press, Oxford.
Whaley, T.P. (1973) *Comprehensive Inorganic Chemistry*, Vol. I (Chapter 8: Sodium, potassium, rubidium, caesium and francium), Pergamon Press, Oxford.

PROBLEMS

1. Why are Group I elements:
   (a) univalent
   (b) largely ionic
   (c) strong reducing agents
   (d) poor complexing agents?
   (e) Why do they have the lowest first ionization energy values in their periods?

2. Why are the Group I metals soft, low melting and of low density? (Refer back to Chapter 5.)

3. Lithium is the smallest ion in Group I. It would therefore be expected to have the highest ionic mobility, and hence solutions of its salts would be expected to have a higher conductivity than solutions of caesium salts. Explain why this is not so.

4. What is the reason for lithium having a greater tendency to form covalent compounds than the other elements in the group?

5. The atomic radius for lithium is 1.23 Å. When the outermost 2*s* electron is ionized off, the ionic radius of Li^+ is 0.76 Å. Assuming that the difference in radii relates to the space occupied by the 2*s* electron, calculate what percentage of the volume of the lithium atom is occupied by the single valence electron. Is this assumption fair? (Volume of a sphere is $\frac{4}{3} \cdot \pi r^3$.) (Answer 76.4%.)

6. Why and in what ways does lithium resemble magnesium?

7. What products are formed when each of the Group I metals is burnt in oxygen? How do these products react with water? Use the molecular

orbital theory to describe the structure of the oxides formed by sodium and potassium.

8. Explain the difference in reactivity of the Group I metals with water.

9. The ionization energies of Group I elements suggest that caesium should be the most reactive, but the standard electrode potentials suggest that lithium is the most reactive. Reconcile these two observations.

10. Describe how you would make lithium hydride. Give equations to show two important properties of lithium hydride. The compound contains the isoelectronic ions Li^+ and H^-. Which ion is the larger and why?

11. Give equations to show the reactions between sodium and: (a) H_2O, (b) H_2, (c) graphite, (d) N_2, (e) O_2, (f) Cl_2, (g) Pb, (h) NH_3.

12. Group I elements generally form very soluble compounds. Name some insoluble or sparingly soluble compounds. How are these elements detected and confirmed in qualitative analysis?

13. Describe the colour and nature of the solutions of Group I metals in liquid ammonia. Give an equation to show how these solutions decompose.

14. Draw the crystal structures of NaCl and CsCl. What is the coordination number of the metal ion in each case? Explain why these two salts adopt different structures.

15. Do the alkali metals form many complexes? Which of the metal ions in the group are best at forming complexes? Which are the best complexing agents?

16. Draw the complexes formed by Li^+, Na^+ and K^+ with acetyl acetone and with salicaldehyde. Why do the coordination numbers differ?

17. What is a crown ether, and what is a crypt? Draw examples of Group I complexes with these molecules. In what way is this type of complex of biological importance?

18. Which of the following methods would you use to extinguish a fire of lithium, sodium or potassium metals? Explain why some of these are unsuitable, and give the reactions involved.
    (a) water
    (b) nitrogen
    (c) carbon dioxide
    (d) asbestos blanket

19. The four general methods of extracting metals are thermal decomposition, displacement of one element by another, chemical reduction, and electrolytic reduction. How are Group I metals obtained and why are the other methods unsuitable?

20. 0.347 g of a metal (A) was dissolved in dilute HNO_3. This solution gave a red coloration to a non-luminous Bunsen burner flame, and on evaporation gave 0.747 g of metal oxide (B). (A) also reacted with nitrogen, forming a compound (C), and with hydrogen, forming (D). On reacting 0.1590 g of (D) with water, a gas (E) was evolved and a sparingly soluble compound (F) formed, which gave a strongly basic reaction and required 200 ml of 0.1000 M hydrochloric acid to neutralize it. Identify the substances (A) to (F) and explain the reactions involved.

The chlor-alkali industry

10

The chlor-alkali industry includes the production of three main chemicals: sodium hydroxide (sometimes called caustic soda), chlorine, and sodium carbonate (sometimes called soda ash). All three chemicals are made from sodium chloride.

NaOH and Cl_2 are produced simultaneously by the electrolysis of an aqueous solution of NaCl. NaOH is the most important alkali used in industry, and Cl_2 is also an extremely important industrial chemical. Sodium carbonate is included with the other two chemicals for two reasons – first because in many applications such as making paper, soap and detergents it can be used interchangeably with sodium hydroxide, and second because Na_2CO_3 can quite easily be converted into NaOH (or vice versa) using the Lime – caustic soda process. In this process, the reaction is reversible, and depending on the relative demands and cost of sodium carbonate and sodium hydroxide it may be used in either direction. Before 1955 Na_2CO_3 was used very extensively for water softening as it prevented the formation of scum when using soap in hard water. Soap is discussed under 'Organic and organometallic compounds' in Chapter 9, and hard water is discussed in Chapter 11. Thus before 1955 it was economic to make Na_2CO_3 from NaOH. More recently the use of soap has declined as detergents have become more widely used, and with this the demand for Na_2CO_3 has declined. Nowadays the reverse reaction is carried out on a limited scale, converting Na_2CO_3 to NaOH.

$$Na_2CO_3 + Ca(OH)_2 \rightleftharpoons CaCO_3 + 2NaOH$$

All three chemicals are classed as 'heavy inorganic chemicals' because of the very large tonnages involved. A list of the chemicals produced in the largest quantities is shown in Table 10.1.

LEBLANC PROCESS

C.W. Scheele discovered chlorine in 1774 by oxidizing hydrochloric acid with manganese dioxide.

$$4HCl + MnO_2 \rightarrow 2Cl_2 + Mn^{2+} + 2H_2O$$

Table 10.1 Tonnes of 'heavy chemicals' produced in 1985

Chemical	Millions of tonnes		
	World	USA	UK
1. H_2SO_4	133.5	36.0	2.5
2. CaO	106.6	14.5	0.85
3. O_2	(100)	16.7	2.5
4. NH_3	80.8	14.3	–
5. NH_4NO_3	(75)	–	–
6. N_2	(60)	–	–
7. ethylene	38.9	13.9	1.4
8. NaOH	34.0	9.8	–
9. HNO_3	30.2	6.9	3.1
10. Na_2CO_3	26.1	7.7	–
11. H_3PO_4	24.3	13.1	0.52
12. Cl_2	23.3	9.4	1.0
13. propylene	20.2	6.8	0.97
14. ethanol	15.8	0.9	0.24
15. benzene	15.6	4.1	0.84
16. vinyl chloride	11.4	3.1	0.31
17. methanol	10.0	2.7	–
18. HCl	9.0	2.5	0.16

He also described the bleaching properties of chlorine, and these eventually led to demand for both chlorine and sodium hydroxide on an industrial scale for use in the textiles industry. At that time there was no chemical industry, so people had to make their own chemicals. The first problem was to make the HCl. This was produced by the Leblanc process. Though the process is now obsolete, it warrants description because it was the first large scale industrial process in Europe; it lasted for most of the nineteenth century, and it illustrates the need to consider what raw materials are needed, how they can be obtained, and the commercial need to sell everything produced. (At this time Europe led the world industrially, and the process was imported into the USA from Europe.)

$$NaCl + \text{concentrated } H_2SO_4 \xrightarrow{\text{heat}} NaHSO_4 + HCl$$

$$NaHSO_4 + NaCl \xrightarrow{\text{heat}} Na_2SO_4 + HCl$$

The HCl was then oxidized to give Cl_2.

$$HCl + MnO_2 \rightarrow Cl_2 + Mn^{2+}$$

The Na_2SO_4 was used either to make glass, or to make Na_2CO_3 and NaOH.

$$Na_2SO_4 + C + CaCO_3 \rightarrow Na_2CO_3 + CaSO_4$$
$$Na_2CO_3 + Ca(OH)_2 \rightarrow 2NaOH + CaCO_3$$

In this process, the chemicals used are H_2SO_4, NaCl, $CaCO_3$ and C, and the products are NaOH and Cl_2 (and to a lesser extent Na_2SO_4). The raw materials were obtained:

$$\text{S or } FeS_2 + O_2 \rightarrow SO_2 \rightarrow SO_3 \rightarrow H_2SO_4$$

NaCl – mined or extracted as brine solution

$CaCO_3$ – mined as limestone

$$CaCO_3 \xrightarrow{\text{heat}} CaO \xrightarrow{H_2O} Ca(OH)_2$$

In 1874 world production of NaOH was 525 000 tonnes, of which 94% was produced by the Leblanc process. Production of NaOH had risen to 1 800 000 tonnes by 1902, but by then only 8% was produced by the Leblanc process. The Leblanc process became obsolete because cheaper methods were found. It was replaced in turn by the Weldon process, the Deacon process, and eventually by electrolysis.

WELDON AND DEACON PROCESSES

The Leblanc process used MnO_2 to oxidize the HCl, but the $MnCl_2$ formed was wasted. The Weldon process (1866) recycled the $MnCl_2$, and was therefore cheaper.

In the Deacon process (1868), air was used to oxidize the HCl instead of using MnO_2. A gas phase reaction was performed between HCl and air on the surface of bricks soaked in a solution of $CuCl_2$, which acted as a catalyst. The reaction is reversible, and a conversion of about 65% is possible.

$$4HCl + O_2 \underset{440\,^{\circ}C}{\overset{CuCl_2 \text{ catalyst}}{\rightleftharpoons}} 2Cl_2 + 2H_2O + \text{heat}$$

Nowadays about 90% of the world supply of chlorine comes from the electrolysis of an aqueous solution of sodium chloride (brine). Most of the remainder is produced by the electrolysis of molten NaCl in the production of sodium metal, electrolysis of aqueous KCl in the production of KOH, and electrolysis of molten $MgCl_2$ in the extraction of magnesium metal. However, a small amount is made by the oxidation of HCl with air, in a slightly modified Deacon process. This started in 1960, and uses a didymium promoted catalyst of Dm_2O_3 and $CuCl_2$ at a slightly lower temperature of 400 °C. (Didymium is an old name and means 'twin'. It was once thought to be an element, but was later resolved into two lanthanide elements, praseodymium and neodymium. The catalyst is a finely powdered mixture of solids which flows like a liquid, and this is termed a fluidized bed.)

ELECTROLYTIC PROCESSES

Electrolysis of brine was first described in 1800 by Cruickshank, but it was not until 1834 that Faraday put forward the Laws of Electrolysis. At that time electrolysis was strictly limited because primary batteries were the only source of electricity. This changed in 1872 when Gramme invented the dynamo. The first commercial electrolytic plant was started in Frankfurt (Germany) in 1891, where the cell was filled, electrolysed, emptied, then refilled. . . and so on. This was therefore a *discontinuous or batch process*. Clearly a cell which could run continuously, and did not need emptying, would produce more and cost less to operate. Many developments and patents attempting to exploit the commercial possibilities appeared over the next twenty years. The first commercially operated plant to use a *continuous diaphragm cell* was probably that designed by Le Seur at Romford (Maine) in 1893, followed by Castner cells at Saltville (Virginia, USA) in 1896. The first in the UK was set up by Hargreaves and Bird in 1897 at Runcorn. In all of these (and also in many modern diaphragm cells), asbestos was used as the diaphragm to separate the anode and cathode compartments. Brine was constantly added, and NaOH and Cl_2 were produced continuously.

About the same time, Castner (who was an American working in Birmingham, England) and Kellner (an Austrian working in Vienna) developed and patented similar versions of the *mercury cathode cell* in 1897. Their combined patents were used by the Castner Kellner Alkali Company, also at Runcorn, and also in 1897.

The same two types of cell, diaphragm and mercury cathode, still remain in use. The early electrolytic plants produced about 2 tonnes of chlorine per day, but modern plants produce 1000 tonnes per day.

In the electrolysis of brine, reactions occur at both the anode and the cathode.

$$\text{Anode:} \quad 2Cl^- \rightarrow Cl_2 + 2e$$

$$\text{Cathode:} \begin{cases} Na^+ + e \rightarrow Na \\ 2Na + 2H_2O \rightarrow 2NaOH + H_2 \end{cases}$$

Side reactions may also occur if the products mix:

$$2NaOH + Cl_2 \rightarrow NaCl + NaOCl + H_2O$$

or

$$2OH^- + Cl_2 \rightarrow \underset{\text{hypochlorite}}{2OCl^-} + H_2$$

and also another reaction may occur to a small extent at the anode:

$$4OH^- \rightarrow O_2 + 2H_2O + 4e$$

DIAPHRAGM CELL

A porous diaphragm of asbestos is used to keep the H_2 and Cl_2 gases (produced at the electrodes) separated from one another. If H_2 and Cl_2 gases mix they react, and the reaction may be explosive. In daylight (and more so in sunlight) a photolytic reaction takes place which produces chlorine atoms. These lead to an explosive chain reaction with hydrogen.

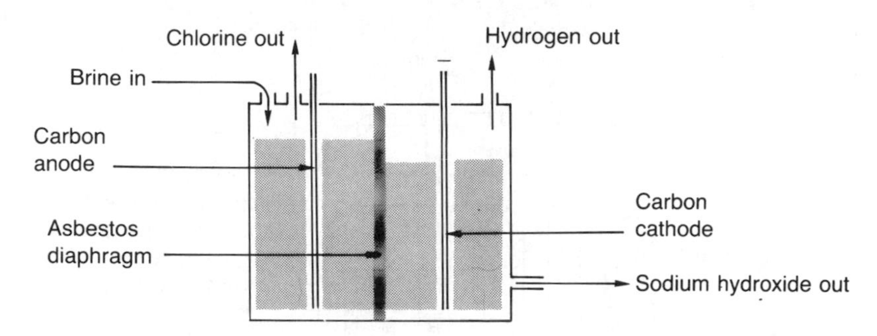

Figure 10.1 A diaphragm cell.

The diaphragm also separates the anode and cathode compartments. This reduces the chance that NaOH produced in the cathode compartment can mix and react with Cl_2 produced in the anode compartment. This reduces the chance of the side reaction producing sodium hypochlorite, NaOCl. However, some sodium hydroxide or OH^- may diffuse into the other compartment, and this is inhibited by maintaining the level of electrolyte higher in the anode compartment than in the cathode compartment, so there is a small positive flow from the anode to the cathode compartment. Traces of oxygen are produced in a side reaction. This reacts with the carbon electrodes, gradually destroying them and forming CO_2.

There is considerable interest in using thin synthetic plastic membranes for the diaphragm instead of asbestos. These membranes are made of a polymer called nafion, supported on a teflon mesh. (Nafion is a copolymer of tetrafluoroethylene and a perfluorosulphonylethoxy ether.) Plastic membranes have a lower resistance than asbestos.

Less than half the NaCl is converted to NaOH, and a mixture of about 11% NaOH and 16% NaCl is usually obtained. This solution is concentrated in a steam evaporator, when a considerable amount of NaCl crystallizes out, giving a final solution containing 50% NaOH and 1% NaCl. *It is important to note that NaOH made in this way always contains some NaCl.* This may or may not matter, depending on how the NaOH is to be used. For most industrial purposes, the product is sold as a solution, as the cost of evaporating it to give the solid exceeds the increased cost of transporting the solution.

MERCURY CATHODE CELL

During the electrolysis of brine, Na^+ ions migrate towards the cathode, and when they get there the ions are discharged.

$$Na^+ + e \rightarrow Na_{(metal)}$$

If the cathode is made of mercury, the Na atoms produced dissolve in the mercury and form an amalgam, or loose alloy. The amalgam is pumped to a different compartment called the denuder, where water trickles over

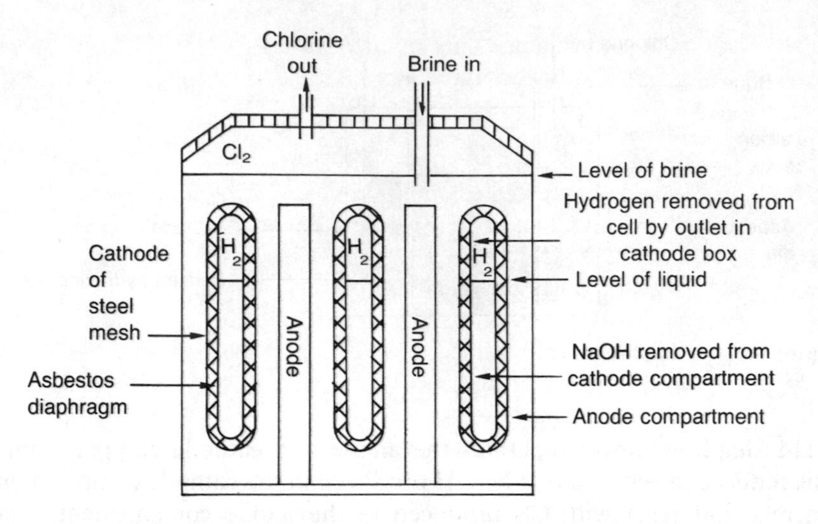

Figure 10.2 Commercial diaphragm cell for Cl_2 and NaOH.

lumps of graphite (here acting as an inert solid). The water and the Na in the amalgam react, and in this way *pure* NaOH at 50% strength is obtained.

$$Na_{(amalgam)} + H_2O \rightarrow NaOH + \tfrac{1}{2}H_2 + Hg$$

The clean mercury is recycled back to the electrolysis tank. Originally the anodes were made of graphite, but because traces of oxygen are produced in a side reaction they become pitted, owing to the formation of CO_2. The anodes are now made of steel coated with titanium. Titanium is very resist-

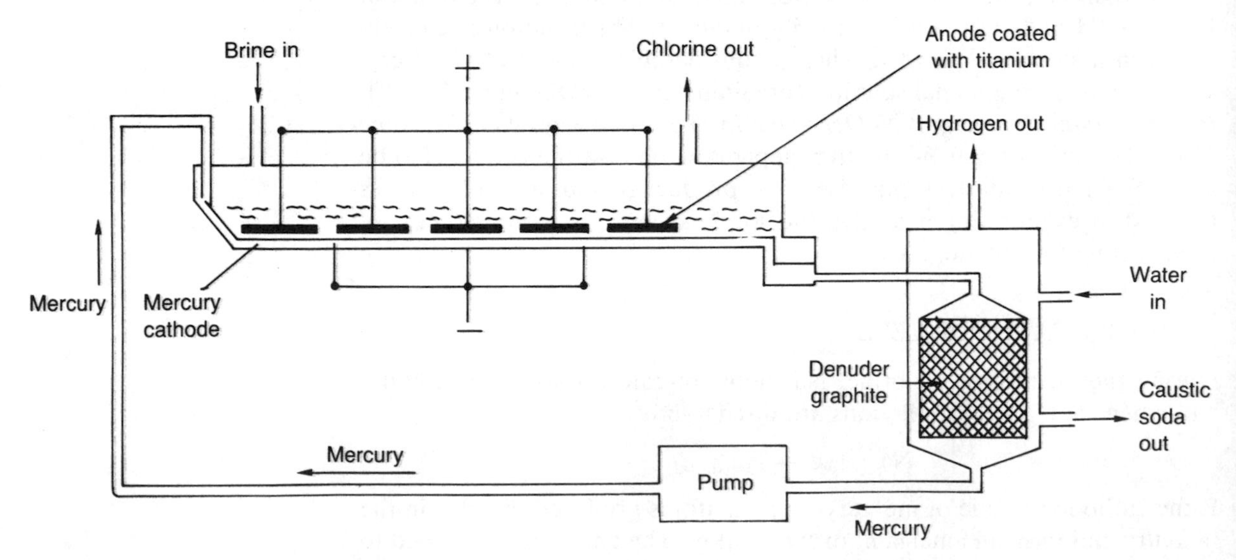

Figure 10.3 Mercury cathode cell for Cl_2 and NaOH.

ant to corrosion, and this not only overcomes the problem of pitting and forming CO_2, but also lowers the electrical resistance.

QUANTITIES

In both electrolytic processes (diaphragm cell and mercury cathode cell) equimolar amounts of Cl and NaOH are produced. Since Cl has an atomic weight of 35.5, and NaOH has a molecular weight of 40, it follows that electrolysis yields 40 parts by weight of NaOH to 35.5 parts of Cl_2. This corresponds to 1.13 tonnes of NaOH for 1 tonne of Cl_2. In 1985 world production of Cl_2 was 23.3 million tonnes, which accounts for 26 million tonnes of NaOH. In fact production of NaOH was 34 million tonnes, so clearly some was made in other ways.

Prior to 1965 demand for NaOH exceeded that for Cl_2, so Cl_2 was cheap. Since then the position has reversed, largely due to the use of large amounts of Cl_2 in making plastics such as polyvinyl chloride. (World production of PVC was 11.5 million tonnes in 1985.)

SODIUM CARBONATE

World production of Na_2CO_3 in 1985 was 26.1 million tonnes, and 45% of this was used in the glass industry. Smaller amounts were used to make various sodium phosphates and polyphosphates which are used for water

Table 10.2 Chlorine production in 1985 (million tonnes)

World production (excluding USSR)	23.3	
USA	9.4	(40%)
West Germany	3.5	(15%)
Canada	1.4	(6%)
France	1.4	(6%)
UK	1.0	(4%)
Japan	0.93	(4%)
Italy	0.92	(4%)
Spain	0.51	(2%)

Table 10.3 Major uses of chlorine

	EEC	USA
Vinyl chloride monomer ($CH_2{=}CHCl$)	31%	18%
Organic intermediates	16%	
Chlorinated solvents (C_2H_5Cl approx. 40 000 tonnes/year, $CH_2Cl \cdot CH_2Cl$ etc.)	14%	22%
Propylene oxide	8%	5%
Bleaching wood pulp and paper		11%
Chloromethanes (CCl_4, $CHCl_3$ etc.)		10%
Inorganic materials (bleaching powder, sodium hypochlorite)		8%
Other uses	31%	26%

Table 10.4 Major uses of caustic soda

	USA
Inorganic chemicals	21%
Organic chemicals	17%
Wood pulp and paper making	14%
Neutralizations	12%
Alumina production	7%
Soap	4%
Rayon	4%
Other uses	21%

Table 10.5 Major uses of sodium carbonate

	USA
Glass – bottles	34%
Sodium phosphates	12%
Glass – sheets and glass fibre	11%
Sodium silicate	5%
Alkaline cleaners	5%
Wood pulp and paper making	4%
Other uses	29%

softening (being added to various cleaning powders), and in wood pulp and paper making. The increased awareness of the effect of 'acid rain' on plants and buildings has led to a new use for Na_2CO_3 in treating the flue gases from coal- and oil-fired power stations, to remove SO_2 and H_2SO_4. This use may eventually account for a large tonnage of Na_2CO_3.

The main producing countries are the USA (30%), the USSR (19%), China (8%), West Germany (5%), Japan (4%), Bulgaria (4%) and Poland (4%). Most of the Na_2CO_3 is produced synthetically by the Solvay (ammonia–soda) process. However, since prehistoric times a natural deposit called Trona, $Na_2CO_3 \cdot NaHCO_3 \cdot 2H_2O$, has been obtained from dried-up lake beds in Egypt. Large amounts are now mined, particularly in the USA and Kenya. In the USA 7.7 million tonnes of Na_2CO_3 were used in 1985. About 5 million tonnes/year of Na_2CO_3 is made from Trona. Trona is sometimes called sodium sesquicarbonate (sesqui means one and a half), and this is converted to sodium carbonate by heating.

$$2(Na_2CO_3 \cdot NaHCO_3 \cdot 2H_2O) \xrightarrow{\text{heat}} 3Na_2CO_3 + CO_2 + 5H_2O$$

In the description of the chlor-alkali industry it was mentioned that sodium carbonate (soda ash) can be used instead of NaOH in applications such as making paper, soap and detergents, and that sodium carbonate can be used to make NaOH by the lime – caustic soda process. However, as NaOH is at present cheap and plentiful, not much sodium carbonate is used for these purposes. With the increased use of detergents, there has been a decline in the use of 'washing soda' $Na_2CO_3 \cdot 10H_2O$ for water softening.

THE SOLVAY (OR AMMONIA–SODA) PROCESS

There have been many attempts to make Na_2CO_3 more cheaply than by the Leblanc process, by using the overall reaction:

$$2NaCl + CaCO_3 \rightarrow Na_2CO_3 + CaCl_2$$

The reaction was first studied by Freshnel in 1811, and several commercial plants were built but were quickly abandoned because they failed to make a profit, or they encountered technical problems such as corrosion of the plant, contamination of the product and blocked pipes. Ernest Solvay was the first to operate a commercial plant successfully, in Belgium (1869).

The process is much more complicated than the overall equation implies, and to make things worse the reaction is reversible and only 75% of the NaCl is converted. The first stage in the process is to purify saturated brine, and then react it with gaseous ammonia. The ammoniated brine is then carbonated with CO_2, forming $NaHCO_3$. This is insoluble in the brine solution because of the common ion effect and so can be filtered off, and on heating to 150 °C it decomposes to anhydrous Na_2CO_3 (called light soda ash in industry because it is a fluffy solid with a low packing density of about $0.5\,g\,cm^{-3}$). Next CO_2 is stripped (removed) by heating the solution, and the CO_2 is reused. Then the NH_3 is driven off by adding alkali (a slurry of lime in water), and the ammonia is reused. Lime (CaO) is obtained by heating limestone ($CaCO_3$), and this also provides the CO_2 required. When lime is mixed with water it gives $Ca(OH)_2$.

$$NH_3 + H_2O + CO_2 \rightarrow NH_4 \cdot HCO_3$$

$$NaCl + NH_4 \cdot HCO_3 \rightarrow NaHCO_3 + NH_4Cl$$

$$2NaHCO_3 \xrightarrow{150^\circ} Na_2CO_3 + CO_2 + H_2O$$

$$CaCO_3 \xrightarrow[\text{lime kiln}]{1100\,^\circ C \text{ in}} CaO + CO_2$$

$$CaO + H_2O \rightarrow Ca(OH)_2$$

$$2NH_4Cl + Ca(OH)_2 \rightarrow 2NH_3 + CaCl_2 + 2H_2O$$

Thus the materials consumed are NaCl and $CaCO_3$, and there is one useful product, Na_2CO_3, and one by-product, $CaCl_2$. There is little requirement for $CaCl_2$, so only a little is recovered from solution, and the rest is wasted. The largest use of Na_2CO_3 is for glass making (Table 10.5), and this requires 'heavy ash', which is $Na_2CO_3 \cdot H_2O$. To obtain this, the 'light ash' produced in the Solvay process (which is anhydrous Na_2CO_3) is recrystallized from hot water.

FURTHER READING

Adam, D.J. (1980) Early industrial electrolysis, *Education in Chemistry*, **17**, 13–14, 16.

Borgstedt, H.U. and Mathews, C.K. (1987) *Applied Chemistry of the Alkali Metals*, Plenum, London.

Boynton, R.S. (1980) *Chemistry and Technology of Lime and Limestone*, 2nd ed., John Wiley, Chichester.

Buchner, W., Schleibs, R., Winter, G. and Buchel, K. H. (1989) *Industrial Inorganic Chemistry*, V.C.H. Publishers, Weinheim.

Grayson, M. and Eckroth, D. (eds), *Kirk-Othmer Concise Encyclopedia of Chemical Technology*, John Wiley.

Kirk-Othmer Encyclopedia of Chemical Technology (1984) (26 volumes), 3rd ed., Wiley-Interscience.

Stephenson, R.M. (1966) *Introduction to the Chemical Process Industries*, Van Nostrand Rienhold, New York.

Thompson, R. (ed.) (1986) *The Modern Inorganic Chemicals Industry* (chapter by Purcell, R.W., The Chlor-Alkali Industry; chapter by Campbell, A., Chlorine and Chlorination, Special Publication No. 31, The Chemical Society, London.

Venkatesh, S. and Tilak, S. (1983) Chlor-alkali technology, *J. Chem. Ed.*, **60**, 276–278.

Production figures for the top 50 chemicals produced in the USA each year are published in *Chemical and Engineering News* in one of the issues in June each year (see Appendix K).

PROBLEMS

1. What chemicals are obtained industrially from sodium chloride? Outline the processes.
2. Describe in detail the industrial electrolysis of sodium chloride. Comment on the purity of the products.
3. What are the main uses of chlorine, sodium and caustic soda? Why has demand for chlorine increased dramatically?
4. What is Na_2CO_3 used for? Why has its use declined? Explain how at different times NaOH has been converted into Na_2CO_3, and at other times Na_2CO_3 has been converted into NaOH.

Group II – the alkaline earth elements

11

Table 11.1 Electronic structures

Element	Symbol	Electronic structure	
Beryllium	Be	$1s^2 2s^2$	or [He] $2s^2$
Magnesium	Mg	$1s^2 2s^2 2p^6 3s^2$	or [Ne] $3s^2$
Calcium	Ca	$1s^2 2s^2 2p^6 3s^2 3p^6 4s^2$	or [Ar] $4s^2$
Strontium	Sr	$1s^2 2s^2 2p^6 3s^2 3p^6 3d^{10} 4s^2 4p^6 5s^2$	or [Kr] $5s^2$
Barium	Ba	$1s^2 2s^2 2p^6 3s^2 3p^6 3d^{10} 4s^2 4p^6 4d^{10} 5s^2 5p^6 6s^2$	or [Xe] $6s^2$
Radium	Ra		[Rn] $7s^2$

INTRODUCTION

The Group II elements show the same trends in properties as were observed with Group I. However, beryllium stands apart from the rest of the group, and differs much more from them than lithium does from the rest of Group I. The main reason for this is that the beryllium atom and Be^{2+} are both extremely small, and the relative increase in size from Be^{2+} to Mg^{2+} is four times greater than the increase between Li^+ and Na^+. Beryllium also shows some diagonal similarities with aluminium in Group III.

The elements form a well graded series of highly reactive metals, but are less reactive than Group I. They are typically divalent, and generally form colourless ionic compounds. The oxides and hydroxides are less basic than those of Group I: hence their oxosalts (carbonates, sulphates, nitrates) are less stable to heat. Magnesium is an important structural metal, and is used in large amounts (393 000 tonnes in 1988). Several compounds are used in vast quantities: limestone ($CaCO_3$) is used to make quicklime CaO (126 million tonnes in 1988) and cement (1172 million tonnes in 1988), and 18 million tonnes of chalk are also used. Other compounds used on a large scale include gypsum $CaSO_4$ (86.5 million tonnes in 1988), fluorite CaF_2 (5.1 million tonnes in 1988), magnesite $MgCO_3$ (11.7 million tonnes in 1988) and barytes $BaSO_4$ (5 million tonnes in 1988).

Mg^{2+} and Ca^{2+} are essential elements in the human body, and Mg^{2+} is an important constituent of chlorophyll.

ELECTRONIC STRUCTURE

All Group II elements have two *s* electrons in their outer shell. Ignoring the filled inner orbitals, their electronic structures may be written $2s^2$, $3s^2$, $4s^2$, $5s^2$, $6s^2$ and $7s^2$.

OCCURRENCE AND ABUNDANCE

Beryllium is not very familiar, partly because it is not very abundant (2 ppm) and partly because it is difficult to extract. It is found in small quantities as

Table 11.2 Abundance of the elements in the earth's crust, by weight

	ppm	Relative abundance
Be	2.0	51
Mg	27 640	6
Ca	46 600	5
Sr	384	15
Ba	390	14
Ra	1.3×10^{-6}	

the silicate minerals beryl $Be_3Al_2Si_6O_{18}$ and phenacite Be_2SiO_4. About 10 000 tonnes of beryl are mined annually, mainly in the USA (70%), the USSR (20%) and Brazil (7%). (The gemstone emerald has the same formula as beryl, but also contains small amounts of chromium which make it green in colour.)

Magnesium is the sixth most abundant element in the earth's crust (27 640 ppm or 2.76%). Magnesium salts occur to about 0.13% in sea water. Entire mountain ranges (e.g. the Dolomites in Italy) consist of the mineral dolomite [$MgCO_3 \cdot CaCO_3$]. Calcined dolomite is used as a refractory for lining furnaces, and dolomite is used for road making. There are also large deposits of magnesite $MgCO_3$: 11.7 million tonnes were mined in 1988. The main sources are China 17%, North Korea and the USSR 16% each, Austria 10%, Turkey 9% and Greece 8%. There are also deposits of sulphates such as epsomite $MgSO_4 \cdot 7H_2O$ and kieserite $MgSO_4 \cdot H_2O$. Carnallite [$KCl \cdot MgCl_2 \cdot 6H_2O$] is mined as a source of potassium. Mg also occurs in a wide range of silicate minerals, including olivine $(Mg, Fe)_2SiO_4$, talc $Mg_3(OH)_2Si_4O_{10}$, chrysotile $Mg_3(OH)_4Si_2O_5$ (asbestos) and micas such as $K^+[Mg_3(OH)_2(AlSi_3O_{10}]^-$.

Calcium is the fifth most abundant element in the earth's crust (46 600 ppm or 4.66%), and it occurs throughout the world in many common minerals. There are vast sedimentary deposits of $CaCO_3$ existing as whole mountain ranges of limestone, marble and chalk (the white cliffs of Dover), and also as coral. These originated from the shells of marine life. Though limestone is typically white, in many places it is coloured yellow, orange or brown owing to traces of iron. There are two crystalline forms of $CaCO_3$, calcite and aragonite. Calcite is the more common: it forms colourless rhombohedral crystals. Aragonite is orthorhombic, and is commonly red–brown or yellow in colour, and this accounts for the colour of the landscape of the Red Sea area, the Bahamas and the Florida keys. Limestone is commercially important as a source of lime CaO. Lime is produced on an enormous scale (126 million tonnes in 1988) second in volume only to H_2SO_4. Limestone chippings are also used for making roads.

Fluoroapatite $[3(Ca_3(PO_4)_2) \cdot CaF_2]$ is commercially important as a source of phosphate. Gypsum $CaSO_4 \cdot 2H_2O$ and anhydrite $CaSO_4$ are major minerals. World production of gypsum was 86.5 million tonnes in 1988. The major sources were the USA 17%, Canada 10%, Iran 9.5%, China 8%, Spain 8%, France 6.5%, the USSR 5.5%, Thailand 5% and the UK 4%. A much smaller amount of anhydrite was mined. Gypsum is used in making Portland cement, plasterboard, and plaster, and in glass making. Its use in plaster is not just recent, since in the Valley of the Kings in Egypt, the walls of the ancient burial tombs of Tutankhamun and those of the other kings were plastered with $CaSO_4$ and then engraved with hieroglyphics. The White Sands National Park and missile range in New Mexico, USA (where the first atomic bomb was tested) have an area of 100 miles by 40 miles of pure white sand dunes composed of gypsum. Fluorite CaF_2 is important as the main source of fluorine.

Strontium (384 ppm) and barium (390 ppm) are much less abundant, but are well known because they occur as concentrated ores which are easy to extract. Strontium is mined as celestite $SrSO_4$ and strontianite $SrCO_3$. World production of Arontium minerals was 171 500 tonnes in 1988. The main producers are Mexico 26%, Turkey 20%, Spain 16%, the UK 15% and Iran 13%. Ba is mined as barytes $BaSO_4$. World production was 5 million tonnes in 1988. It is found throughout the world and the largest producers are China 21%, the USSR 11%, Mexico 8%, India 7%, Turkey 6% and the USA 6%. Radium is extremely scarce and is radioactive. It was first isolated by Pierre and Marie Curie by processing many tonnes of the uranium ore pitchblende. It was used for radiotherapy treatment of cancer at one time: other forms of radiation are now used (^{60}Co, X-rays, or a linear accelerator) for this purpose. Marie Curie was awarded the Nobel Prize for Chemistry in 1911 for isolating and studying radium and polonium.

EXTRACTION OF THE METALS

The metals of this group are not easy to produce by chemical reduction because they are themselves strong reducing agents, and they react with carbon to form carbides. They are strongly electropositive and react with water, and so aqueous solutions cannot be used for displacing them with another metal, or for electrolytic production. Electrolysis of aqueous solutions can be carried out using a mercury cathode, but recovery of the metal from the amalgam is difficult. All the metals can be obtained by electrolysis of the fused chloride, with sodium chloride added to lower

continued overleaf

the melting point, although strontium and barium tend to form a colloidal suspension.

In older processes BeO is extracted from beryl $Be_3Al_2Si_6O_{18}$ by heat treatment or fusion with alkali, followed by treatment with sulphuric acid to give soluble $BeSO_4$. Addition of NH_4OH gives $Be(OH)_2$ which on heating gives BeO. BeO has ceramic properties, and is used in nuclear reactors. Alternatively beryllium is extracted from the silicate minerals by treatment with HF, forming the soluble complex sodium tetrafluoroberyllate $Na_2[BeF_4]$, and converting this to the hydroxide and then to the oxide. Be itself is prepared by converting the hydroxide to $BeCl_2$ by heating with C and Cl_2, followed by electrolysis of the fused chloride. Alternatively beryllium is obtained by reducing BeF_2 with magnesium. Be is used to make alloys with other metals. Addition of about 2% Be to Cu metal increases its strength by a factor of 5 or 6. An alloy of Be with Ni is used to make springs and electrical contacts. Be has a very low cross-section for neutron capture, and is used in the nuclear energy industry. Both Be and BeO have been used in nuclear reactors, and for this purpose they must be of extremely high purity. High purity material is obtained by making basic beryllium acetate, purifying this by distilling it under reduced pressure, and then either decomposing it to the oxide by heating, or converting it to the chloride and then reducing the chloride with Ca or Mg to give the metal. The absorption of electromagnetic radiation by a solid depends on its electron density. Be has a very low electron density and has a smaller absorbing power than any other solid, and for this reason it is used to make the windows of X-ray tubes.

Magnesium is the only Group II metal to be produced on a large scale. World production was 393 000 tonnes in 1988. The largest producers were the USA 36%, the USSR 23% and Norway 13%. Mg is an extremely important light weight structural metal because of its low density ($1.74\,g\,cm^{-3}$ compared with steel $7.8\,g\,cm^{-3}$ or aluminium $2.7\,g\,cm^{-3}$). Mg forms many binary alloys, often containing up to 9% Al, 3% Zn and 1% Mn, traces of the lanthanides praseodymium Pr and neodymium Nd, and traces of thorium. The metal and its alloys can be cast, machined and welded quite easily. It is used to make the bodies of aircraft, aircraft parts and motor car engines. Up to 5% Mg is usually added to Al to improve its properties. Chemically it is important in Grignard reagents such as C_2H_5MgBr.

Mg was formerly prepared by heating MgO and C to 2000 °C, at which temperature C reduces MgO. The gaseous mixture of Mg and CO was then cooled very rapidly to deposit the metal. This 'quenching' or 'shock-cooling' was necessary as the reaction is reversible, and if cooled slowly the reaction will come to equilibrium further to the left.

$$MgO + C \rightleftharpoons Mg + CO$$

Magnesium is now produced by high temperature reduction, and by electrolysis.

1. In the Pidgeon process Mg is produced by reducing calcined dolomite with ferrosilicon at 1150 °C under reduced pressure.

$$[CaCO_3 \cdot MgCO_3] \xrightarrow{\text{heat}} CaO \cdot MgO \xrightarrow{+Fe/Si} Mg + Ca_2SiO_4 + Fe$$

2. Electrolysis may be carried out either on fused $MgCl_2$, or on partially hydrated $MgCl_2$. The $MgCl_2$ is produced in two ways.

Dow sea water process

Sea water contains about 0.13% Mg^{2+} ions, and the extraction of magnesium depends on the fact that $Mg(OH)_2$ is very much less soluble than $Ca(OH)_2$. Slaked lime $Ca(OH)_2$ is added to sea water, and the calcium ions dissolve and $Mg(OH)_2$ is precipitated. This is filtered off, treated with HCl to produce magnesium chloride, and electrolysed.

$$\underset{\text{slaked lime}}{Ca(OH)_2} + \underset{\text{sea water}}{MgCl_2} \rightarrow \underset{\text{precipitate}}{Mg(OH)_2} + CaCl_2$$

$$Mg\,(OH)_2 + HCl \xrightarrow{\text{heat}} MgCl_2$$

Dow natural brine process

Dolomite $[MgCO_3 \cdot CaCO_3]$ is calcined (heated strongly) to give calcined dolomite $MgO \cdot CaO$. This is treated with HCl, giving a solution of $CaCl_2$ and $MgCl_2$. This is treated with more calcined dolomite and CO_2 is bubbled in. $CaCO_3$ is precipitated, leaving a solution of $MgCl_2$ which is then electrolysed.

$$CaCl_2 \cdot MgCl_2 + CaO \cdot MgO + 2CO_2 \rightarrow 2MgCl_2 + \underset{\text{precipitated}}{2CaCO_3}$$

Ca metal is used to make alloys with Al for bearings. It is used in the iron and steel industry to control carbon in cast iron and as a scavenger for P, O and S. Other uses are as a reducing agent in the production of other metals – Zr, Cr, Th and U – and for removing traces of N_2 from argon. Chemically CaH_2 is sometimes used as a source of H_2. World production of Ca is about 1000 tonnes/year. The metal is obtained by electrolysis of the fused $CaCl_2$, which is obtained either as a waste product from the Solvay process, or from $CaCO_3$ and HCl.

The remaining metals Sr and Ba are produced on a very much smaller scale by electrolysis of their fused chlorides, or from their oxides by reduction with aluminium (a thermite reaction).

SIZE OF ATOMS AND IONS

Group II atoms are large, but are smaller than the corresponding Group I elements as the extra charge on the nucleus draws the orbital electrons in. Similarly the ions are large, but are smaller than those of Group I, especially because the removal of two orbital electrons increases the effective nuclear charge even further. Thus, these elements have higher densities than Group I metals.

Table 11.3 Size and density

	Metallic radius (Å)	Ionic radius M^{2+} six-coordinate (Å)	Density ($g\,cm^{-3}$)
Be	1.12	0.31*	1.85
Mg	1.60	0.72	1.74
Ca	1.97	1.00	1.55
Sr	2.15	1.18	2.63
Ba	2.22	1.35	3.62
Ra		1.48	5.5

* Four-coordinate radius, six-coordinate value = 0.45 Å.

Group II metals are silvery white in colour. They have two valency electrons which may participate in metallic bonding, compared with one electron for Group I metals. Consequently Group II metals are harder, have higher cohesive energy, and have much higher melting points and boiling points than Group I elements (see Table 11.4), but the metals are relatively soft. The melting points do not vary regularly, mainly because the metals adopt different crystal structures (see the section on Metallic bonds and metallic structures in Chapter 2).

Table 11.4 Melting and boiling points of Group I and II elements

	Melting point (°C)	Boiling point (°C)		Melting point (°C)	Boiling point (°C)
Be	1287	(2500)	Li	181	1347
Mg	649	1105	Na	98	881
Ca	839	1494	K	63	766
Sr	768	1381	Rb	39	688
Ba	727	(1850)	Cs	28.5	705
Ra	(700)	(1700)			

Figures in brackets are approximate.

IONIZATION ENERGY

The third ionization energy is so high that M^{3+} ions are never formed. The ionization energy for Be^{2+} is high, and its compounds are typically covalent. Mg also forms some covalent compounds. However, the compounds formed by Mg, Ca, Sr and Ba are predominantly divalent and ionic. Since the atoms are smaller than those in Group I, the electrons are more tightly held so that the energy needed to remove the first electron (first ionization energy) is greater than for Group I. Once one electron has been removed, the ratio of charges on the nucleus to orbital electrons is increased, so that the remaining electrons are more tightly held. Hence the energy needed to remove a second electron is nearly double that required for the first. The total energy required to produce gaseous divalent ions for Group II elements (first ionization energy + second ionization energy) is over four times greater than the energy required to produce M^+ from Group I metals. The fact that ionic compounds are formed shows that the energy given out when a crystal lattice is formed more than offsets that used in producing ions.

Table 11.5 Ionization energies and electronegativity

	Ionization energy ($kJ\,mol^{-1}$)			Pauling's electronegativity
	1st	2nd	3rd	
Be	899	1757	14847	1.5
Mg	737	1450	7731	1.2
Ca	590	1145	4910	1.0
Sr	549	1064		1.0
Ba	503	965		0.9
Ra	509	979	(3281)	

Estimated value in brackets.

ELECTRONEGATIVITY

The electronegativity values of Group II elements are low, but are higher than the values for Group I. When Mg, Ca, Sr and Ba react with elements such as the halogens and oxygen at the right hand side of the periodic table, the electronegativity difference is large and the compounds are ionic.

The value for Be is higher than for the others. BeF_2 has the biggest electronegativity difference for a compound of Be and so is the most likely compound of Be to be ionic. BeF_2 has a very low conductivity when fused, and is regarded as covalent.

HYDRATION ENERGIES

The hydration energies of the Group II ions are four or five times greater than for Group I ions. This is largely due to their smaller size and increased charge, and $\Delta H_{\text{hydration}}$ decreases down the group as the size of the ions increases. In the case of Be a further factor is the very strong complex $[Be(H_2O)_4]^{2+}$ that is formed. The crystalline compounds of Group II contain more water of crystallization than the corresponding Group I compounds. Thus NaCl and KCl are anhydrous but $MgCl_2 \cdot 6H_2O$, $CaCl_2 \cdot 6H_2O$ and $BaCl_2 \cdot 2H_2O$ all have water of crystallization. Note that the number of molecules of water of crystallization decreases as the ions become larger.

Since the divalent ions have a noble gas structure with no unpaired electrons, their compounds are diamagnetic and colourless, unless the acid radical is coloured.

Table 11.6 Hydration energies

	Ionic radius (Å)	ΔH hydration ($kJ\,mol^{-1}$)
Be^{2+}	0.31*	−2494
Mg^{2+}	0.72	−1921
Ca^{2+}	1.00	−1577
Sr^{2+}	1.18	−1443
Ba^{2+}	1.35	−1305

* Four-coordinate radius.

ANOMALOUS BEHAVIOUR OF BERYLLIUM

Be differs from the rest of the group for three reasons.

1. It is extremely small, and Fajans' rules state that small highly charged ions tend to form covalent compounds.
2. Be has a comparatively high electronegativity. Thus when beryllium reacts with another atom, the difference in electronegativity is seldom large, which again favours the formation of covalent compounds. Even BeF_2 (electronegativity difference 2.5) and BeO (electronegativity difference 2.0) show evidence of covalent character.
3. Be is in the second row of the periodic table, and the outer shell can hold a maximum of eight electrons. (The orbitals available for bonding are one 2*s* and three 2*p* orbitals.) Thus Be can form a maximum of four conventional electron pair bonds, and in many compounds the maximum coordination number of Be is 4. The later elements can have more than eight outer electrons, and may attain a coordination number of 6 using one *s*, three *p* and two *d* orbitals for bonding. Exceptions occur if multi-centre bonding occurs, as for example in basic beryllium acetate, when higher coordination numbers are obtained.

Thus we should expect Be to form mainly covalent compounds, and commonly have a coordination number of 4. Anhydrous compounds of Be are predominantly two-covalent, and BeX_2 molecules should be linear.

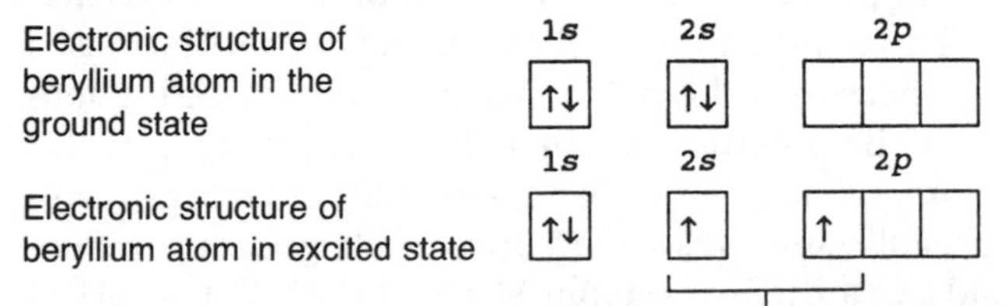

In fact linear molecules exist only in the gas phase, as this electronic arrangement has not filled the outer shell of electrons. In the solid state fourfold coordination is always achieved. There are several ways by which this can be achieved:

1. Two ligands that have a lone pair of electrons may form coordinate bonds using the two unfilled orbitals in the valence shell of Be. Thus two F^- ions might coordinate to BeF_2, forming $[BeF_4]^{2-}$. Similarly diethyl ether can coordinate to Be(+II) in $BeCl_2$, forming $[BeCl_2(OEt_2)_2]$.
2. The BeX_2 molecules may polymerize to form chains, containing bridging halogen groups, for example $(BeF_2)_n$, $(BeCl_2)_n$. Each halogen forms one normal covalent bond, and uses a lone pair to form a coordinate bond.
3. $(BeMe_2)_n$ has essentially the same structure as $(BeCl_2)_n$, but the bonding in the methyl compound is best regarded as three-centre two-electron bonds covering one Me and two Be atoms.
4. A covalent lattice may be formed with a zinc blende or wurtzite structure (coordination number 4), for example by BeO and BeS.

In water beryllium salts are extensively hydrolysed to give a series of hydroxo complexes of unknown structure. They may be polymeric and of the type:

```
⎡HO      OH      OH⎤2-   ⎡HO      OH      OH      OH⎤2-
   \    /  \    /            \    /  \    /  \    /
    Be       Be               Be       Be       Be
   /    \  /    \            /    \  /    \  /    \
⎣HO      OH      OH⎦     ⎣HO      OH      OH      OH⎦
```

If alkali is added to these solutions the polymers break down to give the simple mononuclear berrylate ion $[Be(OH)_4]^{2-}$, which is tetrahedral. Many beryllium salts contain the hydrated ion $[Be(H_2O)_4]^{2+}$ rather than Be^{2+}, and the hydrated ion too is a tetrahedral complex ion. Note that the coordination number is 4. Forming a hydrated complex increases the effective size of the beryllium ion, thus spreading the charge over a larger area. Stable ionic salts such as $[Be(H_2O)_4]SO_4$, $[Be(H_2O)_4](NO_3)_2$ and $[Be(H_2O)_4]Cl_2$ are known.

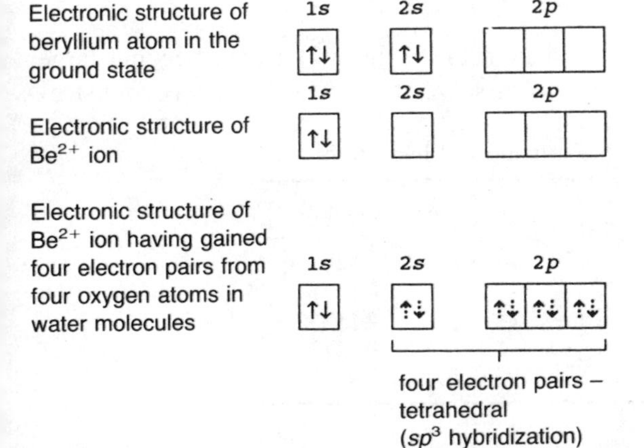

Beryllium salts are acidic when dissolved in pure water because the hydrated ion hydrolyses, producing H_3O^+. This happens because the Be—O bond is very strong, and so in the hydrated ion this weakens the O—H bonds, and hence there is a tendency to lose protons. The initial reaction is

$$H_2O + [Be(H_2O)_4]^{2+} \rightleftharpoons [Be(H_2O)_3(OH)]^+ + H_3O^+$$

but this may be followed by further polymerization, involving hydroxo-bridged structures $[Be_2OH]^{3+}$, $[Be_3(OH)_4]^{3+}$. In alkaline solutions $[Be(OH)_4]^{2-}$ is formed. The other Group II salts do not interact so strongly with water, and do not hydrolyse appreciably.

Beryllium salts rarely have more than four molecules of water of crystallization associated with the metal ion, because there are only four orbitals available in the second shell of electrons, whereas magnesium can have a coordination number of 6 by using some $3d$ orbitals as well as $3s$ and $3p$ orbitals.

SOLUBILITY AND LATTICE ENERGY

The solubility of most salts decreases with increased atomic weight, though the usual trend is reversed with the fluorides and hydroxides in this group. Solubility depends on the lattice energy of the solid, and the hydration energy of the ions as explained below.

Some lattice energy values for Group II compounds are listed in Table 11.7. The lattice energies are much higher than the values for Group I compounds, because of the effect of the increased charge on the ions in the Born–Landé equation. (See Chapter 3.) Taking any one particular negative ion, the lattice energy decreases as the size of the metal increases.

Table 11.7 Lattice energies of some compounds ($kJ\,mol^{-1}$)

	MO	MCO_3	MF_2	MI_2
Mg	−3923	−3178	−2906	−2292
Ca	−3517	−2986	−2610	−2058
Sr	−3312	−2718	−2459	
Ba	−3120	−2614	−2367	

The hydration energy also decreases as the metal ions become larger (Table 11.8). For a substance to dissolve, the hydration energy must ex-

Table 11.8 Enthalpies of hydration

	ΔH ($kJ\,mol^{-1}$)
Be^{2+}	−2494
Mg^{2+}	−1921
Ca^{2+}	−1577
Sr^{2+}	−1443
Ba^{2+}	−1305

ceed the lattice energy. Consider a related group of compounds, such as the chlorides of all the Group II metals. On descending the group the metal ions become larger and so both the lattice energy and the hydration energy decrease. A decrease in lattice energy favours increased solubility, but a decrease in hydration energy favours decreased solubility. These two factors thus change in opposite directions, and the overall effect depends on which of the two has changed most. With most compounds, on descending the group, the hydration energy decreases more rapidly than the lattice energy: hence the compounds become less soluble as the metal gets larger. However, with fluorides and hydroxides the lattice energy decreases more rapidly than the hydration energy, and so their solubility increases on descending the group.

SOLUTIONS OF THE METALS IN LIQUID AMMONIA

The metals all dissolve in liquid ammonia as do the Group I metals. Dilute solutions are bright blue in colour due to the spectrum from the solvated electron. These solutions decompose very slowly, forming amides and evolving hydrogen, but the reaction is accelerated by many transition metals and their compounds.

$$2NH_3 + 2e \rightarrow 2NH_2^- + H_2$$

Evaporation of the ammonia from solutions of Group I metals yields the metal, but with Group II metals evaporation of ammonia gives hexammoniates of the metals. These slowly decompose to give amides.

$$M(NH_3)_6 \rightarrow M(NH_2)_2 + 4NH_3 + H_2$$

Concentrated solutions of the metals in ammonia are bronze coloured, due to the formation of metal clusters.

CHEMICAL PROPERTIES

Reaction with water

The reduction potential of beryllium is much less than those for the rest of the group. (Standard electrode potentials E° of $Be^{2+}|Be$ -1.85, $Mg^{2+}|Mg$ -2.37, $Ca^{2+}|Ca$ -2.87, $Sr^{2+}|Sr$ -2.89, $Ba^{2+}|Ba$ -2.91, $Ra^{2+}|Ra$ -2.92 volts.) This indicates that beryllium is much less electropositive (less metallic) than the others, and beryllium does not react with water. There is some doubt whether it reacts with steam to form the oxide BeO, or fails to react at all.

Ca, Sr and Ba have reduction potentials similar to those of the corresponding Group I metals, and are quite high in the electrochemical series. They react with cold water quite readily, liberating hydrogen and forming metal hydroxides.

$$Ca + 2H_2O \rightarrow Ca(OH)_2 + H_2$$

Table 11.9 Some reactions of Group II elements

Reaction	Comment
$M + 2H_2O \rightarrow M(OH)_2 + H_2$	Be probably reacts with steam, Mg with hot water, and Ca, Sr and Ba react rapidly with cold water
$M + 2HCl \rightarrow MCl_2 + H_2$	All the metals react with acids, liberating hydrogen
$Be + NaOH \rightarrow Na_2[Be(OH)_4] + H_2$	Be is amphoteric
$2M + O_2 \rightarrow 2MO$	Normal oxide formed by all group members
with excess oxygen $Ba + O_2 \rightarrow BaO_2$	Ba also forms the peroxide
$M + H_2 \rightarrow MH_2$	Ionic 'salt-like' hydrides formed at high temperatures by Ca, Sr and Ba
$3M + N_2 \rightarrow M_3N_2$	All form nitrides at high temperatures
$3M + 2P \rightarrow M_3P_2$	All the metals form phosphides at high temperatures
$M + S \rightarrow MS$	All the metals form sulphides
$M + Se \rightarrow MSe$	All the metals form selenides
$M + Te \rightarrow MTe$	All the metals form tellurides
$M + F_2 \rightarrow MF_2$	All the metals form fluorides
$M + Cl_2 \rightarrow MCl_2$	All the metals form chlorides
$M + Br_2 \rightarrow MBr_2$	All the metals form bromides
$M + I_2 \rightarrow MI_2$	All the metals form iodides
$3M + 2NH_3 \rightarrow 2M(NH_2)_2$	All the metals form amides at high temperatures

Magnesium has an intermediate value, and it does not react with cold water but it decomposes hot water.

$$Mg + 2H_2O \rightarrow Mg(OH)_2 + H_2$$

or

$$Mg + H_2O \rightarrow MgO + H_2$$

Mg forms a protective layer of oxide, so despite its favourable reduction potential it does not react readily unless the oxide layer is removed by amalgamating with mercury. In the formation of the oxide film it resembles aluminium.

HYDROXIDES

$Be(OH)_2$ is amphoteric, but the hydroxides of Mg, Ca, Sr and Ba are basic. The basic strength increases from Mg to Ba, and Group II shows the usual trend that basic properties increase on descending a group.

Solutions of $Ca(OH)_2$ and $Ba(OH)_2$ are called lime water and baryta water respectively, and are used to detect carbon dioxide. When CO_2 is bubbled through these solutions, they become 'turbid' or 'milky' due to the formation of a suspension of solid particles of $CaCO_3$ or $BaCO_3$. If excess CO_2 is passed through these 'milky' solutions then the turbidity disappears as soluble bicarbonates form with the excess CO_2. Baryta water is rather too sensitive as it gives a positive test for CO_2 by exhaling breath on it, whereas with lime water, breath (or other gas) must be blown through the solution as bubbles.

$$Ca^{2+}(OH^-)_2 + CO_2 \rightarrow \underset{\substack{\text{insoluble}\\\text{white precipitate}}}{CaCO_3} + H_2O \xrightarrow[]{\substack{\text{Excess}\\CO_2}} \underset{\text{soluble}}{Ca^{2+}(HCO_3^-)_2}$$

The bicarbonates of Group II metals are only stable in solution. Caves in limestone regions often have stalactites growing down from the roof, and stalagmites growing up from the floor. Water percolating through the limestone contains some $Ca^{2+}(HCO_3^-)_2$ in solution. The soluble bicarbonate decomposes slowly into the insoluble carbonate, and this results in the slow growth of the stalactites and stalagmites.

$$Ca^{2+}(HCO_3^-)_2 \rightarrow CaCO_3 + CO_2 + H_2O$$

HARDNESS OF WATER

Hard water contains dissolved salts such as magnesium and calcium carbonates, bicarbonates or sulphates. It is difficult to produce lather from soap with hard water, and an insoluble scum is formed. The metal ions Ca^{2+} and Mg^{2+} react with the stearate ions from the soap, forming an insoluble scum of calcium stearate before any lather is produced.Hard water also produces scale (insoluble deposits) in water pipes, boilers, and kettles.

'Temporary hardness' is due to the presence of $Mg(HCO_3)_2$ and $Ca(HCO_3)_2$. It is called 'temporary' because it can be removed by boiling, which drives off CO_2 and upsets the equilibrium.

$$2HCO_3^- \rightleftharpoons CO_3^{2-} + CO_2 + H_2O$$

Thus the bicarbonates decompose into carbonates, and calcium carbonate is precipitated. If this is filtered off, or allowed to settle, the water is free from hardness. Temporary hardness can also be removed by adding slaked lime to precipitate calcium carbonate. This is called 'lime softening', and by operating at pH 10.5, temporary hardness due to HCO_3^- can be almost completely removed.

$$Ca(HCO_3)_2 + Ca(OH)_2 \rightleftharpoons 2CaCO_3 + 2H_2O$$

'Permanent hardness' is not removed by boiling, and is due mainly to $MgSO_4$ or $CaSO_4$ in solution. Small quantities of pure water are prepared in the laboratory either by distilling the water, or by passing it through an ion-exchange resin, when the Ca^{2+} and Mg^{2+} ions are replaced by Na^+. The sodium salts do not affect the lathering power. Ion-exchange methods are widely used in industry. Water may also be softened by adding various phosphates, such as trisodium phosphate Na_3PO_4, sodium pyrophosphate $Na_4P_2O_7$, sodium tripolyphosphate $Na_5P_3O_{10}$, or Grahams salt (Calgon) $(NaPO_3)_n$. These form a complex with the calcium and magnesium ions and 'sequester' them, that is keep them in solution. At one time large quantities of sodium carbonate were used in the lime–soda process to soften water. The effect of adding Na_2CO_3 is to precipitate $CaCO_3$.

$$CaSO_4 + Na_2CO_3 \rightarrow CaCO_3 + 2NaSO_4$$

REACTION WITH ACIDS AND BASES

The metals all react with acids and liberate H_2, although Be reacts slowly. Be is rendered passive by concentrated HNO_3, i.e. it does not react. This is because concentrated HNO_3 is a strong oxidizing agent and it forms a very thin layer of oxide on the surface of the metal, which protects it from further attack by the acid. Be is amphoteric, as it also reacts with NaOH, giving H_2 and sodium berrylate. Mg, Ca, Sr and Ba do not react with NaOH, and are purely basic. This illustrates that basic properties increase on descending the group.

$$Mg + 2HCl \rightarrow MgCl_2 + H_2$$

$$Be + 2NaOH + 2H_2O \rightarrow Na_2[Be(OH)_4] + H_2$$

$$\text{or } \underset{\text{sodium beryllate}}{NaBeO_2 \cdot 2H_2O} + H_2$$

OXIDES AND PEROXIDES

All the elements in this group burn in O_2 to form oxides MO. Be metal is relatively unreactive in the massive form, and does not react below 600 °C, but the powder is much more reactive and burns brilliantly. The elements also burn in air, forming a mixture of oxide and nitride. Mg burns with dazzling brilliance in air, and evolves a lot of heat. This is used to start a thermite reaction with aluminium, and also to provide light in flash photography using light bulbs, not electronics.

$$Mg + \text{air} \rightarrow MgO + Mg_3N_2$$

BeO is usually made by ignition of the metal, but the other metal oxides are usually obtained by thermal decomposition of the carbonates MCO_3. Other oxosalts such as $M(NO_3)_2$ and MSO_4, and also $M(OH)_2$, all decompose to the oxide on heating. The oxosalts are less stable to heat than the corresponding Group I salts because the metals and their hydroxides are less basic than those of Group I.

CaO (quicklime) is made in enormous quantities (126 million tonnes in 1988) by roasting $CaCO_3$ in a lime kiln.

$$CaCO_3 \xrightarrow{\text{heat}} CaO + CO_2$$

MgO is not very reactive, especially if it has been ignited at high temperatures, and for this reason it is used as a refractory. BeO is also used as a refractory. They combine a number of properties that make them useful for lining furnaces. These factors are:

1. High melting points (BeO approx. 2500 °C, MgO approx. 2800 °C).
2. Very low vapour pressures.
3. Very good conductors of heat.
4. Chemical inertness.
5. Electrical insulators.

All Be compounds should be used with care as they are toxic and dust or smoke cause berylliosis.

CaO, SrO and BaO react exothermically with water, forming hydroxides.

$$CaO + H_2O \rightarrow Ca(OH)_2$$

$Mg(OH)_2$ is extremely insoluble in water (approx. $1 \times 10^{-4}\,g\,l^{-1}$ at 20 °C) but the other hydroxides are soluble and the solubility increases down the group ($Ca(OH)_2$ approx. $2\,g\,l^{-1}$; $Sr(OH)_2$ approx. $8\,g\,l^{-1}$; $Ba(OH)_2$ approx. $39\,g\,l^{-1}$). $Be(OH)_2$ is soluble in solutions containing an excess of OH^{-1}, and is therefore amphoteric. $Mg(OH)_2$ is weakly basic, and is used to treat acid indigestion. The other hydroxides are strong bases. $Ca(OH)_2$ is called slaked lime.

BeO is covalent and has a 4:4 zinc sulphide (wurtzite) structure, but all the others are ionic and have a 6:6 sodium chloride structure.

Attempts to predict the structure using the sizes of the ions and the radius ratio are only partly successful (Table 11.10). The correct structure is predicted for BeO, MgO and CaO, but for SrO and BaO the predicted coordination number is 8, though the structures found are six-coordinate. Crystals adopt the structure that has the most favourable lattice energy, and the failure of the radius ratio concept in this case leads us to examine the assumptions on which it is based. (See Chapter 3 under 'A more critical look at radius ratios'.) Ionic radii are not known with great accuracy and they change with different coordination numbers. Also, ions are not necessarily spherical, or inelastic.

Table 11.10 Radius ratios and coordination numbers

Oxide	Radius ratio M^{2+}/O^{2-}	Predicted coordination number	Coordination number found
BeO	0.32	4	4
MgO	0.51	6	6
CaO	0.71	6	6
SrO	0.84	8	6
BaO	0.96	8	6

As the atoms get larger, the ionization energy decreases and the elements also become more basic. BeO is insoluble in water but dissolves in acids to give salts, and in alkalis to give beryllates, which on standing precipitate as the hydroxide. BeO is therefore amphoteric. MgO reacts with water, forming $Mg(OH)_2$ which is weakly basic. CaO reacts very readily with water, evolving a lot of heat and forming $Ca(OH)_2$ which is a moderately strong base. $Sr(OH)_2$ and $Ba(OH)_2$ are even stronger bases. The oxides are usually prepared by thermal decomposition of the carbonates, nitrates or hydroxides. The increase in basic strength is illustrated by the temperatures at which the carbonates decompose:

$BeCO_3$	$MgCO_3$	$CaCO_3$	$SrCO_3$	$BaCO_3$
<100 °C	540 °C	900 °C	1290 °C	1360 °C

The carbonates are all ionic, but $BeCO_3$ is unusual because it contains the hydrated ion $[Be(H_2O)_4]^{2+}$ rather than Be^{2+}.

$CaCO_3$ occurs as two different crystalline forms, calcite and aragonite. Both forms occur naturally as minerals. Calcite is the more stable: each Ca^{2+} is surrounded by six oxygen atoms from CO_3^- ions. Aragonite is a metastable form, and its standard enthalpy of formation is about $5\,kJ\,mol^{-1}$ higher than that of calcite. In principle aragonite should decompose to calcite, but a high energy of activation prevents this happening. Aragonite can be made in the laboratory by precipitating from a hot solution. Its crystal structure has Ca^{2+} surrounded by nine oxygen atoms. This is a rather unusual coordination number.

Calcium oxide (lime) is prepared on a large scale (126 million tonnes in 1988) by heating $CaCO_3$ in lime kilns.

$$CaCO_3 \rightarrow CaO + CO_2$$

Lime is used:

1. In steel making to remove phosphates and silicates as slag.
2. By mixing with SiO_2 and alumina or clay to make cement.
3. For making glass.
4. In the lime–soda process, which is part of the chlor-alkali industry, converting Na_2CO_3 to NaOH or vice versa.
5. For 'softening' water.
6. To make CaC_2.
7. To make slaked lime $Ca(OH)_2$ by treatment with water.

Bleaching powder is made by passing Cl_2 into slaked lime, and about 90 000 tonnes a year are produced. Though bleaching powder is often written as $Ca(OCl)_2$, it is really a mixture.

$$3Ca(OH)_2 + 2Cl_2 \rightarrow Ca(OCl)_2 \cdot Ca(OH)_2 \cdot CaCl_2 \cdot 2H_2O$$

Soda lime is a mixture of NaOH and $Ca(OH)_2$ and is made from quicklime (CaO) and aqueous sodium hydroxide. It much easier to handle than NaOH.

Peroxides are formed with increasing ease and increasing stability as the metal ions become larger. Barium peroxide BaO_2 is formed by passing air over BaO at 500 °C. SrO_2 can be formed in a similar way but this requires a high pressure and temperature. CaO_2 is not formed in this way, but can be made as the hydrate by treating $Ca(OH)_2$ with H_2O_2 and then dehydrating the product. Crude MgO_2 has been made using H_2O_2, but no peroxide of beryllium is known. The peroxides are white ionic solids containing the $[O—O]^{2-}$ ion and can be regarded as salts of the very weak acid hydrogen peroxide. Treating peroxides with acid liberates hydrogen peroxide.

$$BaO_2 + 2HCl \rightarrow BaCl_2 + H_2O_2$$

SULPHATES

The solubility of the sulphates in water decreases down the group, Be > Mg ≫ Ca > Sr > Ba. Thus $BeSO_4$ and $MgSO_4$ are soluble, but $CaSO_4$ is sparingly soluble, and the sulphates of Sr, Ba and Ra are virtually insoluble. The significantly higher solubilities of $BeSO_4$ and $MgSO_4$ are due to the high enthalpy of solvation of the smaller Be^{2+} and Mg^{2+} ions. Epsom salt $MgSO_4 \cdot 7H_2O$ is used as a mild laxative. Calcium sulphate can exist as a hemihydrate $CaSO_4 \cdot \frac{1}{2}H_2O$ which is important in the building trade as plaster of Paris. This is made by partially dehydrating gypsum.

$$\underset{\text{gypsum}}{CaSO_4 \cdot 2H_2O} \xrightarrow{150\,°C} \underset{\text{plaster of Paris}}{CaSO_4 \cdot \tfrac{1}{2}H_2O} \xrightarrow{200\,°C} \underset{\text{anhydrite}}{CaSO_4} \xrightarrow{1100\,°C} CaO + SO_3$$

When powdered plaster of Paris $CaSO_4 \cdot \frac{1}{2}H_2O$ is mixed with the the correct amount of water it sets into a solid mass of $CaSO_4 \cdot 2H_2O$ (gypsum). Plaster of Paris is used for plastering walls, and also to make plaster casts (moulds) for a variety of purposes, industrial, sculptural, and in hospitals to encase limbs so that broken bones are set straight. Alabaster is a fine grained form of $CaSO_4 \cdot H_2O$ which is shiny like marble, and is used to make ornaments. $CaSO_4$ is slightly soluble in water (2 g per litre), so objects made from alabaster or gypsum cannot be kept outdoors. $BaSO_4$ is both insoluble in water and opaque to X-rays, and is used as a 'barium meal' to provide a shadow of the stomach or duodenum on an X-ray picture, which is useful in diagnosing stomach or duodenal ulcers. The sulphates all decompose on heating, giving the oxides:

$$MgSO_4 \xrightarrow{\text{heat}} MgO + SO_3$$

In the same way as with the stability and thermal decomposition of the carbonates, the more basic the metal is, the more stable the sulphate is. This is shown by the temperatures at which decomposition occurs:

$BeSO_4$	$MgSO_4$	$CaSO_4$	$SrSO_4$
500 °C	895 °C	1149 °C	1374 °C

Heating the sulphates with C reduces them to sulphides. Most barium compounds are made from barium sulphide.

$$BaSO_4 + 4C \rightarrow BaS + 4CO$$

Group II elements also form perchlorates $MClO_4$, which have very similar structures to the sulphates, and the ClO_4^- ion is tetrahedral and similar in size to the SO_2^{2-} ion. However, they differ chemically, since perchlorates are strong oxidizing agents. Magnesium perchlorate is used as a drying agent called anhydrone. Anhydrone should not be used with organic materials because it is a strong oxidizing agent, and accidental contact with organic material could cause an explosion.

NITRATES

Nitrates of the metals can all be prepared in solution and can be crystallized as hydrated salts by the reaction of HNO_3 with carbonates, oxides or hydroxides. Heating the hydrated solids does not give the anhydrous nitrate because the solid decomposes to the oxide. Anhydrous nitrates can be prepared using liquid dinitrogen tetroxide and ethyl acetate. Beryllium is unusual in that it forms a basic nitrate in addition to the normal salt.

$$BeCl_2 \xrightarrow{N_2O_4} Be(NO_3)_2 \cdot 2N_2O_4 \xrightarrow[\text{under vacuum}]{\text{warm to } 50\,^{\circ}C} Be(NO_3)_2 \xrightarrow{125\,^{\circ}C} \underset{\text{basic beryllium nitrate}}{[Be_4O(NO_3)_6]}$$

Basic beryllium nitrate is covalent and has an unusual structure (Figure 11.1b). Four Be atoms are located at the corners of a tetrahedron, with six NO_3^- groups along the six edges of the tetrahedron, and the (basic) oxygen at the centre. The structure is of interest partly because beryllium is unique in forming a series of stable covalent molecules of formula $[Be_4O(R_6)]$ where R may be NO_3^-, $HCOO^-$, CH_3COO^-, $C_2H_5COO^-$, $C_6H_5COO^-$, etc. Thus, basic beryllium nitrate is one of a series of similar molecules (cf. basic beryllium acetate (Figure 11.5 b)). The structure is also of interest because the NO_3^- groups act as bidentate ligands in forming a bridge between two Be atoms. (See 'Chelates' in Chapter 7 for a discussion of multidentate groups.)

HYDRIDES

The elements Mg, Ca, Sr and Ba all react with hydrogen to form hydrides MH_2. Beryllium hydride is difficult to prepare, and less stable than the others. Impure BeH_2 (contaminated with various amounts of ether) was first made by reducing $BeCl_2$ with lithium aluminium hydride $Li[AlH_4]$. Pure samples can be obtained by reducing $BeCl_2$ with lithium borohydride $Li[BH_4]$ to give BeB_2H_8, then heating BeB_2H_8 in a sealed tube with triphenylphosphine PPh_3.

$$2BeCl_2 + LiAlH_4 \rightarrow 2BeH_2 + LiCl + AlCl_3$$
$$BeB_2H_8 + 2PPh_3 \rightarrow BeH_2 + 2Ph_3PBH_3$$

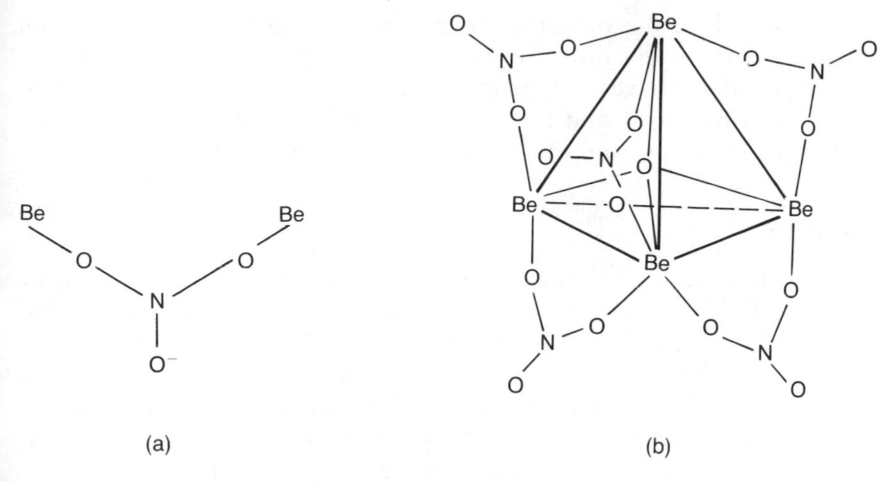

Figure 11.1 (a) A bridging NO_3^- group. (b) Basic beryllium nitrate.

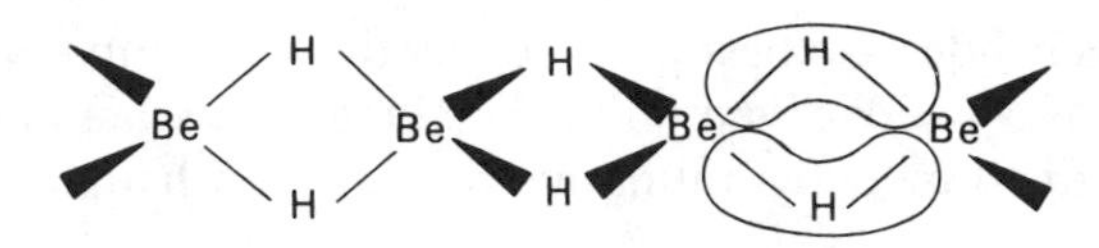

Figure 11.2 Polymeric BeH_2 structure with three-centre bonds.

$$BeCl_2 + 2Li[BH_4] \rightarrow BeB_2H_8 + 2LiCl$$
$$Ca + H_2 \rightarrow CaH_2$$

The hydrides are all reducing agents and are hydrolysed by water and dilute acids with the evolution of hydrogen.

$$CaH_2 + 2H_2O \rightarrow Ca(OH)_2 + 2H_2$$

CaH_2, SrH_2 and BaH_2 are ionic, and contain the hydride ion H^-. Beryllium and magnesium hydrides are covalent and polymeric. $(BeH_2)_n$ presents an interesting structural problem. In the gas phase it seems probable that several species may be present comprising polymeric chains and rings. The solid is polymeric, and its X-ray structure shows that it contains chains with hydrogen bridges between beryllium atoms.

Be is bonded to four H atoms, and the H atoms appear to be forming two bonds. Since Be has two valency electrons, and H only one, it is apparent that there are not enough electrons to form the usual electron pair bonds in which two electrons are shared between two atoms. Instead of this, three-centre bonds are formed in which a 'banana-shaped' molecular orbital (or three-centre bond) covers three atoms Be...H...Be, and contains two electrons. (This is called a three-centre two-electron bond.) This is an example of a cluster compound, where the monomeric molecule BeH_2 (formed with normal bonds) would result in only four electrons in the outer shell of the beryllium atom. This situation is termed 'electron deficient', and, by clustering, each atom shares its electrons with several neighbours and receives a share in their electrons. Clustering is important in metals (Chapter 5) and the boron hydrides (Chapter 12), and in the halides of the second and third row transition metals (Chapter 18).

HALIDES

Halides MX_2 can be made by heating the metals with the halogen, or by the action of halogen acid on either the metal or the carbonate. The beryllium halides are covalent, hygroscopic and fume in air due to hydrolysis. They sublime, and they do not conduct electricity. Anhydrous beryllium halides cannot be obtained from materials made in aqueous solutions because the hydrated ion $[Be(H_2O)_4]^{2+}$ is formed, e.g. $[Be(H_2O)_4]Cl_2$ or $[Be(H_2O)_4]F_2$. Attempts to dehydrate result in hydrolysis.

$$[Be(H_2O)_4]Cl_2 \xrightarrow{\text{heat}} Be(OH)_2 + 2HCl$$

The anhydrous halides are best prepared by the following reactions. Reaction with CCl_4 is a standard method for making anhydrous chlorides which cannot be obtained by dehydrating hydrates. (See Chapter 16.)

$$BeO + NH_4F \rightarrow (NH_4)_2[BeF_4] \xrightarrow{\text{heat}} BeF_2 + 2NH_4F$$

$$BeO + C + Cl_2 \xrightleftharpoons{700\,^\circ C} BeCl_2 + CO$$

$$2BeO + CCl_4 \xrightarrow{800\,^\circ C} 2BeCl_2 + CO_2$$

The anhydrous halides are polymeric. Beryllium chloride vapour contains $BeCl_2$ and $(BeCl_2)_2$, but the solid is polymerized. Though the structure of the $(BeCl_2)_n$ polymer is similar to that for $(BeH_2)_n$, the bonding is different. Both show clustering, but the hydride has three-centre bonding, whereas the halides have halogen bridges, in which a halogen atom bonded to one beryllium atom uses a lone pair of electrons to form a coordinate bond to another beryllium atom.

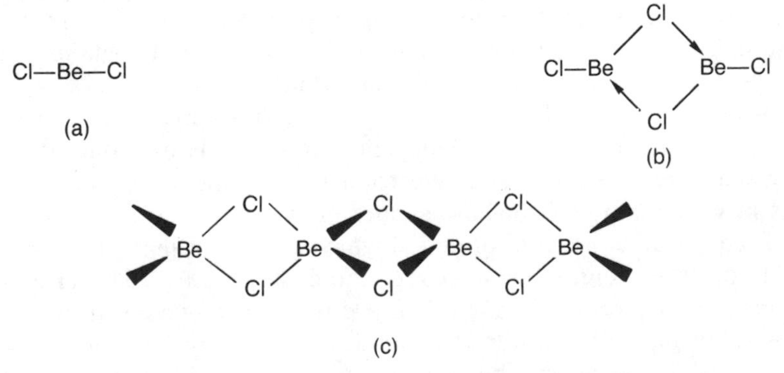

Cl—Be—Cl

(a)

(b)

(c)

Figure 11.3 $BeCl_2$ structures: (a) monomer, (b) dimer and (c) polymer.

Beryllium fluoride is very soluble in water, owing to the high solvation energy of Be in forming $[Be(H_2O)_4]^{2+}$. The other fluorides MF_2 are all almost insoluble.

Fluorides of the other metals are ionic, have high melting points, and are insoluble in water. CaF_2 is a white, insoluble, high melting solid. It is very important industrially, and is the main source of both F_2 and HF.

$$CaF_2 + H_2SO_4 \rightarrow 2HF + CaSO_4$$

$$HF + KF \rightarrow KHF_2 \xrightarrow{\text{electrolysis}} F_2$$

World production of fluorite CaF_2 was 5.1 million tonnes in 1988. The main sources are China 22.5%, Mexico and Mongolia both 16%, the USSR 11% and South Africa 6%. CaF_2 is also used to make prisms and cell windows for spectrophotometers.

The chlorides, bromides and iodides of Mg, Ca, Sr and Ba are ionic, have much lower melting points than the fluorides, and are readily soluble in water. The solubility decreases somewhat with increasing atomic number. The halides all form hydrates, and they are hygroscopic (absorb water vapour from the air). Several million tonnes of $CaCl_2$ are produced each year. Large amounts are discarded in solution as a waste product from the Solvay process because it is not economic to recover the solid and demand is low. $CaCl_2$ is widely used for treating ice on roads, particularly in very cold countries, because a 30% eutectic mixture of $CaCl_2/H_2O$ freezes at $-55\,°C$, compared with $NaCl/H_2O$ at $-18\,°C$. $CaCl_2$ is also used to make concrete set more quickly and to improve its strength, and as 'brine' in refrigeration plants. A minor use is in laboratories as a desiccant (drying agent). Anhydrous $MgCl_2$ is important in the electrolytic method for extracting magnesium.

NITRIDES

The alkaline earth elements all burn in nitrogen and form ionic nitrides M_3N_2. This is in contrast to Group I where Li_3N is the only nitride formed.

$$3Ca + N_2 \rightarrow Ca_3N_2$$

Because the N_2 molecule is very stable, it requires a lot of energy to convert N_2 into N^{3-} nitride ions. The large amount of energy required comes from the very large amount of lattice energy evolved when the crystalline solid is formed. The lattice energy is particularly high because of the high charges on the ions M^{2+} and N^{3-}. Li is alone in Group I in forming a nitride, and here the very small size of Li^+ results in a high lattice energy. (See Born–Landé equation, Chapter 3.)

Be_3N_2 is rather volatile in accord with the greater tendency of Be to covalency, but the other nitrides are not volatile. All the nitrides are all crystalline solids which decompose on heating and react with water, liberating ammonia and forming either the metal oxide or hydroxide, e.g.

$$Ca_3N_2 + 6H_2O \rightarrow 3Ca(OH)_2 + 2NH_3$$

CARBIDES (see also Chapter 13)

When BeO is heated with C at 1900–2000 °C a brick red coloured carbide of formula Be_2C is formed. This is ionic, and has an anti-fluorite structure, i.e. like the CaF_2 structure except that the positions of the metal ions and anions are interchanged. It is unusual because it reacts with water, forming methane, and is thus called a methanide.

$$Be_2C + 4H_2O \rightarrow 2Be(OH)_2 + CH_4$$

Group II metals typically form ionic carbides of formula MC_2. The metals Mg, Ca, Sr and Ba form carbides of formula MC_2, either when the metal is heated with carbon in an electric furnace, or when their oxides are heated with carbon. CaC_2 made in this way is a grey coloured solid, but is colourless when pure. BeC_2 is made by heating Be with ethyne.

On heating, MgC_2 changes into Mg_2C_3. This contains C_3^{4-}, and it reacts with water to form propyne $CH_3C{\equiv}CH$ (methyl acetylene).

$$Ca + 2C \xrightarrow{1100\,°C} CaC_2$$

$$CaO + 3C \xrightarrow{2000\,°C} CaC_2 + CO$$

The MC_2 carbides all have a distorted sodium chloride type of structure. M^{2+} replaces Na^+ and $C{\equiv}C^{2-}$ replaces Cl^-. The C_2^{2-} ions are elongated, not spherical like Cl^-. Thus the axis along which the C_2^{2-} ions lie is lengthened compared with the other two axes, and this is called tetragonal distortion. The structure is shown in Chapter 3 Figure 3.12. At temperatures above 450 °C the C_2^{2-} ions adopt random positions rather than being aligned in this way, and the unit cell then becomes cubic rather than tetragonal.

Calcium carbide is the best known. It reacts with water, liberating ethyne (formerly called acetylene), and is thus called an acetylide.

$$CaC_2 + 2H_2O \rightarrow Ca(OH)_2 + C_2H_2$$

At one time this reaction was the main source of ethyne for oxy-acetylene welding. Production of CaC_2 peaked at 7 million tonnes/year in 1960 but had declined slightly to 6.2 million tonnes in 1985 because ethyne is now obtained from processing oil.

CaC_2 is an important chemical intermediate. When CaC_2 is heated in an electric furnace with atmospheric nitrogen at 1100 °C, calcium cyanamide CaNCN is formed. This is an important reaction because it is one method of fixing atmospheric nitrogen. (The Haber process for NH_3 is another method of fixing nitrogen.)

$$CaC_2 + N_2 \rightarrow CaNCN + C$$

The cyanamide ion $[N{=}C{=}N]^{2-}$ is isoelectronic with CO_2, and has the same linear structure. CaNCN is produced on a large scale, particularly in locations where there is cheap electricity. At one time about 1.3 million tonnes/year was produced, though this has declined appreciably. It is widely used as a slow acting nitrogenous fertilizer, as it hydrolyses slowly over a period of months. CaNCN has the advantage over more soluble nitrogenous fertilizers such as NH_4NO_3 or urea, in that it is not washed away with the first rainstorm.

$$\mathrm{CaNCN + 5H_2O \rightarrow CaCO_3 + 2NH_4OH}$$

Other important industrial uses of CaNCN are the manufacture of cyanamide H_2NCN which is used to make urea and thiourea, and of melamine which forms hard plastics with formaldehyde.

$$\mathrm{CaNCN + H_2SO_4 \rightarrow H_2NCN + CaSO_4}$$

$$\mathrm{CaNCN + CO_2 + H_2O \rightarrow \underset{cyanamide}{H_2NCN} + CaCO_3}$$

$$\mathrm{H_2NCN + H_2O \xrightarrow{pH <2 \ or >12} H_2N \cdot CO \cdot NH_2 \ (urea)}$$

$$\mathrm{H_2NCN + H_2S \longrightarrow H_2N \cdot CS \cdot NH_2 \ (thiourea)}$$

$$\mathrm{\underset{cyanamide}{H_2NCN} \xrightarrow{pH\ 7-9} \underset{dicyanamide}{NCNC(NH_2)_2} \xrightarrow{pyrolysis}}$$

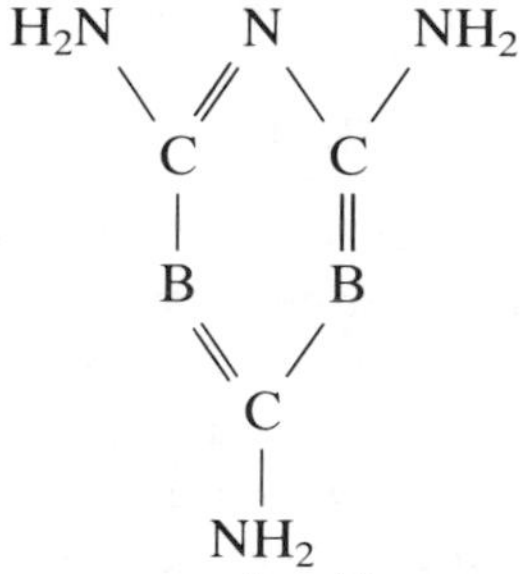

```
H2N       N       NH2
   \    //  \    /
     C        C
     |        ||
     B        B
       \\    /
          C
          |
         NH2
```

cyanuric amide
(melamine)

It is interesting that BaC_2 also reacts with N_2, but it forms a cyanide $Ba(CN)_2$, not a cyanamide $(NCN)^{2-}$.

INSOLUBLE SALTS

The sulphates of calcium, strontium and barium are insoluble, and the carbonates, oxalates, chromates and fluorides of the whole group are insoluble. This is a useful factor in qualitative analysis.

ORGANOMETALLIC COMPOUNDS

Both Be and Mg form an appreciable number of compounds with M—C bonds, but only a few have been isolated for Ca, Sr and Ba.

Victor Grignard, a Frenchman, won the Nobel Prize for Chemistry in 1912 for his work on organomagnesium compounds which are now called Grignard reagents. These are probably the most versatile reagents in organic chemistry, and can be used to make a wide variety of alcohols, aldehydes, ketones, carboxylic acids, esters, amides and alkenes. The use of Grignard reagents and lithium alkyls provide the two general methods for making organometallic compounds, and hence Grignard compounds are very important in inorganic chemistry too.

Grignard reagents are made by the slow addition of an alkyl or aryl halide (Cl, Br or I) to a continuously stirred mixture of magnesium turnings in an absolutely dry organic solvent such as diethyl ether. Water and air must be rigorously excluded. The reaction often has an induction period before it starts, and may require the addition of a crystal of iodine to penetrate the oxide film on the metal to make it start. Very reactive magnesium can be prepared by reducing magnesium halides with potassium in the presence of KI, and the use of this makes it easier to prepare Grignard reagents, and extends their uses.

$$\mathrm{Mg + RBr \xrightarrow{\text{dry ether}} \underset{\text{Grignard reagent}}{RMgBr} \quad (R = alkyl\ or\ aryl)}$$

Grignard reagents are all very reactive. Iodides are the most reactive, and chlorides the least reactive. Alkyl Grignard compounds are usually more reactive than aryl Grignard compounds.

All Grignard reagents are rapidly hydrolysed by water to give the parent hydrocarbon.

$$\mathrm{2RMgBr + 2H_2O \rightarrow 2RH + Mg(OH)_2 + MgBr_2}$$

Grignard reagents are not stored, but are made and used when required without isolating them. They are normally solvated or polymerized with halogen bridges. Their structures have long been the subject of controversy. X-ray structures of solid $PhMgBr \cdot 2Et_2O$ and $EtMgBr \cdot 2Et_2O$ show that the magnesium is tetrahedrally coordinated by bromine, the organic group, and oxygen atoms from ether molecules, but in solution several species may be present. Some typical reactions are:

$$\mathrm{RMgBr + CO_2 \xrightarrow{+\,acid} R \cdot COOH\ (carboxylic\ acid)}$$

$$\mathrm{RMgBr + R_2C{=}O \xrightarrow{+\,H_2O} R_3C \cdot OH\ (tertiary\ alcohol)}$$

$$\mathrm{RMgBr + R \cdot CHO \xrightarrow{+\,H_2O} R_2CHOH\ (secondary\ alcohol)}$$

$$\mathrm{RMgBr + O_2 \xrightarrow{+\,acid} ROH\ (primary\ alcohol)}$$

$$\mathrm{RMgBr + S_8 \longrightarrow RSH\ and\ R_2S}$$

$$\mathrm{RMgBr + HCHO \xrightarrow{+\,acid} RCH_2 \cdot OH}$$

$$\mathrm{RMgBr + I_2 \longrightarrow RI}$$

$$\mathrm{RMgBr + H^+ \longrightarrow RH}$$

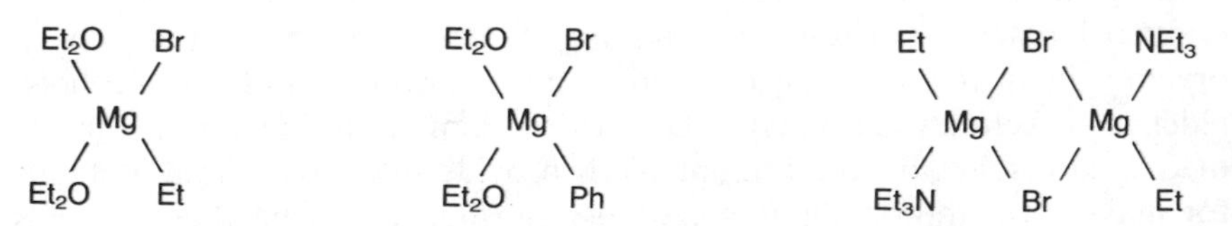

Figure 11.4 Structures of some Grignard compounds.

$$RMgBr + BeCl_2 \longrightarrow BeR_2$$
$$RMgBr + LiR \longrightarrow MgR_2$$
$$RMgBr + BCl_3 \longrightarrow BR_3$$
$$RMgBr + SiCl_4 \longrightarrow \underset{\text{alkyl and aryl chlorosilanes}}{RSiCl_3,\ R_2SiCl_2,\ R_3SiCl,\ R_4Si}$$

The alkyl and aryl chlorosilanes are commercially important in the manufacture of silicones (see Group IV).

$BeCl_2$ reacts with Grignard compounds, forming reactive alkyls and aryls. The compound $Be(Me)_2$ is dimerized in the vapour state, and polymerized in the solid. The structure formed is a chain structure which resembles that in $BeCl_2$. The bonding is, however, very different. The bonding in Be—Me—Be is best described as two electrons forming a three-centre bridge bond involving both Be atoms and the CH_3 group, similar to the three-centre bonds in $(BeH_2)_n$. This is different from the halogen bridging in $(BeCl_2)_n$ where the bridging chlorine atom forms two normal two-electron bonds.

Though much less studied than Grignard compounds, the other Group II metals also form dialkyl and diaryl compounds. These may be prepared using Grignard compounds, lithium alkyls/aryls or mercury alkyls/aryls.

$$BeCl_2 + 2MeMgCl \xrightarrow{Et_2O} BeMe_2 \cdot (Et_2O)_n + 2MgCl_2$$

$$BeCl_2 + 2LiEt \xrightarrow{Et_2O} BeEt_2 \cdot (Et_2O)_n$$

$$Be + HgMe_2 \xrightarrow{\text{warm}} BeMe_2 + Hg$$

Similar reactions may be used to make dialkyls and diaryls of Mg, Ca, Sr and Ba. The Ca, Sr and Ba compounds are much more reactive than the corresponding magnesium compound. The beryllium alkyls react with $BeCl_2$ to form 'beryllium Grignard' compounds.

$$BeMe_2 + BeCl_2 \rightarrow 2MeBeCl$$

These are less reactive than the corresponding magnesium (Grignard) compounds.

COMPLEXES

Group II metals are not noted for their ability to form complexes. The factors favouring complex formation are small highly charged ions with suitable empty orbitals of low energy which can be used for bonding. All the elements in the group form divalent ions, and these are smaller than the corresponding Group I ions: hence Group II elements are better at forming complexes than Group I elements. Be is appreciably smaller than the others, and so Be forms many complexes. Of the others, only Mg and Ca show much tendency to form complexes in solution, and these are usually with oxygen-donor ligands.

Beryllium fluoride BeF_2 readily coordinates two extra F^- ions, forming the $[BeF_4]^{2-}$ complex. The tetrafluoroberyllates $M_2[BeF_4]^{2-}$ are well known complex ions, and resemble the sulphates in properties. In most cases beryllium is four-coordinate in complexes and the tetrahedral arrangement adopted correlates with the orbitals available for complex formation.

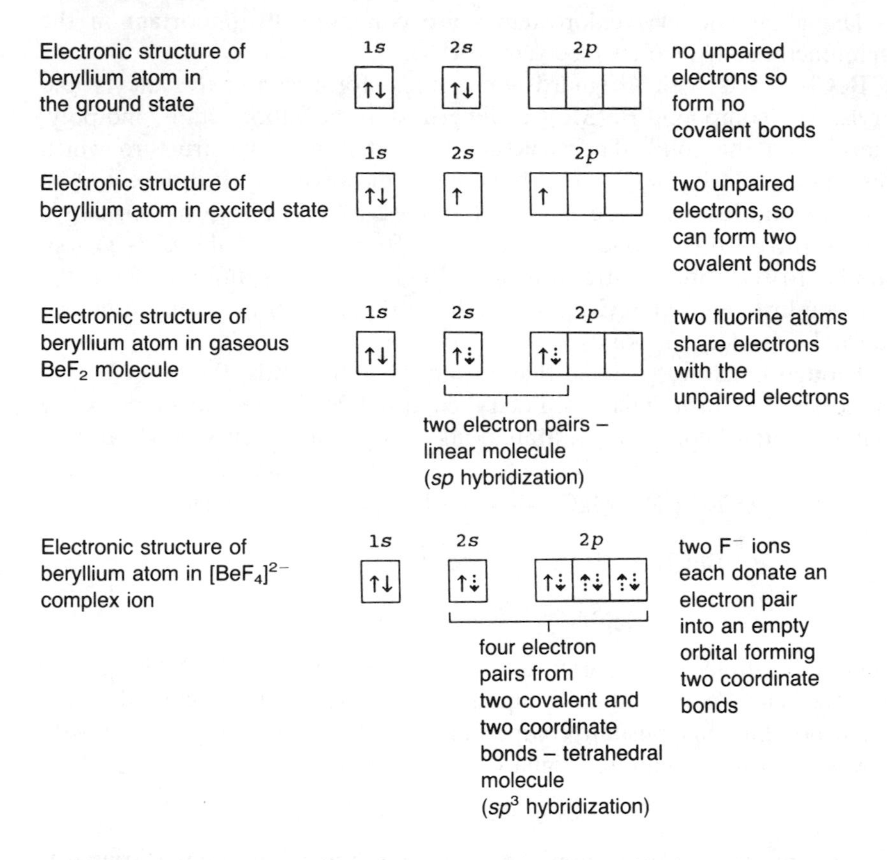

In a similar way complexes of the type $BeCl_2 \cdot D_2$ are formed (where D is an ether, aldehyde or ketone with an oxygen atom which has a lone pair of electrons that can be donated). These complexes, like $[Be(H_2O)_4]^{2+}$, are tetrahedral.

Many stable chelate complexes of Be are known, including beryllium oxalate $[Be(ox)_2]^{2-}$, with β-diketones such as acetylacetone, and with catechol. In all of these the Be^{2+} ion is tetrahedrally coordinated (see Figure 11.5b).

A complex with an unusual structure called basic beryllium acetate $[Be_4O(CH_3COO)_6]$ is formed if $Be(OH)_2$ is evaporated with acetic acid. The structure comprises a central oxygen atom surrounded by four beryllium atoms located at the corners of a tetrahedron, with six acetate groups arranged along the six edges of the tetrahedron (see Figure 11.5a). This structure is one of a series with different organic acids replacing the acetate

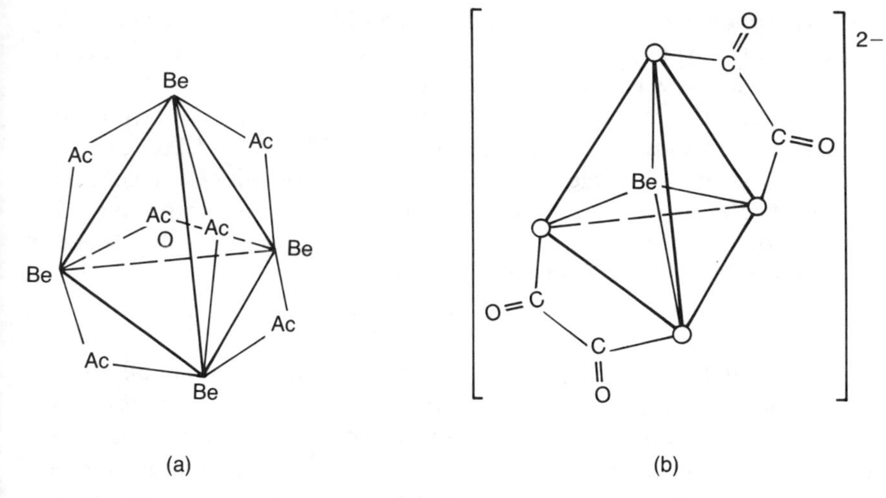

Figure 11.5 (a) Basic beryllium acetate $Be_4O(CH_3COO)_6$. (b) Beryllium oxalate complex $[Be(ox)_2]^{2-}$.

group, and it is the same as the structure of basic beryllium nitrate (Figure 11.1b). Basic beryllium acetate is soluble in organic solvents. It is covalent, and thus has a fairly low melting point (285 °C) and boiling point (330 °C). These are low enough for it to be distilled, which is useful in the purification of beryllium.

Be compounds are said to taste sweet, but do **not** test this for yourself as they are extremely toxic. This is due to their very high solubility and their ability to form complexes with enzymes in the body. Be displaces Mg from some enzymes because it has a stronger complexing ability. Contact with the skin causes dermatitis, and inhaling dust or smoke causes a disease called berylliosis which is rather like silicosis.

Magnesium forms a few halide complexes such as $[NEt_4]_2[MgCl_4]$, but Ca, Sr and Ba do not.

From the viewpoint of life on this planet the most important complex formed by magnesium is chlorophyll. The magnesium is at the centre of a flat heterocyclic organic ring system called a porphyrin, in which four nitrogen atoms are bonded to the magnesium (see Figure 11.6). Chlorophyll is the green pigment in plants which absorbs light in the red region from sunlight, and makes the energy available for photosynthesis. In this process CO_2 is converted into sugars.

$$6CO_2 + 6H_2O \xrightarrow[\text{sunlight}]{\text{chlorophyll}} \underset{\text{glucose}}{C_6H_{12}O_6} + 6O_2$$

Almost all life ultimately depends on chlorophyll and photosynthesis. The oxygen in the atmosphere is a by-product of photosynthesis, and foodstuffs are either parts of plants, or animals which feed on plants. Though photosynthesis is commonly associated with higher plants, about half actually occurs in algae, and certain green, brown, red and purple bacteria. The photosynthetic bacteria are anaerobic and are poisoned by O_2. These bacteria oxidize H_2S to S, or oxidize an organic molecule instead of the usual reaction where H_2O is oxidized to O_2.

Calcium and the rest of the group only form complexes with strong complexing agents. Examples include acetylacetone, $CH_3 \cdot CO \cdot CH_2 \cdot CO \cdot CH_3$ (which has two donor oxygen atoms), and ethylenediaminetetraacetic acid, EDTA, which has four donor oxygen atoms and two donor nitrogen atoms in each molecule. The free acid H_4EDTA is insoluble, and the disodium salt Na_2H_2EDTA is the most used reagent.

$$Ca^{2+} + [H_2EDTA]^{2-} \rightarrow [Ca(EDTA)]^{2-} + 2H^+$$

EDTA will form six-coordinate complexes with most metal ions in solution provided that the pH is suitably adjusted. Since Be is invariably four-coordinate it does not complex appreciably with EDTA. In contrast, calcium and magnesium complex with EDTA, and titrations are performed using EDTA in buffered solutions to estimate the amounts of Ca^{2+} and Mg^{2+} present to determine the 'hardness' of the water. EDTA

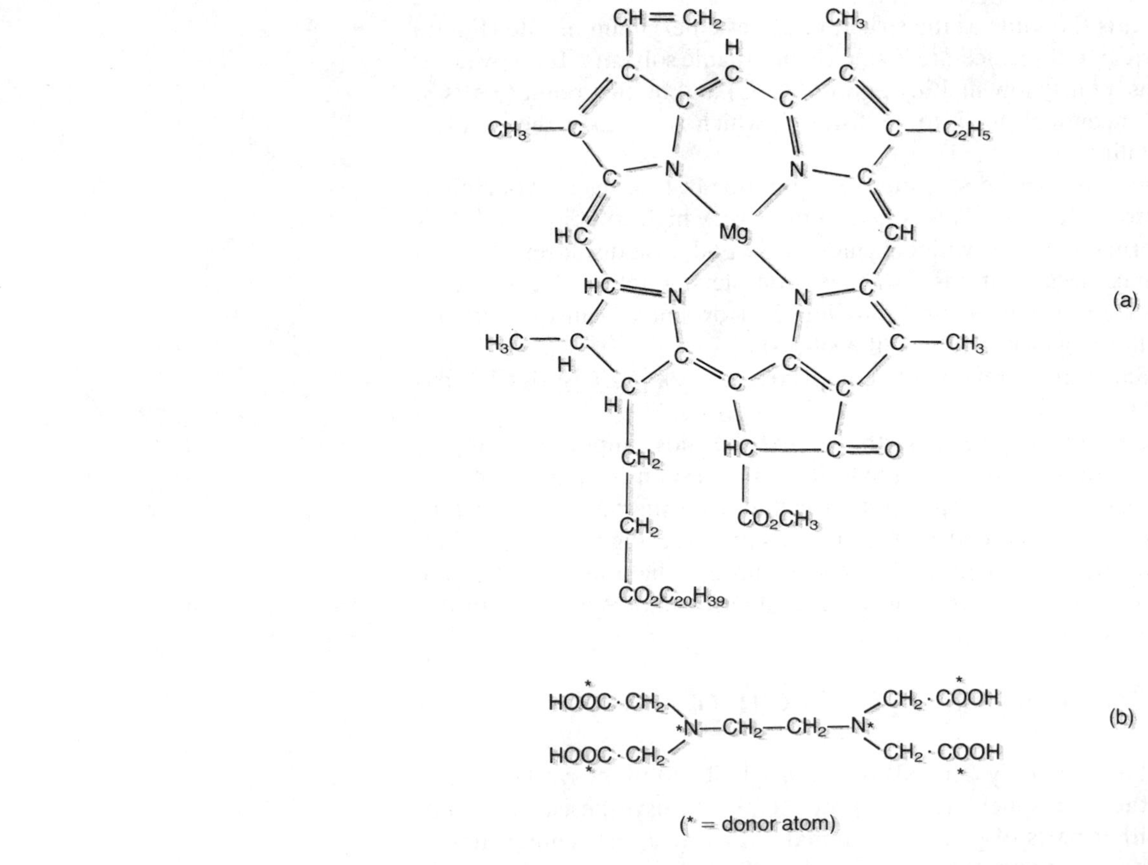

(a)

(b)

(* = donor atom)

Figure 11.6 **(a)** Chlorophyll a. (b) Ethylenediaminetetraacetic acid EDTA.

titrations of Ca^{2+} and Mg^{2+} are carried out at higher pH than those of many other metals (e.g. Zn^{2+}, Cd^{2+} and Pb^{2+}) as their complexes are less stable, and at lower pH values the EDTA is protonated instead of the Ca or Mg complex forming. EDTA is sometimes added to detergents to soften the water. Various polyphosphate ions also form complexes in solution, and this is also important in water softening.

Stable solid complexes with crown ethers and crypts have been isolated. Details of these are given in 'Complexes crowns and crypts' in Chapter 9.

BIOLOGICAL ROLE OF Mg^{2+} AND Ca^{2+}

Mg^{2+} ions are concentrated in animal cells, and Ca^{2+} are concentrated in the body fluids outside the cell, in much the same way that K^+ concentrates inside the cell and Na^+ outside. Mg^{2+} ions form a complex with ATP, and are constituents of phosphohydrolases and phosphotransferases, which are enzymes for reactions involving ATP and energy release. They are also essential for the transmission of impulses along nerve fibres. Mg^{2+} is important in chlorophyll, in the green parts of plants. Ca^{2+} is important in bones and teeth as apatite $Ca_3(PO_4)_2$, and the enamel on teeth as fluoroapatite $[3(Ca_3(PO_4)_2) \cdot CaF_2]$. Ca^{2+} ions are important in blood clotting, and are required to trigger the contraction of muscles and to maintain the regular beating of the heart.

DIFFERENCES BETWEEN BERYLLIUM AND THE OTHER GROUP II ELEMENTS

Beryllium is anomalous in many of its properties and shows a diagonal relationship to aluminium in Group III:

1. Be is very small and has a high charge density so by Fajans' rules it has a strong tendency to covalency. Thus the melting points of its compounds are lower (BeF_2 m.p. 800°C whilst the fluorides of the rest of group melt about 1300°C). The Be halides are all soluble in organic solvents and hydrolyse in water rather like the aluminium halides. The other Group II halides are ionic.
2. Beryllium hydride is electron deficient and polymeric, with multicentre bonding, like aluminium hydride.
3. The halides of beryllium are electron deficient, and are polymeric, with halogen bridges. $BeCl_2$ usually forms chains but also exists as the dimer. $AlCl_3$ is dimeric.
4. Be forms many complexes – not typical of Groups I and II.
5. Be is amphoteric, liberating H_2 with NaOH and forming beryllates. Al forms aluminates.
6. $Be(OH)_2$, like $Al(OH)_3$, is amphoteric.
7. Be, like Al, is rendered passive by nitric acid.
8. The standard electrode potentials for Be and Al, −1.85 volts and −1.66 volts respectively, are much closer than the value for Be is to the values for Ca, Sr and Ba (−2.87, −2.89 and −2.90 volts) respectively.

9. Be salts are extensively hydrolysed.
10. Be salts are among the most soluble known.
11. Beryllium forms an unusual carbide Be_2C, which, like Al_4C_3, yields methane on hydrolysis.

There is plainly a diagonal similarity between beryllium in Group II and aluminium in Group III. Just as was the case with lithium and magnesium, the similarity in atomic and ionic sizes is the main factor underlying this relationship.

FURTHER READING

Bell, N.A. (1972) Beryllium halides and complexes, *Adv. Inorg. Radiochem.*, **14**, 225.

Dietrich, B. (1985) Coordination chemistry of alkali and alkaline earth cations with macrocyclic ligands, *J. Chem. Ed.*, **62**, 954–964. (Crowns and crypts.)

Everest, D.A. (1973) Beryllium, *Comprehensive Inorganic Chemistry*, Vol. I, Pergamon Press, Oxford.

Goodenough, R.D. and Stenger, V.A. (1973) Magnesium, calcium, strontium, barium and radium, *Comprehensive Inorganic Chemistry*, Vol. I, Pergamon Press, Oxford.

Hanusa, T.P. (1987) Re-examining the diagonal relationships, *J. Chem. Ed.*, **64**, 686–687.

Hughes, M.N. (1972) *The Inorganic Chemistry of Biological Processes* (Chapter 8), John Wiley, London. (Group I and II metals in biology.)

Hughes, M.N. and Birch, N.J. (1982) IA and IIA cations in biology, *Chemistry in Britain*, **18**, 196–198.

Parker, D. (1983) Alkali and alkaline earth cryptates, *Adv. Inorg. Radiochem.*, **27**, 1–26.

Sargeson, A.M. (1979) Caged metal ions, *Chemistry in Britain*, **15**, 23–27. (A straightforward account of crown ethers, crypts etc.)

Schubert, J. (1973) Readings from *Scientific American in Chemistry of the Environment* (Chapter 34: Beryllium and berylliosis), W.H. Freeman, San Francisco.

Skilleter, D.N. (1990) To be or not to be – the story of beryllium toxicity, *Chemistry in Britain*, **26**, 26–30.

Spiro, T.G. (ed.) (1983) *Calcium in Biology*, Wiley-Interscience, New York.

Waker, W.E.C. (1980) *Magnesium and Man*, Harvard University Press, London.

PROBLEMS

1. Why are Group II elements smaller than their Group I counterparts?

2. Why are Group II metals harder, and why do they have higher melting points, than Group I metals?

3. What is the reason why compounds of Be are much more covalent than other Group II compounds?

4. What is the structure of $BeCl_2$ in the gaseous state, and as a solid? Why is $BeCl_2$ acidic when dissolved in water?

5. Describe the difference in structure between BeH_2 and CaH_2.

6. Why do the halides and hydrides of Be polymerize?

7. What precautions are necessary when handling beryllium compounds?
8. What are the usual coordination numbers for Be^{2+} and Mg^{2+}? What is the reason for the difference?
9. Compare the extent of hydration of Group I and Group II halides. Why do Be salts seldom contain more than four molecules of water of crystallization?
10. Give equations to show the reactions between Ca and: (a) H_2O, (b) H_2, (c) C, (d) N_2, (e) O_2, (f) Cl_2, (g) NH_3.
11. Compare the reaction with water of Group I and Group II metals. How does the basic strength of Group II hydroxides vary within the group? Is this trend typical of the rest of the periodic table?
12. The hardness of water may be 'temporary' or 'permanent'.
    (a) What causes each of these conditions, and how is each treated?
    (b) Find (from other literature sources) how naturally occurring zeolites, synthetic ion-exchange resins and polyphosphates may be used for softening water.
13. Do the alkaline earth metal ions form many complexes? Are Group II better or worse than Group I at forming complexes? What is the reason for the difference? Which of the metal ions in the group are best at forming complexes? Which are the best complexing agents? Name one complex of a Group II metal which is of biological importance.
14. Outline the preparation, properties, structure and use of basic beryllium acetate.
15. Under what conditions do the Group II metal ions form stable complexes with EDTA? How are the amounts of Ca^{2+} and Mg^{2+} present in water estimated by titration with EDTA? (Consult a practical textbook, e.g. Vogel). Are the EDTA complexes more or less stable than those of most other metal ions? Why is the titration performed at a high pH? What indicator is used?
16. Describe how you would prepare a Grignard reagent from Mg, and list five different uses of the reagent in preparative reactions. (Refer also to the section on silicones.)
17. The four general methods of extracting metals are thermal decomposition, displacement of one element by another, chemical reduction, and electrolytic reduction. How are Group II metals obtained and why are the other methods unsuitable?
18. On treatment with cold water, an element (A) reacted quietly, liberating a colourless, odourless gas (B), and a solution (C). Lithium reacted with (B) yielding a solid product (D) which effervesced with water to give a strongly basic solution (F). When carbon dioxide was

bubbled through solution (C) an initial white precipitate (G) was formed, but this re-dissolved, forming solution (H) when more carbon dioxide was added. Precipitate (G) effervesced when moistened with concentrated hydrochloric acid, and gave a deep red coloration to a Bunsen burner flame. When (G) was heated with carbon at 1000 °C a caustic white compound (I) was formed, which when heated with carbon at 1000 °C gave a solid (J) of some commercial importance. Name the substances (A) to (J) and give balanced chemical equations for each of the reactions.

19. When a white substance (A) was treated with dilute hydrochloric acid, a colourless gas (B) was evolved, which turned moist litmus paper red. On bubbling (B) through lime water a precipitate (C) was formed, but passage of further gas resulted in a clear solution (D). A small sample of (A) was moistened with concentrated hydrochloric acid and placed on a platinum wire, and introduced into a Bunsen burner flame where it caused a green flame coloration. On strong heating (A) decomposed, giving a white solid (E) which turned red litmus paper blue. 1.9735 g of (A) was heated strongly and gave 1.5334 g of (E). The sample of (E) was dissolved in water and made up to 250 ml in a standard flask. 25 ml aliquots were titrated with acid and required 20.30 ml of 0.0985 M hydrochloric acid. Name the compounds (A) to (E) inclusive, and give equations for all the reactions. Calculate the gram molecular weight of (A).

The *p*-Block Elements

Part Three

The group III elements

12

Table 12.1 Electronic structures and oxidation states

Element	Symbol	Electronic configuration	Oxidation states*
Boron	B	[He] $2s^2\ 2p^1$	**III**
Aluminium	Al	[Ne] $3s^2\ 3p^1$	(I) **III**
Gallium	Ga	[Ar] $3d^{10}\ 4s^2\ 4p^1$	I **III**
Indium	In	[Kr] $4d^{10}\ 5s^2\ 5p^1$	I **III**
Thallium	Tl	[Xe] $4f^{14}\ 5d^{10}\ 6s^2\ 6p^1$	**I** III

* The most important oxidation states (generally the most abundant and stable) are shown in bold. Other well-characterized but less important states are shown in normal type. Oxidation states that are unstable, or in doubt, are given in parentheses.

GENERAL PROPERTIES

Boron is a non-metal, and always forms covalent bonds. Normally it forms three covalent bonds at 120° using sp^2 hybrid orbitals. There is no tendency to form univalent compounds. All BX_3 compounds are electron deficient, and may accept an electron pair from another atom, thus forming a coordinate bond. BF_3 is commercially important as a catalyst. Boron also forms a large number of compounds in which the boron atoms form an open basket type of structure, and some which are a closed polyhedron. Other atoms such as carbon may be included in the polyhedron. The bonding in these compounds is of considerable theoretical interest, and involves multi-centre bonds.

The four elements Al, Ga, In and Tl all form trivalent compounds. The heavier members show the 'inert pair effect', and univalent compounds become increasingly important in the order Ga → In → Tl. These four elements (Table 12.1) are more metallic, and more ionic, than B. They are moderately reactive metals. Their compounds are on the borderline between ionic and covalent. Many of their compounds are covalent when anhydrous, but they form ions in solution.

OCCURRENCE AND ABUNDANCE

Boron is a fairly rare element, but it is well known because it occurs as concentrated deposits of borax $Na_2B_4O_7 \cdot 10H_2O$ and kernite $Na_2B_4O_7 \cdot 4H_2O$. The largest deposits are in the Mojave desert (in California), and in Death Valley (in Utah). These are desert regions. Over a long period of time, any rain has leached these alkaline salts from the hills into lakes in the valleys. These lakes have long since dried up leaving solid deposits on the surface from 10 to 50 metres deep. The amount of crude borate minerals mined in 1988 was 2.4 million tonnes. The largest producers are the USA 53%, Turkey 30%, the USSR 8% and Argentina 7%.

Aluminium is the most abundant metal, and the third most abundant element (after oxygen and silicon) by weight in in the earth's crust. It is well known and is commercially important. Aluminium metal is produced on a vast scale. Primary production was 17.6 million tonnes in 1988, and an additional 4 million tonnes is recycled. The most important ore of aluminium is bauxite. This is a generic name for several minerals with formulae varying between $Al_2O_3 \cdot H_2O$ and $Al_2O_3 \cdot 3H_2O$. World production of bauxite was 100 million tonnes in 1988. Aluminium also occurs in large amounts in aluminosilicate rocks such as feldspars, and micas. When these rocks weather they form clay minerals or other metamorphic rocks. There is no easy or economical method of extracting aluminium from feldspars, micas or clays.

Table 12.2 Abundance of the elements in the earth's crust by weight

	ppm	Relative abundance
B	9	38
Al	83 000	3
Ga	19	33 =
In	0.24	63
Tl	0.5	60

Gallium is twice as abundant as boron, but indium and thallium are much less common. All three elements, Ga, In and Tl, occur as sulphides. Ga, In and Tl are not very well known. This is partly because they do not occur as concentrated ores, and partly because there are no major uses for them. Small amounts of Ga are found in the ores of the elements adjacent to it in the periodic table (Al, Zn and Ge). Traces of In and Tl are found in ZnS and PbS ores. Production of In is about 52 tonnes/year and of Ga and Tl about 10 tonnes/year.

EXTRACTION AND USES OF THE ELEMENTS

Extraction of boron

Amorphous boron of low purity (called Moissan boron) is obtained by reducing B_2O_3 with Mg or Na at a high temperature. It is 95–98% pure (being contaminated with metal borides), and is black in colour.

$$\underset{\text{borax}}{Na_2B_4O_7 \cdot 10H_2O} \xrightarrow{\text{acid}} \underset{\text{orthoboric acid}}{H_3BO_3} \xrightarrow{\text{heat}} B_2O_3 \xrightarrow{\text{Mg or Na}} 2B + 3MgO$$

It is difficult to obtain pure crystalline boron, as it has a very high melting point (2180 °C), and the liquid is corrosive. Small amounts of crystalline boron may be obtained:

1. By reducing BCl_3 with H_2. This is done on the kilogram scale.
2. Pyrolysis of BI_3 (Van Arkel method).
3. Thermal decomposition of diborane or other boron hydrides.

$$2BCl_3 + 3H_2 \xrightarrow{\text{red hot W or Ta filament}} 2B + 6HCl$$

$$2BI_3 \xrightarrow[\text{Van Arkel}]{\text{red hot W or Ta filament}} 2B + 3I_2$$

$$B_2H_6 \xrightarrow{\text{heat}} 2B + 3H_2$$

Uses of boron

An important use of boron is to make boron steel or boron carbide control rods for nuclear reactors. Boron has a very high cross-section for capturing neutrons. Control rods made of boron steel or boron carbide may be lowered into a reactor to absorb neutrons and thus slow the reactor down. Boron carbide is also used as an abrasive. Boron is used to make impact resistant steel, as it increases the hardenability (that is the depth to which it will harden) of steel.

Borax $Na_2B_4O_7 \cdot 10H_2O$, orthoboric acid H_3BO_3 and boron sesquioxide B_2O_3 find many uses. World production of borax was roughly 2.4 million tonnes in 1988 (USA 53%, Turkey 30%, USSR 8%, Argentina 7%). Borax is mixed with NaOH and sold as 'Polybor' and 'Timbor' for treating timber and hardboard against attack by wood-boring insects. Other uses are as a food preservative, a mild antiseptic and a flame retardant for fabric and wood. Borax is used as a flux in brazing and in silver soldering. The borax reacts with many metal oxides, forming easily melted borates. The flux removes oxides such as Cu_2O from the surface of hot brass, and thus allows the clean metal surfaces to fuse with the solder. Borax is also used in making enamel, and in leather tanning.

Orthoboric acid is made by hydrolysing many boron compounds. About 170 000 tonnes is made each year by acidifying solutions of borax. Reaction of H_3BO_3 with H_2O_2 gives the mono peroxoboric acid which probably has the structure $[(HO)_3B(OOH)]$. Sodium peroxoborate $Na_2[B_2(O_2)_2(OH)_4] \cdot 6H_2O$ is a constituent of many washing powders. European washing powders may contain 20% of sodium peroxoborate, though it is not used for this

continued overleaf

purpose in the USA. World production is 550 000 tonnes/year. Peroxoborates act as brighteners, because they absorb UV light and emit visible light. This makes both white and coloured fabrics appear extra bright. If used at temperatures over 80 °C, peroxoborates decompose to hydrogen peroxide H_2O_2, which acts as a bleach.

Boron sesquioxide B_2O_3 is used in making borosilicate (heat resistant) glass (e.g. Pyrex which contains 14% B_2O_3). Borosilicate glass has a lower coefficient of thermal expansion and is easier to work than normal 'soda glass'. H_3BO_3, B_2O_3 and calcium borate are used to make soda-free glass fibre, which is used for thermal insulation in houses.

Extraction of aluminium

Aluminium is obtained from the ore bauxite which may be AlO.OH ($Al_2O_3 \cdot H_2O$) or $Al(OH)_3$ ($Al_2O_3 \cdot 3H_2O$). One-hundred million tonnes were mined in 1988. The largest sources are Australia 36%, Guinea 17%, Brazil 8%, Jamaica 7% and the USSR 6%.

The first step is to purify the ore. In the *Bayer process*, waste materials (mainly iron and silicon compounds) are removed because these would spoil the properties of the product. NaOH is added to the ore, and as Al is amphoteric it dissolves, forming sodium aluminate. SiO_2 also dissolves as silicate ions. Any insoluble waste materials, particularly iron oxide, are removed by filtering. Next, aluminium hydroxide is precipitated from the strongly alkaline aluminate solution. This may be done either by bubbling in some CO_2 (an acidic oxide which lowers the pH), or by seeding the solution with Al_2O_3. The silicate ions remain in solution. The $Al(OH)_3$ precipitate is calcined (heated strongly) which converts it to purified Al_2O_3.

Aluminium is usually extracted by the *Hall–Héroult process*. Al_2O_3 is melted with cryolite $Na_3[AlF_6]$, and electrolysed in a graphite lined steel tank, which serves as the cathode. The anodes are also made of graphite. The cell runs continuously, and at intervals molten aluminium (m.p. 660 °C) is drained from the bottom of the cell and more bauxite is added. Some cryolite is mined as a mineral in Greenland, but this is insufficient to meet the demand for it, and most is made synthetically.

$$Al(OH)_3 + 3NaOH + 6HF \rightarrow Na_3[AlF_6] + 6H_2O$$

Cryolite improves the electrical conductivity of the cell as Al_2O_3 is a poor conductor. In addition, the cryolite serves as an added impurity and lowers the melting point of the mixture to about 950 °C. Other impurities such as CaF_2 and AlF_3 may also be added. (A typical electrolyte mixture is 85% $Na_3[AlF_6]$, 5% CaF_2, 5% AlF_3 and 5% Al_2O_3.) Various products are formed at the anodes, including O_2, CO_2, F_2 and carbon compounds of fluorine. These erode the anodes, which must be replaced periodically. The traces of fluorine formed cause serious corrosion. Large amounts of Li_2CO_3 are now used as an alternative impurity, because this reduces corrosion. Energy consumption is very high, and the process is only economic where there is a source of cheap electricity, usually hydroelectric power.

Uses of aluminium

Aluminium metal is moderately soft and weak when pure, but is much stronger when alloyed with other metals. Its main advantage is its lightness (low density $2.73 \, g \, cm^{-3}$). Some alloys are used for special purposes: duralumin which contains about 4% Cu, and several aluminium bronzes (alloys of Cu and Al with other metals such as Ni, Sn and Zn). The metal produced in the largest quantity is iron/steel (776 million tonnes in 1988) but production of aluminium is the second largest. (Total production of Al was 21.6 million tonnes in 1988, made up of 17.6 million tonnes primary production and 4 million tonnes recycled.) The largest producers of Al metal are the USA 22%, the USSR 14%, Canada 9%, Australia 7%, and Brazil and Norway 5% each. There are many uses for aluminium and its alloys:

1. As structural metals in aircraft, ships, cars, and heat exchangers.
2. In buildings (doors, windows, cladding panels and mobile homes).
3. Containers such as cans for drinks, tubes for toothpaste etc. and metal foil.
4. For cooking utensils.
5. To make electric power cables (on a weight for weight basis they conduct twice as well as copper).
6. Finely divided aluminium powder is called 'aluminium bronze', and is used in preparing aluminium paint.

It has been assumed for many years that Al^{3+} is completely harmless and non-toxic to humans. $Al(OH)_3$ is widely used as an anti-acid treatment for indigestion. $Al_2(SO_4)_3$ is used to treat drinking water, and cooking utensils are made of aluminium. However, there are indications that aluminium may not be as harmless as was once thought. Aluminium is acutely toxic to people with kidney failure, who are unable to excrete the element. Patients suffering from Alzheimer's disease (which causes senility) have deposits of aluminium salts in the brain. This element, though toxic, is normally excreted very efficiently. Any abundant element will inevitably be taken up by plants and then by animals, and in general the biosphere either makes use of elements (in which case they are classed as essential) or positively rejects them. Elements are hardly ever toxicologically neutral.

Gallium, indium and thallium

Traces of gallium are found in bauxite, and the ratio of Ga to Al is about 1/5000. During the Bayer process for purifying alumina the concentration of Ga in the alkaline solution gradually increases to about 1/250. Ga is extracted by electrolysis of this solution. Indium and thallium occur in minute quantities in ZnS and PbS ores. These sulphide ores are roasted

continued overleaf

with air in a smelter, to convert them to ZnO and PbO. The Ga and In are recovered from the flue dust, and they are extracted by electrolysis of aqueous solutions of their salts.

The metals are soft, silver coloured and reactive. They dissolve in acids. There are no large scale uses for Ga, In or Tl, but small amounts of Ga are used to dope crystals to make transistors. Semiconductor manufacture requires ultra-high-purity Ga. This is obtained by zone refining. Gallium is also used in other semiconductor devices. Gallium arsenide GaAs is isoelectronic with Ge, and is used in light-emitting diodes (LEDs) and laser diodes. There is a lot or research into using GaAs to make memory chips for computers, since these work 5 to 10 times faster than those made from silicon. Indium is used to dope crystals to make *p–n–p* transistors, and in thermistors (InAs and InSb). It is also used in low melting point solder (commonly used to solder semiconductor chips) and other low melting alloys.

OXIDATION STATES AND TYPE OF BONDS

The (+III) oxidation state

The elements all have three outer electrons. Apart from Tl they normally use these to form three bonds, giving an oxidation state of (+III). Are the bonds ionic or covalent? Covalency is suggested by the following:

1. Fajans' rules – small size of the ions and their high charge of 3+ favours the formation of covalent bonds.
2. The sum of the first three ionization energies is very large, and this also suggests that bonds will be largely covalent.
3. The electronegativity values are higher than for Groups I and II, and when reacting with other elements the difference is not likely to be large.

Boron is considerably smaller than the other elements and thus has a higher ionization energy than the others. The ionization energy is so high that B is always covalent.

Many simple compounds of the remaining elements, such as $AlCl_3$ and $GaCl_3$, *are covalent when anhydrous*. However, Al, Ga, In and Tl *all form metal ions in solution*. The type of bonds formed depends on which is most favourable in terms of energy. This change from covalent to ionic happens because the ions are hydrated, and the amount of hydration energy evolved exceeds the ionization energy. Consider $AlCl_3$: $5137\,kJ\,mol^{-1}$ are required to convert Al to Al^{3+}, $\Delta H_{hydration}$ for Al^{3+} is $-4665\,kJ\,mol^{-1}$ and $\Delta H_{hydration}$ for Cl^- is $-381\,kJ\,mol^{-1}$. Thus the total hydration energy is:

$$-4665 + (3 \times -381) = -5808\,kJ\,mol^{-1}$$

This exceeds the ionization energy, so $AlCl_3$ ionizes in solution.

The hydrated metal ions have six molecules of water which are held very strongly in an octahedral structure $[M(H_2O)_6]^{3+}$. The metal–oxygen bonds are very strong. This weakens the O—H bonds, and protons are released to neighbouring water molecules, forming H_3O^+ (hydrolysis).

$$H_2O + [M(H_2O)_6]^{3+} \rightarrow [M(H_2O)_5(OH)]^{2+} + H_3O^+$$

The (+I) oxidation state – the 'inert pair effect'

In the *s*-block, Group I elements are univalent, and Group II elements are divalent. In Group III we would expect the elements to be trivalent. In most of their compounds this is the case, but some of the elements show lower valency states as well. There is an increasing tendency to form univalent compounds on descending the group. Compounds with Ga(I), In(I) and Tl(I) are known. With Ga and In the (+I) oxidation state is less stable than the (+III) state. However, the stability of the lower oxidation state increases on descending the group. Tl(I) thallous compounds are more stable than Tl(III) thallic compounds.

How and why does monovalency occur? The atoms in this group have an outer electronic configuration of s^2p^1. Monovalency is explained by the *s* electrons in the outer shell remaining paired, and not participating in bonding. This is called the 'inert pair effect'. If the energy required to unpair them exceeds the energy evolved when they form bonds, then the *s* electrons will remain paired. The strength of the bonds in MX_3 compounds decreases down the group. The mean bond energy for chlorides are $GaCl_3$ = 242, $InCl_3$ = 206 and $TlCl_3$ = 153 kJ mol^{-1}. Thus the *s* electrons are most likely to be inert in thallium.

The inert pair effect is not the explanation of why monovalency occurs in Group III. It merely describes what happens, i.e two electrons do not participate in bonding. The reason that they do not take part in bonding is energy. The univalent ions are much larger than the trivalent ions, and (+I) compounds are ionic, and are similar in many ways to Group I elements.

The inert pair effect is not restricted to Group III, but also occurs among the heavier elements in other groups in the *p*-block. Examples from Group IV are Sn^{2+} and Pb^{2+}, and examples from Group V are Sb^{3+} and Bi^{3+}. The lower oxidation state becomes more stable on descending the group. Thus Sn^{2+} is a reducing agent but Pb^{2+} is stable and Sb^{3+} is a reducing agent but Bi^{3+} is stable. When the *s* electrons remain paired, the oxidation state is always two lower than the usual oxidation state for the group.

Thus in the *s*-block, Groups I and II show only the group valency. Groups in the *p*-block show variable valency, differing in steps of two. Variable valency also occurs with elements in the *d*-block. This arises from using different numbers of *d* electrons for bonding, so in this case the valency can change in steps of one (e.g. Cu^+ and Cu^{2+}, Fe^{2+} and Fe^{3+}).

The (+II) oxidation state

Gallium is *apparently* divalent in a few compounds, such as $CaCl_2$. However, Ga is not really divalent, as the structure of $GaCl_2$ has been shown to be $Ga^+[GaCl_4]^-$ which contains Ga(I) and Ga(III).

MELTING POINTS, BOILING POINTS, AND STRUCTURES

The melting points of the Group III elements do not show a regular trend as did the metals of Groups I and II. The Group III values are not strictly comparable because B and Ga have unusual crystal structures.

Boron has an unusual crystal structure which results in the melting point being very high. There are at least four different allotropic forms. Boron has insufficient electrons to fill the valence shell even after forming bonds. The variety and complexity of the allotropic forms illustrates the number of ways in which boron attempts to solve this problem. Other elements solve this problem by metallic bonding, but small size and high ionization energy make this impossible for boron. All four allotropic forms contain icosahedral units with boron atoms at all 12 corners. (Note that an icosahedron is a regular shape with 12 corners and 20 faces.) In these units twelve B atoms form a regular shape, and each B atom is bonded to five equivalent neighbours (at a distance of 1.77 Å). The difference between the allotropic forms arises in the way the icosahedra are bonded together. The simplest form is α-rhombohedral boron. In this, half the atoms are bonded to one atom in another icosahedron (at a distance of 1.71 Å), and half the atoms are bonded to atoms in two different icosahedra (at a distance of 2.03 Å). Plainly this is neither a regular structure nor a metallic structure. Only 37% of space is occupied by the atoms, compared with 74% for a close-packed arrangement. This shows that icosahedra fill up space ineffectively. The other allotropes have even more complicated structures.

The elements Al, In and Tl all have close-packed metal structures. Gallium has an unusual structure. Each metal atom has one close neighbour at a distance of 2.43 Å, and six more distant neighbours at distances between 2.70 Å and 2.79 Å. This remarkable structure tends towards discrete diatomic molecules rather than a metallic structure. This accounts for the incredibly low melting point of gallium of 30 °C. Ga is also unusual because the liquid expands when it forms the solid, i.e. the solid is

Table 12.3 Melting and boiling points

	Melting point (°C)	Boiling point (°C)
B	2180	3650
Al	660	2467
Ga	30	2403
In	157	2080
Tl	303	1457

less dense than the liquid. This property is unique to Ga, Ge and Bi.

Though the melting points decrease from Al to In as expected on descending a group, it increases again for Tl. The boiling point of B is unusually high, but the values for Ga, In and Tl decrease on descending the group as expected. Note that the boiling point for Ga is in line with the others, whereas its melting point is not. The very low melting point is due to the unusual crystal structure, but the structure no longer exists in the liquid.

SIZE OF ATOMS AND IONS

The metallic radii of the atoms do not increase regularly on descending the group (Table 12.4). However, the values are not strictly comparable. Boron is not a metal, and the radius given is half the closest approach in the structure. Ga has an unusual structure, and the value given is half the closest approach. The others have close-packed metal structures.

The ionic radii for M^{3+} increase down the group, though not in the regular way observed in Groups I and II. There are two reasons for this:

1. There is no evidence for the existence of B^{3+} under normal conditions, and the value is an estimate.
2. The electronic structures of the elements are different.Ga, In and Tl follow immediately after a row of ten transition elements. They therefore have ten *d* electrons, which are less efficient at shielding the nuclear charge than the *s* and *p* electrons. (Shielding is in the order $s > p > d > f$.) Poor shielding of the nuclear charge results in the outer electrons being more firmly held by the nucleus. Thus atoms with a d^{10} inner shell are smaller and so have higher ionization energies than would otherwise be expected. This contraction in size is sometimes called the *d*-block contraction. In a similar way Tl follows immediately after 14 *f*-block elements. The size and ionization energy of Tl are affected even more by the presence of 14 *f* electrons, which shield the nuclear charge even less effectively. The contraction in size from these *f*-block elements is called the *lanthanide contraction*.

Table 12.4 Ionic and covalent radii, and electronegativity values

	Metallic radius (Å)	Ionic radius M^{3+} (Å)	Ionic radius M^{+} (Å)	Pauling's electronegativity
B	(0.885)	(0.27)	–	2.0
Al	1.43	0.535	–	1.5
Ga	(1.225)	0.620	1.20	1.6
In	1.67	0.800	1.40	1.7
Tl	1.70	0.885	1.50	1.8

For values in brackets see text.

Table 12.5 Standard reduction potentials (volts)

Oxidation states	
Acid solution +III +II +I 0	Basic solution +III +II +I 0
Al^{3+} —(−1.66)— Al	$Al(OH)_3$ —(−2.31)— Al
Ga^{3+} —(−0.44)— Ga^+ —(−0.79)— Ga Ga^{3+} → Ga: −0.56	$H_2GaO_3^-$ —(−1.22)— Ga
In^{3+} —(−0.44)— In^+ —(−0.18)— In In^{3+} → In: −0.34	$In(OH)_3$ —(−1.0)— In
Tl^{3+} —(+2.06)— Tl^+ —(−0.34)— Tl Tl^{3+} → Tl: +1.26	$Tl(OH)_3$ —(−0.05)— $Tl(OH)$

The large difference in size between B and Al results in many differences in properties. Thus B is a non-metal, has a very high melting point, always forms covalent bonds, and forms an acidic oxide. In contrast, Al is a metal, has a much lower melting point, and its oxide is amphoteric. (It is safe to generalize in this way, but unsafe to argue quantitatively that Al^{3+} is twice the size of B^{3+} or that the metallic radii differ by a factor of 1.6, as B^{3+} does not exist, and B is not a metal.)

ELECTROPOSITIVE CHARACTER

The electropositive or metallic nature of the elements increases from B to Al, but then decreases from Al to Tl. This is shown by the standard electrode potentials for the reaction:

$$M^{3+} + 3e \rightarrow M$$

The increase in metallic character from B to Al is the usual trend on descending a group associated with increasing size. However, Ga, In and Tl do not continue the trend. The elements are less likely to lose electrons (and are thus less electropositive), because of the poor shielding by *d* electrons described previously.

Table 12.6 Standard electrode potentials $E°$

	$M^{3+} \mid M$ (volts)	$M^{+} \mid M$ (volts)
B	(−0.87*)	
Al	−1.66	+0.55
Ga	−0.56	−0.79†
In	−0.34	−0.18
Tl	+1.26	−0.34

* For $H_3BO_3 + 3H^+ + 3e^- \rightarrow B + 3H_2O$)
† Value in acidic solution.

The standard electrode potentials $E°$ for $M^{3+}|M$ become less negative from Al to Ga to In and the potential becomes positive for Tl. Since $\Delta G = -nFE°$ it follows that ΔG, the free energy of formation of the metal, e.g. $Al^{3+} + 3e \rightarrow Al$, is positive. Thus it is difficult to make this reaction work. (The reverse reaction $Al \rightarrow Al^{3+} + 3e$ occurs spontaneously.) The standard potential becomes less negative on descending the group so it becomes less difficult for the reaction $M^{3+} \rightarrow M$ to occur. Thus the (+III) oxidation state becomes less stable in aqueous solution on descending the group. In a similar way, the $E°$ values for $M^+|M$ show that the (+I) state increases in stability. With Tl, the (+I) state is more stable than the (+III) state.

It should be remembered that in this type of argument $E°$ and ΔG refer to the reaction with H_2:

$$Al^{3+} + \tfrac{3}{2}H_2 \rightarrow Al + 3H^+$$

IONIZATION ENERGY

The ionization energies increase as expected (first ionization energy < second ionization energy < third ionization energy). The sum of the first three ionization energies for each of the elements is very high. Thus boron

Table 12.7 Ionization energies

	Ionization energy ($kJ\,mol^{-1}$)			
	1st	2nd	3rd	Sum of three
B	801	2427	3659	6887
Al	577	1816	2744	5137
Ga	579	1979	2962	5520
In	558	1820	2704	5082
Tl	589	1971	2877	5437

has no tendency to form ions, and always forms covalent bonds. The other elements normally form covalent compounds except in solution.

The ionization energy values do not decrease smoothly down the group. The decrease from B to Al is the usual trend on descending a group associated with increased size. The poor shielding by *d* electrons and the resulting *d-block contraction* affect the values for the later elements.

REACTIONS OF BORON

Pure crystalline boron is very unreactive. However, it is attacked at high temperatures by strong oxidizing agents such as a mixture of hot concentrated H_2SO_4 and HNO_3, or by sodium peroxide. In contrast, finely divided amorphous boron (which contains between 2% and 5% of impurities) is more reactive. It burns in air or oxygen, forming the oxide. It also burns at white heat in nitrogen, forming the nitride BN. This is a slippery white solid with a layer structure similar to graphite. Boron also burns in the halogens, forming trihalides. It reacts directly with many elements, forming borides, which are hard and refractory. It reduces strong HNO_3 and H_2SO_4 slowly, and also liberates H_2 from fused NaOH.

Table 12.8 Some reactions of amorphous boron

Reaction	Comment
$4B + 3O_2 \rightarrow 2B_2O_3$	At high temperature
$2B + 3S \rightarrow B_2S_3$	At 1200 °C
$2B + N_2 \rightarrow 2BN$	At very high temperature
$2B + 3F_2 \rightarrow 2BF_3$ $2B + 3Cl_2 \rightarrow 2BCl_3$ $2B + 3Br_2 \rightarrow 2BBr_3$ $2B + 3I_2 \rightarrow 2BI_3$	At high temperature
$2B + 6NaOH \rightarrow 2Na_3BO_3 + 3H_2$	When fused with alkali
$2B + 2NH_3 \rightarrow 2BN + 3H_2$	At very high temperature
$B + M \rightarrow M_xB_y$	Many metals form borides (not group I) often nonstoichiometric

REACTIONS OF THE OTHER ELEMENTS

Reaction with water and air

The metals Al, Ga, In and Tl are silvery white. Thermodynamically Al should react with water and with air, but in fact it is stable in both. The

Table 12.9 Some reactions of the other Group III metals

Reaction	Comment
$4M + 3O_2 \rightarrow M_2O_3$	All react at high temperature Al very strongly exothermic Ga only superficially oxidized Tl forms some Tl_2O as well
$2Al + N_2 \rightarrow 2AlN$	Only Al at high temperature
$2M + 3F_2 \rightarrow 2MF_3$ $2M + 3Cl_2 \rightarrow 2MCl_3$ $2M + 3Br_2 \rightarrow 2MBr_3$	All the metals form trihalides } Tl^+ also formed
$2M + 3I_2 \rightarrow 2MI_3$	Al, Ga, In only
$TlI + I_2 \rightarrow Tl^+[I_3]^-$	thallium(I) triiodide formed
$2M + 6HCl \rightarrow 2MCl_3 + 3H_2$	All react with dilute mineral acids, Al passivated by HNO_3 particularly when concentrated
$Al + NaOH + H_2O \rightarrow NaAlO_2 + H_2$ $Na_3AlO_3 + H_2$	Al and Ga only
$M + NH_3 \rightarrow MNH_2$	All the metals form amides

reason is that a very thin oxide film forms on the surface and protects the metal from further attack. This layer is only 10^{-4} to 10^{-6} mm thick. If the protective oxide covering is removed, for example by amalgamating with mercury, then the metal readily decomposes cold water, forming Al_2O_3 and liberating hydrogen.

Aluminium articles are often 'anodized' to give a decorative finish. This is done by electrolysing dilute H_2SO_4 with the aluminium as the anode. This produces a much thicker layer of oxide on the surface (10^{-2} mm). This layer can take up pigments, thus colouring the aluminium.

Aluminium burns in nitrogen at high temperatures, forming AlN. The other elements do not react.

Reaction with acids and alkalis

Aluminium dissolves in dilute mineral acids liberating hydrogen.

$$2Al + 6HCl \rightarrow 2Al^{3+} + 6Cl^- + 3H_2$$

However, concentrated HNO_3 renders the metal passive because it is an oxidizing agent, and produces a protective layer of oxide on the surface. Aluminium also dissolves in aqueous NaOH (and is therefore amphoteric), liberating hydrogen and forming aluminates. (The nature of aluminates is discussed later.)

$$2Al + 2NaOH + 6H_2O \rightarrow \underset{\text{sodium aluminate}}{NaAl(OH)_4 \text{ or } NaAlO_2 \cdot 2H_2O} + 3H_2$$

Reaction with oxygen

Aluminium burns readily in air or oxygen, and the reaction is strongly exothermic. This is known as the *Thermite reaction*.

$$2Al_{(s)} + \tfrac{3}{2}O_{2(g)} \rightarrow Al_2O_{3(s)} + \text{energy} \qquad \Delta H^\circ = -1670\,\text{kJ}$$

The *Thermite reaction* evolves so much energy that it can be dangerous. The aluminium becomes white hot, and often causes fires. For this precise reason mixtures of Al and an oxide such as Fe_2O_3 or SiO_2 (to provide the oxygen) were used to make incendiary bombs during World War II. Warships are sometimes made of aluminium alloys to reduce their weight. A thermite reaction can be started if the ship is hit by a missile. Such fires on ships caused considerable casualties in the Falklands Islands conflict. The very strong affinity of Al for oxygen is used in the metallurgical extraction of other metals from their oxides.

$$8Al + 3Mn_3O_4 \rightarrow 4Al_2O_3 + 9Mn$$
$$2Al + Cr_2O_3 \rightarrow Al_2O_3 + 2Cr$$

Reaction with the halogens, and sulphate

Aluminium reacts with the halogens quite readily, even when cold, forming trihalides.

Aluminium sulphate is used in large amounts (3.4 million tonnes in 1985). It is made by treating bauxite with H_2SO_4. It is used as a coagulant and precipitant in treating both drinking water and sewage. It is also used in the paper industry, and as a mordant in dyeing.

Alums

Aluminium ions may crystallize from aqueous solutions, forming double salts. These are called *aluminium alums* and have the general formula $[M^I(H_2O)_6][Al(H_2O)_6](SO_4)_2$. M^I is a singly charged cation such as Na^+, K^+ or NH_4^+. The crystals are usually large octahedra, and are extremely pure. Purity is especially important in some applications. Potash alum $[K(H_2O)_6][Al(H_2O)_6](SO_4)_2$ is used as a mordant in dyeing. In this application Al^{3+} is precipitated as $Al(OH)_3$ on cloth to help the dyes bind to the cloth as aluminium complexes. It is essential that Fe^{3+} is absent in order to obtain the 'true' bright colours. Double salts break up in solution, into their constituent ions. Crystals are made up of $[M(H_2O)_6]^+$, $[Al(H_2O)_6]^{3+}$ and two SO_4^- ions. The ions are simply the right size and charge to crystallize together. Some M^{3+} ions other than Al^{3+} also form *alums* of formula $[M^I(H_2O)_6][M^{III}(H_2O)_6](SO_4)_2$. The most common trivalent ions are Fe^{3+} and Cr^{3+}, but others include Ti^{3+}, V^{3+}, Mn^{3+}, Co^{3+}, In^{3+}, Rh^{3+}, Ir^{3+} and Ga^{3+}.

Cement

Aluminium compounds, particularly tricalcium aluminate $Ca_3Al_2O_6$, are very important as constituents of Portland and high alumina cements. The formula of tricalcium aluminate is better written $Ca_9[Al_6O_{18}]$, because it contains 12-membered rings of Si—O—Si—O made by joining six AlO_4 tetrahedra. Portland cement is made by heating the correct mixture of limestone ($CaCO_3$) with sand (SiO_2) and clay (aluminosilicate) at a temperature of 1450–1600 °C in a rotary kiln. When mixed with sand and water Portland cement sets to give concrete, a hard whitish insoluble solid, similar in appearance to Portland stone. (Portland stone is limestone quarried on Portland Bill in Dorset, England.) Between 2% and 5% of gypsum $CaSO_4 \cdot 2H_2O$ is added to slow down the setting process, as slow setting greatly increases the strength. The composition of cement is usually given in terms of the oxides. A typical composition for Portland cement is CaO 70%, SiO_2 20%, Al_2O_3 5%, Fe_2O_3 3%, $CaSO_3 \cdot 2H_2O$ 2%. Total world production of cement was 1172 million tonnes in 1988, about 70% of which was Portland cement.

High alumina cement is made by fusing limestone and bauxite with small amounts of SiO_2 and TiO_2 at 1400–1500 °C in either an open hearth furnace or a rotary kiln. High alumina cement is more expensive than is Portland cement, but has one major advantage over Portland cement – it sets much quicker and develops high strength within one day. It is used to make beams for bridges and buildings. High alumina cement has good resistance to sea water and dilute mineral acids. It withstands temperatures up to 1500 °C and so may be used with refractory bricks in furnaces. A typical analysis of high alumina cement is CaO 40%, Al_2O_3 40%, SiO_2 10%, Fe_2O_3 10%. There has been much publicity over structural failures of beams made of high alumina cement. Failure is due to prolonged exposure to hot wet conditions, or using too much water when mixing the sand and cement. This latter results in it setting too quickly and thus not having time to crystallize properly.

Reactions of Ga, In and Tl

Gallium and indium are stable in air and are not attacked by water unless free oxygen is present. Thallium is a little more reactive and is superficially oxidized by air. All three metals dissolve in dilute acids, liberating hydrogen. Gallium is amphoteric like aluminium, and it dissolves in aqueous NaOH, liberating H_2 and forming gallates. The oxides and hydroxides of Al and Ga are also amphoteric. In contrast, the oxides and hydroxides of In and Tl are purely basic.

All three metals react with the halogens on gentle warming.

SOME PROPERTIES OF THALLIUM(I)

Thallium(I) or thallous compounds are well known. They are typically colourless. They are also extremely poisonous. When ingested, traces turn the hair very black, but larger doses cause loss of hair and death. They are toxic because they upset the enzyme systems in the body.

In aqueous solution the Tl^+ ion is much more stable than Tl(III). The ionic radius of Tl^+ (1.50 Å) is between that of K^+ (1.38 Å) and Rb^+(1.52 Å). For this reason Tl^+ resembles Group I ions in a number of ways. TlOH and Tl_2O are both soluble in water, and are strongly basic. They absorb CO_2 from the air, forming Tl_2CO_3. The solubility of most of the salts is slightly lower than for potassium salts. Tl^+ can replace K^+ in some enzymes, and can thus be used as a biological tracer in the body. There are some differences. TlOH is yellow, and on heating to 100 °C it turns into black Tl_2O. The coordination number of Tl^+ is usually 6 or 8, compared with 6 for Group I ions. TlF is soluble in water, but the other halides are almost insoluble. There is also some similarity with Ag^+, as TlCl is sensitive to light. It darkens when exposed to light rather like AgCl, but TlCl is not soluble in NH_4OH whilst AgCl is soluble.

COMPOUNDS OF BORON AND OXYGEN

Boron sesquioxide and the borates

These are the most important compounds of boron. Sesqui means one and a half, so the oxide should have a formula $MO_{1\frac{1}{2}}$, or M_2O_3. All the elements in the group form sesquioxides when heated in oxygen. B_2O_3 is made more conveniently by dehydrating boric acid:

$$\underset{\text{orthoboric acid}}{H_3BO_3} \xrightarrow{100\,°C} \underset{\text{metaboric acid}}{HBO_2} \xrightarrow{\text{red heat}} \underset{\text{boron sesquioxide}}{B_2O_3}$$

B_2O_3 is a typical non-metallic oxide and is acidic in its properties. It is the anhydride of orthoboric acid, and it reacts with basic (metallic) oxides, forming salts called borates or metaborates. In the borax bead test, B_2O_3 or borax $Na_2B_4O_7 \cdot 10H_2O$ is heated in a Bunsen burner flame with metal oxides on a loop of platinum wire. The mixture fuses to give a glass-like metaborate bead. Metaborate beads of many transition metals have characteristic colours, and so this reaction provides a means of identifying the metal. This simple test provided the first proof that vitamin B_{12} contained cobalt.

$$CoO + B_2O_3 \longrightarrow \underset{\text{cobalt metaborate (blue colour)}}{Co(BO_2)_2}$$

However, it is possible to force B_2O_3 to behave as a basic oxide by reacting with very strongly acidic compounds. Thus with P_2O_5 or As_2O_5, boron phosphate or boron arsenate are formed.

$$B_2O_3 + P_2O_5 \rightarrow 2BPO_4$$

Orthoboric acid H_3BO_3 is soluble in water, and behaves as a weak monobasic acid. It does not donate protons like most acids, but rather it accepts OH^-. It is therefore a Lewis acid, and is better written as $B(OH)_3$.

$$\underset{[H_3BO_3]}{B(OH)_3} + 2H_2O \rightleftharpoons H_3O^+ + [B(OH)_4]^- \qquad pK = 9.25$$

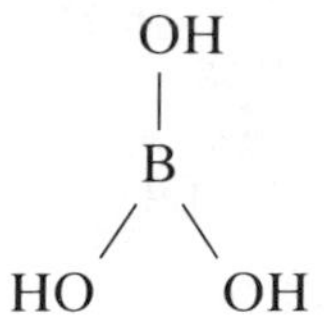

plane triangle

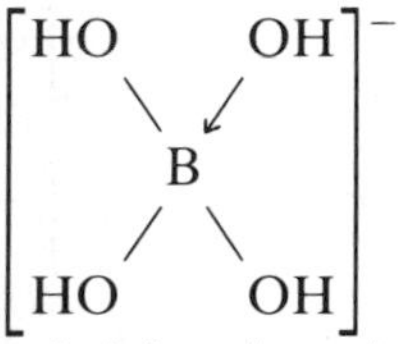

tetrahedral metaborate ion

Polymeric metaborate species are formed at higher concentrations, for example:

$$\underset{[3H_3BO_3]}{3B(OH)_3} \rightleftharpoons H_3O^+ + [B_3O_3(OH)_4]^- + H_2O \qquad pK = 6.84$$

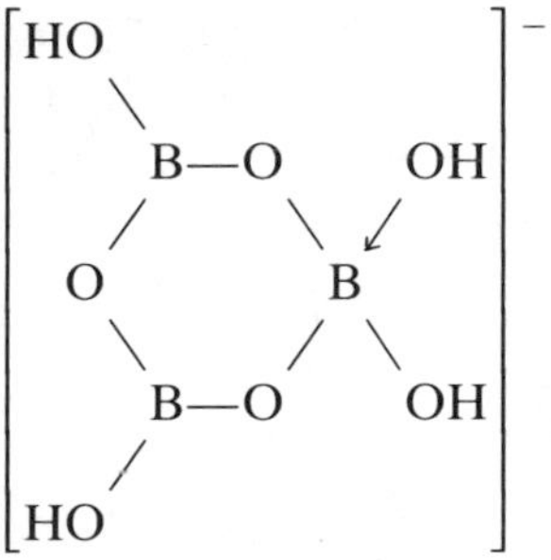

Acidic properties of H_3BO_3 or $B(OH)_3$

Since $B(OH)_3$ only partially reacts with water to form H_3O^+ and $[B(OH)_4]^-$, it behaves as a weak acid. Thus H_3BO_3 or $(B(OH)_3)$ cannot be titrated satisfactorily with NaOH, as a sharp end point is not obtained. If certain organic polyhydroxy compounds such as glycerol, mannitol or sugars are added to the titration mixture, then $B(OH)_3$ behaves as a strong monobasic acid. It can now be titrated with NaOH, and the end point is detected using phenolphthalein as indicator (indicator changes at pH 8.3–10.0).

$$B(OH)_3 + NaOH \rightleftharpoons Na[B(OH)_4]$$
$$\underset{\text{sodium metaborate}}{NaBO_2} + 2H_2O$$

The added compound must be a *cis*-diol, to enhance the acidic properties in this way. (This means that it has OH groups on adjacent carbon atoms in the *cis* configuration.) The *cis*-diol forms very stable complexes with the

$[B(OH)_4]^-$ formed by the forward reaction above, thus effectively removing it from solution. The reaction is reversible. Thus removal of one of the products at the right hand side of the equation upsets the balance, and the reaction proceeds completely to the right. Thus all the $B(OH)_3$ reacts with NaOH: in effect it acts as a strong acid in the presence of the *cis*-diol.

```
   |                                              [  |                ]-       |
—C—OH     [HO      OH]-                           [—C—O      OH   ]    HO—C—
   |      [   \   ↙  ]      −2H2O                 [  |   \  ↙     ]        |
   |    + [     B    ]   ——————→                  [  |    B       ]  +     |
   |      [   /   \  ]                            [  |   /  \     ]        |
—C—OH     [HO      OH]                            [—C—O      OH   ]    OH—C—
   |                                              [  |            ]        |

        [  |           |  ]-
        [—C—O      O—C—   ]
        [  |  \   ↙    |  ]
−2H2O   [  |    B      |  ]
——————→ [  |  /   \    |  ]
        [—C—O      O—C—   ]
        [  |           |  ]
```

Structures of borates

In the simple borates, each B atom is bonded to three oxygen atoms, arranged at the corners of an equilateral triangle. This would be predicted from the orbitals used for bonding.

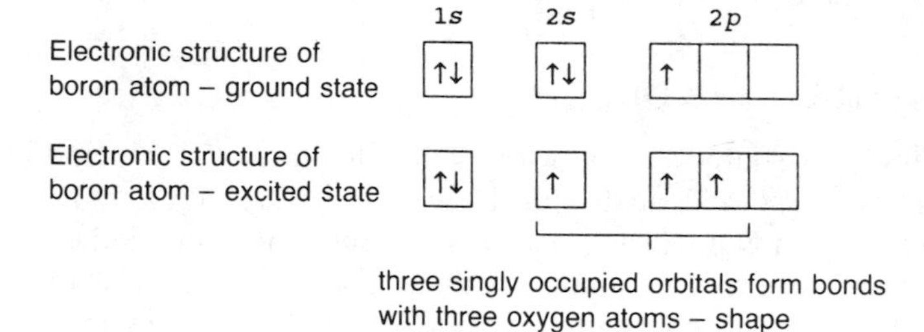

Thus orthoboric acid contains triangular BO_3^{3-} units. In the solid the $B(OH)_3$ units are hydrogen bonded together into two-dimensional sheets with almost hexagonal symmetry. (See Figure 12.1.) The layers are quite a large distance apart (3.18 Å), and thus the crystal breaks quite easily into very fine particles. At one time orthoboric acid was used as a mildly antiseptic talcum powder for babies, because it forms a fine powder. It is no longer used since it sometimes caused a rash.

The orthoborates contain discrete BO_3^{3-} ions, and examples include

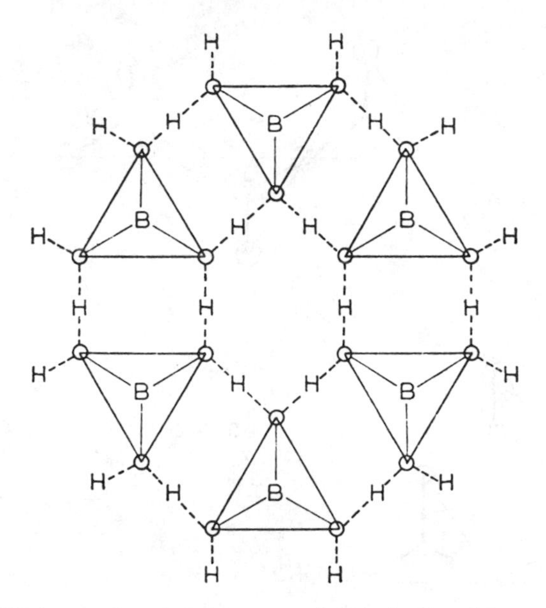

Figure 12.1 Hydrogen bonded structure of orthoboric acid.

$Mg_3(BO_3)_2$ and the lanthanide orthoborates $Ln^{III}BO_3$. In the metaborates simple units (BO_3 planar triangular units or BO_4 tetrahedra) join together to form a variety of polymeric chain and ring structures (see Figure 12.2).

Thus two triangular units join by sharing one corner in $Mg_2[B_2O_5]$ and $Co^{II}[B_2O_5]$. These are called pyroborates by analogy with pyrophosphates. Three triangular units share corners and form a ring in sodium and potassium metaborates $NaBO_2$ and KBO_2 (Figure 12.2b) which are better written $Na_3[B_3O_6]$ and $K_3[B_3O_6]$. Many triangular units may polymerize into an infinite chain, e.g. as calcium metaborate $[Ca(BO_2)_2]_n$ (Figure 12.2a).

In a similar way discrete tetrahedral units are found in $Na_2[B(OH)_4]Cl$ and $Ta^{V}BO_4$. Two tetrahedra may join by sharing one corner, as in $Mg[(HO)_3B \cdot O \cdot B(OH)_3]$. Other structures form rings, chains, sheets and three-dimensional polymers.

Some interesting structures including both triangular and tetrahedral units are formed when polymerization occurs. The spiro compound $K[B_5O_6(OH)_4]$ (Figure 12.2c) contains one tetrahedral unit and four triangular units. Borax is usually written as $Na_2B_4O_7 \cdot 10H_2O$ but is actually made from two tetrahedra and two triangular units joined as shown (Figure 12.2d) and should be written $Na_2[B_4O_5(OH)_4] \cdot 8H_2O$.

Borax

The most common metaborate is borax $Na_2B_4O_7 \cdot 10H_2O$. It is a useful primary standard for titrating against acids.

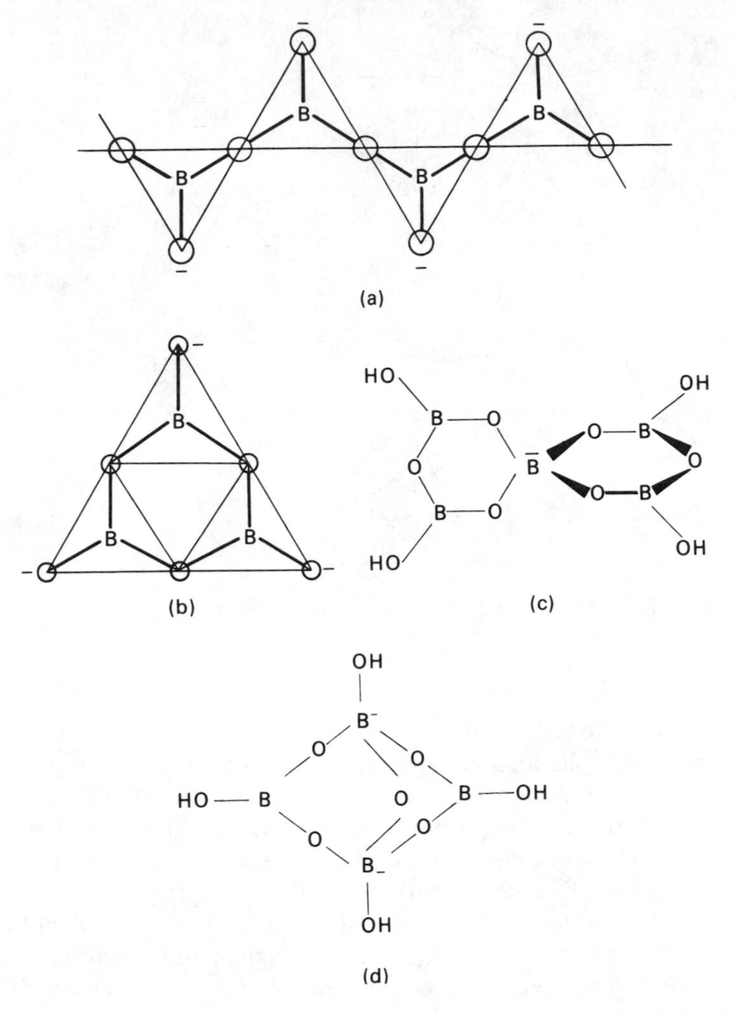

Figure 12.2 Structures of some borates. (a) Metaborate chain $[Ca(BO_2)_2]_n$ made up of triangular BO_3 units. (b) Metaborate ring $K_3[B_3O_6]$ made up of triangular BO_3 units. (c) Complex metaborate $K[B_5O_6(OH)_4]$ called the spiro anion is made up of four triangular BO_3 units and one tetrahedral BO_4 unit. (d) Borax ($Na_2B_4O_7 \cdot 10H_2O$) is made up of two triangular and two tetrahedral units. This ion is $[B_4O_5(OH_4]^{2-}$, and the other water molecules are associated with the metal ions.

$$Na_2B_4O_7 \cdot 10H_2O + 2HCl \rightarrow 2NaCl + 4H_3BO_3 + 5H_2O$$

One of the products H_3BO_3 is itself a weak acid. Thus the indicator used to detect the end point of this reaction must be one that is unaffected by H_3BO_3. Methyl orange is normally used, which changes in the pH range 3.1–4.4.

One mole of borax reacts with two moles of acid. This is because when borax is dissolved in water both $B(OH)_3$ and $[B(OH)_4]^-$ are formed, but only the $[B(OH)_4]^-$ reacts with HCl.

$$[B_4O_5(OH)_4]^{2-} + 5H_2O \rightleftharpoons 2B(OH)_3 + 2[B(OH)_4]^-$$
$$2[B(OH)_4]^- + 2H_3O^+ \rightarrow 2B(OH)_3 + 4H_2O$$

The last reaction will titrate at pH 9.2, so the indicator must have $pK_a <$ ca. 8. Borax is also used as a buffer since its aqueous solution contains equal amounts of weak acid and its salt.

Sodium peroxoborate

Large amounts of sodium peroxoborate are produced, and world production is about 550 000 tonnes/year. There are two main preparative methods:

1. Electrolysis of a solution of sodium borate (containing some Na_2CO_3).
2. By oxidizing boric acid or sodium metaborate with hydrogen peroxide.

$$\underset{\text{sodium metaborate}}{2NaBO_2} + 2H_2O_2 + 6H_2O \rightarrow \underset{\text{sodium peroxoborate}}{Na_2[(OH)_2B(O{-}O)_2B(OH)_2] \cdot 6H_2O}$$

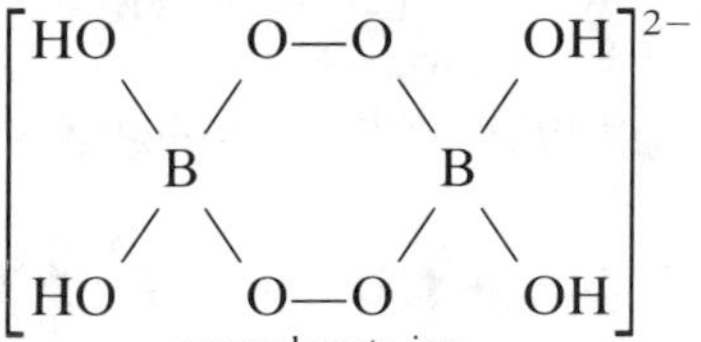

$$\left[\begin{array}{ccccc} HO & & O{-}O & & OH \\ & \diagdown\ \diagup & & \diagdown\ \diagup & \\ & B & & B & \\ & \diagup\ \diagdown & & \diagup\ \diagdown & \\ HO & & O{-}O & & OH \end{array}\right]^{2-}$$

peroxoborate ion

Sodium peroxoborate is used as a brightener in washing powders. It is compatible with enzymes which are added to some 'biological' powders. In very hot water (over 80 °C) the peroxide linkages O—O break down to give H_2O_2.

Isopolyacids of B, Si and P

Other elements form polymeric compounds similar to the borates; notably Si forms silicates and P forms phosphates. These polymeric compounds are called *isopolyacids*. (The name *isopolyacid* means that acidic ions are polymerized together. Iso means 'the same', and indicates that only one type of ion is involved. If two different types of ion polymerize together, for example phosphate and molybdate ions, the phosphomolybdate polymer is called a *heteropolyacid*.)

The principles underlying the structures of borates have been set out by Christ and Clark (see Further Reading). These principles may be summarized as follows:

1. B often forms triangular BO_3 units. Sometimes these remain monomeric, but they may form polynuclear ions by sharing the O atoms at corners. This links the triangular units together into chains, rings and two-dimensional flat sheet-like structures.
2. sometimes forms tetrahedral BO_4 units. More complex polynuclear borates contain both triangular BO_3 units and tetrahedral BO_4 units linked together by sharing corners. These structures are not flat.

3. Hydrated borates may accept protons. These are added in the following order: (i) O^{2-} ions are converted to OH^-, (ii) tetrahedral O are protonated, (iii) the O in planar triangles are protonated and (iv) any free OH^- groups are converted to H_2O.
4. Hydrated borates may polymerize by eliminating H_2O. This may be followed by breaking or rearranging B—O bonds.
5. H_3BO_3 often exists in the presence of more complex polyanions.

These polymeric borate structures tend to break up when dissolved in water.

In contrast, the structures of phosphates and silicates are always based on tetrahedral PO_4 and SiO_4 units. The tetrahedra may polymerize into chains, rings and three-dimensional structures. These structures are rather more stable and do not break up in solution.

Qualitative analysis of boron compounds

When borates are treated with HF (or with concentrated H_2SO_4 and CaF_2) the volatile compound BF_3 is formed. If the BF_3 gas produced is introduced into a flame (for example a bunsen flame) the flame gives a characteristic green coloration.

$$\text{conc. } H_2SO_4 + CaF_2 \rightarrow 2HF + CaSO_4$$
$$H_3BO_3 + 3HF \rightarrow 2BF_3 + 3H_2O$$

An alternative test is to make the ester methyl borate $B(OCH_3)_3$. The suspected borate sample is mixed with concentrated H_2SO_4 to form H_3BO_3, and warmed with methyl alcohol in a small evaporating basin.

$$B(OH)_3 + 3CH_3OH \rightarrow B(OCH_3)_3 + 3H_2O$$

The concentrated H_2SO_4 removes the water formed. The mixture is then set on fire. Methyl borate is volatile, and colours the flame green.

Fluoboric acid

H_3BO_3 dissolves in aqueous HF, forming fluoboric acid HBF_4.

$$H_3BO_3 + 4HF \rightarrow H^+ + [BF_4]^- + 3H_2O$$

Fluoboric acid is a strong acid, and commercial solutions contain 40% acid. The $[BF_4]^-$ ion is tetrahedral, and fluoborates resemble perchlorates ClO_4^- and sulphates in crystal structure and solubility ($KClO_4$ and KBF_4 are both not very soluble in water). The $[BF_4]^-$ and $[ClO_4]^-$ ions have a very low tendency to form complexes in aqueous solution, though a few complexes are formed in non-aqueous media.

Borides

There are over 200 binary compounds between metals and boron. There are many different stoichiometries. The most common are M_2B, MB,

MB_2, MB_4 and MB_6, though formulae as diverse as M_5B and MB_{66} are known. Some of the compounds are nonstoichiometric. The formulae of some of the compounds cannot be rationalized by the application of simple valency rules, and are best explained by multi-centre bonding.

They may be prepared by heating the metal with boron, and by a variety of other methods. The metal-rich borides are mostly with transition metals. They are hard, and have very high melting points: ZrB_2, HfB_2, NbB_2 and TaB_2 all melt over 3000 °C. The melting points and the electrical conductivities are often higher than for the parent metals. Thus TiB_2 conducts five times as well as Ti metal. Borides are often chemically inert, and they have several uses:

1. Boron carbide is commonly written as B_4C. It is produced in quantities of tonnes by reducing B_2O_3 with C at 1600 °C. It is a useful source of B, and is also used as an abrasive for polishing. It is used in brake linings for cars. Fibres of B_4C have an enormous tensile strength, and are used to make bullet-proof clothing. These fibres are made as follows:

$$6H_2 + 4BCl_3 + C_{(fibre)} \xrightarrow{1700-1800\,°C} B_4C_{(fibres)} + 12HCl$$

   Boron carbide should be written $B_{13}C_2$, but its composition varies and may approach $B_{12}C_3$. The structure comprises a series of B_{12} icosahedra. Each icosahedron is linked to six others through either four B—C—B linkages and two B—B linkages, or through six B—C—B linkages. The structure is a cluster compound, and can only be explained by multi-centre bonding.
2. Powder fabrication techniques are used to fabricate parts such as turbine blades and rocket nozzles from powdered borides such as CrB_2, TiB_2, and ZrB_2.
3. Boron steel is used in the nuclear industry for shielding and for control rods in reactors because ^{10}B has a very high absorption cross-section for thermal neutrons.

THE OTHER GROUP III OXIDES

Alumina Al_2O_3 can be made by dehydrating $Al(OH)_3$, or from the elements. Two different forms are known. These are α-Al_2O_3 or corundum, and γ-Al_2O_3.

Corundum is found as a mineral, and α-Al_2O_3 is also made by heating $Al(OH)_3$ or γ-Al_2O_3 above 1000 °C. Corundum is extremely hard (9 on Moh's scale) and is used as 'jewellers' rouge' to polish glass. (See Appendix N). An impure form of corundum, contaminated with iron oxide and silica, is called emery. This is used to make emery paper ('sandpaper' used to polish metals). Corundum is unaffected by acids. It has a high melting point of over 2000 °C. It is used as a refractory to line furnaces and to make containers for high temperature reactions. The crystal structure of corundum is hexagonally close-packed oxygen atoms, with two thirds of the octahedral holes filled by Al^{3+} ions. γ-Al_2O_3 is made by dehydrating $Al(OH)_3$ below 450 °C, and in contrast to α-Al_2O_3 it dissolves in acids, absorbs water, and is used for chromatography.

Alumina is white, but it can be coloured by the addition of Cr_2O_3 or Fe_2O_3. White sapphires are gem quality corundum. Synthetic rubies can be made by strongly heating a mixture of Al_2O_3 and Cr_2O_3, for example in an oxy-hydrogen flame. Rubies are very hard, and are used for jewellery and to make bearings in watches and instruments. Thus ruby is a mixed oxide. Blue sapphires are another mixed oxide containing traces of Fe^{2+}, Fe^{3+} and Ti^{4+}. The mineral spinel $MgAl_2O_4$ is another mixed oxide. It gives its name to the structure adopted by many $M^{II}M_2^{III}O_4$ compounds.

Aluminium has a very strong affinity for oxygen. The enthalpy of formation of Al_2O_3 is $-1670\,kJ\,mol^{-1}$, higher (more negative) than for practically all other metal oxides. Thus Al may be used in the thermite reduction of less stable metal oxides. The overall reaction may be considered as the sum of two reactions:

$$2Al_{(s)} + \tfrac{3}{2}O_{2(g)} \rightarrow Al_2O_{3(s)} \qquad \Delta H^\circ = -1670\,kJ$$

$$Fe_2O_{3(s)} \rightarrow 2Fe_{(s)} + \tfrac{3}{2}O_{2(g)} \qquad \Delta H^\circ = +824\,kJ$$

$$2Al_{(s)} + Fe_2O_{3(s)} \rightarrow Al_2O_{3(s)} + 2Fe_{(s)} \qquad \Delta H^\circ = -846\,kJ$$

Since ΔH° is negative, it is energetically feasible for Al to reduce FeO to Fe. The heat produced is sufficient to melt both Fe and Al_2O_3, and this reaction can be used for welding metals. In this particular case, reduction with C is a cheaper method of extracting iron, but in other cases thermite reduction is used for extraction.

$$3Mn_3O_4 + 8Al \rightarrow 4Al_2O_3 + 9Mn$$

Qualitative analysis of aluminium

In qualitative analysis $Al(OH)_3$ is precipitated as a white gelatinous substance when NH_4OH is added to the solution (after previously removing acid-insoluble sulphides with H_2S). $Fe(OH)_3$, $Cr(OH)_3$ and $Zn(OH)_2$ are also precipitated, but $Fe(OH)_3$ is brown, $Cr(OH)_3$ grey–green or grey–blue. $Zn(OH)_2$ is white, like $Al(OH)_3$, but it is not gelatinous. $Zn(OH)_2$ dissolves in excess NH_4OH, whereas $Al(OH)_3$ does not. A confirmatory test for aluminium is the formation of a red precipitate from $Al(OH)_3$ and the dye aluminon.

Amphoteric behaviour – aluminates

$Al(OH)_3$ is amphoteric. It reacts principally as a base, i.e. it reacts with acids to form salts that contain the $[Al(H_2O)_6]^{3+}$ ion. However, $Al(OH)_3$ shows some acidic properties when it dissolves in NaOH, forming sodium aluminate. (However, $Al(OH)_3$ is reprecipitated by the addition of carbon dioxide, showing that the acidic properties are very weak.)

$$Al(OH)_3 \xrightarrow{\text{excess NaOH}} NaAl(OH) \quad \text{sodium}$$
$$NaAlO_2 \cdot 2H_2O \quad \text{aluminate}$$

The formula of aluminates is often written as $NaAlO_2 \cdot 2H_2O$ (which is equivalent to $[Al(OH)_4]^-$). Raman spectra suggest that the structure of the aluminate ion is more complicated than this implies, and the structure changes with both pH and concentration:

1. Between pH 8 and 12 the ions polymerize using OH bridges and each aluminium is octahedrally coordinated.
2. In dilute solutions above pH values of 13, a tetrahedral $[Al(OH)_4]^-$ ion exists.
3. In concentrated solutions above 1.5 M and at pH values greater than 13 the ion exists as a dimer: $[(HO)_3Al—O—Al(OH)_3]^{2-}$.

Ga_2O_3 and $Ga(OH)_3$ are both amphoteric like the corresponding Al compounds. $Ga(OH)_3$ is white and gelatinous and dissolves in alkali, forming gallates. Tl_2O_3 and In_2O_3 are completely basic, and form neither hydrates nor hydroxides. In contrast, thallous hydroxide TlOH is a strong base, and is soluble in water. Thus TlOH differs from the trivalent hydroxides and resembles the Group I hydroxides. Where an element can exist in more than one valency state, there is a general tendency for the lowest valency state to be the most basic.

Thallium(III) acetate and trifluoroacetate can be made by dissolving the oxide in the appropriate acid. They are used in the synthesis of organometallic thallium compounds.

TETRAHYDRIDOBORATES (BOROHYDRIDES)

Stable complexes containing the group $[BH_4]^-$ are well known. These should be called tetrahydridoborates, though the old name borohydride is still widely used. The tetrahydridoborate ion $[BH_4]^-$ is tetrahedral, and the sodium salt $Na[BH_4]$ is the most important compound. It is usually prepared from trimethylborate. It is ionic, and has a sodium chloride structure.

$$\underset{\text{trimethylborate}}{4B(OMe)_3} + 4NaH \xrightarrow[\text{tetrahydrofuran solvent}]{250\,°C,\ \text{high pressure}} Na[BH_4] + 3Na[B(OMe)_4]$$

Other tetrahydridoborates are made by treating $Na[BH_4]$ with the appropriate metal chloride. The alkali metal tetrahydridoborates are white ionic solids and react with water with varying ease. Thus $Li[BH_4]$ reacts violently with water, but $Na[BH_4]$ may be recrystallized from cold water with only slight decomposition, and $K[BH_4]$ is quite stable.

$$Li[BH_4] + 2H_2O \rightarrow LiBO_2 + 4H_2$$

The alkali metal tetrahydridoborates are valuable reducing agents in both inorganic and organic chemistry. $Na[BH_4]$ is stable in alcoholic and aqueous solutions. This makes it a useful reagent for reducing aldehydes to primary alcohols, and ketones to secondary alcohols. It is a nucleophilic reagent, and attacks sites of low electron density. Thus other functional groups such as C=C, COOH and NO_2 are not normally attacked.

$$R \cdot CHO \xrightarrow{Na[BH_4]} R \cdot CH_2OH \quad \text{primary alcohol}$$

$$\begin{matrix} R \\ \quad \backslash \\ \quad C{=}O \\ \quad / \\ R' \end{matrix} \xrightarrow{Na[BH_4]} \begin{matrix} R \\ \quad \backslash \\ \quad CHOH \\ \quad / \\ R' \end{matrix} \quad \text{secondary alcohol}$$

Not all tetrahydridoborates are ionic. The beryllium, aluminium and transition metal borohydrides become increasingly covalent and volatile. In these the $[BH_4]^-$ group acts as a ligand and forms covalent compounds with metal ions. One or more H atoms in a $[BH_4]^-$ act as a bridge and bond to the metal, forming a three-centre bond with two electrons shared by three atoms. The $[BH_4]^-$ ligand is unusual in that it may form one, two or three such three-centre bonds to the metal ion. Thus $Be(BH_4)_2$, $Al(BH_4)_3$ and $Zr(BH_4)_4$ are covalent, and react strongly with water. In the Al and Zr compounds, each BH_4^- forms two hydrogen bridges, whilst in the Be compound each BH_4^- forms three hydrogen bridges.

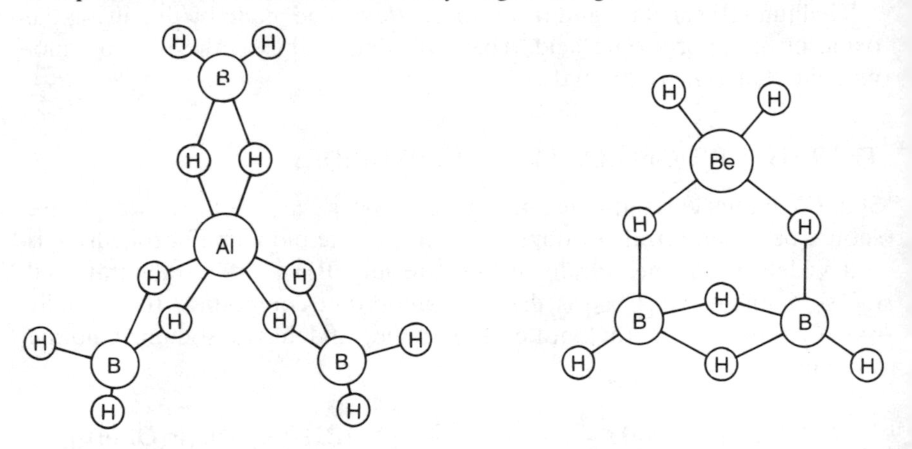

Figure 12.3 Structures of $Al(BH_4)_3$ and $Be(BH_4)_2$. (After H.J. Eméleus and A.G. Sharpe, *Modern Aspects of Inorganic Chemistry*, 4th ed., 1973, Routledge and Kegan Paul.)

The other elements in the group also form electron-deficient hydrides. Thus $(AlH_3)_n$ exists as a white polymer of unknown structure, but it may contain aluminium atoms joined by hydrogen bridges similar to those in diborane. Aluminium hydride can be made from LiH and $AlCl_3$ in ether solution, but with excess LiH lithium aluminium hydride $Li[AlH_4]$ is formed instead.

$$LiH + AlCl_3 \rightarrow (AlH_3)_n \xrightarrow{\text{excess LiH}} Li[AlH_4]$$

$Li[AlH_4]$ is a most useful organic reducing agent because it will reduce functional groups, but in general it does not attack double bonds. It is analogous to the borohydrides but cannot be used in aqueous solutions.

Gallium forms compounds analogous to the borohydrides, e.g. $Li[GaH_4]$. Indium forms a polymeric hydride $(InH_3)_n$, but there is some doubt about the existence of a hydride of thallium.

TRIHALIDES

All the elements form trihalides. The boron halides BX_3 are covalent and gaseous. BF_3 is by far the most important. It is a colourless gas, boiling point −101 °C, and is made in large quantities:

$$B_2O_3 + 3CaF_2 + \text{conc. } 3H_2SO_4 \xrightarrow{\text{heat}} 2BF_3 + 3CaSO_4 + 3H_2O$$

$$B_2O_3 + 6NH_4BF_4 \xrightarrow{\text{heat}} 8BF_3 + 6NH_3 + 3H_2O$$

Both BF_3 gas and its complex with diethyl ether $(C_2H_5)_2O \rightarrow BF_3$ (a viscous liquid) are commercially available.

The shape of the BF_3 molecule is a planar triangle with bond angles of 120 °. This is predicted by VSEPR theory as the most stable shape for three outer electron pairs round B. The valence bond theory also predicts a planar triangle with hybridization of one *s* and two *p*orbitals used for bonding. However, the B atom only has six electrons in its outer shell and this is termed *electron deficient.*

Electronic structure of boron atom – excited state

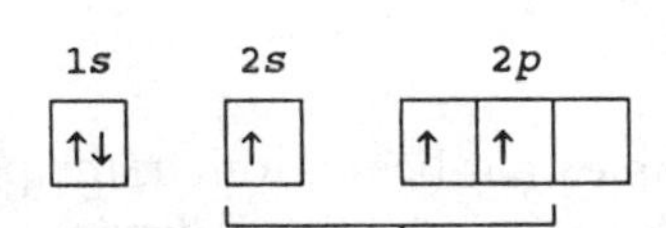

three singly occupied orbitals form bonds with unpaired electrons from three halogen atoms – shape plane triangle (sp^2 hybridization)

The bond lengths in BF_3 are 1.30 Å, and are significantly shorter than the sum of the covalent radii (B = 0.80 Å, F = 0.72 Å). The bond energy is very high: 646 kJ mol^{-1}, which is higher than for any single bond. The shortness and strength of the bonds is interpreted in terms of a $p\pi$–$p\pi$ interaction, that is the bonds possess some double bond character. The empty $2p_z$ atomic orbital on B which is not involved in hybridization is perpendicular to the triangle containing the sp^2 hybrid orbitals. This *pz* orbital may accept an electron pair from a full p_z orbital on any one of the three fluorine atoms. Thus a dative π bond is formed, and the B atom attains an octet of electrons. If one localized double bond existed, then there would be one short bond and two longer ones. However, all measurements show that the three bond lengths are identical. The old valence bond explanation of this was resonance between three structures with the double bond in different positions. The modern explanation is that the double bond is delocalized. The four p_z atomic orbitals from B and the three F atoms form a four-centre π molecular orbital covering all four atoms which contains two bonding electrons. Delocalized π bonding is described more fully in Chapter 4.

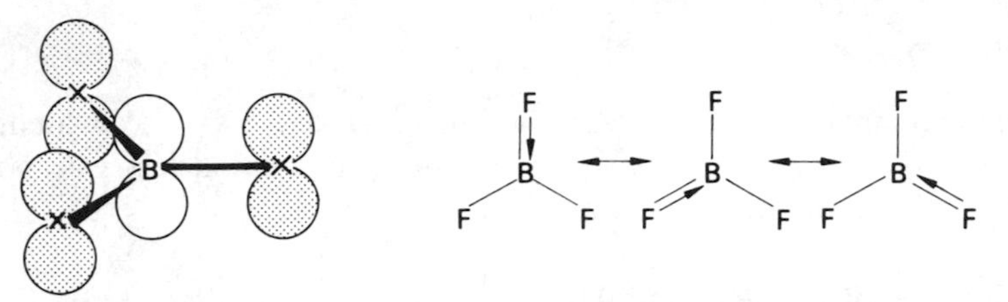

Figure 12.4 Structure of BF_3.

The empty $2p_z$ orbital on the boron atom in BF_3 can also be filled by a lone pair of electrons from donor molecules such as Et_2O, NH_3, $(CH_3)_3N$ or by ions such as F^-. When this occurs, a tetrahedral molecule or ion is formed.

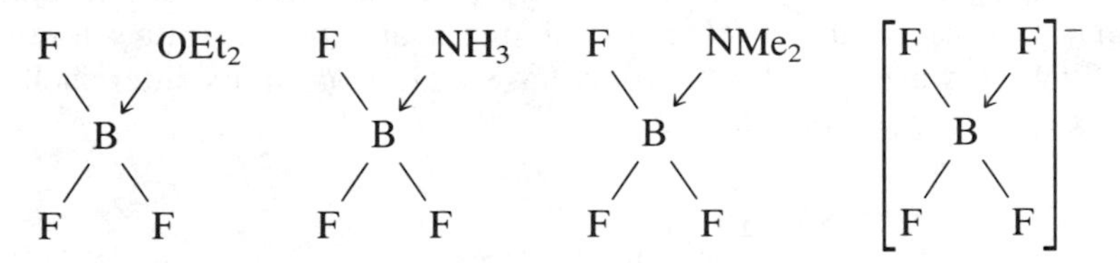

Once a tetrahedral complex has been formed, the possibility for π bonding no longer exists. In $H_3N \rightarrow BF_3$ the B—F distance is 1.38 Å, and in $Me_3N \rightarrow BF_3$ the distance is 1.39 Å, much longer than the 1.30 Å in BF_3. Since the boron halides will accept electron pairs from numerous atoms and ions such as F^-, O, N, P and S, they are acting as strong Lewis acids.

The trihalides are important industrial chemicals, particularly BF_3 and to a lesser extent BCl_3. They are used to prepare elemental boron. They are also very useful for promoting certain organic reactions. In some cases BF_3 is used up in the reaction, and in others it acts as a catalyst by forming a BF_3 complex with one or both reactants. Forming an 'intermediate compound' in this way lowers the activation energy. Examples include:

1. Friedel–Crafts reactions such as alkylations and acylations. In these the BF_3 is used up in the reaction, and so is not strictly catalytic.

$$C_6H_6 + C_2H_5F + BF_3 \rightarrow C_6H_5 \cdot C_2H_5 + H^+ + [BF_4]^-$$

2. It acts as a catalyst in several reactions:

$$\text{Acid} + \text{alcohol} \rightarrow \text{ester} + \text{water}$$
$$\text{Benzene} + \text{alcohol} \rightarrow \text{alkylbenzene} + \text{water}$$

3. Considerable quantities of BF_3 are also used as a polymerization catalyst in the production of polyisobutenes (used to make viscostatic lubricating oils), coumarone–indene resins, and butadiene–styrene rubbers.

About 4000 tonnes a year of BF_3 are produced in the USA from B_2O_3 or borax:

$$B_2O_3 + 6HF + 3H_2SO_4 \rightarrow 2BF_3 + 3H_2SO_4 \cdot H_2O$$

$$Na_2B_4O_7 + 12HF \xrightarrow{-H_2O}$$

$$[Na_2O(BF_3)_4] \xrightarrow{+2H_2SO_4} 4BF_3 + 2NaHSO_4 + H_2O$$

The boron halides are all hydrolysed by water. BF_3 hydrolyses incompletely and forms fluoborates. This is because the HF first formed reacts with the H_3BO_3.

$$4BF_3 + 12H_2O \rightarrow 4H_3BO_3 + 12HF$$
$$12HF + 3H_3BO_3 \rightarrow 3H^+ + 3[BF_4]^- + 9H_2O$$
$$\overline{4BF_3 + 3H_2O \rightarrow H_3BO_3 + 3H^+ + 3[BF_4]^-}$$

The other halides hydrolyse completely, giving boric acid.

$$BCl_3 + 3H_2O \rightarrow H_3BO_3 + 3HCl$$

The fluorides of Al, Ga, In and Tl are ionic and have high melting points. The other halides are largely covalent when anhydrous. $AlCl_3$, $AlBr_3$ and $GaCl_3$ exist as dimers, thus attaining an octet of electrons. The dimeric formula is retained when the halides dissolve in non-polar solvents such as benzene. However, when the halides dissolve in water, the high enthalpy of hydration is sufficient to break the covalent dimer into $[M \cdot 6H_2O]^{3+}$ and $3X^-$ ions.

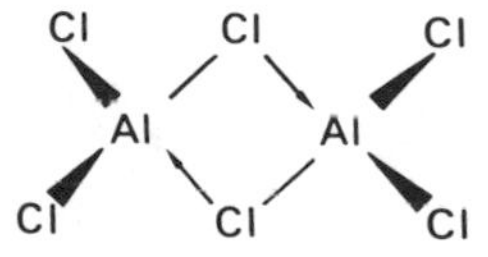

Figure 12.5 Structure of $AlCl_3$ dimer.

Group III elements have only three valency electrons. When these are used to form three covalent bonds, the atom has a share in only six electrons. The compounds are therefore electron deficient. The BX_3 halides attain an octet by π bonding. The other elements in the group have larger atoms and cannot get effective π overlap, so they polymerize to remedy the electron deficiency.

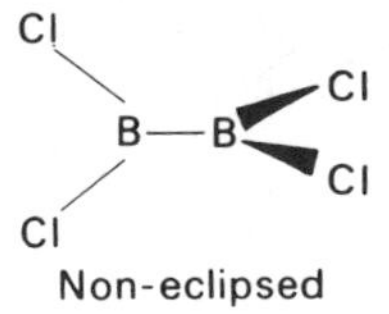

Non-eclipsed

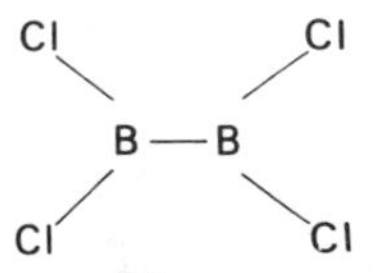

Planar

$AlCl_3$ is an important industrial chemical. Production is about 25 000 tonnes/year in the USA alone. Anhydrous $AlCl_3$ (and to a lesser extent $AlBr_3$) is used as the 'catalyst' in a variety of Friedel–Crafts type of reactions for alkylations and acylations. Large amounts of ethylbenzene are made in this way and are used to make styrene. (Polystyrene production was 6.3 million tonnes in 1985.)

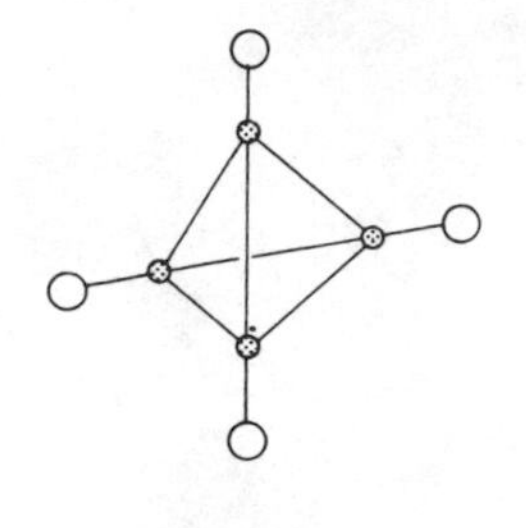

(a) B_4Cl_4

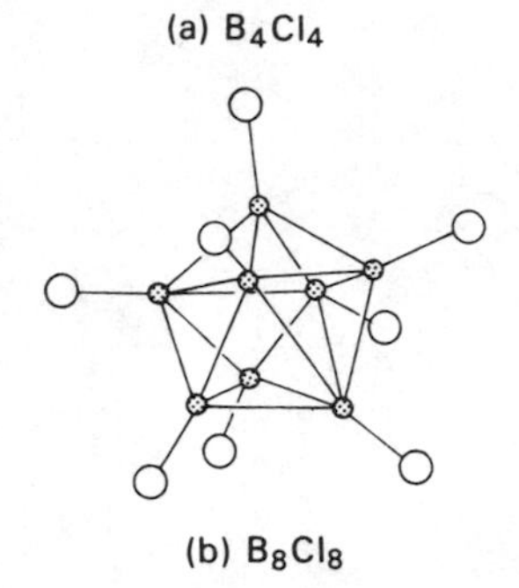

(b) B_8Cl_8

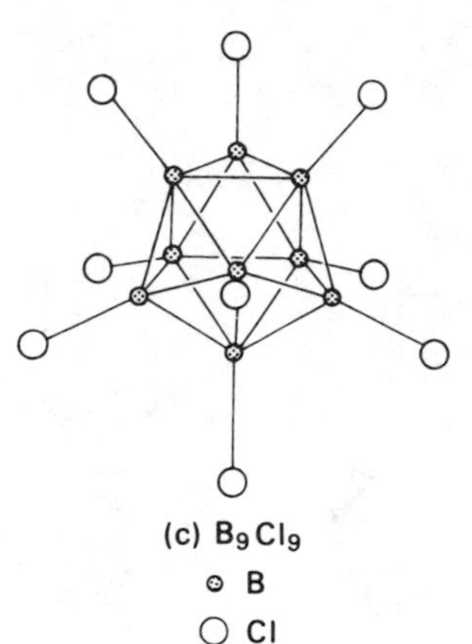

(c) B_9Cl_9

Figure 12.7 Boron monochloride structures $(BCl)_n$ showing polyhedral boron cages. (After A.G. Massey, *The Typical Elements*, Penguin, 1972.)

$$C_6H_5 \cdot H + CH_3CH_2Cl + AlCl_3 \rightarrow C_6H_5 \cdot CH_2CH_3 + H^+ + [AlCl_4]^-$$

This is not true 'catalytic' action, as the $AlCl_3$ is used up, and the formation of $[AlCl_4]^-$ or $[AlBr_4]^-$ is an essential part of the reaction. Acylations are similar:

$$C_6H_5 \cdot H + RCOCl + AlCl_3 \rightarrow RCOC_6H_5 + H^+ + [AlCl_4]^-$$

$AlCl_3$ is also used to catalyse the reaction to make ethyl bromide (which is used to make the petrol additive $PbEt_4$).

$$CH_2{=}CH_2 + HBr \rightarrow C_2H_5Br$$

$AlCl_3$ is also used in the manufacture of anthraquinone (used in the dyestuffs industry), and dodecylbenzene (used to make detergents), and in the isomerization of hydrocarbons (petroleum industry).

TlI_3 is an unusual compound. It is isomorphous with CsI_3 and NH_4I_3, and contains the linear triiodide ion I_3^-. Thus the metal is present as Tl^+, in the oxidation state (+I), not (+III).

DIHALIDES

Boron forms halides of formula B_2X_4. These decompose slowly at room temperature. B_2Cl_4 can be made as follows:

$$2BCl_3 + 2Hg \xrightarrow[\text{low pressure}]{\text{electric discharge}} B_2Cl_4 + Hg_2Cl_2$$

There is free rotation about the B—B bond, and in the gaseous and liquid states the molecule adopts a non-eclipsed conformation. In the solid state the molecule is planar, because of crystal forces and ease of packing. Gallium and indium also form 'dihalides':

$$GaCl_3 + Ga \rightarrow 2GaCl_2$$
$$In + \underset{\text{gas}}{HCl} \rightarrow 2InCl_2$$

These are more properly written $Ga^+[GaCl_4]^-$ and $In^+[InCl_4]^-$ and contain M(I) and M(III) rather than Ga(II) and In(II).

MONOHALIDES

Boron forms several stable polymeric monohalides $(BX)_n$.

$$B_2Cl_4 \xrightarrow{\text{mercury discharge}} B_4Cl_4$$

B_2Cl_4 —(slow decomposition)→ B_8Cl_8, B_9Cl_9, $B_{10}Cl_{10}$, $B_{11}Cl_{11}$, $B_{12}Cl_{12}$

The compounds B_4Cl_4, B_8Cl_8 and B_9Cl_9 are crystalline solids, and their structures (Figure 12.7) have a closed cage or polyhedron of B atoms, where each B atom is bonded to three other B atoms and to one Cl atom. Since B has only three valency electrons there are not enough electrons to form normal electron pair bonds. It is probable that multi-centre σ bonds cover all the B atoms in the cage.

Al, Ga and In all form monohalides MX in the gas phase at elevated temperatures, e.g.

$$AlCl_3 + 2Al \xrightarrow[\text{temperature}]{\text{high}} 3AlCl$$

These compounds are not very stable, and are covalent.

Thallium forms univalent thallous halides which are more stable than the thallium trihalides. This illustrates the inert pair effect, and TlF is ionic.

COMPLEXES

Group III elements form complexes much more readily than the *s*-block elements, because of their smaller size and increased charge. Tetrahedral hydride and halide complexes such as $Li[AlH_4]$ and $H[BF_4]$ have already been mentioned. In addition many octahedral complexes such as $[GaCl_6]^{3-}$, $[InCl_6]^{3-}$ and $[TlCl_6]^{3-}$ are known. The most important octahedral complexes are those with chelate groups, for example β-diketones such as acetylacetone, oxalate ions, dicarboxylic acids, pyrocatechol, and also 8-hydroxyquinoline. (See Figure 12.8.) The latter complex has been used in the gravimetric determination of aluminium.

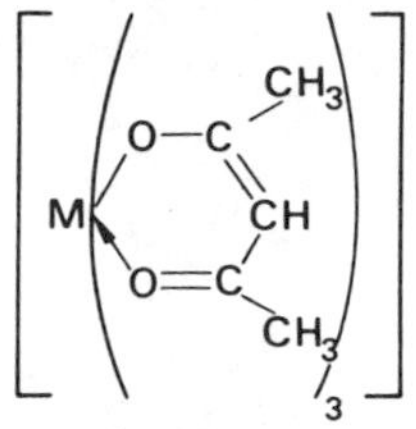

acetyl acetone complex

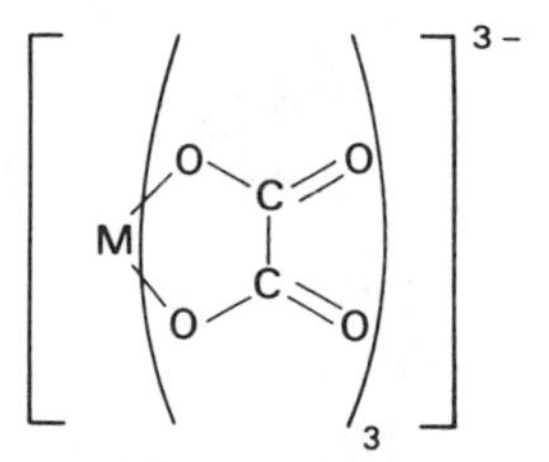

oxalate complex

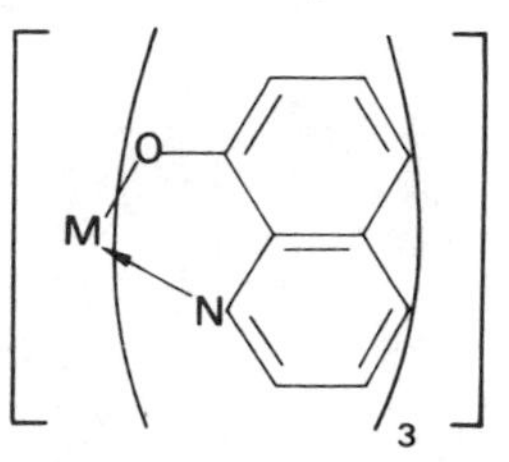

8-hydroxyquinoline complex

Figure 12.8 Some complexes.

DIFFERENCES BETWEEN BORON AND THE OTHER ELEMENTS

Boron differs significantly from the other elements in Group III, mainly because the atoms are very small. It is always covalent, and it is non-metallic. In addition, boron shows a diagonal relationship with silicon in Group IV.

1. B_2O_3 is an acidic oxide, like SiO_2. This is in contrast to Al_2O_3, which is amphoteric.
2. H_3BO_3, which may be written $B(OH)_3$, is acidic, whilst $Al(OH)_3$ is amphoteric.
3. Simple borates and silicate ions can polymerize, forming isopolyacids. Both are built on similar structural principles, namely by sharing oxygen atoms. Complicated chains, rings, sheets and other structures are formed in this way. Aluminium forms no analogous compounds.
4. The hydrides of B are gaseous, readily hydrolysed and spontaneously inflammable. In contrast aluminium hydride is a polymeric solid $(AlH_3)_n$. SiH_4 is gaseous, readily hydrolysed and inflammable.
5. Apart from BF_3, the halides of B and Si hydrolyse readily and vigorously. The aluminium halides are only partly hydrolysed in water.

BORON HYDRIDES

Compounds known

None of the Group III elements react directly with hydrogen, but several interesting hydrides are known. The boron hydrides are sometimes called boranes by analogy with the alkanes (hydrocarbons). Almost 20 boranes have been reported, and 11 are well characterized. They fall into two series:

1. $B_nH_{(n+4)}$ (called *nido*-boranes).
2. A less stable series $B_nH_{(n+6)}$ (called *arachno*-boranes).

In cases where the nomenclature is ambiguous, as for example for pentaborane, it is usual to include the number of hydrogen atoms in the name.

Table 12.10 The two series of boranes

Nido-boranes $B_nH_{(n+4)}$	m.p. (°C)	b.p. (°C)	*Arachno*-boranes $B_nH_{(n+6)}$	m.p. (°C)	b.p. (°C)
B_2H_6 diborane	−165	−93			
			B_4H_{10} tetraborane	−120	18
B_5H_9 pentaborane-9	−47	60	B_5H_{11} pentaborane-11	−122	65
B_6H_{10} hexaborane-10	−62	108	B_6H_{12} hexaborane-12	−82	
B_8H_{12} octaborane-12	dec		B_8H_{14} octaborane-14	dec	
			B_9H_{15} (nonaborane or (enneaborane	3	
$B_{10}H_{14}$ decaborane	−100	213			

dec = decomposes.

Preparation

Diborane is the simplest and most studied of the hydrides. It is used to prepare the higher boranes, and is an important reagent in synthetic organic chemistry. For the latter purpose it is normally generated in situ. It is a versatile reagent for the production of organoboranes, which are useful intermediates in organic synthesis. Alternatively diborane is used as a powerful electrophilic reducing agent for certain functional groups. It attacks sites with a high electron density such as N in cyanides and nitrites, and O in carbonyl compounds.

$$R{-}C{\equiv}N \rightarrow RCH_2NH_2$$
$$R{-}NO_2 \rightarrow RNH_2$$
$$R{-}CHO \rightarrow RCH_2OH$$

Diborane may be prepared by a variety of methods. Boranes were first prepared by Alfred Stock, who pioneered this branch of chemistry between 1912 and 1936. He heated Mg and B to give magnesium boride Mg_3B_2, and then treated this with orthophosphoric acid. The reaction gives a mixture of products. There were enormous difficulties with this early work because the compounds were highly reactive, flammable, and hydrolysed by water. Stock developed vacuum line techniques, which were previously unknown, to handle these reactive compounds. This preparative method has now been superseded except for making B_6H_{10}.

$$\underset{\text{magnesium boride}}{Mg_3B_2} + H_3PO_4 \rightarrow \underset{\text{mainly } B_4H_{10}}{\text{mixture of boranes}} \xrightarrow{\text{heat}} \underset{\text{diborane}}{B_2H_6}$$

Many other methods have been used:

$$B_2O_3 + 3H_2 + 2Al \xrightarrow{750\text{ atmospheres, }150\,°C} B_2H_6 + Al_2O_3$$

$$\underset{\text{gas}}{2BF_3} + 6NaH \xrightarrow{180\,°C} \underset{\text{gas}}{B_2H_6} + 6NaF$$

There are several convenient laboratory preparations:

1. Reducing the etherate complexes of the boron halides with $Li[AlH_4]$.

$$4[Et_2O \cdot BF_3] + 3Li[AlH_4] \xrightarrow{\text{ether}} 2B_2H_6 + 3Li[AlF_4] + 4Et_2O$$

2. Reacting $Na[BH_4]$ and iodine in the solvent diglyme. Diglyme is a polyether $CH_3OCH_2CH_2OCH_2CH_2OCH_3$.

$$2Na[BH_4] + I_2 \xrightarrow{\text{in diglyme solution}} B_2H_6 + H_2 + 2NaI$$

3. Reducing BF_3 with $Na[BH_4]$ in diglyme.

$$4[Et_2O \cdot BF_3] + 3Na[BH_4] \xrightarrow{\text{in diglyme}} 2B_2H_6 + 3Na[BF_4] + 4Et_2O$$

Method (3) is particularly useful when diborane is required as a reaction intermediary. It is produced in situ, and used without the need to isolate or purify it.

Diborane is a colourless gas, and must be handled with care as it is highly reactive. It catches fire spontaneously in air and explodes with oxygen. The heat of combustion is very high. In the laboratory it is handled in a vacuum frame. Since it reacts with the grease used to lubricate taps, special taps must be used. It is instantly hydrolysed by water, or aqueous alkali. At red heat the boranes decompose to boron and hydrogen.

$$B_2H_6 + 3O_2 \rightarrow 2B_2O_3 + 3H_2O \qquad \Delta H = -2165\,kJ\,mol^{-1}$$

$$B_2H_6 + 6H_2O \rightarrow 2H_3BO_3 + 3H_2$$

Most syntheses of higher boranes involve heating B_2H_6, sometimes with hydrogen. A complex reaction occurs when B_2H_6 is heated in a sealed tube. Various higher boranes are formed (B_4H_{10}, B_5H_{11}, B_6H_{12} and $B_{10}H_{14}$). The B_2H_6 molecule probably breaks into the very reactive intermediate $\{BH_3\}$ which has only a transient existence and reacts with B_2H_6, giving another intermediate $\{B_3H_9\}$. This loses hydrogen, forming $\{B_3H_7\}$, which reacts with more $\{BH_3\}$ to give B_2H_{10}. In a similar way a variety of higher boranes are formed depending on the exact conditions. For example:

$$B_2H_6 \xrightarrow{\text{(5 h at 80–90 °C, 200 atmospheres)}} B_4H_{10}$$

$$B_2H_6 + H_2 \xrightarrow{\text{(rapid at 200–250 °C)}} B_5H_9$$

$$B_2H_6 \xrightarrow{\text{(slow pyrolysis in sealed tube 150 °C)}} B_{10}H_{14}$$

Most of the higher boranes are liquids, but B_6H_{10} and $B_{10}H_{14}$ are solids. As the molecular weight increases, they gradually become more stable in air, and less sensitive to water. $B_{10}H_{14}$ is inert in air and can be recovered from aqueous solutions. At one time the boranes were considered as possible high energy rocket fuels. The aim was to replace hydrocarbon fuels in military aircraft and missiles. Over a tonne of $B_{10}H_{14}$ was made for this purpose. Interest in this use disappeared when it was found that combustion to B_2O_3 was incomplete. Because of this the exhaust nozzles of the rocket became partly blocked with an involatile BO polymer.

REACTIONS OF THE BORANES

Hydroboration

A very important reaction occurs between B_2H_4 (or BF_3 + $NaBH_4$) and alkenes and alkynes.

$$B_2H_6 + RCH{=}CHR \rightarrow B(CH_2{-}CH_2R)_3$$

$$B_2H_6 + RC{\equiv}R \rightarrow B(RC{=}CHR)_3$$

The reactions are carried out in dry ether under an atmosphere of nitrogen because B_2H_6 and the products are very reactive. The alkylborane products BR_3 are not usually isolated. They may be converted as follows:

1. to hydrocarbons by treatment with carboxylic acids,
2. to alcohols by reaction with alkaline H_2O_2, or
3. to either ketones or carboxylic acids by oxidation with chromic acid.

The complete process is called hydroboration, and results in *cis*-hydrogenation, or *cis*-hydration. Where the organic molecule is not symmetrical, the reaction follows the anti-Markovnikov rule, that is B attaches to the least substituted C atom.

$$BR_3 + 3CH_3COOH \rightarrow \underset{\text{hydrocarbon}}{3RH} + B(CH_3COO)_3$$

$$B(CH_2\cdot CH_2R)_3 + H_2O_2 \rightarrow \underset{\text{primary alcohol}}{3RCH_2CH_2OH} + H_3BO_3$$

$$\left[\begin{matrix} R \\ \quad\diagdown \\ \quad\quad CH \\ \quad\diagup \\ R' \end{matrix}\right]_3\!\!-B \xrightarrow{H_2CrO_4} \begin{matrix} R \\ \quad\diagdown \\ \quad\quad C{=}O \\ \quad\diagup \\ R' \end{matrix} \quad \text{ketone}$$

$$(CH_3\cdot CH_2)_3{-}B \xrightarrow{H_2CrO_4} \underset{\text{carboxylic acid}}{CH_3COOH}$$

$$(CH_3\cdot CH_2)_3{-}B + CO \xrightarrow{\text{diglyme}} \left[(CH_3\cdot CH_2)_3{-}CBO\right]_2$$

$$\xrightarrow{H_2O_2} \left[CH_3\cdot CH_2\right]_3 COH$$

Hydroboration is a simple and useful process for two main reasons:

1. The mild conditions required for the initial hydride addition.
2. The variety of products which can be produced using different reagents to break the B—C bond.

H.C. Brown won the Nobel Prize for Chemistry in 1979 for work on these organoboron compounds.

Reaction with ammonia

All the boranes act as Lewis acids and can accept electron pairs. Thus they react with amines, forming simple adducts. They also react with ammonia, but the products depend on the conditions:

$$B_2H_6 + 2(Me)_3N \rightarrow 2[Me_3N\cdot BH_3]$$

$$B_2H_6 + NH_3 \xrightarrow[\text{low temperature}]{\text{excess } NH_3} B_2H_6\cdot 2NH_3$$

$$\xrightarrow[\text{higher temperature}]{\text{excess } NH_3} (BN)_x \quad \text{boron nitride}$$

$$\xrightarrow[\text{higher temperature}]{\text{ratio } 2NH_3:1B_2H_6} B_3N_3H_6 \quad \text{borazine}$$

The compound $B_2H_6\cdot 2NH_3$ is ionic, and comprises $[N_3N \rightarrow BH_2 \leftarrow NH_3]^+$ and $[BH_4]^-$ ions. On heating, it forms borazine.

Boron nitride is a white slippery solid. One B atom and one N atom together have the same number of valency electrons as two C atoms. Thus boron nitride has almost the same structure as graphite, with sheets made up of hexagonal rings of alternate B and N atoms joined together. The sheets are stacked one on top of the other, giving a layer structure (Figure 12.9).

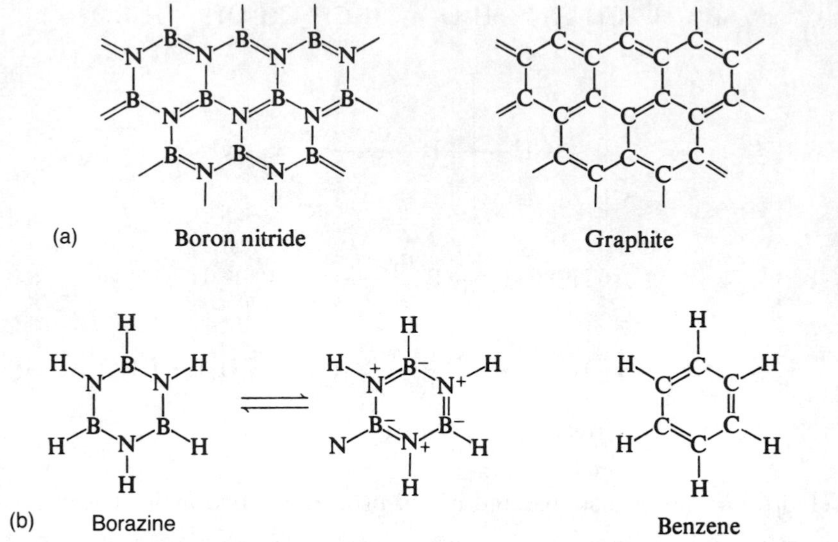

(a) Boron nitride Graphite

(b) Borazine Benzene

Figure 12.9 Similarity in structure between (a) boron nitride and graphite, (b) borazine and benzene.

Borazine $B_3N_3H_6$ is sometimes called 'inorganic benzene' because its structure shows some formal similarity with benzene, with delocalized electrons and aromatic character. Their physical properties are also similar.

Borazine and substituted borazines may be made:

$$3BCl_3 + 3NH_4Cl \xrightarrow{140\,^{\circ}C} B_3N_3H_3Cl_3 \xrightarrow{Na[BH_4]} B_3N_3H_6$$

$$B_3N_3H_3Cl_3 + MeMgBr \rightarrow B_3N_3H_3(Me)_3$$

Borazine forms π complexes such as $B_3N_3H_6—Cr(CO)_3$ with transition metal compounds. Borazine is considerably more reactive than benzene, and substitution reactions occur quite readily:

$$B_3N_3H_6 + 3HCl \rightarrow B_3N_3H_9Cl_3$$

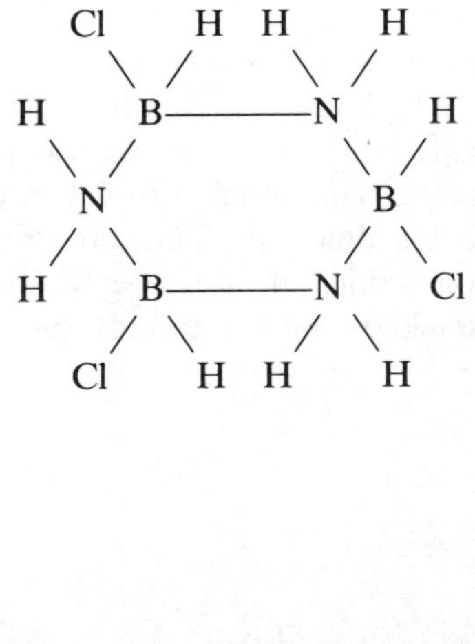

If heated with water, borazine hydrolyses slowly.

$$B_3N_3H_6 + 9H_2O \rightarrow 3NH_3 + 3H_3BO_3 + 3H_2$$

Some other reactions of boranes

$$B_2H_6 + 6H_2O \rightarrow 2B(OH)_3 + 6H_2$$
$$2H_3BO_3 + 6H_2$$
$$B_2H_6 + 6MeOH \rightarrow 2B(OMe)_3 + 6H_2$$
$$B_2H_6 + 2Et_2S \rightarrow 2[Et_2S \rightarrow BH_3]$$
$$B_2H_6 + 2LiH \rightarrow 2Li[BH_4]$$
$$2B_2H_6 + 2Na \rightarrow Na[BH_4] + Na[B_3H_8] \quad \text{(slow)}$$
$$B_2H_6 + HCl \rightarrow B_2H_5Cl + H_2$$
$$B_2H_6 + 3Cl_2 \rightarrow 2BCl_3 + 6HCl$$

STRUCTURES OF THE BORANES

The bonding and structures of the boranes are of great interest. They are different from all other hydrides. There are not enough valency electrons to form conventional two-electron bonds between all of the adjacent pairs of atoms, and so these compounds are termed electron deficient.

In diborane there are 12 valency electrons, three from each B atom and six from the H atoms. Electron diffraction results indicate the structure shown in Figure 12.10.

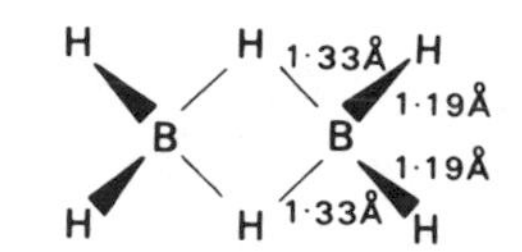

Figure 12.10 The structure of diborane.

The two bridging H atoms are in a plane perpendicular to the rest of the molecule and prevent rotation between the two B atoms. Specific heat measurements confirm that rotation is hindered. Four of the H atoms are in a different environment from the other two. This is confirmed by Raman spectra and by the fact that diborane cannot be methylated beyond $Me_4B_2H_2$ without breaking the molecule into BMe_3.

The terminal B—H distances are the same as the bond lengths measured in non-electron-deficient compounds. These are assumed to be normal covalent bonds, with two electrons shared between two atoms. We can describe these bonds as two-centre two-electron bonds (2c-2e).

Thus the electron deficiency must be associated with the bridge groups. The nature of the bonds in the hydrogen bridges is now well established. Obviously they are abnormal bonds as the two bridges involve only one electron from each boron atom and one from each hydrogen atom, making a total of four electrons. An sp^3 hybrid orbital from each boron atom overlaps with the 1*s* orbital of the hydrogen. This gives a delocalized molecular orbital covering all three nuclei, containing one pair of electrons and making up one of the bridges (see Figure 12.11). This is a three-centre two-electron bond (3c-2e). A second three-centre bond is also formed.

The higher boranes have an open cage structure (Figure 12.12). Both normal and multi-centre bonds are required to explain these structures:

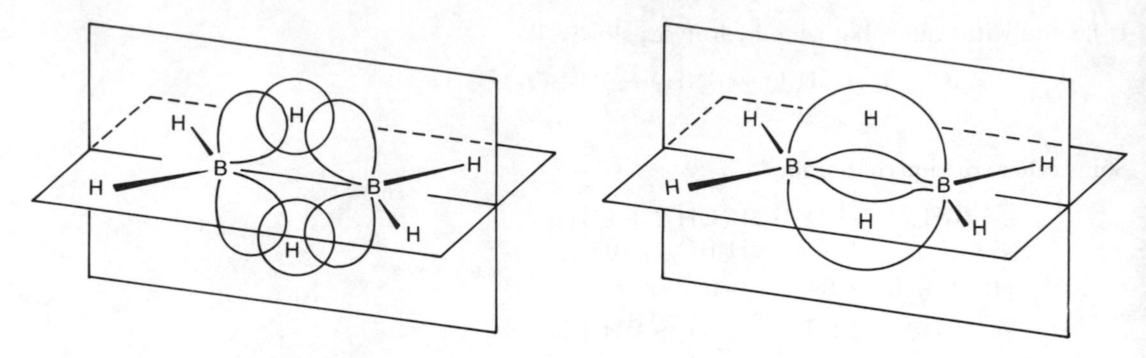

Figure 12.11 Overlap of approximately sp^2 hybrid orbitals from B with an *s* orbital from H to give a 'banana-shaped' three-centre two-electron bond.

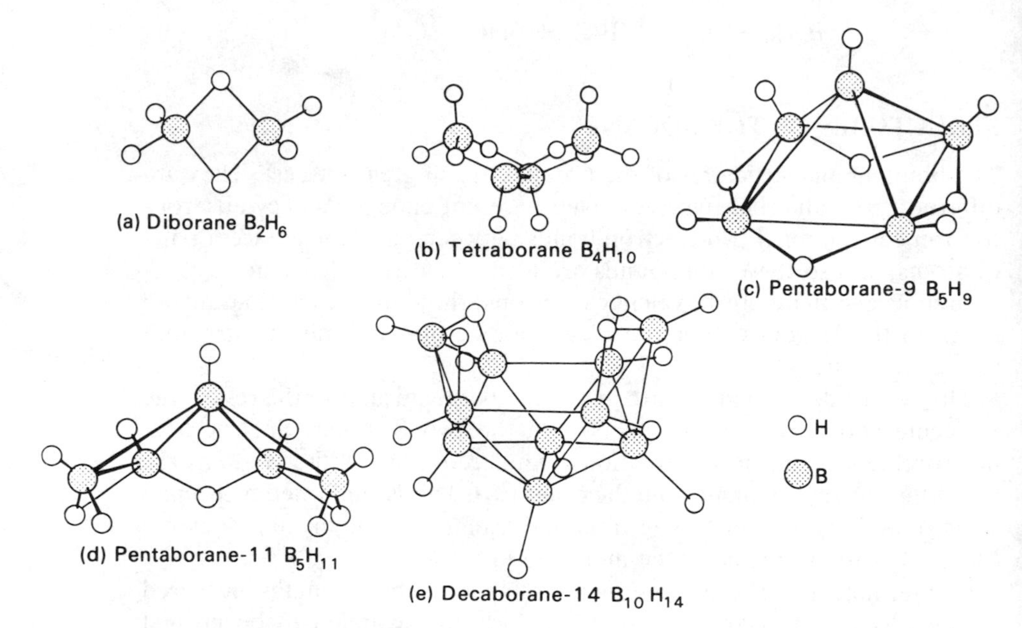

Figure 12.12 Structures of some boranes. (a) Diborane B_2H_6, with two three-centre B...H...B bonds. (b) Tetraborane B_4H_{10}, with four three-centre B...H...B bonds and one B—B bond. (c) Pentaborane-9 B_5H_9, where the boron atoms form a square-based pyramid with four three-centre B...H...B bonds, and multi-centre bonds from the apical B atom to the four B atoms in the square. (d) Pentaborane-11 B_5H_{11}, where the boron atoms form a distorted square-based pyramid, with three three-centre B...H...B bonds and three-centre B...B...B bonds in two of the triangular faces. (e) Decaborane-14. (After A.G. Massey, *The Typical Elements*, Penguin, 1972.)

$$Al—C_2H_5 \rightarrow Al—CH_2 \cdot CH_2 \cdot C_2H_5 \rightarrow Al—(CH_2 \cdot CH_2)_n \cdot C_2H_5$$

1. Terminal B—H bonds. These are normal covalent bonds, that is two-centre two-electron (2c-2e) bonds.
2. B—B bonds. These are also normal 2c-2e bonds.

3. Three-centre bridge bonds including B...H...B as in diborane. These are 3c-2e bonds.
4. Three-centre bridge bonds including B...B...B, similar to the hydrogen bridge in (3). These are called 'open boron bridge bonds', and are of the type 3c-2e.
5. Closed 3c-2e bonds between three B atoms.

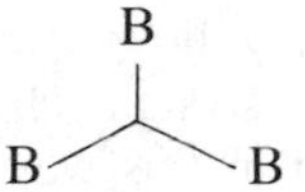

Decaborane-14 has ten B atoms. This number is two short of being able to form a regular icosahedron. An icosahedron has 12 corners and 20 faces. It is a closed cage structure, and is particularly stable. The two extra atoms required to complete the cage may be added by reacting $B_{10}H_{14}$ with an ethyne, forming a carborane (Figure 12.13).

$$B_{10}H_{14} + RC{\equiv}CR \rightarrow B_{10}C_2H_{10}R_2$$

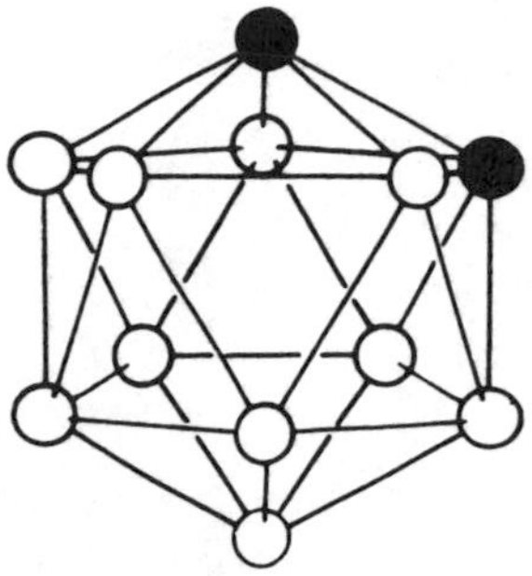

Figure 12.13 Structure of orthocarborane, one of the three isomeric forms of the icosahedral carborane $B_{10}C_2H_{10}R_2$. (After A.G. Massey, *The Typical Elements*, Penguin, 1972.)

Alternatively two B atoms may be added to complete the cage by reacting $B_{10}H_{14}$ with $Me_3N \rightarrow BH_3$.

$$B_{10}H_{14} + 2Me_3N{\rightarrow}BH_3 \rightarrow 2[Me_3NH]^+[B_{12}H_{12}]^{2-}$$

ORGANOMETALLIC COMPOUNDS

Besides the carboranes and the alkylboranes discussed earlier, all the Group III trihalides will react with Grignard reagents and organolithium reagents, forming trialkyl or triaryl compounds:

$$BF_3 + 3C_2H_5MgI \rightarrow B(C_2H_5)_3$$
$$AlCl_3 + 3CH_3MgI \rightarrow Al(CH_3)_3$$
$$GaCl_3 + 3C_2H_5Li \rightarrow Ga(C_2H_5)_3$$
$$InBr_3 + 3C_6H_5Li \rightarrow In(C_6H_5)_3$$

The aluminium compounds are unusual because they have dimeric structures, and appear to have three-centre bonds involving sp^3 hybrid orbitals on Al and C in the Al—C—Al bridges (Figure 12.14).

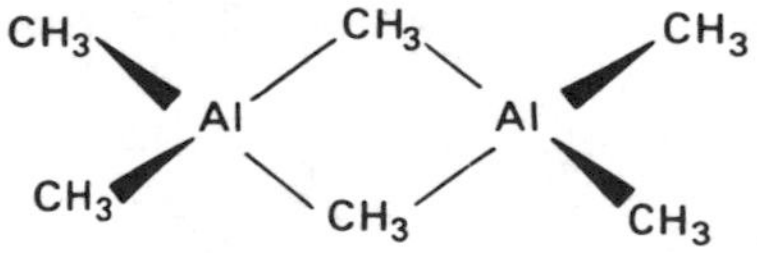

Figure 12.14 Structure of aluminium trimethyl dimer.

Another important route to organoaluminium compounds is from aluminium metal and H_2. The two elements do not react directly to give AlH_3. However, aluminium does take up hydrogen in the presence of aluminium alkyl catalysts (Ziegler catalysts).

$$Al + \tfrac{3}{2}H_2 + 2Et_3Al \rightarrow 3Et_2AlH$$

Alkenes may be added to Al—H bonds.

$$Et_2AlH + \underset{\text{ethene}}{H_2C{=}CH_2} \rightarrow Et_2Al{-}CH_2{-}CH_3 \quad \text{i.e. } Et_3Al$$

At 90–120 °C and 100 atmospheres pressure, ethene molecules are slowly inserted into the Al—C bonds.

$$\mathrm{Al}\begin{matrix} / \; C_2H_5 \\ - \; C_2H_5 \\ \backslash \; C_2H_5 \end{matrix} \rightarrow \mathrm{Al}\begin{matrix} / \; CH_2 \cdot CH_2 \cdot C_2H_5 \\ - \; C_2H_5 \\ \backslash \; C_2H_5 \end{matrix} \rightarrow \mathrm{Al}\begin{matrix} / \; (CH_2 \cdot CH_2)_n \cdot C_2H_5 \\ - \; (CH_2 \cdot CH_2)_n \cdot C_2H_5 \\ \backslash \; (CH_2 \cdot CH_2)_n \cdot C_2H_5 \end{matrix}$$

Long chains up to C_{200} may be grown. Hydrolysis of these longish-chain aluminium alkyls gives straight chain hydrocarbons called polyethylene or polythene. Formed in this way these are low molecular weight polymers (or oligomers), and are of no commercial use. If a transition metal catalyst such as $TiCl_4$ (Natta catalyst) is used then polymerization is much quicker. Furthermore the reaction does not require a high pressure and a much higher molecular weight polymer is produced. These high molecular weight polymers are very important as one route to commercial polythene (see Chapter 20 under 'Organometallic compounds').

Alcohols with chain lengths of about C_{14} are produced from ethene by growing suitable aluminium alkyls, oxidizing with air and then hydrolysing with water. Reaction of these alcohols with SO_3 gives sulphonates R—SO_3H which are then neutralized to make biodegradable detergents CH_3—$(CH_2)_n$—OSO_3^- Na^+. (Detergents are discussed in Group VI under SO_3.)

Aluminium alkyls catalyse the dimerization of propene in the formation of isoprene:

$$\underset{\text{propene}}{2CH_3 \cdot CH{=}CH_2} \xrightarrow{R_3Al} CH_3 \cdot CH_2 \cdot CH_2 \cdot \underset{\displaystyle CH_3}{\underset{|}{C}}{=}CH_2$$

$$\xrightarrow{\text{crack}} \underset{\text{isoprene}}{CH_2{=}CH \cdot \underset{\displaystyle CH_3}{\underset{|}{C}}{=}CH_2} + CH_4$$

The use of aluminium alkyls in producing alcohols and isoprene is of considerable industrial importance. (Commercial polythene is made in two ways, either using titanium catalysts or by a peroxide-induced free radical polymerization, rather than with aluminium alkyls.)

FURTHER READING

Borax Review (1987) Borax Holdings Ltd.

Brown, H.C. (1972) *Boranes in Organic Chemistry*, Cornell University Press.

Brown, H.C. (1975) *Organic Synthesis via Boranes*, Wiley, New York.

Brown, H.C. (1979) Hydride reductions: A 40–year revolution in organic chemistry, *Chem. Eng. News*, March 5, 24–29.

Callahan, K.P. and Hawthorn, M.F. (1976) Ten years of metallocarboranes, *Adv. Organometallic Chem.*, **14**, 145–186.

Christ C.L. and Clark J.R. (1977) A crystal classification of borate structures with emphasis on hydrated borates, *Phys. Chem. Minerals*, **2**, 59.

Greenwood, N.N. (1963) The chemistry of gallium, *Adv. Inorg. Chem. Radiochem.*, **5**, 91.
Greenwood, N.N. (1973) *Comprehensive Inorganic Chemistry*, Vol. I (Chapter 11: Boron), Pergamon Press, Oxford.
Greenwood, N.N. (1975) *Boron*, Pergamon Press, Oxford and Elmsford, NY.
Greenwood, N.N. (1977) The synthesis, structure and chemical reactions of metalloboranes, *Pure Appl. Chem.*, **49**, 791–802.
Greenwood, N.N. and Ward, I.M. (1974) Metalloboranes and metal–boron bonding, *Chem Soc. Rev.*, **3**, 231–271.
Grimes, R.N. (1971) *Carboranes*, Academic Press, New York.
Grimes, R.N. (ed.) (1982) *Metal Interactions with Boron Clusters* (Modern Inorganic Chemistry Series), Plenum Publishing Corp., London.
Grimes, R.N. (1983) Carbon-rich carboranes and their metal derivatives, *Adv. Inorg. Chem. Radiochem.*, **26**, 55–118.
Haupin, W.E. (1983) Electrochemistry of the Hall–Heroult process for aluminium *J. Chem. Ed.*, **60**, 279–282.
Iversen, S. (1988) The chemistry of dementia, *Chemistry in Britain*, **24**, 338–342, 364. (Effects of aluminium in the body.)
James, B.D. and Wallbridge, M.G.H. (1970) Metal tetrahydroborates, *Progr. Inorg. Chem.*, **11**, 97–231. (A comprehensive review with over 600 references.)
Johnson, B.G.F. (ed.) (1980) *Transition Metal Clusters*, Wiley, New York. (Chapters by various leading workers in this field.)
Johnson, B.G.F. and Lewis, J. (1981) *Adv. Inorg. Chem. Radiochem.*, **24**, 225. (A good review of cluster compounds.)
Kennedy, J.D. (1984, 1986) The Polyhedral Metalloboranes, Parts I and II, *Progr. Inorg. Chem.*, **32**, 519–679; **34**, 211–434.
Lancashire, R. (1982) Bauxite and aluminium production, *Education in Chemistry*, **19**, 74–77.
Lee, A.G. (1971) *The Chemistry of Thallium*, Elsevier, Amsterdam.
Lee, A.G. (1972) Coordination chemistry of thallium(I), *Coordination Chem. Rev.*, **8**, 289.
Liebman, J.F., Greenberg, A. and Williams, R.E. (eds) (1988) *Avances in Boron and the Boranes*, VCH, New York.
Massey, A.G., Boron subhalides (1980) *Chemistry in Britain*, **16**, 588–598.
Massey, A.G. (1983) The sub-halides of boron, *Adv. Inorg. Chem. Radiochem.*, **26**, 1–54.
Massey, R.C. (1989) *Aluminium in Food and the Environment* (Special Publication No. 73), Royal Society of Chemistry, London.
Muetterties, E.L. (ed.) (1975) *Boron Hydride Chemistry*, Academic Press, New York.
Sanderson, R.T. (1997) More on complex hydrides, *Chem. Eng. News*, **29**, 3. ($Li[AlH_4]$ etc.)
Stinson, S.C. (1980) Hydride reducing agents, use expanding, *Chem. Eng. News*, Nov 3, 18–20.
Thompson, R. (ed.) (1981) *Speciality Inorganic Chemicals* (Chapter by Wade, K., Sodium borohydride and its uses, Royal Society for Chemistry, London, 1981.
Thompson, R. (ed.) (1986) *The Modern Inorganic Chemicals Industry* (Chapter by Thompson, R., Production and Uses of Inorganic Boron Compounds, (Special Publication No. 31), The Chemical Society, London.
Wade, K. and Bannister, A.J. (1973) *Comprehensive Inorganic Chemistry*, Vol. I (Chapter 12: Aluminium, gallium, indium and thallium), Pergamon Press, Oxford.
Wade, K. (1976) Structural and bonding patterns in cluster chemistry, *Adv. Inorg. Chem. Radiochem.*, **18**, 1–66.
Wade, K. (1971) *Electron Deficient Compounds*, Nelson, London.

Walton, R.A. (1971) Coordination complexes of the thallium(III) halides and their behaviour in non-aqueous media, *Coordination Chem. Rev.*, **6**, 1–25.
Wiberg, E. and Amberger, E. (1971) *Hydrides of the Elements of Main Groups I–IV* (Chapters 5 and 6), Elsevier, Amsterdam.

PROBLEMS

1. The first element in each of the main groups in the periodic table shows anomalous properties when compared with other members of the same group. Discuss this statement with particular reference to the elements Li, Be and B.

2. What is the main source of boron? Outline the steps in the extraction of boron.

3. Draw the structure of the B_{12} unit found in the solid structure of boron. What is this shape called?

4. (a) List features which make borax a useful primary standard, and give a balanced equation to show its use in titrations.
   (b) Work out the shape of the BO_3^{3-} ion and explain why it has this structure.

5. Orthoboric acid may be written as H_3BO_3 or $B(OH)_3$. How does it ionize in water and which way of writing the formula is the most helpful? How strong an acid is it? Why does glycerol enhance its acidic properties? Write a balanced equation for a neutralization reaction with boric acid.

6. Give an account of isopolyacids with special reference to borates.

7. Give one method for the preparation of diborane, B_2H_6. Why is it called an electron-deficient compound? Draw the structure of diborane and give the bond lengths. What is unusual about the bonding in this compound?

8. How does diborane react with (a) ammonia, (b) boron tribromide and (c) trimethylboron?

9. Describe the use of diborane in hydroboration.

10. Give the preparation, structure and uses of sodium borohydride.

11. Compare the structures of BF_3 gas, $AlCl_3$ gas and $AlCl_3$ in aqueous solution.

12. Explain the following:
    (a) BF_3 has no dipole moment, but PF_3 has a substantial dipole
    (b) BF_3 and BrF_3 molecules have different shapes.

13. Describe the preparation and give the structures of the dihalides of B, Ga and In.

14. What is the principal ore of aluminium? How is the ore purified, and

how is the metal extracted? What is the process called? What is the function of cryolite in the process? What is aluminium used for?

15. From the position of Al in the electrochemical series, would you expect it to be stable in water? Why is it stable in air and water?

16. Give equations to show the reactions between Al and: (a) $H^+_{(aq)}$, (b) NaOH, (c) N_2, (d) O_2, (e) Cl_2.

17. Give two examples of alums. What species are present when these materials dissolve in water?

18. How is aluminium chloride prepared? What is its structure when anhydrous and when dissolved in water? Give an example of the use of $AlCl_3$ as a Friedel–Crafts catalyst.

19. How would you prepare LiH and $LiAlH_4$. What are they used for?

20. Give two different preparations for Et_3Al and describe its use as a catalyst for polymerizing C_2H_4. Compare the products formed with those described in the section on titanium in Chapter 20.

21. Compare and contrast the chemistry of boron and aluminium.

22. Give reasons for trivalency and monovalency in Group III elements, and comment on the validity of divalent compounds such as $GaCl_2$.

23. Write notes on the chemistry of thallium in the (+I) oxidation state.

24. Substance (A) is a yellowish-white deliquescent solid which sublimes and has a vapour density of 133. (A) reacted violently with water, forming solution (B). A sample of (B) gave a curdy white precipitate (C) on addition of dilute HNO_3 and $AgNO_3$ solution, but this readily dissolved on the addition of dilute NH_4OH, though a gelatinous white precipitate was formed in its place. (D) was filtered off and dissolved in excess NaOH, forming a clear solution (E). When CO_2 was passed into (E), compound (D) was reprecipitated.
Substance (A) dissolved unchanged in dry ether, and when this solution was reacted with LiH one of two products (F) or (G) was formed, depending on whether the LiH was in excess or not.
Qualitative analysis of solution (B) gave a white gelatinous precipitate in Group III. When 0.1333 g of (A) was dissolved in water and treated with 8-hydroxyquinoline, 0.4594 g of precipitate was obtained.
Identify the compounds (A) to (G) and give equations for all of the reactions.

13 The group IV elements

Table 13.1 Electronic structures and oxidation states

Element		Electronic structure	Oxidation states*
Carbon	C	[He] $2s^2\ 2p^2$	**IV**
Silicon	Si	[Ne] $3s^2\ 3p^2$	(II) **IV**
Germanium	Ge	[Ar] $3d^{10}\ 4s^2\ 4p^2$	II **IV**
Tin	Sn	[Kr] $4d^{10}\ 5s^2\ 5p^2$	II **IV**
Lead	Pb	[Xe] $4f^{14}\ 5d^{10}\ 6s^2\ 6p^2$	**II** IV

* The most important oxidation states (generally the most abundant and stable) are shown in bold. Other well-characterized but less important states are shown in normal type. Oxidation states that are unstable, or in doubt, are given in parentheses.

INTRODUCTION

Carbon is extremely widespread in nature. It is an essential constituent of all living matter, as proteins, carbohydrates and fats. Carbon dioxide is essential in photosynthesis, and is evolved in respiration. Organic chemistry is devoted to the chemistry of carbon-containing compounds. Inorganic compounds produced on a large scale include carbon black, coke, graphite, carbonates, carbon dioxide, carbon monoxide (as a fuel gas), urea, calcium carbide, calcium cyanamide and carbon disulphide. There is great interest in organometallic compounds, carbonyls and π bonding complexes.

The discovery that flint (hydrated SiO_2) had a sharp cutting edge was very important in the development of human technology. Nowadays silicon is important in a number of materials produced in high tonnages. These include cement, ceramics, clays, bricks, glass and the silicone polymers. The very pure element is important in the microelectronics industries (transistors and computer chips).

Germanium is little known, but tin and lead are very well known and have been used as metals since before Biblical times. Lead sheet was used on the floor in the Hanging Gardens of Babylon (one of the wonders of the ancient world) to prevent the water escaping.

OCCURRENCE OF THE ELEMENTS

The elements are all well known, apart from germanium. Carbon is the seventeenth, and silicon the second most abundant element by weight in the earth's crust (Table 13.2). Germanium minerals are very rare. Ge occurs as traces in the ores of other metals and in coal, but it is not well known. Both Si and Ge are important for making semiconductors and transistors. Though the abundances of tin and lead are comparatively low, they occur as concentrated ores which are easy to extract, and both metals have been well known since before Biblical times.

Carbon occurs in large quantities combined with other elements and compounds mainly as coal, crude oil, and carbonates in rocks such as calcite $CaCO_3$ and magnesite $MgCO_3$ and dolomite $[MgCO_3 \cdot CaCO_3]$.

Table 13.2 Abundance of the elements in the earth's crust by weight

	ppm	Relative abundance
C	180	17
Si	272 000	2
Ge	1.5	54
Sn	2.1	49 =
Pb	13	36

Carbon is also found in the native form: large amounts of graphite are mined, and extremely small quantities (in tonnage terms) of diamonds are mined too! Both CO_2 and CO are important industrially. CO_2 occurs in small amounts in the atmosphere, but this is important as CO_2 has a vital role in the carbon cycle with photosynthesis and respiration. CO is an important fuel, and forms some interesting carbonyl complexes. Silicon occurs very widely, as silica SiO_2 (sand and quartz), and in a wide variety of silicate minerals and clays. Germanium is only found as traces in some silver and zinc ores, and in some types of coal. Tin is mined as cassiterite SnO_2, and lead is found as the ore galena PbS.

EXTRACTION AND USES OF THE ELEMENTS

Carbon

Carbon black (soot) is produced in large amounts (4.09 million tonnes in 1985). It is made by the incomplete combustion of hydrocarbons from natural gas or oil. The particle size is very small. Over 90% is used in the rubber industry to make car tyres. Its other main use is in newspaper ink.

Coke is produced in very large amounts (347 million tonnes in 1985). It is produced by high temperature carbonization of coal. The coal is heated in large ovens in the absence of air. Coke is extremely important in the metallurgical extraction of iron and many other metals. The distillation of coal also provides a valuable source of organic chemicals.

continued overleaf

In 1988 605 000 tonnes of natural graphite were mined (China 31%, South Korea 17%, the USSR 14%, Brazil 8% and Mexico 7%). This is usually found as a mixture with mica, quartz and silicates, which contains 10–60% C. Graphite is separated from most of the impurities by flotation. Finally it is purified by heating with HCl and HF in a vacuum to remove the last traces of silicon compounds as SiF_4. Sedimentary deposits of carbon are mined in Mexico. This was once thought to be amorphous carbon, but is now regarded as microcrystalline (very finely divided) graphite. Nearly as much graphite is made synthetically as is mined.

$$3C + SiO_2 \xrightarrow{\text{heat}} SiC + 2CO \xrightarrow{2500\,^\circ C} C_{(graphite)} + Si_{(g)}$$

Graphite is used for making electrodes, in steel making and metal foundries, for crucibles, as a lubricant, and in pencils, brake linings and brushes for electric motors. It is also used as the moderator in the cores of gas cooled nuclear reactors, where it slows down neutrons.

Activated charcoal is made by heating or chemically oxidizing sawdust or peat. World production was 658 000 tonnes in 1985. Active carbon has an enormous surface area, and is used to purify and decolorize sugar and other chemicals. It is also used to absorb poisonous gases in gas masks, in filter beds at sewage plants and as a catalyst for some reactions.

The largest sources of diamonds are Australia 38%, Zaire 20%, Botswana 16.5%, the USSR 12% and South Africa 9%. World production of natural diamonds was 92 400 000 carats or 18.48 tonnes in 1988. Large diamonds are cut as gemstones, and their size is measured in carats (1 g = 5 carats). About 30% are used as gemstones, and 70% are used for industrial purposes, mainly for making drills, or as an abrasive powder for cutting and polishing, as diamond is very hard (10 on Mohs' scale) – see Appendix N. It is economic to make small industrial quality diamonds synthetically, by high temperature and pressure treatment of graphite.

Silicon

More than a million tonnes of Si are produced annually. Most of it is added to steel to deoxidize it. This is important in the manufacture of high silicon corrosion resistant steels. For this purpose it is convenient to use ferrosilicon. This is an alloy of Fe and Si. In 1985 3.2 million tonnes of ferrosilicon were produced. It is made by reducing SiO_2 and scrap iron with coke.

$$SiO_2 + Fe + 2C \rightarrow FeSi + 2CO$$

The element Si is obtained by reducing SiO_2 with high purity coke. There must be an excess of SiO_2, to prevent the formation of the carbide SiC. Si is a shiny blue–grey colour and has an almost metal-like lustre, but it is a semiconductor, not a metal. High purity Si (for the semiconductor industry) is made by converting Si to $SiCl_4$, purifying this by distillation, and reducing the chloride with Mg or Zn.

$$SiO_2 + 2C \rightarrow Si + 2CO$$
$$Si + 2Cl_2 \rightarrow SiCl_4$$
$$SiCl_4 + 2Mg \rightarrow Si + MgCl_2$$

The electronics industry requires small quantities of ultrapure silicon and germanium (with a purity better than $1:10^9$). These materials are insulators when pure, but become *p*-type or *n*-type semiconductors when doped with a Group V or Group III element respectively. These are used as transistors and semiconductor devices. Very pure Si is also used to make computer chips (see Chapter 3 under 'Micro-minaturised semiconductor devices'). To obtain ultrapure Si or Ge, the materials are first purified as much as possible, for example by careful fractional distillation of $SiCl_4$ to get pure Si. For the final stage of purification a process called zone refining is used. This is an excellent method for small quantities. A rod of the element, which has already been purified extensively, is placed in a long quartz tube filled with an inert gas. A heating coil melts a thin disc of the rod. The heater moves slowly from one end to the other, and pure Si or metal crystallizes from the melt. The impurities are more soluble in the liquid, and are carried to the end of the rod, where they are cut off and discarded. Semiconductor quality Si can also be made by sodium reduction of $Na_2[SiF_6]$, which is a by-product from making phosphate fertilizers from fluoroapatite.

$$Na_2[SiF_6] + 4Na \rightarrow Si + 6NaF$$

Germanium

Ge has been recovered from coal ash, but it is now recovered from the flue dust from smelting Zn ores. A number of steps are required in the recovery of Ge from flue dust to concentrate and purify it. These give pure GeO_2 which is reduced by H_2 to Ge at 500 °C. Transistor grade (ultrapure) material is obtained by zone refining. World production was about 49 tonnes in 1988 (USA 43%, USSR 22%, China and Austria 12% each and Japan 10%). It is used mainly for making transistors and semiconductor devices. It is transparent to infra-red light and is therefore also used for making prisms and lenses and windows in infra-red spectrophotometers and scientific apparatus.

Tin

The only important ore is cassiterite SnO_2. Mine production was 205 000 tonnes (metal content) in 1988. The main supplies now come from Brazil 21.5%, China 15%, Indonesia and Malaysia 14% each, the USSR 8%, Thailand 7%, Bolivia 5% and Australia 3%. In the UK, tin was mined in Cornwall from Roman times until this century, but these mines are now uneconomic.

SnO_2 is reduced to the metal using carbon at 1200–1300 °C in an electric

continued overleaf

furnace. The product often contains traces of Fe, which make the metal hard. Fe is removed by blowing air through the molten mixture to oxidize the iron to FeO, which then floats to the surface.

The main uses of Sn are electroplating steel to make tin-plate, and making alloys. Tin-plate is extensively used for making cans for food and drinks. The most important alloy is solder (Sn/Pb), but there are many others, including bronze (Cu/Sn), gun metal (Cu/Sn/Pb/Zn) and pewter (Sn/Sb/Cu). SnO_2 is used as a glaze in ceramics, and is often mixed with other metal oxides as pigments for pottery. $SnCl_4$ and Me_2SnCl_2 are used to produce very thin films of SnO_2 on glass. This toughens the glass, so bottles can be made with thinner walls, and glass can be made scratch resistant (useful for spectacles). Slightly thicker films are put on glass windows, to reduce heat losses. The film allows visible light to pass through, but reflects IR radiation, and thus keeps heat inside a room. A film of SnO_2 is put onto aircraft windows. This conducts electricity and thus produces heat, and prevents the window from frosting up. Large amounts of organotin compounds are used (estimated to be over 40 000 tonnes/year). Compounds R_2SnX_2 (where R is an alkyl group such as n-octyl and X is an organic acid residue such as laurate) are used to stabilize halogenated plastics such as PVC. Without a stabilizer, the plastic degrades in sunlight, air or on heating and becomes brittle and discoloured. The butyl compound Bu_2SnX_2 is used to 'cure' or vulcanize silicone rubber at room temperature. Inorganic tin compounds are used as flame retardants and smoke suppressants. Triorgano compounds such as Bu_3SnOH or Ph_3SnOAc are extensively used in agriculture to control fungi such as potato blight (*Botrytis infestans*) and similar fungal attack of vines, rice and sugar beet. Similar compounds kill red spider mites which attack fruit crops such as apples and pears, and other insects and larvae. They are also used to preserve wood. They make very effective and long-lasting antifouling paint for boats, preventing the build-up of barnacles. This use has now been banned because the heavy metal Sn seems to ha e entered the food chain.

Lead

The main ore is galena PbS. This is black, shiny and very dense.The main sources are the USSR 17%, Australia 14%, the USA 10%, Canada 9%, and Peru, Mexico and China 6% each. Galena is mined and separated from other minerals by froth flotation. There are two methods of extracting the element:

1. Roast in air to give PbO, and then reduce with coke or CO in a blast furnace.

$$\begin{array}{l} 2PbS + 3O_2 \rightarrow 2PbO \xrightarrow{+C} 2Pb_{(liquid)} + CO_{2(gas)} \\ \qquad\qquad\qquad\quad + SO_2 \end{array}$$

2. PbS is partially oxidized by heating and blowing air through it. After

some time the air is turned off and heating is continued, and the mixture undergoes self-reduction.

$$3PbS \xrightarrow[\text{air}]{\text{heat in}} PbS + 2PbO \xrightarrow[\text{absence of air}]{\text{heat in}} 3Pb_{(liquid)} + SO_{2(gas)}$$

The metal contains a number of metallic impurities: Cu, Ag, Au, Sn, As, Sb, Bi and Zn. These are removed by cooling to near the freezing point of Pb, when first Cu and then Zn containing most of the Ag and Au solidify. Preferential oxidation converts As, Sb and Sn to As_2O_3, Sb_2O_3, and SnO_2 which float on the surface of the molten metal and may be skimmed off. World production of Pb was 5.7 million tonnes in 1988. Of this 3.4 million tonnes was primary production from PbS. The main sources of ores are the USSR 15%, Australia 14%, the USA 12%, Canada 11% and China 9%. The recycling of scrap lead yielded 2.3 million tonnes. About 55% of the Pb produced is used to make lead/acid storage batteries. More than 158 million car batteries were produced in 1985. In these the supporting grid for the electrodes is made of an alloy of 91% Pb and 9% Sb. The active anode material is PbO_2, and the cathode material is spongy Pb. Over 80% of 'battery lead' is recovered from worn-out batteries and recycled. About 15% of lead production is used for lead sheets, lead pipes and solder. The manufacture of $PbEt_4$ as an additive to petrol at one time used 10–20%, but this is declining rapidly. About 10% is used in paints and pigments. Red lead paint containing Pb_3O_4 metal and white lead $(PbCO_3)_2 \cdot Pb(OH)_2$ was at one time widely used as an opacifier in paint. Their use has declined because lead is toxic, and TiO_2 is a good alternative opacifier. Calcium plumbate Ca_2PbO_4 is used for rustproofing corrugated steel sheets, and $PbCrO_4$ is used as a strong yellow pigment for road signs and markings. Lead compounds are also included in crown glass and cut glass, and in ceramic glazes.

STRUCTURE AND ALLOTROPY OF THE ELEMENTS

Carbon exists in six allotropic forms. These are diamond, α- and β-graphite, the rare hexagonal form of diamond, and two others produced by very high temperature or radiation treatment, which are thought to contain some triple bonds.

Si, Ge and Sn also exists as a metallic form. Pb exists only in the metallic form. Ge is unusual because the liquid expands when it forms the solid. This property is unique to Ga, Ge and Bi.

$$\underset{\substack{\text{grey tin}\\\text{(diamond structure)}}}{\alpha\text{-Sn}} \xrightleftharpoons{13.2\,^{\circ}\text{C}} \underset{\substack{\text{white tin}\\\text{(metallic)}}}{\beta\text{-Sn}}$$

Diamond is extremely unreactive, and in contrast graphite is quite reactive.

Diamonds are typically colourless, though industrial diamonds are often black. Most naturally occurring diamonds contain a trace of nitrogen, but 'blue diamonds' contain a trace of Al instead. In diamond each C atom is tetrahedrally surrounded by four other C atoms, each at a distance of 1.54 Å. The tetrahedra are linked together into a three-dimensional giant molecule. The unit cell is cubic. Strong covalent bonds extend in all directions. Thus the melting point is abnormally high (about 3930 °C) and the structure is very hard (see Figure 13.1). (In a rare modification of diamond, the tetrahedra are arranged differently to give a wurtzite-like structure and a hexagonal unit cell.)

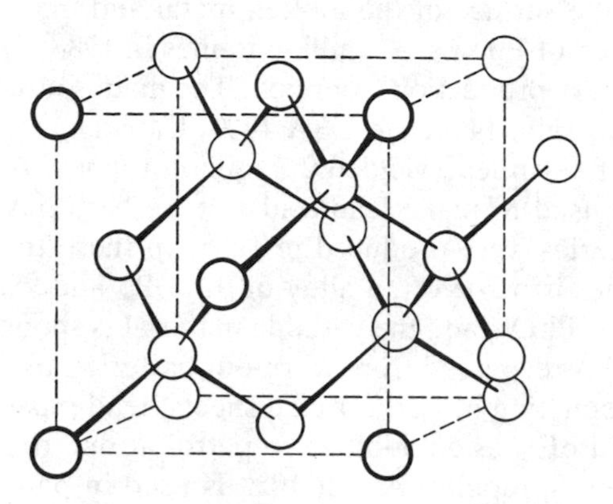

Figure 13.1 The crystal structure of diamond. (Wells, A.F., *Structural Inorganic Chemistry*, Clarendon Press, Oxford.)

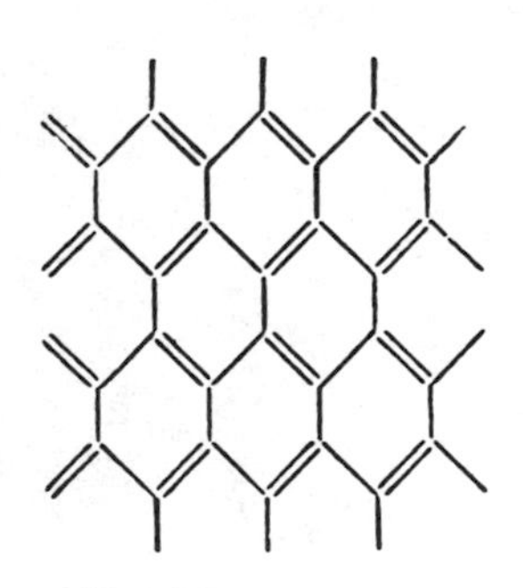

Figure 13.2 The structure of a graphite sheet.

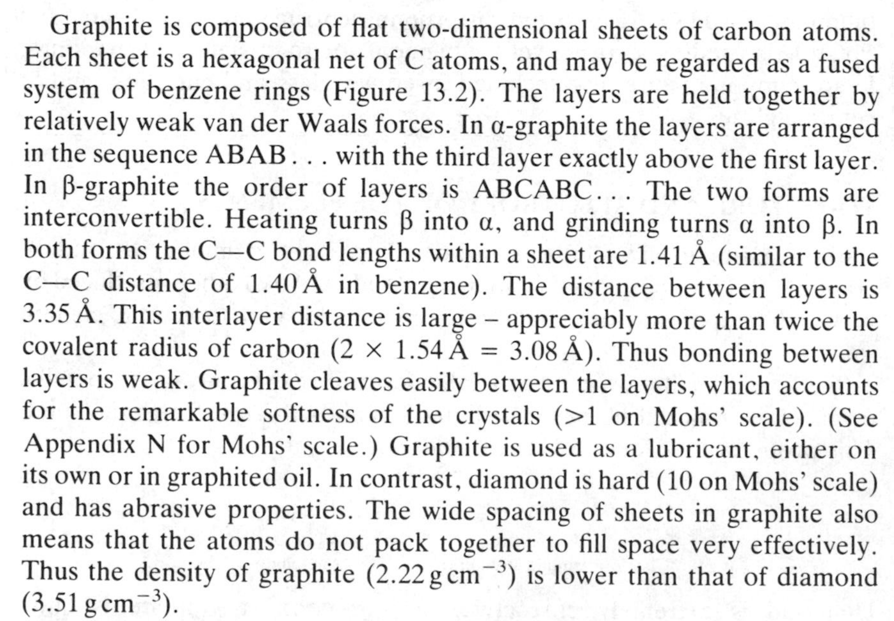

Graphite is composed of flat two-dimensional sheets of carbon atoms. Each sheet is a hexagonal net of C atoms, and may be regarded as a fused system of benzene rings (Figure 13.2). The layers are held together by relatively weak van der Waals forces. In α-graphite the layers are arranged in the sequence ABAB . . . with the third layer exactly above the first layer. In β-graphite the order of layers is ABCABC . . . The two forms are interconvertible. Heating turns β into α, and grinding turns α into β. In both forms the C—C bond lengths within a sheet are 1.41 Å (similar to the C—C distance of 1.40 Å in benzene). The distance between layers is 3.35 Å. This interlayer distance is large – appreciably more than twice the covalent radius of carbon (2 × 1.54 Å = 3.08 Å). Thus bonding between layers is weak. Graphite cleaves easily between the layers, which accounts for the remarkable softness of the crystals (>1 on Mohs' scale). (See Appendix N for Mohs' scale.) Graphite is used as a lubricant, either on its own or in graphited oil. In contrast, diamond is hard (10 on Mohs' scale) and has abrasive properties. The wide spacing of sheets in graphite also means that the atoms do not pack together to fill space very effectively. Thus the density of graphite ($2.22\,g\,cm^{-3}$) is lower than that of diamond ($3.51\,g\,cm^{-3}$).

In graphite only three of the valency electrons of each carbon atom are involved in forming σ bonds (using sp^2 hybrid orbitals). The fourth electron forms a π bond. The π electrons are delocalized over the whole sheet, and as they are mobile, graphite conducts electricity. Conduction can occur in a sheet, but not from one sheet to another.

Graphite is thermodynamically more stable than diamond, and its free energy of formation is $1.9\,kJ\,mol^{-1}$ lower at room temperature and ordinary pressure. Thermodynamically it is favourable for diamonds to turn into graphite. They do not normally do so because there is a high energy of activation for the process. If this activation energy is available, the change does occur, and diamond tipped drills do burn out and form graphite if they get too hot. The reverse process is not thermodynamically possible, and it requires very forcing high energy conditions to convert graphite to diamond. Graphite can be converted to synthetic diamonds at 1600 °C by a pressure of 50 000–60 000 atmospheres.

DIFFERENCES BETWEEN CARBON, SILICON AND THE REMAINING ELEMENTS

In general, the first element in a group differs from the rest of the group because of its smaller size and higher electronegativity. These result in the first element having a higher ionization energy, being more covalent, and being less metallic.

Using the classical theory of bonding, the first atom is limited to forming a maximum of four covalent bonds, because only *s* and *p* orbitals are available for bonding. This would limit the coordination number to 4 in these compounds. The majority of carbon compounds are either three- or four-coordinate. However, multi-centre bonds are now well established, and a number of compounds are known where carbon has higher coordination numbers, as shown in Table 13.3.

Table 13.3 Some carbon compounds with higher coordination numbers

Compound	Coordination number
$Al_2(CH_3)_6$	5
$B_{10}C_2H_{10}R_2$	6
$Li_4(CH_3)_4$	7
$[Co_8C(CO)_{18}]^{2-}$	8

In addition, carbon differs from the other elements in its unique ability to form $p\pi$–$p\pi$ multiple bonds, such as C=C, C≡C, C=O, C=S and C≡N. The later elements do not form $p\pi$–$p\pi$ bonds, principally because the atomic orbitals are too large and diffuse to obtain effective overlap, but they can use *d* orbitals in multiple bonding, particularly between Si and N or O. Thus $N(SiH_3)_3$ is planar and has $p\pi$–$d\pi$ bonding but $N(CH_3)_3$ is pyramidal and has no π bonding.

Carbon also differs from the others in its marked ability to form chains (catenation). This is because the C—C bonds are very strong, and the bonds Si—Si, Ge—Ge and Sn—Sn decrease progressively in strength (Table 13.4).

Table 13.4 Bond energies

Bond	Bond energy ($kJ\,mol^{-1}$)	Remarks
C—C	348	Forms many chains of great length
Si—Si	297	Forms a few chains up to Si_8H_{18} in hydrides and $Si_{16}F_{34}$, Si_6Cl_{14}, Si_4Br_{10} with halogens
Ge—Ge	260	Forms a few chains up to Ge_6H_{14} in hydrides and Ge_2Cl_6 with Cl
Sn—Sn	240	Forms dimer Sn_2H_6 in hydrides

Carbon and silicon have only *s* and *p* electrons, but the other elements follow a completed transition series with ten *d* electrons. Thus some differences are expected, and carbon and silicon differ both from one another and from the rest of the group, while germanium, tin and lead form a graded series.

CARBON DATING

The technique of carbon dating can be used to measure the age of archeological objects. Carbon occurs largely as the isotope ^{12}C, but there is a small amount of ^{13}C, which leads to the average atomic weight of 12.011. In the atmosphere, nitrogen is bombarded by cosmic neutrons, which produces the isotope ^{14}C.

$$^{14}_{7}N + ^{1}_{0}n \rightarrow ^{14}_{6}C + ^{1}_{1}H$$

This carbon reacts with oxygen, forming O_2, which is eventually used by green plants in photosynthesis to make glucose sugar. The glucose may be used by the plant to build starch, proteins, cellulose and other materials in the plant. All plant tissues thus contain traces of ^{14}C. Animals eat plants, so they too contain traces of ^{14}C. This isotope of carbon is weakly radioactive. It undergoes β-decay, and has a half life of 5668 years. While the plant or animal is alive, a natural balance exists between the intake of radiocarbon and that lost by decay. This steady state gives 15.3 ± 0.1 disintegrations per minute per gram of carbon. When the plant or animal dies the intake of radiocarbon ceases, but β-decay continues. Thus a very old sample of wood, cloth, paper, leather etc. will be less radioactive than a recent sample. A very small sample is burnt in oxygen, and the CO_2 produced is introduced into a suitable radiation counter. By carefully measuring the present radioactive decay rate, it is possible to calculate how long ago the plant or animal died. This provides an absolute scale for dating objects of plant or animal origin between 1000 and 20 000 years old. The technique has recently been used to determine the age of the Turin shroud, and many other objects. W.F. Libby was awarded the Nobel Prize for Chemistry in 1960 for developing this technique.

PHYSICAL PROPERTIES

Covalent radii

The covalent radii increase down the group. The difference in size between Si and Ge is less than might be otherwise expected because Ge has a full 3*d* shell which shields the nuclear charge rather ineffectively. In a similar way the small difference in size between Sn and Pb is because of the filling of the 4*f* shell.

Table 13.5 Radii, melting points and electronegativity values

	Covalent radius	Ionization energies ($kJ\ mol^{-1}$)				Melting point	Boiling point	Pauling's electro-negativity values
	(Å)	1st	2nd	3rd	4th	(°C)	(°C)	
C	0.77	1086	2354	4622	6223	4100		2.5
Si	1.17	786	1573	3232	4351	1420	3280	1.8
Ge	1.22	760	1534	3300	4409	945	2850	1.8
Sn	1.40	707	1409	2943	3821	232	2623	1.8
Pb	1.46	715	1447	3087	4081	327	1751	1.8

Ionization energy

The ionization energies decrease from C to Si, but then change in an irregular way because of the effects of filling the *d* and *f* shells. The amount of energy required to form M^{4+} ions is extremely large and hence simple ionic compounds are rare. The only elements which will give a large enough electronegativity difference to give ionic character are F and O. The compounds SnF_2, PbF_2, SnF_4, PbF_4, SnO_2, and PbO_2 are significantly ionic, but the only significant metal ion is Pb^{2+}.

Melting points

C has an extremely high melting point. Si melts appreciably lower than C, but the values for Si and Ge are still high. They all have the very stable diamond type of lattice. Melting involves breaking the strong covalent bonds in this lattice, and so requires a lot of energy. The melting points decrease on descending the group because the M—M bonds become weaker as the atoms increase in size (Table 13.4). Sn and Pb are metallic, and have much lower melting points. They do not use all four outer electrons for metallic bonding.

Metallic and non-metallic character

The change from non-metal to metal with increasing atomic number is well illustrated in Group IV, where C and Si are non-metals, Ge has some metallic properties, and Sn and Pb are metals. The increase in metallic character shows itself in the structures and appearance of the elements, in physical properties such as malleability and electrical conductivity, and in chemical properties such as the increased tendency to form M^{2+} ions and the acidic or basic properties of the oxides and hydroxides.

Four-covalent compounds

The majority of the compounds are four-covalent. In this case all four outer electrons take part in bonding. In the valence bond theory this is explained by promoting electrons from the ground state to an excited state. The energy needed to unpair and promote the electron is more than repaid by the energy released on forming two extra covalent bonds. The distribution of the four orbitals results in a tetrahedral structure, consistent with sp^3 hybridization.

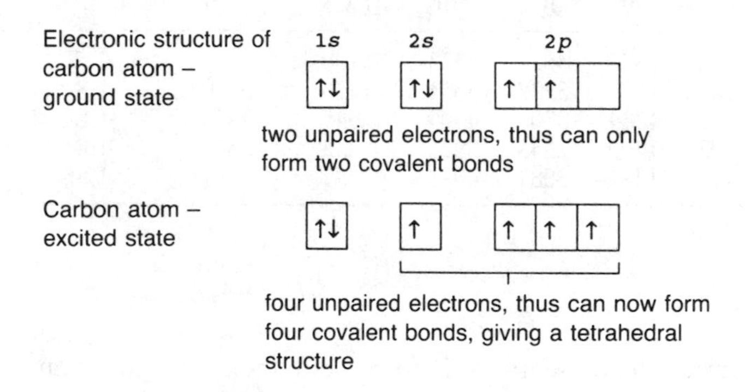

CHEMICAL REACTIVITY

The elements in this group are relatively unreactive, but reactivity increases down the group. The M^{II} oxidation state becomes increasingly stable on descending the group. Pb often appears more noble (unreactive) than expected from its standard electrode potential of -0.13 volts. The unreactiveness is partly due to a surface coating of oxide, and partly due to the high overpotential for the reduction of H^+ to H_2 at a Pb surface. The production of H_2 from H^+ at a lead electrode is kinetically unfavourable, so a much larger potential is required than the standard reduction potential.

C, Si and Ge are unaffected by water. Sn reacts with steam to give SnO_2 and H_2. Pb is unaffected by water, probably because of a protective oxide film.

C, Si and Ge are unaffected by dilute acids. Sn dissolves in dilute HNO_3, forming $Sn(NO_3)_2$. Pb dissolves slowly in dilute HCl, forming the sparingly soluble $PbCl_2$ and quite readily in dilute HNO_3, forming $Pb(NO_3)_2$ and oxides of nitrogen. Pb also dissolves in organic acids (e.g. acetic, citric and oxalic acids). Pb does not dissolve in dilute H_2SO_4 because a surface coating of $PbSO_4$ is formed.

Diamond is unaffected by concentrated acids, but graphite reacts with hot concentrated HNO_3, forming mellitic acid, and with a mixture of hot concentrated HF/HNO_3, forming graphite oxide. Si is oxidized and fluorinated by concentrated HF/HNO_3. Ge dissolves slowly in hot concentrated H_2SO_4 and in HNO_3. Sn dissolves in several concentrated acids. Pb does not dissolve in concentrated HCl because a surface coating of $PbCl_2$ is formed.

C is unaffected by alkalis. Si reacts slowly with cold aqueous solutions of NaOH, and readily with hot solutions, giving solutions of silicates $[SiO_4]^{4-}$. Sn and Pb are slowly attacked by cold alkali, and rapidly by hot alkali, giving stannates $Na_2[Sn(OH)_6]$ and plumbates $Na_2[Pb(OH)_6]$. Thus Sn and Pb are amphoteric.

Diamond does not react with the halogens, but graphite reacts with F_2 at 500 °C, forming intercalation compounds or graphite fluoride $(CF)_n$. Si and Ge react readily with all the halogens, forming volatile halides SiX_4 and GeX_4. Sn and Pb are less reactive. Sn reacts with Cl_2 and Br_2 in the cold, and with F_2 and I_2 on warming, giving SnX_4. Pb reacts with F_2 in the cold, forming PbF_2, and with Cl_2 on heating, giving $PbCl_2$.

INERT PAIR EFFECT

The inert pair effect shows itself increasingly in the heavier members of the group. There is a decrease in stability of the (+IV) oxidation state and an increase in the stability of the (+II) state on descending the group. Ge(+II) is a strong reducing agent whereas Ge(+IV) is stable. Sn(+II) exists as simple ions which are strongly reducing but Sn(+IV) is covalent and stable. Pb(+II) is ionic, stable and more common than Pb(+IV), which is oxidizing. The lower valencies are more ionic because the radius of M^{2+} is greater than that of M^{4+} and according to Fajans' rules, the smaller the ion the greater the tendency to covalency.

STANDARD REDUCTION POTENTIALS (VOLTS)

Acid solution	**Basic solution**
Oxidation state **+IV** **+II** **0**	**+IV** **+II** **0**
$Sn^{4+} \xrightarrow{+0.15} Sn^{2+} \xrightarrow{-0.14} Sn$	$[Sn(OH)]_6^{2-} \xrightarrow{-0.90} HSnO_2^- \xrightarrow{-0.91} Sn$
$PbO_2 \xrightarrow{+1.46} Pb^{2+} \xrightarrow{-0.13} Pb$	$PbO_2 \xrightarrow{+0.28} PbO \xrightarrow{-0.54} Pb$

GRAPHITE COMPOUNDS

The distance between layers in graphite is large: hence the bonding between layers is weak. Thus a large number of substances can invade the space between sheets, forming intercalation compounds of varying composition. When atoms, molecules or ions invade the space between the layers, they cause an increase in the interlayer distance. Provided that the graphite sheets remain flat, the new compound retains its graphite-like character: the π electrons continue to be delocalized over the whole layer, and are thus able to conduct electricity. If the invading atoms add electrons to the π system, the electrical conductivity is increased. Reactions of this kind (i.e. intercalation reactions) are often reversible.

When graphite is heated to about 300 °C with the vapours of the heavier Group I metals K, Rb and Cs, it absorbs metal, forming a bronze coloured compound C_8M. The bronze colour is due to the formation of metal atom clusters at these relatively high metal concentrations, in the same way as clusters are formed in solutions of these metals in liquid ammonia. If C_8M is heated to 350 °C under reduced pressure, metal is lost and a series of intercalation compounds are formed ranging from steel blue to blue or black in colour, depending on the number of layers invaded by the metal (see Table 13.6).

$$C + M \rightarrow C_8M$$
$$C_8M \rightarrow C_{24}M \rightarrow C_{36}M \rightarrow C_{48}M \rightarrow C_{60}M$$

Intercalation compounds of Li and Na are more difficult to make, but a series of compounds is known: C_6Li, $C_{12}Li$, $C_{16}Li$, $C_{18}Li$ and $C_{40}Li$.

The crystal structure of C_8K is known. The graphite sheets remain intact, but the gap between the sheets increases because of the presence of metal atoms. The C atoms in one sheet are arranged vertically above those in the sheet below, rather than in the ABAB... arrangement found in α-graphite. Since the sheets remain flat, they retain their delocalized π electron system. Thus C_8K can conduct electricity, but the electrical resistance is appreciably lower than for α-graphite, i.e. C_8M conducts better than graphite (resistance at 285 K: α-graphite 28.4 ohm cm; C_8K 1.02 ohm cm). Graphite is diamagnetic, and C_8K is paramagnetic. This suggests that bonding between metal and graphite layers involves the transfer of an electron from the alkali metal atom to the π system (that is to the conduction band) of the graphite sheets ($K \rightarrow K^+ + e$). The presence of the invading species forces the graphite sheets apart from their usual distance of 3.35 Å up to a distance as great as 10 Å. These alkali metal graphite compounds are highly reactive. They may explode in water, and react vigorously in air.

$FeCl_3$ reacts with graphite and forms a different type of intercalation compound. Similar behaviour is found with:

1. the halogens Cl_2 and Br_2;
2. HF;

Table 13.6 Idealized representation of graphite compounds showing different numbers of layers invaded by metal

C_8M bronze	$C_{24}M$ steel-blue	$C_{36}M$ blue	$C_{48}M$ black	$C_{60}M$ black
every layer invaded	every second layer invaded	every third layer invaded	every fourth layer invaded	every fifth layer invaded
−C	−C	−C	−C	−C
−M	−M	−M	−M	−M
−C	−C	−C	−C	−C
−M				
−C	−C	−C	−C	−C
−M	−M			
−C	−C	−C	−C	−C
−M		−M		
−C	−C	−C	−C	−C
−M	−M		−M	
−C	−C	−C	−C	−C
−M				−M
−C	−C	−C	−C	−C
−M	−M	−M		
−C	−C	−C	−C	−C
−M				
−C	−C	−C	−C	−C
−M	−M		−M	
−C	−C	−C	−C	−C
−M		−M		
−C	−C	−C	−C	−C
−M	−M			−M
−C	−C	−C	−C	−C

3. a large number of halides including $CdCl_2$, $CuBr_2$, $FeCl_3$, $AlCl_3$, ClF_3, TiF_4, $MoCl_5$, SbF_5, UCl_6, and XeF_6;
4. a number of oxides CrO_3, MoO_3, SO_3, N_2O_5 and Cl_2O_7;
5. and some sulphides FeS_2, PdS, V_2S_3.

Some of the invading compounds can act as electron pair acceptors. In others, for example $FeCl_3$, the compound C_6FeCl_3 is formed in which the $FeCl_3$ forms a layer lattice within the host lattice of graphite. This is almost the same as the layer lattice formed by $FeCl_3$ itself. The presence of this kind of invading species increases the electrical conductivity of graphite by a factor of up to ten times. There seems to be a transfer of electrons from graphite to the invading atoms. With Cl_2 and Br_2 the halogen may remove bonding electrons from graphite ($Cl + e \rightarrow Cl^-$) thus leaving a 'positive hole' in the valence band. The positive hole can migrate, and therefore can carry current. It is not known how conduction occurs in the halide intercalation compounds.

A third class of compounds is formed between O and F with graphite. These compounds are non-conducting. Graphite oxide is formed when graphite is oxidized with strong reagents such as concentrated HNO_3, $HClO_4$ or $KMnO_4$. Graphite oxide is unstable, pale lemon coloured and nonstoichiometric. It decomposes slowly at 70 °C, and catches fire at 200 °C, forming H_2O, CO_2, CO and C. The O : C ratio approaches 1 : 2, but is often short of oxygen and frequently contains hydrogen. The interlayer spacing is increased to 6–7 Å. The oxide absorbs water, alcohols, acetone and a variety of molecules. This may increase the interlayer spacing up to 19 Å. X-ray diffraction shows a layer structure with puckered sheets made up of a hexagonal network of atoms. The C_6 units are mostly in the chair conformation, but a few C=C bonds remain. The oxygen forms bridging (ether-like) linkages C—O—C and C—OH groups which may undergo keto–enol tautomerism ≡C—OH to ⟩C=O. The sheets are buckled because all four electrons on a C atom are now involved in σ bonding. This destroys the delocalized system of mobile π electrons found in the flat sheets in graphite, and this explains the loss of electrical conductivity.

Graphite fluoride is formed by heating graphite in F_2 at 450 °C. The reaction proceeds at a lower temperature in the presence of HF. This can happen in cells producing F_2, and not only will it destroy the electrode, but it may also cause an explosion. The product CF_n is nonstoichiometric, and n varies from 0.7 to 0.98. The colour varies from black through grey to silver and white with increasing fluorine content. The interlayer spacing is about 8 Å. The structure is thought to be a layer structure with buckled sheets. It involves tetrahedral bonding by C atoms. CF is non-conducting, and very unreactive.

CARBIDES

Compounds of carbon and a less electronegative element are called carbides. This excludes compounds with N, P, O, S and the halogens from this section. Carbides are of three main types:

1. ionic or salt-like
2. interstitial or metallic
3. covalent

The formulae of some of the compounds cannot be rationalized by the application of simple valency rules. All three types are prepared by heating the metal or its oxide with carbon or a hydrocarbon at temperatures of 2000 °C.

Salt-like carbides

It is convenient to group these depending on whether the structure contains C, C_2 or C_3 'anions'.

Beryllium carbide Be_2C is a red solid and may be made by heating C and BeO at 2000 °C. Aluminium carbide Al_4C_3 is a pale yellow solid formed by heating the elements in an electric furnace. Be_2C contains individual C atoms/ions, but the structure of Al_4C_3 is complex. It is misleading to formulate the structure as $4Al^{3+}$ and $3C^{4-}$ as such a high charge separation is unlikely. Both Be_2C and Al_4C_3 are called *methanides* because they react with H_2O, yielding methane.

Carbides with a C_2 unit are well known. They are formed mainly by the elements in Group I ($M^I_2C_2$); Group II ($M^{II}C_2$); the coinage metals (Cu, Ag, Au); Zn and Cd; and some of the lanthanides (LnC_2 and $Ln_4(C_2)_3$). These are all colourless ionic compounds and contain the carbide ion $(—C\equiv C—)^{2-}$. By far the most important compound is CaC_2. This is made commercially by strongly heating lime and coke:

$$CaO + 3C \rightarrow CaC_2 + CO \qquad \Delta H = +466\,kJ\,mol^{-1}$$

The reaction is endothermic, and a temperature of 2200 °C is required. These carbides react exothermically with water, liberating ethyne (formerly called acetylene), so they are called *acetylides*.

$$CaC_2 + 2H_2O \rightarrow Ca(OH)_2 + HC\equiv CH$$

Production of CaC_2 reached a maximum of 7 million tonnes/year in 1960, but has since declined slightly to 6.2 million tonnes in 1985. At one time it was the major source of acetylene for oxy-acetylene welding, but ethyne is now obtained mainly from oil. CaC_2 is an important chemical intermediate and is used on an industrial scale to produce calcium cyanamide. Cyanamide is used as a nitrogenous fertilizer, and to make urea and melamine. (See Chapter 11, under 'Carbides'.)

$$CaC_2 + N_2 \xrightarrow{1100\,°C} Ca(NCN) + C$$

The acetylides have a NaCl type of lattice, with Ca^{2+} replacing Na^+ and C_2^{2-} replacing Cl^-. In CaC_2, SrC_2 and BaC_2 the elongated shape of the $(C\equiv C)^{2-}$ ion causes tetragonal distortion of the unit cell, that is it elongates the unit cell in one direction – see Chapter 3 Figure 3.12.

One of the two carbides of magnesium Mg_2C_3 contains a C_3 unit, and on hydrolysis with water it yields propyne $CH_3—C\equiv CH$.

Interstitial carbides

These are formed mostly by transition elements, and some of the lanthanides and actinides. The Cr, Mn, Fe, Co and Ni groups form a large number of carbides with a wide range of stoichiometries. They are typically infusible or are very high melting, and are very hard. For example, TaC has a melting point of 3900 °C, and is very hard (9–10 on Mohs' scale of hardness), and WC is also very hard. Both are used to make cutting tools for lathes. Interstitial carbides retain many of the properties of metals. They conduct electricity by metallic conduction, and have a lustre like a metal.

In these compounds, C atoms occupy octahedral holes in the close-packed metal lattice, and so do not affect the electrical conductivity of the metal. Provided that the size of the metal is greater than 1.35 Å, the octahedral holes are large enough to accommodate C atoms without distorting the metal lattice. (Since we are considering a metal lattice, 12-coordinate radii must be used.) If all the octahedral holes are occupied the formula is MC. Interstitial carbides are generally unreactive. They do not react with H_2O like ionic carbides. Most react slowly with concentrated HF or HNO_3.

Some metals, including Cr, Mn, Fe, Co and Ni, have radii below 1.35 Å: hence the metal lattice is distorted. Thus the structures are more complicated, for compounds such as V_2C, Mn_5C_2, Fe_3C, V_4C_3 and others. Cementite Fe_3C is an important constituent of steel. These carbides are more reactive, and are hydrolysed by dilute acids, and in some cases by water, giving a mixture of hydrocarbons and H_2.

Some carbides are based on the NaCl structure, with C occupying all of the Cl^- positions. These include carbides of some of the early transition metals TiC, ZrC, HfC, VC, NbC, TaC, CrC and MoC, and those of some actinides such as ThC, UC and PuC.

Covalent carbides

SiC and B_4C are the most important. Silicon carbide is hard (9.5 on Mohs' scale), infusible and chemically inert.It is widely used as an abrasive called carborundum, and about 300 000 tonnes are produced annually by heating quartz or sand with an excess of coke in an electric furnace at 2000–2500 °C.

$$SiO_2 + 2C \rightarrow Si + 2CO$$
$$Si + C \rightarrow SiC$$

SiC is very unreactive. It is unaffected by acids (except H_3PO_4), but it does react with NaOH and air, and with Cl_2 at 100 °C.

$$SiC + 2NaOH + 2O_2 \rightarrow Na_2SiO_3 + CO_2 + H_2O$$
$$SiC + 2Cl_2 \rightarrow SiCl_4$$

SiC is often dark purple, black or dark green due to traces of Fe and other impurities, but pure samples are pale yellow to colourless. SiC has a three-dimensional structure of Si and C atoms, each atom tetrahedrally surrounded by four of the other kind. There are a large number of different crystal forms based on either the diamond or wurtzite structures. Boron carbide is even harder than silicon carbide and is used both as an abrasive and as a shield from radiation. It is manufactured in tonne quantities. Its formula is more correctly represented by $B_{13}C_2$ (see Chapter 12, under 'Borides').

OXYGEN COMPOUNDS

Carbon forms more oxides than the other elements, and these oxides differ from those of the other elements because they contain $p\pi - p\pi$ multiple bonds between C and O. Two of these oxides, CO and CO_2, are extremely stable and important. Three are less stable: C_3O_2, C_5O_2 and $C_{12}O_9$. Others which are even less stable include graphite oxide, C_2O and C_2O_3.

Carbon monoxide CO

CO is a colourless, odourless, poisonous gas. It is formed when C is burned in a limited amount of air. In the laboratory it is prepared by dehydrating formic acid with concentrated H_2SO_4.

$$H \cdot COOH + H_2SO_4 \rightarrow CO + H_2O$$

CO can be detected because it burns with a blue flame. It also reduces an aqueous $PdCl_2$ solution to metallic Pd, and when passed through a solution of I_2O_5 it liberates I_2, i.e. it reduces I_2O_5 to I_2. The latter reaction is used to estimate CO quantitatively. The I_2 is titrated with $Na_2S_2O_3$.

$$PdCl_2 + CO + H_2O \rightarrow Pd + CO_2 + 2HCl$$

$$5CO + I_2O_5 \rightarrow 5CO_2 + I_2$$

CO is toxic because it forms a complex with haemoglobin in the blood, and this complex is more stable than oxy-haemoglobin. This prevents the haemoglobin in the red blood corpuscles from carrying oxygen round the body. This causes an oxygen deficiency, leading to unconsciousness and then death. CO is sparingly soluble in water and is a neutral oxide. CO is an important fuel, because it evolves a considerable amount of heat when it burns in air.

$$2CO + O_2 \rightarrow 2CO_2 \qquad \Delta H^\circ = -565\,\text{kJ mol}^{-1}$$

The following are all important industrial fuels:

1. *Water gas*: an equimolecular mixture of CO and H_2.
2. *Producer gas*: a mixture of CO and N_2.
3. *Coal gas*: a mixture of CO, H_2, CH_4 and CO_2, produced at a gasworks by distilling coal, and stored in large gas holders. This was the 'town gas' supplied to peoples' homes for cooking and heating. In the UK it has now been replaced by natural gas (CH_4), but town gas is still used in some countries.

Water gas is made by blowing steam through red or white hot coke.

$$C + H_2O \xrightarrow{\text{red heat}} CO + H_2 \quad \text{(water gas)} \qquad \Delta H^\circ = +131\,\text{kJ mol}^{-1}$$
$$\Delta S^\circ = +134\,\text{kJ mol}^{-1}$$

The water gas reaction is strongly endothermic ($\Delta G = \Delta H - T\Delta S$). Thus the coke cools down, and at intervals the steam must be turned off and air blown through to reheat the coke. It is a particularly good fuel, i.e. it has a high calorific value, because both CO and H_2 burn and evolve heat.

Producer gas is made by blowing air through red hot coke.

$$\underbrace{C + O_2 + 4N_2}_{\text{air}} \rightarrow CO_2 + 4N2$$
$$\downarrow +C$$
$$2CO + 4N_2 \quad \text{(producer gas)}$$

The overall reaction is exothermic, so the coke does not cool down as with water gas.

$$2C + O_2 \rightarrow 2CO \qquad \Delta H^\circ = -221\,\text{kJ mol}^{-1} \text{ and } \Delta S^\circ = +179\,\text{kJ mol}^{-1}$$

Producer gas is a less efficient fuel than water gas, i.e. it has a lower calorific value, as only part of the gas will burn. The approximate composition of producer gas is 70% N_2, 25% CO, 4% CO_2 with traces of CH_4, H_2 and O_2.

CO is a good reducing agent and can reduce many metal oxides to the metal. (See 'The occurrence and isolation of the elements', and 'Thermodynamics of reduction processes', Chapter 6.)

$$Fe_2O_3 + 3CO \xrightarrow{\text{blast furnace}} 2Fe + 3CO_2$$
$$CuO + CO \rightarrow Cu + CO_2$$

CO is an important ligand. It can donate an electron pair to many transition metals, forming carbonyl compounds. The number of CO molecules bonded to the metal in this way is generally in accordance with the effective atomic number rule (see Chapter 7). However, the bonding is more complicated than this implies. A number of different stoichiometries are formed (Table 13.7).

Table 13.7 Binary metal carbonyls formed by the first row transition elements

Sc	Ti	V	Cr	Mn	Fe	Co	Ni	Cu	Zn
		$V(CO)_6$	$Cr(CO)_6$	$Mn_2(CO)_{10}$	$Fe(CO)_5$	$Co_2(CO)_8$	$Ni(CO)_4$		
					$Fe_2(CO)_9$	$Co_4(CO)_{12}$			
					$Fe_3(CO)_{12}$	$Co_6(CO)_{16}$			

Carbonyl compounds may be made by a variety of reactions:

$$Ni + 4CO \xrightarrow{28\,^\circ\text{C}} Ni(CO)_4$$

$$Fe + 5CO \xrightarrow{200\,^\circ\text{C under pressure}} Fe(CO)_5$$

$$2Fe(CO)_5 \xrightarrow{\text{photolysis}} Fe_2(CO)_9 + CO$$

$$CrCl_6 + 3Fe(CO)_5 \xrightarrow{\text{heat}} Cr(CO)_6 + 3FeCl_2 + 9CO$$

In the Mond process (now obsolete) for purifying nickel, nickel carbonyl $Ni(CO)_4$ was made from Ni and CO at 50 °C. (Water gas was used as the source of CO.) $Ni(CO)_4$ is a gas and can be separated from other metals and impurities. The $Ni(CO)_4$ gas was then decomposed at 230 °C. Though the original process is obsolete, a modified process is used in Canada.

The bonding in CO may be represented as three electron pairs shared between the two atoms:

$$:\mathrm{C} \,\vdots\!\vdots\, \mathrm{O}: \quad \text{or} \quad \mathrm{C{\equiv}O}$$

It is better represented using the molecular orbital theory (see Chapter 4).

$$\sigma 1s^2,\ \sigma^* 1s^2,\ \sigma 2s^2,\ \sigma^* 2s^2,\ \begin{cases}\pi 2p_y^2, \\ \pi 2p_z^2,\end{cases} \sigma 2p_x^2,\ \sigma^* 2p_x^0,\ \begin{cases}\pi^* 2p_y^0 \\ \pi^* 2p_z^0\end{cases}$$

$$\xrightarrow{\text{increasing energy}}$$

The carbon–metal bond in carbonyls may be represented as the donation of an electron pair from carbon to the metal $M \leftarrow C{\equiv}O$. This original σ bond is weak. A stronger second bond is formed by back bonding, sometimes called dative π bonding. This arises from sideways overlap of a full d_{xy} orbital on the metal with the empty antibonding $\pi^* 2p_y$ orbital of the carbon, thus forming a π $M \rightarrow C$ bond. The total bonding is thus $M{=}C{=}O$. The filling, or partial filling, of the antibonding orbital on C reduces the bond order of the C—O bond from the triple bond in CO towards a double bond. This is shown by the increase in C—O bond length from 1.128 Å in CO to about 1.15 Å in many carbonyls.

CO is the most studied organometallic ligand. Because of the back bonding it is sometimes called a π acceptor ligand. The drift of π electron density from M to C makes the ligand more negative, which in turn enhances its σ donating power. Thus CO forms weak bonds to Lewis acids (electron pair acceptors) such as BF_3 as only σ bonding is involved. In contrast CO forms strong bonds to transition metals where both σ and π bonding can occur. Other π acceptor ligands include CN^-, RNC, and

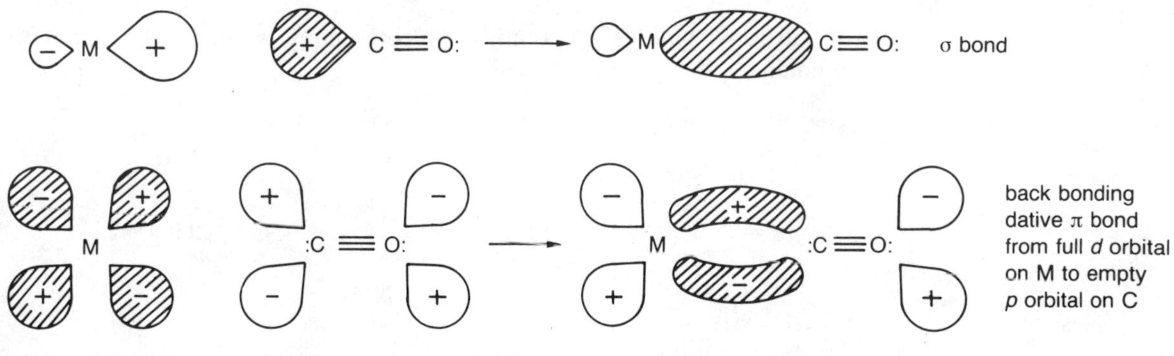

Figure 13.3 Schematic of orbital overlaps in metal carbonyls. (After N.N. Greenwood and A. Earnshaw, *Chemistry of the Elements*, Pergamon, 1984, p. 351.)

NO^+. Comparing these ligands, the strengths of the σ bonds are in the order $CN^- > RNC > CO > NO^+$, whilst their π acceptor properties are in the reverse order.

CO is a very versatile ligand. It may act as a bridging group between the two metal atoms, for example in di-iron ennea carbonyl $Fe_2(CO)_9$. CO may stabilize metal clusters by the C forming a multi-centre bond with three metal atoms, and the π^* orbitals in CO may be involved in bonding to other metal atoms.

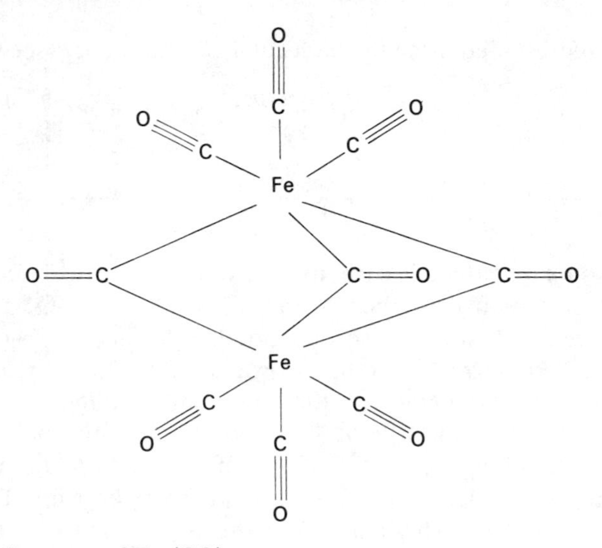

Figure 13.4 Structure of $Fe_2(CO)_9$.

Carbon monoxide is quite reactive, and combines readily with O, S and the halogens F, Cl and Br.

$$CO + \tfrac{1}{2}O_2 \rightarrow CO_2$$

$$CO + S \rightarrow COS \quad \text{carbonyl sulphide}$$

$$CO + Cl_2 \rightarrow COCl_2 \quad \text{carbonyl chloride (phosgene)}$$

The carbonyl halides are readily hydrolysed by water, and react with ammonia to form urea:

$$COCl_2 + H_2O \rightarrow 2HCl + CO_2$$

$$Cl_2C{=}O + 2NH_3 \xrightarrow{\text{gas phase}} (NH_2)_2C{=}O + 2HCl$$

urea

Carbonyl chloride is extremely toxic, and was used as a poisonous gas in World War I. Nowadays it is produced in quite large quantities to make tolylene diisocyanate which is an intermediate in the manufacture of polyurethane plastics.

Carbon dioxide CO_2

CO_2 is a colourless, odourless gas. It is a major industrial chemical, and production in the USA exceeds 33 million tonnes/year. The main industrial source is as a by-product from the manufacture of hydrogen for making ammonia.

$$CO + H_2O \rightleftharpoons CO_2 + H_2$$

$$CH_4 + 2H_2O \rightarrow CO_2 + 4H_2$$

It is also recovered from fermentation processes in breweries, from the gases evolved from calcining limestone in lime kilns and from the flue gases from coal-burning electric power stations. The CO_2 is recovered by absorbing it in either aqueous Na_2CO_3 or ethanolamine.

$$C_6H_{12}O_6 \xrightarrow[\text{anaerobic conditions}]{\text{yeast under}} 2C_2H_5OH + 2CO_2$$

$$CaCO_3 \xrightarrow{\text{strong heat}} CaO + CO_2$$

It is obtained in small amounts by the action of dilute acids on carbonates. It can also be made by burning carbon in excess of air.

$$CaCO_3 + 2HCl \rightarrow CaCl_2 + CO_2 + H_2O$$

$$C + O_2 \rightarrow CO_2$$

Recovery of CO_2

$$Na_2CO_2 + CO_2 + H_2O \underset{\text{hot}}{\overset{\text{cool}}{\rightleftharpoons}} 2NaHCO_3$$

Girbotol process

$$\underset{\text{ethanolamine}}{2HOCH_2CH_2NH_2} + CO_2 + H_2O \underset{100-150\,^\circ C}{\overset{30-60\,^\circ C}{\rightleftharpoons}} HOCH_2CH_2NH_3)_2CO_3$$

CO_2 gas can be liquified under pressure between −57 °C and +31 °C. About 80% is sold in liquid form, and 20% as solid. The solid is produced as white snow by expanding the gas from cylinders. (Expansion causes cooling.) This is compacted into blocks and sold. Solid CO_2 sublimes directly to the vapour state (without going through the liquid state) at −78 °C under atmospheric pressure. Over half the CO_2 produced is used as a refrigerant. Solid CO_2 is called 'dry ice' or 'cardice', and is used to freeze meat, frozen foods and ice cream, and in the laboratory as a coolant. Over a quarter is used to carbonate drinks (Coca-Cola, lemonade, beer etc.). Other uses include the manufacture of urea, as an inert atmosphere, and for neutralizing alkalis. Over 6 million tonnes/year of urea is produced worldwide. (Urea is used as a nitrogenous fertilizer and for making formaldehyde urea resins.)

$$\underset{}{CO_2} + 2NH_3 \xrightarrow[\text{pressure}]{180\,^\circ\text{C}} \underset{\substack{\text{ammonium}\\ \text{carbamate}}}{NH_4CO_2NH_2} \rightarrow \underset{\text{urea}}{CO(NH_2)_2} + H_2O$$

Small scale uses of CO_2 include use in fire extinguishers, blasting in coal mines, as an aerosol propellant, and for inflating life-rafts.

CO_2 gas is detected by its action on lime water $Ca(OH)_2$ or baryta water $Ba(OH)_2$, as a white insoluble precipitate of $CaCO_3$ or $BaCO_3$ is formed. If more CO_2 is passed through the mixture, the cloudiness disappears as the soluble bicarbonate is formed.

$$Ca(OH)_2 + CO_2 \rightarrow \underset{\substack{\text{white}\\ \text{precipitate}}}{CaCO_3} + H_2O$$

$$CaCO_3 + CO_2 + H_2O \rightarrow \underset{\text{soluble}}{Ca(HCO_3)_2}$$

CO_2 is an acidic oxide, and reacts with bases, forming salts. It dissolves in water but it is only slightly hydrated to carbonic acid H_2CO_3, and the solution contains few carbonate or bicarbonate ions. A hydrate $CO_2 \cdot 8H_2O$ can be crystallized at 0 °C under a pressure of 50 atmospheres CO_2.

$$CO_2 + H_2O \rightleftharpoons H_2CO_3$$

Carbonic acid has never been isolated, but it gives rise to two series of salts, hydrogencarbonates (otherwise called bicarbonates), and carbonates.

$$NaOH + (H_2CO_3) \begin{cases} \nearrow NaHCO_3 & \text{sodium bicarbonate (acid salt)} \\ \searrow Na_2CO_3 & \text{sodium carbonate (normal salt)} \end{cases}$$

CO_2 can also act as a ligand, and it form a few complexes such as $[Rh(CO_2)Cl(PR_3)_3]$ and $[Co(CO_2)(PPh_3)_3]$. In the first complex the C atom in CO_2 is bonded to the metal. In the second complex the CO_2 acts as a bidentate ligand with one C atom and one O atom bonded to the metal, and the CO_2 molecule is bent.

The structure of CO_2 is linear O—C—O. Both C—O bonds are the same length. In addition to σ bonds between C and O, there is a three-centre four-electron π bond covering all three atoms. This adds two π bonds to the structure in addition to the two σ bonds. Thus the C—O bond order is two. This is described in more detail in Chapter 4.

Biologically, carbon dioxide is important in the processes of photosynthesis, where the green parts of plants manufacture glucose sugar. Ultimately all animal and plant life depends on this process.

$$6CO_2 + 6H_2O \xrightarrow{\text{sunlight}} \underset{\text{glucose}}{C_6H_{12}O_6} + 6O_2$$

The reverse reaction occurs during the process of respiration, where animals and plants release energy.

$$C_6H_{12}O_6 + 6O_2 \rightarrow 6CO_2 + 6H_2O + \text{energy}$$

Carbon suboxides

Carbon suboxide C_3O_2 is a foul-smelling gas, boiling point 6 °C. It is made by dehydrating malonic acid with P_4O_{10}.

$$\underset{\text{malonic acid}}{HOOC \cdot CH_2 \cdot COOH} \xrightarrow[150\,°C]{P_4O_{10}} O{=}C{=}C{=}C{=}O + 2H_2O$$

It is stable at −78 °C and the molecule is linear. At room temperature the gas polymerizes to a yellow solid, and at higher temperatures to red and purple solids. The oxide reacts with H_2O, giving malonic acid, and with HCl and NH_3 as follows:

$$C_3O_2 + 2HCl \rightarrow CH_2(COCl)_2 \quad \text{(acid chloride)}$$
$$C_3O_2 + 2NH_3 \rightarrow CH_2(CONH_2)_2 \quad \text{(amide)}$$

There are disputed reports that C_5O_2 is formed by thermolysis of C_3O_2. The only other stable suboxide is $C_{12}O_9$. This is a white solid, and is the anhydride of mellitic acid $C_6(COOH)_6$.

CARBONATES

There are two series of salts from carbonic acid H_2CO_3, namely carbonates CO_3^{2-} and hydrogencarbonates HCO_3^-. The CO_3^{2-} ion is flat. The CO_4^{4-} ion does not exist, even though SiO_4^{4-} does. This is probably because C is too small, and the situation is analogous to the formation of NO_3^- and PO_4^{3-} in Group V. The structure of the CO_3^{2-} ion may be represented as follows:

Electronic structure of carbon having gained four electrons by forming four bonds in CO_3^{2-}

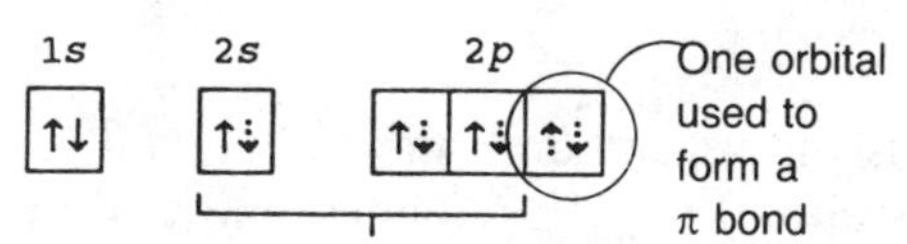

The π bonding in the CO_3^{2-} ion is best described using a delocalized π molecular orbital covering all four atoms. (See Chapter 4.)

Many carbonates of formula $M^{II}CO_3$ have the same structure as calcite, but others have the aragonite structure. The structure adopted is related to the size of the metal ions.

Mn^{2+}	Mg^{2+}	Co^{2+}	Zn^{2+}	Fe^{2+}	Cd^{2+}	Ca^{2+}	Sr^{2+}	Pb^{2+}	Ba^{2+}
0.67	0.72	0.74	0.74	0.78	0.97	1.00	1.18	1.21	1.35
← calcite structure →									
						← aragonite structure →			

Some carbonates are produced in very large amounts: 22 million tonnes/year of Na_2CO_3, 200 000 tonnes/year $NaHCO_3$, and 50 000 tonnes/year of Li_2CO_3 (see Chapters 9 and 10).

Carbonate ions are colourless and hence the carbonates of Group I and II metals are white. Though Ag^+ salts are typically white, Ag_2CO_3 is yellow due to the strong polarizing effect of Ag^+. $(NH_4)CO_3$ and Group I carbonates are readily soluble in water, except Li_2CO_3 which is only slightly soluble. Tl_2CO_3 is moderately soluble, but the other Group III carbonates are sparingly soluble or insoluble. Carbonates all react with acids, liberating CO_2.

$$Na_2CO_3 + 2HCl \rightarrow 2NaCl + CO_2 + H_2O$$

Group I carbonates are stable to heat, and melt without decomposing. Group II carbonates all decompose if heated sufficiently strongly. Their stability increases as the size of the metal ion increases. Most other carbonates decompose easily.

$$CaCO_3 \xrightarrow{\text{heat}} CaO + CO_2$$

	$BeCO_3$	$MgCO_3$	$CaCO_3$	$SrCO_3$	$BaCO_3$
Decomposition temperature	<100 °C	540 °C	900 °C	1290 °C	1360 °C

The only solid bicarbonates known are those of the Group I metals and of NH_4^+. These are colourless solids, and are somewhat less soluble than the corresponding carbonates. They decompose easily on heating. The solid structures of these contain polymeric chains of HCO_3^- groups hydrogen bonded together.

THE CARBON CYCLE

Though carbon is the seventeenth most abundant element in the earth's crust, and totals about 2×10^{16} tonnes, most of this is in the form of coal, oil and various carbonates (limestone and dolomite) which are immobilized.

In contrast there is a rapid turnover of CO_2 in the atmosphere, carbon compounds in living matter, CO_2 dissolved in the oceans, and more slowly with carbonate minerals formed on the sea bed. An equilibrium exists between them. The proportion of CO_2 in the atmosphere is approximately 0.046% by weight, and 0.031% by volume. Though only a small percentage, CO_2 is essential for life, and amounts to 2500 billion tonnes (2500×10^9 tonnes).

Photosynthesis by the green parts of plants and some brown and blue algae removes about 360 billion tonnes of CO_2 from the atmosphere a year – roughly 15%. Glucose sugar is the first product formed. This may be used for respiration and energy release by the plant, or incorporated into plant cells. These may be eaten by animals, and used for respiration or for producing animal cells. Eventually the same amount of CO_2 is returned to the atmosphere either by respiration of the plant or animal, or by death and putrefaction of plant or animal remains.

Combustion of fossil fuels, mainly coal, oil and natural gas, and burning tropical rain forests adds about 25 billion tonnes of CO_2 to the atmosphere each year. (Coal production in 1988 was 4749 million tonnes, crude oil 2944 million tonnes, natural gas $2 \times 10^{12}\,m^3$. Important amounts of CO_2 are released by burning limestone to make lime for making cement. Lime production in 1988 was 126 million tonnes so about 100 million tonnes of CO_2 was produced.

$$CaCO_3 \rightarrow CaO + CO_2$$

It is estimated that in 1988 the USA added 1.2 billion tonnes of CO_2 to the atmosphere, the USSR 1 billion tonnes and western Europe 0.8 billion tonnes. If all of this CO_2 remained in the atmosphere, it is estimated that the CO_2 content will double by the year 2020. With the ever increasing use of fossil fuels the amount of CO_2 could double even sooner.

The CO_2 molecule absorbs strongly in the infrared region, and its presence in the atmosphere decreases the loss of heat from the earth by radiation. This global warming is called the 'greenhouse effect'. (Other gases, including the oxides of nitrogen from car exhausts, Freons from aerosols and refrigerators and methane from bacteria in the soil and in the rumen of cows, also add to the greenhouse effect.) The magnitude of this effect, and whether it exists at all are controversial. The concentration of atmospheric CO_2 has increased by 10% since 1958, and on the basis of measurements of CO_2 from ice cores is some 25% higher than before the industrial revolution. This corresponds to about half the CO_2 produced from burning fossil fuels, and most of the remainder has been absorbed by the oceans. A United Nations report suggests that if nothing is done the mean temperature of the earth will rise by 2.5 °C in the next 30 years. This is an average, varying from 2 °C at the equator to 4 °C at the poles. This could have dramatic effects on the climate. Some fertile areas like the grain belt in the USA would become desert and crops would not grow. The increased temperature would cause more evaporation of water and hence more rain, flooding and tropical storms in certain parts of the world. Part of the polar ice-caps would melt, and this together with the thermal expansion of the sea would flood vast areas of land.

It is by no means certain that these catastrophic changes will occur. It must be emphasized that there is a long timescale for the expected warming due to greenhouse gases. Furthermore in nature one change in the biosphere is usually balanced by another with the opposite effect. Biological and other feedbacks are likely to affect the future concentrations of greenhouse gases. More CO_2 in the atmosphere may enhance plant growth which will use it up. A whole forest can grow in 30 years. Since large amounts of CO_2 dissolve in the sea, this should make the pH of the sea decrease. In the extreme this increase in acidity might dissolve the calcareous shells of molluscs and other sea life, thus destroying them. This is far from certain, since if global warming does occur, the solubility of CO_2 in water will decrease. However, an increase in CO_2 in the surface waters could well lead to an increase in plankton – small marine plants, which use CO_2 in photosynthesis. In addition slow reactions occur on the ocean floor:

$$CO_3^{2-} + CO_2 + H_2O \rightarrow 2HCO_3^-$$

The silicate sediments on the ocean floor play an important role in fixing the composition and pH of the water. An increase in pH is compensated by the dissolution of certain minerals and the precipitation of others. Thus silicate rocks may change to carbonates and SiO_2. If the pH subsequently changes the other way then these processes are reversed.

The greenhouse debate is a reminder of the limited evidence, uncertainty and long timescale of the problems of global warming. The author's opinion is that it must be prudent to conserve energy, to improve efficiency and to reduce wasteful practices, since there is no immediate alternative in sight to using fossil fuels. Nuclear power is the only major alternative energy source, and at present many people find it unacceptable. The hydrogen economy is a long way off, and solar power and the power of wind and waves can at best provide only a very small fraction of our energy requirements.

SULPHIDES

Carbon disulphide CS_2 is the most important sulphide of carbon. It is a colourless volatile liquid, b.p. 46°C. It is dangerous to handle because it is very flammable, it has a very low flash point (30°C) and it ignites spontaneously at 100°C. It is very poisonous, affecting the brain and central nervous system. Pure samples smell like ether, but organic impurities frequently give it an extremely foul smell. It is a commercially important chemical, and world production was 584700 tonnes in 1985. At one time it was produced by heating charcoal and S vapour at about 850°C. Nowadays it is produced mainly from a gas phase reaction between natural gas and sulphur, catalysed by Al_2O_3 or silica gel.

$$CH_4 + 4S \xrightarrow{600\,°C} CS_2 + 2H_2S$$

The main uses of CS_2 are as follows:

1. The manufacture of viscose rayon (artificial silk) and cellophane. CS_2 reacts with cellulose and NaOH to form sodium cellulose dithiocarbonate (cellulose xanthate).

$$CS_2 + \text{cellulose—OH} + NaOH \rightarrow \begin{matrix} \text{cellulose—O} \\ \quad\quad \backslash \\ \quad\quad\quad C{=}S \\ \quad\quad / \\ NaS \end{matrix}$$

sodium cellulose xanthate

This is dissolved in lye (aqueous alkali) to give a viscous solution called 'viscose'. On acidification, 'viscose' is converted back to cellulose in the form of fibres (either rayon or cellulose wool), or as a thin film (cellophane).

2. The manufacture of CCl_4 (see later under 'Halides').
3. Smaller amounts are used as a solvent for S in the cold vulcanization of rubber.

CS_2 reacts with aqueous NaOH, giving a mixture of sodium carbonate and sodium trithiocarbonate:

$$3CS_2 + 6NaOH \rightarrow Na_2CO_3 + 2Na_2CS_3 + 3H_2O$$

CS_2 reacts with NH_3, giving ammonium dithiocarbamate:

$$CS_2 + 2NH_3 \rightarrow NH_4[H_2NCS_2]$$

CS_2 is a linear molecule with a similar structure to CO_2. CS_2 forms complexes more readily than CO_2. The complex $[Pt(CS_2)(PPh_3)]$ is structurally similar to $[Co(CO_2)(PPh_3)_3]$. The CS_2 acts as a bidentate ligand with one C atom and one S atom bonded to the metal, and the CS_2 molecule is bent. The S atoms may bond to other metal atoms, giving more complicated complexes. The bonding cannot be explained by classical localized bonds.

If a high frequency electric discharge is passed through CS_2 vapour then CS is formed.It is unlike CO, and is a highly reactive radical even at the temperature of liquid air. Passing an arc through CS_2 gives C_3S_2. This is thought to have the structure S=C=C=C=S. It is a red liquid that polymerizes slowly (as does C_3O_2).

OXIDES OF SILICON

Two oxides of silicon, SiO and SiO_2, have been reported. Silicon monoxide is thought to be formed by high temperature reduction of SiO_2 with Si, but its existence at room temperature is in doubt.

$$SiO_2 + Si \rightarrow 2SiO$$

Silicon dioxide SiO_2 is commonly called silica, and it is widely found as sand and quartz. Group IV elements typically form four bonds. Carbon can form $p\pi-p\pi$ double bonds and hence CO_2 is a discrete molecule and is a gas. Silicon cannot form double bonds in this way using $p\pi-p\pi$ orbitals. (A substantial number of silicon compounds are now known to contain $p\pi-p\pi$ bonds in which the silicon atom appears to use d orbitals for bonding.) Thus SiO_2 forms an infinite three-dimensional structure, and SiO_2 is a high melting solid. SiO_2 exists in at least 12 different forms. The main ones are quartz, tridymite and cristobalite, each of which has different structures at high and low temperatures. α-Quartz is by far the most common, and is a major constituent of granite and sandstone. Pure SiO_2 is colourless, but traces of other metals may colour it, giving semi-precious gemstones such as amethyst (violet), rose quartz (pink), smoky quartz (brown), citrine (yellow), and non-precious materials such as flint (often black due to C), agate and onyx (banded).

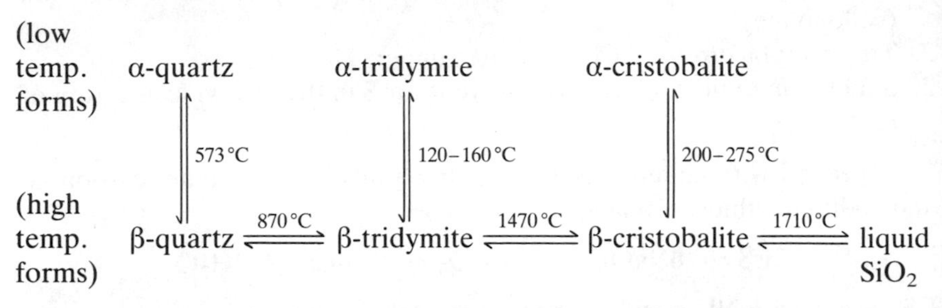

In all of these forms each Si is tetrahedrally surrounded by four O atoms. Each corner is shared with another tetrahedron, thus giving an infinite array. The difference between these structures is the way in which the tetrahedral SiO_4 units are arranged. α-Quartz is the most stable form at room temperature and in this the tetrahedra form helical chains. These are interlinked. Since the helix may be left or right handed, they cannot be superimposed, so it exists as *d* and *l* optical isomers. Individual crystals can be separated by hand. In cristobalite the Si atoms have the same arrangement as the C atoms in diamond, with O atoms midway between them. The relation between tridymite and cristobalite is the same as that between wurtzite and zinc blende. Heating any solid form of SiO_2 to its softening temperature, or slow cooling of molten SiO_2, gives a glass-like solid. This is amorphous, and contains a disordered mixture of rings, chains and three-dimensional units.

Silica in any form is unreactive. It is an acidic oxide and so does not react with acids. However, it does react with HF, forming silicon tetrafluoride SiF_4. This reaction is used in qualitative analysis to detect silicates: when the SiF_4 comes into contact with a drop of water it is hydrolysed to silicic acid. This can be seen as a white solid forming on the surface of the drop of water.

$$H_2SO_4 + CaF_2 \rightarrow HF \xrightarrow{+SiO_2} \underset{+\,H_2O}{SiF_4} \xrightarrow{+H_2O} HF + \begin{cases} Si(OH)_4 \text{ or} \\ SiO_2 \cdot 2H_2O \end{cases}$$

SiO_2 is an acidic oxide: it dissolves slowly in aqueous alkali, and more rapidly in fused alkalis MOH or fused carbonates M_CO_3, forming silicates.

$$SiO_2 + NaOH \rightarrow (Na_2SiO_3)_n \quad \text{and} \quad Na_4SiO_4$$

This reaction accounts for ground glass stoppers sticking in reagent bottles containing NaOH. Of the halogens, only fluorine attacks SiO_2.

$$SiO_2 + 2F_2 \rightarrow SiF_4 + O_2$$

Quartz is important as a piezo-electric material for the crystals in gramophone pickups, for cigarette and gas lighters and for making crystal oscillators for radios and computers. There is insufficient natural quartz of high enough purity, and so it is made synthetically by hydrothermal growth of seed crystals from aqueous NaOH and vitreous silica at 400 °C under pressure.

Vitreous silica has a low coefficient of expansion, is quite resistant to shock, and is very transparent to visible and ultraviolet light. It is used for laboratory glassware, and for optical components such as lenses and prisms and cells to hold samples in UV–visible spectrophotometers.

Silica gel is amorphous and very porous. It is obtained by dehydrating silicic acid, and contains about 4% water. It is widely used as a drying agent, a catalyst, and in chromatography. The mineral opal, which is used as a white or pearl-like gemstone, is hard (amorphous) silica gel. The beginnings of ordered structures are shown by various minerals, many of which are cut and polished as gemstones:

agate (often banded colours)
onyx (often white and black bands)
carnelian (yellow or red)
bloodstone (green with red spots)
jasper (usually red or brown but sometimes green, blue or yellow)
flint (colourless or black if C present)

These are best written $SiO_2 \cdot nH_2O$.

Kieselguhr is another form of SiO_2. It is a fine white powder, and about 2 million tonnes/year are obtained by open cast mining in Europe and North America. It is used in filtration plants, as an abrasive, and as an inert filler. (Gelignite is a mixture of the explosive nitrobenzene (liquid) and inert kieselguhr (solid).)

OXIDES OF GERMANIUM, TIN AND LEAD

The dioxides GeO_2, SnO_2 and PbO_2 normally adopt a TiO_2 structure with 6 : 3 coordination. The basicity of the oxides increases down the group: this is the usual trend. Thus CO_2 and SiO_2 are purely acidic. GeO_2 is not as strongly acidic as SiO_2, and SnO_2 and PbO_2 are amphoteric. GeO_2, SnO_2 and PbO_2 dissolve in alkali to form germanates, stannates and plumbates respectively. The germanates have complicated structures similar to the silicates, but the stannates and plumbates contain $[Sn(OH)_6]^{2-}$ and $[Pb(OH)_6]^{2-}$ complex ions. There is no evidence of the existence of $Ge(OH)_4$, $Sn(OH)_4$ and $Pb(OH)_4$ and these are better represented as $MO_2(H_2O)_n$, where n is about two. All three oxides are insoluble in acids except when a complexing agent such as F^- or Cl^- is present, when complex ions such as $[GeF_6]^{2-}$ and $[SnCl_6]^{2-}$ are formed.

The lower oxides GeO, SnO and PbO have layer lattices rather than the typical ionic structures. They are slightly more basic and ionic than the corresponding higher oxides. GeO is distinctly acidic, whilst SnO and PbO are amphoteric and dissolve in both acids and bases. The increased stability of the lower valence states on descending a group is illustrated by the fact that Ge^{II} and Sn^{II} are quite strong reducing agents whereas Pb^{II} is stable.

PbO is commercially important. It exists as a red form called litharge and a yellow form called massicot. Litharge is used in large amounts to make lead glass, and in ceramic glazes. World production is about 250 000 tonnes/year. 'Black oxide' of lead is a mixture of PbO and Pb, and is extensively used to make the plates in electric storage batteries for motor cars. The anode is oxidized to PbO_2, and the cathode is reduced to spongy lead. About 700 000 tonnes/year are used worldwide.

Lead also forms a mixed oxide Pb_3O_4. This is called red lead and may be represented as $2PbO \cdot PbO_2$: clearly it contains Pb(II) and Pb(IV). Pb_3O_4 is used in paint to prevent the rusting of iron and steel. It is also used to colour and vulcanize plastic and artificial rubber. Smaller amounts are used in ceramics and glassmaking. World production is about 18 000 tonnes/year.

PbO_2 is used as a strong oxidizing agent, and is produced in situ in lead storage batteries.

SILICATES

Occurrence in the earth's crust

About 95% of the earth's crust is composed of silicate minerals, alumino-silicate clays, or silica. These make up the bulk of all rocks, sands, and their breakdown products clays and soil. Many building materials are silicates: granite, slates, bricks, and cement. Ceramics and glass are also silicates.

The three most abundant elements are O, Si and Al. Together they make up 81% of the earth's crust, that is four out of five atoms are one of these. This is a much higher abundance than in the earth as a whole or in the universe. During the cooling of the earth the lighter silicate materials crystallized and floated to the surface, resulting in the concentration of silicates in the earth's crust.

N.L. Bowen has summarized the sequence in which these crystalline minerals appeared as the magma cooled, and this is called Bowen's Reaction Series (Figure 13.5).

Several points arise:

1. The simpler silicate units crystallized first.
2. Hydroxyl groups appear in the later minerals, and F may be substituted instead of OH.
3. Isomorphous replacement, i.e. changing one metal for another without changing the structure, occurs particularly in the later minerals.
4. The orthoclase feldspars, muscovite mica and quartz are the major minerals of granite.
5. As the silicates cooled further, they shrank and cracked. The hydro-

first ↓	olivine	$M^{II}_2SiO_4$
	pyroxenes	$M^{II}_2(SiO_3)_2$
	amphiboles	$M^{II}_7(Al,Si)_4O_{11} \cdot (OH)_2$
	biotite micas	$(K,H)_2(Mg,Fe^{II})_2(Al,Fe^{III})_2(SiO_4)_3$
	orthocase feldspars	$KAlSi_3O_8$
	muscovite micas	$KAl_2(AlSi_3O_{10}) \cdot (OH)_2$
	quartz	SiO_2
last	zeolites	$Na_2(Al_2Si_3O_{10})2H_2O$

(Leaving a small amount of water, SO_2, S, Pb, Cu, Ag, Sn, As, Sb, Bi and other transition metals in solution under a very high temperature and pressure)

Figure 13.5 Sequence in which minerals are thought to have crystallized.

thermal (hot water) solution moved through the cracks nearer the surface to regions of lower temperature and pressure where the elements precipitated and then combined with S, forming veins of sulphides.

Soluble silicates

Silicates can be prepared by fusing an alkali metal carbonate with sand in an electric furnace at about 1400 °C.

$$Na_2CO_3 \xrightarrow{1400\,°C} CO_2 + Na_2O \xrightarrow{+SiO_2} Na_4SiO_4, (Na_2SiO_3)_n \text{ and others}$$

The product is a soluble glass of sodium or potassium silicate. It is dissolved in hot water under pressure, and is filtered from any insoluble material. The composition of the product varies, but is approximately $Na_2Si_2O_5 \cdot 6H_2O$. Nearly 3 million tonnes/year of soluble silicates are produced, mostly sodium compounds. They are used in liquid detergent preparations to keep the pH high, so that grease and fat can be dissolved by forming a soap. Soluble silicates must not be used if the water is hard, or they will react with Ca^{2+} to form insoluble calcium silicate. Sodium silicate is also used as an adhesive (for example for pasting paper, bonding paper pulp and corrugated cardboard), in asbestos roof tiles, in fireproof paint and putty, and in making silica gel.

Principles of silicate structures

The majority of silicate minerals are very insoluble, because they have an infinite ionic structure and because of the great strength of the Si—O bond. This made it difficult to study their structures, and physical properties such as cleavage and the hardness of rocks were originally studied. The structural principles in silicate structures have only become apparent since the structures have been solved by X-ray crystallographic methods.

1. The electronegativity difference between O and Si, $3.5 - 1.8 = 1.7$, suggests that the bonds are almost 50% ionic and 50% covalent.
2. The structure may therefore be considered theoretically by both ionic and covalent methods. The radius ratio $Si^{4+} : O^{2-}$ is 0.29, which suggests that Si is four-coordinate, and is surrounded by four O atoms at

the corners of a tetrahedron. This can also be predicted from the use of the 3*s* and three 3*p* orbitals by Si for bonding. Thus silicates are based on $(SiO_4)^{4-}$ tetrahedral units.

3. The SiO_4 tetrahedra may exist as discrete units, or may polymerize into larger units by sharing corners, that is by sharing O atoms.
4. The O atoms are often close-packed, or nearly close-packed. Close-packed structures have tetrahedral and octahedral holes, and metal ions may occupy either octahedral or tetrahedral sites depending on their size. Most metal ions are the right size to fit one type of hole, though Al^{3+} can fit into either. Thus Al can replace either a metal in one of the holes, or a silicon atom in the lattice. This is particularly important in the aluminosilicates.

Occasionally Li may occupy sites with a coordination number of 6 rather than the usual 4, and K and Ca may have a coordination number of 8 rather than the usual 6. The radius ratio principle is a useful guide, but it is only strictly applicable to ionic compounds, and silicates are partly covalent. In these compounds the full charge separation to give Si^{4+} and O^{2-} does not occur, and empirical 'effective ionic radii' may be used instead of normal ionic radii (see Further Reading, R.D. Shannon.)

CLASSIFICATION OF SILICATES

The way in which the $(SiO_4)^{4-}$ tetrahedral units are linked together provides a convenient classification of the many silicate minerals.

Orthosilicates (neso-silicates)

A wide variety of minerals contain discrete $(SiO_4)^{4-}$ tetrahedra, that is they share no corners (see Figure 13.6). They have the formula $M_2^{II}[SiO_4]$, where M may be Be, Mg, Fe, Mn or Zn, or $M^{IV}[SiO_4]$, for example $ZrSiO_4$. Different structures are formed depending on the coordination number adopted by the metal.

In willemite $Zn_2[SiO_4]$, and phenacite $Be_2[SiO_4]$, the Zn and Be atoms have a coordination number of 4, and occupy tetrahedral holes.

In forsterite $Mg_2[SiO_4]$, the Mg has a coordination number of 6 and

Table 13.8 Types of holes occupied in close-packed structures

Oxide	Radius ratio	Coordination number	Type of hole occupied
$Be^{2+} : O^{2-}$	0.25	4	Tetrahedral
$Si^{4+} : O^{2-}$	0.29	4	Tetrahedral
$Al^{3+} : O^{2-}$	0.42	4 or 6	Tetrahedral or Octahedral
$Mg^{2+} : O^{2-}$	0.59	6	Octahedral
$Fe^{2+} : O^{2-}$	0.68	6	Octahedral

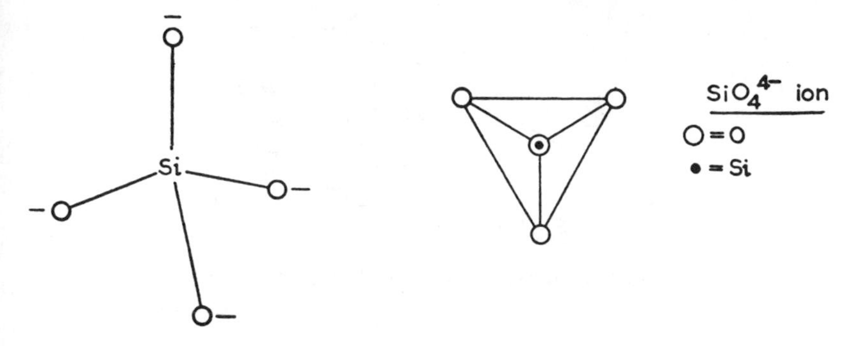

Figure 13.6 Structure of orthosilicates. (After T. Moeller.)

occupies octahedral holes. When octahedral sites are occupied, it is quite common to get isomorphous replacement of one divalent metal ion by another of similar size, without changing the structure. The mineral olivine $(Mg, Fe)_2[SiO_4]$ has the same structure as forsterite, but about one tenth of the Mg^{2+} ions in forsterite are replaced by Fe^{2+} ions. The ions have the same charge and similar radii (Mg^{2+} 0.72 Å, Fe^{2+} 0.78 Å), and occupy the same type of hole. Thus substitution of one metal for another does not change the structure. This mineral may also have Mn^{II} in some octahedral sites, thus giving $(Mg, Fe, Mn)_2[SiO_4]$. These structures are all related to hexagonal close-packing.

Zircon $ZrSiO_4$ is used as a gemstone as it can be cut to look like a diamond, but is much cheaper. Zircon is much softer than diamond, and the cut edges which make the gem attractive eventually wear and spoil the look of the stone. Zircon has a coordination number of 8. The structure is not close-packed.

The garnets are another important group of minerals with discrete tetrahedra. Large crystals of garnet are cut and polished and used as a red gemstone. Much larger amounts (100 975 tonnes in 1988) are used to make 'sandpaper'. The formula is $M_3^{II}M_2^{III}[(SiO_4)_3]$. M^{II} may be Mg, Ca or Fe^{II}, and these are six-coordinate. M^{III} may be Fe^{III}, Cr or Al and these are eight-coordinate.

Pyrosilicates (soro-silicates, disilicates)

Two tetrahedral units are joined by sharing the O at one corner, thus giving the unit $(Si_2O_7)^{6-}$. This is the simplest of the condensed silicate ions. The name pyro comes from the similarity in structure with pyrophosphates such as $Na_4P_2O_7$, and these were named because they can be made by heating orthophosphates (see Figure 13.7).

Pyrosilicates are rare. One example is thortveitite $Sc_2[Si_2O_7]$. A number of lanthanide disilicates have similar formulae $Ln_2[Si_2O_7]$. These are not quite the same, as the Si—O—Si angle is not 180° as in the Sc compound, but varies down to 133°, and the coordination number of the metal changes from 6 to 7 and then to 8 as the size of the metal increases. Hemimorphite

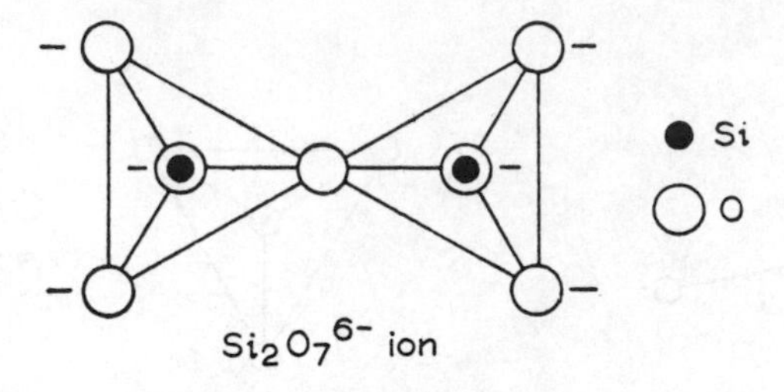

Figure 13.7 Structure of pyrosilicates $Si_2O_7^{6-}$. (After T. Moeller.)

$Zn_4(OH)_2[Si_2O_7] \cdot H_2O$ is another example, but structural studies show no difference in the lengths of the bridging and terminal Si—O bonds. Thus the representation as a disilicate ion may be misleading, and the structure may be better considered as $[SiO_4]$ and $[ZnO_3(OH)]$ tetrahedra linked to give a three-dimensional network.

Cyclic silicates

If two oxygen atoms per tetrahedron are shared, ring structures may be formed of general formula $(SiO_3)_n^{2n-}$ (Figure 13.8). Rings containing three, four, six and eight tetrahedral units are known, but those with three and six are the most common. The cyclic ion $Si_3O_9^{6-}$ occurs in wollastonite $Ca_3[Si_3O_9]$ and in benitoite $BaTi[Si_3O_9]$. The $Si_6O_{18}^{12-}$ unit occurs in beryl $Be_3Al_2[Si_6O_{18}]$. In beryl the Si_6O_{18} units are aligned one above the other, leaving channels. Na^+, Li^+ and Cs^+ are commonly found in these channels, and because of the channels the mineral is permeable to gases

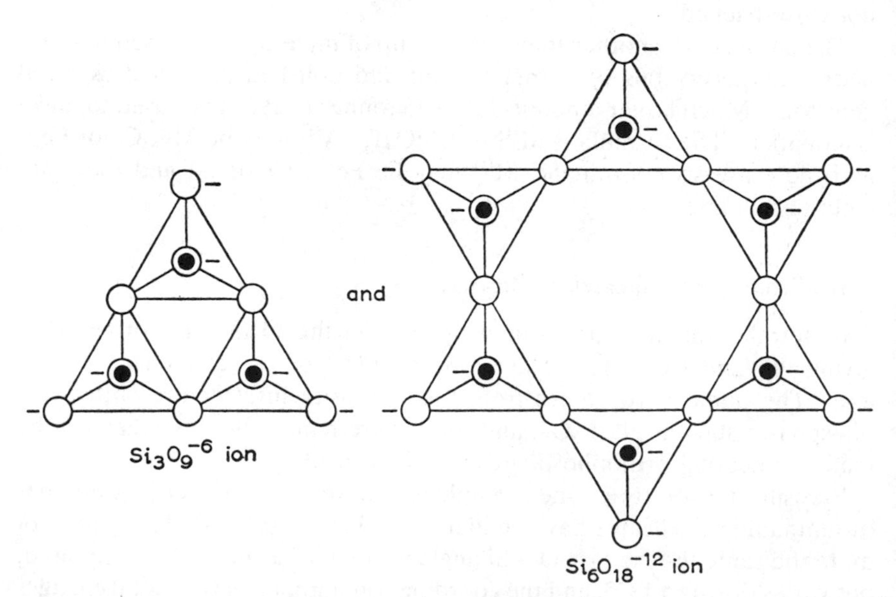

Figure 13.8 Structure of cyclic silicates $Si_3O_9^{6-}$ and $Si_6O_{18}^{12-}$. (After T. Moeller.)

consisting of small atoms or molecules, e.g. helium. Beryl and emerald are both gemstones. Beryl is found with granite and usually forms pale green crystals which are six-sided prisms. Emerald has the same formula as beryl except that it contains 1–2% Cr which gives it a strong green colour.

Chain silicates

Simple chain silicates or pyroxenes are formed by the sharing of the O atoms on two corners of each tetrahedron with other tetrahedra. This gives the formula $(SiO_3)_n^{2n-}$ (see Figures 13.9 and 13.10). A large number of important minerals form chains, but there are a variety of different structures formed because the arrangement of the tetrahedra in space may vary

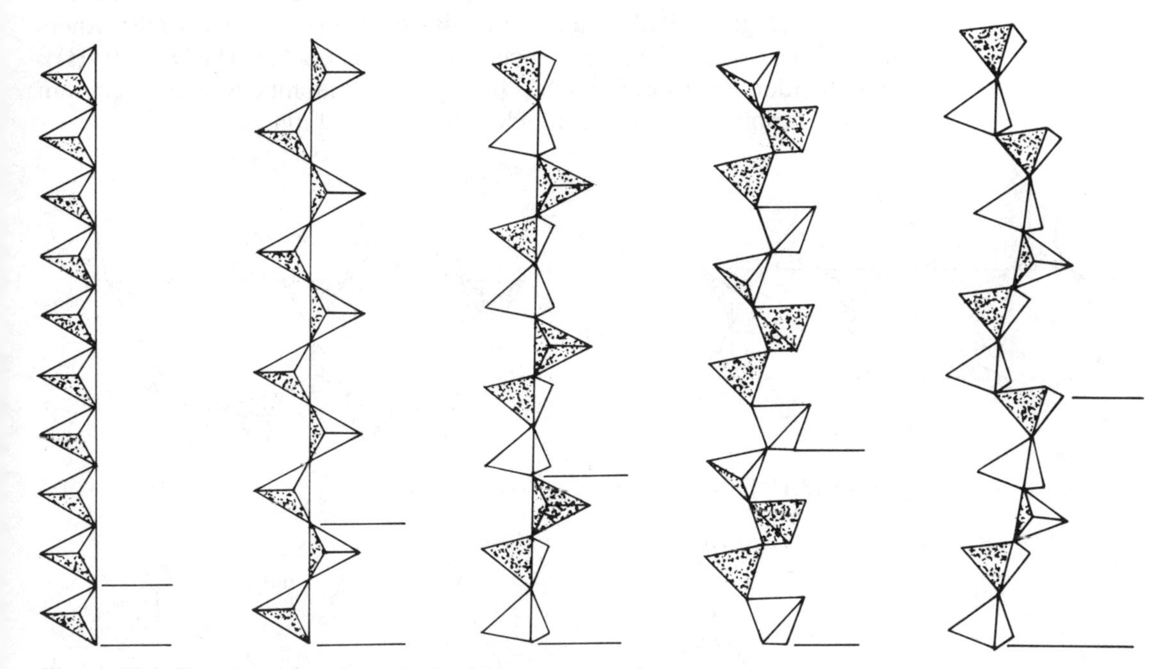

Figure 13.9 Structure of various single chains.

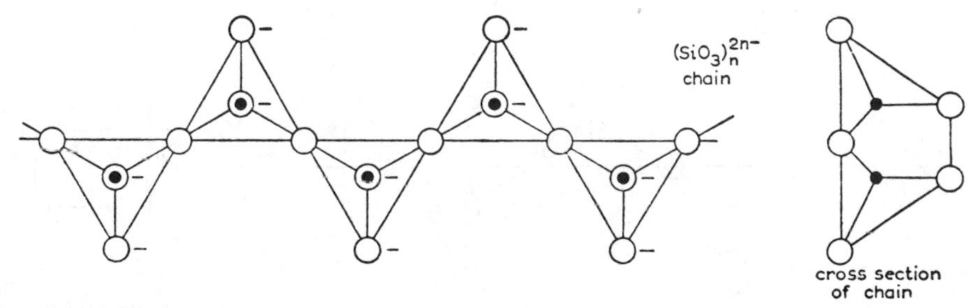

Figure 13.10 Structure of pyroxenes $(SiO_3)_n^{2n-}$. (After T. Moeller.)

and thus affect the repeat distance along the chain. The most common arrangement repeats after every second tetrahedron, for example in spodumene $LiAl[(SiO_3)_2]$ (which is the main source of Li), enstatite $Mg_2[(SiO_3)_2]$, and diopsite $CaMg[(SiO_3)_2]$. Wollastonite $Ca_3[(SiO_3)_3]$ has a repeat unit of three tetrahedra, and others are known with repeat units of 4, 5, 6, 7, 9 and 12.

Double chains can be formed when two simple chains are joined together by shared oxygens. These minerals are called amphiboles, and they are well known. There are several ways of forming double chains, giving formulae $(Si_2O_5)_n^{2n-}$, $(Si_4O_{11})_n^{6n-}$, $(Si_6O_{17})_n^{10-}$ and others. (See Figure 13.11.)

The most numerous and best known amphiboles are the asbestos minerals. These are based on the structural unit $(Si_4O_{11})_n^{6n-}$. In this structure (Figure 13.12) some tetrahedra share two corners, whilst others share three corners. Examples include tremolite, $Ca_2Mg_5[(Si_4O_{11})_2](OH)_2$, and crocidolite $Na_2Fe_3^{II}Fe_2^{III}[(Si_4O_{11})_2](OH)_2$. Amphiboles always contain hydroxyl groups, which are attached to the metal ions.

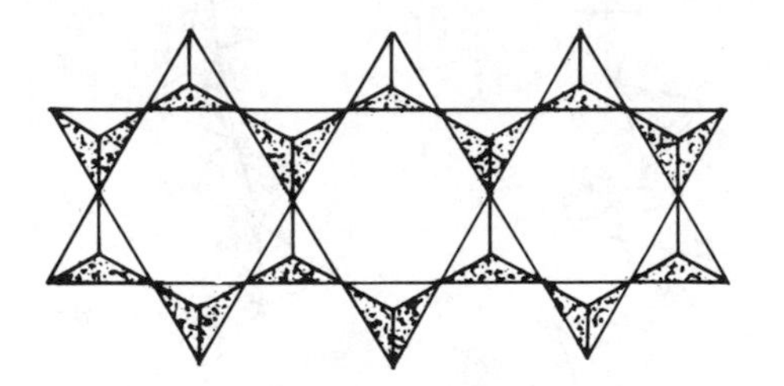

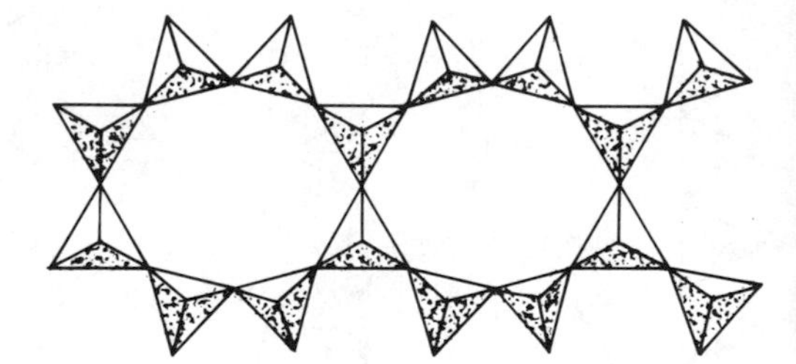

Figure 13.11 Structure of various double chains.

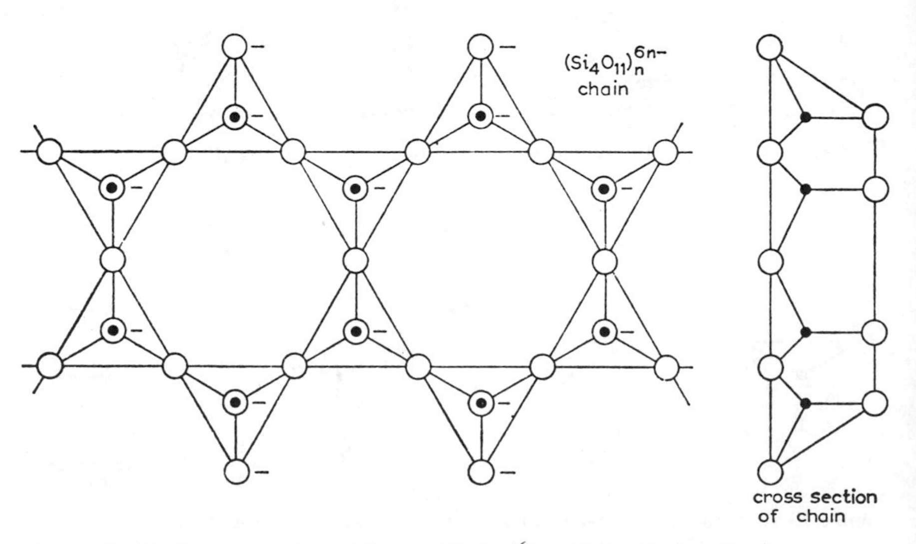

Figure 13.12 Structure of amphiboles $(Si_4O_{11})_n^{6n-}$. (After T. Moeller.)

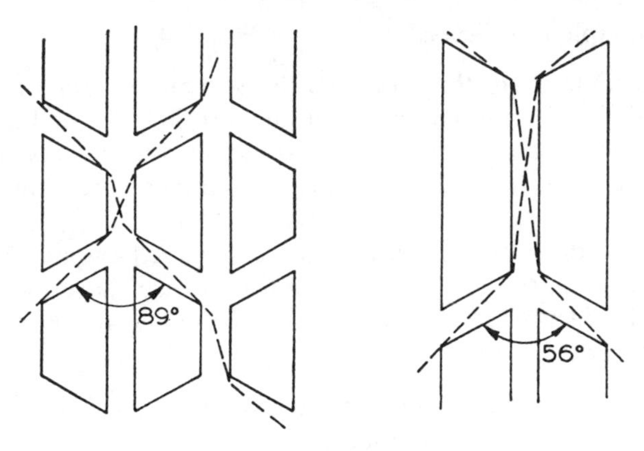

Figure 13.13 Cleavage angles in pyroxenes and amphiboles.

The Si—O bonds in the chains are strong and directional. Adjacent chains are held together by the metal ions present. Thus pyroxenes and amphiboles cleave readily parallel to the chains, forming fibres. For this reason they are called fibrous minerals. The cleavage angle for pyroxenes is 89°, and for amphiboles 56°. This is used as a means of identifying the minerals. These angles are related to the size of the cross-sectional trapezium of the chains and the way in which they are packed together (see Figure 13.13).

Asbestos is of considerable commercial importance, and 4.3 million tonnes were mined in 1988 (USSR 61%, Canada 16%, Brazil 5% and Zimbabwe 4%). It is useful because it is strong, cheap, resistant to heat and flames, and also resistant to acids and alkalis. Most is used to make asbestos reinforced cement and roofing sheets. Smaller amounts are used for brake linings, clutch linings, asbestos paper, adding to vinyl floor coverings, and thermal insulation for pipes, and fibres may be woven into asbestos cloth to make fire fighting suits.

Asbestos minerals come from two different groups of silicates:

1. The amphiboles.
2. The sheet silicates.

The amphiboles include crocidolite $Na_2Fe_3^{II}Fe_2^{III}[Si_8O_{22}](OH)_2$, which is called blue asbestos, and others derived from it by isomorphous replacement, for example amosite or brown asbestos $(Mg, Fe^{II})_7[Si_8O_{22}](OH)_2$. Together these make up about 5% of the asbestos used.

The mineral chrysotile $Mg_3(OH)_4[Si_2O_5]$ is called white asbestos, and this is derived from serpentine, and is a sheet silicate. This constitutes 95% of the asbestos used.

Though asbestos is chemically inert, it presents a serious health hazard. Inhaling asbestos dust causes asbestosis or scarring of the lungs. It also causes lung cancer. Blue asbestos appears to be the worst hazard. The disease may have a latent period of 20–30 years. The best control is to minimize asbestos dust, and to handle asbestos wet if possible.

Sheet silicates (phyllo-silicates)

When SiO_4 units share three corners the structure formed is an infinite two-dimensional sheet of empirical formula $(Si_2O_5)_n^{2n-}$ (see Figure 13.14). There are strong bonds within the Si—O sheet, but much weaker forces hold each sheet to the next one. Thus these minerals tend to cleave into thin sheets.

Structures with simple planar sheets are rare. A large number of sheet silicates are important and well known. These have slightly more complicated structures, and are made up of either two or three layers joined together. These include:

1. Clay minerals (kaolinite, pyrophyllite, talc)
2. White asbestos (chrysotile, biotite)
3. Micas (muscovite and margarite)
4. Montmorillonites (Fullers earth, bentonite and vermiculite)

Consider how a two-layer structure may be formed. If we start with a simple silicate sheet, then one side of the sheet contains all the unshared (singly bound) O atoms. The pure hydroxides $Al(OH)_3$ and $Mg(OH)_2$ both crystallize with layer structures (and Al and Mg are six-coordinate and occupy octahedral sites). The unshared O in a silicate sheet have almost the same relative positions as two thirds of the OH groups on each side of these hydroxide layers. If a Si_2O_5 layer is placed alongside a layer of

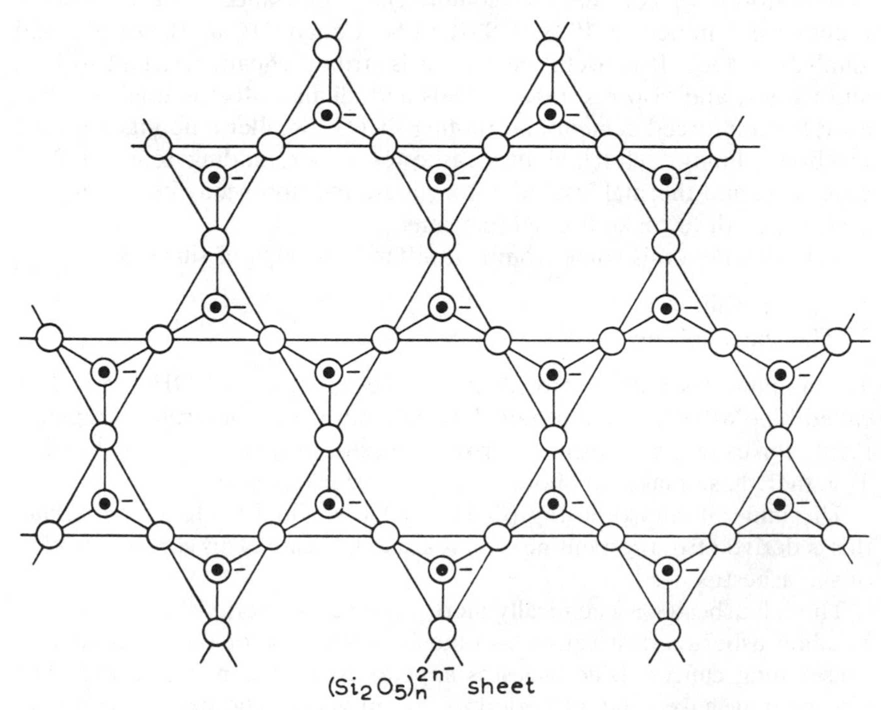

Figure 13.14 Structure of sheet silicates $(Si_2O_5)_n^{2n-}$. (After T. Moeller.)

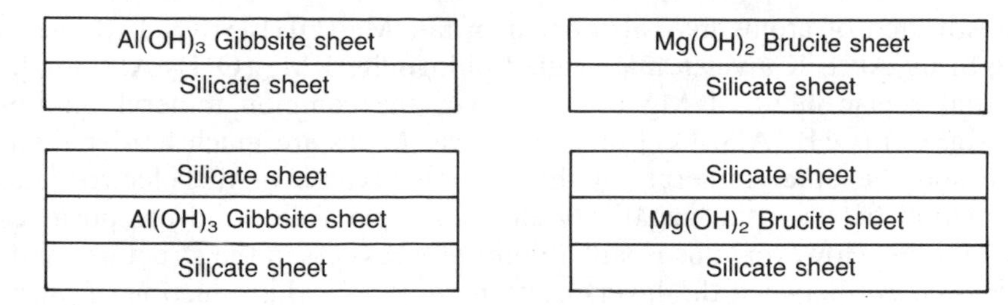

Figure 13.15 Two- and three-layer structures.

γ-gibbsite ($Al(OH)_3$, then many of the O atoms will coincide. The OH groups in $Al(OH)_3$ can be removed and an electrically neutral two-layer structure is formed. These double layers stacked parallel give the mineral kaolinite, which has the formula $Al_2(OH)_4[Si_2O_5]$. It is a white solid and is formed by the decomposition of granite. Large amounts are used for filling paper, and as a refractory. World production of kaolinite was 23 million tonnes in 1988 (USA 37%, UK 14%, USSR 13%). China clay or kaolin is mined in Cornwall. It is a high grade of kaolinite, and when small amounts of SiO_2 and mica are removed it can be mixed with water to give a white or nearly white plastic clay. Small amounts are used for making porcelain, china cups and plates, sanitary ware, and other ceramics, and for chromatography, as treatment for indigestion and for poultices.

One $Al(OH)_3$ layer has two equivalent sides. A Si_2O_5 layer was combined with one side in kaolinite, but a second Si_2O_5 sheet can combine with the other side thus giving a three-layer structure. This is made up of silicate, $Al(OH)_3$. Pyrophyllite $Al_2(OH)_2[(Si_2O_5)_2]$ has this three-layer structure.

A layer of brucite $Mg(OH)_2$ may be combined with a Si_2O_5 layer to form a two-layer structure. This gives the mineral chrysotile $Mg_3(OH)_4[Si_2O_5]$ (white asbestos), which is of considerable commercial importance. Alternatively brucite may combine with two Si_2O_5 layers, forming a triple-sheet structure called talc or soapstone $Mg_3(OH)_2[(Si_2O_5)_2]$. These triple sheets are electrically neutral, and there are no metal ions holding one sheet to the next, so the structure breaks very easily, and is very soft. Soapstone has a slippery feeling (hence its name). It acts as a dry lubricant. World production of talc was 7.6 million tonnes in 1988. It is used mainly in making ceramics, paper and paint. Small amounts are used in cosmetics, as talc can be ground into a fine powder and used as talcum powder.

Substitution of atoms may occur in the triple-layer structures of the pyrophyllite types. If Si is partly replaced by Al (in tetrahedral holes) then the sheet becomes negatively charged. These charges are balanced by positively charged metal ions which are placed between the layers. This gives rise to the mica minerals. These are a group of minerals characterized by the fact that they are readily split into glistening transparent flexible sheets of varied colour. Muscovite has the formula $KAl_2(OH)_2[AlSi_3O_{10}]$ and is called white mica. Margarite has the formula $CaAl_2(OH)_2[lSi_3O_{10}]$.

Substitution of atoms may also occur in talc $Mg_3(OH)_2[Si_4O_{10}]$. Replacing Si by Al + K gives a mica called phlogopite $KMg_3(OH)_2[AlSi_3O_{10}]$. Partial replacement of Mg by Fe^{II} gives the common mineral biotite $K(MgFe^{II})_3(OH)_2[AlSi_3O_{10}]$ or black mica. Micas are much harder than talc and the other minerals in this section because of the electrostatic attraction between the negatively charged triple sheets and the positive metal ions. However, this is still a point of weakness in the structure, and micas cleave between the layers quite readily. World production of mica was 249 000 tonnes in 1988. The main sources are the USA 52%, the USSR 20%, Canada 6%, India 5% and France 4%. Sheet mica is used as an electrical insulator and as a former on which to wind electric heating elements (e.g. in electric irons). It is also used in electrical capacitors and for windows in furnaces. Finely ground mica is used as a filler in plastics and rubber, in insulating board, and in polychromatic and glitter paint (e.g. for motor cars).

The clay minerals are formed from other silicates by weathering, or by hydrothermal processes, i.e. by the action of water under heat and pressure. They contain electrically neutral layers, e.g. kaolinite, and pyrophyllite. The clay minerals also include the montmorillonites, which have negatively charged layers, but the number of charges is much lower than in the micas. Some, but not all, of the octahedral Al^{III} in pyrophyllite $Al_2(OH)_2[(Si_2O_5)_2]$ are replaced by Mg^{II}, giving $(Mg_{0.33}, Al_{1.67})(OH)_2[(Si_2O_5)_2]^{0.33-}$. The triple sheets thus have a small negative charge, and 1/3 M^+ or 1/6 M^{2+} must be incorporated between the layers. These metal ions can be hydrated, and the minerals are sometimes called hydromicas. Finely divided particles suspended in water have thixotropic properties. The particles are small plates with negative charges on the surface, and positive charges on the edge. The particles are free to move in water, and they arrange themselves + to − and thus give a semi-solid gel-like mass. If stirred the +/− attractions are broken, and the suspension becomes watery, that is of lower viscosity. They are used in thixotropic non-drip emulsion paints. These minerals can also act as ion exchangers. Fuller's Earth is a calcium montmorillonite. It is very absorbant: about 3.7 million tonnes were produced in 1988, mainly to decolorize and deodorize vegetable and mineral oils, fats and waxes. It is also used to mop up oil spills, and as litter for pet animals. It can act as an ion exchanger for Ca^{2+}, and replacement of Ca^{2+} by Na^+ gives the mineral bentonite. This has marked thixotropic properties and is used as drilling mud, and in water based emulsion paints (6.5 million tonnes of bentonite were produced in 1988).

If in talc $Mg_3(OH)_2[(Si_2O_5)_2]$ substitution of Mg^{2+} in the brucite sheet occurs, and if also replacement of Si^{4+} with Al^{3+} occurs in the silicate sheet, then vermiculites are formed. A typical formula is $Na_x(Mg, Al, Fe)_3(OH)_2[((Si, Al)_2O_5)_2] \cdot H_2O$. If vermiculites are heated they dehydrate in an unusual way by extruding little worms: hence the name. These materials are porous and light in weight, and are used for packing and insulation, and as 'soil' for growing plants by ring culture. Over half a million tonnes are produced each year.

Three-dimensional silicates

Sharing all four corners of a SiO_4 tetrahedron results in a three-dimensional lattice of formula SiO_2 (quartz, tridymite, cristobalite etc.). These contain no metal ions, but three-dimensional structures can form the basis of silicate structures if there is isomorphous replacement of some of the Si^{4+} by Al^{3+} plus an additional metal ion. This gives an infinite three-dimensional lattice, and the additional cations occupy holes in the lattice. Replacing one quarter of the Si^{4+} in SiO_2 with Al^{3+} gives a framework ion $AlSi_3O_8^-$. The cations are usually the larger metal ions such as K^+, Na^+, Ca^{2+} or Ba^{2+}. The smaller ions Fe^{3+}, Cr^{3+} and Mn^{2+} which were common in the chain and sheet silicates do not occur in the three-dimensional silicates because the cavities in the lattice are too large. Replacements of one quarter or one half of the Si atoms are quite common, giving structures $M^I[AlSi_3O_8]$ and $M^{II}[Al_2Si_2O_8]$. Such replacements result in three groups of minerals:

1. feldspars
2. zeolites
3. ultramarines.

The feldspars are the most important rock forming minerals and constitute two thirds of the igneous rocks. For example, granite is made up of feldspars, with some micas and quartz. Feldspars are divided into two classes:

Orthoclase feldspars		*Plagioclase feldspars*	
orthoclase	$K[AlSi_3O_8]$	albite	$Na[AlSi_3O_8]$
celsian	$Ba[Al_2Si_2O_8]$	anorthite	$Ca[Al_2Si_2O_8]$

The orthoclases are more symmetrical than are plagioclases as K^+ and Ba^{2+} are just the right size to fit into the lattice whilst Na^+ and Ca^{2+}, being smaller, allow distortion.

Zeolites have a much more open structure than the feldspars. The anion skeleton is penetrated by channels, giving a honeycomb-like structure. These channels are large enough to allow them to exchange certain ions. They can also absorb or lose water and other small molecules without the structure breaking down. Zeolites are often used as ion-exchange materials, and as molecular sieves. Natrolite $Na_2[Al_2Si_3O_{10}]2H_2O$ is a natural ion exchanger. Permutit water softeners use sodium zeolites. Zeolites take Ca^{2+} ions from hard water and replace them by Na^+, thereby softening the water. The sodium zeolite natrolite gradually becomes a calcium zeolite, and eventually has to be regenerated by treatment with a strong solution of NaCl, when the reverse process takes place. In addition to naturally occurring minerals, many synthetic zeolites have been made. Zeolites also act as molecular sieves by absorbing molecules which are small enough to enter the cavities, but not those which are too big to enter. They can absorb water, CO_2, NH_3 and EtOH, and they are useful for separating straight chain hydrocarbons from branched chain compounds. Some other zeolites are heulandite $Ca[Al_2Si_7O_{18}]6H_2O$, chabazite $Ca[Al_2Si_4O_{12}]6H_2O$, and analcite $Na[AlSi_2O_6]H_2O$. Molecular sieves can be made with pores of appropriate size to remove small molecules selectively.

The mineral lapis lazuli is a splendid blue colour and was highly prized as a pigment for oil paintings in the middle ages. It contains ultramarine $Na_8[(AlSiO_4)_6]S_2$, in which the colour is produced by the polysulphide ion. The ultramarines are a group of related compounds, which contain no water, but do contain cations such as Cl^-, SO_4^{2-} and S_2^{2-}. Some examples of ultramarines are:

ultramarine	$Na_8[(AlSiO_4)_6]S_2$
sodalite	$Na_8[(AlSiO_4)_6]Cl_2$
nosean	$Na_8[(AlSiO_4)_6]SO_4$

The ultramarines are now produced synthetically. Ultramarine itself is made by igniting kaolinite, sodium carbonate and sulphur in the absence of air. The product may be green, blue or red depending on the particular polysulphide species present. Typically it is used as a blue pigment in oil based paints and ceramics. Before the days of detergents with artificial brighteners, synthetic ultramarine was used as a blueing agent (called dolly blue) to make domestic washing appear white by masking any yellowness.

SILICATES IN TECHNOLOGY

Many silicates are used as a direct result of their physical properties. For example, clay minerals are used for absorbing chemicals, micas are used for electrical insulation, asbestos is used for thermal insulation, agate and flint are used as hard or sharp surfaces, and a variety of gemstones are used for ornaments and jewellery.

Silicates are extremely important because the cement, ceramic and glass industries are based on their chemistry. Metallurgical extraction processes often produce silicates as waste products or slag, either because the minerals are silicates, or because the minerals contain silicate impurities. Some of the main technological applications are as follows.

Alkali silicates

These are used mainly as glue, as described earlier.

Cement

Both Portland and high alumina cement are described in Chapter 12.

Ceramics

Ceramics are inorganic materials that can be made into a paste and shaped at normal temperatures: the shape is then 'fired' at a high temperature.

Firing gives the product strength, either by sintering the crystallites together, or partly melting the paste. A number of carbides, oxides, and, in particular, clays, are treated in this way. The process is important for making bricks, tiles and pottery.

On heating, kaolinite $Al_2(OH)_4[Si_2O_5]$ loses water at 500–600 °C, giving $Al_2O_3 \cdot 2SiO_2$, and at about 950 °C forms a solid solution of mullite (approximate formula $3Al_2O_3 \cdot 2SiO_2$) and SiO_2. This remains solid till at least 1595 °C, and then softens only slowly. Since this is an iron-free clay, the product is white.

The porosity of the product depends on the temperature reached. If the firing temperature is high, the product does not absorb liquids. The presence of Fe^{3+} in the clay imparts a red or purple colour, and clays with a high proportion of CaO give a yellow or buff coloured product. Most ceramics except bricks and floor tiles are covered with a glassy coating called a glaze. This is done by dipping the article in an aqueous suspension of a heavy metal oxide such as SnO_2 or PbO_2 before firing. Pigments may be applied either before glazing (underglaze colour), or after firing and on top of the glaze, when a second firing is required.

Glass

A small amount of glass is made of silica. This has excellent properties, but very high temperatures are needed to produce it. *Silica glass* is too expensive for general use, but is used in scientific instruments.

The temperature required for melting can be reduced by adding various oxides to the melt, thus obtaining *silicate glass*. A number of oxides may be used including Na_2O, K_2O, MgO, CaO, BaO, B_2O_3, Al_2O_3, PbO and ZnO. Glass is a solid solution, and so its composition may vary. The amount of oxide added is not very large, and so the SiO_4 tetrahedra have a significant role in the structure. If only Na_2O or K_2O were used, the glass would be water soluble. Normal domestic glass for windows is a *calcium-alkali silicate glass* made by fusing the alkali metal carbonate, $CaCO_3$ and SiO_2. (The carbonates decompose to oxides on heating.) If Na_2CO_3 is used we obtain *soda glass*, which is also used for cheap laboratory glassware. Using K_2CO_3 *potash glass*. Most of the CaO may be replaced by PbO, giving *lead glass*, which has a higher refractive index, and is used for making optical parts and glass ornaments. If Al_2O_3 is used, Al^{3+} may be present in the structure as a free metal ion, or it may replace Si^{4+} in SiO_4 tetrahedra. If B_2O_3 is used, B^{3+} replaces some Si^{4+} in the tetrahedral skeleton. *Borosilicate glasses* containing B, and sometimes Al as well, are important. They have a low coefficient of expansion, and can withstand heat changes without cracking. They contain less alkali and so are less prone to chemical attack. Such glasses are widely used for laboratory equipment as in Pyrex glassware.

Glass is produced in very large amounts. In 1985 26.5 million tonnes of glass for bottles was produced, and in addition 4.9 million tonnes of sheet glass was produced. This was used mainly as windows.

Additives for fining, for decolorizing and for colouring may be used when making glass. Fining agents such as $NaNO_3$ or As_2O_3 are added to remove bubbles. The fining agent decomposes and gives off large bubbles of gases in the melt, which sweep out the small bubbles that are always formed. Decolorizing agents may be added to eliminate impurities and to obtain colourless glass. Fe^{3+} gives a yellow–brown colour, a mixture of Fe^{3+} and Fe^{2+} gives a green colour and Fe^{2+} gives a light blue colour. Other colouring agents may be added – Co^{2+} gives a deep blue, and colloidal particles of Cu give ruby-red colours. CaF_2 is sometimes added as a clouding agent to make *opal glass*.

ORGANOSILICON COMPOUNDS AND THE SILICONES

Organosilicon compounds

Si—C bonds are almost as strong as C—C bonds. Thus silicon carbide SiC is extremely hard and stable. Many thousands of organosilicon compounds containing Si—C bonds have been made, most since 1950. Many of these are inert, and stable to heat (e.g. $SiPh_4$ can be distilled in air at 428 °C). However, the vast range of organic compounds is not replicated by silicon for three main reasons:

1. Silicon has little tendency to bond to itself (catenate) whilst carbon has a strong tendency to do so. The largest chains formed by Si are contained in $Si_{16}F_{34}$ and Si_8H_{18}, but these compounds are exceptional. This is related to the weakness of Si—Si bonds in contrast to the strength of C—C bonds (see Table 13.4).
2. Silicon does not form $p\pi-p\pi$ double bonds, whilst carbon does so readily. (Note that a disilene $Me_2SI{=}SiMeH$ has been isolated, but only by using matrix isolation methods with solid argon. Various transient reaction species with Si=C and Si=N bonds are known, and $(Me_3Si)_2Si{=}C(OSiMe_3)(C_{10}H_{15})$ exists as crystals at room temperature, and is stable in the absence of air. These are rare exceptions.)
3. Silicon forms a number of compounds containing $p\pi-p\pi$ double bonds in which the silicon atom uses *d* orbitals (see later).

Preparation of organosilicon compounds

There are several ways of forming Si—C bonds:

1. By a Grignard reaction

$$SiCl_4 + CH_3MgCl \rightarrow CH_3SiCl_3 + MgCl_2$$
$$CH_3SiCl_3 + CH_3MgCl \rightarrow (CH_3)_2SiCl_2 + MgCl_2$$
$$(CH_3)_2SiCl_2 + CH_3MgCl \rightarrow (CH_3)_3SiCl + MgCl_2$$
$$(CH_3)_3SiCl + CH_3MgCl \rightarrow (CH_3)_4Si + MgCl_2$$

This is useful in the laboratory, or on a small scale.

2. Using an organolithium compound

$$4LiR + SiCl_4 \rightarrow SiR_4 + 4LiCl$$

   This also is useful in the laboratory, and R may be alkyl or aryl.
3. By the Rochow 'Direct Process'. Alkyl or aryl halides react directly with a fluidized bed of silicon in the presence of large amounts (10%) of a copper catalyst.

$$Si + 2CH_3Cl \xrightarrow{\text{Cu catalyst 280–300°C}} (CH_3)_2SiCl_2$$

   This is the main industrial method for making methyl and phenyl chlorosilanes which are of considerable commercial importance in the production of silicones. The yield is about 70%, with varying amounts of other products: $MeSiCl_3$ (10%) and Me_3SiCl (5%), and smaller amounts of Me_4Si and $SiCl_4$ and others such as $MeSiHCl_2$. Both Grignard and direct methods yield a mixture of products, and very careful fractionation is important as the boiling points are close: Me_3SiCl (57.7°C), $MeSiCl_3$ (66.4°C) and Me_2SiCl_2 (69.6°C).
4. Catalytic addition of Si—H to an alkene. This is a useful general method, but is not applicable to making the methyl and phenyl silanes required by the silicone industry.

Except for Ph_3SiCl, which is solid, the products are volatile liquids. They are highly reactive and flammable and the reaction with water is strongly exothermic.

Silicones

The silicones are a group of organosilicon polymers. They have a wide variety of commercial uses as fluids, oils, elastomers (rubbers) and resins. Annual production is estimated as about 300 000 tonnes/year. They are now produced on a larger scale than any other group of organometallic compounds.

The complete hydrolysis of $SiCl_4$ yields SiO_2, which has a very stable three-dimensional structure. The fundamental research of F. S. Kipping on the hydrolysis of alkyl substituted chlorosilanes led, not to the expected silicon compound analogous to a ketone, but to long-chain polymers called silicones.

$$R_2SiCl_2 \xrightarrow[-2HCl]{+2H_2O} R_2Si(OH)_2 \xrightarrow{-H_2O} \cancel{R_2Si{=}O}$$

$$HO{-}SiR_2{-}OH + HO{-}SiR_2{-}OH \xrightarrow{-H_2O} HO{-}SiR_2{-}O{-}SiR_2{-}OH$$

$$\begin{array}{ccccccccccccc} & & R & & & & & & R & & R & & \\ & & | & & & & & & | & & | & & \\ HO & - & Si & - & OH & + & HO & - & Si & - & O - Si & - & OH \rightarrow \\ & & | & & & & & & | & & | & & \\ & & R & & & & & & R & & R & & \end{array}$$

$$\begin{array}{ccccccccc} & & R & & R & & R & & \\ & & | & & | & & | & & \\ HO & - & Si & -O- & Si & -O- & Si & - & OH \text{ etc.} \\ & & | & & | & & | & & \\ & & R & & R & & R & & \end{array}$$

The starting materials for the manufacture of silicones are alkyl or aryl substituted chlorosilanes. Methyl compounds are mainly used, though some phenyl derivatives are used as well. Hydrolysis of dimethyldichlorosilane $(CH_3)_2SiCl_2$ gives rise to straight chain polymers and, as an active OH group is left at each end of the chain, polymerization continues and the chain increases in length. $(CH_3)_2SiCl_2$ is therefore a chain building unit. Normally, high polymers are obtained.

$$\begin{array}{ccccccccccccccccc} & & CH_3 & & CH_3 & & CH_3 & & CH_3 & & CH_3 & & CH_3 & & CH_3 & & \\ & & | & & | & & | & & | & & | & & | & & | & & \\ HO & - & Si & -O- & Si & -O- & Si & -O- & Si & -O- & Si & -O- & Si & -O- & Si & - & OH \\ & & | & & | & & | & & | & & | & & | & & | & & \\ & & CH_3 & & CH_3 & & CH_3 & & CH_3 & & CH_3 & & CH_3 & & CH_3 & & \end{array}$$

Hydrolysis under carefully controlled conditions can produce cyclic structures, with rings containing three, four, five or six Si atoms:

$$\begin{array}{ccccccc} Me & & & O & & & Me \\ & \diagdown & \diagup & & \diagdown & \diagup & \\ & Si & & & & Si & \\ \diagup & | & & & & | & \diagdown \\ Me & O & & & & O & Me \\ & & \diagdown & & \diagup & & \\ & & & Si & & & \\ & & \diagup & & \diagdown & & \\ & & Me & & Me & & \end{array} \qquad \begin{array}{ccccccc} & & Me & & Me & & \\ & & | & & | & & \\ Me & - & Si & -O- & Si & - & Me \\ & & | & & | & & \\ & & O & & O & & \\ & & | & & | & & \\ Me & - & Si & -O- & Si & - & Me \\ & & | & & | & & \\ & & Me & & Me & & \end{array}$$

tris cyclo-dimethylsiloxane   tetrakis cyclo-dimethylsiloxane

Hydrolysis of trimethylmonochlorosilane $(CH_3)_3SiCl$ yields $(CH_3)_3SiOH$ trimethylsilanol as a volatile liquid, which can condense, giving hexamethyldisiloxane. Since this compound has no OH groups, it cannot polymerize any further.

$$\begin{array}{ccccccccccccccc} & & CH_3 & & & & & & CH_3 & & & & CH_3 & & CH_3 & & \\ & & | & & & & & & | & & & & | & & | & & \\ CH_3 & - & Si & - & OH & + & HO & - & Si & - & CH_3 \rightarrow CH_3 & - & Si & -O- & Si & - & CH_3 \\ & & | & & & & & & | & & & & | & & | & & \\ & & CH_3 & & & & & & CH_3 & & & & CH_3 & & CH_3 & & \end{array}$$

hexamethyldisiloxane

If some $(CH_3)_3SiCl$ is mixed with $(CH_3)_2SiCl_2$ and hydrolysed, the $(CH_3)_3SiCl$ will block the end of the straight chain produced by $(CH_3)_2SiCl_2$. Since there is no longer a functional OH group at this end of the chain, it cannot grow any more at this end. Eventually the other end will be blocked in a similar way. Thus $(CH_3)_3SiCl$ is a chain stopping unit, and the ratio of $(CH_3)_3SiCl$ and $(CH_3)_2SiCl_2$ in the starting mixture will determine the average chain size.

$$HO-\overset{\displaystyle CH_3}{\underset{\displaystyle CH_3}{\overset{|}{\underset{|}{Si}}}}-O-\overset{\displaystyle CH_3}{\underset{\displaystyle CH_3}{\overset{|}{\underset{|}{Si}}}}-O-\overset{\displaystyle CH_3}{\underset{\displaystyle CH_3}{\overset{|}{\underset{|}{Si}}}}-O-\overset{\displaystyle CH_3}{\underset{\displaystyle CH_3}{\overset{|}{\underset{|}{Si}}}}-OH + HO-\overset{\displaystyle CH_3}{\underset{\displaystyle CH_3}{\overset{|}{\underset{|}{Si}}}}-CH_3 \rightarrow$$

$$HO-\overset{\displaystyle CH_3}{\underset{\displaystyle CH_3}{\overset{|}{\underset{|}{Si}}}}-O-\overset{\displaystyle CH_3}{\underset{\displaystyle CH_3}{\overset{|}{\underset{|}{Si}}}}-O-\overset{\displaystyle CH_3}{\underset{\displaystyle CH_3}{\overset{|}{\underset{|}{Si}}}}-O-\overset{\displaystyle CH_3}{\underset{\displaystyle CH_3}{\overset{|}{\underset{|}{Si}}}}-O-\overset{\displaystyle CH_3}{\underset{\displaystyle CH_3}{\overset{|}{\underset{|}{Si}}}}-CH_3$$

The hydrolysis of methyl trichlorosilane $RSiCl_3$ gives a very complex cross-linked polymer.

$$\begin{array}{ccccccccc}
 & & | & & & & & & \\
 & & O & & R & & & & \\
 & & | & & | & & & & \\
 & R- & Si & -O- & Si & -O- & & & \\
 & & | & & | & & & & \\
 & & O & & O & & R & & \\
 & & | & & | & & | & & \\
-O- & & Si & -O- & Si & -O- & Si & -O- & \\
 & & | & & | & & | & & \\
 & & R & & R & & O & & \\
 & & & & & & | & &
\end{array}$$

In a similar way addition of a small amount of CH_3SiCl_3 to the hydrolysis mixture produces a few cross-links, or provides a site for attaching other molecules. By controlled mixing of the reactants, any given type of polymer can be produced.

Silicones are fairly expensive but have many desirable properties. They were originally developed as electrical insulators, because they are more stable to heat than are organic polymers, and if they do break down they do not produce conducting materials as carbon does. They are resistant to heat, oxidation and most chemicals. They are strongly water repellent, are good electrical insulators, and have non-stick properties and anti-foaming properties. Their strength and inertness are related to two factors:

1. Their stable silica-like skeleton of Si—O—Si—O—Si. The Si—O bond energy is very high ($502\,kJ\,mol^{-1}$).
2. The high strength of the Si—C bond.

Their water repellency arises because a silicone chain is surrounded by organic side groups, and looks like an alkane from the outside. Silicones may be liquids, oils, greases, elastomers (rubbers) or resins.

Straight chain polymers of 20 to 500 units are used as silicone fluids. They make up 63% of the silicones used. If they are made by hydrolysing a mixture of $(CH_3)_2SiCl_2$ and $(CH_3)_3SiCl$, then the chain lengths vary considerably. Commercially they are made by treating a mixture of tetrakis cyclodimethylsiloxane $(Me_2SiO)_4$ and hexamethyl disiloxane $(Me)_3SiOSi(Me)_3$ with 100% H_2SO_4. The cyclo compound provides chain building units, and the hexamethylsiloxane provides chain stopping groups. The average chain length is determined by the ratio of these reactants. The H_2SO_4 splits Si—O—Si bonds, forming Si—O—SO_4H esters and Si—OH. The esters hydrolyse back to give Si—O—Si bonds. This process goes on repeatedly, and results in the size of chains all becoming similar. The boiling point and viscosity increase with chain length, giving compounds ranging from watery liquids to viscous oils and greases. The fluids are used as water repellents for treating masonry and buildings, glassware and fabrics. They are also included in car polish and shoe polish. Silicone fluids are non-toxic and have a low surface tension. Addition of a few parts per million of a silicone greatly reduces foaming in sewage disposal, textile dyeing, beer making (fermentation) and frothing of cooking oil in making potato crisps or chips. Silicone oils are used as dielectric insulating material in high voltage transformers. They are also used as hydraulic fluids. Methyl silicones can be used as light duty lubricating oil, but are not suitable for heavy duty applications like gearboxes, because the oil film breaks down under high pressure. Silicones with some phenyl groups are better lubricants. These oils can be mixed with lithium stearate soaps to give greases.

Silicone rubbers are made of long, straight chain polymers, (dimethylpolysiloxanes) between 6000 and 600 000 Si units long, mixed with a filler – usually finely divided SiO_2 or occasionally graphite. They are usually produced by hydrolysis of dimethyldichlorosilane with KOH. Great care must be taken to exclude chain blocking and cross-linking groups. Rubbers make up about 25% of the silicones produced. Silicone rubbers are useful because they retain their elasticity from $-90\,°C$ to $+250\,°C$, which is a much wider range than for natural rubber. They are also good electrical insulators. They may be vulcanized to give hard rubber as follows:

1. By oxidizing with a small amount of benzoyl peroxide which produces occasional cross-links (up to 1% of the Si atoms may be cross-linked).
2. By building a cross-linking unit into the chain.

The most heat-stable side groups are phenyl groups, followed by the methyl, ethyl and propyl groups in descending order of stability. On heating in air to 350–400 °C, silicones are rapidly oxidized and cross-links are formed. The polymer becomes brittle and cracks, and low molecular weight polymers and cyclic structures are evolved. Strong heating in the absence of air causes silicones to soften and form volatile products, but oxidation and cross-linking do not occur.

Silicone resins are rigid polymers rather like bakelite. They are made by dissolving a mixture of $PhSiCl_3$ and $(Ph)_2SiCl_2$ in toluene and hydrolysing with water. The partly polymerized product is washed to remove HCl, and can then be shaped or moulded. Finally the product is heated with a quaternary ammonium salt as catalyst to condense any remaining OH groups in the structure. The final product is extensively cross-linked. About 12% of silicones produced are resins. These resins are used as electrical insulators, often mixed with glass fibre for additional strength. They are used to make printed circuit boards and to encapsulate integrated circuit chips and resistors. They are also used as non-stick coatings for pans, and for moulds for car tyres and bread.

HYDRIDES

All the elements form covalent hydrides, but the number of compounds formed and the ease with which they form differs greatly. Carbon forms a vast number of chain and ring compounds including:

1. The alkanes (paraffins) C_nH_{2n+2}
2. The alkenes (olefines) C_nH_{2n}
3. The alkynes (acetylenes) C_nH_{2n-2}
4. Aromatic compounds

These are the basis of organic chemistry. There is a strong tendency to catenation (forming chains) because the C—C bond is very strong.

Silicon forms a limited number of saturated hydrides, Si_nH_{2n+2}, called the silanes. These may exist as straight chains or branched chains, containing up to eight Si atoms. Ring compounds are very rare. No analogues of alkenes or alkynes are known. Monosilane SiH_4 is the only silicon hydride of importance. SiH_4 and $SiHCl_3$ were first made by treating an Al/Si alloy with dilute HCl. A mixture of silanes was prepared by hydrolysing magnesium silicide, Mg_2Si, with sulphuric or phosphoric acid. These compounds are colourless gases or volatile liquids. They are highly reactive, and catch fire or explode in air. Apart from SiH_4 they are thermally unstable. It only became possible to study them when A. Stock invented a method of handling reactive gases in a vacuum frame.

$$2Mg + Si \xrightarrow{\text{heat in absence of air}} Mg_2Si$$

$$\begin{aligned} Mg_2Si + H_2SO_4 \rightarrow\ & SiH_4 \quad (40\%) \\ & Si_2H_6 \quad (30\%) \\ & Si_3H_8 \quad (15\%) \\ & Si_4H_{10} \quad (10\%) \\ & \left.\begin{matrix} Si_5H_{12} \\ Si_6H_{14} \end{matrix}\right\} (5\%) \end{aligned}$$

More recently monosilane has been prepared by reducing $SiCl_4$ with $Li[AlH_4]$, LiH or NaH in ether solution at low temperatures. This is a much better method, as it gives one product rather than a mixture and it gives a quantitative yield.

$$SiCl_4 + Li[AlH_4] \rightarrow SiH_4 + AlCl_3 + LiCl$$
$$Si_2Cl_6 + 6LiH \rightarrow Si_2H_6 + 6LiCl$$
$$Si_3Cl_8 + 8NaH \rightarrow Si_3H_8 + 8NaCl$$

Silanes may also be prepared by direct reaction by heating Si or ferrosilicon with anhydrous HX or RX in the presence of a copper catalyst.

$$Si + 2HCl \rightarrow SiH_2Cl_2$$
$$Si + 3HCl \rightarrow SiHCl_3 + H_2$$
$$Si + 2CH_3Cl \rightarrow CH_3SiHCl_2 + C + H_2$$

The silanes are much more reactive than the alkanes. The alkanes are chemically unreactive apart from reaction with the halogens, and burning in air. In contrast the silanes are strong reducing agents, ignite in air and explode in Cl_2. Pure silanes do not react with dilute acids or pure water in silica apparatus, but they hydrolyse readily in alkaline solutions, or even with the trace of alkali which leaches out from glass apparatus.

$$Si_2H_6 + (4 + n)H_2O \xrightarrow{\text{trace of alkali}} 2SiO_2 \cdot nH_2O + 7H_2$$

Compounds with Si—H bonds undergo an important hydrosilation reaction with alkenes, in the presence of a platinum catalyst. The reaction is similar to hydroboration, and the products may be used to make silicones.

$$RCH{=}CH_2 + SiHCl_3 \rightarrow RCH_2CH_2SiCl_3$$

The difference in behaviour between alkanes and silanes is attributed to several factors:

1. Pauling's electronegativity values are: C = 2.5, Si = 1.8, and H = 2.1. Thus the bonding electrons between C and H or Si and H are not equally shared, leaving a δ^- charge on C and a δ^+ charge on Si. Thus Si is vulnerable to attack by nucleophilic reagents.

$$\overset{\delta^-}{C}\text{—}\overset{\delta^+}{H} \qquad \overset{\delta^+}{Si}\text{—}\overset{\delta^-}{H}$$

2. The larger size of Si makes it easier to attack.
3. Si has low energy *d* orbitals which may be used to form an intermediate compound, and thus lower the activation energy of the process.

Several germanium hydrides or germanes Ge_nH_{2n+2} are known up to $n = 5$. They are straight chain compounds and are colourless gases or volatile liquids. They are similar to the silanes, but are less volatile, less flammable and are unaffected by water or aqueous acids or alkalis. GeH_4 can be made:

$$GeCl_4 + Li[AlH_4] \xrightarrow{\text{dry ether}} GeH_4 + LiCl + AlCl_3$$

$$GeO_2 + Na[BH_4] \xrightarrow{\text{aqueous solution}} GeH_4 + NaBO_2$$

Stannane SnH_4 is much less stable. It can be made by reducing $SnCl_4$ with $Li[AlH_4]$ or $Na[BH_4]$. It is a strong reducing agent. It is unaffected by water and dilute acids and alkali, but it reacts slowly with concentrated solutions. Distannane Sn_2H_6 is known and is even less stable. No higher stannanes are known. Plumbane PbH_4 is even less stable and even more difficult to prepare. The preparative methods used for the other hydrides fail. It has been made in trace amounts and at low concentrations by cathodic reduction and detected using a mass spectrometer.

CYANIDES

The alkali metal cyanides, particularly NaCN, are made in quantity, about 120 000 tonnes/year. Until about 1965 it was made by the *Castner process* by high temperature reactions from sodamide.

$$Na + NH_3 \rightarrow NaNH_2 + \tfrac{1}{2}H_2$$

$$NaNH_2 + C \xrightarrow{750\,^{\circ}C} NaCN + H_2$$

Since 1965 hydrogen cyanide HCN has become commercially available (currently 300 000 tonnes/year): NaCN is made from this. HCN was formerly prepared by acidification of NaCN or $Ca(CN)_2$.

$$2NaCN + H_2SO_4 \rightarrow 2HCN + Na_2SO_4$$
$$Ca(CN)_2 + H_2SO_4 \rightarrow 2HCN + CaSO_4$$

In the modern industrial processes, a gas phase reaction occurs between CH_4 and NH_3 at about 1200 °C in the presence of a catalyst.

Degussa process $CH_4 + NH_3 \rightarrow HCN + 3H_2$ (Pt catalyst)

Andrussow process $2CH_4 + 2NH_3 + 3O_2 \rightarrow 2HCN + 6H_2O$ (Pt/Rh catalyst)

HCN is extremely poisonous. It has an abnormally high boiling point of 26 °C, because of hydrogen bonding. It is one of the weakest acids known, weaker than HF. Over half the HCN produced is used to make the polymer methyl methacrylate, and the remainder is used to make NaCN, $(ClCN)_3$, and various cyanide complexes such as $K_4[Fe(CN)_6]$ and $K_3[Fe(CN)_6]$, and in the extraction of Ag and Au.

$$(ClCN)_3 + 3NH_3 \rightarrow 3HCl + C_3N_3(NH_2)_3 \quad \text{(melamine)}$$
$$4Au + 8NaCN + 2H_2O + O_2 \rightarrow 4Na[Au(CN)_2] + 4NaOH$$

HCN has been used as a non-aqueous ionizing solvent.

The cyanide ion is important in forming stable complexes, particularly with the metals of the Cr, Mn, Fe, Co, Ni, Cu and Zn groups. Two common complexes are ferrocyanides $[Fe(CN)_6]^{4-}$ and ferricyanides $[Fe(CN)_6]^{3-}$. The later transition elements form stable cyanide complexes because they can use filled *d* orbitals for $d\pi - p\pi$ backbonding in addition to the original σ coordinate bond $M \leftarrow (CN)^-$. The bonding is similar to that in the carbonyls, and the CN^- ion acts as a π acceptor. The negative charge on CN^- makes it a stronger σ donor than CO, but the charge also weakens the effectiveness of CN^- as a π acceptor.

The extreme toxicity of cyanides is due to CN^- complexing with metals in enzymes and haemoglobin in the body, thus preventing normal metabolism. Besides forming many complexes analogous to halide complexes, the cyanide ion often brings out the maximum coordination number of a metal. Thus Fe^{3+} forms $[FeCl_4]^-$ with chloride ions, but $[Fe(CN)_6]^{3-}$ with cyanide ions. Many metal ions such as Cu^+, Ni^+, Mn^+, Au^+ and Mn^{3+}, which are too unstable to exist in solution, are quite stable when complexed with cyanide ions. The formation of complexes is important in the extraction of silver and gold, as the metals dissolve in a solution of NaCN in the presence of air, and form sodium argentocyanide or sodium aurocyanide, from which the metal is recovered by reduction with zinc.

$$4Ag + 8NaCN + 2H_2O + O_2 \rightarrow 4Na[Ag(CN)_2] + 4NaOH$$
$$4Au + 8NaCN + 2H_2O + O_2 \rightarrow 4Na[Au(CN)_2] + 4NaOH$$

Cyanide ions may act as both complexing and reducing agents:

$$2Cu^{2+} + 4CN^- \rightarrow (CN)_2 + 2CuCN \xrightarrow{+CN^-} [Cu(CN)_4]^{3-}$$

In this reaction the cyanide ion is itself oxidized to cyanogen $(CN)_2$ in much the same way as I^- is oxidized to I_2 by Cu^{2+}. In alkaline solution, cyanogen disproportionates into cyanide and cyanate ions.

$$(CN)_2 + 2OH^- \rightarrow H_2O + CN^- + NCO^-$$

The cyanate ion is isoelectronic with carbon dioxide; hence they have similar structures and are both linear.

$$O{=}C{=}O \qquad {}^-N{=}C{=}O$$

COMPLEXES

The ability to form complexes is favoured by a high charge, small size and availability of empty orbitals of the right energy. Carbon is in the second period and has a maximum of eight electrons in its outer shell. In four-covalent compounds of carbon, the second shell contains the maximum of eight electrons. Because this structure resembles that of a noble gas, these compounds are stable, and carbon does not form complexes. Four-covalent compounds of the subsequent elements can form complexes due to the availability of *D* orbitals, and they generally increase their coordination number from 4 to 6.

$$SiF_4 + 2F^- \rightarrow [SiF_6]^{2-}$$
$$GeF_4 + 2NMe_3 \rightarrow [GeF_4 \cdot (NMe_3)_2]$$
$$SnCl_4 + 2Cl^- \rightarrow [SnCl_6]^{2-}$$

The VSEPR theory suggests that because there are six outer electron pairs these complexes will be octahedral. The valence bond theory requires that four covalent and two coordinate bonds are formed and give an octahedral structure. For example, $[SiF_6]^{2-}$:

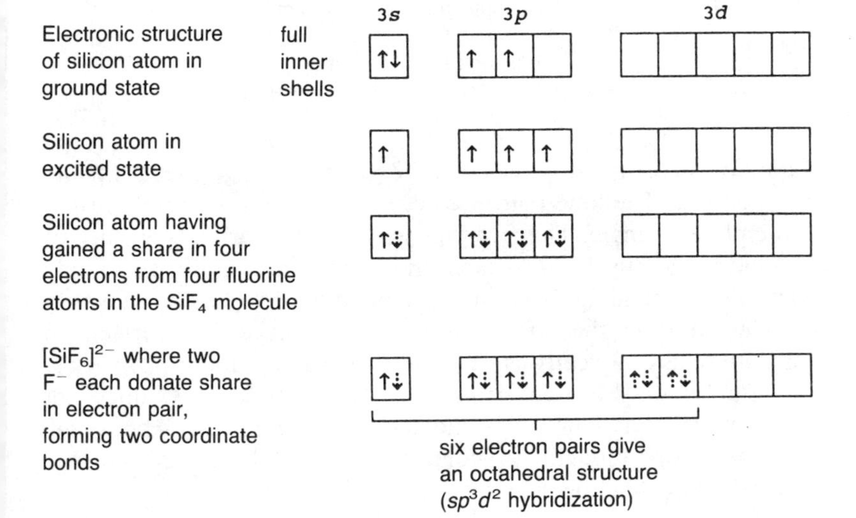

The arguments over hybridizing *d* orbitals have been discussed in Chapter 4. The $[SiF_6]^{2-}$ ion is usually formed from SiO_2 and aqueous HF.

$$SiO_2 + 6HF \rightarrow 2H^+ + [SiF_6]^{2-} + 2H_2O$$

The $[SiF_6]^{2-}$ complex is stable in water and alkali, but the others in the group are less stable. $[GeF_6]^{2-}$ and $[SnF6]^{2-}$ are hydrolysed by alkali, and $[PbF_6]^{2-}$ is hydrolysed by both alkali and water. Ge, Sn and Pb also form chloride complexes such as $[PbCl_6]^{2-}$, and oxalate complexes such as $[Pb(ox)_3]^{2-}$.

Lead tetraacetate $Pb(CH_3COO)_4$ can be obtained as a colourless, solid by treating Pb_3O_4 with glacial acetic acid. It is water sensitive, and is widely used as a selective oxidizing agent in organic chemistry. Its best known application is in the cleavage of 1,2-diols (glycols), as present, for example, in carbohydrates.

$$\begin{matrix} -\mathrm{C}-\mathrm{OH} \\ | \\ -\mathrm{C}-\mathrm{OH} \end{matrix} \xrightarrow{Pb(CH_3COO)_4} \begin{matrix} -\mathrm{C}-\mathrm{O} \\ | \\ -\mathrm{C}-\mathrm{O} \end{matrix}\!\!> Pb\,(CH_3COO)_2 \rightarrow \begin{matrix} -\mathrm{C}{=}\mathrm{O} \\ \\ -\mathrm{C}{=}\mathrm{O} \end{matrix} + Pb(CH_3COO)_2$$

INTERNAL π BONDING USING *d* ORBITALS

The compounds trimethylamine $(CH_3)_3N$ and trisilylamine $(SiH_3)_3N$ have similar formulae, but have totally different structures (Figure 13.16).In trimethylamine, the arrangement of electrons is as follows:

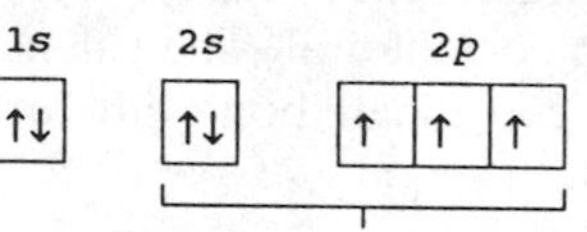

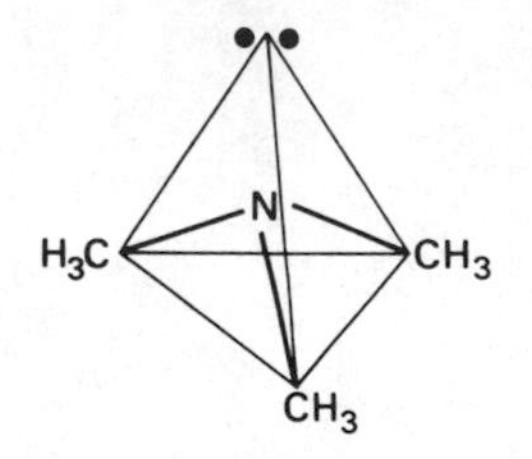

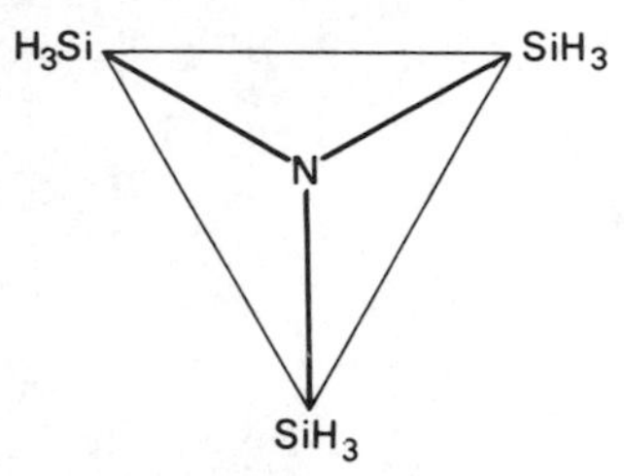

Figure 13.16 Trimethylamine $N(CH_3)_3$ and trisilylamine $N(SiCH_3)_3$.

In trisilylamine, three sp^2 orbitals are used for σ bonding, giving a plane triangular structure. The lone pair of electrons occupy a *p* orbital at right angles to the plane triangle. This overlaps with empty *d* orbitals on each of the three silicon atoms, and results in π bonding, more accurately described as *pπ–dπ* bonding, because it is from a full *p* orbital to an empty *d* orbital. This shortens the bond lengths N—Si. Since the nitrogen no longer has a lone pair of electrons, the molecule has no donor properties. Similar *pπ–dπ* bonding is impossible in $(CH_3)_3N$ because C does not possess *d* orbitals and hence this molecule is pyramidal. About 200 compounds are now thought to contain *pπ–dπ* bonds (see Further Reading, Raabe and Michl).

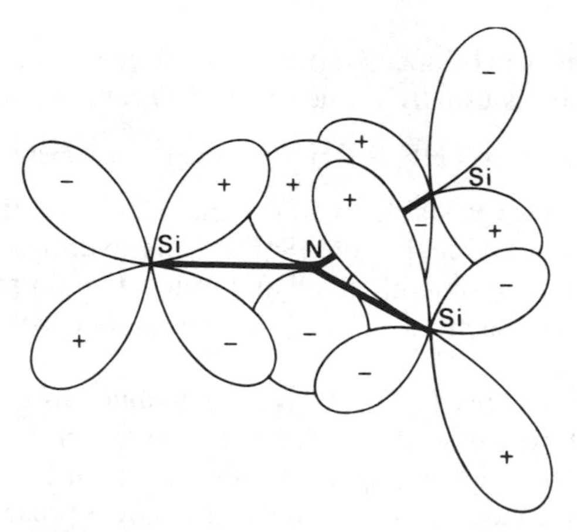

Figure 13.17 pπ–dπ bonding in trisilylamine. (From Mackay and Mackay, *Introduction to Modern Inorganic Chemistry*, 4th ed., Blackie, 1989.)

TETRAHALIDES

All the tetrahalides are known except PbI_4. They are typically covalent, tetrahedral, and very volatile. The exceptions are SnF_4 and PbF_4, which have three-dimensional structures and are high melting (SnF_4 sublimes at 705 °C, PbF_4 melts at 600 °C).

Carbon

Tetrafluoromethane (carbon tetrafluoride) CF_4 is an exceptionally unreactive gas. It can be made as follows:

$$CO_2 + SF_4 \rightarrow CF_4 + SO_2$$
$$SiC + 2F_2 \rightarrow SiF_4 + C$$
$$CF_2Cl_2 + F_2 \rightarrow CF_4 + Cl_2 \quad \text{(industrial method)}$$

Other fluorine compounds such as hexafluoroethane C_2F_6 and tetrafluoroethylene C_2F_4 are known. Under pressure C_2F_4 polymerizes to $(C_2F_4)_n$, giving polytetrafluoroethylene or PTFE. This is a hard, white solid plastic with a greasy feel to the touch, and is much heavier (more dense) than one would expect. It is a good electrical insulator, and is chemically inert. It is expensive, and is used in the laboratory because of its inertness. It has a very low coefficient of friction and is used for coating non-stick pans and razor blades. Fluorocarbons are useful lubricants, solvents and insulators.

$$CHCl_3 + HF \xrightarrow{SbFCl_4 \text{ catalyst}} CF_2ClH \xrightarrow{\text{heat}} C_2F_4$$

$$C_2F_4 \xrightarrow{\text{pressure}} (C_2F_4)_n$$

Tetrachloromethane (carbon tetrachloride) CCl_4 is manufactured mainly from carbon disulphide.

$$CS_2 + 3Cl_2 \xrightarrow{FeCl_3 \text{ catalyst } 30\,^\circ C} CCl_4 + S_2Cl_2$$

$$CS_2 + 2S_2Cl_2 \xrightarrow{FeCl_3 \text{ catalyst } 60\,^\circ C} CCl_4 + 6S$$

CCl_4 is extensively used as a solvent, and for the preparation of Freons. It is also used in fire extinguishers, where the heavy vapour excludes oxygen and thus puts the fire out.

$$CCl_4 + 2HF \xrightarrow[SbCl_5 \text{ catalyst}]{\text{anhydrous conditions}} CCl_2F_2 + 2HCl$$

The carbon halides are not hydrolysed *under normal conditions* because they have no *d* orbitals, and cannot form a five-coordinate hydrolysis intermediate. In contrast the silicon halides hydrolyse readily. Silicon has $3d$ orbitals available, and these may be used to coordinate OH^- ions or water as a first step in hydrolysis. In any atom there are always empty orbitals, but these are usually too high in energy to be used. If sufficient energy is provided by using superheated steam then CCl_4 will hydrolyse:

$$CCl_4 + H_2O \xrightarrow[\text{steam}]{\text{superheated}} \underset{\text{carbonyl chloride (phosgene)}}{COCl_2} + 2HCl$$

Phosgene is highly toxic, and was used as a poisonous gas in World War I. It is now made by combining CO and Cl_2 with a C catalyst in sunlight, and is used to make isocyanates for the manufacture of polyurethanes.

Freons

Mixed chlorofluorohydrocarbons such as $CFCl_3$, CF_2Cl_2 and CF_3Cl are known as Freons. They are unreactive and non-toxic and are widely used as refrigeration fluids and as the propellant in aerosols. At one time, nearly 700 000 tonnes of Freons were produced annually (half in the USA). There are several Freons, $CFCl_3$, CF_2Cl_2 and CF_3Cl, they are sometimes called CFCs, which is short for chlorinated fluorocarbons. Despite being so inert they have been found to cause environmental damage. Their use in aerosols was banned in the USA from 1980, and in Europe from 1990. They are still used in refrigerators, and a total ban on Freons in Europe is planned by the year 2000. Cheaper and environmentally friendlier aerosol propellants such as CO_2 and butane are replacing Freons. There are difficulties, as butane is flammable, and cannot be used with food. CO_2 has a low vapour pressure when cold, and is therefore no use for windscreen de-icers.

Freons are very much more effective 'greenhouse gases' in the atmosphere than is CO_2, though the amount of Freons present is extremely small. Much more seriously, the Freons have penetrated the upper atmosphere (5–20 miles high), and are causing damage to the ozone layer. There has been a loss of about 6% of the ozone between 1980 and 1990. A hole in the ozone layer has appeared over the South Pole, and a similar hole seems to be developing over the North Pole. The ozone layer is important as it filters the radiation from the sun and prevents most of the harmful UV radiation from reaching the earth. Excessive exposure to UV radiation should be avoided as it causes skin cancer (melanoma) in humans.

In the upper atmosphere Freons undergo a photolytic reaction and produce free chlorine atoms (which are radicals). These react readily with ozone. The ClO radicals formed decompose slowly, re-forming chlorine radicals, which react with more ozone . . . and so on. The chlorine radicals do not recombine to form Cl_2, because they need a three-body collision to dissipate the energy, and such collisions are extremely rare in the upper atmosphere. There is no effective sink for chlorine radicals. Once formed they are used again and again, so a small number of radicals make a very effective scavenger for ozone.

$$\left.\begin{matrix} CFCl_3 \\ CF_2Cl_2 \\ CF_3Cl \end{matrix}\right] \xrightarrow{\text{photolysis}} Cl$$

$$Cl + O_3 \xrightarrow{\text{rapid}} O_2 + ClO$$

$$ClO \longrightarrow Cl + O$$

$$ClO + O \rightarrow Cl + O_2$$

(Cl formed is returned to react with further O_3.)

Overall reaction: $2O_3 \rightarrow 3O_2$

Silicon

The silicon halides can be prepared as follows:

$$Si \quad or \quad SiC + 2X_2 \xrightarrow{heat} SiX_4$$

In marked contrast to the inertness of CF_4, CCl_4 and the Freons, SiF_4 is readily hydrolysed by alkali.

$$SiF_4 + 8OH^- \rightarrow SiO_4^{4-} + 4F^- + 4H_2O$$

The silicon halides are rapidly hydrolysed by water to give silicic acid.

$$SiCl_4 + 4H_2O \rightarrow Si(OH)_4 + 4HCl$$

In the case of the tetrafluoride, a secondary reaction occurs between the resultant HF and the unchanged SiF_4, forming the hexafluorosilicate ion $[SiF_6]^{2-}$.

$$SiF_4 + 2HF \rightarrow 2H^+ + [SiF_6]^{2-}$$

Other members of the series Si_nX_{2n+2} can be produced by pyrolysis (strong heating). These are either volatile liquids or solids. The longest chains known are $Si_{16}F_{34}$, Si_6Cl_{14} and Si_4Br_{10}. The chains are longer than those formed in the hydrides. This is due to dative π back bonding from full p orbitals on the halogen with empty d orbitals on Si.

$$SiCl_4 + Si \rightarrow Si_2Cl_6 + \text{higher members of the series}$$
$$5Si_2Cl_6 \rightarrow Si_6Cl_{14} + 4SiCl_4$$

Germanium, tin and lead

Ge, Sn and Pb form two series of halides, MX_4 and MX_2. With Ge the (+IV) oxidation state is the most stable, but with Sn the (+II) state is the most stable.

The tetrahalides are all colourless volatile liquids except for GeI_4 and SnI_4 which are bright orange solids. Compounds formed by the main group elements are normally white. Colour is associated with electrons being promoted from one energy level to another, and absorbing or emitting the energy difference between the two levels. This is common in the transition elements, where there are often unfilled energy levels in the d shell, allowing promotion from one d level to another. In the main groups, the s and p electron shells are normally filled when a compound is formed, so promotion within the same shell is not possible. Promotion from one shell to another, for example from the $2p$ to the $3p$ level, involves so much energy that absorption lines would appear in the ultraviolet rather than in the visible region. Thus the tetrahalides would be expected to be white in colour. The orange colour of SnI_4 is caused by the absorption of blue light, the reflected light thus containing a higher proportion of red and orange. The energy absorbed in this way causes the transfer of an electron from I to Sn. (This corresponds to the temporary reduction of Sn(IV) to Sn(III).)

Since transferring an electron to another atom is transferring a charge, such spectra are called *charge transfer spectra*. This occurs in SnI_4 and GeI_4 because the atoms have similar energy levels. This would be expected because they are close in the periodic table, and have similar sizes. Charge transfer spectra do not occur with the other halides.

$GeCl_4$ and $GeBr_4$ are hydrolysed less readily. $SnCl_4$ and $PbCl_4$ hydrolyse in dilute solutions, but hydrolysis is often incomplete and can be repressed by the addition of the appropriate halogen acid.

$$Sn(OH)_4 \underset{H_2O}{\overset{HCl}{\rightleftharpoons}} SnCl_4 \underset{H_2O}{\overset{HCl}{\rightleftharpoons}} [SnCl_6]^{2-}$$

In the presence of excess acid, halides of Si, Ge, Sn and Pb increase their coordination number from 4 to 6, and form complex ions, such as $[SiF_6]^{2-}$, $[GeF_6]^{2-}$, $[SnCl_6]^{2-}$ and $[SnCl_5]^-$. PbI_4 is not known, probably because of the oxidizing power of Pb(+IV) and the reducing power of I^-, which results in PbI_2 always being formed.

Catenated halides

Carbon forms a number of catenated halides, perhaps the best known being Teflon or polytetrafluoroethylene, which is described above. The polymers formed have chain lengths of several hundred carbon atoms

Silicon forms polymers $(SiF_2)_n$ and $(SiCl_2)_n$ by passing the tetrahalide over heated silicon. These polymers decompose on heating into low molecular weight polymers (or oligomers) of formula Si_nX_{2n+2}. The longest chains known are $Si_{16}F_{34}$, Si_6Cl_{14} and Si_4Br_{10}.

Germanium forms the dimer Ge_2Cl_6, but Sn and Pb do not form any catenated halides.

DIHALIDES

There is a steady increase in the stability of dihalides:

$$CX_2 \ll SiX_2 < GeX_2 < SnX_2 < PbX_2$$

SiF_2 can be made by high temperature reactions, and can be trapped by cooling in liquid N_2. When the product warms up, polymerization occurs giving a range of compounds up to $Si_{16}F_{34}$.

$$SiF_4 + Si \rightleftharpoons 2SiF_2$$

GeF_2 is a white solid made either by heating Ge with anhydrous HF, or from GeF_4 and Ge. It has an unusual fluorine bridged polymeric structure, based on a trigonal bipyramid. GeF_3 units share two F atoms (giving the formula GeF_2), and the Ge also forms a weaker interaction to another F, with a lone pair in the fifth position. These units are linked into infinite spiral chains. SnF_2 and $SnCl_2$ are white solids, and are obtained by heating Sn or SnO with gaseous HF or HCl. Stannous fluoride SnF_2 was used together with tin pyrophosphate $Sn_2P_2O_7$ in the original 'Crest' fluoride toothpaste. This is surprising as Sn is toxic, and NaF is now used instead. The crystal structure of SnF_2 is made of Sn_4F_8 tetramers. These form an eight-membered puckered ring —Sn—F—Sn—F—, with weaker interactions linking the rings together. $SnCl_2$ partly hydrolyses in water, forming the basic chloride Sn(OH)Cl. SnF_2 and $SnCl_2$ both dissolve in solutions containing halide ions.

$$SnF_2 + F^- \rightarrow [SnF_3]^- \qquad pK \sim 1$$
$$SnCl_2 + Cl^- \rightarrow [SnCl_3]^- \qquad pK \sim 2$$

Sn^{2+} ions do occur in perchlorate solutions, but the stannous ion is readily oxidized by air to Sn^{IV} unless precautions are taken. Sn^{2+} ions are hydrolysed by water mainly to $[Sn_3(OH)_4]^{2+}$, with small amounts of $[SnOH]^+$ and $[Sn_2(OH)_2]^{2+}$. The $[Sn_3(OH)_4]^{2+}$ ion is probably cyclic and the compounds $[Sn_3(OH)_4SO_4$ and $[Sn_3(OH)_4](NO_3)_2$ are known. The compounds PbX_2 are much more stable than PbX_4. Pb is the only element in the group with well defined cations. The salts PbX_2 can all be made from a water soluble Pb^{2+} salt and the corresponding halide ion or halogen acid. The plumbous ion is partially hydrolysed by water.

$$Pb^{2+} + 2H_2O \rightarrow [PbOH]^+ + H_3O^+$$

CLUSTER COMPOUNDS

There is a well established tendency for the heavier members of Groups IV, V and VI to form polyatomic ions. These may be chains, rings or clusters.

Reduction of Ge, Sn and Pb by Na in liquid ammonia gives metal ions containing several atoms. These have been shown to be metal clusters. Crystalline compounds containing such ions can be isolated by forming complexes with ethylenediamine, or with crypt ligands. Examples include $[Na(crypt)]_2^+$ $[Sn_5]^{2-}$, $[Na(crypt)]_2^+$ $[Pb_5]^{2-}$, $[Na_4(en)_5Ge_9]$ and $[Na_4(en)_5Sn_9]$. The shape of M_5 clusters is a trigonal bipyramid, and M_9 clusters are unicapped square anti-prisms. (The latter consists of a square anti-prism, i.e. a cube with the top face with four corners rotated 45° relative to the bottom face. Unicapped means an extra atom projects from one of the faces.)

REACTION MECHANISMS

Many inorganic reactions, such as double decomposition, involve only ions, and these occur virtually instantaneously. Typically, organic reactions are slower because they involve breaking covalent bonds, and they occur either by substituting one group for another, or by adding on an extra group to give an intermediate which then eliminates another group to give the product.

The hydrolysis of $SiCl_4$ is rapid because Si can use a *d* orbital to form a

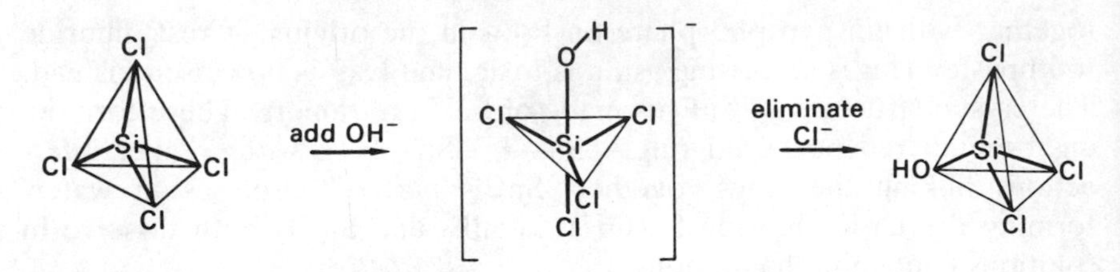

Figure 13.18 Hydrolysis of $SiCl_4$.

five-coordinate intermediate, and the reaction occurs by an S_N2 mechanism (Figure 13.18). A lone pair of electrons from the oxygen is donated to an empty *d* orbital on Si, forming a five-coordinate intermediate which has a trigonal bipyramidal structure.

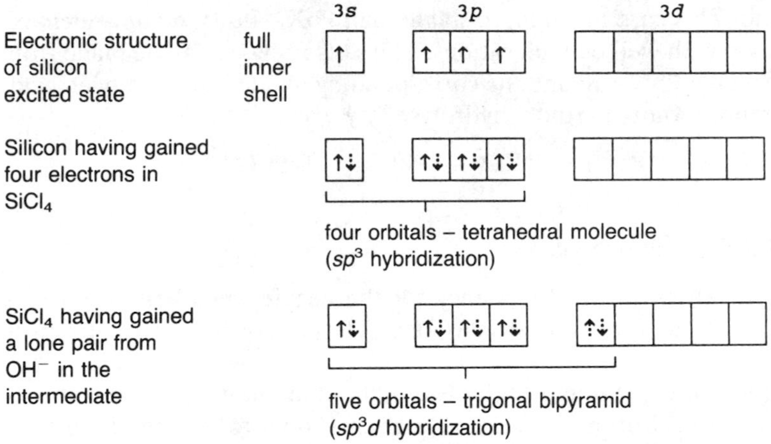

		3s	3p	3d
Electronic structure of silicon excited state	full inner shell	↑	↑ ↑ ↑	
Silicon having gained four electrons in $SiCl_4$		↑↓	↑↓ ↑↓ ↑↓	

four orbitals – tetrahedral molecule (sp^3 hybridization)

		3s	3p	3d
$SiCl_4$ having gained a lone pair from OH^- in the intermediate		↑↓	↑↓ ↑↓ ↑↓	↑↓

five orbitals – trigonal bipyramid (sp^3d hybridization)

If the hydrolysis is performed on an asymmetrically substituted, and consequently optically active, silicon compound such as MeEtPhSi*Cl, then Walden inversion will occur, resulting in inversion of the structure from *d* to *l* or vice versa (Figure 13.19). In a similar way, the reduction of $R_1R_2R_3$ Si*Cl with $Li[AlH_4]$ to give $R_1R_2R_3$ Si*H also involves inversion of structure.

Other mechanisms are possible because the conversion of $R_1R_2R_3$Si*H

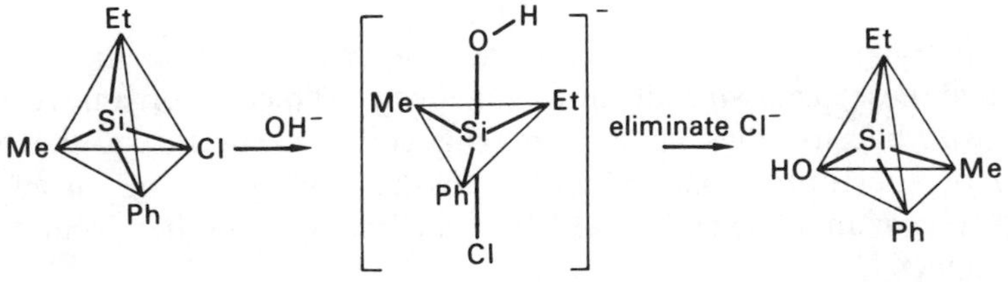

Figure 13.19 Walden inversion of structure.

to $R_1R_2R_3Si^*Cl$ occurs with the retention of structure. If $R_1R_2R_3Si^*Cl$ is dissolved in ether or CCl_4 it is recovered unchanged, but dissolving in CH_3CN results in racemization.

ORGANIC DERIVATIVES

The elements of this group have an extensive organometallic chemistry. The divalent state becomes increasingly stable and important on descending the group (the inert pair effect), yet rather surprisingly the organometallic derivatives of Sn and Pb all contain M^{IV} and not M^{II}.

The alkyl silicon chlorides are important as the starting materials for the manufacture of silicones. The silicone polymers have already been described. Tetra organic derivatives of Si, Ge, Sn and Pb may be prepared from the halides using Grignard or organolithium reagents.

$$SiCl_4 + MeMgCl \rightarrow MeSiCl_3, Me_2SiCl_2, Me_3SiCl, Me_4Si$$

$$PbCl_2 + LiEt \rightarrow PbEt_2 \rightarrow Pb + PbEt_4$$

Tetraethyl lead is produced in large amounts and used an an 'anti-knock' additive to increase the octane number of petrol. The commercial preparation uses a sodium/lead alloy.

$$Na/Pb + 4EtCl \rightarrow PbEt_4 + 4NaCl$$

Lead is poisonous to man, and burning petrol containing $PbEt_4$ releases lead into the atmosphere. In 1974 about 230 000 tonnes of $PbEt_4$ was produced in the USA, 55 000 tonnes in the UK, and an estimated total of 500 000 tonnes worldwide. At that time, $PbEt_4$ was produced in larger tonnage than any other organometallic compound. Its use is declining sharply due to legislation that new cars must run on lead-free petrol.

About 40 000 tonnes/year of organotin compounds R_2SnX_2 and R_3SnX are used. About two thirds are used to stabilize PVC plastics and the remainder are used in agriculture to control fungi, and pests such as insects and larvae.

FURTHER READING

Abel, E.W. (1973) *Comprehensive Inorganic Chemistry*, Vol. 2, (Chapter 17: Tin; Chapter 18: Lead), Pergamon Press, Oxford.

Abel, E.W. and Stone, F.G. (1969, 1970) The chemistry of transition metal carbonyls, *Q. Rev. Chem. Soc.*, Part I – Structural considerations, **23**, 325; Part II – Synthesis and reactivity, **24**, 498.

Ainscough, E.W. and Brodie, A.M. (1984) Asbestos – structures, uses and biological activities, *Education in Chemistry*, **21**, 173–175.

Barrer, R.M. (1978) *Zeolites and Clay Minerals as Sorbents and Molecular Sieves*, Academic Press, London.

Bode, H. (1977) *Lead-Acid Batteries*, Wiley, New York.

Bolin B. (1970) The carbon cycle, *Scientific American*, September.

Breck, D.W. (1973) *Molecular Sieves*, Wiley, New York.

Cusak, P.A. (1986) *Investigations Into Tin-Based Flame Retardants and Smoke Suppressants*, International Tin Institute, London.

Donovan, R.J. (1978) Chemistry and pollution of the stratosphere, *Education in Chemistry*, **15**, 110–113.
Drake, J.E. and Riddle, C. (1970) Volatile compounds of the hydrides of silicon and germanium with the elements of Group V and VI, *Q. Rev. Chem. Soc.*, **24**, 263.
Ebsworth, E.A.V. (1963) *Volatile Silicon Compounds*, Pergamon Press, Oxford.
Elliott, S. and Rowland, F.S. (1987) Chlorofluorocarbons and stratospheric ozone, *J. Chem. Ed.*, **64**, 387–390.
Emeléus H.J. and Sharpe, A.G. (1973) *Modern Aspects of Inorganic Chemistry*, 4th ed. (Complexes of Transition Metals: Chapter 20, Carbonyls), Routledge and Kegan Paul.
Fleming, S. (1976) *Dating in Archaeology: a Guide to Scientific Techniques*, Dent, London. (Carbon dating etc.)
Glasser, L.S.D. (1982) Sodium silicates, *Chemistry in Britain*, 33–39.
Greniger, D. *et al.* (1976) *Lead Chemicals*, International Lead Zinc Research Organization, New York. (Comprehensive data on lead and its compounds.)
Griffith, W.P. (1973) *Comprehensive Inorganic Chemistry*, Vol. 4 (Chapter 46: Carbonyls, cyanides, isocyanides and nitrosyls), Pergamon Press, Oxford.
Harrison, P.G. (ed.) (1989) *Chemistry of Tin*, Blackie, Glasgow.
Harrison, R.M. and Laxen, D.P.H. (1981) *Lead Pollution*, Chapman and Hall, London.
Holliday, A.K., Hughes, G. and Walker, S.M. (1973) *Comprehensive Inorganic Chemistry*, Vol. I (Chapter 13: Carbon), Pergamon Press, Oxford.
Hunt, C. (1987) Silicones reinvestigated – 50 years ago, *Education in Chemistry*, **24**, 7–11.
Johansen, H.H. (1977) Recent developments in the chemistry of transition metal carbides and nitrides, *Survey Progr. Chem.*, **8**, 57–81.
Johnson, B.F.G. (1976) The structures of simple binary carbonyls, *JCS Chem. Commun.*, 211–213.
Molina, M.J. and Rowland, F.S. (1974) Stratospheric sink for chlorofluoromethanes: chlorine catalyzed destruction of ozone, *Nature*, **249**, 810–812.
Nicholson, J.W. (1989) The early history of organotin chemistry, *J. Chem. Ed.*, **66**, 621–622.
Noll, W. with contributions by Glenz, O. and Hecht, G. (1968) *Chemistry and Technology of Silicones*, Academic Press, New York.
Pearce, C.A. (1972) *Silicon Chemistry and Applications* (Monograph for Teachers, No. 20), Chemical Society, London.
Pizey, J.S. (ed.) (1977) *Synthetic Reagents* (Chapter 4 by Butler R.N., Lead tetraacetate), John Wiley, Chichester.
Raabe, G. and Michl, J. (1985) Multiple bonding to silicon, *Chem. Rev.*, **85**, 419–509.
Rochow, E.G. (1973) *Comprehensive Inorganic Chemistry*, Vol. I (Chapter 15: Silicon), Pergamon Press, Oxford.
Segal, D. (1989) Making advanced ceramics, *Chemistry in Britain*, **25**, 151, 154–156.
Selig, H. and Ebert, L.B. (1980) Graphite intercalation compounds, *Adv. Inorg. Chem. Radiochem.*, **23**, 281–327. (A comprehensive review with over 300 references.)
Shannon R.D. (1976) Revised effective ionic radii. *Acta Cryst.*, **A32**, 751–767. (The most up to date and generally accepted values for ionic radii.)
Sharpe, A.G. (1976) *Chemistry of Cyano Complexes of the Transition Metals*, Academic Press, London.
Thompson, R. (ed.) (1986) *The Modern Inorganic Chemicals Industry* (Chapter by Farmer, J.B., Inorganic cyanogen compounds; chapter by Barby, D., Soluble silicates and their derivatives), Special Publication No. 31, The Chemical Society, London.

Thrush, B.A. (1977) The chemistry of the stratosphere and its pollution, *Endeavour*, **1**, 3–6.
Toth, L.E. (1971) *Transition Metal Carbides and Nitrides*, Academic Press, London.
Turner, D. (1980) Lead in petrol, *Chemistry in Britain*, **16**, 312–314.
Wells, A.F. (1984) *Structural Inorganic Chemistry*, 5th ed. (Chapter 23: Silicon), Oxford University Press, Oxford.
Wiberg, E. and Amberger, E. (1971) *Hydrides of the Elements of Main Groups I–IV*, (Chapter 7: Silicon hydrides; Chapter 10: Lead hydrides), Elsevier, Amsterdam. (Comprehensive review, over 700 references.)
Zuckerman, J.J. (ed.) (1976) Organotin compounds: New chemistry and applications, *Advances in Chemistry Series*, No. 157, American Chemical Society.

PROBLEMS

1. What are the most common oxidation states of carbon and tin? Why is there any difference?
2. (a) Draw the structures of diamond and graphite.
   (b) Explain the difference in density between diamond and graphite.
   (c) Explain the difference in electrical conductivity between diamond and graphite.
   (d) Which allotropic form of C has the lowest energy?
   (e) Why does the less stable form exist when the other form is thermodynamically favoured?
3. List as many ways as possible of making CO and CO_2. What are they used for, and how are they detected?
4. Explain the bonding in CO and CO_2.
5. List the advantages and limitations of CO as a reducing agent in the extraction of metals from their oxides.
6. Give equations to explain what reaction occurs when the following are heated: (a) $CaCO_3$, (b) $CaCO_3$ and SiO_2, (c) $CaCO_3$ and C, (d) CaC_2 and N_2.
7. Explain what happens when CO_2 is passed into a solution of $Ca(OH)_2$. What happens with excess CO_2?
8. Give equations to show what reactions occur between CO and: (a) O_2, (b) S, (c) Cl_2, (d) Ni, (f) Fe, (g) Fe_2O_3.
9. Write the formulae for the mononuclear carbonyls formed by V, Cr, Fe and Ni. What is the effective atomic number rule? Which of these complexes obey the rule?
10. Explain with the aid of suitable diagrams how CO forms σ and π bonds in $Ni(CO)_4$.
11. Draw the structures of the polynuclear carbonyls $Mn_2(CO)_{10}$, $Fe_3(CO)_{12}$, $Ru_3(CO)_{12}$ and $Rh_4(CO)_{12}$.

12. Why does $CaCO_3$ dissolve slightly when excess CO_2 is passed into an aqueous slurry? Write equations to explain the effect of a small amount of CO_2 on lime water, and of an excess of CO_2.
13. Give reasons why CO_2 is a gas and SiO_2 is a solid.
14. Explain the π bonding in CO_2, CO_3^-, SO_2 and SO_3.
15. Why is CCl_4 unaffected by water whilst $SiCl_4$ is rapidly hydrolysed? Is CCl_4 unreactive towards superheated steam?
16. Compare the shapes of the following pairs of molecules or ions, and suggest reasons for the differences in each pair:
    (a) CCl_4 and $TeCl_4$
    (b) CO_2 and NO_2
    (c) SiF_4 and ICl_4^-
17. Why is SnI_4 an orange coloured solid when CCl_4 and $SiBr_4$ are colourless liquids?
18. Starting with labelled $BaCO_3$ (containing ^{14}C), how would you prepare labelled Na_2CO_3, $CaCO_3$, CaC_2, CaNCN, C_2H_2, CH_3OH, CS_2 and $Ni(CO)_4$.
19. How is CS_2 made, and what is it used for?
20. How are NaCN and $(CN)_2$ made? What is NaCN used for?
21. Give two preparations for monosilane and compare its chemical properties with those of CH_4. Explain any differences.
22. Give three examples of Freons. How are they made, what are they used for, and how do they damage the environment?
23. Compare and contrast the structures of trimethylamine and trisilylamine.
24. Draw the structures of six different types of silicate and give the name and formula of one example of each type.
25. Describe the uses of soluble sodium silicates.
26. Describe the use of zeolites as water softeners.
27. How are silicate impurities removed in the extraction of Al and Fe?
28. Describetwo commercial methods for manufacturing alkyl substituted chlorosilanes. How are the reaction products separated and how may polymers with almost any specified properties be prepared from them?
29. How can silicates be detected in qualitative analysis?
30. What are the main ores of Sn and Pb, and how are the metals extracted?
31. Give equations to show the reactions between Sn and: (a) $H^+_{(aq)}$, (b) NaOH, (c) HNO_3, (d) O_2, (e) Cl_2.

32. What are the main uses of lead?
33. What is red lead, and what is it used for?
34. What are the main sources of lead pollution, what can be done about them, and what are the effects of lead on humans?
35. How would you make lead tetra-acetate? What is its structure, and what is it used for?
36. Describe the changes in physical and chemical properties which may be observed in the elements C, Si, Ge, Sn and Pb. Give reasons for these changes.

14 The group V elements

Table 14.1 Electronic structures and oxidation states

Element		Electronic structure	Oxidation states*
Nitrogen	N	[He] $2s^2\ 2p^3$	**−III** −II −I 0 I II **III** IV **V**
Phosphorus	P	[Ne] $3s^2\ 3p^3$	**III** **V**
Arsenic	As	[Ar] $3d^{10}\ 4s^2\ 4p^3$	**III** **V**
Antimony	Sb	[Kr] $4d^{10}\ 5s^2\ 5p^3$	**III** **V**
Bismuth	Bi	[Xe] $4f^{14}\ 5d^{10}\ 6s^2\ 6p^3$	**III** V

* The most important oxidation states (generally the most abundant and stable) are shown in bold. Other well-characterized but less important states are shown in normal type. Oxidation states that are unstable, or in doubt, are given in parentheses.

ELECTRONIC STRUCTURE AND OXIDATION STATES

The elements of this group all have five electrons in their outer shell. They exhibit a maximum oxidation state of five towards oxygen by using all five outer electrons in forming bonds. The tendency for the pair of *s* electrons to remain inert (the inert pair effect) increases with increasing atomic weight. Thus, only the *p* electrons are used in bonding and trivalency results. Valencies of three and five are shown with the halogens and with sulphur. The hydrides are trivalent. Nitrogen exhibits a very wide range of oxidation states: (−III) in ammonia NH_3, (−II) in hydrazine N_2H_4, (−I) in hydroxylamine NH_2OH, (0) in nitrogen N_2, (+I) in nitrous oxide N_2O, (+II) in nitric oxide NO, (+III) in nitrous acid HNO_2, (+IV) in nitrogen dioxide NO_2 and (+V) in nitric acid HNO_3. The negative oxidation states arise because the electronegativity of H = 2.1 and that for N = 3.0.

OCCURRENCE, EXTRACTION AND USES

Nitrogen

Though nitrogen comprises 78% of the earth's atmosphere, it is not a very abundant element in the earth's crust (only the thirty-third equal). Nitrates are all very soluble in water so they are not widespread in the earth's crust, though deposits are found in a few desert regions. The largest is a 450-mile-long belt along the coast of northern Chile, where $NaNO_3$ (Chile saltpetre) is found together with small amounts of KNO_3, $CaSO_4$ and $NaIO_3$ under a thin layer of sand or soil. This provided the main source of nitrates prior to World War I, when synthetic processes were developed for the manufacture of nitrates from atmospheric nitrogen. A major deposit of saltpetre KNO_3 occurs in India.

Nitrogen is an essential constituent of proteins and amino acids. (An average composition of a protein is C = 50%, O = 25%, N = 17%, H = 7%, S = 0.5%, P = 0.5% by weight.) Nitrates and other nitrogen compounds are extensively used in fertilizers and explosives. Earlier this century $NaNO_3$ was very important as a fertilizer. Bat guano was also

Table 14.2 Abundance of the elements in the earth's crust, by weight

	ppm	Relative abundance
N	19	33 =
P	1120	11
As	1.8	52
Sb	0.20	64
Bi	0.008	71

important. (This is bat droppings, which were found in large amounts in limestone caverns in Kentucky, Tennessee and Carlsbad, New Mexico.) In the last 50 years these sources have largely been replaced by NH_3 and NH_4NO_3 from the huge synthetic ammonia and nitric acid industries.

Nitrogen gas is used in large amounts as an inert atmosphere. This is mainly in the iron and steel and other metallurgical industries, and in oil refineries for purging catalytic cracking vessels, re-forming vessels and pipes. Liquid nitrogen is used as a refrigerant. Large amounts of N_2 are used in the manufacture of ammonia and calcium cyanamide. N_2 is obtained commercially by condensing air to the liquid state, and then fractionally distilling the liquid air. N_2 has a lower boiling point than O_2 and distils off first. Six industrial gases are obtained in this way: N_2, O_2, Ne, Ar, Kr and Xe. A typical analysis of air is shown in Table 14.3.

World production of N_2 is growing rapidly and exceeds 60 million tonnes/year. (This is largely because liquid O_2 is essential for modern steel making processes, and N_2 is produced at the same time.) About two thirds of the N_2 is sold as gas. This may either be compressed in steel cylinders, or piped to where it is used. One third is sold as liquid N_2. Nitrogen obtained in this way always contains traces of oxygen and the noble gases. Commercial N_2 normally contains up to 20 ppm of O_2, 'oxy-free' N_2 contains up to 2 ppm O_2, and ultrapure N_2 has no O_2 but may have up to 10 ppm Ar.

A cylinder of N_2 is the usual source of N_2 in the laboratory, but samples of the gas may be obtained by making ammonium nitrite and then warming it. N_2 is also obtained by oxidizing ammonia, for example with calcium

continued overleaf

Table 14.3 Abundance of different gases in dry air

	% by volume	b.p. of gas (°C)
N_2	78.08	−195.8
O_2	20.95	−183.1
Ar	0.934	−186.0
CO_2	0.025–0.050	−78.4 (sublimes)
Ne	0.0015	−246.0
H_2	0.0010	−253.0
He	0.000 52	−269.0
Kr	0.000 11	−153.6
Xe	0.000 008 7	−108.1

hypochlorite, bromine water or CuO. Small quantities of very pure N_2 may be obtained by carefully warming sodium azide NaN_3 to about 300 °C. Thermal decomposition of NaN_3 is used to inflate the air bags used as safety devices in some cars.

$$NH_4Cl + NaNO_2 \rightarrow NaCl + NH_4NO_2 \xrightarrow{\text{warm}} N_2 + 2H_2O$$

$$4NH_3 + 3Ca(OCl)_2 \rightarrow 2N_2 + 3CaCl_2 + 6H_2O$$

$$8NH_3 + 3Br_2 \rightarrow N_2 + 6NH_4Br$$

$$2NaN_3 \xrightarrow{300\,°C} 3N_2 + 2Na$$

Phosphorus

Phosphorus is the eleventh most abundant element in the earth's crust. Phosphorus is essential for life, both as a structural material in higher animals, and in the essential metabolism of both plants and animals. About 60% of bones and teeth are $Ca_3(PO_4)_2$ or $[3(Ca_3(PO_4)_2) \cdot CaF_2]$, and an average person has 8 lbs (3.5 kg) of calcium phosphate in his body. Nucleic acids such as DNA and RNA contain the genetic material for each cell. These nucleic acids are made up of polyester chains of phosphates and sugars with organic bases (adenine, cytosine, thymine and guanine). Phosphorus, in the form of adenosine triphosphate ATP and adenosine diphosphate ADP, is of vital importance for the production of energy in cells. When water splits a phosphate group off ATP, forming ADP, 33 kJ mole^{-1} of energy is released.

Vast amounts of phosphates are used in fertilizers. World production of phosphate rock was 159 million tonnes in 1988 (USA 29%, USSR 22%, Morocco and Senegal 16% each, China 9%, Tunisia and Jordan 4% each). Most is mined as fluoroapatite $[Ca_3(PO_4)_2 \cdot CaF_2]$, but some is found as hydroxyapatite $[Ca_3(PO_4)_2 \cdot Ca(OH)_2]$ and chloroapatite $[Ca_3(PO_4)_2 \cdot CaCl_2]$ where OH^- and Cl^- are substituted for F^- in the crystal structure. About 90% of phosphate rock is used directly to make fertilizers, and the remainder is used to make phosphorus and phosphoric acid.

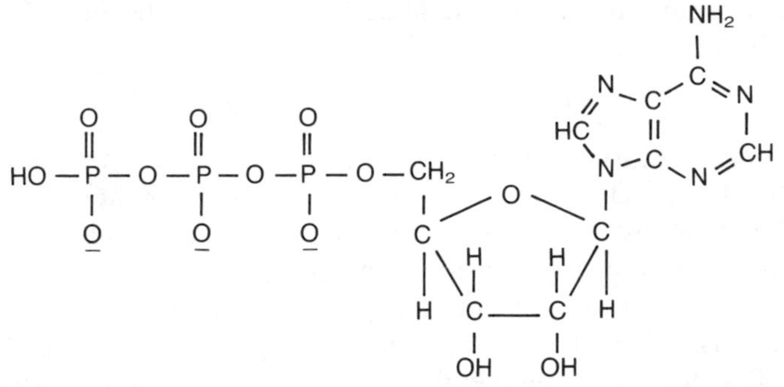

Figure 14.1 Structure of adenosine triphosphate ATP.

World production of elemental P is about 1.2 million tonnes/year, but is declining. It is obtained by the reduction of calcium phosphate with C in an electric furnace at 1400–1500 °C. Sand (silica SiO_2) is added to remove the calcium as a fluid slag (calcium silicate) and to drive off the phosphorus as P_4O_{10}. The P_4O_{10} is reduced to phosphorus by C. At this temperature gaseous phosphorus distils off, mainly as P_4 but with some P_2. This is condensed to white phosphorus P_4 by passing the gas through water.

$$2Ca_3(PO_4)_2 + 6SiO_2 \rightarrow 6CaSiO_3 + P_4O_{10}$$
$$P_4O_{10} + 10C \rightarrow P_4 + 10CO$$

About 85% of the elemental P produced is used to make very pure phosphoric acid H_3PO_4. About 10% is used to make P_4S_{10} (used making organo P—S compounds) and P_4S_3 (which is used to make matches).

$$P_4 + 5O_2 \rightarrow P_4O_{10} + 6H_2O \rightarrow 4H_3PO_4$$
$$P_4 + 10S \rightarrow P_4S_{10}$$

Other uses are for making $POCl_3$ and phosphor bronze.

Arsenic, antimony and bismuth

The elements As, Sb and Bi are not very abundant. Their most important source is as sulphides occurring as traces in other ores. They are well known because they are obtained as metallurgical by-products from roasting sulphide ores in a smelter. Care should be taken since As and Sb compounds are poisonous. The colours of the sulphide ores are distinctive.

Arsenic is obtained as As_2O_3 in the flue dust from roasting CuS, PbS, FeS, CoS and NiS in air. World production of As_2O_3 was about 50 000 tonnes in 1988. The oxide may be converted to As by reduction with C. The only common ores are arsenopyrites FeAsS (white–grey colour with metallic lustre), realgar As_4S_4 (red–orange colour) and orpiment As_2S_3 (yellow colour). The last two are found in volcanic areas. The element As is obtained commercially by heating arsenides such as NiAs, $NiAs_2$ or

continued overleaf

$FeAs_2$, or arsenopyrites FeAsS, to about 700 °C in the absence of air, when the As sublimes out.

$$4FeAsS \rightarrow As_{4(g)} + 4FeS$$

There are few uses for As metal, but it is used to alloy with lead to make the lead harder. Small amounts are used to dope semiconductors and make light emitting diodes. Arsenic compounds are generally made from As_2O_3. Their main uses, e.g. rat poison, in medicine to kill parasites, and for preventing wood rot, arise from their poisonous nature.

Antimony is obtained as Sb_2O_3 in the flue dust from roasting ZnS ores. This is easily reduced to the metal with carbon. The most important ore is stibnite Sb_2S_3 (iridescent metal-like needles). The metal is obtained by fusing with iron:

$$Sb_2S_3 + 3Fe \rightarrow 2Sb + 3FeS$$

Antimony metal is used in alloys with Sn and Pb. It is also used to electroplate steel to prevent rusting. World production of Sb was 63 900 tonnes in 1988.

Bi_2O_3 is obtained from the flue dust from roasting PbS, ZnS and CuS, and can be reduced to the metal with carbon. It also occurs as the minerals bismuthinite Bi_2S_3 and bismite Bi_2O_3. Because of its low melting point bismuth metal can be cast in much the same way as lead. As, Sb and Bi metals are all too brittle to work. Bi is used in low melting alloys. (One use of these alloys is as a low melting plug for automatic fire sprinkler systems.) Other uses are in batteries, bearings, solder and ammunition. World production was 3900 tonnes in 1988.

GENERAL PROPERTIES AND STRUCTURES OF THE ELEMENTS

Nitrogen

The first element differs from the rest as was the case in the previous groups. Thus nitrogen is a colourless, odourless, tasteless gas which is diamagnetic and exists as diatomic molecules N_2. The other elements are solids and exist as several allotropic forms. The N_2 molecule contains

Table 14.4 Melting and boiling points

	Melting point (°C)	Boiling point (°C)
N_2	−210	−195.8
P_4	44	281
α-As	816*	615 (sublimes)
α-Sb	631	1587
α-Bi	271	1564

* At 38.6 atmospheres pressure.

a triple bond N≡N with a short bond length of 1.09 Å. This bond is very stable, and the dissociation energy is consequently very high ($945.4\,kJ\,mol^{-1}$). Thus N_2 is inert at room temperature, though it does react with Li, forming the nitride Li_3N. Other isoelectronic species such as CO, CN^- and NO^+ are much more reactive than N_2, and this is because the bonds are partly polar, whilst in N_2 they are not. At elevated temperatures N_2 becomes increasingly reactive, and reacts directly with elements from Groups II, III and IV, with H_2 and with some of the transition metals.

Active nitrogen can be made by passing an electric spark through N_2 gas at a low pressure. This forms atomic nitrogen, and the process is associated with a yellow–pink afterglow. Active nitrogen will react with a number of elements, and breaks many normally stable molecules.

The nitrogen cycle

There is a continual turnover of nitrogen between the atmosphere, the soil, the sea and living organisms, which is estimated to be between 10^8 and 10^9 tonnes/year. This is called the nitrogen cycle. Consider the combined nitrogen in the soil: this is present as nitrates, nitrites and ammonium compounds. Losses of combined nitrogen from the soil occur for several reasons:

1. Plants absorb these compounds, and use them to make protoplasm, in order to grow. Plants may be eaten by animals, and animals may eat animals. Animals excrete nitrogenous wastes usually as urea or uric acid, which is returned to the soil. Death and decay eventually return all the nitrogen to the soil.
2. A group of denitrifying bacteria called *Denitrificans* convert nitrates into N_2 or NH_3 gases, which escape into the atmosphere. (Horse stables often smell of ammonia for this reason.) NH_3 is returned to the soil by the first rainstorm, but N_2 is not. Examples of denitrifying bacteria include *Pseudomonas* and *Achromobacter*.

$$\text{nitrates} \rightarrow \text{nitrites} \rightarrow NO_2 \rightarrow N_2 \rightarrow NH_3$$

3. A net loss of nitrogen compounds in the soil occurs through the drainage of surface water into the sea. There it supports marine plant life.
4. There is a small loss of NO and NO_2 into the atmosphere from the burning of plants and coal, and from car exhausts. Though this may be unpleasant and produce smog locally, the amounts are small, and the nitrogen is returned to the soil when it rains.

There are net gains to the supply of combined nitrogen in the soil:

1. The largest gain is from nitrifying bacteria which fix N_2 gas and turn it into nitrates or ammonium salts. This produces over 60% of the nitrogen gain. It is estimated that approximately 175 million tonnes of N_2 are fixed annually by bacteria. This may be compared with 98 million

tonnes of NH_3 produced in 1988 by man (mainly by the Haber–Bosch process, but some from the distillation of coal). The most important genus of bacteria is *Rhizobium*. These live symbiotically in the nodules of roots of plants in the family Leguminosae, i.e. peas, beans, clover and alder trees. Other nitrifying bacteria live free in the soil, for example the blue-green bacteria *Anabena* and *Nostoc*, aerobic bacteria such as *Azotobacter* and *Bei-jerinckia*, and anaerobic bacteria such as *Clostridium pastorianum*. These bacteria require traces of certain transition metals such as Mo, Fe, Co and Cu from the soil, and also B. The nitrogen fixing enzyme 'nitrogenase' was isolated from *Clostridium pastorianum* in 1960: the same enzyme system is responsible for nitrogen fixation in the other bacteria too.

Nitrogenase contains two components. One is a Mo–Fe protein of molecular weight 220 000, containing 24–32 Fe, two Mo and a labile sulphide group, and the other is an Fe protein of molecular weight 60 000, containing four Fe and four S. Nitrogenase reduces N_2 to NH_3 and the N_2 is thought to form a complex with the Mo–Fe protein. Bonding in N_2 complexes favours end-on coordination to the metal. σ donation from N to the metal is more important than π back bonding from the metal to N. Nitrogenase also reduces N_2O, RCN and N_3^- to NH_3 and it also reduces ethyne C_2H_2 to ethene C_2H_4.

Blue-green bacteria can also fix N_2, and this is important in growing rice.

The number of kilograms of nitrogen fixed in one acre of fertile soil in one year by different organisms is: *Rhizobium* 120, blue-green bacteria 10, *Azotobacter* 0.1, *Clostridium* 0.1.

2. About 20% comes from NH_4NO_3, which is used in vast quantities as an artificial fertilizer. The Haber–Bosch process is used to fix atmospheric N_2 and to make NH_3, and the Ostwald process is used to convert NH_3 to HNO_3. Reacting NH_3 and HNO_3 gives NH_4NO_3.
3. Relatively small amounts come from deposits such as Chile saltpetre $NaNO_3$ which is mined and used as fertilizer.
4. Minor gains come from the effects of lightning, and from photochemical changes. Lightning may cause N_2 and O_2 in the air to form NO and NO_2. This is essentially similar to the obsolete Birkeland–Eyde process. The strong UV radiation in the upper atmosphere may cause similar photochemical changes, giving oxides of nitrogen. The NO_2 formed forms a very dilute solution of HNO_3 in rain water.

Phosphorus

Phosphorus is solid at room temperature. White phosphorus is soft, waxy and reactive. It reacts with moist air and gives out light (chemiluminescence). It ignites spontaneously in air at about 35 °C, and is stored under water to prevent this. It is highly toxic. It exists as tetrahedral P_4 molecules, and the tetrahedral structure remains in the liquid and gaseous states. Above 800 °C P_4 molecules in the gas begin to dissociate into P_2

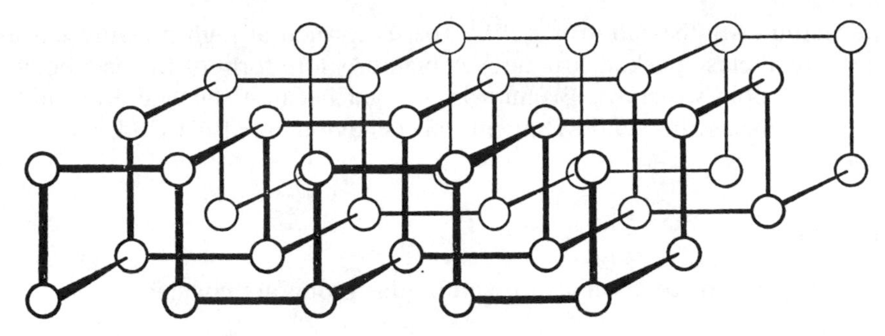

Figure 14.2 The structure of black phosphorus. The atoms are arranged in corrugated planes in crystalline black phosphorus (Van Wazer, J.R., Phosphorus and Its Compounds, Vol.I, Interscience, New York – London, 1958, p.121).

molecules, which have a bond energy of 489.6 kJ mol^{-1}. (This is only half the value for N_2 because the orbitals in the third shell are much larger and give relatively poor $p\pi$–$p\pi$ overlap.) If white phosphorus is heated to about 250 °C, or a lower temperature in the presence of sunlight, then red phosphorus is formed. This is a polymeric solid, which is much less reactive than white phosphorus. It is stable in air and does not ignite unless it is heated to 400 °C. It need not be stored under water. It is insoluble in organic solvents. Heating white phosphorus under high pressure results in a highly polymerized form of P called black phosphorus. This is thermodynamically the most stable allotrope. It is inert and has a layer structure (Figure 14.2). Other more doubtful allotropes have been reported.

Arsenic, Antimony and Bismuth

Solid As, Sb and Bi each exist in several allotropic forms. Arsenic vapour contains tetrahedral As_4 molecules. A reactive yellow form of the solid resembles white phosphorus and is thought to contain tetrahedral As_4 units. Sb also has a yellow form. All three elements have much less reactive metallic or α-forms. These have layer structures, but the layers are

Table 14.5 Radii, ionization energies and electronegativity

	Covalent radius (Å)	Ionization energies (kJ mol^{-1}) 1st	2nd	3rd	Pauling's electronegativity
N	0.74	1403	2857	4578	3.0
P	1.10	1012	1897	2910	2.1
As	1.21	947	1950	2732	2.0
Sb	1.41	834	1590	2440	1.9
Bi	1.52	703	1610	2467	1.9

puckered. Another allotrope of Sb that is formed at high pressure has a hexagonal close-packed structure. A high pressure form of Bi has a body-centred cubic structure. Bismuth is unusual because the liquid expands when it forms the solid. This unusual behaviour is also found with Ga and Ge.

BOND TYPE

The majority of compounds formed by this group are covalent.

Outer electronic structure of Group V element

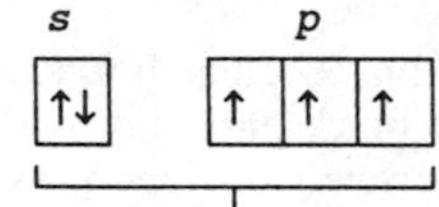

three unpaired electrons form σ bonds with three other atoms
four electron pairs give a tetrahedral shape with one position occupied by a lone pair

A coordination number of 4 is obtained if the lone pair is donated (that is used to form a coordinate bond) to another atom or ion. An example is the ammonium ion $[H_3N \rightarrow H]^+$ (Figure 14.4).

It requires too much energy to remove all five outer electrons so M^{5+} ions are not formed. However, Sb and Bi can lose just three electrons, forming M^{3+} ions, but the ionization energy is too high for the other elements to do so. Both SbF_3 and BiF_3 exist as ionic solids. The M^{3+} ions are not very stable in solution. They can exist in fairly strong acid solutions, but are rapidly hydrolysed in water to give the antimony oxide ion or bismuth oxide ion SbO^+ and BiO^+. This change is reversed by adding 5 M HCl.

$$Bi^{3+} \underset{HCl}{\overset{H_2O}{\rightleftharpoons}} [BiO]^+$$

$$BiCl_3 + H_2O \rightleftharpoons BiOCl + 2HCl$$

Nitrogen atoms may gain three electrons and so attain a noble gas configuration, forming ionic nitrides containing the N^{3-} ion. It takes

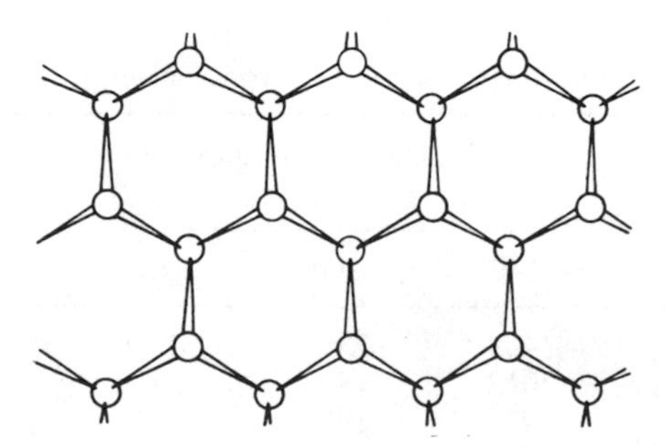

Figure 14.3 Layer structure of bismuth.

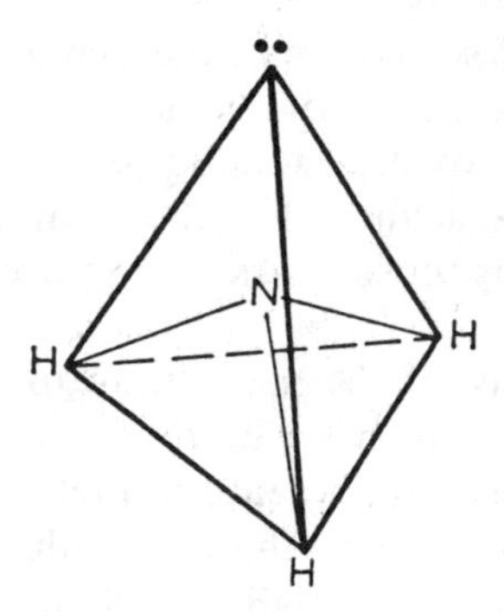

Figure 14.4 Structure of ammonia.

2125 kJ mol^{-1} of energy to form N^{3-}. Thus ionic nitrides are formed only by metals which have low ionization energies and can form nitrides with high lattice energies (Li_3N, Be_3N_2, Mg_3N_2, Ca_3N_2). Though compounds such as Na_3P and Na_3Bi are known, these are not ionic.

Nitrogen cannot extend its coordination number beyond 4 because there are only four orbitals available in the second shell of electrons. Thus nitrogen cannot form complexes by accepting electron pairs from other ligands, but the subsequent elements can form such complexes. Thus the other elements may have coordination numbers of 5 or 6 as, for example, in PCl_5 gas and $[PCl_6]^-$. The formation of complexes may be explained by involving one or two *d* orbitals in bonding. Thus sp^3d or sp^3d^2 hybridization may occur. The 3*d* orbitals of an isolated phosphorus atom are much larger than the 3*s* and 3*p* orbitals. This might at first sight suggest that the use of the 3*d* orbitals for bonding is improbable. However, when electronegative ligands are placed round the phosphorus atom, the 3*d* orbitals contract to nearly the same size as the 3*s* and 3*p* orbitals. (The extent of *d* orbital participation in σ bonding is controversial and is discussed in Chapter 4.)

Nitrogen also differs from the other elements in that it can form strong $p\pi$–$p\pi$ multiple bonds. Because of this it forms several compounds which have no counterparts in the other elements. These include nitrates NO_3^-, nitrites NO_2^-, azides N_3^-, nitrogen N_2, oxides of nitrogen N_2O, NO, NO_2, N_2O_4, cyanides CN^-, and azo and diazo compounds. Because nitrogen can form multiple bonds, the oxides N_2O_3 and N_2O_5 are monomeric, whilst the trioxides and pentoxides of the other elements are dimeric.

METALLIC AND NON-METALLIC CHARACTER

Group V shows the usual trend, that metallic character increases on descending the group. Thus N and P are non-metals, As and Sb are metalloids, which show many metallic properties, and Bi is a true metal. The increasing metallic character is shown by the following:

1. In the appearance and structures of the elements.
2. By their tendency to form positive ions.
3. By the nature of their oxides. Metallic oxides are typically basic, and non-metallic oxides are acidic. Thus the normal oxides of N and P are strongly acidic, whereas those of As and Sb are amphoteric and that of Bi is largely basic.
4. The electrical resistivity of the metallic forms (α-As 33, α-Sb 39 and α-Bi 106 μohm cm) are much lower than for white phosphorus (1×10^{17} μohm cm), indicating an increase in metallic properties. However, the resistivity values are higher than the values for a good conductor such as Cu, 1.67 μohm cm, and higher than Sn, 11, and Pb, 20 μohm cm, in the adjacent group. (See Appendix J.)

REACTIVITY

Nitrogen is relatively unreactive, which is why it has accumulated in such large amounts in the atmosphere.

White phosphorus catches fire when exposed to air, burning to form P_4O_{10}. It is stored under water to prevent this. Red phosphorus is stable in air at room temperature, though it reacts on heating.

Arsenic is stable in dry air, but tarnishes in moist air, giving first a bronze then a black tarnish. When heated in air it sublimes at 615 °C and forms As_4O_6, not As_4O_{10}. Strong heating in oxygen can give either of these oxides, depending on the amount of oxygen present. This reluctance to attain the maximum oxidation state for the group is found in the elements Ga, As, Se and Br, that is in the elements immediately following the filling of the first *d* shell. As_4O_{10} and H_3AsO_4 are used as oxidizing agents in volumetric analysis.

Sb is less reactive, and is stable towards water and to air at room temperature. On heating in air it forms Sb_4O_6, Sb_4O_8 or Sb_4O_{10}. Bi forms Bi_2O_3 on heating.

HYDRIDES

The elements all form volatile hydrides of formula MH_3, which are all poisonous, foul smelling gases. On descending the group from NH_3 to BiH_3:

1. The hydrides become increasingly difficult to prepare.
2. Their stability decreases.
3. Their reducing power increases.
4. The ease of replacing the hydrogen atoms by other groups such as Cl or Me decreases.
5. Their ability to act as electron donors, using the lone pair of electrons for coordinate bond formation, decreases.

Ammonia NH_3

NH_3 is a colourless gas with a pungent odour. The gas is quite poisonous. It dissolves very readily in water with the evolution of heat. At 20 °C and one atmosphere pressure 53.1 g NH_3 dissolves in 100 g water. This corresponds to 702 volumes of NH_3 dissolving in 1 volume of H_2O. In solution ammonia forms ammonium hydroxide NH_4OH, and behaves as a weak base.

$$NH_3 + H_2O \rightleftharpoons NH_4^+ + OH^- \qquad K = 1.8 \times 10^5 \, mol\, l^{-1}$$

NH_3 and NH_4OH both react with acids, forming ammonium salts. These salts resemble potassium salts in solubility and in their crystal structures. Like the Group I salts, ammonium salts are typically colourless. There are some differences. Ammonium salts are usually slightly acidic if they have been formed with strong acids such as HNO_3, HCl and H_2SO_4, since NH_4OH is only a weak base. Ammonium salts decompose quite readily on heating. If the anion is not particularly oxidizing (e.g. Cl^-, CO_3^{2-} or SO_4^{2-}) then ammonia is evolved:

$$NH_4Cl \xrightarrow{\text{heat}} NH_3 + HCl$$

$$(NH_4)_2SO_4 \xrightarrow{\text{heat}} 2NH_3 + H_2SO_4$$

If the anion is more oxidizing (e.g. NO_2^-, NO_3^-, ClO_4^-, $Cr_2O_7^{2-}$) then NH_4^+ is oxidized to N_2 or N_2O.

$$\overset{-III}{N}H_4NO_2 \xrightarrow{\text{heat}} \overset{0}{N_2} + 2H_2O$$

$$\overset{-III}{N}H_4NO_3 \xrightarrow{\text{heat}} \overset{+I}{N_2}O + 2H_2O$$

$$(NH_4)_2Cr_2O_7 \xrightarrow{\text{heat}} N_2 + 4H_2O + Cr_2O_3$$

NH_3 burns in oxygen with a pale yellow flame:

$$4NH_3 + 3O_2 \rightarrow 2N_2 + 3H_2O$$

The same reaction occurs in air, but the heat of reaction is insufficient to maintain combustion unless heat is supplied, for example in a gas flame. Certain mixtures of NH_3/O_2 and NH_3/air are explosive.

NH_3 is prepared in the laboratory by heating an ammonium salt with NaOH. This is a standard test in the laboratory for NH_4^+ compounds.

$$NH_4Cl + NaOH \rightarrow NaCl + NH_3 + H_2O$$

The NH_3 evolved may be detected:

1. By its characteristic smell.
2. By turning moist litmus paper blue.
3. By forming dense white clouds of NH_4Cl with the stopper from a bottle of HCl.
4. By forming a yellow–orange–brown precipitate with Nessler's solution.

World production of NH_3 was 98 million tonnes in 1988. Most was manufactured synthetically from H_2 and N_2 by the Haber–Bosch process (see later) but some was obtained from coal gas purification and during coke production from coal. Ammonia can also be obtained from the hydrolysis of calcium cyanamide, CaNCN. Calcium cyanamide is usually used as a fertilizer, and this reaction occurs slowly in the soil.

$$CaNCN + 3H_2O \rightarrow 2NH_3 + CaCO_3$$

(CaNCN is also used to make melamine, urea and thiourea – See Chapter 11.) In the past when town gas was made as a fuel by dry distilling coal in the absence of air, any nitrogenous compounds in coal were converted into NH_3. This NH_3 was obtained as a by-product.

Ammonium salts

Ammonium salts are all very soluble in water. They all react with NaOH, liberating NH_3. The NH_4^+ ion is tetrahedral. Several ammonium salts are important.

NH_4Cl is well known. At one time it was obtained by heating camel dung: NH_4Cl is easily purified by sublimation! It can be recovered as a by-product from the Solvay process. It is used in 'dry batteries' of the Leclanché type. It is also used as a flux when tinning or soldering metals, since many metal oxides react with NH_4Cl, forming volatile chlorides, thus leaving a clean metal surface.

NH_4NO_3 is used in enormous amounts as a nitrogenous fertilizer. It is deliquescent. Because it can cause explosions it is often mixed with $CaCO_3$ or $(NH_4)_2SO_4$ to make it safe. It is also used as an explosive, since on strong heating (above 300 °C), or with a detonator, very rapid decomposition occurs. The solid has almost zero volume and it produces seven volumes of gas: this causes the explosion:

$$2NH_4NO_3 \rightarrow 2N_2 + O_2 + 4H_2O$$

Smaller amounts of $(NH_4)_2SO_4$ are also used as a fertilizer. At one time $(NH_4)_2SO_4$ was obtained as a by-product from making coal gas (town gas). Since natural gas has become available in developed countries, town gas is no longer made. $(NH_4)_2SO_4$ is made by passing NH_3 and CO_2 gases into a slurry of $CaSO_4$ in water:

$$2NH_3 + CO_2 + H_2O \rightarrow (NH_4)_2CO_3$$
$$(NH_4)_2CO_3 + CaSO_4 \rightarrow CaCO_3 + (NH_4)_2SO_4$$

Small amounts of diammonium hydrogen phosphate $(NH_4)_2HPO_4$ and ammonium dihydrogen phosphate $NH_4H_2PO_4$ are used as fertilizers. They are also used for fireproofing wood, paper and textiles. NH_4ClO_4 is used as an oxidizing agent in solid fuel rocket propellants.

Phosphine PH_3

Phosphine PH_3 is a colourless and extremely toxic gas, which smells slightly of garlic or bad fish. It is highly reactive. It can be formed either by hydrolysing metal phosphides such as Na_3P or Ca_3P_2 with water, or by hydrolysing white phosphorus with NaOH solution.

$$Ca_3P_2 + 6H_2O \rightarrow 2PH_3 + 3Ca(OH)_2$$
$$P_4 + 3NaOH + 3H_2O \rightarrow PH_3 + 3NaH_2PO_2$$

PH_3, unlike NH_3, is not very soluble in water: aqueous solutions are neutral. It is more soluble in CS_2 and other organic solvents. Phosphonium salts such as $[PH_4]^+Cl^-$ can be formed, but require PH_3 and anhydrous HCl (in contrast to the ready formation of NH_4X in aqueous solution). Pure PH_3 is stable in air, but it catches fire when heated to about 150 °C.

$$PH_3 + 2O_2 \rightarrow H_3PO_4$$

PH_3 frequently contains traces of diphosphine P_2H_6 which cause it to catch fire spontaneously. This is the origin of the flickering light called will-o'-the-wisp, which is sometimes seen in marshes.

Arsine AsH_3, stibine SbH_3 and bismuthine BiH_3

The bond energy (Table 14.6) and the stability of the hydrides both decrease on descending the group. Consequently, arsine AsH_3, stibine SbH_3 and bismuthine BiH_3 are only obtained in small amounts. AsH_3 and SbH_3 are both very poisonous gases. AsH_3, SbH_3 and BiH_3 can be prepared by hydrolysing binary metal compounds such as Zn_3As_2, Mg_3Sb_2 or Mg_3Bi_2 with water or dilute acid. AsH_3 and SbH_3 are formed in Marsh's test for As and Sb compounds. Before the use of instruments for analysis, this test was used as a forensic test. Practically all As or Sb compounds can be reduced with Zn and acid, forming AsH_3 or SbH_3. The gaseous hydrides are passed through a glass tube heated with a Bunsen burner. SbH_3 is less stable than AsH_3: hence it decomposes before passing through the flame, and gives a metallic mirror on the glass tube. AsH_3 is more stable, and requires stronger heating to make it decompose. Thus AsH_3 gives a mirror after the flame.

Structure of the hydrides

The structure of ammonia may either be described as pyramidal, or tetrahedral with one position occupied by a lone pair (Figure 14.4). This shape is predicted using the VSEPR theory since there are four electron pairs in the outer shell. These comprise three bonding pairs and one lone pair. The repulsion between a lone pair and a bond pair of electrons always exceeds that between two bond pairs. Thus the bond angles are reduced from 109°27′ to 107°48′, and the regular tetrahedral shape is slightly distorted.

Electronic structure of nitrogen atom – ground state

Nitrogen having gained a share in three electrons from three hydrogen atoms in NH_3 molecule

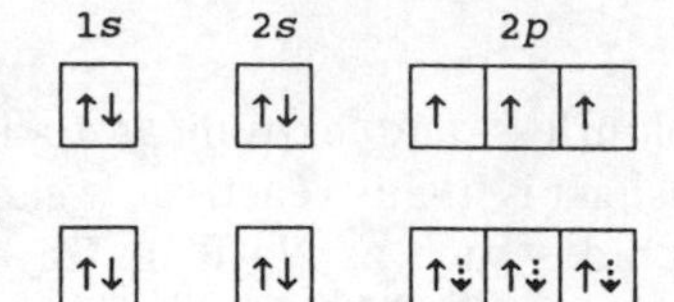

four orbitals in the outer shell (three bond pairs and one lone pair) tetrahedral arrangement with one position occupied by a lone pair

The hydrides PH_3, AsH_3 and SbB_3 would be expected to be similar. However, the bond pairs of electrons are much further away from the central atom than they are in NH_3. Thus the lone pair causes even greater distortion in PH_3, AsH_3 and SbH_3. The bond angle decreases to 91°18′ (Table 14.6). These bond angles suggest that in PH_3, AsH_3, SbH_3 and BiH_3 the orbitals used for bonding are close to pure *p* orbitals.

The melting and boiling points of the hydrides increase from PH_3 through AsH_3 to SbH_3. The values for NH_3 seem out of line with this trend: one might have expected the boiling point of NH_3 to be −110 °C or −120 °C. The reason why NH_3 has a higher boiling point and is much less volatile than expected is that it is hydrogen bonded in the liquid state. The other hydrides do not form hydrogen bonds.

These hydrides are strong reducing agents and react with solutions of metal ions to give phosphides, arsenides and stibnides. They are flammable and extremely poisonous.

Table 14.6 Some properties of the hydrides

	m.p. (°C)	b.p. (°C)	Bond energy ($kJ\,mol^{-1}$)	Bond angle	Bond length (Å)
NH_3	−77.8	−34.5	N–H = 389	H–N–H = 107°48′	1.017
PH_3	−133.5	−87.5	P–H = 318	H–P–H = 93°36′	1.419
AsH_3	−116.3	−62.4	As–H = 247	H–As–H = 91°48′	1.519
SbH_3	−88	−18.4	Sb–H = 255	H–Sb–H = 91°18′	1.707

Donor properties

NH_3 can donate its lone pair of electrons quite strongly to form complexes. Thus ammonia forms ammonium NH_4^+ salts, and also coordination complexes with metal ions from the Co, Ni, Cu and Zn groups, for example the $[Co(NH_3)_6]^{3+}$ ion, very readily.

PH_3 acts as an electron donor and forms numerous complexes such as $[F_3B \leftarrow PH_3]$, $[Cl_3Al \leftarrow PH_3]$ and $[Cr(CO)_3(PH_3)_3]$. A variety of other trivalent phosphorus compounds such as PF_3, PCl_3, PEt_3, $P(OR)_3$ and PPh_3 also form complexes, which in some ways resemble complexes with CO. Thus the lone pair on P is used to form the coordinate bond to an empty orbital on the B or metal (a σ bond). In the case of metals, this original coordinate bond may be reinforced by back bonding from π overlap of a filled *d* orbital on the metal with an empty *d* orbital on P.

The donor properties of the other hydrides are very weak, and they have little or no tendency to form coordinate bonds.

In NH_3 the lone pair occupies an sp^3 hybrid orbital. In AsH_3 and SbH_3 the bond angles become close to 90° which suggests that the orbitals used for M—H bonding are almost pure *p* orbitals. If the three *p* orbitals are used for M—H bonding, the lone pair must occupy a spherical *s* orbital. This is larger, and less directional, and hence less effective for forming a coordinate bond. This means that any σ bond will be very weak. In addition the 4*d* and 5*d* orbitals are too large for effective π back bonding. These two factors account for the difference in complexing power between the hydrides. Nitrogen forms several hydrides (see Table 14.7).

Table 14.7 Hydrides of nitrogen

Formula	Name	Oxidation state
NH_3	Ammonia	−III
N_2H_4	Hydrazine	−II
NH_2OH	Hydroxylamine	−I

Hydrazine N_2H_4

Hydrazine is a covalent liquid, which fumes in air, and smells similar to NH_3. Pure hydrazine burns readily in air with the evolution of a large amount of heat.

$$N_2H_{4(l)} + O_{2(g)} \rightarrow N_{2(g)} + 2H_2O \qquad \Delta H = -621\,\text{kJ mol}^{-1}$$

The methyl derivatives $MeNHNH_2$ and Me_2NNH_2 are mixed with N_2O_4 and used as a rocket fuel in the space shuttle, in guided missiles, and (earlier) in the Apollo lunar modules.

N_2H_4 is a weak base and reacts with acids, forming two series of salts. The salts are white ionic crystalline solids, and are soluble in water.

$$N_2H_4 + HX \rightarrow N_2H_5^+ + X^-$$
$$N_2H_4 + 2HX \rightarrow N_2H_6^{2+} + 2X^-$$

When dissolved in water (in neutral or basic solutions) hydrazine or its salts are powerful reducing agents. They are used to produce silver and copper mirrors and to precipitate the platinum metals. Hydrazine also reduces I_2 and O_2.

$$N_2H_4 + 2I_2 \rightarrow 4HI + N_2$$
$$N_2H_4 + 2O_2 \rightarrow 2H_2O_2 + N_2$$
$$N_2H_4 + 2CuSO_4 \rightarrow Cu + N_2 + 2H_2SO_4$$

In acidic solutions, hydrazine usually behaves as a mild reducing agent, though powerful reducing agents can reduce N_2H_4 to NH_3, thus causing N_2H_4 to be oxidized.

$$\underset{(-II)}{N_2H_4} + Zn + 2HCl \rightarrow \underset{(-III)}{2NH_3} + ZnCl_2$$

Hydrazine may act as an electron donor. The N atoms have a lone pair of electrons, which can form coordinate bonds to metal ions such as Ni^{2+} and Co^{2+}.

World production of hydrazine is nearly 20 000 tonnes/year. Most is used as rocket fuel. Other uses are the manufacture of 'blowing agents' (for producing blown plastics), as agricultural chemicals, and to treat the boiler feed water in power stations to prevent oxidation of the boiler and pipes. In the laboratory phenylhydrazine is used to characterize carbonyl compounds and sugars by forming crystalline derivatives called osazones. Osazones can be identified by microscopic examination of the shape of the crystals, or by melting point determination.

$$\underset{\text{glucose}}{\begin{array}{l} CH_2\cdot OH \\ | \\ (CH\cdot OH)_3 \\ | \\ CH\cdot OH \\ | \\ CHO \end{array}} + \underset{\text{phenylhydrazine}}{H_2N\text{—}NHC_6H_5} \rightarrow \underset{\text{glucose phenylhydrazone (an osazone)}}{\begin{array}{l} CH_2\cdot OH \\ | \\ (CH\cdot OH)_3 \\ | \\ CH\cdot OH \\ | \\ CH{=}N\text{—}NHC_6H_5 \end{array}} + H_2O$$

Hydrazine is still manufactured by the Raschig process, in which ammonia is oxidized by sodium hypochlorite in dilute aqueous solution:

$$NH_3 + NaOCl \rightarrow NH_2Cl + NaOH \quad \text{(fast)}$$
$$2NH_3 + NH_2Cl \rightarrow NH_2NH_2 + NH_4Cl \quad \text{(slow)}$$

A side reaction between chloramine and hydrazine may destroy some or all of the product.

$$N_2H_4 + 2NH_2Cl \rightarrow N_2 + 2NH_4Cl$$

This reaction is catalysed by heavy metal ions present in solution. For this reason distilled water is used (rather than tap water), and glue or gelatin is added to mask (i.e. complex with) the remaining metal ions. The use of excess of ammonia reduces the incidence of chloramine reacting with hydrazine. The use of a dilute solution of the reactants is necessary to minimize another side reaction:

$$3NH_2Cl + 2NH_3 \rightarrow N_2 + 3NH_4Cl$$

A 2% solution of hydrazine can be made by this method. This solution is concentrated either by distillation, or by adding H_2SO_4 to precipitate the salt hydrazine sulphate $N_2H_4 \cdot H_2SO_4$.

Electron diffraction and infrared data indicate that the structure of hydrazine is related to that of ethane. Each N atom is tetrahedrally surrounded by one N, two H and a lone pair. The two halves of the molecule are rotated 95° about the N—N bond and adopt a gauche (non-eclipsed) conformation. The N—N bond length is 1.45 Å.

```
H         :
 \       /
: —N—N—H
 /       \
H         H
```

Phosphorus forms an unstable hydride P_2H_4, which has very little chemical similarity to N_2H_4.

Hydroxylamine NH_2OH

Hydroxylamine forms colourless crystals that melt at 33 °C. It is thermally unstable and decomposes into NH_3, N_2, HNO_2 and N_2O easily. It explodes if heated strongly. It is usually handled in aqueous solution, or as one of its salts, since these are more stable than free NH_2OH.

Hydroxylamine is a weaker base than is ammonia or hydrazine. Salts contain the hydroxylammonium ion $[NH_3OH]^+$.

$$NH_2OH + HCl \rightarrow [NH_3OH]^+Cl^-$$
$$NH_2OH + H_2SO_4 \rightarrow [NH_3OH]^+HSO_4^-$$

The appropriate reduction potentials suggest that hydroxylamine should disproportionate. It disproportionates slowly in acidic solutions:

$$4[\overset{-I}{N}H_2OH \cdot H]^+ \rightarrow \overset{+I}{N_2}O + 2\overset{-III}{N}H_4^+ + 2H^+ + 3H_3O$$

and rapidly in alkaline solutions:

$$3\overset{-I}{N}H_2OH \rightarrow \overset{0}{N_2} + \overset{-III}{N}H_3 + 3H_2O$$

Both NH_2OH and its salts are very poisonous, and they are also strong reducing agents.

Hydroxylamine is manufactured by reducing nitrites, or from nitromethane:

$$NH_4NO_2 + NH_4HSO_3 + SO_2 + 2H_2O \rightarrow [NH_3OH]^+HSO_4^- + (NH_4)_2SO_4$$
$$CH_3NO_2 + H_2SO_4 \rightarrow [NH_3OH]^+HSO_4^- + CO$$

Hydroxylamine has donor properties (like NH_3 and N_2H_4): the N atom can form coordinate bonds, and can complex with metals. In addition it adds

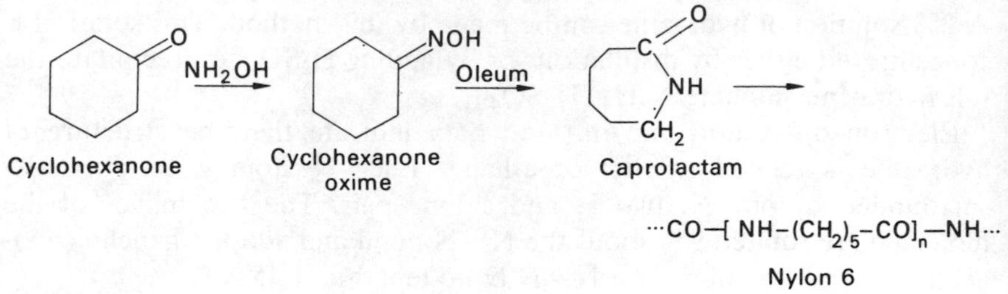

NOH
NH_2OH
Oleum
NH
CH_2
Cyclohexanone
Cyclohexanone oxime
Caprolactam
$\cdots CO\text{—}[NH\text{–}(CH_2)_5\text{–}CO]_n\text{—}NH\cdots$
Nylon 6

Figure 14.5 Nylon-6.

easily to double bonds in organic molecules, and thus provides an easy way of introducing N atoms into molecules.

NH_2OH is manufactured in large quantities to make cyclohexanone oxime, which is converted to caprolactam and then polymerized to give nylon-6.

LIQUID AMMONIA AS A SOLVENT

Ammonia gas is easily condensed (boiling point −33 °C) to give liquid ammonia. Liquid ammonia is the most studied non-aqueous solvent and it resembles the aqueous system quite closely. Liquid ammonia, like water, will dissolve a wide variety of salts. Both water and ammonia undergo self-ionization:

$$2H_2O \rightleftharpoons H_3O^+ + OH^-$$
$$2NH_3 \rightleftharpoons NH_4^+ + NH_2^-$$

Thus, substances which produce H_3O^+ ions in water are acids, and ammonium salts are acids in liquid ammonia. Similarly, substances producing OH^- in water or NH_2^- in liquid ammonia are bases in that solvent.

Thus acid–base neutralization reactions occur in both solvents, and phenolphthalein may be used to detect the end point in either:

$$\underset{\text{acid}}{HCl} + \underset{\text{base}}{NaOH} \rightarrow \underset{\text{salt}}{NaCl} + \underset{\text{solvent}}{H_2O} \quad \text{(in water)}$$

$$NH_4Cl + NaNH_2 \rightarrow NaCl + 2NH_3 \quad \text{(in ammonia)}$$

In a similar way, precipitation reactions occur in both solvents. However, the direction of the reaction is a function of the solvent:

$$(NH_4)_2S + Cu^{2+} \rightarrow 2NH_4^+ + Cu_2S\downarrow \quad \text{(in water)}$$
$$(NH_4)_2S + Cu^{2+} \rightarrow 2NH_4^+ + Cu_2S\downarrow \quad \text{(in ammonia)}$$

$$BaCl_2 + 2AgNO_3 \rightarrow Ba(NO_3)_2 + 2AgCl\downarrow \quad \text{(in water)}$$
$$Ba(NO_3)_2 + 2AgCl \rightarrow BaCl_2\downarrow + 2AgNO_3 \quad \text{(in ammonia)}$$

Amphoteric behaviour is observed in both solvents; for example,

$Zn(OH)_2$ is amphoteric in water and $Zn(NH_2)_2$ is amphoteric in ammonia:

$$Zn^{2+} + NaOH \rightarrow \underset{\text{insoluble}}{Zn(OH)_2} + \overset{\text{excess}}{NaOH} \rightarrow \underset{\text{soluble}}{Na_2[Zn(OH)_4]} \quad \text{(in water)}$$

$$Zn^{+} + KNH_2 \rightarrow \underset{\text{insoluble}}{Zn(NH_2)_2} + \overset{\text{excess}}{KNH_2} \rightarrow \underset{\text{soluble}}{K_2[Zn(NH_2)_4]} \quad \text{(in ammonia)}$$

Liquid ammonia is an extremely good solvent for the alkali metals and the heavier Group II metals Ca, Sr and Ba. The metals are very soluble and solutions in liquid ammonia have a conductivity comparable to that of pure metals. The ammonia solvates the metal ions, but is resistant to reduction by the free electrons. These solutions of metals in liquid ammonia are very good reducing agents because of the presence of free electrons.

$$Na \xrightarrow{\text{liquid ammonia}} [Na(NH_3)_n]^+ + e$$

Solutions of ammonium salts in liquid ammonia are used to clean the cooling systems in some nuclear reactors. Liquid sodium is used to cool fast breeder nuclear reactors, such as that at Dounreay in Scotland. Liquid ammonia is a good solvent for metals, but the surfaces are left wet with NH_3. When this evaporates it may leave a trace of finely divided sodium which is pyrophoric. Thus it is necessary to destroy the sodium by using an acid such as an ammonium salt in liquid ammonia.

$$2NH_4Br + 2Na \xrightarrow{\text{in } NH_3} 2NaBr + H_2 + 2NH_3$$

Because liquid ammonia accepts protons readily, it enhances the ionization of so-called weak acids such as acetic acid.

$$NH_3 + CH_3 \cdot COOH \rightleftharpoons CH_3 \cdot COO^- + NH_4^+$$

The NH_3 removes H^+ and thus causes the reaction to proceed in the forward direction. Thus acetic acid has a pK_a value of 5 in water but is almost completely ionized in liquid ammonia. Ammonia thus reduces the difference between the strengths of acids. In this respect ammonia is called a levelling solvent (see Chapter 8, under Acids and bases).

HYDROGEN AZIDE AND THE AZIDES

Hydrogen azide HN_3 (formerly called hydrazoic acid) is a colourless liquid b.p. 37 °C, which is highly poisonous and has an irritating odour. Both the liquid and the gas explode on heating or with a violent shock.

$$2HN_3 \rightarrow H_2 + 3N_2$$

HN_3 is slightly more stable in aqueous solution, but should be treated with care. It dissociates slightly in aqueous solution ($pK_a \cong 5$). It behaves as a weak acid, of similar strength to acetic acid. It reacts with electropositive metals, forming salts called azides, but unlike other acid + metal reactions, no hydrogen is evolved.

$$6HN_3 + 4Li \rightarrow \underset{\text{lithium azide}}{4LiN_3} + 2NH_3 + 2N_2$$

Covalent azides are used as detonators and explosives. Ionic azides are usually much more stable, and some are used as organic intermediates and dyestuffs.

The most important method of making azides is by passing nitrous oxide gas into fused sodamide at 190 °C under anhydrous conditions. The water vapour produced reacts with more sodamide. Alternatively nitrous oxide can be passed into a solution of sodamide in liquid ammonia as a solvent.

$$N_2O + NaNH_2 \rightarrow NaN_3 + H_2O$$
$$H_2O + NaNH_2 \rightarrow NH_3 + NaOH$$
$$\overline{N_2O + 2NaNH_2 \rightarrow NaN_3 + NH_3 + NaOH}$$

The sodium azide so obtained may be converted to hydrogen azide by treatment with H_2SO_4 followed by distillation. Lead azide $Pb(N_3)_2$ can be precipitated from a solution of sodium azide and a soluble lead salt such as $Pb(NO_3)_2$. $Pb(N_3)_2$ is sensitive to shock and is used as a detonator to set off a high explosive charge. It is particularly reliable, and works even in damp conditions. Numerous other metal azides are known.

Cyanuric triazide is a powerful explosive (Figure 14.6).

The $(N_3)^-$ ion is considered as a pseudohalide ion (see Chapter 16). It forms the extremely unstable and explosive compounds fluorazide FN_3, chlorazide ClN_3, bromazide BrN_3 and iodazide IN_3, but the dimer $N_3—N_3$ is unknown.

Analysis of N_3^- is by reduction with H_2S.

$$NaN_3 + H_2S + H_2O \rightarrow NH_3 + N_2 + S + NaOH$$

The N_3^- ion has 16 outer electrons and is isoelectronic with CO_2. The N_3^- ion is linear (N—N—N) as is CO_2. Four electrons are used for the two σ bonds. Each of the end N atoms has one non-bonding pair of electrons. This leaves $16 - 4 - (2 \times 2) = 8$ electrons for π bonding. If the bonding

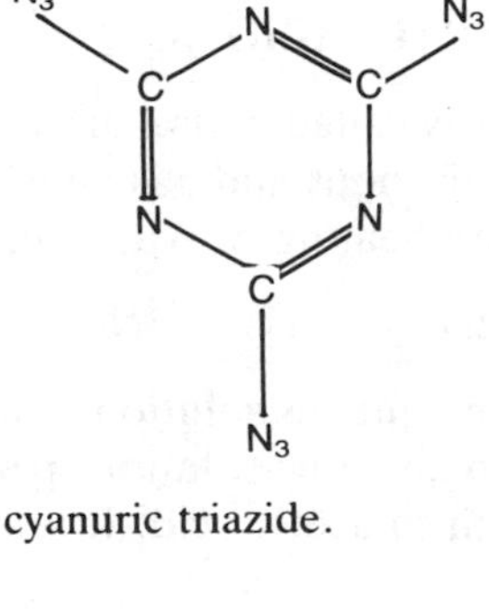

Figure 14.6 Structure of cyanuric triazide.

and non-bonding electrons are assumed to occupy the $2s$ and $2p_x$ orbitals, this leaves six atomic orbitals for π bonding. These are three $2p_y$ AOs and three $2p_z$ AOs. The three $2p_y$ orbitals form three three-centre π molecular orbitals. The lowest MO in energy is bonding, the highest is antibonding, and the remaining MO is non-bonding. In a similar way the three $2p_z$ atomic orbitals give bonding, non-bonding and antibonding MOs. The eight π electrons fill both of the bonding MOs and both of the non-bonding MOs. Thus there are two σ and two π bonds, giving a bond order of 2. Also both N—N bonds are the same length, 1.16 Å.

The hydrogen azide molecule has a bent structure. The addition of the extra electron from H means that one electron must now occupy an antibonding MO, and hence the two N–N bond lengths are different:

```
H
 \
  N———————N———————N
   1.24 Å   1.13 Å
```

The bond angle H—N—N is 112°, and the two N—N bonds are of significantly different lengths, and the bond orders are probably 1.5 and 2 respectively.

FERTILIZERS

Plant fertilizers normally contain three main ingredients:

1. *Nitrogen* in a combined form (commonly as ammonium nitrate, other ammonium salts or nitrates, or as urea). Nitrogen is essential for plant growth, particularly of leaves, since it is a constituent of amino acids and proteins, which must be made to make new cells.
2. *Phosphorus* for root growth, usually as a slightly soluble form of phosphate such as 'superphosphate' or 'triple superphosphate'. Phosphate rocks such as fluoroapatite $[3Ca_3(PO_4)_2 \cdot CaF_2]$ are mined. Basic slag, which is a by-product from the steel industry, is also used as a phosphate fertilizer.
3. *Potassium ions* for flowering, often provided as K_2SO_4.

NITROGEN FIXATION

There is a large amount of N_2 gas in the atmosphere, but plants are unable to utilize this because N_2 gas is so stable and unreactive. Fertile soil contains combined nitrogen, mainly in the form of nitrates, nitrites, ammonium salts or urea $CO(NH_2)_2$. These compounds are absorbed from the soil water by the roots of the plants. This reduces the fertility of the soil, though much of the nitrogen is eventually returned to the soil due to death and decay of the plants. It has long been known that using a plot of land to grow different crops in rotation gives a better yield than growing the same crop repeatedly. Furthermore, growing clover one year greatly

STANDARD REDUCTION POTENTIALS (VOLTS)

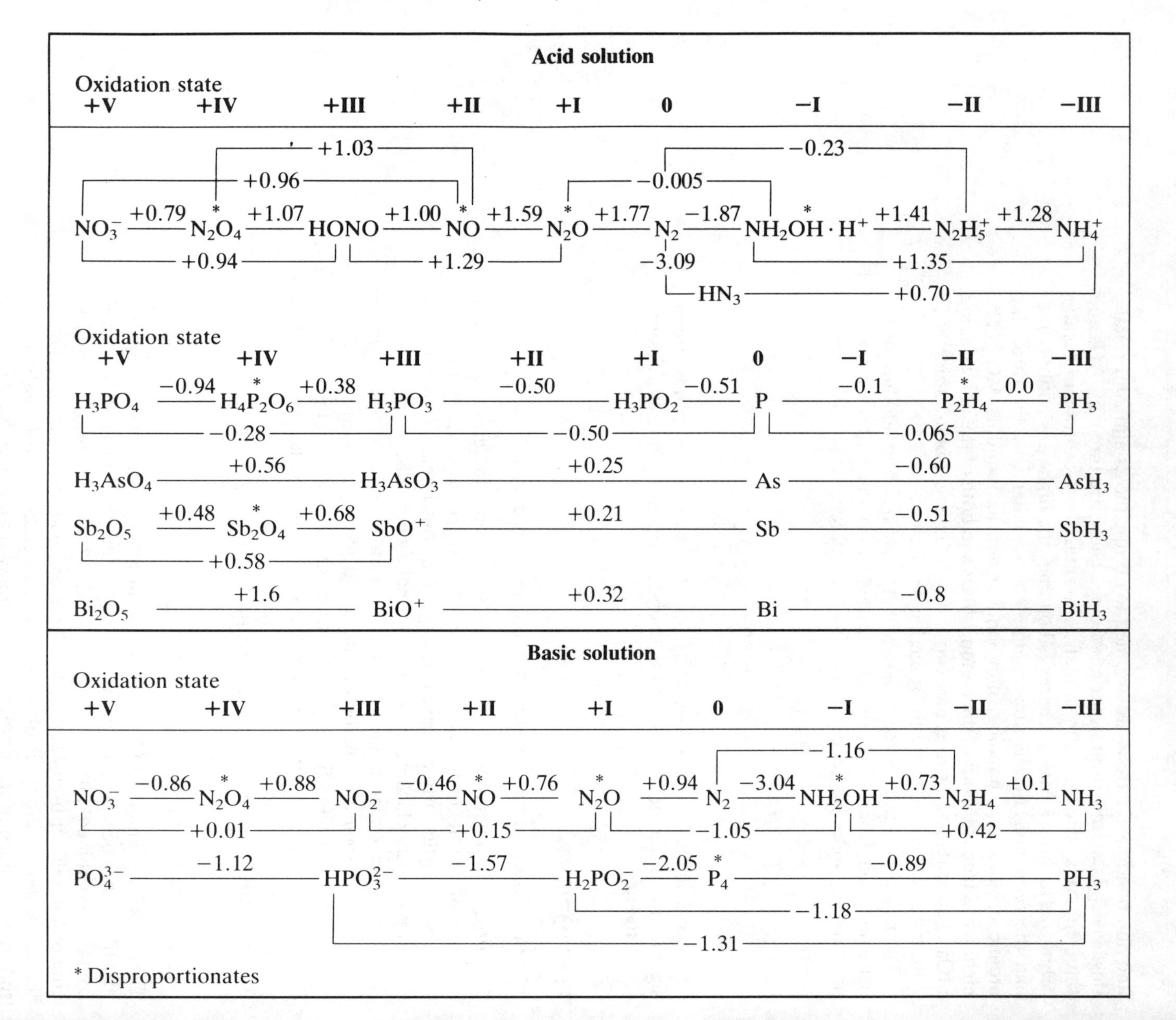

increases the yield of corn the following year. A few species of bacteria and cyanobacteria can 'fix' atmospheric nitrogen, that is can convert N_2 gas into combined forms. These bacteria can have a great effect on the fertility of soil by producing 'combined nitrogen'. The most important nitrogen fixing genus of bacteria is called *Rhizobium*. It lives symbiotically in the nodules on the roots of plants in the family Leguminosae, e.g. peas, beans, clover and alder trees. Other bacteria exist in the soil near roots, and are also able to fix nitrogen, but in smaller amounts (see 'Nitrogen cycle').

Though plants require nitrates, bacteria in the soil will readily convert other nitrogenous compounds into nitrates.

$$NH_4^+ \xrightarrow[\text{and } Nitrobacter]{Nitrosomonas} NO_2^- \xrightarrow{Nitrobacter} NO_3^-$$

Chemical processes involving the fixation of atmospheric nitrogen include the Haber–Bosch process for ammonia, and the formation of calcium cyanamide, which both involve the use of high temperatures and pressure. Bacteria can fix nitrogen easily at room temperature and atmospheric pressure, yet man requires expensive plant with high temperatures and pressures to do the same.

There is considerable research interest into finding transition metal catalysed systems which will absorb nitrogen and produce ammonia for fertilizers cheaply and without the necessity for high temperatures or pressure. The first dinitrogen complex, the pentaammine(dinitrogen)ruthenium cation, was made in 1965 by reducing ruthenium trichloride with hydrazine. Other methods have now been found, e.g. replacement of a labile ligand in a complex by N_2. Dinitrogen complexes have now been made for almost all the transition elements.

$$[Ru(NH_3)_5H_2O]^{2+} + N_2 \xrightarrow{\text{aqueous solution}} [Ru(NH_3)_5N_2]^{2+}$$

The formation of this stable nitrogen complex led to studies with other metals. Complexes with titanium(II) are the most promising, and reduction of titanium alkoxides yields either ammonia or hydrazine. A complete cycle of reactions for fixing atmospheric nitrogen to ammonia has been reported:

$$Ti^{IV}(OR)_4 \xrightarrow{Na} Ti^{II}(OR)_2 + 2NaOR$$

$$\rightarrow Ti(OR)_2 \xrightarrow{N_2} [Ti(OR_2)N_2]$$

$$[Ti(OR)_2N_2] \xrightarrow{\text{reduce}} [Ti(OR_2)N_2]^{6-}$$

$$[Ti(OR_2)N_2]^{6-} \xrightarrow{+6H^+} 2NH_3 + \boxed{Ti(OR)_2}$$

or

$$[Ti(OR)_2N_2] \xrightarrow{\text{reduce}} [Ti(OR_2)N_2]^{4-} \xrightarrow{+4H^+} N_2H_4 + Ti(OR)_2$$

Cyanamide process

Production of calcium cyanamide reached a maximum of 1.3 million tonnes/year, but has now declined a little. It is used in large amounts as a nitrogenous fertilizer, and as a source of organic chemicals such as melamine.

$$CaC_2 + N_2 \xrightarrow{1100°C} CaNCN + C$$

$$CaNCN + 5H_2O \rightarrow CaCO_3 + 2NH_4OH$$

Haber–Bosch process

The most important commercial process is the Haber–Bosch process. Fritz Haber discovered how to make N_2 and H_2 combine directly in the laboratory. He was awarded the Nobel Prize for Chemistry in 1918. Carl Bosch was a chemical engineer who developed the plant to make ammonia using this reaction on an industrial scale. He too was awarded the Nobel Prize for Chemistry in 1931 for his work on high pressure reactions.

$$\underbrace{N_2 + 3H_2}_{\text{4 volumes}} \rightleftharpoons \underbrace{2NH_3}_{\text{2 volumes}} + \text{heat}$$

The reaction is reversible, and Le Chatelier's principle suggests that a high pressure and low temperature are required to drive the reaction to the right, and thus form NH_3. A low temperature gives a higher percentage conversion to NH_3, but the reaction is slow in reaching equilibrium, and a catalyst is required. In practice the conditions used are 200 atmospheres pressure, a temperature of 380–450 °C and a catalyst of *promoted iron*. It is more economic to use a higher temperature, so that equilibrium will be reached much faster, even though this gives a lower percentage conversion. At a temperature of about 400 °C a 15% conversion is obtained with a single pass over the catalyst. The gas mixture is cooled to condense liquid NH_3, and the unchanged mixture of N_2 and H_2 gases is recycled. The plant is made of steel alloyed with Ni and Cr.

The catalyst is made by fusing Fe_3O_4 with KOH and a refractory material such as MgO, SiO_2 or Al_2O_3. This is broken into small lumps and put into the ammonia convertor, where the Fe_3O_4 is reduced to give small crystals of iron in a refractory matrix. This is the active catalyst.

The actual plant is more complicated than this one-stage reaction implies, since the N_2 and H_2 must be made before they can be converted to NH_3. The cost of H_2 is of great importance for the economy of the process. Originally the H_2 required was produced by electrolysis of water. This was expensive, and a cheaper method using coke and water was then used (water gas, producer gas). Nowadays the H_2 is produced from hydrocarbons, either naptha or CH_4. All traces of S must be removed since these poison the catalyst.

$$CH_4 + 2H_2O \rightleftharpoons CO_2 + 4H_2$$
$$CH_4 + H_2O \rightleftharpoons CO + 3H_2$$

Some air is added. The O_2 burns with some of the H_2, thus leaving N_2 to give the required reaction ratio $N_2 : H_2$ of 1:3.

$$\underset{\text{air}}{(4N_2 + O_2)} + 2H_2 \rightleftharpoons 4N_2 + 2H_2O$$

CO must also be removed as it too poisons the catalyst.

$$CO + H_2O \rightleftharpoons CO_2 + H_2$$

Finally the CO_2 is removed in a scrubber by means of a concentrated solution of K_2CO_3, or ethanolamine.

World production of NH_3 has risen from about 1 million tonnes/year in 1950 to 98 million tonnes in 1988. This is not quite the largest tonnage of any chemical produced, but since NH_3 has a very low molecular weight it constitutes a larger amount of substance (moles) than any other chemical. The largest producers are the USSR 27%, China 21%, the USA 18%, Canada 4%, Romania 4%, the Netherlands 3%, Mexico 3%, West Germany, Poland, Italy and East Germany 2% each.

About 75% of the ammonia is used as a fertilizer (30% direct application of NH_3 gas or NH_4OH to the soil, 20% NH_4NO_3, urea 15%, ammonium phosphate 10%, $(NH_4)_2SO_4$ 3%). Other uses include the following:

1. Making HNO_3, which can be used to make NH_4NO_3 (fertilizer), or explosives such as nitroglycerine, nitrocellulose and TNT. HNO_3 can be used for many other purposes.
2. Making caprolactam, which on polymerization forms nylon-6 (see hydroxylamine).
3. Making hexamethylenediamine which is used in making nylon-6-6, polyurethanes and polyamides.
4. Making hydrazine and hydroxylamine.
5. Liquid NH_3 is often used as a cheaper and more convenient way of transporting H_2 than cylinders of compressed H_2 gas. The H_2 is obtained from NH_3 by heating over a catalyst of finely divided Ni or Fe.
6. Ammonia has been used as the cooling liquid in refrigerators. It has a very high heat of vaporization, and convenient boiling and freezing points. With the environmental concern over using Freons in refrigerators, this use of NH_3 could increase.

The widespread use of nitrates as fertilizers greatly boosts crop yields. Since nitrates are soluble, the run-off water into lakes and rivers also contains nitrates. This causes several problems.

1. It produces increased growth of algae and other aquatic plants, which may clog up rivers and lakes, and may make mudbanks in estuaries turn green.

2. There is concern that nitrates are harmful in drinking water. They cause a disease in babies called methaemoglobinaemia, which reduces the amount of oxygen in the baby's blood. In extreme forms this causes the 'blue baby syndrome'. There is also concern that nitrates could be linked with stomach cancer. Because of this, the EEC have set a safety limit of 25 ppm for nitrates in drinking water.
3. There is some concern that denitrification to oxides of nitrogen, particularly N_2O, may harm the ozone layer.

UREA

Urea is widely used as a nitrogenous fertilizer. It is very soluble, and hence quick acting, but it is easily washed away. It has a very high nitrogen content (46%). It is manufactured from NH_3, and the reaction proceeds in two stages.

$$2NH_3 + CO_2 \xrightarrow[\text{high pressure}]{180\text{–}200\,^\circ C} \underset{\text{ammonium carbamate}}{NH_2COONH_4} \rightarrow \underset{\text{urea}}{NH_2 \cdot CO \cdot NH_2} + H_2O$$

In the soil, urea slowly hydrolyses to ammonium carbonate.

$$NH_2CONH_2 + 2H_2O \rightarrow (NH_4)_2CO_3$$

PHOSPHATE FERTILIZERS

Phosphate rocks such as fluoroapatite $[Ca_3(PO_4)_2 \cdot CaF_2]$ are very insoluble, and thus are of no use to plants. Superphosphate is made by treating phosphate rock with concentrated H_2SO_4. The acid salt $Ca(H_2PO_4)_2$ is more soluble, and over a period of weeks the superphosphate will dissolve in the soil water.

$$3[Ca_3(PO_4)_2 \cdot CaF_2] + 7H_2SO_4 \rightarrow \underbrace{3Ca(H_2PO_4)_2 + 7CaSO_4}_{\text{superphosphate}} + 2HF$$

The $CaSO_4$ is an insoluble waste product, and is of no value to plants, but is not removed from the product sold.

'Triple superphosphate' is made in a similar way, using H_3PO_4 to avoid the formation of the waste product $CaSO_4$.

$$[Ca_3(PO_4)_2 \cdot CaF_2] + 6H_3PO_4 \rightarrow \underbrace{4Ca(H_2PO_4)_2}_{\text{triple superphosphate}} + 2HF$$

HALIDES

Trihalides

All the possible trihalides of N, P, As, Sb and Bi are known. The nitrogen compounds are the least stable. Though NF_3 is stable, NCl_3 is explosive.

It was formerly sold as 'agene' to bleach flour to make white bread. This use declined rapidly when it was suspected that bread made from flour bleached in this way sent dogs mad! NBr_3 and NI_3 are known only as their unstable ammoniates $NBr_3 \cdot 6NH_3$ and $NI_3 \cdot 6NH_3$. The latter compound can be made by dissolving I_2 in 0.880 NH_4OH. It detonates unless excess ammonia is present, and students are warned **not** to prepare this compound. The other 16 trihalides are stable.

The trihalides are predominantly covalent and, like NH_3, have a tetrahedral structure with one position occupied by a lone pair. The exceptions are BiF_3 which is ionic and the other halides of Bi and SbF_3 which are intermediate in character.

The trihalides typically hydrolyse readily with water, but the products vary depending on the element:

$$NCl_3 + 4H_2O \rightarrow NH_4OH + 3HOCl$$
$$PCl_3 + 3H_2O \rightarrow H_3PO_3 + 3HCl$$
$$AsCl_3 + 3H_2O \rightarrow H_3AsO_3 + 3HCl$$
$$SbCl_3 + H_2O \rightarrow SbO^+ + 3Cl^- + 2H^+$$
$$BiCl_3 + H_2O \rightarrow BiO^+ + 3Cl^- + 2H^+$$

They also react with NH_3.

e.g. $$PCl_3 + 6NH_3 \rightarrow P(NH_2)_3 + 3NH_4Cl$$

NF_3 behaves differently from the others. It is unreactive, rather like CF_4, and does not hydrolyse with water, dilute acids or alkali. It does react if sparked with water vapour.

$$2NF_3 + 3H_2O \rightarrow 6HF + N_2O_3$$

The trihalides, particularly PF_3, can act as donor molecules using their lone pair to form a coordinate bond. PF_3 is rather less reactive towards water and is more easily handled than the other halides. It is very similar to CO as a ligand, and $Ni(PF_3)_4$ can be made from nickel carbonyl $Ni(CO)_4$.

$$Ni(CO)_4 + 4PF_3 \rightarrow Ni(PF_3)_4 + 4CO$$

Many trifluorophosphine complexes of the transition metals are known. Much of the work was done by J. Chatt and his group at ICI. Though most of the trihalides are made from the elements, PF_3 is made by the action of CaF_2 (or other fluoride) on PCl_3. PF_3 is a colourless, odourless gas, which is very toxic because it forms a complex with haemoglobin in the blood, thus starving the body of oxygen.

NF_3 has little tendency to act as a donor molecule. The molecule is tetrahedral with one position occupied by a lone pair, and the bond angle F—N—F is 102°30′. However, the dipole moment is very low (0.23 Debye units) compared with 1.47D for NH_3. The highly electronegative F atoms attract electrons, and these moments partly cancel the moment from the lone pair, and this reduces both the dipole moment and its donor power.

The trihalides also show acceptor properties, and can accept an electron pair from another ion such as F^-, forming complex ions such as $[SbF_5]^{2-}$ and $[Sb_2F_7]^-$. They also react with a variety of organometallic reagents, forming compounds MR_3.

PCl_3 is the most important trihalide, and 250000 tonnes/year are produced commercially from the elements. Some PCl_3 is used to make PCl_5.

$$PCl_3 + Cl_2 \text{ (or } S_2Cl_2) \rightarrow PCl_5$$

PCl_3 is widely used in organic chemistry to convert carboxylic acids to acid chlorides, and alcohols to alkyl halides.

$$PCl_3 + 3RCOOH \rightarrow 3RCOCl + H_3PO_3$$
$$PCl_3 + 3ROH \rightarrow 3RCl + H_3PO_3$$

PCl_3 can be oxidized by O_2 or P_4O_{10} to give phosphorus oxochloride $POCl_3$.

$$2PCl_3 + O_2 \rightarrow 2POCl_3$$
$$6PCl_3 + P_4O_{10} + 6Cl_2 \rightarrow 10POCl_3$$

$POCl_3$ is used in large amounts in the manufacture of trialkyl and triaryl phosphates $(RO)_3PO$.

$$O{=}PCl_3 + 3EtOH \rightarrow O{=}P(OEt)_3 \quad \text{Triethyl phosphate}$$

$$O{=}PCl_3 + 3HO{-}C_6H_4{-}CH_3 \rightarrow O{=}P(O\cdot C_6H_4\cdot CH_3)_3 \quad \text{Tritollyl phosphate}$$

Several of these phosphate derivatives are commercially important:

1. Triethyl phosphate is used in producing systemic insecticides.
2. Tritolyl phosphate is a petrol additive.
3. Triaryl phosphates and trioctyl phosphate are used as plasticizers for polyvinyl chloride.
4. Tri-n-butyl phosphate is used for solvent extraction.

Pentahalides

Nitrogen is unable to form pentahalides because the second shell contains a maximum of eight electrons, i.e. four bonds. The subsequent elements have suitable *d* orbitals, and form the following pentahalides:

PF_5	PCl_5	PBr_5	PI_5
AsF_5	$(AsCl_5)$		
SbF_5	$SbCl_5$		
BiF_5			

$AsCl_5$ is highly reactive and unstable, and has only a temporary existence. BiF_5 is highly reactive, and explodes with water, forming O_3 and F_2O. It oxidizes UF_4 to UF_6, and BrF_3 to BrF_5, and fluorinates hydrocarbons.

The pentahalides are prepared as follows:

$$3PCl_5 + 5AsF_3 \rightarrow 3PF_5 + 5AsCl_3$$
$$PCl_3 + Cl_2 \text{ (in } CCl_4) \rightarrow PCl_5$$
$$2As_2O_3 + 10F_2 \rightarrow 4AsF_5 + 3O_2$$
$$2Sb_2O_3 + 10F_2 \rightarrow 4SbF_5 + 3O_2$$
$$2Bi + 5F_2 \rightarrow 2BiF_5$$

These molecules have a trigonal bipyramid shape in the gas phase (see Figure 14.7), as expected from the VSEPR theory for five pairs of electrons. The valence bond explanation of the shape is:

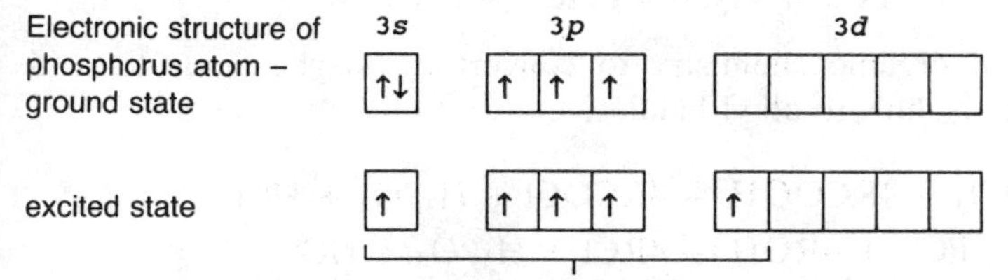

The trigonal bipyramid is not a regular structure. Electron diffraction on PF_5 gas shows that some bond angles are 90° and others are 120°, and the axial P—F bond lengths are 1.58 Å whilst the equatorial P—F lengths are 1.53 Å. In contrast nmr studies suggest that all five F atoms are equivalent. This paradox may be explained quite simply. Electron diffraction gives an instantaneous picture of the molecule, whilst nmr gives the picture averaged over several milliseconds. The axial and equatorial F atoms are thought to interchange their positions in less time than that needed to take the nmr. The interchange of axial and equatorial positions is called 'pseudorotation'.

PF_5 remains covalent and keeps this structure in the solid state. However, PCl_5 is close to the ionic–covalent borderline, and it is covalent in the gas and liquid states, but is ionic in the solid state. PCl_5 solid exists as $[PCl_4]^+$ and $[PCl_6]^-$: the ions have tetrahedral and octahedral structures respectively. In the solid, PBr_5 exists as $[PBr_4]^+Br^-$, and PI_5 appears to be $[PI_4]^+$ and I^- in solution.

PCl_5 is the most important pentahalide, and it is made by passing Cl_2 into a solution of PCl_3 in CCl_4. World production is about 20 000 tonnes/year. Complete hydrolysis of the pentahalides yields the appropriate -ic acid. Thus PCl_5 reacts violently with water:

$$PCl_5 + 4H_2O \rightarrow \underset{\text{phosphoric acid}}{H_3PO_4} + 5HCl$$

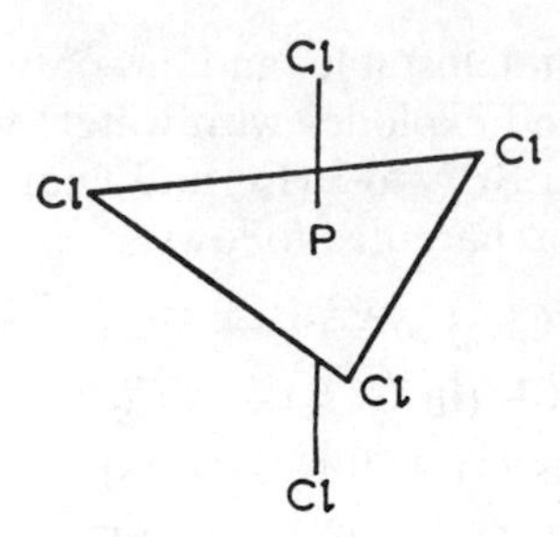

Figure 14.7 Structure of gaseous phosphorus pentachloride.

If equimolar amounts are used, the reaction is more gentle and yields phosphorus oxochloride $POCl_3$.

$$PCl_5 + H_2O \rightarrow POCl_3 + 2HCl$$

PCl_5 is used in organic chemistry to convert carboxylic acids to acid chlorides, and alcohols to alkyl halides.

$$PCl_5 + 4RCOOH \rightarrow 4RCOCl + H_3PO_4 + HCl$$
$$PCl_5 + 4ROH \rightarrow 4RCl + H_3PO_4 + HCl$$

It reacts with P_4O_{10}, forming $POCl_3$, and with SO_2, forming thionyl chloride $SOCl_2$.

$$6PCl_5 + P_4O_{10} \rightarrow 10POCl_3$$
$$PCl_5 + SO_2 \rightarrow POCl_3 + SOCl_2$$

PCl_5 also reacts with NH_4Cl, forming a variety of phosphonitrilic chloride polymers (see later).

$$nPCl_5 + nNH_4Cl \rightarrow (NPCl_2)_n + 4nHCl \quad \text{(ring compounds } n = 3\text{–}8)$$
$$\text{and} \quad Cl_4P \cdot (NPCl_2)_n \cdot NPCl_3 \quad \text{(chain compounds)}$$

Despite the existence of pentahalides, no hydrides MH_5 are known. To attain the five-valent state, *d* orbitals must be used. Hydrogen is not sufficiently electronegative to make the *d* orbitals contract sufficiently, though PHF_4 and PH_2F_3 have been isolated.

OXIDES OF NITROGEN

The oxides and oxoacids of nitrogen all exhibit $p\pi$–$p\pi$ multiple bonding between the nitrogen and oxygen atoms. This does not occur with the heavier elements in the group, and consequently nitrogen forms a number of compounds which have no P, As, Sb or Bi analogues. Nitrogen forms a very wide range of oxides, exhibiting all the oxidation states from (+I) to (+VI). The lower oxides are neutral, and the higher ones are acidic (Table 14.8).

Table 14.8 Oxides of nitrogen

Formula	Oxidation number	Name
N_2O	+I	Nitrous oxide
NO	+II	Nitric oxide
N_2O_3	+III	Nitrogen sesquioxide
NO_2, N_2O_4	+IV	Nitrogen dioxide, dinitrogen tetroxide
N_2O_5	+V	Dinitrogen pentoxide
(NO_3, N_2O_6) very unstable	+VI	Nitrogen trioxide, dinitrogen hexoxide

Nitrous oxide N_2O

N_2O is a stable, relatively unreactive colourless gas. It is prepared by careful thermal decomposition of molten ammonium nitrate at about 280 °C. If heated strongly it explodes. N_2O can also be made by heating a solution of NH_4NO_3 acidified with HCl.

$$NH_4NO_3 \rightarrow N_2O + 2H_2O$$

N_2O is a neutral oxide and does not form hyponitrous acid $H_2N_2O_2$ with water nor hyponitrites with alkali. It is important in the preparation of sodium azide, and hence also of the other azides:

$$N_2O + 2NaNH_2 \rightarrow NaN_3 + NH_3 + NaOH$$

The largest use of N_2O is as a propellant for whipped ice-cream. Because it has no taste, and is non-toxic, it meets the strict food and health regulations.

N_2O is used as an anaesthetic, particularly by dentists. It is sometimes called 'laughing gas', because small amounts cause euphoria. It requires a partial pressure of 760 mm Hg of N_2O to anaesthetize a patient completely. Thus if oxygen is also supplied, the patient may not be completely unconscious. If deprived of oxygen for long, the patient will die. Plainly N_2O is unsuitable for long operations. Usually N_2O is administered to put the patient 'to sleep', and O_2 to make him recover consciousness.

The molecule is linear as would be expected for a triatomic molecule with 16 outer shell electrons (see also N_3^- and CO_2). However, CO_2 is symmetrical (O—C—O), whereas in N_2O the orbital energies favour the formation of the asymmetrical molecule N—N—O rather then the symmetrical molecule N—O—N. The bond lengths are short, and the bond orders have been calculated as N—N 2.73 and N—O 1.61.

$$N \overset{1.126\,\text{Å}}{\text{———}} N \overset{1.186\,\text{Å}}{\text{———}} O$$

Nitric oxide NO

NO is a colourless gas and is an important intermediate in the manufacture of nitric acid by the catalytic oxidation of ammonia (Ostwald process). It was also important in the obsolete Birkeland–Eyde process which involved sparking nitrogen and oxygen. NO is prepared in the laboratory by the reduction of dilute HNO_3 with Cu, or reduction of HNO_2 with I^-:

$$3Cu + 8HNO_3 \rightarrow 2NO + 3Cu(NO_3)_2 + 4H_2O$$
$$2HNO_2 + 2I^- + 2H^+ \rightarrow 2NO + I_2 + 2H_2O$$

NO is a neutral oxide and is not an acid anhydride.

NO has 11 valency electrons. It is impossible for them all to be paired, and hence this is an odd electron molecule and the gas is paramagnetic. It is diamagnetic in the liquid and solid states, because the molecule dimerizes, forming O—N—N—O. The asymmetrical dimer O—N—O—N has been observed to be formed as a red solid in the presence of HCl or other Lewis acids.

The bond length N—O is 1.15 Å, which is intermediate between a double and a triple bond. Bonding is best described using the molecular orbital theory (see Chapter 4). The bonding is similar to that in N_2 and CO which both have 10 outer electrons. NO has 11 outer electrons, and the extra unpaired electron occupies an antibonding π^*2p orbital. This reduces the bond order from 3 in N_2 to $2\frac{1}{2}$ in NO. If this electron is removed by oxidizing NO, the nitrosonium ion NO^+ is formed. In NO^+ the bond order is 3, and the N—O bond length contracts from 1.15 Å in NO to 1.06 Å in NO^+.

Odd electron molecules are usually highly reactive and tend to dimerize. NO is unusually stable for an odd electron molecule. Nevertheless it reacts instantly with oxygen to give NO_2, and with the halogens it gives nitrosyl halides, e.g. NOCl.

$$2NO + O_2 \rightarrow 2NO_2$$
$$2NO + Cl_2 \rightarrow 2NOCl$$

NO readily forms coordination complexes with transition metal ions. These complexes are called nitrosyls. Fe^{2+} and NO form the complex $[Fe(H_2O)_5NO]^{2+}$, which is responsible for the colour in the 'brown-ring test' for nitrates. Most nitrosyl complexes are coloured. Another example is sodium nitroprusside $Na_2[Fe(CN)_5NO] \cdot 2H_2O$. NO often acts as a three-electron donor, in contrast to most ligands which donate two electrons. Thus three CO groups may be replaced by two NO groups:

$$[Fe(CO)_5] + 2NO \rightarrow [Fe(CO)_2(NO)_2] + 3CO$$
$$[Cr(CO)_6] + 4NO \rightarrow [Cr(NO)_4] + 6CO$$

In these complexes the M—N—O atoms are linear, or close to linear. However, in 1968 the M—N—O angle in $[Ir(CO)(Cl)(PPh_3)(NO)]^+$ was found to be 123°, and since then a number of other complexes have been found with bond angles in the range 120–130°. These bent bonds, which are weaker than straight bonds, are of considerable theoretical interest. NO may also act as a bridging ligand between two or three metal atoms in a similar way to CO.

Nitrogen sesquioxide N_2O_3

N_2O_3 can only be obtained at low temperatures. It can be made by condensing equimolar amounts of NO and NO_2 together, or by reacting NO with the appropriate amount of O_2. This gives a blue liquid or solid, which is unstable and dissociates into NO and NO_2 at −30 °C.

$$NO + NO_2 \rightarrow N_2O_3$$
$$4NO + O_2 \rightarrow 2N_2O_3$$

It is an acidic oxide and is the anhydride of nitrous acid HNO_2. With alkali it forms nitrites.

$$N_2O_3 + H_2O \rightarrow 2HNO_2$$
$$N_2O_3 + NaOH \rightarrow 2NaNO_2 + H_2O$$

N_2O_3 reacts with the concentrated acids, forming nitrosyl salts:

$$N_2O_3 + 2HClO_4 \rightarrow 2NO[ClO_4] + H_2O$$
$$N_2O_3 + 2H_2SO_4 \rightarrow 2NO[HSO_4] + H_2O$$

The oxide exists in two different forms. These may be interconverted by irradiation with light of the appropriate wavelength. The N—N bond length from microwave spectra is 1.864 Å in the asymmetrical form. This is exceptionally long and thus the bond is exceptionally weak compared with the N—N bond found in hydrazine (length of 1.45 Å).

```
O          O
 \\       //
   N----N              O=N      N=O
          \\              \    /
           O                O
asymmetrical form      symmetrical form
                  (has a two fold rotation axis)
```

asymmetrical form

symmetrical form (has a two fold rotation axis)

Nitrogen dioxide NO_2 and dinitrogen tetroxide N_2O_4

NO_2 is a red–brown poisonous gas and is produced on a large scale by oxidizing NO in the Ostwald process for the manufacture of nitric acid. In the laboratory it is prepared by heating lead nitrate:

$$2Pb(NO_3)_2 \rightarrow 2PbO + 4NO_2 + O_2$$

The gaseous products O_2 and NO_2 are passed through a U-tube cooled in ice. The NO_2 (b.p. 21 °C) condenses. The $Pb(NO_3)_2$ must be carefully dried, since NO_2 reacts with water. The NO_2 is obtained as a brown liquid which turns paler on cooling, and eventually becomes a colourless solid. This is because NO_2 dimerizes into colourless N_2O_4. NO_2 is an odd electron molecule, and is paramagnetic and very reactive. It dimerizes to N_2O_4, pairing the previously unpaired electrons. N_2O_4 has no unpaired electrons and is diamagnetic.

$$\underset{\substack{\text{paramagnetic}\\\text{brown}}}{2NO_2} \rightleftharpoons \underset{\substack{\text{diamagnetic}\\\text{colourless}}}{N_2O_4}$$

N_2O_4 is a mixed anhydride, because it reacts with water to give a mixture of nitric and nitrous acids:

$$N_2O_4 + H_2O \rightarrow HNO_3 + HNO_2$$

The HNO_2 formed decomposes to give NO.

$$2HNO_2 \rightarrow NO_2 + NO + H_2O$$
$$2NO_2 + H_2O \rightarrow HNO_3 + HNO_2$$

Thus moist NO_2 or N_2O_4 gases are strongly acidic.

The NO_2 molecule is angular with an O—N—O angle of 132°. The bond length O—N of 1.20 Å is intermediate between a single and a double bond. X-ray diffraction on solid N_2O_4 shows the structure to be planar.

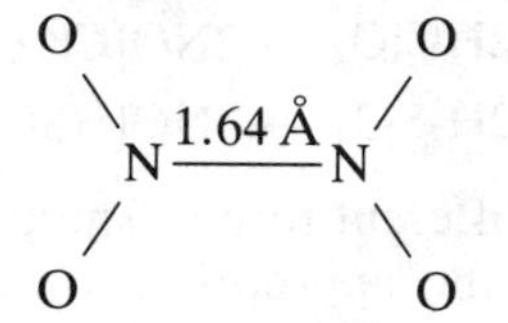

The N—N bond is very long (1.64 Å), and is therefore weak. It is much longer than the single bond N-N distance of 1.47 Å in N_2H_4, but there is no satisfactory explanation of why it is long.

Liquid N_2O_4 is useful as a non-aqueous solvent. It self-ionizes:

$$N_2O_4 \rightleftharpoons \underset{\text{acid}}{NO^+} + \underset{\text{base}}{NO_3^-}$$

In N_2O_4 substances containing NO^+ are acids and those containing NO_3^- are bases. A typical acid–base reaction is:

$$\underset{\text{acid}}{NOCl} + \underset{\text{base}}{NH_4NO_3} \rightarrow \underset{\text{salt}}{NH_4Cl} + \underset{\text{solvent}}{N_2O_4}$$

Liquid N_2O_4 is particularly useful as a solvent for preparing anhydrous metal nitrates and also nitrato complexes.

$$ZnCl_2 + N_2O_4 \rightarrow Zn(NO_3)_2 + 2NOCl$$
$$TiBr_4 + N_2O_4 \rightarrow Ti(NO_3)_4 + 4NO + 2I_2$$

The NO_2–N_2O_4 system is a strong oxidizing agent. NO_2 reacts with fluorine and chlorine, forming nitryl fluoride NO_2F and nitryl chloride NO_2Cl. It oxidizes HCl to Cl_2 and CO to CO_2.

$$2NO_2 + F_2 \rightarrow 2NO_2F$$
$$2NO_2 + Cl_2 \rightarrow 2NO_2Cl$$
$$2NO_2 + 4HCl \rightarrow 2NOCl + Cl_2 + 2H_2O$$
$$NO_2 + CO \rightarrow CO_2 + NO$$

Dinitrogen pentoxide N_2O_5

N_2O_5 is prepared by carefully dehydrating HNO_3 with P_2O_5 at low temperatures. It is a colourless deliquescent solid, which is highly reactive, is a strong oxidizing agent, and is light sensitive. It is the anhydride of HNO_3.

$$N_2O_5 + H_2O \rightarrow 2HNO_3$$
$$N_2O_5 + Na \rightarrow NaNO_3 + NO_2$$
$$N_2O_5 + NaCl \rightarrow NaNO_3 + NO_2Cl$$
$$N_2O_5 + 3H_2SO_4 \rightarrow H_3O^+ + 2NO_2^+ + 3HSO_4^-$$

In the gas phase N_2O_5 decomposes into NO_2, NO and O_2. Nitrogen trioxide NO_3 may be formed by treating N_2O_5 with O_3.

X-ray diffraction shows that solid N_2O_5 is ionic $NO_2^+\ NO_3^-$: it should in reality be called nitronium nitrate. It is covalent in solution and in the gas phase, and probably has the structure:

```
O           O
 \         /
  N — O — N
 /         \
O           O
```

OXOACIDS OF NITROGEN

Nitrous acid HNO_2

Nitrous acid is unstable except in dilute solution. It is easily made by acidifying a solution of a nitrite. Barium nitrite is often used with H_2SO_4, since the insoluble $BaSO_4$ can be filtered off easily.

$$Ba(NO_2)_2 + H_2SO_4 \rightarrow 2HNO_2 + BaSO_4$$

Group I metal nitrites can be made by heating nitrates, either on their own or with Pb.

$$2NaNO_3 \xrightarrow{\text{heat}} 2NaNO_2 + O_2$$

$$NaNO_3 + Pb \xrightarrow{\text{heat}} NaNO_2 + PbO$$

Nitrous acid and nitrites are weak oxidizing agents and will oxidize Fe^{2+} to Fe^{3+}, and I^- to I_2: they themselves are reduced to N_2O or NO. However, HNO_2 and nitrites are oxidized by $KMnO_4$ and Cl_2, forming nitrates NO_3^-.

Large amounts of nitrites are used to make diazo compounds, which are converted into azo dyes, and also pharmaceutical products.

$$PhNH_2 + HNO_2 \rightarrow \underset{\text{phenyldiazonium chloride}}{PhN_2Cl} + 2H_2O$$

Nitrites are important in the manufacture of hydroxylamine:

$$NH_4NO_2 + NH_4HSO_3 + SO_2 + 2H_2O \rightarrow [NH_3OH]^+HSO_4^- + (NH_4)_2SO_4$$

Sodium nitrite is used as a food additive in cured meat, sausages, hot dogs, bacon and tinned ham. Though an approved additive, its use is controversial. $NaNO_2$ is slightly poisonous. The tolerance limit for humans is 5–10 g per day depending on body weight. NO_2^- ions inhibit the growth of bacteria, particularly *Clostridium botulinum*, which causes botulism (a particularly unpleasant form of food poisoning). Reductive decomposition of NO_2^- gives NO, which forms a red complex with haemoglobin, and improves the look of meat. There is concern that during the cooking of meat, the nitrites may react with amines and be converted into nitrosamines $R_2N—N=O$, which are thought to cause cancer. Certainly secondary and tertiary aliphatic amines form nitrosamines with nitrites:

$$Et_2NH + HNO_2 \rightarrow Et_2NNO + H_2O$$

$$Et_3N + HNO_2 \rightarrow [Et_3NH][NO_2] \xrightarrow{\text{heat}} Et_2NNO + EtOH$$

The nitrite ion is a good ligand and forms many coordination complexes. Since lone pairs of electrons are present on both N and O atoms, either N or O can form a coordinate bond. This gives rise to isomerism between nitro complexes $M \leftarrow NO_2$ and nitrito complexes $M \leftarrow ONO$, for example $[Co(NH_5)_5(NO_2)]^{2+}$ and $[Co(NH_3)_5(ONO)]^{2+}$. This is discussed in Chapter 7, under 'Isomerism'. If a solution of Co^{2+} ions is treated with NO_2^- ions, first Co^{2+} ions are oxidized to Co^{3+}, then NO_2^- ions form the complex $[Co(NO_2)_6]^{3-}$. Precipitation of potassium cobaltinitrite $K_3[Co(NO_2)_6]$ is used to detect K^+ qualitatively. The NO_2^- ion may act as a chelating ligand, and bond to the same metal twice, or it may act as a bridging ligand joining two metal atoms.

The nitrite ion NO_2^- has a plane triangular structure, with N at the centre, two corners occupied by O atoms, and the third corner occupied by the lone pair. A three-centre bond covers the N and the two O atoms and the bond order is 1.5 for the N—O bonds, which have bond lengths in between those for a single and double bond. (More details are given in Chapter 4, under 'Examples of molecular orbital treatment involving delocalised π bonding'.)

Nitric acid HNO_3

HNO_3 is the most important oxoacid of nitrogen. (The three most important industrial acids in order of tonnages produced are (1) H_2SO_4, (2) HNO_3 and (3) HCl.) Pure nitric acid is a colourless liquid, but on exposure to light it turns slightly brown because of slight decomposition into NO_2 and O_2.

$$4HNO_3 \rightarrow 4NO_2 + O_2 + 2H_2O$$

It is a strong acid and is 100% dissociated in dilute aqueous solutions into H_3O^+ and NO_3^-. It forms a large number of salts called nitrates, which are typically very soluble in water.

The shape of the NO_3^- ion is a planar triangle, like the CO_3^{2-} ion. The later elements in both groups form tetrahedral oxoacid ions such as PO_4^{3-} and SiO_4^{4-}. This difference in shape is probably due to the small size of the N and C atoms and their restriction to eight electrons in their outer shell.

HNO_3 is an excellent oxidizing agent particularly when hot and concentrated. H^+ ions are oxidizing, but the NO_3^- ion is an even stronger oxidizing agent in acid solution. Thus metals like copper and silver which are insoluble in HCl dissolve in HNO_3. Some metals such as gold are insoluble even in HNO_3, but will dissolve in aqua regia, a mixture of 25% concentrated HNO_3 and 75% concentrated HCl. The enhanced ability to dissolve metals shown by aqua regia arises from the oxidizing power of HNO_3 coupled with the ability of Cl^- to form complexes with the metal ions.

HNO_3 was originally made from $NaNO_3$ or KNO_3 and concentrated H_2SO_4. The first synthetic method was the Birkeland–Eyde process. This sparked N_2 and O_2 together in an electric arc furnace, and passed the gas into water. The process was started in Norway in 1903, but is now obsolete, because of the high cost of electricity.

$$N_2 + O_2 \xrightarrow{\text{spark}} NO \xrightarrow{+O_2} NO_2 \xrightarrow{H_2O} 4HNO_3$$

The Ostwald process depends on the catalytic oxidation of ammonia to NO, followed by oxidation of NO to NO_2, and conversion of NO_2 with water to HNO_3. The first plant was set up in Germany in 1908, and Ostwald was awarded the Nobel Prize in 1909. The method is still used and about 20 million tonnes/year of HNO_3 are produced. The overall process is:

$$4NH_{3(g)} + 5O_{2(g)} \xrightarrow[\text{5 atmospheres } 850\,^\circ\text{C}]{\text{platinum/rhodium catalyst}} 4NO_{(g)} + 6H_2O_{(g)}$$

The NO and air are cooled and the mixture of gases is absorbed in a countercurrent of water.

$$2NO_{(g)} + O_{2(g)} \rightleftharpoons 2NO_{2(g)}$$
$$2NO_{2(g)} + H_2O_{(l)} \rightarrow HNO_3 + HNO_2$$
$$2HNO_2 \rightarrow H_2O + NO_2 + NO$$
$$3NO_2 + H_2O \rightarrow 2HNO_3 + NO$$

(the NO produced is recycled to the first step)

overall $$NH_3 + 2O_2 \rightarrow HNO_3 + H_2O$$

This gives a HNO_3 solution of concentration 60% by weight. Distillation only increases the concentration to 68% since a constant boiling mixture is formed. 'Concentrated' HNO_3 contains 98% acid and is produced by dehydrating with concentrated sulphuric acid, or by mixing with a 72% magnesium nitrate solution, followed by distillation.

When nitric acid is mixed with concentrated sulphuric acid, the nitronium ion NO_2^+ is formed. This is the active species in the nitration of benzene $\xrightarrow{NO_2^+}$ nitrobenzene (NO_2) $\xrightarrow{\text{reduce}}$ aniline (NH_2) (used for dyestuffs)

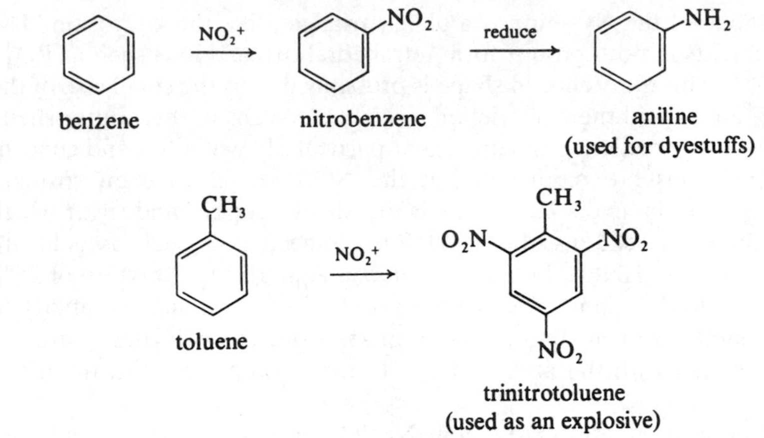

toluene (CH_3) $\xrightarrow{NO_2^+}$ trinitrotoluene (CH_3, O_2N, NO_2, NO_2) (used as an explosive)

Figure 14.8 Nitration of benzene and toluene.

aromatic organic compounds. This is an important step in making explosives, or the nitro compounds may be reduced to aniline and used for making dyestuffs:

$$HNO_3 \xrightarrow{-H_2O} NO_2^+$$

Covalent nitrates are less stable than ionic nitrates. (This is a similar behaviour to that of the azides.) Nitroglycerine, nitrocellulose, trinitrotoluene (TNT) and fluorine nitrate (FNO_3) are all explosive (Figure 14.9).

World production of explosives is quoted as 2.9 million tonnes for 1985, but the true value may be higher than this.

HNO_3 is a strong oxidizing agent, and is used to oxidize cyclohexanol/cyclohexanone mixtures to adipic acid (which reacts with hexamethylenediamine in the manufacture of nylon-66).

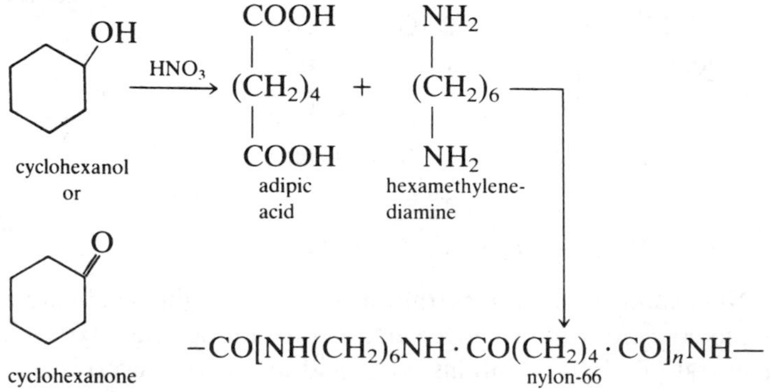

cyclohexanol (OH) or cyclohexanone (O) $\xrightarrow{HNO_3}$ $COOH-(CH_2)_4-COOH$ (adipic acid) + $NH_2-(CH_2)_6-NH_2$ (hexamethylenediamine)

$$-CO[NH(CH_2)_6NH\cdot CO(CH_2)_4\cdot CO]_nNH-$$

nylon-66

HNO_3 is also used to oxidize *p*-xylene to terephthalic acid for the manufacture of terylene.

The structure of the nitrate ion is a planar triangle. All three oxygen

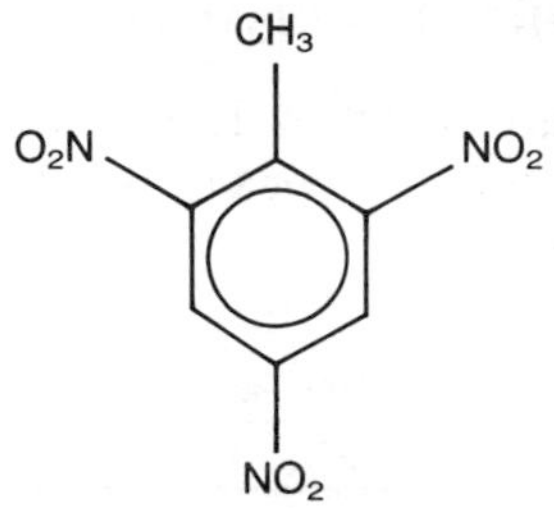

methyl-2,4,6-trinitrobenzene
(trinitrotoluene, TNT)

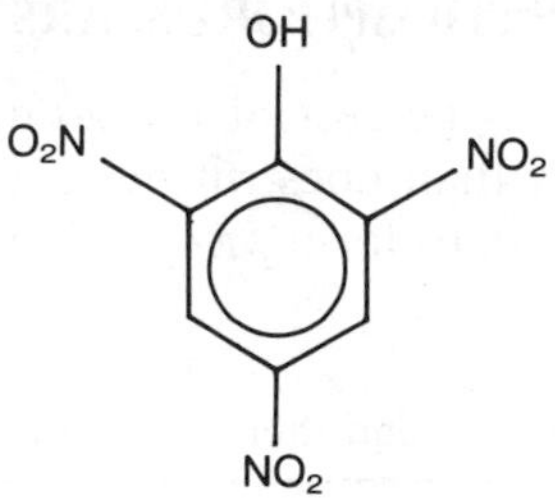

2,4,6-trinitrophenol
(picric acid)

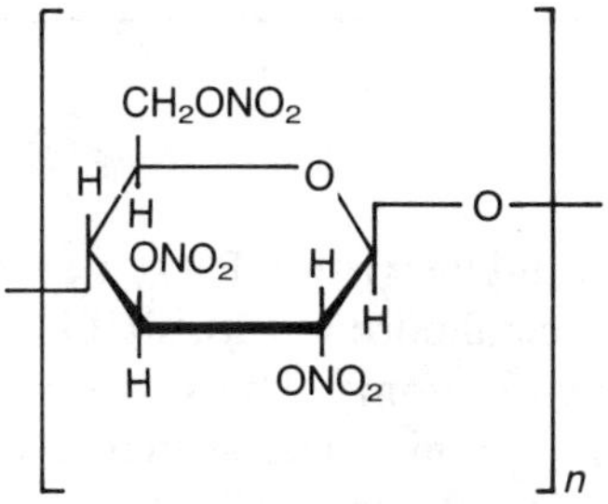

cellulose nitrate
(nitrocellulose-gun cotton)

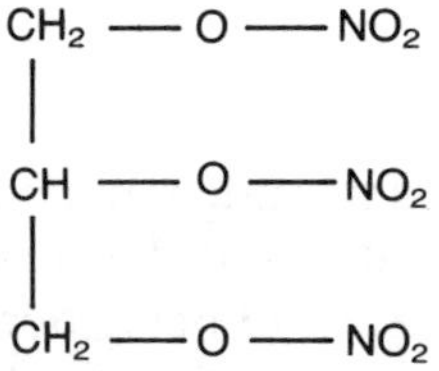

propane-1,2,3-triyl trinitrate
(nitroglycerine)

Figure 14.9 Some explosives.

atoms are equivalent. In addition to the σ bonds, four-centre π molecular orbitals cover the N and the three O atoms. Each of the N—O bonds has a bond order of $1\frac{1}{3}$, 1 from the σ bond and $\frac{1}{3}$ from the π bond. (This is described more fully in Chapter 4, under 'Examples of molecular orbital treatment involving delocalised π bonding'.)

Reduction of nitrates in acid solution gives either NO_2 or NO, but in alkaline solutions with metals such as Devarda's alloy (Cu/Al/Zn), ammonia is produced.

$$3Cu + 8HNO_3 \xrightarrow{\text{cold dilute} < 1\,M} 2NO + Cu(NO_3)_2 + 4H_2O$$

$$Cu + 3HNO_3 \xrightarrow{\text{stronger acid}} NO_2 + Cu(NO_3)_2 + H_2O$$

$$\text{Devarda's alloy (Cu/Al/Zn)} + NaOH \rightarrow H$$

$$NO_3^- + 9H \rightarrow NH_3 + 3H_2O$$

$$NO_2^- + 7H \rightarrow NH_3 + 2H_2O$$

	NO_3^-	NO_2	NO	NH_3
oxidation state of N	(+V)	(+IV)	(+II)	(−III)

OXIDES OF PHOSPHORUS, ARSENIC AND BISMUTH

The oxides of the rest of the group are listed in Table 14.9. They form fewer oxides than does nitrogen, presumably because of the inability of these elements to form $p\pi-p\pi$ double bonds.

Table 14.9 Oxides and their oxidation states

P_4O_6 III	As_4O_6 III	Sb_4O_6 III	Bi_2O_3 III
P_4O_7 III V		$(SbO_2)_n$ III V	
P_4O_8 III V			
P_4O_9 III V			
P_4O_{10} V	As_4O_{10} V	Sb_4O_{10} V	

Trioxides

Phosphorus trioxide is dimeric and should be written P_4O_6, not P_2O_3. P_4O_6 has four P atoms at the corners of a tetrahedron, with six O atoms along the edges, each O being bonded to two P atoms. The structures of As_4O_6 and Sb_4O_6 are similar to this. Bi_2O_3 is ionic. The structure of P_4O_{10} is shown in Figure 14.10. Since the P—O—P angle is 127° the O atoms are strictly above the edges, but it is more convenient to draw them on the edges.

Because yellow phosphorus is more reactive than is N_2, phosphorus oxides (unlike nitrogen oxides) can all be obtained by burning phosphorus in air.

$$P_4 + 3O_2 \xrightarrow{\text{limited supply of air}} P_4O_6$$

P_4O_6 is formed by burning phosphorus in a limited supply of air. It is a soft white solid (m.p. 24 °C, b.p. 175 °C). It is removed from the reaction mixture and is purified by distillation. (Higher oxides are formed in a plentiful supply of air.) P_4O_6 will burn in air, forming P_4O_{10}.

$$P_4O_6 + 2O_2 \rightarrow P_4O_{10}$$

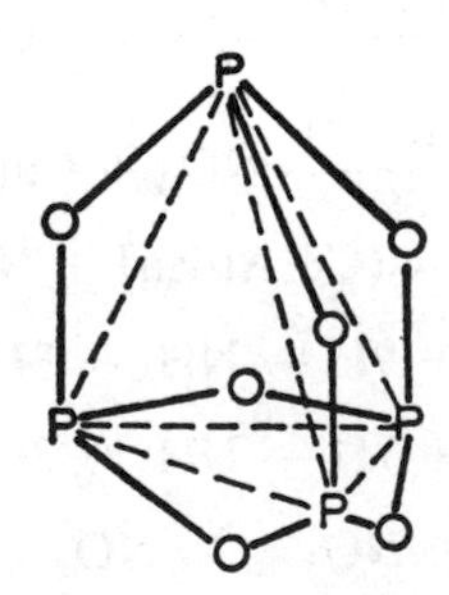

Figure 14.10 Structure of phosphorus trioxide P_4O_6.

As_4O_6 and Sb_4O_6 are obtained by burning the metals in air or oxygen, since they have less tendency to form higher oxides. Heating the sulphide minerals As_4S_4 (realgar) or As_2S_3 (orpiment) in air also gives As_4O_6. Both As_4O_6 and Sb_4O_6 are very poisonous. Bi_2O_3 is not dimeric like the others.

The basicity of oxides and hydroxides usually increases on descending a group. P_4O_6 is acidic and hydrolyses in water, forming phosphor*ous* acid. (This is considered in more detail later.) Arsenious oxide As_4O_6 is sparingly soluble in water, and Sb_4O_6 is insoluble. As_4O_6 and Sb_4O_6 are both amphoteric since they react with alkali, forming arsenites and antimonites, and with concentrated HCl, forming arsenic and antimony trichlorides. In the past, various copper arsenites were used as brilliant green pigments. The best known are Scheele's green $Cu_2As_2O_5$ and Paris green $[(CH_3COO)Cu_2(AsO_3)]$. They are seldom used nowadays because they are toxic, and, even worse, in damp places bacteria and moulds can produce poisonous volatile substances such as AsH_3 and $As(CH_3)_3$. Bi_2O_3 is wholly basic.

$$P_4O_6 + 6H_2O \rightarrow 4H_3PO_3$$

$$As_4O_6 + 12NaOH \rightarrow 4Na_3AsO_3 + 6H_2O$$

$$As_4O_6 + 12HCl \rightarrow 4AsCl_3 + 6H_2O$$

Pentoxides

Phosphorus pentoxide is the most important oxide, and is quite common. It is dimeric and has the formula P_4O_{10}, not P_2O_5. Its structure is derived from that of P_4O_6. Each P atom in P_4O_6 forms three bonds to O atoms. There are five electrons in the outer shell of a P atom. Three electrons have been used in bonding, and the other two comprise a lone pair, which is situated on the outside of the tetrahedral unit. In P_4O_{10} the lone pairs on each of the four P atoms form a coordinate bond to an oxygen atom (Figure 14.11a).

Measurement of the P—O bond lengths shows that the bridging bonds on the edges are 1.60 Å but the coordinate bonds on the corners are 1.43 Å. The bridging bonds compare with those in P_4O_6 (1.65 Å) and are normal single bonds. The bonds on the corners are much shorter than a single bond, and are in fact double bonds. These double bonds are different in origin from the 'usual' double bonds such as that in ethene which arises from $p\pi - p\pi$ overlap with one electron coming from each C atom. The second bond in P=O is formed by $p\pi - d\pi$ back bonding. A full *p* orbital on the O atom overlaps sideways with an empty *d* orbital on the P atom. Thus it differs from the double bond in ethene in two respects:

1. A *p* orbital overlaps with a *d* orbital, rather than *p* with *p*.
2. Both electrons come from one atom, and hence the bond is a 'dative bond'.

A similar type of back bonding is found in the carbonyls.

As_4O_{10} has a similar structure to P_4O_{10} in the gas phase. However, the

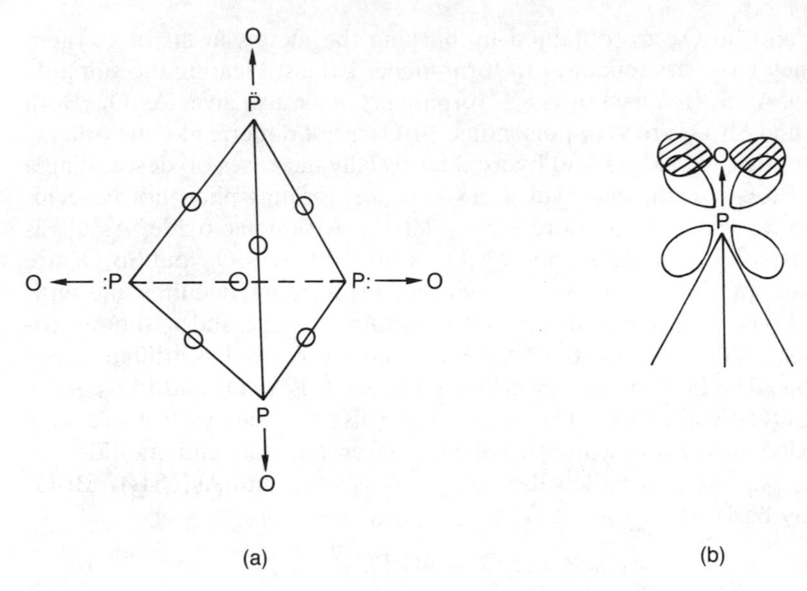

Figure 14.11 Structure of phosphorus pentoxide P_4O_{10}. (a) Formation of σ bonds. (b) Orbitals involved in back bonding.

crystal contains equal numbers of $[AsO_4]$ tetrahedra and $[AsO_6]$ octahedra joined together by sharing corners. As_4O_{10} is a strong oxidizing agent, and oxidizes HCl to Cl_2. It is deliquescent, and very soluble in water.

P_4O_{10} is formed by burning P in an excess of air or oxygen, but As and Sb require more drastic oxidation by concentrated HNO_3 to form the pentoxides. As_4O_{10} and Sb_4O_{10} lose oxygen when they are heated, and form the trioxides.

P_4O_{10} absorbs water from the air or from other compounds, and becomes sticky. Because of this strong affinity for water, P_4O_{10} is used as a drying agent. Finely powdered P_4O_{10} is sometimes spread over glass wool and used for drying purposes. This provides a large drying surface, which is not easily covered by solid hydrolysis products. P_4O_{10} hydrolyses violently in water, forming phosphor*ic* acid H_3PO_4. The manufacture of pure H_3PO_4 by this route is the largest use of P_4O_{10}.

$$P_4O_{10} + 6H_2O \rightarrow 4H_3PO_4$$

P_4O_{10} reacts with alcohols and ethers, forming phosphate esters. (The relation of these esters to phosphor*ic* acid is shown by writing H_3PO_4 as $O{=}P(OH)_3$.)

$$P_4O_{10} + 6EtOH \rightarrow 2O{=}P(OEt)(OH)_2 + 2O{=}P(OEt)_2(OH)$$
$$P_4O_{10} + 6Et_2O \rightarrow 4O{=}P(OEt)_3$$

As_4O_{10} dissolves slowly in water, forming arsenic acid H_3AsO_4. This is tribasic, and is a much stronger acid than is arsenious acid. Salts such as lead arsenate $PbHAsO_4$ and calcium arsenate $Ca_3(AsO_4)_2$ are used as insecticides against locusts, cotton weevils, and fruit moths. Sb_4O_{10} is insoluble in water and antimonic acid is not known. Antimonates containing $[Sb(OH)_6]^-$, however, are known.

Bi does not form a pentoxide, showing that the stability of the highest oxidation states decreases on descending the group. The usual trend that higher oxidation states are more acidic is also observed.

Other oxides

The oxides P_4O_7, P_4O_8 and P_4O_9 are very uncommon. In these oxides one molecule contains P atoms in both the oxidation states (+III) and (+V). P_4O_7 is best made by dissolving P_4O_6 in tetrahydrofuran and reacting it with the correct amount of oxygen. Heating P_4O_6 under vacuum in a sealed tube gives a mixture of red phosphorus and the oxides P_4O_7, P_4O_8 and P_4O_9. These oxides have structures in between those of P_4O_6 and P_4O_{10}, in that they have one, two or three apical O atoms attached to P atoms. Hydrolysis with water thus yields a mixture of oxoacids in both oxidation states, phosphor*ic* acid P(+V) and phosphor*ous* acid P(+III).

$$\left.\begin{matrix} P_4O_8 \\ P_4O_9 \end{matrix}\right] \xrightarrow{+H_2O} \underset{\text{orthophosphoric acid}}{H_3PO_4} + \underset{\text{orthophosphorous acid}}{H_3PO_3}$$

OXOACIDS OF PHOSPHORUS

Phosphorus forms two series of oxoacids:

1. The phosphor*ic* series of acids, in which the oxidation state of P is (+V), and in which the compounds have oxidizing properties.
2. The phosphor*ous* series of acids, which contain P in the oxidation state (+III), and which are reducing agents.

In all of these, P is four-coordinate and tetrahedrally surrounded wherever possible. $p\pi-p\pi$ back bonding gives rise to P=O bonds. The hydrogen atoms in OH groups are ionizable and are acidic, but the P—H bonds found in the phosphor*ous* acids have reducing, not acidic, properties. Simple phosphate ions can condense (polymerize) together to give a wide range of more complicated isopolyacids or their salts.

THE PHOSPHORIC ACID SERIES

Orthophosphoric acids

The simplest phosphor*ic* acid is H_3PO_4 orthophosphor*ic* acid (Figure 14.12).The acid contains three replaceable H atoms, and is tribasic. It undergoes stepwise dissociation:

$$H_3PO_4 \rightleftharpoons H^+ + H_2PO_4^- \qquad K_{a1} = 7.5 \times 10^{-3}$$
$$H_2PO_4^- \rightleftharpoons H^+ + HPO_4^{2-} \qquad K_{a2} = 6.2 \times 10^{-8}$$
$$HPO_4^{2-} \rightarrow H^+ + PO_4^{3-} \qquad K_{a3} = 1 \times 10^{-12}$$

Three series of salts can be formed:

1. Dihydrogen phosphates, for example sodium dihydrogen phosphate NaH_2PO_4, which is slightly acidic in water.
2. Monohydrogen phosphates, for example disodium hydrogen phosphate Na_2HPO_4, which is slightly basic in water.
3. Normal phosphates such as trisodium phosphate Na_3PO_4, which are appreciably basic in solution.

NaH_2PO_4 and Na_2HPO_4 are made industrially by neutralizing H_3PO_4 with 'soda ash' (Na_2CO_3), but NaOH is required to make Na_3PO_4. All three salts exist in the anhydrous state and also in a number of hydrated forms, and they are used extensively.

Phosphor*ic* acid also forms esters with alcohols:

$$\underset{\text{acid}}{(HO)_3P{=}O} + \underset{\text{alcohol}}{3EtOH} \rightarrow \underset{\substack{\text{ester}\\\text{(triethyl phosphate)}}}{(EtO)_3P{=}O} + \underset{\text{water}}{3H_2O}$$

Phosphates are detected analytically by mixing a solution of the salt with dilute HNO_3 and ammonium molybdate solution. A yellow precipitate of a complex ammonium 12-molybdophosphate forms slowly, confirming the presence of phosphates. Arsenates form a similar precipitate, but only on heating the mixture.

The orthophosphates of Group I metals (except Li) and NH_4^+ are soluble in water. Most of the other metal orthophosphates are soluble in dilute HCl or acetic acids. Titanium, zirconium and thorium phosphates are insoluble even in acids. Thus in qualitative analysis a solution of zirconyl nitrate is commonly added to remove any phosphate present in solution.

Phosphates can be estimated quantitatively by adding a solution containing Mg^{2+} and NH_4OH solution to a solution of the phosphate. Magnesium ammonium phosphate is precipitated quantitatively, and this is filtered, washed, ignited, and weighed as magnesium pyrophosphate $Mg_2P_2O_7$.

$$Mg^{2+} + NH_4^+ + PO_4^{3-} \rightarrow MgNH_4PO_4$$
$$2MgNH_4PO_4 \rightarrow Mg_2P_2O_7 + 2NH_3 + H_2O$$

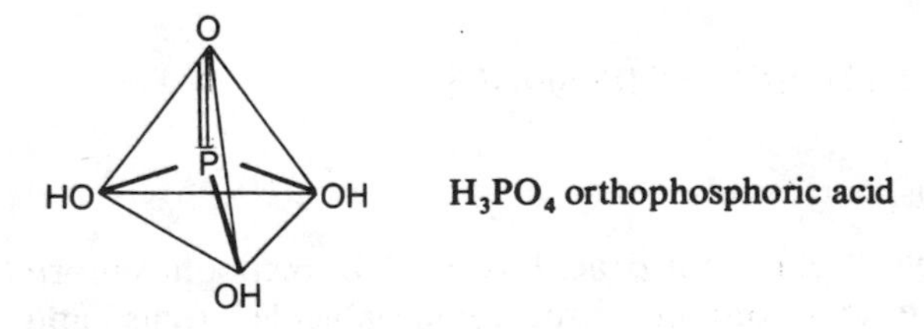

Figure 14.12 Structure of orthophosphoric acid H_3PO_4.

Impure orthophosphor*ic* acid H_3PO_4 is prepared in large amounts by treating phosphate rock with H_2SO_4. This is called the 'wet process'. The $CaSO_4$ is hydrated to gypsum $CaSO_4 \cdot 2H_2O$, which is filtered off, and the F^- is converted to $Na_2[SiF_6]$ and removed. The H_3PO_4 is concentrated by evaporation. Most of the H_3PO_4 made in this way is used to make fertilizer.

$$Ca_3(PO_4)_2 + 3H_2SO_4 \rightarrow 2H_3PO_4 + 3CaSO_4$$

$$[3(Ca_3(PO_4)_2) \cdot CaF_2] + 10H_2SO_4 + 16H_2O \rightarrow 6H_3PO_4 + 10CaSO_4 + 2HF$$

Pure H_3PO_4 is made by the 'furnace process'. Molten P is burnt in a furnace with air and steam. First P_4O_{10} is formed by reaction between P and O, and then this is immediately hydrolysed.

$$P_4 + 5O_2 \rightarrow P_4O_{10}$$

$$P_4O_{10} + 6H_2O \rightarrow 4H_3PO_4$$

Phosphor*ic* acid is hydrogen bonded in aqueous solution, and because of this the 'concentrated acid' is syrupy and viscous. Concentrated acid is widely used and contains about 85% by weight of H_3PO_4 (100% pure (anhydrous) H_3PO_4 is seldom used, but it can be prepared as colourless deliquescent crystals by evaporation at low pressure). Most of the acid (solution) made in this way is used in the laboratory, and in food and pharmaceutical preparations.

H_3PO_4 may also be made by the action of concentrated HNO_3 on P.

$$P_4 + 20HNO_3 \rightarrow 4H_3PO_4 + 20NO_2 + 4H_2O$$

Orthophosphor*ic* acid loses water steadily on heating:

$$\underset{\text{orthophosphoric acid}}{H_3PO_4} \xrightarrow[220\,^\circ C]{\text{gentle heat}} \underset{\text{pyrophosphoric acid}}{H_4P_2O_7} \xrightarrow[320\,^\circ C]{\text{strong heat}} \underset{\text{metaphosphoric acid}}{(HPO_3)_n}$$

Polyphosphates

A very large number of polyphosphor*ic* acids and their salts, the polyphosphates, arise by polymerizing acidic $[PO_4]$ units forming isopoly-acids. These consist of chains of tetrahedra, each sharing the O atoms at one or two corners of the $[PO_4]$ tetrahedron, giving simple unbranched chains, in a similar way to the formation of pyroxenes by the silicates.

The hydrolysis of P_4O_{10} proceeds in stages, and an understanding of these stages leads to an understanding of the wide range of phosphor*ic* acids (Figure 14.14).

$$P_4O_{10} + 6H_2O \rightarrow 4H_3PO_4 \quad \text{(overall reaction)}$$

Polyphosphates are straight chain compounds. The basicity of the various acids, that is the number of replaceable H atoms, can be found by drawing the structure and counting the number of OH groups. Thus

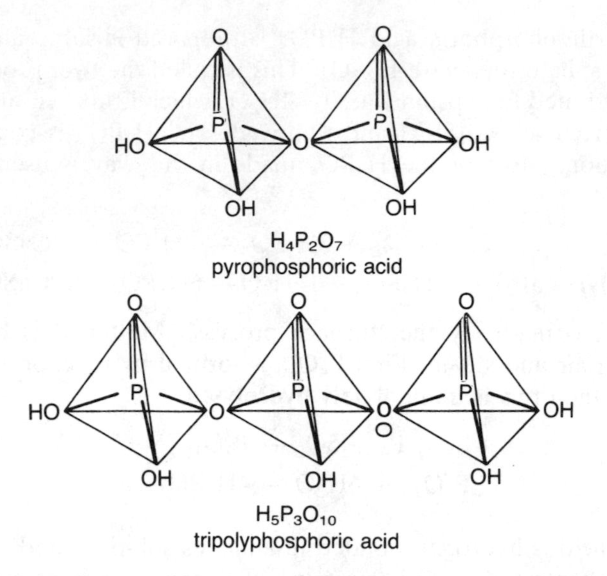

Figure 14.13 Pyrophosphoric acid $H_4P_2O_7$ and tripolyphosphoric acid $H_5P_3O_{10}$.

orthophosphor*ic* acid is tribasic, pyrophosphor*ic* acid is tetrabasic, tripolyphosphor*ic* acid is pentabasic, tetrapolyphosphor*ic* acid is hexabasic, and tetrametaphosphor*ic* acid is tetrabasic.

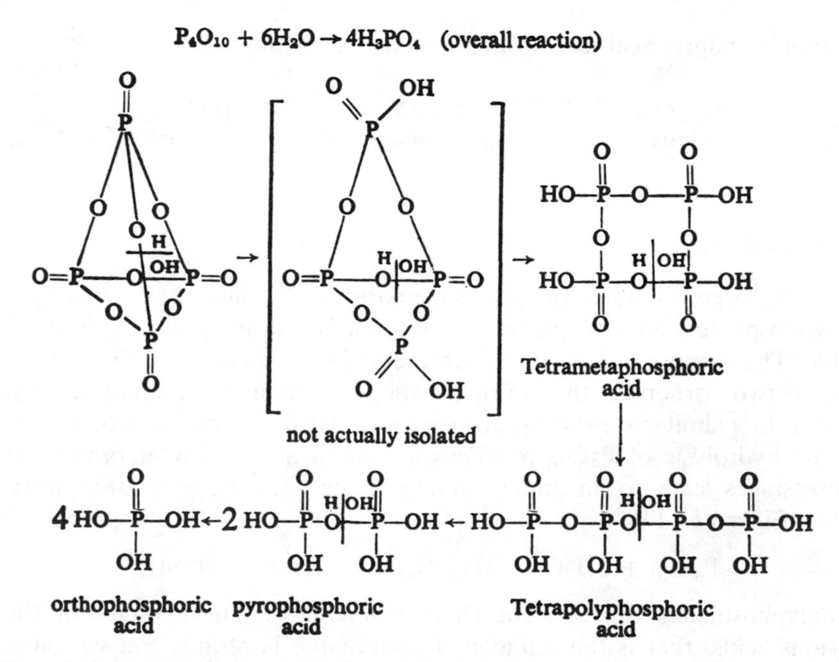

Figure 14.14 Scheme for the hydrolysis of P_4O_{10}.

Many polyphosphates are known. Chains of different lengths with up to ten [PO_4] units have been isolated, but the first four in the series are well known.

H_3PO_4	orthophosphor*ic* acid
$H_4P_2O_7$	dipolyphosphor*ic* acid (pyrophosphor*ic* acid)
$H_5P_3O_{10}$	tripolyphosphor*ic* acid
$H_6P_4O_{13}$	tetrapolyphosphor*ic* acid

Some very long chain polymers called Graham's salt, Kurrol salt and Maddrell's salt are also known. These are named after the person who first reported them, and they are discussed later.

Disodium dihydrogenpyrophosphate $Na_2H_2P_2O_7$ is mixed with $NaHCO_3$ and used in bread making to leaven the bread, that is to make it rise. They react and evolve CO_2 when heated together. This is an easier way of making batches of bread than using yeast, and is used commercially.

$$Na_2H_2P_2O_7 + 2NaHCO_3 \rightarrow Na_4P_2O_7 + 2CO_2 + 2H_2O$$

$Ca_2P_2O_7$ is used as the abrasive/polishing agent in fluoride toothpaste, and $Na_4P_2O_7$ is mixed with starch and flavouring to make 'instant pudding' mixtures.

At one time sodium pyrophosphate $Na_4P_2O_7$ was added to soap powders and solutions as a water softener, to prevent the formation of scum in hard water. For many purposes detergents, e.g. anionic and non-ionic surfactants (surface active agents), have replaced soap. Also $Na_4P_2O_7$ has been replaced by sodium tripolyphosphate $Na_5P_3O_{10}$. Between 20% and 45% of $Na_5P_3O_{10}$ is added to solid detergent powders and liquid detergents (washing up fluids etc.) used in the home and in industry. (The lower figure applies to the USA, where there have been bad experiences of extensive pollution of rivers and lakes.) Sodium tripolyphosphate is called a 'filler' because it increases the quantity of material in the packet. Its main usefulness, however, is in serving as a water softener. It does this by forming a stable soluble complex with Ca^{2+} and Mg^{2+}. This is called sequestration, and results in the effective removal (masking) of these ions which are responsible for hardness in water. Thus Ca^{2+} and Mg^{2+} ions do not form precipitates with CO_3^{2-} ions or with soap in its presence. $Na_5P_3O_{10}$ also makes the solution alkaline which helps dissolve grease and improves the action of the detergent. $Na_5P_3O_{10}$ can be prepared in the following ways:

1. The most common method of preparation is to fuse the correct quantities of Na_2HPO_4 and NaH_2PO_4. Recrystallization from water gives the hexahydrate $Na_5P_3O_{10} \cdot 6H_2O$:

$$2Na_2HPO_4 + NaH_2PO_4 \xrightarrow{450\,°C} Na_5P_3O_{10} + 2H_2O$$

2. In Germany it is largely made by fusing Na_2O and P_4O_{10}. On cooling, the pyrophosphate $Na_2P_2O_7$ crystallizes out first, but with slow cooling this changes into $Na_5P_3O_{10}$:

$$10Na_2O + 3P_4O_{10} \xrightarrow[\text{cool slowly}]{1000\,°C} 4Na_5P_3O_{10}$$

Long chain polyphosphates – linear metaphosphates

The very long chain polyphosphates have caused confusion in the past since they were originally called metaphosphates, a name used for ring compounds. When the number of units in the polymer n becomes very large, the formula of a chain polyphosphate $(PO_3)_n \cdot PO_4$ becomes indistinguishable from that of a true metaphosphate, that is a ring compound with a formula $(PO_3)_n$. The long chains are sometimes called linear metaphosphates.

Graham's salt is the best known of these long chain polyphosphates, and is formed by quenching molten $NaPO_3$. It forms a glassy solid instead of crystallizing. In industry it is incorrectly called sodium hexametaphosphate. This is wrong because it does not contain six $[PO_4]$ units and is a high molecular weight polymer $(NaPO_3)_n$, which usually has a mean molecular weight of 12 000–18 000, and up to 200 $[PO_4]$ units in the chain. Though mainly made up of long chains, it does contain up to 10% of ring metaphosphates and a little cross-linked material. (Molecular weights of these long chain polymeric species can be determined by titrating the end groups, and also from osmotic pressure, diffusion, viscosity, electrophoresis, and ultracentrifuge measurements.) Graham's salt is soluble in water. These solutions give precipitates with metal ions such as Pb^{2+} and Ag^+, but not with Ca^{2+} and Mg^{2+}. Graham's salt is sold commercially under the trade name Calgon. It is widely used for softening water. It sequesters Ca^{2+} and Mg^{2+} in a similar way to $Na_5P_3O_{10}$. Many of these polyphosphates are used for water softening, and also for descaling boilers and pipes.

Heating $Na_2H_2P_2O_7$ results in dehydration, but three different products are formed depending on the vapour pressure of water. If heated in air (an open system), where water can escape, then cyclic sodium trimetaphosphate is formed. Heating in a closed system, where the water cannot escape, yields either Maddrell's high temperature or low temperature salt. These are crystalline, as is Kurrol's salt. They consist of chains of tetrahedral $[PO_4]$ units, and they differ in the way the tetrahedra are oriented in the chains. Thus Kurrol's salt is made up of helical chains of $[PO_4]$ units, and the structure contains an equal number of left handed and right handed helices. Thus chains may differ in their length and they may also have different repeat units, as was found in the chain silicates.

These and other relationships are shown in (Figure 14.15).

When (cyclic) sodium trimetaphosphate melts at about 625 °C, long chain polyphosphates are formed. If the liquid is cooled rapidly these chains remain (Graham's salt). Annealing Graham's salt above 550 °C gives Kurrol's salt. This exists in two forms, one fibrous and the other plate-like. They have different densities. The two different forms are

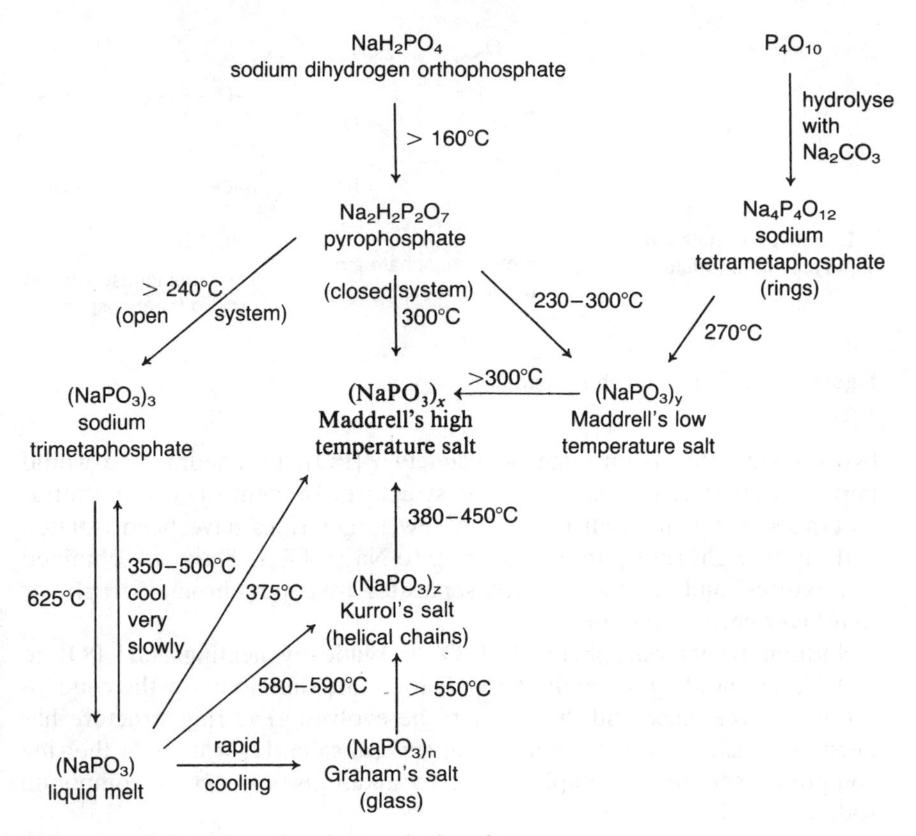

Figure 14.15 Relation of various polyphosphates.

similar to the asbestos minerals in the silicates, some of which are made of chains, and others of sheets. Annealing one form of Kurrol salt at 400 °C gives sodium trimetaphosphate, and the other gives Maddrell's high temperature salt. All forms of sodium polyphosphate revert to (cyclic) sodium trimetaphosphate near the melting point 625 °C, or on annealing (prolonged heating) at 400 °C. This is presumably because the trimetaphosphate has the most stable crystal structure.

Metaphosphates – cyclophosphates

The metaphosphates form a family of ring compounds. The old name of metaphosphates is still widely used even though according to IUPAC nomenclature cyclo- should be used to indicate the formation of rings. They can be prepared by heating orthophosphates:

$$nH_3PO_4 \xrightarrow{\text{heat } 316\,°C} (HPO_3)_n + nH_2O$$

There is no evidence for the existence of free monometaphosphate ions PO_3^-, or of dimetaphosphate ions. The latter would involve the sharing of

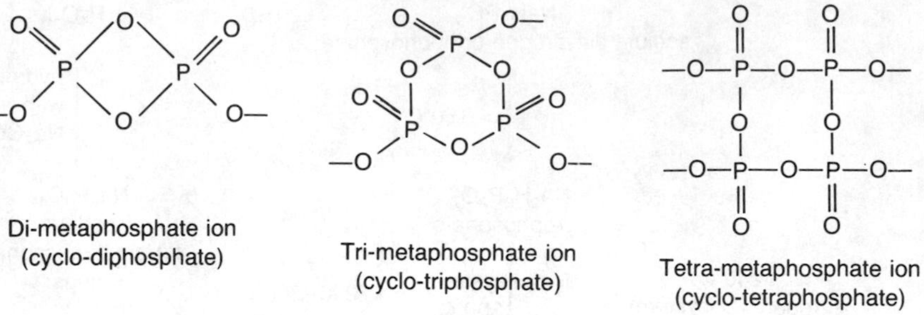

Di-metaphosphate ion (cyclo-diphosphate)

Tri-metaphosphate ion (cyclo-triphosphate)

Tetra-metaphosphate ion (cyclo-tetraphosphate)

Figure 14.16 Some polyphosphate ions.

two corners, that is an edge between two $[PO_4]$ tetrahedra, and would impose a great deal of strain on the structure. In contrast, tri- and tetra-metaphosphates are well known. A few larger rings have been isolated with up to eight $[PO_4]$ units, that is up to $Na_8[P_8O_{24}]$. These are obtained as mixtures, and are conveniently separated by paper chromatography or thin layer chromatography.

Sodium trimetaphosphate $Na_3P_3O_9$ is made by heating NaH_2PO_4 to 640 °C, and holding the melt at 500 °C for some time to allow the condensation to take place and the water to be evolved. The ring structure has been established by X-ray analysis of several salts. Hydrolysis of the ring compound sodium trimetaphosphate by alkali gives the chain compound sodium tripolyphosphate.

$$3NaH_2PO_4 \xrightarrow{\text{heat}} Na_3P_3O_9 + 3H_2O$$

$$Na_3P_3O_9 + 2NaOH \rightarrow Na_5P_3O_{10} + H_2O$$

Sodium tetrametaphosphate $Na_4P_4O_{12} \cdot 4H_2O$ is formed when P_4O_{10} is treated with a solution of cold NaOH or $NaHCO_3$.

Hypophosphoric acid $H_4P_2O_6$

This contains P in the oxidation state (+IV) and has one less O atom than pyrophosphor*ic* acid $H_4P_2O_7$. It is prepared by hydrolysis and oxidation of

$4H_2O$ → 2[H(OH)P–P(OH)H] $\xrightarrow{4O_2}$ 2O=P(OH)2–P(OH)2=O

hypophosphoric acid

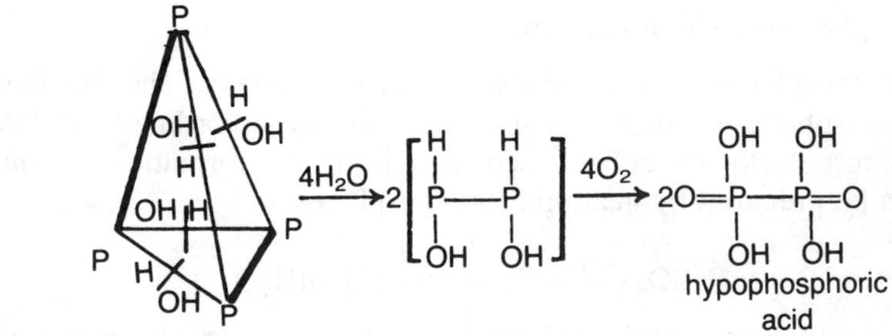

Figure 14.17 Hydrolysis and oxidation of yellow phosphorus.

red phosphorus by NaOCl, or yellow phosphorus by water and air. There are no P—H bonds, and so this acid is not a reducing agent. There are four acidic hydrogens, and hence the acid is tetrabasic and can form four series of salts, though usually two hydrogens are replaced. It is unusual in that it contains a P—P bond. This is much stronger than the P—O—P bond, so hydrolysis is slow.

$$(HO)_2P(=O)—P(=O)(OH)_2 + H|OH \rightarrow O{=}P(OH)_2—H + HO—P(=O)(OH)_2$$

H_3PO_3 orthophosphor*ous* acid; H_3PO_4 orthophosphor*ic* acid

THE PHOSPHOROUS ACID SERIES

The phosphor*ous* acids are less well known. They all contain phosphorus in the oxidation state (+III). They have P—H bonds and are therefore reducing agents.

Hydrolysis of P_4O_6 in a manner analogous to the hydrolysis of P_4O_{10} already described yields pyro- and orthophosphor*ous* acids, which are both dibasic and reducing agents.

$$HO—P(H)(=O)—O—P(H)(=O)—OH \qquad H—P(H)(=O)—OH$$

pyrophosphor*ous* acid; orthophosphor*ous* acid

Orthophosphorous acid H_3PO_3

H_3PO_3 contains two acidic H atoms (the OH groups), and one reducing H (the P—H hydrogen atom). Consequently only two of the three H atoms can ionize, and the acid is dibasic.

$$H_3PO_3 \rightleftharpoons H^+ + H_2PO_3^- \qquad K_{a1} = 1.6 \times 10^{-2}$$
$$H_2PO_3^- \rightleftharpoons H^+ + HPO_3^{2-} \qquad K_{a2} = 7 \times 10^{-7}$$

Thus H_3PO_3 can form two series of salts:

1. Dihydrogen phosphites, for example NaH_2PO_3.
2. Monohydrogen phosphites, for example Na_2HPO_3.

The phosphites are very strong reducing agents in basic solutions. In acid solutions they are converted to H_3PO_3, which is still a moderately strong reducing agent.

Metaphosphorous acid $(HPO_2)_n$

This can be prepared from phosphine at low pressure.

$$PH_3 + O_2 \xrightarrow{25\,\text{mm Hg}} H_2 + HPO_2$$

If the formula were HPO_2, the P atom would only form three bonds or else form double bonds. In fact, it polymerizes rather than form double bonds. The structure is not known, but by analogy with metaphosphor*ic* acid it may well be a ring structure.

Hypophosphorous acid H_3PO_2

H_3PO_2 contains P in the oxidation state (+I), and has one O atom less than the orthophosphor*ous* acid. It is prepared by alkaline hydrolysis of phosphorus.

$$P_4 + 3OH^- + 3H_2O \rightarrow PH_3 + 3H_2PO_2^-$$

The acid is monobasic and a very strong reducing agent. Salts of this acid are called hypophosphites, and sodium hypophosphite NaH_2PO_2 is used industrially to bleach wood and to make paper.

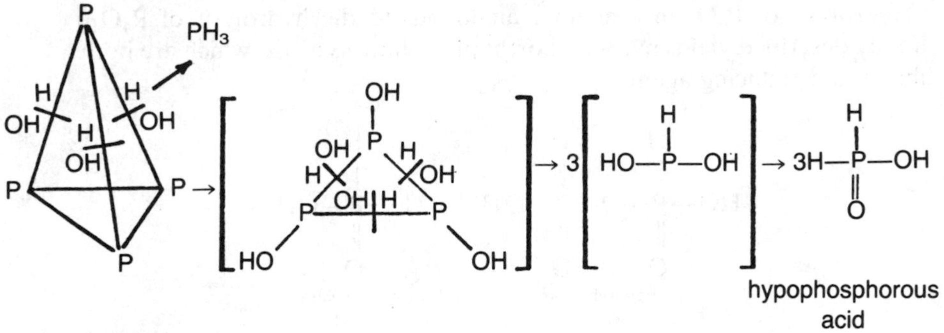

Figure 14.18 Alkaline hydrolysis of phosphorus.

MAJOR USES OF PHOSPHATES

Phosphate rock is mined on a vast scale (159 million tonnes in 1988). The minerals vary both in purity and in composition. Industry expresses the production of phosphates in terms of the P_2O_5 content. On this basis world production of phosphates is about 49 million tonnes per year. (This is equivalent to 69 million tonnes of H_3PO_4.) The major commercial uses are as follows:

85% For fertilizers such as superphosphate, triple superphosphate, and ammonium phosphate. These do not need to be especially pure.

5% Added to detergents (builders, i.e fillers) mainly sodium tripolyphosphate in powders and sodium pyrophosphate in liquid preparations.

3% Used in the food industry to give the acid taste in drinks such as cola (pH 2), sarsaparilla and root beer, and as an emulsifier in processed cheese, dried milk, sausages etc.

2½% For treating metals.

(a) Rustproofing, by dipping the hot metal into phosphoric acid, or heating the acid to 90–95 °C (sometimes with Zn^{2+}, Mn^{2+} Cu^{2+} or other ions present) in processes such as Parkerizing and Bonderizing. Small metal parts such as nuts, bolts and screws are treated in this way, and also motor car bodies, refrigerators etc. before they are painted.

(b) Pickling metals, that is removing scale and oxide from the surface of iron and steel by dipping in an acid bath.

(c) 'Bright dipping' of aluminium parts. The parts are connected to the anode and electrolysed in a bath of H_3PO_4 with a small amount of HNO_3 and a trace of $Cu(NO_3)_2$. This gives a highly polished Al surface protected by a clear layer of Al_2O_3.

1% For industrial uses such as water softening (particularly calgon, and trisodium phosphate Na_3PO_4), buffers (NaH_2PO_4 and Na_2HPO_4), paint strippers (Na_3PO_4), and removing H_2S from gases particularly in the petroleum industry (K_3PO_4).

1% For making phosphorus sulphides (for matches).

1% For making organophosphorus compounds: plasticizers (triaryl phosphates), insecticides (triethyl phosphate) and petrol additives (tritollyl phosphate).

1% For pharmaceutical products such as toothpaste ($CaHPO_4 \cdot 2H_2O$, or $Ca_2P_2O_7$ in fluoride toothpaste), and combined baking powder ($Ca(H_2PO_4)_2$ which is slightly acidic, mixed with $NaHCO_3$).

½% Flameproofing fabrics (ammonium phosphates and urea phosphate $NH_2CONH_2 \cdot H_3PO_4$).

The excessive use of phosphates as water softeners is criticized by environmentalists, since it contributes to water pollution. The phosphates in domestic waste water pass through sewage disposal systems into rivers and lakes. There they nourish bacteria, which grow excessively and deplete the water of dissolved oxygen, thus killing the fish. The phosphates may also produce a massive overgrowth of water plants. When this crop of plants dies, there will be excessive decay and putrefaction which may also kill the fish. The tendency of the acidic ions of P to condense and give isopolyacids is quite strong. The phosphates and phosphites are similar to the arsenates and arsenites. Condensed As anions are much less stable than the corresponding P polyanions and they are rapidly hydrolysed in water. Antimonates and antimonites are known, but Sb has a coordination number of 6, and these salts contain the octahedral $[Sb(OH)_6]^-$ ion.

SULPHIDES OF PHOSPHORUS

When P and S are heated together to a temperature over 100 °C, P_4S_3, P_4S_5, P_4S_7 and P_4S_{10} may be formed depending on the relative amounts of

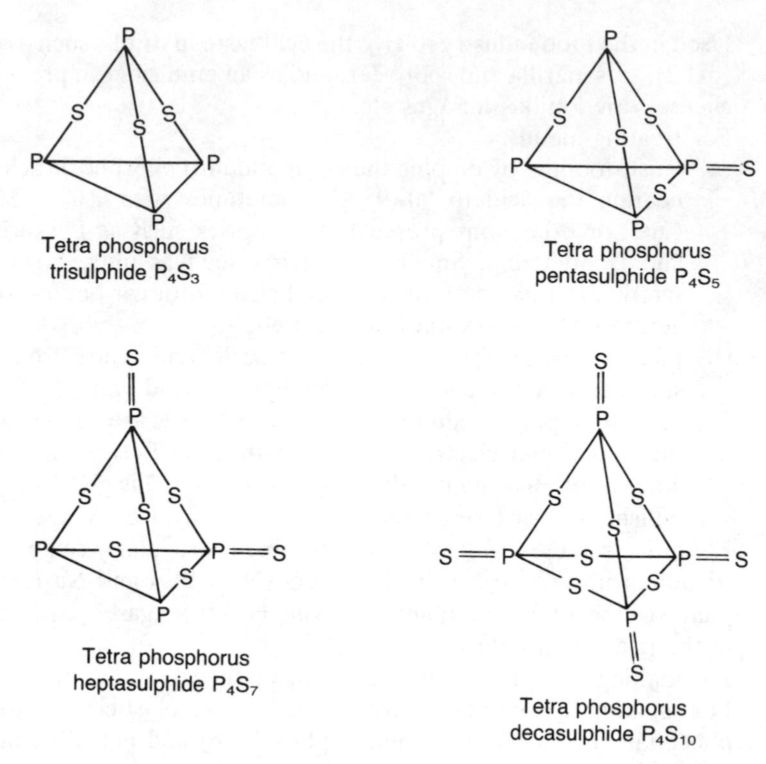

Figure 14.19 Structures of phosphorus sulphides.

reactants present. Two more compounds P_4S_4 and P_4S_9 have been made using other reactions.

P_4S_{10} is structurally the same as P_4O_{10}, but the absence of P_4S_6 is surprising. The structures of the other sulphides have no counterparts in the oxides. They are loosely related to the structures of the oxides P_4O_6 and P_4O_{10}, with a tetrahedron of P atoms, with some S atoms bridging between P atoms and others occupying apical positions attached to P atoms.

P_4S_3

Phosphorus trisulphide P_4S_3 is the most stable sulphide. It is made by heating red phosphorus and a limited amount of sulphur to 180 °C in an inert atmosphere. It is soluble in organic solvents such as toluene and carbon disulphide: traces of unreacted P can be removed either by recrystallization from toluene or by distillation. P_4S_3 is used commercially for making matches. Matches contain P_4S_3, $KClO_3$, fillers and glue. The friction between the match and the sandpaper on the side of the box initiates a violent reaction between the P_4S_3 and $KClO_3$. This generates enough heat to make the match burst into flame.

P_4S_{10}

P_4S_{10} is the most important sulphide. It is made by reacting liquified white phosphorus at 300 °C with a slight excess of sulphur. World production is about 250 000 tonnes/year. It hydrolyses in water, forming phosphor*ic* acid in a similar way to P_4O_{10}.

$$P_4S_{10} + 16H_2O \rightarrow 4H_3PO_4 + 10H_2S$$

The most important reaction of P_4S_{10} is hydrolysis by alcohols and phenols to give dialkyl or diaryl dithiophosphor*ic* acids.

$$P_4S_{10} + 8EtOH \rightarrow 4(EtO)_2 \cdot P \cdot (S) \cdot SH + 2H_2S$$

The Zn salts of dialkyl and diaryl thiophosphates $[(RO)_2 \cdot P \cdot (S)]_2Zn$ are used as extreme pressure additives in high pressure lubricants such as gearbox oil. $(EtO)_2 \cdot P \cdot (S) \cdot Na$ and $(EtO)_2 \cdot P \cdot (S) \cdot NH_4$ are used as flotation agents for concentrating sulphide ores such as PbS and ZnS before smelting. The methyl and ethyl derivatives are used in the manufacture of pesticides such as malathion and parathion.

$$(EtO)_2 \cdot P \cdot (S) \cdot SH + Cl_2 \rightarrow (EtO)_2 \cdot P \cdot (S) \cdot Cl + HCl$$

$$(EtO)_2 \cdot P \cdot (S) \cdot Cl + NaO \cdot C_6H_4 \cdot NO_2 \rightarrow \underset{\text{parathion}}{(EtO)_2 \cdot P \cdot (S) \cdot O \cdot C_6H_4 \cdot NO_2}$$

These organophosphorus esters are very effective insecticides. They prevent the nervous system of insects from working properly, thus killing the insects very rapidly. Acetylcholine is a chemical neurotransmitter, produced to transmit nerve impulses across a synaptic junction. Normally the enzyme acetylcholinesterase destroys the acetylcholine once the im-

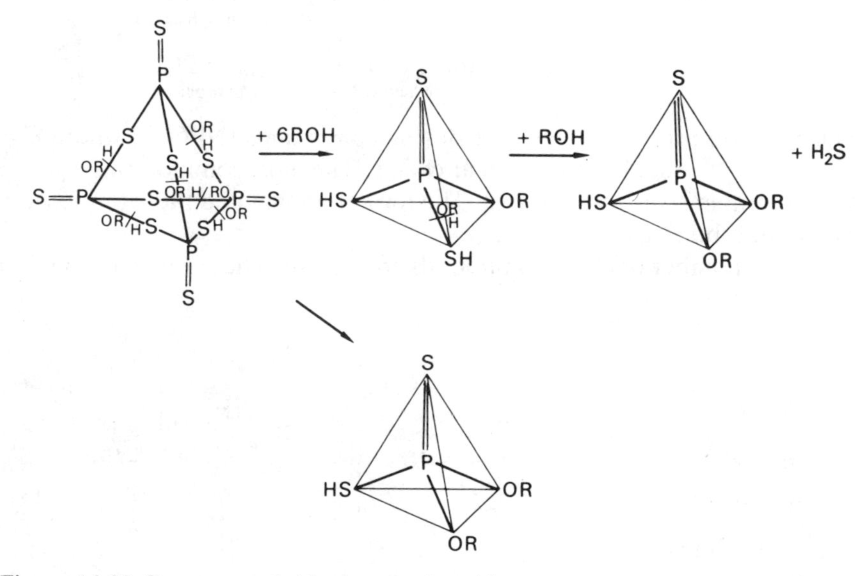

Figure 14.20 Structures of thiophosphoric acids.

pulse has been transmitted. These organophosphorus esters inhibit the action of acetylcholinesterase. Malathion and parathion are not toxic if eaten by mammals, since the digestive system breaks the molecule down before it enters the body.

PHOSPHAZENES AND CYCLOPHOSPHAZENES (PHOSPHONITRILIC COMPOUNDS)

Nitrogen and phosphorus show only a slight tendency to catenate *by themselves*. The maximum chain length for nitrogen is three in the azide N_3^- ion, and two for phosphorus in a few compounds such as P_2H_4 and $(Me_2)(S)P—P(S)(Me_2)$. A few ring compounds exist with four, five or six P or As atoms joined together.

In contrast to this, N and P may bond *together*, forming a large number of phosphazenes. In these the P atom is in the oxidation state (+V) and N is in the (+III) state. The compounds are formally unsaturated. Thus monophosphazines may be made by reacting an azide with PCl_3, POR_3 or $P(C_6H_5)_3$:

$$PCl_3 + C_6H_5N_3 \rightarrow Cl_3P{=}NC_6H_5 + N_2$$
$$P(C_6H_5)_3 + C_6H_5N_3 \rightarrow (C_6H_5)_3P{=}NC_6H_5 + N_2$$

Diphosphazenes can be made as follows:

$$3PCl_5 + 2NH_4Cl \rightarrow [Cl_3P{=}N{-}PCl_2{=}N{-}PCl_3]^+Cl^- + 8HCl$$

However, N and P catenate together, forming an interesting series of polymers.

$$nPCl_5 + nNH_4Cl \xrightarrow{120-150\,°C} \underset{\text{(ring compounds cyclophosphazenes)}}{(NPCl_2)_n + 4nHCl}$$

$$\text{and} \quad \underset{\text{(chain compounds polyphosphazenes)}}{Cl_4P\cdot(NPCl_2)_n\cdot NPCl_3}$$

This reaction produces a mixture of ring compounds $(NPCl_2)_n$ where n = 3, 4, 5, 6. . . , and fairly short linear chains. The most common rings (n = 3 and 4) contain six or eight atoms. The former are flat and the latter exist in 'chair' and 'boat' conformations.

A large number of chain compounds are known. These range from short

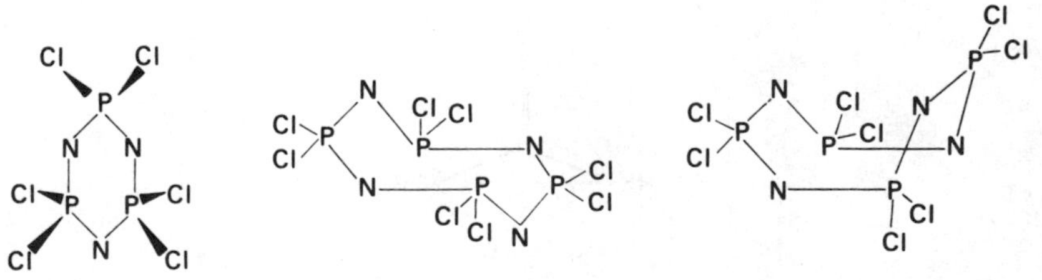

Figure 14.21 Some cyclophosphazene compounds.

chains P_2NCl_7, $P_3N_2Cl_9$, $P_4N_3Cl_{11}$ to those with up to 10^4 units $[—N{=}PCl_2^-]$ linked together. These compounds were originally called phosphonitrilic halides, but are now named systematically poly(chlorophosphazenes).

The chlorine atoms are reactive, and most reactions of chlorophosphazenes involve replacement of Cl by groups such as alkyl, aryl, OH, OR, NCS or NR_2. Alkyl or aryl groups may be introduced using lithium or Grignard reagents. Substitution may be complete, or partial, and in the latter case many different isomers are possible.

$$[NPCl_2]_3 + 6CH_3MgI \rightarrow [NP(CH_3)_2]_3 + 3MgCl_2 + 3MgI_2$$
$$[NPCl_2]_3 + 6C_6H_5Li \rightarrow [NP(C_6H_5)_2]_3 + 6LiCl$$
$$[NPCl_2]_3 + 6NaOR \rightarrow [NP(OR)_2]_3 + 6NaCl$$
$$[NPCl_2]_3 + 6NaSCN \rightarrow [NP(SCN)_2]_3 + 6NaCl$$

Similar compounds are formed with Br and F. The largest rings formed contain 34 atoms in the chlorides and 12 atoms in the bromides. Some of the long chain polymers are rubber-like, and those with perfluoroalkoxy side groups $[NP(OCH_2CF_3)_2]_n$ resemble polythene.

There are many potential uses for the high molecular weight phosphazenes, as rigid plastics, expanded foam, and fibres, since they are waterproof and fireproof, and are unaffected by petrol, oil, and solvents. They also form flexible plastics which are useful for fuel hoses and gaskets since they retain their elasticity at low temperatures. The phosphazenes are at present far too expensive for general use. Thin films of poly(aminophosphazene) are used in hospitals to cover severe burns and other extensive wounds since they prevent the loss of body fluids and keep germs out.

There are two main points of interest in these P—N compounds:

1. The nature of the bonding is not understood. In all of these phosphazene compounds the (apparent) P—N and P=N bonds are equivalent. Their bond lengths are 1.56–1.59 Å, which is much shorter than the usual single bond distance of 1.77 Å. Thus the bonding in these compounds is not adequately represented by a system of alternate single

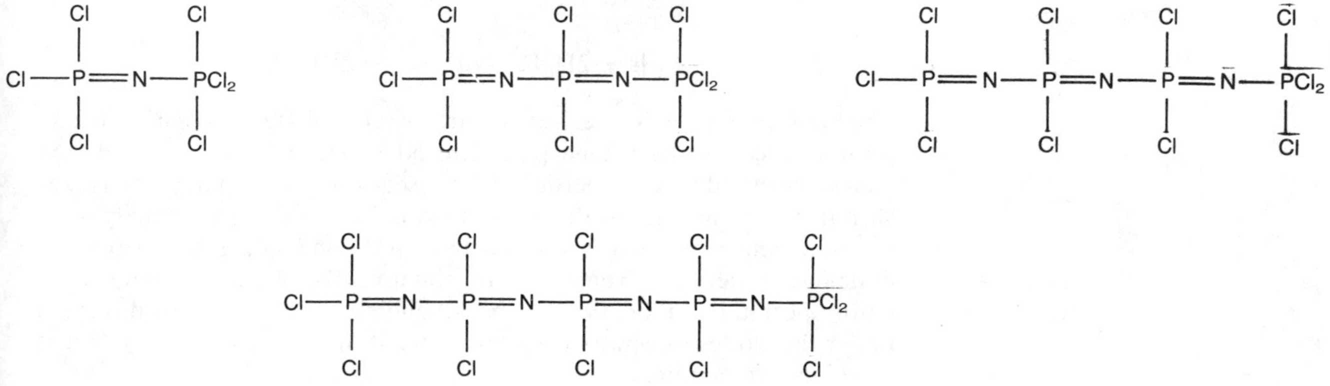

Figure 14.22 Some polyphosphazene chain compounds.

and double bonds, nor can it be explained by $p\pi-p\pi$ bonding and delocalization similar to that in benzene or graphite. It has been suggested that a coordinate bond is formed between a filled sp^2 orbital on N and the empty $3d_{x^2-y^2}$ orbital on P. This is similar to the $p\pi-d\pi$ bonding in the oxides of phosphorus except that in the phosphazenes the π bonds are delocalized over the whole molecule, giving pseudo-aromatic character. There are objections to this because of the size and energy of the d orbitals. Alternatively the singly occupied p_z orbital on N may form a three-centre bond with the d_{xz} and d_{yz} orbitals on the two adjacent P atoms.

2. The polyphosphazenes form a very extensive series of polymers. Carbon based polymers are the most extensive, then the silicones, then the phosphazenes.

ORGANOMETALLIC COMPOUNDS

Nitrogen forms primary, secondary and tertiary amines RNH_2, R_2NH and R_3N which are described fully in organic chemistry texts.

Many organophosphorus compounds are toxic. Some have been used as pesticides, herbicides, and nerve gases. Others play an essential part in life.

The halides of P, As, Sb and Bi react readily with lithium reagents and Grignard reagents, forming alkyl and aryl compounds. The best known are the tertiary phosphine compounds such as triphenyl phosphine.

$$PCl_3 + 3LiEt \rightarrow \underset{\text{triethyl phosphine}}{PEt_3} + 3LiCl$$

$$PCl_3 + 3PhMgCl \rightarrow \underset{\text{triphenyl phosphine}}{PPh_3} + 3MgCl_2$$

The trimethyl derivatives of P, As, Sb and Bi are all attacked by air, but the triaryl compounds are stable. It is not necessary to substitute all three halogen atoms, and mixed halo organo species can be made either by using an excess of PCl_3, or by using a weaker alkylating or arylating agent.

$$\underset{\text{excess}}{PCl_3} + LiEt \rightarrow EtPCl_2 + LiCl$$

$$PCl_3 + 2HgR_2 \rightarrow R_2PCl + 2RHgCl$$

The structures of MR_3 derivatives are pyramidal (tetrahedral with one position occupied by a lone pair) like NH_3. The trialkyls of P and As possess strong donor properties and consequently form many complexes with transition metals. In these a σ bond is formed using the lone pair of electrons and a π bond is formed by 'back bonding', arising from the donation of electrons from a full d orbital on the transition metal to an empty d orbital on P or As. This 'back bonding' is similar to that discussed under the oxides of phosphorus, but since it involves two d orbitals it is called $d\pi-d\pi$ bonding.

Some MR_5 derivatives can be made in a similar way, and their structures are similar to that of PCl_5, that is a trigonal bipyramidal structure. It is rare to have five organic groups bonded to P.

$$PCl_5 + C_6H_5Li \rightarrow P(C_6H_5)Cl_4 + LiCl$$
$$PCl_5 + 2C_6H_5Li \rightarrow P(C_6H_5)_2Cl_3 + 2LiCl$$
$$PCl_5 + 3C_6H_5Li \rightarrow P(C_6H_5)_3Cl_2 + 3LiCl \quad \text{etc.}$$

Several ions NR_4^+, PR_4^+, AsR_4^+ and SbR_4^+ exist which are tetrahedral like the ammonium ion.

Treatment of $POCl_3$ with lithium or Grignard reagents yields trialkyl and triaryl phosphine oxides.

$$POCl_3 + 3LiR \rightarrow POR_3 + 3LiCl$$

Phosphate esters play a vital role in many life processes:

1. The release of energy in living matter by adenosine triphosphate (ATP → ADP + energy) has been described earlier. Nicotinomide adenine dinucleotide (NAD) is important in the degradation of citric acid in the Krebs' cycle for the release of energy. Another ester, phosphocreatine, is important in the regeneration of ATP, and others control the synthesis and storage of carbohydrates such as glycogen in animals.
2. Phosphate esters are also important in the synthesis of proteins and nucleic acids. Deoxyribonucleic acids, DNA, are responsible for the storage and transfer of genetic information. The sequence of organic bases is specific for each nucleic acid. The DNA molecules comprise two strands which are hydrogen bonded together and form a double helix. Ribonucleic acids, RNA, are similar, but are usually single strands, and form a single helix. Their function is to act as a template to produce identical nucleic acids, with the same sequence of bases, and the same orientation in space.
3. Phosphate esters also play a part in photosynthesis, and the conversion of surplus sugar into starch in plants.
4. Phosphate esters are also involved in nitrogen fixation.

FURTHER READING

Addison, C.C. (1980) Dinitrogen tetroxide, nitric acid, and their mixtures as media for inorganic reactions, *Chem. Rev.*, **80**, 21–39.

Addison, C.C., Logan, N., Wallwork, S.C. and Garner, C.D. (1971) Structural aspects of coordinated nitrate groups, *Q. Rev. Chem. Soc.*, **25**, 289–322.

Allcock, H.R. (1972) *Phosphorus Nitrogen Compounds*, Academic Press, London.

Allcock, H.R. (1985) Inorganic macromolecules, *Chem. Eng. News*, March 18 issue, 22–36.

Arena, B.J. (1986) Ammonia: confronting a primal trend, *J. Chem. Ed.*, **63**, 1040–1043.

Aylett, B.J. (1979) Arsenic, antimony and bismuth, *Organometallic Compounds*, 4th ed., Vol. I, Part 2, Chapman and Hall, London.

Baudler, M. (1987) Polyphosphorus compounds – new results and new perspectives, *Angewandte Chemie*, International Edition, **26**, 419–441.

Bergersen, F.J. and Postgate, J.R. (eds) (1987) *A Century of Nitrogen Fixation Research*, Royal Society, London.

Bossard, G.E. *et al.* (1983) Reactions of coordinated dinitrogen, *Inorg. Chem.*, **22**, 1968–1970.

Bottomley, F. and Burns, R.C. (1979) *Treatise on Dinitrogen Fixation*, Wiley, New York.

Broughton, W.J. and Phler, A. (1986) *Nitrogen Fixation*, Clarendon, Oxford.

Cardulla, F. (1983) Hydrazine, *J. Chem. Ed.*, **60**, 505–508.

Chatt, J., Dilworth, J.R. and Richards, R.L. (1978) Recent advances in the chemistry of nitrogen fixation, *Chem. Rev.*, **78**, 589–625.

Chatt, J., da C. Pina, L.M. and Richards, R.L. (eds) (1980) *New Trends in the Chemistry of Nitrogen Fixation*, Academic Press, London and New York. (Conference papers).

Coates, G.E. and Wade, K. (1967) Antimony and bismuth, *Organometallic Compounds*, 3rd ed., Vol. I (Chapter 5), Methuen, London.

Colburn, C.B. (ed.) (1966) *Developments in Inorganic Nitrogen Chemistry*, Vol. I (Chapter 2 by Yoffe, A.D., The Inorganic Azides; Chapter 5 by Nielsen, M.L., Phosphorus-nitrogen compounds), Elsevier, Amsterdam.

Corbridge, D.E. (1985) *Phosphorus: An Outline of Its Chemistry, Biochemistry and Technology*, 3rd ed. (No. 6, Studies in Inorganic Chemistry), Elsevier, Oxford.

Emsley, J. and Hall, D. (1976) *The Chemistry of Phosphorus*, Harper and Row, New York.

Evans, H.J., Bottomley, P.J. and Newton, W.E. (eds) (1985) *6th International Symposium on Nitrogen Fixation* (held at Corvallis, Or.), Dordrecht, Lancaster, Nijhoff.

Eysseltov, J. and Dirkse, T.P. (eds) (1988) *Alkali Metal Orthophosphates*, Pergamon.

Gallon, J.R. and Chaplin, A.E. (1987) *An Introduction to Nitrogen Fixation*, Cassell.

Glidewell, C. (1990) The nitrate/nitrite controversy, *Chemistry in Britain*, **26**, 26–30. (On the harmful effects of nitrates in drinking water.)

Goldwhite, H. (1981) *Introduction to Phosphorus Chemistry (Cambridge texts in Chemistry and Biochemistry), Cambridge University Press, Cambridge.*

Griffith, E.J. (1975) The chemical and physical properties of condensed phosphates, Pure Appl. Chem., **44**, 173–200.

Griffith, W.P. (1968) Organometallic nitrosyls, *Adv. Organometallic Chem.* **7**, 211.

Griffith, W.P. (1973) *Comprehensive Inorganic Chemistry*, Vol. 4 (Chapter 46: Carbonyls, cyanides, isocyanides and nitrosyls), Pergamon Press, Oxford.

Hamilton, C.L. (ed.) (1973) *Chemistry in the Environment* (Chapter 5 by Delwiche, C.C., The Nitrogen Cycle), Readings from Scientific American, W.H. Freeman, San Francisco.

Heal, H.G. (1980) *The Inorganic Heterocyclic Chemistry of Sulphur, Nitrogen and Phosphorus*, Academic Press, London.

Henderson, R.A., Leigh, G.J. and Pickett, C.J. (1983) The chemistry of nitrogen fixation and models for the reactions of nitrogenase, *Adv. Inorg. Chem. Radiochem.*, **27**, 197–292.

Hoffmann, H. and Becke-Goehring, M. (1976) Phosphorus sulphides, *Topics Phosphorus Chem.*, **8**, 193–271. (A comprehensive review with over 400 references.)

Holm, R.H. (1981) Metal clusters in biology: quest for synthetic representation of the catalytic site of nitrogenase, *Chem. Soc. Rev.*, **10**, 455–490.

Jander, J. (1976) Recent chemistry and structure investigation of NI_3, NBr_3, NCl_3 and related compounds, *Adv. Inorg. Chem. Radiochem.*, **19**, 1–63.

Johnson, B.F.G. and McCleverty, J.A. (1966) Nitric oxide compounds of transition metals, *Progr. Inorg. Chem.*, **7**, 277.

Jolly, W.L. (1972) *Metal Ammonia Solutions*, Dowden, Hutchinson and Row, Stroudburg, PA.

Jones, K. (1973) *Comprehensive Inorganic Chemistry*, Vol. II (Chapter 19: Nitrogen), Pergamon Press, Oxford.
Kanazawa, T. (ed.) (1989) *Inorganic Phosphate Materials*, Kodansha, Tokyo.
Kulaev, I.S. (1980) *The Biochemistry of Inorganic Polyphosphates*, John Wiley, Chichester.
Lagowski, J. (ed.) (1967) *The Chemistry of Non-aqueous Solvents*(Chapter 4 by Lee, W.H., Nitric Acid; Chapter 7 by Lagowski, J.J. and Moczygemba, G.A., Liquid ammonia, Academic Press, New York.
Lee, J.A., Rorison, I.H. and McNeill, S. (eds) (1983) *Nitrogen as an Ecological Factor* (22nd symposium of the British Ecological Society, Oxford, 1981), Blackwell Scientific Publications, Oxford.
Lieu, N.H. et al. *(1984) Reduction of molecular nitrogen in molybdenum (III–V) hydroxide/titanium(III), Inorg. Chem.*, **23**, 2772–2777.
McAuliffe, C.A. and Levason, W. (1979) *Phosphine, Arsine and Stibine Complexes of the Transition Metals*, Elsevier, Amsterdam.
Nicholls, D. (1979) *Inorganic Chemistry in Liquid Ammonia* (Topics in Inorganic and General Chemistry, Monograph 17), Elsevier.
Postgate, J.R. (1982) *The Fundamentals of Nitrogen Fixation*, Cambridge University Press, Cambridge.
Richards, R.L. (1979) Nitrogen fixation, *Education in Chemistry*, **16**, 66–69.
Richards, R.L. (1988) Biological nitrogen fixation, *Chemistry in Britain*, **24**, 133–134, 136.
Schmidt, E.W. (1984) *Hydrazine and its Derivatives: Preparation, Properties Applications*, John Wiley, New York and Chichester.
Smith, J.D. (1973) *Comprehensive Inorganic Chemistry*, Vol. II (Chapter 21: Nitrogen), Pergamon Press, Oxford.
Toy, A.D.F. (1973) *Comprehensive Inorganic Chemistry*, Vol. II (Chapter 20: Chemistry of phosphorus), Pergamon Press, Oxford.
Toy, A.D.F. (1975) *Chemistry of Phosphorus*, Pergamon Press, New York.
Thompson, R. (ed.) (1986) *The Modern Inorganic Chemicals Industry* Chapter by Grant, W.J. and Redfearn, S.L., Industrial gases; chapter by Andrew, S.P., Modern processes for the production of ammonia, nitric acid and ammonium nitrate; chapter by Childs, A.F., Phosphorus, phosphoric acid and inorganic phosphates), Special Publication No. 31, The Chemical Society, London.
Waddington, T.C. (ed.) (1965) *Non Aqueous Solvent Systems* (Chapter 1 by Jolly, W.L. and Hallida, C.J., Liquid ammonia), Nelson.
Wright, A.N. and Winkler, C.A. (1968) *Active Nitrogen*, Academic Press, New York.
Yamabe, T., Hori, K., Minato, T. and Fukui, K. (1980) Theoretical study on the bonding nature of transition metal complexes of molecular nitrogen, *Inorg. Chem.*, **19**, 2154–2159.

PROBLEMS

1. Use the molecular orbital theory to describe the bonding in N_2 and NO. What is the bond order in each case?

2. Explain why nitrogen molecules have the formula N_2, whilst phosphorus has the formula P_4.

3. Outline how nitrogen and phosphorus are obtained commercially.

4. Write balanced equations to show the effect of heat on (a) $NaNO_3$, (b) NH_4NO_3, (c) a mixture of NH_4Cl and $NaNO_2$, (d) $Cu(NO_3)_2 \cdot 2H_2O$, (e) $Pb(NO_3);_2$ and (f) NaN_3.

5. Give equations to show how the following materials react with water: (a) Li_3N, (b) CaNCN, (c) AlN, (d) NO_2, (e) N_2O_5, (f) NCl_3.
6. Describe the commercial methods for manufacturing NH_3 and HNO_3. How are the starting materials obtained? What are their main uses? How is HNO_3 concentrated?
7. Explain the π bonding in NO_3^-.
8. Write an account of the chemistry of the oxides of nitrogen. Describe and give equations for the preparation of each, and discuss their properties, reactivity structures and bonding.
9. Describe the conditions under which the following react, and give the products in each case:
   (a) copper and nitric acid
   (b) nitrous oxide and sodamide
   (c) calcium carbide and nitrogen
   (d) cyanide ions and cupric sulphate
   (e) ammonia and an acidified solution of sodium hypochlorite
   (f) nitrous acid and iodide ions.
10. Describe the production of hydrazine and hydrazine sulphate. What practical difficulties are involved? What are they used for?
11. Explain what happens and give equations for the reaction of an aqueous solution of hydrazine sulphate with:
    (a) an aqueous solution of I_2 in KI
    (b) an alkaline solution of copper sulphate
    (c) an aqueous solution of potassium ferricyanide $K_3[Fe(CN)_6]$
    (d) an ammoniacal solution of silver nitrate.
12. Why is NF_3 stable whilst NCl_3 and NI_3 are explosive?
13. Why is it that NF_3 has no donor properties, but PF_3 forms many complexes with metals? Give examples of such complexes.
14. Give a preparation of NH_2OH and describe one of its major uses.
15. What are the main ingredients in fertilizers? How are they made, and what use do the plants make of them?
16. Compare the oxides of nitrogen and phosphorus.
17. Substance (A) is a gas of vapour density 8.5. On oxidation at high temperature with a platinum catalyst it gave a colourless gas (B), which rapidly turned brown in air, forming a gas (C). (B) and (C) were condensed together to give substance (D), which reacted with water, forming an acid (E). On treatment of (E) with an acidified solution of KI, gas (B) was evolved, but when (E) was treated with a solution of NH_4Cl, a stable colourless gas (F) was evolved. (F) did not support combustion, but magnesium continued to burn in it. However, (F) reacted with calcium carbide in an electric furnace, forming a solid

(G), which was slowly hydrolysed by water, forming a solution of substance (A), which turned Nessler's reagent yellow. Identify substances (A) to (G) and explain the reactions involved.

18. Compare the structures of the oxides and sulphides of phosphorus.

19. Give equations for the reactions of the following compounds with water: (a) P_4O_6, (b) P_4O_{10}, (c) PCl_3, (d) PCl_5, (e) Na_3P.

20. What do you understand by $p\pi-d\pi$ bonding in the oxides and oxoacids of phosphorus? Give examples to show how this may explain some of the differences in the chemistry of nitrogen and phospho?

21. Explain why the P—O bond length in $POCl_3$ is 1.45 Å whereas the sum of the single bond covalent radii of phosphorus and oxygen is 1.83 Å.

22. Discuss the uses of phosphates in analysis and in industry.

23. Compare and contrast the structures and behaviour of phosphates, silicates and borates.

24. Suggest reasons why PF_5 is known but NF_5 is not.

25. Give examples of phosphazenes. How are they made, and what are their structures?

26. Give equations for the reactions of the following compounds with water: (a) As_4O_6, (b) As_4O_{10}, (c) $SbCl_3$, (d) Mg_3Bi, (e) Na_3As.

15 Group VI – the chalcogens

Table 15.1 Electronic structures and oxidation states

Elements		Electronic structure	Oxidation states*			
Oxygen	O	[He] $2s^2\ 2p^4$	**−II** (−I)			
Sulphur	S	[Ne] $3s^2\ 3p^4$	−II	(II)	IV	**VI**
Selenium	Se	[Ar] $3d^{10}\ 4s^2\ 4p^4$	(−II)	II	**IV**	VI
Tellurium	Te	[Kr] $4d^{10}\ 5s^2\ 5p^4$		II	**IV**	VI
Polonium	Po	[Xe] $4f^{14}\ 5d^{10}\ 6s^2\ 6p^4$		II	**IV**	

* The most important oxidation states (generally the most abundant and stable) are shown in bold. Other well-characterized but less important states are shown in normal type. Oxidation states that are unstable, or in doubt, are given in parentheses.

GENERAL PROPERTIES

The first four elements are non-metals. Collectively they are called 'the chalcogens' or ore-forming elements, because a large number of metal ores are oxides or sulphides.

Several chemicals in this group are industrially important. H_2SO_4 is the most important chemical in the chemical industry. A staggering figure of 163 million tonnes was produced in 1988. One hundred million tonnes/year of O_2 is produced, and most is used in iron and steel making. In 1988 60.5 million tonnes of S were produced in 1988, most of which is used to make H_2SO_4. One million tonnes/year of Na_2SO_3 is used, mostly to bleach wood pulp and paper. Worldwide, 824 400 tonnes of H_2O_2 were produced in 1985.

The elements show the usual increase in metallic character on descending the group. This is shown by their reactions, the structures of the elements, and an increased tendency to form M^{2+} ions together with a decrease in stability of M^{2-} ions. O and S are totally non-metallic. Non-metallic character is weaker in Se and Te. Po is markedly metallic, and is also radioactive and short-lived.

Oxygen is a very important element in inorganic chemistry, since it reacts with almost all the other elements. Most of its compounds are covered under the other elements.

S, Se and Te are moderately reactive and burn in air to form dioxides. They combine directly with most elements, both metals and non-metals, though less readily than with O. As expected for non-metals, S, Se and Te are not attacked by acids except those which are oxidizing agents. Po shows metallic properties since it dissolves in H_2SO_4, HF, HCl and HNO_3, forming pink solutions of Po^{II}. However, Po is strongly radioactive, and the α-emission decomposes the water, and the Po^{II} solution is quickly oxidized to yellow solutions of Po^{IV}.

Oxygen shows several differences from the rest of the group. These are associated with its smaller size, higher electronegativity, and the lack of suitable *d* orbitals. Oxygen can use $p\pi$ orbitals to form strong double bonds. The other elements can also form double bonds, but these become weaker as the atomic number increases. Thus CO_2 (O=C=O) is stable, CS_2 less stable, CSe_2 polymerizes rather than form double bonds and CTe_2 is unknown. Oxygen also forms strong hydrogen bonds which greatly affect the properties of water and other compounds.

Sulphur shows a much greater tendency to form chains and rings than the other elements (see Allotropic Forms). Sulphur forms an extensive and unusual range of compounds with nitrogen which are not matched by the other elements.

Whereas O and S have only *s* and *p* electrons, Se follows after the first transition series and has *d* electrons too. The filling of the 3*d* shell affects the properties of Ge, As, Se and Br. The atoms are smaller, and the electrons are held more tightly. This is the reason why Se is reluctant to attain the highest oxidation state of (+VI) shown by S. Thus HNO_3 oxidizes S to H_2SO_4 (S +VI) but only oxidizes Se to H_2SeO_3 (Se +IV).

All compounds of Se, Te and Po are potentially toxic, and should be handled with care. Organo derivatives, and volatile compounds such as H_2Se and H_2Te, are 100 times more toxic than HCN.

ELECTRONIC STRUCTURE AND OXIDATION STATES

The elements all have the electronic structure S^2P^4. They may attain a noble gas configuration either by gaining two electrons, forming M^{2-} ions, or by sharing two electrons, thus forming two covalent bonds. The electronegativity of O is very high – second only to F. The electronegativity difference between M and O is large. Thus most metal oxides are ionic and contain O^{2-} ions, and the oxidation state of O is (−II). Sulphides, selenides and tellurides are formed with the more electronegative metals in Groups I and II and the lanthanides, and these compounds are some of the most stable formed. Compounds are often written as containing S^{2-}, Se^{2-} and Te^{2-}. The electronegativity differences suggest that these compounds are close to the 50% ionic, 50% covalent borderline. In the same way as for PCl_5, these compounds may be covalent in the solid but ionic in aqueous solution.

The elements also form compounds containing two covalent (electron-pair) bonds such as H_2O, F_2O, Cl_2O, H_2S and SCl_2. Where the chalcogen atom is the least electronegative atom in the molecule (e.g. in SCl_2 where the electronegativity of S = 2.5 and Cl = 3.5) the S shows an oxidation state of (+II).

In addition, the elements S, Se and Te show oxidation states of IV and VI, and these are more stable than the +II state.

ABUNDANCE OF THE ELEMENTS

Oxygen is the most abundant of all elements. It exists in the free form as O_2 molecules and makes up 20.9% by volume and 23% by weight of the atmosphere. Most of this has been produced by photosynthesis, the process where the chlorophyll in the green parts of plants uses the sun's energy to make foodstuffs such as glucose sugar.

$$6CO_2 + 6H_2O + \text{energy from the sun} \rightarrow C_6H_{12}O_6 + 6O_2$$

Oxygen makes up 46.6% by weight of the earth's crust, where it is the major constituent of silicate minerals. Oxygen also occurs as many metal oxide ores, and as deposits of oxosalts such as carbonates, sulphates, nitrates and borates. Oceans cover three quarters of the earth's surface, and oxygen makes up 89% by weight of the water in the oceans. Ozone O_3 exists in the upper atmosphere, and is of great importance. This is discussed later.

Sulphur is the sixteenth most abundant element and constitutes 0.034% by weight of the earth's crust. It occurs mainly in the combined form as numerous sulphide ores, and as sulphates (particularly gypsum $CaSO_4 \cdot 2H_2O$). It is not economic to mine these to obtain S, although gypsum is mined for other uses. The native element can be obtained from volcanic ources in many places, but these sources are little used now except in Japan and Mexico. From Biblical times up till the present century volcanic sources provided the major source of S. In early times S in the form of brimstone (burning rock) was used for fumigation. From the thirteenth century until the middle of the nineteenth century it was used to make gunpowder. In the present century the major use has been to make H_2SO_4.

The other elements Se, Te and Po are very scarce.

Table 15.2 Abundance of the elements in the earth's crust, by weight

	ppm	Relative abundance
O	455 000	1
S	340	16
Se	0.05	68
Te	0.001	74 =
Po	trace	

EXTRACTION AND USES OF THE ELEMENTS

Extraction and separation of oxygen

Oxygen is produced industrially by the fractional distillation of liquid air. (See Chapter 14, under 'Occurrence, extraction and uses of Nitrogen'.) Most of the O_2 is used in the steel making industry. Gas produced in this way commonly contains traces of N_2 and the noble gases, particularly Ar. Steel cylinders of compressed O_2 are used for many purposes, including oxy-acetylene welding, and in the laboratory. O_2 is administered together with an anaesthetic for surgical operations. It is sometimes prepared on a small scale in the laboratory by thermal decomposition of $KClO_3$ (with MnO_2 as catalyst), though the product often contains traces of Cl_2 or ClO_2. Small amounts of O_2 as an emergency breathing supply in aircraft are produced by heating $NaClO_3$:

$$2KClO_3 \xrightarrow{150\,°C\ MnO_2\ catalyst} 2KCl + 3O_2$$

by the catalytic decomposition of hypochlorites:

$$2HOCl \xrightarrow{Co^{2+}} 2HCl + O_2$$

or by the electrolysis of water with a trace of H_2SO_4 or barium hydroxide solution.

Uses of oxygen

Practically all the elements react with oxygen, either at room temperature or on heating. (The only exceptions are a few noble metals such as Pt, Au and W, and the noble gases.) Even though the bond energy of O_2 is high ($493\,kJ\,mol^{-1}$) the reactions are generally strongly exothermic, and once started often continue spontaneously.

Oxygen is essential for respiration (for the release of energy in the body) by both animals and plants. It is therefore essential for life.

$$\underset{glucose}{C_6H_{12}O_6} + 6O_2 \xrightarrow{respiration} 6CO_2 + 6H_2O + energy$$

The complex formed between oxygen and haemoglobin (the red pigment in blood) is of vital importance since it is the method by which higher animals transport oxygen round the body to the cells which actually use it.

World production of liquid and gaseous O_2 is about 100 million tonnes/year. By far the largest consumption of oxygen (60–80%) is in the iron and steel industry. Here pure oxygen is used to convert pig iron into steel in the *basic oxygen process* which originated as the *Kaldo* and *LD* processes. Since the late 1950s these have replaced the Bessemer process (which used air). Plants to produce oxygen are frequently located adjacent to, or are part of, modern steel making plants, and the O_2 is piped from one plant to the other. There are three advantages to the modern methods using O_2:

continued overleaf

1. The conversion to steel is quicker.
2. Larger ingots of pig iron can be used (Bessemer 6 tonnes, BOP 100 tonnes).
3. The metal does not form nitrides, which can occur when air is used.

In some places oxygen is introduced with air into blast furnaces used for the reduction of iron oxides to impure pig iron by coke. The reason for using oxygen is largely to allow the use of some heavy hydrocarbons (naptha) as fuels as a partial replacement for expensive metallurgical coke. Oxygen is also used for oxy-acetylene welding and metal cutting. Other important chemical uses of oxygen include the following:

1. The preparation of TiO_2 from $TiCl_4$. TiO_2 is used as a white pigment in paint and paper and as a filler in plastics.
2. To oxidize NH_3 in the manufacture of HNO_3.
3. In the manufacture of oxirane (ethylene oxide) from ethene.
4. As the oxidant in rockets.

Extraction of sulphur

World production of S was 60.5 million tonnes in 1988. The main producers are the USA and USSR 18% each, Canada 15%, Poland 9%, China 8%, and Japan and Mexico 4% each. There are several methods of extracting S:

Recovered from natural gas and petroleum	48%
Mined by the Frasch process	19%
From pyrites	17%
Recovered from smelter gases	12%
Mined as sulphur ore	4%
Made from $CaSO_4$	0.03%

Large amounts of sulphur are obtained from natural gas plants. In Canada these plants are the major source (90%) of S since the natural gas there may contain up to 20% H_2S. It is essential that all traces of sulphur compounds are removed from natural gas, since H_2S has an objectionable smell. Furthermore, burning sulphur compounds forms SO_2 which has an acrid smell and is corrosive. In a similar way, large amounts of S are obtained from oil refineries (60% of the USA total and 37% of the USSR total). After cracking long chain hydrocarbons, H_2S and other sulphur derivatives are removed because of their objectionable smell. About a third of the H_2S is oxidized in air to give SO_2, which is subsequently reacted with the remaining H_2S. This provides a second major source of S in the USA and Japan. With the enormous increase in the use of natural gas and oil, more S is now obtained from gas and petroleum than is mined by the Frasch process (see later).

Major deposits of native S are found in the USA (the Gulf of Mexico States; Louisiana, Texas, and Mexico), and in the upper Vistula region of Poland and the Ukraine. These deposits of S were formed by anaerobic bacteria which metabolize $CaSO_4$ to form H_2S and S.

$$2H_2S + 3O_2 \rightarrow 2SO_2 + 2H_2O$$
$$SO_2 + 2H_2S \rightarrow 2H_2O + 3S$$

These underground deposits of S are mined by the Frasch process, and yield S in a very high state of purity. In this process, three concentric pipes are sunk in a borehole down to the underground deposit. Superheated steam is passed down the outer pipe, and this melts the sulphur. Compressed air is blown down the inner pipe and forces molten sulphur up the middle pipe. One bore hole can cover an area of about half an acre. This technique was developed to overcome difficulties of mining in swamp areas or through quicksand, and for offshore mining in Louisiana. Mining in the USA started in the 1890s, and in Poland in the 1950s.

SO_2 is obtained as a by-product in the extraction of metals from sulphide ores. The most important metal sulphide is iron pyrites (fool's gold) FeS_2. This is mined in large amounts in the USSR, Spain, Portugal, Japan and many other places. Non-ferrous metals such as wurtzite ZnS, galena PbS, several forms of copper sulphide and NiS all yield SO_2 in smelters. The SO_2 is used to make H_2SO_4. Because there are a large number of metal production processes, this method of producing S currently yields more S than the other two methods. However, it is produced as SO_2 rather than S. The metals in the *p*-block and about half of the transition metals form sulphide minerals: all these metals are collectively called chalcophiles. Some of the most important sulphide minerals are listed in Table 15.3.

There are vast amounts of S in the form of sulphates dissolved in the oceans, and as mineral deposits such as $CaSO_4$. There are smaller deposits of other metals such as $FeSO_4$ and $Al_2(SO_4)_3$. In Poland SO_2 is obtained by heating $CaSO_4$ with coke in a rotary kiln. Production is about 20000 tonnes/year. The SO_2 is used for the Contact process for the manufacture of H_2SO_4. Production of elemental S from sulphates is not much used since other sources of S are at present cheaper.

Table 15.3 Some important sulphide ores

MoS_2	Molybdenite
FeS_2	Pyrites (fool's gold)
FeS_2	Marcasite
FeAsS	Arsenopyrites
$(Fe,Ni)_9S_8$	Pentlandite
Cu_2S	Copper glance or chalcocite
$CuFeS_2$	Copper pyrites or chalcopyrite
Cu_5FeS_4	Bornite or peacock mineral
Ag_2S	Silver glance or argentite
ZnS	Zinc blende or sphalerite
ZnS	Wurtzite
HgS	Cinnabar
PbS	Galena or lead glance
As_2S_3	Orpiment
As_4S_4	Realgar
Sb_2S_3	Stibnite
Bi_2S_3	Bismuthinite

continued overleaf

$$2CaSO_4 + C \xrightarrow{1200\,^\circ C} 2SO_2 + 2CaO + CO_2$$

At one time elemental S was obtained in large amounts as a by-product in the production of coal gas. Since natural gas has replaced coal gas (town gas) in many developed areas, this source has largely disappeared except in less developed areas.

Acid rain and SO_2

Coal typically contains 2% S, and may contain up to 4%. This represents a huge potential source of S, which could be extracted as SO_2 from the flue gases. Worldwide about 4749 million tonnes of coal were produced in 1988. The largest use is in coal burning electricity generating plants. Worldwide this produces about 90 million tonnes of S, that is 180 million tonnes of SO_2. (In the UK 108 million tonnes of coal are used in power stations, producing about 3 million tonnes per year of S or 6 million tonnes of SO_2.) Because it is uneconomic to remove the SO_2, only about 1% of this total tonnage is recovered as H_2SO_4. The majority is discharged into the atmosphere, where it causes acid rain.

The atmospheric chemistry of acid rain is not fully understood. SO_2 is oxidized by ozone or hydrogen peroxide to SO_3. This reacts with water or hydroxyl radicals to give H_2SO_4. Ammonium sulphate is also formed, and can be seen as an atmospheric haze (sometimes described as an aerosol of fine particles). *Wet deposition* occurs after raindrops become nucleated with aerosol particles of SO_3 or $(NH_4)_2SO_4$, but SO_2 does not dissolve in significant amounts. Instead SO_2 is deposited by *dry deposition*, and is absorbed directly on both solid and liquid ground surfaces. In 1982 UK deposition was about 50 units (kg hectare^{-1} year^{-1}) of dry and 5 units of wet. The name 'acid rain' is misleading since it refers to both wet and dry deposition.

Inevitably power stations are located (and SO_2 is emitted) in densely populated regions. Using high chimney stacks to disperse the gas merely moves the problem on to someone else. For example, 10% of SO_2 pollution in Sweden actually comes from Sweden, but 80% comes from the industrial regions of Europe (East and West Germany, Poland, Czechoslovakia) and 10% from Britain. Acid rain causes damage to trees, plants, fish and buildings, and causes respiratory ailments in man and animals. About 60% of atmospheric SO_2 comes from coal fired power stations. Most of the rest comes from oil refineries, oil fired power stations and smelters.

Total elimination of SO_2 pollution is not possible for both economic and technical reasons. However, we have the technology to reduce pollution to a low figure. The methods used are scrubbing the flue gases with a slurry of $Ca(OH)_2$, or reducing the SO_2 to S using H_2S and an activated alumina catalyst.

$$Ca(OH)_2 + SO_2 \rightarrow CaSO_3 + H_2O$$
$$2H_2S + SO_2 \rightarrow 3S + 2H_2O$$

Uses of sulphur

S is an essential though minor constituent of certain proteins. It is present in the amino acids cystine, cysteine and methionine.

World production of S was 60.5 million tonnes in 1988. Almost 90% of this is converted to SO_2, then to SO_3 and finally to H_2SO_4. Sixty per cent of the H_2SO_4 produced is used to make fertilizers. The remainder is used to make a variety of other chemicals. Sulphites SO_3^{2-}, hydrogen sulphites HSO_3^- and SO_2 are important for bleaching.

The 10% of S for non-acid purposes is used as elemental S. Some is used to make carbon disulphide CS_2, which is used to make CCl_4 and viscose rayon. Sulphur reacts with alkenes and forms sulphur cross-links between molecules. This is important in the vulcanizing of rubber. Sulphur and selenium will dehydrogenate saturated hydrocarbons. Other uses of sulphur are in the manufacture of fungicides, insecticides and gunpowder. Gunpowder is an intimate mixture of saltpetre $NaNO_3$ (75%), charcoal (15%) and sulphur (10%). It was discovered by Roger Bacon in 1245, and was the first explosive which could propel a bullet or cannon ball. It was first used for this purpose at the Battle of Crécy in 1346. It was then used in land and sea warfare for 500 years until better explosives such as guncotton, nitroglycerine and cordite were discovered.

Extraction and uses of selenium and tellurium

Se and Te occur among sulphide ores and are obtained in concentrated form from anode sludge after the electrolytic refining of copper. This sludge also contains the platinum metals, and Ag and Au. Se and Te are also obtained from flue dust produced during the roasting of sulphide ores such as PbS, CuS or FeS_2. The dust is trapped by means of an electrostatic precipitator. Both elements also occur in the native form together with S.

World production of Se metal was 1478 tonnes in 1988. Most is used to decolorize glass, though Cd(S,Se) is used to make pink and red coloured glass. Se is used in Xerox-type photocopiers to make the photoreceptor to capture the image. The photoreceptor is a thin film of Se on an Al support. The photoreceptor is sensitized electrostatically by a high voltage, and then an image is focussed on it as on the film in a camera. Areas exposed to light lose their electrostatic charge. Toner powder sticks to the areas still charged, and the powder is transferred to a sheet of paper and heated to fuse the powder to the paper. Thus a copy of the original image is obtained. The photoreceptor is then wiped clean, sensitized again, and reused. Selenium is an essential element in the body in trace amounts, and is a component in a number of important enzymes, e.g. glutathione peroxidase, which protects cells against attack by peroxide. However, Se is toxic in larger quantities.

continued overleaf

World production of Te metal was 144 tonnes in 1988. The major part of this is used in making steel and non-ferrous alloys, for example to harden lead.

Both Se and Te compounds are absorbed by the human body and are excreted as foul smelling organic derivatives in breath and sweat.

Discovery and production of polonium

Polonium was discovered by Marie Curie by processing very large amounts of thorium and uranium minerals, and separating the decay products. Polonium is one of the decay products (see 'Radioactive decay series' in Chapter 31). The separation was observed by studying the radioactivity.

$$^{210}_{82}Pb \xrightarrow[\text{half life 22.3 years}]{\beta} {}^{210}_{82}Bi \xrightarrow[\text{half life 5.0 days}]{\beta} {}^{210}_{84}Po \xrightarrow[\text{half life 138.4 days}]{\beta} {}^{206}_{82}Pb$$

Marie Curie shared the award of a Nobel Prize for Physics with H. A. Becquerel and Pierre Curie in 1903 for work on radioactivity, which was then a new technique. In 1911 she was awarded a second Nobel Prize (this time for Chemistry), for the discovery of polonium and radium. Polonium is named after Marie Curie's home country Poland. Polonium is now made artificially in gram quantities from bismuth by neutron irradiation in a nuclear reactor. The metal is extracted by sublimation.

$$^{209}_{83}Bi + {}^{1}_{0}n \rightarrow {}^{210}_{83}Bi \rightarrow {}^{210}_{84}Po + {}^{0}_{-1}e$$

All isotopes of polonium are highly radioactive. The most stable isotope is $^{210}_{84}Po$, but this is an intense α emitter and has a half life of 138 days. The α emission decomposes water, which complicates any studies of polonium compounds in aqueous solutions. Thus the chemistry of polonium is not well known.

STRUCTURE AND ALLOTROPY OF THE ELEMENTS

All the elements except Te are polymorphic, that is they exist in more than one allotropic form.

Oxygen

Oxygen occurs as two non-metallic forms, dioxygen O_2 and ozone O_3. Dioxygen O_2 is stable as a diatomic molecule, which accounts for it being a gas. (S, Se, Te and Po have more complicated structures, e.g. S_8, and are solids at normal temperatures.) The bonding in the O_2 molecule is not as simple as it might at first appear. If the molecule had two covalent bonds, then all the electrons would be paired and the molecule should be diamagnetic.

$$:\ddot{O}\cdot + \cdot\ddot{O}: \rightarrow :\ddot{O}\vdots\ddot{O}: \quad \text{or} \quad O{=}O$$

Oxygen is paramagnetic and therefore contains unpaired electrons. The explanation of this phenomenon was one of the early successes of the molecular orbital theory. The structure is described (see 'Examples of molecular orbital treatment' in Chapter 4).

Liquid oxygen is pale blue in colour, and the solid is also blue. The colour arises from electronic transitions which excite the ground state (a triplet state) to a singlet state. This transition is 'forbidden' in gaseous oxygen. In liquid or solid oxygen a single photon may collide with two molecules simultaneously and promote both to excited states, absorbing red–yellow–green light, so O_2 appears blue. The origin of the excited singlet states in O_2 lies in the arrangement of electrons in the antibonding π^*2p_y and π^*2p_z molecular orbitals, and is shown below.

	π^*p_y π^*p_z		State	Energy/kJ
Second excited state (electrons have opposite spins)	↑ | ↓	singlet	$^1\Sigma_g^+$	157
First excited state (electrons paired)	↑↓ |	singlet	$^1\Delta_g$	92
Ground state (electrons have parallel spins)	↑ | ↑	triplet	$^3\Sigma_g^-$	0

Singlet O_2 is excited, and is much more reactive than normal ground state triplet oxygen. Singlet oxygen can be generated photochemically by irradiating normal oxygen in the presence of a sensitizer such as fluorescein, methylene blue or some polycyclic hydrocarbons. Singlet oxygen can also be made chemically:

$$H_2O_2 + OCl^- \xrightarrow{EtOH} O_2\ (^1\Delta_g) + H_2O + Cl^-$$

Singlet oxygen can add to a diene molecule in the 1,4 positions, rather like a Diels–Alder reaction. It may add 1,2 to an alkene which can be cleaved into two carbonyl compounds.

$CH_2{=}CH{-}CH{=}CH_2$ + singlet O_2 → (cyclic peroxide, O—O ring)

Singlet oxygen may be involved in biological oxidations.

Ozone O_3 is the triatomic allotrope of oxygen. It is unstable, and decomposes to O_2. The structure of O_3 is angular, with an O—O—O bond angle of 116°48′. Both O—O bond lengths are 1.28 Å, which is intermediate between a single bond (1.48 Å in H_2O_2) and a double bond (1.21 Å in O_2). (The structure of O_3 is described near the end of Chapter 4.) The older valence bond representation as resonance hybrids is now seldom used. The structure is described as the central O atom using sp^2 hybrid orbitals to bond to the terminal O atoms. The central atom has one lone pair, and the terminal O atoms have two lone pairs. This leaves four electrons for π bonding. The p_z atomic orbitals from the three atoms form three delocalized molecular orbitals covering all three atoms. One MO is bonding, one non-bonding, and one antibonding. The four π electrons fill the bonding and non-bonding MOs and thus contribute one delocalized π bond to the molecule in addition to the two σ bonds. Thus the bond order is 1.5, and the π system is described as a four-electron three-centre bond.

Sulphur

Sulphur has more allotropic forms than any other element. These different forms arise partly from the extent to which S has polymerized, and partly from the crystal structures adopted. The two common crystalline forms are α or rhombic sulphur which is stable at room temperature, and β or monoclinic sulphur which is stable above 95.5 °C. These two forms change reversibly with slow heating or slow cooling. Rhombic sulphur occurs naturally as large yellow crystals in volcanic areas. A third modification known as γ-monoclinic sulphur is nacreous (looks like mother-of-pearl). It can be made by chilling hot concentrated solutions of S in solvents such as CS_2, toluene or EtOH. All three forms contain puckered S_8 rings with a crown conformation (Figure 15.1), and differ only in the overall packing of the rings in the crystal. This affects their densities:

α-rhombic	$2.069\,g\,cm^{-3}$
β-monoclinic	$1.94–2.01\,g\,cm^{-3}$
γ-monoclinic	$2.19\,g\,cm^{-3}$

Engel's sulphur (ε-sulphur) is unstable and contains S_6 rings arranged in the chair conformation. It is made by pouring $Na_2S_2O_3$ solution into concentrated HCl and extracting the S with toluene. It can also be made as follows:

$$H_2S_4 + S_2Cl_2 \rightarrow S_6 + 2HCl$$

Several other rings S_7, S_9, S_{10}, S_{11}, S_{12}, S_{18} and S_{20} have been made by Schmidt and his group. They are usually obtained by 1 : 1 reactions in dry ether between hydrogen polysulphides and polysulphur dichlorides with the required number of S atoms, for example:

$$H_2S_8 + S_2Cl_2 \rightarrow S_{10} + 2HCl$$
$$H_2S_8 + S_4Cl_2 \rightarrow S_{12} + 2HCl$$

In all of these ring compounds the S—S distance is 2.04 – 2.06 Å, and the bond angle S—S—S is in the range 102–108°. They are all soluble in CS_2.

Plastic or χ-sulphur is obtained by pouring liquid sulphur into water. Several other forms can be produced by quenching molten S. These may be fibrous, laminar, or rubber-like, and a commercial form is called Crystex. These are all metastable, and revert to the α (cyclo-S_8) form on standing. Their structures contain spiral chains, and sometimes S_8 and other rings.

Sulphur melts to form a mobile liquid. As the temperature is raised the colour darkens. At 160 °C the S_8 rings break, and the diradicals so formed

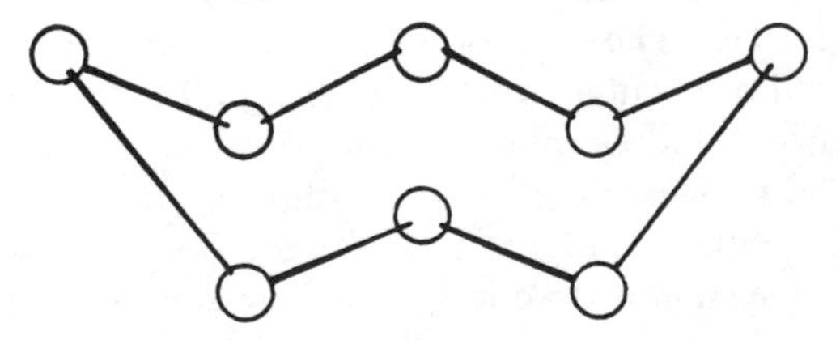

Figure 15.1 Structure of S_8 molecule. (Heslop and Robinson, *Inorganic Chemistry*, Elsevier, Amsterdam, 1963.)

polymerize, forming long chains of up to a million atoms. This makes all the physical properties change discontinuously. The viscosity increases sharply, and continues to rise up to 200 °C. At higher temperatures chains break, and shorter chains and rings are formed, which makes the viscosity decrease up to 444 °C, the boiling point. The vapour at 200 °C consists mostly of S_8 rings, but contains 1–2% of S_2 molecules. At 600 °C the gas mainly consists of S_2 molecules.

The S_2 molecule is paramagnetic and blue coloured like O_2, and presumably has similar bonding. S_2 gas is stable up to 2200 °C. The stability of S_2 is used in the quantitative analysis of S compounds. They are burnt in a reducing flame, and the colour from excited S_2 is measured spectrophotometrically. S_3 and S_4 are also known.

Selenium, tellurium and polonium

Six allotropes of selenium are known. Interest in these is because Se is used in electronic devices. These include capturing the image in Xerox-type photocopiers, as rectifiers (to convert alternating current into direct current), and as light emitting diodes (LEDs). There are four red forms. Three different red non-metallic forms are known containing Se_8 rings. They differ in the way the rings are packed in the crystal. An 'amorphous' red form contains polymeric chains. There are in addition two grey forms. The most stable is the grey metallic form, which contains infinite spiral chains of Se atoms with weak metallic interaction between adjacent chains. A black vitreous form of Se is commercially available, and is made of large irregular rings with up to 1000 atoms.

Tellurium has only one crystalline form, which is silvery white and semi-metallic. This is similar to grey Se, but has stronger metallic interaction.

Polonium is a true metal. It exists as an α-form which is cubic and a β-form which is rhombohedral. Both forms are metallic.

Thus there is a marked decrease in the number of allotropic forms from S to Se to Te. There is an increase in metallic character down the group. The electrical properties also change from insulators (O and S), to semiconductors (Se and Te), to metallic conduction (Po). The structures change from simple diatomic molecules, to rings and chains, to a simple metallic lattice.

CHEMISTRY OF OZONE

O_3 is an unstable, dark blue diamagnetic gas, b.p. −112 °C. The colour is due to intense absorption of red light (λ 557 and 602 nm). It also absorbs strongly in the UV region (λ 255 nm). This is particularly important since there is a layer of O_3 in the upper atmosphere which absorbs harmful UV radiation from the sun, thus protecting people on the earth. The use of chlorofluorocarbons in aerosols and refrigerators, and their subsequent escape into the atmosphere, is blamed for making holes in the ozone layer over the Antarctic and Arctic. It is feared that this will allow an excessive amount of UV light to reach the earth which will cause skin cancer (melanoma) in humans. Oxides of nitrogen (from car exhausts) and the halogens can also damage the O_3 layer. (See Chapter 13.)

O_3 has a characteristic sharp smell, often associated with sparking electrical equipment. The gas is toxic, and continuous exposure to concentrations of 0.1 ppm must be avoided.

O_3 is usually prepared by the action of a silent electric discharge upon oxygen between two concentric metallized tubes in an apparatus called an ozonizer. Concentrations of up to 10% of O_3 are obtained in this way. Higher concentrations or pure O_3 can be obtained by fractional liquifaction of the mixture. The pure liquid is dangerously explosive. Low concentrations of O_3 can be made by UV irradiation of O_2. This occurs in the atmosphere when photochemical smog is formed over some cities, for example over Los Angeles or Tokyo. The photochemical change is useful for producing low concentrations to sterilize food, particularly for cold storage. O_3 can also be made by heating O_2 to over 2500 °C and quenching. In all of these preparations oxygen atoms are produced, and these react with O_2 molecules to form O_3.

O_3 is also used as a disinfectant. For example, it is used to purify drinking water, since it destroys bacteria and viruses. Its advantage over chlorine for this purpose is that it avoids the unpleasant smell and taste of chlorine, since any excess O_3 soon decomposes to O_2. For similar reasons it is used to treat water in swimming pools.

The amount of O_3 in a gas mixture may be determined by passing the gas into a KI solution buffered with a borate buffer (pH 9.2). The iodine that is liberated is titrated with sodium thiosulphate.

$$O_3 + 2K^+ + 2I^- + H_2O \rightarrow I_2 + 2KOH + O_2$$

Alternatively the gas may be decomposed catalytically, and the change in volume measured.

$$\underset{\text{(2 volumes)}}{2O_3} \rightarrow \underset{\text{(3 volumes)}}{3O_2} \qquad \Delta G = -163\,\text{kJ mol}^{-1}$$

O_3 is thermodynamically unstable, and decomposes to O_2. The decomposition is exothermic, and is catalysed by many materials. The solid and liquid often decompose explosively. The gas decomposes slowly, even when warmed, *providing catalysts and UV light are absent*. O_3 is an extremely powerful oxidizing agent, second only to F_2 in oxidizing power, and reacts much more readily than oxygen.

$$3PbS + 4O_3 \rightarrow 3PbSO_4$$
$$2NO_2 + O_3 \rightarrow N_2O_5 + O_2$$
$$S + H_2O + O_3 \rightarrow H_2SO_4$$
$$2KOH + 5O_3 \rightarrow 2KO_3 + 5O_2 + H_2O$$

Potassium ozonide KO_3 is an orange coloured solid and contains the paramagnetic O_3^- ion. O_3 adds to unsaturated organic compounds at room

Table 15.4 Physical properties of the elements

	Covalent radius (Å)	Ionic radius M^{2-} (Å)	First ionization energy ($kJ\ mol^{-1}$)	Pauling's electro-negativity	Melting point (°C)	Boiling point (°C)
O	0.74	1.40	1314	3.5	−229	−183
S	1.04	1.84	999	2.5	114	445
Se	1.14	1.98	941	2.4	221	685
Te	1.37	2.21	869	2.1	452	1087
Po			813	2.0	254	962

Values for covalent radii are for two-coordination.

temperature, forming ozonides. These are not usually isolated, but can be cleaved to aldehydes and ketones in solution, or oxidized by air to give carboxylic acids.

The difference in free energy between the different oxidation states of S is not very great. PoO_3 (Po(+VI)) is appreciably more oxidizing than the (+VI) states of the other elements, and selenate SeO_4^{2-} and tellurate TeO_4^{2-} have a much greater oxidizing power than SO_4^{2-}. The differences between the oxidizing power of the (+IV) states of polonite, tellurite, selenite and sulphite are smaller. The potentials also show the decrease in stability from H_2O to H_2S to H_2Se to H_2Te to H_2Po: the last three are thermodynamically unstable.

OXIDATION STATES (+II), (+IV) AND (+VI)

Oxygen is never more than divalent because when it has formed two covalent bonds it has attained a noble gas configuration, and there are no low energy orbitals which can be used to form further bonds. However, the elements S, Se, Te and Po have empty *d* orbitals which may be used for bonding, and they can form four or six bonds by unpairing electrons.

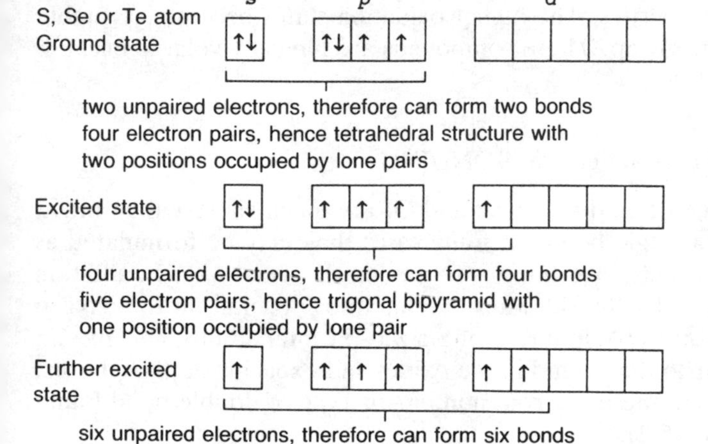

STANDARD REDUCTION POTENTIALS (VOLTS)

Acid solution

Oxidation state: +VI, +V, +IV, +III, $+II\frac{1}{2}$, +II, +I, 0, −I, −II

$$O_2 \xrightarrow{+0.69} H_2O_2{}^{*} \xrightarrow{+1.78} H_2O$$

$O_2 \rightarrow H_2O$: +1.23

$$SO_4^{2-} \xrightarrow{-0.22} S_2O_6^{2-} \xrightarrow{+0.57} SO_2 \xrightarrow{+0.51} S_4O_6^{2-} \xrightarrow{+0.08} S_2O_3^{2-} \xrightarrow{+0.47} S_8 \xrightarrow{+0.14} H_2S$$

$SO_4^{2-} \rightarrow SO_2$: +0.17; $SO_2 \rightarrow S_8$: +0.45

$$SeO_4^{2-} \xrightarrow{+1.15} H_2SeO_3 \xrightarrow{+0.74} Se \xrightarrow{-0.40} H_2Se$$

$$H_6TeO_6 \xrightarrow{+0.92} TeO_2 \xrightarrow{+0.57} Te \xrightarrow{-0.72} H_2Te$$

$H_6TeO_6 \rightarrow Te^{4+}$: +1.02; $Te^{4+} \rightarrow Te$: +0.53

$$PoO_3 \xrightarrow{+1.52} PoO_2 \xrightarrow{+0.80} Po^{2+} \xrightarrow{+0.65} Po \xrightarrow{-1.00} H_2Po$$

$PoO_2 \rightarrow Po$: +0.72; $PoO_3 \rightarrow Po^{2+}$: +1.16

* Disproportionates

Compounds of S, Se and Te with O are typically tetravalent. The (+IV) state shows both oxidizing and reducing properties. Fluorine brings out the maximum oxidation state of (+VI). Compounds in the (+VI) state show oxidizing properties. The higher oxidation states become less stable on descending the group. These compounds are typically volatile because they are covalent.

BOND LENGTHS AND $p\pi-d\pi$ BONDING

The bonds between S and O, or Se and O, are much shorter than might be expected for a single bond. In some cases they may be formulated as localized double bonds. A σ bond is formed in the usual way. In addition a π bond is formed by the sideways overlap of a p orbital on the oxygen with a d orbital on the sulphur, giving a $p\pi-d\pi$ interaction. This $p\pi-d\pi$ bonding is similar to that found in the oxides and oxoacids of phosphorus, and is in contrast to the more common $p\pi-p\pi$ type of double bond found in ethene (Figure 15.2).

To obtain effective $p\pi-d\pi$ overlap, the size of the d orbital must be similar to the size of the *p* orbital. Thus sulphur forms stronger π bonds than the larger elements in the group. On crossing a period in the periodic table, the nuclear charge is increased and more *s* and *p* electrons are added. Since these *s* and *p* electrons shield the nuclear charge incompletely, the size of the atom and the size of the *d* orbitals decreases from Si to P to S to Cl. The decrease in the size of the 3*d* orbitals in this series of elements leads to progressively stronger *p*π–*d*π bonds. Thus in the silicates there is hardly any *p*π–*d*π bonding. Thus SiO4 units polymerize into an enormous variety of structures linked by Si—O—Si σ bonds. In the phosphates, π bonding is stronger, but a large number of polymeric phosphates exist. In the oxoacids of sulphur, π bonding is even stronger and has become a dominant factor. Thus only a small amount of polymerization occurs, and only a few polymeric compounds are known with S—O—S linkages. For chlorine, *p*π–*d*π bonding is so strong that no polymerization of oxoanions occurs.

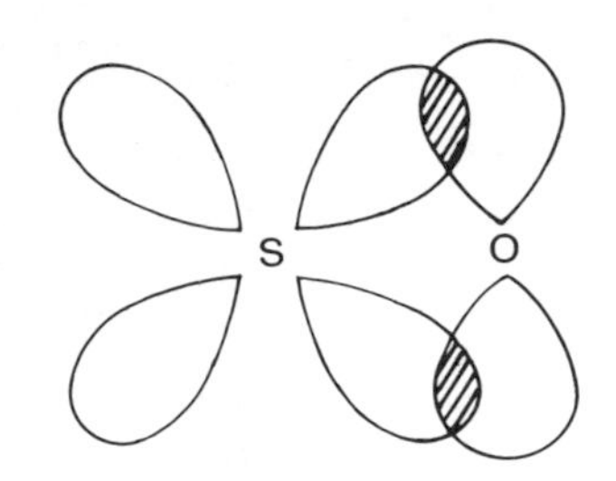

Figure 15.2 pπ–dπ overlap.

In cases where there is more than one π bond in the molecule it may be more appropriate to explain the π bonding in terms of delocalized molecular orbitals covering several atoms.

DIFFERENCES BETWEEN OXYGEN AND THE OTHER ELEMENTS

Oxygen differs from the rest of the group in that it is more electronegative and therefore more ionic in its compounds.

Hydrogen bonding is very important for O compounds, but it is only recently that weak hydrogen bonds involving S have been proved to exist.

The absence of higher valency states and the limitation to a coordination number of 4 are a consequence of the limitation of the second shell to eight electrons. The other elements can have a coordination number of 6 by using *d* orbitals.

Oxygen can use *p*π orbitals to form strong double bonds. The other elements can also form double bonds, but these become weaker as the atomic number increases.

GENERAL PROPERTIES OF OXIDES

Practically all of the elements react with oxygen to form oxides. There are several ways in which oxides may be classified, depending on their structure or their chemical properties. First consider the classification according to their geometric structure. In this way oxides are classified as normal oxides, peroxides or suboxides.

Normal oxides

In these, the oxidation number of M can be deduced from the empirical formula M_xO_y, taking the oxidation number of oxygen as (−II). These oxides, for example H_2O, MgO and Al_2O_3, contain only M—O bonds.

Peroxides

These contain more oxygen than would be expected from the oxidation number of M. Some are ionic and contain the peroxide ion O_2^{2-}, for example those of Group I and II metals (Na_2O_2 and BaO_2). Others are covalently bound and contain —O—O— in the structure, for example H_2O_2 (H—O—O—H), peroxomonosulphuric acid and peroxodisulphuric acid.

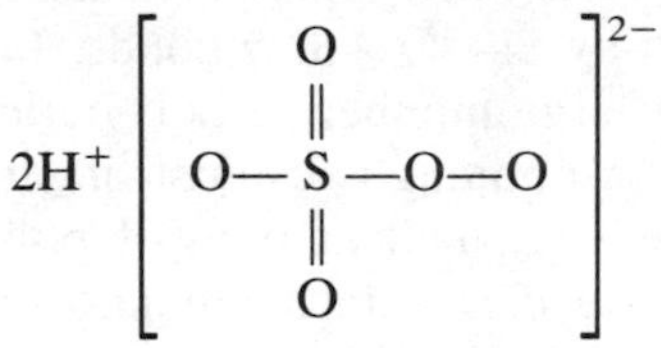

$$2H^+ \left[\begin{array}{c} O \\ \| \\ O{-}S{-}O{-}O \\ \| \\ O \end{array} \right]^{2-} \qquad 2H^+ \left[\begin{array}{c} O \qquad\quad O \\ \| \qquad\quad \| \\ O{-}S{-}O{-}O{-}S{-}O \\ \| \qquad\quad \| \\ O \qquad\quad O \end{array} \right]^{2-}$$

peroxomonosulphuric acid 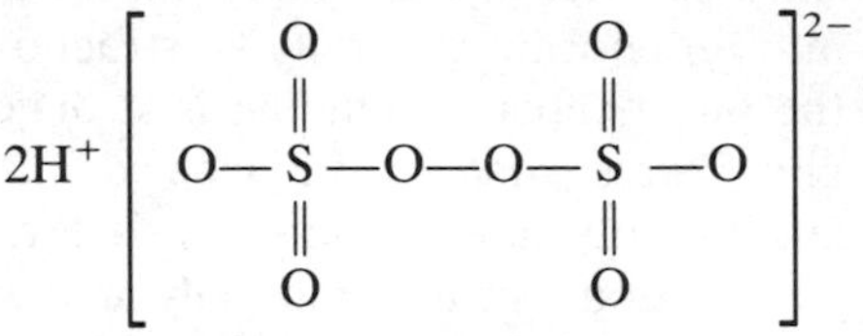 — peroxodisulphuric acid

Peroxo compounds are strong oxidizing agents, and are hydrolysed by water to give H_2O_2.

$$H_2SO_5 + H_2O \rightarrow H_2SO_4 + H_2O_2$$

Superoxides, e.g. KO_2, contain more oxygen than would be expected (see Chapter 9).

Suboxides

Here the formula contains less oxygen than would be expected from the oxidation number of M. They involve M—M bonds in addition to M—O bonds, for example O=C=C=C=O.

A second method of classifying oxides depends on their acid–base properties. Thus oxides may be acidic, basic, amphoteric or neutral, depending on the products formed when they react with water.

Basic oxides

Metallic oxides are generally basic. Most metal oxides are ionic and contain the O^{2-} ion. The oxides of the more electropositive metals, Groups I and II, and the lanthanides are typical. A large amount of energy is required to form an ionic oxide. This is because the O_2 molecule must first be broken into atoms, and then the energy (the electron affinity) required to add two electrons to form O^{2-} is also large. Thus ionic oxides are formed by compounds with a high lattice energy to offset this. Thus ionic oxides typically have high melting points (Na_2O 1275 °C, MgO 2800 °C, La_2O_3 2315 °C). When they react with water the O^{2-} ion is converted into OH^-.

$$Na_2O + H_2O \rightarrow 2NaOH$$

However, many metal oxides with formulae M_2O_3 and MO_2, though ionic, do not react with water. Examples include $Tl?_2O_3$, Bi_2O_3 and ThO_2. These react with acids to form salts, and so are basic. Where a metal can exist in more than one oxidation state, and thus form more than one oxide, e.g. CrO, Cr_2O_3, CrO_3, PbO, PbO_2, and Sb_4O_6, Sb_4O_{10}, the lowest oxidation state is the most ionic and the most basic. Thus CrO is basic, Cr_2O_3 amphoteric and CrO_3 acidic.

Amphoteric oxides

Many metals yield oxides which are amphoteric, and react with both strong acids and strong bases. Examples include BeO, Al_2O_3, Ga_2O_3, SnO, PbO and ZnO. Solutions of these oxides may undergo acidic or basic dissociation:

$$2Al^{3+} + 6OH^- \rightleftharpoons Al_2O_3 + 3H_2O \rightleftharpoons 6H^+ + 2[Al(OH)_6]^{3-}$$
$$Pb^{2+} + 2OH^- \rightleftharpoons PbO + H_2O \rightleftharpoons H^+ + [PbO \cdot OH]^-$$

Acidic oxides

Non-metallic oxides are usually covalent. Many occur as discrete molecules (CO_2, NO, SO_2, Cl_2O) and have low melting and boiling points, though some, such as B_2O_3 and SiO_2, form infinite 'giant molecules' and have high melting points. They are all acidic. There are many are the anhydrides of acids.

$$B_2O_3 + 3H_2O \rightarrow 2H_3BO_3$$
$$N_2O_5 + H_2O \rightarrow 2HNO_3$$
$$P_4O_{10} + 6H_2O \rightarrow 4H_3PO_4$$
$$SO_3 + H_2O \rightarrow H_2SO_4$$

Others which do not react with water such as SiO_2 do react with NaOH, thus showing their acidic properties. In cases where the element exists in more than one oxidation state, e.g. N_2O_3 and N_2O_5, SO_2 and SO_3, the higher oxidation state is the most acidic.

$$N_2O_3 + H_2O \rightarrow 2HNO_2$$
$$N_2O_5 + H_2O \rightarrow 2HNO_3$$

N_2O_3 contains N(+III) and N_2O_5 contains N(+V). HNO_3 is a stronger acid than HNO_2. This may be rationalized since the higher the oxidation state of the central atom the more it will attract electrons, thus weakening any O—H bonds and facilitating the release of H^+.

Neutral oxides

A few covalent oxides have no acidic or basic properties (N_2O, NO, CO).

Reactions between oxides

More important than the reaction of an oxide with water is its relation to, and its reaction with, other oxides. If oxides are arranged in a series from the most basic to the most acidic, then the further apart two oxides are in the series, the more stable the compound formed when they react together. This can be put on a quantitative basis by considering the changes in standard free energy.

$$\begin{array}{lllll} & Na_2O_{(s)} & +\ H_2O_{(l)} & \rightarrow 2NaOH_{(s)} & \\ \Delta G^\circ & -376 & \quad -234 & \quad 2(-376) & \Delta G = -142\,\text{kJ mol}^{-1} \end{array}$$

$$\begin{array}{lllll} & CaO_{(s)} & +\ H_2O_{(l)} & \rightarrow Ca(OH)_{2(s)} & \\ \Delta G^\circ & -602 & \quad -234 & \quad -895 & \Delta G = -59\,\text{kJ mol}^{-1} \end{array}$$

$$\begin{array}{lllll} & Al_2O_{3(s)} & +\ 3H_2O_{(l)} & \rightarrow 2Al(OH)_{3(s)} & \\ \Delta G^\circ & -1572 & \quad 3(-234) & \quad 2(-1138) & \Delta G = -2\,\text{kJ mol}^{-1} \end{array}$$

From the ΔG values, Na_2O is the most basic and Al_2O_3 the least basic. In fact Na_2O is strongly basic and Al_2O_3 is amphoteric. From the free energy values for the hydroxides, NaOH is chemically the most reactive and $Al(OH)_3$ the most stable. This is also true in the following reactions:

$$\begin{array}{lllll} & CaO_{(s)} & +\ CO_{2(g)} & \rightarrow CaCO_{3(s)} & \\ \Delta G^\circ & -602 & \quad -393 & \quad -1129 & \Delta G = -134\,\text{kJ mol}^{-1} \end{array}$$

$$\begin{array}{lllll} & CaO_{(s)} & +\ N_2O_{5(g)} & \rightarrow Ca(NO_3)_{2(s)} & \\ \Delta G^\circ & -602 & \quad +134 & \quad -740 & \Delta G = -272\,\text{kJ mol}^{-1} \end{array}$$

$$\begin{array}{lllll} & CaO_{(s)} & +\ SO_{3(g)} & \rightarrow CaSO_{4(s)} & \\ \Delta G^\circ & -602 & \quad -368 & \quad -1317 & \Delta G = -347\,\text{kJ mol}^{-1} \end{array}$$

From the ΔG values, SO_3 is the most strongly acidic oxide, N_2O_5 is the next strongest, and CO_2 is the weakest. From the free energy values of the salts formed, $Ca(NO_3)_2$ is the least stable and the most reactive, whilst $CaSO_4$ is the most stable and the least reactive.

The order of acidic strength of oxides can be obtained as follows:

$$K_2O,\ CaO,\ MgO,\ CuO,\ H_2O,\ SiO_2,\ CO_2,\ N_2O_5,\ SO_3$$

most basic ⟷ most acidic

It is possible to predict if a reaction is possible or impossible. If CaO is added to a mixture of H_2O and SO_3 (H_2SO_4), the more stable $CaO \cdot SO_3$ ($CaSO_4$) will form, that is:

$$H_2SO_4 + CaO \rightarrow CaSO_4 + H_2O$$

but

$$CuSO_4 + CO_2 \rightarrow \text{no reaction}$$

Free energy data and acidic power are clearly related. Thermodynamics allows us to predict if a reaction is possible in terms of energy. This use of ΔG values is not limited to oxides, but applies to all reactions. Thermodynamics does not allow us to predict the rate of a reaction. For example, the reaction below is thermodynamically possible:

$$CaO + SiO_2 \rightarrow CaSiO_3$$

The reaction is very slow at normal temperatures, though it is more rapid at high temperatures in a blast furnace.

Much inorganic chemistry consists of remembering which compounds react, and comparing the different stabilities of hydroxides, silicates, carbonates, nitrates, sulphates etc. The use of a series such as the one above minimizes this memory work.

For comprehensive lists of standard free energy data see Bard, A.J., Parsons, R. and Jordan, J.; and Latimer, W.M. in Further Reading.

OXIDES OF SULPHUR, SELENIUM, TELLURIUM AND POLONIUM

Table 15.5 Oxides

Element	MO_2	MO_3	Other oxides
S	SO_2	SO_3	S_2O (S_2O_2) (SO) (S—O—O) (SO_4) S_6O, S_7O, S_8O, S_9O, $S_{10}O$
Se	SeO_2	SeO_3	
Te	TeO_2	TeO_3	TeO
Po	PoO_2		PoO

Dioxides MO_2

SO_2 is produced commercially on a vast scale:

1. By burning S in air.
2. By burning H_2S in air.
3. By roasting various metal sulphide ores with air in smelters (particularly FeS_2, and to a smaller extent CuS and ZnS).
4. Large amounts are produced as a waste product by burning coal and, to a lesser extent, other fossil fuels, oil and gas. This undoubtedly harms the environment.

SO_2 is a colourless gas (b.p. $-10\,°C$, m.p. $-75.5\,°C$), which has a choking smell, and is very soluble in water ($39\,cm^{-3}$ of SO_2 gas will dissolve in $1\,cm^{-3}$ of water). The SO_2 in solution is almost completely present as various hydrated species such as $SO_2 \cdot 6H_2O$ and the solution contains only a minute amount of sulphurous acid H_2SO_3. SO_2 levels above 5 ppm are poisonous to man, but plants are harmed at appreciably lower levels.

SO_2 may be detected in the laboratory:

1. By its smell.
2. Because it turns a filter paper moistened with acidified potassium dichromate solution green, due to the formation of Cr^{3+}.

$$K_2Cr_2O_7 + 3SO_2 + H_2SO_4 \rightarrow Cr_2(SO_4)_3 + K_2SO_4 + H_2O$$

3. Because it turns starch iodate paper blue (due to starch and I_2).

$$2KIO_3 + 5SO_2 + 4H_2O \rightarrow I_2 + 2KHSO_4 + 3H_2SO_4$$

Quantitative methods for measuring SO_2 in the atmosphere are highly developed because of environmental concern over 'acid rain'. Methods include:

1. Oxidation to H_2SO_4, followed by determination of the H_2SO_4 by titration.

$$SO_2 + H_2O_2 \rightarrow H_2SO_4$$

2. Reaction with $K_2[HgCl_4]$ to give a mercury complex which reacts with the dye pararosaniline, and is estimated colorimetrically.

$$K_2[HgCl_4] + 2SO_2 + 2H_2O \rightarrow K_2[Hg(SO_3)_2] + 4HCl$$

3. Burning in a hydrogen flame in a flame photometer, and measuring the spectrum of S_2. (See also the discussion of singlet oxygen.)

Most of the SO_2 produced is oxidized to SO_3 by the Contact process, and used to manufacture H_2SO_4. Smaller amounts of SO_2 are used to make sulphites SO_3^{2-}, for bleaching, and for preserving food and wine.

$$2SO_{2(g)} + O_{2(g)} \rightleftharpoons 2SO_{3(g)} \quad \Delta H = -98\,\text{kJ mol}^{-1}$$

The forward reaction is exothermic, and is favoured by a low temperature. Since there is a decrease in the number of moles of gas, the process is favoured by a high pressure. In practice the reaction is carried out at atmospheric pressure. The formation of SO_3 is favoured by an excess of O_2, and removing the SO_3 from the reaction mixture. A catalyst is used to obtain a reasonable conversion in a reasonable time. In the Contact process a platinum gauze and platinized asbestos were both used at one time. Pt is an excellent catalyst, and it works at moderately low temperatures. However, it is very expensive and is susceptible to poisoning, particularly by metals such as As. Nowadays a V_2O_5 catalyst activated with K_2O is used instead, and is supported on kieselguhr or silica. This is much cheaper, and is resistant to poisoning. The catalyst is inactive below 400 °C and breaks down between 600 °C and 650 °C. Dust may clog the catalyst surface, and impair its efficiency. To prevent this the gases are passed through an electrostatic precipitator. The catalyst may last for over 20 years. Most commercial plants are four-stage convertors. The gases are passed over four beds of catalyst in turn, and are cooled in between each catalyst bed. SO_3 is removed after passing over the first three beds, and again after the fourth bed.

SO_2 is used to make other products:

$$2SO_2 + Na_2CO_3 + H_2O \rightarrow \underset{\substack{\text{sodium hydrogen sulphite} \\ \text{(sodium bisulphite)}}}{2NaHSO_3} + CO_2$$

$$2NaHSO_3 + Na_2CO_3 \rightarrow \underset{\text{sodium sulphite}}{2Na_2SO_3} + H_2O + CO_2$$

$$Na_2SO_3 + S \xrightarrow{\text{heat}} \underset{\text{sodium thiosulphate}}{Na_2S_2O_3}$$

SO_2 has also been used as a non-aqueous solvent. A wide range of covalent compounds, both inorganic and organic, are soluble in liquid SO_2, and it is a useful reaction medium. At one time it was thought that SO_2 underwent self-ionization, with a system of 'acid–base' reactions in this solvent, but this is now known to be incorrect.

$$2SO_2 \rightleftharpoons SO^{2+} + SO_3^{2-}$$

SO_2 gas forms discrete V-shaped molecules, and this structure is retained in the solid state (Figure 15.3). The bond angle is 119°30′. The bonding in SO_2 is described in Chapter 4.

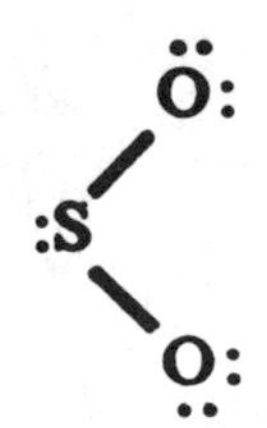

Figure 15.3 Structure of SO_2.

The dioxides SeO_2, TeO_2 and PoO_2 are made by burning the element in air. SeO_2 is solid at room temperature. The gas has the same structure as SO_2, but the solid forms infinite chains which are not planar (Figure 15.4). TeO_2 and PoO_2 both crystallize in two ionic forms.

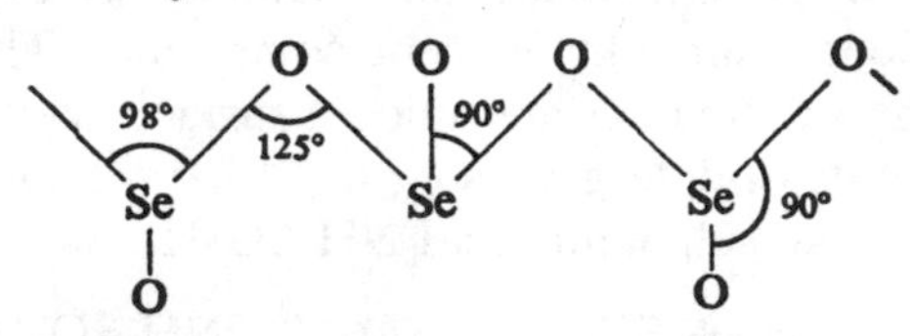

Figure 15.4 Structure of SeO_2.

The reaction of the dioxides with water also differs. SO_2 is very soluble, but largely forms hydrated SO_2 with only minute amounts of sulphurous acid H_2SO_3. (Dissolving SO_2 in alkali gives sulphites – salts of H_2SO_3.) Pure H_2SO_3 cannot be isolated. SeO_2 reacts with water and forms selenious acid, H_2SeO_3, which may be obtained in a crystalline state. TeO_2 is almost insoluble in water, but dissolves in alkali to form tellurites. It also dissolves in acids to form basic salts: this illustrates the amphoteric character of TeO_2. Thus there is the usual increase in basic character on descending a group. SeO_2 is used to oxidize aldehydes and ketones.

$$R{-}CH_2{-}CO{-}R + SeO_2 \rightarrow R{-}CO{-}CO{-}R + Se + H_2O$$

Trioxides MO_3

SO_3 is the only important trioxide. It is manufactured on a huge scale by the Contact process in which SO_2 reacts with O_2 in the presence of a catalyst (Pt or V_2O_5). (See under SO_2.) The SO_3 is not usually isolated, and practically all of it is converted to H_2SO_4. SO_3 reacts vigorously with water, evolving a large amount of heat and forming H_2SO_4. Commercially it is not possible just to react SO_3 with water. The SO_3 reacts with water vapour and causes the formation of a dense mist of H_2SO_4 droplets, which are difficult to condense and pass out of the absorber into the atmosphere. To avoid this, it has been found best to dissolve SO_3 in 98–99% H_2SO_4 in ceramic packed towers, to give oleum or fuming sulphuric acid. This is mainly pyrosulphuric acid $H_2S_2O_7$. Water is continuously added to keep the concentration of H_2SO_4 constant.

$$H_2S_2O_7 + H_2O \rightarrow 2H_2SO_4$$

In the gas phase SO_3 has a plane triangular structure (Figure 15.5). The bonding is best described as S forming three σ bonds, giving rise to a plane triangle, and three delocalized π bonds. (See Chapter 4.)

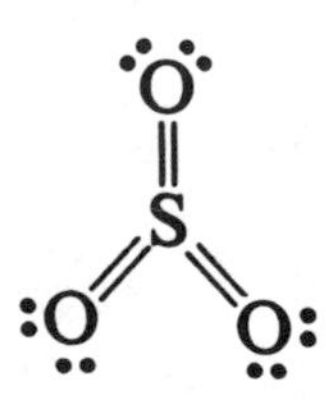

Figure 15.5 Structure of SO_3 gas.

At room temperature SO_3 is solid and exists in three distinct forms. γ-SO_3 is ice-like and is a cyclic trimer $(SO_3)_3$, m.p. 16.8 °C. If SO_3 is kept

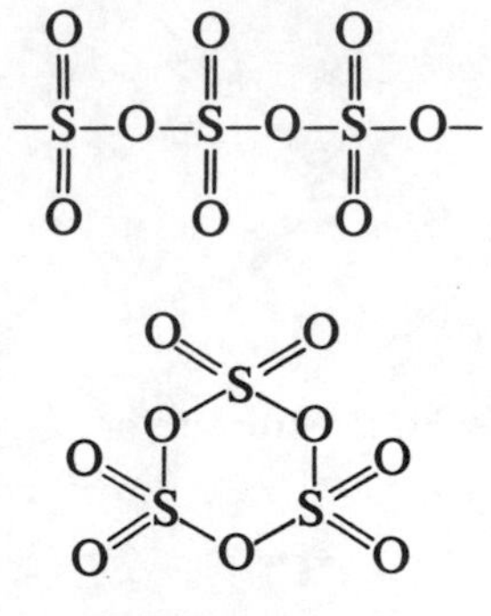

Figure 15.6 Structure of SO_3 chains and SO_3 cyclic trimer.

for a long time, or if a trace of water is present, either β-SO_3 or α-SO_3 is formed. Both look like asbestos, and comprise bundles of white silky needles. β-SO_3 (m.p. 32.5 °C) is made up of infinite helical chains of tetrahedral $[SO_4]$ units each sharing two corners. This structure is similar to that of chain phosphates. α-SO_3 (m.p. 62.2 °C) is the most stable form, and is made of chains cross-linked into sheets (Figure 15.6).

SO_3 is a powerful oxidizing agent, especially when hot. It oxidizes HBr to Br_2 and P to P_4O_{10}.

Commercially SO_3 is important in the manufacture of H_2SO_4, and also for sulphonating long chain alkylbenzene compounds. The sodium salts of these alkylbenzene sulphonates are anionic surface active agents, and are the active ingredients of detergents.

SO_3 is used to make sulphamic acid NH_2SO_3H

$$\underset{\text{urea}}{NH_2 \cdot CO \cdot NH_2} + SO_3 + H_2SO_4 \rightarrow \underset{\text{sulphamic acid}}{2NH_2SO_3H} + CO_2$$

Sulphamic acid is the only strong acid that exists as a solid at room temperature. It is used for cleaning the plant at sugar refineries and breweries, for desalination evaporators, and for destroying any excess nitrites present after dyeing with diazo dyestuffs.

SeO_3 is formed by the action of a silent electric discharge on Se and O_2 gases, and TeO_3 is formed from telluric acid H_6TeO_6 by strong heating. Both trioxides are acid anhydrides.

$$SeO_3 + H_2O \rightarrow H_2SeO_4 \quad \text{selenic acid}$$
$$TeO_3 + 3H_2O \rightarrow H_6TeO_6 \quad \text{telluric acid}$$

Other oxides

S_2O is formed when S and SO_2 are subjected to a silent electric discharge. It is very reactive, attacks metals and KOH, and will polymerize. It can exist for a few days at low pressures. It is of interest spectroscopically because of its similarity in structure to O_2. Until recently it was incorrectly formulated as SO.

A range of oxides from S_6O to $S_{10}O$ have been made by dissolving the cyclo forms of S_6, S_7, S_8, S_9 and S_{10} in CS_2 or CH_2Cl_2 and oxidizing with trifluoroperoxoacetic acid at −10 °C to −30 °C.

$$\text{cyclo-}S_8 + CF_3C(O\text{—}O)OH \rightarrow S_8O + CF_3COOH$$

These compounds are all orange–yellow in colour, and retain the original ring of S atoms, but have an oxygen atom double bonded to one of the ring S atoms.

TeO and PoO are obtained by thermal decomposition of a sulphite.

$$TeSO_3 \rightarrow TeO + SO_2$$

Detergents

Soap is the original detergent. It is excellent for cleaning, and is 100% biodegradable (i.e. it is completely broken down by bacteria in sewage disposal plants or rivers). It has a typical formula of $C_{17}H_{35}COO^-Na^+$. It has two disadvantages:

1. It forms an insoluble precipitate or 'scum' when hard water containing Ca^{2+} or Mg^{2+} is used.
2. It cannot be used for industrial purposes in acidic solutions, since the fatty acids are precipitated.

$$RCOO^- + H_2O^+ \rightleftharpoons RCOOH + H_2O$$

Detergents are surface active agents. They consist of molecules with a non-polar organic part, and a polar group. If a non-polar material is placed in an ionic solvent containing a detergent, then the detergent molecules will arrange themselves on the surface so that the non-polar part of the molecule points towards the non-polar material, and the polar group points towards the solvent. Thus particles of dirt become surrounded by detergent molecules, forming a micelle. This effectively 'dissolves' the dirt. The first new detergents, introduced in about 1950, were branched chain alkylbenzene sulphonates (ABS). They are called 'hard detergents' and a typical formula is:

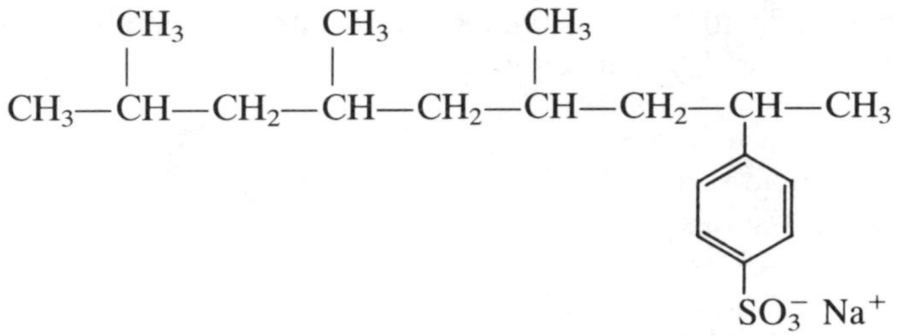

$$CH_3-\underset{\overset{|}{CH_3}}{CH}-CH_2-\underset{\overset{|}{CH_3}}{CH}-CH_2-\underset{\overset{|}{CH_3}}{CH}-CH_2-\underset{\underset{SO_3^-\ Na^+}{C_6H_4}}{CH}-CH_3$$

These are excellent detergents, with good surface active and cleaning properties. The problem is that waste water containing the detergent passes through the drains to the sewage treatment plants. The detergents cause problems with frothing in the sewage plants. Bacteria in the treatment plant break down various waste products, both sewage and the detergents. However, the detergent is only about 50–60% biodegradable. The bacteria cannot degrade the benzene ring, and they have difficulty and are slow at degrading branched chains. Thus much of the detergent is discharged into rivers and foams badly, especially if protein is present in the water.

'Soft' detergents also contain a non-polar and a polar part of the molecule, but they are linear alkylbenzene sulphonates (LAS) and have an unbranched aliphatic chain. They are 90% biodegradable. The straight alkyl chain is completely degraded, but the aromatic ring is not. Thus soft detergents cause considerably less pollution than do hard detergents.

If straight chain alcohols such as lauryl alcohol (available from coconut oil or tallow) are sulphonated, the product is an excellent detergent, which is rapidly and completely biodegraded. Alcohols of chain length C_{14} can be produced by the Ziegler–Natta process (see under 'Organometallic compounds', Chapter 12) and are useful for making biodegradable detergents such as Teepol.

At first sight it seems highly desirable to use biodegradable detergents. However, their use can create other problems. The bacteria responsible for degrading will grow rapidly, feeding on the degradable detergent. In growing they may use up all the oxygen dissolved in the water. The lack of oxygen kills all other forms of aquatic life such as fish or plants. Under extreme anaerobic conditions SO_4^{2-} ions may be reduced to H_2S, giving unpleasant smells.

OXOACIDS OF SULPHUR

The oxoacids of sulphur are more numerous and more important than those of Se and Te. Many of the oxoacids of sulphur do not exist as free acids, but are known as anions and salts. Acids ending in -ous have S in the oxidation state (+IV), and form salts ending in -ite. Acids ending in -ic have S in the oxidation state (+IV) and form salts ending in -ate.

As discussed previously under bond lengths and $p\pi - d\pi$ bonding, the oxoanions have strong π bonds and so they have little tendency to polymerize compared with the phosphates and silicates. To emphasize structural similarities the acids are listed in four series:

1. sulphurous acid series
2. sulphuric acid series
3. thionic acid series
4. peroxoacid series.

1. Sulphurous acid series

H_2SO_3 sulphurous acid	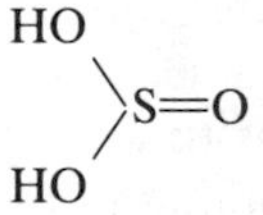 $(HO)_2S{=}O$	S(+IV)
$H_2S_2O_5$ di- or pyrosulphurous acid	$HO{-}S(=O)_2{-}S(=O){-}OH$	S(+V), S(+III)
$H_2S_2O_4$ dithionous acid	$HO{-}S(=O){-}S(=O){-}OH$	S(+III)

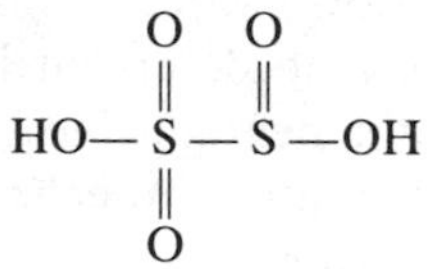

2. Sulphuric acid series

H_2SO_4 sulphuric acid	$HO{-}S(=O)_2{-}OH$	S(+IV)

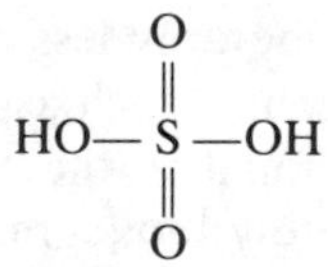

$H_2S_2O_3$ thiosulphuric acid

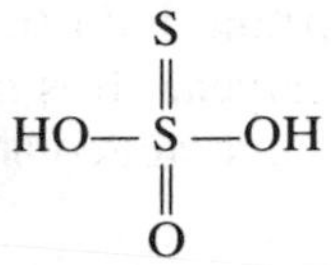

$$\begin{array}{c} S \\ \| \\ HO{-}S{-}OH \\ \| \\ O \end{array}$$

S(+IV), S(−II)

$H_2S_2O_7$ di- or pyrosulphuric acid

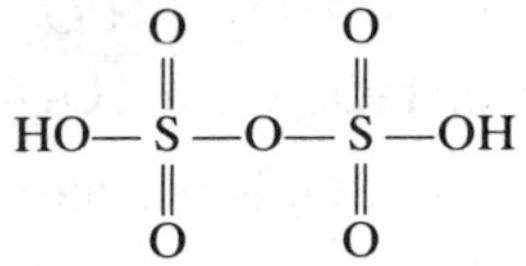

$$\begin{array}{ccccc} & O & & O & \\ & \| & & \| & \\ HO{-} & S & {-}O{-} & S & {-}OH \\ & \| & & \| & \\ & O & & O & \end{array}$$

S(+VI)

3. Thionic acid series

$H_2S_2O_6$ dithionic acid

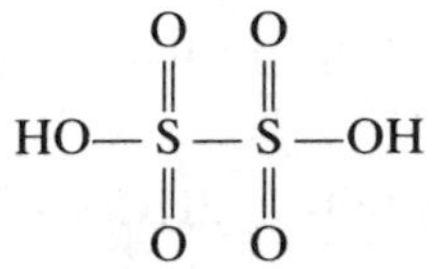

$$\begin{array}{ccccc} & O & & O & \\ & \| & & \| & \\ HO{-} & S & {-} & S & {-}OH \\ & \| & & \| & \\ & O & & O & \end{array}$$

S(+V)

$H_2S_nO_6$ polythionic acid (n = 1–12)

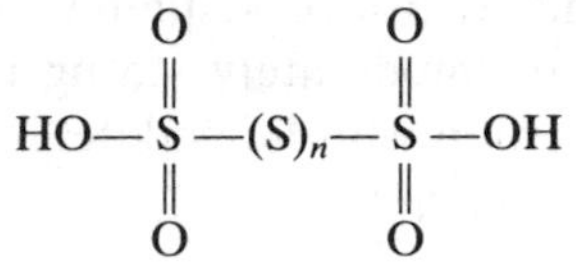

$$\begin{array}{ccccc} & O & & O & \\ & \| & & \| & \\ HO{-} & S & {-}(S)_n{-} & S & {-}OH \\ & \| & & \| & \\ & O & & O & \end{array}$$

S(+V), S(0)

4. Peroxoacid series

H_2SO_5 peroxomonosulphuric acid

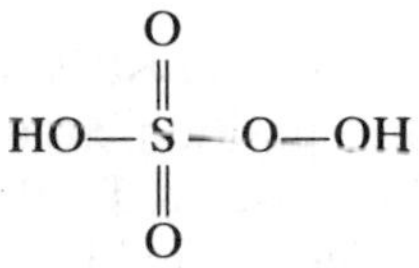

$$\begin{array}{ccc} & O & \\ & \| & \\ HO{-} & S & {-}O{-}OH \\ & \| & \\ & O & \end{array}$$

S(+VI)

$H_2S_2O_8$ peroxodisulphuric acid

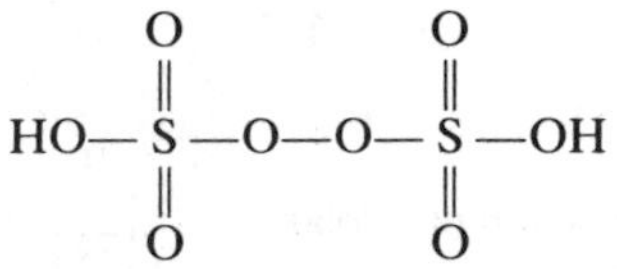

$$\begin{array}{ccccc} & O & & O & \\ & \| & & \| & \\ HO{-} & S & {-}O{-}O{-} & S & {-}OH \\ & \| & & \| & \\ & O & & O & \end{array}$$

S(+VI)

Sulphurous acid series

Though SO_2 is very soluble in water, most is present as hydrated SO_2 ($SO_2 \cdot H_2O$). Sulphurous acid H_2SO_3 may exist in the solution in minute amounts, or not at all, though the solution is acidic. Its salts, the sulphites SO_3^{2-}, form stable crystalline solids. Many sulphites are insoluble or are sparingly soluble in water, e.g. $CaSO_3$, $BaSO_3$ or Ag_2SO_3. However, those of the Group I metals and ammonium are soluble in water, and in dilute solutions the hydrogen sulphite (bisulphite) ion HSO_3^- is the predominant species. Crystals of hydrogen sulphites have only been formed with a few large metal ions, e.g. $RbHSO_3$ and $CsHSO_3$. Most attempts to isolate hydrogen sulphites lead to internal dehydration with the formation of disulphites $S_2O_5^{2-}$:

$$2NaHSO_3(aq) \rightleftharpoons Na_2S_2O_5 + H_2O$$

Na_2SO_3 is an important industrial chemical, and world production exceeds 1 million tonnes/year. It is made by passing SO_2 into an aqueous solution of Na_2CO_3 to give aqueous $NaHSO_3$, then treating the solution with more Na_2CO_3.

$$Na_2CO_3 + 2SO_2 + H_2O \rightarrow 2NaHSO_3 + CO_2$$
$$2NaHSO_3 + Na_2CO_3 \rightarrow 2Na_2SO_3 + H_2O + CO_2$$

The main use of Na_2SO_3 is as a bleach for wood pulp in the paper making industry. Some is used to treat boiler feed water (it removes O_2 and thus reduces corrosion of pipes and boilers). Small amounts are used in photographic developer.

Sulphites and hydrogen sulphites liberate SO_2 on treatment with dilute acids:

$$Na_2SO_3 + 2HCl \rightarrow 2NaCl + SO_2 + H_2O$$

Sulphites and hydrogen sulphites both contain S in the oxidation state (+IV) and are moderately strong reducing agents. Sulphites are determined by reaction with I_2, and determination of the excess I_2 with sodium thiosulphate.

$$NaHSO_3 + I_2 + H_2O \rightarrow NaHSO_4 + 2HI$$
$$2Na_2S_2O_3 + I_2 \rightarrow Na_2S_4O_6 + 2Na^+ + 2I^-$$

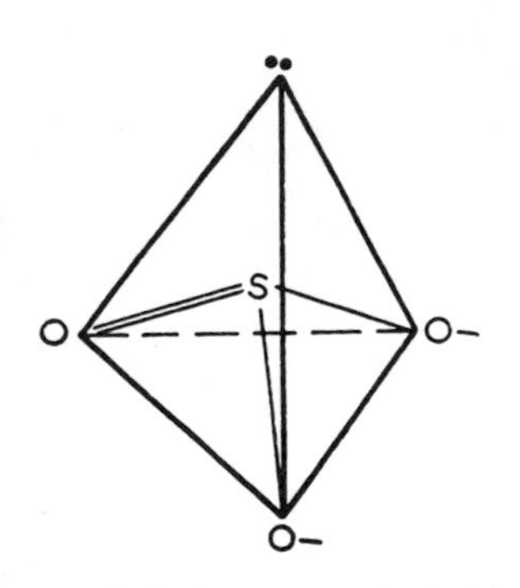

Figure 15.7 Structure of sulphite ion SO_3^{2-}.

The sulphite ion exists in crystals and has a pyramidal structure, that is tetrahedral with one position occupied by alone pair.The bond angles O—S—O are slightly distorted (106°) due to the lone pair, and the bond lengths are 1.51 Å. The π bond is delocalized, and hence the S—O bonds have a bond order of 1.33.

Electronic structure of sulphur atom – excited state

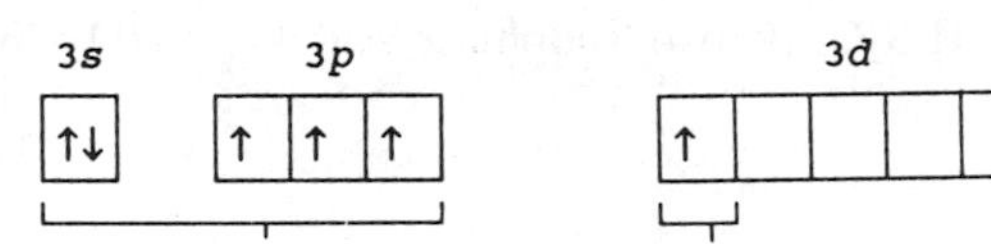

three unpaired electron form σ bonds forms π bond with three oxygen atoms
four electron pairs, hence tetrahedral with one position occupied by a lone pair

On heating solid hydrogen sulphites, or treating their solutions with SO_2, disulphites are formed. These contain an S—S linkage.

$$2RbHSO_3 \xrightarrow{\text{heat}} Rb_2S_2O_5 + H_2O$$
$$Na_2SO_3(aq) + SO_2 \rightarrow Na_2S_2O_5$$

$Na_2S_2O_5$ is called sodium disulphite, but in the past it has been called sodium pyrosulphite and sodium metabisulphite. The free acid $H_2S_2O_5$ is not known. Adding acid to disulphites gives SO_2.

$$Na_2S_2O_5 + HCl \rightarrow NaHSO_3 + NaCl + SO_2$$

On oxidation sulphites form sulphates, and with sulphur they form thiosulphates.

$$SO_3^{2-} + H_2O_2 \rightarrow SO_4^{2-} + H_2O$$
$$SO_3^{2-} + S \rightarrow S_2O_3^{2-} \quad \text{thiosulphate}$$

Reduction of sulphite solution plus SO_2 with Zn dust, or electrolytically, yields dithionites. These contain S in the oxidation state (+III).

$$2HSO_3^- + SO_2 \xrightarrow{Zn} \underset{\text{dithionite}}{S_2O_4^{2-}} + SO_3^{2-} + H_2O$$

$$2Na^+ \left[\begin{array}{c} \text{O} \quad \text{O} \\ \| \quad \| \\ \text{O—S—S—O} \end{array} \right]^{2-}$$

sodium dithionite

The dithionite ion has an eclipsed conformation, with a very long S—S bond (2.39 Å) and S—O bond lengths of 1.51 Å. Sodium dithionite $Na_2S_2O_4$ crystallizes out on adding NaCl to the mixture. The parent acid does not exist. $Na_2S_2O_4$ is a powerful reducing agent, which has a variety of industrial uses. These include bleaching paper pulp and making dyestuffs. It is used to treat water since it reduces many heavy metal ions (Pb^{2+}, Cu^+, Bi^{3+}) to the metal. In NaOH solution $Na_2S_2O_4$ is used to absorb oxygen in gas analysis. It is also used to preserve foodstuffs and fruit squashes.

Sulphuric acid series

H_2SO_4 is the most important acid used in the chemical industry. World production was 163 million tonnes in 1988. The main producers in millions tonnes/year are: USA 39; USSR 20; Japan 6; West Germany 4.5; France, UK, Poland and Canada all over 3. By far the most important commercial process for its manufacture is the Contact process, in which SO_2 is oxidized by air to SO_3, using a catalytic surface. Formerly a platinum gauze or platinized asbestos was used as catalyst. This has now been replaced by vanadium pentoxide, which is slightly less efficient but is cheaper and less easily poisoned. The SO_3 could be mixed with water to give H_2SO_4, but the reaction is violent and produces a dense chemical mist which is difficult to condense. Instead, the SO_3 is passed into 98% H_2SO_4, forming pyrosulphuric acid $H_2S_2O_7$, sometimes called oleum or fuming sulphuric acid. (Some trisulphuric acid $H_2S_3O_{10}$ is also formed.) This solution may be sold as oleum, or diluted with water to give concentrated sulphuric acid which is a 98% mixture with water (an 18 M solution).

$$H_2S_2O_7 + H_2O \rightarrow 2H_2SO_4$$
$$H_2S_3O_{10} + 2H_2O \rightarrow 3H_2SO_4$$

(The older lead chamber process is now obsolete. This used NO_2 as a homogeneous catalyst, to oxidize SO_2 in the presence of water. The NO formed reacts with air to reform NO_2.

Table 15.6 Uses of H_2SO_4 and oleum in the USA and UK

	USA	UK
Fertilizers	65%	32%
Manufacturing chemicals	5%	16%
Paint/pigments	2%	15%
Detergents	2%	11%
Fibres/cellulose film	2%	9%
Pickling metals	5%	2.5%
Refining petrol	5%	1%

$$NO_2 + SO_2 + H_2O \rightarrow H_2SO_4 + NO$$
$$2NO + O_2 \rightarrow 2NO_2$$

The disadvantages of the lead chamber process were that it only produced 78% H_2SO_4, not concentrated H_2SO_4, and the acid was less pure than that from the Contact process.)

The main uses of H_2SO_4 in the USA and the UK are compared in Table 15.6. The largest use is in converting calcium phosphate into superphosphate, which is used as a fertilizer. Fatty acids are sulphonated to make detergents. TiO_2 is the most widely used white pigment, and large amounts of H_2SO_4 are used to purify the mineral ilmenite ($FeTiO_3$). Pickling is the removal of oxides and scale from the surface of metals. H_2SO_4 is used as a catalyst in the production of high octane fuels by alkylating unsaturated hydrocarbons. H_2SO_4 has an imporant electrochemical use as the electrolyte in lead storage batteries.

Pure sulphuric acid melts at 10.5°C, forming a viscous liquid. It is strongly hydrogen bonded, and in the absence of water it does not react with metals to produce H_2. Many metals reduce H_2SO_4 (S +VI) to SO_2 (S +IV), especially if heated. If pure H_2SO_4 is heated, a little SO_3 is evolved, and an azeotropic mixture of 98.3% H_2SO_4 and 1.7% water is produced. This boils at 338°C. Pure H_2SO_4 is used as a non-aqueous solvent and as a sulphonating agent.

Anhydrous H_2SO_4 and concentrated H_2SO_4 mix with water in all proportions, and evolve a great deal of heat (880 kJ mol^{-1}). If water is poured into concentrated acid, the heat evolved leads to boiling of the drops of water and causes violent splashing. *The safe way to dilute strong acids is to carefully pour the acid into the water with stirring.*

Concentrated H_2SO_4 has quite strong oxidizing properties. Thus when NaBr is dissolved in concentrated H_2SO_4, HBr is formed but in addition some Br^- ions are oxidized to Br_2. Cu does not react with acids because it is lower than H in the electrochemical series. However, several noble metals such as Cu dissolve in concentrated H_2SO_4 due to its oxidizing properties. The oxidizing properties of SO_4^{2-} convert Cu into Cu^{2+}.

Concentrated H_2SO_4 absorbs water avidly, and is an effective drying agent for gases. It is sometimes used as a drying agent in desiccators. It dehydrates HNO_3, forming the nitronium ion NO_2^+, which is very important in then it ration of organic compounds.

$$HNO_3 + 2H_2SO_4 \rightarrow NO_2^+ + H_3O^+ + 2HSO_4^-$$

H_2SO_4 can also remove the elements of water, for example in the preparation of ethers.

$$2C_2H_5OH + H_2SO_4 \rightarrow C_2H_5 \cdot O \cdot C_2H_5 + H_2SO_4 \cdot H_2O$$

It removes water so strongly from some organic compounds that they char, and only the carbon remains. Paper and cloth are completely destroyed.

In dilute aqueous solution H_2SO_4 acts as a strong acid. The first proton dissociates very readily, and hence hydrogensulphates HSO_4^- formed. The second proton dissociates much less readily, to form sulphates SO_4^{2-}. Because of this, solutions of hydrogensulphates are acidic.

The SO_4^{2-} ion is tetrahedral. The bond lengths are all equal (1.49 Å) and are all rather short. The bond order of the S—O bonds is approximately 1.5. The bonding is best explained as four σ bonds between S and the O atoms, with two π bonds delocalized over the S and the four O atoms.

Thiosulphuric acid $H_2S_2O_3$ cannot be formed by adding acid to a thiosulphate because the free acid decomposes in water into a mixture of S, H_2S, H_2S_n, SO_2 and H_2SO_4. It can be made in the absence of water (e.g. in ether) at low temperatures (−78 °C).

$$H_2S + SO_3 \xrightarrow{\text{ether}} H_2S_2O_3 \cdot (Et_2O)_n$$

In contrast, the salts which are called thiosulphates are stable and numerous. Thiosulphates are made by boiling alkaline or neutral sulphite solutions with S, and also by oxidizing polysulphides with air.

$$Na_2SO_3 + S \xrightarrow{\text{boiling water}} Na_2S_2O_3$$

$$2Na_2S_3 + 3O_2 \xrightarrow{\text{heat in air}} 2Na_2S_2O_3 + 2S$$

The thiosulphate ion is structurally similar to the sulphate ion (Figure 15.8).

Hydrated sodium thiosulphate $Na_2S_2O_3 \cdot 5H_2O$ is called 'hypo'. It forms very large colour less hexagonal crystals, m.p. 48 °C. It is readily soluble in water and solutions are used for iodine titrations in volumetric analysis. Iodine very rapidly oxidizes thiosulphate ions $S_2O_3^{2-}$ to tetrathionate ions $S_4O_6^{2-}$, and the I_2 is reduced to I^- ions.

$$2Na_2S_2O_3 + I_2 \rightarrow \underset{\text{sodium tetrathionate}}{Na_2S_4O_6} + 2NaI$$

$Na_2S_2O_3$ is used in the bleaching industry to destroy any excess Cl_2 on fabrics after they have been through a bleach bath. Similarly $Na_2S_2O_3$ is sometimes used to remove the taste from heavily chlorinated drinking

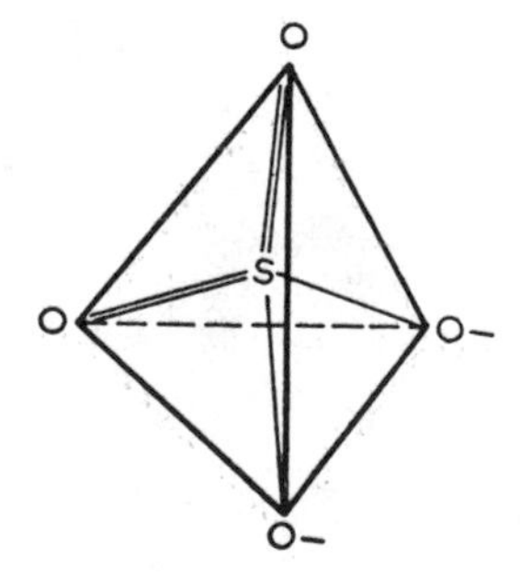

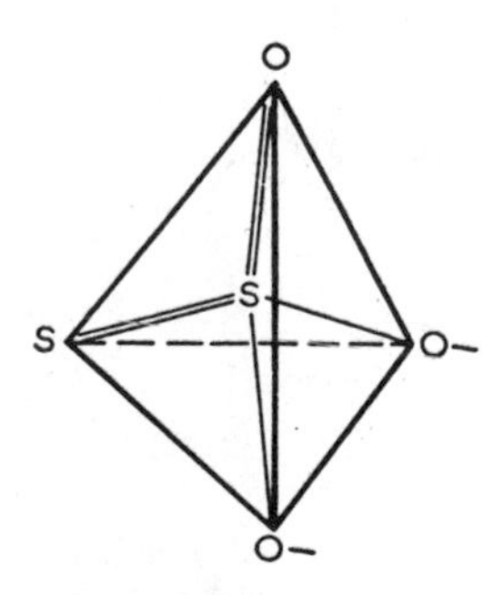

Figure 15.8 Structure of sulphate and thiosulphate ions.

water. Since Cl_2 is a stronger oxidizing agent than I_2, hydrogensulphate ions are formed rather than tetrathionate ions.

$$Na_2S_2O_3 + 4Cl_2 + 5H_2O \rightarrow 2NaHSO_4 + 8HCl$$

Hypo is used in photography for 'fixing' films and prints. Photographic emulsions are made of $AgNO_3$, AgCl and AgBr. Parts of the film exposed to light begin to decompose to Ag, thus forming a negative image. The process is enhanced by the developer solution. After developing, the film or print is put in a solution of hypo ($Na_2S_2O_3$). This forms a soluble complex with silver salts, thus dissolving any unchanged silver salts in the photographic emulsion. When there is no photographic emulsion left, the film or print can safely be exposed to light.

$$Na_2S_2O_3 + AgBr \rightarrow Ag_2S_2O_3 \xrightarrow{+2Na_2S_2O_3} Na_5[Ag(S_2O_3)_3]$$

Pyrosulphates can be made by heating hydrogensulphates strongly, or by dissolving SO_3 in H_2SO_4. Some trisulphuric acid $H_2S_3O_{10}$ is also formed, but polysulphuric acids higher than this are not known.

$$2NaHSO_4 \rightarrow Na_2S_2O_7 + H_2O$$

$$H_2SO_4 + SO_3 \rightarrow H_2S_2O_7$$

$$H_2SO_4 + 2SO_3 \rightarrow H_2S_3O_{10}$$

Thionic acid series

Dithionic acid $H_2S_2O_6$ is known only in solution. The acid is dibasic, and salts called dithionates are known, e.g. $Na_2S_2O_6$. No acid salts exist. The acid and its salts contain S in the oxidation state (+V). Dithionates can be made by oxidizing a sulphite, but on a larger scale they are made by oxidizing a cooled aqueous solution of SO_2 with MnO_2 or Fe_2O_3.

$$2MnO_2 + 3SO_2 \rightarrow Mn^{II}S_2O_6 + MnSO_4$$

Most dithionates are readily soluble in water. They can be isolated as BaS_2O_6, and converted into other salts by double decomposition reactions. Dithionates are stable to mild oxidizing agents, though strong oxidizing agents, such as $KMnO_4$ or the halogens, oxidize them to sulphate. Similarly, mild reducing agents have no effect but strong reducing agents reduce dithionates to dithionites and sulphites.

$$2Na_2S_2O_6 + 2Na/Hg \rightarrow Na_2S_2O_4 + 2Na_2SO_3 + Hg$$

$$\begin{array}{c} \quad\ \ O \quad O \\ \quad\ \ \| \quad\ \| \\ HO—S—S—OH \\ \quad\ \ \| \quad\ \| \\ \quad\ \ O \quad O \end{array}$$

dithionic acid

The dithionate ion has a structure similar to that of ethane, but the two SO_3 groups adopt an almost eclipsed conformation. The S—S length is

2.15 Å and the S—O bonds are 1.43 Å – again rather short. The bond angles S—S—O are close to tetrahedral (103°).

A range of polythionates have been known as salts since the early work of Wakenroder on the effect of H_2S on aqueous solutions of SO_2. Ions such as trithionate $S_3O_6^{2-}$, tetrathionate $S_4O_6^{2-}$, pentathionate $S_5O_6^{2-}$ and hexathionate $S_6O_6^{2-}$ are named according to the total number of S atoms present. It is only in recent times that the parent acids have been made.

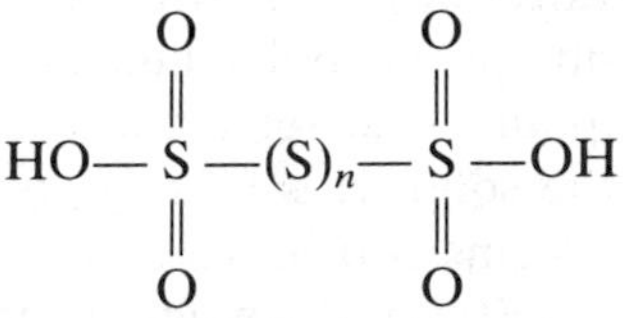

$$\mathrm{HO{-}\overset{\overset{\displaystyle O}{\|}}{\underset{\underset{\displaystyle O}{\|}}{S}}{-}(S)_n{-}\overset{\overset{\displaystyle O}{\|}}{\underset{\underset{\displaystyle O}{\|}}{S}}{-}OH}$$

polythionic acid

Peroxoacid series

The name peroxo indicates that the compound contains an —O—O— linkage. Two peroxoacids of sulphur are known: peroxomonosulphuric acid H_2SO_5 and peroxodisulphuric acid $H_2S_2O_8$. No peroxoacids of Se and Te are known. $H_2S_2O_8$ is a colourless solid, m.p. 65 °C. It is obtained by electrolysis of sulphates at high current density. It is soluble in water, and is a powerful and useful oxidizing agent. It will convert Mn^{2+} to permanganate and Cr^{3+} to chromate. The most important salts are $(NH_4)_2S_2O_8$ and $K_2S_2O_8$. $(NH_4)_2S_2O_8$ is used to initiate the polymerization of vinyl acetate in the production of synthetic rayon, and of tetrafluoroethylene in the formation of PTFE. $K_2S_2O_8$ is used as an initiating agent in the polymerization of vinyl chloride to PVC and styrene—butadiene copolymer rubbers.

Hydrolysis of peroxodisulphuric acid gives peroxomonosulphuric acid, H_2SO_5, which is often called Caro's acid.

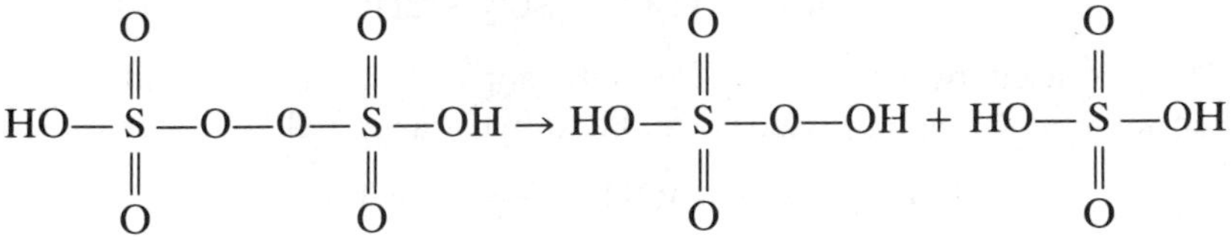

$$\mathrm{HO{-}\overset{\overset{\displaystyle O}{\|}}{\underset{\underset{\displaystyle O}{\|}}{S}}{-}O{-}O{-}\overset{\overset{\displaystyle O}{\|}}{\underset{\underset{\displaystyle O}{\|}}{S}}{-}OH \rightarrow HO{-}\overset{\overset{\displaystyle O}{\|}}{\underset{\underset{\displaystyle O}{\|}}{S}}{-}O{-}OH + HO{-}\overset{\overset{\displaystyle O}{\|}}{\underset{\underset{\displaystyle O}{\|}}{S}}{-}OH}$$

H_2SO_5 can also be made from chlorosulphuric acid:

$$(HO)(Cl)SO_2 + H_2O_2 \rightarrow (HO)(HOO)SO_2 + HCl$$

It forms colourless crystals, m.p. 45 °C, but must be handled with care since it may explode.

OXOACIDS OF SELENIUM AND TELLURIUM

Selenium forms two oxoacids, selenious acid H_2SeO_3 and selenic acid H_2SeO_4. Selenious acid is formed when SeO_2 dissolves in water. The solid acid can be isolated and two series of salts (the normal and acid selenites) containing SeO_3^{2-} and $HSeO_3^-$ are known. The acid is converted to selenic acid by refluxing with H_2O_2. Pyroselenates containing $Se_2O_7^{2-}$ can be made by heating selenates, but the free acid is not known.

H_2SeO_4, like H_2SO_4, is a strong acid, and selenates are isomorphous with sulphates. Both H_2SeO_4 and H_2SeO_3 are moderately strong oxidizing agents.

TeO_2 is almost insoluble in water so that tellurous acid has not been characterized. The dioxide does react with strong bases and forms tellurites, acid tellurites and various polytellurites. Telluric acid H_6TeO_6 is quite different from sulphuric and selenic acids and exists as octahedral $Te(OH)_6$ molecules in the solid. It is a fairly strong oxidizing agent, but a weak dibasic acid and forms two series of salts, examples of which are $NaTeO(OH)_5$ and $Li_2TeO_2(OH)_4$. The acid is prepared by the action of powerful oxidizing agents such as $KMnO_4$ on Te or TeO_2.

OXOHALIDES

Thionyl compounds

Only S and Se form oxohalides. These are called thionyl and selenyl halides, and the following are known.

SOF_2	$SOCl_2$	$SOBr_2$
$SeOF_2$	$SeOCl_2$	$SeOBr_2$

Thionyl chloride $SOCl_2$ is a colourless fuming liquid, b.p. 78 °C, and is usually prepared as follows:

$$PCl_5 + SO_2 \rightarrow SOCl_2 + POCl_3$$

Most thionyl compounds are readily hydrolysed by water, though SOF_2 only reacts slowly.

$$SOCl_2 + H_2O \rightarrow SO_2 + 2HCl$$

$SOCl_2$ is used by organic chemists to convert carboxylic acids to acid chlorides, and it is also used to make anhydrous metal chlorides.

$$SOCl_2 + R{-}COOH \rightarrow R{-}COCl + SO_2$$

Thionyl bromide is prepared from the chloride and HBr, and thionyl fluoride is obtained from the chloride by reacting with SbF_3.

The structure of these oxohalides is tetrahedral with one position occupied by a lone pair.

Sulphuryl compounds

The following sulphuryl halides are known:

SO_2F_2	SO_2Cl_2	SO_2FCl	SO_2FBr
SeO_2F_2			

Sulphuryl chloride SO_2Cl_2 is a colourless fuming liquid, b.p. 69 °C, and is made by direct reaction of SO_2 and Cl_2 in the presence of a catalyst. It is used as a chlorinating agent. Sulphuryl fluoride is a gas and is not hydrolysed by water, but the chloride fumes in moist air and is hydrolysed by water. The sulphuryl halides have a distorted tetrahedral structure. They may be regarded as derivatives of H_2SO_4, where both OH groups have been replaced by halogens. If only one group is replaced, halosulphuric acids are obtained.

$$FSO_3H \qquad ClSO_3H \qquad BrSO_3H$$

Fluorosulphuric acid forms many salts, but chlorosulphuric acid forms none and is used as a chlorinating agent in organic chemistry.

HYDRIDES

The elements all form covalent hydrides. These are water H_2O, hydrogen sulphide H_2S, hydrogen selenide H_2Se, hydrogen telluride H_2Te and hydrogen polonide H_2Po. Water is liquid at room temperature, but the others are all colourless, foul smelling toxic gases. All but H_2Te can be made from the elements. It is easier to make H_2S, H_2Se and H_2Te by the action of mineral acids on metal sulphides, selenides and tellurides, or hydrolysis:

$$FeS + H_2SO_4 \rightarrow H_2S + FeSO_4$$

$$FeSe + 2HCl \rightarrow H_2Se + FeCl_2$$

$$Al_2Se_3 + 6H_2O \rightarrow 3H_2Se + 2Al(OH)_3$$

$$Al_2Te_3 + 6H_2O \rightarrow 3H_2Te + 2Al(OH)_3$$

H_2Te can also be made by electrolysing cooled dilute H_2SO_4 with a Te cathode. H_2Po has only been obtained in trace amounts from a mixture of Mg, Po and dilute acid.

H_2S, H_2Se and H_2Te are all soluble in water and all burn in air with a blue flame.

$$2H_2S + 3O_2 \rightarrow 2H_2O + 2SO_2$$

H_2S is about twice as soluble in water as CO_2, and 1 volume of water can absorb 4.6 volumes of H_2S at 0 °C and 2.6 volumes at 20 °C. A saturated solution is used as a laboratory reagent, but it does not keep well since air slowly oxidizes it and sulphur is deposited. H_2S is a very weak dibasic acid. Most metal sulphides can be regarded as salts of H_SS, and since it is dibasic two series of salts can be derived from it, the hydrogen sulphides, e.g. NaHS, and the normal sulphides, e.g. Na_2S.

$$H_2S + NaOH \rightarrow NaHS + H_2O$$

$$H_2S + NaHS \rightarrow Na_2S + H_2O$$

The alkali metal sulphides are all soluble in water and hydrolyse strongly (a 1 M solution is about 90% hydrolysed), so they are strongly basic.

$$Na_2S + H_2O \rightarrow NaHS + NaOH$$

Most of the sulphides of the heavy metals are insoluble in water and so do not hydrolyse. If a dilute solution of ammonia is saturated with H_2S then ammonium hydrogen sulphide NH_4HS is formed, not $(NH_4)_2S$. The latter only exists at low temperatures in the absence of water. Solutions of NH_4HS are colourless, and when mixed with an equimolar amount of NH_3 it is used as a laboratory reagent. More commonly 'yellow ammonium sulphide' is used as a laboratory reagent, for example to precipitate metal sulphides in qualitative analysis. Yellow ammonium sulphide is really a mixture of ammonium polysulphides, and is made by dissolving sulphur in colourless NH_4HS/NH_3 solution.

Water

Water is the most abundant chemical compound, and the oceans cover almost 71% of the earth's surface. For this reason we have no need to prepare water. However, sea water contains many dissolved salts, and less than 3% of the earth's water is fresh water, and most of that is present as polar ice. The preparation of pure water for drinking and laboratory use is a major industry. The human body is more tolerant of some impurities than is industry. The EEC have set limits for impurities in drinking water.

Table 15.7 EEC limits for contaminants in drinking water

	($\mu g\ litre^{-1}$)	Source of contaminant and suspected problems
Al	200	$Al_2(SO_4)_3$ is added to water in some places to clarify it. Al may be related to Alzheimer's disease (senile dementia)
Pb	50	Water pipes made of lead are the problem. Pb damages the brains of children
NO_3^-	50	Nitrates are added as fertilizers to make crops grow. This gets into the water supply. Nitrates affect the level of O_2 in the blood of babies, and cause 'blue baby syndrome'. Nitrates may be related to stomach cancer
Trihalomethanes	100	Water supplies are treated with chlorine to kill bacteria. Over-chlorination may allow chloroform to be formed by reaction with peat. This is a possible cause of cancer of the gut and bladder
Pesticides	0.1	For individual pesticides
	0.5	For all pesticides together DDT is now banned in many places. Its harmful effects occur because it accumulates and undergoes biological amplification in the food chain, and UV light converts DDT into polychlorinated biphenyls which are toxic

Very pure water is required for laboratory and industrial use. The only way to remove all solid solutes is distillation. This is expensive since water has a high boiling point and high latent heat of evaporation. During distillation the water tends to dissolve appreciable amounts of CO_2 from the atmosphere, which make it acidic. A cheaper method is to produce deionized water. This is done by passing the water down two different ion-exchange columns one after the other. (Alternatively a 'mixed bed' may be used, i.e. a single column made up of two different ion-exchange materials.) Ion-exchange resins are insoluble polymeric solids, containing a reactive group. They are manufactured in bead form, and are permeable to water. The first column contains a sulphonic acid resin, that is an organic resin with acidic groups $—SO_3H$. This removes all metal ions from solution and replaces them with H^+:

$$\text{resin-SO}_3\text{H} + \text{Na}^+ \rightarrow \text{resin-SO}_3\text{Na} + \text{H}^+$$

$$2\ \text{resin-SO}_3\text{H} + \text{Ca}^{2+} \rightarrow (\text{resin-SO}_3)_2\text{Ca} + 2\text{H}^+$$

The second column contains a resin with basic groups $-NR_4^+OH^-$, which removes negative ions.

$$\text{resin-N(CH}_3)_4^+\ \text{OH}^- + \text{Cl}^- \rightarrow \text{resin-N(CH}_3)_4^+\ \text{Cl}^- + \text{OH}^-$$

Water produced in this way usually contains soluble silicates and CO_2. When all the reactive sites on the resin have been used, the resins can be regenerated by treating the first one with dilute H_2SO_4, and the second one with a Na_2CO_3 solution.

Drinking water is usually much less pure, and water with no dissolved salts does not taste very nice. The World Health Organization recommends a maximum desirable content of not more than 0.5 grams of dissolved solids per litre, though the maximum permissible level is three times this. If the fresh water source contains silt, this is allowed to settle out. Light suspended particles and colloidal particles which discolour the water are removed by treating with $Al(OH)_3$ or $Fe(OH)_3$. These coagulate suspended particles, thus clarifying the water. (Alum is the most widely used coagulating agent.) If necessary some water softening may be performed by ion-exchange, or by mixing water from different sources. The water is then chlorinated, or treated with ozone to kill bacteria. These are present because of drainage from fields into rivers and lakes, and also from the disposal of partly treated and untreated sewage. Failure to treat drinking water is the major cause of enteritis. In some underdeveloped parts of the world up to half the children under the age of five die from this cause or other waterborne diseases.

Sea water has a high salt content. The production of drinking water and water for crops from sea water is called desalination. It requires a large amount of energy, and is therefore expensive. It is only carried out when the shortage of fresh water is severe, but it has become increasingly important in arid regions like the Persian Gulf. Distillation, ion-exchange, electrodialysis, reverse osmosis and the freezing out of ice have all been used.

Table 15.8 Some properties of H_2O, H_2S, H_2Se and H_2Te

	Enthalpies of formation ($kJ\,mol^{-1}$)	Bond angle	Boiling point (°C)
H_2O	−242	H—O—H = 104°28′	100
H_2S	−20	H—S—H = 92°	−60
H_2Se	+81	H—Se—H = 91°	−42
H_2Te	+154		−2.3

Apart from water, the other hydrides are all poisonous and have unpleasant odours. The hydrides decrease in stability from H_2O to H_2S to H_2Se to H_2Te. (This is shown by the decrease in their enthalpies of formation – Table 15.8.) They become less stable because the bonding orbitals become larger and more diffuse: hence overlap with the hydrogen 1*s* orbital is less effective.

The H—O—H bond angle in water is 104°28′, in accordance with the VSEPR prediction of slightly less than tetrahedral due to the presence of lone pairs of electrons. Thus the orbitals used for bonding by O are close to sp^3 hybrids. In H_2S, H_2Se and H_2Te the bond angles become close to 90°. This suggests that almost pure *p* orbitals on Se and Te are used for bonding to hydrogen.

In a series of similar compounds, the boiling points usually increase as the atoms become larger and heavier. If the boiling points increase, then the volatility decreases. This trend is shown by the boiling points of H_2S, H_2Se, H_2Te and H_2Po, but the boiling point of water is anomalous.

Water has an abnormally low volatility because its molecules are associated with each other by means of hydrogen bonds in both the solid and liquid states. The structure of liquid water is not known for certain, but probably consists of groups of two or three molecules hydrogen bonded together. The structure of ordinary hexagonal ice is known. At high pressures other more dense structures are formed. A total of nine different forms of ice are known. X-ray studies do not often reveal the positions of H atoms. In this case the H positions were found by neutron diffraction on solid deuterium oxide D_2O. The structure is similar to wurtzite ZnS (see Chapter 3), with O atoms occupying both the Zn^{2+} and the S^{2-} positions. The H atoms are located just off the line joining two O atoms, and the O—H. . .O angle is 104°28′. The strength of a hydrogen bond is about 20 $kJ\,mol^{-1}$. This association is responsible for the abnormally high boiling point and melting point of water.

The H bonding is the main reason why covalent compounds have a very low solubility in water. When two substances mix there is an increase in entropy since the order decreases. Thus mixing is always favoured. However, in the case of water, dissolving something means that hydrogen bonds must be broken. Unless there is an interaction between the dissolved material and water greater than the energy lost through breaking hydrogen bonds, then the material will not dissolve. Covalent materials have little interaction with water, and so are insoluble. Ionic materials become hydrated, and polar materials take part in the hydrogen bonding, so they are soluble.

A unique property of water is that the solid is less dense than the liquid. This is why lakes and the sea freeze from the top downwards. The ice at the top makes it difficult to cool the water underneath, so even at the North Pole there is water underneath the ice. But for this the sea would freeze solid and the polar ice-caps would cover much more of the earth's surface. The maximum density of water occurs at 4 °C. On melting, the hydrogen bonded network in the solid partly breaks down. Ice has a rather open structure, with quite large cavities. On partial melting some 'free' water molecules occupy some of these cavities, and hence the density increases. This effect outweighs the effect of thermal expansion up to 4 °C, but above this temperature expansion is the larger effect so the density decreases.

An unusual form of water called 'polywater' was reported and extensively studied between 1966 and 1973. Polywater was reported to have a freezing point of −40 °C and a very high density of $1.4\,g\,cm^{-3}$. It was obtained when water was formed in glass or quartz capillary tubes. This caused excitement at first because polywater was thought to comprise a larger number of water molecules polymerized together. It is now known to be a colloidal mixture of silicates, and a variety of ions Na^+, K^+, Ca^{2+}, BO_3^{3-}, NO_3^-, SO_4^{2-} and Cl^- which had been leached from the glass!

Other hydrides

The hydrides dissociate to a varying degree, forming H^+ ions. They are all very weak acids and there is an increase in acidic strength from H_2O to H_2Te. The large difference in electronegativity taken in conjunction with Fajans' rules (the larger the negative ion the greater the tendency to covalency) suggest that H_2Te *gas* should be the most covalent. Acidic behaviour *in solution* is discussed for the halogen acids (see Chapter 16) and depends on the enthalpy of formation of the molecule, the ionization energy, electron affinity and enthalpies of hydration. In the compounds H_2O, H_2S, H_2Se and H_2Te the most important factor is the enthalpy of formation, the values being −120, −10, +43 and $+77\,kJ\,mol^{-1}$. The stability decreases (the last two are in fact thermodynamically unstable), thus accounting for the greater dissociation of H_2Te.

$$H_2Te_{(hydrated)} + H_2O \rightleftharpoons H_3O^+ + HTe^-_{(hydrated)}$$

The more acidic the hydrogen atom in the hydrides, the more stable will be the salts formed from them, i.e. oxides, sulphides, selenides and tellurides.

Peroxides and polysulphides

Oxygen, and to a greater extent sulphur, differ from the remainder in their ability to catenate and form polyoxides and polysulphides, which are less stable than the normal salts. Unbranched polysulphane chains containing up to eight sulphur atoms have been prepared.

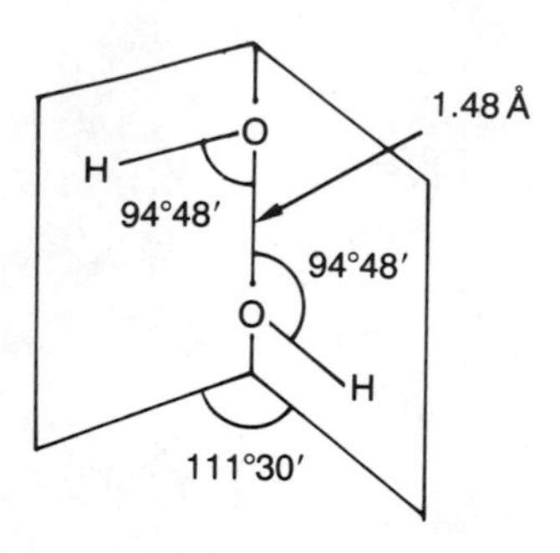

Figure 15.9 Structure of H_2O_2 in the gas phase.

H_2O_2 H—O—O—H H_2S_2 H—S—S—H
H_2S_3 H—S—S—S—H
H_2S_4 H—S—S—S—S—H

H_2O_2 and H_2S_2 have similar 'skew' structures. The dimensions of H_2O_2 gas are shown in Figure 15.9. H_2O_2 is the smallest molecule known to show restricted rotation, in this case about the O—O bond, and this is presumably due to repulsion between the OH groups. A similar structure is retained in the liquid and solid, but the bond lengths and angles are slightly changed because of hydrogen bonding.

H_2O_2 and H_2S_2 can be prepared by the addition of acid to a peroxide or a persulphide salt.

$$BaO_2 + H_2SO_4 \rightarrow H_2O_2 + BaSO_4$$
$$Na_2S_2 + H_2SO_4 \rightarrow H_2S_2 \text{ (also } H_2S_3) + Na_2SO_4$$

In most of its reactions H_2O_2 acts as a strong oxidizing agent. In acidic solutions these reactions are often slow, but in basic solution they are usually fast. H_2O_2 will oxidize Fe^{2+} to Fe^{3+}, $[Fe^{II}(CN)_6]^{4-}$ (ferrocyanide) to $[Fe^{III}(CN)_6]^{3-}$ (ferricyanide), NH_2OH to HNO_3 and SO_3^{2-} to SO_4^{2-}.
Ionic peroxides such as Na_2O_2 give H_2O_2 with water or dilute acids. Na_2O_2 reacts with gaseous CO_2.

$$2Na_2O_2 + 2CO_2 \rightarrow 2Na_2CO_3 + O_2$$

Heating Na_2O_2 with many organic compounds results in their oxidation to carbonates. Fusing Na_2O_2 with Fe^{2+} salts gives sodium ferrate $Na_2[FeO_4]$ which contains Fe(+VI).

With stronger oxidizing agents H_2O_2 is oxidized, that is H_2O_2 is forced to act as a reducing agent, and in such cases O_2 is always evolved.

$$2KMnO_4 + 5H_2O_2 + 3H_2SO_4 \rightarrow 2MnSO_4 + K_2SO_4 + 5O_2 + 8H_2O$$
$$KIO_4 + H_2O_2 \rightarrow KIO_3 + O_2 + H_2O$$
$$2Ce(SO_4)_2 + H_2O_2 \rightarrow Ce_2(SO_4)_3 + 2H_2SO_4 + O_2$$

H_2S_2 is not oxidizing. H_2O_2 is fairly stable and decomposes only slowly in the absence of catalysts. H_2S_2 is less stable, and its decomposition is catalysed by hydroxyl ions.

$$H_2O_2 \rightarrow H_2O + \tfrac{1}{2}O_2$$
$$H_2S_2 \rightarrow H_2S + S$$

Hydrogen polyselenides and polytellurides do not exist, but some of their salts are known.

Hydrogen peroxide

Pure H_2O_2 is a colourless liquid which resembles water quite closely. It is more hydrogen bonded than is water and so has a higher boiling point (b.p. 152 °C, m.p. −0.4 °C). It is more dense than water (density 1.4 g cm^{-3}). Though it has a high dielectric constant, it is of little use as an ionizing solvent because it is decomposed by many metal ions, and it oxidizes many compounds.

H_2O_2 is a major industrial chemical, and over 824 400 tonnes were produced in 1985. It is used extensively as a mild bleaching agent for textiles and paper/wood pulp. It is used for several environmental purposes: to restore aerobic conditions in sewage waters, and to oxidize cyanides and sulphides. It is an important rocket fuel. It is also used for making other chemicals, particularly sodium peroxoborate $Na_2[B_2(O_2)_2(OH)_4] \cdot 6H_2)$ (annual production 550 000 tonnes/year), which is used as a brightener in washing powders (see Chapter 12). Organic peroxides are used to initiate addition polymerization reactions (PVC, polyurethanes and epoxy resins), and sodium chlorite $NaClO_2$ is used for bleaching. Smaller amounts of H_2O_2 are used to bleach hair, feathers, fats and waxes. It is used as an oxidizing agent in the laboratory, and as an antiseptic to treat wounds. It is useful to counteract chlorine, and in this reaction H_2O_2 behaves as a reducing agent.

$$H_2O_2 + Cl_2 \rightarrow 2HCl + O_2$$

H_2O_2 is unstable, and the rate at which it decomposes (disproportionates) depends on the temperature and concentration. Many impurities catalyse the decomposition, which may become very violent, especially with concentrated solutions. Catalysts include metal ions Fe^{2+}, Fe^{3+}, Cu^{2+}, Ni^{2+}, metal surfaces such as Pt or Ag, MnO_2, charcoal or alkali – even the small amount leached from glass.

$$2H_2O_2 \rightarrow 2H_2O + O_2$$

At one time H_2O_2 was obtained by electrolysis of H_2SO_4 or $(NH_4)_2SO_4$ at a high current density to form peroxosulphates, which were then hydrolysed.

$$2SO_4^{2-} \xrightarrow{\text{electrolysis}} S_2O_8^{2-} + 2e$$

$$\underset{\text{peroxodisulphuric acid}}{H_2S_2O_8} + H_2O \rightarrow \underset{\text{peroxomonosulphuric acid}}{H_2SO_5} + H_2SO_4$$

$$H_2SO_5 + H_2O \rightarrow H_2SO_4 + H_2O_2$$

H_2O_2 is now produced on an industrial scale by a cyclic process (Figure 15.10). 2-Ethyl anthroquinol is oxidized by air to the corresponding quinone and H_2O_2. The anthraquinone is reduced back to anthraquinol with hydrogen at a moderate temperature using platinum, palladium or Raney nickel as catalyst. The cycle is then repeated. The reaction is carried out in a mixture of organic solvents (ester/hydrocarbon or octanol/methylnaphthalene). The solvent must:

1. dissolve the quinol and quinone
2. resist oxidation
3. be immiscible with water.

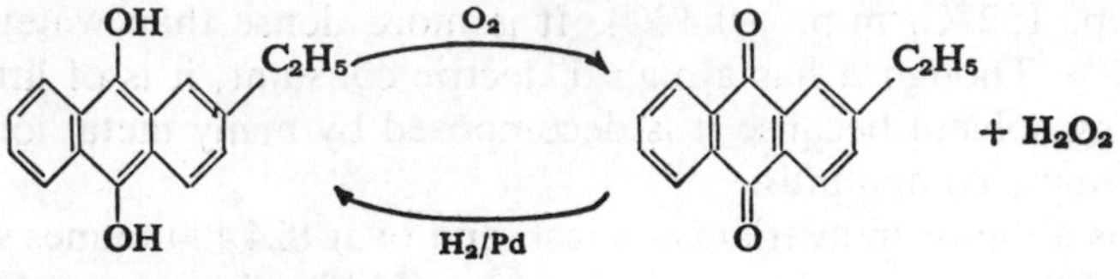

Figure 15.10 Production of H_2O_2.

The H_2O_2 is extracted with water as a 1% solution. This is concentrated by distillation under reduced pressure, and sold as a 30% (by weight) solution which has a pH of about 4.0 (85% solutions are also produced). H_2O_2 solutions are stored in plastic or wax coated glass vessels, often with negative catalysts such as urea or sodium stannate added as stabilizers. Solutions keep quite well, but must be handled with care since they may explode with traces of organic material or specs of dust.

HALIDES

Compounds with the halogens are listed in Table 15.9. Since F is more electronegative than O, binary compounds are oxygen fluorides, whereas similar chlorine compounds are chlorine oxides. Some of these compounds including the oxides of iodine, are therefore described in Chapter 16 under 'Halogen oxides'.

Fluorine brings out the maximum valency of six with S, Se and Te; and

Table 15.9 Compounds with the halogens

	MX_6	MX_4	MX_2	M_2X_2	M_2X	Others
O			OF_2	O_2F_2		O_3F_2, O_4F_2
			Cl_2O		ClO_2	Cl_2O_6, Cl_2O_7
			Br_2O		BrO_2	BrO_3
						I_2O_4, I_4O_9, I_2O_5
S	SF_6	SF_4	SF_2	S_2F_2		SSF_2, S_2F_4, S_2F_{10}
		SCl_4	SCl_2	S_2Cl_2		
				S_2Br_2		
Se	SeF_6	SeF_4				
		$SeCl_4$		Se_2Cl_2		
		$SeBr_4$		Se_2Br_2		
Te	TeF_6	TeF_4				
		$TeCl_4$	$TeCl_2$			
		$TeBr_4$	$TeBr_2$			
		TeI_4	TeI_2			
Po		$PoCl_4$	$PoCl_2$			
		$PoBr_4$	$PoBr_2$			
		PoI_4	(PoI_2)			

Compounds shown in parentheses are unstable.

SF_6, SeF_6 and TeF_6 are all formed by direct combination. They are all colourless gases and have an octahedral structure as predicted by the VSEPR theory. The low boiling point indicates a high degree of covalency.

Electronic structure of sulphur – excited state

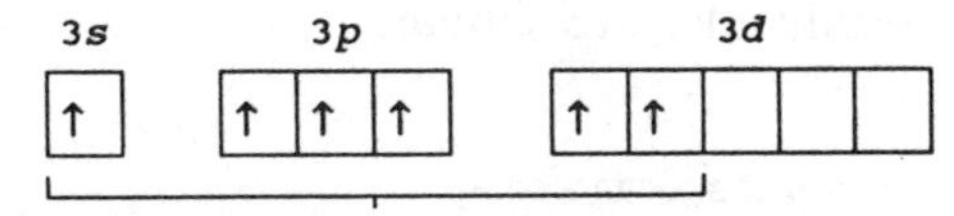

six unpaired electrons form bonds with six fluorine atoms, hence octahedral shape

SF_6 is a colourless, odourless, non-flammable gas, which is insoluble in water and extremely inert. It is used as a gaseous dielectric (insulator) in high voltage transformers and switchgear. SeF_6 is slightly more reactive and TeF_6 is hydrolysed by water. This is possibly due to the larger size of Te which permits the larger coordination number necessary in the first stage of hydrolysis.

$$TeF_6 + 6H_2O \rightarrow 6HF + H_6TeO_6$$

Coordination numbers greater than 6 are not common, but TeF_6 does add F^- ions, forming $[TeF_7]^-$ and $[TeF_8]^{2-}$.

Many tetrahalides are known. It is difficult to prepare tetrafluorides by direct combination even with diluted F_2, because they readily change to hexafluorides. SF_4 is gaseous, SeF_4 liquid and TeF_4 solid. They have been prepared:

$$S + F_2 \text{ (diluted with } N_2) \rightarrow SF_4 \text{ and } SF_6$$
$$3SCl_2 + 4NaF \rightarrow SF_4 + S_2Cl_2 + 4NaCl$$
$$S + 4CoF_3 \rightarrow SF_4 + 4CoF_2$$
$$SeCl_4 + 4AgF \rightarrow SeF_4 + 4AgCl$$
$$TeO_2 + 2SeF_4 \rightarrow TeF_4 + 2SeOF_2$$

SF_4 is highly reactive, but is more stable than the lower fluorides. In contrast to the relatively stable hexafluorides, the tetrahalides are very sensitive to water.

$$SF_4 + 2H_2O \rightarrow SO_2 + 4HF$$

SF_4 is a powerful fluorinating agent.

$$3SF_4 + 4BCl_3 \rightarrow 4BF_3 + 3Cl_2 + 3SCl_2$$
$$5SF_4 + I_2O_5 \rightarrow 2IF_5 + 5OSF_2$$

It is a useful selective fluorinating agent for organic chemicals, for example:

$$R{-}COOH \rightarrow R{-}CF_3$$
$$R_2C{=}O \rightarrow R_2CF_2$$
$$R{-}CHO \rightarrow R{-}CHF_2$$
$$R{-}OH \rightarrow RF$$

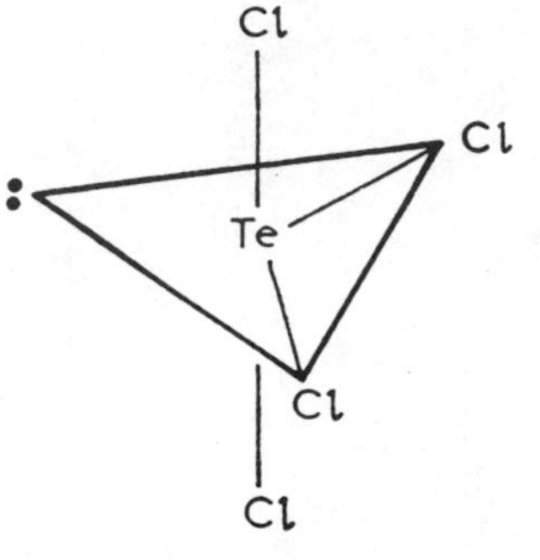

Figure 15.11 Structure of $TeCl_4$.

S, Se, Te and Po form tetrachlorides by direct reaction with chlorine. SCl_4 is a rather unstable liquid, but the other tetrachlorides are solids. The structure of $TeCl_4$ is a trigonal bipyramid with one equatorial position occupied by a lone pair (Figure 15.11). It is probable that the other tetrahalides are similar.

Electronic structure of tellurium atom – excited state

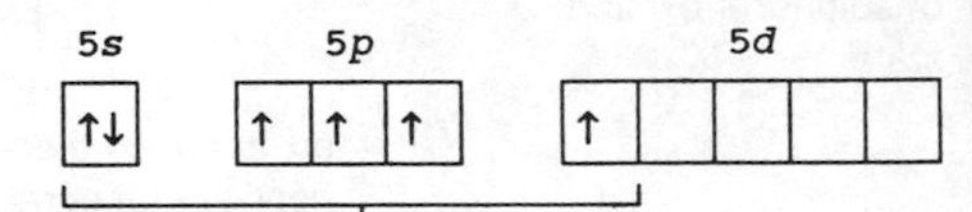

four unpaired electrons can form bonds with four chlorine atoms
five electron pairs, hence trigonal bipyramid with one position occupied by a lone pair

$TeCl_4$ reacts with hydrochloric acid and gives the complex ion $[TeCl_6]^{2-}$, which is isomorphous with $[SiF_6]^{2-}$ and $[SnCl_6]^{2-}$.

$$TeCl_4 + 2HCl \rightarrow H_2[TeCl_6]$$

Po also forms complex halide ions and a series of compounds $(NH_4)_2[PoX_6]$ and $Cs[PoX_6]$ and $Cs_2[PoX_6]$ are known where X is Cl, Br or I.

Tetrabromides of Se, Te and Po are known, but $SeBr_4$ is unstable and hydrolyses readily.

$$2SeBr_4 \rightarrow Se_2Br_2 + 3Br_2$$

$$SeBr_4 + 4H_2O \rightarrow \underset{\text{unstable}}{[Se(OH)_4]} + 4HBr$$

$$\downarrow$$

$$H_2SeO_3 + H_2O$$

Te and Po are the only elements which form tetraiodides.

SCl_2 is the best known dihalide. It is a foul smelling red liquid (m.p. −122 °C, b.p. 59 °C). Heating S and Cl_2 gives S_2Cl_2, and if this is saturated with chlorine SCl_2 is formed. Reacting SCl_2 with hydrogen polysulphides at low temperatures yields a range of dichlorosulphanes.

$$H_2S_2 + 2SCl_2 \rightarrow S_4Cl_2 + 2HCl$$

$$H_2S_n + 2SCl_2 \rightarrow S_{(n+2)}Cl_2 + 2HCl$$

SCl_2 is commercially important since it readily adds across double bonds in alkenes. It has been used to produce the notorious 'mustard gas', first used in World War I, and more recently in 1988 in the Iran–Iraq war.

$$SCl_2 + 2CH_2{=}CH_2 \rightarrow \underset{\substack{\text{di(2-chloroethyl) sulphide} \\ \text{or mustard gas}}}{S(CH_2CH_2Cl)_2}$$

Mustard gas is not a gas but a volatile liquid (m.p. 13 °C, b.p. 215 °C). It was sprayed as a mist that stayed close to the ground, and was blown by gentle winds onto the enemy. It causes severe blistering of the skin and death. In living cells it is converted into the divinyl compound $(CH_2CH)_2S$ which reacts with and disrupts proteins in the cell.

The dihalides form angular molecules, based on a tetrahedron with two positions occupied by lone pairs. The lone pairs distort the tetrahedral angle of 109°28′ to 103° in SCl_2, 101.5° in F_2O and 98° in $TeBr_2$.

Electronic structure of sulphur atom – ground state

3s	3p		
↑↓	↑↓	↑	↑

two unpaired electrons can form bonds with two chlorine atoms
four electron pairs, hence structure is tetrahedral with two lone pairs

Dimeric monohalides such as S_2F_2, S_2Cl_2, Se_2Cl_2 and Se_2Br_2 are formed by direct action between S and Se and the halogens. These monohalides are hydrolysed slowly and tend to disproportionate.

$$2S_2F_2 + 2H_2O \rightarrow 4HF + SO_2 + 3S$$

$$\overset{+II}{2SeCl_2} \rightarrow \overset{+IV}{SeCl_4} + \overset{0}{Se}$$

S_2Cl_2 is a toxic yellow liquid (m.p. −76 °C, b.p. 138 °C), with a revolting smell. It is commercially important in vulcanizing rubber, and in preparing chlorohydrins. The use of S_2Cl_2 has been described earlier for making rings of sulphur atoms with 7–20 atoms.

$$H_2S_8 + S_2Cl_2 \rightarrow S_{10} + 2HCl$$

It can also be used to make dichlorosulphanes.

$$H_2S_n + 2S_2Cl_2 \rightarrow S_{(n+4)}Cl_2 + 2HCl$$

The structure of S_2Cl_2 and the other monohalides is similar to that of H_2O_2, with a bond angle of 104° which is due to distortion by two lone pairs.

S_2F_2 is an unstable compound. It is formed by the action of a mild fluorinating agent such as AgF on S. (Direct reaction of S with F gives SF_6, and even when the F is diluted with N_2 it gives SF_4.) S_2F_2 exists in two different isomeric forms, F—S—S—F (like Cl—S—S—Cl and H—O—O—H), and thiothionyl fluoride S═SF_2.

The compound S_2F_{10} has an unusual structure, of two octahedra joined together.

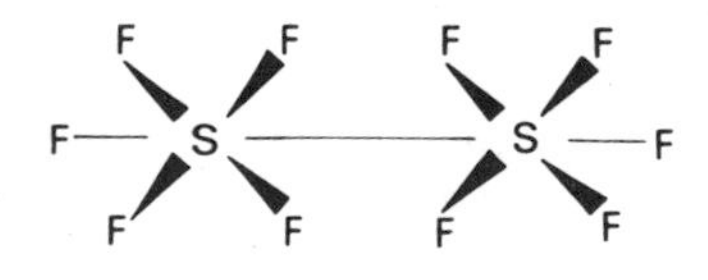

COMPOUNDS OF SULPHUR AND NITROGEN

A number of ring and chain compounds containing S and N exist. The elements N and S are diagonally related in the periodic table, and have similar charge densities. Their electronegativities are close (N 3.0, S 2.5) so covalent bonding is expected. The compounds formed have unusual structures which cannot be explained by the usual bonding theories. Attempting to work out oxidation states is unhelpful or misleading.

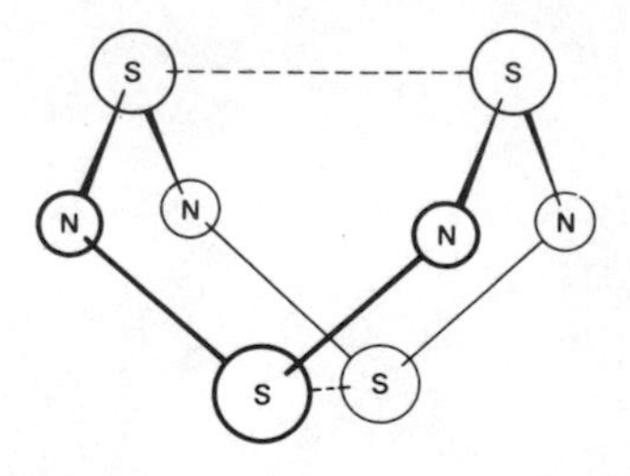

Figure 15.12 Structure of N_4S_4.

The best known is tetrasulphur tetranitride S_4N_4, and this is the starting point for many other S—N compounds. S_4N_4 may be made as follows:

$$6SCl_2 + 16NH_3 \rightarrow S_4N_4 + 2S + 14NH_4Cl$$

$$6S_2Cl_2 + 16NH_3 \xrightarrow{CCl_4} S_4N_4 + 8S + 12NH_4Cl$$

$$6S_2Cl_2 + 4NH_4Cl \rightarrow S_4N_4 + 8S + 16HCl$$

S_4N_4 is a solid, m.p. 178 °C. It is *thermochromic*, that is it changes colour with temperature. At liquid nitrogen temperatures it is almost colourless, but at room temperature it is orange–yellow, and at 100 °C it is red. It is stable in air, but may detonate with shock, grinding or sudden heating. The structure is a heterocyclic ring. This is cradle shaped, and differs structurally from the S_8 ring, which is crown shaped. The X-ray structure (Figure 15.12) shows that the average S—N bond length is 1.62 Å. Since the sum of the covalent radii for S and N is 1.78 Å, the S—N bonds seem to have some double bond character. The fact that the bonds are of equal length suggests that this is delocalized. The S . . . S distances at the top and bottom of the cradle are 2.58 Å. The van der Waals (non-bonded) distance S . . . S is 3.30 Å, and the single bond distance S—S is 2.08 Å. This indicates weak S—S bonding, and S_4N_4 is thus a cage structure.

Many different sizes of rings exist, for example cyclo-S_2N_2, cyclo-S_4N_2, cyclo-S_4N_3Cl, cyclo-$S_3N_3Cl_3$. In addition bicyclo compounds $S_{11}N_2$, $S_{15}N_2$, $S_{16}N_2$, $S_{17}N_2$ and $S_{19}N_2$ are known. The last four may be regarded as two heterocyclic S_7N rings, with the N atoms joined through a chain of 1–5 S atoms.

S_4N_4 is very slowly hydrolysed by water, but reacts rapidly with warm NaOH with the break-up of the ring:

$$S_4N_4 + 6NaOH + 3H_2O \rightarrow Na_2S_2O_3 + 2Na_2SO_3 + 4NH_3$$

If S_4N_4 is treated with Ag_2F in CCl_4 solution then $S_4N_4F_4$ is formed. This has an eight-membered S—N ring, with the F atoms bonded to S. This results from breaking the S—S bonds across the ring. Similarly the formation of adducts such as $S_4N_4 \cdot BF_3$ or $S_4N_4 \cdot SbF_5$ (in which the extra group is bonded to N) breaks the S—S bonds and increases the mean S—N distance from 1.62 Å to 1.68 Å. This is presumably because the electron attracting power of BF_3 or SbF_5 withdraws some of the π electron density.

Reduction of S_4N_4 with $SnCl_2$ in MeOH gives tetrasulphur tetraimide $S_4(NH)_4$. Several imides can be made by reacting S_4N_4 with S, or S_2Cl_2 with NH_3. These imides are related to an S_8 ring in which one or more S atoms have been replaced by imide NH groups, for example in S_7NH, $S_6(NH)_2$, $S_5(NH)_3$ and $S_4(NH)_4$.

If S_4N_4 is vaporized under reduced pressure and passed through silver wool, then disulphur dinitrogen S_2N_2 is formed.

$$S_4N_4 + 4Ag \rightarrow S_2N_2 + 2Ag_2S + N_2$$

S_2N_2 is a crystalline solid, which is insoluble in water but soluble in many organic solvents. It explodes with shock or heat. The structure is cyclic and the four atoms are very nearly square planar.

The most important reaction of S_2N_2 is the slow polymerization of the solid or vapour to form polythiazyl $(SN)_x$. This is a bronze coloured shiny solid that looks like a metal. It conducts electricity and conductivity increases as the temperature decreases, which is typical of a metal. It becomes a superconductor at 0.26 K. The crystal structure shows that the four-membered rings in S_2N_2 have opened and polymerized into a long chain polymer. The atoms have a zig-zag arrangement, and the chain is almost flat. Conductivity is much greater along the chains than in other directions, and so the polymer behaves as a one-dimensional metal. The resistivity is quite high at room temperature (about 1×10^9 μohm cm along a chain), but this falls to about 1×10^6 μohm cm at 4 K. (See Appendix J for values for other elements.)

ORGANO DERIVATIVES

Oxygen forms many organo derivatives R_2O, which are called ethers. Similar derivatives of S, Se and Te may be prepared using Grignard or organolithium reagents:

$$SCl_2 + 2LiR \xrightarrow{\text{ether}} R_2S + 2LiCl$$
$$SCl_4 + 4RMgCl \rightarrow R_4S + 4MgCl_2$$

Dialkyl sulphides R_2S have a similar structure to water (tetrahedral with two positions occupied by lone pairs), and the lone pairs make them useful donor molecules.

Haemoglobin is the pigment in the blood of most animals. It is red when oxygen is present and blue when there is no oxygen. Haemoglobin is essential for absorbing molecular oxygen in the lungs, where it forms oxyhaemoglobin. Oxyhaemoglobin releases the oxygen in the parts of the body where it is needed, forming (reduced) haemoglobin. Haemoglobin

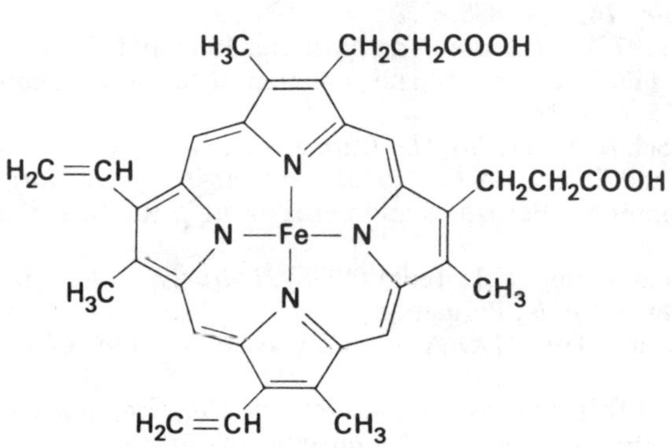

Figure 15.13 Structure of haem.

has a molecular weight of about 65 000. It is made up of four haem groups, which are flat heterocyclic porphyrin ring systems containing iron and a globular protein (Figure 15.13). (See 'Bioinorganic chemistry of iron', Chapter 24.) Haemoglobin reacts with O_2 and forms oxyhaemoglobin. In this complex the O–O axis of the O_2 molecule lies parallel to the plane of the porphyrin ring, with two equal Fe...O distances. It is thought that molecular oxygen is π bonded to the iron.

A few transition metal complexes can also form π bonded complexes with molecular oxygen.

$$L_4Pt + O_2 \rightarrow L_2PtO_2 \qquad (L = P(C_6H_5)_3)$$

$$L_2Ir(Cl)(CO) + O_2 \rightleftharpoons L_2Ir(Cl)(CO)(O_2)$$

FURTHER READING

Bagnall, K.W. (1973) *Comprehensive Inorganic Chemistry*, Vol. II (Chapter 24: Selenium, tellurium and polonium), Pergamon Press, Oxford.

Bailey, P.S. (1978) *Ozonation in Organic Chemistry*, Academic Press, New York.

Bard, A.J., Parsons, R. and Jordan, J. (1985) *Standard Potentials in Aqueous Solution* (Monographs in Electroanalytical Chemistry and Electrochemistry Series, Vol. 6), Marcel Dekker. (Commissioned by IUPAC to replace the earlier values in Latimer's book.)

Bevan, D.J.M. (1973) *Comprehensive Inorganic Chemistry*, Vol. 3, (Chapter 49: Nonstoichiometric compounds), Pergamon Press, Oxford. (A good introduction to nonstoichiometric compounds.)

Bland, W.J. (1984) Sulphuric acid – modern manufacture and uses, *Education in Chemistry*, **21**, 7–10.

Campbell, I.M. (1977) *Energy and the Atmosphere*, Wiley, London. (Acid rain etc.)

Clive, D.L.J. (1978) *Modern Organo-Selenium Chemistry*, Pergamon Press, New York.

Cocks, A. and Kallend, T. (1988) The chemistry of atmospheric pollution, *Chemistry in Britain*, **24**, 884–885.

Cooper, W.C. (1972) *Tellurium*, Van Nostrand-Reinhold, New York.

Donovan, R.J. (1978) Chemistry and pollution of the stratosphere, *Education in Chemistry*, **15**, 110–113.

Dotto, L. and Schiff, H. (1978) *The Ozone War*, Doubleday, New York.

Emeléus, H.J. and Sharpe, A.G. (1973) *Modern Aspects of Inorganic Chemistry*, 4th ed. (Chapter 12: Peroxides and peroxy-acids), Routledge and Kegan Paul, London.

Fogg, P.G.T. and Young, C.L. (eds) (1988) *Hydrogen Sulfide, Deuterium Sulfide and Hydrogen Selenide*, Pergamon.

Greenwood, G. and Hill, H.O.A. (1982) Oxygen and life, *Chemistry in Britain*, **18**, 194.

Govindgee, R. (1977) Photosynthesis, *McGraw Hill Encyclopedia of Science and Technology*, 4th ed., Vol. 10. (An introductory article.)

Heal, H.G. (1980) *The Inorganic Heterocyclic Chemistry of Sulphur, Nitrogen and Phosphorus*, Academic Press, London.
Heicklen, J. (1976) *Atmospheric Chemistry*, Academic Press, New York. (Acid rain etc.)
Holloway, J.H. and Laycock, D. (1983) Preparations and reactions of inorganic main-group oxide fluorides, *Adv. Inorg. Chem. Radiochem.*, **27**, 157–195.
Horvath, M., Bilitzky, L. and Huttner, J. (1985) *Ozone* (Topics in Inorganic and General Chemistry Series No. 20), Elsevier.
Hynes, H.B.N. (1973) *The Biology of Polluted Waters*, Liverpool University Press. (A good textbook on water pollution.)
Lagowski, J. (ed.) (1967) *The Chemistry of Non-aqueous Solvents*, Vol. III (Chapter by Burow, D.F., Liquid sulphur dioxide), Academic Press, New York.
Latimer, W.M. (1959) *The Oxidation States of the Elements and Their Potentials in Aqueous Solution*, 2nd ed., Prentice Hall. (Old, but until very recently the standard source of oxidation potential data.)
Murphy, J.S. and Orr, J.R. (1975) *Ozone Chemistry and Technology*, Franklin Institute Press, Philadelphia.
Nickless, G. (ed.) (1968) *Inorganic Sulfur Chemistry*, Elsevier, Amsterdam.
Ochiai, E.I. (1975) Bioinorganic chemistry of oxygen, *J. Inorg. Nuclear Chem.*, **37**, 1503–1509.
Ogryzlo, E.A. (1965) Why liquid oxygen is blue, *J. Chem. Ed.*, **42**, 647–648.
Oxygen in the Metal and Gaseous Fuel Industries (1978) Special Publication No. 32, Royal Society for Chemistry, London. (Proceedings of first BOC Priestley Conference.)
Oxygen and Life (1981) Special Publication No. 39, Royal Society for Chemistry, London. (Proceedings of second BOC Priestley Conference.)
Patai, S. (ed.) (1983) *The Chemistry of Peroxides*, John Wiley Chichester.
Phillips, A. (1977) The modern sulphuric acid process, *Chemistry in Britain*, **13**, 471.
Roesky, H.W. (1979) Cyclic sulphur–nitrogen compounds, *Adv. Inorg. Chem. Radiochem.*, **22**, 240–302.
Roy, A.B. and Trudinger, P.A. (1970) *The Biochemistry of Inorganic Compounds of Sulphur*, Cambridge University Press, Cambridge.
Schaap, A.P. (ed.) (1976) *Singlet Molecular Oxygen*, Wiley, New York.
Schmidt, M. and Siebert, W. (1973) *Comprehensive Inorganic Chemistry*, Vol. II (Chapter 23: Oxyacids of sulfur), Pergamon Press, Oxford.
Thompson, R. (ed.) (1976) *The Modern Inorganic Chemicals Industry* (Chapter by Grant, W.J. and Redfearn, S.L., Industrial gases; chapter by Arden, T.V., Water purification and recycling; chapter by Crampton, C.A. *et al.*, Manufacture, properties and uses of hydrogen peroxide and inorganic peroxy compounds; chapter by Phillips, A., The modern sulphuric acid process), Special Publication No. 31, The Chemical Society, London.
Thrush, B.A. (1977) The chemistry of the stratosphere and its pollution, *Endeavour*, **1**, 3–6.
Vaska, L. (1976) Dioxygen–metal complexes: towards a unified view, *Acc. Chem. Res.*, **9**, 175–183.
Waddington, T.C. (ed.) (1965) *Non Aqueous Solvents* (Chapter 4 by Gillespie, R.J. and Robinson, E.A., Sulfuric acid; chapter 6 by Waddington, T.C., Liquid sulfur dioxide), Nelson.
Wasserman, H.H. and Murray, R.W. (eds) (1979) *Singlet Oxygen*, Academic Press, New York.
West, J.R. (1975) *New Uses of Sulfur*, ACS Advances in Chemistry series, No. 140, American Chemical Society.
Zurer, P.S. (1987) The Antarctic ozone hole, *Chem. Eng. News*, 7 August, 7–13; 2 November, 22–26.

PROBLEMS

1. Write equations for the preparation of oxygen from: (a) H_2O, (b) H_2O_2, (c) Na_2O_2, (d) $NaNO_3$, (e) $KClO_3$, (f) HgO.
2. How is oxygen obtained commercially, and what are its main uses?
3. Compare the oxides of Na and Ca with those of S and N. Comment on their melting points, the nature of the bonding, and their reactions with water, acids and bases.
4. In what ways and on what basis may oxides be classified?
5. Use the molecular orbital theory to describe the bonding in each of the following and give the bond order and magnetic properties (paramagneticor diamagnetic) in each case: (a) O_2 (b) superoxide ion O_2^-, (c) peroxide ion O_2^{2-}.
6. Explain the following facts:
   (a) Liquid oxygen sticks to the poles of a magnet but liquid nitrogen does not.
   (b) The $N{-}O^+$ ion has a shorter bond length than does NO, even though the latter has an extra electron.
7. How is ozone prepared in the laboratory? What is its structure and what are its main uses? There is a layer of ozone in the upper atmosphere: why is this important to man?
8. Give equations for the reaction between O_2 and (a) Li, (b) Na, (c) K, (d) C, (e) CH_4, (f) N_2, (g) S, (h) Cl_2, (i) PbS, (j) CuS.
9. Why have oxygen molecules the formula O_2 whilst sulphur is S_8?
10. Describe one method by which hydrogen peroxide is prepared. Give the structure of H_2O_2 in the gas phase. Write balanced equations for the reaction of H_2O_2 with:
    (a) an acidified solution of $KMnO_4$
    (b) aqueous HI
    (c) an acidic solution of potassium hexacyanoferrate(II)
11. What are the main sources of sulphur? Which are the two most common allotropic forms?
12. Describe the Frasch process for mining sulphur.
13. Describe the changes which occur on heating sulphur.
14. Explain the differences in bond angles and boiling points of H_2O and H_2S.
15. Explain how π bonding occurs in O_2, O_3, SO_3 and SO_4^{2-}.
16. Describe how sulphuric acid is manufactured on an industrial scale. List its main uses.

17. Describe the preparation, properties and structure of SO_2, SO_3, H_2SO_5, $H_2S_2O_8$.

18. (a) Describe the differences in structure between gaseous and solid SO_3.
    (b) What reaction occurs between SO_3 and H_2SO_4? Give the structure of the product.
    (c) Describe the action of heat on $NaHSO_3$.
    (d) Compare the structures of the SO_4^{2-} and $S_2O_3^{2-}$ ions.
    (e) What reaction occurs between $Na_2S_2O_3$ and I_2?
    (f) Why are sulphurous acid and sulphites reducing?

19. Compare and contrast sulphuric acid, selenic acid and telluric acid.

20. How is $Na_2S_2O_3$ made? Explain its uses in photography and volumetric analysis.

21. What are the main fluorides of sulphur? How are they made, what are their structures and what are their uses?

22. Explain why SF_6 is unreactive towards water but TeF_6 reacts.

23. Suggest reasons why SF_6 is known but OF_6 is not.

16 Group VII – the halogens

Table 16.1 Electronic structures and oxidation states

Element		Electronic configuration	Oxidation states*
Fluorine	F	[He] $2s^2\ 2p^5$	−I
Chlorine	Cl	[Ne] $3s^2\ 3p^5$	−I +I +III +IV +V +VI +VII
Bromine	Br	[Ar] $3d^{10}\ 4s^2\ 4p^5$	−I +I +III +IV +V +VI
Iodine	I	[Kr] $4d^{10}\ 5s^2\ 5p^5$	−I +I +III +V +VII
Astatine	At	[Xe] $4f^{14}\ 5d^{10}\ 6s^2\ 6p^5$	

* The most important oxidation states (generally the most abundant and stable) are shown in bold. Other well-characterized but less important states are shown in normal type. Oxidation states that are unstable, or in doubt, are given in parentheses.

INTRODUCTION

The name halogen comes from the Greek, and means salt former. The elements all react directly with metals to form salts, and they are also very reactive with non-metals. Fluorine is the most reactive element known.

The elements all have seven electrons in their outer shell. The s^2p^5 configuration is one *p* electron less than that of the next noble gas. Thus atoms complete their octet either by acquiring an electron (i.e. through forming an ionic bond, giving X^-), or by sharing an electron with another atom (thus forming a covalent bond). Compounds with metals are typically ionic, whilst those with non–metals are covalent.

The halogens show very close group similarities. Fluorine (the first element in the group) differs in several ways from the rest of the group. The first element of each of the main groups all show differences from the subsequent elements. The reasons for the difference are:

1. The first element is smaller than the rest, and holds its electrons more firmly.
2. It has no low-lying *d* orbitals which may be used for bonding.

The properties of chlorine and bromine are closer than those between the other pairs of elements because their sizes are closer. The ionic radius of Cl^- is 38% larger than that of F^-, but the radius of Br^- is only 6.5% larger than that of Cl^-. The relatively small change in size occurs because Br^- contains ten $3d$ electrons, which shield the nuclear charge ineffectively. This also results in the electronegativity values being particularly close for these two elements. Thus there is little difference in polarity of the bonds formed by Cl and Br with other elements.

The oxidation states (±I) are by far the most common. Whether it is (−I) or (+I) depends on whether the halogen is the most electronegative element. Higher oxidation states exist for all of the elements except F. The lack of low-lying empty *d* orbitals in the second shell prevents F from forming more than one normal covalent bond.

Fluorine is a very strong oxidizing agent, and this together with its small size allows it to form compounds that bring out the highest oxidation state of other elements. Examples include IF_7, PtF_6, SF_6 and many hexafluorides, BiF_5, SF_5, TbF_4, AgF_2, and $K[Ag^{III}F_4]$.

The elements all exist as diatomic molecules, and they are all coloured. Gaseous F_2 is light yellow, Cl_2 gas is yellow–green, Br_2 gas and liquid are dark red–brown, and I_2 gas is violet. The colours arise from the absorption of light on promoting an electron from the ground state to a higher state. On descending the group the energy levels become closer, so the promotion energy becomes less and the wavelength of the band becomes longer.

I_2 solid crystallizes as black flakes, and has a slightly metallic lustre. Though the X-ray structure shows discrete I_2 molecules, the colour is reminiscent of charge transfer compounds, and the properties are different from those of other molecular solids. The solid conducts electricity to a small extent, and the conductivity increases when the temperature is raised. This behaviour is like that of an intrinsic semiconductor, and different from metals. However, liquid I_2 conducts very slightly. This is ascribed to self-ionization:

$$3I_2 \rightleftharpoons I_3^+ + I_3^-$$

The stable isotopes of the halogens all have a nuclear spin. This is used in nmr spectroscopy. Chemical shifts are conveniently measured using the isotope ^{19}F.

Several chemicals are of commercial importance and are produced on a vast scale. These include Cl_2 (23 million tonnes in 1985), anhydrous HCl and hydrochloric acid (9 million tonnes in 1985), anhydrous HF and hydrofluoric acid (1 million tonnes/year), Br_2 (398 500 tonnes in 1988) and ClO_2 (200 000 tonnes/year).

OCCURRENCE AND ABUNDANCE

The halogens are all very reactive, and do not occur in the free state. However, all except astatine are found in combined form in the earth's crust. (Astatine is radioactive and has a short half life.) Fluorine is the thirteenth most abundant element by weight in the earth's crust, and

continued overleaf

Table 16.2 Abundance of the elements in the earth's crust, by weight

	ppm	Relative abundance
F	544	13
Cl	126	20
Br	2.5	47
I	0.46	62

chlorine the twentieth. These two elements are reasonably abundant, but bromine and iodine are comparatively rare.

The main source of fluorine is the mineral CaF_2 called fluorspar or fluorite. (The name fluorspar was given because the mineral fluoresces, that is it emits light, when it is heated.) World production was 5.1 million tonnes in 1988. The largest producers are China 22.5%, Mexico and Mongolia 16% each, the USSR 11% and South Africa 6%. Another well known fluorine containing mineral is fluoroapatite $[3(Ca_3(PO_4)_2 \cdot CaF_2]$. This is used primarily as a source of phosphorus. It is not used to produce HF and F_2 because the mineral contains appreciable amounts of SiO_2. The HF produced reacts with the SiO_2 in the mineral to form fluorosilicic acid, $H_2[SiF_6]$. Some $H_2[SiF_6]$ is made in this way and is used as an alternative to NaF for fluoridizing drinking water. The mineral cryolite $Na_3[AlF_6]$ is rather rare. It is found only in Greenland, and is used in the electrolytic extraction of aluminium.

The most abundant compound of chlorine is NaCl, and it is used to produce virtually all the Cl_2 and HCl made. World consumption of NaCl was 179.6 million tonnes in 1988. Some salt is mined, and some is obtained by solar evaporation of sea water. Chlorides and bromides are leached from the land by rain, and are washed into the sea. Sea water usually contains about 15 000 ppm (1.5%) of NaCl. Certain inland seas contain much more (the Dead Sea contains 8% and the Great Salt Lake, Utah, contains 23%). The dried-up beds of inland lakes and seas contain large deposits of NaCl, mixed with smaller amounts of $CaCl_2$, KCl and $MgCl_2$. In contrast, the fluoride content of sea water is very low (1.2 ppm). This is because the water contains a large concentration of Ca^{2+}, and CaF_2 is insoluble.

Bromides occur in sea water. Iodides only occur in low concentration in sea water, but they are absorbed and concentrated by seaweed. At one time iodine was extracted from seaweed. There are now better sources. Natural brines have higher concentrations of I^-. Sodium iodate $NaIO_3$ and sodium periodate $NaIO_4$ occur as impurities in $NaNO_3$ deposits in Chile.

EXTRACTION AND USES OF THE ELEMENTS

Fluorine

Fluorine is extremely reactive, and this causes great difficulties in the preparation and handling of the element. The first preparation of fluorine was by Moissan in 1886. He was subsequently awarded the Nobel Prize for Chemistry in 1906 for this work. Fluorine is obtained by treating CaF_2 with concentrated H_2SO_4 to give an aqueous mixture of HF. This is distilled, yielding anhydrous liquid HF. Then a cooled solution of KHF_2 in anhydrous HF is electrolysed, giving F_2 and H_2.

$$CaF_2 + H_2SO_4 \rightarrow CaSO_4 + 2HF$$

$$KF + HF \rightarrow K[HF_2]$$

$$HF + K[HF_2] \xrightarrow{\text{electrolyse}} H_2 + F_2$$

There are many difficulties in obtaining fluorine.

1. HF is corrosive, and etches glass and also causes very painful skin wounds. These arise partly by dehydrating the tissue, and partly from the acidic nature of HF. The wounds are slow to heal because F^- ions remove Ca^{2+} ions from the tissues.
2. Gaseous HF is also very toxic (3 ppm) compared with HCN (10 ppm).
3. Anhydrous HF is only slightly ionized and is therefore a poor conductor of electricity. Thus a mixture of KF and HF is electrolysed to increase the conductivity. Moissan used a solution of KF in HF with a mole ratio of 1 : 13. This has the disadvantage that the vapour pressure of HF is high, and this gives problems with toxicity and corrosion, even when the reaction mixture is cooled to −24 °C. Modern methods use medium temperature fluorine generators. These use a mole ratio of 1 : 2 of KF : HF so that the vapour pressure of HF is much lower. This mixture melts at about 72 °C which is a much easier temperature to maintain. Note that KF and HF react to form the acid salt $K^+[F{-}H{-}F]^-$.
4. Water must be rigorously excluded or the fluorine produced will oxidize it to oxygen.
5. The hydrogen liberated at the cathode must be separated from the fluorine liberated at the anode by a diaphragm, otherwise they will react explosively.
6. Fluorine is extremely reactive. It catches fire, for example with traces of grease or with crystalline silicon. Glass and most metals are attacked. It is difficult to find suitable materials from which to make the reaction vessels. Moissan used a platinum U-tube, since platinum is very unreactive (but it is very expensive). Copper or Monel metal (Cu/Ni alloy) are now used instead, because they cost less. A protective fluoride film forms on the surface of the metal and slows down further attack.
7. The cathodes are made of steel, the anodes are carbon, and teflon is used for electrical insulation. Graphite anodes must not be used, since graphite reacts with fluorine, forming graphite compounds CF. In these, fluorine atoms progressively invade the space between the sheets of graphite, forcing them apart, and buckling them. This gradually stops the graphite from conducting, the current needed increases, more heat is produced, and eventually an explosion may occur. Ungraphitized carbon is used to avoid this. It is made from powdered coke compacted and impregnated with copper.

continued overleaf

Cylinders of F_2 are now commercially available. However, for many purposes F_2 is converted to ClF_3 (b.p. 12°C), which though very reactive is less unpleasant and easier to transport.

$$3F_2 + Cl_2 \xrightarrow{200-300\,°C} 2ClF_3$$

Production of fluorine first became important for the manufacture of inorganic fluorides such as AlF_3 and synthetic $Na_3[AlF_6]$. Both are used in the extraction of aluminium. The natural mineral cryolite is only found in Greenland, and this source is largely exhausted.

It was discovered in the 1940s that the isotopes of uranium could be separated by gaseous diffusion of UF_6. This was important in preparing enriched uranium to make the first atomic bomb. Gaseous diffusion is still used to make enriched uranium fuel for nuclear reactors. The nuclear industry uses about 75% of the fluorine produced. UF_6 is made as follows:

$$U \text{ or } UO_2 + HF \rightarrow UF_4$$
$$UF_4 + F_2 \rightarrow UF_6$$
$$UF_4 + ClF_3 \rightarrow UF_6$$

The fluorocarbons are a very interesting and useful group of compounds, derived from hydrocarbons by substituting F for H. Tetrafluoromethane CF_4 is the fluorocarbon corresponding to methane. Completely fluorinated compounds C_nF_{2n+2} are called perfluoro compounds. Thus CF_4 is perfluoromethane. Perfluoro compounds have very low boiling points for their molecular weight: this is associated with very weak intermolecular forces. Fluorocarbons are extremely inert. Unlike methane, CF_4 can be heated in air without burning. Fluorocarbons are inert to concentrated HNO_3 and H_2SO_4, to strong oxidizing agents such as $KMnO_4$ or O_3, and to strong reducing agents such as $Li[AlH_4]$ or C at 1000°C. They are attacked by molten Na. When pyrolysed at very high temperatures the C—C bonds break rather than the C—F bonds. Tetrafluoroethene $F_2C{=}CF_2$ (b.p. −76.6°C) can be made:

$$2CHClF_2 \xrightarrow{500-1000\,°C} CF_2{=}CF_2 + 2HCl$$

Fluoroalkenes of this type can be polymerized either thermally, or using a free radical initiator. Depending on the degree of polymerization, that is on the molecular weight produced, the products may be oils, greases or a solid of high molecular weight called polytetrafluoroethylene. This is similar to ethene (formerly called ethylene) polymerizing to give polyethylene (polythene). Polytetrafluoroethylene is known commercially as PTFE or Teflon. It is a very inert solid plastic material, and is useful because it is completely resistant to chemical attack and is an electrical insulator. Though expensive, it is used in laboratories. It is also used as a coating for non-stick pans.

Freons are mixed chlorofluorocarbons. Compounds such as $CClF_3$, CCl_2F_2 and CCl_3F are important as non-toxic refrigerating fluids and aerosol propellants. They too are very inert, and are discussed later. $CF_3CHBrCl$ is used as an anaesthetic called Fluothane.

Another use of F_2 is to make SF_6, which is a very inert gas used as a dielectric (insulating) medium for high voltage equipment. F_2 is also used to make other fluorinating agents ClF_3, BrF_3 and IF_5 and SbF_5. The earlier use of liquid F_2 as an oxidizing agent in rocket motors has now been discontinued. Anhydrous HF has many uses.

Traces of fluoride ions F^- in drinking water (about 1 ppm) greatly reduce the incidence of dental caries (tooth decay). The F^- ions make the enamel on teeth much harder, by converting hydroxyapatite $[3(Ca_3(PO_4)_2 \cdot Ca(OH)_2]$ (the enamel on the surface of teeth) into the much harder fluoroapatite $[3(Ca_3(PO_4)_2 \cdot CaF_2]$. However, F^- concentrations above 2 ppm cause discoloration, the brown mottling of teeth, and higher concentrations are harmful. In some places NaF and $H_2[SiF_6]$ are added to drinking water, where the natural water is very soft and contains insufficient naturally occurring F^- ions. NaF is now used in fluoride toothpaste. (The original fluoride toothpaste contained SnF_2 and $Sn_2P_2O_7$.)

Chlorine

Chlorine was first prepared by Scheele by oxidizing HCl with MnO_2. This method was used as a laboratory preparation, but chlorine is now readily available in cylinders.

$$H_2SO_4 + NaCl \rightarrow HCl + NaHSO_4$$
$$4HCl + MnO_2 \rightarrow MnCl_2 + 2H_2O + Cl_2$$

Gas prepared from MnO_2 in this way must be purified. First it is passed through water to remove HCl, and then through concentrated H_2SO_4 to remove water. It may be further dried by passing it over CaO and P_4O_{10}.

Chlorine is produced commercially on a vast scale by two main methods. About 23 million tonnes per year are produced annually:

1. By the electrolysis of aqueous NaCl solutions in the manufacture of NaOH.
2. By electrolysis of fused NaCl in the manufacture of sodium. (See Chapter 10.)

Before 1960 chlorine was a by-product from these processes. Since then there has been a great increase in the use of chlorine, mainly in the manufacture of plastics such as polyvinyl chloride (11.5 million tonnes of PVC were made in 1985). Thus chlorine is now the major product.

$$2NaCl + 2H_2O \xrightarrow{\text{electrolyse}} 2NaOH + Cl_2 + 2H_2$$
$$2NaCl \xrightarrow{\text{electrolyse}} 2Na + Cl_2$$

At one time chlorine was produced by oxidizing HCl with air, using the Deacon process. This process became obsolete. (See Chapter 10.) How-

continued overleaf

ever, a modified Deacon process is now used to a small extent. It utilizes HCl obtained as a by-product from the pyrolysis of 1,2-dichloroethane to vinyl chloride, and uses an improved catalyst ($CuCl_2$ with didymium oxide as promoter; didymium is an old name meaning 'twin', and it consists of two lanthanide elements praseodymium and neodymium). This works at a slightly lower temperature than the original process.

$$CH_2Cl—CH_2Cl \xrightarrow{400-450\,°C} CH_2{=}CHCl + HCl$$

Chlorine gas is toxic. It was used as a poison gas in World War I. The gas is detectable by smell at a concentration of 3 ppm, and 15 ppm causes a sore throat and running eyes. Higher concentrations cause coughing, lung damage, and death.

World production of chlorine excluding the USSR is about 23 million tonnes per year (USA 40%, West Germany 15%, Canada 6%, France 6%, UK, Japan and Italy 4% each). About two thirds of this is used to make organic chloro compounds, one fifth for bleaching, and the rest for the manufacture of a variety of inorganic chemicals. The main two organic compounds produced are:

1,2-dichloroethane
vinyl chloride monomer

Both are used in the plastics industry. Other uses include the production of:

chlorinated solvents including methyl and ethyl chlorides
perchloro- and dichloroethene
mono-, di- and trichlorobenzene
benzene hexachloride
the insecticide DDT
chlorinated phenols
plant growth hormones (2,4-dichlorophenoxyacetic acid and 2,4,6-trichlorophenoxyacetic acid are used as selective weedkillers)

Large amounts of chlorine are used for bleaching textiles, wood, pulp and paper. Chlorine is widely used throughout the developed world to purify drinking water, because it kills bacteria. It is also used to make a wide variety of inorganic chemicals including:

bleaching powder
sodium hypochlorite NaOCl
chlorine dioxide ClO_2
sodium chlorate $NaClO_3$
many metal and non-metal chlorides

Bromine

Bromine is obtained from sea water and brine lakes. Sea water contains about 65 ppm Br^-. Thus 15 tonnes of sea water contain about 1 kg of bromine. Bromine is extracted from sea water, but it is more economic to use more concentrated brine sources such as the Dead Sea, or brine from wells in Arkansas and Michigan (USA) and Japan, which contain 2000–5000 ppm of Br^-. First H_2SO_4 is added to adjust the pH to about 3.5. Then Cl_2 gas is passed through the solution to oxidize the Br^- to Br_2. This is an example of displacement of one element by another higher in the electrochemical series.

$$Cl_2 + 2Br^- \rightarrow 2Cl^- + Br_2$$

The Br_2 is removed by a stream of air, because Br_2 is quite volatile. The gas is passed through a solution of Na_2CO_3, when the Br_2 is absorbed, forming a mixture of NaBr and $NaBrO_3$. Finally the solution is acidified and distilled to give pure bromine.

$$3Br_2 + 3Na_2CO_3 \rightarrow 5NaBr + NaBrO_3 + 3CO_2$$

$$5NaBr + NaBrO_3 + 3H_2SO_4 \rightarrow 5HBr + HBrO_3 + 3Na_2SO_4$$

$$5HBr + HBrO_3 \rightarrow 3Br_2 + 3H_2O$$

World production of bromine was 398 500 tonnes per year in 1988 (USA 41%, Israel 26, USSR 16%, UK 7% and France 5%). In 1955 about 90% was used to make 1,2-dibromoethane, $CH_2Br \cdot CH_2Br$, but the figure is now under 50%. 1,2-dibromoethane is added to petrol to act as a lead scavenger. Tetraethyl lead is added to petrol to improve its octane rating, but when it burns it forms lead deposits. 1,2-dibromoethane is added to prevent the build-up of lead deposits on the sparking plug and in the engine. The lead passes out with the exhaust gases, mainly as PbClBr. The use of $PbEt_4$ as an anti-knock additive to petrol has already declined, and will decline further because of legislation against its use, and environmental concern over the toxic effects of lead. Therefore the use of 1,2-dibromoethane has also declined.

Almost 20% of the bromine produced is used to make organic derivatives such as methyl bromide, ethyl bromide and dibromochloropropane. These are used in agriculture: MeBr acts as a nematocide (kills earthworms) and as a pesticide against insects and fungi. The other compounds are used as pesticides.

Nearly 10% is used to make flame retardants. Bromo compounds may be included in the polymerization when making acrylic and polyester fibres. It is more common to 'fireproof' fabrics and carpets by treating them with tris(dibromopropyl)phosphate, $(Br_2C_3H_5O)_3PO$. This may be done either when spinning the thread, or after manufacture.

Other uses include the manufacture of photographic emulsions and pharmaceuticals. AgBr is light sensitive and is used for photographic films, and also for water sanitation and dyestuffs. KBr is used as a sedative, and as an anticonvulsant in treating epilepsy.

Iodine

There are two different commercial methods of obtaining iodine. The method used depends on whether the source is Chile saltpetre or natural

continued overleaf

brines (for example from wells in Oklahoma or Michigan USA, and Japan).

Chile saltpetre is mainly $NaNO_3$, but it contains traces of sodium iodate $NaIO_3$ and sodium periodate $NaIO_4$. Pure $NaNO_3$ is obtained by dissolving saltpetre in water and crystallizing $NaNO_3$. The iodate residues thus accumulate and concentrate in the mother liquor. Eventually, this concentrate is divided into two parts. One part is reduced with $NaHSO_3$ to give I^-. This is mixed with the untreated part, giving I_2, which is filtered off as a solid and then purified by sublimation.

$$2IO_3^- + 6HSO_3^- \rightarrow 2I^- + 6SO_4^{2-} + 6H^+$$
$$5I^- + IO_3^- + 6H^+ \rightarrow 3I_2 + 3H_2O$$

Sea water contains only about 0.05 ppm of I^-, which is too low for commercial recovery. Natural brine, which may contain 50–100 ppm, is treated with Cl_2, to oxidize I^- ions to I_2. This is blown out with air in the same way as bromine. Alternatively, after oxidation with Cl_2, the solution may be passed through an ion-exchange resin. The I_2 is adsorbed on the column as the triiodide ion I_3^-, and finally is removed from the resin by treatment with alkali.

World production of I_2 was 15 300 tonnes per year in 1988 (Japan 49%, Chile 26%, USSR 13% and USA 9%). There is no one dominant use. Half is used to make a variety of organic compounds including iodoform CHI_3 (used as an antiseptic), and methyl iodide CH_3I. AgI is used for photographic films, and for seeding clouds to produce rain. Small amounts of iodine are required in the human diet, so traces (10 ppm) of NaI are added to table salt. KI is added to animal and poultry feeds. The thyroid gland produces a growth regulating hormone called thyroxine which contains iodine. Deficiency of iodine causes the disease goitre. Iodine has limited use as an antiseptic; tincture of iodine is an aqueous solution of I_2 in KI, and French iodine is a solution in alcohol. In the laboratory iodides and iodates are used in volumetric analysis, and Nessler's reagent $K_2[HgI_4]$ is used to detect ammonia.

Astatine

Astatine does not occur in nature, but over twenty artificial isotopes have been made. All of these are radioactive. The most stable isotopes are ^{210}At (half life 8.3 hours), and ^{211}At (half life 7.5 hours). The latter was first made in 1940 by a nuclear reaction in which bismuth was bombarded with high energy α particles.

$$^{209}_{83}Bi + ^{4}_{2}He \rightarrow ^{211}_{85}At + 2^{1}_{0}n$$

Tracer methods were used to study the chemistry of ^{211}At, using minute quantities of about 10^{-14} mole. This isotope decays by orbital electron capture and by α-emission (see Chapter 31 under 'Modes of decay'). Astatine appears to resemble iodine quite closely.

SIZE OF ATOMS AND IONS

Table 16.3 Ionic and covalent radii

	Covalent radius (Å)	Ionic radius X^- (Å)
F	0.72	1.33
Cl	0.99	1.84
Br	1.14	1.96
I	1.33	2.20

IONIZATION ENERGY

The ionization energies of the halogens show the usual trend to smaller values as the atoms increase in size. The values are very high, and there is little tendency for the atoms to lose electrons and form positive ions.

Table 16.4 Ionization and hydration energies, electron affinity

	First ionization energy ($kJ\,mol^{-1}$)	Electron affinity ($kJ\,mol^{-1}$)	Hydration energy X^- ($kJ\,mol^{-1}$)
F	1681	−333	−515
Cl	1256	−349	−381
Br	1143	−325	−347
I	1009	−296	−305
At	–	−270	–

The ionization energy for F is appreciably higher than for the others, because of its small size. F always has an oxidation state of (−I) except in F_2. It forms compounds either by gaining an electron to form F^-, or by sharing an electron to form a covalent bond.

Hydrogen has an ionization energy of $1311\,kJ\,mol^{-1}$, and it forms H^+ ions. It is at first surprising that the halogens Cl, Br and I have lower ionization energies than H, yet they do not form simple X^+ ions. The ionization energy is the energy required to produce an ion from a single isolated gaseous atom. Usually we have a crystalline solid, or a solution, so the lattice energy or hydration energy must also be considered. Because H^+ is very small, crystals containing H^+ have a high lattice energy, and in solution the hydration energy is also very high ($1091\,kJ\,mol^{-1}$). The negative ions also have a hydration energy. Thus H^+ ions are formed because the lattice energy, or the hydration energy, exceeds the ionization energy. In contrast the halide ions X^+ would be large and thus have low hydration and lattice energies. Since the ionization energy would be larger than the lattice energy or hydration energy, these ions are not normally

Table 16.5 Electronegativity and electrode potential

	Pauling's electronegativity	Standard electrode potential $E°$ (volts)
F	4.0	+2.87
Cl	3.0	+1.40
Br	2.8	+1.09
I	2.5	+0.62
At	2.2	+0.3

formed. However, a few compounds are known where I^+ is stabilized by forming a complex with a Lewis base, for example $[I(pyridine)_2]^+ NO_3^-$. These are discussed later under 'Basic properties of the halogens'.

The electron affinities for the halogens are all negative. This shows that energy is evolved when a halogen atom gains an electron, and $X \rightarrow X^-$. Thus, the halogens all form halide ions.

TYPE OF BONDS FORMED, AND OXIDATION STATES

Most compounds formed by the halogens and metals are ionic. However, covalent halides are formed in a few cases where the metal ions are very small and have a high charge. (The structures of $BeCl_2$ and $AlCl_3$ are unusual – see Chapters 11 and 12.)

The halogens all have very high electronegativity values (see Table 16.5). When they react with metals there will be a large electronegativity difference: hence they form ionic bonds. Halide ions are produced quite easily. This is shown by the large electron affinity values (Table 16.4). Note that energy is evolved when a gaseous halogen atom gains an electron, and also because of their large positive standard electrode potentials for $X_2|2X^-$ (Table 16.5). (The standard electrode potentials may be converted to an energy term using the relationship $\Delta G° = -nFE°$, where n is the number of electrons (2 in this case), and F is the Faraday constant $96\,486\,kJ\,mol^{-1}$.) The $E°$ values decrease down the group and thus the energy evolved on forming halide ions also decreases down the group. Many iodides are partly covalent. For example, CdI_2 forms a layer structure, and all the iodides have much lower melting points than the fluorides.

When two halogen atoms form a molecule, they form a covalent bond. Most compounds between the halogens and non-metals are also covalent. Fluorine is always univalent, and since it is the most electronegative element it always has the oxidation number (−I). With Cl, Br and I, a covalence of one is the most common. The oxidation state may be either (−I) or (+I) depending on which atom in the molecule has the greater electronegativity.

Cl, Br and I also exhibit higher valencies, with oxidation numbers of

(+III), (+V) and (+VII). These higher valency states are covalent, and arise quite logically by promoting electrons from filled *p* and *s* levels to empty *d* levels. The unpaired electrons then form three, five or seven covalent bonds. There are numerous examples of higher valency states in the interhalogens and halogen oxides.

		ns	*np*	*nd*
Electronic structure of halogen atom – ground state	full inner shell	↑↓	↑↓ ↑↓ ↑	(empty)

(Only one unpaired electron, so can only form one covalent bond)

		ns	*np*	*nd*
Electronic structure of halogen atom – excited state	full inner shell	↑↓	↑↓ ↑ ↑	↑

(Three unpaired electrons, so can form three covalent bonds)

		ns	*np*	*nd*
Electronic structure of halogen atom – further excited state	full inner shell	↑↓	↑ ↑ ↑	↑ ↑

(Five unpaired electrons, so can form five covalent bonds)

		ns	*np*	*nd*
Electronic structure of halogen atom – still further excited	full inner shell	↑	↑ ↑ ↑	↑ ↑ ↑

(Seven unpaired electrons, so can form seven covalent bonds)

The oxidation states (+IV) and (+VI) occur in the oxides ClO_2, BrO_2, Cl_2O_6 and BrO_3.

MELTING AND BOILING POINTS

The melting and boiling points of the elements increase with increased atomic number. At room temperature, fluorine and chlorine are gases, bromine is liquid, and iodine is a solid. In temperate climates, only two elements are liquid at room temperature, bromine and mercury. (In very hot climates caesium and thallium are also liquid.) At atmospheric pressure I_2 solid sublimes without melting.

Table 16.6 Melting and boiling points

	Melting point (°C)	Boiling point (°C)
F_2	−219	−188
Cl_2	−101	−34
Br_2	−7	60
I_2	114	185

Table 16.7 Bond energy and bond lengths of X_2

	Bond energy (free energy of dissociation) ($kJ\,mol^{-1}$)	Bond length X_2 (Å)
F	126	1.43
Cl	210	1.99
Br	158	2.28
I	118	2.66

BOND ENERGY IN X_2 MOLECULE

The elements all form diatomic molecules. It would be expected that the bond energy in the X_2 molecules would decrease as the atoms become larger, since increased size results in less effective overlap of orbitals. Cl_2, Br_2 and I_2 show the expected trend (Table 16.7), but the bond energy for F_2 does not fit the expected trend.

The bond energy in F_2 is abnormally low ($126\,kJ\,mol^{-1}$), and this is largely responsible for its very high reactivity. (Other elements in the first row of the periodic table also have weaker bonds than the elements which follow in their respective groups. For example in Group V the N—N bond in hydrazine is weaker than P—P, and in Group VI the O—O bond in peroxides is weaker than S—S.) Two different explanations have been suggested for the low bond energy:

1. Mulliken postulated that in Cl_2, Br_2 and I_2 *some pd* hybridization occurred, allowing *some* multiple bonding. This would make the bonds stronger than in F_2 in which there are no *d* orbitals available.
2. Coulson suggested that since fluorine atoms are small, the F—F distance is also small (1.48 Å), and hence internuclear repulsion is appreciable. The large electron–electron repulsions between the lone pairs of electrons on the two fluorine atoms weaken the bond.

It seems unnecessary to invoke multiple bonding to explain these facts, and the simpler Coulson explanation is widely accepted.

OXIDIZING POWER

Electron affinity is the tendency of the atoms to gain electrons. This reaches a maximum at chlorine. (See Table 16.4.) Oxidation may be regarded as the removal of electrons, so that an oxidizing agent gains electrons. Thus the halogens act as oxidizing agents. The strength of an oxidizing agent (that is, its oxidation potential) depends on several energy terms and is best represented by a Born–Haber type of energy cycle (Figure 16.1).

The oxidation potential is the energy change between the element in its standard state, and in its hydrated ions. Thus for iodine the change is from

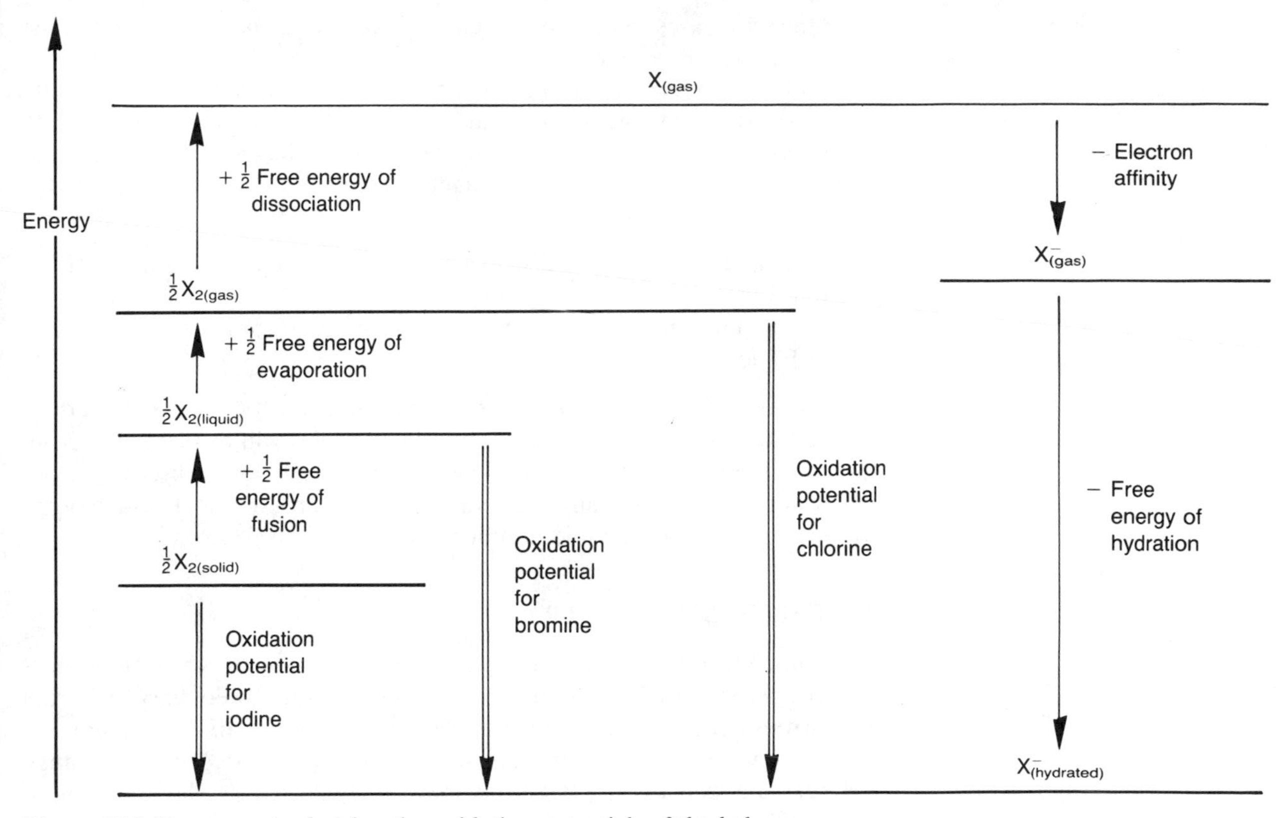

Figure 16.1 Energy cycle showing the oxidation potentials of the halogens.

$\frac{1}{2}I_{2(solid)}$ to $I^-_{(hydrated)}$. Thus the oxidation potential is equal to the sum of the energy put in as the enthalpies of fusion, evaporation and dissociation, less the energy evolved as the electron affinity and free energy of hydration.

In a similar cycle the oxidation potential for bromine can be calculated for the change from $\frac{1}{2}Br_{2(liquid)}$ to $Br^-_{(hydrated)}$. (Note that since in its standard state bromine is liquid, the free energy of fusion must be omitted. Similarly in calculating the oxidation potential for chlorine and fluorine,

Table 16.8 Free energy (ΔG°) values for $\frac{1}{2}X_2 \rightarrow X_{(hydrated)}$ (all values in $kJ\,mol^{-1}$)

	$\frac{1}{2}$ Free energy of fusion	$\frac{1}{2}$ Free energy of evaporation	$\frac{1}{2}$ Free energy of dissociation	Electron affinity	Free energy of hydration	Sum of ΔG°
F_2	–	–	+126/2	−333	−460	−730
Cl_2	–	–	+210/2	−349	−348	−592
Br_2	–	+31/2	+158/2	−325	−318	−548.5
I_2	+15/2	+44/2	+118/2	−296	−279	−486.5

since they are gases, both the free energies of fusion and evaporation must be omitted.)

Though the electron affinity of chlorine is the highest, it is not the strongest oxidizing agent (see Table 16.8). When all the terms in the energy cycle are summed, fluorine has the most negative $\Delta G°$ value, so fluorine is the strongest oxidizing agent. There are two main reasons for this:

1. F_2 has a low enthalpy of dissociation (arising from the weakness of the F—F bond).
2. F_2 has a high free energy of hydration (arising from the smaller size of the F^- ion).

Fluorine is a very strong oxidizing agent, and it will replace Cl^- both in solution and also when dry. Similarly, chlorine gas will displace Br^- from solution. (This is the basis of the commercial extraction of bromine from sea water.) In general any halogen of low atomic number will oxidize halide ions of higher atomic number.

REACTION WITH WATER

The halogens are all soluble in water, but the extent to which they react with the water, and the reaction mechanism that is followed, vary. Fluorine is so strong an oxidizing agent that it oxidizes water to oxygen. The reaction is spontaneous and strongly exothermic. (The free energy change is large and negative.) Oxidation may be regarded as the removal of electrons, so that an oxidizing agent gains electrons. Thus the fluorine atoms are reduced to fluoride ions.

$$F_2 + 3H_2O \rightarrow 2H_3O^+ + 2F^- + \tfrac{1}{2}O_2 \qquad \Delta G° = -795\ \text{kJ mol}^{-1}$$

A similar reaction between chlorine and water is thermodynamically possible, but the reaction is very slow because the energy of activation is high.

$$Cl_2 + 3H_2O \rightarrow 2H_3O^+ + 2Cl^- + \tfrac{1}{2}O_2$$

With chlorine an alternative disproportionation reaction occurs rapidly:

$$\begin{array}{lccccccc} & Cl_2 & + & H_2O & \rightarrow & HCl & + & HOCl \\ \text{Oxidation state of chlorine} & (0) & & & & (-I) & & (+I) \end{array}$$

Table 16.9 Concentrations in saturated aqueous solutions at 25 °C

	Solubility (mol l^{-1})	Concentration X_2 (hydrated) (mol l^{-1})	Concentration HOX (mol l^{-1})
Cl_2	0.091	0.061	0.030
Br_2	0.21	0.21	1.1×10^{-3}
I_2	0.0013	0.0013	6.4×10^{-6}

A similar disproportionation reaction occurs to a very limited extent with Br_2 and I_2. Thus a saturated aqueous solution of Cl_2 at 25 °C contains about two thirds hydrated X_2 and one third OCl^-. Solutions of Br_2 and I_2 contain only a very small amount of OBr^-, and a negligible amount of OI^-, respectively.

Iodine is an even weaker oxidizing agent. The free energy change is positive, which shows that energy must be supplied to make it oxidize water.

$$I_2 + H_2O \rightarrow 2H^+ + 2I^- + \tfrac{1}{2}O_2 \qquad \Delta G^\circ = +105\,\text{kJ mol}^{-1}$$

It follows that for the reverse reaction ΔG° would be $-105\,\text{kJ mol}^{-1}$, so the reverse reaction should occur spontaneously. This is the case. Atmospheric oxygen oxidizes iodide ions to iodine. At the end point of an iodine titration with sodium thiosulphate, the iodine originally resent is all converted to iodide ions. Thus the bluish colour produced by the starch indicator with iodine disappears, and the solution becomes colourless.

$$I_2 + 2S_2O_3^{2-} \rightarrow 2I^- + S_4O_6^{2-}$$

If the titration flask is allowed to stand for two or three minutes, the indicator turns blue again. This is because some atmospheric oxidation has taken place, forming I_2, which reacts with the starch to give the blue colour again.

$$2I^- + \tfrac{1}{2}O_2 + 2H^+ \rightarrow I_2 + H_2O$$

The end point of the titration is usually taken as being when the colour disappears *and the solution remains colourless for half a minute.*

REACTIVITY OF THE ELEMENTS

Fluorine is the most reactive of all the elements in the periodic table. It reacts with all the other elements except the lighter noble gases He, Ne and Ar. It reacts with xenon under mild conditions to form xenon fluorides. (See Chapter 17.) Reactions with many elements are vigorous, and often explosive. In the massive form a few metals such as Cu, Ni, Fe and Al acquire a protective fluoride coating. However, if these metals are in powdered form (with a large surface area), or if the reaction mixture is heated, then the reaction is vigorous. The reactivity of the other halogens decreases in the order Cl > Br > I. Chlorine and bromine react with most of the elements, though less vigorously than does fluorine. Iodine is less reactive and does not combine with some elements such as S and Se. Fluorine and chlorine often oxidize elements further than do bromine and iodine, by this means bringing out higher valencies, for example in PBr_3 and PCl_5, and in S_2Br_2, SCl_2 and SF_6.

The great reactivity of fluorine is attributable to two factors:

1. The low dissociation energy of the F—F bond (which results in a low activation energy for the reaction).
2. The very strong bonds which are formed.

Table 16.10 Some bond energies of halogen compounds (all values in kJ mol^{-1})

	HX	BX_3	AlX_3	CX_4	NX_3	X_2
F	566	645	582	439	272	159
Cl	431	444	427	347	201	243
Br	366	368	360	276	243	193
I	299	272	285	238	*	151

* Unstable and explosive.

Both of these properties arise from the small size of fluorine. The weak F—F bond arises because of repulsion between the lone pairs of electrons on the two atoms. Strong bonds arise because of the high coordination number and high lattice energy.

Some bond energies are shown in Table 16.10. These explain why the halogens form very strong bonds. Many are stronger than the C—C bond, which is itself regarded as a very strong bond. (The C—C bond energy is 347 kJ mol^{-1}.)

Table 16.11 Some reactions of the halogens

Reaction	Comment
$2F_2 + 2H_2O \rightarrow 4H^+ + 4F^- + O_2$ $2I_2 + 2H_2O \leftarrow 4H^+ + 4X^- + O_2$ $X_2 + H_2O \rightarrow H^+ + X^- + HOX$	Vigorous reaction with F I reaction in reverse direction Cl > Br > I (F not at all)
$X_2 + H_2 \rightarrow 2HX$	All the halogens
$nX_2 + 2M \rightarrow 2MX_n$	Most metals form halides F the most vigorous
$X_2 + CO \rightarrow COX_2$	Cl and Br form carbonyl halides
$3X_2 + 2P \rightarrow PX_3$ $5X_2 + 2P \rightarrow PX_5$	All the halogens form trihalides As, Sb and Bi also form trihalides F, Cl and Br form pentahalides AsF_5, SbF_5, BiF_5, $SbCl_5$
$X_2 + 2S \rightarrow S_2X_2$ $2Cl_2 + S \rightarrow SCl_4$ $3F_2 + S \rightarrow SF_6$	Cl and Br Cl only F only
$X_2 + H_2S \rightarrow 2HX + S$	All the halogens oxidize S^{2-} to S
$X_2 + SO_2 \rightarrow SO_2X_2$	F and Cl
$3X_2 + 8NH_3 \rightarrow N_2 + 6NH_4X$	F, Cl and Br
$X_2 + X'_2 \rightarrow 2XX'$ $X_2 + X'X \rightarrow X'X_3$	Interhalogen compounds formed higher interhalogen compounds

HYDROGEN HALIDES HX

It is usual to refer to pure anhydrous HX compounds as hydrogen halides, and their aqueous solutions as hydrohalic acids or simply halogen acids.

The halogens all react with hydrogen and form hydrides HX, though except for HCl this is not the usual way of preparing them. Reactivity towards hydrogen decreases down the group. Hydrogen and fluorine react violently. The reaction with chlorine is slow in the dark, faster in daylight, and explosive in sunlight. The reaction with iodine is slow at room temperature.

HF

Industrially HF is made by heating CaF_2 with strong H_2SO_4. The reaction is endothermic: hence the need for heating. It is important that SiO_2 impurities are removed from the CaF_2, as otherwise they consume much of the HF produced.

$$SiO_2 + 4HF \rightarrow SiF_4 + 2H_2O$$
$$SiF_4 + 2HF_{(aq)} \rightarrow H_2[SiF_6]$$

The HF is purified by successive washing, cooling and fractional distillation, giving a product that is 99.95% pure. World production of HF is about 1 million tonnes per year, with over 80000 tonnes a year being produced in the UK.

$$CaF_2 + H_2SO_4 \rightarrow CaSO_4 + 2HF$$

Gaseous HF is very toxic, and should be handled only in a good fume cupboard. Solutions of HF are called hydrofluoric acid, and this is very corrosive. Hydrofluoric acid is normally handled in metal apparatus made of copper or Monel, because hydrofluoric acid attacks glass with the formation of fluorosilicate ions $[SiF_6]^{2-}$.

$$SiO_2 + 6HF \rightarrow [SiF_6]^{2-} + 2H^+ + 2H_2O$$

Surprisingly little corrosion occurs at concentrations above 80%. The main uses of HF are as follows:

1. For making chlorofluorocarbons (Freons). These are sometimes called CFCs. They are used as refrigerating fluids and as the propellant in aerosols. The use of CFCs is being phased out because they damage the ozone layer in the upper atmosphere. (See Chapter 13, under 'Tetrahalides'.)

$$CCl_4 + 2HF \xrightarrow[+\ SbCl_5]{\text{anhydrous conditions}} \underset{\text{Freon}}{CCl_2F_2} + 2HCl$$

2. For making AlF_3 and synthetic cryolite used in the electrolytic extraction of aluminium. (See Chapter 3.)
3. Smaller amounts are used for uranium processing (through intermediates UF_4 and UF_6).
4. Anhydrous HF is used as an alkylation catalyst in the petrochemical industry, for making long chain alkylbenzene compounds. These are then converted into alkylbenzene sulphonates and used as detergents.
5. Aqueous HF is used for etching glass, pickling steel and making many fluorides such as BF_3.

$$B_2O_3 + 6HF \xrightarrow{\text{conc. } H_2SO_4} 2BF_3 + 3H_2SO_4 \cdot H_2O$$

$$Al_2O_3 + 6HF \rightarrow 2AlF_3 + 3H_2O$$

HCl

HCl is produced on a very large scale. World production was 9 million tonnes in 1985 (USA 28%, China 21%, West Germany 11%, Japan 7%, France 3%, Spain, Canada, Mexico, UK, Romania and East Germany 2% each). There are several different preparative methods:

1. At one time HCl was made exclusively by the 'salt cake' method. In this method, concentrated H_2SO_4 was added to rock salt (NaCl). The reaction was endothermic, and was performed in two stages at different temperatures. The first of the reactions was carried out at about 150 °C. The solid NaCl reacted with H_2SO_4 and became coated with insoluble $NaHSO_4$. This prevented further reaction, and accounts for the name 'salt cake'. In the second stage, the mixture was heated to about 550 °C, when further reaction with H_2SO_4 occurred and Na_2SO_4 was formed. This by-product was sold for paper making and glass making.

$$NaCl + H_2SO_4 \xrightarrow{150\,°C} HCl_{(g)} + NaHSO_4$$

$$NaCl + NaHSO_4 \xrightarrow{550\,°C} HCl_{(g)} + Na_2SO_4$$

2. Large amounts of impure HCl have become available in recent years as a by-product from the heavy organic chemical industry. For example, HCl is produced in the conversion of 1,2-dichloethane $CH_2Cl—CH_2Cl$ to vinyl chloride $CH_2═CHCl$, and in the manufacture of chlorinated ethanes and chlorinated fluorocarbons. This is now the largest source of HCl.
3. High purity HCl is made by direct combination of the elements. A gaseous mixture of H_2 and Cl_2 is explosive. However, the reaction proceeds quietly if the gases are burnt in a hydrogen–chlorine flame in a special combustion chamber. The process is strongly exothermic.
4. HCl is conveniently made in the laboratory by treating NH_4Cl with concentrated H_2SO_4. NH_4Cl costs more than NaCl (which was used in the 'salt cake' process). However, NH_4Cl is preferred because

NH_4HSO_4 is soluble, and the reaction does not stop at the halfway stage.

$$2NH_4Cl + H_2SO_4 \rightarrow 2HCl + (NH_4)_2SO_4$$

Hydrogen chloride gas is very soluble in water. Aqueous solutions of HCl are sold as *hydrochloric acid*. A saturated solution at 20 °C contains 42% HCl by weight, and 'concentrated' acid normally contains about 38% HCl by weight (approx. 12 M). Pure hydrochloric acid is colourless, but technical grades are sometimes yellow because of contamination by Fe(III). The largest use is for 'pickling' metals, that is removing oxide layers from the surface. It is also used to make metal chlorides, in the manufacture of dyestuffs, and in the sugar industry.

Gaseous HCl is conveniently prepared in the laboratory from concentrated HCl and concentrated H_2SO_4.

HBr and HI

HBr and HI are made by the reaction of concentrated phosphoric acid H_3PO_4 on metal bromides or iodides, in a similar reaction to the 'salt cake' process for HCl. Note that a non-oxidizing acid such as phosphoric acid must be used. Concentrated H_2SO_4 is a strong oxidizing agent and would oxidize HBr to Br_2 and HI to I_2.

$$H_3PO_4 + NaI \rightarrow HI + NaH_2PO_4$$

The usual laboratory preparation involves reducing bromine or iodine with red phosphorus in water. Thus HBr is made by adding bromine to a mixture of red phosphorus and water. For HI, water is added to a mixture of phosphorus and iodine.

$$H_3PO_4 + NaBr \rightarrow HBr + NaH_2PO_4$$

$$\overset{\text{red}}{2P} + 3Br_2 \rightarrow 2PBr_3 \xrightarrow{+6H_2O} 6HBr + 2H_3PO_3$$

$$\overset{\text{red}}{2P} + 3I_2 \rightarrow 2PI_3 \xrightarrow{+6H_2O} 6HI + 2H_3PO_3$$

HF is only just liquid at room temperature (b.p. 19.9 °C), and HCl to HBr and HI are gases. The boiling points increase regularly from HCl,

Table 16.12 Some properties of HX compounds

	Melting point (°C)	Boiling point (°C)	Density ($g\,cm^{-1}$)	pK_a values	Composition of azeotrope (weight %)
HF	−83.1	19.9	0.99	3.2	35.37
HCl	−114.2	−85.0	1.19	−7	20.24
HBr	−86.9	−66.7	2.16	−9	47.0
HI	−50.8	−35.4	2.80	−10	57.0

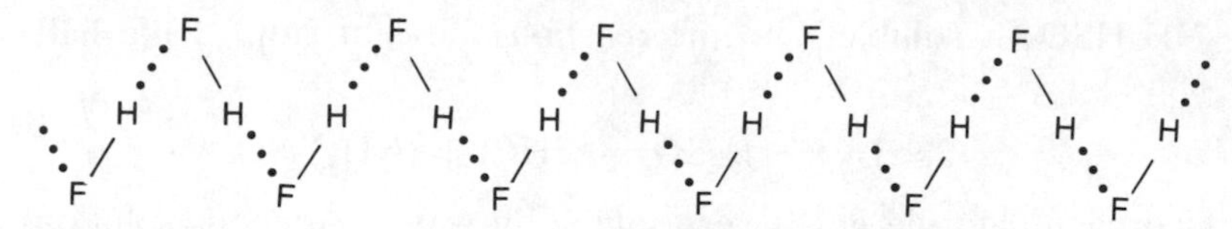

Figure 16.2 Hydrogen bonded chain in solid HF.

HBr to HI, but the value for HF is completely out of line with the others.

The unexpectedly high boiling point of HF arises because of the hydrogen bonds formed between the F atom of one molecule and the H atom of another molecule. This links the molecules together as $(HF)_n$, and they form zig-zag chains in both the liquid and the solid. Some hydrogen bonding also occurs in the gas, which consists of a mixture of cyclic $(HF)_6$ polymers, dimeric $(HF)_2$, and monomeric HF. HCl, HBr and HI are not hydrogen bonded in the gas and liquid, though HCl and HBr are weakly hydrogen bonded in the solid.

Hydrogen bonds are generally weak ($5–35\,kJ\,mol^{-1}$) compared with normal covalent bonds (C—C $347\,kJ\,mol^{-1}$), but their effect is highly significant. The most electronegative elements fluorine and oxygen (and to a lesser extent chlorine) form the strongest hydrogen bonds. (The bond energy of the hydrogen bond in F—H. . . F is $29\,kJ\,mol^{-1}$ in $HF_{(g)}$).

In the gaseous state the hydrides are essentially covalent. However, in aqueous solutions they ionize. H^+ are not produced since the proton is transferred from HCl to H_2O, thus giving $[H_3O]^+$. HCl, HBr and HI ionize almost completely and are therefore strong acids. HF only ionizes slightly and is therefore a weak acid.

$$HCl + H_2O \rightarrow [H_3O]^+ + Cl^-$$

The aqueous solutions form azeotropic mixtures with maximum boiling points, because of a negative deviation from Raoult's law. Azeotropic mixtures are sometimes used as standards for volumetric analysis, because the azeotrope always has the same composition.

Though HCl, HBr and HI completely ionize in water, the degree of ionization is much less in poorer ionizing solvents such as anhydrous acetic acid. HCl ionizes less than HI in glacial acetic acid as solvent. Thus in acetic acid, HI is the strongest acid, followed by HBr and HCl, and HF is the weakest.

It is at first paradoxical that HF is the weakest acid in water, since HF has a greater electronegativity difference than the other hydrides, and therefore has more ionic character. However, acidic strength is the tendency of hydrated molecules to form hydrogen ions:

$$HX_{(hydrated)} \rightarrow H^+_{(hydrated)} + X^-_{(hydrated)}$$

This may be represented in stages: dissociation, ionization and hydration in an energy cycle.

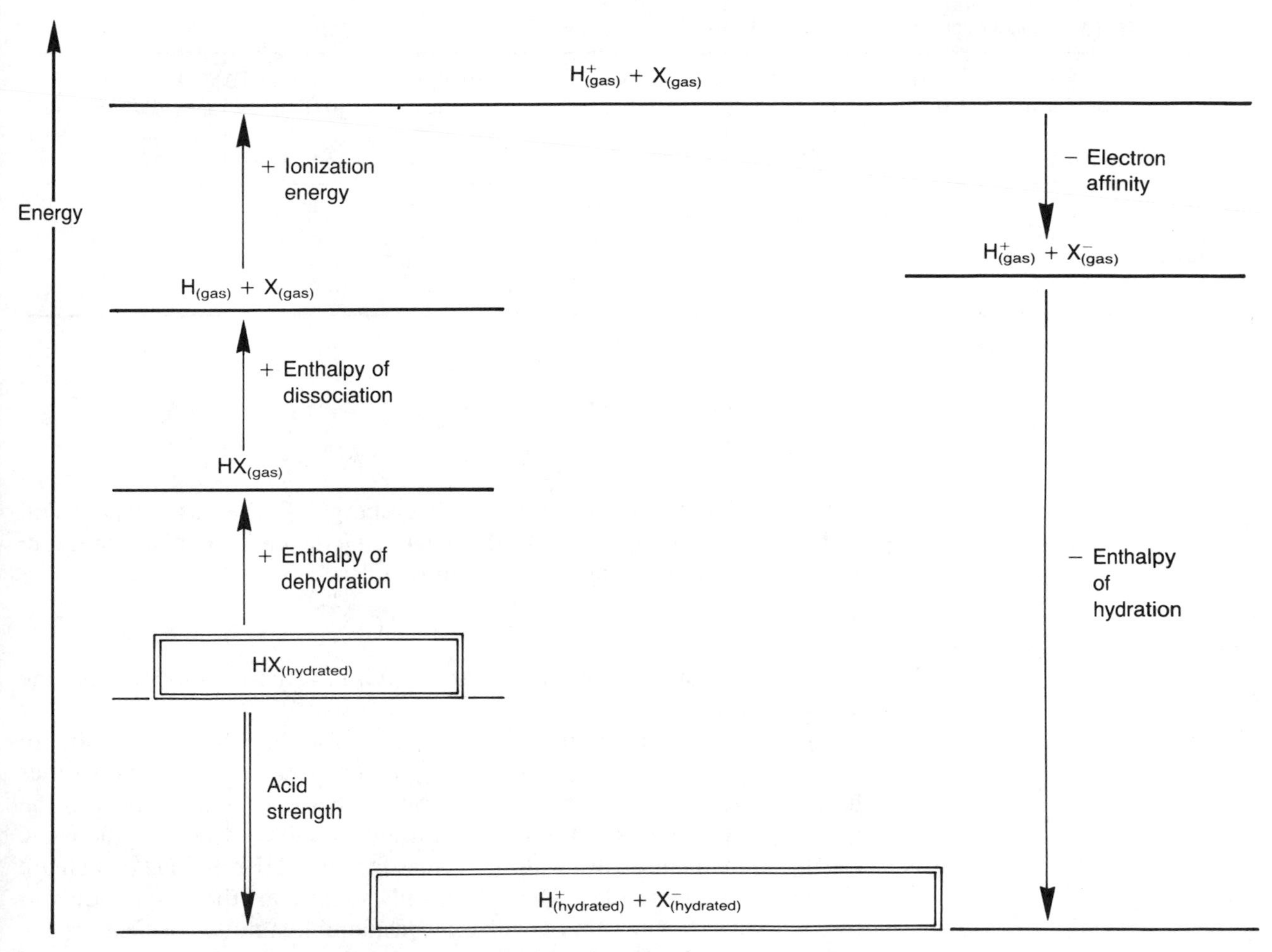

Figure 16.3 Energy cycle showing the acid strengths of the halogens.

The acid strength is equal to the sum of all the energy terms round the energy cycle in Figure 16.3.

acid strength = free energy of dehydration
+ free energy of dissociation
+ ionization energy of H^+
+ electron affinity X^-
+ free energy of hydration of H^+ and X^-

The factors which make HF the weakest halogen acid in water become apparent if the various thermodynamic terms are examined in more detail. The dissociation constant *k* for the change

$$HX_{(hydrated)} \rightleftharpoons H^+_{(hydrated)} + X^-_{(hydrated)}$$

Table 16.13 Energy cycle. (All values in kJ mol^{-1}.)

	Enthalpy dehydration	Enthalpy dissociation	Ionization energy $H \rightarrow H^+$	Electron affinity of X	Enthalpy hydration		Total ΔH	$T\Delta S$	$\Delta G = (\Delta H - T\Delta S)$
					H^+	X^-			
HF	48	566	1311	−333	−1091	−151	−14	−29	+15
HCl	18	431	1311	−348	−1091	−381	−60	−13	−47
HBr	21	366	1311	−324	−1091	−347	−64	−4	−60
HI	23	299	1311	−295	−1091	−305	−58	+4	−62

is given by the equation:

$$\Delta G^\circ = -RT \ln k$$

(where ΔG° is the Gibbs standard free energy, R the gas constant and T the absolute temperature). However, ΔG depends on the change in enthalpy ΔH and the change in entropy ΔS

$$\Delta G = \Delta H - T\Delta S$$

Table 16.13 shows the enthalpy changes (ΔH) for the various stages in the above energy cycle.

Consider first the total enthalpy change ΔH for the dissociation of $HX_{(hydrated)}$ into $H^+_{(hydrated)}$ and $X^-_{(hydrated)}$. The ΔH values for the various halogen acids are all negative which means that energy is evolved in the process, so the change is thermodynamically possible. However, the value for HF is small compared with the values for HCl, HBr and HI (which are all similar in magnitude). Thus HF is only slightly exothermic in aqueous solution whereas the others evolve a considerable amount of heat.

The low total ΔH value for HF is the result of several factors.

1. The enthalpies of dissociation show that the H—F bond is much stronger than the H—Cl, H—Br or H—I bonds. Thus the dissociation energy of HF is nearly twice that required to dissociate HI. (The strength of the HF bond is also shown by the short bond length of 1.0 Å compared with 1.7 Å in HI.)
2. The heat of dehydration for the step $HX_{(hydrated)} \rightarrow HX_{(gas)}$ is much higher for HF than for the others. This is because of the strong hydrogen bonding which occurs in aqueous HF solutions.
3. The unexpectedly low value for the electron affinity of F^- also contributes, and though the enthalpy of hydration of F^- is very high, it is not enough to offset these other terms.

If allowance is made for the $T\Delta S$ term, the ΔH values can be converted into corresponding ΔG values. From these the dissociation constants are obtained: HF $k = 10^{-3}$, HCl $k = 10^8$, HBr $k = 10^{10}$ and HI $k = 10^{11}$.

The dissociation constants show quite clearly that HF is only very slightly ionized in water, and is therefore a weak acid. In a similar way, the others are almost totally dissociated, and are therefore strong acids.

Liquid HF has been used as a non-aqueous solvent. It undergoes self-ionization:

$$2HF \rightleftharpoons [H_2F]^+ + F^-$$

Acid–base reactions occur in this solvent system. However, the solvent itself has a very strong tendency to donate protons. Thus when the familiar mineral acids HNO_3, H_2SO_4 and HCl are dissolved in HF, the mineral acids are forced to accept protons from the HF. Thus the so-called mineral acids are actually behaving as bases *in this solvent*. The very strong proton donating powers of HF mean that very few substances act as acids in HF. Perchloric acid is an exception, and it does behave as an acid. The only other known acids in liquid HF are fluoride acceptors such as SbF_5, NbF_5, AsF_5 and BF_3. Many compounds react with HF, thus limiting its usefulness as a solvent. It is a useful medium for preparing fluoro complexes such as $[SbF_6]^-$, and fluorides.

HALIDES

Ionic halides

Most halides where the metal has an oxidation state of (+I), (+II), or +(III) are ionic. This includes Group I, Group II (except Be), the lanthanides, and *some* of the transition metals. Most ionic halides are soluble in water, giving hydrated metal ions and halide ions. A few are insoluble: LiF, CaF_2, SrF_2, BaF_2, and the chlorides, bromides and iodides of Ag(+I), Cu(+I), Hg(+I) and Pb(+II). The solubilities usually increase from F^- to Br^- to Cl^- to I^- (provided that they are all ionic), because the lattice energy decreases as the ionic radii increase.

Molecular (covalent) halides

Among the metals which show variable valency, the highest oxidation state is usually found with the fluorides. Thus osmium forms OsF_6, but only $OsCl_4$, $OsBr_4$ and OsI_4. High oxidation states are covalent. For a metal with variable oxidation states, the higher oxidation states will be covalent and the lower ones ionic. For example, UF_6 is covalent and gaseous, whereas UF_4 is an ionic solid. Similarly $PbCl_4$ is covalent and $PbCl_2$ is ionic. Most of the more electronegative elements also form covalent halides, sometimes called molecular halides. A large number of these are hydrolysed quite readily by water:

$$BCl_3 + 3H_2O \rightarrow H_3BO_3 + 3H^+ + 3Cl^-$$
$$SiCl_4 + 4H_2O \rightarrow Si(OH)_4 + 4H^+ + 4Cl^-$$
$$PCl_3 + 3H_2O \rightarrow H_3PO_3 + 3H^+ + 3Cl^-$$
$$PCl_5 + 4H_2O \rightarrow H_3PO_4 + 5H^+ + 5Cl^-$$

Sometimes when the maximum covalency is obtained, the halides are inert to water. Thus CCl_4 and SF_6 are stable. This is because of kinetic rather than thermodynamic factors, and CCl_4 does hydrolyse with superheated steam to form phosgene $COCl_2$. Molecular halides are usually gases or volatile liquids. This is because there are strong bonds within the molecule, but only weak van der Waals forces holding the molecules together. A number of fluorides show multiple bonding when the central atom has suitable vacant orbitals. This contributes to the high strength and shortness of many bonds of fluorine (B—F, C—F, N—F and P—F).

Bridging halides

Halide bridges are sometimes formed between two atoms. (Less commonly they are formed between three atoms.) Thus $AlCl_3$ forms a dimeric structure whereas BeF_2 and $BeCl_2$ form infinite chains. The bridges are depicted as the halogen forming one normal covalent bond, and donating a lone pair of electrons to form a second (coordinate) bond. Both bonds are identical. The bridge may be described in molecular orbital terms as a three-centre four-electron bond. Halogen bridges involving chlorine and bromine are typically bent, but those involving fluorine may be either bent or linear. Several pentafluorides such as NbF_5 and TaF_5 form cyclic tetramers with linear bridges.

Preparation of anhydrous halides

There are several general methods for making anhydrous halides:

Direct reaction of the elements.

Most metals react vigorously with F_2, and give fluorides in the highest oxidation states. Some non-metals such as P and S explode. Elevated temperatures are usually required to prepare chlorides, bromides and iodides.

$$2Fe + 3F_2 \rightarrow 2FeF_3$$
$$Fe + Br_2 \rightarrow FeBr_2$$
$$Fe + I_2 \rightarrow FeI_2$$

Reactions are easier in a solvent such as tetrahydrofuran, though the products are often solvated.

Reacting the oxide with carbon and the halogen

It is assumed that the carbon first reduces the oxide to the metal, followed by reaction of the metal with the halogen.

$$TiO_2 + C + 2Cl_2 \rightarrow TiCl_4 + CO_2$$

Reaction of metal with anhydrous HX

Many metals will react with HF, HCl, HBr or HI gases.

$$2Al + 6HCl \rightarrow 2AlCl_3 + 3H_2$$
$$Cr + 2HF \rightarrow CrF_2 + H_2$$
$$Fe + 2HCl \rightarrow FeCl_2 + H_2$$

Reaction of oxides with halogen compounds

Heating halogen compounds such as NH_4Cl, CCl_4, ClF_3, BrF_3, S_2Cl_2 or $SOCl_2$ with oxides often gives halides.

$$Sc_2O_3 + 6NH_4Cl \xrightarrow{300\,°C} 2ScCl_3 + 6NH_3 + 3H_2O$$
$$2BeO + CCl_4 \xrightarrow{800\,°C} 2BeCl_2 + CO_2$$
$$3UO_2 + 4BrF_3 \rightarrow 3UF_4 + 3BrO_2 + \tfrac{1}{2}Br_2$$
$$3NiO + 2ClF_3 \rightarrow 3NiF_2 + Cl_2 + 1\tfrac{1}{2}O_2$$

Halogen exchange

Many halides will react either with the halogens, or with excess of another halide and replace one halogen atom by another. Thus several metal fluorides such as AgF_2, ZnF_2, CoF_3, AsF_3, SbF_3 and SbF_5 can be used to make fluorides, and also HF.

$$PCl_3 + SbF_3 \rightarrow PF_3 + SbCl_3$$
$$CoCl_2 + 2HF \rightarrow CoF_2 + 2HCl$$

Chlorides may be converted to iodides by treatment with KI in acetone. Similarly KBr can be used for bromides.

$$TiCl_4 + 4KI \rightarrow TiI_4 + 4KCl$$

Dehydrating hydrated halides

Hydrated halides may be prepared in a variety of ways, such as dissolving carbonates, oxides or metals in the appropriate halogen acid. Evaporation gives hydrated halides. Some of these may be dehydrated simply by heating, or by heating in a vacuum, but oxohalides are often produced. Chlorides may be dehydrated by distilling with thionyl chloride. Other halides may be dehydrated by treatment with 2,2-dimethoxypropane.

$$VCl_3 \cdot 6H_2O + 6SOCl_2 \rightarrow VCl_3 + 12HCl + 6SO_2$$
$$CrF_3 \cdot 6H_2O + 6CH_3C(OCH_3)_2CH_3 \rightarrow CrF_3 + 12CH_3OH + 6(CH_3)_2CO$$

HALOGEN OXIDES

The compounds with oxygen probably show greater differences between the different halogens than any other class of compound. Differences

Table 16.14 Compounds of the halogens with oxygen (and their oxidation states)

Oxidation state	(−I) (+I)	(+IV)	(+V)	(+VI)	(+VIII)	Others
Fluorides	$OF_2(-I)$ $O_2F_2(-I)$					O_4F_2
Oxides	$Cl_2O(+I)$	ClO_2		Cl_2O_6 ClO_3	Cl_2O_7	$ClClO_4$
	$Br_2O(+I)$	BrO_2				
			I_2O_5			I_4O_9 I_2O_4

between F and the others arise for the usual reasons (small size, lack of *d* orbitals and high electronegativity). In addition oxygen is less electronegative than F, but more electronegative than Cl, Br and I. Thus binary compounds of F and O are fluorides of oxygen rather than oxides of fluorine. The other halogens are less electronegative than oxygen and thus form oxides. There is only a small difference in electronegativity between the halogens and oxygen, so the bonds are largely covalent. Rather surprisingly I_2O_4 and I_4O_9 are stable and ionic.

Most of the halogen oxides are unstable, and tend to explode when subjected to shock, or sometimes even when exposed to light. The iodine oxides are the most stable, then the chlorine oxides, but the bromine oxides all decompose below room temperature. The higher oxidation states are more stable than the lower states. Of the compounds shown in Table 16.14, ClO_2, Cl_2O, I_2O_5 and OF_2 are the most important.

OF_2 Oxygen difluoride

OF_2 is a pale yellow gas, formed by passing F_2 into dilute (2%) NaOH. OF_2 is a strong oxidizing agent, and has been used as a rocket fuel. It reacts vigorously with metals, S, P and the halogens, giving fluorides and oxides. It dissolves in water and gives a neutral solution, so it is not an acid anhydride. With NaOH it gives fluoride ions and oxygen.

$$2F_2 + 2NaOH \rightarrow 2NaF + H_2O + OF_2$$

O_2F_2 Dioxygen difluoride

O_2F_2 is an unstable orange–yellow solid, and is a violent oxidizing and fluorinating agent. It is formed by passing an electric discharge through a mixture of F_2 and O_2 at very low pressure and at liquid air temperature. It decomposes at $-95\,^{\circ}C$. Its structure is similar to that of H_2O_2, except that the O—O bond length of 1.22 Å is much shorter than the O—O distance of 1.48 Å in H_2O_2. The O—F bond lengths are 1.58 Å, which is much longer than in OF_2. O_4F_2 is made in a similar way, and apparently contains a chain of four oxygen atoms. O_5F_2 and O_6F_2 have been reported.

Cl_2O Dichlorine monoxide

Cl_2O is a yellow–brown gas. It is commercially important. Both laboratory and commercial preparations are by heating freshly precipitated (yellow) mercuric oxide with the halogen gas diluted with dry air.

$$2Cl_2 + 2HgO \xrightarrow{300\,°C} HgCl_2 \cdot HgO + Cl_2O$$

Cl_2O explodes in the presence of reducing agents, or NH_3, or on heating.

$$3Cl_2O + 10NH_3 \rightarrow 6NH_4Cl + 2N_2 + 3H_2O$$

Cl_2O gas is very soluble in water (144 g Cl_2O dissolves in 100 g H_2O at −9 °C), forming hypochlorous acid, and the two are in equilibrium.

$$Cl_2O + H_2O \rightleftharpoons 2HOCl$$

Cl_2O dissolves in NaOH solution, forming sodium hypochlorite.

$$Cl_2O + 2NaOH \rightarrow 2NaOCl + H_2O$$

Most of the Cl_2O produced is used to make hypochlorites. NaOCl is sold in aqueous solution. $Ca(OCl)_2$ is a solid commonly known as 'bleaching powder'. The latter is also made by passing Cl_2 into $Ca(OH)_2$. These are used to bleach wood pulp and fabrics, and as disinfectants. Some Cl_2O is used to make chlorinated solvents.

The structures of OF_2, Cl_2O and Br_2O are all related to a tetrahedron with two positions occupied by lone pairs of electrons.

	1s	2s	2p
Electronic structure of oxygen atom in its ground state	↑↓	↑↓	↑↓ ↑ ↑
Electronic structure of oxygen atom having gained two electrons by forming bonds to two halogen atoms	↑↓	↑↓	↑↓ ↑↓ ↑↓

four electron pairs – tetrahedral with two positions occupied by lone pairs

Repulsion between the lone pairs reduces the bond angle in F_2O from the tetrahedral angle of 109°28′ to 105° (Figure 16.4). In Cl_2O (and presumably Br_2O) the bond angle is increased because of steric crowding of the larger halogen atoms.

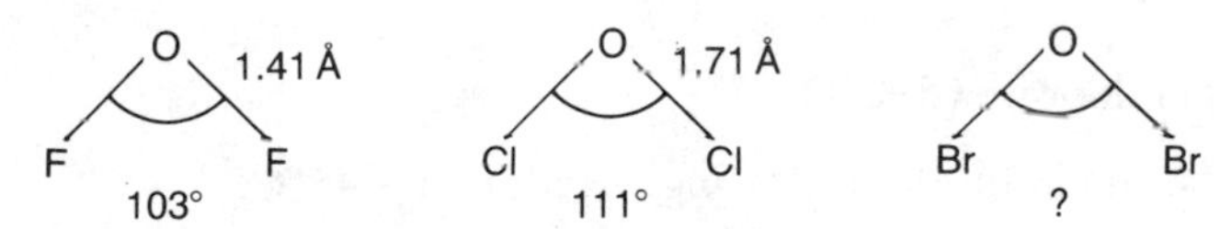

Figure 16.4 Bond angles in F_2O, Cl_2O and Br_2O.

XO_2 Chlorine dioxide

ClO_2 is a yellow gas which condenses to a deep red liquid, b.p. 11 °C. In spite of its high reactivity (or perhaps because of this) ClO_2 is of commercial importance, and is the most important of the oxides. ClO_2 is a powerful oxidizing and chlorinating agent. Large quantities are used for bleaching wood pulp and cellulose, and for purifying drinking water. It is 30 times as effective as chlorine in bleaching flour (to make white bread).

ClO_2 liquid explodes above −40 °C. The gas detonates readily when concentrated above 50 mmHg partial pressure. It explodes when mixed with reducing agents. Because of this, it is made in situ, and is used diluted with air or CO_2. The safest laboratory preparation is from sodium chlorate and oxalic acid, as this automatically dilutes the gas with CO_2.

$$2NaClO_3 + 2(COOH)_2 \xrightarrow{H_2O\ 90\,°C} 2ClO_2 + 2CO_2 + (COONa)_2 + 2H_2O$$

The gas is made commercially from $NaClO_3$. It is difficult to obtain total production figures, since for safety reasons ClO_2 is produced where it is used, and is always diluted (for safety). An estimate is 200 000 tonnes per year, half in the USA. A pure product is formed using SO_2. Using HCl causes contamination with Cl_2, but this may be unimportant or even useful for bleaching and sterilization.

$$2NaClO_3 + SO_2 + H_2SO_4 \xrightarrow{\text{trace of NaCl}} 2ClO_2 + 2NaHSO_4$$

$$2HClO_3 + 2HCl \rightarrow 2ClO_2 + Cl_2 + 2H_2O$$

It is also used to manufacture sodium chlorite $NaClO_2$, which is also used for bleaching textiles and paper.

$$2ClO_2 + 2NaOH + H_2O_2 \rightarrow 2NaClO_2 + O_2 + 2H_2O$$

Some other reactions are:

$$2ClO_2 + 2NaOH \rightarrow NaClO_2 + NaClO_3 + H_2O$$

$$2ClO_2 + 2O_3 \rightarrow Cl_2O_6 + 2O_2$$

The ClO_2 molecule is paramagnetic and contains an odd number of electrons. Odd electron molecules are generally highly reactive and ClO_2 is typical. Odd electron molecules often dimerize in order to pair the electrons, but ClO_2 does not. This is thought to be because the odd electron is delocalized. The molecule is angular with an O—Cl—O angle of 118°. The bond lengths are both 1.47 Å and are shorter than for single bonds.

Chlorine perchlorate $Cl \cdot ClO_4$

This can be made by the following reaction at −45 °C.

$$CsClO_4 + ClOSO_2F \rightarrow Cs(SO_3)F + ClOClO_3$$

It is less stable than ClO_2, and decomposes to O_2, Cl_2 and Cl_2O_6 at room temperature.

Cl_2O_6 Dichlorine hexoxide

Cl_2O_6 is a dark red liquid, which freezes to give a yellow solid at $-180\,^\circ C$. Cl_2O_6 is in equilibrium with the monomer ClO_3, and is made from ClO_2 and O_3. The structure of neither the liquid nor the solid is known. Both are diamagnetic, and so have no unpaired electrons. Possible structures are shown in Figure 16.5.

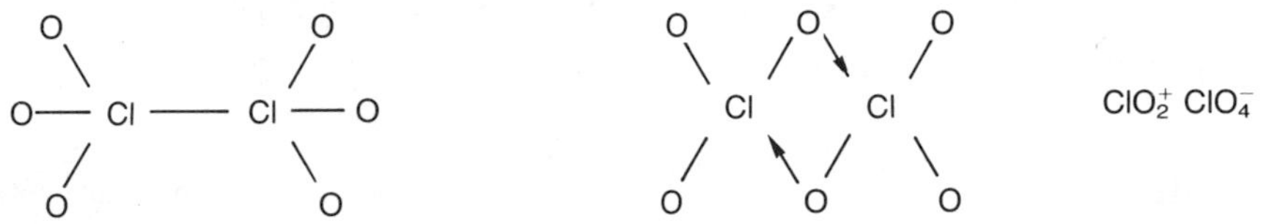

$ClO_2^+ \; ClO_4^-$

Figure 16.5 Possible structures of Cl_2O_6.

Cl_2O_6 is a strong oxidizing agent and explodes on contact with grease. Hydrolysis of Cl_2O_6 with water or alkali gives chlorate and perchlorate. Reaction with anhydrous HF is reversible.

$$Cl_2O_6 + 2NaOH \rightarrow \underset{\text{chlorate}}{NaClO_3} + \underset{\text{perchlorate}}{NaClO_4} + H_2O$$

$$Cl_2O_6 + H_2O \rightarrow \underset{HOClO_2}{HClO_3} + \underset{HOClO_3}{HClO_4}$$

$$Cl_2O_6 + HF \rightleftharpoons FClO_2 + HClO_4$$

$$Cl_2O_6 + N_2O_4 \rightarrow ClO_2 + [NO_2]^+[ClO_4]^-$$

Dichlorine heptoxide Cl_2O_7

Cl_2O_7 is a colourless oily liquid. It is moderately stable and is the only exothermic oxide of chlorine, but it is shock sensitive. It is made by carefully dehydrating perchloric acid with phosphorus pentoxide, or H_3PO_4. Its structure is $O_3Cl—O—ClO_3$, with a bond angle of 118°36′ at the central oxygen. It is less reactive than the lower oxides, and does not ignite organic materials. It reacts with water, forming perchloric acid.

$$2HClO_4 \underset{H_2O}{\overset{P_4O_{10}}{\rightleftharpoons}} Cl_2O_7$$

Oxides of bromine

Less is known about the oxides of bromine, and they are much less important than those of chlorine. Br_2O is a dark brown liquid, prepared either by reacting Br_2 gas with HgO (in the same way as Cl_2O is made), or by carefully decomposing BrO_2. It does not form HOBr to any appreciable extent by reaction with water, but with NaOH it gives OBr^-. It is a strong oxidizing agent, and oxidizes I_2 to I_2O_5.

Bromine dioxide BrO_2 is a pale yellow solid. It may be prepared by the action of an electric discharge on Br_2 and O_2 gas at low temperature and pressure, or by reacting bromine and ozone at −78 °C.

$$Br_2 + 2O_3 \rightarrow 2BrO_2 + O_2$$

It is only stable below −40 °C. It has a similar structure to ClO_2, but it is much less important than ClO_2. BrO_2 hydrolyses in alkaline solutions, giving bromide and bromate.

$$6BrO_2 + 6NaOH \rightarrow NaBr + 5NaBrO_3 + 3H_2O$$

With F_2 it gives $FBrO_2$.

Oxides of iodine

The oxides of iodine are much more stable than those of the other elements. Iodine pentoxide I_2O_5 forms stable white hygroscopic crystals. It is formed by heating iodic acid HIO_3 to 170 °C.

$$2HIO_3 \rightarrow I_2O_5 + H_2O$$

It is very soluble in water, and is the anhydride of iodic acid. Because it is hygroscopic, commercial I_2O_5 has usually picked up some water, and has the formula $I_2O_5 \cdot HIO_3$. I_2O_5 decomposes to I_2 and O_2 on heating to 300 °C. It is also an oxidizing agent, which leads to its use analytically for the detection and estimation of carbon monoxide. It oxidizes CO to CO_2 quantitatively at room temperature, liberating iodine, which can be titrated with sodium thiosulphate.

$$I_2O_5 + 5CO \rightarrow 5CO_2 + I_2$$

This is useful in analysing gases, such as the exhaust gas from car engines or gases from blast furnaces, for CO. I_2O_5 also oxidizes H_2S to SO_2 and NO to NO_2. With fluorinating agents such as F_2, BrF_3 or SF_4 it forms IF_5.

$$2I_2O_5 + 10F_2 \rightarrow 4IF_5 + 5O_2$$

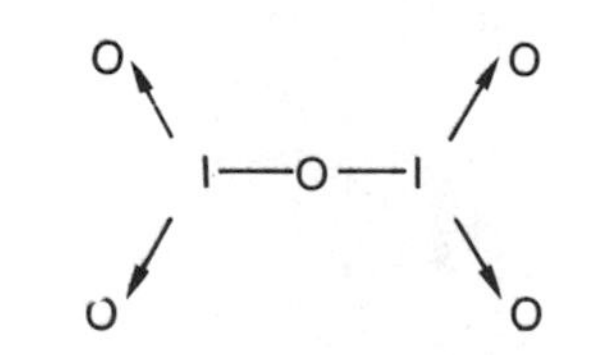

Figure 16.6 Structure of I_2O_5.

The structure of I_2O_5 is shown in Figure 16.6. The solid is a three-dimensional network, with strong intermolecular I...O interactions linking molecules together.

The oxides I_2O_4 and I_4O_9 are moderately stable, though less stable than I_2O_5. I_2O_4 is a yellow hygroscopic solid, which can be made as follows:

$$2I_2 + 3O_3 \rightarrow I_4O_9$$

$$HIO_3 \xrightarrow{P_2O_5 \text{ or } H_3PO_4\ (-H_2O)} I_4O_9$$

When heated above 75 °C it decomposes to I_2O_5.

$$4I_4O_9 \rightarrow 6I_2O_5 + 2I_2 + 3O_2$$

Dehydrating HIO_3 with concentrated H_2SO_4 gives I_2O_4. This is a lemon yellow solid. When heated above 135 °C it decomposes to I_2O_5.

$$5I_2O_4 \rightarrow 4I_2O_5 + I_2$$

The structures of these oxides are not known, but I_2O_4 is probably $IO^+ \cdot IO_3^-$ and I_4O_9 is probably $I^{3+} \cdot (IO_3^-)_3$.

STANDARD REDUCTION POTENTIALS (VOLTS)

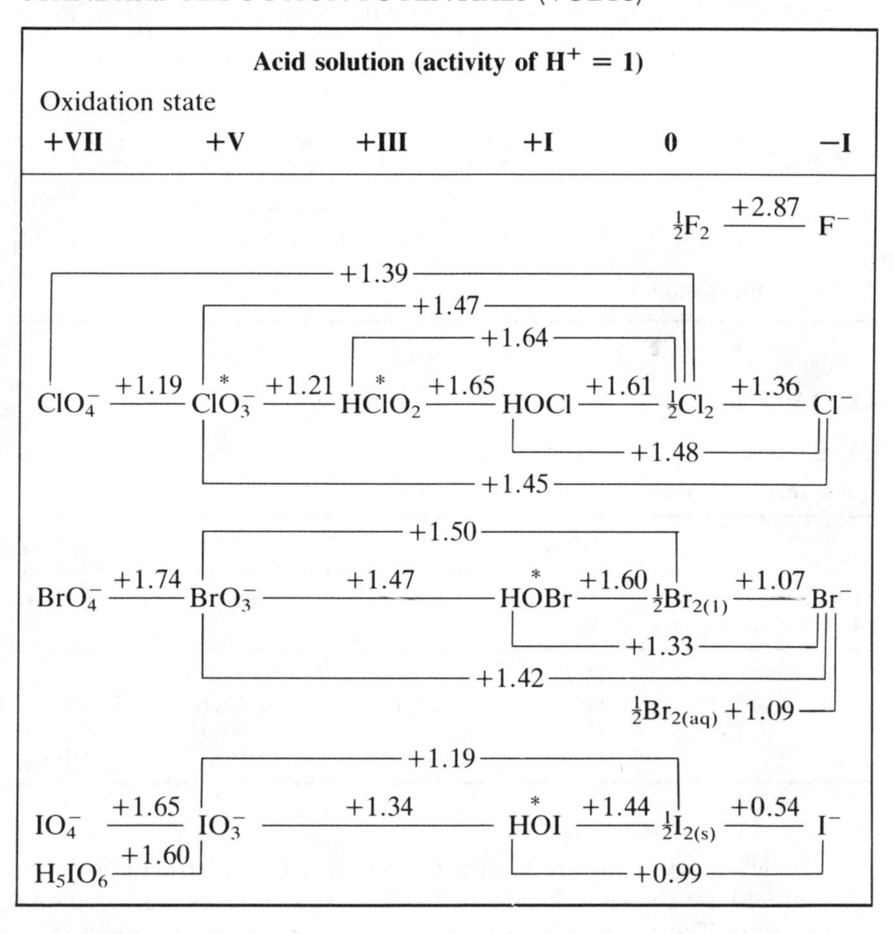

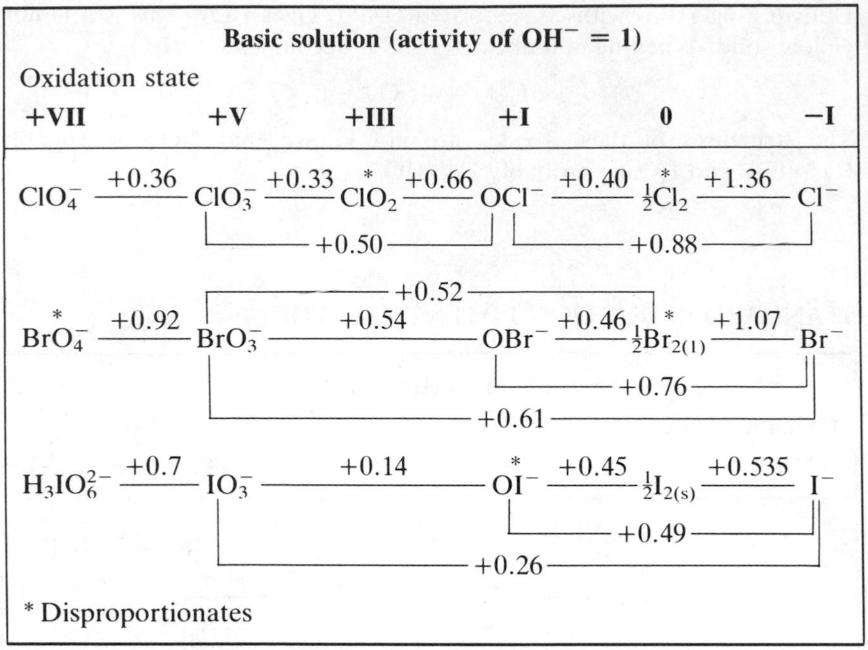

Basic solution (activity of OH^- = 1)

Oxidation state

+VII	**+V**	**+III**	**+I**	**0**	**−I**

$$ClO_4^- \xrightarrow{+0.36} ClO_3^- \xrightarrow{+0.33} \overset{*}{ClO_2} \xrightarrow{+0.66} OCl^- \xrightarrow{+0.40} \overset{*}{\tfrac{1}{2}Cl_2} \xrightarrow{+1.36} Cl^-$$

ClO_3^- to OCl^-: +0.50; OCl^- to Cl^-: +0.88

$$\overset{*}{BrO_4^-} \xrightarrow{+0.92} BrO_3^- \xrightarrow{+0.54} OBr^- \xrightarrow{+0.46} \overset{*}{\tfrac{1}{2}Br_{2(l)}} \xrightarrow{+1.07} Br^-$$

BrO_3^- to $\tfrac{1}{2}Br_{2(l)}$: +0.52; OBr^- to Br^-: +0.76; BrO_3^- to Br^-: +0.61

$$H_3IO_6^{2-} \xrightarrow{+0.7} IO_3^- \xrightarrow{+0.14} \overset{*}{OI^-} \xrightarrow{+0.45} \tfrac{1}{2}I_{2(s)} \xrightarrow{+0.535} I^-$$

OI^- to I^-: +0.49; IO_3^- to I^-: +0.26

* Disproportionates

OXOACIDS

Table 16.15 The oxoacids

	HOX	HXO_2	HXO_3	HXO_4
Oxidation states of the halogens	(+I)	(+III)	(+V)	(+VII)
	HOF			
	HOCl	$HClO_2$	$HClO_3$	$HClO_4$
	HOBr		$HBrO_3$	$HBrO_4$
	HOI		HIO_3	HIO_4

Four series of oxoacids are known (Table 16.15). The structures of the ions formed are shown in Figure 16.7. All these structures are based on a tetrahedron. The sp^3 hybrid orbitals used for bonding form only weak σ bonds, because the *s* and *P* levels differ appreciably in energy. The ions are stabilized by strong $p\pi-d\pi$ bonding between full 2*p* orbitals on oxygen with empty *d* orbitals on the halogen atoms. Even so, many of the oxoacids are known only in solution, or as their salts.

Fluorine has no *d* orbitals, and thus cannot form $p\pi-d\pi$ bonds. For a long time it was thought that F could not form any oxoacids. It is now known that HOF can be made under special conditions, but it is very unstable. No other oxoacids of F are known.

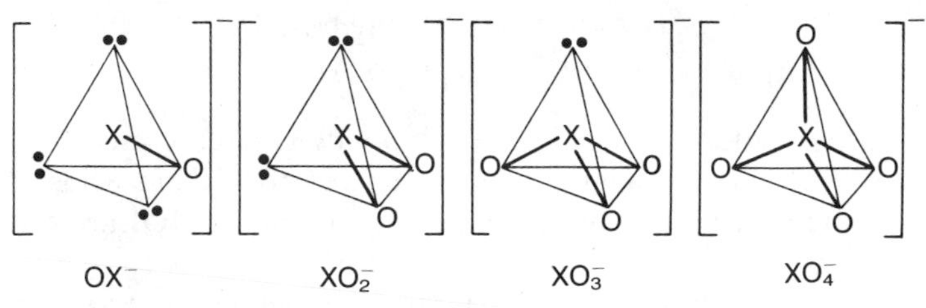

Figure 16.7 Structures of the oxoacids.

Hypohalous acids HOX

The hypohalous acids HOF, HOCl, HOBr and HOI are all known, and the halogen has the oxidation state (+I).

HOF is a colourless unstable gas. It was first made in 1968, using the matrix isolation technique. F_2 and H_2O were trapped in an unreactive matrix of solid nitrogen. (This requires a very low temperature.) The gases were photolysed, and the HOF formed. Since this too was trapped in the solid nitrogen, it was unable to collide and react with any other molecules such as H_2O, F_2 or O_2. Thus a product was obtained. More recently HOF has been made by passing F_2 over ice at 0 °C, and removing the product into a cold trap.

$$F_2 + H_2O \xrightleftharpoons{-40\,°C} HOF + HF$$

HOF is unstable, and decomposes on its own to HF and O_2. It is a strong oxidizing agent and oxidizes H_2O to H_2O_2 quite readily. The OF group occurs in $F_3C—OF$, $O_2N—OF$, $F_5S—OF$ and $O_3Cl—OF$ and these are all strong oxidizing agents. HOF should be a stronger acid than HOCl.

HOCl, HOBr and HOI are not very stable, and are known only in aqueous solutions. They are very weak acids, but they are good oxidizing agents especially in acidic solutions. They can be prepared by shaking the halogen with freshly precipitated HgO in water, for example:

$$2HgO + H_2O + 2Cl_2 \rightarrow HgO \cdot HgCl_2 + 2HOCl$$

Hypochlorous acid is the most stable. Sodium hypochlorite NaOCl is well known, and is used extensively for bleaching cotton fabric, and as a domestic bleach (sold under various trade names: Parazone, Lanry, Domestos, Chlorox etc.). It is also used as a disinfectant and sterilizing agent. NaOCl is produced commercially by electrolysing cold brine whilst stirring vigorously. During electrolysis hydrogen is liberated at the cathode. This increases the concentration of OH^- in the solution. The stirring mixes the Cl_2 formed at the anode with the OH^- so they can react together.

$$\text{anode} \quad \begin{cases} 2Cl^- \rightarrow Cl_2 \\ Cl_2 + 2OH^- \rightarrow OCl^- + Cl^- + H_2O \end{cases}$$

$$\text{cathode} \quad 2H^+ \rightarrow H_2$$

The halogens Cl_2, Br_2 and I_2 all dissolve to some extent in water, forming hydrated X_2 molecules and X^- and OX^- ions.

$$H_2O + X_2 \rightarrow X_2 \text{ (hydrated)}$$
$$\hookrightarrow HX + HOX$$

In a saturated solution of chlorine, about two thirds exists as hydrated molecules, and the rest as hydrochloric and hypochlorous acids. A much smaller amount of HOBr and a negligible amount of HOI are formed by similar means.

Dissolving the halogens in NaOH can in principle be used to make all the hypohalite ions.

$$X_2 + 2NaOH \rightarrow NaX + NaOX + H_2O$$

However, the hypohalite ions tend to disproportionate, particularly in basic solutions. The rate of the disproportionation reaction increases with temperature. Thus when Cl_2 dissolves in NaOH at or below room temperature, a reasonably pure solution of NaOCl and NaCl is obtained. However, in hot solutions (80 °C) the sodium hypochlorite disproportionates rapidly, and a good yield of sodium chlorate is obtained.

$$\overset{(+I)}{3OCl^-} \xrightarrow{\text{hot}} \overset{(-I)}{2X^-} + \overset{(+V)}{XO_3^-}$$

Hypobromites can only be made at about 0 °C: at temperatures above 50 °C quantitative yields of BrO_3^- are obtained.

$$Br_2 + 2OH^- \xrightarrow{0\,°C} Br^- + OBr^- + H_2O$$
$$3Br_2 + 6OH^- \xrightarrow{>50\,°C} 5Br^- + BrO_3^- + 3H_2O$$

Hypoiodites disproportionate rapidly at all temperatures, and IO_3^- is produced quantitatively.

Thus the hypohalites all tend to disproportionate. The reduction potentials show that OBr^- and OI^- are unstable to disproportionation, since their reduction potentials do not decrease progressively from oxidation state (+V) to (+I) to (0). However, the standard reduction potentials suggest that OCl^- should just be stable under standard conditions.

$$\overset{(+III)}{HClO_2} \xrightarrow{+1.65 \text{ volt}} \overset{(+I)}{HOCl} \xrightarrow{+1.61 \text{ volt}} \overset{(+V)}{\tfrac{1}{2}Cl_2} \quad \text{oxidation states}$$

However, the values +1.65 volts and +1.61 volts are almost the same. These are standard potentials, measured under standard conditions. Differences from standard conditions of temperature and concentration change the potentials sufficiently for disproportionation to occur.

$$\underset{\text{hypochlorite}}{\overset{(+I)}{3OCl^-}} \rightarrow \underset{\text{chloride}}{\overset{(-I)}{2Cl^-}} + \underset{\text{chlorate}}{\overset{(+V)}{ClO_3^-}} \quad \text{Oxidation states}$$

Halous acids HXO_2

The only halous acid known for certain is chlorous acid $HClO_2$. This only exists in solution. It is a weak acid, but is stronger than HOCl. The chlorine atom exists in the oxidation state (+III). $HClO_2$ is made by treating barium chlorite with H_2SO_4, and filtering off the $BaSO_4$.

$$Ba(ClO_2)_2 + H_2SO_4 \rightarrow 2HClO_2 + BaSO_4$$

Salts of $HClO_2$ are called chlorites, and are made either from ClO_2 and sodium hydroxide, or ClO_2 and sodium peroxide.

$$2ClO_2 + 2NaOH \rightarrow \underset{\text{chlorite}}{NaClO_2} + \underset{\text{chlorate}}{NaClO_3} + H_2O$$

$$2ClO_2 + Na_2O_2 \rightarrow 2NaClO_2 + O_2$$

Chlorites are used as bleaches. They are stable in alkaline solution even when boiled, but in acid solution they disproportionate, particularly when heated.

$$\overset{(+III)}{5HClO_2} \rightarrow \overset{(+IV)}{4ClO_2} + \overset{(-I)}{NaCl} \quad \text{oxidation states}$$

Halic acids HXO_3

Three halic acids are known: $HClO_3$, $HBrO_3$ and HIO_3. The halogen has the oxidation state (+V). $HClO_3$ and $HBrO_3$ are not very stable, but are known in solution, and as salts. $HClO_3$ and $HBrO_3$ detonate if attempts are made to evaporate them to dryness. The main reaction is:

$$4HClO_3 \rightarrow 4ClO_2(gas) + 2H_2O(gas) + O_2(gas)$$

In contrast, iodic acid HIO_3 is reasonably stable, and exists as a white solid. The halic acids all behave as strong oxidizing agents and strong acids.

HIO_3 can be made by oxidizing I_2 with concentrated HNO_3 or O_3. $HClO_3$ and $HBrO_3$ are made by treating the barium halates with H_2SO_4, and filtering off the $BaSO_4$.

$$Ba(ClO_3)_2 + H_2SO_4 \rightarrow 2HClO_3 + BaSO_4$$

Chlorates may be made in two ways:

1. Passing Cl_2 into a hot solution of NaOH.
2. Electrolysing hot chloride solutions that are vigorously stirred.

Only one sixth of the chlorine is converted to ClO_3^-, which appears very inefficient. However, the NaCl produced is electrolysed again, and is thus not wasted.

$$6NaOH + 3Cl_2 \xrightarrow{80\,°C} NaClO_3 + 5NaCl + 3H_2O$$

$$2Cl^- + 2H_2O \xrightarrow{\text{electrolyse}} Cl_2 + H_2 + 2OH^-$$
$$6NaOH + 3Cl_2 \longrightarrow NaClO_3 + 5NaCl + 3H_2O$$

Chlorates and bromates decompose on heating, but the way they decompose is complex and is not fully understood. $KClO_3$ may decompose in two different ways, depending on the temperature.

1. Heating $KClO_3$ to 400–500 °C is the well known laboratory experiment to produce oxygen. It also gives a trace of Cl_2 or ClO_2 (though this is seldom mentioned). Decomposition occurs at 150 °C if a catalyst such as MnO_2 or powdered glass is present to provide a surface from which O_2 can escape.

$$2KClO_3 \rightarrow 2KCl + 3O_2$$

   When $Zn(ClO_3)_2$ is heated it decomposes to O_2 and Cl_2.

$$2Zn(ClO_3)_2 \rightarrow 2ZnO + 2Cl_2 + 5O_2$$

2. In the absence of a catalyst, especially at a lower temperature, $KClO_3$ tends to disproportionate to perchlorate and chloride.

$$4KClO_3 \rightarrow 3KClO_4 + KCl$$

Chlorates are much more soluble than bromates and iodates. The iodates of Ce^{4+}, Zr^{4+}, Hf^{4+} and Th^{4+} are precipitated from 6 M HNO_3, and this is a useful means of separation for these metals.

Chlorates are used to make fireworks and matches. Sodium chlorate is widely used as a powerful weedkiller. Its effects remain for some time, and it prevents growth for one growing season. Solid chlorates, bromates and iodates should be handled with care. Chlorates can explode on grinding, on heating, or if they come into contact with easily oxidized substances such as organic matter or sulphur. They are particularly dangerous in the solid form, but are much safer in solution. Solid sodium chlorate has been used by terrorists in making bombs. The solid must be finely powdered (a dangerous process), and mixed intimately with something it can reduce, such as sugar. Mixing is highly dangerous. Such bombs are notoriously unreliable and dangerous.

Perhalic acids HXO_4

Perchloric and periodic acids and their salts are well known. Perbromates were unknown until 1968, and are not common.

World consumption of perchlorates is about 30 000 tonnes per year. $NaClO_4$ is made by electrolysing aqueous $NaClO_3$ using smooth platinum anodes in a steel container which also acts as the cathode. The platinum electrode gives a high oxygen overpotential, and thus prevents the electrolysis of water.

$$NaClO_3 + H_2O \xrightarrow{\text{electrolysis}} NaClO_4 + H_2$$

All other perchlorates and perchloric acid $HClO_4$ are made from $NaClO_4$.

1. NH_4ClO_4 is a white solid and was formerly used as a blasting compound in mining. It is now used in the booster rockets in the Challenger Space Shuttle. NH_4ClO_4 oxidizes the fuel (Al powder). A Shuttle launch uses nearly 700 tonnes of NH_4ClO_4, and this accounts for half the per-

chlorates used. NH_4ClO_4 will absorb sufficient ammonia to liquefy itself.

2. Nearly 500 tonnes of $HClO_4$ are used annually, mostly to make other perchlorates. $HClO_4$ is a colourless liquid and can be made from NH_4ClO_4 and dilute nitric acid, or from $NaClO_4$ and concentrated hydrochloric acid.

$$NH_4ClO_4 + HNO_3 \rightarrow HClO_4 + NH_4NO_3$$

In principle perchlorates could be made from the disproportionation of chlorates, but the reaction is slow and of little use.

$$4ClO_3^- \rightarrow 3ClO_4^- + Cl^-$$

$HClO_4$ is commercially available as 70% $HClO_4$, which has almost the composition of the dihydrate $HClO_4 \cdot 2H_2O$. It is the only oxoacid of chlorine that can be isolated in the anhydrous state. It is made by dehydrating $HClO_4 \cdot 2H_2O$ with fuming sulphuric acid, then removing $HClO_4$ by vacuum distillation.

$$HClO_4 \cdot 2H_2O + 2H_2S_2O_7 \rightarrow HClO_4 + 2H_2SO_4$$

$HClO_4$ is one of the strongest acids known. The anhydrous compound is a powerful oxidizing agent which explodes on contact with organic material (wood, paper, cloth, grease, rubber or chemicals), and sometimes on its own. The cold concentrated (70% aqueous) solution is a much weaker oxidizing agent. Hot concentrated solutions have been used for 'wet ashing', where all organic materials in the sample are oxidized to CO_2, leaving only the inorganic constituents for analysis. Alcohols must not be present since perchlorate esters are explosive. Often a mixture of $HClO_4$ and HNO_3 is used for wet ashing, since the HNO_3 oxidizes alcohols and removes this risk.

3. Magnesium perchlorate $MgClO_4$ is used as the electrolyte in so-called 'dry batteries'. It is very hygroscopic, and is a very effective desiccant called 'anhydrone'.
4. $KClO_4$ is used in fireworks and flares. Those with a bang and flash usually use $KClO_4$, Al and S, and those that are flares have $KClO_4$ and Mg. A red colour is obtained by adding some $SrCO_3$ or Li_2CO_3, and $CuCO_3$ gives blue. Making fireworks is dangerous – do not attempt it!

Virtually all metal perchlorates, except those with the larger Group I ions K^+, Rb^+ and Cs^+, are soluble in water. The sparing solubility of $KClO_4$ is used to detect potassium in qualitative analysis. (A solution of $NaClO_4$ is added to the solution containing K^+, and $KClO_4$ is precipitated.) The ClO_4^- ion has only a very slight tendency to form complexes with metal ions. Thus perchlorates are often used as an inert ion in the study of metal ions in aqueous solution. However, in the absence of other ligands, ClO_4^- ions may act as unidentate or bidentate ligands. The perchlorate ion is tetrahedral.

For a long time efforts to make perbromates were unsuccessful, and it was thought that they could not exist, until traces were obtained from β decay of $^{83}SeO_4^{2-}$. They can be made from bromates by the action of powerful oxidizing agents such as F_2 or XeF_2, or by electrolysis of an aqueous solution.

$$KBrO_3 + F_2 + 2KOH \rightarrow KBrO_4 + 2KF + H_2O \quad \text{(20\% yield)}$$

$$RbBrO_3 + XeF_2 + H_2O \rightarrow RbBrO_4 + 2HF + Xe \quad \text{(10\% yield)}$$

$$LiBrO_3 \xrightarrow{\text{electrolytic oxidation}} LiBrO_4 \quad \text{(1\% yield)}$$

Solid perbromates are stable. $KBrO_4$ is stable up to 275 °C, and is isomorphous with $KClO_4$. $HBrO_4$ is stable in solution up to a concentration of 6 M, but the concentrated acid is a vigorous oxidizing agent. In dilute solutions perbromates only oxidize slowly, and Cl^- is not oxidized.

Periodates can be made by oxidizing I_2 or I^- in aqueous solution. Commercially they are made by oxidizing iodates in alkaline solution, either with Cl_2 or electrolytically.

$$IO_3^- + 6OH^- + Cl_2 \longrightarrow IO_6^{5-} + 3H_2O + 2Cl^-$$

$$IO_3^- + 6OH^- \xrightarrow{-2\text{ electrons}} IO_6^{5-} + 3H_2O$$

The common form of periodic acid is $HIO_4 \cdot 2H_2O$ or H_5IO_6. This is called paraperiodic acid, and exists as white crystals which melt with decomposition at 128.5 °C. Water is lost on heating to 100 °C under reduced pressure, yielding periodic acid HIO_4. On strong heating this eventually decomposes losing O_2 and forming I_2O_5.

$$\underset{\text{paraperiodic acid}}{2H_5IO_6} \xrightarrow[-4H_2O]{100\,°C} \underset{\text{periodic acid}}{2HIO_4} \xrightarrow{200\,°C} \underset{\text{iodine pentoxide}}{I_2O_5} + O_2 + H_2O$$

Periodates are of two structural types: tetrahedral IO_4^- and octahedral $(OH)_5IO$. In aqueous solutions at room temperature, the main ion is IO_4^-. The structures of periodates are much more complicated than this implies. A wide range of isopolyacids exist with octahedral units (based on one I and six O atoms), linked together by sharing the O atoms at two corners (that is sharing an edge of the octahedron), or sharing three corners (that is a face of the octahedron).

Chemically, periodates are important as oxidants, and they will oxidize Mn^{2+} to MnO_4^-. They are also used to oxidize organic compounds. Solutions of periodic acid are used to determine the structure of organic compounds by degradative methods. HIO_4 is called a glycol splitting agent since it splits (oxidizes) 1 : 2 glycols into aldehydes.

$$IO_4^- + R{-}\underset{OH}{\underset{|}{CH}}{-}\underset{OH}{\underset{|}{CH}}{-}R \rightarrow R{-}CHO + R{-}CHO + IO_3^- + H_2O$$

Strength of the oxoacids

$HClO_4$ is an extremely strong acid, whilst HOCl is a very weak acid. The dissociation of an oxoacid involves two energy terms:

1. Breaking an O—H bond to produce a hydrogen ion and an anion.
2. Hydrating both ions.

Plainly the ClO_4^- ion is larger than the OCl^- ion, so the hydration energy of ClO_4^- is less than that for OCl^-. This would suggest that HOCl should ionize more readily than $HClO_4$. Since we know the reverse to be true, the reason must be the energy required to break the O—H bond.

Oxygen is more electronegative than chlorine. In the series of oxoacids HOCl, $HClO_2$, $HClO_3$, $HClO_4$, an increasing number of oxygen atoms are bonded to the chlorine atom. The more oxygen atoms that are bonded, the more the electrons will be pulled away from the O—H bond, and the more this bond will be weakened. Thus $HClO_4$ requires the least energy to break the O—H bond and form H^+. Hence $HClO_4$ is the strongest acid.

In general, for any series of oxoacids, the acid with the most oxygen (that is the one with the highest oxidation number) is the most dissociated. Thus the acid strengths decrease $HClO_4 > HClO_3 > HClO_2 >$ HOCl. In exactly the same way, H_2SO_4 is a stronger acid than H_2SO_3, and HNO_3 is a stronger acid than HNO_2.

INTERHALOGEN COMPOUNDS

The halogens react with each other to form interhalogen compounds. These are divided into four types AX, AX_3, AX_5 and AX_7.

They can all be prepared by direct reaction between the halogens, or by the action of a halogen on a lower interhalogen. The product formed depends on the conditions.

Table 16.16 Interhalogen compounds, and their physical state at 25 °C

AX	AX_3	AX_5	AX_7
ClF(g) colourless			
BrF(g) pale brown	ClF_3(g) colourless		
BrCl(g) red–brown	BrF_3(l) pale yellow	ClF_5(g) colourless	
ICl(s) ruby red	$(ICl_3)_2$(s) bright yellow	BrF_5(l) colourless	IF_7(g) colourless
IBr(s) black	(IF_3)(s) (unstable) yellow	IF_5(l) colourless	
(IF)* (unstable)			

* Disproportionates rapidly into IF_5 and I_2.

$$Cl_2 + F_2 \text{ (equal volumes)} \xrightarrow{200\,°C} 2ClF$$
$$Cl_2 + 3F_2 \text{ (excess } F_2) \xrightarrow{300\,°C} 2ClF_3$$
$$I_2 + Cl_2 \text{ liquid (equimolar)} \longrightarrow ICl$$
$$I_2 + Cl_2 \text{ liquid (excess } Cl_2) \longrightarrow (ICl_3)_2$$
$$Br_2 + F_2 \text{ (diluted with nitrogen)} \rightarrow BrF_3$$
$$Br_2 + F_2 \text{ (excess } F_2) \longrightarrow BrF_5$$
$$I_2(s) + 5F_2 \xrightarrow{20\,°C} IF_5$$
$$I_2(g) + 7F_2 \xrightarrow{250-300\,°C} IF_7$$

There are never more than two different halogens in a molecule. The bonds are essentially covalent because of the small electronegativity difference, and the melting and boiling points increase as the difference in electronegativity increases.

The compounds formed in the AX and AX_3 groups are those where the electronegativity difference is not too great. The higher valencies AX_5 and AX_7 are shown by large atoms such as Br and I associated with small atoms such as F. This is because it is possible to pack more small atoms round a large one.

The interhalogens are generally more reactive than the halogens (except F_2). This is because the A—X bond in interhalogens is weaker than the X—X bond in the halogens. The reactions of interhalogens are similar to those of the halogens. Hydrolysis gives halide and oxohalide ions. Note that the oxohalide ion is always formed from the larger halogen present.

$$BrF_5 \xrightarrow{H_2O} 5F^- + BrO_3^- \text{ bromate ion}$$
$$ICl \xrightarrow{H_2O} Cl^- + OI^- \text{ hypoiodite ion}$$
$$\longrightarrow Cl^- + IO_3^- \text{ iodate ion}$$

Interhalogen compounds will fluorinate many metal oxides, metal halides and metals.

$$3UO_2 + 4BrF_3 \rightarrow 3UF_4 + 2Br_2 + 3O_2$$
$$UF_4 + ClF_3 \rightarrow UF_6 + ClF$$

AX compounds

All six compounds can be made by controlled reaction of the elements. ClF is very reactive. ICl and IBr are the most stable and can be obtained pure at room temperature. The compounds have properties intermediate between those of the constituent halogens.

ClF fluorinates many metals and non-metals:

$$6ClF + 2Al \rightarrow 2AlF_3 + 3Cl_2$$
$$6ClF + U \rightarrow UF_6 + 3Cl_2$$
$$6ClF + S \rightarrow SF_6 + 3Cl_2$$

It can simultaneously chlorinate and fluorinate a compound either by oxidizing the element or adding to a double bond.

$$ClF + SF_4 \rightarrow SF_5Cl$$
$$ClF + CO \rightarrow COFCl$$
$$ClF + SO_2 \rightarrow ClSO_2F$$

Iodine monochloride ICl is well known. It is used as Wij's reagent in the estimation of the iodine number of fats and oils. The iodine number is a measure of the number of double bonds, i.e. the degree of unsaturation of the fat. The ICl adds to double bonds in the fat. The ICl solution is brown coloured, and when it is added to an unsaturated fat the colour disappears until all the double bonds have reacted. The iodine number is simply the volume (ml) of a standard solution of ICl which reacts with a fixed weight of fat.

$$—CH{=}CH— + ICl \rightarrow —\underset{\displaystyle I}{\underset{|}{CH}}—\underset{\displaystyle Cl}{\underset{|}{CH}}—$$

When ICl reacts with organic compounds it often iodinates them, though chlorination may occur depending on the conditions.

$$\text{salicylic acid} \begin{cases} \xrightarrow{\text{+ ICl vapour}} \text{chlorination} \\ \xrightarrow[\text{+ ICl in nitrobenzene}]{} \text{iodination} \end{cases}$$

It is thought that the attacking species is I^+, since the I atoms substitute in positions where there is an excess of electrons. Both ICl and IBr are partially ionized in the fused state. Conductivity measurements show that ICl ionizes to the extent of about 1%, and electrolysis suggests that the method of ionization is:

$$2ICl \rightleftharpoons I^+ + [ICl_2]^-$$

The I^+ ion is solvated, giving $[I_2Cl]^+$. Both ICl and IBr can be used as non-aqueous ionizing solvents.

The interhalogens also form addition compounds with alkali halides. These compounds are ionic, and are called polyhalides.

$$NaBr + ICl \rightarrow Na^+ [BrICl]^-$$
$$KI + ICl \rightarrow K^+ [I_2Cl]^-$$

AX_3 compounds

The compounds ClF_3, BrF_3, IF_3 and $(ICl_3)_2$ can all be made by direct combination of the elements if the conditions are chosen with care. ClF_3 is a liquid (b.p. 11.8 °C), and is commercially available. It is produced by direct action:

$$Cl_2 + 3F_2 \xrightarrow{200-300\,°C} 2ClF_3$$

or

$$ClF + F_2 \rightarrow ClF_3$$

ClF_3 reacts with excess Cl_2, forming ClF. BrF_3 behaves similarly.

$$ClF_3 + Cl_2 \rightarrow 3ClF$$

IF_3 is only stable below −30 °C, and tends to form the more stable IF_5. It can also be made using XeF_2 to fluorinate I_2.

$$3XeF_2 + I_2 \rightarrow 2IF_3 + 3Xe$$

I_2Cl_6 is easily made by adding I_2 solid to liquid Cl_2, but on warming to room temperature it dissociates:

$$I_2Cl_6 \rightarrow 2ICl + 2Cl_2$$

IF_3 is unstable above −30 °C. Direct reaction of F_2 and I_2 gives the more stable IF_5. IF_3 can be made by diluting the F_2 with CCl_3F and keeping the reaction cold, or by reaction with XeF_2.

$$3XeF_2 + I_2 \rightarrow 2IF_3 + 3Xe$$

Both ClF_3 and BrF_3 are well known as covalent liquids. ClF_3 is one of the most reactive compounds known, and its properties are aggressive. It catches fire spontaneously with wood and most building materials – even asbestos. It was used in incendiary bombs in World War II. Despite the dangers, peace time production of ClF_3 runs into hundreds of tonnes per year. It is available in steel cylinders. It is mainly used by the nuclear industry for fuel processing. It is used to make gaseous UF_6, which is useful in making enriched ^{235}U fuel. It is also important in separating the fission products from spent fuel rods. Pu and most of the fission products form involatile tetrafluorides like PuF_4, whilst U forms volatile UF_6.

$$2ClF_3 + U \rightarrow UF_6 + Cl_2$$
$$4ClF_3 + 3Pu \rightarrow 3PuF_4 + 2Cl_2$$

ClF_3 reacts explosively with water, stopcock grease and many organic compounds, including cotton and paper. It is a powerful fluorinating agents for inorganic compounds.

$$4ClF_3 + 6MgO \rightarrow 6MgF_2 + 2Cl_2 + 3O_2$$
$$4ClF_3 + 2Al_2O_3 \rightarrow 4AlF_3 + 2Cl_2 + 3O_2$$
$$2ClF_3 + 2AgCl \rightarrow 2AgF_2 + Cl_2 + 2ClF$$
$$2ClF_3 + 2NH_3 \rightarrow 6HF + Cl_2 + N_2$$
$$ClF_3 + BF_3 \rightarrow [ClF_2]^+ [BF_4]^-$$
$$ClF_3 + SbF_5 \rightarrow [ClF_2]^+ [SbF_6]^-$$
$$ClF_3 + PtF_5 \rightarrow [ClF_2]^+ [PtF_6]^-$$

ClF_3 can be used to fluorinate organic compounds provided it is diluted with nitrogen to moderate the reaction. ClF_3 has been used as fuel in short range rockets, reacting with hydrazine. This is technically easier than using liquid O_2 or F_2, since the reactants can be stored without refrigeration, and they ignite spontaneously on mixing.

$$4ClF_3 + 3N_2H_4 \rightarrow 12HF + 3N_2 + 2Cl_2$$

Bromine trifluoride BrF_3 is a red liquid, and can be made from the elements at or near room temperature. It is manufactured in multi-tonne quantities. It is less violent in its reactions than ClF_3, but it reacts in an analogous manner. The order of reactivity of the interhalogens is:

$$ClF_3 > BrF_5 > IF_7 > ClF > BrF_3 > IF_5 > BrF > IF_3 > IF$$

Like ClF_3, BrF_3 is used in nuclear processing and reprocessing to make UF_6, and it is used to make many other fluorides. It liberates oxygen quantitatively from many oxides (B_2O_3, SiO_2, As_2O_5, I_2O_5, CuO, TiO_2), and also from oxosalts such as carbonates and phosphates. Measuring the O_2 evolved is used as a method of analysis.

$$4BrF_3 + 3SiO_2 \rightarrow 3SiF_4 + 2Br_2 + 3O_2$$

$$4BrF_3 + 3TiO_2 \rightarrow 3TiF_4 + 2Br_2 + 3O_2$$

The interhalogens are all potential non-aqueous ionizing solvents. BrF_3 has been more widely used as a solvent than the others. This is for three main reasons:

1. It has a convenient liquid range (m.p. 8.8 °C, b.p. 126 °C).
2. It is a good, but not too violent, fluorinating agent.
3. It self-ionizes considerably, and much more than ClF_3.

$$2BrF_3 \rightleftharpoons [BrF_2]^+ + [BrF_4]^-$$

Thus substances producing $[BrF_2]^+$ ions are acids and $[BrF_4]^-$ ions are bases in this solvent.

The structure of the AX_3 type of interhalogen molecule is of interest. In ClF_3, Cl is the central atom (Figure 16.8).

Electronic structure of chlorine atom – excited state

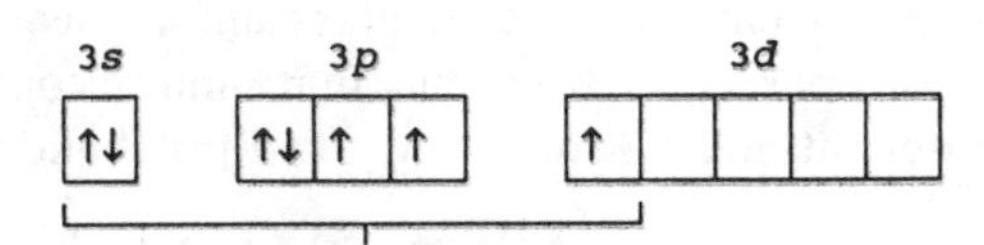

Three unpaired electrons form bonds with three fluorine atoms, plus two lone pairs, giving a total of five electron pairs. Shape trigonal bipyramid with two positions occupied by lone pairs

The way to predict which of the three possible arrangements will be formed is described in Chapter 4.

A structural study of ClF_3 by microwave spectroscopy shows that the molecule is T-shaped, with bond angles of 87°40′. This is close to 90°, and suggests structure 3. The distortion from 90° is because of repulsion between the lone pairs. Note that two of the bond lengths are the same

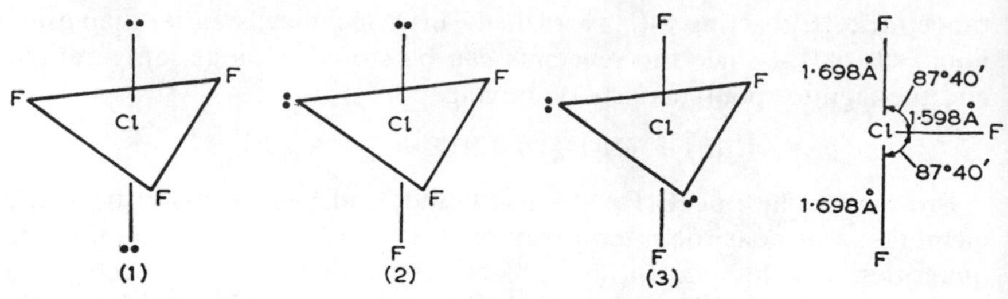

Figure 16.8 Possible structures for chlorine trifluoride ClF_3.

and are different from the third. This is to be expected because a trigonal bipyramid is not a regular shape. Equatorial bonds (those in the triangle) are different from apical bonds (those pointing up and down). The X-ray structure of crystalline ClF_3 shows that the molecule is T-shaped, with a bond angle of 87°0′ and bond lengths 1.716 Å and 1.621 Å. The structure of BrF_3 (by microwave) is also T-shaped, with a bond angle of 86°12′ and bond lengths of 1.810 Å and 1.721 Å. ICl_3 gas decomposes into ICl and Cl_2, so its structure is not known. The liquid has an appreciable electrical conductivity due to self-ionization:

$$2ICl_3 \rightleftharpoons ICl_2^+ + ICl_4^-$$

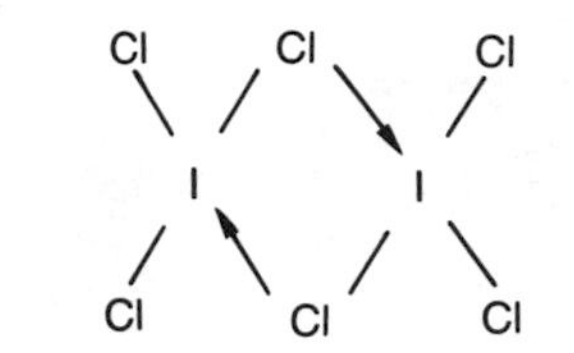

Figure 16.9 Structure of I_2Cl_6.

In the solid state, two T-shaped ICl_3 molecules join together, forming a planar dimeric molecule $(ICl_3)_2$. The terminal I—Cl bonds are normal single bonds of length 2.38 Å and 2.39 Å. The bridging I—Cl bonds are appreciably longer (2.68 Å and 2.72 Å).

AX_5 compounds

Three compounds are known: ClF_5, BrF_5 and IF_5. ClF_5 and BrF_5 react extremely vigorously, but they are exceeded in violence by ClF_3. (See the order of reactivity given earlier.) IF_5 is a little less reactive, and unlike the others it can be used in glass apparatus. Production is several hundred tonnes per year. They fluorinate many compounds, react explosively with water, attack silicates, and form polyhalides.

$$ClF_5 + 2H_2O \rightarrow FClO_2 + 4HF$$
$$BrF_5 + 3H_2O \rightarrow HBrO_3 + 5HF$$
$$2BrF_5 + SiO_2 \rightarrow SiF_4 + 2BrF_3 + O_2$$
$$BrF_5 + CsF \rightarrow Cs^+[BrF_6]^-$$
$$IF_5 + KI \rightarrow K^+ [IF_6]^-$$

Liquid IF_5 self-ionizes, and therefore conducts electricity.

$$2IF_5 \rightleftharpoons IF_4^+ + IF_6^-$$

The AX_5 compounds all have structures based on a square based pyramid, that is octahedral with one position unoccupied. The central atom is displaced slightly below the plane. The structure may be understood by considering IF_5. I is the central atom in the molecule (Figure 16.10).

Electronic structure of iodine atom – excited state

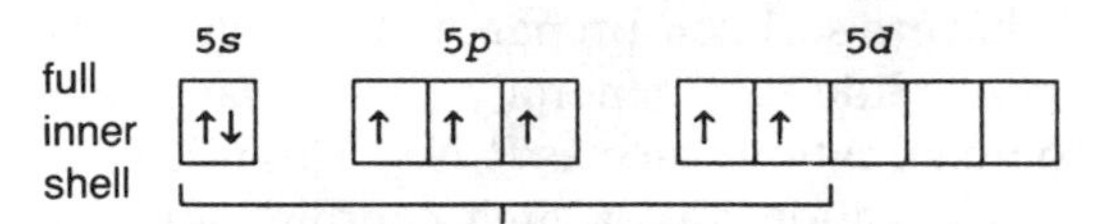

five unpaired electrons form bonds with five fluorine atoms, plus one lone pair giving a total of six electron pairs structure is octahedral with one position occupied by a lone pair (alternatively described as square based pyramid)

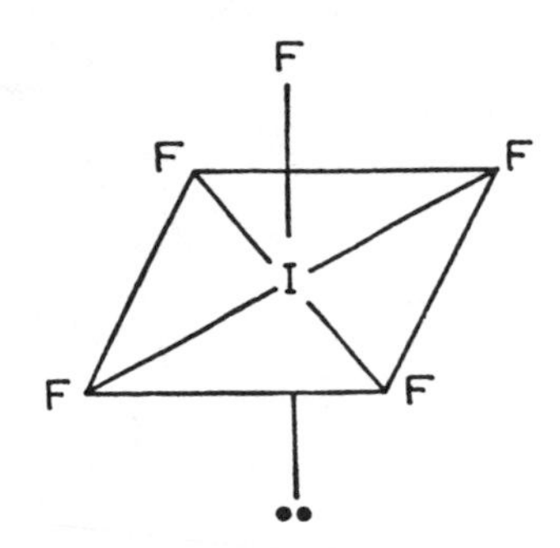

Figure 16.10 Structure of IF_5.

AX_7 compounds

IF_7 is formed by direct combination of the elements at 250–300 °C, by heating IF_5 with F_2, or by treating iodides with F_2.

$$KI + 4F_2 \rightarrow IF_7 + KF$$
$$PdI_2 + 8F_2 \rightarrow 2IF_7 + PdF_2$$

IF_7 is a violent fluorinating agent, and reacts with most elements. It also reacts with water, SiO_2 and CsF.

$$IF_7 + H_2O \rightarrow IOF_5 + 2HF$$
$$2IF_7 + SiO_2 \rightarrow 2IOF_5 + SiF_4$$
$$IF_7 + CsF \rightarrow Cs^+[IF_8]^-$$

The structure of IF_7 is unusual – a pentagonal bipyramid (Figure 16.11). It is probably the only known example of a non-transition element using three *d* orbitals for bonding.

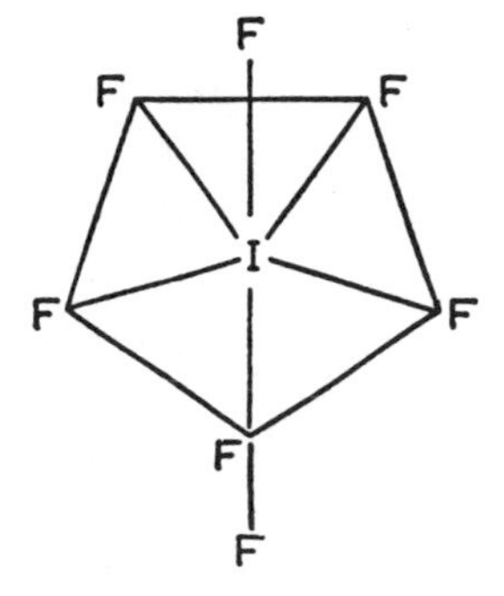

Figure 16.11 Structure of IF_7.

Electronic structure of iodine atom – excited state

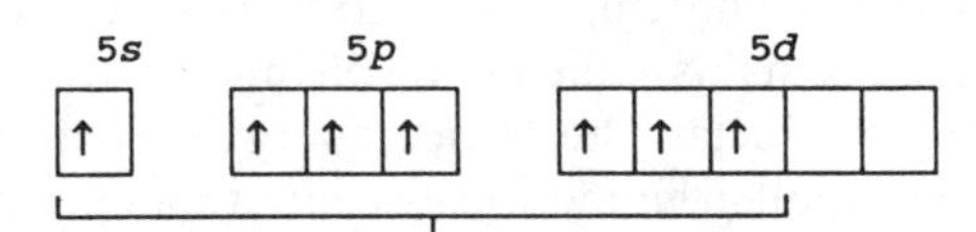

seven unpaired electrons form bonds with seven fluorine atoms
seven pairs of electrons form a pentagonal bipyramid

POLYHALIDES

Halide ions often react with molecules of halogens or interhalogens and form polyhalide ions. Iodine is only slightly soluble in water ($0.34\,g\,l^{-1}$). Its solubility is greatly increased if some iodide ions are present in the solution. The increase in solubility is due to the formation of a polyhalide ion, in this case the triiodide ion I_3^-. This is stable both in aqueous solution and in ionic crystals.

$$I_2 + I^- \rightarrow I_3^-$$

More complex ions such as pentaiodide I_5^-, heptaiodide I_7^- and enneaiodide I_9^- have also been prepared. Crystalline compounds containing the larger polyiodide ions generally contain large metal ions such as Cs^+ or large complex cations such as R_4N^+. This is because a large anion together with a large cation give a high coordination number and hence a high lattice energy. Polyhalides such as $KI_3 \cdot H_2O$, RbI_3, NH_4I_5, $[(C_2H_5)_4N]I_7$ and $RbI_9 \cdot 2C_6H_6$ may be formed by the direct addition of I_2 to I^-, either with or without a solvent.

The Br_3^- ion is much less stable and less common than I_3^-. A few unstable Cl_3^- compounds are known, and the ion is formed in concentrated solution.

$$Cl^-_{(aq)} + Cl_2 \rightarrow Cl^-_{3(aq)}$$

No F_3^- compounds are known, presumably because fluorine has no available *d* orbitals and therefore cannot expand its octet.

Many polyhalides are known which contain two or three different halogens, for example $K[ICl_2]$, $K[ICl_4]$, $Cs[IBrF]$ and $K[IBrCl]$. These are formed from interhalogens and metal halides.

$$ICl + KCl \rightarrow K^+[ICl_2]^-$$
$$ICl_3 + KCl \rightarrow K^+[ICl_4]^-$$
$$IF_5 + CsF \rightarrow Cs^+[IF_6]^-$$
$$ICl + KBr \rightarrow K^+[BrICl]^-$$
$$2ClF + AsF_5 \rightarrow [FCl_2]^+ [AsF_6]^-$$

Polyhalides are typical ionic compounds (crystalline, stable and soluble in water, conduct electricity when in solution), though they tend to decompose on heating. The products of the decomposition (that is which halogen remains attached to the metal) are governed by the lattice energy of the products. The lattice energy of the alkali metal halides is highest for the smaller halide ions, so the smaller halogen remains bonded to the metal.

$$Cs[I_3] \xrightarrow{\text{heat}} CsI + I_2$$
$$Rb[ICl_2] \xrightarrow{\text{heat}} RbCl + ICl$$
$$K[BrICl] \xrightarrow{\text{heat}} KCl + IBr$$

The structures of the polyhalides are known. The trihalides $K[I_3]$, $K[ICl_2]$ and $Cs[IBrF]$ all contain a linear trihalide ion. This may be explained by considering the orbitals used. For example, see $[ICl_2]^-$ in Figure 16.12.

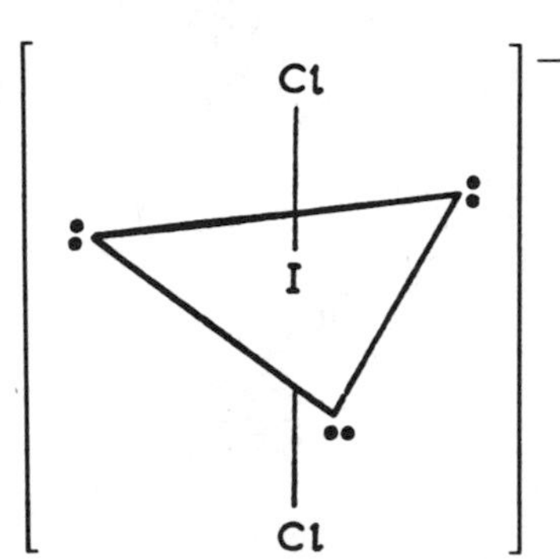

Figure 16.12 Structure of $[ICl_2]^-$ ion.

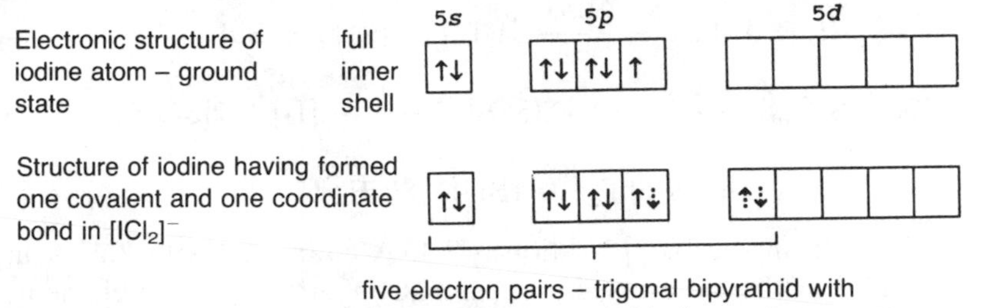

Similarly the structures of the pentahalide ions $[ICl_4]^-$ and $[BrF_4]^-$ are square planar (see Figure 16.13).

Figure 16.13 Structure of $[ICl_4]^-$ ion. (The structure of the $[I_5]^-$ ion is different.)

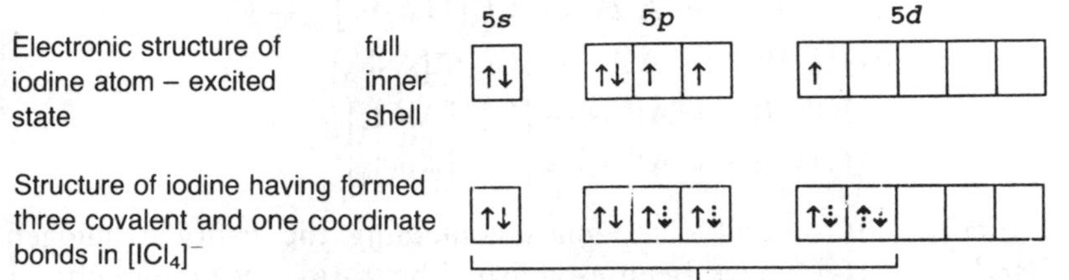

BASIC PROPERTIES OF THE HALOGENS

Elements typically become more metallic or basic on descending a main group. Thus in Groups IV, V and VI the first elements C, N and O are non-metals, but the heavier members Sn, Pb, Bi and Po are metals. Metallic properties decrease on crossing a period. Little is known about astatine, though it would be expected to show more tendency to form cations than the other members of the group. Thus the trend to metallic properties is less obvious in Group VII. The increasing stability of positive ions indicates an increasing tendency to basic or metallic character. It must be emphasized that iodine is not a metal.

Fluorine is the most electronegative element and has no basic properties (that is, it has no tendency to form positive ions).

Iodine dissolves in oleum and other strongly oxidizing solvents, forming bright blue solutions which are paramagnetic. For a long time these solutions were thought to contain I^+, but they are now known to contain the $[I_2]^+$ cation. (The Group VI elements S, Se and Te behave in a similar way. Thus S dissolves, forming blue coloured paramagnetic solutions containing various $[S_n]^{2+}$ cations such as $[S_4]^{2+}$, $[S_8]^{2+}$ and $[S_{19}]^{2+}$.) $[Br_2]^+$ is also formed in oleum and is bright red coloured and paramagnetic. $[Cl_2]^+$ has been observed spectroscopically in discharge tubes. Several crystalline compounds containing $[I_2]^+$ or $[Br_2]^+$ have been prepared.

$$2I_2 + 5SbF_5 \xrightarrow{SO_2\ \text{solvent}} [I_2]^+[Sb_2F_{11}]^- + SbF_3$$

$$2I_2 + S_2O_6F_2 \xrightarrow{H_2SO_3F} 2[I_2]^+[SO_3F]^- \xrightarrow{-80\,°C} [I_4]^{2+}\ 2[SO_3F]^-$$

$$Br_2 + SbF_5 \xrightarrow{BrF_5} [Br_2]^+[Sb_3F_{16}]^-$$

The bond length in the $[Br_2]^+$ cation is 2.15 Å compared with 2.27 Å in Br_2. The bond in the cation is shorter and stronger than in the element, and shows that the electron removed came from an antibonding orbital. In a similar way the bond in $[I_2]^+$ is stronger than in I_2.

Many other compounds containing cations such as $[Cl_3]^+$, $[Br_3]^+$, $[I_3]^+$, $[Br_5]^+$ and $[I_5]^+$ have been prepared.

$$Cl_2 + ClF_3 + AsF_5 \rightarrow [Cl_3]^+[AsF_6]^- + F_2$$
$$Br_2 + BrF_3 + AsF_5 \rightarrow [Br_3]^+[AsF_6]^- + F_2$$
$$I_2 + ICl + AlCl_3 \rightarrow [I_3]^+[AlCl_4]^-$$
$$2I_2 + ICl + AlCl_3 \rightarrow [I_5]^+[AlCl_4]^-$$

The structures of several compounds containing the cationic halogen ions $[Br_3]^+$ and $[I_3]^+$ have been established by X-ray crystallography or Raman spectroscopy. These ions are always bent, in contrast to the triiodide ion $[I_3]^-$ which is linear. The structures of the other ions are not known with certainty. A compound I_7SO_3F has been reported (as a maximum on a phase diagram). It is a black solid but it is not known whether it contains $[I_7]^+$.

Positive bromine also exists in other complexes such as $Br(pyridine)_2NO_3$ and BrF_3. These ionize and give:

$$Br(pyridine)_2NO_3 \rightleftharpoons [Br(pyridine)_2]^+ + NO_3^-$$
$$2BrF_3 \rightleftharpoons [BrF_2]^+ + [BrF_4]^-$$

Complexes containing positive iodine are more numerous. ICl and IBr both conduct electricity when molten. Electrolysis of ICl liberates I_2 and Cl_2 at both electrodes. This is consistent with the following ionization:

$$3ICl \rightleftharpoons [I_2Cl]^+ + [ICl_2]^-$$

Molten ICN behaves in a similar way to ICl.

$$3ICN \rightleftharpoons [I_2CN]^+ + [I(CN)_2]^- \quad \text{in melt}$$

However, electrolysis of ICl dissolved in pyridine gives I_2 only at the cathode. This suggests the more simple type of ionization:

$$2ICl \xrightleftharpoons{pyridine} [I(pyridine)_2]^+ + [ICl_2]^-$$

The addition of $AlCl_3$ to a melt of ICl greatly increases its conductivity, and is explained by the formation of complex ions:

$$AlCl_3 + 2ICl \rightarrow [I_2Cl]^+ + [AlCl_4]^-$$

ICl behaves as an electrophilic iodinating agent. It converts acetanilide to 4-iodoacetanilide, and salicylic acid to 3,5-diiodosalicylic acid. Because the attacked sites have an electron excess, the iodine must be positive.

If a solution of iodine in an inert solvent is passed down a cationic ion-exchange column, some iodine is retained in the resin.

$$H^+Resin^- + I_2 \rightarrow I^+Resin^- + HI$$

The positive ion retained may be eluted with KI, to estimate the amount of I^+, or it may be allowed to react with various reagents.

$$I^+Resin^- + KI \rightarrow I_2 + K^+Resin^-$$
$$I^+Resin^- + \text{anhydrous } H_2SO_4 \rightarrow I_2SO_4 + H^+Resin^-$$
$$I^+Resin^- + \text{alcoholic } HNO_3 \rightarrow INO_3 + H^+Resin^-$$

I^+ reacts with OH^- in aqueous solutions.

$$I^+ + OH^- \rightarrow HOI$$
$$2HOI + OI^- \rightarrow IO_3^- + 2I^- + 2H^+$$

For this reason I^+ will only exist in water if it is stabilized by coordination to some other molecule. A large number of compounds are known which contain I^+ stabilized in a complex ion. Many pyridine complexes are known such as $[I(pyridine)_2]NO_3$, $[I(pyridine)_2]ClO_4$, [I(pyridine)]acetate and [I(pyridine)]benzoate.

Molten ICl_3 has a high conductivity ($8.4 \times 10^{-3} ohm^{-1} cm^{-1}$). When ICl_3 is electrolysed both I_2 and Cl_2 are liberated at both electrodes. This suggests ionization:

$$2ICl_3 \rightleftharpoons [ICl_2]^+ + [ICl_4]^-$$

Treatment of I_2 with fuming HNO_3 and acetic anhydride gives the ionic compound $I(acetate)_3$. If a saturated solution of $I(acetate)_3$ in acetic anhydride is electrolysed using silver electrodes, one equivalent of AgI is formed at the cathode for every three Faradays of electricity passed. This would seem to indicate ionization giving I^{3+}:

$$I(acetate)_3 \rightleftharpoons I^{3+} + 3(acetate^-)$$

There is no structural evidence for the presence of I^{3+}. Other ionic compounds which may contain I^{3+} are iodine phosphate IPO_4 and iodine fluosulphonate $I(SO_3F)_3$.

Pseudohalogens and pseudohalides

A few ions are known, consisting of two or more atoms of which at least one is N, that have properties similar to those of the halide ions. They are therefore called pseudohalide ions. Pseudohalide ions are univalent, and these form salts resembling the halide salts. For example, the sodium salts are soluble in water, but the silver salts are insoluble. The hydrogen compounds are acids like the halogen acids HX. Some of the pseudohalide

Table 16.17 The important pseudohalogens

Anion		Acid		Dimer	
CN^-	cyanide ion	HCN	hydrogen cyanide	$(CN)_2$	cyanogen
SCN^-	thiocyanate ion	HSCN	thiocyanic acid	$(SCN)_2$	thiocyanogen
$SeCN^-$	selenocyanate ion			$(SeCN)_2$	selenocyanogen
OCN^-	cyanate ion	HOCN	cyanic acid		
NCN^{2-}	cyanamide ion	H_2NCN	cyanamide		
ONC^-	fulminate ion	HONC	fulminic acid		
N_3^-	azide ion	HN_3	hydrogen azide		

ions combine to form dimers comparable with the halogen molecules X_2. These include cyanogen $(CN)_2$, thiocyanogen $(SCN)_2$ and selenocyanogen $(SeCN)_2$.

The best known pseudohalide is CN^-. This resembles Cl^-, Br^- and I^- in the following respects:

1. It forms an acid HCN.
2. It can be oxidized to form a molecule cyanogen $(CN)_2$.
3. It forms insoluble salts with Ag^+, Pb^{2+} and Hg^+.
4. 'Interpseudohalogen' compounds ClCN, BrCN and ICN can be formed.
5. AgCN is insoluble in water but soluble in ammonia, as is AgCl.
6. It forms a large number of complexes similar to halide complexes, e.g. $[Cu(CN)_4]^{2-}$ and $[CuCl_4]^{2-}$, and $[Co(CN)_6]^{3-}$ and $[CoCl_6]^{3-}$

FURTHER READING

Banks, R.E. (ed.) (1982) *Preparation, Properties and Industrial Applications of Organofluorine Compounds*, Ellis Horwood, Chichester.

Banks, R.E., Sharp, D.W.A. and Tatlow, J.C. (1987) *Fluorine the First 100 Years (1886–1986)*, Elsevier.

Brown, D. (1968) *Halides of the Transition Elements*, Vol. I (Halides of the lanthanides and actinides), Wiley, London and New York.

Brown, I. (1987) Astatine: organonuclear chemistry and biomedical applications, *Adv. Inorg. Chem.*, **31**, 43–88.

Canterford, J.H. and Cotton, R. (1968) *Halides of the Second and Third Row Transition Elements*, Wiley, London.

Canterford, J.H. and Cotton, R. (1969) *Halides of the First Row Transition Elements*, Wiley, London.

Colton, R. and Canterford, J.H. (1968) *Halides of the Second and Third Row Transition Elements*, Wiley, London.

Colton, R. and Canterford, J.H. (1969) *Halides of the First Row Transition Elements*, Wiley, London.

Donovan, R.J. (1978) Chemistry and pollution of the stratosphere, *Education in Chemistry*, **15**, 110–113.

Downs, A.J. and Adams, C.J. (1973) *Comprehensive Inorganic Chemistry*, Vol. II (Chapter on Chlorine, bromine, iodine and astatine), Pergamon Press, Oxford.

Downs, A.J. and Adams, C.J. (1975) *The Chemistry of Chlorine, Bromine, Iodine and Astatine*, Pergamon, New York.

Fogg, P.G.T. and Gerrard, W. (eds) (1989) *Hydrogen Halides in Non-aqueous Solvents*, Pergamon.

Gillespie, R.J. and Morton, M.J. (1971) Halogen and interhalogen cations, *Q. Rev. Chem. Soc.*, **25**, 553.
Golub, A.M., Kohler, H. and Skopenko, V.V. (eds) (1986) *Chemistry of Pseudohalides* (Topics in Inorganic and General Chemistry Series No. 21), Elsevier, Oxford.
Gutman, V. (ed.) (1967) *Halogen Chemistry*, 3 Volumes, Academic Press, London, (Vol. I: Halogens, interhalogens, polyhalides, noble gas halides, Vol II: Halides of non-metals, Vol. 3: Complexes), (Old but comprehensive.)
Lagowski, J. (ed.) (1978) *The Chemistry of Non-aqueous Solvents*, Vol. II (Chapter 1 by Klanberg, F., Liquid HCl, HBr and HI; Chapter 2 by Kilpatrick, M. and Jones, J.G., Anhydrous hydrogen fluoride as a solvent medium; Chapter 3 by Martin, D.M., Rousson, R. and Weulersse, J.M., The Interhalogens), Academic Press, New York.
Langley, R.H. and Welch, L. (1983) Fluorine (Chemical of the month), *J. Chem. Ed.*, **60**, 759–761.
O'Donnell, T.A. (1973) *Comprehensive Inorganic Chemistry*, Vol. II (Chapter 25: Fluorine), Pergamon Press, Oxford.
Price, D., Iddon, B. and Wakefield, B.J. (eds) (1988) *Bromine Compounds*, Elsevier.
Sconce, J.S. (1962) *Chlorine: Its Manufacture, Properties and Uses*, Reinhold, New York.
Shamir, J. (1979) Polyhalogen cations, *Structure and Bonding*, **37**, 141–210.
Tatlow, J.C. *et al.* (eds), *Advances in Fluorine Chemistry*, Butterworths, London. (A continuing series.)
Thompson, R. (ed.) (1986) *The Modern Inorganic Chemicals Industry* (Chapter by Fielding, H.C. and Lee, B.E., Hydrogen fluoride, Inorganic fluorides and fluorine; Chapter by Purcell, R.W., The chlor-alkali industry, Chapter by Campbell, A., Chlorine and chlorination; Chapter by McDonald, R.B. and Merriman, W.R., Bromine and the bromine-chemicals industry), Special Publication No. 31, The Chemical Society, London, (reprinted 1986).
Thrush, B.A. (1977) The chemistry of the stratosphere and its pollution, *Endeavour*,
1, 3–6.
Waddington, T.C. (ed.) (1969) *Nonaqueous Solvent Systems* (Chapter 2 by Hyman, H.H. and Katz, J.J., Liquid hydrogen fluoride; Chapter 3 by Peach, M.E. and Waddington, T.C, The higher hydrogen halides as ionizing solvents), Nelson.

PROBLEMS

1. Describe how fluorine is produced, the apparatus used and any precautions you consider necessary. Is elemental fluorine widely used in reactions?

2. Why is it not possible to obtain F_2 by electrolysis of aqueous NaF, aqueous HF or anhydrous HF?

3. What fluorinating agents are often used instead of F_2? Give equations to show their use.

4. What are the main uses of fluorine?

5. Write balanced equations to show the reactions between HF and (a) SiO_2, (b) CaO, (c) KF, (d) CCl_4, (e) U, (f) graphite.

6. What are the common sources of chlorine, bromine and iodine in salts? Where do they occur and how are the elements extracted from the appropriate salts?

7. What are the main uses of Cl_2?

8. Give equations to show how the halogen acids HF, HCl, HBr and HI may be prepared in aqueous solution. Why is HF a weak acid compared with HI in water?

9. $HCl_{(g)}$ can be prepared from NaCl and H_2SO_4. $HBr_{(g)}$ and $HI_{(g)}$ cannot be made in a similar way from NaBr and NaI. Explain why this is so.

10. Write equations for the reactions between Cl_2 and (a) H_2, (b) CO, (c) P, (d) S, (e) SO_2, (f) $Br^-_{(aq)}$, (g) NaOH.

11. Explain what happens when an aqueous solution of $AgNO_3$ is added to solutions of NaF, NaCl, NaBr and NaI. What change (if any) occurs when ammonia is added to each of these mixtures?

12. (a) Draw the structures of OF_2, Cl_2O, O_2F_2 and I_2O_5.
    (b) Explain the bond angle in OF_2 and give a reason why it is different in Cl_2O.
    (c) Why are the O—F bonds in O_2F_2 longer than in OF_2, whereas the O—O bond in O_2F_2 is short compared with that in H_2O_2?

13. Describe the preparation and an analytical use of I_2O_5.

14. (a) Give the names of four different types of oxoacid of the halogens and give the formula of either an acid or a salt derived from the acid for each type.
    (b) Describe the preparation of the following compounds and give one use of each: NaOCl, $NaClO_2$, $NaClO_3$, HIO_4.

15. Iodine is almost insoluble in water, but it dissolves readily in an aqueous solution of KI. Explain why this is so.

16. (a) Explain the shape of the I_3^{-14} ion.
    (b) Explain why solid Cs_3 is stable but solid NaI_3 is not.

17. (a) Give the formulae of 11 interhalogen compounds.
    (b) Draw the shapes of the following molecules and ions, showing the positions of the lone pairs of electrons:

$$ClF,\ BrF_3,\ IF_5,\ IF_7,\ I_3^-,\ ICl_4^-\ \text{and}\ I_5^-.$$

18. List differences between the chemistry of fluorine and the other halogens and give reasons for these differences.

Group 0 – the noble gases

17

Table 17.1 Electronic structures

Element	Symbol		Electronic structure
Helium	He		$1s^2$
Neon	Ne	[He]	$2s^2\ 2p^6$
Argon	Ar	[Ne]	$3s^2\ 3p^6$
Krypton	Kr	[Ar]	$3d^{10}\ 4s^2\ 4p^6$
Xenon	Xe	[Kr]	$4d^{10}\ 5s^2\ 5p^6$
Radon	Rn	[Xe]	$4f^{14}\ 5d^{10}\ 6s^2\ 6p^6$

NAME OF GROUP, AND THEIR ELECTRONIC STRUCTURES

The elements of Group 0 have been called 'the inert gases' and 'the rare gases'. Both are misnomers, since the discovery of the xenon fluorides in 1962 shows that xenon is not inert, and argon makes up 0.9% by volume of the atmosphere. The name 'noble gases' implies that they tend to be unreactive, in the same way that the noble metals are often reluctant to react and are the least reactive metals.

Helium has two electrons which form a complete shell $1s^2$. The other noble gases have a closed octet of electrons in their outer shell ns^2np^6. This electronic configuration is very stable and is related to their chemical inactivity. These atoms have an electron affinity of zero (or slightly negative), and have very high ionization energies – higher than any other elements. Under normal conditions the noble gas atoms have little tendency to gain or lose electrons. Thus they have little tendency to form bonds, and so they exist as single atoms.

OCCURRENCE AND RECOVERY OF THE ELEMENTS

The gases He, Ne, Ar, Kr and Xe all occur in the atmosphere. A mixture of the noble gases was first obtained by Cavendish in 1784. Cavendish removed N_2 from air by adding excess O_2 and sparking. The NO_2 formed was absorbed in NaOH solution. The excess O_2 was removed by burning

continued overleaf

with S, and absorbing the SO_2 in NaOH solution. This gave a small volume of unreactive gas.

Ar is quite abundant and can be recovered by fractional distillation of liquid air (see under 'Nitrogen', Chapter 14). Ar constitutes 0.93% by volume of air (i.e. 9300 ppm). It originates in the air mostly from electron capture (β+ decay) of potassium:

$$^{40}_{19}K + ^{0}_{-1}e \rightarrow ^{40}_{18}Ar$$

World production of Ar is over 700 000 tonnes/year.

The other noble gases are much less abundant. The abundance of He in the atmosphere is only about 5 ppm by volume and recovery from air would be very expensive. A cheaper source is from natural gas deposits, where the hydrocarbons are liquified, leaving He gas. The He has been produced by radioactive decay, and trapped underground. The richest source is in southwest USA, where the natural gas contains 0.5–0.8% He. This provides most of the world's supply of He. Other natural gas deposits containing appreciable amounts of He have been found in Algeria, Poland, the USSR and Canada. World production is about 5000 tonnes/year.

The non-radioactive noble gases are all produced industrially by fractional distillation of liquid air. This gives large amounts of nitrogen and oxygen, and only a small amount of the noble gases. (The oxygen is mainly used for steel making.) Of the noble gases, Ar is obtained in the largest amounts, and it is the cheapest.

Rn is radioactive and is produced by the decay of radium and thorium minerals. A convenient source is ^{226}Ra, and 100 g of radium yields about 2 ml of radon per day:

$$^{226}_{88}Ra \rightarrow ^{222}_{86}Rn + ^{4}_{2}He$$

The most stable isotope ^{222}Rn is itself α active and has a half life of only 3.8 days, so only tracer studies have been made.

USES OF THE ELEMENTS

The largest use of Ar is to provide an inert atmosphere for metallurgical processes. This includes welding stainless steel, titanium, magnesium and aluminium, and in the production of titanium (Kroll and IMI processes). Smaller amounts are used in growing silicon and germanium crystals for transistors, and in electric light bulbs, fluorescent lamps, radio valves and Geiger–Müller radiation counters.

Helium has the lowest boiling point of any liquid, and it is used in cryoscopy to obtain the very low temperatures required for superconductivity, and lasers. It is used as the cooling gas in one type of gas cooled nuclear reactor, and as the flow-gas in gas–liquid chromatography. It is also used in weather balloons and airships. Though H_2 has a lower density and is cheaper and more readily available than He, H_2 is highly flammable. Thus on safety grounds He is used in preference to H_2 in airships. He is much less dense than air. One cubic metre of He gas at atmospheric

continued overleaf

pressure can lift 1 kg. Helium is used in preference to nitrogen to dilute oxygen in the gas cylinders used by divers. This is because nitrogen is quite soluble in blood, so a sudden change in pressure causes degassing and gives bubbles of N_2 in the blood. This causes the painful (or fatal) condition called 'bends'. Helium is only slightly soluble so the risk of 'bends' is reduced.

Small amounts of Ne are used in neon discharge tubes which give the familiar reddish orange glow of 'neon' signs. The other gases are also used in discharge tubes to give different colours.

PHYSICAL PROPERTIES

The elements are all colourless, odourless monatomic gases. The enthalpy of vaporization is a measure of the forces holding the atoms together. The values are very low because the only forces between the atoms are very weak van der Waals forces. The enthalpy of vaporization increases down the group as the polarizability of the atoms increases.

Table 17.2 Physical properties of the noble gases

	First ionization energy ($kJ\,mol^{-1}$)	Enthalpy of vaporization ($kJ\,mol^{-1}$)	Melting point (°C)	Boiling point (°C)	Atomic radii (Å)	Abundance in atmosphere (% volume)
He	2372	0.08		−269.0	1.20	5.2×10^{-4}
Ne	2080	1.7	−248.6	−246.0	1.60	1.5×10^{-3}
Ar	1521	6.5	−189.4	−186.0	1.91	0.93
Kr	1351	9.1	−157.2	−153.6	2.00	1.1×10^{-4}
Xe	1170	12.7	−111.8	−108.1	2.20	8.7×10^{-6}
Rn	1037	18.1	−71	−62		

Because the interatomic forces are very weak, the melting points and boiling points are also very low. The boiling point of He is the lowest of any element, only four degrees above absolute zero.

The atomic radii of the elements are all very large, and increase on descending the group. It must be noted that these are non-bonded radii, and should be compared with the van der Waals radii of other elements rather than with covalent (bonded) radii.

The noble gases are all able to diffuse through glass, rubber and plastic materials, and some metals. This makes them difficult to handle in the laboratory, particularly since glass Dewar flasks cannot be used for low temperature work.

SPECIAL PROPERTIES OF HELIUM

Helium is unique. It has the lowest boiling point of any substance known. All other elements become solids on cooling, but cooling only produces helium liquid. It only forms a solid under high pressure (about 25 atmospheres). There are two different liquid phases. Helium I is a normal liquid, but helium II is a superfluid. A superfluid is a most unusual state of matter. Normally atoms are free to move in a gas, can move in a more restricted way in a liquid, and can only vibrate about fixed positions in a solid. As the temperature decreases, the amount of thermal motion of atoms decreases, and gases become liquids, and eventually solids. When the temperature of helium gas is lowered to 4.2 K it liquifies as helium I. Rather surprisingly the liquid continues to boil vigorously. At 2.2 K, the liquid suddenly stops boiling (which with normal materials is when a solid is formed). In this case helium II is formed. This is still a liquid because the interatomic forces are not strong enough to form a solid, but thermal motion of the atoms has actually stopped. Helium I is a normal liquid, and when it changes to helium II at the λ-point temperature, many physical properties change abruptly. The specific heat changes by a factor of 10. The thermal conductivity increases by 10^6 and becomes 800 times greater than for copper. It becomes a superconductor (i.e. shows zero electrical resistance). The viscosity becomes effectively zero and 1/100th of that of gaseous hydrogen. It spreads to cover all surfaces at temperatures below the λ-point. Thus *the liquid can actually flow up the sides of the vessel* and over the edge until the levels on both sides are the same. The surface tension and compressibility are also anomalous.

CHEMICAL PROPERTIES OF THE NOBLE GASES

The noble gases were isolated and discovered because of their lack of reactivity. For a long time it was thought that they really were chemically inert. Before 1962, the only evidence for compound formation by the noble gases was some molecular ions formed in discharge tubes, and clathrate compounds.

Molecular ions formed under excited conditions

Several molecular ions such as He_2^+, HeH^+, HeH^{2+} and Ar_2^+ are formed under high energy conditions in discharge tubes. They only survive momentarily and are detected spectroscopically. Neutral molecules such as He_2 are unstable.

Clathrate compounds

Clathrate compounds of the noble gases are well known. Normal chemical compounds have ionic or covalent bonds. However, in the clathrates atoms or molecules of the appropriate size are trapped in cavities in the crystal lattice of other compounds. *Though the gases are trapped, they do not form bonds.*

If an aqueous solution of quinol (1,4-dihydroxybenzene) is crystallized under a pressure of 10–40 atmospheres of Ar, Kr or Xe, the gas becomes trapped in cavities of about 4 Å diameter in the β-quinol structure. When the clathrate is dissolved, the hydrogen bonded arrangement of β-quinol breaks down and the noble gas escapes. Other small molecules such as O_2, SO_2, H_2S, MeCN and CH_3OH form clathrates as well as Ar, Kr and Xe. The smaller noble gases He and Ne do not form clathrate compounds because the gas atoms are small enough to escape from the cavities. The composition of these clathrate compounds corresponds to 3 quinol : 1 trapped molecule, though normally all the cavities are not filled.

The gases Ar, Kr and Xe may be trapped in cavities in a similar way when water is frozen under a high pressure of the gas. These are clathrate compounds, but are more commonly called 'the noble gas hydrates'. They have formulae approximating to $6H_2O$: 1 gas atom. He and Ne are not trapped because they are too small. The heavier noble gases can also be trapped in cavities in synthetic zeolites, and samples have been obtained containing up to 20% of Ar by weight. Clathrates provide a convenient means of storing radioactive isotopes of Kr and Xe produced in nuclear reactors.

CHEMISTRY OF XENON

The first real compound of the noble gases was made in 1962. Bartlett and Lohman had previously used the highly oxidizing compound platinum hexafluoride to oxidize oxygen.

$$PtF_6 + O_2 \rightarrow O_2^+[PtF_6]^-$$

The first ionization energy for $O_2 \rightarrow O_2^+$ is $1165\,kJ\,mol^{-1}$, which is almost the same as the value of $1170\,kJ\,mol^{-1}$ for $Xe \rightarrow Xe^+$. It was predicted that xenon should react with PtF_6. Experiments showed that when deep red PtF_6 vapour was mixed with an equal volume of Xe, the gases combined immediately at room temperature to produce a yellow solid. They (incorrectly) thought the product obtained was xenon hexafluoroplatinate(V), $Xe^+[PtF_6]^-$. The reaction has since been shown to be more complicated, and the product is really $[XeF]^+[Pt_2F_{11}]^-$.

$$Xe[PtF_6] + PtF_6 \xrightarrow{25\,°C} [XeF]^+[PtF_6]^- + PtF_5 \xrightarrow{\text{heat } 60\,°C} [XeF]^+[Pt_2F_{11}]^-$$

Soon after this it was found that Xe and F_2 reacted at 400 °C to give a colourless volatile solid XeF_4. This has the same number of valency electrons as, and is isostructural with, the polyhalide ion $[ICl_4]^-$. Following these discoveries there was a rapid extension of the chemistry of the noble gases, and in particular of xenon.

The ionization energies of He, Ne and Ar are much higher than for Xe, and are too high to allow the formation of similar compounds. The ionization energy for Kr is a little lower than for Xe, and Kr does form KrF_2. The ionization energy of Rn is less than for Xe, and Rn might be expected to form compounds similar to those of Xe. Rn is radioactive, has no stable isotopes, and all the isotopes have short half lives. This has limited work on radon compounds, and only RnF_2 and a few complexes are known.

Table 17.3 Structures of some xenon compounds

Formula	Name	Oxidation state	m.p. (°C)	Structure
XeF_2	xenon difluoride	(+II)	129	linear (RnF_2 and $XeCl_2$ are similar)
XeF_4	xenon tetrafluoride	(+IV)	117	square planar ($XeCl_4$ is similar)
XeF_6	xenon hexafluoride	(+VI)	49.6	distorted octahedron
XeO_3	xenon trioxide	(+VI)	explodes	pyramidal (tetrahedral with one corner unoccupied)
XeO_2F_2		(+VI)	30.8	trigonal bipyramid (with one position unoccupied)
$XeOF_4$		(+VI)	−46	square pyramidal (octahedral with one position unoccupied)
XeO_4	xenon tetroxide	(+VIII)	−35.9	tetrahedral
XeO_3F_2		(+VIII)	−54.1	trigonal bipyramid
$Ba_2[XeO_6]^{4-}$	barium perxenate	(+VIII)	dec. >300	octahedral

Xe reacts directly only with F_2. However, oxygen compounds can be obtained from the fluorides. There is some evidence for the existence of $XeCl_2$ and $XeCl_4$, and one compound is known with a Xe—N bond. Thus there is quite an extensive chemistry of Xe. The principal compounds are listed in Table 17.3.

Xenon reacts directly with fluorine when the gases are heated at 400 °C in a sealed nickel vessel, and the products depend on the F_2/Xe ratio.

$$Xe + F_2 \rightarrow \begin{cases} 2:1 \text{ mixture} \rightarrow XeF_2 \\ 1:5 \text{ mixture} \rightarrow XeF_4 \\ 1:20 \text{ mixture} \rightarrow XeF_6 \end{cases}$$

The compounds XeF_2, XeF_4 and XeF_6 are all white solids. They can be sublimed at room temperature, and can be stored indefinitely in nickel or Monel containers. The lower fluorides form higher fluorides when heated with F_2 under pressure. The fluorides are all extremely strong oxidizing and fluorinating agents. They react quantitatively with hydrogen as follows:

$$XeF_2 + H_2 \rightarrow 2HF + Xe$$

$$XeF_4 + 2H_2 \rightarrow 4HF + Xe$$
$$XeF_6 + 3H_2 \rightarrow 6HF + Xe$$

They oxidize Cl^- to Cl_2, I^- to I_2 and cerium(III) to cerium(IV):

$$XeF_2 + 2HCl \rightarrow 2HF + Xe + Cl_2$$
$$XeF_4 + 4KI \rightarrow 4KF + Xe + 2I_2$$
$$SO_4^{2-} + XeF_2 + Ce_2^{III}(SO_4)_3 \rightarrow 2Ce^{IV}(SO_4)_2 + Xe + F_2$$

They fluorinate compounds:

$$XeF_4 + 2SF_4 \rightarrow Xe + 2SF_6$$
$$XeF_4 + Pt \rightarrow Xe + PtF_4$$
$$XeF_4 + C_6H_6 \rightarrow Xe + C_6H_5F + HF$$

The fluorides differ in their reactivity with water. XeF_2 is soluble in water, but undergoes slow hydrolysis. Hydrolysis is more rapid with alkali.

$$2XeF_2 + 2H_2O \rightarrow 2Xe + 4HF + O_2$$

XeF_4 reacts violently with water, giving xenon trioxide XeO_3.

$$3XeF_4 + 6H_2O \rightarrow 2Xe + XeO_3 + 12HF + 1\tfrac{1}{2}O_2$$

XeF_6 also reacts violently with water, but slow hydrolysis by atmospheric moisture gives the highly explosive solid XeO_3.

$$XeF_6 + 6H_2O \rightarrow XeO_3 + 6HF$$

With small quantities of water, partial hydrolysis occurs, giving a colourless liquid xenon oxofluoride $XeOF_4$. The same product is formed when XeF_6 reacts with silica or glass:

$$XeF_6 + H_2O \rightarrow XeOF_4 + 2HF$$
$$2XeF_6 + SiO_2 \rightarrow XeOF_4 + SiF_4$$

XeO_3 is an explosive white hygroscopic solid. It reacts with XeF_6 and $XeOF_4$.

$$XeO_3 + 2XeF_6 \rightarrow 3XeOF_4$$
$$XeO_3 + XeOF_4 \rightarrow 2XeO_2F_2$$

XeO_3 is soluble in water, but does not ionize. However, in alkaline solution above pH 10.5 it forms the xenate ion $[HXeO_4]^-$.

$$XeO_3 + NaOH \rightarrow \underset{\text{sodium xenate}}{Na^+ [HXeO_4]^-}$$

Xenates contain Xe(+VI) and they slowly disproportionate in solution to perxenates (which contain Xe(+VIII)) and Xe.

$$2[HXeO_4]^- + 2OH^- \rightarrow \underset{\text{perxenate ion}}{[XeO_6]^{4-}} + Xe + O_2 + 2H_2O$$

Several perxenates of Group I and II metals have been isolated, and the crystal structures of $Na_4XeO_6 \cdot 6H_2O$ and $Na_4XeO_6 \cdot 8H_2O$ have been

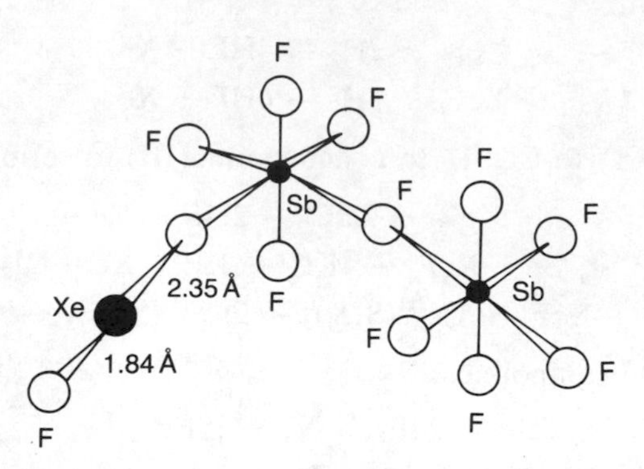

Figure 17.1 Structure of $XeF_2 \cdot 2SbF_5$. (From Mackay and Mackay, *Introduction to Modern Inorganic Chemistry*, 4th ed., Blackie, 1989.)

determined by X-ray crystallography. The solubility of sodium perxenate in 0.5 M NaOH is only 0.2 grams per litre, so precipitation of sodium perxenate could be used as a gravimetric method of analysis for sodium. Perxenates are extremely powerful oxidizing agents, which will oxidize HCl to Cl_2, H_2O to O_2, and Mn^{2+} to MnO_4^-. With concentrated H_2SO_4 they give xenon tetroxide XeO_4, which is volatile and explosive.

Xenon fluoride complexes

XeF_2 acts as a fluoride donor and forms complexes with covalent pentafluorides including PF_5, AsF_5, SbF_5 and the transition metal fluorides NbF_5, TaF_5, RuF_5, OsF_5, RhF_5, IrF_5 and PtF_5. These are thought to have the structure

$$XeF_2 \cdot MF_5 \qquad [XeF]^+ [MF_6]^-$$
$$XeF_2 \cdot 2MF_5 \qquad [XeF]^+ [M_2F_{11}]^-$$

and

$$2XeF_2 \cdot MF_5 \qquad [Xe_2F_3]^+ [MF_6]^-$$

The structures of some of the XeF_2 complexes in the solid state are known. In the complex $XeF_2 \cdot 2SbF_5$ (Figure 17.1) the two Xe–F distances differ greatly (1.84 Å and 2.35 Å). This suggests the formulation $[XeF]^+$ $[Sb_2F_{11}]^-$. However, the Xe–F distance of 2.35 Å is much less than the van der Waals (non-bonded) distance of 3.50 Å. This suggests that one fluorine atom forms a fluorine bridge between Xe and Sb. In fact the structure is intermediate between that expected for the ionic structure, and that for the fully covalent bridge structure.

XeF_4 forms only a few complexes, for example those with PF_5, AsF_5 and SbF_5. XeF_6 can act as a fluoride donor, forming complexes such as:

$$XeF_6 \cdot BF_3$$
$$XeF_6 \cdot GeF_4$$

$$XeF_6 \cdot 2GeF_4$$
$$XeF_6 \cdot 4SnF_4$$
$$XeF_6 \cdot AsF_5$$
$$XeF_6 \cdot SbF_5$$

XeF_6 may also act as a fluoride acceptor. With RbF and CsF it reacts as follows:

$$XeF_6 + RbF \rightarrow Rb^+[XeF_7]^-$$

On heating, the $[XeF_7]^-$ ion decomposes:

$$2Cs^+[XeF_7]^- \xrightarrow{50\,^\circ C} XeF_6 + Cs_2[XeF_8]$$

STRUCTURE AND BONDING IN XENON COMPOUNDS

The structures of the more common xenon halides, oxides and oxoions are given in Table 17.4. The nature of the bonds and the orbitals used for bonding in these compounds are of great interest and have been the subject of considerable controversy.

XeF_2

XeF_2 is a linear molecule with both Xe–F distances 2.00 Å. The bonding may be explained quite simply by promoting an electron from the $5p$ level of Xe to the $5d$ level. The two unpaired electrons form bonds with fluorine atoms. The five electron pairs point to the corners of a trigonal bipyramid. Of these, three are lone pairs and occupy the equatorial positions, and two are bond pairs and occupy the apical positions. The atoms thus form a linear molecule (Figure 17.2).

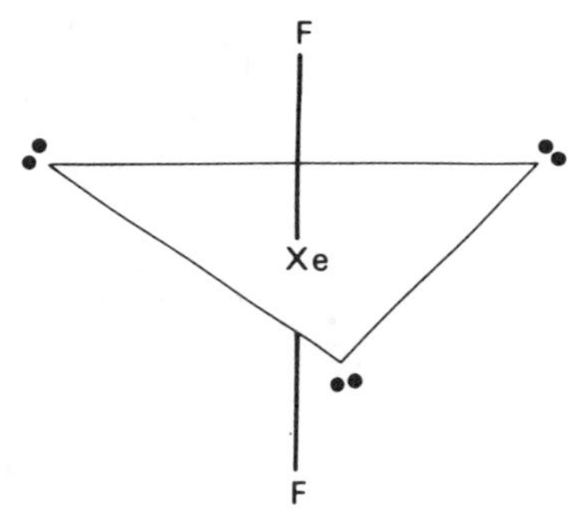

Figure 17.2 XeF_2 molecule.

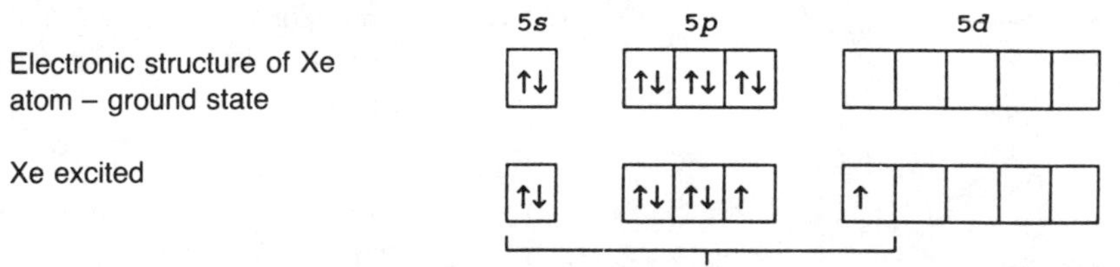

This explains the observed structure, but an objection is that the $5d$ orbitals of Xe appear to be too large for effective overlap of orbitals. The maximum in the radial electron distribution function for a $5d$ orbital in a Xe atom occurs at a distance of 4.9 Å from the nucleus. It has been noted in Chapter 4 in the section 'The extent of d orbital participation in molecular bonding' that highly electronegative atoms like fluorine cause a large contraction in the size of d orbitals. If this contraction is big enough, the valence bond explanation will suffice.

A second objection is over the mixing of orbitals (sp^3d hybridization). Mixing is only effective between orbitals of similar energy, and the Xe $5d$ orbitals would seem too high in energy to contribute to such a scheme of hybridization. (The difference in energy between a $5p$ and a $5d$ level is about 960 kJ mol^{-1}.)

The molecular orbital explanation involving three-centre bonds is more acceptable. The outer electronic configurations of the atoms are

	$5s$	$5p$ (p_x p_y p_z)		$2s$	$2p$ (p_x p_y p_z)
Xe	↑↓	↑↓ ↑↓ ↑↓	F	↑↓	↑↓ ↑↓ ↑

Assume that bonding involves the $5p_z$ orbital of Xe and the $2p_z$ orbitals of the two F atoms. For bonding to occur, orbitals with the same symmetry must overlap. These three atomic orbitals combine to give three molecular orbitals, one bonding, one non-bonding and one antibonding. This is represented in a simple way in Figure 17.3. The original three atomic orbitals contained four electrons (two in the Xe $5p_z$ and one in each of the F $2p_z$). These electrons will occupy the molecular orbitals of lowest energy. The order of energy is:

$$\text{bonding MO} < \text{non-bonding MO} < \text{antibonding MO}$$

Thus two electrons occupy the bonding MO, and this pair of electrons is responsible for binding all three atoms. The remaining two electrons occupy the non-bonding MO. These electrons are situated mainly on the F atoms, and confer some ionic character. The bonding may be described

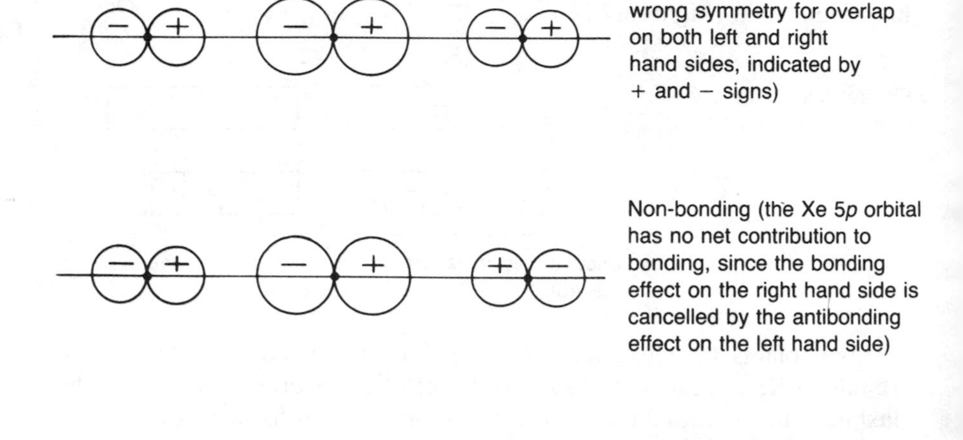

Figure 17.3 Possible combinations of atomic orbitals in XeF_2.

as three-centre, four-electron σ bonding. A linear arrangement of the atoms gives the best overlap of orbitals, in agreement with the observed structure. These bonds should be compared with the three-centre two-electron bonds described for B_2H_6 (see Chapter 12).

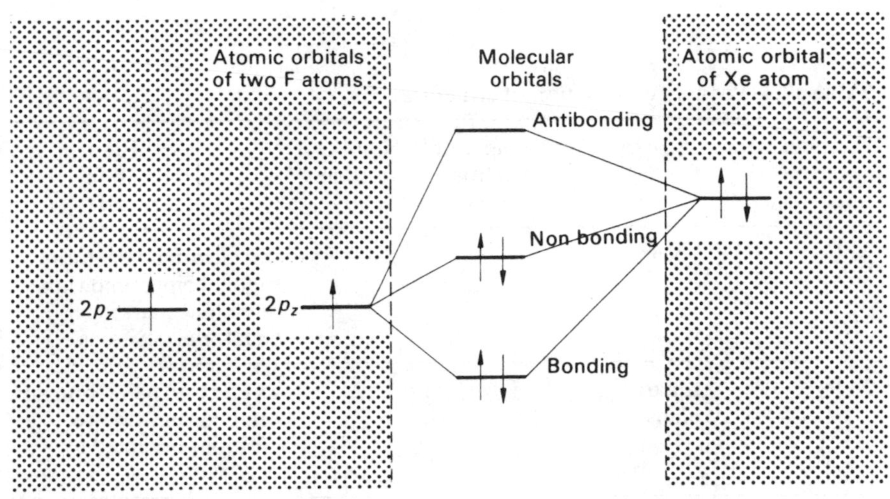

Figure 17.4 Molecular orbitals in XeF_2.

XeF_4

The structure of XeF_4 is square planar. The valence bond theory explains this by promoting two electrons as shown:

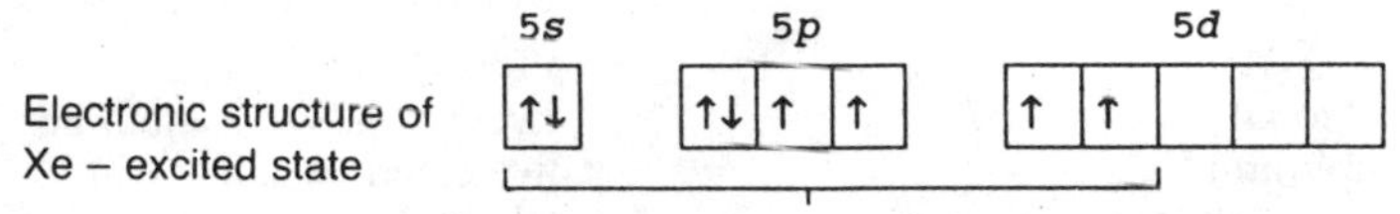

four unpaired electrons form bonds to four fluorine atoms
six electron pairs form octahedral structure
with two positions occupied by lone pairs

The problem of whether the size of the xenon 5*d* orbitals will allow effective overlap, or their energy will allow mixing and hybridization, is the same as in XeF_2. The molecular orbital explanation of XeF_4 is similar to that for XeF_2. The Xe atom bonds to four F atoms. The xenon $5p_x$ orbital forms a three-centre MO with 2*p* orbitals from two F atoms just as in XeF_2. The $5p_y$ orbital forms another three-centre MO involving two more F atoms. The two three-centre orbitals are at right angles to each other, thus giving a square planar molecule.

XeF_6

The structure of XeF_6 is a distorted octahedron. The bonding in XeF_6 has caused considerable controversy which is not completely resolved. The

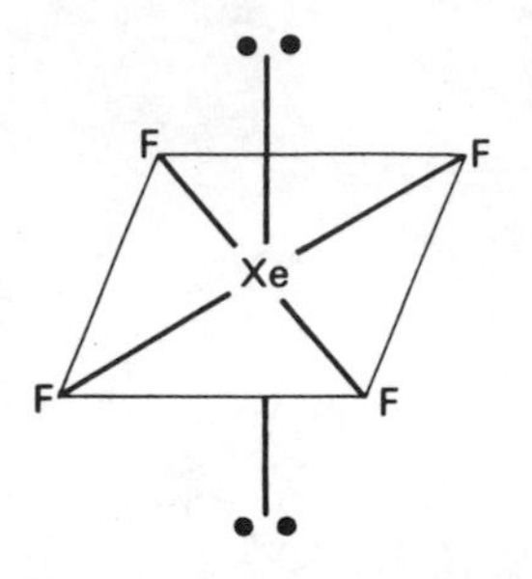

Figure 17.5 XeF_4 molecule.

structure may be explained in valence bond terms by promoting three electrons in Xe:

	5s	5p	5d
Electronic structure of xenon – excited state	↑↓	↑ ↑ ↑	↑ ↑ ↑

Table 17.4 Possible explanation of structures

Formula	Structure	Number of electron pairs	Number of lone pairs	VSEPR explanation of structure
XeF_2	linear	5	3	five electron pairs form trigonal bipyramid with three lone pairs in equatorial positions
XeF_4	square planar	6	2	six electron pairs form an octahedron with two positions occupied by lone pairs
XeF_6	distorted octahedron	7	1	pentagonal bipyramid, or capped octahedron with one lone pair
XeO_3	pyramidal	7	1	three π bonds so the remaining four electron pairs form a tetrahedron with one corner occupied by a lone pair
XeO_2F_2	trigonal bipyramid	7	1	two π bonds so remaining five electron pairs form trigonal bipyramid with one equatorial position occupied by a lone pair
$XeOF_4$	square pyramidal	7	1	one π bond so remaining six electron pairs form an octahedron with one position occupied by a lone pair
XeO_4	tetrahedral	8	0	four π bonds so remaining four electron pairs form a tetrahedron
XeO_3F_2	trigonal bipyramid	8	0	three π bonds so remaining five electron pairs form a trigonal bipyramid
$Ba_2[XeO_6]^{4-}$	octahedral	8	0	two π bonds so remaining six electron pairs form an octahedron

The six unpaired electrons form bonds with fluorine atoms. The distribution of seven orbitals gives either a capped octahedron or a pentagonal bipyramid (as in IF_7). (A capped octahedron has a lone pair pointing through one of the faces of the octahedron.) Since there are six bonds and one lone pair, a capped octahedron would give a distorted octahedral molecule. The molecular orbital approach fails with XeF_6, since three three-centre molecular orbital systems mutually at right angles would give a regular octahedral shape.

The structure of XeF_6 is most unusual in that it rapidly fluctuates with time. The time-averaged shape measured by X-ray crystallographic methods is a distorted octahedron, but the structure continually moves through the regular shape. The fact that XeF_6 forms adducts (such as $XeF_6 \cdot SbF_5$ discussed earlier) implies the presence of a lone pair of electrons on Xe which is donated to SbF_5. This lends some support to the valence bond approach and possibly to the presence of sp^3d^3 hybrid orbitals.

The shapes of oxygen containing compounds of Xe are correctly predicted by the valence bond method. (Electrons in π bonds (double bonds) must be subtracted before counting the number of electron pairs which determine the primary shape of the molecule.)

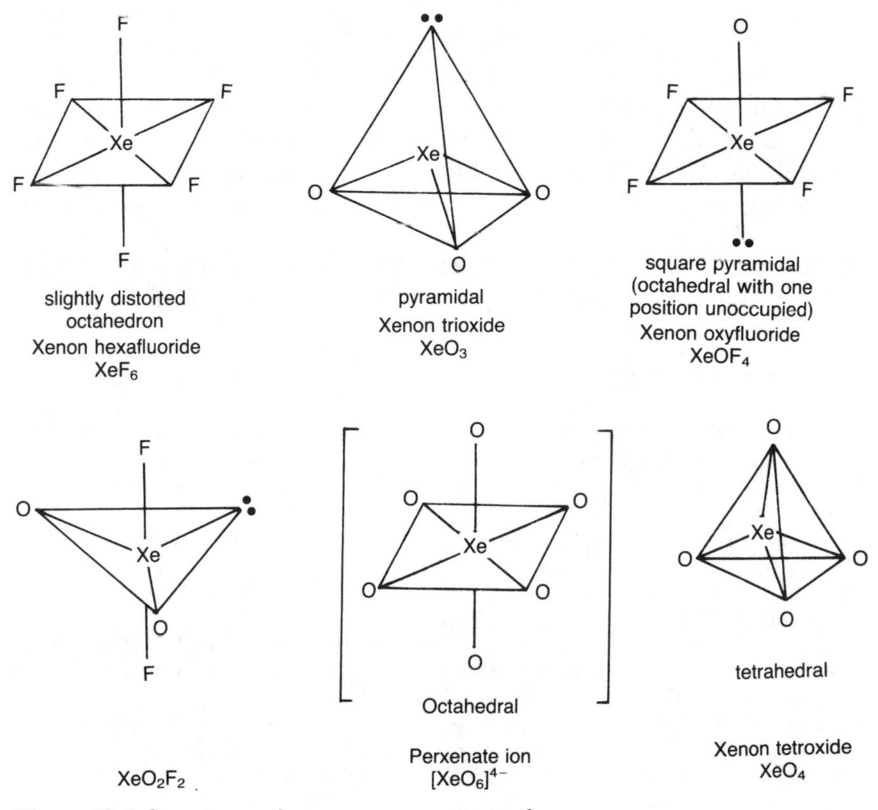

Figure 17.6 Structures of some xenon compounds.

VALEDICTION

For many years the noble gases were thought to be completely unreactive. This was associated with the concept that an octet of electrons is the only stable arrangement. The octet rule has done much to help the understanding of why atoms react, how many bonds they will form, and the shape of the periodic table. The discovery of the noble gas compounds has shown that though the 'octet' arrangement is very stable, it can be broken, and that there are other stable arrangements of electrons.

Two important points emerge:

1. Only the heavier noble gases (Kr, Xe and Rn) form these compounds. This is related to their lower ionization energies.
2. Compounds are only formed with electronegative ligands.

The discovery of the noble gas compounds led to a flurry of practical work attempting the synthesis of new compounds. There was also much theoretical work attempting to explain the structure and bonding in these compounds. This involved calculations on large computers on the extent of *d* orbital participation in bonding by elements in the *s*- and *p*-blocks. These conclusions may be summarized as follows:

1. In compounds of high coordination number with elements of high electronegativity, such as PF_5, SF_6, IF_5 and XeF_6, the *d* orbitals appear to be significantly involved in σ bonding. (These compounds may all be described without using *d* orbitals if three-centre bonds are formed.)
2. In compounds with elements of low electronegativity such as H_2S and PH_3, the *d* orbital population is very low (1–2%). However, this small contribution considerably improves the agreement between the observed and calculated values for the dipole moments and the energy levels.
3. The use of *d* orbitals makes a very significant contribution to π bonds, for example $p\pi - d\pi$ bonding in the phosphates and oxoacids of sulphur, and in PF_3.

FURTHER READING

Bartlett, N. (1962) Xenon hexafluoroplatinate(V), $Xe^+[PtF]^-$, *Proc. Chem. Soc.*, 218. (The original paper on the first true noble gas compound.)

Bartlett, N. (1971) *The Chemistry of the Noble Gases*, Elsevier, Amsterdam.

Bartlett, N. and Sladky, F.E. (1973) *Comprehensive Inorganic Chemistry*, Vol. I (Chapter 6: The chemistry of the krypton, xenon and radon), Pergamon Press, Oxford.

Berecz, E. and Balla-Ach, S.M. (1984) *Gas Hydrates* (Studies in Inorganic Chemistry, Vol. 4), Elsevier, Oxford.

Cockett, A.H. and Smith, K.C. (1973) *Comprehensive Inorganic Chemistry*, Vol. I (Chapter 5: The monatomic gases: physical properties and production), Pergamon Press, Oxford.

Hawkins, D.T., Falconer, W.E. and Bartlett, N. (1978) *Noble Gas Compounds*, Plenum Press, London. (Contains references from 1962 to 1976.)

Holloway, J.H. (1987) Twenty-five years of noble gas chemistry, *Chemistry in Britain*, **23**, 658–672.

Holloway, J.H. and Laycock, D. (1983) Preparations and reactions of inorganic main-group oxide fluorides, *Adv. Inorg. Chem. Radiochem.*, **27**, 157–195.
Huston, J.L. (1982) Chemical and physical properties of some xenon compounds, *Inorg. Chem.*, **21**, 685–688.
Moody, G.J. (1974) A decade of xenon chemistry, *J. Chem. Ed.*, **51**, 628.
Selig, H. and Holloway, J.H. (1984) *Topics in Current Chemistry*, Cationic and anionic complexes of the noble gases, (ed. by Bosche, F.L.), Springer Verlag, **124**, 33.
Seppelt, K. and Lentz, D. (1982) Novel developments in noble gas chemistry, *Progr. Inorg. Chem.*, **29**, 167–202.
Thompson, R. (ed.) (1986) *The Modern Inorganic Chemicals Industry*, (Chapter by Grant W.J. and Redfearn, S.L., Industrial Gases), Special Publication No. 31, The Chemical Society, London. (reprinted 1986).
Wilks, J. and Betts, D.S. (1987) *An Introduction to Liquid Helium*, 2nd ed., Clarendon, Oxford.

PROBLEMS

1. Where did the helium present in the earth and its atmosphere come from? How is He obtained commercially, and what is it used for? What is the boiling point of He in °C and K?

2. How much argon is present in the earth's atmosphere? How is Ar obtained commercially, and what is it used for?

3. Explain why Ar is used in the Kroll process to extract Ti, for welding, and in electric light bulbs.

   (a) Draw the structures of XeF_2, XeF_4 and XeF_6.
   (b) How may these compounds be prepared from Xe?
   (c) Give balanced equations to show how these three compounds react with water.

4. How did Bartlett interpret the reaction between Xe and PtF_6, and how is this reaction now interpreted?

5. (a) Draw the structures of XeF_2, XeF_4 and XeF_6.
   (b) How may these compounds be prepared from Xe?
   (c) Give balanced equations to show how these three compounds react with water.

6. How may the compounds XeO_3, $XeOF_4$ and Ba_2XeO_6 be prepared, and what are their structures?

7. 'In some ways the discovery of the noble gas compounds has created more problems than it has solved.' Discuss this statement with particular reference to the stability of a closed electron shell and the participation of *d* orbitals in bonding by elements of the *s*- and *p*-blocks.

8. Suggest reasons why the only binary compounds of the noble gases are fluorides and oxides of Kr, Xe and Rn.

The *d*-Block Elements

Part Four

An introduction to the transition elements

18

Table 18.1 The elements

Sc	Ti	V	Cr	Mn	Fe	Co	Ni	Cu	Zu
Scan-dium	Tita-nium	Vana-dium	Chro-mium	Man-ganese	Iron	Cobalt	Nickel	Copper	Zinc
Y	Zr	Nb	Mo	Tc	Ru	Rh	Pd	Ag	Cd
Yttrium	Zir-conium	Nio-bium	Molyb-denum	Tech-netium	Ruthe-nium	Rho-dium	Palla-dium	Silver	Cad-mium
La	Hf	Ta	W	Re	Os	Ir	Pt	Au	Hg
Lan-thanum	Haf-nium	Tan-talum	Tungsten	Rhenium	Osmium	Iridium	Platinum	Gold	Mer-cury
Ac									
Acti-nium									

INTRODUCTION

Three series of elements are formed by filling the $3d$, $4d$ and $5d$ shells of electrons. Together these comprise the d-block elements. They are often called 'transition elements' because their position in the periodic table is between the s-block and p-block elements. Their properties are transitional between the highly reactive metallic elements of the s-block, which typically form ionic compounds, and the elements of the p-block, which are largely covalent. In the s- and p-blocks, electrons are added to the outer shell of the atom. In the d-block, electrons are added to the penultimate shell, expanding it from 8 to 18 electrons. Typically the transition elements have an incompletely filled d level. The zinc group has a d^{10} configuration and since the d shell is complete, compounds of these elements are not typical and show some differences from the others. The elements make up three complete rows of ten elements and an incomplete fourth row. The position of the incomplete fourth series is discussed with the f-block elements.

METALLIC CHARACTER

In the *d*-block elements the penultimate shell of electrons is expanding. Thus they have many physical and chemical properties in common. Thus all the transition elements are metals. They are therefore good conductors of electricity and heat, have a metallic lustre and are hard, strong and ductile. They also form alloys with other metals.

VARIABLE OXIDATION STATE

One of the most striking features of the transition elements is that the elements usually exist in several different oxidation states. Furthermore, the oxidation states change in units of one, e.g. Fe^{3+} and Fe^{2+}, Cu^{2+} and Cu^{+}.

The oxidation states shown by the transition elements may be related to their electronic structures. Calcium, the *s*-block element preceding the first row of transition elements, has the electronic structure:

$$\text{Ca} \quad 1s^2 2s^2 2p^6 3s^2 3p^6 4s^2$$

It might be expected that the next ten transition elements would have this electronic arrangement with from one to ten *d* electrons added in a regular way: $3d^1$, $3d^2$, $3d^3 \ldots 3d^{10}$. This is true except in the cases of Cr and Cu. In these two cases one of the *s* electrons moves into the *d* shell, because of the additional stability when the *d* orbitals are exactly half filled or completely filled (Table 18.2).

Table 18.2 Oxidation states

	Sc	Ti	V	Cr	Mn	Fe	Co	Ni	Cu	Zn
Electronic structure	d^1s^2	d^2s^2	d^3s^2	~~d^4s^2~~ d^5s^1	d^5s^2	d^6s^2	d^7s^2	d^8s^2	~~d^9s^2~~ $d^{10}s^1$	$d^{10}s^2$
Oxidation states				I					I	
	II	II	II	II	II	II	II	II	II	II
	III	III	III	III	III	III	III	III	III	
		IV	IV	IV	IV	IV	IV	IV		
			IV	V	V	V	V			
				VI	VI	VI				
					VII					

Thus Sc could have an oxidation number of (+II) if both *s* electrons are used for bonding and (+III) when two *s* and one *d* electrons are involved. Ti has an oxidation state (+II) when both *s* electrons are used for bonding, (+III) when two *s* and one *d* electrons are used and (+IV) when two *s* and two *d* electrons are used. Similarly, V shows oxidation numbers (+II), (+III), (+IV) and (+V). In the case of Cr, by using the single *s* electron for bonding, we get an oxidation number of (+I): hence by using varying numbers of d electrons oxidation states of (+II), (+III), (+IV), (+V) and (+VI) are possible. Mn has oxidation states (+II), (+III), (+IV), (+V), (+VI) and (+VII). Among these first five elements, the correlation between electronic structure and minimum and maximum oxidation states in simple compounds is complete. In the highest oxidation states of these first five elements, all of the s and d electrons are being used for bonding. Thus the properties depend only on the size and valency, and consequently show some similarities with elements of the main groups in similar oxidation states. For example, SO_4^{2-} and CrO_4^{2-} are isostructural, as are $SiCl_4$ and $TiCl_4$.

Once the d^5 configuration is exceeded, i.e. in the last five elements, the tendency for all the d electrons to participate in bonding decreases. Thus Fe has a maximum oxidation state of (+VI). However, the second and third elements in this group attain a maximum oxidation state of (+VIII), in RuO_4 and OsO_4. This difference between Fe and the other two elements Ru and Os is attributed to the increased size.

These facts may be conveniently memorized, because the oxidation states form a regular 'pyramid' as shown in Table 18.2. Only Sc(+II) and Co(+V) are in doubt. The oxidation number of all elements in the elemental state is zero. In addition, several of the elements have zero-valent and other low-valent states in complexes. Low oxidation states occur particularly with π bonding ligands such as carbon monoxide and dipyridyl.

Similar but not identical pyramids of oxidation states are found in the second and third rows of transition elements. The main differences are as follows:

1. In the iron group the second and third row elements show a maximum oxidation state of (+VIII) compared with (+VI) for Fe.
2. The electronic structures of the atoms in the second and third rows do not always follow the pattern of the first row. The structures of the nickel group are:

Ni	$3d^8$	$4s^2$
Pd	$4d^{10}$	$5s^0$
Pt	$5d^9$	$6s^1$

Since a full shell of electrons is a stable arrangement, the place where this occurs is of importance.

The d levels are complete at copper, palladium and gold in their respective series.

Ni			Cu	$3d^{10}$	$4s^1$	Zn	$3d^{10}$	$4s^2$
Pd	$4d^{10}$	$5s^0$	Ag			Cd	$3d^{10}$	$4s^2$
Pt			Au	$5d^{10}$	$6s^1$	Hg	$3d^{10}$	$4s^2$

Even though the ground state of the atom has a d^{10} configuration, Pd and the coinage metals Cu, Ag and Au behave as typical transition elements. This is because in their most common oxidation states Cu(II) has a d^9

configuration and Pd(II) and Au(III) have d^8 configurations, that is they have an incompletely filled d level. However, in zinc, cadmium and mercury the ions Zn^{2+}, Cd^{2+} and Hg^{2+} have a d^{10} configuration. Because of this, these elements do not show the properties characteristic of transition elements.

Stability of the various oxidation states

Compounds are regarded as stable if they exist at room temperature, are not oxidized by the air, are not hydrolysed by water vapour and do not disproportionate or decompose at normal temperatures. Within each of the transition groups, there is a difference in stability of the various oxidation states that exist. In general the second and third row elements exhibit higher coordination numbers, and their higher oxidation states are more stable than the corresponding first row elements. This can be seen from Table 18.3. This gives the known oxides and halides of the first, second and third row transition elements. Stable oxidation states form

Table 18.3 (a) Oxides and halides of the first row

		Sc	Ti	V	Cr	Mn	Fe	Co	Ni	Cu	Zn
+II	O		TiO	VO^m	CrO	**MnO**	FeO	**CoO**	**NiO**	**CuO**	**ZnO**
	F			VF_2	**CrF_2**	**MnF_2**	FeF_2	**CoF_2**	**NiF_2**	**CuF_2**	**ZnF_2**
	Cl		$TiCl_2$	VCl_2	$CrCl_2$	**$MnCl_2$**	**$FeCl_2$**	**$CoCl_2$**	**$NiCl_2$**	$CuCl_2$	**$ZnCl_2$**
	Br		$TiBr_2$	VBr_2	$CrBr_2$	**$MnBr_2$**	**$FeBr_2$**	**$CoBr_2$**	**$NiBr_2$**	$CuBr_2$	**$ZnBr_2$**
	I		TiI_2	VI_2	CrI_2	**MnI_2**	**FeI_2**	**CoI_2**	**NiI_2**		**ZnI_2**
+III	O	**Sc_2O_3**	Ti_2O_3	V_2O_3	**Cr_2O_3**	Mn_2O_3	**Fe_2O_3**	$(Co_2O_3)^h$	$(Ni_2O_3)^h$		
	F	**ScF_3**	TiF_3	VF_3	CrF_3	MnF_3	**FeF_3**	CoF_3			
	Cl	**$ScCl_3$**	$TiCl_3$	**VCl_3**	**$CrCl_3$**		$FeCl_3$				
	Br	**$ScBr_3$**	$TiBr_3$	**VBr_3**	**$CrBr_3$**		$FeBr_3$				
	I	**ScI_3**	TiI_3	**VI_3**	**CrI_3**						
+IV	O		**TiO_2**	VO_2	CrO_2	MnO_2		$(CoO_2)^h$	$NiO_2{}^h$		
	F		**TiF_4**	**VF_4**	CrF_4	MnF_4					
	Cl		**$TiCl_4$**	VCl_4	$CrCl_4{}^g$						
	Br		**$TiBr_4$**	VBr_4	$CrBr_4{}^g$						
	I		**TiI_4**		CrI_4						
+V	O			**V_2O_5**							
	F			VF_5	CrF_5						
	Cl										
	Br										
	I										
+VI	O				CrO_3						
	F				(CrF_6)						
	Cl										
	Br										
	I										
+VII	O					Mn_2O_7					
	F										
	Cl										
	Br										
	I										
Other compounds						Mn_3O_4	**Fe_3O_4**	Co_3O_4		Cu_2O	
										$CuCl$	
										$CuBr$	
										CuI	

(b) Oxides and halides of the second row

		Y	Zr	Nb	Mo	Tc	Ru	Rh	Pd	Ag	Cd
+II	O			NbO				RhO	**PdO**	AgO[x]	CdO
	F								**PdF_2**	AgF_2	CdF_2
	Cl		$ZrCl_2$		$[Mo_6Cl_8]Cl_4$[c]				**$PdCl_2$**		$CdCl_2$
	Br		$ZrBr_2$?		$[Mo_6Br_8]Br_4$[c]				**$PdBr_2$**		$CdBr_2$
	I		ZrI_2?		$[Mo_6I_8]I_4$[c]				**PdI_2**		CdI_2
+III	O	Y_2O_3					Ru_2O_3[h]	**Rh_2O_3**	(Pd_2O_3)[h]?	(Ag_2O_3)	
	F	YF_3		(NbF_3)[c]	MoF_3		**RuF_3**	**RhF_3**	**$Pd[PdF_6]$**		
	Cl	YCl_3	$ZrCl_3$	$NbCl_3$[c]	$MoCl_3$[m]		**$RuCl_3$**	**$RhCl_3$**			
	Br	YBr_3	$ZrBr_3$	$NbBr_3$[c]	$MoBr_3$		**$RuBr_3$**	**$RhBr_3$**			
	I	YI_3	ZrI_3	NbI_3[c]	MoI_3		**RuI_3**	**RhI_3**			
+IV	O		**ZrO_2**	NbO_2	MoO_2[m]	TcO_2[m]	**RuO_2**	RhO_2	(PdO_2)[h]		
	F		**ZrF_4**	NbF_4	MoF_4		RuF_4	RhF_4	PdF_4		
	Cl		**$ZrCl_4$**	$NbCl_4$[m]	$MoCl_4$[m]	$TcCl_4$	$RuCl_4$				
	Br		**$ZrBr_4$**	$NbBr_4$[m]	$MoBr_4$						
	I		**ZrI_4**	NbI_4[m]	MoI_4?						
+V	O			**Nb_2O_5**	Mo_2O_5						
	F			**NbF_5**	MoF_5	TcF_5	**RuF_5**	(RhF_5)			
	Cl			**$NbCl_5$**	$MoCl_5$						
	Br			**$NbBr_5$**							
	I			**NbI_5**							
+VI	O				**MoO_3**	TcO_3	(RuO_3)[h]				
	F				**MoF_6**	**TcF_6**	RuF_6	RhF_6			
	Cl				$(MoCl_6)$	$(TcCl_6)$?					
	Br										
	I										
+VII	O					**Tc_2O_7**					
	F										
	Cl										
	Br										
	I										
Other compounds				Nb_6F_{14}[c]			RuO_4			**Ag_2O**	
				Nb_6I_{14}[c]						**AgF**	
										$AgCl$	
										$AgBr$	
										AgI	

oxides, fluorides, chlorides, bromides and iodides. Strongly reducing states probably do not form fluorides and/or oxides, but may well form the heavier halides. Conversely, strongly oxidizing states form oxides and fluorides, but not iodides.

COMPLEXES

A The transition elements have an unparalleled tendency to form coordination compounds with Lewis bases, that is with groups which are able to donate an electron pair. These groups are called ligands. A ligand may be a neutral molecule such as NH_3, or an ion such as Cl^- or CN^-. Cobalt forms more complexes than any other element, and forms more compounds than any other element except carbon.

$$Co^{3+} + 6NH_3 \rightarrow [Co(NH_3)_6]^{3+}$$
$$Fe^{2+} + 6CN^- \rightarrow [Fe(CN)_6]^{4-}$$

Table 18.3 (*continued*) (c) Oxides and halides of the third row

		La	Hf	Ta	W	Re	Os	Ir	Pt	Au	Hg
+II	O			(TaO)					$(PtO)^h$		HgO
	F										HgF_2
	Cl		$HfCl_2$?		$W_6Cl_{12}^c$	$(ReCl_2)$		$(IrCl_2)$?	**$PtCl_2$**		$HgCl_2$
	Br		$HfBr_2$?		$W_6Br_{12}^c$	$(ReBr_2)$			**$PtBr_2$**		$HgBr_2$
	I				$W_6I_{12}^c$	(ReI_2)			**PtI_2**		HgI_2
+III	O	La_2O_3				$Re_2O_3^h$		$Ir_2O_3^h$	$(Pt_2O_3)^h$?	Au_2O_3	
	F	LaF_3		$(TaF_3)^c$				IrF_3		AuF_3	
	Cl	$LaCl_3$	$HfCl_3$	$TaCl_3^c$	$W_6Cl_{18}^c$	$Re_3Cl_9^c$		**$IrCl_3$**	$PtCl_3$?	$AuCl_3$	
	Br	$LaBr_3$	$HfBr_3$	$TaBr_3^c$	$W_6Br_{18}^c$	$Re_3Br_9^c$		**$IrBr_3$**	$PtBr_3$?	$AuBr_3$	
	I	LaI_3	HfI_3		$W_6I_{18}^c$	$Re_3I_9^c$		**IrI_3**	PtI_3?		
+IV	O		**HfO_2**	TaO_2	WO_2^m	ReO_2^m	**OsO_2**	**IrO_2**	**PtO_2**		
	F		**HfF_2**		WF_4	ReF_4	**OsF_4**	**IrF_4**	**PtF_4**		
	Cl		**$HfCl_2$**	$TaCl_4^m$	WCl_4	**$ReCl_4^m$**	**$OsCl_4$**	$(IrCl_4)$	**$PtCl_4$**		
	Br		**$HfBr_2$**	$TaBr_4^m$	WBr_4	**$ReBr_4$**	$OsBr_4$		**$PtBr_4$**		
	I		**HfI_2**	TaI_4^m	WI_4?	**ReI_4**	OsI_4		**PtI_4**		
+V	O			**Ta_2O_5**	(W_2O_5)	(Re_2O_5)					
	F			**TaF_5**	WF_5^d	ReF_5	OsF_5	(IrF_5)	$(PtF_5)_4$		
	Cl			**$TaCl_5$**	WCl_5	$ReCl_5$	$OsCl_5$				
	Br			**$TaBr_5$**	WBr_5	$ReBr_5$					
	I			**TaI_5**							
+VI	O				**WO_3**	ReO_3	$(OsO_3)^h$	(IrO_3)	$(PtO_3)^h$		
	F				**WF_6**	ReF_6	**OsF_6**	IrF_6	PtF_6		
	Cl				WCl_6	$(ReCl_6)$?					
	Br				WBr_6						
	I										
+VII	O					**Re_2O_7**					
	F					**ReF_7**	(OsF_7)				
	Cl										
	Br										
	I										
Other compounds							**OsO_4** $OsCl_{3.5}$		Pt_3O_4	Au_2O $AuCl$ AuI	$Hg_2F_2^m$ $Hg_2Cl_2^m$ $Hg_2Br_2^m$ $Hg_2I_2^m$

In Table 18.3 the most stable compounds are bold, unstable compounds are in parentheses, h indicates hydrated oxides, g indicates that it occurs only as a gas, m indicates metal–metal bonding, c indicates cluster compounds, x indicates mixed oxide and d indicates that it disproportionates.

This ability to form complexes is in marked contrast to the *s*- and *p*-block elements which form only a few complexes. The reason transition elements are so good at forming complexes is that they have small, highly charged ions and have vacant low energy orbitals to accept lone pairs of electrons donated by other groups or ligands. Complexes where the metal is in the (+III) oxidation state are generally more stable than those where the metal is in the (+II) state.

Some metal ions form their most stable complexes with ligands in which the donor atoms are N, O or F. Such metal ions include Group I and II elements, the first half of the transition elements, the lanthanides and actinides, and the *p*-block elements except for their heaviest member. These metals are called class-a acceptors, and correspond to 'hard' acids (see 'Acids and bases', Chapter 8). In contrast the metals Rh, Ir, Pd, Pt, Ag, Au and Hg form their most stable complexes with the heavier elements of Groups V, VI and VII. These metals are called class-b acceptors, and correspond to 'soft' acids. The rest of the transition metals, and the heaviest elements in the *p*-block, form complexes with both types of donors, and are thus 'intermediate' in nature. These are shown (a/b) in Table 18.4.

Table 18.4 Class-a and class-b acceptors

Li (a)	Be (a)											B (a)	C (a)	N (a)	O
Na (a)	Mg (a)											Al (a)	Si (a)	P (a)	S (a)
K (a)	Ca (a)	Sc (a)	Ti (a)	V (a)	Cr (a)	Mn (a)	Fe (a/b)	Co (a/b)	Ni (a/b)	Cu (a/b)	Zn (a)	Ga (a)	Ge (a)	As (a)	Se (a)
Rb (a)	Sr (a)	Y (a)	Zr (a)	Nb (a)	Mo (a)	Tc (a/b)	Ru (a/b)	Rh (b)	Pd (b)	Ag (b)	Cd (a/b)	In (a)	Sn (a)	Sb (a)	Te (a)
Cs (a)	Ba (a)	La (a)	Hf (a)	Ta (a)	W (a)	Re (a/b)	Os (a/b)	Ir (b)	Pt (b)	Au (b)	Hg (b)	Tl (a/b)	Pb (a/b)	Bi (a/b)	Po (a/b)
Fr (a)	Ra (a)	Ac (a)													
			Ce (a)	Pr (a)	Nd (a)	Pm (a)	Sm (a)	Eu (a)	Gd (a)	Tb (a)	Dy (a)	Ho (a)	Er (a)	Tm (a)	Yb (a)
			Th (a)	Pa (a)	U (a)	Np (a)	Pu (a)	Am (a)	Cm (a)	Bk (a)	Cf (a)	Es (a)	Fm (a)	Md (a)	Mo (a)

The nature of coordination complexes and the important crystal field theory of bonding are discussed in Chapter 7.

SIZE OF ATOMS AND IONS

The covalent radii of the elements (Table 18.5) decrease from left to right across a row in the transition series, until near the end when the size increases slightly. On passing from left to right, extra protons are placed in the nucleus and extra orbital electrons are added. The orbital electrons shield the nuclear charge incompletely (*d* electrons shield less efficiently than *p* electrons, which in turn are shielded less effectively than *s* electrons). Thus the nuclear charge attracts all of the electrons more strongly: hence a contraction in size occurs.

Atoms of the transition elements are smaller than those of the Group I or II elements in the same horizontal period. This is partly because of the usual contraction in size across a horizontal period discussed above, and

Table 18.5 Covalent radii of the transition elements (Å)

K	Ca	Sc	Ti	V	Cr	Mn	Fe	Co	Ni	Cu	Zn
1.57	1.74	1.44	1.32	1.22	1.17	1.17	1.17	1.16	1.15	1.17	1.25
Rb	Sr	Y	Zr	Nb	Mo	Tc	Ru	Rh	Pd	Ag	Cd
2.16	1.91	1.62	1.45	1.34	1.29	–	1.24	1.25	1.28	1.34	1.41
Cs	Ba	La *	Hf	Ta	W	Re	Os	Ir	Pt	Au	Hg
2.35	1.98	1.69	1.44	1.34	1.30	1.28	1.26	1.26	1.29	1.34	1.44

* 14 Lanthanide elements

partly because the orbital electrons are added to the penultimate *d* shell rather than to the outer shell of the atom.

The transition elements are divided into vertical groups of three (triads) or sometimes four elements, which have similar electronic structures. On descending one of the main groups of elements in the *s*- and *p*-blocks, the size of the atoms increases because extra shells of electrons are present. The elements in the first group in the *d*-block show the expected increase in size Sc → Y → La. However, in the subsequent groups there is an increase in radius of 0.1 → 0.2 Å between the first and second member, but hardly any increase between the second and third elements. This trend is shown both in the covalent radii (Table 18.5) and in the ionic radii (Table 18.6). Interposed between lanthanum and hafnium are the 14 lanthanide elements, in which the antepenultimate 4*f* shell of electrons is filled.

Table 18.6 The effect of the lanthanide contraction on ionic radii

Ca^{2+}	1.00	Sc^{3+}	0.745	Ti^{4+}	0.605	V^{3+}	0.64
Sr^{2+}	1.18	Y^{3+}	0.90	Zr^{4+}	0.72	Nb^{3+}	0.72
Ba^{2+}	1.35	La^{3+}	1.032 *	Hf^{4+}	0.71	Ta^{3+}	0.72

* 14 Lanthanides

There is a gradual decrease in size of the 14 lanthanide elements from cerium to lutetium. This is called the lanthanide contraction, and is discussed in Chapter 29. The lanthanide contraction cancels almost exactly the normal size increase on descending a group of transition elements. The covalent radius of Hf and the ionic radius of Hf^{4+} are actually smaller than the corresponding values for Zr. The covalent and ionic radii of Nb are the same as the values for Ta. Therefore the second and third row transition elements have similar radii. As a result they also have similar lattice energies, solvation energies and ionization energies. Thus the differences in properties between the first row and second row elements are much greater than the differences between the second and third row elements. The effects of the lanthanide contraction are less pronounced towards the right of the *d*-block. However, the effect still shows to a lesser degree in the *p*-block elements which follow.

DENSITY

The atomic volumes of the transition elements are low compared with elements in neighbouring Groups I and II. This is because the increased nuclear charge is poorly screened and so attracts all the electrons more strongly. In addition, the extra electrons added occupy inner orbitals. Consequently the densities of the transition metals are high. Practically all have a density greater than $5\,g\,cm^{-3}$. (The only exceptions are Sc $3.0\,g\,cm^{-3}$ and Y and Ti $4.5\,g\,cm^{-3}$.) The densities of the second row are high and third row values are even higher. (See Appendix D.) The two elements with the highest densities are osmium $22.57\,g\,cm^{-3}$ and iridium $22.61\,g\,cm^{-3}$. To get some feel for how high this figure really is, a football made of osmium or iridium measuring 30 cm in diameter would weigh 320 kg or almost one third of a tonne!

MELTING AND BOILING POINTS

The melting and boiling points of the transition elements are generally very high (see Appendices B and C). Transition elements typically melt above 1000 °C. Ten elements melt above 2000 °C and three melt above 3000 °C (Ta 3000 °C, W 3410 °C and Re 3180 °C). There are a few exceptions. The melting points of La and Ag are just under 1000 °C (920 °C and 961 °C respectively). Other notable exceptions are Zn (420 °C), Cd (321 °C) and Hg which is liquid at room temperature and melts at −38 °C. The last three behave atypically because the *d* shell is complete, and *d* electrons do not participate in metallic bonding. The high melting points are in marked contrast to the low melting points for the *s*-block metals Li (181 °C) and Cs (29 °C).

REACTIVITY OF METALS

Many of the metals are sufficiently electropositive to react with mineral acids, liberating H_2. A few have low standard electrode potentials and remain unreactive or noble. Noble character is favoured by high enthalpies of sublimation, high ionization energies and low enthalpies of solvation. (See 'Born–Haber cycle', Chapter 6.) The high melting points indicate high heats of sublimation. The smaller atoms have higher ionization energies, but this is offset by small ions having high solvation energies. This tendency to noble character is most pronounced for the platinum metals (Ru, Rh, Pd, Os, Ir, Pt) and gold.

IONIZATION ENERGIES

The ease with which an electron may be removed from a transition metal atom (that is, its ionization energy) is intermediate between those of the *s*- and *p*-blocks. Values for the first ionization energies vary over a wide range from $541\,kJ\,mol^{-1}$ for lanthanum to $1007\,kJ\,mol^{-1}$ for mercury.

These are comparable with the values for lithium and carbon respectively. This would suggest that the transition elements are less electropositive than Groups I and II and may form either ionic or covalent bonds depending on the conditions. Generally, the lower valent states are ionic and the higher valent states covalent. The first row elements have many more ionic compounds than elements in the second and third rows.

COLOUR

Many ionic and covalent compounds of transition elements are coloured. In contrast compounds of the *s*- and *p*-block elements are almost always white. When light passes through a material it is deprived of those wavelengths that are absorbed. If absorption occurs in the visible region of the spectrum, the transmitted light is coloured with the complementary colour to the colour of the light absorbed. Absorption in the visible and UV regions of the spectrum is caused by changes in electronic energy. Thus the spectra are sometimes called electronic spectra. (These changes are often accompanied by much smaller changes in vibrational and rotational energy.) It is always possible to promote an electron from one energy level to another. However, the energy jumps are usually so large that the absorption lies in the UV region. Special circumstances can make it possible to obtain small jumps in electronic energy which appear as absorption in the visible region.

Polarization

NaCl, NaBr and NaI are all ionic, and are all colourless. AgCl is also colourless. Thus the halide ions Cl^-, Br^- and I^-, and the metal ions Na^+ and Ag^+, are typically colourless. However, AgBr is pale yellow and AgI is yellow. The colour arises because the Ag^+ ion polarizes the halide ions. This means that it distorts the electron cloud, and implies a greater covalent contribution. The polarizability of ions increases with size: thus I^- is the most polarized, and is the most coloured. For the same reason Ag_2CO_3 and Ag_3PO_4 are yellow, and Ag_2O and Ag_2S are black.

Incompletely filled *d* or *f* shell

Colour may arise from an entirely different cause in ions with incomplete *d* or *f* shells. This source of colour is very important in most of the transition metal ions.

In a free isolated gaseous ion the five *d* orbitals are degenerate, that is they are identical in energy. In real life situations the ion will be surrounded by solvent molecules if it is in solution, by other ligands if it is in a complex, or by other ions if it is in a crystal lattice. The surrounding groups affect the energy of some *d* orbitals more than others. Thus the *d* orbitals are no longer degenerate, and at their simplest they form two groups of orbitals of different energy. Thus in transition element ions with a partly filled *d* shell it is possible to promote electrons from one *d* level to another *d* level of higher energy. This corresponds to a fairly small energy difference, and so light is absorbed in the visible region. The colour of a transition metal complex is dependent on how big the energy difference is between the two *d* levels. This in turn depends on the nature of the ligand, and on the type of complex formed. Thus the octahedral complex $[Ni(NH_3)_6]^{2+}$ is blue, $[Ni(H_2O)_6]^{2+}$ is green and $[Ni(NO_2)_6]^{4-}$ is brown–red. The colour changes with the ligand used. The colour also depends on the number of ligands and the shape of the complex formed.

The source of colour in the lanthanides and the actinides is very similar, arising from $f \rightarrow f$ transitions. With the lanthanides the 4*f* orbitals are deeply embedded inside the atom, and are well shielded by the 5*s* and 5*p* electrons. The *f* electrons are practically unaffected by complex formation: hence the colour remains almost constant for a particular ion regardless of the ligand. The absorption bands are also very narrow.

Some compounds of the transition metals are white, for example $ZnSO_4$ and TiO_2. In these compounds it is not possible to promote electrons within the *d* level. Zn^{2+} has a d^{10} configuration and the *d* level is full. Ti^{4+} has a d^0 configuration and the *d* level is empty. In the series Sc(+III), Ti(+IV), V(+V), Cr(+VI) and Mn(+VII), these ions may all be considered to have an empty *d* shell: hence *d–d* spectra are impossible and they should be colourless. However, as the oxidation number increases these states become increasingly covalent. Rather than form highly charged simple ions, oxoions are formed TiO^{2+}, VO_2^+, VO_4^{3-}, CrO_4^{2-} and MnO_4^-. $VO_4^- \cdot VO_2^+$ is pale yellow, but CrO_4^{2-} is strongly yellow coloured, and MnO_4^- has an intense purple colour in solution though the solid is almost black. The colour arises by charge transfer. In MnO_4^- an electron is momentarily transferred from O to the metal, thus momentarily changing O^{2-} to O^- and reducing the oxidation state of the metal from Mn(VII) to Mn(VI). Charge transfer requires that the energy levels on the two different atoms are fairly close. Charge transfer always produces intense colours since the restrictions of the Laporte and spin selection rules do not apply to transitions between atoms.

The *s*- and *p*-block elements do not have a partially filled *d* shell so there cannot be any *d–d* transitions. The energy to promote an *s* or *p* electron to a higher energy level is much greater and corresponds to ultraviolet light being absorbed. Thus the compound will not be coloured.

MAGNETIC PROPERTIES

When a substance is placed in a magnetic field of strength *H*, the intensity of the magnetic field in the substance may be greater than or less than *H*.

If the field in the substance is greater than *H*, the substance is paramagnetic. It is easier for magnetic lines of force to travel through a paramagnetic material than through a vacuum. Thus paramagnetic materials attract lines of force, and, if it is free to move, a paramagnetic material will move from a weaker to a stronger part of the field. Paramagnetism arises as a result of unpaired electron spins in the atom.

If the field in the substance is less than H, the substance is diamagnetic. Diamagnetic materials tend to repel lines of force. It is harder for magnetic lines of force to travel through diamagnetic materials than through a vacuum, and such materials tend to move from a stronger to a weaker part of a magnetic field. In diamagnetic compounds all the electron spins are paired. The paramagnetic effect is much larger than the diamagnetic effect.

It should be noted that Fe, Co and Ni are ferromagnetic. Ferromagnetic materials may be regarded as a special case of paramagnetism in which the moments on individual atoms become aligned and all point in the same direction. When this happens the magnetic susceptibility is greatly enhanced compared with what it would be if all the moments behaved independently. Alignment occurs when materials are magnetized, and Fe, Co and Ni can form permanent magnets. Ferromagnetism is found in several of the transition metals and their compounds. It is also possible to get antiferromagnetism by pairing the moments on adjacent atoms which point in opposite directions. This gives a magnetic moment less than would be expected for an array of independent ions. It occurs in several simple salts of Fe^{3+}, Mn^{2+} and Gd^{3+}. Since ferromagnetism and antiferromagnetism depend on orientation, they disappear in solution.

Many compounds of the transition elements are paramagnetic, because they contain partially filled electron shells. If the magnetic moment is measured, the number of unpaired electrons can be calculated. The magnetochemistry of the transition elements shows whether the d electrons are paired. This is of great importance in distinguishing between *high-spin* and *low-spin* octahedral complexes.

There are two common methods of measuring magnetic susceptibilities: the Faraday and the Gouy methods. The Faraday method is useful for measurements on a very small single crystal, but there are practical difficulties because the forces are very small. The Gouy method is more often used. Here the sample may be presented as a long rod of material, a solution, or a glass tube packed with powder. One end of the sample is placed in a uniform magnetic field and the other end in a very low or zero field. The forces observed here are much larger, and can be measured using a modified laboratory balance.

The volume susceptibility κ_1 of a compound is measured using a magnetic balance (Gouy balance). κ_1 is dimensionless, and is readily converted into the molar susceptibility χ_M, which has units $m^3\,mol^{-1}$. From this the magnetic moment of the compound μ can be calculated if the small diamagnetic contribution is ignored:

$$\mu^2 = \frac{3kT\chi_M}{N^\circ\mu_o}$$

where k is the Boltzmann constant ($1.3805 \times 10^{-23}\,J\,K^{-1}$)
μ_0 is the permeability of free space ($4\pi \times 10^{-7}\,H\,m^{-1}$)
T is the absolute temperature and N° is the Avogadro constant.

Thus μ has SI units JT^{-1}, and

$$\mu = \sqrt{(3k)/(N°\mu_o)} \, . \, \sqrt{\chi_M T}$$

It is convenient to express the magnetic moment μ in units of Bohr magnetons μ_B, where

$$\mu_B = \frac{eh}{4\pi m_e} = 9.273 \times 10^{-24}\, JT^{-1}$$

where e is the electronic charge, h is Planck's constant and m_e is the mass of an electron.

Thus the magnetic moment in Bohr magnetons becomes:

$$\frac{\mu}{\mu_B} = \text{constant} \, . \, \sqrt{\chi_M \, . \, T} \qquad (18.1)$$

where *constant* $= \sqrt{(3k)/(N°\mu_o)}/\mu_B = 797.5\, m^{3/2}\, mol^{-1/2}\, K^{1/2}$

The magnetic moment μ of a transition metal can give important information about the number of unpaired electrons present in the atom and the orbitals that are occupied, and sometimes indicates the structure of the molecule or complex. If the magnetic moment is due entirely to the spin of unpaired electrons μ_S, then

$$\mu_S = \sqrt{4S(S + 1)} \, . \, \mu_B$$

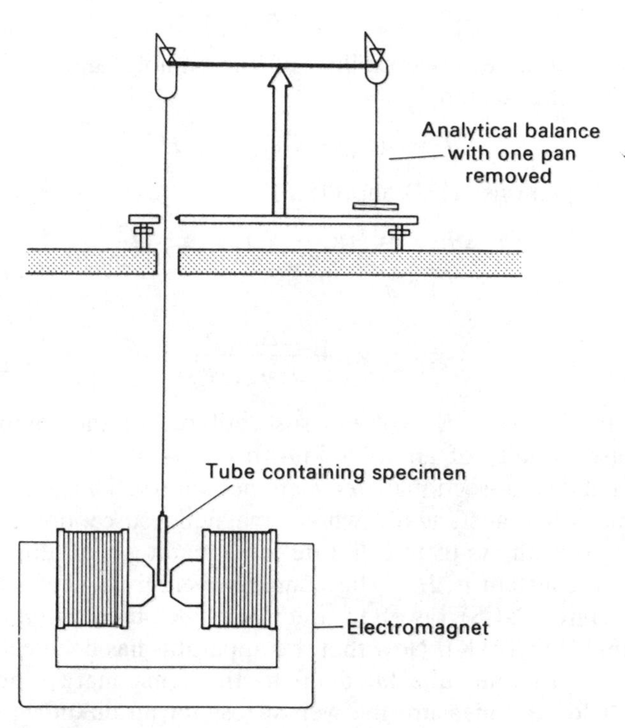

Figure 18.1 Diagram of a Gouy magnetic balance.

where S is the total spin quantum number. This gives the magnetic moment in SI units of JT^{-1}, and the magnetic moment in Bohr magnetons is given by $\sqrt{(4S(S+1))}$. This equation is related to the number of unpaired electrons n by the equation:

$$\mu_S = \sqrt{n(n+2)} \, . \, \mu_B$$

The object of the experiment is to determine the volume susceptibility $\varkappa$, by weighing a sample in and out of a magnetic field, and calculate in turn the molar susceptibility χ_M, the magnetic moment μ, the total spin quantum number S and eventually n the number of electrons responsible for the paramagnetism.

Measurement of magnetic moments

Let us examine how μ is obtained. First χ_M can be determined experimentally. A sample tube which has a narrow diameter and a flat bottom is filled with sample up to the calibration mark. The sample may be a finely divided solid, or a solution. The cross-sectional area of the tube is a. The sample and tube are weighed in the usual way. Then the apparent mass is measured again by weighing in the presence of a strong magnetic field of force H. A difference in weight occurs. If g is the acceleration due to gravity then the force F acting on the sample is given by:

$$F = \Delta m \, . \, g \tag{18.2}$$

If $\varkappa_1$ is the volume susceptibility of the sample and $\varkappa_2$ the volume susceptibility of air, then

$$F = \tfrac{1}{2}(\varkappa_1 - \varkappa_2) \, . \, a \, . \, \mu_o \, . \, H^2 \tag{18.3}$$

Combining equations (18.2) and (18.3)

$$\Delta m \, . \, g = \tfrac{1}{2}(\varkappa_1 - \varkappa_2) \, . \, a \, . \, \mu_o \, . \, H^2$$

hence

$$\varkappa_1 = \varkappa_2 + \frac{2\Delta m \, . \, g}{a \, . \, \mu_o \, . \, H^2}$$

We wish to calculate the volume susceptibility of the sample $\varkappa_1$. The volume susceptibility of air $\varkappa_2$ is known (0.364×10^{-12}), but the field strength H and cross-sectional area a are not known. Thus we carry out the experiment using a standard whose magnetic susceptibility is known accurately. This allows us to calibrate the apparatus, and thus deduce the value of the constant $a \, . \, H^2$. The complex mercury(II) tetrathiocyanatocobalt(II) $Hg[Co(NCS)_4]$ is often used as a solid standard ($\chi_M = 206.6 \times 10^{-12} \, m^3 \, mol^{-1}$ at 293 K). Now that the apparatus has been calibrated, we use the same sample tube filled up to the same mark, and the same magnetic field, and measure the weight loss for an unknown compound. The volume susceptibility $\varkappa_1$ of the unknown compound can thus be obtained. This is readily converted into the molar susceptibility of the compound χ_M as follows:

$$\chi_M = \frac{\varkappa_1 . M}{D} = \frac{M}{D}\left[\frac{2\Delta m . g}{a . \mu_o . H^2}\right]$$

where M is the molar mass of the compound, and D is its density.

Diamagnetic materials have no unpaired electrons, and have a magnetic moment $\mu = 0$. The external magnetic field induces a small magnetic moment which is in opposition to the external field. Thus diamagnetic materials repel lines of force, and show a slight decrease in weight. In contrast paramagnetism arises where there is one or more unpaired electron in a compound. Paramagnetic materials attract lines of force and increase in weight, because the sample is pulled down into the gap between the pole-pieces of the magnetic balance (see Figure 18.1). For a transition metal complex the weight change measured with the Gouy balance is the sum of the effects from the paramagnetic metal ion and the diamagnetic ligands and ions present. Thus the value of χ_M derived from this weight change is the net magnetism, which is the sum of $\chi_{\text{paramagnetic}} + \chi_{\text{diamagnetic}}$. Since we wish to measure the paramagnetism of the metal ion, we must make a diamagnetic correction. The easiest way to make the correction for $\chi_{\text{diamagnetic}}$ is to add up the diamagnetic corrections (sometimes called Pascal's constants) for the atoms and ions in the molecule and contributions from multiple bonds. These data are known (Table 18.7).

$$\chi_{\text{diamagnetic}} = \Sigma\chi_{(\text{atom corrections})} + \Sigma\chi_{(\text{multiple bonds})}$$

Thus $\chi_{\text{paramagnetic}}$ can be obtained:

$$\chi_{\text{paramagnetic}} = \chi_{\text{measured}} - \chi_{\text{diamagnetic}}$$

Finally $\chi_{\text{paramagnetic}}$ is converted into magnetic moment of the metal ion in Bohr magnetons using equation 18.1:

$$\frac{\mu}{\mu_B} = 797.5 \cdot \sqrt{\chi_M . T}$$

The unpaired electron gives rise to a magnetic field because of its spin, and also because of the orbital angular momentum. The general equation for the magnetic moments of the first row of transition metal ions is:

$$\mu(S + L) = \sqrt{4S(S + 1) + L(L + 1)} . \mu_B$$

where S is the total of the spin quantum numbers, and L is the resultant of the orbital angular momentum quantum numbers of all the electrons in the molecule. (See coupling of orbital angular momenta and coupling of spins in Chapter 32.) For an electron the spin quantum number m_s has a value of $\pm\frac{1}{2}$: hence $S = m_s \cdot n$ where n is the number of unpaired electrons.

In many compounds of the first row transition elements, the orbital contribution is quenched by the electric fields of the surrounding atoms. As a slight approximation the orbital contributions can be ignored and the

Table 18.7 Some diamagnetic corrections (Pascal's constants) (All table values must be multiplied by 10^{-12}. Units are $m^3 mol^{-1}$)

NH_4^+	−167	OH^-	−151	H	−36	C=C	+30
Li^+	−12	O^{2-}	−151	C	−75	C=N	+105
Na^+	−85	F^-	−114	C (aromatic)	−79	C=O	+81
K^+	−187	Cl^-	−294	N	−70	N=N	+23
Rb^+	−269	Br^-	−435	N (aromatic)	−59	N=O	+22
Cs^+	−422	I^-	−636	P(+V)	−331	C≡C	+10
Mg^{2+}	−63	CO_3^{2-}	−370	O ether/ROH	−58	C≡N	+10
Ca^{2+}	−131	NO_2^-	−126	O_2 (in COOH)	−42		
Cr^{2+}	−188	NO_3^-	−326	S	−189		
Cr^{3+}	−138	SO_3^{2-}	−478	F	−79		
Mn^{2+}	−176	SO_4^{2-}	−503	Cl	−253		
Mn^{3+}	−126	BF_4^-	−490	Br	−395		
Fe^{2+}	−161	CN^-	−163	I	−561		
Fe^{3+}	−126	CNO^-	−264	acetate	−402		
Co^{2+}	−161	CNS^-	−390	oxalate	−427		
Co^{3+}	−126	ClO_3^-	−380	ethylenediamine	−578		
Ni^{2+}	−161	ClO_4^-	−402	NH_3	−226		
Cu^{2+}	−161	H_2O	−163	dipyridyl	−1319		
Zn^{2+}	−189			PPh_3	−2098		

observed magnetic moment may be considered to arise only from unpaired spins. The *spin-only* magnetic moment μ_S may be written:

$$\mu_S = \sqrt{4S(S+1)} \cdot \mu_B$$

The magnetic moment μ (measured in Bohr magnetons, BM) is related to the number of unpaired spins n by the equation:

$$\mu_S = \sqrt{n(n+2)} \cdot \mu_B$$

The spin-only results are shown in Table 18.8. The simple spin-only formula gives good agreement with many *high-spin* complexes of first row transition metals, shown in Table 18.9.

Table 18.8 Spin only magnetic moments for numbers of unpaired electrons

Number of unpaired electrons n	Magnetic moment μ_S (BM)	Total spin quantum number S
1	1.73	1/2
2	2.83	2/2 = 1
3	3.87	3/2
4	4.90	4/2 = 2
5	5.92	5/2

Table 18.9 Magnetic moments of some first row complexes

Ion	Number of unpaired electrons	Experimental magnetic moment (BM)	Calculated magnetic moment spin only formula μ_S (BM)
Ti^{3+}	1	1.7–1.8	1.73
V^{3+}	2	2.8–3.1	2.83
Cr^{3+}	3	3.7–3.9	3.87
Cr^{2+}, Mn^{3+}	4	4.8–4.9	4.90
Mn^{2+}, Fe^{3+}	5	5.7–6.0	5.92
Fe^{2+}	4	5.0–5.6	4.90
Co^{2+}	3	4.3–5.2	3.87
Ni^{2+}	2	2.9–3.9	2.83
Cu^{2+}	1	1.9–2.1	1.73

An example

This is best explained by an example. Magnetic measurements on $CuSO_4 \cdot 5H_2O$ at 293 K using a Gouy balance gave a value $\varkappa_1 = 1.70 \times 10^{-4}$. The molecular weight is 250.18, so the molar mass M is 0.250 kg mol^{-1}. The density D is 2.29 g cm^{-3} = 2.29×10^3 kg m^{-3}. The value for χ_M (not corrected for diamagnetic effects) is:

$$\chi_M = \frac{1.70 \times 10^{-4} \times 0.250\,\text{kg mol}^{-1}}{2.29 \times 10^3\,\text{kg m}^{-3}} = 1.858 \times 10^{-8}\,\text{mol}^{-1}\,\text{m}^3$$

The diamagnetic correction is obtained by adding together the contributions from each of the constituent ions and molecules given in Table 18.7:

$$\begin{array}{lr} Cu^{2+} & -161 \times 10^{-12} \\ SO_4^{2-} & -503 \times 10^{-12} \\ 5H_2O \quad 5 \times (-163 \times 10^{-12}) = & -815 \times 10^{-12} \\ \hline \chi_{\text{diamagnetic}} & -1479 \times 10^{-12} \\ & \text{or } -0.148 \times 10^{-8} \end{array}$$

The corrected value for χ_M is thus:

$$(1.856 + 0.148) \times 10^{-8}\,\text{mol}^{-1}\,\text{m}^3 = 2.004 \times 10^{-8}\,\text{mol}^{-1}\,\text{m}^3$$

Using equation (18.1) the true magnetic moment in Bohr magnetons is:

$$\frac{\mu}{\mu_B} = 797.5\sqrt{2.004 \times 10^{-8} \times 293}$$

$$= 1.93 \text{ Bohr magnetons}$$

Assuming the spin-only formula

$$\mu = \sqrt{n(n+2)}\,.\,\mu_B$$

If $n = 1$ $\quad \mu = \sqrt{1(1+2)}\,.\,\mu_B = 1.73$ BM

Thus Cu^{2+} in $CuSO_4 \cdot 5H_2O$ has one unpaired electron

Though the agreement for first row transition metal complexes given in Table 18.9 is generally very good, in some cases, for example Co^{2+}, the observed value for μ is higher than calculated by the spin-only formula. This suggests that there is also an orbital contribution. To have an orbital angular moment it must be possible to transform an orbital into an equivalent (degenerate) orbital by rotation. It is possible to transform the t_{2g} orbitals (d_{xy}, d_{xz} and d_{yz}) into each other by rotating 90°. It is not possible to transform the e_g orbitals in this way (e.g. the $d_{x^2-y^2}$ into the d_{z^2}), since they have different shapes. If the t_{2g} orbitals are all singly occupied, then it is not possible to transform the d_{xy} into d_{xz} or d_{yz} since they already contain an electron with the same spin. Similarly it is not possible to transform the t_{2g} orbitals if they are all doubly occupied. Thus configurations with $(t_{2g})^3$ or $(t_{2g})^6$ have no orbital contribution, but the others all have an orbital contribution. Thus in octahedral complexes the following arrangements have an orbital contribution:

$$(t_{2g})^1(e_g)^0 \quad (t_{2g})^2(e_g)^0 \quad (t_{2g})^4(e_g)^2 \quad (t_{2g})^5(e_g)^2$$

Co^{2+} has the $(t_{2g})^5$ $(e_g)^2$ configuration: hence the high value of μ is due to the orbital contribution. In a similar way an orbital contribution arises in tetrahedral cases with the following configurations:

$$(e)^2(t_2)^1 \quad (e)^2(t_2)^2 \quad (e)^4(t_2)^4 \quad (e)^4\ (t_2)^5$$

In the second and third row transition elements, and particularly in the lanthanide elements (where unpaired electrons occupy 4*f* orbitals), the orbital motion is not prevented or quenched. Thus the orbital contribution L must be included in the calculations. In some cases there is coupling between the spin contribution S and the orbital contribution L (spin orbit coupling, or Russell–Saunders coupling) to give a new quantum number J. In this case a more complicated formula is used. This is described in Chapter 29. The equations are.

$$\mu = g\sqrt{J(J+1)}\,.\,\mu_B$$

where

$$g = 1 + \frac{S(S+1) - L(L+1) + J(J+1)}{2J(J+1)}$$

Rearranging, this becomes

$$g = 1\tfrac{1}{2} + \frac{S(S+1) - L(L+1)}{2J(J+1)}$$

Using this equation the agreement between the observed and the calculated magnetic moments of the trivalent lanthanide elements is very close. For further details see Chapter 29. Spin orbit coupling gives rise to fine structure in absorption spectra. It splits the degenerate lower levels into a set of different levels. These levels may be populated by using thermal energy, giving rise to a temperature dependent magnetic moment.

Pierre Curie found that the measured magnetic susceptibility χ_M for paramagnetic materials varied with temperature. He put forward the Curie law that the paramagnetic susceptibility χ_M varies inversely with the absolute temperature, and

$$\chi_M = \frac{C}{T}$$

where C is a constant characteristic for the particular substance, called the Curie constant. Thus the magnetic field tends to align the moments of the paramagnetic atoms or ions, and thermal agitation tends to randomize them. Applying a statistical treatment:

$$\chi_M = \frac{\mu_o(N^\circ\mu^2)/(3k)}{T}$$

Thus the magnetic moment in Bohr magnetons is given by:

$$\frac{\mu}{\mu_B} = \sqrt{(3k/N^\circ)}/\mu_B \,.\, \sqrt{\chi_M \,.\, T}$$

hence

$$\frac{\mu}{\mu_B} = 797.5\sqrt{\chi_M \,.\, T}$$

The Curie law is obeyed with great accuracy by some systems such as $[FeF_6]^{3-}$. However, many paramagnetic materials deviate slightly from this ideal behaviour, and obey the Curie–Weiss law:

$$\chi_M = \frac{C}{T + \theta} \quad \text{and} \quad \mu = 797.5\sqrt{\chi_M \,.\, (T + \theta)} \,.\, \mu_B$$

(θ is called the Weiss constant and is an empirical quantity.)

CATALYTIC PROPERTIES

Many transition metals and their compounds have catalytic properties. Some of the more important ones are listed here:

$TiCl_4$	Used as the Natta catalyst in the production of polythene.
V_2O_5	Converts SO_2 to SO_3 in the Contact process for making H_2SO_4.
MnO_2	Used as a catalyst to decompose $KClO_3$ to give O_2.
Fe	Promoted iron is used in the Haber–Bosch process for making NH_3.
$FeCl_3$	Used in the production of CCl_4 from CS_2 and Cl_2.
$FeSO_4$ and H_2O_2	Used as Fenton's reagent for oxidizing alcohols to aldehydes.
$PdCl_2$	Waker process for converting $C_2H_4 + H_2O + PdCl_2$ to $CH_3CHO + 2HCl + Pd$.
Pd	Used for hydrogenation (e.g. phenol to cyclohexanone).
Pt/PtO	Adams catalyst, used for reductions.

Pt	Formerly used for $SO_2 \rightarrow SO_3$ in the Contact process for making H_2SO_4.
Pt	Is increasingly being used in three stage-convertors for cleaning car exhaust fumes.
Pt/Rh	Formerly used in the Ostwald process for making HNO_3 to oxidize NH_3 to NO.
Cu	Is used in the direct process for manufacture of $(CH_3)_2SiCl_2$ used to make silicones.
Cu/V	Oxidation of cyclohexanol/cyclohexanone mixtures to adipic acid which is used to make nylon-66.
$CuCl_2$	Deacon process of making Cl_2 from HCl.
Ni	Raney nickel, numerous reduction processes (e.g. manufacture of hexamethylenediamine, production of H_2 from NH_3, reducing anthraquinone to anthraquinol in the production of H_2O_2).
Ni complexes	Reppe synthesis (polymerization of alkynes, e.g. to give benzene or cyclooctatetraene).

In some cases the transition metals with their variable valency may form unstable intermediate compounds. In other cases the transition metal provides a suitable reaction surface.

Enzymes are catalysts that enhance the rates of specific reactions. They are proteins and are produced by living cells from amino acids. They work under mild conditions, often give 100% yields and may speed a reaction by 10^6 or 10^{12} times. Some enzymes require the presence of metal ions as *cofactors*, and these are called *metalloenzymes*. Many (but not all) metalloenzymes contain a transition metal. Some metalloenzymes are listed in Table 18.10.

NONSTOICHIOMETRY

A further feature of the transition elements is that they sometimes form nonstoichiometric compounds. These are compounds of indefinite structure and proportions. For example, iron(II) oxide FeO .sr4should be written with a bar over the formula $\overline{FeO}$ to indicate that the ratio of Fe and O atoms is not exactly 1 : 1. Analysis shows that the formula varies between $Fe_{0.94}O$ and $Fe_{0.84}O$. Vanadium and selenium form a series of compounds ranging from $VSe_{0.98}$ to VSe_2. These are given the formulae:

$$\overline{VSe} \quad (VSe_{0.98} \rightarrow VSe_{1.2})$$
$$\overline{V_2Se_3} \quad (VSe_{1.2} \rightarrow VSe_{1.6})$$
$$\overline{V_2Se_4} \quad (VSe_{1.6} \rightarrow VSe_2)$$

Nonstoichiometry is shown particularly among transition metal compounds of the Group VI elements (O, S, Se, Te). It is mostly due to the variable valency of transition elements. For example copper is precipitated from a solution containing Cu^{2+} by passing in H_2S. The sulphide is completely insoluble, but this is not used as a gravimetric method for

Table 18.10 Metalloenzymes and metalloproteins (metalloproteins in brackets)

Metal	Enzyme/metalloprotein	Biological function
Mo	Xanthine oxidase	Metabolism of purines
	Nitrate reductase	Utilization of nitrates
Mn^{II}	Arginase	Urea formation
	Phosphotransferases	Adding or removing PO_4^{3-}
Fe^{II} or Fe^{III}	Aldehyde oxidase	Oxidation of aldehydes
	Catalase	Decomposes H_2O_2
	Peroxidase	Decomposes H_2O_2
	Cytochromes	Electron transfer
	Ferredoxin	Photosynthesis
	(Haemoglobin)	O_2 transport in higher animals
	Succinic dehydrogenase	Aerobic oxidation of carbohydrates
Fe and Mo	Nitrogenase	Nitrogen fixation
Co	Glutamic mutase	Metabolism of amino acids
	Ribonucleotide reductase	Biosynthesis of DNA
Cu^{I} or Cu^{II}	Amine oxidases	Oxidation of amines
	Ascorbate oxidase	Oxidation of ascorbic acid
	Cytochrome oxidase	Principal terminal oxidase
	Galactose oxidase	Oxidation of galactose
	Lysine oxidase	Elasticity of aortic walls
	Dopamine hydroxylase	Producing noradrenaline to generate nerve impulses in the brain
	Tyrosinase	Skin pigmentation
	Ceruloplasmin	Utilization of Fe
	(Haemocyanin)	O_2 transport in invertebrates
	Plastocyanin	Photosynthesis
Zn^{II}	Alcohol dehydrogenase	Metabolism of alcohol
	Alkaline phosphatase	Releasing PO_4^{3-}
	Carbonic anhydrase	Regulation of pH and CO_2 formation
	Carboxypeptidase	Digestion of proteins

analysing for Cu because the precipitate is a mixture of CuS and Cu_2S. Sometimes nonstoichiometry is caused by defects in the solid structures.

ABUNDANCE

Three of the transition metals are very abundant in the earth's crust. Fe is the fourth most abundant element by weight, Ti the ninth and Mn the twelfth. The first row of transition elements largely follow Harkins' rule that elements with an even atomic number are in general more abundant than their neighbours with odd atomic numbers. Manganese is an exception. The second and third row elements are much less abundant than the first row. Tc does not occur in nature. Of the last six elements in the second and third rows (Tc, Ru, Rh, Pd, Ag, Cd; Re, Os, Ir, Pt, Au, Hg) none occurs to an extent of more than 0.16 parts per million (ppm) in the earth's crust.

Table 18.11 Abundance of the transition elements in the earth's crust, in ppm by weight

Sc	Ti	V	Cr	Mn	Fe	Co	Ni	Cu	Zn
25	6320	136	122	1060	60 000	29	99	68	76
Y	Zr	Nb	Mo	Tc	Ru	Rh	Pd	Ag	Cd
31	162	20	1.2	–	0.0001	0.0001	0.015	0.08	0.16
La	Hf	Ta	W	Re	Os	Ir	Pt	Au	Hg
35	2.8	1.7	1.2	0.0007	0.005	0.001	0.01	0.004	0.08

DIFFERENCES BETWEEN THE FIRST ROW AND THE OTHER TWO ROWS

Metal–metal bonding and cluster compounds

Metal–metal (M–M) bonding occurs not only in the metals themselves, but also in some compounds. M–M bonding is quite rare in the first row transition elements. It occurs only in a few carbonyl compounds such as $Mn_2(CO)_{10}$, $Fe_2(CO)_9$, $Co_2(CO)_8$, $Fe_3(CO)_{12}$ and $Co_4(CO)_{12}$, and in carboxylate complexes such as chromium(II) acetate $Cr_2(CH_3COO)_4(H_2O)_2$, and in solid nickel dimethylglyoxime.

In the second and third row elements M–M bonds are much more common:

1. They form carbonyls with M–M bonds similar to those from the first row such as $Ru_3(CO)_{12}$, $Os_3(CO)_{12}$, $Rh_4(CO)_{12}$ and $Ir_4(CO)_{12}$, and a type not formed by the first row $Rh_6(CO)_{16}$.
2. The metals Mo, Ru and Rh form binuclear carboxylate complexes such as $Mo_2(CH_3COO)_4(H_2O)_2$ which are similar to chromium(II) acetate.
3. The halide ions $[Re_2Cl_8]^{2-}$ and $[Mo_2Cl_9]^{3-}$ also have M–M bonds.
4. The lower halides of several elements have a group of three or six metal atoms bonded together and are called cluster compounds. The elements are:

   Nb Mo
   Ta W Re

   $[Nb_6Cl_{12}]^{2+}$ and $[Ta_6Cl_{12}]^{2+}$ have unusual structures. Both contain six metal atoms arranged in a cluster at the corners of an octahedron, with 12 bridging halogen atoms across the corners. There is extensive M–M bonding within the octahedron. The so-called 'dihalides' of Mo and W are really Mo_6Cl_{12} and W_6Br_{12}, and these contain the $[M_6X_8]^{4+}$ ion. This too has a remarkable structure. Six metal atoms are arranged in a cluster at the corners of an octahedron, with eight halogen atoms located above each of the eight faces of the octahedron and 'bonded' to three metal atoms. $ReCl_3$ is really trimeric Re_3Cl_9, and comprises a triangle of three Re atoms with three bridging halogen atoms across the three corners, and six halogen atoms that bridge to other Re_3Cl_9 units.

Stability of oxidation states

The (+II) and (+III) states are important for all the first row transition elements. Simple ions M^{2+} and M^{3+} are common with the first row but are less important for second and third row elements, which have few ionic compounds. Similarly the first row form a large number of extremely stable complexes such as $[Cr^{III}Cl_6]^{3-}$ and $[Co^{III}(NH_3)_6]^{3+}$. No equivalent complexes of Mo or W, or Rh or Ir, are known.

The higher oxidation states of the second and third row elements are more important and much more stable than those of the first row elements. Thus the chromate ion $[CrO_4]^{2-}$ is a strong oxidizing agent but molybdate $[MoO_4]^{2-}$ and tungstate $[WO_4]^{2-}$ are stable. Similarly the permanganate ion $[MnO_4]^-$ is a strong oxidizing agent but pertechnate $[TcO_4]^-$ and perrhenate $[ReO_4]^-$ ions are stable.

Some compounds exist in high oxidation states which have no counterparts in the first row, for example WCl_6, ReF_7, RuO_4, OsO_4 and PtF_6.

Complexes

The coordination number 6 is widespread in the transition elements, giving an octahedral structure. The coordination number 4 is much less common, giving tetrahedral and square planar complexes. Coordination numbers of 7 and 8 are uncommon for the first row but are much more common in the early members of the second and third rows. Thus in $Na_3[ZrF_7]$ the $[ZrF_7]^{3-}$ is a pentagonal bipyramid, and in $(NH_4)_3[ZrF_7]$ it is a capped trigonal prism. In $Cu_2[ZrF_8]$ the Zr is at the centre of a square antiprism.

Size

The second row elements are all larger than the first row elements. Because of the lanthanide contraction the radii of the third row are almost the same as those for the second row.

Magnetism

When transition elements form octahedral complexes, the d levels are split into t_{2g} and e_g sub-levels. Consider a first row element. If the ligands possess a strong field they cause a large difference in energy between these two sub-levels. Thus the electrons occupying the d level fill the lower t_{2g} level even if this means they must be paired. If pairing of electrons occurs, the complex is called *low spin* or *spin paired*. Alternatively if the ligands have only a weak field the splitting is small, and only when each of the t_{2g} and e_g levels contains one electron does pairing of spins occur. Such complexes are called *high spin* or *spin free*. Thus with a first row element the strength of the ligand field determines whether a low-spin or a high-spin complex is formed.

The second and third row transition elements tend to give low-spin complexes, that is it is more favourable in terms of energy to pair electrons in the lower energy *d* levels rather than use the higher levels, regardless of the ligand.

The spin only formula gives reasonable agreement relating the observed magnetic moment of first row transition metal complexes to the number of unpaired electrons. For the second and third row transition elements the orbital contribution is significant, and in addition spin orbit coupling may occur. Thus the spin only approximation is no longer valid, and more complicated equations must be used. Thus the simple interpretation of magnetic moments in terms of the number of unpaired electrons cannot be extended from the first row of transition elements to the second and third rows. The second and third rows also show extensive temperature dependent paramagnetism. This is explained by the spin orbit coupling removing the degeneracy from the lowest energy level in the ground state.

Abundance

The ten first row transition elements are reasonably common and make up 6.79% of the earth's crust. The remaining transition elements are mostly very scarce. Even though the abundance of Zr is 162 ppm, La 31 ppm, Y 31 ppm and Nb 20 ppm, the 20 elements in the second and third rows elements together make up only 0.025% of the earth's crust. Tc does not occur in nature.

FURTHER READING

Bevan, D.J.M. (1973) *Comprehensive Inorganic Chemistry*, Vol. 3 (Chapter 49: Nonstoichiometric compounds), Pergamon Press, Oxford. (A good introduction to nonstoichiometric compounds.)

Brown, D. (1968) *Halides of the Transition Elements*, Vol. I (Halides of the lanthanides and actinides), Wiley, London and New York.

Canterford, J.H. and Cotton, R. (1968) *Halides of the Second and Third Row Transition Elements*, Wiley, London.

Canterford, J.H. and Cotton, R. (1969) *Halides of the First Row Transition Elements*, Wiley, London.

Corbett, J.D. (1981) Extended metal–metal bonding in halides of the early transition metals, *Acc. Chem. Res.* **14**, 239.

Cotton, F.A. (1983) Multiple metal–metal bonds, *J. Chem. Ed.*, **60**, 713–720.

Crangle, J. (1977) *The Magnetic Properties of Solids*, Arnold, London.

Diatomic metals and metallic clusters (1980) (Conference papers of the Faraday Symposia of the Royal Society of Chemistry, No. 14), Royal Society of Chemistry, London.

Earnshaw, A. (1968) *Introduction to Magnetochemistry*, Academic Press, London.

Emeléus, H.J. and Sharpe, A.G. (1973) *Modern Aspects of Inorganic Chemistry*, 4th ed. (Chapters 14 and 15: Complexes of Transition Metals; Chapter 16: Magnetic properties), Routledge and Kegan Paul, London.

Figgis, B.N. and Lewis, J. (1960) The magnetochemistry of complex compounds, Chapter 6 in *Modern Inorganic Chemistry* (ed. Lewis, J. and Wilkins, R.G.), Interscience, New York and Wiley, London, 1960. (A good account of measuring magnetic moments).

Greenwood, N.N. (1968) *Ionic Crystals, Lattice Defects and Non-Stoichiometry*, Butterworths, London.

Griffith, W.P. (1973) *Comprehensive Inorganic Chemistry*, Vol. 4 (Chapter 46: Carbonyls, cyanides, isocyanides and nitrosyls), Pergamon Press, Oxford.
Johnson, B.F.G. (1976) The structures of simple binary carbonyls, *JCS Chem. Comm.*, 211–213.
Johnson, B.F.G. (ed.) (1980) *Transition Metal Clusters*, John Wiley, Chichester.
Lewis, J. (1988) Metal clusters revisited, *Chemistry in Britain*, **24**, 795–800.
Lewis, J. and Green, M.L. (eds) (1983) *Metal Clusters in Chemistry* (Proceedings of the Royal Society Discussion Meeting May 1982), The Society, London.
Lever, A.B.P. (1984) *Inorganic Electronic Spectroscopy*, Elsevier, Amsterdam. (Up-to-date and comprehensive, and a good source of spectral data.)
Moore, P. (1982) Colour in transition metal chemistry, *Education in Chemistry*, **19**, 10–11, 14.
Muetterties, E.L. (1971) *Transition Metal Hydrides*, Marcel Dekker, New York.
Muetterties, E.L. and Wright, C.M. (1967) High coordination numbers, *Q. Rev. Chem. Soc.*, **21**, 109.
Nyholm, R.S. (1953) Magnetism and inorganic chemistry, *Q. Rev. Chem. Soc.*, **7**, 377.
Nyholm, R.S. and Tobe, M.L. (1963) The stabilization of oxidation states of the transition metals, *Adv. Inorg. Radiochem.*, **5**, 1.
Shriver, D.H., Kaesz, H.D. and Adams, R.D. (eds) (1990) *Chemistry of Metal Cluster Complexes*, VCH, New York.
Sharpe, A.G. (1976) *Chemistry of Cyano Complexes of the Transition Metals*, Academic Press, London, 1976.

PROBLEMS

1. How would you define a transition element? List the properties associated with transition elements.

2. How do the following properties vary in the transition elements:
   (a) ionic character
   (b) basic properties
   (c) stability of the various oxidation states
   (d) ability to form complexes?

3. Give examples of, and suggest reasons for, the following features of transition metal chemistry:
   (a) The lowest oxide of a transition metal is basic whereas the highest oxide is usually acidic.
   (b) A transition metal usually exhibits higher oxidation states in its fluorides than in its iodides.
   (c) The halides become more covalent with increasing oxidation state of the metal and are more susceptible to hydrolysis.

4. Write notes on the following:
   (a) The effective atomic number rule
   (b) Ligands which stabilize low oxidation states
   (c) Back bonding in metal carbonyls

5. Describe the methods by which extremely pure samples of the metals may be prepared.

6. Which of the M^{2+} and M^{3+} ions of the first row transition elements are

stable in aqueous solution, which are oxidizing and which are reducing?

7. Explain why certain ligands such as F^- tend to bring out the maximum oxidation state of an element, whilst others such as CO and dipyridyl bring out the lowest oxidation states.
8. Give reasons why carbonyl and cyanide complexes of the later transition elements Cr, Mn, Fe, Co, Ni are more stable, more common, and more likely to exist than similar compounds of the *s*-block or early transition elements.
9. Why do the second and third rows of transition elements resemble each other much more closely than they resemble the first row?
10. What do you understand by the terms paramagnetism and diamagnetism? Predict the magnetic moment for octahedral complexes of Fe^{2+} with strong field ligands and with weak field ligands.

The scandium group

19

Table 19.1 Electronic structures and oxidation states

Element		Electronic structure	Oxidation states
Scandium	Sc	[Ar] $3d^1\ 4s^2$	III
Yttrium	Y	[Kr] $4d^1\ 5s^2$	III
Lanthanum	La	[Xe] $5d^1\ 6s^2$	III
Actinium	Ac	[Rn] $6d^1\ 7s^2$	III

INTRODUCTION

These four elements are sometimes grouped with the 14 lanthanides and called collectively the 'rare earths'. This is a misnomer because the scandium group are *d*-block elements and the lanthanides are *f*-block elements. In addition the scandium group is by no means rare, except for actinium which is radioactive. The trends in properties in the family Sc, Y, La and Ac are quite regular, and similar to the trends in Groups I and II. There are few important industrial uses of the elements or their compounds apart from Mischmetal, which is used in the metallurgical industries.

OCCURRENCE, SEPARATION, EXTRACTION AND USES

Sc is the thirty-first most abundant element by weight in the earth's crust. It is thinly distributed. It occurs as the rather rare mineral thortveitite $Sc_2[Si_2O_7]$. It is available as a by-product from the extraction of U. There is very little use for Sc or its compounds.

Y and La are the twenty-ninth and twenty-eighth most abundant elements. They are found together with the lanthanide elements in bastnaesite $M^{III}CO_3F$, monazite $M^{III}PO_4$ and other minerals. It is extremely difficult to separate the individual elements. This is covered in some detail in Chapter 29. It is also difficult to extract the metals from their compounds. The metals are electropositive and react with water. Their oxides are very stable so that a thermite reaction cannot be used (Al_2O_3

continued overleaf

Table 19.2 Abundance of the elements in the earth's crust, by weight

	ppm	Relative abundance
Sc	25	31
Y	31	29
La	35	28
Ac	trace	

enthalpy of formation 1675 kJ mol^{-2}, La_2O_3 enthalpy of formation 1884 kJ mol^{-2}). The metals can be obtained by reduction of the chlorides or fluorides with calcium at 1000 °C, under an atmosphere of argon.

Small amounts of Y are used to make the red phosphor for TV tubes. Some is used to make synthetic garnets used in radar and as gemstones. In contrast about 5000 tonnes of La is produced annually, mostly as Mischmetal. This is an unseparated mixture of La and the lanthanide metals (50% Ce, 40% La, 7% Fe, 3% other metals). It is used extensively to improve the strength and workability of steel, and also in Mg alloys. Small amounts are used as 'lighter flints'. La_2O_3 is used in optical glass such as Crooke's lenses, which give protection against UV light.

Traces of Ac are always found associated with U and Th, as it is formed as a decay product in both the thorium and actinium natural radioactive decay series (Chapter 31).

$$^{232}_{90}Th \xrightarrow{\alpha} {}^{228}_{88}Ra \xrightarrow{\beta} {}^{228}_{89}Ac$$

$$^{235}_{92}U \xrightarrow{\alpha} {}^{231}_{90}Th \xrightarrow{\beta} {}^{231}_{91}Pa \xrightarrow{\alpha} {}^{227}_{89}Ac \xrightarrow{\beta} {}^{227}_{90}Th$$

Ac can be made by irradiating Ra with neutrons in a nuclear reactor:

$$^{226}_{88}Ra + {}^{1}_{0}n \rightarrow {}^{227}_{88}Ra \xrightarrow{\beta} {}^{227}_{89}Ac$$

At best this produces milligram quantities: separation from other elements is performed by ion exchange. $^{228}_{89}Ac$ and $^{227}_{89}Ac$ are the only naturally occurring isotopes, and both are intensely radioactive. The half lives of $^{228}_{89}Ac$ and $^{227}_{89}Ac$ are 6 hours and 21.8 years respectively. Thus any Ac present when the earth was formed will have long since decayed. Any found now must have been produced fairly recently by radioactive decay of another element. This explains the scarcity of naturally occurring Ac. The high activity has limited the study of the chemistry of the element.

OXIDATION STATE

The elements always exist in the oxidation state (+III), and occur as M^{3+} ions. The formation of M^{3+} requires the removal of the two *s* and one *d* electron. Thus the ions have a d^0 configuration, and *d*–*d* spectra are impossible. As a result the ions, and their compounds, are colourless and diamagnetic. The sum of the first three ionization energies for scandium is a little less than the sum for aluminium. The properties of scandium are similar in some ways to those of aluminium.

SIZE

The covalent and ionic radii of the elements increase regularly on descending the group as in the *s*-block. In the groups of transition elements shown in Table 19.3 the second and third row elements are almost identical in size because of the lanthanide contraction. However, this happens after La.

Table 19.3 Some physical properties

Element	Covalent radius	Ionic radius M^{3+}	Ionization energies ($kJ\,mol^{-1}$)			Standard electrode potential	Melting point	Pauling's electro-negativity
	(Å)	(Å)	1st	2nd	3rd	$E°$ (V)	(°C)	
Sc	1.44	0.745	631	1235	2393	−2.08	1539	1.3
Y	1.62	0.900	616	1187	1968	−2.37	1530	1.2
La	1.69	1.032	541	1100	1852	−2.52	920	1.1
Ac		1.12				−2.6	817	1.1

CHEMICAL PROPERTIES

The metals have moderately high standard electrode potentials. They are quite reactive, and reactivity increases with increased size. They tarnish in air and burn in oxygen, giving oxides M_2O_3. Y forms a protective oxide coating in air, which makes it unreactive.

$$2La + 3O_2 \rightarrow 2La_2O_3$$

The metals react slowly with cold water, but more rapidly with hot water, liberating hydrogen and forming either the basic oxide or the hydroxide.

$$2La + 6H_2O \rightarrow 2La(OH)_3 + 3H_2$$

$$La(OH)_3 \rightarrow \underset{\text{basic oxide}}{LaO \cdot OH} + H_2O$$

$Sc(OH)_3$ appears not to exist as a definite compound, but the basic oxide $ScO \cdot OH$ is well established, and is amphoteric like $Al(OH)_3$. Since scandium is amphoteric, it dissolves in NaOH, liberating H_2.

$$Sc + 3NaOH + 3H_2O \rightarrow Na_3[Sc(OH)_6]^{3-} + 1\tfrac{1}{2}H_2$$

Basic properties of the oxides and hydroxides increase on descending the group, and $Y(OH)_3$ and $La(OH)_3$ are basic. Thus the oxides and hydroxides form salts with acids, and $Y(OH)_3$ and $La(OH)_3$ react with carbon dioxide:

$$2Y(OH)_3 + 3CO_2 \rightarrow Y_2(CO_3)_3 + 3H_2O$$

$La(OH)_3$ is a strong enough base to liberate NH_3 from ammonium salts. Because the oxides (and hydroxides) are either amphoteric or weak bases, their oxosalts can be decomposed to oxides by heating. This is similar to the behaviour of Group II elements, but the decomposition occurs more easily, i.e. at a lower temperature.

$$2Y(OH)_3 \xrightarrow{\text{heat}} Y_2O_3 + 3H_2O$$
$$Y_2(CO_3)_3 \longrightarrow Y_2O_3 + 3CO_2$$
$$2Y(NO_3)_3 \longrightarrow Y_2O_3 + 6NO_2 + 1\tfrac{1}{2}O_2$$
$$Y_2(SO_4)_3 \longrightarrow Y_2O_3 + 3SO_2 + 1\tfrac{1}{2}O_2$$

The metals react with the halogens, forming trihalides MX_3. These resemble the halides of Ca. The fluorides are insoluble (like CaF_2) and the other halides are deliquescent and very soluble (like $CaCl_2$). If the chlorides are prepared in solution, they crystallize as hydrated salts. Heating these hydrated salts does not yield anhydrous halides. On heating, $ScCl_3(H_2O)_7$ decomposes to the oxide, whilst the others give oxohalides.

$$2ScCl_3 \cdot (H_2O)_7 \xrightarrow{\text{heat}} Sc_2O_3 + 6HCl + 4H_2O$$
$$YCl_3 \cdot (H_2O)_7 \longrightarrow YOCl + 2HCl + 6H_2O$$
$$LaCl_3 \cdot (H_2O)_7 \longrightarrow LaOCl + 2HCl + 6H_2O$$

Anhydrous chlorides can be made from the oxides with NH_4Cl:

$$Sc_2O_3 + 6NH_4Cl \xrightarrow{300\,^\circ C} 2ScCl_3 + 6NH_3 + 3H_2O$$

Anhydrous $ScCl_3$ differs from $AlCl_3$ as $ScCl_3$ is monomeric whilst $(AlCl_3)_2$ is dimeric. In addition $ScCl_3$ shows no Friedel–Crafts catalytic properties.

The salts generally resemble those of calcium, and the fluorides, carbonates, phosphates and oxalates are insoluble.

The elements all react with hydrogen on heating to 300 °C, forming highly conducting compounds of formula MH_2. These do not contain M^{2+}, but probably contain M^{3+} and $2H^-$, and have the extra electron in a conduction band. Except for Sc they can absorb more H_2 and lose their conducting power, forming compounds which are not quite stoichiometric but approach the composition MH_3. The exact composition depends on the temperature and the pressure of the hydrogen. The hydrides react with water, liberating hydrogen, and are salt-like (ionic) hydrides containing the hydride ion H^-.

Scandium forms a carbide ScC_2 when the oxide is heated with carbon in an electric furnace. The carbide reacts with water, liberating ethyne.

$$Sc_2O_3 + C \xrightarrow{1000\,^\circ C} 2ScC_2 \xrightarrow{H_2O} C_2H_2 + ScO \cdot OH$$

At one time the carbide was thought to contain Sc(+II) and to contain Sc^{2+} and $(C{\equiv}C)^{2-}$. Magnetic measurements suggest that it contains Sc^{3+}

ions and C_2^{2-} ions and the free electrons are delocalized into a conduction band, thus giving some metallic conduction.

COMPLEXES

Despite the charge of +3, the metal ions in this group do not have a strong tendency to form complexes. This is because of their fairly large size. Sc^{3+} is the smallest ion in the group and forms complexes more readily than the other elements. These include $[Sc(OH)_6]^{3-}$ and $[ScF_6]^{3-}$, both of which have a coordination number of 6 and are octahedral in shape.

$$ScF_3 + 3NH_4F \rightarrow 3NH_4^+ + [ScF_6]^{3-}$$

By far the most common donor atom in complexes is O, and complexes occur particularly with chelating ligands. Thus complexes are formed with strong complexing agents such as oxalic acid, citric acid, acetylacetone and EDTA. The complexes of the larger metals Y and La often have higher coordination numbers than 6, and coordination number 8 is the most common. Thus [Sc(acetylacetone)$_3$] has a coordination number of 6 and is octahedral. Y(acetylacetone)$_3$(H_2O)] has a coordination number of 7 and the structure is a capped trigonal prism. In [La(acetylacetone)$_3 \cdot (H_2O)_2$] the coordination number is 8 and the structure is a distorted square antiprism. In $[La \cdot EDTA \cdot (H_2O)_4] \cdot 3H_2O$ the EDTA forms four bonds from O atoms and two from N atoms, and the O atoms in the four water molecules all form bonds to La, giving a coordination number of 10. The ions NO_3^- and SO_4^{2-} also act as bidentate ligands and form complexes with high coordination numbers. In $[Sc(NO_3)_5]^{2-}$ four NO_3^- groups are bidentate and one is unidentate giving a coordination number of 9. In $[Y(NO_3)_5]^{2-}$ all five NO_3^- groups are bidentate and the coordination number is 10. In $La_2(SO_4)_3 \cdot 9H_2O$ half of the La atoms are 12-coordinate.

Lanthanum salts have been used as a biological tracer. La^{3+} appears to replace Ca^{2+} in the conduction of nerve impulses along the axons of nerve cells, and also in structure promoting in cell membranes. La^{3+} is readily detected by electron spin resonance. The similarity in biological role may arise because the ions are similar in size (La^{3+} = 1.032 Å, Ca^{2+} = 1.00 Å).

FURTHER READING

Callow, R.J. (1967) *The Industrial Chemistry of the Lanthanons, Yttrium, Thorium and Uranium*, Pergamon Press, New York.

Horovitz, C.T. (ed.) (1975) *Scandium: Its Occurrence, Chemistry, Physics, Metallurgy, Biology and Technology*, Academic Press, London.

Melson, G.A. and Stotz, R.W. (1971) The coordination chemistry of scandium, *Coordination Chem. Rev.*, **7**, 133.

Vickery, R.C. (1973) *Comprehensive Inorganic Chemistry*, Vol. 3 (Chapter 31: Scandium, yttrium, lanthanum), Pergamon Press, Oxford.

20 The titanium group

Table 20.1 Electronic structures and oxidation states

Element		Electronic structure	Oxidation states*
Titanium	Ti	[Ar] $3d^2\ 4s^2$	(−I) (0) (II) III **IV**
Zirconium	Zr	[Kr] $4d^2\ 5s^2$	(II) (III) **IV**
Hafnium	Hf	[Xe] $4f^{14}\ 5d^2\ 6s^2$	(II) (III) **IV**

* The most important oxidation states (generally the most abundant and stable) are shown in bold. Other well-characterized but less important states are shown in normal type. Oxidation states that are unstable, or in doubt, are given in parentheses.

INTRODUCTION

Titanium is a commercially important element. Vast quantities of TiO_2 are used as a pigment and filler, and Ti metal is important for its strength, low density, and corrosion resistance. $TiCl_3$ is important as a Ziegler–Natta catalyst for making polythene and other polymers. Zirconium is used to make the cladding for fuel rods in water cooled nuclear reactors. Hafnium is used to make control rods for certain reactors.

OCCURRENCE AND ABUNDANCE

Ti is the ninth most abundant element by weight in the earth's crust. The main ores are ilmenite $FeTiO_3$ and rutile TiO_2. In 1988 world production was 7.8 million tonnes of ilmenite and 456 000 tonnes of rutile. These had a TiO_2 content of 4.3 million tonnes. The main producers are Canada 32%, Australia 24%, Norway 11.5%, and the USSR and Malaysia 6% each. Zr is the eighteenth most abundant element, and is found mainly as zircon $ZrSiO_4$ and small amounts of baddeleyite ZrO_2. World production of zirconium minerals was 918 000 tonnes in 1988. Hf is very similar in size and properties to Zr because of the lanthanide contraction. Thus Hf occurs to the extent of 1–2% in Zr ores. It is particularly difficult to separate Zr and Hf – even more difficult than separating the lanthanides.

Table 20.2 Abundance of the elements in the earth's crust, by weight

	ppm	Relative abundance
Ti	6320	9
Zr	162	18
Hf	2.8	45

EXTRACTION AND USES

Ti has been called 'the wonder metal' because of its unique and useful properties. It is very hard, high melting (1667 °C) and is stronger and much lighter than steel (densities Ti = $4.4\,g\,cm^{-3}$, Fe = $7.87\,g\,cm^{-3}$). However, even traces of non-metal impurities, for example H, C, N or O, make Ti, and also the other two metals Zr and Hf, brittle. Ti has better corrosion resistance than stainless steel. It is a better conductor of heat and electricity than the Sc group metals. Ti metal and alloys of Ti with Al are used extensively in the aircraft industry in jet and gas turbine engines, and in airframes. Supersonic aircraft like Concorde can use Al as the structural

Table 20.3 Some physical properties

Element	Covalent radius (Å)	Ionic radius M^{4+} (Å)	Melting point (°C)	Boiling point (°C)	Density ($g\,cm^{-3}$)	Pauling's electro-negativity
Ti	1.32	0.605	1667	3285	4.50	1.5
Zr	1.45	0.72	1857	4200	6.51	1.4
Hf	1.44	0.71	2222	4450	13.28	1.3

skin (m.p. 660 °C) by limiting the speed to Mach 2.2 (2.2 times the speed of sound). If and when SST (supersonic transport) planes are made which fly at three times the speed of sound, it is probable that they will be made of Ti (m.p. 1660 °C). Ti is also used in marine equipment and in chemical plant. Small amounts of Ti alloyed with steel harden and toughen the steel. World production of Ti metal is almost 100 000 tonnes per year.

The metal is difficult to extract because of its high melting point and because it reacts readily with air, oxygen, nitrogen, and hydrogen at elevated temperatures. The oxide cannot be reduced by C or CO because it forms carbides. Since TiO_2 is very stable, the first stage is to form $TiCl_4$ by heating it with C and Cl_2 at 900 °C.

$$TiO_2 + 2C + 2Cl_2 \rightarrow TiCl_4 + 2CO$$

$$2FeTiO_3 + 6C + 7Cl_2 \rightarrow 2TiCl_4 + 6CO + 2FeCl_3$$

$TiCl_4$ is a liquid (b.p. 137 °C) and is removed from $FeCl_3$ and other impurities by distillation. One of the following methods is then used.

continued overleaf

Kroll process

Originally Wilhelm Kroll produced Ti by reducing $TiCl_4$ with Ca in an electric furnace. Later Mg was used, and Imperial Metal Industries (IMI) use Na instead. At this temperature Ti is highly reactive and reacts readily with air or N_2. It is therefore necessary to perform the reaction under an atmosphere of argon.

$$TiCl_4 + 2Mg \xrightarrow{1000-1150\,°C} Ti + 2MgCl_2$$

The $MgCl_2$ formed can be removed by leaching with water, or better with dilute HCl as this dissolves any excess Mg. Alternatively the $MgCl_2$ is removed by vacuum distillation. This leaves a sponge of Ti rather than a solid block. The Ti is converted to the massive form by melting in an electric arc under a high vacuum or an atmosphere of argon. The IMI process is almost the same. $TiCl_4$ is reduced by Na under an atmosphere of argon, and NaCl is leached out with water. The Ti is in the form of small granules. These can be fabricated into metal parts using 'powder forming' techniques and sintering in an inert atmosphere. The high fuel costs, the necessity to use argon, the cost of Mg or Na, and the need to reheat a second time all make Ti expensive. The high cost prevents its being more widely used. Zirconium is also produced by the Kroll process.

The van Arkel–de Boer method

Small amounts of very pure metal can be produced by this process. Impure Ti or Zr are heated in an evacuated vessel with I_2. TiI_4 or ZrI_4 is formed, and volatilizes (thus separating it from any impurities). At atmospheric pressure TiI_4 melts at 150 °C and boils at 377 °C; ZrI_4 melts at 499 °C and boils at 600 °C. Under reduced pressure, however, the boiling points are lower. The gaseous MI_4 is decomposed on a white hot tungsten filament. As more metal is deposited on the filament it conducts electricity better. Thus more electric current must be passed to keep it white hot.

$$\text{impure } Ti + 2I_2 \xrightarrow{50-250\,°C} TiI_4 \xrightarrow[\text{tungsten filament}]{1400\,°C} Ti + 2I_2$$

Zr is produced on a much smaller scale than Ti. Zr is even more corrosion resistant than is Ti, and is used in chemical plants. Its most important use is to make the cladding (that is the casing) for UO_2 fuel in water cooled nuclear reactors. Its corrosion resistance, high melting point (1857 °C) and low absorption of neutrons make it very suitable. (Of the metallic elements only Be and Mg have lower neutron absorption cross-sections than Zr. They are unsuitable for use as cladding as Be is brittle and Mg corrodes too easily.) Hf always occurs with Zr. Their chemical properties are almost identical, and for most purposes there is no need to separate them. However, Hf absorbs neutrons very strongly, and Zr used for cladding must be free from Hf.

The similarity in size of the ions makes separation exceedingly difficult.

The same problem is encountered in separating the lanthanide elements (see Chapter 29). Zr and Hf are separated by solvent extraction of their nitrates into tri-n-butyl phosphate or thiocyanates into methylisobutyl ketone. Alternatively the elements can be separated by ion exchange of an alcoholic solution of the tetrachlorides on silica gel columns. On eluting the column with an alcohol/HCl mixture, the Zr comes off first.

Zr is also used to make alloys with steel, and a Zr/Nb alloy is an important superconductor. The very high absorption of thermal neutrons by Hf is turned to good use. Hf is used to make control rods for regulating the free neutron levels in the nuclear reactors used in submarines.

OXIDATION STATES

The most common and most stable oxidation state for all the elements is (+IV). Anhydrous compounds such as $TiCl_4$ are covalent and the molecules are tetrahedral in the gaseous state. Most of the halides retain their tetrahedral structure in the solid. TiF_4 has the largest electronegativity difference and is the most likely compound to be ionic. It is a volatile white powder, which sublimes at 284 °C – not behaviour typical of an ionic salt. Its crystal structure is a polymeric F bridged structure in which each Ti is octahedrally surrounded by six F atoms.

In the oxidation state (+IV) the elements have a d^0, configuration with no unpaired electrons: hence their compounds are typically white or colourless and diamagnetic.

Ti^{4+} ions do not exist in solution but oxoions are formed instead. The titanyl ion TiO^{2+} is found in solution but it usually polymerizes in crystalline salts.

The oxidation state (+III) is reducing, and Ti^{3+} ions are more strongly reducing than Sn^{2+}. They are reasonably stable, and exist as solids and in solution. Since the M^{3+} ions have a d^1 configuration they have one unpaired electron and are paramagnetic. The magnetic moments of their compounds are close to the spin only value of 1.73 Bohr magnetons. With only one d electron, there is only one possible d–d electronic transition: hence there is only one band in the visible spectrum, and nearly all the compounds are a pale purple–red colour.

The (+II) state is very unstable and is so strongly reducing that it reduces water. Thus few compounds are known and these exist only in the solid state. The (0), (−I) and (−II) states are found in the dipyridyl complexes $[Ti^{0}(dipy)_3]$, $Li[Ti^{-I}(dipy)_3]^{3.5}$ tetrahydrofuran and $Li_2[Ti^{-II}(dipy)_3]^{5}$ tetrahydrofuran. The lower oxidation states tend to disproportionate.

$$2Ti^{III}Cl_3 \xrightarrow{\text{heat}} Ti^{IV}Cl_4 + Ti^{II}Cl_2$$
$$2Ti^{II}Cl_2 \rightarrow Ti^{IV}Cl_4 + Ti^{0}$$

SIZE

The covalent and ionic radii increase normally from Ti to Zr, but Zr and Hf are almost identical in size. The reason why Hf does not show the expected increase in size is that between La and Hf the 4*f* level is filled with the 14 lanthanide elements. There is a small decrease in size from one lanthanide element to the next. The overall decrease in size across the 14 elements is called the *lanthanide contraction* (see Chapter 29). The decrease in size caused by the lanthanide contraction cancels out the expected size increase from Zr to Hf. Since the sizes of Hf and La are almost identical and they have a similar outer electronic structure, their chemical properties are almost identical. Separation of these two elements is difficult as noted above under 'Extraction and uses'.

REACTIVITY AND PASSIVE BEHAVIOUR

The compact (massive) forms of the metals are unreactive or passive at low and moderate temperatures. This results from a thin impermeable oxide film which forms on the surface and prevents further attack. This is particularly effective with Ti. (Note, however, that the finely divided metals are pyrophoric.) At room temperature the metals are unaffected by either acids or alkali. However, Ti dissolves slowly in hot concentrated HCl, giving Ti^{3+} and H_2. Ti is oxidized by hot HNO_3, giving the hydrated oxide $TiO_2 \cdot (H_2O)_n$. Zr dissolves in hot concentrated H_2SO_4 and aqua regia. The best solvent for all the metals is HF, because they form hexafluoro complexes.

$$Ti + 6HF \rightarrow H_2[TiF_6] + 2H_2$$

At about 450 °C all three metals begin to react with many substances. At temperatures over 600 °C they are highly reactive. They form oxides MO_2, halides MX_4, interstitial nitrides MN and interstitial carbides MC by direct combination. Like the scandium group, the powdered metals absorb H_2. The amount absorbed depends on the temperature and pressure, and interstitial compounds are formed of limiting composition MH_2. These interstitial hydrides are stable in air, and are unaffected by water. This is in contrast to the behaviour of the ionic hydrides of the scandium group and *s*-block elements.

STANDARD REDUCTION POTENTIALS (VOLTS)

Acid solution

Oxidation state

+IV		+III		+II		+I	0
TiO^{2+}	$\xrightarrow{+0.10}$	Ti^{3+}	$\xrightarrow{-0.37}$	Ti^{2+}	$\xrightarrow{-1.63}$		Ti

(+IV) STATE

Oxides

World production of TiO_2 was 4.3 million tonnes in 1988. More than half is used as a white pigment in paint, and as an opacifier to make coloured paint opaque. It has replaced the earlier pigments white lead ($2PbCO_3 \cdot Pb(OH)_2$), $BaSO_4$ and $CaSO_4$ for this purpose. It has three advantages over Pb: it covers better, it is non-toxic, and it does not blacken if exposed to H_2S. The other major uses of TiO_2 are for whitening paper, and as a filler in plastics and rubber. Some is used for delustering and whitening nylon.

Naturally occurring TiO_2 is invariably coloured by impurities. There are two commercial processes for obtaining pure TiO_2, the older sulphate process and the more recent chloride process. In the chloride process rutile (TiO_2) is heated with chlorine and coke at 900 °C, forming $TiCl_4$. This is volatile and can thus be separated from any impurities. The $TiCl_4$ is heated with O_2 at about 1200 °C, forming pure TiO_2 and Cl_2. The chlorine is reused.

$$TiO_2 + 2C + 2Cl_2 \rightarrow TiCl_4 + 2CO$$

$$TiCl_4 + 2O_2 \rightarrow TiO_2 + 2Cl_2$$

In the sulphate process, ilmenite $FeTiO_3$ is digested with concentrated H_2SO_4: $Fe^{II}SO_4$, $Fe_2^{III}(SO_4)_3$ and titanyl sulphate $TiO \cdot SO_4$ are formed as a sulphate cake. This latter is leached with water and any insoluble material is removed. Fe^{III} in the solution is reduced to Fe^{II} using scrap iron, and then $FeSO_4$ is crystallized out by vacuum evaporation and cooling. The $TiOSO_4$ solution is hydrolysed by boiling, and then the solution is seeded with rutile or anatase crystals as required.

The crystal structures of the oxides suggest that they are ionic. However, the sum of the first four ionization energies is so high (8800 kJ mol^{-1} for Ti^{4+}) that this seems improbable. TiO_2 occurs in three different crystalline forms: rutile, anatase and brookite. Rutile is the most common: here each Ti is surrounded octahedrally by six O atoms (see Chapter 3, under 'Ionic compounds of the type AX_2'). The structures of the other two forms are distorted octahedral arrangements. The oxides are insoluble in water. M^{4+} ions do not exist in solution, but MO^{2+} ions are formed, giving basic salts such as titanyl sulphate $TiO \cdot SO_4$. Either TiO^{2+} ions or $[Ti(OH)_2]^{2+}$ are present in solution, but in the solid they polymerize into oxygen bridged $(MO)_n^{2+}$ chains. The X-ray structure of $TiO \cdot SO_4$ shows that each Ti is surrounded octahedrally by six O atoms: two chain O atoms, O atoms from three SO_4^- groups and O from one H_2O.

In a similar way zirconyl ions ZrO^{2+} exist in solution, and they form polymeric species in crystals. Zirconyl nitrate $ZrO(NO_3)_2$ forms an oxygen bridged chain structure similar to $TiO \cdot SO_4$. $ZrO(NO_3)_2$ is soluble in water and dilute HNO_3, and is used to remove phosphate in qualitative analysis. Otherwise it would interfere with the systematic analysis for metals. The phosphates of Ti, Zr and Hf are all insoluble.

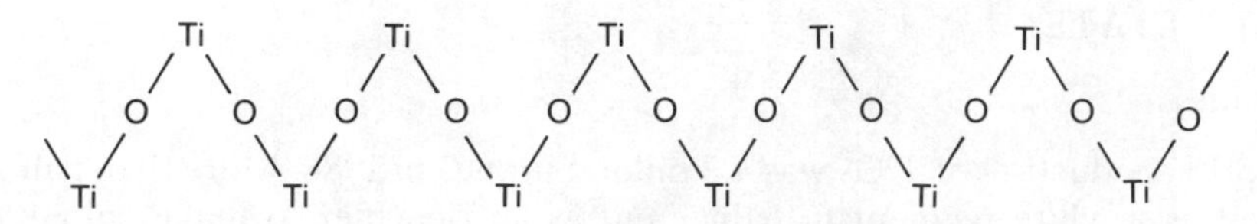

Figure 20.1 Polymeric $(TiO)_n^{2+}$ chain.

TiO_2, ZrO_2 and HfO_2 are all very stable white solids, are non-volatile, and are rendered refractory by strong ignition. On strong heating ZrO_2 becomes very hard and its high melting point of 2700 °C and its resistance to chemical attack make it useful for making high temperature crucibles and furnace linings. If the solids have been prepared dry, or have been heated, they do not react with acids. If they are prepared in solution, for example by hydrolysing $TiCl_4$, the oxide dissolves in HCl, HF and H_2SO_4, forming complexes $[TiCl_6]^{2-}$, $[TiF_6]^{2-}$ or $[Ti(SO_4)_3]^{2-}$.

The basic properties of the oxides increase with atomic number: TiO_2 is amphoteric, and ZrO_2 and HfO_2 are increasingly basic. TiO_2 dissolves in both bases and acids, forming titanates and titanyl compounds:

$$\underset{\text{titanyl sulphate}}{TiO \cdot SO_4} \xleftarrow{\text{conc. } H_2SO_4} TiO_2 \cdot (H_2O)_n \xrightarrow{\text{conc. NaOH}} \underset{\text{sodium titanate}}{Na_2TiO_3 \cdot (H_2O)_n}$$

$Ti(OH)_4$ is not known because it dehydrates to give the hydrated oxide.

Mixed oxides

If the oxides TiO_2, ZrO_2 or HfO_2 are fused (at temperatures of 1000–2500 °C) with the appropriate quantities of other metal oxides, titanates, zirconates and hafnates are formed. These are mixed oxides, and typically do not contain discrete ions. Thus anhydrous sodium titanate Na_2TiO_3 can be made by fusing TiO_2 with Na_2O, Na_2CO_3 or NaOH. Reduction of Na_2TiO_3 with H_2 at high temperatures gives titanium bronzes, which are nonstoichiometric materials of formula $Na_{0.2-0.25}TiO_2$. These have a high electrical conductivity, and have a blue–black metal-like appearance, and are similar to the tungsten bronzes. Calcium titanate $CaTiO_3$ occurs naturally as perovskite, and ilmenite (iron titanate) $Fe^{II}TiO_3$ provides the largest source of Ti.

The ilmenite structure consists of a lattice of hexagonal close-packed O atoms, with Ti atoms occupying one third of the octahedral sites, and Fe (or the other metal) occupying another one third of the octahedral sites. This structure is formed when the other metal is about the same size as Ti. The structure is the same as corundum Al_2O_3, except that corundum has two Al^{3+} ions rather than one Ti^{4+} and one Fe^2.

When the two metals differ appreciably in size the *perovskite structure* (Figure 20.2) is formed. This is a cubic close-packed array of O and Ca (so that the Ca has a coordination number of 12) with Ti occupying one quarter of the octahedral holes. The holes occupied are those bounded completely by O atoms, thus keeping Ca and Ti as far apart as possible.

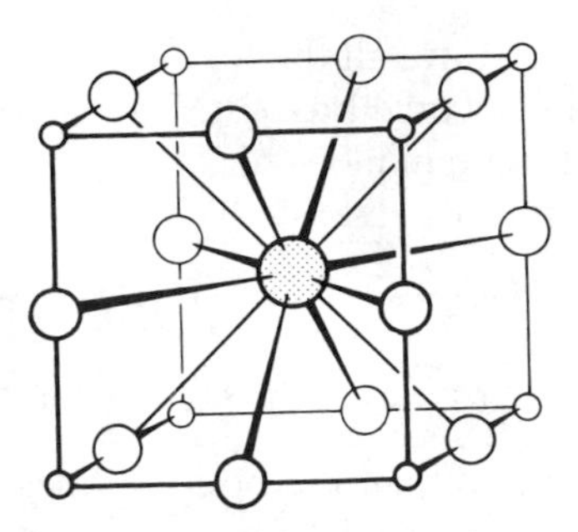

Figure 20.2 Perovskite structure.

$BaTiO_3$ has a perovskite structure. The Ba^{2+} ion is too large to fit into the close-packed oxide lattice without expanding it. This increases the size of the octahedral holes, so that Ti can 'rattle' in its octahedral hole. In an electric field the Ti atoms are drawn to one side of the hole, thus causing some polarization and making the crystal strongly ferroelectric. It is also piezoelectric (pressure produces an electric current, and vice versa). This makes it useful as a transducer for crystals in gramophone pickups and microphones, and for ceramic capacitors and other electronic uses.

Other titanates have the formula $M_2^{II}TiO_4$, where M may be Mg, Mn, Fe, Co or Zn. Mg_2TiO_4 has a *spinel structure* (like $MgAl_2O_4$). The oxide ions form a cubic close-packed array and the Mg ions occupy one half of the octahedral holes and Ti occupies one eighth of the tetrahedral holes. Thus these compounds contain discrete $[TiO_4]^{4-}$ ions.

Peroxides

A characteristic property of Ti(IV) solutions is that they form an intense yellow–orange colour on addition of H_2O_2. This reaction can be used for the colorimetric determination of either Ti(IV) or H_2O_2. The colour is thought to be due to the formation of a peroxo complex. Below pH 1 the main species is $[Ti(O_2) \cdot OH \cdot (H_2O)_n]^+$, in which the peroxo group is bidentate.

Halides

$TiCl_4$ is the best known halide and is made commercially by passing Cl_2 over heated TiO_2 and C. The other halides MX_4 can be made in a similar way. To avoid handling F_2, the fluorides can be prepared from $TiCl_4$ by reaction with anhydrous HF.

$$TiCl_4 + 4HF \rightarrow TiF_4 + 4HCl$$

The iodides can also be made by heating the halogens and metal. They are important in the van Arkel–de Boer process for purifying the metals.

$TiCl_4$ is a colourless, diamagnetic, covalent fuming liquid. $ZrCl_4$ is a white solid. In the gaseous state, all the halides are tetrahedral, but in the solid zig-zag chains of MX_6 octahedra exist. The halides are all hydrolysed vigorously by water, and fume in moist air, giving TiO_2, though hydrolysis with aqueous HCl gives the oxochloride $TiOCl_2$.

$$TiCl_4 + 2H_2O \longrightarrow TiO_2 \cdot (H_2O)_n + 4HCl$$

$$TiCl_4 + H_2O \xrightarrow{HCl} TiOCl_2 + 2HCl$$

The fluorides are more stable than the other halides.

The tetrahalides act as Lewis acids (electron pair acceptors) with a wide variety of donors, forming a large number of octahedral complexes.

$$TiF_4 \xrightarrow{\text{conc. HF}} [TiF_6]^{2-} \quad \text{stable}$$
$$TiCl_4 \xrightarrow{\text{conc. HCl}} [TiCl_6]^{2-} \quad \text{very unstable}$$

Other ligands include phosphines R_3P, arsines R_3As, oxygen donors R_2O, and nitrogen donors such as pyridine, ammonia and trimethylamine. The complexes formed have the formula $[TiX_4 \cdot L_2]$ and are octahedral. In most cases the added ligands are *cis* to each other.

Other coordination numbers are found in complexes. A few unusual five-coordinate complexes exist such as $Et_4N[Ti^{IV}Cl_5]^-$ and $[TiCl_4 \cdot AsH_3]$. A few seven-coordinate complexes are known: $Na_3[ZrF_7]$ and $Na_3[HfF_7]$. These have a pentagonal bipyramidal shape like IF_7. However, the structure of $(NH_4)_3[ZrF_7]$ is a capped trigonal prism: the Zr is at the centre of a trigonal prism of six F atoms, with an extra F in the middle of one face. Another unusual compound is $Ti(NO_3)_4$. The NO_3^- groups are bidentate, that is two O atoms from each NO_3^- are bonded to Ti. Thus the coordination number of Ti is 8, and the shape is a nearly regular triangulated dodecahedron called a bisdisphenoid. $Na_4[ZrF_8]$ and $Na_4[HfF_8]$ also have a bisdisphenoid configuration. However, in $[Cu(H_2O)_6]_2^{2+}[ZrF_8]^{4-}$ the Zr is eight-coordinate but has a square antiprism structure. (This may be visualized as a cube in which the top face has been rotated 45°.) The structures of the seven- and eight-coordinate complexes are given in Figure 20.3.

(+III) STATE

All the (+III) compounds have a d^1 configuration and are coloured and paramagnetic. Ti(III) is much more basic than Ti(IV), and the addition of alkali to Ti^{3+} solutions precipitates $Ti_2O_3 \cdot (H_2O)_n$, which is purple in colour, and insoluble in excess alkali.

The halides TiX_3 are readily formed by reducing TiX_4 compounds. Thus anhydrous $TiCl_3$ can be obtained as a violet powder by reducing $TiCl_4$ with H_2 at 600 °C. $TiCl_3$ is important as the Ziegler–Natta catalyst (see later). Reduction of aqueous solutions containing Ti(+IV) with Zn gives the purple aqua ion $[Ti(H_2O)_6]^{3+}$. This is a powerful reducing agent, and is more powerful than Sn^{2+}. It is oxidized directly by air, and must be kept

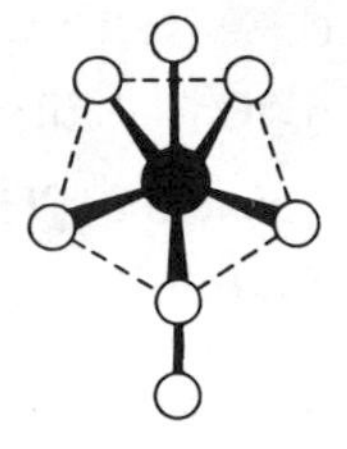

(a) $[ZrF_7]^{3-}$ ion in $Na_3[ZrF_7]$ pentagonal bipyramid

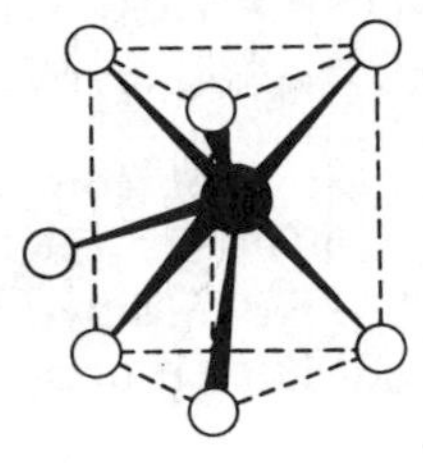

(b) $[ZrF_7]^{3-}$ ion in $(NH_4)_3[ZrF_7]$ trigonal prism with an extra atom in the middle of a rectangular face

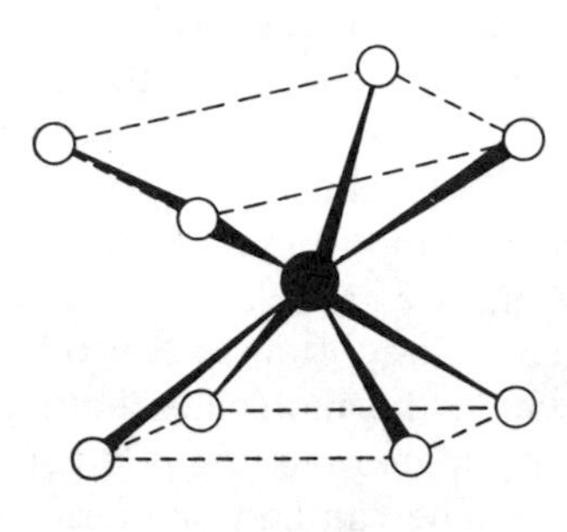

(c) $[ZrF_7]^{4-}$ ion in $[Cu(H_2O)_6]_2{}^{2+}[ZrF_8]^{4-}$ square antiprism (like a cube with the top face rotated 45°)

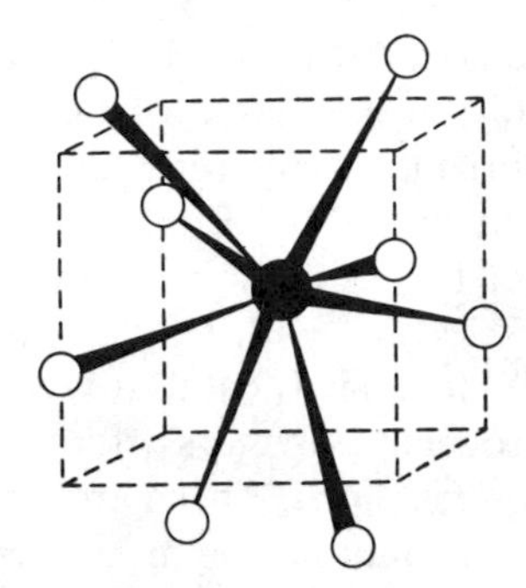

(d) $[ZrF_8]^{4-}$ ion in $Na_4[ZrF_8]$ bisdisphenoid

Figure 20.3 Structures of some fluoride complexes.

out of direct sunlight. There are two different hydrated forms of $Ti^{III}Cl_3$. These have different colours. In one complex the Ti^{III} is surrounded by six H_2O molecules, $[Ti(H_2O)_6]^{3+}$ $3Cl^-$, and in the other it is surrounded by five H_2O molecules and one Cl^-, giving $[TiCl(H_2O)_5]^{2+}$ $2Cl^-$. These two environments give rise to a different degree of crystal field splitting of the *d* levels: hence the energy jump for the single *d* electron is different in the two cases, and their colours are different.

Ti^{3+} is used in volumetric analysis for the determination of Fe^{3+} and also organic nitro compounds. In the iron titration the end point may be detected with ammonium thiocyanate which remains red whilst any Fe^{3+} is present, or by methylene blue which is reduced and decolorized as soon as Ti^{3+} is in excess.

$$TiCl_4 \xrightarrow{650°C} \underset{\text{Violet}}{TiCl_3} \xrightarrow{H_2O} \left\{ \begin{array}{c} [Ti(H_2O)_6]^{3+}Cl_3^- \\ \text{Violet} \\ [Ti(H_2O)_5Cl]^{2+}Cl_2^- \\ \text{Green} \end{array} \right\} \xrightarrow{\text{Disproportionate on heating}} \left\{ \begin{array}{c} TiCl_4 \\ \text{and} \\ TiCl_2 \end{array} \right.$$

$$Ti \xrightarrow{\text{Hot HCl}} TiCl_3$$

$$FeCl_3 + TiCl_3 + H_2O \rightarrow FeCl_2 + TiO \cdot Cl_2 + 2HCl$$
$$R \cdot NO_2 + 6Ti^{3+} + 4H_2O \rightarrow R \cdot NH_2 + 6TiO^{2+} + 6H^+$$

A wide variety of complexes are formed, for example $[Ti^{III}F_6]^{3-}$, $[Ti^{III}Cl_6]^{3-}$, $[Ti^{III}Cl_5 \cdot H_2O]^{2-}$, $[TiBr_3 \cdot (dipyridyl)_2]$, and $[TiBr_2 \cdot (dipyridyl)_2]^+$ $[TiBr_4 \cdot dipyridyl]^-$.

Zr(III) and Hf(III) are unstable in water, and exist only as solid compounds.

ORGANOMETALLIC COMPOUNDS

Ziegler–Natta catalysts

Solutions of $AlEt_3$ and $TiCl_4$ in a hydrocarbon solvent react exothermically to form a brown solid. This is the important Ziegler–Natta catalyst for polymerizing ethene (ethylene) to form polythene. Ziegler and Natta were awarded the Nobel Prize for Chemistry in 1963 for this work. (See also Chapter 12.) (Similar catalytic activity has been found from mixtures of Li, Be or Al alkyls with halides from the Ti, V and Cr groups.)

The $AlCl_3/TiCl_4$ catalyst is of great commercial importance. It produces stereoregular polymers (that is polymers where the molecules have the same orientation). These are stronger and have higher melting points than atactic or random polymers. Practically any alkene can be polymerized.

A large amount of work has been directed to how the catalyst works. The active species is Ti^{III}, and the $AlEt_3$ can reduce $TiCl_4$ to $TiCl_3$ in situ, or $TiCl_3$ may be added instead. Then one of the Cl atoms is replaced by an ethyl group. A possible mechanism is that the double bond in ethene attaches itself to a vacant site on a Ti atom on the surface of the catalyst. A carbon shift reaction occurs, and the ethene migrates and is inserted between Ti and C in the Ti—Et bond. This extends the C chain from two to four atoms, leaving a vacant site on Ti. The process is repeated, and the C chain grows in length. A similar reaction occurs with other alkenes such as propylene $CH_3—CH=CH_2$. When the double bond attaches to Ti the CH_3 group must always point away from the surface simply because the reaction occurs on a surface. Thus when the molecule migrates and is inserted into the Ti—C bond it always has the same orientation. This is called *cis insertion* of the alkene, and explains why the polymers produced are stereoregular.

Polymerization of ethene was originally carried out using a high temperature and pressure. By using a Ziegler–Natta catalyst polymerization can be carried out under relatively mild conditions from room temperature to 93 °C, and from atmospheric pressure to 100 atmospheres. Eventually the product is hydrolysed with water or alcohol and the catalyst is removed. The polymer produced is called high-density polythene and has a density 0.95–0.97 g cm^{-3} and a melting point of 135 °C. It has a molecular weight of 20 000–30 000, and consists of straight chains with very little branching. This form of polythene is relatively hard and stiff. Another form of polythene called low-density polythene is much softer and has a lower density, 0.91–0.94 g cm^{-3}, and a lower melting point of about

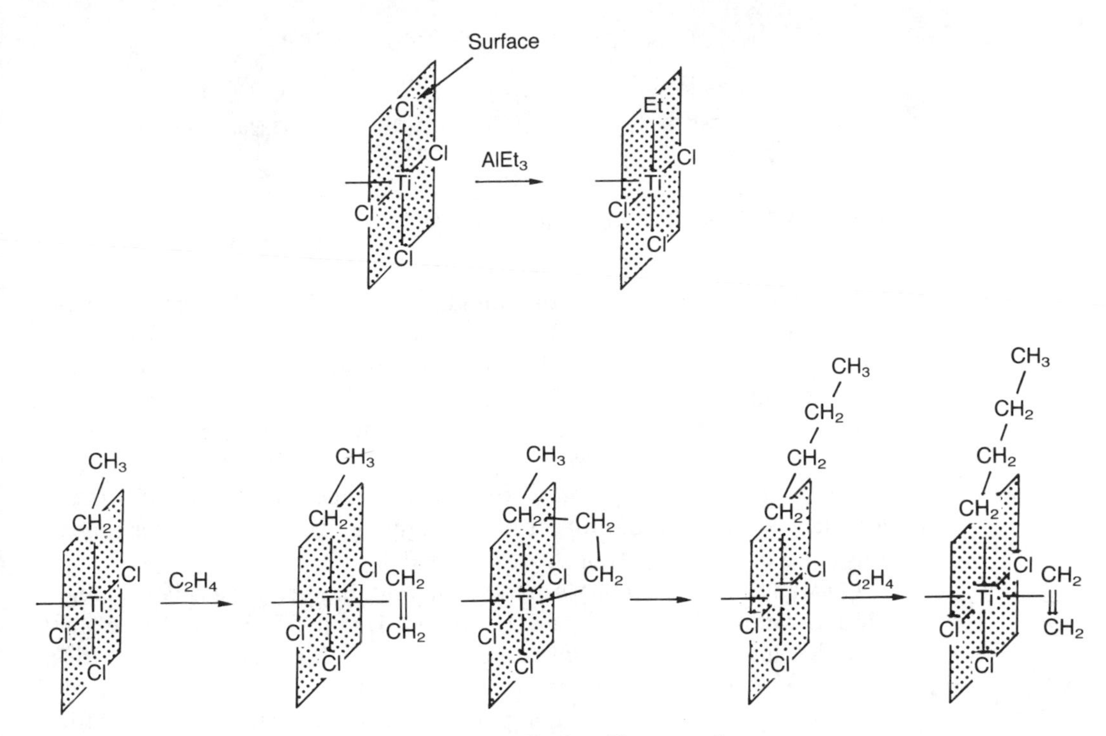

Figure 20.4 Suggested polymerization with a Ziegler–Natta catalyst.

115 °C. This consists of much branched chains, and is produced by a free radical polymerization of ethene and a promoter under much more severe conditions (190–210 °C and 1500 atmospheres pressure). Production of all forms of polythene was 17.9 million tonnes, and of polypropylene 6.5 million tonnes, both in 1985.

Other compounds

No stable carbonyl compounds are known. $Ti(CO)_6$ has been formed using the matrix isolation technique by condensing Ti vapour and CO in a solid matrix of inert gas at very low temperatures. It has been studied spectroscopically. The absence of carbonyls is probably due to the shortage of *d* electrons for back bonding.

Cyclopentadienyl compounds are much more stable and better known. In describing organometallic compounds it is convenient to describe the hapticity η of a group as the number of C atoms associated with the metal. Several stable bis(cyclopentadienyls) are known such as $[Ti(\eta^5\text{-}C_5H_5)_2(CO)_2]$, $[Ti(\eta^5\text{-}C_2H_5)_2(NR_2)_2]$ and $[Ti(\eta^5\text{-}C_5H_5)_2(SCN)_2]$. The structures are roughly tetrahedral, but the cyclopentadiene molecules are pentahaptic, that is five C atoms in each ring are attached to Ti. Reduction of these compounds gives $[Ti(C_5H_5)_2 \cdot X]$ or $[Ti(C_5H_5)_2]$. The latter formula resembles ferrocene $[Fe(C_5H_5)_2]$, but the structure of the Ti compound is dimeric, and hence is different.

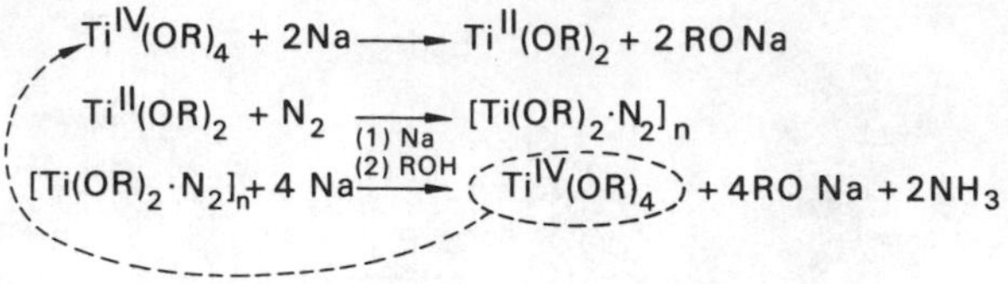

$$Ti^{IV}(OR)_4 + 2Na \longrightarrow Ti^{II}(OR)_2 + 2RONa$$

$$Ti^{II}(OR)_2 + N_2 \longrightarrow [Ti(OR)_2 \cdot N_2]_n$$

$$[Ti(OR)_2 \cdot N_2]_n + 4Na \xrightarrow[(1)\,Na]{(2)\,ROH} Ti^{IV}(OR)_4 + 4RONa + 2NH_3$$

Figure 20.5 Cycle for nitrogen fixation.

Tetra(cyclopentadienyl) compounds such as $Ti(C_5H_5)_4$ can be made from $TiCl_4$ and NaC_5H_5. The formula may be written $[Ti(\eta^5\text{-}C_5H_5)_2(\eta^1\text{-}C_5H_5)_2]$, where two cyclopentadiene rings are attached by five C atoms (π bonded) and two rings are attached by one C atom (σ bonded). Nuclear magnetic resonance studies on these tetra(cyclopentadienyl) compounds suggest that in the η^1 rings the C bonded to Ti continually changes, and also the η^5 and η^1 rings interchange their roles. Hf forms an identical compound, but rather surprisingly the Zr compound has three η^5 rings and only one η^1 ring.

Only a few alkyls and aryls are known, and they are generally unstable in air and water. $C_6H_5 \cdot Ti(C_3H_7O)_3$, CH_3TiCl_3 and $Ti(CH_2 \cdot Ph)_4$ are stable at 10 °C, and $Ti(CH_3)_4$ is stable below −20 °C. Most compounds with an alkyl group attached to Ti will polymerize alkenes.

Some organometallic Ti^{II} compounds are able to fix N_2 gas and produce NH_3. One example is $(C_{10}H_{10}Ti)_2$. This cycle could be similar to the nitrogen fixation process in nature. (Dinitrogen complexes are discussed under ruthenium(II) complexes in Chapter 24.)

FURTHER READING

Allen, A.D. (1973) Complexes of dinitrogen, *Chem Rev.*, **73**, 11.

Boor, J. (1979) *Ziegler–Natta Catalysts and Polymerizations*, Academic Press, New York.

Canterford, J.H. and Cotton, R. (1968) *Halides of the Second and Third Row Transition Elements*, Wiley, London.

Canterford, J.H. and Cotton, R. (1969) *Halides of the First Row Transition Elements*, Wiley, London.

Corbett, J.D. (1981) Extended metal–metal bonding in halides of the early transition metals, *Acc. Chem. Res.*, **14**, 239.

Davis, K.A. (1982) Titanium dioxide (Chemical of the month), *J. Chem. Ed.*, **59**, 158–159.

Jones, D.J. (1988) The story of titanium, *Chemistry in Britain*, **24**, 1135–1138.

Kepert, D.L. (1972) *The Early Transition Metals* (Chapter 2), Academic Press, London.

Kettle, S.F.A. (1969) *Coordination Compounds*, Nelson, London. (Spectra Ti^{3+}.)

Shriver, D.H., Kaesz, H.D. and Adams, R.D. (eds) (1990) *Chemistry of Metal Cluster Complexes*, VCH, New York.

Sinn, H. and Kaminsky, W. (1980) Ziegler–Natta Catalysts, *Adv. Organometallic Chem.*, **18**, 99–143. (A good account.)

Thompson, R. (ed.) (1986) *The Modern Inorganic Chemicals Industry*, (Chapter by Darby, R.S. and Leighton, J. Titanium dioxide pigments), Special Publication No. 31, The Chemical Society, London.

The vanadium group

21

Table 21.1 Electronic structures and oxidation states

Element		Electronic structure	Oxidation states*
Vanadium	V	[Ar] $3d^3\ 4s^2$	(−I) (0) (I) (II) III **IV** V
Niobium	Nb	[Kr] $4d^3\ 5s^2$	(−I) (0) (I) (II) III (IV) **V**
Tantalum	Ta	[Xe] $4f^{14}\ 5d^3\ 6s^2$	(−I) (0) (I) (II) III (IV) **V**

* The most important oxidation states (generally the most abundant and stable) are shown in bold. Other well-characterized but less important states are shown in normal type. Oxidation states that are unstable, or in doubt, are given in parentheses.

INTRODUCTION

Vanadium is commercially important as the alloy ferrovanadium which is used to make alloy steels. V_2O_5 is well known and is an important catalyst, and V metal is also used as a catalyst. The vanadates have an extensive solution chemistry. Niobium and tantalum are only used in small quantities. However, there is great theoretical interest in the cluster compounds they form in their low oxidation states.

ABUNDANCE, EXTRACTION AND USES

V, Nb and Ta have odd atomic numbers and are less abundant than their neighbours. However, V is the nineteenth most abundant element by weight in the earth's crust, and is the fifth most abundant transition element. It is widely spread, and there are few concentrated deposits. Much is obtained as a by-product from other processes. It occurs in lead ores as vanadinite $PbCl_2 \cdot 3Pb_3(VO_4)_2$, in uranium ores as carnotite $K_2(UO_2)_2(VO_4)_2 \cdot 3H_2O$ and in some crude oil from Venezuela and Canada. Vanadate residues are heated with Na_2CO_3 or NaCl at 800 °C. The sodium vanadate $NaVO_3$ formed is then leached out with water, acidified with H_2SO_4 to precipitate red coloured sodium polyvanadate, and heated to 700 °C to give V_2O_5. (This is black, possibly because of

continued overleaf

impurities: usually V_2O_5 is red or orange.) Over 75% of the V_2O_5 is converted to an iron/vanadium alloy called ferrovanadium. This contains about 50% Fe. It is made by heating V_2O_5 with iron or iron oxide and a reducing agent such as C, Al or ferrosilicon. Ferrovanadium is used commercially to make steel alloys for springs and high speed cutting tools. Vanadium metal is seldom used on its own. It is difficult to prepare because at the elevated temperatures used in metallurgy it reacts with O_2, N_2 and C. Pure vanadium can be obtained by reducing VCl_5 with H or Mg, by reducing V_2O_5 with Ca, or by electrolysing a fused halide complex. World production of V in alloys and pure metal was 31 600 tonnes in 1988.

V_2O_5 is extremely important as the catalyst in the conversion of SO_2 into SO_3 in the Contact process for making H_2SO_4. It has replaced Pt as catalyst

Table 21.2 Abundance of the elements in the earth's crust, by weight

	ppm	Relative abundance
V	136	19
Nb	20	32
Ta	1.7	53

because it is much cheaper, and is less susceptible to poisoning by impurities such as arsenic. Vanadium is an important catalyst in oxidation reactions such as naphthalene $\rightarrow$ phthalic acid, and toluene $\rightarrow$ benzaldehyde. A Cu/V alloy is used as a catalyst in the oxidation of cyclohexanol/cyclohexanone mixtures to adipic acid. (Adipic acid is used to make nylon-66.) Vanadium is also used as a catalyst for the reduction (hydrogenation) of alkenes and aromatic hydrocarbons.

Niobium and tantalum occur together. The most important mineral is pyrochlorite $CaNaNb_2O_6F$. Much smaller amounts of columbite $(Fe,Mn)Nb_2O_6$ and tantalite $(Fe,Mn)Ta_2O_6$ are also mined. However, 60% of Ta is recovered from the slag from extracting Sn. The ores are dissolved either by fusion with alkali or in acid. Formerly separation of Nb and Ta was achieved by treatment with a solution of HF, when Nb formed soluble $K_2[NbOF_5]$ and Ta formed insoluble $K_2[TaF_7]$. Separation is now performed by solvent extraction from dilute HF to methyl isobutyl ketone. The metals are obtained either by reducing the pentoxides with Na, or by electrolysis of molten fluoro complexes such as $K_2[NbF_7]$.In 1988 world production was 17 900 tonnes of Nb and 400 tonnes of Ta.

Nb is used in various stainless steels, and Nb/steel is used to encapsulate the fuel elements for some nuclear reactors. A Nb/Zr alloy is a superconductor at low temperatures, and is used to make wire for very powerful electromagnets. Ta is used to make capacitors for the electronics industry. Because it is not rejected by the human body it is valuable for making metal plates, screws and wire for repairing badly fractured bones. Tantalum carbide TaC is one of the highest melting solids known (about 3800 °C).

OXIDATION STATES

The maximum oxidation state for this group is (+V). All three elements show the full range of oxidation states from (−I) to (+V). For vanadium the (+II) and (+III) states are reducing, (+IV) is stable, and (+V) slightly oxidizing. For Nb and Ta the (+V) state is by far the most stable and the best known, although lower oxidation states are known.

V(+V) is reduced by zinc and acid to V^{2+}, Nb(+V) is reduced to Nb^{3+} but Ta(+V) is not reduced. This illustrates the increasing stability of the (+V) state on descending the group. At the same time the lower oxidation states become less stable. *This is the opposite trend to that in the main groups*.

SIZE

The atoms are smaller than those of the Ti group due to the poor shielding of the nucleus by *d* electrons. The covalent and ionic radii of Nb and Ta are identical because of the lanthanide contraction (Table 21.3). Consequently these two elements have very similar properties, occur together, and are very difficult to separate.

GENERAL PROPERTIES

V, Nb and Ta are silvery coloured metals with high melting points. V has the highest melting point in the first row transition elements. This is associated with the maximum participation of *d* electrons in metallic bonding. The melting points of Nb and Ta are high, but the maximum melting point in the second and third row transition elements occurs in the next group with Mo and W.

The pure metals V, Nb and Ta are moderately soft and ductile, but traces of impurities make them harder and brittle. They are extremely resistant to corrosion due to the formation of a surface film of oxide. At room temperature they are not affected by air, water or acids, other than HF with which they form complexes. V also dissolves in oxidizing acids such as hot concentrated H_2SO_4, HNO_3 and aqua regia. V is unaffected by alkali, showing that it is completely basic, but Nb and Ta dissolve in fused alkali.

Table 21.3 Some physical properties

	Covalent radius (Å)	Ionic radius (Å)		Melting point (°C)	Boiling point (°C)	Density ($g\,cm^{-3}$)	Pauling's electro-negativity
		M^{2+}	M^{3+}				
V	1.22	0.79	0.640	1915	3350	6.11	1.6
Nb	1.34	–	0.72	2468	4758	8.57	1.6
Ta	1.34	–	0.72	2980	5534	16.65	1.5

At high temperatures all three metals react with many non-metals. The products are often interstitial compounds which are non stoichiometric.

V forms many different positive ions, but Nb and Ta form virtually none. Thus though Nb and Ta are metals, their compounds in the (+V) state are mostly covalent, volatile, and readily hydrolysed – properties associated with non-metals.

The tendency to form simple ionic compounds decreases as the oxidation state increases. Even though V^{2+} and V^{3+} are reducing, they exist both in the solid and in solution (as hexahydrate ions). They have an extensive aqueous chemistry. The oxidation state (+IV) is dominated by the VO^{2+} ion. This is very stable and exists in a wide range of compounds both as solids and in solution (as the hydrated ion). Some covalent (+IV) compounds such as VCl_4 also exist. The (+V) state may be covalent as in VF_5, or form VO_2^+ or VO_4^{3-} hydrated ions. The chemistry of Nb and Ta is largely confined to the (+V) state.

The basic properties of the oxides M_2O_5 increase down the group. V_2O_5 is amphoteric but mainly acidic. It dissolves slightly in water, giving a pale yellow acidic solution. It dissolves readily in NaOH, forming colourless solutions which contain a wide range of vanadate ions. The ions formed depend on the pH: various isopolyvanadates at intermediate pH and orthovanadate VO_4^{3-} at high pH. The aqueous chemistry of the polymerized vanadates is quite complex. V_2O_5 also dissolves slightly in concentrated H_2SO_4, forming the pale yellow VO_2^+ ion. Nb_2O_5 and Ta_2O_5 are rather unreactive but are amphoteric. They have only very weak acidic properties. Niobates and tantalates are only formed by fusing with NaOH. They are decomposed by weak acids or CO_2, and only partially imitate the behaviour of the isopolyvanadates.

Whilst V^{2+} and V^{3+} are well known, Nb(II), Ta(II), Nb(III) and Ta(III) are not ionic but exist as cluster compounds M_6X_{12}, in which groups of metal atoms are bonded together.

COLOUR

Colour in transition metal compounds very commonly arises from *d–d* electronic transitions. It can also arise from defects in the solid state (see Chapter 3) and from charge transfer spectra. (Charge transfer spectra are discussed under SnI_4 in Chapter 13.) The oxidation states below (+V) are coloured because they have an incomplete *d* shell of electrons and give *d–d* spectra. However, the (+V) state has a d^0 configuration and so colourless compounds would be expected. NbF_5, TaF_5 and $TaCl_5$ are white, but V_2O_5 is red or orange, $NbCl_5$ is yellow, $NbBr_5$ is orange and NbI_5 is brass coloured. The colours arise because of charge transfer.

COMPOUNDS WITH NITROGEN, CARBON AND HYDROGEN

At high temperatures the metals react with N_2, forming interstitial nitrides MN, and with C, forming two series of carbides MC and MC_2. Carbides such as NbC and TaC are interstitial, refractory and very hard, like TiC and HfC in the previous group. TaC has the highest melting point of any compound, about 3800 °C. In contrast, carbides such as VC_2 are ionic and react with water, liberating ethyne.

All three elements react with H_2 on heating, forming nonstoichiometric hydrides. The amount of hydrogen absorbed depends on the temperature and pressure. Here, as in the titanium group, the metal lattice expands as hydrogen enters interstitial positions. Thus the density of the hydride is less than that of the metal. It is difficult to decide if these are true compounds or solid solutions, as the maximum hydrogen contents are $VH_{0.71}$, $Nb_{0.86}$ and $Ta_{0.76}$.

HALIDES

When V is heated with the halogens, halides of different oxidation states are formed: VF_5, VCl_4, VBr_3 and VI_3. Nb and Ta react with all of the halogens on heating to give pentahalides MX_5. The range of halides which have been formed is summarized in Table 21.4. All the halides are volatile, covalent and hydrolysed by water.

(+V) halides

V forms only a pentafluoride, but Nb and Ta form the full range of halides. These may be formed by direct reaction of the elements or by the reactions

$$M_2O_5 + F_2 \rightarrow MF_5 \qquad NbCl_5 \text{ or } TaCl_5 + F_2 \rightarrow MF_5$$

$$2VF_4 \xrightarrow[\text{disproportionates}]{\text{heat } 600\,°C} VF_5 + VF_3 \qquad NbCl_5 \text{ or } TaCl_5 + HF \rightarrow MF_5$$

Table 21.4 Halides

Oxidation states			
(+II)	(+III)	(+IV)	(+V)
VF_2 blue	VF_3 yellow–green	**VF_4** green	VF_5 colourless (liquid)
VCl_2 pale green	**VCl_3** red–violet	VCl_4 reddish brown	–
VBr_2 orange–brown	**VBr_3** brown	VBr_4 magenta	–
VI_2 red–violet	**VI_3** black–brown	–	–
–	(NbF_3) *blue	NbF_4 black	**NbF_5** white
–	$NbCl_3$ black	$NbCl_4$ violet	**$NbCl_5$** yellow
–	$NbBr_3$ brown	$NbBr_4$ brown	**$NbBr_5$** orange
–	NbI_3	NbI_4 grey	**NbI_5** brass
–	(TaF_3) *blue	–	**TaF_5** white
–	$TaCl_3$ black	$TaCl_4$ black	**$TaCl_5$** white
–	$TaBr_3$	$TaBr_4$ blue	**$TaBr_5$** yellow
–	–	TaI_4	**TaI_5** black

The most stable oxidation states are shown in bold.
* may be oxide fluorides.

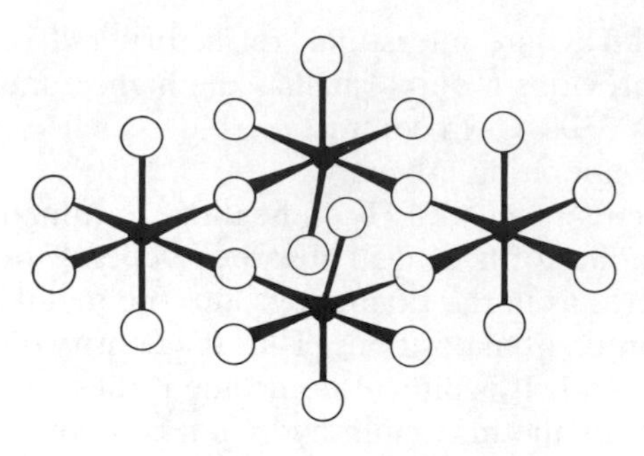

Figure 21.1 Tetrameric structure of NbF_5 and TaF_5.

VF_5 is a colourless liquid, but the other pentahalides are solids. They adopt several different structures. The fluorides are built of octahedral MF_6 units with two *cis* fluorine atoms in each octahedron acting as bridging groups V—F—V with other octahedra. VF_5 forms long chains of octahedra in this way, but NbF_5 and TaF_5 form cyclic tetramers with four octahedra joined in this way (Figure 21.1). This structure is also found in other pentahalides, e.g. MoF_5, RuF_5 and OsF_5. Solid $NbCl_5$ and $TaCl_5$ are dimeric with two octahedra joined by sharing two corners, i.e. one edge (Figure 21.2).

All of the pentahalides can be sublimed under an atmosphere of the appropriate halogen. In the vapour phase they probably exist as monomeric trigonal bipyramids.

The pentafluorides all react with F^- ions, forming octahedral $[MF_6]^-$ complexes. As in the titanium group, the heavier elements can form complexes with higher coordination numbers. With high concentrations of F^- the complexes $[NbOF_5]^{2-}$, $[NbF_7]^{2-}$ and $[TaF_7]^{2-}$ are formed. The structures of the seven-coordinate species are capped trigonal prisms, i.e. a trigonal prism with one extra atom in a rectangular face (see Figure 19.2b). With an even higher concentration of F^-, Ta forms $[TaF_8]^{3-}$ with a square antiprism structure (Figure 19.2c) while Nb forms $[NbOF_6]^{3-}$ which is an octahedron with an extra atom in one of the faces. This inability of Ta to form oxohalides has been used to separate Nb and Ta.

On warming the halides MF_5 and MCl_5 with donors such as dimethyl ether $(CH_3)_2O$ or dimethyl sulphoxide $(CH_3)_2SO$, the halides extract O and form oxochlorides $MOCl_3$. Oxohalides are also formed when the pentahalides are heated in air. $VOCl_3$ can also be made by heating any V_2O_5 with Cl_2 (or sometimes C and Cl_2). The oxohalides are all readily hydrolysed by water to the hydrated pentoxide. The oxohalides are tetrahedral in shape.

The pentahalides react with N_2O_4 in giving solvated nitrates such as $NbO_2 \cdot NO_3 \cdot 0.67MeCN$, and anhydrous nitrates such as $NbO(NO_3)_3$.

The (+V) halides have a d^0 configuration, and cannot give d–d spectra. The fluorides are white but the other halides are coloured due to charge transfer spectra.

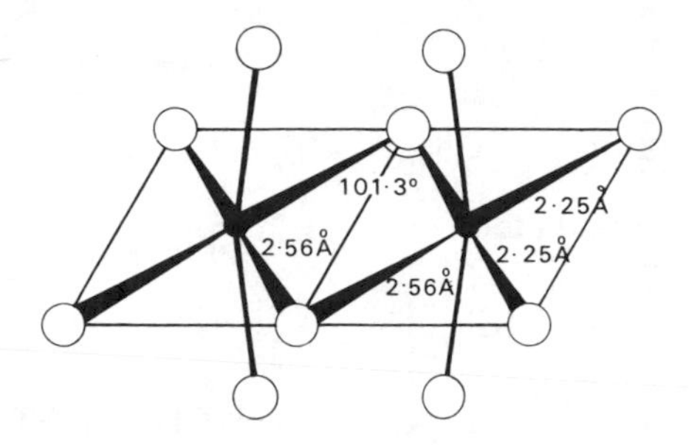

Figure 21.2 Dimeric structure of $NbCl_5$.

(+IV) halides

All the tetrahalides are known except TaF_4. These may be prepared as follows:

$$V + Cl_2 \rightarrow VCl_4 \quad NbX_5 \text{ or } TaX_5 \xrightarrow[H_2,\text{ Al, Nb or Ta}]{\text{reduce with}} MX_4$$

$$V + HF \rightarrow VF_4$$

VCl_4 is tetrahedral in the gas. The d^1 configuration of V(+IV) would be expected to make this unstable and to cause distortion. In the liquid it is dimeric. NbF_4 is a black, paramagnetic involatile solid made up of regular octahedra joined in a chain by their edges. The tetrachlorides, tetrabromides and tetraiodides of Nb and Ta are also brown–black solids and are diamagnetic. This suggests extensive metal–metal interaction. In NbI_4, the structure is a chain of octahedra joined by their edges. The Nb atoms are displaced from the centre of the NbI_6 octahedron and occur in pairs, thus permitting weak Nb—Nb bonds of length 3.20 Å, and pairing the previously unpaired electron spins (Figure 21.3). $NbCl_4$ is similar and has M—M bonds of length 3.06 Å.

The tetrahalides tend to disproportionate:

$$2VCl_4 \xrightarrow{\text{room temperature}} 2VCl_3 + Cl_2 \text{ (}VCl_5\text{ does not exist)}$$

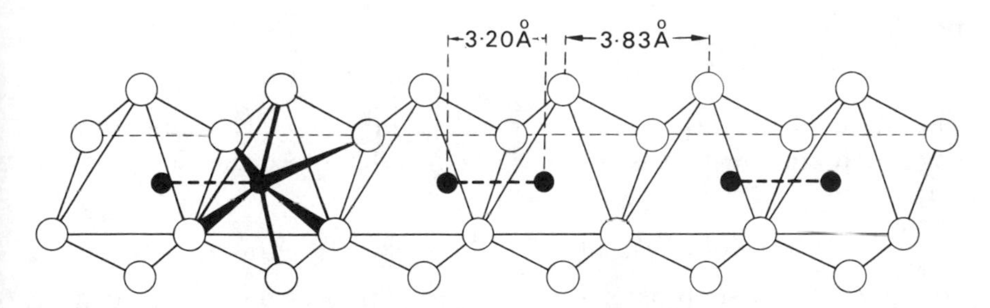

Figure 21.3 Polymeric structure of NbI_4, with metal–metal bonds.

$$2VF_4 \xrightarrow{600\,°C} VF_3 + VF_5$$

$$2TaCl_4 \xrightarrow{400\,°C} TaCl_3 + TaCl_5$$

They also hydrolyse with water:

$$VCl_4 \xrightarrow{H_2O} VOCl_2$$

$$4Ta^{IV}Cl_4 \xrightarrow{5H_2O} Ta_2^{V}O_5 + 2Ta^{III}Cl_3 + 10HCl$$

(+III) halides

All the trihalides are known except for TaI_3. They are reducing, have a d^2 configuration, and are brown or black in colour. The VX_3 compounds are all polymeric compounds in which V is octahedrally surrounded. VCl_3 and VBr_3 can be made from the elements, and VF_3 is made from VCl_3 and HF. VF_3 can be crystallized from water, giving $[VF_3 \cdot (H_2O)_3]$, and, for the other halides, $[V(H_2O)_6]^{3+}$ and three X^- ions. VCl_3 forms complexes such as $VCl_3 \cdot (NMe_3)_2$, which has a trigonal bipyramid shape. VI_3 disproportionates as follows:

$$VI_3 \rightarrow VI_2 + VI_4$$

Spectra

The V^{3+} ion has a d^2 configuration. The two d electrons occupy two of the t_{2g} orbitals, i.e. any two from d_{xy}, d_{xz}, and d_{yz}. The ground state is triply degenerate and has the symbol $^3T_{1g}(F)$. At first sight it might be expected that by promoting these electrons to the e_g level two d–d absorption bands would occur in the electronic spectrum. Under suitable conditions three bands are observed. If one electron is promoted from the t_{2g} level to the e_g level then the most stable arrangement (i.e. that with the lowest energy) will be when the two electrons occupy orbitals as far apart as possible, i.e. at right angles to each other. Thus if one electron occupies the d_{xy} orbital, the structure would be more stable if the other electron occupied the d_{z^2} orbital rather than the $d_{x^2-y^2}$ orbital. There are three degenerate ways of arranging these two electrons in orbitals perpendicular to each other:

$$(d_{xy})^1(d_{z^2})^1 \quad (d_{xz})^1(d_{x^2-y^2})^1 \quad \text{and} \quad (d_{yz})^1(d_{x^2-y^2})^1$$

This state is the $^3T_{2g}$. Somewhat higher in energy is another triply degenerate state written as $^3T_{2g}$ in which the orbitals occupied are at 45° to each other:

$$(d_{xy})^1(d_{x^22y^2})^1 \quad (d_{xz})^1(d_{z^2})^1 \quad \text{and} \quad (d_{yz})^1(d_{z^2})^1$$

If both electrons are promoted to the e_g level then assuming they remain unpaired the only arrangement possible is:

$$(d_{x^2-y^2})^1 \quad (d_{z^2})^1$$

This state has the symbol $^3A_{2g}$. Thus there are three possible transitions from the ground state to excited states with the same multiplicity:

$$^3T_{1g}(F) \rightarrow {}^3T_{2g}$$
$$^3T_{1g}(F) \rightarrow {}^3T_{1g}(P)$$
and
$$^3T_{1g}(F) \rightarrow {}^3A_{2g}$$

Thus three bands are possible. (See Chapter 32, particularly Figure 32.15). In the spectrum of $[V(H_2O)_6]^{3+}$ only two bands are actually observed, at about $17\,000\,cm^{-1}$ and $24\,000\,cm^{-3}$. The third band arising from the transition to $^3A_{2g}$ is not observed experimentally. Since this transition is a two electron transition it is less probable than the others and so will have a low intensity. Furthermore this band overlaps, and is hidden by, the very intense charge transfer band in the UV region. All three bands are observed when V^{3+} is incorporated into an Al_2O_3 lattice.

The discussion above indicates that in addition to the crystal field splitting Δ_o the interelectronic repulsion must also be taken into account when explaining the spectra. The repulsion terms are described by the Racah parameters B and C (see Chapter 32).

Nb and Ta halides

The trihalides of Nb and Ta are typically nonstoichiometric. In $NbCl_3$ the Nb ions occupy octahedral holes in a distorted hexagonal close-packed array of Cl^- ions in such a way that niobium atoms in three adjacent octahedra are close enough to be bonded together into a metal cluster. Compounds where three or more metal atoms are held together by multi-centre bonding are called cluster compounds.

(+II) halides

All the vanadium dihalides are known. The VX_2 compounds are prepared by reducing the trihalides with Zn/acid in aqueous solution. VF_2 has a rutile TiO_2 structure, and the others have a CdI_2-type layer structure. They are soluble in water, giving violet solutions containing $[V(H_2O)_6]^{2+}$. Addition of NaOH precipitates $V(OH)_2$, and addition of H_2SO_4 and ethanol precipitates violet crystals of $VSO_4 \cdot 6H_2O$. The compounds are strongly reducing, and are hygroscopic. Their solutions are readily oxidized by air to $[V(H_2O)_6]^{3+}$, and they are often used to remove traces of O_2 from the noble gases. They also reduce H_2O with the liberation of H_2. Nb and Ta behave very differently. High temperature reduction of the pentahalides NbX_5 and TaX_5 with sodium or aluminium yields a series of lower halides such as M_6Cl_{14}, M_6I_{14}, Nb_6F_{15}, Ta_6Cl_{15}, Ta_6Br_{15} and Ta_6Br_{17}. These are all based on the $[M_6X_{12}]^{n+}$ unit. For example, if Nb_6Cl_{14} is dissolved in water and alcohol, and treated with $AgNO_3$

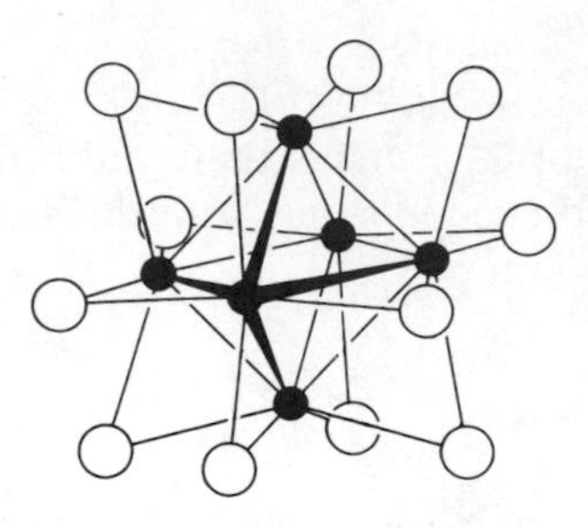

Figure 21.4 Structure of $[Nb_6Cl_{12}]^{2+}$ ion, showing the octahedral cluster of metal atoms bonded together, and the 12 bridging chlorine atoms.

solution, only two chlorine atoms are precipitated as AgCl. This indicates the presence of $[Nb_6Cl_{12}]^{2+}$ ions and two Cl^- ions, and this structure has been confirmed by X-ray crystallography (Figure 21.4). The Nb atoms are bonded together, forming an octahedral cluster of six metal atoms. The halogen atoms are situated above each edge of the octahedron and are bonded to two Nb atoms. Thus the halogens form bridges along the 12 edges of the octahedron. The structure is held together by both multicentre bonding over the six metal atoms and by the halogen bridges.

Cluster compounds are currently attracting a lot of attention. If one of these clusters is linked by halogen bridges to four other clusters, the composition is $[M_6X_{12}]$ (the cluster) + $\frac{1}{2}X_4$ (the halogen bridges), i.e. M_6X_{14} or $MX_{2.3}$. Structures of this type are sheet-like, and diamagnetic because of the metal–metal bonding. Alternatively a cluster may be linked to six other clusters by halogen bridges, giving a three-dimensional structure and a formula of $[M_6X_{12}] + \frac{1}{2}X_6$, i.e. M_6X_{15} or $MX_{2.5}$. These are

STANDARD REDUCTION POTENTIALS (VOLTS)

Acid solution

Oxidation state

+V		IV		III		+II	+I	0
$V(OH)_4^+$	+1.0	VO^{2+}	+0.34	V^{3+}	−0.26	V^{2+}	−1.18	V
$V(OH)_4^+$ → V: −0.26								
Nb_2O_5	0.05			Nb^{3+}	−1.10			Nb
Nb_2O_5 → Nb: −0.64								
Ta_2O_5	−0.75							Ta

paramagnetic and the magnetic moment corresponds to one unpaired electron. Cluster compounds of this type are characteristic of the lower oxidation states of Nb and Ta. Many of these cluster compounds are soluble in water, and the cluster remains intact during reactions. The clusters can be oxidized:

$$[M_6X_{12}]^{2+} \rightarrow [M_6X_{12}]^{3+} \rightarrow [M_6X_{12}]^{4+}$$

A very unusual structure is found in Nb_6I_{11}, where the six Nb atoms form an octahedral cluster as before, but eight iodine atoms are situated above the eight faces of the octahedron. Each of these I atoms act as bridging group to three metal atoms within the octahedron. The rest of the I atoms are bonded to the six corners of the metal octahedron. These act as halogen bridges to other octahedra. The formula is thus $[Nb_6I_8] + \frac{1}{2}I_6$ which is Nb_6I_{11}.

OXIDES

The metals all react with oxygen at elevated temperatures and give pentoxides M_2O_5. V_2O_5 is orange or red depending on the state of division, but the others are white. When made this way, VO_2 can also be formed. Pure V_2O_5 is obtained either by acidifying ammonium metavanadate, or simply by heating it:

$$2NH_4VO_3 + H_2SO_4 \rightarrow V_2O_5(\text{hydrated}) + (NH_4)_2SO_4 + H_2O$$

$$2NH_4VO_3 \xrightarrow{\text{heat}} V_2O_5 + 2NH_3 + H_2O$$

The main oxides formed are shown in Table 21.5. The lower oxides often exist over a wide range of composition.

(+V) oxides

The oxides M_2O_5 can all be made by heating the metal in oxygen. However, V_2O_5 is best made by heating ammonium metavanadate NH_4VO_3. Nb_2O_5 and Ta_2O_5 are commonly made by ignition of other Nb or Ta compounds in air. In the pentoxides the metal atoms all have a d^0 configuration, and might be expected to be colourless. Nb_2O_5 and Ta_2O_5 are white, but V_2O_5 is orange or red coloured due to charge transfer.

Table 21.5 Oxides

Oxidation state			
(+II)	(+III)	(+IV)	(+V)
VO	V_2O_3	VO_2	**V_2O_5**
NbO	–	NbO_2	**Nb_2O_5**
(TaO)	–	TaO_2	**Ta_2O_5**

The most stable oxidation states are shown in bold.

V_2O_5 is amphoteric, but is mainly acidic. With very strong NaOH it forms colourless orthovanadate ions VO_4^{3-}. At slightly higher pH these polymerize to form a wide range of isopolyacids called polyvanadates. These are described later. V_2O_5 dissolves in very strong acid, forming eventually the pale yellow dioxovanadium(V) ion VO_2^+. This ion has an angular shape. Some reactions of V_2O_5 are as follows:

$$V_2O_5 + NaOH \rightarrow \text{various vanadates}$$
$$V_2O_5 + H_2O_2 \rightarrow \text{peroxovanadates (red colour)}$$
$$V_2O_5 + Cl_2 \rightarrow VOCl_3$$
$$V_2O_5 + SO_2 \rightarrow VO_2 + SO_3$$
$$V_2O_5 + H_2 \rightarrow VO_2 + V_2O_3$$

Though Nb_2O_5 and Ta_2O_5 react with HF, and form niobates and tantalates when fused with NaOH, they are better described as unreactive rather than amphoteric.

The structure of V_2O_5 is unusual and consists of distorted trigonal bipyramids of VO_5 units sharing edges with other units to form zig-zag double chains. Its use as a catalyst in the Contact process has been mentioned previously. The catalytic activity may be because it can reversibly lose or gain oxygen when heated.

(+IV) oxides

VO_2 can be made from V_2O_5 with mild reducing agents such as Fe^{2+}, SO_2, or oxalic acid. V(+IV) has a d^1 configuration and VO_2 is dark blue in colour. The oxide is amphoteric, but is more basic than acidic. In acids it forms blue solutions containing the oxovanadium(IV) ion VO^{2+}. This is commonly called the vanadyl ion. A large number of vanadyl compounds are known: vanadyl sulphate $VOSO_4$ and vanadyl halides VOX_2. Several vanadyl complexes are also known, $[VOX_4]^{2-}$ where X is a halogen, $[VO(oxalate)_2]^{2-}$, $[VO(bipyridyl)_2Cl]^+$, $[VO(NCS)_4]^{2-}$ and $[VO(acetylacetone)_2]$.

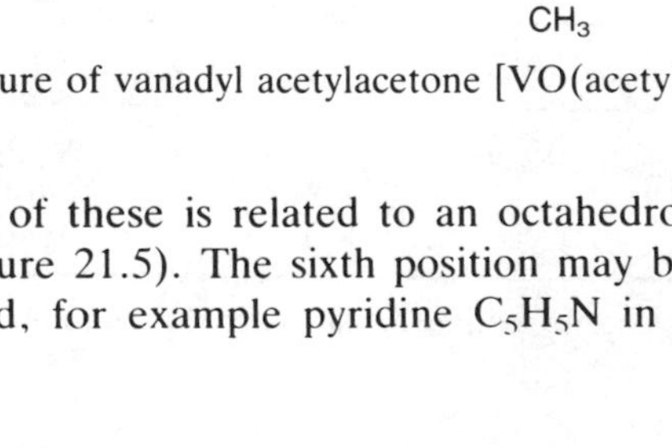

Figure 21.5 Structure of vanadyl acetylacetone $[VO(acetylacetone)_2]$.

The structure of these is related to an octahedron with one position unoccupied (Figure 21.5). The sixth position may be filled quite readily by a sixth ligand, for example pyridine C_5H_5N in $[VO(acetylacetone)_2$

(pyridine)]. In these compounds the V—O bonds are 1.56–1.59 Å. This is shorter than a single bond, and the bond is better represented as V=O. The π bond arises from back bonding from a filled *p* orbital on O with an empty *d* orbital on V, similar to the oxides and oxoacids of phosphorus.

(+III) oxides

V_2O_3 is nonstoichiometric ($VO_{1.35-1.5}$) and contains simple ions arranged in the corundum Al_2O_3 type of structure. It can be produced by high temperature reduction of V_2O_5 with carbon or hydrogen, or by electrolytic reduction of a vanadate. The oxide is completely basic. It dissolves in acids, forming blue or green hydrated ions $[V(H_2O)_6]^{3+}$, and these solutions are quite strongly reducing. Addition of NaOH precipitates hydrated $V(OH)_3$, but this has no tendency to dissolve in excess NaOH. The oxidation state (+III) has a d^2 configuration, and the oxide is black and the hydrated ion is blue. A considerable number of octahedral complexes are known such as $[V(H_2O)_6]^{3+}$, $[VF_6]^{3-}$, $[V(CN)_6]^{3-}$ and $[V(oxalate)_3]^{3-}$. Vanadium also forms a triacetate complex, which is dimeric $[V^{III}{}_2(acetate)_6]$. This has an unusual structure. Four of the acetate groups act as bridging groups through O atoms to the two V atoms. The remaining two acetate groups act as unidentate ligands and one is attached to each V. The structure is related to the structures of chromium(II) acetate $[Cr_2(acetate)_4 \cdot (H_2O)_2]$ and copper acetate $[Cu_2(acetate)_4 \cdot (H_2O)_2]$. Both are dimeric, and both have four bridging acetate groups, but the terminal acetate groups in the V compound are replaced by two molecules of water. In water, $[V(H_2O)_6]^{3+}$ partially hydrolyses to $V(OH)^{2+}$ and VO^+.

(+II) oxides

VO is nonstoichiometric, of composition $VO_{0.94-1.12}$. The solid is ionic and has a defect NaCl type of structure. The (+II) state has a d^3 configuration, and the oxide is grey–black and has a metallic lustre. It is made by reducing V_2O_5 with hydrogen at 1700 °C. It has a fairly high electrical conductivity, which is probably due to metal–metal bonding in the structure. The oxide is completely basic, and is soluble in water. Addition of NaOH to this solution precipitates $V(OH)_2$. VO dissolves in acids, forming violet coloured $[V(H_2O)_6]^{2+}$ ions. These solutions are strongly reducing, and are very readily oxidized both by air and water. Thus the violet solution soon turns green through oxidation to $[V(H_2O)_6]^{3+}$. A few octahedral complexes are known such as $K_4[V(CN)_6] \cdot 7H_2O$ and $[V(ethylenediamine)_3]Cl_2$. $K_4[V(CN)_7]2H_2O$ is also known, and the structure of the anion is a pentagonal bipyramid.

VANADATES

Though V_2O_5 is amphoteric, it is mainly acidic. It dissolves in very strong NaOH, forming a colourless solution containing orthovanadate ions

VO_4^{3-}. These are tetrahedral in shape. If acid is added gradually to lower the pH, the ions add protons and polymerize. Thus a very large number of different isopolyacids are formed in solution. The oxoions polymerize to form dimers, trimers and pentamers. When the solution becomes acidic the hydrated oxide $V_2O_5(H_2O)_n$ is precipitated. This dissolves in very strong acid, forming various complex ions, until finally the dioxovanadium ion VO_2^+ is formed. Various solids have been crystallized out at different pH values, but these do not necessarily have the same structure, and are unlikely to be hydrated to the same extent as the species in solution. The following scheme could explain the observations, though the extent to which the various species are hydrated is unknown.

$$\underset{\text{colourless}}{[VO_4]^{3-}} \xrightarrow{\text{pH } 12} \underset{\text{colourless}}{[VO_3 \cdot OH]^{2-}} \xrightarrow{\text{pH } 10} \underset{\text{colourless}}{[V_2O_6 \cdot OH]^{3-}} \xrightarrow{\text{pH } 9} \underset{\text{orange}}{[V_3O_9]^{3-}} \xrightarrow{\text{pH } 7}$$

$$\underset{\text{red}}{[V_5\overset{*}{O}_{14}]^{3-}} \xrightarrow{\text{pH } 6.5} \underset{\text{brown precipitate}}{V_2\overset{*}{O}_5 \cdot (H_2O)_n} \xrightarrow{\text{pH } 2.2} [V_{10}O_{28}]^{6-} \xrightarrow{\text{pH } <1} \underset{\text{pale yellow}}{[VO_2]^+}$$

*These species are included because solids of this formula have been precipitated.

Nb_2O_5 and Ta_2O_5 are white and chemically inert. They are hardly attacked by acids, except HF which forms fluoro complexes. If the pentoxides are fused with NaOH, niobates and tantalates are formed. These precipitate the hydrated oxides at pH 7 and 10 respectively, and the only isopolyion found in solution is $[M_6O_{19}]^{8-}$.

In the main groups, the phosphates, silicates and borates all show a strong tendency for the oxoions to polymerize, forming a very large number of isopolyacids. In a similar way in the *d*-block, molybdates and tungstates also polymerize to form a large number of isopolyacids. This tendency is shown to a lesser extent by $(TiO^{2+})_n$ and $CrO_4^{2-} \rightarrow Cr_2O_7^{2-}$.

Vanadate ions also form complexes with the ions of other acids. Because there is more than one type of acid unit which condenses, these are called heteropolyacids. They always contain vanadate, molybdate or tungstate ions together with one or more acidic ions (such as phosphate, arsenate or silicate) from about 40 elements. The ratio between the numbers of the different types of units is commonly 12 : 1 or 6 : 1, giving the 12-polyacids and the 6-polyacids. However, other ratios are also found. A study of heteropolyacids is very difficult because:

1. The molecular weight is high, often 3000 or more.
2. The water content is variable.
3. The ions present change with the pH.
4. The species present in solution are probably different from those which crystallize out.

LOW OXIDATION STATES

Only a few compounds are known. The (−I) state occurs with $[M(CO)_6]^-$ for all three elements, and for the dipyridyl complex $Li[V(dipyridyl)_3] \cdot$

ether. The zero-valent state occurs with $[V(CO)_6]$. This is not very stable. Since V has an odd atomic number, it follows that in $V(CO)_6$ it has an unpaired electron and hence an incomplete shell of electrons. Other first row metal carbonyls with an unpaired electron dimerize and form a M—M bond, thus pairing the electrons. $V(CO)_6$ is unusual in that it remains monomeric. [V(dipyridyl)$_3$] is another example of V(0). The (+I) state is found in [V(dipyridyl)$_3$]$^+$.

ORGANOMETALLIC COMPOUNDS

This group does not form many compounds with M—C σ bonds. The main examples are $V(CO)_6$, which, though pyrophoric and not very stable, can be prepared in much larger quantities than $Ti(CO)_6$. An unusual compound hexakis(dinitrogen) $[V(N_2)_6]$ is also known, and is thought to be isoelectronic and isostructural with the carbonyl.

V forms bis(cyclopentadienyl) compounds such as $[V(\eta^5-C_5H_5)_2Cl_2]$, $[V(\eta^5-C_5H_5)_2Cl_2]$ and $[V(\eta^5-C_5H_5)_2Cl_2]$. The last is a simple sandwich compound and is called vanadocene. This is like ferrocene in the next chapter, and unlike the titanium compounds. The hapticity η^5 indicates that in each ring five carbon atoms are 'bonded' to V. Vanadocene is extremely air sensitive, and is a dark violet paramagnetic solid. Nb and Ta also form cyclopentadienyl compounds such as $[Nb(\eta^5\text{-}C_5H_5)_2(\eta^1\text{-}C_5H_5)_2]$, $[Nb(\eta^5\text{-}C_5H_5)_2(\eta^1\text{-}C_5H_5)_2Cl_3]$ and $[Nb(\eta^5\text{-}C_5H_5)_2(\eta^1\text{-}C_5Hd5)_2Cl_3]$, in which two rings are η^5, and two rings are η^1 bonded.

FURTHER READING

Brown, D. (1973) *Comprehensive Inorganic Chemistry*, Vol. 3 (Chapter 35: Niobium and tantalum), Pergamon Press, Oxford.

Clark, R.J.H. (1973) *Comprehensive Inorganic Chemistry*, Vol. 3 (Chapter 34: Vanadium), Pergamon Press, Oxford.

Canterford, J.H. and Cotton, R. (1968) *Halides of the Second and Third Row Transition Elements*, Wiley, London.

Canterford, J.H. and Cotton, R. (1969) *Halides of the First Row Transition Elements*, Wiley, London.

Corbett, J.D. (1981) Extended metal–metal bonding in halides of the early transition metals, *Acc. Chem. Res.*, **14**, 239.

Cotton, F.A. (1975) Compounds with multiple metal to metal bonds, *Chem. Soc. Rev.*, **27**, 27–53.

Diatomic metals and metallic clusters (1980) (Conference papers of the Faraday Symposia of the Royal Society of Chemistry, No. 14), Royal Society of Chemistry, London.

Emeléus, H.J. and Sharpe, A.G. (1973) *Modern Aspects of Inorganic Chemistry*, 4th ed. (Chapters 14 and 15: Complexes of Transition Metals: Chapter 20: Carbonyls) Routledge and Kegan Paul, London.

Fairbrother, F. (1967) *The Chemistry of Niobium and Tantalum*, Elsevier, Amsterdam.

Hagenmuller, P. (1973) *Comprehensive Inorganic Chemistry*, Vol. 4 (Chapter 50: Tungsten bronzes, vanadium bronzes), Pergamon Press, Oxford.

Hunt, C.B. (1977) Metallocenes – the first 25 years, *Education in Chemistry*, **14**, 110–113.

Johnson, B.F.G. (ed.) (1980) *Transition Metal Clusters*, John Wiley, Chichester.

Kepert, D.L. (1972) *The Early Transition Metals* (Chapter 3: V, Nb, Ta), Academic Press, London.
Kepert, D.L. (1973) *Comprehensive Inorganic Chemistry*, Vol. 4 (Chapter 51: Isopolyanions and heteropolyanions), Pergamon Press, Oxford.
Lewis, J. and Green, M.L. (eds) (1983) *Metal Clusters in Chemistry* (Proceedings of the Royal Society Discussion Meeting May 1982), The Society, London.
Muetterties, E.L. (1971) *Transition Metal Hydrides*, Marcel Dekker, New York.
Ophard, C.E. and Stupgia, S. (1984) Synthesis and spectra of vanadium complexes, *J. Chem. Ed.*, **61**, 1102–1103.
Pope, M.T. (1983) *Heteropoly and Isopoly Oxo Metalates*, Springer-Verlag. (A good account of polyacids).
Shriver, D.H., Kaesz, H.D. and Adams, R.D. (eds) (1990) *Chemistry of Metal Cluster Complexes*, VCH, New York, 1990.

The chromium group

22

Table 22.1 Electronic structures and oxidation states

Element		Electronic structure	Oxidation states*
Chromium	Cr	[Ar] $3d^5$ $4s^1$	(−II) (−I) 0 (I) **II** **III** (IV) (V) **VI**
Molybdenum	Mo	[Kr] $4d^5$ $5s^1$	(−II) (−I) 0 I (II) III IV V **VI**
Tungsten	W	[Xe] $4f^{14}$ $5d^4$ $6s^2$	(−II) (−I) 0 I (II) (III) IV V **VI**

* The most important oxidation states (generally the most abundant and stable) are shown in bold. Other well-characterized but less important states are shown in normal type. Oxidation states that are unstable, or in doubt, are given in parentheses.

INTRODUCTION

Chromium metal is produced on a large scale, and is used extensively in ferrous and non-ferrous alloys, and for electroplating. The metals molybdenum and tungsten are produced in appreciable amounts. Sodium dichromate is also used in large amounts. CrO_3 and Cr_2O_3 are both used commercially.

Tungstate and molybdate ions both form extensive series of iso- and heteropolyacids. Chromium(II) acetate has an unusual structure with a quadruple bond. The lower halides MoX_2 and WX_2 form interesting cluster compounds based on the octahedral $[M_6X_8]^{4+}$ metal cluster. Mo is important in nitrogen fixation.

ABUNDANCE, EXTRACTION AND USES

Chromium is the twenty-first most abundant element by weight in the earth's crust. This is about as common as chlorine. Molybdenum and tungsten are quite rare.

The only commercially important ore of Cr is chromite $FeCr_2O_4$. This is the chromium analogue of magnetite Fe_3O_4, which is better written as $Fe^{II}Fe_2^{III}O_4$. Chromite has a spinel structure. In this structure the O atoms are arranged in a cubic close-packed lattice with Fe^{II} in one eighth of the available tetrahedral holes and Cr^{III} in one quarter of the octahedral holes.

continued overleaf

Chromite has a slight lustre and looks like pitch, with a brownish cast to the colour. It may be slightly magnetic. World production of chromite was 11.7 million tonnes in 1988, with a Cr content of 3.4 million tonnes. The largest sources of chromite ore are South Africa 36%, the USSR 28%, Turkey 7%, India 6.5%, Albania 6%, and Finland and Zimbabwe 5% each. Small amounts of crocoite $PbCrO_4$ and chrome ochre Cr_2O_3 are also mined.

Chromium is produced in two forms: ferrochrome and pure Cr metal, depending on what it is to be used for. Ferrochrome is an alloy containing Fe, Cr and C. It is produced by reducing chromite with C. In 1988, 3.3 million tonnes of ferrochrome were produced. It is used to make many ferrous alloys, including stainless steel and hard 'chromium' steel.

$$FeCr_2O_4 + C \xrightarrow[\text{furnace}]{\text{electric}} \underbrace{Fe + 2Cr}_{\text{ferrochrome}} + 4CO$$

Several steps are required to obtain pure chromium. First chromite is fused with NaOH in air, when the Cr is oxidized to sodium chromate.

$$2FeCr_2^{III}O_4 + 8NaOH + 3\tfrac{1}{2}O_2 \xrightarrow{1100\,^\circ C} 4Na_2[Cr^{VI}O_4] + Fe_2O_3 + 4H_2O$$

Fe_2O_3 is insoluble but sodium chromate is soluble. Thus the $Na_2[CrO_4]$ is removed by dissolving it in water, and is then acidified to give sodium dichromate. This is less soluble, and can be precipitated. The sodium dichromate is reduced to Cr_2O_3 by heating with C.

$$Na_2[Cr_2O_7] + 2C \rightarrow Cr_2O_3 + Na_2CO_3 + CO$$

Finally Cr_2O_3 is reduced to the metal by Al or Si.

$$Cr_2O_3 + 2Al \rightarrow 2Cr + Al_2O_3$$

Since the metal is brittle, it is seldom used on its own. It is used to make non-ferrous alloys. Alternatively Cr_2O_3 is dissolved in H_2SO_4, and deposited electrolytically on the surface of a metal. This both protects the metal from corrosion and gives it a shiny appearance.

Molybdenum occurs as the mineral molybdenite MoS_2. World production of ores in 1988 had a molybdenum content of 92 000 tonnes. The largest sources are the USA 47%, Chile 15.5%, Canada 13.5% and the USSR 12.5%. Some MoS_2 is also obtained as a by-product from CuS ores.

Table 22.2 Abundance of the elements in the earth's crust, by weight

	ppm	Relative abundance
Cr	122	21
Mo	1.2	56 =
W	1.2	56 =

MoS_2 is roasted in air, converting it to MoO_3. This may be added to steel directly, or MoO_3 may be heated with Fe and Al to give ferromolybdenum, which is then added to steel. Almost 90% of Mo is used to make cutting steel or stainless steel. Pure Mo is obtained by dissolving MoO_3 in dilute NH_4OH and precipitating ammonium molybdate, dimolybdate or paramolybdate. This is reduced with hydrogen to give the metal. Mo metal is used as a catalyst in the petrochemical industry.

Tungsten occurs as tungstates, the most common being wolframite $FeWO_4 \cdot MnWO_4$ and scheelite $CaWO_4$. World production in 1988 had 43 000 tonnes metal content. The largest sources are China 49%, the USSR 21%, Mongolia 5%, South Korea 4%, and Portugal and Australia 3% each. Different processes are used to extract W from wolframite and scheelite. Wolframite is fused with Na_2CO_3, forming sodium tungstate, which is leached out and acidified to give 'tungstic acid' (the hydrated oxide). Scheelite is acidified with HCl when 'tungstic acid' is precipitated and other materials dissolve. 'Tungstic acid' is then heated to give the anhydrous oxide, which is reduced to give the metal by heating with hydrogen at 850 °C.

Mo and W are obtained by this method in the form of powders. Their melting points are high, so melting to give the massive metal would be expensive. Instead metal objects are obtained by fabricating the powder into the required shape and sintering (heating but not melting) under an atmosphere of H_2. Both Mo and W are alloyed with steel, giving very hard alloys, which are used to make 'cutting steel'. This is used to make machine tools. Cutting steel retains its cutting edge even when the metal becomes red hot. About half the W produced is used to make tungsten carbide WC, which is extremely hard (10 on Moh's scale) and will cut glass. WC is used to make the tips for drills. W metal is used to make the filaments in electric light bulbs. Molybdenum disulphide MoS_2 has a layer lattice and is an excellent lubricant, either on its own or when added to hydrocarbon oil.

OXIDATION STATES

The ground state electronic configuration of Cr and Mo is d^5s^1, with a stable half-filled d^5 configuration, whilst W has a d^4s^2 arrangement.

From the electronic structures, Cr and Mo might be expected to form compounds with oxidation states from (+I) to (+VI), and W from (+II) to (+VI) inclusive. They form these states, and in addition some lower states occur as dipyridyl complexes, carbonyl complexes and carbonyl ions.

For Cr the (+II), (+III) and (+VI) states are well known. Cr(+II) is reducing, Cr(+III) is the most stable and important and Cr(+VI) is strongly oxidizing. The most important states for Mo and W are (+V) and (+VI). Whilst Cr(+VI) is strongly oxidizing, Mo(+VI) and W(+VI) are stable. Similarly Cr(+III) is stable but Mo(+III) and W(+III) are strongly reducing. This fits the usual trend that on descending a group the higher

Table 22.3 Oxides and halides

Oxidation states				
(+II)	(+III)	(+IV)	(+V)	(+VI)
–	**Cr_2O_3**	CrO_2	–	CrO_3
–	–	MoO_2	Mo_2O_5	**MoO_3**
–	–	WO_2	(W_2O_5)	**WO_3**
CrF_2	CrF_3	CrF_4	CrF_5	(CrF_6)
$CrCl_2$	**$CrCl_3$**	$CrCl_4$	–	–
$CrBr_2$	**$CrBr_3$**	$CrBr_4$	–	–
CrI_2	**CrI_3**	CrI_4	–	–
–	MoF_3	MoF_4	MoF_5	MoF_6
$MoCl_2$	$MoCl_3$	$MoCl_4$	$MoCl_5$	$(MoCl_6)$
$MoBr_2$	$MoBr_3$	$MoBr_4$	–	–
MoI_2	MoI_3	MoI_4?	–	–
–	–	WF_4	WF_5	**WF_6**
WCl_2	WCl_3	WCl_4	WCl_5	WCl_6
WBr_2	WBr_3	WBr_4	WBr_5	WBr_6
WI_2	WI_3	WI_4?	–	–

The most stable oxidation states are bold, unstable are bracketed.

oxidation states become more stable and the lower states become less stable. A list of known oxides and halides is given in Table 22.3.

GENERAL PROPERTIES

The metals are hard and have very high melting points and low volatility (Table 22.4). The melting point of W is the next highest to carbon.

Cr is unreactive or passive at low temperatures because it is protected by a surface coating of oxide, thus resembling Ti and V in previous groups. It is because of this passive behaviour that Cr is extensively used for electroplating onto iron and other metals to prevent corrosion. Cr dissolves

Table 22.4 Some physical properties

	Covalent radius (Å)	Ionic radius (Å)		Melting point (°C)	Boiling point (°C)	Density ($g\,cm^{-3}$)	Pauling's electro-negativity
		M^{2+}	M^{3+}				
Cr	1.17	0.80[h] 0.73[l]	0.615	1900	2690	7.14	1.6
Mo	1.29	–	0.69	2620	4650	10.28	1.8
W	1.30	–	–	3380	5500	19.3	1.7

h = high spin value, l = low spin radius.

in HCl and H_2SO_4, but is passivated by HNO_3 or aqua regia. Mo and W are relatively inert, and are only slightly attacked by aqueous acids and alkalis. Mo reacts initially with HNO_3, but then becomes passive. Both Mo and W dissolve in HNO_3/HF mixtures, and also in fused Na_2O_2 and fused KNO_3/NaOH. Cr reacts with HCl gas, forming anhydrous $CrCl_2$ and H_2.

The metals do not react with O_2 at normal temperatures (apart from the surface coating). However, on strong heating Cr forms α-Cr_2O_3, which is green coloured and has a corundum structure. In contrast, Mo and W form MO_3. Similarly, on heating Cr with the halogens, trivalent halides CrX_3 are formed. In contrast Mo and W form MCl_6 on heating with Cl_2, and MF_6 is formed at room temperature.

$$2Cr + 3O_2 \rightarrow Cr_2O_3$$
$$2Mo + 3O_2 \rightarrow 2MoO_3$$
$$2Cr + 3Cl_2 \rightarrow 2CrCl_3$$
$$Mo + 3Cl_2 \rightarrow MoCl_6$$

As a result of the lanthanide contraction, there is a close similarity in the size and the properties of Mo and W. The difference between these two elements is greater than in the titanium group between Zr and Hf and in the vanadium group between Nb and Ta. Thus Mo and W can be easily separated in the conventional scheme for qualitative analysis of metals: $WO_3(H_2O)_n$ is precipitated with the insoluble chlorides in Group 1, and molybdates are reduced by H_2S in Group 2 and MoS_2 and S are precipitated.

STANDARD REDUCTION POTENTIALS (VOLTS)

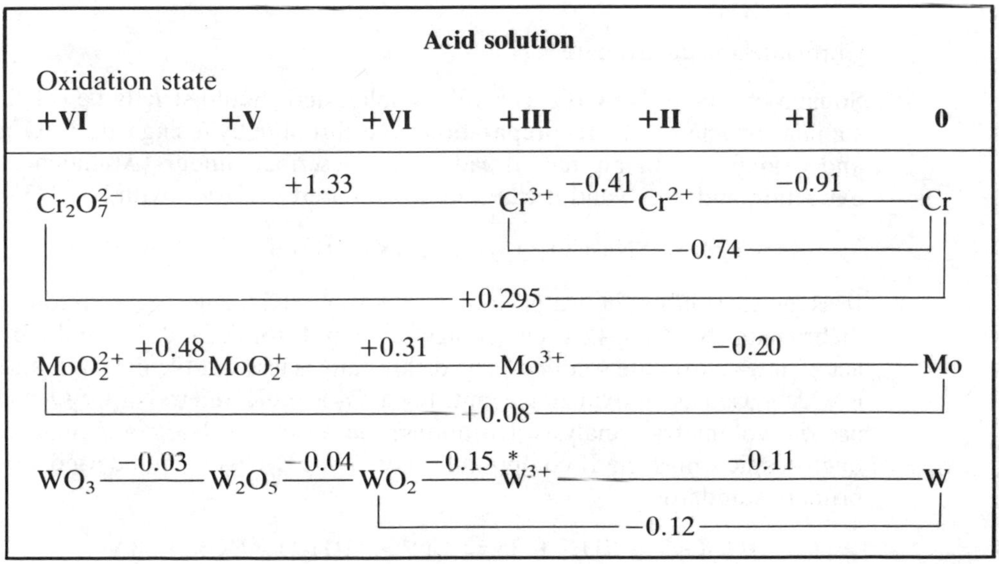

Acid solution

Oxidation state

+VI **+V** **+VI** **+III** **+II** **+I** **0**

$Cr_2O_7^{2-}$ —— +1.33 —— Cr^{3+} —— −0.41 —— Cr^{2+} —— −0.91 —— Cr

Cr^{3+} —— −0.74 —— Cr

$Cr_2O_7^{2-}$ —— +0.295 —— Cr

MoO_2^{2+} —— +0.48 —— MoO_2^{+} —— +0.31 —— Mo^{3+} —— −0.20 —— Mo

MoO_2^{2+} —— +0.08 —— Mo

WO_3 —— −0.03 —— W_2O_5 —— −0.04 —— WO_2 —— −0.15 * —— W^{3+} —— −0.11 —— W

WO_2 —— −0.12 —— W

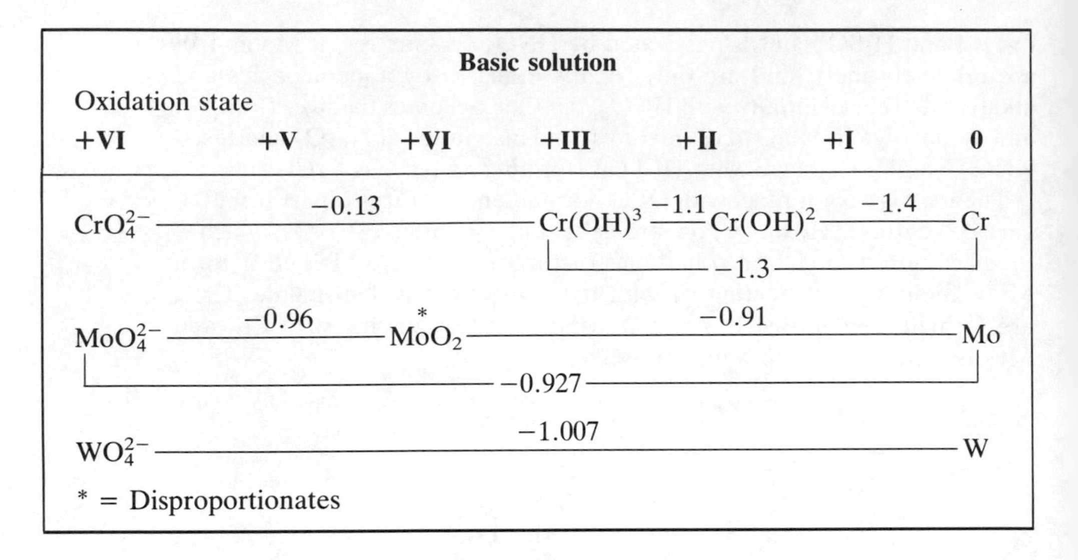

Many of these potentials have been calculated from thermodynamic data, and the existence of species such as Mo^{3+}, W^{3+} and W_2O_5 is questionable.

(+VI) STATE

A limited number of Cr(+VI) compounds are known. These are very strong oxidizing agents and include chromates $[CrO_4]^{2-}$, dichromates $[Cr_2O_7]^{2-}$, chromium trioxide CrO_3, oxohalides CrO_3X^- and CrO_2X_2 (X = F, Cl, Br or I), and $CrOX_4$ (X = F or Cl) and CrF_6.

Chromate and dichromate

Sodium chromate Na_2CrO_4 is a yellow solid, and should strictly be called sodium chromate(VI). Its preparation from chromite by fusing with NaOH and oxidizing with air has already been described under 'Abundance, extraction and uses', and it can also be prepared by fusion with Na_2CO_3.

$$4FeCr_2O_4 + 8Na_2CO_3 + 7O_2 \rightarrow 8Na_2CrO_4 + 2Fe_2O_3 + 8CO_2$$

It is quite soluble in water and is a strong oxidizing agent. Sodium dichromate $Na_2Cr_2O_7$ is an orange coloured solid, and is made by acidifying a chromate solution. The dichromate is less soluble in water, and is widely used as an oxidizing agent. $K_2Cr_2O_7$ is preferred to $Na_2Cr_2O_7$ for use in volumetric analysis (titrations) because the Na compound is hygroscopic whilst the K compound is not. Thus $K_2Cr_2O_7$ can be used as a primary standard.

$$\tfrac{1}{2}Cr_2O_7^{2-} + 7H^+ + 3e \rightleftharpoons Cr^{3+} + 3\tfrac{1}{2}H_2O \quad E^\circ = 1.33\,V$$

Peroxo compounds

When hydrogen peroxide is added to an acidified solution of a dichromate (or any other Cr(+VI) species), a complicated reaction occurs. The products depend on the pH and the concentration of Cr.

$$Cr_2O_7^{2-} + 2H^+ + 4H_2O_2 \rightarrow 2CrO(O_2)_2 + 5H_2O$$

A deep blue–violet coloured peroxo compound $CrO(O_2)_2$ is formed. This decomposes rapidly in aqueous solution into Cr^{3+} and oxygen. The peroxo compound can be extracted into ether, where it reacts with pyridine, forming the adduct py · $CrO(O_2)_2$ (Figure 22.1a). The structure of this is approximately a pentagonal pyramid. Cr is at the centre of a pentagon of four O atoms (from the two peroxo groups) and the N from pyridine, and one O above the pentagon in an apical position.

In less acidic solutions $K_2Cr_2O_7$ and H_2O_2 give salts which are violet coloured and diamagnetic. These are thought to contain $[CrO(O_2)(OH)]^-$, but the structures are not known as the compounds are explosive. In alkaline solution with 30% H_2O_2, a red–brown compound K_3CrO_8 is formed which is a tetraperoxo species $[Cr(O_2)_4]^{3-}$, and contains Cr(+V). In ammonia solution the dark red–brown compound $(NH_3)_3CrO_4$ is formed which contains Cr(+IV) (Figure 22.1b).

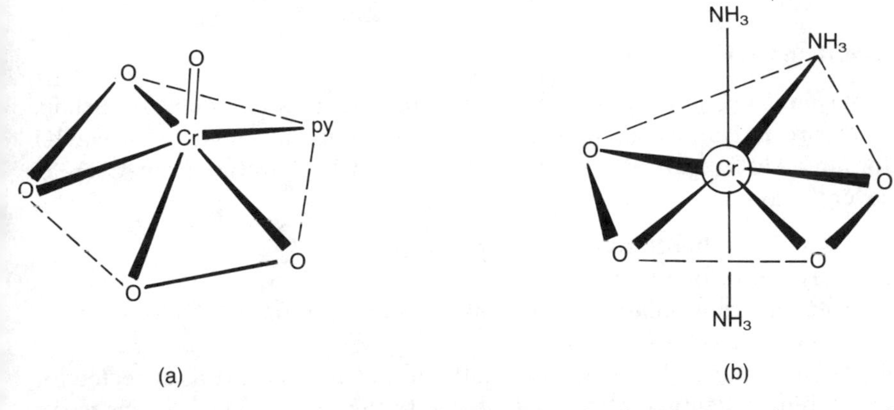

Figure 22.1 Structures of (a) py · $CrO(O_2)_2$, (b) $(NH_3)_3CrO_4$ (pentagonal bipyramid).

Chromium trioxide (chromic acid)

CrO_3 is a bright orange solid, and is commonly called 'chromic acid'. It is usually prepared by adding concentrated H_2SO_4 to a saturated solution of sodium dichromate.

$$Na_2Cr_2O_7 + H_2SO_4 \rightarrow 2CrO_3 + Na_2SO_4 + H_2O$$

The colour arises from charge transfer (not *d–d* spectra as Cr(+VI) has a d^0 configuration). CrO_3 is toxic, and corrosive. The crystal structure consists of chains of fused tetrahedra. CrO_3 dissolves readily in water, and is both a very strong acid and an oxidizing agent. It is an acidic oxide, and dissolves in NaOH solutions, forming the chromate ion CrO_4^{2-}. On heating above 250 °C it loses oxygen in stages, eventually forming green coloured Cr_2O_3.

$$2CrO_3 \rightarrow 2CrO_2 + O_2$$
$$2CrO_2 \rightarrow Cr_2O_3 + \tfrac{1}{2}O_2$$

CrO_3 reacts with F_2 at normal pressures, forming the oxofluorides CrO_2F_2 and $CrOF_4$, but at 170 °C and 25 atmospheres CrF_6 is formed.

$$CrO_3 + F_2 \xrightarrow{150\,^\circ C} CrO_2F_2 + \tfrac{1}{2}O_2$$

$$CrO_3 + 2F_2 \xrightarrow{220\,^\circ C} CrOF_4 + O_2$$

$$CrO_3 + 3F_2 \xrightarrow{170\,^\circ C,\ 25\ \text{atmospheres}} CrF_6 \text{ lemon yellow solid}$$

CrO_3 is widely used to make chromium plating solutions. It can be dissolved in acetic acid and used in this form as an oxidant in organic chemistry, though reactions may be explosive. Chromic acid solutions are used to clean laboratory glassware.

MoO_3 and WO_3

MoO_3 and WO_3 are formed by heating the metal in air. They are acidic. They are not attacked by acids except HF, but they dissolve in NaOH forming MoO_4^{2-} and WO_4^{2-} ions. MoO_3 and WO_3 differ from CrO_3 in several ways:

1. They have almost no oxidizing properties.
2. They are insoluble in water.
3. Their melting points are much higher (CrO_3 m.p. 197 °C, MoO_3 m.p. 795 °C and WO_3 m.p. 1473 °C).
4. Their colour and structures are different. MoO_3 is white as expected for d^0, but on heating it turns yellow due to the formation of defects in the solid. The structure is a layer lattice. WO_3 is lemon yellow in colour, and has a slightly distorted rhenium trioxide ReO_3 structure of WO_6 octahedra sharing corners in three dimensions. (See Figure 23.4a.)

Mixed oxides

Several mixed oxides can be made by fusing MoO_3 or WO_3 with Group I or II oxides. These comprise chains or rings of MoO_6 or WO_6 octahedra. Moist WO_3 turns slightly blue on exposure to UV light. Mild reduction of aqueous suspensions of MoO_3 and WO_3 or acidic solutions of molybdates or tungstates also gives a blue colour. The 'blue oxides' so produced are thought to have Mo or W in oxidation states of (+IV) and (+V), and contain some OH^- instead of O^{2-} to balance the charges.

Oxohalides

Oxohalides of the type MO_2Cl_2 may be formed by dissolving the trioxide in strong acid, or in some cases by the action of strong acids on salts such as dichromates, or by direct addition of the halogens to the dioxide.

$$CrO_3 + 2HCl \xrightarrow{\text{conc. } H_2SO_4} CrO_2Cl_2 + H_2O$$

$$K_2Cr_2O_7 + 6HCl \xrightarrow{\text{conc. } H_2SO_4} 2CrO_2Cl_2 + 2KCl + 3H_2O$$

Chromyl chloride CrO_2Cl_2 is a deep red coloured liquid. It is formed in qualitative analysis to confirm the presence of chloride ions. The suspected chloride is mixed with solid $K_2Cr_2O_7$ and gently warmed with concentrated H_2SO_4. Deep red vapours of CrO_2Cl_2 are formed, and if these are passed into aqueous NaOH the solution turns yellow due to the formation of Na_2CrO_4.

Chromyl and molybdenyl chlorides are covalent acid chlorides and are readily decomposed by water. Tungstenyl chloride hydrolyses less readily.

Halides

CrF_6 is a yellow solid made by heating the elements under pressure in a bomb, and cooling rapidly. The product is unstable and decomposes into CrF_5 and F_2. In contrast MoF_6 and WF_6 are very stable. They are both low melting (MoF_6 17.4 °C, WF_6 1.9 °C), volatile, and easily hydrolysed. They are diamagnetic and colourless as expected for a d^0 configuration. However, $MoCl_6$ and WCl_6 are black and WBr_6 is dark blue. WCl_6 is made by heating the metal in Cl_2. It reacts with water, forming tungstic acid. WCl_6 is soluble in EtOH, ether and CCl_4, and is used as the starting point for making other compounds.

(+V) STATE

There are few Cr(+V) compounds, and they are unstable and decompose to Cr(+III) and Cr(+VI). One example is K_3CrO_8, a red–brown compound formed from $NaCrO_4$ and H_2O_2 in alkaline solution (see above). K_3CrO_8 contains the tetraperoxo species $[Cr(O_2)_4]^{3-}$. Another example is CrF_5 which is made by heating the elements at 500 °C or heating CrO_3 with F_2. It is a red solid, based on CrF_6 octahedra linked to give a *cis*-bridged polymer.

MoF_5 has a tetrameric structure of four octahedra joined into a ring, like NbF_5 and TaF_5 (see Figure 21.1). Heating Mo and Cl_2 gives Mo_2Cl_{10}. This is soluble in benzene and other organic solvents. It exists as monomeric $MoCl_5$ in solution, but dimerizes to Mo_2Cl_{10} in the solid. Mo_2Cl_{10} is used as the starting point for making other Mo compounds. It is rapidly hydrolysed by water, and removes O from oxygenated solvents, forming oxochlorides. Mo_2Cl_{10} is paramagnetic (μ = 1.6 BM), indicating that there is one unpaired electron and thus no metal–metal bonding.

(+IV) STATE

Cr(+IV) compounds are also rare. CrF_4 is formed by heating the elements at 350 °C. $MoCl_4$ exists in two polymeric forms, one like $NbCl_4$ (see Figure 21.3) comprising chains of octahedra with the metal atoms displaced in pairs, forming metal–metal bonds, and the other form without metal–metal bonds.

CrO_2 is made from CrO_3 by hydrothermal reduction, and has a rutile (TiO_2) structure. The oxide is black in colour and has some metallic conductivity. It is also ferromagnetic, and is widely used to make high quality magnetic recording tapes. MoO_2 and WO_2 are both made by reducing the trioxide with hydrogen. They are brown–violet in colour and are insoluble in non-oxidizing acids, but dissolve in concentrated HNO_3, forming MoO_3 or WO_3. The dioxides have a copper-like lustre and have distorted rutile structures with strong metal–metal bonds. The oxohalide $CrOF_2$ is also known.

(+III) STATE

Chromium

Chromic compounds Cr^{3+} are the most important and most stable compounds of chromium. Although this oxidation state is very stable in acidic solution, it is easily oxidized to Cr(+VI) in alkaline solution.

Cr_2O_3 is a green solid which is used as a pigment. The most convenient preparation is by heating ammonium dichromate $(NH_4)_2Cr_2O_7$ in the well known volcano experiment used in some fireworks. (Once the reaction is started, it produces enough heat to continue on its own. The green coloured Cr_2O_3 powder is blown in the air by the large volume of N_2 and water vapour produced, and settles like dust from a volcano.) Cr_2O_3 is also formed by burning the metal in air, or by heating CrO_3. It has a corundum Al_2O_3 structure.

$$(NH_4)_2Cr_2O_7 \rightarrow Cr_2O_3 + N_2 + 4H_2O$$

$$4Cr + 3O_2 \rightarrow 2Cr_2O_3$$

$$4CrO_3 \rightarrow 2Cr_2O_3 + 3O_2$$

The addition of NaOH to Cr^{3+} solutions does *not* precipitate the hydroxide, but the hydrous (hydrated) oxide is precipitated instead.

$$Cr^{3+} + 3OH^- \rightarrow Cr(OH)_3 \rightarrow Cr_2O_3(H_2O)_n$$

The oxide becomes inert to acids and bases if heated strongly, but otherwise it is amphoteric, giving $[Cr(H_2O)_6]^{3+}$ with acids, and 'chromites' with concentrated alkali. The species present in 'chromite' solutions is probably $[Cr^{III}(OH)_6]^{3-}$ or $[Cr^{III}(OH)_5H_2O]^{2-}$. Cr_2O_3 is commercially important. It is formed as one step in the extraction of chromium. It is used as a pigment in paint, rubber and cement, and as a catalyst for a wide variety of reactions including the manufacture of polythene and butadiene.

All the anhydrous CrX_3 halides are known. $CrCl_3$ is a solid which forms red–violet flakes. The flakiness is related to the layer lattice structure of the solid. The chloride ions are cubic close-packed, and to maintain stoichiometry one third of the octahedral holes must be occupied by Cr^{3+} ions. Two thirds of the holes are occupied in one layer, and none in the next layer, and consequently only weak van der Waals forces hold some chloride layers together. In aqueous solution the halides form the violet coloured hexaaqua ion $[Cr(H_2O)_6]^{3+}$, and halogen complexes such as $[Cr(H_2O)_5Cl]^{2+}$, $[Cr(H_2O)_4Cl_2]^+$ and $[Cr(H_2O)_3Cl_3]$. The hexaaqua ion also occurs in many crystalline compounds such as $[Cr(H_2O)_6]Cl_3$ and the alums. The alums are double salts, for example chrome alum $K_2SO_4 \cdot Cr_2(SO_4)_3 \cdot 24H_2O$, which crystallizes from mixed solutions of $Cr_2(SO_4)_3$ and K_2SO_4. The structure is better shown if the formula is written $[K(H_2O)_6]\ [Cr^{III}(H_2O)_6]\ [SO_4]_2$. In solution the alums dissociate completely into simple ions.

The hexaaqua ion is acidic, and it may form a dimer by means of two hydroxo bridges.

$$[Cr(H_2O)_6]^{3+} \rightleftharpoons H^+ + [Cr(H_2O)_5OH]^{2+} \rightleftharpoons \left[(H_2O)_4Cr \begin{array}{c} H \\ O \\ / \quad \backslash \\ \\ \backslash \quad / \\ O \\ H \end{array} Cr(H_2O)_4\right]^{4+} + 2H_2O$$

Cr^{3+} ions form an enormous number and variety of complexes. These are typically six-coordinate with octahedral structures, and are very stable both as solids and in aqueous solution. The stability is related to the high crystal field stabilization energy from its d^3 electronic configuration. The magnetic moments of these complexes are close to the spin only value of 3.87 BM expected for three unpaired electrons. Complexes include the hexaaqua $[Cr(H_2O)_6Cl]^{2+}$ and halogen complexes $[Cr(H_2O)_6]^{3+}$, mentioned above. The ammine and oxalate complexes show many different forms of isomerism, for example the ammine complexes:

$[Cr(NH_3)_6]^{3+}$	only one form
$[Cr(NH_3)_5Cl]^{2+}$	only one form
$[Cr(NH_3)_4Cl_2]^+$	*cis* and *trans* isomers
$[Cr(NH_3)_3Cl_3]$	*mer* and *fac* isomers

and in the oxalate complexes:

$[Cr(oxalate)_3]^{3-}$ *d* and *l* isomers

Isomerism is discussed more fully in Chapter 7 (see Figure 7.3). There are also many cyanide and thiocyanate complexes. Reinecke's salt $NH_4[Cr(NH_3)_2(NCS)_4] \cdot H_2O$ is often used to precipitate large positive ions. This is because when the anion and cation are similar in size, the crystal has a high coordination number and thus a higher lattice energy.

Cr(+III) forms an unusual basic acetate $[Cr_3O(CH_3COO)_6L_3]^+$ (where L is water or some other ligand). The structure consists of a triangle of three Cr atoms with an O atom at the centre. The six acetate groups act as bridges between the Cr atoms – two acetate groups across each edge of the triangle. Each Cr atom is octahedrally surrounded by six atoms: O atoms from four acetate groups, the central O, and L in the sixth position. L may be water or another ligand. Cr^{3+} has a d^3 arrangement and should have a magnetic moment of 3.87 BM. The magnetic moment of the complex at room temperature is only 2 BM. It is thought that the smaller value is due to partial pairing of d electrons on the three metal atoms by means of $d\pi - p\pi$ bonding through O. This type of carboxylate complex is formed by the trivalent ions of Cr, Mn, Fe, Ru, Rh and Ir. (Partial spin pairing and a reduced magnetic moment are also found in other complexes such as $[(NH_3)_5Cr—OH—Cr(NH_3)_5]^{5-}$ and $[(NH_3)_5Cr—O—Cr(NH_3)_5]^{4-}$ where the magnetic moment is temperature dependent, and is about 1.3 BM at room temperature, but almost zero at −200 °C.)

Spectra

Cr(+III) has a d^3 electronic configuration. In the ground state these electrons occupy the t_{2g} orbitals, i.e. $(t_{2g})^3$. The two e_g orbitals are empty, providing two 'holes' into which electrons can be promoted. The situation is analagous to that for d^2 described for V^{3+} in Chapter 21. The electronic spectra of Cr(+III) complexes exhibit three absorption bands. In the ground state, the d_{xy}, d_{xz} and d_{yz} orbitals each contain one electron, giving the singly degenerate state ${}^4A_{2g}(F)$. The first excited state corresponds to promoting one electron, i.e. $(t_{2g})^2\,(e_g)^1$, and gives two terms ${}^4T_{2g}(F)$ and ${}^4T_{1g}(F)$. The second excited state corresponds to promoting two electrons, i.e. $(t_{2g})^1\,(e_g)^2$. This gives only one quartet that is triply degenerate, so the term symbol is ${}^4T_{1g}(P)$. The transitions are ${}^4T_{2g}(F) \rightarrow {}^4T_{2g}(F)$, ${}^4T_{2g}(F) \rightarrow {}^4T_{1g}(F)$ and ${}^4A_{2g}(F) \rightarrow {}^4T_{1g}(P)$. In the hexaaqua ion $[Cr(H_2O)_6]^{3+}$ bands are found at 17 400 cm^{-1} and 24 700 cm^{-1}, and there is a shoulder on the charge transfer band at 37 800 cm^{-1}. (See also Chapter 32.)

Molybdenum and tungsten

Mo(+III) and W(+III) do not exist as oxides, but all the halides are known except WF_3 (Table 22.3). These compounds do not contain simple ions. Mo(+III) compounds are fairly stable, but slowly oxidize in air and slowly hydrolyse in water. They form octahedral complexes with halide ions in solution.

$$MoCl_3 + 3Cl^- \rightarrow [MoCl_6]^{3-}$$

Two solid forms of $MoCl_3$ are known, one with cubic close packing of chlorine atoms, the other based on hexagonal close packing. In both forms the Mo atoms are displaced from the centres of adjacent octahedra, and form metal–metal bonds of length 2.76 Å. In contrast W(+III) compounds are unstable. WCl_3 is really W_6Cl_{18} and forms a cluster compound $[W_6Cl_{12}]^{6+}$ structurally like $[Nb_6Cl_{12}]^{2+}$ (see Figure 21.4). W_6Br_{18} also forms a cluster compound, but its structure contains $[W_6Br_8]^{6+}$ and has the same structure as $[Mo_6Br_8]^{4+}$ (Figure 22.3).

(+II) STATE

Chromous compounds are well known, are ionic and contain Cr^{2+}. Solutions containing $[Cr(H_2O)_6]^{2+}$ can be produced either by electrolytically reducing solutions containing Cr^{3+}, or by reducing them with zinc amalgam. They can also be produced from Cr metal and acids. The $[Cr(H_2O)_6]^{2+}$ ion is sky blue coloured. It is one of the strongest reducing agents known in aqueous solution.

$$Cr^{3+} + e \rightarrow Cr^{2+} \quad E^\circ = -0.41\,V$$

If the solution is acidic Cr^{2+} slowly reduces water to H_2. Cr(+II) compounds are oxidized by air to Cr^{3+}. Cr^{2+} is used to remove the last trace of oxygen from nitrogen, and has other uses as a reducing agent. Cr(+II) may be stabilized by forming coordination compounds, such as $[Cr(NH_3)_6]^{2+}$ or $[Cr(dipyridyl)_3]^{2+}$. Though Cr^{2+} is stable to disproportionation, the dipyridyl complex disproportionates.

$$2[Cr(dipyridyl)_3]^{2+} \rightarrow [Cr(dipyridyl)_3]^{+} + [Cr(dipyridyl)_3]^{3+}$$

Hydrated salts such as $CrSO_4 \cdot 7H_2O$, $Cr(ClO_4)_2 \cdot 6H_2O$ and $CrCl_2 \cdot 4H_2O$ can be isolated, but they cannot be dehydrated as they decompose on heating.

Anhydrous Cr(+II) halides can be made either by reducing the trihalides with hydrogen at 500 °C, or from the metal and HF, HCl, HBr or I_2 at 600 °C. The dihalides are all readily oxidized in air to the (+III) state unless protected by an inert atmosphere such as N_2. $CrCl_2$ is the most important, and it dissolves in water, giving the sky blue coloured $[Cr(H_2O)_6]^{2+}$ ion.

Chromium forms many complexes, especially with N ligands and chelating groups. These are easily oxidized, particularly if moist. Almost all the complexes are octahedral, and both high-spin and low-spin complexes are known. The high-spin complexes have the electronic configuration $(t_{2g})^3\,(e_g)^1$. The asymmetrical filling of the e_g orbitals causes Jahn–Teller distortion similar to that found in complexes of Cu^{2+}.

Chromium(II) acetate dihydrate $Cr_2(CH_3COO)_4 \cdot 2H_2O$ is one of the most stable chromous compounds. It is easily prepared by adding sodium acetate to a solution containing Cr^{2+} under an atmosphere of N_2. Hydrated chromium(II) acetate is precipitated as a red solid. It is a good starting material for the preparation of other Cr(+II) salts. It has an unusual dimeric bridge structure (Figure 22.2). Each Cr^{2+} is surrounded by a distorted octahedron made up of four O atoms from four bidentate acetate groups, one O from a H_2O molecule, and the other Cr^{2+} ion. The four

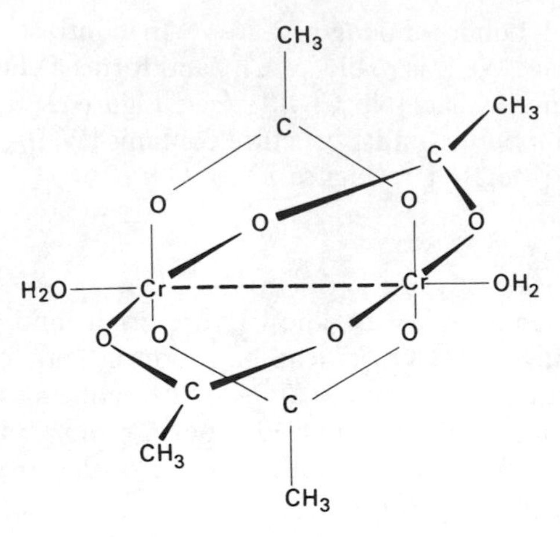

Figure 22.2 Structure of $Cr_2(CH_3COO)_4 \cdot 2H_2O$.

acetate groups bridge between the two Cr atoms. The very short distance of 2.36 Å between the two Cr atoms is evidence of a strong metal–metal bond. Cr^{2+} has a d^4 configuration and has four unpaired electrons, but chromium(II) acetate is diamagnetic. This suggests that all four unpaired electrons take part in M–M bonding. Assuming that the ligands are bonded using one s, three p and the $d_{x^2-y^2}$ orbitals, the d_{z^2} orbital can form a σ bond to the other Cr^{2+}. In addition the d_{xz} and d_{yz} orbitals can form π bonds between the Cr atoms, and the d_{xy} orbital forms a δ bond. Quadruple bonds of this type are found with other heavy transition metals Mo, W, Tc and Re, e.g. $[Mo_2(CH_3COO)_4]$ (note there are no axial H_2O ligands), $[Mo_2Cl_8]^{4-}$, $[W_2Cl_4(PR_3)_4]$, $[W_2(CH_3)_8]^{4-}$, $[W_2(C_8H_8)_3]$, $[Re_2Cl_8]^{2-}$ and $[Re_2Br_8]^{2-}$. At one time quadruple bonds were regarded as anomalies, but they may be more common than originally thought. It is suggested that once a M–M multiple bond has been formed, the metal can easily be reduced to give a bond of higher order.

Mo and W do not form difluorides, but the other six (+II) halides are known. They are usually made by reduction or thermal decomposition of higher halides. They do not exist as simple ions, but form 'cluster compounds' instead. $MoBr_2$ is really $[Mo_6Br_8]Br_4 \cdot 2H_2O$, and all six of these so-called 'dihalides' have the same structure based on a cluster $[M_6X_8]^{4+}$ with four halide ions and two H_2O acting as electron donors. The structure of the $[M_6X_8]^{4+}$ unit is an octahedral cluster of six metal atoms. There is extensive M–M bonding, and eight face-bridging halogen atoms occupy the eight triangular faces of the octahedron (Figure 22.3). Thus each halogen is bonded to three metal atoms. Each metal atom has an unoccupied coordination position. The $M_6X_8^{4+}$ units can thus accept six coordinate bonds to the metal atoms at the corners of the octahedron from the remaining four X^- ions and two H_2O molecules, or from any other suitable electron pair donor. The six ligands on the corners are labile, and undergo replacement reactions quite readily, forming for example

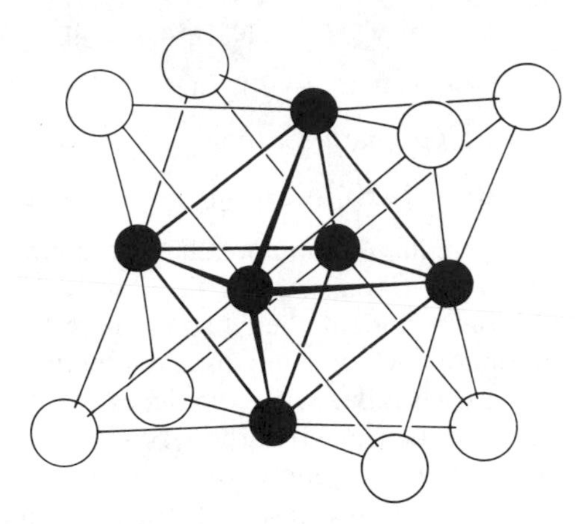

Figure 22.3 Structure of $[Mo_6Br_8]^{4+}$ ion showing the octahedral M_6 cluster.

$[Mo_6Cl_8Br_4]$ or $[Mo_6Cl_8(Me_2SO)_6]^{4+}$. In contrast the bridging halogens in the cluster undergo replacement very slowly. Despite the low oxidation state, Mo_6Cl_{12} is not reducing, though W_6Cl_{12} is reducing in solution.

The bonding in these compounds is not settled. The compounds are diamagnetic; therefore Mo must use all six outer electrons d^5s^1 or W d^4s^2 for bonding. Since there are six M atoms there are 36 valence electrons. It seems probable that eight electrons are used to bond the eight Cl atoms on the faces, and four electrons are transferred to form four X^- ions. This leaves 24 electrons to form M–M bonds along the 12 edges of the M_6 octahedron.

(+I) STATE

The oxidation state (+I) expected for the atoms with a d^5s^1 configuration is very uncommon. It is doubtful if Cr^+ exists except when stabilized in a complex. Trisdipyridyl chromium(I) perchlorate $[Cr(dipyridyl)_3]^+ ClO_4^-$ is known. Mo and W form sandwich-type structures such as $(C_6H_6)_2Mo^+$ and $C_5H_5MoC_6H_6$ where the metal is in the (+I) state.

ZERO STATE, (−I) AND (−II)

The zero oxidation state arises in metal carbonyls such as $M(CO)_6$, where the σ bonding electrons are donated by the CO group to the metal, and strong $d\pi$–$p\pi$ back bonding occurs from the filled metal orbitals. All three metals form octahedral carbonyl compounds of this type. They are stable and may be sublimed under reduced pressure. They are soluble in organic solvents. The bipyridyl complex $[Cr(bipyridyl)_3]$ is also octahedral.

An unusual complex dibenzene chromium $[Cr(\eta^6\text{-}C_6H_6)_2]$ was made by E.O. Fischer in 1955. It forms dark brown crystals and has a sandwich structure similar to ferrocene, though it is much more air sensitive than ferrocene. Cr has a coordination number of 12. It was made as follows:

$$3CrCl_3 + 2Al + AlCl_3 + 6C_6H_6 \rightarrow 3[Cr(\eta^6\text{-}C_6H_6)_2]^+ + 3[AlCl_4]^-$$

$$2[Cr(\eta^6\text{-}C_6H_6)_2]^+ + Na_2S_2O_4 + 4OH^- \rightarrow 2[Cr(\eta^6\text{-}C_6H_6)_2] + 2Na_2SO_3 + 2H_2O$$

It may also be made by a Grignard reaction:

$$CrCl_3 + 2C_6H_5MgBr \rightarrow [Cr(\eta^6\text{-}C_6H_6)_2]^+ Cl^- + MgBr_2 + MgCl_2$$

For work on this and similar organometallic compounds, Fischer was awarded the Nobel Prize for Chemistry in 1973, jointly with G. Wilkinson who did parallel work on cyclopentadienyl compounds.

The η^5-C_5H_5 compounds include complexes containing only one cyclopentadienyl ring, and in a similar way complexes are formed with one benzene ring. For example, benzene tricarbonyl complexes for Cr, Mo and W, e.g. $[Cr(\eta^6\text{-}C_6H_6)(CO)_3]$, are yellow solids; the metal has a coordination number of 9. The benzene complexes are more reactive and less thermally stable than their η^5-C_5H_5 counterparts.

The lower oxidation states occur in carbonyl ions: (−I) in $[M_2(CO)_{10}]^{2-}$, and (−II) in $[M(CO)_5]^{2-}$.

CHROMATES, MOLYBDATES AND TUNGSTATES

The oxides CrO_3, MoO_3 and WO_3 are strongly acidic, and dissolve in aqueous NaOH forming discrete tetrahedral chromate CrO_4^{2-}, molybdate MoO_4^{2-} and tungstate WO_4^{2-} ions.

$$CrO_3 + 2NaOH \rightarrow 2Na^+ + CrO_4^{2-} + H_2O$$

Chromates, molybdates and tungstates exist both in solution and as solids. Chromates are strong oxidizing agents, but molybdates and tungstates have only weak oxidizing powers. Molybdates and tungstates can be reduced to form the blue oxides.

On acidifying, chromates CrO_4^{2-} form $HCrO_4^-$ and orange–red dichromates $Cr_2O_7^{2-}$, in which two tetrahedral units join together by sharing the oxygen atom at one corner (Figure 22.4). $HCrO_4^-$ and $Cr_2O_7^{2-}$ exist in equilibrium over a wide range of pH from 2–6.

CrO_3 is precipitated from very concentrated acid (below pH 1).

$$\underset{\text{yellow}}{CrO_4^{2-}} \rightleftharpoons \underset{\text{orange}}{Cr_2O_7^{2-}} \rightleftharpoons CrO_3$$

$Na_2Cr_2O_7$ is the most important chromium compound, and is produced as one step in the extraction of chromium. Apart from that used in the extraction of the metal, 304 700 tonnes were used in 1985 for chrome tanning of leather, making various lead chromes, for 'anodizing' aluminium, and as an oxidizing agent. There is some evidence for further polymerization giving a limited polychromate series. Trichromates $Cr_3O_{10}^{2-}$ and tetrachromates $Cr_4O_{13}^{2-}$ have been found.

When molybdate and tungstate solutions are acidified they condense and give an extensive range of polymolybdates and polytungstates. Below a pH of 1 the hydrated oxides are precipitated. $MoO_3 \cdot 2H_2O$ is yellow and $WO_3 \cdot 2H_2O$ is white. The formation of polyacids is a prominent feature of the chemistry of Mo and W. Other transition elements V, Nb, Ta and U

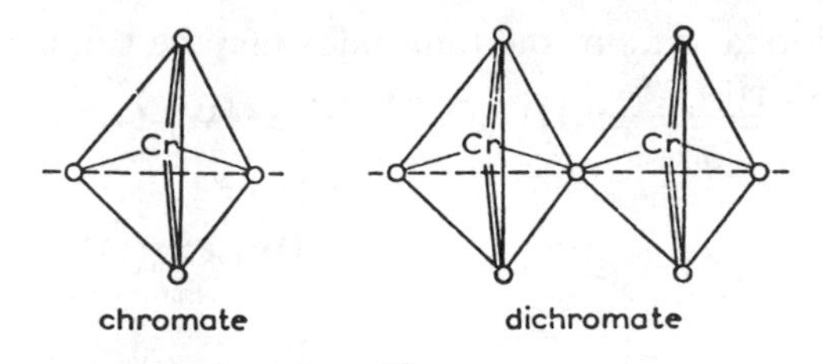

Figure 22.4 Chromate and dichromate ions.

also form polyacids, but to a lesser extent. The polyanions contain MoO_6 or WO_6 octahedra, which are joined together in a variety of ways by sharing corners or edges, but not faces. The polyacids of Mo and W are divided into two main types:

1. Isopolyacids, where the anions which condense together are all of the same type – for example all MoO_6 groups or all WO_6 groups.
2. Heteropolyacids, where two or more different types of anion condense together – for example molybdate or tungstate groups with phosphate, silicate or borate groups.

The isopolyacids of Mo and W are not completely understood. They are very difficult to study because the extent of hydration and protonation of the various species in solution is not known. The fact that a solid can be crystallized from solution does not prove that the ion has that structure or even exists in solution. The first step in polyacid formation as the pH is lowered must be to increase the coordination number of Mo or W from 4 to 6 by adding water molecules. The relationship between the stable species so far known is:

$$\underset{\text{normal molybdate}}{[MoO_4]^{4-}} \xrightarrow{\text{pH } 6} \underset{\text{paramolybdate}}{[Mo_7O_{24}]^{6-}} \xrightarrow{\text{pH } 1.5\text{–}2.9} \underset{\text{octamolybdate}}{[Mo_8O_{26}]^{4-}} \xrightarrow{\text{pH} < 1} \underset{\text{hydrated oxide}}{MoO_3 \cdot 2H_2O}$$

The structures of the paramolybdate and octamolybdate ions have been confirmed by X-ray crystallographic studies of their crystalline salts (Figure 22.5).

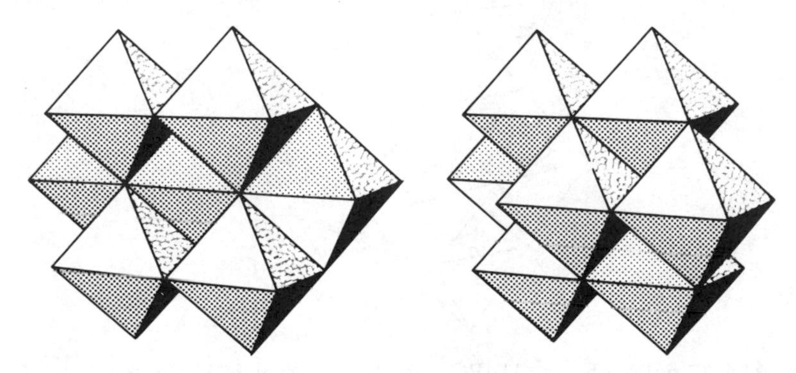

Figure 22.5 Some molybdate ions. (From H.J. Eméleus and A.G. Sharpe, *Modern Aspects of Inorganic Chemistry*, 4th ed., Routledge and Kegan Paul, 1973.)

The present understanding of the tungstates may be summarized:

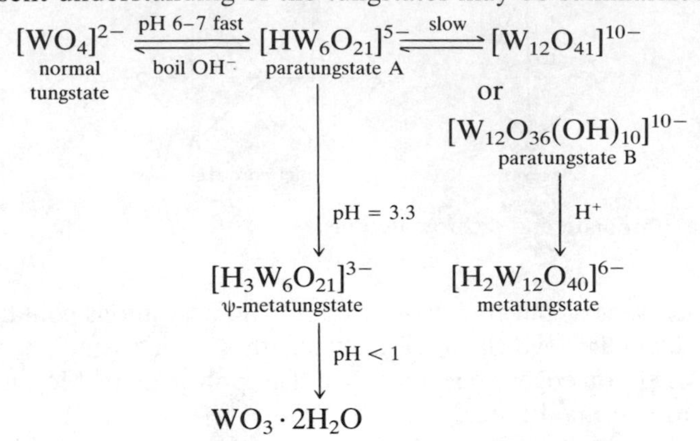

Heteropolyions are formed if a molybdate or tungstate solution is acidified in the presence of phosphate, silicate or metal ions. The second anion provides a centre round which the MoO_6 or WO_6 octahedra condense, by sharing oxygen atoms with other octahedra and with the central group. The central groups are often oxoanions such as PO_4^{3-}, SiO_4^{4-} compounds, and BO_4^{3-}, but other elements including Al, Ge, Sn, As, Sb, Se, Te, I and many of the transition elements will serve as the second group. The ratio of MoO_6 or WO_6 octahedra to P, Si, B or other central atom is usually 12 : 1, 9 : 1 or 6 : 1, although other ratios occur less commonly. A well known example of heteropolyacid formation is the test for phosphates. A phosphate solution is warmed with ammonium molybdate and nitric acid, and a yellow precipitate of ammonium phosphomolybdate $(NH_4)_3[PO_4 \cdot Mo_{12}O_{36}]$ is formed.

The structures of several heteropolyacids have been established. In the 12-heteropolyacids, for example 12-phosphotungstic acid, 12 WO_6 octahedra surround a PO_4 tetrahedron. This ion may be considered as four groups of three WO_6 octahedra (Figure 22.6).

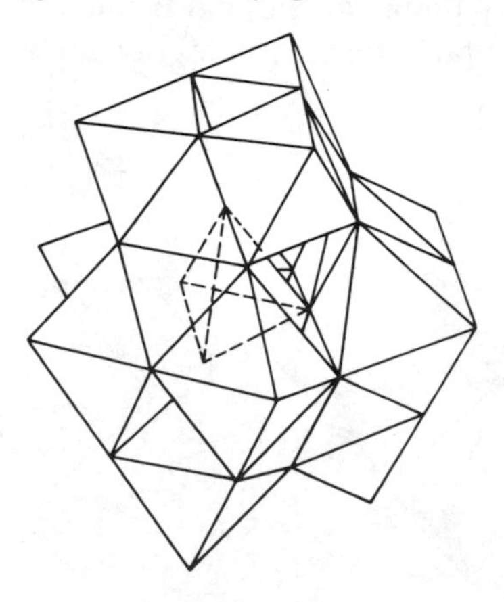

Figure 22.6 12-polyacid, e.g. $H_3[PO_4 \cdot W_{12}O_{36}]$. (From H.J. Eméleus and A.G. Sharpe, *Modern Aspects of Inorganic Chemistry*, 4th ed., RKP, 1973.)

The 6-heteropolyacids accommodate larger central atoms, which have a coordination number of 6. The arrangement of six MoO_6 octahedra as shown in Figure 22.7 leaves a central cavity large enough to accept the octahedron from the hetero atom, and has been found in $K_6[TeMo_6O_{24}]$.

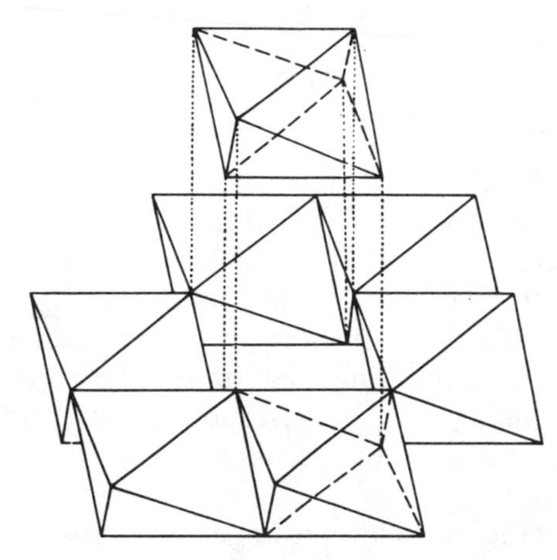

Figure 22.7 6-polyacid, e.g. $K_6[TeMo_6O_{24}]$. (From H.J. Eméleus and A.G. Sharpe, *Modern Aspects of Inorganic Chemistry*, 4th ed., RKP, 1973.)

TUNGSTEN BRONZES

Tungsten bronzes were originally made by strongly heating sodium tungstate Na_2WO_4 with WO_3 and H_2. They are now made by heating Na_2WO_4 with W metal, when blue, purple, red or yellow tungsten bronzes are formed. These are semi-metallic solids which have a lustre and conduct electricity. They are very inert to both strong acids and strong alkali. They are used in the production of 'bronze' and 'metallic' paints.

Tungsten bronzes are nonstoichiometric compounds of formula M_xWO_3, where M is Na, K, a Group II metal or a lanthanide, and x is always less than one. The Na^2 or other metal ions occupy interstitial positions. The colour depends on the proportion of M present, and for the sodium compounds: $x \sim 0.9$ yellow or gold, $x \sim 0.7$ orange, $x \sim 0.5$ red, $x \sim 0.3$ blue–black. The variable amount of Na^+ produces a defective lattice and some of the sites which should be occupied by alkali metals are vacant. It might be thought that for each Na^+ removed from $NaWO_3$, one tungsten would change from W(+V) to W(+VI). The properties of the tungsten bronzes are better explained by assuming that all the tungsten atoms are in the (+VI) state. The valency electrons from the alkali metals are free to move throughout the lattice, giving metallic conduction. The conductivity decreases with increased temperature, as in a metal. The structure is WO_6 octahedra joined together by sharing all their corners with other octahedra. Further details of tungsten bronzes are given in Further Reading.

Molybdenum also forms bronzes similar to those of W, but a high pressure is needed to form them, and the Mo compounds are less stable. This may be because Mo(+V) is less stable than W(+V), or because the solid structure is different and contains MoO_6 octahedra joined by a mixture of corner and edge sharing with other octahedra. Lithium also forms bronzes, but these do not conduct electricity.

BIOLOGICAL IMPORTANCE

Trace amounts of Cr and Mo are necessary in the diet of mammals. Cr(+III) and insulin are both involved in maintaining the correct level of glucose in the blood. In cases of Cr deficiency, glucose is only removed from the blood half as fast as normally. Some cases of diabetes may reflect faulty metabolism of Cr. The most important medical aspect of Cr salts is that larger amounts either ingested or on the skin are carcinogenic. Compounds containing Cr(VI), e.g. dichromate and chromate, are particularly so. Thus care should be taken when performing titrations using $K_2Cr_2O_7$ or $KCrO_4$.

Mo is present in the catalysts of nitrogen fixing bacteria. (The amount of nitrogen fixed biologically is estimated at 175 million tonnes per year, compared with a total of 81 million tonnes of NH_3 produced by the Haber–Bosch process and the distillation of coal.) The best known nitrogen fixing bacterium is *Rhizobium*. This contains the metallo-enzyme nitrogenase.

Nitrogenase contains two proteins, molybdoferredoxin and azoferredoxin. Molybdoferredoxin is brown, air sensitive, contains two Mo atoms, 24–36 Fe atoms and 24–36 S atoms together with a protein, and has a molecular weight of about 225 000. Azoferredoxin is yellow, air sensitive, is a derivative of ferredoxin $Fe_4S_4(SR)_4$ and has a molecular weight in the range 50 000–70 000. It is not certain exactly how nitrogen fixation occurs. It is thought that N_2 bonds to the Mo in molybdoferredoxin (whether end-on or sideways-on is not known). If the Mo is sufficiently reduced it can probably bond to the antibonding orbitals on N_2. Then the Fe in azoferredoxin is reduced by free ferredoxin $Fe_4S_4(SR)_4$. An electron is transferred from reduced azoferredoxin to molybdoferredoxin, possibly via the Fe, then from the Mo to the N_2. Protons are then added to N_2, eventually giving NH_3. Adenosine triphosphate ATP is required to provide the necessary energy for the process. Nitrogenase is not specific for the reaction $N_2 \rightarrow NH_3$, but also reduces alkynes to alkenes, and cyanides.

FURTHER READING

Abel, E.W. and Stone, F.G. (1969, 1970) The chemistry of transition metal carbonyls, *Q. Rev. Chem. Soc.*, Part I – Structural considerations, **23**, 325; Part II – Synthesis and reactivity, **24**, 498.

Bevan, D.J.M. (1973) *Comprehensive Inorganic Chemistry*, Vol. 4 (Chapter 49: Non-stoichiometric compounds), Pergamon Press, Oxford.
Burgmayer, S.J.N. and Stiefel, E.I. (1985) Molybdenum enzymes, *J. Chem. Ed.*, **62**, 943–953.
Canterford, J.H. and Cotton, R. (1968) *Halides of the Second and Third Row Transition Elements*, Wiley, London.
Canterford, J.H. and Cotton, R. (1969) *Halides of the First Row Transition Elements*, Wiley, London.
Corbett, J.D. (1981) Extended metal–metal bonding in halides of the early transition metals, *Acc. Chem. Res.*, **14**, 239.
Cotton, F.A. (1975) Compounds with multiple metal to metal bonds, *Chem. Soc. Rev.*, **4**, 27–53.
Cotton, F.A. (1978) Discovering and understanding multiple metal to metal bonds, *Acc. Chem. Res.*, **11**, 225–232.
Cotton, F. A. and Chisholm, M.H. (1978) Chemistry of compounds containing metal–metal bonds, *Acc. Chem. Res.*, **11**, 356–362.
Coughlin, M. (ed.) (1980) *Molybdenum and Molybdenum-containing Enzymes*, Pergamon Press, Oxford.
Diatomic metals and metallic clusters (Conference papers of the Faraday Symposia of the Royal Society of Chemistry, No. 14), Royal Society of Chemistry, London.
Dilworth, J.R. and Lappert, M.F. (eds) (1983) *Some Recent Developments in the Chemistry of Chromium, Molybdenum and Tungsten*, Royal Society of Chemistry, Dalton Division. (Proceedings of a conference at the University of Sussex.)
Emeléus, H.J. and Sharpe, A.G. (1973) *Modern Aspects of Inorganic Chemistry*, 4th ed. (Chapters 14 and 15: Complexes of Transition Metals: Chapter 20: Carbonyls) Routledge and Kegan Paul, London.
Greenwood, N.N. (1968) *Ionic Crystals, Lattice Defects and Non-Stoichiometry*, Butterworths, London. (Defect structures.)
Hagenmuller, P. (1973) *Comprehensive Inorganic Chemistry*, Vol. 4 (Chapter 50: Tungsten bronzes, vanadium bronzes), Pergamon Press, Oxford.
Hudson, M. (1982) Tungsten: its sources, extraction and uses, *Chemistry in Britain*, **18**, 438–442.
Hunt, C.B. (1977) Metallocenes – the first 25 years, *Education in Chemistry*, **14**, 110–113.
Johnson, B.F.G. (ed.) (1980) *Transition Metal Clusters*, John Wiley, Chichester.
Kepert, D.L. (1972) *The Early Transition Metals* (Chapter 4: Cr, Mo, W), Academic Press, London.
Kepert, D.L. (1973) *Comprehensive Inorganic Chemistry*, Vol. 4 (Chapter 51: Isopolyanions and heteropolyanions), Pergamon Press, Oxford.
Lewis, J. and Green, M.L. (eds) (1983) *Metal Clusters in Chemistry* (Proceedings of the Royal Society Discussion Meeting May 1982), The Society, London.
Mitchell, P.C.H. (ed.) (1974) Chemistry and uses of molybdenum, *J. Less Common Metals*, **36**, 3–11.
Pope, M.T. (1983) *Heteropoly and Isopoly Oxo Metalates*, Springer-Verlag. (A good account of poly acids.)
Richards, R.L. (1979) Nitrogen fixation, *Education in Chemistry*, **16**, 66–69.
Rollinson, C.L. (1973) *Comprehensive Inorganic Chemistry*, Vol. 3 (Chapter 36: Chromium, molybdenum and tungsten), Pergamon Press, Oxford.
Shriver, D.H., Kaesz, H.D. and Adams, R.D. (eds) (1990) *Chemistry of Metal Cluster Complexes*, VCH, New York.
Templeton, J.L. (1979) Metal–metal bonds of order four, *Progr. Inorg. Chem.*, **26**, 211–300.
Toth, L.E. (1971) *Transition Metal Carbides and Nitrides*, Academic Press, London.

23 The manganese group

Table 23.1 Electronic structures and oxidation states

Element		Electronic structure	Oxidation states*
Manganese	Mn	[Ar] $3d^5\ 4s^2$	(−I) 0 (I) **II** (III) IV (V) (VI) VII
Technetium	Tc	[Kr] $4d^5\ 5s^2$	0 (II) (III) **IV** (V) VI **VII**
Rhenium	Re	[Xe] $4f^{14}\ 5d^5\ 6s^2$	0 (I) (II) **III** **IV** (V) VI **VII**

* The most important oxidation states (generally the most abundant and stable) are shown in bold. Other well-characterized but less important states are shown in normal type. Oxidation states that are unstable, or in doubt, are given in parentheses.

INTRODUCTION

Manganese is produced in very large amounts, most of which is used in the steel industry. Large quantities of MnO_2 are also produced, and are used mainly for making 'dry' batteries and in the brick industry. $KMnO_4$ is an important oxidizing agent. Mn^{III} forms a basic acetate with an unusual structure. Mn is biologically important and is necessary for photosynthesis. The elements technetium and rhenium are rarely encountered. They differ from manganese in that they have little cationic chemistry, their high oxidation states are much more stable, and the oxidation states (+II), (+III) and (+IV) form cluster compounds and compounds with metal–metal bonds.

ABUNDANCE, EXTRACTION AND USES

Manganese is the twelfth most abundant element by weight in the earth's crust, and is mined as the ore pyrulsite MnO_2. This is a secondary material, which has been formed by alkaline waters leaching Mn from igneous rocks and depositing it as MnO_2. World production of Mn ores was 22.7 million tonnes in 1988, containing about 9 million tonnes of Mn. The largest producers are the USSR 40.5%, South Africa 15%, Gabon 10%, Australia 9%, Brazil 8% and China 7%. Pure Mn is now obtained by electrolysis of aqueous $MnSO_4$ solutions. (Formerly it was made by reducing MnO_2 or Mn_3O_4 with Al in a thermite reaction, but the reaction with MnO_2 was particularly violent.) There is little use for the pure metal. Ninety-five per cent of the Mn ores mined are used in the steel industry to produce alloys. Ferromanganese is the most important, and contains 80% Mn. World production of ferromanganese was 2.7 million tonnes in 1985. It is made by reducing the appropriate mixture of Fe_2O_3 and MnO_2 with carbon in a blast furnace, or an electric-arc furnace, with some limestone added to remove silicate impurities as calcium silicate slag. Alloys with a lower Mn content include silicomanganese (approximately 65% Mn, 20% Si, 15% Fe) and spiegeleisen which is similar to cast iron and contains 5–25% Mn. Mn is an important additive in making steel. It acts as a scavenger (removing both oxygen and sulphur and thus preventing bubbles and brittleness) and in addition it forms a very hard steel alloy. (Hadfield steel contains about 13% Mn and 1.25% C. It is very hard wearing and resistant to shock, and is used for rock crushing machinery and excavators.) Smaller amounts of Mn are also used in non-ferrous alloys. For example, manganin is an alloy containing 84% Cu, 12% Mn and 4% Ni. It is widely used in electrical instruments because its electrical resistance is almost unaffected by temperature.

Table 23.2 Abundance of the elements in the earth's crust, by weight

	ppm	Relative abundance
Mn	1060	12
Tc	0	
Re	0.0007	76

Technetium does not occur in nature, and was the first man-made element. All its isotopes are radioactive, and its chemistry has only recently been studied. ^{99}Tc is one of the fission products of uranium. It is a β emitter with a half life of 2.1×10^5 years. It is obtained in kilogram quantities from spent fuel rods from reactors at nuclear power stations. The rods may contain 6% Tc. These rods must be stored for several years to allow the short-lived radioactive species to decay. Tc can be extracted by oxidation to Tc_2O_7 which is volatile. Alternatively solutions can be separated by ion exchange and solvent extraction. The Tc_2O_7 can be dissolved in water, forming the pertechnate ion TcO_4^-, and crystallized as ammonium or potassium pertechnate. Ammonium pertechnate NH_4TcO_4 can be reduced with H_2 to give the metal. Tc metal has no commercial uses. ^{97}Tc and ^{98}Tc can be made by neutron bombardment of Mo. Small amounts of Tc compounds are sometimes injected into patients to allow radiographic scanning of the liver and other organs.

Rhenium is a very rare element, and occurs in small amounts in molybdenum sulphide ores. Re is recovered as Re_2O_7 from the flue dust from roasting these ores. This is dissolved in NaOH, giving a solution containing perrhenate ions ReO_4^-. The solution is concentrated and then KCl added

continued overleaf

to precipitate potassium perrhenate $KReO_4$. The metal is obtained by reducing $KReO_4$ or NH_4ReO_4 with hydrogen. World production was only 2.4 tonnes in 1988. Most of it is used to make Pt–Re alloys which are used as catalysts for making low-lead or lead-free petrol for cars. Small amounts are used as a catalyst for hydrogenation and dehydrogenation reactions. Because of its very high melting point (Table 23.4) it is used in thermocouples, electric furnace windings and mass spectrometer filaments.

OXIDATION STATES

The electronic structure for this group of elements is d^5s^2. The highest oxidation state of (+VII) is obtained when all these electrons are used for bonding. Mn shows the widest range of oxidation states of all the elements, ranging from (−III) to (+VII). The (+II) state is the most stable and most common and Mn^{2+} ions exist in the solid, in solution and as complexes. However, in alkaline solution Mn^{2+} is readily oxidized to MnO_2. The (+IV) state is found in the main ore pyrulsite MnO_2. Mn(+VII) is well known as $KMnO_4$. This is one of the strongest oxidizing agents known in solution, and is stronger than Cr(+VI) in the previous group. Mn(+III) and Mn(+VI) tend to disproportionate. The lower oxidation states exist as carbonyl compounds or as substituted carbonyl complexes.

In contrast to the highly oxidizing properties of Mn(+VII), the (+VII) state is the most common and most stable for Tc and Re. Tc(+VII) and Re(+VII) show only slight oxidizing properties. The (+VI) state tends to disproportionate and is not well known. The (+V) and (+IV) states for Tc

Table 23.3 Oxides and halides

Oxidation states						
(+II)	(+III)	(+IV)	(+V)	(+VI)	(+VII)	Others
MnO	Mn_2O_3	MnO_2	–	–	Mn_2O_7	Mn_3O_4
–	–	TcO_2	–	TcO_3	**Tc_2O_7**	
–	Re_2O_3^h	ReO_2	(Re_2O_5)	ReO_3	**Re_2O_7**	
MnF_2	MnF_3	MnF_4	–	–	–	
$MnCl_2$	–	–	–	–	–	
$MnBr_2$	–	–	–	–	–	
MnI_2	–	–	–	–	–	
–	–	–	TcF_5	**TcF_6**	–	
–	–	$TcCl_4$	–	($TcCl_6$?)	–	
–	–	–	–	–	–	
–	–	–	–	–	–	
–	–	ReF_4	ReF_5	ReF_6	**ReF_7**	
($ReCl_2$)	$ReCl_3$	**$ReCl_4$**	$ReCl_5$	($ReCl_6$?)	–	
($ReBr_2$)	$ReBr_3$	**$ReBr_4$**	$ReBr_5$	–	–	
(ReI_2)	ReI_3	**ReI_4**	–	–	–	

The most stable oxidation states are shown in bold, unstable ones in brackets.
h = hydrous oxide.

and Re have an extensive chemistry. Re(+III) is also stable and the halides form cluster compounds with metal–metal bonds. The (+II) and lower oxidation states are uncommon and are strongly reducing.

Thus on descending the group there is an increase in stability of the highest oxidation state and also a decrease in stability of the lower states. The stability of the various states is shown in Table 23.3. The tendency to disproportionate is shown in the reduction potentials.

STANDARD REDUCTION POTENTIALS (VOLTS)

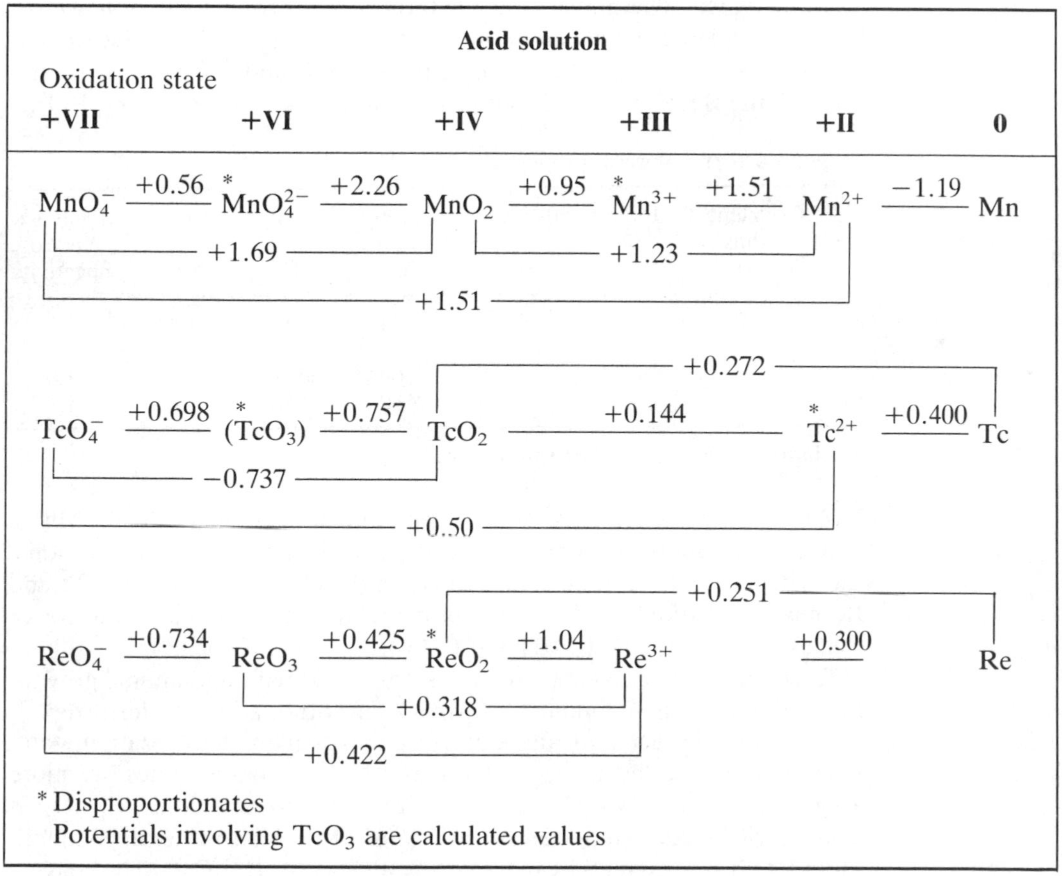

Acid solution

Oxidation state

+VII		+VI		+IV		+III		+II		0
MnO_4^-	+0.56	* MnO_4^{2-}	+2.26	MnO_2	+0.95	* Mn^{3+}	+1.51	Mn^{2+}	−1.19	Mn
TcO_4^-	+0.698	* (TcO_3)	+0.757	TcO_2	+0.144			* Tc^{2+}	+0.400	Tc
ReO_4^-	+0.734	ReO_3	+0.425	* ReO_2	+1.04	Re^{3+}	+0.300			Re

Linking potentials:

- MnO_4^- → MnO_2: +1.69
- MnO_2 → Mn^{2+}: +1.23
- MnO_4^- → Mn^{2+}: +1.51
- TcO_2 → Tc: +0.272
- TcO_4^- → TcO_2: −0.737
- TcO_4^- → Tc^{2+}: +0.50
- ReO_2 → Re: +0.251
- ReO_3 → Re^{3+}: +0.318
- ReO_4^- → Re^{3+}: +0.422

* Disproportionates

Potentials involving TcO_3 are calculated values

GENERAL PROPERTIES

Mn is more reactive than its neighbours in the periodic table. It reacts slowly with H_2O, liberating H_2, and it dissolves readily in dilute acids. The finely divided metal is pyrophoric in air, but the massive metal does not react unless heated. When strongly heated the massive metal reacts with many non-metals such as O_2, N_2, Cl_2 and F_2, forming Mn_3O_4, Mn_3N_2, $MnCl_2$ and a mixture of MnF_2 and MnF_3. The melting point of the metal is appreciably lower than for the earlier first row elements Ti, V and Cr. It is unusual in that it has four different solid structures (α or body-centred cubic, cubic close-packed, β, and χ). The α form is stable at room temperature and has a body-centred cubic structure. (See Chapter 5.)

Tc and Re metals are less reactive than Mn. They do not react with H_2O, or non-oxidizing acids. They do not dissolve in HCl and HF, but they react with oxidizing acids such as concentrated HNO_3 and H_2SO_4, forming pertechnic acid $HTcO_4$ and perrhenic acid $HReO_4$. (This is not the usual reaction of acid + metal to give a salt + H_2. For example, with concentrated HNO_3, the NO_3^- ion is a stronger oxidizing agent than H_3O^+, and NO_2 is evolved.) Tc and Re undergo similar reactions with H_2O_2 and bromine water. The massive metals tarnish (oxidize) slowly in moist air, but the powdered metals are more reactive. Heating with O_2 gives Tc_2O_7 and Re_2O_7 which are both low melting (119.5 °C and 300 °C respectively) and volatile. Heating with F_2 gives TcF_5 and TcF_6, and ReF_6 and ReF_7.

Table 23.4 Some physical properties

	Covalent radius (Å)	Ionic radius (Å)		Melting point (°C)	Boiling point (°C)	Density (g cm^{-3})	Pauling's electro-negativity
		M^{2+}	M^{3+}				
Mn	1.22	0.67	0.645^h 0.58^l	1244	2060	7.43	1.5
Tc	1.34	–	–	2200	4567	11.5	1.9
Re	1.34	–	–	3180	5650	21.0	1.9

h = high spin value, l = low spin radius.

Many ionic compounds of Mn are known including Mn^{2+}, Mn^{3+}, MnO_4^{2-} and MnO_4^-. In contrast Tc and Re have virtually no aqueous ionic chemistry apart from the oxoions TcO_4^- and ReO_4^-. The elements Tc and Re have a marked tendency to form metal–metal bonds in the lower oxidation states (+II), (+III) and (+IV).

Trace amounts of manganese are essential for plant and animal growth. For this reason small amounts of $MnSO_4$ are often added to fertilizers.

The basic character of any element changes with the oxidation state. Low oxidation states are more basic and high oxidation states are more acidic. MnO and Mn_2O_3 are basic oxides and are ionic. MnO_2 is amphoteric and does not exist as Mn^{4+} ions. Mn(+V) is rather uncommon. Mn(+VI) is represented by manganates such as $Na_2[MnO_4]]$. This may be regarded as a salt of the unstable acidic oxide MnO_3, which does not exist in the free state. Mn(+VII) occurs as Mn_2O_7 which is strongly acidic. The corresponding acid, permanganic acid $HMnO_4$, is a very strong acid.

Almost all manganese compounds are coloured. Mn^{2+} is pale pink, and MnO_2 is black, both because of *d–d* transitions. The (+VII) oxidation state has a d^0 configuration and would be expected to be colourless. Whilst perrhenates ReO_4^- containing Re(+VII) are colourless, permanganates MnO_4^- containing Mn(+VII) are intensely coloured. The purple–black colour arises from charge transfer spectra.

Mn resembles iron in its physical and chemical properties. Mn is harder and more brittle than iron, but melts at a lower temperature (Mn = 1245 °C, Fe = 1537 °C). All three metals Mn, Tc and Re are usually obtained as grey powders, but in the massive form they look like platinum. Re has the second highest melting point of all the metals (W = 3410 °C; Re = 3180 °C).

Manganese is quite electropositive and dissolves in cold dilute non-oxidizing acids. It is not very reactive towards non-metals at room temperature, but reacts readily on heating.

Mn is much more reactive than Re. Similar behaviour is observed in adjacent groups: Cr is more reactive than W, and Fe more reactive than Os. This decrease in electropositive character is opposite to the trend shown in the main groups of the periodic table. In addition, Re tends to attain a higher oxidation state than Mn when they both react with the same element. A comparison of some of the reactions of Mn and Re is given in Table 23.5.

Table 23.5 Some reactions of manganese and rhenium

Reagent	Mn	Re
N_2	Mn_3N_2 formed at 1200 °C	No reaction
C	Mn_3C	No reaction
H_2O	$Mn^{2+} + H_2$	No reaction
Dilute acid	$Mn^{2+} + H_2$	No reaction
Strong acid	$Mn^{2+} + H_2$	Dissolves slowly
Halogens	MnX_2 and MnF_3	ReF_6, $ReCl_5$, $ReBr_3$
S	MnS	ReS_2
O_2	Mn_3O_4	Re_2O_7

The manganese group elements have seven outer electrons, but the similarity to Group VII elements, the halogens, is very slight except in the highest oxidation state. Mn_2O_7 and Cl_2O_7 may be compared; MnO_4^- and ClO_4^- are isomorphous and have similar solubilities, and IO_4^- and ReO_4^- are quite similar. There are much closer similarities between Mn and its horizontal neighbours Cr and Fe. The chromates CrO_4^{2-}, manganates MnO_4^{2-} and ferrates FeO_4^{2-} are similar. The solubilities of the lower oxides are also similar, which accounts for the occurrence of iron and manganese together.

LOWER OXIDATION STATES

The (−I) oxidation state is found in the carbonylate anion $[Mn(CO)_5]^-$. The zero-valent state exists as the carbonyls $[Mn_2(CO)_{10}]$ and $[Re_2(CO)_{10}]$ and as a coordination complex $K_6[Mn(CN)_6] \cdot 2NH_3$ which is unstable and highly reducing.

Mn(+I) and Re(+I) are only obtained with difficulty and are strongly reducing.

$$K_3[Mn^{III}(CN)_6] \xrightarrow{\text{K in liquid } NH_3} K_5[Mn^{I}(CN)_6]$$

$$[Re_2(CO)_{10}] \xrightarrow{Cl_2 \text{ under pressure}} 2[Re^{I}(CO)_5Cl]$$

The cyano complex $Na_5[Mn^{I}(CN)_6]$ is formed by dissolving powdered Mn in aqueous NaCN in the complete absence of air.

(+II) STATE

Mn(+II) salts can easily be made from MnO_2.

$$MnO_2 + 4HCl \rightarrow MnCl_2 + Cl_2 + 2H_2O$$
$$2MnO_2 + 2H_2SO_4 \rightarrow 2MnSO_4 + O_2 + 2H_2O$$

Most manganous salts are soluble in water and form hydrated Mn^{2+} ions, but $Mn_3(PO_4)_2$ and $MnCO_3$ are sparingly soluble. $[Mn(H_2O)_6]^{2+}$ ions are pink. They are also formed by dissolving the metal in acid, or by reducing higher oxidation states. Small amounts of $MnSO_4$ are added to fertilizers, as Mn is an essential trace element for plants. Adding NaOH or NH_4OH to a solution of Mn^{2+} ions gives a very pale pink gelatinous precipitate of $Mn(OH)_2$, which turns brown–black due to oxidation to MnO_2. The standard electrode potentials show that this cannot occur in acid solution, but occurs easily in basic solutions.

Oxidation state

+VII **+VI** **+V** **+IV** **+III** **+II** **0**

Acid solution

$$MnO_4^- \xrightarrow{+0.56\,*} MnO_4^{2-} \xrightarrow{+2.26} MnO_2 \xrightarrow{+0.95\,*} Mn^{3+} \xrightarrow{+1.51} Mn^{2+} \xrightarrow{-1.19} Mn$$

Basic solution

$$MnO_4^- \xrightarrow{+0.56} MnO_4^{2-} \xrightarrow{+0.27\,*} MnO_4^{3-} \xrightarrow{+0.93} MnO_2 \xrightarrow{+0.1} Mn^{3+} \xrightarrow{-0.2} Mn(OH)_2 \xrightarrow{-1.55} Mn$$

* = Disproportionates

Mn^{2+} has the electronic configuration $3d^5$ which corresponds to a half filled *d* shell. Thus Mn^{2+} is more stable than other divalent ions of transition metals and is more difficult to oxidize than Cr^{2+} or Fe^{2+}. Most Mn(+II) complexes are octahedral, and have a high-spin arrangement with five unpaired electrons (Figure 23.1). This arrangement gives zero crystal field stabilization energy (see Chapter 7). Thus complexes such as $[Mn(NH_3)_6]^{2+}$ and $[MnCl_6]^{4-}$ are not stable except in solution. Complexes with chelating ligands are more stable and $[Mn(ethylenediamine)_3]^{2+}$, $[Mn(oxalate)_3]^{4-}$ and $[Mn.EDTA]^{4-}$ can be isolated as solids.

These complexes all have very pale colours. This is because a *d–d* transition in a high-spin d^5 complex requires not only the promotion of an

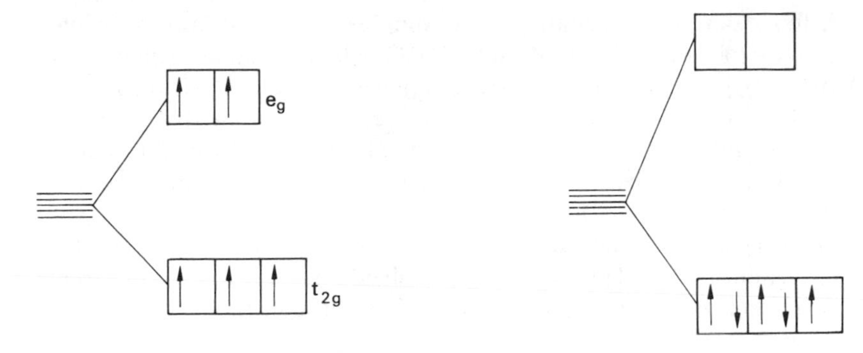

Figure 23.1 High- and low-spin arrangements in octahedral crystal field.

electron from the t_{2g} level to the e_g level, but also the reversing of its spin. The spin selection rule states that when promoting an electron its spin may not be changed. The rule is only partially obeyed, but the probability of a transition where the spin is changed is low. Such transitions are termed *spin forbidden*. Because the probability of a spin forbidden transition is small, the colour is only about one hundredth the intensity of those observed in most normal *spin allowed* transitions. (It must also be remembered that all d–d transitions are forbidden by the Laporte selection rule. This states that when promoting an electron the change in the subsidiary quantum number l must be ± 1, so $s \rightarrow p$ or $p \rightarrow d$ transitions are allowed but not $d \rightarrow d$. This selection rule is less restrictive as it is possible to get round it if the complex is not symmetrical, or by the mixing of orbitals, or by thermal motion of the ligands. These are discussed in Chapter 32.) Details of crystal field splitting of the d levels are given in Chapter 7.

Strong field ligands may force the electrons to pair up, giving a *spin paired* complex. The only common spin paired complex of Mn is $[Mn(CN)_6]^{4-}$ which has only one unpaired electron (Figure 23.1). $K_4[Mn(CN)_6] \cdot 3H_2O$ is blue in colour, and has the same structure as $K_4[Fe(CN)_6]$. $[Mn(CNR)_6]^{2+}$ and $[Mn(CN)_5 \cdot NO]^{3-}$ also have a spin paired or low-spin arrangement with only one unpaired electron. The low-spin arrangement is a slightly more stable arrangement than the high-spin case. In the low-spin case the crystal field stabilization energy is five times Δ_o from filling five electrons into the low energy t_{2g} orbitals. This is partly cancelled out by the energy needed to pair two electron spins. The low-spin complexes are more reactive. They can be oxidized very readily because removing an electron removes one unit of pairing energy and thus increases the crystal field stabilization energy. These complexes can also be reduced fairly readily, since adding an electron fills the orbitals symmetrically.

$$[Mn^{I}(CN)_6]^{5-} \xleftarrow[\text{reduction}]{\text{Zn}} [Mn^{II}(CN)_6]^{4-} \xrightarrow[\text{oxidation}]{\text{air}} [Mn^{III}(CN)_6]^{3-}$$

In low-spin d^5 complexes, d–d electronic transitions are *spin permitted* and the compounds are quite strongly coloured.

A few examples of square planar complexes are also known, including Mn phthalocyanine, and $MnSO_4 \cdot 5H_2O$ which contains square planar $[Mn(H_2O)_4]^{2+}$ units. The halide complexes $[MnCl_4]^{2-}$, $[MnBr_4]^{2-}$ and $[MnI_4]^{2-}$ are tetrahedral and are green–yellow in colour. In solution they add two molecules of water or two halide ions to form pink coloured octahedral complexes. These octahedra can polymerize by means of halide bridges.

Re(+II) is not ionic, and is found only in a few complexes such as $[Re(pyridine)_2Cl_2]$. This has been resolved into *cis* and *trans* isomers, showing that it is square planar rather than tetrahedral.

(+III) STATE

Mn_3O_4 is black in colour and is the most stable oxide at high temperatures. It is formed by heating any oxide or hydroxide of Mn to 1000 °C. It contains both Mn(+II) and Mn(+III), i.e. $(Mn^{II} \cdot Mn_2^{III}O_4)$, and has a spinel structure. The O atoms are close packed with Mn(+III) in octahedral holes and Mn(+II) in tetrahedral holes.

The hydrated manganic ion $[Mn(H_2O)_6]^{3+}$ can be obtained in solution by electrolysis, by oxidizing Mn^{2+} with potassium peroxodisulphate $K_2S_2O_8$, or by reducing MnO_4^-. It cannot be obtained in strong concentrations partly because water reduces Mn^{3+} to the very stable Mn^{2+} (which has enhanced stability due to the d^5 configuration). Mn^{3+} disproportionates in acid solution (see the standard reduction potentials). A small number of Mn^{3+} salts are known, e.g. Mn_3, $Mn_2(SO_4)_3$ and the oxide Mn_2O_3. These disproportionate in acid and hydrolyse in water.

$$2Mn^{3+} + 2H_2O \xrightarrow{\text{acid}} Mn^{2+} + Mn^{IV}O_2 + 4H^+$$

$$Mn^{3+} + 2H_2O \longrightarrow MnO \cdot OH + 3H^+$$

Mn(+III) complexes are stable in aqueous solution. Most complexes are octahedral and high spin, with magnetic moments close to the spin only value of 4.90 BM for four unpaired electrons. These include $[Mn(H_2O)_6]^{3+}$ which is found in alums such as $Cs^IMn^{III}(SO_4)_2 \cdot 12H_2O$, and in the acetylacetone complex $[Mn(acac)_3]$ whose crystal structure is a slightly distorted octahedron. The Jahn–Teller theorem predicts a slight distortion of the shape, as the electronic structure is $(t_{2g})^3\,(e_g)^1$, and the e_g level is not symmetrically filled. This distortion is similar to that found in Cr^{2+} and Cu^{2+} complexes. The oxalate complex $[Mn^{III}(oxalate)_3]^{3-}$ may be formed during permanganate–oxalate titrations. This would upset analytical calculations which are based on the reduction of MnO_4^- to Mn^{2+}. However, the oxalate complex is thermally unstable so the titration is performed at about 60 °C to decompose the complex.

The complex $K_3[Mn^{III}(CN)_6]$ is formed when air is bubbled through a solution containing Mn^{2+} and KCN. Since CN^- acts as a strong field ligand, it causes spin pairing, and the complex is low spin. The electronic structure is $(t_{2g})^4\ (e_g)^0$, the e_g level is symmetrically filled, and the complex ion is a regular octahedron as expected.

Manganese forms an unusual basic acetate when Mn^{2+} is oxidized with $KMnO_4$ in glacial acetic acid. The compound has the formula $[Mn_3O(CH_3COO)_6]^+\ [CH_3 \cdot COO]^-$ and is a deep red–brown coloured solid. It is used industrially for the oxidation of toluene $C_6H_5 \cdot CH_3$ to phenol C_6H_5OH. It will also oxidize alkenes to lactones, and it is used as the starting material for making many Mn(+III) compounds. The structure is unusual and consists of a triangle of three Mn atoms with an O atom at the centre. The six acetate groups act as bridges between the Mn atoms – two groups across each edge of the triangle. Thus each Mn atom is linked to four acetate groups and the central O, and the sixth position of the octahedron is occupied by water or another ligand giving $[Mn_3O(CH_3COO)_6L_3]^+$. This type of carboxylate complex is formed by the trivalent ions of Cr, Mn, Fe, Ru, Rh and Ir. A similar preparation in sulphuric acid gives an intensely red coloured solution. This is thought to contain a very similar sulphate complex with SO_4^{2-} groups acting as bridging groups. The solution is as strongly oxidizing as $KMnO_4$ and at one time it was used as an alternative oxidizing agent for use in sulphate solutions.

Mn(+III) and Mn(+IV) complexes are involved in the release of O_2 during photosynthesis.

Tc(+III) is unstable but $Re_2O_3 \cdot (H_2O)_n$ and the heavier halides are known. The halides are trimeric and $(ReCl_{33}$ and $(ReBr_3)_3$ are dark red solids and $(ReI_3)_3$ is black. Their structure consists of a metal atom cluster in which three Re atoms form an equilateral triangle (with formal double bonds between the metal atoms). Three halogen atoms act as bridging groups along the edges of the triangle, and the remaining six halogen atoms are coordinated two to each Re. An isolated Rc_3X_9 unit is shown in Figure 23.2a. There is one unfilled coordination position on each Re atom, marked L. In the solid this is filled by further halogen bridging with other Re_3X_9 units, giving a polymer (Figure 23.2b). The trimeric structure is very stable, with Re–Re distances of 2.48 Å, and this structure remains even in the gas phase at 600 °C. This structure also forms the basis of the structure of many Re(+III) complexes, where three additional groups L are added to the isolated Re_3X_9 unit (Figure 23.2c).

The simplest explanation of the double bonds between Re atoms is that each Re has nine atomic orbitals available for bonding (five *d*, one *s*, and three *p*). The metal is surrounded by five ligands, leaving four unused orbitals. Assuming the unused orbitals are pure *d* or mainly *d* in character, there are 12 atomic orbitals for Re–Re bonding. If these are delocalized over the three atoms there will be six bonding MOs and six antibonding MOs. Each Re(+III) has a d^4 configuration: hence the 12 electrons can enter the six bonding MOs, corresponding to double bonds between each of the three Re atoms. Since all the electrons are paired, the clusters should be diamagnetic, and this has been found experimentally.

If Re_3Cl_9 or Re_3Br_9 is dissolved in concentrated HCl or HBr, one, two

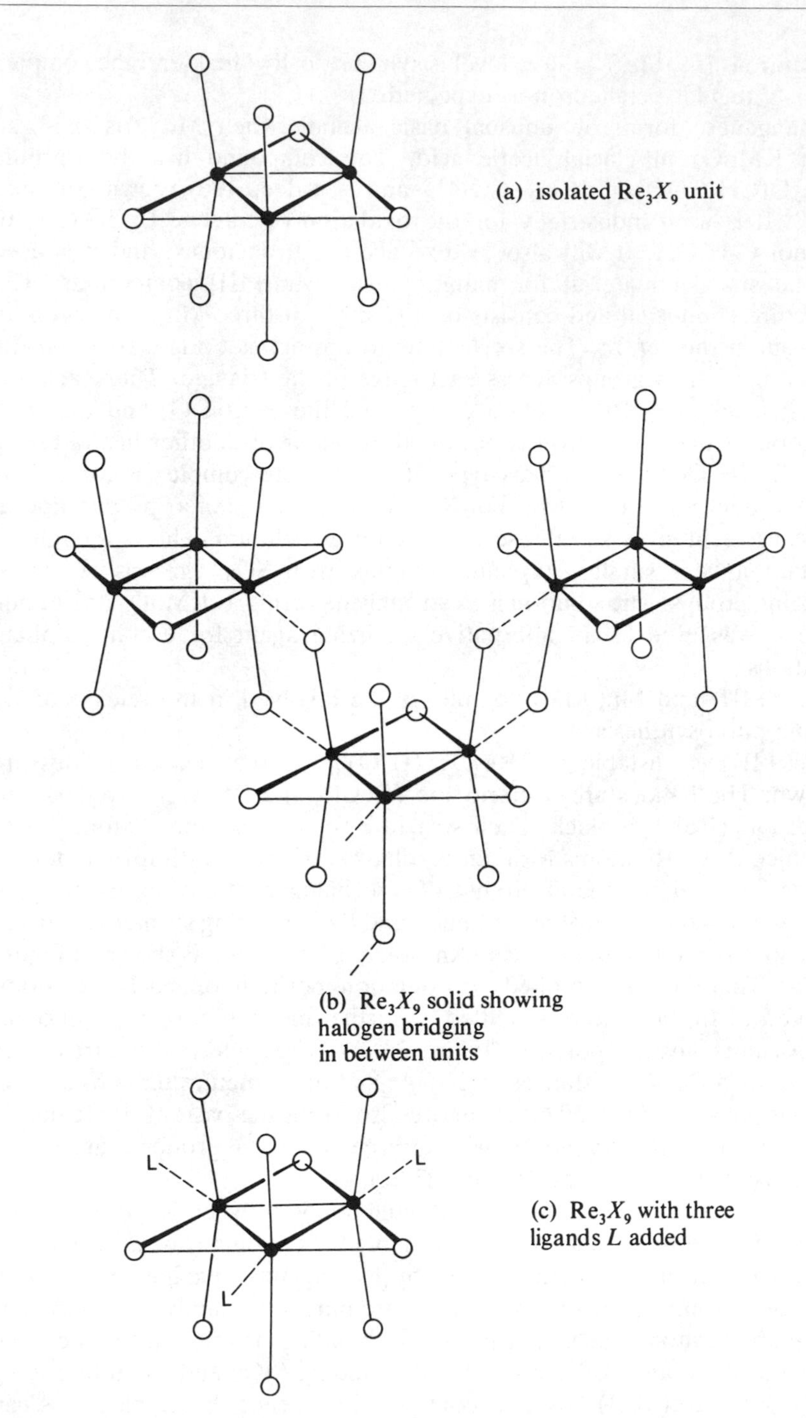

Figure 23.2 Various Re_3X_9 structures.

or three halide ions may be added to the isolated Re_3X_9 unit. Halide complexes such as $K^+[Re_3Br_{10}]^-$, $K_2^+[Re_3Br_{11}]^{2-}$ and $K_3^+[Re_3Br_{12}]^{3-}$ can be obtained from solution. It is also possible to obtain $[Re_3X_9 \cdot 3H_2O$. The extra ligands L are more labile, i.e. more reactive, than the bridging groups.

A quite different type of halide complex is formed when perrhenates are reduced by H_2 or sodium hypophosphite NaH_2PO_2 in acid solution. These contain a dimeric unit $[Re_2Cl_8]^{2-}$ or $[Re_2Br_8]^{2-}$ which is isostructural with $[Mo_2Cl_8]^{4-}$. It is made up of two approximately square planar ReX_4 units linked by a Re—Re bond of length 2.24 Å (Figure 23.3).

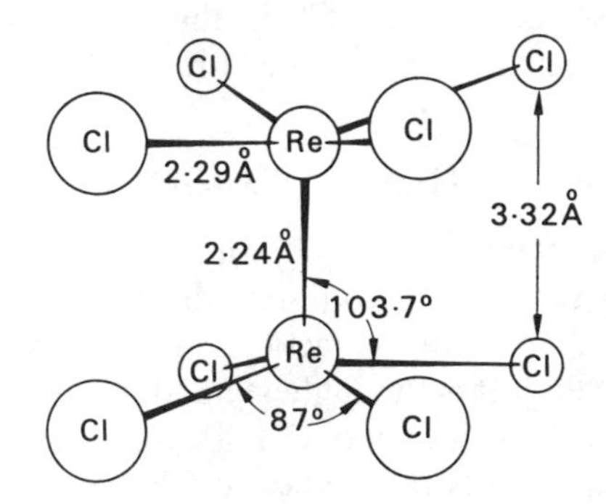

Figure 23.3 Structure of $[Re_2Cl_8]^{2-}$ ion.

The metal–metal bond length is very short, and is interpreted as a quadruple bond. If the Re—Re bond points along the z axis, the square planar ReX_4 unit uses the s, p_x, p_y and $d_{x^2-y^2}$ orbitals for σ bonding to the X atoms. A σ bond between the two Re atoms will have contributions from the p_z and d_{z^2} orbitals which lie along the axis. With the eclipsed conformation the d_{xz} and d_{yz} orbitals on the two Re atoms overlap sideways, forming two π bonds. Finally the d_{xy} orbitals which lie in the two ReX_4 planes overlap with each other, giving a δ bond.

(+IV) STATE

Very few Mn(+IV) compounds are known. However, MnO_2 is the most important oxide in the group, and is commercially important. It is not the most stable oxide, as it is an oxidizing agent, and decomposes to Mn_3O_4 on heating to 530 °C. MnO_2 occurs naturally as the black coloured mineral pyrolusite. It has the rutile structure which is also found for many other oxides of formula MO_2. It can be made as follows:

1. By heating Mn in O_2.
2. By oxidizing Mn^{2+}, for example by heating $Mn(NO_3)_2 \cdot 6H_2O$ in air.
3. Very pure MnO_2 is made by electrolytic oxidation of $Mn^{II}SO_4$.
4. In the laboratory hydrated MnO_2 is precipitated from solution when performing permanganate titrations in alkaline solution.

$$MnO_4^- + 2H_2O + 3e \rightarrow MnO_2 + 4OH^- \quad E^\circ = 1.23\,V$$

Hydrated MnO_2 has some cation-exchange properties, and if it is dehydrated it loses some O_2, forming a nonstoichiometric product.

MnO_2 does not react with most acids unless heated. It dissolves in concentrated HCl, but does not form Mn(+IV) in solution. Instead it oxidizes Cl^- to Cl_2, and is itself reduced to Mn^{2+}. Scheele discovered Cl_2 using this reaction.

$$MnO_2 + 4H^+ + 4Cl^- \rightarrow Mn^{2+} + 2Cl^- + Cl_2 + 2H_2O$$

This was the commercial method of producing Cl_2 until electricity became commercially available and electrolysis became the preferred method. This reaction is still used to produce Cl_2 in the laboratory if cylinders are not available. MnO_2 oxidizes hot concentrated H_2SO_4 in a similar way, liberating O_2.

$$2MnO_2 + 2H_2SO_4 \rightarrow 2MnSO_4 + O_2 + 2H_2O$$

Fusing MnO_2 with NaOH gives sodium manganate $Na_2[MnO_4]$. This is a dark green compound, which contains Mn(+VI) and is oxidizing.

MnF_4 is formed by direct reaction between Mn and F_2. It is blue coloured and unstable. It is the highest halide formed by Mn. A few (+IV) complexes are also known including $K_2[MnF_6]$, $K_2[MnCl_6]$, $K_2[Mn(CN)_6]$ and $K_2[Mn(IO_3)_6]$. This is the highest oxidation state in which Mn forms complexes.

Over half a million tonnes per year of MnO_2 is used in 'dry batteries' (Leclanché cells). This must be very pure, and is produced electrolytically (see above). Large amounts of MnO_2 are also used in the brick industry to colour bricks red or brown. It is used to make red or purple glass. MnO_2 is used in the production of potassium permanganate:

$$MnO_2 + KNO_2 \xrightarrow{\text{fuse in NaOH}} K_2MnO_4 + NO$$

$$K_2MnO_4 + H_2O \xrightarrow{\text{electrolytic oxidation}} KMnO_4 + KOH + H_2$$

MnO_2 is also used as an oxidizing agent in organic chemistry, for oxidizing alcohols and other compounds:

$$\underset{\text{toluene}}{C_6H_5 \cdot CH_3} + 2MnO_2 + 2H_2SO_4 \rightarrow \underset{\text{benzaldehyde}}{C_6H_5 \cdot CHO} + 2Mn_2SO_4 + 3H_2O$$

$$2C_6H_5 \cdot NH_2 + 4MnO_2 + 5H_2SO_4 \rightarrow \underset{\text{quinhydrone}}{O{=}C_6H_4{=}O} + (NH_4)_2SO_4 + 4MnSO_4 + 4H_2O$$

MnO_2 is used as a catalyst in the preparation of oxygen from $KClO_3$. If $KClO_3$ is heated to a temperature of 400–500 °C it decomposes and evolves O_2. With MnO_2 present decomposition occurs at 150 °C. The product is contaminated with Cl_2 or ClO_2.

$$2KClO_3 \rightarrow 2KCl + 3O_2$$

The (+IV) state is the second most stable state for both Tc and Re. The oxides TcO_2 and ReO_2 are black and brown respectively, and can be made in a variety of ways:

1. By burning the metals in a limited amount of oxygen.
2. By heating the heptoxides M_2O_7 with M.
3. By thermal decomposition of the ammonium salts NH_4MO_4.
4. The hydrated oxides are conveniently made by reducing solutions of TcO_4^- or ReO_4^- with Zn/HCl. These may be dehydrated by heating.

TcO_2 is insoluble in alkali, but ReO_2 reacts with fused alkali, forming rhenites ReO_3^{2-}. Both oxides have a distorted rutile structure. The metal atoms in adjacent octahedra are displaced from the centre, giving a substantial metal–metal interaction like that in MoO_2 and WO_2.

The sulphides TcS_2 and ReS_2 are both known. Rhenium sulphides are effective catalysts for organic hydrogenation reactions, and have the advantage over Pt that they are not 'poisoned' by sulphur compounds.

$TcCl_4$ has a solid structure of $TcCl_6$ octahedra linked into a zig-zag chain, similar to $ZrCl_4$. It is paramagnetic, and there are no metal–metal bonds. All four ReX_4 halides are known. $ReCl_4$ can be prepared as follows:

$$2ReCl_5 + SbCl_3 \rightarrow 2ReCl_4 + SbCl_5$$

It is metastable and reactive, and has a structure based on cubic close-packed chlorine atoms. The Re atoms occupy one quarter of the octahedral holes but occur in pairs of adjacent octahedra, forming metal–metal bonds with a Re–Re distance of 2.73 Å.

Many complexes are known. $[ReCl_6]^{2-}$ is octahedral, and is obtained by reduction of $KReO_4$:

$$ReO_4^- \text{ or } TcO_4^- \xrightarrow[+KI]{\text{conc. HCl}} [M^{IV}Cl_6]^{2-}$$

$[ReCl_6]^{2-}$ is hydrolysed in water:

$$[ReCl_6]^{2-} + H_2O \rightarrow ReO_2 \cdot (H_2O)_n$$

The other halide complexes $[MF_6]^{2-}$, $[MBr_6]^{2-}$ and $[MI_6]^{2-}$ are made from the hexachloride and the appropriate halogen acid. $[ReF_6]^{2-}$ is stable in water. Cyanide complexes are formed by treating $[MI_6]^{2-}$ with KCN. Tc forms $[Tc^{IV}(CN)_6]^{2-}$. Re is oxidized by CN^- and forms $[Re^{V}(CN)_8]^{3-}$. In this complex Re has an oxidation state of (+V) and a coordination number of 8, and the structure is probably dodecahedral.

(+V) STATE

Mn(+V) is little known except as the hypomanganate ion MnO_4^{3-}. This can be obtained as a bright blue salt K_3MnO_4 by reducing an aqueous solution of $KMnO_4$ with an excess of sodium sulphite. It is not stable, and tends to disproportionate. Tc is reluctant to form the (+V) state, and Re(+V) compounds are readily hydrolysed by water and at the same time they disproportionate.

$$3Re^{V}Cl_5 + 8H_2O \rightarrow \underset{\text{perrhenic acid}}{HRe^{VII}O_4} + Re^{IV}O_2 \cdot (H_2O)_n$$

$ReCl_5$ is dimeric, and has a structure like $(NbCl_5)_2$ (Figure 21.2) in which two octahedra are joined by an edge. The Re–Re distance is 3.74 Å, and therefore there is no metal–metal bonding. The oxide Re_2O_5 is also known.

(+VI) STATE

The only compounds of Mn(+VI) are the dark green manganates which contain the MnO_4^{2-} ion. Manganates are formed by oxidizing MnO_2 in fused KOH with air, KNO_3, PbO_2, $NaBiO_3$ or other oxidizing agents. They can also be obtained by treating $KMnO_4$ with alkali:

$$4MnO_4^- + 4OH^- \rightarrow 4MnO_4^{2-} + O_2 + H_2O$$

MnO_4^{2-} is quite strongly oxidizing and is only stable in very strong alkali. In dilute alkali, water or acidic solutions it disproportionates:

$$\underset{\text{manganate}}{3Mn^{VI}O_4^{2-}} + 4H^+ \rightarrow \underset{\text{permanganate}}{2Mn^{VII}O_4^-} + Mn^{IV}O_2 + 2H_2O$$

Re(+VI) is known as the red coloured oxide ReO_3, but the existence of TcO_3 is uncertain. The structure of ReO_3 is shown in Figure 23.4a, and the same structure is adopted by other oxides such as WO_3. Each metal is octahedrally surrounded by oxygen atoms. The structure is closely related to the perovskite structure for compounds ABO_3 by the insertion of a large cation at the centre of the cube as shown in Figure 23.4b.

The halides TcF_6, ReF_6 and $ReCl_6$ are known. The fluorides are made by direct reaction of the elements, and the chloride by treating the fluoride with BCl_3. They have a d^1 configuration, and the fluorides are yellow and the chloride is green–black. The magnetic moment is lower than the spin only value for d^1 because of strong spin orbital coupling. The compounds have low melting points ranging from 18 °C to 33 °C, and they are sensitive to water.

$$ReF_6 + H_2O \rightarrow HRe^{VII}O_4 + Re^{IV}O_2 + HF$$

(+VII) STATE

Mn(+VII) is not common, but is very well known as the permanganate ion MnO_4^-. The potassium salt $KMnO_4$ is widely used as an oxidizing agent in

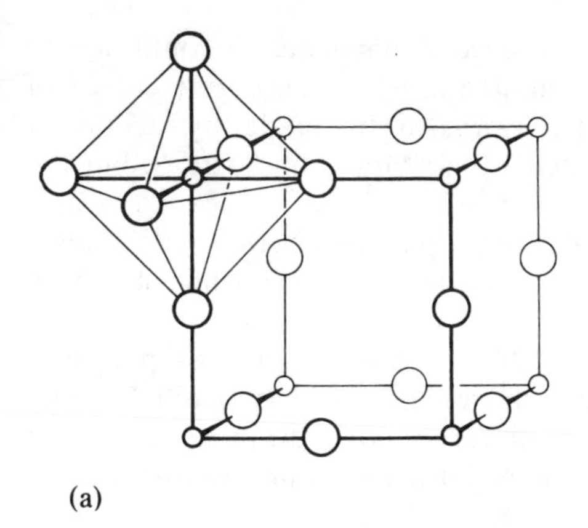

(a)

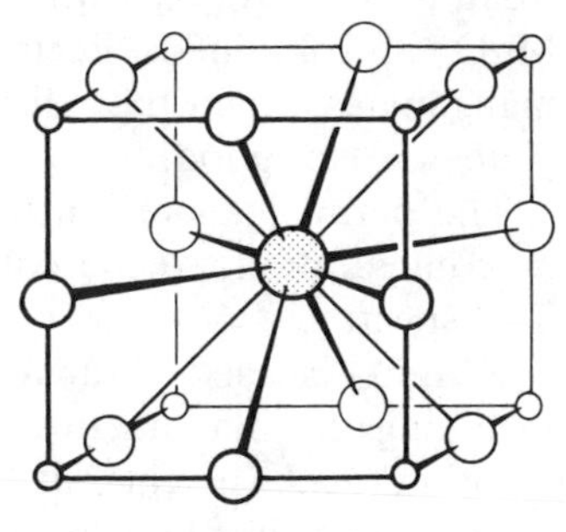

(b) perovskite structure ABO_3, e.g. $CaTiO_3$

Figure 23.4 Structures of (a) rhenium trioxide ReO_3 and (b) perovskite ABO_3, e.g. $CaTiO_3$. (From A.F. Wells, *Structural Inorganic Chemistry*, Oxford, Clarendon Press, 1950.)

both preparative and analytical chemistry. Titrations are normally carried out in acidic solutions, and MnO_4^- is reduced to Mn^{2+}, a change of five electrons.

$$MnO_4^- + 8H^+ + 5e \rightarrow Mn^{2+} + 4H_2O \quad E° = 1.51\,V$$

In alkaline solution $Mn^{IV}O_2$ is formed, involving a three-electron change.

$$MnO_4^- + 2H_2O + 3e \rightarrow MnO_2 + 4OH^- \quad E° = 1.23\,V$$

Thus the reaction which occurs, and hence the stoichiometry, depend on the pH. The purple colour of MnO_4^- acts as its own indicator in titrations.

Permanganate solutions are intrinsically unstable in acidic solution, and decompose slowly. Decomposition is catalysed by sunlight, so $KMnO_4$ solutions should be stored in dark bottles and they must be standardized frequently.

$$4MnO_4^- + 4H^+ \rightarrow 4MnO_2 + 3O_2 + 2H_2O$$

If a small quantity of $KMnO_4$ is added to concentrated H_2SO_4, a green solution containing MnO_3^+ ions is formed.

$$KMnO_4 + 3H_2SO_4 \rightarrow K^+ + MnO_3^+ + 3HSO_4^- + H_3O^+$$

With larger amounts of $KMnO_4$, an explosive oil Mn_2O_7 is formed. (Do **not** try this.)

$KMnO_4$ is manufactured on a large scale:

$$\underset{}{MnO_2} \xrightarrow[\text{oxidize with air or } KNO_3]{\text{fuse with KOH}} \underset{\textit{manganate}}{MnO_4^{2-}} \xrightarrow[\text{in alkaline solution}]{\text{electrolytic oxidation}} \underset{\textit{permanganate}}{MnO_4^-}$$

Permanganates can also be prepared by treating a solution of Mn^{2+} ions with very strong oxidizing agents such as PbO_2 or sodium bismuthate

$NaBiO_3$. In qualitative analysis the presence of manganese is confirmed by treating the solution with sodium bismuthate, when the purple colour of MnO_4^- is obtained. Permanganates can also be made by acidifying manganates, when they disproportionate into MnO_4^- and MnO_2, but the yields are not good.

The permanganate ion has an intense purple colour. Mn(+VII) has a d^0 configuration, so the colour arises from charge transfer and **not** from *d–d* spectra.

$KMnO_4$ is used as an oxidizing agent in many organic preparations including the manufacture of saccharin, ascorbic acid (vitamin C), and nicotinic acid (niacin). It is also used for treating drinking water. (It oxidizes and thus kills bacteria, but does not leave an unpleasant taste as does Cl_2.)

In contrast to the scarcity of Mn(+VII) compounds, Tc(+VII) and Re(+VII) occur in several compounds. These include the heptoxides M_2O_7, heptasulphides M_2S_7, MO_4^- ions, oxohalides, a hydride complex MH_9^{2-} and ReF_7. These are only slightly oxidizing and are relatively stable compounds.

The oxides Tc_2O_7 and Re_2O_7 are formed when the metals are heated in air or oxygen. They are both stable yellow solids. Tc_2O_7 melts at 120 °C and Re_2O_7 at 220 °C, in contrast to Mn_2O_7 which is an explosive oil. Tc_2O_7 is more oxidizing than Re_2O_7.

Both oxides dissolve in water and form colourless solutions of pertechnic acid $HTcO_4$ or perrhenic acid $HReO_4$. A second form of perrhenic acid exists as H_3ReO_5 (compare with periodic acid HIO_4 and H_3IO_5). The per acids are strong acids, i.e. completely ionized in aqueous solution. The solubilities of perrhenates are similar to those of perchlorates. The ions MnO_4^-, TcO_4^- and ReO_4^- are all tetrahedral. The MnO_4^- ion is a powerful oxidizing agent. TcO_4^- and ReO_4^- show mild oxidizing properties. The difference is illustrated by their reaction with H_2S. $KMnO_4$ oxidizes H_2S to S, and is itself reduced to Mn^{2+}, whereas $KTcO_4$ and $KReO_4$ precipitate sulphides Tc_2S_7 and Re_2S_7.

Whilst MnO_4^- is unstable in alkaline solution, TcO_4^- and ReO_4^- are stable from bases.

$KMnO_4$ is a very dark purple–black solid. The MnO_4^- ion is deep purple coloured due to charge transfer spectra. In contrast solutions of TcO_4^- and ReO_4^- are colourless, as the charge transfer band occurs at higher energy in the UV region. However solutions of $HReO_4$ become yellow–green when they are concentrated, and $HTcO_4$ has been isolated as a red solid. These colours arise because the tetrahedral ReO_4^- ion becomes less symmetrical when undissociated $HO—ReO_3$ is formed, and the Raman spectrum of the concentrated solution shows lines due to the acid.

The elements in this group do not form binary hydrides. However, when potassium pertechnate $KTcO_4$ or potassium perrhenate $KReO_4$ is treated with potassium in ethylenediamine or ethanolic solution, hydrido complexes $K_2[TcH_9]$ and $K_2[ReH_9]$ are formed. The structure is a tri-capped trigonal prism, that is a trigonal prism with an extra group pointing out of

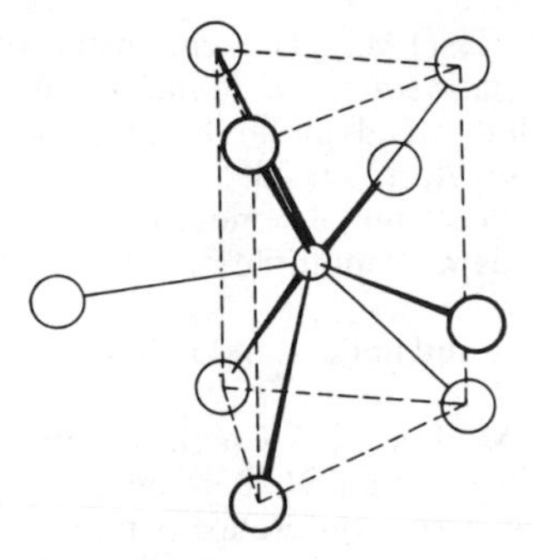

Figure 23.5 Structure of the enneahydridiorhenium(VII) ion $[ReH_9]^{2-}$.

each of the three rectangular faces (Figure 23.5). The proton magnetic resonance spectrum of this ion rather surprisingly shows a single sharp line. This suggests that the H atoms are equivalent. This suggests that the H atoms undergo a very rapid intramolecular site exchange.

When Re is heated with fluorine ReF_7 is formed, but Tc only forms TcF_6. ReF_7 and IF_7 are the only heptahalides known throughout the periodic table, and both have a pentagonal bipyramid structure. Several oxohalides can be made by treating the oxides with the appropriate halogen, or ReF_7 with oxygen or water. These include $ReOF_5$, ReO_2F_3, ReO_3F, ReO_3Cl, TcO_3F and TcO_3Cl. The oxohalides are all pale yellow or colourless compounds which are either low-melting solids or liquids.

BIOLOGICAL IMPORTANCE

Mn^{II} is important in both animal and plant enzymes. In mammals the enzyme arginase is produced in the liver. This is important because it converts nitrogenous waste products into urea in the ornithine–arginine–citrulline cycle (discovered by Hans Krebs who also discovered the tricarboxylic acid cycle). The urea is carried by the blood to the kidneys, where it is excreted in urine.

Mn is an essential trace element for plant growth. It is added to fertilizers in parts of the world where there is a deficiency in the soil. It is essential in a group of enzymes called phosphotransferases.

FURTHER READING

Abel, E.W. and Stone, F.G. (1969, 1970) The chemistry of transition metal carbonyls, *Q. Rev. Chem. Soc.*, Part I – Structural considerations, **23**, 325; Part II – Synthesis and reactivity, **24**, 498.

Canterford, J.H. and Cotton, R. (1968) *Halides of the Second and Third Row Transition Elements*, Wiley, London.

Canterford, J.H. and Cotton, R. (1969) *Halides of the First Row Transition Elements*, Wiley, London.

Clarke, M.J. and Fackler, P.H. (1982) The chemistry of technetium: toward improved diognostic agents, *Structure and Bonding*, **50**, 57–78.

Corbett, J.D. (1981) Extended metal–metal bonding in halides of the early transition metals, *Acc. Chem. Res.*, **14**, 239.

Cotton, F.A. (1983) Multiple metal–metal bonds, *J. Chem. Ed.*, **60**, 713–720.

Cotton, F.A. (1975) Compounds with multiple metal to metal bonds, *Chem. Soc. Rev.*, **4**, 27–53.

Cotton, F.A. (1978) Discovering and understanding multiple metal to metal bonds, *Acc. Chem. Res.*, **11**, 225–232.

Cotton, F.A. and Chisholm, M.H. (1978) Chemistry of compounds containing metal–metal bonds, *Acc. Chem. Res.*, **11**, 356–362.

Diatomic metals and metallic clusters (Conference papers of the Faraday Symposia of the Royal Society of Chemistry, No. 14), Royal Society of Chemistry, London.

Emeléus, H.J. and Sharpe, A.G. (1973) *Modern Aspects of Inorganic Chemistry*, 4th ed. (Chapters 14 and 15: Complexes of Transition Metals; Chapter 20: Carbonyls), Routledge and Kegan Paul, London.

Griffith, W.P. (1973) *Comprehensive Inorganic Chemistry*, Vol. 4 (Chapter 46: Carbonyls, cyanides, isocyanides and nitrosyls), Pergamon Press, Oxford.

Johnson, B.F.G. (ed.) (1980) *Transition Metal Clusters*, John Wiley, Chichester.

Kemmitt, R.D.W. (1973) *Comprehensive Inorganic Chemistry*, Vol. 3 (Chapter 37: Manganese), Pergamon Press, Oxford.

Levason, W. and McAuliffe, C.A. (1972) Higher oxidation state chemistry of manganese, *Coordination Chem. Rev.*, **7**, 353–384.

Lewis, J. and Green, M.L. (eds) (1983) *Metal Clusters in Chemistry* (Proceedings of the Royal Society Discussion Meeting May 1982), The Society, London.

Peacock, R.D. (1973) *Comprehensive Inorganic Chemistry*, Vol. 3 (Chapter 38: Technetium; Chapter 39: Rhenium), Pergamon Press, Oxford.

Pinkerton, T.C. *et al.* (1985) Bioinorganic activity of technetium radiopharmaceuticals, *J. Chem. Ed.*, **62**, 965–973.

Rard, J.A. (1985) Inorganic aspects of ruthenium chemistry, *Chem. Rev.*, **81**, 1.

Rouschias, G. (1974) Recent advances in the chemistry of rhenium, *Chem. Rev.*, **74**, 531–566.

Shriver, D.H., Kaesz, H.D. and Adams, R.D. (eds))1990) *Chemistry of Metal Cluster Complexes*, VCH, New York.

Templeton, J.L. (1979) Metal–metal bonds of order four, *Progr. Inorg. Chem.*, **26**, 211–300.

The iron group

24

IRON, COBALT AND NICKEL GROUPS

Iron	Fe	Cobalt	Co	Nickel	Ni
Ruthenium	Ru	Rhodium	Rh	Palladium	Pd
Osmium	Os	Iridium	Ir	Platinum	Pt

These nine elements together make up Group VIII in the old Mendeleev periodic table. They will be considered in vertical groups or *triads* in the same way as the other transition elements:

Fe	Co	Ni
Ru	Rh	Pd
Os	Ir	Pt

However, the horizontal similarities between these elements are greater than anywhere else in the periodic table except among the lanthanides. As a consequence of the lanthanide contraction, the second and third rows of transition elements are much alike. Because of this, the horizontal similarities are sometimes emphasized by considering these nine elements as two horizontal groups: the three ferrous metals Fe, Co and Ni, and the six platinum metals Ru, Rh, Pd, Os, Ir and Pt.

Fe	Co	Ni
Ru	Rh	Pd
Os	Ir	Pt

INTRODUCTION TO THE IRON GROUP

Iron is used in larger quantities than any other metal, and steel making is of immense importance throughout the world. Iron is also the most important

Table 24.1 Electronic structures and oxidation states

Element	Electronic structure	Oxidation states*
Fe	[Ar] $3d^6\ 4s^2$	0 **II** **III** (IV) (V) (VI)
Ru	[Kr] $4d^7\ 5s^1$	0 II **III** IV (V) VI (VII) VIII
Os	[Xe] $4f^{14}\ 5d^6\ 6s^2$	0 (I) (II) (III) **IV** (V) VI (VII) VIII

* The most important oxidation states (generally the most abundant and stable) are shown in bold. Other well-characterized but less important states are shown in normal type. Oxidation states that are unstable, or in doubt, are given in parentheses.

transition element in plants and animals. Its biological importance is as an electron carrier in plants and animals (cytochromes and ferredoxins), as haemoglobin, the oxygen carrier in the blood of mammals, as myoglobin for oxygen storage, for iron scavenging and storage (ferretin and transferrin) and in nitrogenase (the enzyme in nitrogen fixing bacteria). Iron forms several unusual complexes, including ferrocene.

ABUNDANCE, EXTRACTION AND USES

Iron is the fourth most abundant element in the earth's crust, after O, Si and Al. It makes up 62 000 ppm or 6.2% by weight of the earth's crust, where it is the second most abundant metal (Appendix A). In addition iron and nickel make up most of the earth's core.

The chief ores are haematite Fe_2O_3, magnetite Fe_3O_4, limonite FeO(OH) and siderite $FeCO_3$. (Smaller amounts of pyrites FeS_2 are also found. This has a yellow metallic appearance and is called 'fool's gold' because it has been mistaken for gold.) World production of iron ores was 970 million tonnes in 1988. The largest sources are the USSR 26%, China 17%, Brazil 15%, Australia 10%, the USA 6%, India 5% and Canada 4%. This yielded 538 million tonnes of pig iron in 1988.

Ruthenium and osmium are very rare. They are found in the metallic state together with the platinum metals and the coinage metals (Cu, Ag and Au). The main sources are traces found in the NiS/CuS ores mined in South Africa, Canada (Sudbury, Ontario), and the USSR (the river sand from the Ural mountains). World production of all six platinum group metals was only 267 tonnes in 1988. The largest sources were South Africa 45%, the USSR 44%, Canada 4%, the USA 2% and Japan 0.8%.

Table 24.2 Abundance of the elements in the earth's crust, by weight

	ppm	Relative abundance
Fe	62 000	4
Ru	0.0001	77 =
Os	0.005	72

EXTRACTION OF IRON

The extraction of iron has played a considerable part in the development of modern civilization. The iron age began when man found how to use the charcoal formed by burning wood to extract iron from iron ores, and how to use it to make implements. The industrial revolution began when Abraham Darby developed a process which used coke instead of charcoal at Coalbrookdale in Shropshire, England, in 1773. It is much cheaper and easier to produce coke from coal rather than make charcoal by partly burning wood. Furthermore the greater mechanical strength of coke made it possible to blow air through a mixture of coke and iron ore in a blast furnace and thus extract iron on a much larger scale. These two factors increased the availability of iron and greatly reduced its price. It became possible for the first time to make bridges, ships, steam engines and railway lines using iron. The scale on which iron and steel are produced and their widespread use justify detailed examination of the processes involved.

Blast furnace

Iron is extracted from its oxides in a blast furnace. This is an almost cylindrical furnace, lined with fire bricks. It runs continuously and works on the counter-current principle. It is charged from the top with iron ore, a reducing agent (coke) and slag forming substances (calcium carbonate). The amount of $CaCO_3$ is varied, depending on the amount of silicate materials in the ore. Air is blown in at the bottom. The coke burns producing heat and CO. The temperature of the furnace is nearly 2000 °C at the point where the air enters, but is about 1500 °C at the bottom and 200 °C at the top. The iron oxide is reduced to iron mainly by CO (though perhaps some reduction by C takes place). The molten iron dissolves 3–4% of C from the coke, resulting in the formation of pig iron. The melting point of pure iron is 1535 °C. The impurities in pig iron lower the melting point, possibly to as low as 1015 °C (the eutectic temperature) by the presence of 4.3% C. Molten iron collects in the hearth at the bottom of the blast furnace.

The temperature of the furnace decomposes $CaCO_3$ to CaO, which then reacts with any silicate impurities present (such as sand or clay), forming iron silicate or slag. This is also molten and also drips to the bottom. The slag floats on the molten iron, thus protecting it from oxidation. Molten slag and molten iron are drawn off through separate openings at intervals. The molten iron is run into moulds made of sand and is allowed to solidify into ingots called 'pigs'. Pig iron or cast iron is hard but is very brittle.

A small amount of pig iron is remelted and cast into metal parts, but the brittleness makes it almost impossible to machine the metal on a lathe, though it can be shaped by grinding. Pig iron contains up to 4.3% carbon and possibly other impurities such as Si, P, S and Mn. The non-metallic elements must be removed to reduce the brittleness. All Fe/C alloys

continued overleaf

containing less than 2% C are called steel. Steel is ductile and can be rolled or machined into shape. The hardness and strength increase with increasing C content. The most common forms are mild steel (otherwise called soft steel or low-carbon steel) and hard steel (otherwise called high-carbon steel).

The source of the impurities in pig iron are as follows: C arises from the coke used in the blast furnace. Si, P and S arise from the reduction by C of silicates, phosphates or sulphates present in the ore, or from S in the coke. Small amounts of Mn are often found in iron ore, but for most purposes this is beneficial rather than harmful, as small amounts of Mn actually enhance the physical properties of the metal. Typical analyses of pig iron and mild steel are shown in Table 24.3.

The slag is used as a building material, for example to make breeze blocks, or cement.

Reactions involved

The overall process for the extraction of Fe is:

$$3C + Fe_2O_3 \rightarrow 4Fe + 3CO_2$$

$$CaCO_3 + SiO_2 \rightarrow CaSiO_3 + CO_2$$

The reaction proceeds in several stages at different temperatures. Since the air passes through in a few seconds, the individual reactions do not reach equilibrium.

400 °C $3Fe_2O_3 + CO \rightarrow 2Fe_3O_4 + CO_2$

$Fe_3O_4 + CO \rightarrow 2FeO + CO_2$

500–600 °C $2CO \rightarrow C + CO_2$

The C is deposited as soot and reduces FeO to Fe but it also reacts with the refractory lining of the furnace, and is harmful

800 °C $FeO + CO \rightarrow Fe + CO_2$

900 °C $CaCO_3 \rightarrow CaO + CO_2$

Table 24.3 Typical impurities that may be present in pig iron and mild steel

	Pig iron (%)	Mild steel (%)
C	3–4.3	0.15
Si	1–2	0.03
P	0.05–2	0.05
S	0.05–1	0.05
Mn	0.5–2	0.5

1000 °C $FeO + CO \rightarrow Fe + CO_2$
$CO_2 + C \rightarrow 2CO$
(Together these two reactions appear to be
$FeO + C \rightarrow Fe + CO$)

1800 °C $CaO + SiO_2 \rightarrow CaSiO_3$
$FeS + CaO + C \rightarrow Fe + CaS + CO$
$MnO + C \rightarrow Mn_{(in\ Fe)} + CO$
$SiO_2 + 2C \rightarrow Si_{(in\ Fe)} + 2CO$

A modern blast furnace may be 120 feet (40 metres) high, and 50 feet (15 metres) in diameter at the bottom, and can produce 10 000 tonnes of pig iron per day.

STEEL MAKING

Steel is made by removing most of the C and other impurities from pig iron. The process involves melting, and oxidizing the C, Si, Mn and P in the pig iron so that the impurities are given off as gases or converted into slag.

Puddling

Originally steel was made by 'puddling', which involved mixing molten pig iron with haematite Fe_3O_4, and burning off all the C and other impurities to give wrought iron. Wrought iron has had all the C removed and is fairly soft, and is extremely malleable. It can be converted into steel by adding the required amount of C or other metals to give the type of steel required, but this process is now entirely obsolete.

Bessemer and Thomas processes

In 1855 H. Bessemer (a Frenchman living in England) patented the process named after him which uses a Bessemer converter. This is a large pear-shaped furnace lined with silica. The furnace can be tilted and whilst it is in the horizontal position molten pig iron is poured in. It is then tilted to the vertical position and a blast of compressed air is blown through holes in the bottom of the furnace and through the molten metal.This burns the Si and Mn in the pig iron to give SiO_2 (then iron silicate) and manganese oxide slag. These reactions produce a lot of heat which leads to stronger oxidation of C to CO or CO_2. The course of the reaction can be followed by burning the waste gases and observing the colour or the spectrum of the flame produced. When the C content is sufficiently low, the converter is tipped and the molten steel is poured into cast iron moulds. This gives an ingot of steel, which can be rolled or forged. The process typically takes 20 minutes for a 6 tonne ingot of metal.

Iron ores typically contain either Si or P as impurities, which are oxidized to SiO_2 or P_4O_{10}. A phosphorus content of over 0.05% in steel

continued overleaf

leads to a low tensile strength and 'cold brittleness'. The Bessemer process does not remove phosphorus, so satisfactory steel or wrought iron can only be obtained from pig iron containing little or no P. Furthermore the phosphorus damages the lining of the convertor, and this can only be replaced by taking the convertor out of use.

$$P_4O_{10} + 6Fe + 3O_2 \rightarrow 2Fe_3(PO_4)_2$$

$$Fe_3(PO_4)_2 + 2Fe_3C + 3Fe \rightleftharpoons 2Fe_3P + 6FeO + 2CO$$

$$FeO + \underset{\text{(furnace lining)}}{SiO_2} \longrightarrow FeSiO_3$$

In some countries steel making is based on phosphorus-rich ores. In this case the 'basic Bessemer process' (often called the Thomas and Gilchrist process, patented by S.G. Thomas in 1879) is used instead. There are two differences from the normal Bessemer process:

1. The Thomas convertor is lined with a basic material such as calcined (strongly heated) dolomite or limestone. This reduces reaction between the iron phosphate slag and the lining of the convertor. This prolongs the life of the convertor lining.
2. Limestone $CaCO_3$ or lime CaO are added as slag formers. These are basic and react with P_4O_{10}, forming 'basic slag' $Ca_3(PO_4)_2$, thus removing P from the steel. *Basic slag* is a valuable by-product and is finely ground and sold as a phosphate fertilizer.

Thus a Bessemer convertor with its silicate lining was used for iron ore containing silicate impurities, and Thomas convertor lined with limestone or dolomite was used with ores which had a high P content. A small amount of Mn is added to the molten metal to remove S and O. The Bessemer process reduced the price of steel by a factor of 5 in the UK. It was adopted by Andrew Carnegie in the USA and reduced the price by a factor of 10. The process dominated world steel production until World War I, and was still used in England until the 1960s.

Siemens open hearth process

The open hearth process was invented by Sir William Siemens soon after the Bessemer process was invented. The furnace required external heating by burning gas or oil in air. Molten pig iron was put in a shallow hearth, and the impurities were oxidized by air. The lining of the furnace was acidic or basic depending on the impurities in the pig iron. This process replaced the Bessemer process in many places. The open hearth process is slow and it took about 10 hours to convert 350 tonnes. For this reason it has largely been replaced by the basic oxygen process.

Siemens electric arc furnace

Siemens also patented an electric arc furnace in 1878. Heat is provided either by having an electric arc just above the metal, or by passing an electric current through the metal. This process is still used to produce steel alloys and other high quality steel. Stainless steel typically contains 12–15% of Ni, but that used for cutlery may contain 20% Cr and 10% Ni. High speed cutting steel used as the cutting edges on lathes may contain 18% W and 5% Cr. Alloys with 0.4–1.6% Mn give steel with a high tensile strength. Hadfield steel, which is very tough and is used for rock breaking machinery and excavators, contains about 13% Mn and 1.25% C. Spring steel contains 2.5% Si.

Basic oxygen process

The main process nowadays is the more modern *basic oxygen process* (BOP). It originated in 1952 as the Kaldo and LD processes in Austria. One problem with the older Bessemer and Thomas processes was that the molten metal took up small amounts of nitrogen. In concentrations above 0.01% nitrogen makes steel brittle, and nitriding of the surface makes the metal more difficult to weld. The remedy is to use pure O_2 instead of air to oxidize the pig iron. If O_2 was blown through the bottom of the furnace in the same way as air was blown through in the Bessemer process, the bottom of the furnace would melt. The Kaldo and LD processes both use pure O_2. In the LD process strong convection currents were set up in the melt to obtain an effective reaction, whilst in the Kaldo process the convertor was rotated to ensure mixing and hence an effective reaction. The BOP process is now very widely used. The furnace is initially charged with molten pig iron and lime, and pure O_2 is blown onto the surface of the liquid metal at great speed through retractable water cooled lances. The O_2 penetrates into the melt and oxidizes the impurities rapidly. The heat evolved in oxidizing the impurities keeps the contents of the furnace molten despite the rise in melting point as impurities are removed. No external heat is required. Eventually the furnace is tipped and the molten steel is poured either into moulds to give steel castings, or into ingots, which in turn may be passed to rolling mills. The advantages of using O_2 rather than air are:

1. There is a faster conversion, so a given plant can produce more in a day.
2. Larger quantities can be handled. (A 300 tonnes charge can be converted in 40 minutes compared with 6 tonnes in 20 minutes by the Bessemer process.)
3. It gives a purer product, and the surface is free from nitrides.

Table 24.4 Composition of various steels

%C	name
0.15–0.3	Mild steel
0.3–0.6	Medium steel
0.6–0.8	High-carbon steel
0.8–1.4	Tool steel

continued overleaf

Phase diagram

The iron–carbon phase diagram is very complicated and is described under 'Interstitial alloys and related compounds' in Chapter 5. The production of alloys containing small amounts of V, Cr, Mo, W or Mn gives steels with special properties for particular purposes.

Production figures and uses

World production of steel ingots and castings was 776 million tonnes in 1988. The largest steel producing countries were the USSR 21%, Japan 14%, the USA 12%, China 7%, West Germany 5%, Brazil and Italy 3% each, and the UK, France, South Korea, Poland, Canada, Czechoslovakia and Romania 2% each (2% amounts to over 15 million tonnes!). The largest use is as mild steel for ship building, girders and motor car bodies. Mild steel is malleable and can be bent or machined. It can also be hardened (tempered) by heating to red heat and quenching (cooling rapidly) by plunging into water or oil. In a year, 11.7 million tonnes of tin plate, i.e. thin sheets of mild steel electroplated with a very thin protective layer of tin are used for packaging food and other materials as 'tin cans'. Several ferrous alloys are produced in large amounts: ferrosilicon 3.2 million tonnes/year, ferrochrome 3 million tonnes/year, ferromanganese 2.6 million tonnes/year and ferronickel 385 000 tonnes/year.

Promoted iron is used as the catalyst in the Haber–Bosch process for making NH_3.

EXTRACTION OF RUTHENIUM AND OSMIUM

Ru and Os are obtained from the anode slime which accumulates in the electrolytic refining of Ni. This contains a mixture of the platinum metals together with Ag and Au. The elements Pd, Pt, Ag and Au are dissolved in aqua regia and the residue contains Ru, Os, Rh and Ir. After a complex separation Ru and Os are obtained as powders, and powder forming techniques are used to give the massive metal. These elements are both scarce and expensive. Ru is used to alloy with Pd and Pt, and Os is also used to make hard alloys. All of the platinum metals have specific catalytic properties.

OXIDATION STATES

In the previous groups of transition elements, the maximum oxidation number was attained when all of the valency electrons in the *d* and *s* levels were used for valency purposes. If this trend continued then the maximum oxidation number for this group would be (+VIII). However, this trend is not continued in the second half of the *d*-block and the highest oxidation state for Fe is (+VI). This state is rare and is of little importance.

The main oxidation states for Fe are (+II) and (+III). Fe(+II) is the most stable, and exists in aqueous solution. Fe(+III) is slightly oxidizing, but the Fe(+II) and Fe(+III) states are much closer in stability than has been found in previous groups. In marked contrast the elements Ru and Os form RuO_4 and OsO_4 which are in the (+VIII) state. Ru(+III) and Os(+IV) are the most stable states, though Ru(+V), Os(+VI) and Os(+VIII) are also reasonably stable. Thus the usual trend is observed that on descending a group, the higher oxidation states become more stable. The stability of the various oxidation states is shown by the range of oxides and halides which are known (see Table 24.5).

Table 24.5 Oxides and halides

Oxidation states							
(+II)	(+III)	(+IV)	(+V)	(+VI)	(+VII)	(+VIII)	Others
FeO	Fe_2O_3	–	–	–	–	–	**Fe_3O_4**
–	$Ru_2O_3{}^h$	**RuO_2**	–	$(RuO_3)^h$	–	RuO_4	
–	–	**OsO_2**	–	$(OsO_3)^h$	–	**OsO_4**	
FeF_2	**FeF_3**	–	–	–	–	–	
$FeCl_2$	$FeCl_3$	–	–	–	–	–	
$FeBr_2$	$FeBr_3$	–	–	–	–	–	
FeI_2	–	–	–	–	–	–	
–	**RuF_3**	RuF_4	**RuF_5**	(RuF_6)	–	–	
–	**$RuCl_3$**	$RuCl_4$	–	–	–	–	
–	**$RuBr_3$**	–	–	–	–	–	
–	**RuI_3**	–	–	–	–	–	
–	–	**OsF_4**	OsF_5	**OsF_6**	(OsF_7)	–	
–	–	**$OsCl_4$**	$OsCl_5$	–	–	–	$OsCl_{3.5}$
–	–	$OsBr_4$	–	–	–	–	
–	–	OsI_4	–	–	–	–	

The most stable oxidation states are shown in bold, unstable ones in brackets.
h = hydrous oxide.

GENERAL PROPERTIES

Pure iron is silvery in colour, is not very hard, and is quite reactive (see Table 24.6). The finely divided metal is pyrophoric. Dry air has little effect on massive Fe, but moist air quite quickly oxidizes the metal to hydrous ferric oxide (rust). This forms a non-coherent layer which flakes off, and exposes more metal to attack. Iron dissolves in cold dilute non-oxidizing acids, forming Fe^{2+} and liberating hydrogen. If the acid is warm and if air is present some Fe^{3+} ions are formed as well as Fe^{2+}, whilst oxidizing acids give only Fe^{3+}. Strong oxidizing agents such as concentrated HNO_3 or $K_2Cr_2O_7$ passivate the metal because of the formation of a protective coat of oxide. If this layer is scratched, the exposed metal is once again vulnerable to chemical attack. Fe is slightly amphoteric. It is not affected by dilute NaOH, but it is attacked by concentrated NaOH.

Table 24.6 Some reactions of Fe, Ru and Os

Reagent	Fe	Ru	Os
O_2	Fe_3O_4 at 500 °C Fe_2O_3 at higher temp.	RuO_2 at 500 °C	OsO_4 at 200 °C
S	FeS FeS_2 with excess	RuS_2	OsS_2
F_2	FeF_3	RuF_5	OsF_6
Cl_2	$FeCl_3$	$RuCl_3$	$OsCl_4$
H_2O	Rusts slowly Fe_3O_4 formed at red heat	No reaction	No reaction
Dilute HCl	$Fe^{2+} + H_2$	No reaction	No reaction
Dilute HNO_3	$Fe^{3+} + H_2$	No reaction	No reaction
Aqua regia	Passive	No reaction	OsO_4

In contrast Ru and Os are noble, and are very resistant to attack by acids. However, Os is oxidized to OsO_4 by aqua regia.

The rusting of iron is a special case of corrosion, and is of great practical importance. The process is very complex, but a simplified explanation is that Fe atoms are converted to Fe^{2+} ions and electrons. The electrons move to a more noble metal which may be present as an impurity in the iron, or in contact with it. The electrons discharge H^+ ions present in the water, forming hydrogen, which reacts with atmospheric oxygen to give water:

$$Fe \rightarrow Fe^{2+} + 2e$$

$$2H^+ + 2e \rightarrow 2H \xrightarrow{\frac{1}{2}O_2} H_2O$$

The iron becomes positive and forms the anode, and the noble metal serves as the cathode, i.e. small local electrochemical cells are formed in the surface. The Fe^{2+} ions are subsequently oxidized to Fe(+III), either $FeO \cdot OH$, Fe_2O_3 or Fe_3O_4. Since the oxide does not form a coherent protective film, corrosion continues.

To prevent corrosion, O_2, H_2O and the impurities must be excluded. In practice Fe is often given a protective coating to exclude the water. Electroplating Fe with a thin layer of Sn is widely used and 11.7 million tonnes of 'tin plate' were produced in 1985. Other methods include 'hot dipping' the Fe in molten zinc, galvanizing (electroplating with Zn), Sherardizing, and painting with red lead. Another effective treatment is to convert the outer layer of iron into iron phosphate. This may be done by treating with phosphoric acid or acid solutions of $Mn(H_2PO_4)_2$ or $Zn(H_2PO_4)_2$ in the Parkerizing and Bonderizing processes. Alternatively a sacrificial anode may be used which makes the iron the cathode in the electrolytic cell. Ru and Os are noble, and do not react with water.

The effect of the lanthanide contraction is less pronounced in this part of

Table 24.7 Some physical properties

	Covalent radius (Å)	Ionic radius (Å)		Melting point (°C)	Boiling point (°C)	Density ($g\,cm^{-3}$)	Pauling's electro-negativity
		M^{2+}	M^{3+}				
Fe	1.17	0.78^h 0.61^l	0.645^h 0.55^l	1535	2750	7.87	1.8
Ru	1.24	–	0.68	2282	(4050)	12.41	2.2
Os	1.26	–	–	3045	(5025)	22.57	2.2

h = high spin value, l = low spin radius.

the periodic table: hence the similarities between the second and third row elements are not so close as are found in the earlier transition groups (see Table 24.7). The close horizontal similarities in the ferrous metals and in the platinum metals are largely due to the similarity of their atoms and ions in size (e.g. Fe^{2+} 0.61 Å (low spin), Co^{2+} 0.65 Å (low spin) and Ni^{2+} 0.69 Å). Because osmium is only a little larger than ruthenium, it would be expected to have a much higher density and indeed Os is the second most dense element known (density 22.57 $g\,cm^{-3}$) exceeded only by a small margin by Ir (density 22.61 $g\,cm^{-3}$) (see Appendix D).

LOW OXIDATION STATES

The (−II) state is rare, but occurs in the carbonyl ions $[Fe(CO)_4]^{2-}$ and $[Ru(CO)_4]^{2-}$. The zero-valent state occurs in the carbonyls, which may be mononuclear, e.g. $Fe(CO)_5$, $Ru(CO)_5$ and $Os(CO)_5$, dinuclear, e.g. $Fe_2(CO)_9$ and $Os_2(CO)_9$, or trinuclear, e.g. $Fe_3(CO)_{12}$, $Ru_3(CO)_{12}$, $Os_3(CO)_{12}$ (Figure 24.1).

$Fe(CO)_5$, $Ru(CO)_5$ and $Os(CO)_5$ are liquids at room temperature. The other carbonyls are volatile solids. $Fe(CO)_5$ is available commercially. The di- and trinuclear carbonyls contain M—M bonds. $Fe_2(CO)_9$ has three bridging CO groups joining the two metal atoms, but in $Os_2(CO)_9$ there is probably only one bridging group. $Fe_3(CO)_{12}$ has two CO bridges between one pair of Fe atoms, whilst $Os_3(CO)_{12}$ has no bridging groups. $Fe_2Ru(CO)_{12}$ has the same structure as $Fe_3(CO)_{12}$, but $FeRu_2(CO)_{12}$ has the same structure as $Os_3(CO)_{12}$.

$Fe(CO)_5$ is oxidized quite easily, and the gas forms an explosive mixture with air. These carbonyls react with aqueous alkali or water, forming carbonylate anions:

$$Fe(CO)_5 + 3NaOH \rightarrow Na[HFe(CO)_4] + Na_2CO_3 + H_2O$$

They undergo substitution reactions with other ligands such as PF_3, PCl_3, PPh_3, $AsPh_3$ and unsaturated organic molecules such as benzene. In the polynuclear carbonyls the metal–metal bond and the metal cluster are often retained.

$$Fe(CO)_5 + PF_3 \rightarrow (PF_3)Fe(CO)_4 + CO$$

$$Fe_2(CO)_9 + 6PPh_3 \rightarrow Fe_2(CO)_3(PPh)_6 + 6CO$$

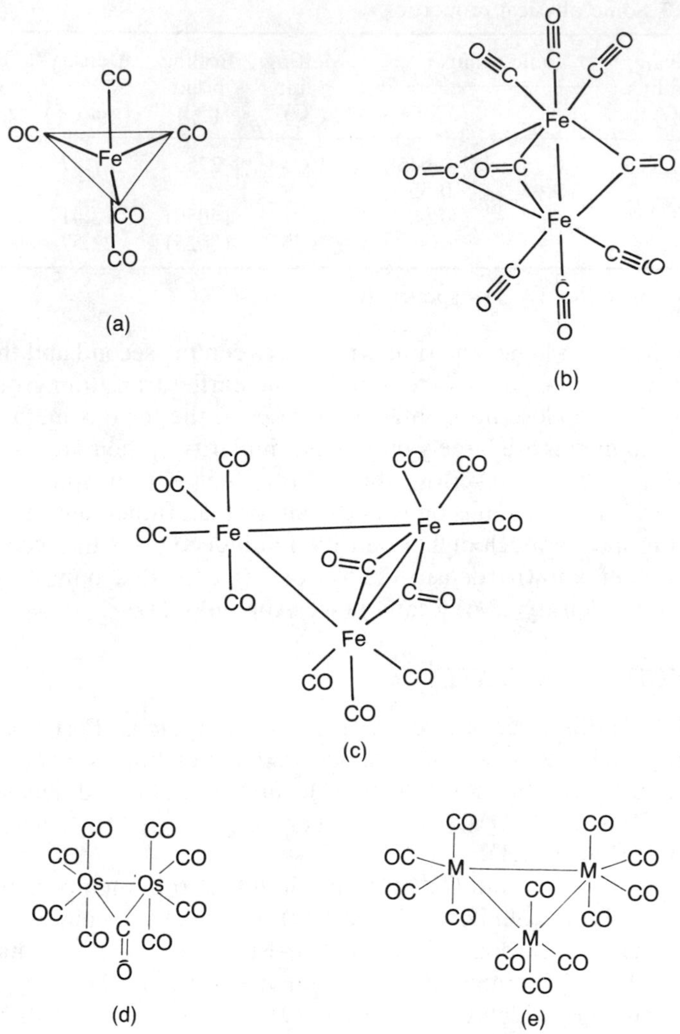

Figure 24.1 Structures of (a) $Fe(CO)_5$, (b) $Fe_2(CO)_9$, (c) $Fe_3(CO)_{12}$, (d) suggested structure of $Os_2(CO)_9$ and (e) structure of $Os_3(CO)_{12}$ and $Ru_3(CO)_{12}$.

These carbonyl compounds are used as catalysts in various reactions:

$$\underset{\text{ethyne}}{2HC{\equiv}CH} + 3CO + H_2O \xrightarrow{Fe(CO)_5} \underset{p\text{-quinol}}{HO{-}C_6H_4{-}OH} + CO_2$$

$$\underset{\text{nitrobenzene}}{C_6H_5NO_2} + 2CO + H_2 \xrightarrow{Ru_3(CO)_{12}} \underset{\text{aniline}}{C_6H_5NH_2} + 2CO_2$$

Nitrosyl compounds are formed by the action of NO on the dinuclear carbonyls:

$$Fe_2(CO)_9 + 4NO \rightarrow 2Fe(CO)_2(NO)_2 + 5CO$$

An unusual derivative $Ru_6C(CO)_{17}$ contains a distorted octahedral cluster of six Ru atoms with a carbon atom at the centre (similar to the arrangement in interstitial metal carbides), but four of the Ru atoms have three terminal CO groups and the remaining two Ru atoms have two terminal CO groups and one bridging CO.

The complex $[Fe(H_2O)_5NO]^{2+}$ is formed in the 'brown ring' test for nitrates. The colour is due to charge transfer. This complex formally contains Fe(+I) and NO^+. Its magnetic moment is approximately 3.9 BM, confirming the presence of three unpaired electrons.

(+II) STATE

Iron(+II) is one of the most important oxidation states, and salts are often called *ferrous salts*. They are well known as crystalline compounds. Most are pale green and contain the $[Fe(H_2O)_6]^{2+}$ ion, for example $FeSO_4 \cdot 7H_2O$, $FeCl_2 \cdot 6H_2O$ and $Fe(ClO_4)_2 \cdot 6H_2O$. The $[Fe(H_2O)_6]^{2+}$ ion exists in aqueous solutions. Ferrous compounds are easily oxidized, and so are difficult to obtain pure. However, the double salt $FeSO_4 \cdot (NH_4)_2SO_4 \cdot 6H_2O$ is used as a standard compound in volumetric analysis for titrations with oxidizing agents such as dichromate, permanganate and ceric solutions. $FeSO_4$ and H_2O_2 are used as Fenton's reagent for producing hydroxyl radicals and, for example, oxidizing alcohols to aldehydes.

Oxides

$\overline{FeO}$ is nonstoichiometric and is metal deficient. It commonly has the formula $Fe_{0.95}O$. It may be formed as a black powder by strongly heating iron(+II) oxalate $Fe^{II}(COO)_2$ in a vacuum and then quenching to prevent its disproportionation:

$$Fe(C_2O_4) \rightarrow FeO + CO_2 + CO$$

$$2Fe^{II}O \rightarrow \underset{Fe^{II}Fe_2^{III}O_4}{Fe_3O_4} + Fe^0$$

$\overline{FeO}$ dissolves in acids and is completely basic. It has a sodium chloride type of lattice, comprising a cubic close-packed arrangement of O^{2-} ions with Fe^{2+} ions occupying all (or almost all) of the octahedral holes. $Fe(OH)_2$ is precipitated from solutions containing Fe(+II) as a white solid, but it rapidly absorbs O_2 from the air and turns dark green and then brown. This is because it oxidizes first to a mixture of $Fe(OH)_2$ and $Fe(OH)_3$, and then to hydrous $Fe_2O_3 \cdot (H_2O)_n$. $Fe(OH)_2$ dissolves in acids. It also dissolves in strong solutions of NaOH, giving a blue–green complex $Na_4[Fe(OH)_6]$ which can be crystallized.

Halides

Iron dissolves in the halogen acids in the absence of air, and from these solutions the hydrated dihalides $FeF_2 \cdot 8H_2O$ (white), $FeCl_2 \cdot 4H_2O$ (pale green) and $FeBr_2 \cdot 4H_2O$ (pale green) are obtained. Hydrated ferrous chloride contains the $[FeCl_2 \cdot 4H_2O]$ octahedral unit in which the two chlorine atoms occupy *trans* positions.

Anhydrous FeF_2 and $FeCl_2$ are made by heating the metal with gaseous HF or HCl, as heating the elements gives FeF_3 and $FeCl_3$. $FeBr_2$ and FeI_2 are made by heating the elements. The anhydrous Fe(+II) halides react with gaseous NH_3, forming salts containing the octahedral complex ion $[Fe(NH_3)_6]^{2+}$. This complex decomposes in water to give $Fe(OH)_2$.

Complexes

Ferrous ions form many complexes. The most important one is haemoglobin (the red pigment in blood) which is discussed later in the section 'Bioinorganic chemistry of iron'. Most of the complexes are octahedral, though a few tetrahedral halide complexes $[FeX_4]^{2-}$ are formed.

The best known complex is the hexacyanoferrate(II) or ferrocyanide ion $[Fe^{II}(CN)_6]^{4-}$. Potassium hexacyanoferrate(II) $K_4[Fe(CN)_6]^{4-}$ is a yellow coloured solid and can be made by the action of CN^- on a ferrous salt in solution. Potassium ferrocyanide is used to test for iron in solution. Fe^{2+} ions give a white precipitate of $K_2Fe^{II}[Fe^{II}(CN)_6]$, but Fe^{3+} ions give deep blue $KFe^{III}[Fe^{II}(CN)_6]$ known as Prussian blue. A deep blue colour is also produced by Fe^{2+} with hexacyanoferrate(III) ions (ferricyanides) $[Fe^{III}(CN)_6]^{3-}$, and this is known as Turnbull's blue $KFe^{II}[Fe^{III}(CN)_6]$. Both have been used as pigments in ink and paint. Recent X-ray work, infrared and Mossbauer spectroscopy have shown that Turnbull's blue is identical to Prussian blue. The intense colour arises from electron transfer between Fe(+II) and Fe(+III). The standard reduction potentials show that it is easier to oxidize the $[Fe^{II}(CN)_6]^{4-}$ ion in aqueous solution to $[Fe^{III}(CN)_6]^{3-}$, than it is to oxidize $[Fe^{II}(H_2O)_6]^{2+}$ to $[Fe^{III}(H_2O)_6]^{3+}$.

$$[Fe^{III}(H_2O)_6]^{3+} + e \rightarrow [Fe^{II}(H_2O)_6]^{2+} \qquad E° = +0.77\,V$$
$$[Fe^{III}(CN)_6]^{3-} + e \rightarrow [Fe^{II}(CN)_6]^{4-} \qquad E° = +0.36\,V$$

This means that Fe(+III) forms a more stable complex with CN^- than does Fe(+II). The CN^- ion is known to form π bonds by accepting electrons from metal ions. Since Fe(+II) has more electrons than Fe(+III) it follows that Fe(+II) should have stronger π bonding and hence shorter bonds than Fe(+III). The Fe—C bond lengths of 1.95 Å in $[Fe^{III}(CN)_6]^{3-}$ and 1.92 Å in $[Fe^{II}(CN)_6]^{4-}$ confirm the difference in π bonding but this does not fit the observed $E°$ values. There must be another significant energy factor involved, and this is the large negative entropy of hydration resulting from the high charge on $[Fe^{II}(CN)_6]^{4-}$.

Cu^{2+} ions may be precipitated from solution as the red–brown complex $Cu_2^{II}[Fe(CN)_6]$ in gravimetric analysis.

The cyanide groups in $K_4[Fe(CN)_6]$ are kinetically inert, and are not easily removed or substituted. Though the salt is said not to be poisonous, care must be taken as it evolves HCN with dilute acids. It is possible to obtain mono-substitution products by replacing one CN^- by H_2O, CO, NO_2^- or NO^+. The best known is sodium nitrosopentacyanoferrate(II) $Na_2[Fe(CN)_5(NO)] \cdot 2H_2O$, which is usually called sodium nitroprusside. This complex has NO^+ as a ligand, and is formed as brown–red crystals by reacting a ferrocyanide with either 30% HNO_3 or with a nitrite:

$$[Fe(CN)_6]^{4-} + NO_3^- + 4H^+ \rightarrow [Fe(CN)_5(NO)]^{2-} + NH_4^+ + CO_2$$

$$Na_4[Fe(CN)_6] + NO_2^- + H_2O \rightarrow Na_2[Fe(CN)_5(NO)] + 2NaOH + CN^-$$

Sodium nitroprusside reacts with sulphide ions to give a purple complex $[Fe(CN)_5(NOS)]^{4-}$. This is used as a sensitive qualitative test for sulphides.

$$2[Fe(CN)_5(NO)]^{2-} + S^{2-} \rightarrow 2[Fe(CN)_5(NOS)]^{4-}$$

Other stable complexes are formed with bidentate ligands such as ethylenediamine, 2,2′-dipyridyl and 1,10-phenanthroline (formerly called *ortho*-phenanthroline (Figure 24.2). The latter complex $[Fe^{II}(phen)_3]^{2+}$ is bright red, and is used for the colorimetric determination of iron, and also as the redox indicator 'ferroin' in titrations. It is easier to oxidize $[Fe(H_2O)_6]^{2+}$ to $[Fe(H_2O)_6]^{3+}$ than it is to oxidize $[Fe(phen)_3]^{2+}$ (red colour) to $[Fe(phen)_3]^{3+}$ (blue colour). Thus the red colour persists until there is excess oxidizing agent present. The greater stability of the iron(II) phenanthroline complex is due to π bonding between the metal and low energy π* antibonding orbitals on the ligand. Similar stabilization also occurs in the dipyridyl complex.

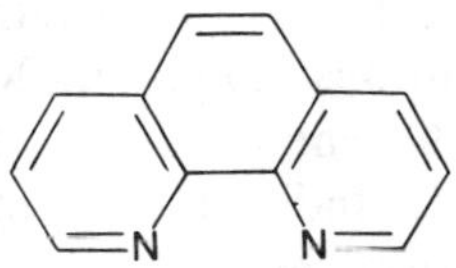

N N

Figure 24.2 1,10-phenanthroline.

Fe^{2+} has a d^6 electronic configuration, and octahedral complexes with weak field ligands have a high-spin arrangement with four unpaired electrons (Figure 24.3a). Strong field ligands such as CN^- and 1,10-phenanthroline cause spin pairing. This makes them more stable because they have a larger crystal field stabilization energy (Figure 24.3b). Spin pairing also results in the complexes being diamagnetic.

The brown ring test for nitrates and nitrites depends on forming a brown complex $[Fe(H_2O)_5 \cdot NO]^{2+}$. In this test a freshly prepared solution of $FeSO_4$ is mixed with the solution containing NO_2^- or NO_3^- ions in a test tube. Concentrated H_2SO_4 is run down the side of the tube so that the acid forms a layer at the bottom. The H_2SO_4 reacts with NO_3^-, forming NO, which combines with Fe^{2+}, slowly forming the brown complex $[Fe(H_2O)_5 \cdot NO]^{2+}$ at the interface between the two liquids. If the mixture gets hot or is shaken the brown colour disappears, NO is evolved and a yellow solution of $Fe^{III}{}_2(SO_4)_3$ remains. Nitrites give the brown colour before H_2SO_4 is added.

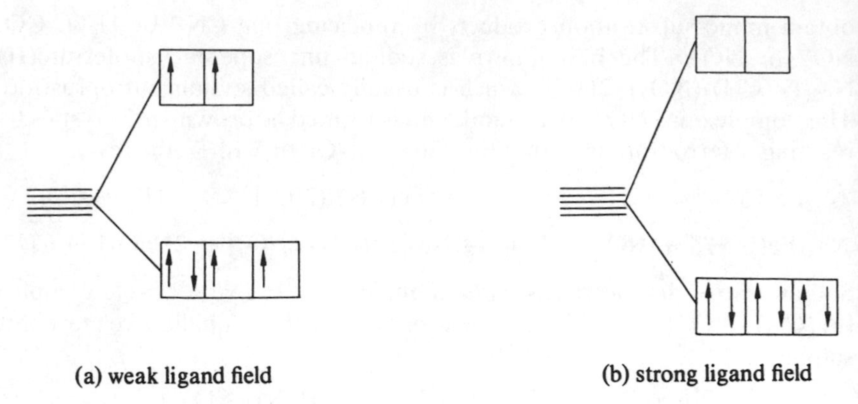

Figure 24.3 Electronic arrangements of the d^6 ion Fe^{2+}.

In addition to these octahedral complexes, the heavier halides form crystalline complexes $[Fe^{II}X_4]^{2-}$. These are tetrahedral complexes and are formed only with large cations. The presence of a large anion and a large cation gives a high lattice energy.

There are few simple compounds of Ru(+II) and Os(+II), though $[Ru^{II}(H_2O)_6]^{2+}$ is known. Except in the presence of non-complexing anions such as BF_4^- and ClO_4^-, the (+II) state exists as complexes. A large number of complexes are formed with ligands capable of back bonding, such as CO and phosphine ligands PR_3. Other complexes are formed with CN^-, Cl^-, NH_3 and amines. These are all tetrahedral. These complexes are formed by reducing solutions containing M(+III) or M(+IV) in the presence of the ligand. The metal has a d^6 configuration, and the complexes are all diamagnetic, indicating a spin paired arrangement, with a large crystal field stabilization energy.

Ruthenium forms a binuclear carboxylate $Ru_2(CH_3COO)_4(H_2O)_2$ similar to chromium(II) acetate and containing a M—M bond. Similar complexes are also formed by Mo and Rh.

Ruthenium also forms an interesting series of complexes $Ru^{II}NO \cdot L_5$, where the ligand L may be: Cl^-, NH_3, H_2O, NO_3^-, OH^-, CN^- and many others. The NO is strongly bonded to the metal by σ and π bonds, and the Ru—NO bond persists through a wide variety of substitution and oxidation–reduction reactions. X-ray studies show that in some of the complexes the Ru—N—O atoms are linear, whilst in others they are bent.

```
Ru—N—O     Ru—N
               \
                O
```

Similar osmium complexes are known, though they have been studied less.

Carbonyl complexes of Fe, Ru and Os have been known for a long time. CO and N_2 are isoelectronic, and for a long time it was speculated that M—NN bonds might be formed analogous to M—CO. The first example of a dinitrogen complex $[Ru(NH_3)_5 \cdot N_2]Cl_2$ was reported in 1965. The complex can be made by reducing an aqueous solution of $RuCl_3$ with hydrazine sulphate or by treating $[Ru^{III}(NH_3)_5 \cdot H_2O]^{3+}$ with NaN_3. The reaction where H_2O is replaced directly by N_2 is of greatest interest, since this might be closer to how bacteria fix atmospheric nitrogen.

$$[Ru(NH_3)_5H_2O]^{2+} + N_2 \rightarrow [Ru(NH_3)_5 \cdot N_2]^{2+}$$

Terminal N_2 ligands have a strong IR band in the range 1930–2230 cm^{-1}. This compares with a band at 2331 cm^{-1} in N_2 gas. The IR band may be used to show that a dinitrogen complex has been formed, and the lowering of the wavenumber shows π bonding has occurred from Ru to N, thus reducing the N—N bond order. Thus N_2 is acting as a π acceptor ligand. N_2 is a weaker σ donor and a weaker π acceptor than CO, and hence dinitrogen complexes are not very stable. Complexes of other metals including Mo, Fe and Co can also take up N_2 at atmospheric pressure, particularly with tertiary phosphine ligands:

$$FeCl_2 + N_2 + 3PEtPh_2 \xrightarrow{NaBH_4} FeH_2(N_2)(PEtPh_2)_3$$

The N_2 ligand may also act as a bridging group:

$$[Ru(NH_3)_5 \cdot N_2]^{2+} + [Ru(NH_3)_5 \cdot H_2O]^{2+}$$
$$\rightarrow [(NH_3)_5 \cdot Ru—N—N—Ru \cdot (NH_3)_5]^{4+}$$

In the dinuclear complex the N—N bond is 1.24 Å, only slightly longer than the distance of 1.098 Å in N_2. The interest in complexes which can coordinate nitrogen arises from the possibility of nitrogen fixation (see also Chapter 20). Nitrogen complexes have also been made for all of the metals in the Mn, Fe and Co groups, together with Mo and Ni. The complexes typically contain the metal in a low oxidation state, often with phosphine or hydride ligands such as $(R_3P)_2Ni \cdot N_2 \cdot Ni(PR_3)_2$ and $(R_3P)_3 \cdot Fe \cdot N_2 \cdot H_2$. Since the N_2 molecule is symmetrical and has zero dipole moment, σ bonding from N to the metal will be very slight (hence the very small change in N—N bond length), and back bonding from the metal to π orbitals on the nitrogen is the main interaction.

(+III) STATE

Fe(+III) is a very important oxidation state, and ferric salts are obtained by oxidizing the corresponding ferrous salts. Ferric solutions are frequently yellow–brown, but the colour is due to the presence of colloidal iron oxide, or $FeO \cdot OH$. Fe(+III) forms crystalline salts with all the common anions except I^-, and many of the salts exist in both anhydrous and hydrated forms. These have a variety of colours: $FeCl_3 \cdot 6H_2O$, $FeF_3 \cdot 4\frac{1}{2}H_2O$ and $Fe_2(SO_4)_3 \cdot 9H_2O$ are yellow; $FeBr_3 \cdot 6H_2O$ and FeF_3 are green; $Fe(NO_3)_3 \cdot 6H_2O$ is colourless; and $Fe(NO_3)_3 \cdot 9H_2O$ is pale purple. Several salts contain the $[Fe(H_2O)_6]^{3+}$ ion. The most common is $Fe_2(SO_4)_3$. This exists as six different hydrates, and is quite widely used as a coagulant to clarify drinking water and also in the treatment of industrial effluent and sewage.

The alums are double salts, and crystallize easily. They also contain $[Fe(H_2O)]^{3+}$. The best known are ammonium ferric alum $(NH_4)[Fe^{III}(H_2O)_6][SO_4]_2 \cdot 6H_2O$ and potash alum $[K(H_2O)_6][Fe^{III}(H_2O)_6][SO_4]_2$. Like the chrome alums, these are used as mordants in the dyeing industry.

Oxides and hydrated oxides

There is no evidence that $Fe(OH)_3$ exists. Hydrolysis of $FeCl_3$ does not give the hydroxide but gives a red–brown gelatinous precipitate of the hydrous oxide $Fe_2O_3(H_2O)_n$. At least part of this precipitate consists of $FeO \cdot OH$. Heating the hydrous oxide to 200 °C gives red–brown α-Fe_2O_3. The structure comprises a hexagonally close-packed lattice of O^{2-} ions with Fe^{3+} ions in two thirds of the octahedral holes. However, if Fe_3O_4 is oxidized, γ-Fe_2O_3 is formed, which has a cubic close-packed arrangement of O^{2-} ions with Fe^{3+} ions randomly distributed in both the octahedral and tetrahedral sites.

Fe_3O_4 is formed as a black solid by igniting Fe_2O_3 at 1400 °C. Fe_3O_4 is a mixed oxide $Fe^{II}Fe_2^{III}O_4$ and has an inverse spinel structure. The O^{2-} ions are cubic close-packed, with the larger Fe^{2+} in one quarter of the octahedral holes. Half of the Fe^{3+} occupy octahedral holes and half occupy tetrahedral holes.

$\overline{Fe_2O_3}$ and $\overline{Fe_3O_4}$, like $\overline{FeO}$, all tend to be nonstoichiometric. The nonstoichiometry is related to their similarity in structure. The cubic close-packed forms differ only in the arrangement of Fe^{2+} and Fe^{3+} in the octahedral and tetrahedral holes.

Basic properties decrease with increased oxidation number. $Fe^{III}{}_2O_3$ is largely basic. The ignited form is difficult to dissolve in acids. However, the freshly precipitated hydrous form dissolves in acids, giving the very pale violet $[Fe(H_2O)_6]^{3+}$ ion, and in concentrated NaOH, forming $[Fe(OH)_6]^{3-}$. This shows that the oxide is slightly amphoteric. Fusion with LiOH, NaOH or Na_2CO_3 gives $LiFeO_2$ or $NaFeO_2$.

$$Fe_2O_3 + Na_2CO_3 \rightarrow 2NaFeO_2 + CO_2$$

At one time these compounds were called ferrites, but they are better represented as mixed oxides, as $LiFeO_2$ has a NaCl structure, and Li^+ and Fe^{3+} are approximately the same size. The average charge on the metal ions is +2, which matches the charge of −2 on the oxide ions. Ferrites hydrolyse with water, forming NaOH. At one time this was used in the obsolete Lowig process for manufacturing caustic soda.

$$Fe_2O_3 + Na_2CO_3 \rightarrow 2NaFeO_2 + CO_2$$
$$2NaFeO_2 + H_2O \rightarrow 2NaOH + Fe_2O_3$$

If Cl_2 is passed into an alkaline solution of hydrated ferric oxide, a red–purple solution is formed containing the ferrate ion $Fe^{VI}O_4^{2-}$. Ferrates are also made by oxidizing Fe(+III) with NaOCl or electrolytically. They contain Fe(+VI) and are stronger oxidizing agents than is $KMnO_4$.

Halides

The halides FeF_3, $FeCl_3$ and $FeBr_3$ can all be made by direct reaction of halogen and metal. The iodide does not exist in a pure state because Fe^{3+} oxidizes I^-. In solution this oxidation is complete but some FeI_3 may exist with FeI_2 in the solid. Fe(+III) compounds are less ionized than those of Fe(+II). FeF_3 is a white solid. It is sparingly soluble in water but the solution does not give a positive test for Fe^{3+} or F^- ions. It combines with alkali metal fluorides, forming complexes, e.g. $Na_3[FeF_6]$. Fe^{3+} has a d^5 configuration. Since F^- is a weak field ligand, each of the *d* orbitals will be singly occupied, giving a high-spin octahedral complex (Figure 24.4a). As a result of the small size of Fe^{3+}, Fe(+III) complexes are generally more stable than those of Fe(+II).

Any *d–d* electronic transition breaks the Laporte selection rule. (This states that Δl, the change in the secondary quantum number for the transition, must be ± 1. For a *d–d* transition $\Delta l = 0$.) In $[FeF_6]^{3-}$ the electronic transition is 'spin forbidden' too, as it involves both promoting an electron and reversing its spin. There is only a low probability of doing this: hence the lack of colour. A similar effect is observed with Mn^{2+} spectra (see also Chapter 32).

In contrast to the absence of colour in FeF_3, $FeCl_3$ solid is almost black. It sublimes at about 300 °C, giving a dimeric gas. $FeBr_3$ is red–brown. The colours arise from charge transfer spectra. The solid structures are layer lattices, consisting of close-packed halide ions with Fe^{3+} occupying two thirds of the octahedral holes in one layer, and none in the next layer. $FeCl_3$ dissolves in both ether and water, giving solvated monomeric species (Figure 24.5). Iron(III) chloride is usually obtained as yellow–brown lumps of the hydrate $FeCl_3 \cdot 6H_2O$. This is very soluble in water and is used both as an oxidizing agent, and as a mordant in dyeing. $FeCl_3$ is also used in the manufacture of CCl_4.

$$CS_2 + 3Cl_2 \xrightarrow{FeCl_3 \text{ catalyst } 30\,°C} CCl_4 + S_2Cl_2$$

$$CS_2 + 2S_2Cl_2 \xrightarrow{FeCl_3 \text{ catalyst } 60\,°C} CCl_4 + 6S$$

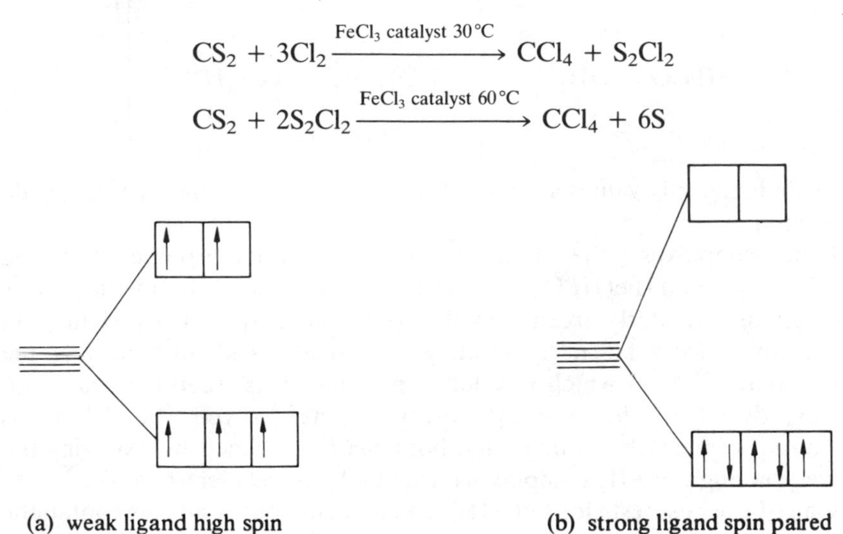

Figure 24.4 d^5 electronic arrangement for Fe^{3+}.

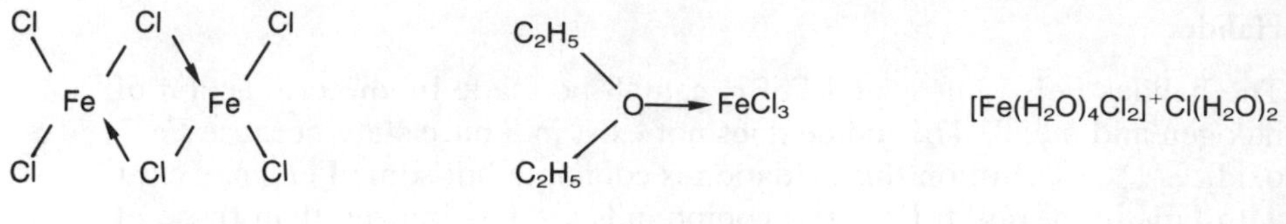

(a) $(FeCl_3)_2$ gas dimer

(b) solvated $FeCl_3$

(c) hydrated $FeCl_3$ $(FeCl_3 \cdot 6H_2O)$

Figure 24.5 Structures of $FeCl_3$.

The structure of $FeCl_3 \cdot 6H_2O$ is unusual. It normally consists of *trans* $[Fe(H_2O)_4Cl_2]Cl \cdot 2H_2O$, though in strong HCl it forms tetrahedral $[FeCl_4]^-$ ions.

Complexes

Fe^{3+} shows a preference for forming complexes with ligands which coordinate through O as opposed to N. Complexes with NH_3 are unstable in water. Complexes with chelating N ligands such as dipyridyl and 1,10-phenanthroline are formed, but are less stable than their Fe(+II) counterparts. These ligands cause spin pairing.

The most common complex is the hydrated ion $[Fe(H_2O)_6]^{3+}$. This is pale purple in very strongly acid solutions, and tends to hydrolyse to yellow solutions at pH 2–3.

$$[Fe(H_2O)_6]^{3+} \rightleftharpoons [Fe(H_2O)_5 \cdot OH]^{2+} + H^+$$

At pH 4–5 the hydroxo species polymerizes to form a dimer which forms a brownish solid. Polymerization of aqua ions is quite common, especially if they have a high charge.

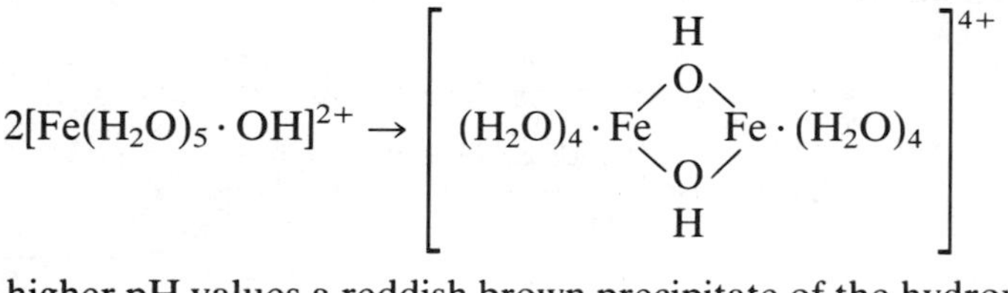

$$2[Fe(H_2O)_5 \cdot OH]^{2+} \rightarrow \left[(H_2O)_4 \cdot Fe \begin{matrix} H \\ O \\ \\ O \\ H \end{matrix} Fe \cdot (H_2O)_4 \right]^{4+}$$

At even higher pH values a reddish brown precipitate of the hydrous oxide is formed.

Other complexes with O donors include those with phosphate ions giving $[Fe(PO_4)_3]^{6-}$ and $[Fe(HPO_4)_3]^{3-}$ (which are colourless), and with oxalate ions giving the dark green complex $[Fe(oxalate)_3]^{3-}$. Rust stains can be removed from fabric by treating with oxalic acid and forming the $[Fe(oxalate)_3]^{3-}$ ion which is soluble in water. This treatment may also remove dyes from the fabric. $[Fe(acetylacetone)_3]$ is red. $[Fe(CN)_6]^{3-}$ is red and $[Fe(phen)_3]^{3+}$ is blue, and both can be obtained by oxidizing the corresponding Fe(+II) complex with $KMnO_4$, or by electrolysis.

One of the best tests for Fe(+III) is to mix aqueous solutions containing Fe^{3+} and SCN^- ions. A blood red colour is produced, which is due to a mixture of $[Fe(SCN)(H_2O)_5]^{2+}$ and also some $Fe(SCN)_3$ and $[Fe(SCN)_4]^-$. The colour may also be used for the estimation of ferric iron. This colour is destroyed by the addition of F^- ions, because $[FeF_6]^{3-}$ is formed.

Fe^{3+} has a d^5 electronic configuration. Thus complexes with weak field ligands will have a high-spin arrangement with five unpaired electrons. Thus *d–d* spectra will be 'spin forbidden': hence the absorption will be very weak. Mn^{2+} also has a d^5 electronic arrangement and also has very weak *d–d* spectra. However, Fe(+III) has an extra charge and is more able to polarize the ligands, thus giving intense charge transfer spectra not found in Mn^{2+}. The only complex which actually shows the weak *d–d* bands is $[Fe(H_2O)_6]^{3+}$. The hydrolysed species have charge transfer bands which mask the *d–d* spectra. Strong field ligands such as CN^-, SCN^- and oxalate form complexes with Fe^{3+} which have a spin paired arrangement (Figure 24.4b). These would be expected to show reasonably intense colours from *d–d* spectra, but these colours too are masked by charge transfer.

Fe(+III) and Ru(+III) form basic acetates of the type $[Fe_3O(CH_3COO)_6L_3]^+$. The structure consists of a triangle of three Fe atoms with an O atom at the centre. The six acetate groups act as bridges between the Fe atoms, two groups across each edge of the triangle. Thus each Fe atom is linked to four acetate groups and the central O, and the sixth position of the octahedron is occupied by water or another ligand. This type of carboxylate complex is formed by the trivalent ions of Cr, Mn, Fe, Ru, Rh and Ir.

Ruthenium and osmium

In the (+III) state, Ru complexes are more numerous than those of Os. Both elements form $[M^{III}(NH_3)_6]^{3+}$, and Ru forms a range of mixed halogen–ammonia complexes. If RuO_4 is added to concentrated HCl and evaporated, a dark red material formulated $RuCl_3 \cdot 3H_2O$ is formed. This is readily soluble in water, and is the starting point of many preparations. It seems to contain not only Ru^{III} species but also some polynuclear Ru^{IV} species. Ru(+III) chloro species catalyse the hydration of alkynes. Ru(+III) forms a basic acetate similar to $[Fe_3O(CH_3COO)_6L_3]^+$ described above.

(+IV) AND HIGHER STATES

Burning Ru or Os metals in air gives RuO_2 as a blue–black solid and OsO_2 as a coppery coloured solid. Both oxides have rutile (TiO_2) structures. Os forms a stable tetrafluoride and tetrachloride.

(+V) STATE

The (+V) state is unstable, and is found only as the pentafluorides, which are tetrameric with bridging F groups as in NbF_5 and TaF_5 (see Figure 21.1a). Fluoride complexes are also known.

$$RuCl_3 + KCl + F_2 \rightarrow K[Ru^{V}F_6]$$

(+VI) STATE

Ferrates $Fe^{VI}O_4^{2-}$ can be made by passing Cl_2 into an alkaline solution of hydrated ferric oxide, by oxidizing Fe(+III) with NaOCl, or electrolytically. They contain Fe(+VI), are purple coloured and are stronger oxidizing agents than $KMnO_4$. The stability of the (+VI) state decreases across the periodic table:

$$CrO_4^{2-} > MnO_4^{2-} > FeO_4^{2-} >> CoO_4^{2-}$$

Ferrates are only stable in strongly alkaline solution, and decompose in water or acid, liberating oxygen.

$$2[Fe^{VI}O_4]^{2-} + 5H_2O \rightarrow Fe^{3+} + 1\tfrac{1}{2}O_2 + 10(OH)^-$$

Sodium and potassium ferrates are very soluble, but $BaFeO_4$ is precipitated. The ferrate ion is tetrahedral like the chromate ion CrO_4^{2-}.

RuF_6 is the highest halide of Ru. It is made by heating the elements and quenching (i.e. cooling very rapidly). RuF_6 is unstable, but in contrast OsF_6 is stable.

(+VIII) STATE

RuO_4 and OsO_4 are yellow coloured volatile solids with melting points of 25 °C and 40 °C respectively. OsO_4 is made either by burning finely divided metal in O_2, or by treating it with concentrated HNO_3. RuO_4 is less stable and is formed by oxidation with permanganate or bromate in H_2SO_4. Both oxides are toxic, smell like ozone and are strongly oxidizing. They have tetrahedral structures. Both are slightly soluble in water but very soluble in CCl_4. Aqueous solutions of OsO_4 are used as a biological stain because the organic matter reduces it to black OsO_2 or Os. OsO_4 vapour is harmful to the eyes for this reason. OsO_4 is used in organic chemistry to add to double bonds and give *cis* glycols. The tetroxides do not show basic properties and HCl reduces OsO_4 to *trans*-$[Os^{VI}O_2Cl_4]^{2-}$, $[Os^{IV}Cl_6]^{2-}$ and $[Os^{IV}{}_2OCl_{10}]^{2-}$. RuO_4 dissolves in NaOH solutions and liberates O_2. Ru(+VIII) is reduced forming first the perruthenate (+VII) ion, then the ruthenate (+VI) ion:

$$4Ru^{VIII}O_4 + 4OH^- \rightarrow \underset{\text{perruthenate}}{4Ru^{VII}O_4^-} + 2H_2O + O_2$$

$$4Ru^{VII}O_4^- + 4OH^- \rightarrow \underset{\text{ruthenate}}{4Ru^{VI}O_4^{2-}} + 2H_2O + O_2$$

The ruthenate ion contains Ru(+VI) and is analogous to the ferrate ion FeO_4^{2-}. OsO_4 is more stable and is not reduced by NaOH. Instead, OH^- ligands are added forming the octahedral *trans*-osmate(VIII) ion.

$$Os^{VIII}O_4 + 2OH^- \rightarrow \underset{\text{perosmate or osmate (VIII)}}{[Os^{VIII}O_4\cdot(OH)_2]^{2-}}$$

The osmate(VIII) ion reacts with NH_3 giving an unusual nitrido complex $[OsO_3N]^-$, and is reduced by EtOH to form *trans*-$[Os^{VI}O_2(OH)_4]^{2-}$.

BIOINORGANIC CHEMISTRY OF IRON

Iron is essential in small amounts for both plant and animal life. However, like Cu and Se it is toxic in larger quantities. Biologically iron is the most important transition element. It is involved in several different processes:

1. As an oxygen carrier in the blood of mammals, birds and fish (haemoglobin).
2. For oxygen storage in muscle tissue (myoglobin).
3. As an electron carrier in plants, animals and bacteria (cytochromes) and for electron transfer in plants and bacteria (ferredoxins).
4. For storage and scavenging of Fe in animals (ferretin and transferrin).
5. As nitrogenase (the enzyme in nitrogen fixing bacteria).
6. As a number of other enzymes: aldehyde oxidase (oxidation of aldehydes), catalase and peroxidase (decomposition of H_2O_2) and succinic dehydrogenase (the aerobic oxidation of carbohydrates).

Haemoglobin

The human body contains about 4 g of iron. About 70% of this is found as haemoglobin, the red pigment in the erythrocytes (red blood cells). Most of the rest is stored as ferretin. The function of haemoglobin is to pick up oxygen at the lungs. The arteries carry blood to parts of the body where oxygen is required such as the muscles. Here the oxygen is transferred to a myoglobin molecule, and stored until the oxygen is required to release energy from glucose sugar. When O_2 is removed from haemoglobin, it is replaced by a water molecule. Next the protein part of haemoglobin absorbs H^+. Indirectly this helps remove CO_2 from the tissues, since CO_2 is converted to HCO_3^- and H^+. The blood removes the more soluble HCO_3^- ions and the reduced haemoglobin removes the H^+. The blood returns to the heart through the veins. It is then pumped to the lungs, where the HCO_3^- ions are converted back to CO_2. This is excreted into the air in the lungs and exhaled. The blood picks up oxygen again, and the process is repeated.

The important factor in haemoglobin acting as an oxygen carrier is the reversibility of the process. If too stable an oxygen complex were formed then too much energy would be released in the lungs leaving less energy when oxygen is released in the muscles. The oxygenated form is called *oxyhaemoglobin* and the reduced form is called *deoxyhaemoglobin*. This transfer of O_2 is remarkable because it involves only Fe(+II), and **not** Fe(+III). Other groups such as CO, CN^- and PF_3 can occupy the O_2 site. Coordination is still reversible, but is much stronger. In the case of CN^-, coordination is irreversible. These ligands reduce or prevent oxygen transfer and may result in death. However, CN^- also interferes with the cytochrome enzyme system, which is the principal reason for its extreme toxicity.

Haemoglobin has a molecular weight of nearly 65 000, and is made up of four subunits. Each subunit comprises a porphyrin complex haem (Figure 24.6) which contains Fe^{2+} bonded to four N atoms, and a globular protein called globin. The globular protein is coordinated to the Fe^{2+} in haem through a fifth N atom from a histidine molecule in the protein. The sixth position round the Fe^{2+} is occupied either by an oxygen molecule or a water molecule.

In oxyhaemoglobin the Fe^{2+} is in the low-spin state and is diamagnetic. It is just the right size to fit in the hole at the centre of the porphyrin ring. The porphyrin is both planar and rigid. In deoxyhaemoglobin the Fe^{2+} is in the high-spin state and is paramagnetic. The size of Fe^{2+} increases by 28% when it changes from low spin to high spin, i.e. from 0.61 Å to 0.78 Å (see Table 24.7). Thus in deoxyhaemoglobin the Fe^{2+} is too large to fit in the hole at the centre of the porphyrin, and is situated 0.7–0.8 Å above the ring, thus distorting the bonds round the Fe. Thus the presence of O_2 changes the electronic arrangement of Fe^{2+} and also distorts the shape of the complex. The globular protein appears to be essential, since if it is removed oxidation of Fe(+II) to Fe(+III) occurs which is not reversible.

Haemoglobin is made up of four subunits, and when one subunit picks up an O_2 molecule, the Fe^{2+} contracts and moves into the plane of the ring. In doing so, it moves the histidine molecule attached to it, and causes conformational changes in the globin chain. Since this chain is hydrogen bonded to the other three units, it changes their conformations too, and enhances their ability to attract O_2. This phenomenon is called the *cooperative effect*. The affinity of haemoglobin for O_2 decreases as the pH decreases, but as blood is well buffered this has only a slight effect. In a similar way when the blood reaches the muscles, once one O_2 has been released the others are released even more easily due to the cooperative effect in reverse.

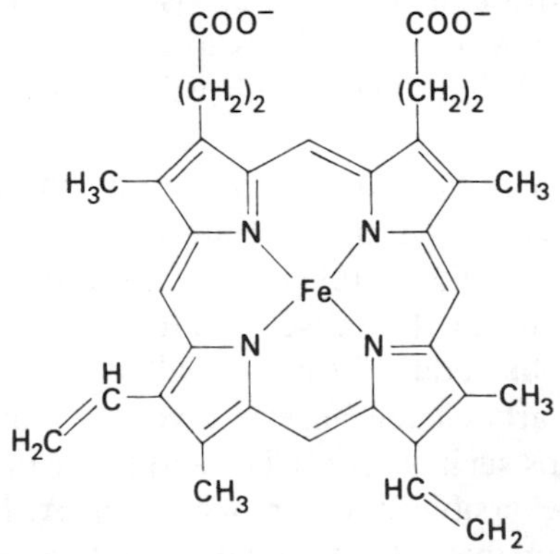

Figure 24.6 Structure of haem b.

CO_2 is the end product from the breakdown of glucose to release energy. There is an appreciable build-up of CO_2 in the muscles. This is removed from the tissue and converted into soluble HCO_3^- ions.

$$CO_2 + H_2O \rightarrow H^+ + HCO_3^-$$

This process is facilitated by the terminal amine groups of deoxyhaemoglobin which pick up the protons produced and thus act as a buffer. The reverse process occurs at the lungs.

The porphyrin ring is conjugated and planar. The characteristic red colour arises from charge transfer between stable π and low-lying π^* orbitals on the ring and Fe.

Myoglobin

Haem is also important biologically in myoglobin which is used to store oxygen in muscles. Myoglobin is similar to one of the units in haemoglobin. It contains only one Fe atom, has a molecular weight of about 17 000, and binds O_2 more strongly than haemoglobin.

Cytochromes

There are many cytochromes, which differ in slight detail, but these are broadly grouped together as cytochrome a, cytochrome b and cytochrome c. The prosthetic group in all cytochromes comprises four haem units, and cytochromes have a molecular weight of about 12 400. As in haemoglobin, Fe is bonded to four N atoms in each porphyrin ring, and the fifth site is occupied by a N atom from the associated protein. The big difference is that the sixth position is usually occupied by a S atom from an amino acid such as methionine, which is part of a protein.

Cytochromes are involved in the release of energy by oxidizing glucose with molecular O_2 in the mitochondria inside living cells. The cytochromes are reversibly oxidized (and thus act as electron carriers). The Fe is in the low-spin state, and it changes reversibly between the (+II) and (+III) states. Cytochromes a, b and c have slightly different reduction potentials and reactions involve all three one after the other in the order b, c, a. In this way the energy from oxidizing glucose is released gradually. The energy is stored in the form of adenosine triphosphate ATP, which is used when required by the cell.

Table 24.8 Reduction potentials E° for cytochrome

Cytochrome b	0.04 V
Cytochrome c	0.26 V
Cytochrome a	0.28 V

Ferretin

Animals, including man, absorb iron as Fe(+II) from food in their digestive systems. Their requirements for Fe are very small as the existing Fe in the body is recycled. Any iron absorbed immediately reacts to form transferrin. A human body contains about 4 g of Fe, roughly 3 g as haemoglobin, and 1 g as ferretin. Ferretin is found in the spleen, liver and bone marrow. When a red blood cell has become aged after an average of 16 weeks, the haemoglobin is broken down and the iron is recovered by transferrin, a non-haem protein. This is a single chain polypeptide of molecular weight 76 000–80 000, which contains two Fe atoms. This transports the Fe to the bone marrow where it is converted to ferretin, which is brown and water soluble. Ferretin contains about 23% Fe. It consists of a roughly spherical sheath of protein called apoferretin which is approximately 120 Å in diameter and which encloses a micelle of Fe(+III) hydroxide, oxide and phosphate. The micelle is an aggregate of particles whose surface bears a charge. The micelle contains 2000–5000 atoms of Fe.

Catalases and peroxidases

Catalases are enzymes which promote the disproportionation of H_2O_2. They also catalyse the oxidation of substrates by H_2O_2. The catalase molecule contains four Fe(+III) haem b groups, and has a molecular weight of about 240 000.

$$2H_2O_2 \rightarrow 2H_2O + O_2$$

The peroxidases also catalyse the decomposition of H_2O_2 but they are associated with a coenzyme AH_2, which is oxidized in the reaction.

$$H_2O_2 + AH_2 \rightarrow 2H_2O + A$$

Ferredoxins

These are a group of non-haem iron proteins which are responsible for electron transfer in plants and bacteria. They serve the same function that cytochromes perform in animals, but ferredoxins have much lower molecular weights (6000–12 000), and they may contain one, two, four or eight Fe atoms. Rubedoxin is the simplest and contains only one Fe atom. This is surrounded by four S atoms from the amino acid cysteine which is linked into protein chains, and may be represented $Fe(S\text{-cysteine})_4$. The four S atoms are roughly tetrahedral.

Other ferredoxins include inorganic sulphide ions as well as cysteine S atoms, e.g. $Fe_2(S_2)(S\text{-cysteine})_4$. Since they contain inorganic sulphide ions, treatment with dilute acids liberates H_2S.

Some bacterial ferredoxins form cluster compounds. The simplest is $Fe_4S_4(S\text{-cysteine})_4$. The cluster may be imagined as a cube with two Fe atoms at diagonally opposite corners on the top face and two Fe atoms occupying the other two diagonally opposite corners on the bottom face. S atoms fill the four unoccupied positions. In addition, each Fe is bonded to a S atom in a cysteine molecule.

Haemerythrin

Despite its name, haemerythrin is a non-haem iron protein which serves as the oxygen carrier in some marine worms. It has a molecular weight of about 108 000, and is made up of eight subunits. Other marine worms have myohaemerythrin in their muscles which comprises just one of these subunits. This is analogous to haemoglobin and myoglobin in higher animals. However, deoxyhaemerythrin contains two high-spin Fe^{II} atoms and oxyhaemerythrin contains two Fe^{III} atoms and oxygen bound as O_2^{2-}.

CYCLOPENTADIENYL AND RELATED COMPOUNDS

Interest in organometallic chemistry began in 1951 when G. Wilkinson *et al.* reported making an astonishing iron–hydrocarbon derivative called di-π-cyclopentadienyliron. Rather surprisingly two research groups working independently prepared the same compound about the same time.

$$2C_5H_5MgBr + FeCl_2 \rightarrow Fe(C_5H_5)_2 + MgBr_2 + MgCl_2$$

This compound is now called ferrocene and has the formula $(\pi\text{-}C_5H_5)_2Fe$. It is stable in air, forms orange crystals, and is diamagnetic. Ferrocene is soluble in organic solvents (alcohol, ether and benzene), and insoluble in water, NaOH solution and concentrated HCl. It is thermally stable up to 500 °C. The X-ray structure shows that this has a sandwich structure in which the metal atom is sandwiched between two parallel planar cyclopentadienyl rings. This gave the clue that organic ligands could use their π system to bond to metals, and this opened up the study of compounds with metal–carbon bonds.

For ferrocene the symmetry of the space group requires the two five-membered rings to have the staggered conformation. In contrast, in the corresponding Ru and Os compounds ruthenocene and osmocene the rings adopt the eclipsed conformation (Figure 24.7). The exact arrangement in ferrocene is not simple. The barrier to internal rotation of the rings is only about 4 kJ mol^{-1}, and electron diffraction (on the gas) suggests an eclipsed arrangement. It is possible that crystal packing forces result in the staggered arrangement. More recent X-ray and neutron diffraction studies on the solid suggest the space group symmetry is met by a disordered arrangement of nearly eclipsed molecules with an angle of 9° between the rings rather than 0° for eclipsed and 36° for a staggered conformation.

Since all five C atoms in the cyclopentadienyl ring are equidistant from Fe, the ring has a hapticity η of 5, i.e. $\eta^5\text{-}C_5H_5$. All the first row transition elements form similar compounds, but they are much less stable than ferrocene.

The perpendicular distance between the rings is 3.25 Å compared with

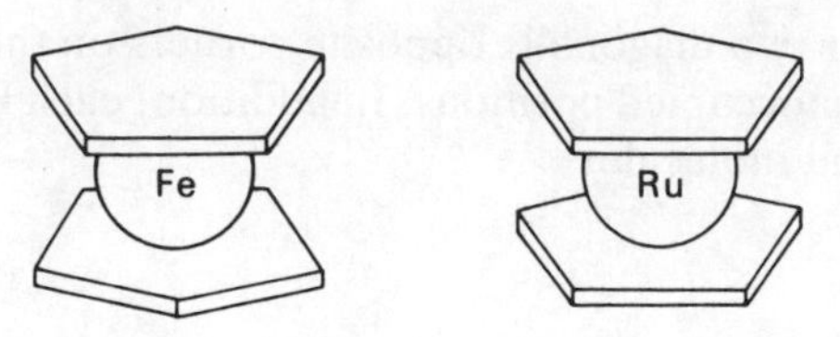

Figure 24.7 Ferrocene and ruthenocene.

3.35 Å in graphite. The C—C bond lengths are all equal (1.39 Å ± 0.06 Å). This is exactly the same bond length as in benzene, and the bond orders are also similar to those in benzene. The cyclopentadiene rings do not undergo reactions of dienes such as the Diels–Alder reaction or catalytic hydrogenation, but they do display aromatic character. Thus ferrocene undergoes Friedel–Crafts acylation. With an equimolar amount of acetyl chloride the following reaction occurs:

$$(\eta^5\text{-}C_5H_5)_2Fe + CH_3COCl \rightarrow (\eta^5\text{-}C_5H_4 \cdot CO \cdot CH_3)(\eta^5\text{-}C_5H_5)Fe + HCl$$

With excess acetyl chloride, disubstitution occurs. (Other cyclopentadienyl compounds are decomposed.) This suggests that the ligand is really $C_5H_5^-$.

$$2C_5H_6 + 2Et_2NH + FeCl_2 \rightarrow Fe(C_5H_5)_2 + 2Et_2NH_2Cl$$

The general method for making cyclopentadienyl compounds is from cyclopentadiene in tetrahydrofuran solution:

$$C_5H_6 + Na \rightarrow Na^+ + C_5H_5^- + \tfrac{1}{2}H_2$$
$$2C_5H_5^- + FeCl_2 \rightarrow (\eta^5\text{-}C_5H_5)_2Fe + 2Cl^- \quad \text{ferrocene}$$
$$2C_5H_5^- + NiCl_2 \rightarrow (\eta^5\text{-}C_5H_5)_2Ni + 2Cl^- \quad \text{nickelocene}$$

Another preparative method uses C_5H_5Tl, which is stable, and insoluble in water:

$$C_5H_6 + TlOH \xrightarrow{H_2O} C_5H_5Tl + H_2O$$
$$2C_5H_5Tl + FeCl_2 \xrightarrow{THF} (\eta^5\text{-}C_5H_5)_2Fe + 2TlCl$$

Another convenient preparative method is to use a strong base to remove a proton from C_5H_6:

$$2C_5H_6 + FeCl_2 + 2Et_2NH \rightarrow Fe(C_5H_5)_2 + 2Et_2NH_2Cl$$

A large number of η^5-C_5H_5 complexes are now known.

Other ring systems are now known to form similar sandwich complexes, including C_6H_6, C_8H_8, C_3Ph_3, C_4H_4 and C_7H_7. Examples include $[Cr(\eta^6\text{-}C_6H_6)_2]$ and $[U(\eta^8\text{-}C_8H_8)_2]$.

Some cyclopentadienyl compounds have two rings at an angle, rather than forming a sandwich. For example, $[(\eta^5\text{-}C_5H_5)_2TiCl_2]$, $[Ti(\eta^5\text{-}C_5H_5)_2(CO)_2]$, $[Ti(\eta^5\text{-}C_5H_5)_2(NR_2)_2]$, $[Ti(\eta^5\text{-}C_5H_5)_2(SCN)_2]$ and $[V(\eta^5\text{-}C_5H_5)_2Cl_2]$ have roughly tetrahedral structures, but the cyclopentadiene mole-

Table 24.9 Some di-η^5-cyclopentadienyl sandwich compounds

$[(\eta^5\text{-}C_5H_5)_2V^{II}]$	Vanadocene	Dark violet solid, air sensitive
$[(\eta^5\text{-}C_5H_5)_2Cr^{II}]$	Chromocene	scarlet crystals, m.p. 173 °C; very air sensitive
$[(\eta^5\text{-}C_5H_5)_2Fe^{II}]$	Ferrocene	Orange crystals, m.p. 174 °C; stable to >500 °C
$[(\eta^5\text{-}C_5H_5)_2Co^{III}]^+$	Cobaltocene	Yellow salts, stable in water, stable to about 400 °C
$[(\eta^5\text{-}C_5H_5)_2Ni^{II}]$	Nickelocene	Bright green solid, m.p. 173 °C (dec); oxidizes in air to $[(\eta^5\text{-}C_5H_5)_2Ni]^+$

cules are pentahaptic, with five C atoms in each ring attached to Ti. Reduction of these compounds gives $[Ti(C_5H_5)_2 \cdot X]$ or $[Ti(C_5H_5)_2]_2$. Note that the latter compound is dimeric, and thus has a different structure to ferrocene.

$Ti(C_5H_5)_4$ is unusual because two cyclopentadienyl rings are pentahaptic (π bonding) and two are monohaptic (σ bonding), i.e. $[Ti(\eta^5\text{-}C_5H_5)_2(\eta^1\text{-}C_5H_5)_2]$. A similar arrangement is found in $[Nb(\eta^5\text{-}C_5H_5)_2(\eta^1\text{-}C_5H_5)_2]$, $[Nb(\eta^5\text{-}C_5H_5)_2(\eta^1\text{-}C_5H_5)_2Cl_2]$ and $[Nb(\eta^5\text{-}C_5H_5)_2(\eta^1\text{-}C_5H_5)_2Cl_3]$.

Other compounds have only one ring, e.g. $[Cr(\eta^5\text{-}C_5H_5)(CO)_3]$, $[Mn(\eta^5\text{-}C_5H_5)(CO)_3]$, $[Cr(\eta^6\text{-}C_6H_6)(CO)_3]$, $[Mo(\eta^7\text{-}C_7H_7)(CO)_3]$ and $[Fe(\eta^4\text{-}C_4H_4)(CO)_3]$.

Ferrocene is sometimes regarded as a compound of Fe^{2+} and two $C_5H_5^-$ ions. The bonding in these aromatic sandwich-type structures is better considered as c bonding involving the lateral overlap of the d_{xz} and d_{yz} orbitals on Fe with the delocalized $p\pi$ aromatic orbital from each cyclopentadienyl ring. The bonding is too complicated to be described in detail here.

The sandwich compounds were discovered and studied independently by E.O. Fischer in Munich and G. Wilkinson at Imperial College, London. For this work they were jointly awarded the Nobel Prize for Chemistry in 1973.

FURTHER READING

Abel, E.W. and Stone, F.G. (1969, 1970) The chemistry of transition metal carbonyls, *Q. Rev. Chem. Soc.*, Part I – Structural considerations, **23**, 325; Part II – Synthesis and reactivity, **24**, 498.

Allen, A.D. (1973) Complexes of dinitrogen, *Chem Rev.*, **73**, 11.

Canterford, J.H. and Cotton, R. (1968) *Halides of the Second and Third Row Transition Elements*, Wiley, London.

Canterford, J.H. and Cotton, R. (1969) *Halides of the First Row Transition Elements*, Wiley, London.

Corbett, J.D. (1981) Extended metal–metal bonding in halides of the early transition metals, *Acc. Chem. Res.*, **14**, 239.

Diatomic metals and metallic clusters, (1980) (Conference papers of the Faraday Symposia of the Royal Society of Chemistry, No. 14), Royal Society of Chemistry, London.

Emeléus, H.J. and Sharpe, A.G. (1973) *Modern Aspects of Inorganic Chemistry*, 4th ed. (Chapters 14 and 15: Complexes of Transition Metals: Chapter 20: Carbonyls), Routledge and Kegan Paul, London.

Griffith, W.P. (1968) Organometallic nitrosyls, *Adv. Organometallic Chem.*, **7**, 211.

Griffith, W. P. (1973) *Comprehensive Inorganic Chemistry*, Vol. 4 (Chapter 46: Carbonyls, cyanides, isocyanides and nitrosyls), Pergamon Press, Oxford.

Hunt, C.B. (1977) Metallocenes – the first 25 years, *Education in Chemistry*, **14**, 110–113.

Johnson, B.F.G. and McCleverty, J.A. (1966) Nitric oxide compounds of transition metals, *Progr. Inorg. Chem.*, **7**, 277.

Johnson, B.F.G. (1976) The structures of simple binary carbonyls, *JCS Chem. Commun.*, 211–213.

Johnson, B.F.G. (ed.) (1980) *Transition Metal Clusters*, John Wiley, Chichester.

Kauffman, G.B. (1983) The discovery of ferrocene, the first sandwich compound, *J. Chem. Ed.*, **60**, 185–186.

Levason, W. and McAuliffe, C.A. (1974) Higher oxidation state chemistry of iron, cobalt and nickel, *Coordination Chem. Rev.*, **12**, 151–184.

Lever, P. and Gray, H.B. (eds) (1989) *Iron Porphyrins* (Physical Bioinorganic Chemistry Series), V.C.H. Publishers.

Lewis, J. and Green, M.L. (eds) (1983) *Metal Clusters in Chemistry* (Proceedings of the Royal Society Discussion Meeting May 1982), The Society, London.

Nicholls, D. (1973) *Comprehensive Inorganic Chemistry*, Vol. 3 (Chapter 40: Iron), Pergamon Press, Oxford.

Perutz, M. (1970) Stereochemistry of cooperative effects in haemoglobin, *Nature*, **228**, 726–739.

Seddon, E.A. and Seddon, K.R. (1984) *Chemistry of Ruthenium* (Topics in Inorganic Chemistry Series, Vol. 19), Elsevier.

Suslick, K.S. and Reinert, T.J. (1985) The synthetic analogs of O_2-Binding heme proteins, *J. Chem. Ed.*, **62**, 974–982.

Toth, L.E. (1971) *Transition Metal Carbides and Nitrides*, Academic Press.

Williams, R.V. (1969) Carbon determination in modern steel making, *Chemistry in Britain*, **5**, 213–216.

Wilkinson, G., Rosenblum, M., Whiting, M.C. and Woodward, R.B. (1952) The structure of iron *bis*-cyclopentadienyl, *J. Am. Chem. Soc.*, **74**, 2125–2126. (The first report of a sandwich type of organometallic compound.)

The cobalt group

25

Table 25.1 Electronic structure and oxidation states

Element		Electronic structure	Oxidation states*
Cobalt	Co	[Ar] $3d^7\ 4s^2$	(−I) 0 (I) **II** **III** (IV)
Rhodium	Rh	[Kr] $4d^8\ 5s^1$	(−I) 0 (I) II **III** IV (VI)
Iridium	Ir	[Xe] $4f^{14}\ 5d^9$	(−I) 0 (I) (II) **III** **IV** (V) (VI)

* The most important oxidation states (generally the most abundant and stable) are shown in bold. Other well-characterized but less important states are shown in normal type. Oxidation states that are unstable, or in doubt, are given in parentheses.

OCCURRENCE, EXTRACTION AND USES

The elements have odd atomic numbers and have low abundance in the earth's crust. Co occurs to the extent of 23 ppm by weight, whilst rhodium and iridium are extremely rare.

Table 25.2 Abundance of the elements in the earth's crust, by weight

	ppm	Relative abundance
Co	30	29
Rh	0.0001	77 =
Ir	0.001	74

There are many ores which contain Co. The commercially important ones are cobaltite CoAsS, smaltite $CoAs_2$ and linnaeite Co_3S_4. These are always found together with Ni ores, often with Cu ores and sometimes with Pb ores. Co is obtained as a by-product from the extraction of the other metals. In 1988 world production of Co ores was 31 200 tonnes of contained metal. The main producers were Zaire 32.5%, Zambia 16%, Australia 11%, the USSR 10% and Canada 9%.

The ore is roasted to convert it to a mixture of oxides called 'speisses'. As_4O_{10} and/or SO_2 come off as gases and are valuable by-products. The

continued overleaf

oxides are treated with H_2SO_4, when Fe (which is often present as an impurity), Co and Ni dissolve and can be separated from Cu or Pb. Lime is added to the solution to precipitate Fe as the hydrous oxide $Fe_2O_3 \cdot (H_2O)_n$. Then NaOCl is added to precipitate $Co(OH)_3$. The hydroxide is ignited to give Co_3O_4 which is reduced by heating with H_2 or charcoal to give Co metal.

Co forms important high temperature alloys with steel, and about one third of the metal produced is used for this purpose. These alloys find important uses in gas turbine engines, and in high speed steel which is used to make cutting tools for lathes. High working speeds can be used as these tools retain their hardness and cutting edge even at red heat. Exceptionally hard alloys can be made which can be used instead of diamonds in rock drills, e.g. stellite (50% Co, 27% Cr, 12% W, 5% Fe and 2.5% C) and widia metal (tungsten carbide WC with 10% Co).

A third of the Co produced is used to make pigments for the ceramic, glass and paint industries. Historically the oxide was used as a blue pigment in the ceramic industry. It is used to make blue glass. Nowadays it is mainly used to counteract the yellow colour of Fe and give a white colour.

Co is ferromagnetic (i.e. it can be magnetized permanently) like Fe and Ni. One fifth of the Co produced is used to make magnetic alloys such as Alnico (containing Al, Ni and Co). These alloys make powerful permanent magnets which are 20–30 times more powerful than magnets of Fe.

Small amounts of the Co salts of fatty acids from linseed oil and naphthenic acid are used as 'driers' to speed the drying of oil paints.

Co is an essential constituent of fertile soil and is present in some enzymes and in vitamin B_{12}.

The artificial isotope ^{60}Co is radioactive, and undergoes β decay (half life 5.2 years). At the same time it gives out intense high energy γ radiation, which is used in hospitals for radiotherapy of cancerous tumours. ^{60}Co is prepared by neutron irradiation of the only naturally occurring isotope ^{59}Co in a nuclear reactor.

$$^{60}_{27}\text{Co} \rightarrow {}^{60}_{28}\text{Ni} + {}_{-1}^{2}\text{e} + \nu + \gamma$$

Trace amounts of rhodium and iridium are found in the metallic state together with the platinum metals and the coinage metals in the NiS/CuS ores mined in South Africa, Canada (Sudbury, Ontario), and the USSR (the river sand from the Ural mountains). World production of all six platinum group metals was only 267 tonnes in 1988. The largest sources were South Africa 45%, the USSR 44%, Canada 4%, the USA 2% and Japan 0.8%. Rh and Ir are obtained from the anode slime which accumulates in the electrolytic refining of Ni. This contains a mixture of the platinum metals together with Ag and Au. The elements Pd, Pt, Ag and Au are dissolved in aqua regia and the residue contains Ru, Os, Rh and Ir. After a complex separation Rh and Ir are obtained as powders. Their melting points are very high and powder forming techniques are used to fabricate metal components. (The powder is formed into the required shape, then sintered, i.e. heated in hydrogen until it congeals. It does not melt.) These elements are both scarce and expensive, and have a limited number of specialist uses.

All the platinum metals have specific catalytic properties. A Pt/Rh alloy was formerly used in the Ostwald process (for making HNO_3) to oxidize NH_3 to NO. Rh is an important catalyst in the control of car exhaust emissions. Rh–phosphine complexes are used as catalysts for hydrogenation reactions. Ir (like Os in the previous group) is used to make very hard alloys which are used to make pivots for instruments. A Pt/Ir alloy is used to make the electrodes for long life sparking plugs. These are expensive but have important military uses, for example in helicopters. The USA used a lot of this alloy during the Vietnam war.

OXIDATION STATES

The trend for the elements in the second half of the *d*-block not to use all their outer electrons for bonding in the maximum oxidation state is continued. A possible report of Co(+V) has been disproved, and even Co(+IV) is unstable. The maximum oxidation state for Rh and Ir is (+VI). For Co, the (+II) and (+III) states are by far the most important. The trend in the later elements of the first row for the (+II) state to be more stable than (+III) is also observed. Co^{2+} ions and the hydrated ion $[Co(H_2O)_6]^{2+}$ exist in many simple compounds and the hydrated ion is stable in water. In contrast simple compounds containing Co(+III) are oxidizing and are relatively unstable. However, Co(+III) is stable and is very important in complexes.

The most stable states for the other elements are Rh(+III), Ir(+III) and Ir(+IV). Simple ionic compounds of these elements are uncommon. The oxides and halides formed are shown in Table 25.3.

GENERAL PROPERTIES

Co resembles iron and is very tough. It is harder and has a higher tensile strength than steel. Co is bluish white and lustrous in appearance. Like iron it is ferromagnetic, but on heating above 1000 °C it changes to a non-magnetic form.

Co is relatively unreactive, and does not react with H_2O, H_2, or N_2, though it reacts with steam, forming CoO. It is oxidized when heated in air and burns at white heat to Co_3O_4. Co dissolves slowly in dilute acids, but like Fe it is rendered passive by concentrated HNO_3. Co combines readily with the halogens, and at elevated temperatures with S, C, P, As, Sb and Sn.

Rh and Ir are also hard metals. In common with the other platinum metals they are much more noble and unreactive. Ir has the highest density of any element, $22.61\,g\,cm^{-3}$. Rh and Ir are resistant to acids, but react with O_2 and the halogens at high temperatures (Table 25.5). All three elements form a large number of coordination compounds.

Table 25.3 Oxides and halides

Oxidation states					
(+II)	(+III)	(+IV)	(+V)	(+VI)	Others
CoO	$(Co_2O_3)^h$	$(CoO_2)^h$	–	–	Co_3O_4
RhO	**Rh_2O_3**	RhO_2	–	–	
–	Ir_2O_3	**IrO_2**	–	(IrO_3)	
CoF_2	CoF_3	–	–	–	
$CoCl_2$	–	–	–	–	
$CoBr_2$	–	–	–	–	
CoI_2	–	–	–	–	
–	**RhF_3**	RhF_4	(RhF_5)	RhF_6	
–	**$RhCl_3$**	–	–	–	
–	**$RhBr_3$**	–	–	–	
–	**RhI_3**	–	–	–	
–	IrF_3	**IrF_4**	(IrF_5)	IrF_6	
$(IrCl_2?)$	**$IrCl_3$**	$(IrCl_4)$	–	–	
–	**$IrBr_3$**	–	–	–	
–	**IrI_3**	–	–	–	

The most stable oxidation states are shown in bold, unstable ones in brackets.
h = hydrous oxide.

Table 25.4 Some physical properties

	Covalent radius (Å)	Ionic radius (Å)		Melting point (°C)	Boiling point (°C)	Density ($g\,cm^{-3}$)	Pauling's electronegativity
		M^{2+}	M^{3+}				
Co	1.16	0.745^h 0.65^l	0.61^h 0.545^l	1495	3100	8.90	1.8
Rh	1.25	–	0.665	1960	3760	12.39	2.2
Ir	1.26	–	0.68	2443	(4550)	22.61	2.2

h = high spin value, l = low spin radius.

Table 25.5 Some reactions of Co, Rh and Ir

Reagent	Co	Rh	Ir
O_2	Co_3O_4	Rh_2O_3 at 600 °C	IrO_2 at 1000 °C
F_2	CoF_2 and CoF_3	RhF_3 at 600 °C	IrF_6
Cl_2	$CoCl_2$	$RhCl_3$ at 400 °C	$IrCl_3$ at 600 °C
H_2O	No reaction	No reaction	No reaction
Dilute HCl or HNO_3	$[Co(H_2O)_6]^{2+} + H_2$	No reaction	No reaction
Conc. HNO_3	Passive	No reaction	No reaction

LOWER OXIDATION STATES

The oxidation states (−I) and (0) occur in a few compounds with π bonding ligands such as CO, PF_3, NO and CN^-. The (−I) state is found in the tetrahedral complexes $[Co(CO)_4]^-$, $[Rh(CO)_4]^-$, $[Co(CO)_3NO]$ and $K[Ir(PF_3)_4]$. The zero oxidation state occurs in $Co_2(CO)_8$, though there is some doubt about the corresponding Ru compounds (Figure 25.1). Other zero-valent compounds are $K_4[Co(CN)_4]$ and $[Co(PMe_3)_4]$.

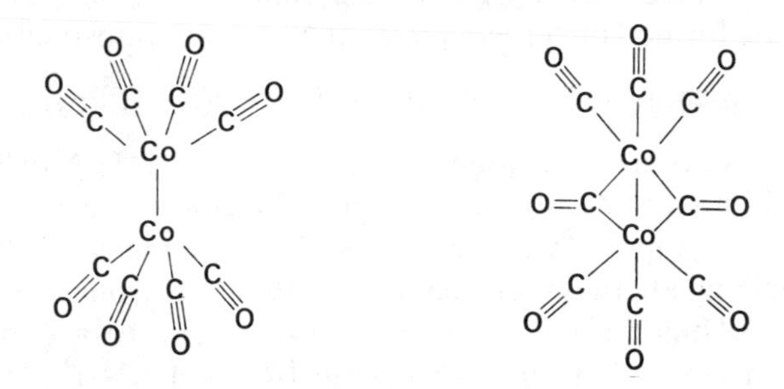

Figure 25.1 Two isomeric forms of $Co_2(CO)_8$, both with metal–metal bonding.

The carbonyls $Co_4(CO)_{12}$, $Rh_4(CO)_{12}$ and $Ir_4(CO)_{12}$ all have M—M bonds and contain a cluster of four metal atoms. The CO groups may be either apical (terminal) or bridging. There are slight differences between compounds of the metals. The Co and Rh compounds have three bridging CO groups but the Ir compound has none (Figure 25.2).

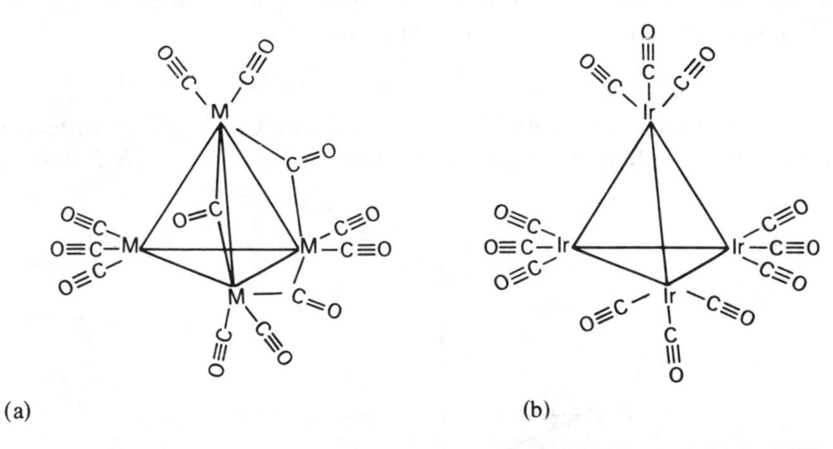

Figure 25.2 The structures of (a) $Co_4(CO)_{12}$ and $Rh_4(CO)_{12}$ and (b) $Ir_4(CO)_{12}$.

The carbonyl ion from $Na_3[Co_6(CO)_{14}]$ and the carbonyls $Co_6(CO)_{16}$ and $Rh_6(CO)_{16}$ have unusual structures, comprising an octahedral cluster of six metal atoms with a CO group bridging three metal atoms on each triangular face of the octahedron. The remaining CO groups are normal terminal groups.

(+I) STATE

Co(+I) exists in many complexes with π bonded ligands. The (+I) state is better known for Co than for any other first row transition metal except Cu. Compounds are usually made by reducing $CoCl_2$ with Zn or N_2H_4, in the presence of the ligand. Their structures are typically a trigonal bipyramid or tetrahedral.

The ion $[Co^{-I}(CO)_4]^-$ reacts with organic isonitriles R—NC, giving $[Co^{I}(CNR)_5]^+$ which has a trigonal bipyramid structure. A dinitrogen complex can be formed by direct uptake of N_2 gas at atmospheric pressure:

$$Co(acac)_3 + N_2 + 3Ph_3P \rightarrow [Co^{I}(H)(N_2)(PPh_3)_3]$$

The ligand (acac) is acetylacetone. The complex $[Co^{I}(H)(N_2)(PPh_3)_3]$ also has a trigonal bipyramidal structure (Figure 25.3) which has a N≡N bond length of 1.11 Å compared with 1.098 Å in N_2. Since the N≡N bond length is almost unchanged, this indicates that σ bonding from N to Co is extremely weak. Thus the N—Co bond is mainly due to π bonding (back bonding) from Co to N similar to that in $[Ru(NH_3)_5N_2]^{2+}$. (Dinitrogen complexes are discussed in Chapter 24 under '+II state – complexes').

The reduced form of vitamin B_{12} also appears to contain Co(+I).

There is a fairly extensive chemistry of Rh(+I) and Ir(+I) complexes with π bonding ligands such as CO, phosphines PR_3 and alkenes. These normally have either a square planar structure, for example *trans*-$[Ir(Cl)(CO)(PPh_3)_2]$ (called Vaska's compound) and $[Rh(Cl)(PPh)_3]$ (called Wilkinson's catalyst), or a trigonal bipyramid structure as in $[Rh(H)(CO)(PPh_3)_3]$. The square planar (+I) compounds undergo an unusual type of reaction called *oxidative addition*. In this a neutral molecule is added to the (+I) complex to give a (+III) octahedral complex.

$$[Ir^{I}(Cl)(CO)(PPh_3)_2] + HCl \rightarrow [Ir^{III}(Cl)_2(CO)(PPh_3)_2H]$$

A similar reaction occurs with H_2, H_2S, CH_3I, and Cl—HgCl. A different reaction occurs when other molecules such as O_2, SO_2, CS_2, RNCS,

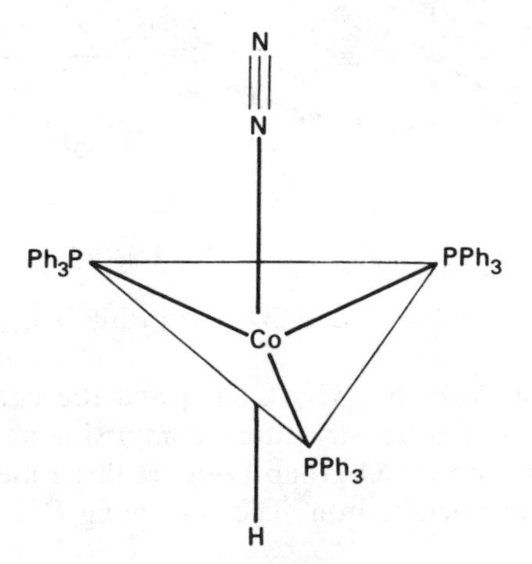

Figure 25.3 The structure of $[Co^{I} \cdot H(N_2)(PPh_3)_3]$.

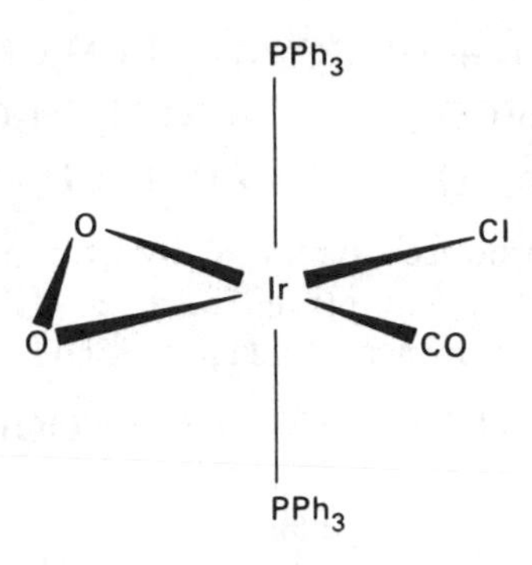

Figure 25.4 $[Ir^{III}Cl(CO)(O_2)(PPh_3)_2]$ complex.

RNCO and RC≡CR are added to the (+I) square planar compounds. (The added molecules all contain multiple bonds.) Here the added molecule acts as a bidentate ligand, thus forming a cyclic structure (Figure 25.4).

Vaska's compound is yellow, and it readily absorbs O_2 and becomes orange coloured. The O_2 may be removed by flushing with N_2. This reversible oxygenation has been studied as a model for the oxygen carrying ability of haemoglobin (see Chapter 24). Oxidative addition reactions have been observed for complexes where the central metal has a d^8 or d^{10} configuration involving Rh^I, Ir^I, Ni^0, Pd^0, Pt^0, Pd^{II} and Pt^{II}. There must be non-bonding **d** electrons available on the metal, and also two vacant coordination sites.

Wilkinson's catalyst $[Rh(Cl)(PPh)_3]$ is red–violet in colour, and is made by refluxing $RhCl_3 \cdot 3H_2O$ with triphenyl phosphine. It has a square planar structure. It is very effectivc for selective hydrogenation of organic molecules *at room temperature and pressure*. Alkene groups at the end of a chain (alk-1-enes) are hydrogenated but double bonds elsewhere in the chain are not affected. It is of importance in the pharmaceutical industry.

Wilkinson's catalyst and various Co compounds such as the carbonyl hydride $HCo^I(CO)_4$ have been used as catalysts in the OXO process. In this process, CO and H_2 are added to an alkene, thus forming an aldehyde. A temperature of 150 °C and 200 atmospheres pressure are required. The OXO process is of considerable industrial importance, as the aldehydes produced can be converted into alcohols. About 3 million tonnes of C_6–C_9 alcohols are produced annually in this way. These are a mixture of straight chain and branched chain molecules depending on the position of the double bond in the hydrocarbon. The straight chain alcohols are used to make polyvinyl chloride and detergents. Efforts to improve the yield of the straight chain products involve the use of triphenylphosphine substituted carbonyls of Co. The complex *trans*-$[Rh(CO)(H)(PPh_3)_3]$ has also been used in the OXO process, and is an important catalyst in the hydrogenation of alkenes. It is active at 25 °C and 1 atmosphere pressure, and for steric reasons it is specific to terminal alkenes rather than double bonds elsewhere in the chain.

$$RCH{=}CH_2 + HCo(CO)_4 \rightarrow RCH_2CH_2Co(CO)_4$$
$$RCH_2CH_2Co(CO)_4 + CO \rightarrow RCH_2CH_2CO \cdot Co(CO)_4$$
$$RCH_2CH_2CO \cdot Co(CO)_4 + H_2 \rightarrow RCH_2CH_2CHO + HCo(CO)_4$$

Acetic acid is also produced synthetically from methyl alcohol, and this reaction is catalysed by complexes such as $[Rh(Cl)(CO)(PPh_3)_2]$ or $[Rh(Cl)(CO)_2]_2$ in the presence of CH_3I, I_2 or HI as activator.

$$CH_3OH + CO \rightarrow CH_3COOH$$

(+II) STATE

The (+II) state is the most important for simple compounds of Co (though the (+III) state is the most important in complexes). Rh(+II) and Ir(+II) are only of minor importance.

A wide range of simple Co(+II) compounds are known including CoO, $Co(OH)_2$, CoS, and salts of the common acids such as $CoCl_2$, $CoBr_2$, $CoSO_4$, $Co(NO_3)_2$, and $CoCO_3$. The hydrated salts are all pink or red and contain the hexahydrate ion $[Co(H_2O)_6]^{2+}$. Most Co(+II) compounds except the carbonate are soluble in water.

If NaOH is added to a solution containing Co^{2+} then $Co(OH)_2$ is first obtained as a blue precipitate which turns pale pink on standing. This is mainly basic, but it is weakly amphoteric as it dissolves in very strong NaOH, giving a blue coloured solution which contains $[Co(OH)_4]^{2-}$. $Co(OH)_2$ slowly oxidizes in air to brown $Co^{III}O \cdot OH$.

CoO is olive green and is formed by heating $Co(OH)_2$ or by heating many Co(+II) salts such as $CoCO_3$ in the absence of air. If CoO is melted with SiO_2 and K_2CO_3 a deep blue glass potassium cobalt(II) silicate is formed. Commercially this blue glass is ground up and the powder, which is called *smalt*, is used as a pigment to give a blue colour to glass, enamels and glazes. Smalt was known to the ancient Egyptians and the Romans. In the laboratory 'cobalt glass' is used to observe the flame test for potassium in the presence of sodium. The blue glass absorbs the intense yellow coloration from sodium, thus allowing the colour from potassium to be seen.

$CoCl_2$ is used as a test for water, both as 'cobalt chloride paper' and as an indicator added to the drying agent 'silica gel'. Hydrated $CoCl_2 \cdot 6H_2O$ is pink coloured and contains octahedral $[Co(H_2O)_6]^{2+}$ ions. If this is partially dehydrated by heating, then blue coloured tetrahedral ions $[Co(H_2O)_4]^{2+}$ are formed. Addition of water produces the reverse change. Thus when the indicator in silica gel is blue the drying agent is effective, but when it is pink the drying agent needs changing.

$$\underset{\text{pink}}{[Co(H_2O)_6]^{2+}} \rightleftharpoons \underset{\text{blue}}{[Co(H_2O)_4]^{2+}} + 2H_2O$$

In a similar way the octahedral aqua ion reacts with excess Cl^- to give the blue coloured tetrahedral ion $[CoCl_4]^{2-}$.

$$\underset{\text{pink}}{[Co(H_2O)_6]^{2+}} + 4Cl^- \rightarrow \underset{\text{blue}}{[Co(Cl)_4]^{2-}} + 6H_2O$$

Both $[Co(H_2O)_6]^{2+}$ and $[Co(Cl)_4]^{2-}$ and most Co(+II) complexes are low spin and have a d^7 configuration. The tetrahedral complexes have more intense colours than the octahedral complexes. This is because a tetrahedron lacks a centre of symmetry and thus easily overcomes the Laporte selection rule (that $\Delta l = 1$), whereas the octahedral complexes have to rely on asymmetric vibrations of ligands to destroy the centre of symmetry. The magnetic moments of both octahedral and tetrahedral complexes are higher than predicted using the spin only formula which would give $\mu = 3.87$ BM. In the octahedral case this is because there is an orbital contribution since with a $t_{2g}^5\ e_g^2$ arrangement it is possible to transform one t_{2g} orbital into another. In tetrahedral complexes the electronic arrangement is $e^4\ t_2^3$ so transformation of the t_2 orbitals is not possible and the orbital contribution is zero. However, in this case spin orbit coupling occurs. This accounts for the higher than expected value of μ (see 'Measurement of magnetic moments', Chapter 18).

Cobalt(II) acetate $Co(CH_3COO)_2 \cdot 4H_2O$ is formed by dissolving $CoCO_3$ in acetic acid. It forms red crystals which are very soluble, and is used as a drying agent for varnish and lacquers.

Anhydrous Co salts cannot be made by heating the hydrated salts, because they decompose to the oxide. Thus dry preparative methods are used. Anhydrous CoF_2 is pink and is obtained from the reaction of HF with $CoCl_2$. Anhydrous $CoCl_2$ (blue) and $CoBr_2$ (green) are made by heating the elements. Anhydrous CoI_2 is blue–black and is made by heating the metal with HI. They all have solid structures in which Co^{2+} is octahedrally coordinated.

Co(+II) forms a number of complexes, but these are less stable than those of Co(+III). Co(+II) complexes may be tetrahedral or octahedral. Since there is only a small difference in stability between them the two forms sometimes exist in equilibrium. The large monodentate ligands Cl^-, Br^-, I^-, OH^- and SCN^- commonly form tetrahedral complexes. Co(+II) forms more tetrahedral complexes than any other transition metal ion. This is associated with the fairly small loss of crystal field stabilization energy of $0.27\Delta_o$ with a d^7 ion in a weak ligand field (see Table 7.15).

The blue coloured complex $Hg[Co(NCS)_4]$ is unusual. The Co^{2+} is tetrahedrally coordinated by N atoms and Hg^{2+} is tetrahedrally coordinated by S atoms, giving a polymeric solid. This compound is often used to calibrate a magnetic balance when measuring magnetic moments.

Most Co(+II) complexes are high spin, but the CN^- ligand produces low-spin complexes. If a solution of a Co^{2+} salt is treated with excess CN^- a green coloured complex $[Co(CN)_5]^{3-}$ is formed. This can be isolated as the barium salt. It is a good catalyst for the hydrogenation of alkenes. The complex is paramagnetic with one unpaired electron, and its shape is a square pyramid. It may dimerize to give purple $[Co_2(CN)_{10}]^{6-}$, which is diamagnetic and has the structure $(CN)_5Co—Co(CN)_5$ similar to the carbonyl $Mn_2(CO)_{10}$. It is interesting that $[Co^{II}(CN)_5]^{3-}$ is formed instead of an octahedral complex $[Co^{II}(CN)_6]^{4-}$. A low-spin octahedral complex would have the configuration $(t_{2g})^6$ $(e_g)^1$, and as the e_g level is not symmetrically filled it would suffer from Jahn–Teller distortion. The CN^- ligand is a strong π acceptor, i.e. it accepts electrons from the metal in back bonding. Back bonding increases the crystal field splitting Δ_o, which makes the e_g orbitals very high in energy and thus strongly 'antibonding'. If an octahedral complex was formed, these high energy e_g orbitals must contain one electron. This makes the octahedral complex too unstable to exist. (In marked contrast Co(+III) has the configuration $(t_{2g})^6$ $(e_g)^0$ and the octahedral complex $[Co^{III}(CN)_6]^{3-}$ is extremely stable).

The $[Co^{II}(CN)_5]^{3-}$ complex is oxidized by air to give a brown coloured peroxo complex $K_6[(CN)_5Co^{III}—O—O—Co^{III}(CN)_5]$ which is discussed under (+III) complexes.

Less commonly Co(+II) forms square planar complexes with bidentate ligands such as dimethylglyoxime, and with tetradentate ligands such as porphyrins. Magnetic measurements can be used to distinguish between tetrahedral and square planar arrangements. Tetrahedral complexes have three unpaired electrons and square planar only one.

Co^{2+} ions are very stable and are difficult to oxidize. Co^{3+} ions are less stable and are reduced by water. In contrast, many Co(+II) complexes are readily oxidized to Co(+III) complexes, and Co(+III) complexes are very stable.

$$[Co^{II}(NH_3)_6]^{2+} \xrightarrow{\text{bubble in air}} [Co^{III}(NH_3)_6]^{3+}$$

This happens because the crystal field stabilization energy of Co(+III) with a d^6 configuration is higher than for Co(+II) with a d^7 arrangement (Figure 25.5).

Certain porphyrin complexes of Co(+II) are structurally similar to haemoglobin (Figure 25.6).

The complex (Figure 25.7) is a Schiff's base and is capable of reversible

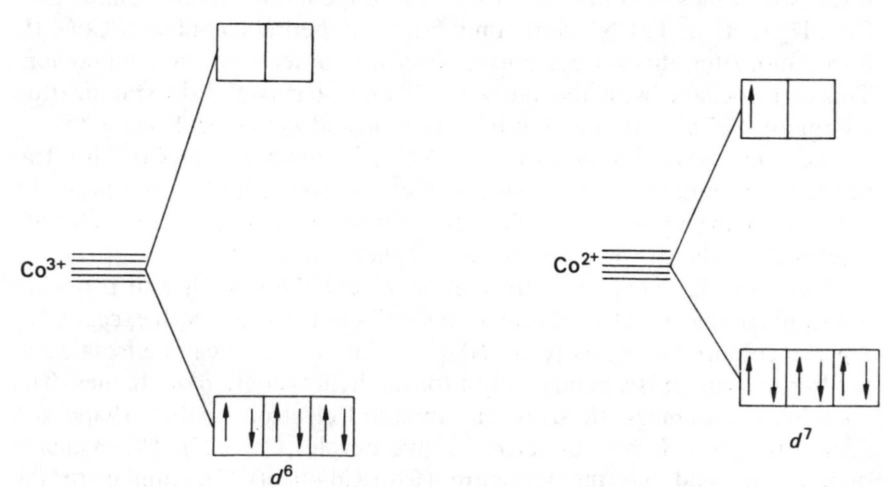

Figure 25.5 Electronic arrangements for d^6 and d^7 ions in a strong octahedral field.

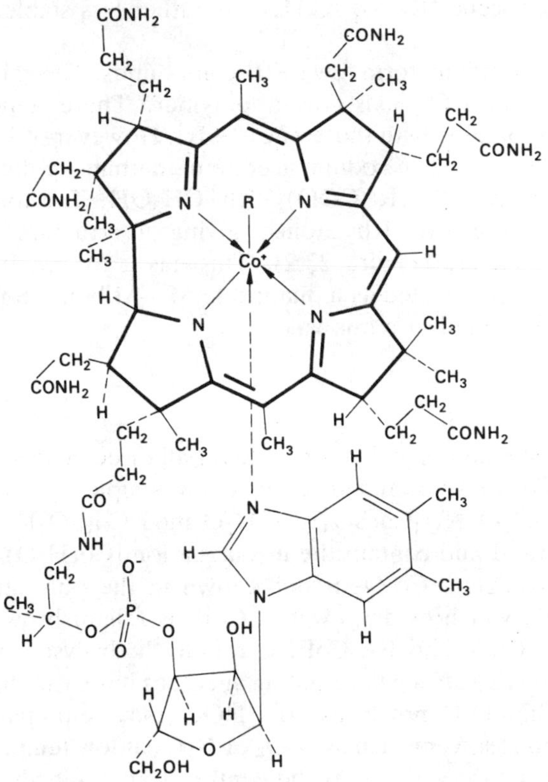

Figure 25.6 Vitamin B_{12}. The corrin ring is shown in heavy type.

oxygenation and deoxygenation in pyridine solution at room temperature. Though Co complexes are not involved in oxygen metabolism in the body, they serve as useful models for metal–oxygen binding in biological systems.

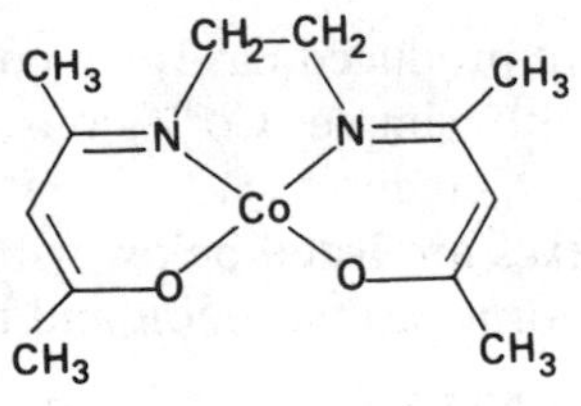

Figure 25.7 A complex of Co(+II) with both N and O donors.

Ferrocene-like complexes are formed by Co and Rh. Cobaltocene $[Co^{II}(\eta^5\text{-}C_5H_5)_2]$ is formed by reacting sodium cyclopentadiene with anhydrous $CoCl_2$ in tetrahydrofuran. It is dark purple, and is air sensitive. It is easily oxidized (i.e. loses an electron) to form the very stable yellow coloured ion $[Co^{III}(\eta^5\text{-}C_5H_5)_2]^+$. The latter is not oxidized even by concentrated HNO_3 , but like ferrocene the rings are attacked by nucleophilic reagents. Rhodocene $[Rh^{II}(\eta^5\text{-}C_5H_5)_2]$ is rather less stable, and tends to dimerize.

Rhodium and iridium form few (+II) compounds. The existence of RhO is uncertain, and $IrCl_2$ exists as a polymer. There appear to be no complexes comparable with those of Co(+II). However, if $RhCl_3 \cdot 3H_2O$ is warmed with a solution of sodium acetate in methanol a dimeric diacetate is formed $HOH_3C \cdot Rh \cdot (R \cdot COO)_4 \cdot Rh \cdot CH_3OH$. The four carboxylate groups bridge the two Rh atoms, giving a structure like that for chromium(II) acetate (Figure 22.2). This has a M—M bond length of 2.39 Å which is interpreted as a quadruple M—M bond. Some complexes with phosphine ligands are known.

(+III) STATE

This is the most common oxidation state for all three metals, particularly in complexes. Co(+III) occurs in only a few simple compounds such as $Co_2(SO_4)_3 \cdot 18H_2O$, $NH_4Co(SO_4)_2 \cdot 12H_2O$ and $KCo(SO_4)_2 \cdot 12H_2O$. These are blue coloured and contain the hexaaqua ion $[Co(H_2O)_6]^{3+}$. They are all strongly oxidizing. Co_2O_3 is not known in the pure state, only as a hydrated oxide which oxidizes water. CoF_3 is a light brown solid, which is made from CoF_2 and F_2. CoF_3 is rapidly hydrolysed by water. It is commonly used as a strong fluorinating agent, as it is easier to handle and is less reactive than F_2. Anhydrous $Co^{III}(NO_3)_3$ may be prepared from CoF_3 in a non-aqueous solvent such as N_2O_4 or N_2O_5 at low temperatures. It has an unusual structure with Co at the centre of an octahedron of O atoms from three bidentate NO_3^- groups. The chemistry of Co(+III) is largely that of its coordination compounds.

The oxide Co_3O_4 is black and is formed by heating the metal in air at 500 °C. It has a spinel structure like Fe_3O_4. These are better written $Fe^{II}Fe^{III}{}_2O_4$ and $Co^{II}Co^{III}{}_2O_4$. In this structure the O atoms are approximately close packed. The larger low-spin Co(+III) ions occupy half of the octahedral holes and the smaller high-spin Co(+II) are in one eighth of the tetrahedral holes.

Co(+III) complexes are produced easily, in contrast to the difficulty in preparing compounds with simple Co^{3+} ions. Co(+III) forms more complexes than any other element.

Some common complexes are listed below with their colours. It can be seen that the cobalt can form cation, anion and neutral complexes.

$[Co(NH_3)_6]^{3+}$	yellow
$[Co(NH_3)_5 \cdot H_2O]^{3+}$	pink
$[Co(NH_3)_5Cl]^{2+}$	purple
$[Co(NH_3)_4CO_3]^+$	purple
$[Co(NH_3)_3(NO_2)_3]$	yellow
$[Co(CN)_6]^{3-}$	violet
$[Co(NO_2)_6]^{3-}$	orange

Practically all complexes of Co(+III) have six ligands in an octahedral arrangement. The metal has a d^6 configuration, and most of the ligands are strong enough to cause spin pairing, giving the electronic arrangement $(t_{2g})^6\ (e_g)^0$. This arrangement has a very large crystal field stabilization energy. Such complexes are diamagnetic. The one exception is $[CoF_6]^{3-}$ which is a high-spin complex, and is paramagnetic. Complexes with nitrogen donor ligands (ammonia and amines) are the most common.

These complexes may be prepared by oxidation of a solution containing Co^{2+} with air or H_2O_2 in the presence of appropriate ligands and a catalyst such as activated charcoal. It is also possible to substitute ligands in an existing complex. The complexes are very stable, and ligand exchange (substitution) reactions occur only slowly. This is the reason why complexes of Co(+III) have been so extensively studied since the 1890s by Werner and others. Much of our knowledge on the stereochemistry, isomerism and general properties of octahedral complexes has come from these studies.

Co^{3+} has an affinity for N donors such as NH_3, ethylenediamine, amines, EDTA and the nitrite ion NO_2^-. The salt sodium cobaltinitrite $Na_3[Co(NO_2)_6]$ is an orange coloured solid. It is used in both qualitative and quantitative analysis to precipitate K^+ as $K_3[Co(NO_2)_6]$. The complex $[Co(CN)_6]^{3-}$ is extremely stable and is not decomposed even by alkalis. The CN^- ligands are very firmly bonded by π back bonding, and the crystal field stabilization energy is very high. The complex is claimed to be non-toxic.

An aqueous solution containing $[Co^{II}(CN)_5]^{3-}$ and KCN can be oxidized by air to give a brown coloured complex $K_6[(CN)_5Co^{III}—O—O—Co^{III}(CN)_5]$. The peroxo bond length O—O is 1.45 Å compared with 1.48 Å in H_2O_2. This complex can be oxidized by air, or better by Br_2, to give a red complex $K_5[(CN)_5Co—O—O—Co(CN)_5]$. Whilst this might contain Co^{III} and Co^{IV}, the X-ray structure shows that the O—O bond length is very much shorter than before at 1.26 Å. The reason for this shortening is that an antibonding electron has been removed from the $O—O^{2-}$ ion and this has now become a superoxide linkage with a bond order of 1.5 (see Chapter 4). If solutions of the peroxo or superoxo complexes are boiled, yellow coloured $K_3[Co(CN)_6]$ is formed.

Several different isomers are found in complexes with the bidentate ligands such as ethylenediamine (en), acetylacetone or oxalate ions:

$$4Co^{2+} + 12en + 4H^+ + O_2 \rightarrow 4[Co^{III}(en)_3]^{3+} + 2H_2O$$

The complex potassium tris(ethylenediamine)cobalt(III) contains the $[Co^{III}(en)_3]^{3+}$ ion which is optically active and exists in *d* and *l* forms (see Chapter 7 under 'Isomerism'). A similar preparation in the presence of HCl gives the dark green salt *trans*-$[Co^{III}(en)_2(Cl)_2]^{2+}$ which on careful evaporation of a neutral solution gives the purple *cis* isomer. Both isomers undergo substitution reactions on heating with water, giving first $[Co(en)_2(Cl)(H_2O)]^{2+}$ then $[Co(en)_2(H_2O)]^{3+}$. Similar substitution reactions occur with other ligands such as NCS^-, giving $[Co(en)_2(NCS)_2]^+$.

Complexes with O donors are generally less stable, but those with chelating ligands such as [Co(acetylacetone)$_3$] and [Co(oxalate)$_3$]$^{3-}$ are stable, and are optically active.

Halogen complexes are rare and [CoF_6]$^{3-}$ is the only hexahalide complex known. This is blue, as is [$CoF_3(H_2O)_3$], and both are unusual because they are high spin, and hence are paramagnetic with a magnetic moment of about 5.8 Bohr magnetons.

Vitamin B_{12} is an important Co complex. The vitamin was isolated from liver after it was found that eating large quantities of raw liver was an effective treatment for pernicious anaemia. Injections of vitamin B_{12} are now used for treatment (more pleasant than eating raw liver!). Vitamin B_{12} is a coenzyme, and serves as a prosthetic group which is tightly bound to several enzymes in the body, but the precise role of vitamin B_{12} is not fully understood. Dorothy Crowfoot Hodgkin was awarded the Nobel Prize for Chemistry in 1964 for X-ray crystallographic work including solving the structure of this enzyme. The complex contains a Co(+III) ion at the centre of a corrin ring system (Figure 25.6). This is similar to the arrangement of Fe(+II) in a porphyrin ring in haemoglobin except that the corrin ring is less conjugated and rings A and D are joined directly. The Co atom is bonded to four ring N atoms. The fifth position is occupied by another N from a side chain (α-5,6-dimethylbenzimadazole) and this is also attached to the corrin ring. The sixth group which makes up the octahedron is the active site, and is occupied by a CN^- group in cyanocobalamin. The CN^- is introduced in isolating the coenzyme, and is not present in the active form in living tissue. This position is occupied by OH^- in hydroxocobalamin, by water or by an organic group such as CH_3 (methylcobalamin), or adenosine. This shows that a metal to carbon σ bond can be formed. The cobalamins can be reduced from Co^{III} to Co^{II} and Co^{I} in neutral or alkaline solutions both in the laboratory and in vivo (in the living body). The Co^{I} complex is strongly reducing. Its structure is five-coordinate, i.e. the site usually occupied by CN^- or OH^- is vacant.

Methylcobalamin is important in the metabolism of certain bacteria which produce methane. These bacteria can also transfer a methyl group CH_3 to a few metals such as Pt^{II}, Au^{I} and Hg^{II}. The latter poses a considerable ecological problem as the bacteria can transform elemental Hg or inorganic Hg salts into highly toxic methyl mercury CH_3Hg^+ or dimethyl mercury $(CH_3)_2Hg$ at the bottom of lakes.

Cobalt is also biologically important in some enzymes. Glutamic mutase is involved in the metabolism of amino acids and ribonucleotide reductase in the biosynthesis of DNA. Traces of cobalt are essential in the diet of animals. Some sheep raised in Australia, New Zealand, Florida and Britain suffered from a deficiency disease which was traced to them grazing on cobalt deficient soil. This can be remedied either by treating the soil periodically, or by forcing the animals to swallow a pellet of cobalt. This pellet remains in the rumen, and slowly releases cobalt into the gut. (Sometimes the animals are made to swallow a metal screw as well. This too remains in the rumen, and its purpose is to scrape any coating off the cobalt pellet. The pellet is recovered and reused when the animals are slaughtered.) Larger amounts of cobalt appear to be harmful. Traces of cobalt (1–1.5 ppm) are added to beer to make it froth better. This has been linked with an increased rate of heart failure among heavy beer drinkers who have a dietary deficiency of protein or thiamine.

All the (+III) halides RhX_3 and IrX_3 are known. RhF_3 is prepared by fluorinating $RhCl_3$, IrF_3 by reducing IrF_6 with Ir, and the others by direct reaction. They are all insoluble in water, unreactive and probably have layer lattices. The oxide Rh_2O_3 is obtained by burning the metal in air. Ir_2O_3 is only obtained with difficulty as the hydrated oxide, by adding alkali to Ir^{III} solutions under an inert atmosphere, as it oxidizes easily to $Ir^{IV}O_2$. In contrast to the oxidizing properties of $[Co^{III}(H_2O)_6]^{3+}$, $[Rh^{III}(H_2O)_6]^{3+}$ exists as a stable yellow coloured ion.

A considerable number of Rh(+III) and Ir(+III) complexes are known. Like the complexes of Co(+III) they are typically octahedral, stable, low spin and diamagnetic, e.g. $[RhCl_6]^{3-}$, $[Rh(H_2O)_6]^{3+}$, and $[Rh(NH_3)_6]^{3+}$. The chloride complexes are made by heating finely divided Rh or Ir with a Group I metal chloride and chlorine.

$$2Rh + 6NaCl + 3Cl_2 \rightarrow 2Na_3[RhCl_6]$$

The complex $Na_3[RhCl_6] \cdot 12H_2O$ is red coloured and is the best known compound of rhodium. On boiling with water it gives $[Rh(H_2O)_6]^{3+}$, and with NaOH it gives $Rh_2O_3 \cdot H_2O$. The yellow coloured hydrated ion is converted back to the chloro complex with HCl. If $Rh_2O_3 \cdot H_2O$ is treated with a limited amount of HCl then $[RhCl_3 \cdot 3H_2O]$ is formed, but with excess acid $[RhCl_6]^{3-}$ is formed instead. $[RhCl_3 \cdot 3H_2O]$ is octahedral, and should exist as two different isomeric forms *fac* and *mer* (see Figure 7.3).

A small number of complexes are known which are not octahedral, e.g. $[RhBr_5]^{2-}$ and $[RhBr_7]^{4-}$. Metal–metal bonds are found in a few complexes:

$[(R_3As)_3Rh^{III}(HgCl)]^+Cl^-$ contains Rh-Hg bond

$[Ir_2Cl_6(SnCl_3)_4]^{4-}$ contains Ir—Sn bonds

Rh(+III) and Ir(+III) form basic acetates $[Rh_3O(CH_3COO)_6L_3]^+$ which have unusual structures. The Rh atoms form a triangle with an O atom at the centre. The six acetate groups act as bridges between the Rh atoms – two acetate groups across each edge of the triangle. Thus each Rh atom is linked to four acetate groups and the central O, and the sixth position of the octahedron is occupied by water or another ligand. The magnetic moment is reduced due to partial pairing of *d* electrons on the three metal atoms by means of $d\pi$–$p\pi$ bonding through O. This type of carboxylate complex is also formed by the trivalent ions of Cr, Mn, Fe, Ru, Rh and Ir.

Several hydride complexes are also known: $[Rh(R_3P)_3 \cdot H \cdot Cl_2]^{2+}$ and $[Ir(R_3P)_3 \cdot H \cdot Cl_2]^{2+}$, $[Ir(R_3P)_3 \cdot H_2Cl]^{4+}$ and $[Ir(R_3P)_3H_3]^{6+}$. Reduction of Rh(+III) and Ir(+III) complexes gives the metal, whereas reduction of Co(+III) gives Co(+II).

(+IV) STATE

This is the highest oxidation state normally obtained for cobalt. Oxidation of alkaline Co^{2+} solutions gives an ill-defined product thought to be hydrated CoO_2, and a complex $Ba_2Co^{IV}O_4$ has been reported.

If the activated charcoal catalyst is omitted from the preparation of $[Co(NH_3)_6]^{3+}$ by the air oxidation of Co^{2+}, then a brown compound containing $[(NH_3)_5Co^{III}—O—O—Co^{III}(NH_3)_5]^{4+}$ can be isolated. This is stable in concentrated NH_4OH solution, or as the solid, but can be oxidized by strong oxidizing agents such as persulphate $(S_2O_8)^{2-}$ to give a green peroxo complex which formally contains Co(+IV).

$$[(NH_3)_5Co^{III}—O—O—Co^{III}(NH_3)_5]^{4+} \xrightarrow{\text{oxidize}} [(NH_3)_5Co^{III}—O—O—Co^{IV}(NH_3)_5]^{5+}$$

The magnetic moment of the green complex is about 1.7 Bohr magnetons. This is in agreement with the presence of Co(+III) (d^6 low spin, diamagnetic) and Co(+IV) (d^5 low spin, one unpaired electron). However, electron spin resonance indicates that both Co atoms are identical. Thus an electron must be able to move across the peroxo bridge and spend an equal amount of time on both metal atoms.

Several other binuclear complexes are known which use $—O—O—^{2-}$, OH^-, NH_2^- or NH^{2-} as bridging groups:

$$[(NH_3)_5Co—NH_2—Co(NH_3)_5]^{5+} \quad \text{(blue)}$$

$$[(NH_3)_4Co(\mu\text{-}NH_2)(\mu\text{-}O_2)Co(NH_3)_4]^{3+} \quad \text{(brown)}$$

$$[(NH_3)_4Co(\mu\text{-}OH)_2Co(NH_3)_4]^{4+} \quad \text{(red)}$$

Ir(+IV) is one of the most stable states, but Rh(+IV) is unstable and forms few compounds. Both metals form tetrafluorides. RhF_4 can be made from $RhCl_3$ and BrF_3. $IrCl_4$ is not very stable. IrO_2 is formed by burning the metal in air, but RhO_2 is only formed by strongly oxidizing Rh(+III) in alkaline solution, for example with sodium bismuthate. Rhodium forms only a few complexes, e.g. $K_2[RhF_6]$ and $K_2[RhCl_6]$, but these react with water, liberating O_2 and eventually forming RhO_2. Iridium forms a variety of halide and aqua complexes $[IrCl_6]^{2-}$, $[IrCl_3(H_2O)_3]^+$, $[IrCl_4(H_2O)_2]$ and $[IrCl_5 \cdot H_2O]^-$. The oxalate complex $[Ir(oxalate)_3]^{2-}$ can be resolved into *d* and *l* optical isomers.

(+V) AND (+VI) STATES

Co(+V) compounds do not exist under normal conditions. Rh(+V) and Ir(+V) exist as pentafluorides $(RhF_5)_4$ and $(IrF_5)_4$. These are very reactive and are readily hydrolysed. They have a tetrameric structure with M—F—M bridges similar to Nb, Ta, Mo, Ru and Os (Figure 21.1). The only complexes known are $Cs[RhF_6]$ and $Cs[IrF_6]$.

Rh(+VI) and Ir(+VI) occur only in RhF_6 and IrF_6, which are made by direct reaction. Neither is stable, though the heavier IrF_6 is more stable than RhF_6.

FURTHER READING

Abel, E.W. and Stone, F.G. (1969, 1970) The chemistry of transition metal carbonyls, *Q. Rev. Chem. Soc.*, Part I – Structural considerations, **23**, 325; Part II – Synthesis and reactivity, **24**, 498.

Allen, A.D. (1973) Complexes of dinitrogen, *Chem Rev.*, **73**, 11.

Canterford, J.H. and Cotton, R. (1968) *Halides of the Second and Third Row Transition Elements*, Wiley, London.

Canterford, J.H. and Cotton, R. (1969) *Halides of the First Row Transition Elements*, Wiley, London.

Diatomic metals and metallic clusters (1980) (Conference papers of the Faraday Symposia of the Royal Society of Chemistry, No. 14), Royal Society of Chemistry, London.

Emeléus, H.J. and Sharpe, A.G. (1973) *Modern Aspects of Inorganic Chemistry*, 4th ed. (Chapters 14 and 15: Complexes of Transition Metals: Chapter 20: Carbonyls), Routledge and Kegan Paul, London.

Golding, B.T. (1983) Cobalt and vitamin B_{12}, *Education in Chemistry*, **20**, 204–207.

Griffith, W. P. (1973) *Comprehensive Inorganic Chemistry*, Vol. 4 (Chapter 46: Carbonyls, cyanides, isocyanides and nitrosyls), Pergamon Press, Oxford.

Johnson, B.F.G. (1976) The structures of simple binary carbonyls, *JCS Chem. Commun.*, 211–213.

Johnson, B.F.G. (ed.) (1980) *Transition Metal Clusters*, John Wiley, Chichester.

Levason, W. and McAuliffe, C.A. (1974) Higher oxidation state chemistry of iron, cobalt and nickel, *Coordination Chem. Rev.*, **12**, 151–184.

Lewis, J. and Green, M.L. (eds) (1983) *Metal Clusters in Chemistry* (Proceedings of the Royal Society Discussion Meeting May 1982), The Society, London.

Nicholls, D. (1973) *Comprehensive Inorganic Chemistry*, Vol. 3 (Chapter 41: Cobalt), Pergamon Press, Oxford.

Phipps, D.A. (1976) *Metals and Metabolism*, Oxford University Press, Oxford.

26 The nickel group

Table 26.1 Electronic structures and oxidation states

Element		Electronic structure	Oxidation states*
Nickel	Ni	[Ar] $3d^8\ 4s^2$	−I 0 (I) **II** (III) (IV)
Palladium	Pd	[Kr] $4d^{10}$	0 (I)? **II** IV
Platinum	Pt	[Xe] $4f^{14}\ 5d^9\ 6s^1$	0 (I) **II** (III)? **IV** (V) (VI)

* The most important oxidation states (generally the most abundant and stable) are shown in bold. Other well-characterized but less important states are shown in normal type. Oxidation states that are unstable, or in doubt, are given in parentheses.

INTRODUCTION

Nickel is moderately abundant and is produced in large quantities. It is used in large quantities in a wide variety of alloys, both ferrous and non-ferrous. It is predominantly divalent and ionic in simple compounds, and exists as Ni(+II) in the most of its complexes. These are commonly square planar or octahedral. Palladium and platinum are both rare and expensive. They are noble and not very reactive, but are slightly more reactive than the other platinum group metals. They are both used as catalysts. The most common oxidation states are Pd(+II) and Pt(+II) and Pt(+IV). These are not ionic.

OCCURRENCE, EXTRACTION AND USES

Nickel

Nickel is the twenty-second most abundant element by weight in the earth's crust. Commercially important Ni ores include sulphides, which are usually mixed with Fe or Cu sulphides, and alluvial deposits of silicates and oxides/hydroxides. Pentlandite $(Fe,Ni)_9S_8$ is the most important ore. It always has a Fe : Ni ratio of 1 : 1. It usually occurs with a form of FeS called pyrrhotite – both are bronze coloured, and are found in the USSR, Canada, and South Africa. Several other sulphide and arsenide ores such as millerite NiS, niccolite NiAs and nickel glance NiAsS were once important, but are now little used. Important alluvial deposits include garnierite, a magnesium–nickel silicate of variable composition $(Mg,Ni)_6Si_4O_{10}(OH)_8$, and nickeliferous limonite $(Fe,Ni)O(OH)(H_2O)_n$. Mine production of ore contained 860 000 tonnes of Ni in 1988. The main sources of ore were Canada 25%, the USSR 24%, and Australia, New Caledonia and Indonesia 7% each.

The extraction of Ni is complicated by the presence of other metals. Sulphide ores now provide most of the nickel produced. The ore is concentrated by flotation and magnetically, then heated with SiO_2. FeS decomposes to FeO, which reacts with SiO_2 to form slag $FeSiO_3$, which

Table 26.2 Abundance of the elements in the earth's crust, by weight

	ppm	Relative abundance
Ni	99	22
Pd	0.015	69
Pt	0.01	70

is easily removed. The remaining sulphide matte is cooled slowly giving an upper silvery layer of Cu_2S and a lower black layer of Ni_2S_3 which can be separated mechanically. (A small amount of metallic Cu/Ni alloy is also formed. This dissolves any of the platinum group metals present, and is used as a source of these rare and expensive elements.) The Ni_2S_3 is then roasted with air and converted to NiO. This may be used directly in steel making. Alternatively NiO may be reduced to the metal by carbon in a smelter. The metal is cast into electrodes which are purified by electrolysis in an aqueous solution of $NiSO_4$.

The Mond process provides an alternative method for producing high purity Ni. This process was patented by L. Mond and was used in South Wales from 1899 until about 1970. NiO and water gas (H_2 and CO) were warmed under atmospheric pressure to 50 °C. The H_2 reduced NiO to Ni, which in turn reacted with CO, forming volatile nickel carbonyl $Ni(CO)_4$. (This is highly toxic and flammable.) Any impurities remained solid. The gas was heated to 230 °C, when it decomposed to give pure metal and the CO was recycled. A new plant in Canada now uses CO and impure metal, but runs at 150 °C and 20 atmospheres pressure to form $Ni(CO)_4$.

$$Ni + 4CO \xrightarrow{50\,°C} Ni(CO)_4 \xrightarrow{230\,°C} Ni + 4CO \qquad \text{(Mond process)}$$

Nickel silicate ores such as garnierite are mixed with gypsum ($CaSO_4$) and smelted with coke. The silicates form $CaSiO_3$ slag, and the Ni forms a sulphide matte which is treated as above.

Most of the Ni produced is used to make ferrous and non-ferrous alloys. Ni improves both the strength of steel and its resistance to chemical attack. In 1985, 385 000 tonnes of ferronickel were produced. Stainless steel may

continued overleaf

contain 12–15% Ni and steel for cutlery contains 20% Cr and 10% Ni. Very strong permanent magnets are made from 'Alnico' steel. Monel metal is very resistant to corrosion and is used in apparatus to handle F_2 and other corrosive fluorides. It contains 68% Ni, 32% Cu and traces of Fe and Mn. Several non-ferrous alloys are important. The Nimonic series of alloys (75% Ni with Cr, Co, Al and Ti) are used in gas turbine and jet engines where they are subjected to high stresses and high temperatures. Others such as Hastelloy C are used for their corrosion resistance. Nichrome contains 60% Ni and 40% Cr and is used to make the wire which gets red hot in electric radiators. Cupro-nickel (80% Cu and 20% Ni) is used to make 'silver' coins. The so called 'nickel–silver' contains roughly 60% Cu, 20% Ni and 20% Zn. This is used to make imitation silver articles and can be electroplated on other metals to give EPNS (electroplated nickel–silver). The name nickel–silver is confusing as it contains no silver. Often steel is electroplated with Ni before electroplating with Cr. Some Ni is used in Ni/Fe storage batteries, which have the advantage that they can be charged at very fast rates without damaging the battery plates. Small amounts of very finely divided Ni (Raney Ni) are used for many reduction processes. Examples include the manufacture of hexamethylenediamine, the production of H_2 from NH_3, and the reduction of anthraquinone to anthraquinol in the production of H_2O_2.

Palladium and platinum

Pd and Pt are rare elements, but they are appreciably more abundant than the other platinum group metals (Ru, Os, Rh and Ir). World production of all six platinum group metals was only 267 tonnes in 1988. Nearly 100 tonnes of this was Pt. Even though Pd is slightly more abundant than Pt, production of Pt is greater than that of Pd. The largest sources were South Africa 45%, the USSR 44%, Canada 5%, the USA 2% and Japan 0.8%). South African sources yield more Pt than Pd, but USSR sources yield more Pd than Pt.

The platinum group metals occur as traces in the sulphide ores of Cu and Ni. They are obtained as concentrates as anode sludge from electrolytic processes for the major metals. The platinum group metals are also obtained from the Cu/Ni alloy produced in the separation of the sulphide matte of Cu_2S and Ni_2S_3 in the process outlined for Ni above. Separation of the platinum metals is complex, but in the last stages $(NH_4)_2[PtCl_6]$ and $[Pd(NH_3)_2Cl_2]$ are ignited to give the respective metals. The metals are obtained as powders or sponges, and are fabricated into solid objects by sintering.

Roughly one third of the Pt produced is used in jewellery, one third in cars and one third for investment and for industrial uses. Pt has been used in jewellery since several centuries BC. The earliest users were the ancient Egyptians and the Indians of Peru and Ecuador. Nowadays it is often used to make the mountings for diamonds in rings and other jewellery. It resembles silver and has been called 'white gold'. Rather confusingly this name is now used for a Pd/Au alloy.

A new and increasing use of Pt is in 'three-way catalytic convertors'. These convertors are fitted to many new cars to reduce pollution from the exhaust gases. It is essential that lead-free petrol is used in the car. The main component of the convertor is a ceramic honeycomb which is coated with Pt, Pd and Rh. The exhaust gases from the engine exhaust pass through the honeycomb at about 300 °C. The precious metals convert unburnt fuel, CO and oxides of nitrogen into harmless CO_2 and N_2. (Leaded petrol must **not** be used as Pb poisons the catalyst.)

Both Pd and Pt find extensive chemical uses as catalysts. $PdCl_2$ is used in the Waker process for converting C_2H_4 to CH_3CHO. Pd is used for hydrogenations such as phenol to cyclohexanone, and also for dehydrogenations. Pt is very important as a catalyst in the oil industry in the reforming of hydrocarbons. Pt/PtO is used as Adam's catalyst for reductions. At one time Pt was used in the Contact process in the manufacture of H_2SO_4 (to convert SO_2 to SO_3); V_2O_5 is now used as catalyst instead of Pt as it is cheaper and less susceptible to poisoning. A Pt/Rh alloy was formerly used to oxidize NH_3 to NO in the Ostwald process for making HNO_3.

In the laboratory Pt crucibles are sometimes used and Pt is also used to make apparatus to handle HF. Pt is also used to seal into soda glass to allow electrical connections to pass through the glass. This is important in making electrodes, thermionic valves etc. Soda glass and Pt have almost the same coefficient of expansion, so the glass does not crack on cooling.

OXIDATION STATES

Ni shows a range of oxidation states from (−I) to (+IV), but its chemistry is predominantly that of the (+II) state. $[Ni(H_2O)_6]^{2+}$ ions are green coloured and are stable both in solution and in many simple compounds. Ni(+II) also forms many complexes, which are mainly square planar or octahedral. The higher oxidation states of all three metals are unstable.

Pd(+II) is the most important state, and occurs as the hydrated ion $[Pd(H_2O)_4]^{2+}$ and in complexes. Pt does not form an aqua ion. Both Pt(+II) and Pt(+IV) are important, but these are not ionic. The (+II) complexes are square planar and (+VI) complexes are octahedral.

Zero-valent states occur for all three elements with π bonding ligands such as CO. The maximum oxidation state of (+VI) is only attained in PtF_6, and Pt(+V) occurs in $[PtF_6]^-$. The highest oxidation state attained by Ni and Pd is (+IV) in NiF_4 and PdF_4. The so-called PdF_3 does not contain Pd^{III} but is really $Pd^{2+}[Pd^{IV}F_6]^{2-}$. The oxides and halides formed are shown in Table 26.3.

GENERAL PROPERTIES

Ni is a silvery white metal, and Pd and Pt are both grey–white. All three elements are unreactive in the massive state. They do not tarnish or react with air or water at normal temperatures.

Ni is often electroplated on to other metals to provide a protective

Table 26.3 Oxides and halides

Oxidation states					
(+II)	(+III)	(+IV)	(+V)	(+VI)	Other
NiO	$(Ni_2O_3)^h$	$NiO_2{}^h$	–	–	
PdO	–	$(PdO_2)^h$	–	–	
$(PtO)^h$	$(Pt_2O_3)^h$	**PtO_2**	–	$(PtO_3)^h$	Pt_3O_4
NiF_2	–	–	–	–	
$NiCl_2$	–	–	–	–	
$NiBr_2$	–	–	–	–	
NiI_2	–	–	–	–	
PdF_2	**$Pd[PdF_6]$**	PdF_4	–	–	
$PdCl_2$	–	–	–	–	
$PdBr_2$	–	–	–	–	
PdI_2	–	–	–	–	
–	–	**PtF_4**	$(PtF_5)_4$	PtF_6	
$PtCl_2$	$PtCl_3$?	**$PtCl_4$**	–	–	
$PtBr_2$	$PtBr_3$?	**$PtBr_4$**	–	–	
PtI_2	PtI_3?	**PtI_4**	–	–	

The most stable oxidation states are shown in bold, unstable ones in brackets.
h = hydrous oxide.

Table 26.4 Some physical properties

	Covalent radius (Å)	Ionic radius (Å) M^{2+}	Ionic radius (Å) M^{3+}	Melting point (°C)	Boiling point (°C)	Density ($g\,cm^{-3}$)	Pauling's electro-negativity
Ni	1.15	0.69	0.60^h 0.56^l	1455	2920	8.91	1.8
Pd	1.28	0.86	0.76	1552	2940	11.99	2.2
Pt	1.29	0.80	–	1769	4170	21.41	2.2

h = high spin value, l = low spin radius.

coating. However, Ni does tarnish when heated in air. Raney Ni is a very finely divided form of Ni used as a catalyst. It is readily oxidized by air and is pyrophoric. Red hot Ni reacts with steam.

Ni dissolves readily in dilute acids, giving hydrated $[Ni(H_2O)_6]^{2+}$ ions and H_2. Like Fe and Co it is rendered passive by concentrated HNO_3 and aqua regia. Pd and Pt are more noble (less reactive) than Ni, but are more reactive than the other elements in the platinum group. Pd dissolves slowly in concentrated HCl in the presence of O_2 or Cl_2, and fairly readily in concentrated HNO_3, giving $[Pd^{IV}(NO_3)_2(OH)_2]$. Pt is the most resistant to acids, but dissolves in aqua regia, giving chloroplatinic acid $H_2[PtCl_6]$ (see Table 26.5).

Ni is unaffected by aqueous alkalis and is therefore used to make the

Table 26.5 Some reactions of Ni, Pd and Pt

Reagent	Ni	Pd	Pt
O_2	NiO	PdO at red heat	PtO at high temp. and pressure
F_2	NiF_2	$Pd^{II}[Pd^{IV}F_6]$ at 500 °C	PtF_4 at red heat
Cl_2	$NiCl_2$	$PdCl_2$	$PtCl_2$
H_2O	No action	No action	No action
Dilute HCl or dilute HNO_3	$Ni^{2+} + H_2$	Dissolves very slowly	No action
Concentrated HNO_3	Passive	Dissolves	No action
Aqua regia	Passive	Dissolves	$H_2[PtCl_6]$

apparatus for manufacturing NaOH. Pd and Pt are both rapidly attacked by fused alkali metal oxides and peroxides, e.g Na_2O and Na_2O_2.

Ni reacts with the halogens on heating. It reacts only slowly with fluorine, so Ni and alloys such as Monel are often used to handle F_2 and corrosive fluorides. Ni also reacts with S, P, Si and B on heating. Red hot Pd reacts with F_2, Cl_2 and O_2. Pt is less reactive but at red heat it reacts with F_2, and at a high temperature and pressure it reacts with O_2.

All three metals absorb gaseous H_2. The amount absorbed depends on the physical state of the metal. However, Pd absorbs very large volumes of H_2, more than any other metal. If red hot Pd is cooled in H_2 it can absorb 935 times its own volume of H_2. The hydrogen is mobile and diffuses through the metal lattice. The conductivity of the metal falls as H_2 is absorbed. The H_2 is evolved on heating. Other gases, including He, are not absorbed, so this process is used to purify H_2.

Pt(+II) and Pt(+IV) form an extremely large number of complexes. (The two most prolific complex forming elements are Co and Pt.)

LOW VALENCY STATES (−I), (0), (+I)

Ni(−I) is found in the carbonyl anion $[Ni_2(CO)_6]^{2-}$.

The zero-valent state is formed by all three metals. $[Ni^0(CO)_4]$ is formed by warming Ni and CO. Its formation and subsequent pyrolysis was important in the Mond process for the purification of the metal. Though the original process became obsolete in about 1970, a modified process is used in Canada. $[Ni(CO)_4]$ is perhaps the best known carbonyl, but its stability is much lower than that of carbonyls in earlier transition metal groups. The $[Ni(CO)_4]$ molecule is tetrahedral, volatile, very poisonous, easily oxidized and flammable. A phosphine derivative $[Ni^0(PF_3)_4]$, and mixed compounds such as $[Ni(CO)_2(PF_3)_2]$, are also known. Reduction of $[Ni^{II}(CN)_4]^{2-}$ by potassium in liquid ammonia gives $K_4[Ni^0(CN)_4]$, whilst reduction with hydrazine sulphate in aqueous media gives $K_4[Ni^I_2(CN)_6]$. Pd and Pt do not form simple carbonyls like $[Ni(CO)_4]$, but they do form phosphine complexes such as $[Pt^0(PPh_3)_4]$ and $[Pt^0(PPh_3)_3]$:

$$2K_2[Pt^{II}Cl_4] + N_2H_4 + 8PPh_3 \xrightarrow{EtOH} 2[Pt^0(PPh_3)_4] + 4KCl + 4HCl + N_2$$

$Pd^0(CO)(PPh_3)_3]$ and $[Pt^0(CO)_2(PPh_3)_2]$ are also known. The ligand CO is a weak σ donor and a strong π acceptor, whilst PPh_3 is a stronger σ donor but a weaker π acceptor. The absence of simple carbonyls for Pd and Pt suggests that they have less tendency to form π bonds than Ni, possibly because they have high ionization energies. $[Ni(CO)_4]$ is reduced by sodium in liquid ammonia to give a carbonyl hydride $[\{Ni(CO)_H\}_2] \cdot (NH_3)_4$. This is red coloured and is dimeric. Reduction may also yield cluster compounds such as $[Ni_5(CO)_{12}]^{2-}$ and $[Ni_6(CO)_{12}]^{2-}$. A series of cluster compounds such as $[Pt_3(CO)_6]_n^{2-}$ are formed by reducing $[PtCl_6]^{2-}$ in alkaline solution under an atmosphere of CO. No similar Pd compounds have so far been observed.

(+II) STATE

The (+II) state is very important for all three elements. A wide variety of simple Ni^{2+} compounds exist. These include all the halides, the oxide, sulphide, selenide and telluride, salts of all the common acids and also some less stable ones such as $NiCO_3$ and salts of oxidizing ions like $Ni(ClO_4)_2$. The hydrated ion $[Ni(H_2O)_6]^{2+}$ gives rise to the green colour characteristic of many hydrated Ni salts. Many anhydrous Ni salts are yellow. Double salts are formed with alkali metals and NH_4^+, for example $NiSO_4(NH_4)_2SO_4 \cdot 6H_2O$. These are isomorphous with the corresponding double salts of Fe^{2+}, Co^{2+} and Mg^{2+}.

Though the chemistry of Ni is simplified by the dominance of the (+II) state, the Ni(+II) complexes are quite complicated. Octahedral and square planar complexes are commonly formed, and a few tetrahedral, trigonal bipyramidal and square-based pyramidal structures are also formed. Pd(+II) and Pt(+II) complexes are all square planar.

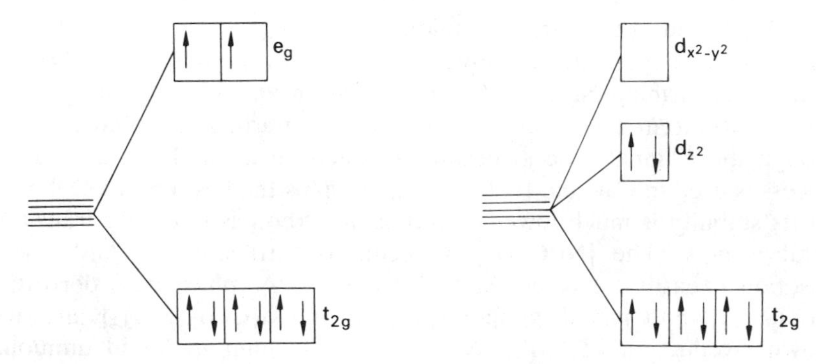

Figure 26.1 d^8 arrangement in weak and strong octahedral fields.

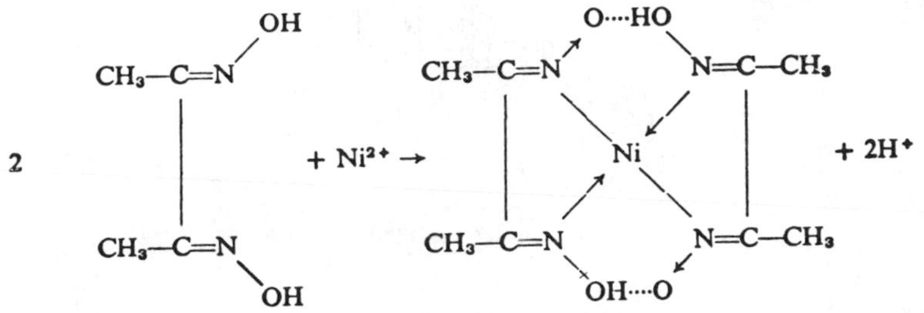

Figure 26.2 Nickel dimethylglyoxime complex.

The complexes formed with ammonia $[Ni(NH_3)_6]^{2+}$, $[Ni(H_2O)_4(NH_3)]^{2+}$, and the complex ethylenediamine $[Ni(\text{ethylenediamine})_3]^{2+}$ are all octahedral. These octahedral complexes are usually blue in colour, and they are paramagnetic as the d^8 ion has two unpaired electrons (Figure 26.1a). In complexes with strong field ligands such as CN^-, the electrons are forced to pair up and diamagnetic square planar complexes such as $[Ni(CN)_4]^{2-}$ are formed (Figure 26.1b).

The red coloured complex precipitated by Ni^{2+} and dimethylglyoxime from slightly ammoniacal solution is also square planar. However, in the solid the square planar molecules are stacked on top of each other and a Ni–Ni interaction occurs. The Ni–Ni distance is 3.25 Å. This was one of the earliest examples of metal to metal bonding, and in the solid Ni should be regarded as octahedrally coordinated rather than square planar. The formation of this complex is used both for the detection and quantitative estimation of Ni. The dimethylglyoxime loses a proton, and forms a stable complex molecule. The complex is stabilized because two five-membered chelate rings are formed, and also by internal hydrogen bonding, shown by dotted lines in Figure 26.2.

The square planar complexes are generally red, brown or yellow in colour. The reason for the formation of square planar complexes is discussed further in Chapter 7.

Several tetrahedral Ni(+II) complexes are known. These generally contain halide ligands, and often phosphine, phosphine oxide or arsine ligands as well, as in $[Ph_4As]_2^+[NiCl_4]^{2-}$, $[Ph_3P)_2 \cdot NiCl_2$ and $[Ph_3AsO)_2NiBr_2]$. These complexes are typically intensely blue coloured, and can be easily distinguished from square planar complexes both by the colour, and because they are paramagnetic (Figure 26.3).

When nickel cyanide is crystallized from a mixture containing ammonia and benzene, benzene ammino nickel cyanide is formed. The benzene molecules are not bonded, but are trapped in the cagework of the crystal. Such compounds are called clathrates and other molecules of a similar size may be trapped in a similar way.

Pd(+II) and Pt(+II) exist as oxides, halides, nitrates and sulphates. The anhydrous solids are generally not ionic. PdO exists in the anhydrous state, but PtO is only known as an unstable hydrated oxide. All the dihalides MX_2 are known except for PtF_2. Unlike the other halides, PdF_2 is ionic.

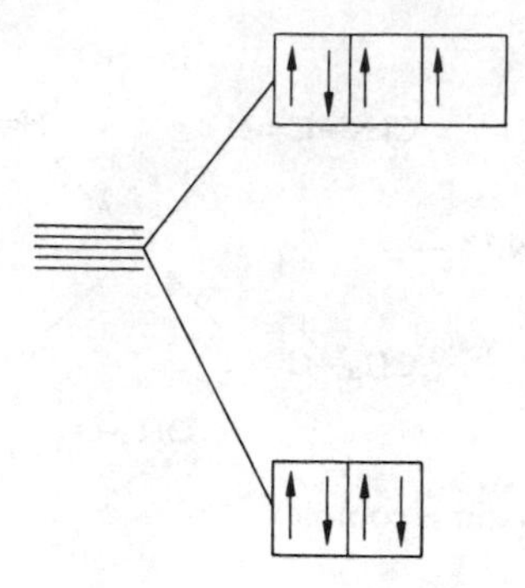

Figure 26.3 d^8 arrangement in tetrahedral field.

The Pd^{2+} ion has a d^8 configuration and is paramagnetic. The $[Pd(H_2O)4]^{2+}$ ion exists in water, and is diamagnetic. Because this complex is spin paired it is presumed to have a square planar structure. All Pd(+II) and Pt(+II) complexes are diamagnetic. In hydrochloric acid the diamagnetic $[PdCl_4]^{2-}$ ion is formed. All the other dihalides are molecular or polymeric, and are diamagnetic. $PdCl_2$ and $PtCl_2$ are made from the elements and both exist in α and β forms. Which is formed depends on the exact conditions used. The α forms are the more common.

$$Pd + Cl_2 \xrightarrow{>550\,^\circ C} \alpha\text{-}(PdCl_2)_n \xrightarrow{\text{slow}} \beta\text{-}(PdCl_2)$$

$$Pd + Cl_2 \xrightarrow{<550\,^\circ C} \beta\text{-}(PdCl_2)$$

α-$PdCl_2$ is a dark red solid, whilst α-$PtCl_2$ is olive green. α-$PdCl_2$ has a flat chain polymeric structure (Figure 26.4a), in which the Pd atoms are in a square planar arrangement. It is hygroscopic and is soluble in water. The structure of α-$PtCl_2$ is not known, but it is insoluble in water, and dissolves in HCl, giving $[PtCl_4]^{2-}$ ions. The β forms of $PdCl_2$ and $PtCl_2$ have an unusual molecular structure. This is based on a Pd_6Cl_{12} or Pt_6Cl_{12} unit. The structure is best described as the metal surrounded by four Cl atoms in a square planar environment, with six such units linked by halogen bridges (Figure 26.4b). This is remarkably similar in shape to the $[Nb_6Cl_{12}]^{2+}$ cluster compound shown in Figure 21.3d. In this the six Nb atoms were linked in an octahedral cluster with halogen atoms bridging all 12 edges. β-$PdCl_2$ is soluble in benzene and retains its structure. Despite the similarity in shape, the β-$PdCl_2$ structure appears to be covalent and stabilized largely by halogen bridges rather than by metal–metal bonding as in $[Nb_6Cl_{12}]^{2+}$.

An important reaction occurs between $PdCl_2$ and alkenes. With ethene, complexes such as $[Pd(C_2H_4)Cl_3]^-$, $[Pd(C_2H_4)Cl_2]_2$ and $[Pd(C_2H_4)_2Cl_2]$ are formed.

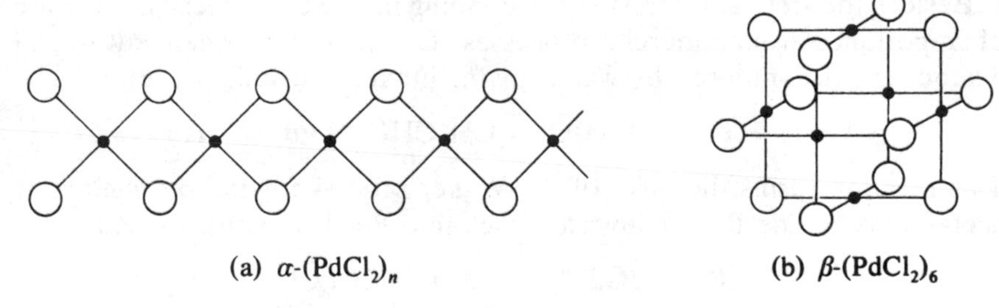

Figure 26.4 Structures of α- and β-$PdCl_2$.

```
[Cl      Cl ]-  [C2H4      Cl       Cl  ]  [C2H4      Cl ]
   \    /           \    /   \    /             \    /
     Pd               Pd       Pd                 Pd
   /    \           /    \   /    \             /    \
[Cl    C2H4 ]   [  Cl       Cl     C2H4 ]  [  Cl     C2H4]
```

Similar compounds are known for Pt: for example, Zeise's salt $K[Pt(C_2H_4)(Cl)_3] \cdot H_2O$ forms yellow crystals and has been known since 1825. The structure of these alkene complexes is unusual. In Zeise's salt the $[Pt(C_2H_4)(Cl)_3]^-$ ion is essentially square planar with Cl at three corners and $H_2C{=}CH_2$ at the other corner. However, the $H_2C{=}CH_2$ molecule is perpendicular to the $PtCl_3$ plane, and the two C atoms are almost equidistant from the Pt. (The Pt—C distances are 2.128 Å and 2.135 Å.) The C=C distance in the complex is 1.375 Å compared with 1.337 Å in ethene and a C—C distance of 1.54 Å in ethane. Thus the double bond is only lengthened slightly in forming the complex. The double bond occupies the coordination position rather than a single C atom, and C_2H_4 acts as a dihapto ligand. Thus the complex should be written $K[Pt(\eta^2\text{-}C_2H_4)(Cl)_3] \cdot H_2O$.

The bonding in these alkene complexes was not understood until 1951 when Dewar suggested that the π bond donated electrons to a vacant σ orbital on the metal, rather than involving bonding an individual C atom. This idea was extended by Chatt in 1953, and current thinking is that bonding is in two parts:

1. A dative bond in which the electron pair in the filled π orbital on ethene overlaps with an empty hybrid orbital on the metal, giving a σ bond.
2. π overlap also occurs between a filled metal *d* orbital and an empty antibonding orbital on ethene. This is π back donation or back bonding. Most of the transition elements form complexes with alkenes. The exceptions are the first few elements where the *d* orbitals on the metal are not sufficiently populated to allow back bonding. The extent of back bonding varies from one complex to another, and is related to the C=C bond length.

Besides theoretical interest in the bonding in these complexes, some are of importance in commercial processes. Complexes between $PdCl_2$ and alkenes are decomposed by water, giving ethanal (acetaldehyde):

$$C_2H_4 + PdCl_2 + H_2O \rightarrow CH_3CHO + Pd + 2HCl$$

This reaction forms the basis of the Waker process for the production of acetaldehyde. The Pd is converted back into $PdCl_2$ in situ by $CuCl_2$:

$$Pd + 2CuCl_2 \rightarrow PdCl_2 + 2CuCl$$

The solution contains HCl and the $CuCl_2$ is regenerated by passing in O_2.

$$2CuCl + 2HCl + \tfrac{1}{2}O_2 \rightarrow 2CuCl_2 + H_2O$$

Thus the overall reaction is:

$$H_2C{=}CH_2 + \tfrac{1}{2}O_2 \rightarrow CH_3CHO$$

This process is practicable because the reaction between Pd and $CuCl_2$ is quantitative, so the catalyst is recycled and only small amounts of Pd are required for replenishment.

With propene $CH_3 \cdot CH{=}CH_2$, the product is acetone. This reaction is also of commercial importance. If the reaction is carried out in acetic acid, ethene is converted to vinyl acetate. Though this is not a commercial process because of corrosion problems and difficulty in catalyst recovery, it has led to a study of palladium(II) acetate $[Pd(CH_3COO)_2]_3$. This has an unusual structure, comprising three metal atoms in a triangle, held together by six bridging acetate groups.

$PdCl_2$ catalyses the reaction between ethene, CO and H_2O:

$$CH_2{=}CH_2 + CO + H_2 \rightarrow CH_3CH_2COOH$$

Magnus' green salt has the formula $[Pt(NH_3)_4]^{2+}$ $[PtCl_4]^{2+}$, and the square planar anions and cations are stacked on top of each other. This structure also occurs in other complexes such as $[Pd(NH_3)_4]^{2+}$ $[Pd(SCN)_4]^{2-}$ and $[Cu(NH_3)_4]^{2+}$ $[PtCl_4]^{2-}$. The metal atoms in adjacent units may interact with each other, giving weak metal–metal bonds. Evidence for this is that if the anion and cation contain Pt(+II) they are colourless, pale yellow or pale red individually, but when stacked together they show an unusual iridescent green colour. They also show increased electrical conductivity. [Pt(ethyenediamine)Cl_2] is stacked in a similar way (Figure 26.5).

$K_2[Pt(CN)_4] \cdot 3H_2O$ is a well known complex and is colourless and stable. In the crystal structure the square planar $[Pt(CN)_4]^{2-}$ units are stacked on top of each other, but the solid does not conduct electricity. However, several complexes can be derived from it which show electrical conduction in one dimension, and they are also dichroic. (Dichroic materials have a different refractive index in different directions, so when they are viewed from different directions they appear differently coloured.) If this compound is oxidized it is possible to obtain bronze coloured compounds which are cation deficient such as $K_2[Pt(CN)_4]Br_{0.3} \cdot 3H_2O$ and

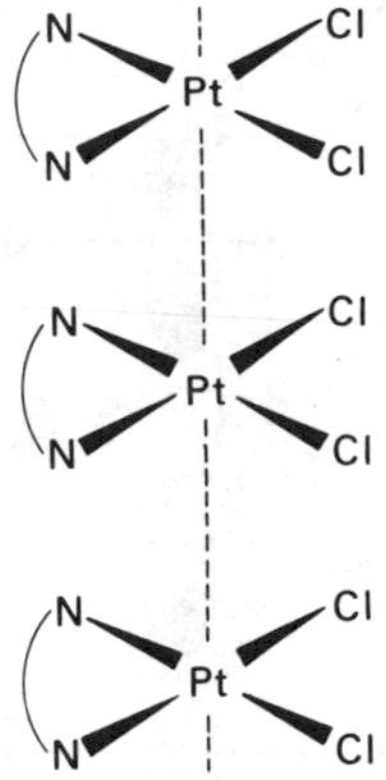

Figure 26.5 Stacks of square planar [Pt(ethylenediamine)Cl_2] molecules.

$K_2[Pt(CN)_4]Cl_{0.3} \cdot 3H_2O$. The filled d_z^2 orbitals on the Pt atoms overlap, giving a delocalized band along the Pt chain. In $K_2[Pt(CN)_4] \cdot 3H_2O$ this band is full: hence it cannot conduct. In $K_2[Pt(CN)_4]Br_{0.3} \cdot 3H_2O$ the Br act as electron acceptors, removing on average 0.3 electrons from each $[Pt(CN)_4]^{2-}$ unit. Thus the d_z^2 band is only five sixths filled, and hence the solid conducts electricity by a metallic mechanism in one dimension (Figure 26.6). In $K_2[Pt(CN)_4] \cdot 3H_2O$ the Pt—Pt distance is 3.48 Å, but the strong overlap of the d_z^2 orbitals in $K_2[Pt(CN)_4]Br_{0.3} \cdot 3H_2O$ reduces the Pt—Pt distance to 2.8–3.0 Å.

A very important medical use of Pt(+II) compounds is the use of the *cis* isomer of $[Pt(NH_3)_2(Cl)_2]$ as an anti-cancer drug for treating several types of malignant tumours. The *trans* isomer is ineffective. The *cis* isomer is called cisplatin, and is highly toxic. It is injected into the bloodstream, and the more reactive Cl groups bond to a N atom in guanosine (part of the DNA molecule). The cisplatin molecule can bond to two different guanosine units, and by bridging between them it upsets the normal reproduction of DNA. Those cells which are undergoing cell division are attacked by cisplatin. Tumours are usually growing rapidly, but so also are the bone marrow cells (producing red and white blood cells), and cells in the testes (producing sperms), so these are also affected. Dramatic results are possible, and a large number of patients are completely cured. There is a critical balance between giving enough cisplatin to kill the tumour and leaving sufficient white blood cells to protect the body from attack by bacteria and viruses.

(+III) STATE

This state in not important for any of the three metals. Few Ni(+III) compounds are known. Oxidation of $Ni(OH)_2$ in alkaline solution with Br_2 gives $Ni_2O_3 \cdot 2H_2O$ as a black solid, which decomposes to NiO on dehydration. If Ni is fused in NaOH and oxygen bubbled through, sodium

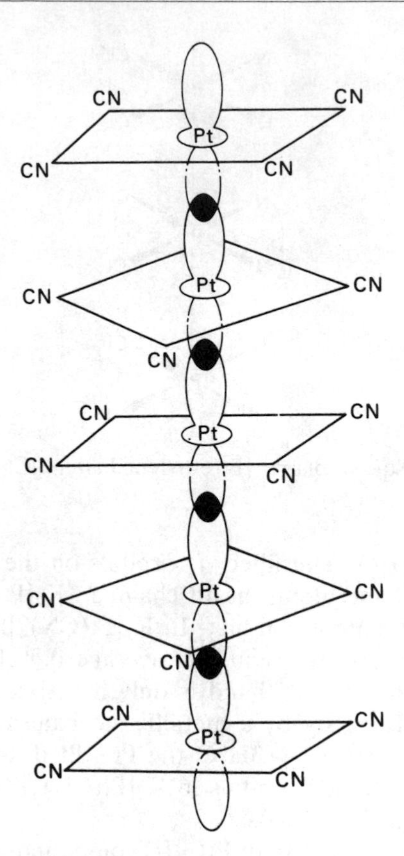

Figure 26.6 Structure of $K_2[Pt(CN)_4]Br_{0.3} \cdot 3H_2O$.

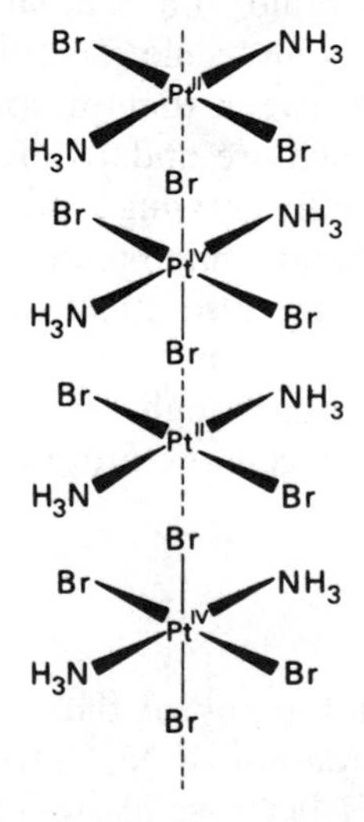

Figure 26.7 Structure of $[Pt(NH_3)_2Br_3]$.

nickelate(III) $Na[Ni^{III}O_2]$ is formed. The (+III) state can be stabilized in complexes. $K_3[NiF_6]$ can be prepared by fluorinating $NiCl_2$ and KCl at a high temperature and pressure. It is a violet solid but is strongly oxidizing and reacts with H_2O, evolving O_2. The structure is octahedral, but is slightly elongated as expected from Jahn–Teller distortion due to its asymmetrically filled $t_{2g}^6\ e_g^1$ electronic arrangement. $[Ni^{III}(ethylenediamine)_2Cl_2]Cl$ is also octahedral. The structure of $[Ni^{III}(PEt_3)_2Br_3]$ is a trigonal bipyramid.

Pd(+III) compounds are very rare and it is doubtful if Pt(+III) exists. Hydrated oxides may be known. The complexes $Na^+[PdF_4]^-$ and $NaK_2[PdF_6]$ have been reported. The $[PdF_6]^{3-}$ ion has four short bonds and two long bonds as expected for a low-spin d^7 octahedral complex. Heating Pd and F_2 gives a stable solid which was once thought to be PdF_3. This has since been shown to be a mixed valency compound $Pd^{2+}[Pd^{IV}F_6]^{2-}$ containing Pd(+II) and Pd(+IV). Complexes which are apparently in the (+III) state such as $[Pt(ethylenediamine)Br_3]$ and $[Pt(NH_3)_2Br_3]$ consist of chains of alternate square planar Pt(+II) units and octahedral Pt(+IV) units (Figure 26.7).

(+IV) STATE

Ni(+IV) is rare. The hydrated oxide is made by powerful oxidation of Ni^{2+} in fused alkali, and the product oxidizes Mn^{2+} to MnO_4^- and decomposes water. Fluorination of $NiCl_2$ and KCl gives the red complex $K_2[Ni^{IV}F_6]$ which is strongly oxidizing and liberates O_2 with water.

PdO_2 is only known in the hydrated form. In contrast PtO_2 is the most stable oxide of Pt and exists in both anhydrous and hydrated forms. The anhydrous oxide is insoluble, but the hydrated form dissolves in acids and alkalis.

PdF_4 is the only known halide of Pd, but all four Pt halides PtX_4 are known. Direct reaction of Pd and F_2 gives PdF_3 (really $Pd^{II}[Pd^{IV}F_6]$) and PdF_4, whilst Pt gives PtF_4, PtF_5 and PtF_6. $PtCl_4$ is formed either by direct reaction, or by heating $H_2[PtCl_6]$.

$$Pt \xrightarrow{\text{aqua regia}} H_2[PtCl_6] \xrightarrow{\text{heat}} PtCl_4 + 2HCl$$

Pd(+IV) forms a few octahedral complexes $[PdX_6]^{2-}$ where X = F, Cl or Br, and $[PdX_4(NH_3)_2]$. These are generally reactive. $[PdF_6]^{2-}$ hydrolyses rapidly in water whilst the other halide complexes are decomposed by hot water, giving $[Pd^{II}X_4]^{2-}$ and halogen.

In contrast Pt(+IV) forms a very large number of very stable octahedral complexes. These range from $[Pt(NH_3)_6]^{4+}$, $[Pt(NH_3)_5Cl]^{3+}$, $[Pt(NH_3)_4Cl_2]^{2+}$...to $[PtCl_6]^{2-}$. Similar series of complexes exist with a wide range of ligands including F^-, Cl^-, Br^-, I^-, OH^-, acetyl acetone, NO_2^-, SCN^-, $SeCN^-$, and CN^-. Some of these were studied by Werner in the early studies on coordination complexes (see Chapter 7).

Chloroplatinic acid is commercially the most common Pt compound. It is formed when a Pt metal sponge dissolves in aqua regia or concentrated HCl saturated with Cl_2. It forms red crystals of formula $H_2[PtCl_6] \cdot 2H_2O$. The sodium or potassium salts are a common starting material for making other Pt(+IV) compounds. Platinized asbestos is used as a catalyst. It is made by soaking asbestos in a solution of chloroplatinic acid, followed by strong heating to decompose the complex to Pt metal. This leaves a small amount of Pt spread over a very large surface.

Platinized platinum, or platinum black, electrodes are often used for conductivity measurements, and these are made by electrolysing hexachloroplatinates $[PtCl_6]^{2-}$.

Platinum is unusual in that it forms alkyl derivatives by a Grignard reaction.

$$4PtCl_4 + 12CH_3MgI \rightarrow [(CH_3)_3PtI]_4 + 8MgCl_2 + 4MgI_2$$

There were reports of $[(CH_3)_4Pt]_4$, but these were incorrect, and the compound formed is actually $[(CH_3)_3Pt \cdot OH]_4$. These complexes exist as tetrameric solids in which Pt is six-coordinate. The Pt—C bond is very stable. These organo derivatives are soluble in organic solvents.

(+V) AND (+VI) STATES

These are only found for Pt, and are rare. The (+V) state is represented by PtF_5, which is tetrameric and has the same structure as many transition metal pentafluorides (Figure 21.3a). The $[PtF_6]^-$ ion also contains Pt(+V) and was first formed by reacting PtF_6 and O_2 to give the compound $O_2^+[PtF_6]^-$. A similar reaction between Xe and PtF_6 led to reports by Bartlett in 1962 of the formation of $Xe^+[PtF_6]^-$, the first reported compound of the noble gases. (This compound was subsequently shown to be $[XeF]^+[Pt_2F_{11}]^-$.) The only examples of the (+VI) state which are known for certain are PtO_3 and PtF_6.

HORIZONTAL COMPARISONS IN THE IRON, COBALT AND NICKEL GROUPS

The ferrous metals Fe, Co and Ni show horizontal similarities and differ from the platinum metals in that the ferrous metals are much more reactive. Within the ferrous metals the reactivity decreases from Fe to Co to Ni. Although the maximum oxidation states are Fe(+VI), Co(+IV) and Ni(+IV), these elements rarely exceed an oxidation state of (+III). The tendency to trivalency decreases across the period. Fe^{3+} is the usual state but Co^{3+} is a strong oxidizing agent unless complexed, and nickel is divalent in all its simple compounds. The lower valency states exist as simple ions. The elements are relatively abundant.

The platinum metals Ru, Rh, Pd, Os, Ir and Pt are much more noble than the ferrous metals, and are little affected by acids. The reactivity of the metals increases from Ru to Rh to Pd and from Os to Ir to Pt, which is the opposite of the trend in the ferrous metals. The halogens react with the metals only at high temperatures, and bring out the higher valencies, e.g. OsF_6, IrF_6, PtF_6. The lower valency states are unstable except in complexes. Few simple ions exist. Because of the lanthanide contraction, the radii of the second and third rows of transition elements are very similar. Thus their atomic volumes are almost the same, so the densities of Os, Ir and Pt are almost double those of Ru, Rh and Pd. All six elements are rare.

Both the ferrous metals and the platinum metals are typical transition elements, and are characterized by:

1. coloured compounds
2. variable valency
3. catalytic properties
4. an ability to form coordination compounds

The differences between the two groups are:

1. increased stability of higher oxidation states
2. disappearance of simple ionic forms
3. increased nobility

These are the normal changes expected in a vertical group.

FURTHER READING

Abel, E. (1989) The Mond connection, *Chemistry in Britain*, **25**, 1014–1016. (History of nickel carbonyl.)

Abel, E.W. and Stone, F.G. (1969, 1970) The chemistry of transition metal carbonyls, *Q. Rev. Chem. Soc.*, Part I – Structural considerations, **23**, 325; Part II – Synthesis and reactivity, **24**, 498.

Canterford, J.H. and Cotton, R. (1968) *Halides of the Second and Third Row Transition Elements*, Wiley, London.

Canterford, J.H. and Cotton, R. (1969) *Halides of the First Row Transition Elements*, Wiley, London.

Emeléus H.J. and Sharpe, A.G. (1973) *Modern Aspects of Inorganic Chemistry*, 4th ed. (Chapters 14 and 15: Complexes of Transition Metals; Chapter 20: Carbonyls), Routledge and Kegan Paul, London.

Hartley, F.R. (1973) *The Chemistry of Platinum and Palladium*, Applied Science Publishers, London.

Hunt, C. (1984) Platinum drugs in cancer chemotherapy, *Education in Chemistry*, **21**, 111–115, 171.

Levason, W. and McAuliffe, C.A. (1974) Higher oxidation state chemistry of iron, cobalt and nickel, *Coordination Chem. Rev.*, **12**, 151–184.

Nicholls, D. (1973) *Comprehensive Inorganic Chemistry*, Vol. 3 (Chapter 42: Nickel), Pergamon Press, Oxford.

Shaw, B.L. and Tucker, N.I. (1973) *Comprehensive Inorganic Chemistry*, Vol. 4 (Chapter 53: Organo-transition metal compounds and related aspects of homogeneous catalysis), Pergamon Press, Oxford.

Wiltshaw, E. (1979) Cisplatin in the treatment of cancer, *Platinum Metals Rev.*, **23**, 90–98.

27 The copper group – coinage metals

Table 27.1 Electronic structures and oxidation states

Element		Electronic structure	Oxidation states*
Copper	Cu	[Ar] $3d^{10} 4s^1$	I **II** (III)
Silver	Ag	[Kr] $4d^{10} 5s^1$	**I** II (III)
Gold	Au	[Xe] $4f^{14} 5d^{10} 6s^1$	I **III** V

* The most important oxidation states (generally the most abundant and stable) are shown in bold. Other well-characterized but less important states are shown in normal type. Oxidation states that are unstable, or in doubt, are given in parentheses.

INTRODUCTION

The elements all have one *s* electron outside a completed *d* shell. They show only slight similarities in properties and considerable differences. All three metals have the same crystal structure (cubic close-packed). They conduct electricity and heat particularly well, and they tend to be noble (unreactive). The only ions which exist in solution (apart from complexes) are Cu^{2+} and Ag^+. The most stable oxidation state varies; Cu(+II), Ag(+I) and Au(+III). Copper is produced on a large scale and 10.8 million tonnes were used in 1988, mostly as the metal and in alloys. Copper is biologically important in various oxidase enzymes, as an oxygen carrier in invertebrates and in photosynthesis. There is great interest in various mixed oxides of copper which act as superconductors.

ABUNDANCE, EXTRACTION AND USES OF THE ELEMENTS

Copper is moderately abundant and is the twenty-fifth most abundant element in the earth's crust. It occurs to the extent of 68 ppm by weight. Silver and gold are quite rare (Table 27.2).

Table 27.2 Abundance of the elements in the earth's crust, by weight

	ppm	Relative abundance
Cu	68	25
Ag	0.08	66 =
Au	0.004	73

Copper

Copper nuggets (i.e. pieces of metal) have been found in the earth, but this source is largely exhausted. The most common ore is chalcopyrites $CuFeS_2$. This has a metallic lustre and is similar in appearance to pyrites FeS_2 (fool's gold) but is more copper coloured. Other ores include Cu_2S (called copper glance or chalcocite; dark grey coloured), basic copper carbonate $CuCO_3 \cdot Cu(OH)_2$ (which is called malachite and is green), cuprous oxide Cu_2O (which is called cuprite and is ruby red coloured) and Cu_5FeS_4 (called bornite or 'peacock ore' because it has a mixture of iridescent colours like a peacock's feathers (blue, red, brown and purple)). Turquoise $CuAl_6(PO_4)_4(OH)_8 \cdot 4H_2O$ is a popular gemstone because of its blue colour and delicate veining.

The sulphide ores are often lean and may contain only 0.4–1% Cu. These are crushed and concentrated by froth-flotation, giving a concentrate with 15% Cu. This is then roasted with air.

$$2CuFeS_2 \xrightarrow[1400-1450\,°C]{O_2} Cu_2S + Fe_2O_3 + 3SO_2$$

Sand is added to remove the iron as iron silicate slag $Fe_2(SiO_3)_3$ which floats on the surface. Air is blown through the liquid matte of Cu_2S with some FeS and silica, causing partial oxidation:

$$2FeS + 3O_2 \rightarrow 2FeO + 2SO_2$$
$$FeO + SiO_2 \rightarrow Fe_2(SiO_3)_3$$
$$Cu_2S + O_2 \rightarrow Cu_2O + SO_2$$

After some time the air is turned off and self-reduction of the oxide and sulphide occurs, giving impure 'blister copper' which is 98–99% pure.

$$Cu_2S + 2Cu_2O \rightarrow 6Cu + SO_2$$

The 'blister copper' is cast into blocks and refined by electrolysis using Cu electrodes with an electrolyte of dilute H_2SO_4 and $CuSO_4$.

It is not economic to treat very lean ores in this way, so these are dug up and left exposed to the air to weather. The CuS oxidizes slowly to $CuSO_4$ which is leached (dissolved out) with water or dilute H_2SO_4. Copper is displaced from the resulting copper sulphate solution by adding scrap iron which is sacrificed.

$$Fe + Cu^{2+} \rightarrow Fe^{2+} + Cu$$

continued overleaf

World production of mined Cu was 8.7 million tonnes in 1988. The largest sources of copper ores are Chile 17%, the USA 16%, the USSR 11%, Canada 8%, Zambia 5.5%, and Zaire and Poland 5% each. In addition about 2.1 million tonnes of scrap metal were recycled, giving a total of 10.8 million tonnes.

The metal is used in the electrical industry because of its high conductivity. It is also used for water pipes because of its inertness. Over 1000 different alloys of copper exist. These include brass (Cu/Zn with 20–50% Zn), so-called 'nickel silver' (55–65% Cu, 10–18% Ni, 17–27% Zn), phosphor bronze (Cu with 1.25–10% Sn and 0.35% P) and various alloys for making coins. Copper sulphate is produced in moderately large amounts (146 500 tonnes in 1985). Several copper compounds are used in agriculture. For example, Bordeaux mixture is basic copper hydroxide, and is made from $CuSO_4$ and $Ca(OH)_2$. It is an important spray for preventing fungus attack on the leaves of potatoes (potato blight) which caused the potato famine in Ireland in 1845–1846. It is also used to spray vines to prevent fungal attack. Basic copper carbonate, copper acetate and copper oxochloride have also been used. Paris Green is an insecticide made from basic copper acetate, arsenious oxide and acetic acid.

There has been enormous interest in a variety of mixed oxides of copper such as $La_{(2-x)}Ba_xCuO_{(4-y)}$ since these behave as superconductors at temperatures below 50 K. G. Bednorz and A. Müller were awarded the Nobel Prize for Physics in 1987 for work on these compounds. Other superconductors which work at higher temperatures (up to 125 K) are based on $YBa_2Cu_3O_{7-x}$. These are described in Chapter 5 under 'Superconductivity'.

Silver

Silver is found as sulphide ores Ag_2S (argentite), as the chloride AgCl (horn silver) and as the native metal. There are three extraction processes:

1. Most is now obtained as a by-product from the extraction of Cu, Pb or Zn. It can be obtained from the anode slime formed in the electrolytic refining of Cu or Zn.
2. Zinc is used to extract silver by solvent extraction from molten lead in Parke's process.
3. Silver and gold are extracted by making soluble cyanide complexes.

World production of silver was 14 861 tonnes in 1988. The main producers were Mexico 16%, the USA and USSR 11% each, Peru and Canada 10% each, Australia 7.5% and Poland 7%. The main uses of silver are as AgCl and AgBr in photographic emulsions, for jewellery and silver ornaments, for batteries and for silvering mirrors.

Gold

Historically gold has been found as lumps of metal in the ground called nuggets. Finds of this kind have started 'gold rushes' in the USA. However, gold occurs mainly as grains of metal disseminated in quartz veins. Many of these rocks have weathered with time. The gold and powdered rock are washed away in streams and accumulate as sediments in river beds. The grains of gold can be separated from silica by 'panning', i.e. swirling them both with water. Gold is very dense ($19.3\,g\,cm^{-3}$) and rapidly settles to the bottom, but the SiO_2, with a density of $2.5\,g\,cm^{-3}$, settles more slowly and is thrown away with the water. This method is little used nowadays since the sources are largely exhausted.

Nowadays rocks containing traces of gold are crushed and extracted either with mercury or with sodium cyanide. Water and powdered rock are passed over mercury, in which the gold dissolves, forming an amalgam. The gold is recovered by distilling the amalgam, when the mercury distils off and is reused. This process is also used with river water and silt in Brazil. Losses of mercury have poisoned considerable stretches of the River Amazon, giving environmental problems. In the cyanide process the crushed rock is treated with a 0.1–0.2% solution of NaCN and aerated.

$$4Au + 8NaCN + 2H_2O + O_2 \rightarrow 4Na[Au(CN)_2] + 4NaOH$$

The sodium argentocyanide complex is soluble, thus separating gold from the rest of the rock. The gold is precipitated from this solution by adding Zn powder. World production of gold was 1785 tonnes in 1988, and the main producers were South Africa 35%, the USSR 16%, the USA 11%, Australia 9% and Canada 7%. The major uses are as gold bullion (which is used as international currency) and for jewellery. Gold used in jewellery gold is usually alloyed with a mixture of Cu and Ag. These alloys retain the golden colour, but are harder. The proportion of gold in the alloy is expressed in *carats*. Pure gold is 24 carats. The alloys commonly used are 9 carat, 18 carat and 22 carat, and these contain 9/24, 18/24 and 22/24 pure gold respectively. Small amounts of gold are used to make corrosion-free electrical contacts, for example on computer boards. A thin film $10^{-11}\,m$ thick is sometimes deposited on glass windows in prestigious skyscraper buildings (e.g. a bank in Toronto). This thin metal film reflects unwanted heat from the sun in the summer, thus keeping the building cool. The film keeps heat in during the winter.

OXIDATION STATES

The elements Cu, Ag and Au show oxidation states of (+I), (+II) and (+III). However, the only simple hydrated ions found in solution are Cu^{2+} and Ag^+. The univalent ions Cu^+ and Au^+ disproportionate in water, and as a result they only exist as insoluble compounds or complexes. Cu(+III), Ag(+III) and Ag(+II) are so strongly oxidizing that they reduce water. Thus they only occur when stabilized in complexes, or as insoluble compounds. The oxides and halides formed are shown in Table 27.3.

Table 27.3 Oxides and halides

Oxidation states					
(+I)	(+II)	(+III)	(+IV)	(+V)	Others
Cu_2O	**CuO**	–	–	–	
Ag_2O	AgO	(Ag_2O_3?)	–	–	
Au_2O	–	Au_2O_3	–	–	
–	**CuF_2**	–	–	–	
CuCl	$CuCl_2$	–	–	–	
CuBr	$CuBr_2$	–	–	–	
CuI	–	–	–	–	
AgF	AgF_2	–	–	–	Ag_2F
AgCl	–	–	–	–	
AgBr	–	–	–	–	
AgI	–	–	–	–	
–	–	**AuF_3**	–	(AuF_5)	
AuCl	–	$AuCl_3$	–	–	
–	–	$AuBr_3$	–	–	
AuI	–	–	–	–	

The most stable oxidation states are shown in bold, unstable ones in brackets.

STANDARD REDUCTION POTENTIALS (VOLTS)

Acid solution

Oxidation state

+III **+II** **+I** **0**

$$CuO^+ \xrightarrow{+1.8} Cu^{2+} \xrightarrow{+0.15} {}^{*}Cu^+ \xrightarrow{+0.52} Cu$$

$$Cu^{2+} \xrightarrow{+0.34} Cu$$

$$AgO^+ \xrightarrow{+2.1} Ag^{2+} \xrightarrow{+1.98} Ag^+ \xrightarrow{+0.80} Ag$$

$$Au^{3+} \xrightarrow{<+1.29} Au^{2+} \xrightarrow{>+1.29} {}^{*}Au^+ \xrightarrow{+1.68} Au$$

$$Au^{3+} \xrightarrow{+1.50} Au$$

* Disproportionates

GENERAL PROPERTIES

The metals in this group have the highest electrical and thermal conductivities known. They are the most malleable and ductile structural metals. This is associated with their cubic close-packed structure. When sufficient force is applied, one plane may be forced to slip over another plane. The structure is very simple, so that when it slips it remains a regular cubic close-packed structure.

Atoms of Cu, Ag and Au (Table 27.1) have one *s* electron in their outer orbital. This is the same outer electronic arrangement as for the Group I metals. In spite of this, there are few similarities apart from the formal stoichiometry of compounds in the (+I) state and the high electrical conductivity of both groups of metals.

The Cu group differ from Group I elements in that the penultimate shell contains ten *d* electrons. The poor screening by the *d* electrons makes the atoms of the copper group much smaller in size. As a result the Cu group have higher densities and are harder. Their ionization energies are higher (Table 27.4), and their compounds are more covalent.

In the Cu group the *d* electrons are involved in metallic bonding. Thus the melting points and enthalpies of sublimation are much higher than for Group I metals.

The higher enthalpy of sublimation and higher ionization energy are the reasons why Cu, Ag and Au tend to be unreactive, i.e. show noble character. Group I metals have large negative standard reduction potentials (E° values) and are at the top of the electrochemical series. They are the most reactive metals in the periodic table. In contrast the currency metals have positive E° values and are thus below hydrogen in the electrochemical series. Thus they do not react with water or liberate H_2 with acids. The nobility increases from Cu to Ag to Au, whereas on descending Group I the reactivity increases. The inertness of Au resembles that of the platinum metals. Cu is inert towards non-oxidizing acids, but reacts with concentrated HNO_3 and H_2SO_4.

$$3Cu + \underset{\text{dilute}}{8HNO_3} \rightarrow 2NO + 3Cu(NO_3)_2 + 4H_2O$$

$$Cu + \underset{\text{concentrated}}{4HNO_3} \rightarrow 2NO_2 + Cu(NO_3)_2 + 2H_2O$$

However, Cu is very slowly oxidized on the surface in moist air, giving a green coating of verdigris. This is basic copper carbonate $CuCO_3 \cdot Cu(OH)_2$ and is familiar on the roofs of buildings covered with copper sheet, and also on copper statues such as the Statue of Liberty in New York.

Ag will dissolve in concentrated HNO_3 and in hot concentrated H_2SO_4. Au is inert to all acids except aqua regia (a 3:1 mixture of concentrated HCl and HNO_3). The HNO_3 acts as an oxidizing agent and the chloride ions as a complexing agent.

Cu reacts with oxygen, but Ag and Au are inert.

Table 27.4 Some physical properties

	Covalent radius (Å)	Ionic radius (Å)			Melting point (°C)	Boiling point (°C)	Density ($g\,cm^{-3}$)	Pauling's electro-negativity
		M^+	M^{2+}	M^{3+}				
Cu	1.17	0.77	0.73	0.54[l]	1083	2570	8.95	1.9
Ag	1.34	1.15	0.94	0.75	961	2155	10.49	1.9
Au	1.34	1.37	–	0.85	1064	2808	19.32	2.4

l = low spin radius.

$$Cu + O_2 \xrightarrow{\text{red heat}} CuO \xrightarrow{\text{higher temperature}} Cu_2O + O_2$$

Cu and Ag metals react with H_2S and S, but Au does not. Silver objects tarnish slowly in air (i.e. polished silver articles gradually blacken). This is due to traces of H_2S in the air which react with Ag, forming black Ag_2S.

$$2Ag + H_2S \rightarrow \underset{\text{black}}{Ag_2S} + H_2$$

In a similar way passing H_2S into solutions containing Cu^{2+} or Ag^+ gives black precipitates of CuS and Ag_2S. All three metals react with the halogens. Simple compounds of Au decompose to the metal quite readily, those of Ag can be reduced fairly easily, and those of Cu less readily.

For metals to react, an atom must first be isolated from the crystal structure and then be ionized. A high enthalpy of sublimation and a high ionization energy will reduce reactivity, though this may be partly offset by the energy gained when the ion is hydrated. Comparing Cu and K, Cu has a much higher melting point (and hence a higher enthalpy of sublimation). Because of the increased nuclear charge of copper, the orbital electrons are more tightly held (and hence the ionization energy is higher). The enthalpy of hydration is not large enough to offset these large amounts of energy, and so potassium is much more reactive than copper.

The oxides and hydroxides of Group I are strongly basic, and are soluble in water. In contrast the oxides of copper are insoluble and weakly basic. Group I compounds all contain simple colourless univalent ions and only form complexes with very strong complexing agents. In contrast the copper group elements show variable valency. The most common oxidation states are Cu(+II), Ag(+I) and Au(+III), and the three elements differ widely in their chemistries. Their compounds are mainly coloured and they show a strong tendency to form coordination complexes.

Copper is important in several catalysts. Cu is used in the direct process for manufacture of alkylchlorosilanes such as $(CH_3)_2SiCl_2$, which is used to make silicones. Cu and V catalyse the oxidation of cyclohexanol/cyclohexanone mixtures to adipic acid which is used to make nylon-66. $CuCl_2$ was used as the catalyst in the Deacon process for making Cl_2 from HCl.

(+I) STATE

In the (+I) state most of the simple compounds and complexes are diamagnetic and colourless because the ions have a d^{10} configuration. There are a few coloured compounds. For example, Cu_2O is yellow or red, Cu_2CO_3 is yellow and CuI is brown. In these cases the colour arises from charge transfer bands and not from d–d spectra.

It might be expected that the (+I) state would be the most common and most stable because of the extra stability resulting from a full d shell. Surprisingly this is not so. Although Ag^+ is stable in both the solid state and solution, Cu^+ and Au^+ disproportionate in water.

$$2Cu^+ \rightleftharpoons Cu^{2+} + Cu \qquad K = \frac{[Cu^{2+}]}{[Cu^+]^2} = 1.6 \times 10^6$$

$$3Au^+ \rightleftharpoons Au^{3+} + 2Au \qquad K = \frac{[Au^{3+}]}{[Au^+]^3} = 1 \times 10^{10}$$

The equilibrium constant for the disproportionation of Cu^+ in solution is high, showing that the equilibrium is largely to the right. Thus the concentration of Cu^+ is very low in solution and typically a Cu^+ ion exists in aqueous solution for less than a second. Similarly Au^+ is virtually non-existent in solution. The only cuprous and aurous compounds that are stable to water are either insoluble or present as complexes. Examples of insoluble Cu(+I) compounds include CuCl, CuCN and CuSCN. Cuprous thiocyanate CuSCN is used to estimate copper gravimetrically.

$$2Cu^{2+} + SO_3^{2-} + 2SCN^- + H_2O \rightarrow 2Cu^ISCN + H_2SO_4$$

Cu^{2+} is reduced to cuprous oxide Cu_2O by mild reducing agents. This is the basis of Fehling's test for reducing sugars (monosaccharides such as glucose). Equal quantities of two different solutions, Fehling's A and Fehling's B, are mixed immediately before adding the sugar and warming. The solution is deep blue coloured and if a reducing agent is present a yellow or red precipitate of Cu_2O is formed. (Fehling's A solution is a solution of copper(II) tartrate, made from $CuSO_4$ and Rochelle salt (sodium potassium tartrate). Fehling's B solution is NaOH.)

Cu_2O is a basic oxide and reacts with the halogen acids HCl, HBr and HI, giving insoluble CuCl, CuBr and CuI. CuF is unknown. CuCl and CuBr are usually made by boiling an acidic solution of $CuCl_2$ or $CuBr_2$ with excess Cu. This gives a solution containing the complex ions $[CuCl_2]^-$ and $[CuBr_2]^-$ which are linear in shape. Diluting these solutions gives white CuCl or yellow CuBr.

Addition of KI to a solution containing Cu^{2+} results in the I^- ions reducing Cu^{2+} to cuprous iodide Cu^II and at the same time I^- is oxidized to I_2. This reaction is used to estimate Cu^{2+} in solution by volumetric analysis. Excess of KI is added to an acidified solution and the I_2 produced is estimated by titrating with sodium thiosulphate.

$$2Cu^{2+} + 4I^- \rightarrow 2CuI + I_2$$
$$2Na_2S_2O_3 + I_2 \rightarrow Na_2S_4O_6 + 2NaI$$

The cuprous halides are partly covalent and have zinc blende structures with tetrahedrally coordinated Cu^+ ions. In the vapour CuCl and CuBr are polymeric and the main species is a six-membered ring.

The cuprous halides are insoluble in water. However, they dissolve in solutions containing an excess of halide ions by forming soluble halide complexes such as $[CuCl_2]^-$, $[CuCl_3]^{2-}$ and $[CuCl_4]^{3-}$. These and other Cu^I complexes are tetrahedral in the solid. (In $KCuCl_2$ and K_2CuCl_3 there are chains in which Cu is tetrahedrally surrounded by Cl, with simple units joined into a chain by shared halide ions.)

The cuprous halides also dissolve in strong HCl, HNO_3 and aqueous solutions of ammonia. Solutions of CuCl in concentrated HCl and CuCl in NH_4OH are important because they absorb carbon monoxide. Three points arise from this:

1. A solution of CuCl in NH_4OH is often used to measure the amount of CO in gas samples, simply by measuring the change in volume of the gas.
2. Though the metals of this group do not form neutral carbonyl compounds, an unstable carbonyl halide [Cu(CO)Cl] is formed by bubbling CO through a solution of CuCl. Both Cu and Au form carbonyl halides [M(CO)Cl] when CO is passed over the heated halide.
3. Several complexes with alkenes and alkynes can be made in a similar way by bubbling the hydrocarbon through a solution of Cu^{I} or Ag^{I}. Alkene complexes can also be made by passing the hydrocarbon over the heated halide. These have the formula [MRX] where R is an unsaturated hydrocarbon and X a halogen. These complexes are very reactive, and are often polymeric. The M—C bonds are not symmetrical, suggesting σ rather than π bonding. Au^{I} forms complexes less readily, and only with high molecular weight alkenes.

Cyanide complexes are well known, and are used to extract Ag and Au as soluble complexes. The metals are recovered from the complex by reduction with zinc.

$$4Au + 8CN^- + 2H_2O + O_2 \rightarrow 4[Au^{I}(CN)_2]^- + 4OH^-$$

Two-coordinate complexes such as $[Au(CN)_2]^-$ have a linear structure. However, in solid $K[Cu^{I}(CN)_2]$ the Cu is bonded to three CN^-, giving a planar triangular arrangement. The CN^- are bonded in the usual way through C, but the third CN^- acts as a bridging group to another Cu atom.

Cyanide ions may react with metal ions in two ways:

1. as a reducing agent
2. as complexing agent

Thus adding KCN to a $CuSO_4$ solution first causes reduction and precipitates cuprous cyanide. This reacts with excess CN^-, forming a soluble four-coordinate complex $[Cu(CN)_4]^{3-}$ which is tetrahedral in shape.

$$2Cu^{2+} + 4CN^- \rightarrow 2Cu^+CN^- + \underset{\text{cyanogen}}{(CN)_2}$$

$$CuCN + 3CN^- \rightarrow [Cu(CN)_4]^{3-}$$

Cu(+I) forms several different polymeric complexes which involve a cluster of four Cu atoms at the corners of tetrahedron, but which do not involve metal–metal bonding. The phosphine and arsine complexes $Cu_4I_4(PR_3)_4$ and $Cu_4I_4(AsR_3)_4$ are examples of such clusters. In these the four Cu form a tetrahedron. The four phosphine or arsine ligands are attached to the four corners, and the I atoms are located above the four faces of the tetrahedron, with each I triply bridged to three Cu atoms.

In $[Cu_4(SPh)_6]^{2-}$ the Cu atoms form a tetrahedron and the six S atoms bridge the six edges of the tetrahedron.

Ag(+I) is the most important state for silver and many simple ionic compounds are known containing Ag^+. Practically all Ag^I salts are insoluble in water. Exceptions include $AgNO_3$, AgF and $AgClO_4$ which are soluble. The salts are typically anhydrous except for $AgF \cdot 4H_2O$. The Ag^+ ion is hydrated in solution but only as the dihydrate $[Ag(H_2O)_2]^+$. Ag^I commonly forms two-coordinate complexes rather than four-coordinate complexes as in Cu^I.

$AgNO_3$ is one of the most important salts. Ag_2O is mainly basic, dissolving in acids. Moist Ag_2O absorbs carbon dioxide and forms Ag_2CO_3. Since Ag_2O dissolves in NaOH it must have slight acidic properties too.

The silver halides are used in photography (see later). AgF is soluble in water but the other silver halides are insoluble. In qualitative analysis solutions containing the halide ions Cl^-, Br^- and I^- are tested by adding $AgNO_3$ solution and dilute HNO_3. A white precipitate of AgCl indicates the presence of a chloride, a pale yellow precipitate of AgBr indicates a bromide and a yellow precipitate of AgI indicates an iodide. The presence of these halide ions may be confirmed by testing the solubility of the silver halide precipitates in ammonium hydroxide. AgCl is soluble in dilute NH_4OH, AgBr dissolves in strong 0.880 ammonia, and AgI is insoluble even in 0.880 ammonia. When AgCl and AgBr dissolve, they form the ammine complex $[H_3N{\rightarrow}Ag{\rightarrow}NH_3]^+$, which is linear.

A few silver compounds with colourless anions are coloured. For example, AgI, Ag_2CO_3 and Ag_3PO_4 are yellow and Ag_2S is black. This is because the Ag^+ ion is small and highly polarizing and the anion, e.g. I^-, is large and highly polarizable. This leads to some covalent character.

Ag^+ forms a variety of complexes. Most simple ligands result in two-coordination and a linear structure, for example $[Ag(NH_3)_2]^+$, $[Ag(CN)_2]^-$ and $[Ag(S_2O_3)_2]^{3-}$. Bidentate ligands form polynuclear complexes. The halide complexes of Cu^+ and Ag^+ are unusual because the stability sequence is $I > Br > Cl > F$, whilst for most metals the sequence is the reverse. Ligands capable of π bonding, such as phosphine derivatives, may form both two- and four-coordinate complexes.

Au(+I) is less stable. It is known as the oxide Au_2O. The halides AuCl and AuBr can be obtained by gently heating the corresponding AuX_3 halide. AuI is precipitated by adding I^- to AuI_3. These Au(+I) compounds disproportionate in water to Au metal and Au(+III).

Au(+I) also exists in linear complexes such as $[NC{-}Au{-}CN]^-$, $[Cl{-}Au{-}Cl]^-$, $[R_3P{-}Au{-}Cl]$ and $[R_3P{-}Au{-}CH_3]$. The cyanide complex is soluble in water and is formed in the cyanide extraction processes by dissolving Au in an aqueous solution containing CN^- in the presence of air or H_2O_2. Phosphine complexes are made by treating Au_2Cl_6 with R_3P in ether solution, giving R_3PAuCl. The Cl may be substituted by other groups such as I, SCN or Me. If R_3PAuCl is strongly reduced, e.g. with $NaBH_4$, then a cluster compound $Au_{11}Cl_3(R_3P)_7$ is formed. The cluster has a structure related to an incomplete icosahedron. There is an Au at the centre and Au at ten of the 12 corners of the icosahedron.

An important use of Au(+I) is in drugs to treat rheumatoid arthritis. The drugs are thought to be linear complexes of the type RS→Au→SR or R_3P→Au—PR_3. Changing the organic group R varies the solubility of the drug in lipids and affects how readily the drug is spread round the body.

There is some evidence that gold can exist as the Au^- auride ion. This has an electronic configuration $d^{10}s^2$ which is a stable arrangement. The compound CsAu has no metallic lustre and the solid does not conduct electricity like a metal. This compound is ionic Cs^+Au^- and is not an alloy. The Au^- ion has been confirmed in liquid ammonia solutions. Au^- is large and has been isolated in solids with several large cations.

Photography

The silver halides are of great importance in photography. A photographic film consists of a light sensitive emulsion of fine particles (grains) of silver salts in gelatine spread on a clear celluloid strip or a glass plate. The grain size is very important to photographers, as this affects the quality of the pictures produced. AgBr is mainly used as the light sensitive material. Some AgI is used in 'fast' emulsions (i.e. ones which can take photographs when the intensity of the light is low, or are used for photographing moving objects like racing cars where the exposure must be very short). AgCl may also be present in the emulsion.

The film is placed in a camera. When the photograph is exposed, light from the subject enters the camera and is focussed by the lens to give a sharp image on the film. The light starts a photochemical reaction by exciting a halide ion, which loses an electron. The electron moves in a conduction band to the surface of the grain, where it reduces a Ag^+ ion to metallic silver. In the early days of photography, exposures were long, often 10 or 20 minutes, so that sufficient silver was produced to give a negative picture. Parts of the subject which are light become black on the film. In modern photography only a short exposure of perhaps 1/100th of a second is used. In this short time, only a few atoms of silver (perhaps 10–50) are produced in each grain exposed to light. Parts of the film which have been exposed to the bright parts of the subject contain a lot of grains with some silver. Parts exposed to paler parts of the subject contain a few grains with some silver, whilst parts not exposed contain none. Thus the film contains a latent image of the subject. However, the number of silver atoms produced is so small that the image is not visible to the eye.

Next the film is placed in a developer solution. This is a mild reducing agent, usually containing quinol. Its purpose is to reduce more silver halide to Ag metal. Ag is deposited mainly where there are already some Ag atoms. Thus the developing process intensifies the latent image on the film so it becomes visible. The correct conditions for processing must be used to obtain an image of the required blackness. The important factors are the concentration of the developer solution, the pH, the temperature and the length of time that the film is kept in the developer solution.

If the film was brought out into daylight at this stage, the unexposed parts of the emulsion would turn black and thus destroy the picture. To prevent this happening any unchanged silver halides are removed by placing the film in a fixer solution. In the early days, strong ammonia was used to dissolve the unchanged AgBr. Though this was effective, the smell of ammonia in the confined space of a darkroom must have been unpleasant and harmful to the photographer. Nowadays a solution of sodium thiosulphate is used as fixer. It forms a soluble complex with silver halides.

$$AgBr + 2Na_2S_2O_3 \rightarrow Na_3[Ag(S_2O_3)_2] + NaBr$$

After fixing, the film can safely be brought out into daylight. Parts blackened by silver represent the light parts of the original picture. This is therefore a negative.

To obtain an image with light and dark the right way round, a print must be made. Light is passed through the negative onto a piece of paper coated with AgBr emulsion. This is then developed and fixed in the same way as before.

(+II) STATE

The (+II) state is the most stable and important for Cu. The cupric ion Cu^{2+} has the electronic configuration d^9 and has an unpaired electron. Its compounds are typically coloured due to d–d spectra and the compounds are paramagnetic. $CuSO_4 \cdot 5H_2O$ and many hydrated cupric salts are blue.

On strong heating, oxosalts such as $Cu(NO_3)_2$ decompose into CuO which is black. Very strong heating (>800 °C) gives Cu_2O. The addition of NaOH to a solution containing Cu^{2+} gives a blue precipitate of the hydroxide. The hydrated ion $[Cu(H_2O)_6]^{2+}$ is formed when the hydroxide or carbonate are dissolved in acid, or when $CuSO_4$ or $Cu(NO_3)_2$ are dissolved in water. This ion has the characteristic blue colour associated with copper salts, and has a distorted octahedral shape. There are two long bonds *trans* to each other, and four short bonds (Figure 27.1). This is called tetragonal distortion and is a consequence of the d^9 configuration. The octahedral arrangement causes crystal field splitting of the d orbitals on Cu into low energy e_g and higher energy t_{2g} levels. The nine d electrons are arranged $(e_g)^6$ and $(t_{2g})^3$. The three t_{2g} electrons occupy the $d_{x^2-y^2}$ and d_z^2 orbitals, two in one orbital and one in the other. Since the t_{2g} level is not symmetrically filled, Jahn–Teller distortion occurs. This removes the degeneracy of the e_g and t_{2g} orbitals (i.e. they are no longer the same energy). Thus the complex is distorted. The $d_{x^2-y^2}$ orbital is under the influence of four ligands approaching along the directions $+x$, $-x$, $+y$ and $-y$. The d_z^2 orbital is under the influence of only two ligands approaching along $+z$ and $-z$. Thus the energy of the $d_{x^2-y^2}$ orbital is raised more than that of the d_z^2 orbital. Because of this, the three electrons in the t_{2g} level are arranged $(d_z^2)^2\ (d_{x^2-y^2})^1$. Because the d_z^2 orbital contains two electrons the

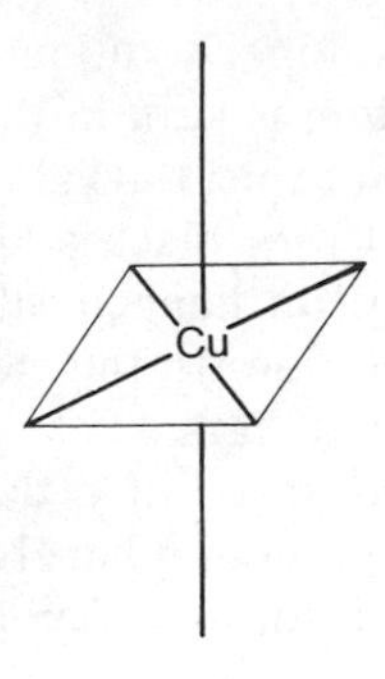

Figure 27.1 Tetragonally distorted octahedron, with four short bonds in a square planar arrangement, and two long *trans* bonds.

ligands approaching along the $+z$ and $-z$ directions are prevented from coming as close to the copper as those approaching along $+x$, $-x$, $+y$ and $-y$ (see Chapter 7.)

This distorted octahedral arrangement is common in copper compounds. For example, the halides CuX_2 have a distorted rutile (TiO_2) structure with a coordination number of 6. The bond lengths in the $Cu^{II}X_2$ halides are:

CuF_2	4 bonds 1.93 Å and 2 bonds 2.27 Å
$CuCl_2$	4 bonds 2.30 Å and 2 bonds 2.95 Å
$CuBr_2$	4 bonds 2.40 Å and 2 bonds 3.18 Å

Similar reasoning to that for the distortion of octahedral Cu^{2+} complexes also applies to octahedral complexes of any other metal ion where the e_g orbitals are not symmetrically filled, i.e. when they contain one or three electrons. Distortions of this kind are found in:

Co^{2+} and Ni^{3+} high-spin $(e_g)^6\,(t_{2g})^1$

and Cr^{2+} and Mn^{3+} low-spin $(e_g)^3\,(t_{2g})^1$

Aqueous Cu^{2+} solutions form many complexes with ammonia and amines, such as $[Cu(H_2O)_5NH_3]^{2+}$, $[Cu(H_2O)_4(NH_3)_2]^{2+}$, $[Cu(H_2O)_3(NH_3)_3]^{2+}$ and $[Cu(H_2O)_2(NH_3)_4]^{2+}$. It is difficult to add a fifth or sixth NH_3, though it is possible to make $[Cu(NH_3)_6]^{2+}$ using liquid ammonia as solvent. The reason for this is the Jahn–Teller effect which distorts the octahedral complex and makes the fifth and sixth bonds long and weak. Similarly with the bidentate ligand ethylenediamine (en) the complexes $[Cu(en)(H_2O)_4]^{2+}$ and $[Cu(en)_2(H_2O)_2]^{2+}$ are formed quite easily but $[Cu(en)_3]^{2+}$ is formed only with a large excess of ethylenediamine.

The halide complexes show two different stereochemistries. In $(NH_4)_2[CuCl_4]$ the $[CuCl_4]^{2-}$ ion is square planar, but in $Cs_2[CuCl_4]$

and $Cs_2[CuBr_4]$ the $[CuX_4]^{2-}$ ions have a slightly squashed tetrahedral shape.

Most Cu(+II) complexes and compounds have a distorted octahedral structure and are blue or green. The metal ion has the d^9 electronic configuration. This leaves only one 'hole' into which an electron may be promoted, so the spectra should be similar to the d^1 case (e.g. Ti^{3+}) and have a single broad absorption band. This is observed, and these complexes absorb in the region 11 000–16 000 cm^{-1}. However, octahedral complexes of Cu^{II} are appreciably distorted (Jahn–Teller effect) so there is more than one peak. These overlap so the band is not symmetrical. It is not possible to assign the peaks unambiguously. The anhydrous salts of CuF_2 and $CuSO_4$ are, perhaps surprisingly, white.

Tetrahedral $[CuCl_4]^{2-}$ ions are orange and square planar $[CuCl_4]^{2-}$ ions are yellow.

For many years no anhydrous nitrates of the transition metals were known. Heating the hydrated salts simply decomposed them. Water is a stronger ligand than the nitrate group, and so a hydrated nitrate prepared in aqueous solution will lose nitrate groups rather than water on heating. The remedy is to prepare anhydrous nitrates without water rather than attempting to remove water from hydrated salts. Many anhydrous nitrates have now been prepared using a non-aqueous solvent such as liquid N_2O_4.

Anhydrous copper(II) nitrate can be made by dissolving Cu metal in a solution of N_2O_4 in ethyl acetate. The reaction is vigorous, and $Cu(NO_3)_2 \cdot N_2O_4$ is crystallized from the solution. This has the structure $NO^+[Cu(NO_3)_3]^-$. Heating to 90 °C drives off N_2O_4 and gives blue anhydrous $Cu(NO_3)_2$. Anhydrous $Cu(NO_3)_2$ can be sublimed at 150–200 °C, and has an unusual crystal structure. There are two different crystalline forms. The structure of one of the two solid forms shows infinite chains of Cu and NO_3^- groups. Each Cu forms two short Cu—O bonds (1.9 Å), and the Cu is coordinated to six more oxygen atoms in other chains, forming longer bonds (2.5 Å), making Cu eight-coordinate (Figure 27.2). In the vapour, $Cu(NO_3)_2$ is monomeric.

Copper(II) acetate is dimeric and hydrated, $Cu_2(CH_3COO)_4 \cdot 2H_2O$. The structure (Figure 27.3) is similar to that of the carboxylate complexes of Cr^{II}, Mo^{II}, Rh^{II} and Ru^{II}, but there is an important difference. The structure consists of two Cu atoms each with a roughly octahedral structure. The four acetate groups act as bridging ligands between the two Cu atoms. Thus O atoms from the acetate groups occupy four positions (in a square plane) round each Cu. The fifth position round each Cu is occupied by the O atom from a water molecule. The other Cu atom occupies the sixth of the octahedral positions. So far the structure is the same as that of chromous acetate. The difference is that the Cu—Cu distance is 2.64 Å, which is significantly longer than the distance of 2.55 Å found in metallic copper. Thus Cu does **not** form a M—M bond, whereas Cr and the other metals which form carboxylate complexes do form M—M bonds. The Cu^{II} ions in this complex have a d^9 configuration with one unpaired electron. If the two Cu ions form a metal–metal bond these electrons will be paired

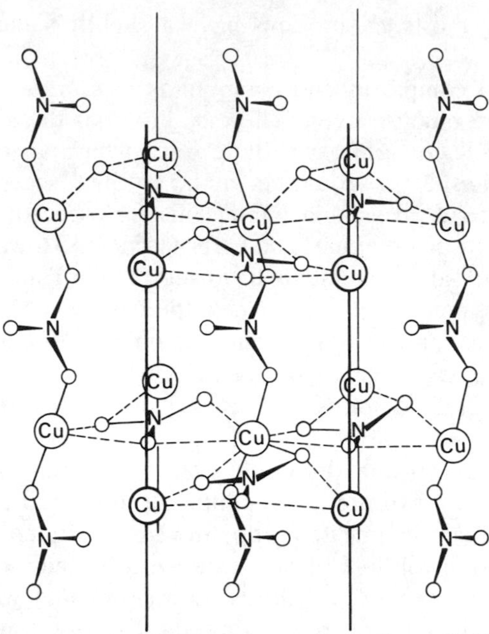

Figure 27.2 The crystal structure of one form of anhydrous $Cu(NO_3)_2$. (S.C. Wallwork, *Proc. Chem. Soc. (London)*, **34**, 1959.)

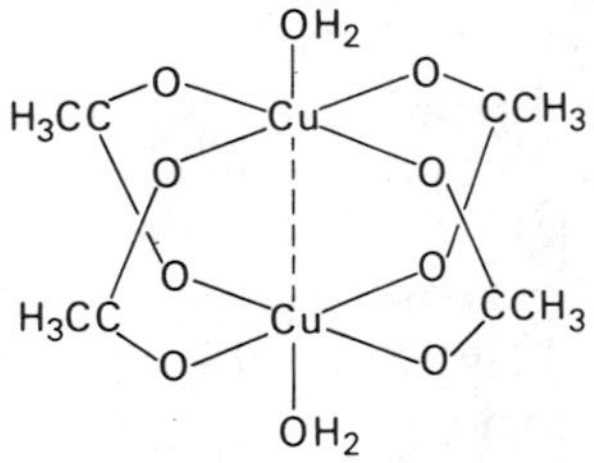

Figure 27.3 Structure of $(Cu(CH_3 \cdot COO)_2 \cdot H_2O)_2$.

and the complex will be diamagnetic. If they do not form a bond then the complex will be paramagnetic. However, the magnetic moment measured at 25 °C is 1.4 BM per Cu atom, rather than the spin only value of 1.73 BM. This suggests that there is an 'interaction' or weak coupling of the unpaired spins on the two Cu atoms in this complex. This is thought to involve lateral overlap of the $3d_{x^2-y^2}$ orbitals on the copper atoms, and is sometimes called δ bonding.

Ag(+II) is known as the fluoride AgF_2. This is a brown solid and is made by heating Ag in F_2. AgF_2 is a strong oxidizing agent, and a good fluorinating agent like CoF_3. It decomposes on heating:

$$AgF_2 \xrightarrow{\text{heat}} AgF + \tfrac{1}{2}F_2$$

Ag^{II} is more stable in complexes such as $[Ag(pyridine)_4]^{2+}$, $[Ag(dipyridyl)_2]^{2+}$ and $[Ag(\textit{ortho}\text{-phenanthroline})_2]^{2+}$ which form stable salts with non-reducing anions such as NO_3^- and ClO_4^-. They are usually prepared by oxidizing a solution of Ag^+ and the ligand with potassium persulphate. The complexes are square planar and paramagnetic.

A black oxide of formula AgO is formed by strong oxidation of Ag_2O in alkaline solution. Ag^{II} has a d^9 configuration and must be paramagnetic, but AgO is diamagnetic. AgO does not contain Ag(+II), but is a mixed oxide $Ag^IAg^{III}O_2$.

Au(+II) probably occurs in dithiolene compounds and in $[Au(B_9C_2H_{11})_2]^{2-}$, but otherwise it only exists as a transient intermediate.

R–C–S
‖ Au
R–C–S

(+III) STATE

The (+III) state is uncommon for Cu and Ag. In alkaline solution Cu^{2+} can be oxidized to $KCu^{III}O_2$, and if a fused mixture of $KCl/CuCl_2$ is fluorinated, $K_3[Cu^{III}F_6]$ is formed. Strong oxidation with periodic acid (H_5IO_6) gives $K_7Cu^{III}(IO_6)_2 \cdot 7H_2O$. Fluorination of a fused mixture of alkali metal halide and silver halide gives $M^+[Ag^{III}F_4]^-$, and electrolytic oxidation of Ag^+ can give impure Ag_2O_3. Oxidation of alkaline Ag_2O with persulphate gives the mixed oxide $Ag^IAg^{III}O_2$. Persulphate in the presence of periodate or tellurate ions gives compounds such as $K_6H[Ag^{III}(IO_6)_2]$ and $Na_6H_3[Ag^{III}(TeO_6)_2]$. These compounds are all unstable and are strong oxidizing agents.

In contrast, Au(+III) is the most common state for gold. There are few simple compounds, and these do not contain Au^{3+} ions. Au^{III} has a d^8 configuration like Pt^{II}, and like Pt it forms square planar complexes. These compounds decompose to the metal quite readily on heating.

$$[Au^{III}Cl_4]^- + OH^- \rightarrow Au(OH)_3 \xrightarrow{\text{dehydrate}} Au_2O_3 \xrightarrow{150\,^\circ C} Au + Au_2O + O_2$$

All the halides AuX_3 are known. $AuCl_3$ is made from the elements, or by dissolving gold in aqua regia and evaporating.

$$Au + HNO_3 + HCl \rightarrow H_3O^+[AuCl_4]^- \cdot 3H_2O \rightarrow AuCl_3$$

$AuBr_3$ is made from the elements and AuI_3 is made from the bromide. AuF_3 can only be formed with a strong fluorinating agent:

$$Au + BrF_3 \rightarrow AuF_3$$

The fluoride is made up of square planar AuF_4 units linked into a chain by *cis* fluoride bridges. The chloride and bromide are dimeric.

```
Cl      Cl      Cl
  \    /  \    /
    Au      Au
  /    \  /    \
Cl      Cl      Cl
```

The 'liquid gold' used to decorate picture frames, glass and ceramic ornaments is a chloro complex of Au^{III} dissolved in an organic solvent. The complex is probably polymeric, and possibly a cluster compound. When 'liquid gold' is heated it decomposes, leaving a film of metallic gold.

Hydrated gold oxide is amphoteric. It dissolves in alkalis to give salts such as sodium aurate $NaAuO_2 \cdot H_2O$, and in strong acids to give $H[AuCl_4]$, $H[Au(NO_3)_4]$ and $H[Au(SO_4)_2]$. Cationic Au(+III) complexes are also known, e.g. $[Au(NH_3)_4](NO_3)_3$ which is square planar. The complexes are generally square planar, and more stable than the simple compounds. The dialkyls R_2AuX where $X = Cl^-$, Br^-, CN^-, SO_3^{2-} are stable organometallic compounds with strong Au—C bonds. The halides are dimeric and the cyanides are tetrameric.

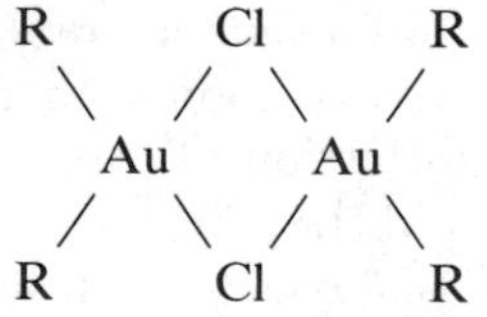

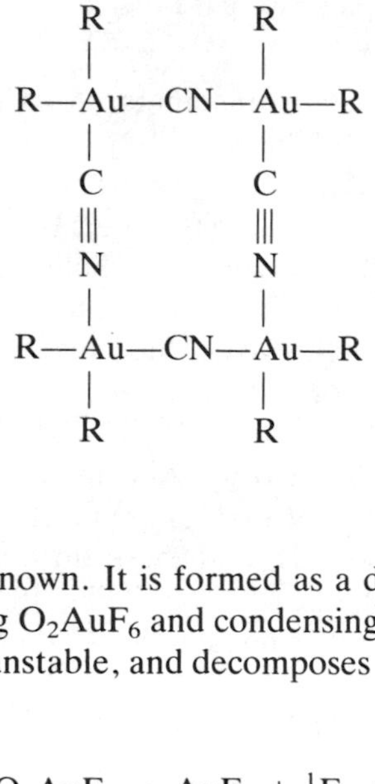

```
                          R           R
                          |           |
                        R—Au—CN—Au—R
R      Cl      R          |           |
  \   /  \   /            C           C
   Au      Au             |||         |||
  /   \  /   \            N           N
R      Cl      R          |           |
                        R—Au—CN—Au—R
                          |           |
                          R           R
```

(+V) STATE

AuF_5 is the only (+V) compound known. It is formed as a dark red diamagnetic polymeric solid by warming O_2AuF_6 and condensing the product on a 'cold finger'. The compound is unstable, and decomposes to AuF_3 and F_2 above 60 °C.

$$Au + 3F_2 + O_2 \xrightarrow[\text{high pressure}]{370\,°C} O_2AuF_6 \rightarrow AuF_5 + \tfrac{1}{2}F_2 + O_2$$

BIOLOGICAL ROLE OF COPPER

Copper is essential to life and adult humans contain about 100 mg. This is the third largest amount of a transition metal, after Fe (4 g) and Zn (2 g). Though small amounts of Cu are essential, larger amounts are toxic. About 4–5 mg of Cu are required daily in the diet, and deficiency in animals results in the inability to use iron stored in the liver. Thus the animal suffers from anaemia. The Cu is bound to proteins in the body either as metalloproteins or as enzymes. Examples include various oxidases and blue proteins. These include:

1. Amine oxidases (oxidation of amines).
2. Ascorbate oxidase (oxidation of ascorbic acid).
3. Cytochrome oxidase (acts with haem as the terminal oxidase step).
4. Galactose oxidase (oxidation of an OH group to CHO in the monosaccharide galactose).

Copper is also important in:

1. Lysine oxidase, which affects the elasticity of aortic walls.
2. Dopamine hydroxylase, which affects brain function.
3. Tyrosinase, which affects skin pigmentation.
4. Ceruloplasmin, which plays a role in Fe metabolism.

Wilson's disease is a hereditary shortage of ceruloplasmin, causing Cu accumulation in the liver, kidneys and brain. This disease is treated by feeding the patient with a chelating agent such as EDTA. The Cu forms a complex and is excreted. At the same time many other essential metals involved in other enzyme systems form EDTA complexes and are excreted. The treatment would thus upset many different enzyme systems, so the required metals must be replaced in the diet and treatment must be very carefully monitored.

Haemocyanin is a copper containing protein which is important as an oxygen carrier in some invertebrate animals. Despite its name it is a non-haem protein. The molecular weight is roughly one million, and the molecule contains two Cu^{II} ions. Despite the d^9 configuration of Cu^{2+} the molecule is diamagnetic because of strong antiferromagnetic coupling between the copper ions. Haemocyanin is found in the blood of snails, crabs, lobsters, octapuses, and scorpions. The oxygenated haemocyanins are blue coloured (unlike human blood), and have one oxygen molecule attached to two Cu atoms. Deoxygenated haemocyanin contains Cu^{I} and is also blue.

There are several blue proteins which contain Cu. They act as electron transfer agents by means of a Cu^{2+}/Cu^{+} couple. Their colour is much more intense than would be expected for d–d spectra. The colour is thought to be caused by charge transfer between Cu and S. Examples include plastocyanin and azurin. Plastocyanin occurs in the chloroplasts of green plants, has a molecular weight of about 10 500 and contains one Cu atom. It is important in photosynthesis as an electron carrier. Azurin is found in bacteria. It contains one Cu per molecule and is structurally similar to plastocyanin, but it has a molecular weight of about 16 000.

FURTHER READING

Ainscough, E.W. and Brodie, A.M. (1985) Gold chemistry and its medical applications, *Education in Chemistry*, **22**, 6–8.

Beinert, H. (1977) Structure and function of copper proteins, *Coordination Chem. Rev.*, **23**, 119–129.

Canterford, J.H. and Cotton, R. (1968) *Halides of the Second and Third Row Transition Elements*, Wiley, London.

Canterford, J.H. and Cotton, R. (1969) *Halides of the First Row Transition Elements*, Wiley, London.

Dirkse, T.P. (ed.) (1986) *Copper, Silver, Gold, Zinc, Cadmium, Mercury Oxides and Hydroxides*, Pergamon, Oxford.

Gerloch, M. (1981) The sense of Jahn–Teller distortions in octahedral copper(II) and other transition metal complexes, *Inorg. Chem.*, **20**, 638–640.

Gernscheim, H. (1977) The history of photography, *Endeavour*, **1**, 18–22.

Hathaway, B.J. and Billing, D.E. (1970) The electronic properties and stereochemistry of mono-nuclear complexes of the copper(II) ion, *Coordination Chem. Rev.*, **5**, 143.

Jardine, F.H. (1975) Copper I complexes, *Adv. Inorg. Chem. Radiochem.*, **17**, 115.

Johnson, B.G.F. and Davis, R. (1973) *Comprehensive Inorganic Chemistry*, Vol. 3 (Chapter 29: Gold), Pergamon Press, Oxford.

Karlin, K.D. and Gultneh, Y. (1985) Bioinorganic chemical modelling of dioxygen-activation copper proteins, *J. Chem. Ed.*, **62**, 983–987.

Karlin, K.D. and Zubieta, J. (eds) (1986) *Biological and Inorganic Copper Chemistry*, Vol. 1, Adenine Press.

Knutton, S. (1986) Copper – the enduring metal, *Education in Chemistry*, **23**, 135–137.

Massey, A.G. (1973) *Comprehensive Inorganic Chemistry*, Vol. 3 (Chapter 27: Copper), Pergamon Press, Oxford.

Siegl, H. (ed.) (1981) *Metal Ions in Biological Systems*, Vol. 13, Copper proteins, Marcel Dekker, New York.

Sorenson, R.J. (1989) Copper complexes as 'radiation recovery' agents, *Chemistry in Britain*, **25**, 169–170, 172.

Thompson, N.R. (1973) *Comprehensive Inorganic Chemistry*, Vol. 3 (Chapter 28: Silver), Pergamon Press, Oxford.

Valentine, J.S. and de Freitas, D.M. (1985) Copper–zinc dismutase, *J. Chem. Ed.*, **62**, 990–996.

West, E.G. (1969) Developments in copper smelting, *Chemistry in Britain*, **5**, 199–202.

The zinc group

28

Table 28.1 Electronic structures and oxidation states

Element		Electronic structure	Oxidation states*
Zinc	Zn	[Ar] $3d^{10}\,4s^2$	**II**
Cadmium	Cd	[Kr] $4d^{10}\,5s^2$	**II**
Mercury	Hg	[Xe] $4f^{14}\,5d^{10}\,6s^2$	I **II**

* The most important oxidation states (generally the most abundant and stable) are shown in bold. Other well-characterized but less important states are shown in normal type.

INTRODUCTION

These elements all have a $d^{10}\,s^2$ electronic arrangement and they typically form M^{2+} ions. However, many of their compounds are appreciably covalent. Hg(+II) compounds are more covalent and its complexes are more stable than is the case for Zn and Cd. Because these ions have a complete *d* shell, they do not behave as typical transition metals. Though the ions are divalent, they show only slight similarity with Group II elements. Thus Zn shows some similarities to Mg. However, Zn is more dense and less reactive due to its smaller radius and higher nuclear charge. Also Zn has a much stronger tendency to form covalent compounds. Zn and Cd are broadly similar in most of their properties. The behaviour of Hg differs appreciably from that of Zn and Cd. In many ways Hg is unique. It is a liquid at room temperature, it is noble, and it forms 'apparently univalent' mercurous compounds. Mercury is the only element in the group with a well established (+I) oxidation state. The increased stability of the lowest oxidation state in the heaviest element in the group is more typical of the *p*-block elements than the transition elements.

Zn is produced on a large scale. In 1988, 7.1 million tonnes of the metal were produced. It is mostly used as a metal for rustproofing, for casting and for making alloys. ZnO is also important commercially.

Zinc has an important role in several enzymes. Biologically it is the second most important transition metal.

ABUNDANCE AND OCCURRENCE

Zn occurs in the earth's crust to the extent of 132 ppm by weight. It is the twenty-fourth most abundant element. World production of Zn was 7.1 million tonnes in 1988. The main regions where it is mined are Canada 19%, the USSR 13.5%, Australia 11%, China and Peru 7%, and Spain· Mexico and the USA 4% each. Cd production in 1988 was 22 000 tonnes and Hg production 5700 tonnes. Cd and Hg are quite rare. In spite of this the elements are familiar because their extraction and purification are simple.

Table 28.2 Abundance of the elements in the earth's crust, by weight

	ppm	Relative abundance
Zn	76	24
Cd	0.16	65
Hg	0.08	66 =

When the earth was formed Zn was deposited as sulphides. ZnS is mined and is called sphaelerite in the USA and zinc blende in Europe. The structure is like that of diamond with half the positions occupied by S and half by Zn or some other metal. Sphaelerite almost always contains iron, and the formula may be written (ZnFe)S. This commonly occurs with galena PbS. Hydrothermal weathering of the sulphides gave deposits of carbonates and silicates. $ZnCO_3$ is another important ore. It is called smithsonite in the USA (after James Smithson the founder of the Smithsonian Institution in Washington DC), but $ZnCO_3$ is called calamine in Europe. Hemimorphite $Zn_4(OH)_2(Si_2O_7) \cdot H_2O$ is less important commercially but is an interesting example of a pyrosilicate.In 1988 the main sources of ores were Canada 19%, the USSR 13.5%, Australia 11%, and China and Peru 7% each. Cadmium ores are very rare. Cd is found as traces in Zn ores and it is extracted from these. Hg is mined as the rather scarce ore cinnabar HgS mainly in the USSR, Spain, Mexico and Algeria.

EXTRACTION AND USES

Extraction of zinc

Zinc ores (mainly ZnS) are concentrated by flotation, then roasted in air to give ZnO and SO_2. (The SO_2 is used to make H_2SO_4.) is extracted from the oxide by two different processes.

1. ZnO may be reduced by carbon monoxide at 1200 °C in a smelter. The reaction is reversible, and the high temperature is required to move the equilibrium to the right. At this temperature the Zn is gaseous. If

the gaseous mixture of Zn and CO_2 was simply removed from the furnace and cooled, then some reoxidation of Zn would occur. Thus the zinc powder obtained would contain large amounts of ZnO.

$$ZnO + CO \rightleftharpoons Zn + CO_2$$

Modern smelters minimize the reoxidation in two ways:
(a) by having excess carbon, so the CO_2 formed is converted to CO.
(b) by shock cooling the gases leaving the smelter so they do not have time to attain equilibrium. This rapid cooling is achieved by spraying the hot gas with droplets of molten lead. This gives 99% pure Zn. Any cadmium present can be separated by distillation.

2. Alternatively ZnS is heated in air at a lower temperature, yielding ZnO and $ZnSO_4$. These are dissolved in H_2SO_4. Zn dust is added to precipitate Cd, and then the $ZnSO_4$ solution is electrolysed to give pure Zn. The electrolytic process is expensive and is not used in the UK.

Extraction of cadmium

Cd is found as traces (2–3 parts per thousand) in most Zn ores, and is extracted from these. The ore is treated and yields a solution of $ZnSO_4$ containing a small amount of $CdSO_4$. Cd is recovered by adding a more electropositive metal (i.e. one higher in the electrochemical series) to displace it from solution. Zn powder is added to the $ZnSO_4/CdSO_4$ solution, when the Zn dissolves and Cd metal is precipitated. Zn is higher in the electrochemical series than Cd, and elements high in the series displace elements lower in the series.

$$Zn_{(solid)} + Cd^{2+}_{(solution)} \rightarrow Zn^{2+}_{(solution)} + Cd_{(solid)} \qquad E^\circ = 0.36\,V$$

The Cd concentrate so obtained is then dissolved in H_2SO_4, and purified by electrolysis. The Zn is recovered from the $ZnSO_4$ solution by electrolysis.

Extraction of mercury

Hg is also scarce. It is mined as the bright red coloured ore cinnabar HgS mainly in the USSR, Spain, Mexico and Algeria. Sometimes droplets of Hg are found in these ores. The ore is crushed and as HgS has a very high density ($8.1\,g\,cm^{-3}$) it is separated from other rocks and concentrated by sedimentation. If the ore is lean, it is heated in air. The Hg vapour formed is condensed, and the SO_2 is used to make H_2SO_4.

$$HgS + O_2 \xrightarrow{600\,°C} Hg + SO_2$$

Rich ores are heated with scrap iron or quicklime.

$$HgS + Fe \rightarrow Hg + FeS$$
$$4HgS + CaO \rightarrow 4Hg + CaSO_4 + 3CaS$$

continued overleaf

Hg obtained in this way may contain traces of other metals dissolved in it, particularly Pb, Zn and Cd. Very pure Hg may be obtained by blowing air through the metal at 250 °C, when traces of other metals form oxides. These oxides float on the surface and are easily removed:

1. By scraping them off the surface as scum
2. By treatment with dilute HNO_3, when the oxides dissolve
3. Hg can be purified and separated from the other metals and oxides by distillation. The boiling points are: Pb, 1751 °C; Zn, 908 °C; Cd, 765 °C; Hg, 357 °C.

Uses of zinc

Zn is used in large amounts for coating iron to prevent it from rusting. A thin coating of Zn may be applied electrolytically (galvanizing). Thicker layers may be applied by hot-dipping (dipping the metal in molten zinc). The latter process is misleadingly called 'hot galvanizing'. Alternatively the object may be coated with powdered Zn and heated (Sherardizing), or sprayed with molten Zn. Large amounts of Zn are used to make alloys. The most common alloy is brass (a Cu/Zn alloy with 20–50% Zn). Zinc is the most widely used metal for casting metal parts. Zinc is also used as the negative electrode in sealed 'dry' batteries (Leclanché cells, mercury cells and alkaline manganese cells). ZnO is sometimes used as a white pigment in paint. It is particularly bright as it absorbs UV light and re-emits it as white light.

Uses of cadmium

Most of the Cd produced is used for protecting steel from corrosion. It is applied electrolytically by Cd plating. Cd absorbs neutrons very well, and is used to make control rods for nuclear reactors. Cd is also used for alkaline Ni/Cd storage batteries used both in diesel locomotives, and also as the '*nicad*' rechargeable 'dry batteries' used in radios and electrical appliances. CdS is an important but expensive yellow pigment. This is used in paint.

Uses of mercury

The largest use of mercury is in electrolytic cells for the production of NaOH and Cl_2. The electrical industry uses Hg in mercury vapour street lights, switches and rectifiers. Historically Hg has been, and still is, used in the extraction of precious metals (particularly silver and gold) as amalgams. Phenyl mercuric acetate and other oganomercury compounds have fungicidal and germicidal properties. They are sometimes used in agriculture for treating seeds. Mercurous chloride $[Hg_2]Cl_2$ is used for treating club-root, a disease in brassicas (the cabbage family of plants). HgO has been used in antifouling paints for ships, etc. $HgCl_2$ is used to make organo derivatives.

All Hg compounds are toxic, but the organo compounds are extremely dangerous and have lasting ecological effects. Small scale uses of mercury include thermometers, barometers and manometers, amalgams, as a detonator (mercury fulminate) and in some medicines.

OXIDATION STATES

The elements in this group all have two *s* electrons beyond a completed *d* shell. Removal of the *s* electrons results in divalent compounds, and the (+II) oxidation state is characteristic of the group.

Hg(+I) mercurous compounds are important. The univalent ion Hg^+ does not exist, as mercurous compounds are dimerized. Mercurous chloride is therefore Hg_2Cl_2 and contains $[Hg—Hg]^{2+}$ ions. In these, the two Hg^+ species which have a $6s^1$ configuration are bonded together using their *s* electrons. Thus mercurous compounds are diamagnetic. Unstable species Zn_2^{2+} abd Cd_2^{2+} have been detected in fused mixtures such as $Cd/CdCl_2$, and $Cd/CdCl_2/AlCl_3$. A yellow compound $[Cd_2][AlCl_4]_2$ has been isolated from such melts, but it disproportionates instantly in water.

$$[Cd_2]^{2+} \rightarrow Cd^{2+} + Cd$$

The normal trend is that on descending a group the bonds become weaker because the orbitals are larger and more diffuse. Thus overlap is less effective. The reverse trend is observed in this group. The Hg—Hg bond is much stronger than the Cd—Cd bond. This is because the first ionization energy of Hg is much higher than for Cd (Table 28.4), and hence Hg shares electrons more readily.

Oxidation states higher than (+II) do not occur. This is because removal of more electrons would destroy the symmetry of a completed *d* shell.

SIZE

The ionic radii of the M^{2+} ions are appreciably smaller than for the corresponding elements in Group II. This is because Zn, Cd and Hg have 10 *d* electrons which shield the nuclear charge rather poorly. The difference in size accounts for the lack of similarity in properties between the two groups.

$$\begin{array}{ll} Ca^{2+} = 1.00\,\text{Å} & Zn^{2+} = 0.74\,\text{Å} \\ Sr^{2+} = 1.18\,\text{Å} & Cd^{2+} = 0.95\,\text{Å} \\ Ba^{2+} = 1.35\,\text{Å} & Hg^{2+} = 1.02\,\text{Å} \end{array}$$

The lanthanide contraction results in the size of the second and third row transition elements being nearly the same. With earlier pairs of transition elements, Zr/Hf and Nb/Ta, the sizes were particularly close. With Cd/Hg the effect of the lanthanide contraction is still felt, but to a much smaller degree. Hg^{2+} is larger than Cd^{2+} but the increase is size is

Table 28.3 Some physical properties

	Covalent radius (Å)	Ionic radius M^{2+} (Å)	Melting point (°C)	Boiling point (°C)	Density ($g\,cm^{-3}$)	Pauling's electro-negativity
Zn	1.25	0.740	420	907	7.14	1.6
Cd	1.41	0.95	321	765	8.65	1.7
Hg	1.44	1.02	−39	357	13.534	1.9

smaller than that between Zn^{2+} and Cd^{2+}. The chemical evidence is that Cd and Hg are dissimilar.

IONIZATION ENERGIES

The first ionization energy for these elements (Table 28.4) is considerably higher than for the corresponding Group II elements. This is because the atoms are smaller and the filled *d* level is poorly shielding. The filled 4*f* shell in Hg further increases the binding energy of the outer electrons, and the first ionization energy for Hg is greater than for any other metal. The second ionization energies are high, but M^{2+} ions are known for all three elements as the solvation or lattice energy is sufficient to offset this. Mercury tends to form covalent compounds. The third ionization energies are so high that (+III) compounds do not exist.

Table 28.4 Promotion and ionization energies

	Promotion energy $s^2 \rightarrow s^1p^1$	Ionization energy ($kJ\,mol^{-1}$)		
	($kJ\,mol^{-1}$)	1st	2nd	3rd
Zn	433	906	1733	3831
Cd	408	876	1631	3616
Hg	524	1007	1810	3302

GENERAL PROPERTIES

Zn, Cd and Hg show few of the properties associated with typical transition elements. This is because they have a complete *d* shell, which is not available for bonding.

1. Zn and Cd do not show variable valency.
2. They have a d^{10} electronic configuration and so cannot produce *d*–*d* spectra. Thus many of their compounds are white. However, some compounds of Hg(+II) and a smaller number of Cd(+II) are highly coloured due to charge transfer from the ligands to the metal. (The

metals of this group are smaller and thus have a greater polarizing power than Group II metals. This increases the chance of covalency and also the chance of charge transfer.)

3. The metals are relatively soft compared with the other transition metals. This is probably because the *d* electrons do not participate in metallic bonding.
4. The melting and boiling points are very low. This explains why the metals are more reactive than the copper group, even though the ionization energies for the two groups suggest the reverse. (Nobility is favoured by a high heat of sublimation, a high ionization energy and a low heat of hydration.)

Mercury is the only metal which is liquid at room temperature. The reason for this is that the very high ionization energy makes it difficult for electrons to participate in metallic bonding. The liquid has an appreciable vapour pressure at room temperature. Thus exposed mercury surfaces should always be covered (for example with toluene), to prevent vaporization, and hence poisoning. The gas is unusual because it is monatomic like the noble gases.

Similarities between Group II elements with an outer electronic structure s^2 and the zinc group an outer electronic structure $d^{10}s^2$ are slight. Both groups are divalent. The hydrated sulphates are isomorphous, and double salts such as $K_2SO_4 \cdot HgSO_4 \cdot 6H_2O$ are analogous to $K_2SO_4 \cdot MgSO_4 \cdot 6H_2O$. However, the zinc group is more noble, more covalent, has a much greater ability to form complexes and is less basic.

Zn and Cd are silvery solids which tarnish rapidly in moist air. Hg is a silvery liquid and does not tarnish readily. Zn and Cd dissolve in dilute non-oxidizing acids, liberating H_2, but Hg does not. All three metals react with oxidizing acids such as concentrated HNO_3 and concentrated H_2SO_4, forming salts and evolving a mixture of oxides of nitrogen and SO_2. Under these conditions Hg forms mercuric Hg^{II} salts, but with dilute HNO_3a mercurous salt $Hg_2(NO_3)_2$ containing Hg^I is slowly formed.

Zn is the only element in the group which shows any amphoteric properties, and it is soluble in alkalis,forming zincates. These are formulated $Na_2[Zn(OH)_4]$, $Na[Zn(OH)_3 \cdot H_2O]$ or $Na[Zn(OH)_3 \cdot (H_2O)_3]$ similar to aluminates.

All three metals form oxides MO, sulphides MS and halides MX_2 by heating the elements. On stronger heating HgO decomposes, and this has been used as a preparation of O_2.

$$Hg + O_2 \xrightarrow{350\,°C} HgO \xrightarrow{400\,°} Hg + O_2$$

All three elements form insoluble sulphides. In qualitative analysis CdS (yellow) and HgS (black) are precipitated by passing H_2S into acidified solutions. ZnS (white) is more soluble and is precipitated by H_2S only from alkaline solutions. $ZnCl_2$ and $CdCl_2$ are ionic, but $HgCl_2$ is covalent. The energy to promote an *s* electron to a *p* level prior to forming two covalent bonds decreases from Zn to Cd, but rather surprisingly increases from Cd to Hg. Zn and Cd react with P to give phosphides, but Hg does not.

All three metals form alloys with several other metals. Those formed by Cu and Zn are called brass, and different brasses have from 20% to 50% Zn. These are commercially important. Alloys of other metals with Hg are called amalgams. Sodium amalgam is produced in the mercury cathode cell in the manufacture of NaOH. Both zinc and sodium amalgams are used as strong reducing agents in the laboratory. Many of the first row transition elements do not form amalgams, and Mn, Cu and Zn are the only ones to do so. The heavier transition elements form amalgams quite readily.

In a vertical group in the *d*-block the second and third row elements are very similar in both size and chemical properties and differ from the first row element. In this group Zn and Cd are very similar and differ considerably from Hg.

STANDARD REDUCTION POTENTIALS (VOLTS)

Acid solution

Oxidation state	+II	+I	0
Zn	$Zn^{2+} \xrightarrow{-0.76} Zn$		
Cd	$Cd^{2+} \xrightarrow{-0.40} Cd$		
Hg	$Hg^{2+} \xrightarrow{+0.91} Hg_2^{2+} \xrightarrow{+0.79} Hg$; $Hg^{2+} \rightarrow Hg$ +0.85		

The reactivity decreases from Zn to Cd to Hg. This is shown by their standard reduction potentials. Zn and Cd are electropositive metals. Hg has a positive potential and is therefore quite noble. The large difference between Cd and Hg may be partly explained by Hg having the highest first ionization energy of any metal, and by the higher solvation energy of Cd^{2+}.

$$Zn^{2+} + 2e \rightarrow Zn \qquad E^\circ = -0.76\,V$$

$$Cd^{2+} + 2e \rightarrow Cd \qquad E^\circ = -0.40\,V$$

but

$$Hg^{2+} + 2e \rightarrow Hg \qquad E^\circ = +0.85\,V$$

OXIDES

ZnO is the only oxide of commercial importance. Its main use is in the production of rubber since it shortens the time taken for vulcanization to occur. ZnO is also used as a white pigment in paint. It is much less used for this purpose than is TiO_2, which has a higher refractive index and hence a better covering power. ZnO is the starting point for making other Zn compounds such as zinc stearate and zinc palmitate. These two compounds are 'soaps' and are used to stabilize plastics and to make paint dry. World production of ZnO was 353 300 tonnes in 1985.

Basic properties increase down the group. ZnO is amphoteric. It dissolves in acids, forming salts, and in alkali, forming zincates such as $[Zn(OH)_4]^{2-}$ and $[Zn(OH)_3 \cdot H_2O]^-$. Addition of alkali to an aqueous solution of a Zn^{2+} salt first gives a white gelatinous precipitate of $Zn(OH)_2$, which redissolves in excess alkali, forming zincates. CdO is largely basic, but with very strong alkali $Na_2[Cd(OH)_4]$ is formed. Both $Zn(OH)_2$ and $Cd(OH)_2$ dissolve in excess ammonia, giving ammine complexes. HgO is completely basic.

All three oxides are formed by direct combination of the elements or by heating the nitrates. ZnO and CdO both sublime, showing that they are appreciably covalent. HgO does not sublime as it decomposes on heating.

$$2HgO \xrightarrow{\text{heat} >400\,^{\circ}C} 2Hg + O_2$$

The thermal stability of the oxides thus decreases from Zn to Cd to Hg.

ZnO is white when cold but turns yellow on heating. It returns to white on cooling. CdO may be yellow, green or brown at room temperature, depending on its previous heat treatment. However, it is white at liquid air temperature. These divalent compounds have a complete *d* shell, so the colour is not from *d*–*d* spectra. Their colour is due to defects in the solid structure. (On heating ZnO it loses O.) The number of defects increases with temperature and is zero at absolute zero (see Chapter 3). HgO exists in red and yellow forms.

Addition of a base to a solution of the salts precipitates $Zn(OH)_2$, $Cd(OH)_2$ and a yellow form of HgO, not $Hg(OH)_2$. The yellow form of HgO has the same crystal structure as the more common red form which is formed by heating the elements or $Hg(NO_3)_2$. The difference in colour is due to the particle size.

When $Zn(OH)_2$ and $Cd(OH)_2$ are treated with H_2O_2 they form peroxides which have variable compositions.

DIHALIDES

All 12 dihalides MX_2 are known. The fluorides have appreciably higher melting points than the other halides, and the fluorides are ionic. The melting points of the chlorides, bromides and iodides are fairly low. This suggests that they are partly covalent.

ZnF_2, CdF_2 and HgF_2 are white solids and are considerably more ionic

Table 28.5 Dihalides and their melting points

ZnF_2	872 °C	CdF_2	1049 °C	HgF_2	645 °C (decomp.)
$ZnCl_2$	283 °C	$CdCl_2$	568 °C	$HgCl_2$	276 °C
$ZnBr_2$	394 °C	$CdBr_2$	567 °C	$HgBr_2$	236 °C
ZnI_2	446 °C	CdI_2	387 °C	HgI_2	259 °C

and have higher melting points than the other halides. They are not very soluble in water. This is partly because of their higher lattice energies, and partly because they do not form halogen complexes in solution. ZnF_2 has the rutile (TiO_2) structure in which Zn^{2+} is octahedrally surrounded by six F^- ions. The larger Cd^{2+} and Hg^{2+} ions are eight-coordinate with a fluorite (CaF_2) structure.

$ZnCl_2$, $ZnBr_2$ and ZnI_2 may be considered as close-packed arrays of halide ions, with Zn^{2+} ions occupying one quarter of the tetrahedral holes. The CdX_2 compounds are close-packed arrays of halide ions with Cd^{2+} occupying half the octahedral holes. There is a considerable amount of polarization, and the crystal lattices are not completely regular as expected for ionic compounds. $CdCl_2$ and CdI_2 form slightly different layer lattices, in which Cd^{2+} ions occupy all the octahedral holes in one layer, and none in the next. This illustrates that they are partly covalent (see Chapter 3). In contrast $HgCl_2$, $HgBr_2$ and HgI_2 are covalent with low melting points. $HgCl_2$ solid contains linear Cl—Hg—Cl molecules with a bond length Hg—Cl of 2.25 Å. There is little interaction between the Hg and Cl atoms other than between the Hg and the two Cl atoms it is closely associated with (interatomic distance 3.34 Å). $HgBr_2$ and HgI_2 form layer lattices.

The halides are all white except for $CdBr_2$ which is pale yellow, and HgI_2 which exists in red and yellow forms. The colour is due to charge transfer.

The chlorides, bromides and iodides of Zn and Cd are hygroscopic, and are very soluble in water. The solubilities are:

$ZnCl_2$	432 g in 100 g water at 25 °C
$CdCl_2$	140 g in 100 g water at 20 °C

The high solubility is partly because the crystal lattice is not very strong (hence the low melting points). Another reason for the high solubility is that the metal ions form a variety of complexes in solution such as $[Zn(H_2O)_6]^{2+}$, $ZnCl^+_{(hydrated)}$, $ZnCl_{2(hydrated)}$ and $[ZnCl_4(H_2O)_2]^{2-}$. Solutions of Zn^{2+} and Cd^{2+} salts are acidic, because of hydrolysis:

$$H_2O + [Zn(H_2O)_6]^{2+} \rightarrow [Zn(H_2O)_5OH]^+ + H_3O^+$$

Concentrated solutions of $ZnCl_2$ are corrosive, and dissolve paper. $ZnCl_2$ is commercially important and is used for treating textiles. $ZnCl_2$ is also used as a flux for soldering. This is sometimes called *killed salts* and it dissolves metal oxides, thus allowing the solder to stick to a clean metal surface.

Zn salts are usually hydrated. Cd salts are less hydrated and when the halides dissolve they do not ionize completely, and may undergo self-complexing. Thus CdI_2 may give a mixture of hydrated Cd^{2+}, CdI^+, CdI_2^- and CdI_4^- in solution, the proportions depending on the concentration. Mercuric salts are usually anhydrous and do not ionize appreciably on dissolution in water. $HgCl_2$ is called *corrosive sublimate*. It was made by heating $HgSO_4$ and NaCl and has been used as an antiseptic since the middle ages. It is very poisonous. However, calomel Hg_2Cl_2 is used in medicine as a powerful laxative.

COMPLEXES

Zn^{2+} and Cd^{2+} form complexes with O donor ligands and also with N and S donor ligands and with halide ions. Hg(II) forms complexes with N, P and S donor ligands, but is reluctant to bond to O. The stability of Hg(+II) complexes is much greater than that of the other two elements. This is unusual because smaller ions usually complex best. No complexes are known with π bonding ligands such as CO, NO or alkenes. However, Zn and Cd do form complexes with CN^-, e.g. $[Zn(CN)_4]^{2-}$. Zn complexes are usually colourless, but Hg complexes (and to a lesser extent Cd complexes) are often coloured because of charge transfer. Coordination numbers from 2 to 8 are known. Since the elements have a d^{10} configuration there is no crystal field stabilization energy.

Zn^{II} and Cd^{II} occur largely in four-coordination as tetrahedral complexes. Many tetrahedral complexes are known (e.g. $[MCl_4]^{2-}$, $[M(H_2O)_4]^{2+}$, $[M(NH_3)_4]^{2+}$, $[M(NH_3)_2(Cl)_2]$, $[Zn(CN)_4]^{2-}$, $[Zn(pyridine)_2Cl_2]$ and $[Cd(pyridine)_2Cl_2]$). In $[Zn(NCS)_4]^{2+}$ the ligand is bonded through N, but in $[Cd(SCN)_4]^{2+}$ the ligands bond through S.

Zn and Cd form several six-coordinate octahedral complexes such as $[M(H_2O)_6]^{2+}$, $[M(NH_3)_6]^{2+}$, $[M(ethylenediamine)_3]^{2+}$ and $[Cd(\textit{ortho}\text{-}phenanthroline)_3]^{2+}$. The octahedral complexes of Zn are not very stable, but Cd forms octahedral complexes more readily and they are more stable than those of Zn because Cd is larger.

Most Hg^{II} complexes are octahedral. These are appreciably distorted with two short bonds and four long bonds. In the extreme this distortion results in only two bonds. Examples of this are the compounds $Hg(CN)_2$ and $Hg(SCN)_2$ and the complex $[Hg(NH_3)_2]Cl_2$. The latter contains the linear $[H_3N—Hg—NH_3]^{2+}$ ion. Hg^{II} also forms some tetrahedral complexes (e.g. $[Hg(SCN)_4]^{2-}$ and halide complexes such as $K_2[HgI_4]$). The latter is used as Nessler's reagent for the detection and quantitative determination of ammonia in solution. Nessler's reagent gives a yellow colour or brown precipitate with concentrations as low as 1 part per million of NH_3. This test is used on drinking water. The presence of NH_4^+ ions in water at this concentration is not in itself harmful, but it may indicate contamination of the water with sewage. Mercury(II) chloride in solution is mainly $HgCl_2$, but $[HgCl_4]^{2-}$ can be formed if Cl^- are in excess. In the Hg complexes, two ligands are held more strongly than the others, and some ammine complexes lose ammonia from the solid.

$$[Hg(NH_3)_4](NO_3)_2 \rightarrow [Hg(NH_3)_2](NO_3)_2 + 2NH_3$$

Two-coordinate complexes are rare for Zn and Cd.

Three-coordination is rare but an example is $[HgI_3]^-$. Five-coordination is not common, but examples include $[CdCl_5]^{3-}$ and $[Zn(terpyridyl)Cl_2]$, which have a trigonal bipyramid shape. There are a few complexes with coordination numbers of 7 and 8. Examples of coordination number 8 are $[Zn(NO_3)_4]^{2-}$ and $[Hg(NO_2)_4]^{2-}$. These involve bidentate O donor ligands with a small 'bite' such as NO_3^- and NO_2^-, in which two O atoms may occupy coordination positions.

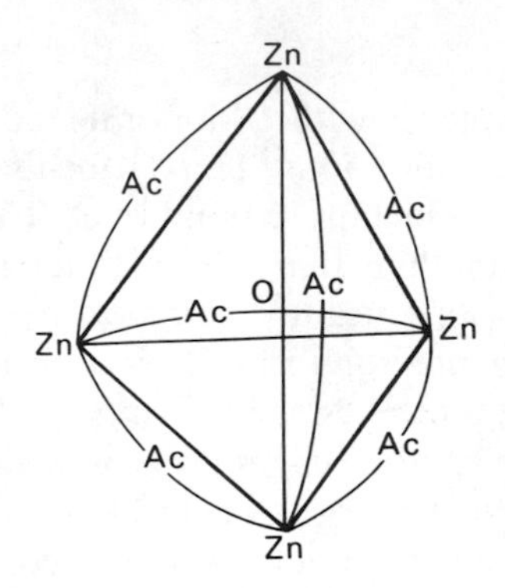

Figure 28.1 Structure of basic zinc acetate $(CH_3COO)_6 \cdot Zn_4O$.

Zinc forms basic zinc acetate $(CH_3COO)_6 \cdot Zn_4O$, a complex very like basic beryllium acetate both in structure and properties (see Figure 28.1). The zinc complex hydrolyses more readily than the beryllium complex, as zinc can increase its coordination number to 6.

MERCURY(+I) COMPOUNDS

Only a few Hg(I) mercurous compounds are known. They contain the mercurous ion $(Hg{-}Hg)^{2+}$, not Hg^+. The two Hg atoms are bonded together using the 6*s* orbitals. Mercury is unique in forming stable dinuclear metal ions. The only other metals which form dinuclear ions are $(Zn{-}Zn)^2$ and $(Cd{-}Cd)^{2+}$. These ions are unstable and have only been detected spectroscopically in melts of $Zn/ZnCl_2$ and $Cd/CdCl_2$.

Mercury(I) compounds can be made by reducing the mercury(II) salt with the metal. Alternatively mercurous nitrate can be made by dissolving Hg in dilute HNO_3. Other salts are made from this by adding $NaHCO_3$ to precipitate Hg_2CO_3, and treating this with HCl, HF, H_2SO_4, etc. to produce the salt required.

All four mercurous halides are known. $[Hg_2]F_2$ is hydrolysed by water, and then disproportionates.

$$[Hg_2]F_2 + 2H_2O \rightarrow 2HF + \underset{\text{unstable}}{[Hg_2](OH)_2}$$

$$[Hg_2](OH)_2 \xrightarrow{\text{disproportionates}} Hg^{II}O + Hg^0 + H_2O$$

Hg_2Cl_2, Hg_2Br_2 and Hg_2I_2 are insoluble in water. The nitrate $[Hg_2](NO_3)_2 \cdot 2H_2O$ is soluble in water, and contains the linear $[H_2O{-}Hg{-}Hg{-}OH_2]^{2+}$ ion, and $[Hg_2](ClO_4)_2 \cdot 4H_2O$ is also soluble. No oxide, hydroxide or sulphide is known.

The standard reduction potentials are so close that oxidizing agents like HNO_3 will convert Hg to Hg^{2+} rather than Hg(I) if the oxidizing agent is present in excess.

$$Hg^{2+} \rightarrow Hg \qquad E^\circ = +0.85\,V$$

and

$$[Hg_2]^{2+} \rightarrow Hg \qquad E^\circ = +0.79\,V$$

The reduction potential diagram shows that $[Hg_2]^{2+}$ is stable to disproportionation by a small margin under standard conditions.

$$Hg^{2+} + Hg \rightleftharpoons Hg_2^{2+} \qquad E^\circ = +0.13\,V$$

The equilibrium constant K for the reaction can be calculated from the potential E°:

$$E^\circ = \frac{RT}{nF} \ln K$$

$$K = \frac{\text{concentration } Hg_2^{2+}}{\text{concentration } Hg^{2+}} = \text{approx. } 170$$

Thus solutions of mercurous compounds contain one Hg^{2+} ion for every 170 $[Hg_2]^{2+}$ ions. The equilibrium is finely balanced. If a reagent is added which removes Hg^{2+} from this mixture (by forming either an insoluble compound or a complex) then the equilibrium moves to the left. Under these conditions mercurous ions disproportionate completely into Hg^{2+} and Hg. For example, the addition of OH^- or S^{2-} ions gives a precipitate of HgO and Hg, or HgS and Hg. The absence of several mercurous compounds such as hydroxides, sulphides and cyanides is because they precipitate Hg^{2+}, and allow disproportionation to occur. Conversely, mercurous compounds can be made by heating mercuric compounds with at least a 50% excess of Hg.

$$HgCl_2 + Hg \rightarrow [Hg_2]^{2+}Cl_2$$

In qualitative analysis the presence of Hg_2Cl_2 in solution is confirmed by the formation of a black precipitate when treated with NH_4OH. The dark residue may be $Hg(NH_3)_2Cl_2$, $HgNH_2Cl$ or $Hg_2NCl \cdot H_2O$ or a mixture of all three together with Hg.

Mercurous ions form few complexes. This is because of the large size of the binuclear ion, and because most ligands cause disproportionation.

Mercury is unique in the Hg(+I) state in that it consists of two directly linked metal atoms. The mercurous ion thus has the structure $[Hg—Hg]^{2+}$. Evidence for this comes from several sources:

X-ray diffraction

The crystal structures of several mercurous compounds have been determined by X-ray diffraction. For example, consider $Hg^{I}Cl$. If the compounds comprised Hg^+ and Cl^- the structure should contain alternate Hg^+ and Cl^- ions. It does not have this structure, but contains a linear arrangement Cl—Hg—Hg—Cl. The other Hg^{I} compounds also have Hg—Hg pairs rather than discrete Hg^+ ions. The Hg—Hg bond length varies in different compounds: Hg_2F_2, 2.51 Å; Hg_2Cl_2, 2.53 Å; Hg_2Br_2,

2.49 Å; Hg_2I_2, 2.69 Å; $Hg_2(NO_3)_2 \cdot 2H_2O$, 2.54 Å; Hg_2SO_4, 2.50 Å. These are all much shorter than the Hg—Hg distance of 3.00 Å in solid mercury.

Equilibrium constant

Equilibrium constants can be measured and provide evidence about the species present. Mercurous compounds can often be made from the corresponding mercuric compound by treatment with mercury. If the reaction is:

$$Hg(NO_3)_2 + Hg \rightarrow 2HgNO_3$$
$$Hg^{2+} + Hg \rightarrow 2Hg^{+}$$

then by the law of mass action

$$\frac{[Hg^{+}]^2}{[Hg^{2+}]} = \text{constant}$$

Experiments have shown this to be untrue. If, however:

$$Hg^{2+} + Hg \rightarrow (Hg_2)^{2+} \qquad \text{then} \qquad \frac{[Hg_2]^{2+}}{[Hg^{2+}]} = \text{constant}$$

This has been verified experimentally, thus proving that mercurous ions are $(Hg_2)^{2+}$, that is $(Hg—Hg)^{2+}$.

Concentration cell

Measurement of the e.m.f. of a concentration cell of mercurous nitrate shows that the mercurous ion carries two positive charges. For the cell below, the potential $E°$ was measured and found to be 0.029 V at 25 °C.

Hg	0.005 M mercurous nitrate in M/10 HNO_3 (*1*)	0.05 M mercurous nitrate in M/10 HNO_3 (*2*)	Hg

$$E = \frac{2.303RT}{nF} \log \frac{c_2}{c_1}$$

Substituting values in this equation:

$$0.029 = \frac{0.059}{n} \log \frac{0.05}{0.005}$$

hence $n = 2.0$

The number of charges on the ion n can be calculated since R the gas constant, T the absolute temperature, F the Faraday, and c_1 and c_2 the concentrations of the solutions, are all known. The value of n was 2, confirming $(Hg_2)^{2+}$.

Raman spectra

The Raman spectrum for mercurous nitrate contains the lines characteristic of the NO_3^- group. Similar lines appear in the spectra of many other nitrates. Mercurous nitrate has an extra line in the spectrum at $171.7\,cm^{-1}$ which is attributed to the Hg—Hg bond. Homonuclear stretching of a diatomic species is Raman active but not infra-red active. This was the first example of Raman spectra identifying a new species.

Magnetic properties

All mercurous compounds are diamagnetic both in the solid state and in solution. Hg^+ would have an unpaired electron and would be paramagnetic, but in $[Hg—Hg]^{2+}$ the electrons are all paired and it should therefore be diamagnetic.

Cryoscopic measurements

The depression of freezing point produced depends on the number of particles dissolved in the liquid. The observed depression fits for mercurous nitrate ionizing into Hg_2^{2-} and two NO_3^- ions, not into Hg^+ and NO_3^-.

POLYCATIONS

The compounds $Hg_3(AlCl_4)_2$ and $Hg_4(AsF_6)_2$ contain cations with three and four mercury atoms joined together in an almost linear chain. They can be made by reactions such as:

$$2Hg + HgCl_2 + 2AlCl_3 \rightarrow Hg_3(AlCl_4)_2$$
$$3Hg + 3AsF_5 \rightarrow Hg_3(AsF_6)_2 + AsF_3$$
$$4Hg + 3AsF_5 \rightarrow Hg_4(AsF_6)_2 + AsF_3$$

Their crystal stuctures give the following bond lengths in the cations:

$$\left[Hg \xrightarrow{2.55\,Å} Hg \xrightarrow{2.55\,Å} Hg\right]^{2+} \qquad \left[Hg \xrightarrow{2.57\,Å} Hg \xrightarrow{2.70\,Å} Hg \xrightarrow{2.57\,Å} Hg\right]^{2+}$$

The bond lengths, though not identical, are within the range shown in most mercurous compounds. The nature of the bonding in these ions is of interest. In the mercurous Hg_2^{2+} ion the two Hg atoms are bonded together using the 6*s* orbital. This idea cannot be extended to three or four Hg atoms. If the mercurous ion was made from a Hg^{2+} ion with a Hg atom coordinated to it $Hg^{2+} \leftarrow Hg$, then this might be extended to the trimer $Hg \rightarrow Hg^{2+} \leftarrow Hg$. However, it is hard to see how a linear tetramer could be formed. The answer may lie in multi-centre bonding. An unusual solid $[Hg_{2.85}(AsF_6)]_n$ can be made in liquid SO_2:

$$6n(Hg) + 3n(AsF_5) \rightarrow [Hg_{2.85}(AsF_6)]_n + n(AsF_3)$$

The crystal structure contains chains of Hg atoms in the *a* and *b* directions. The compound conducts electricity almost as well as mercury, and it becomes a superconductor at low temperatures.

ORGANOMETALLIC COMPOUNDS

The first organometallic compound was Zeise's salt $K[Pt(C_2H_4)Cl_3]H_2O$. This was isolated in 1825, but remained a chemical curiosity. The first useful organometallic compounds were prepared by Sir Edward Frankland in 1849, and they found use in organic synthesis. These were of two types:

1. zinc alkyls ZnR_2 and
2. alkyl zinc halides RZnX.

They were originally prepared as follows:

$$\mathrm{EtI + Zn \xrightarrow[N_2]{\text{inert atmosphere}} EtZnI \xrightarrow{\text{heat}} Et_2Zn + ZnI_2}$$

The reaction works best with alkyl or aryl iodides, but it is cheaper to use RBr with a Zn/Cu alloy. CO_2 may also be used as the inert atmosphere. The products ZnR_2 are either covalent liquids or low melting solids. The alkyl zinc halides RZnX apparently have the structure:

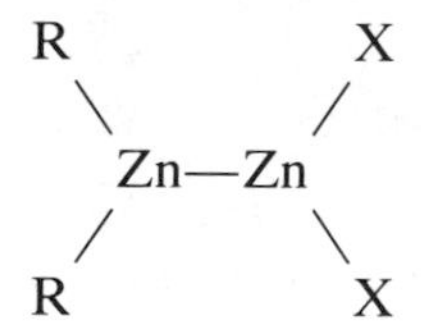

These were used in organic syntheses before Grignard reagents RMgX were discovered. Grignard compounds are usually more convenient to make and use.

Organo compounds of Zn and Cd (and also Li and Mg) all decompose rapidly on contact with water and air. The Zn and Cd alkyls are conveniently prepared from Grignard or alkyl lithium reagents, or alkyl mercury compounds.

$$\mathrm{CdCl_2 + 2RMgCl \rightarrow CdR_2 + 2MgCl_2}$$

$$\mathrm{ZnCl_2 + 2RLi \rightarrow ZnR_2 + 2LiCl}$$

$$\mathrm{Zn + HgR_2 \rightarrow ZnR_2 + Hg}$$

If required the Grignard-type compounds can be made as follows:

$$\mathrm{CdR_2 + CdI_2 \rightarrow 2RCdI}$$

R_2Zn and R_2Cd react in a similar way to Grignard reagents and lithium alkyls, but are less reactive and are unaffected by CO_2. Their lesser reactivity allows selective alkylation:

$$\mathrm{R_2Zn + EtOH \rightarrow RZnOEt + RH}$$

The most important use of organocadmium compounds is to produce ketones. This also exploits the lower reactivity of the Cd alkyls, as Grignard reagents would react further with the ketone produced.

$$R_2Cd + 2CH_3COCl \rightarrow 2\ \begin{matrix} R \\ \diagdown \\ CO \\ \diagup \\ CH_3 \end{matrix} + CdCl_2$$

A large number of organomercury compounds are known of types R_2Hg and monomeric RHgX. They are much more stable to air and water than the zinc compounds. They are easily made by Grignard reactions with $HgCl_2$ in tetrahydrofuran, or by reacting HgX_2 with a hydrocarbon:

$$HgCl_2 + RMgBr \rightarrow RHgCl + \tfrac{1}{2}MgCl_2 + \tfrac{1}{2}MgBr_2$$
$$RHgCl + RMgBr \rightarrow R_2Hg + \tfrac{1}{2}MgCl_2 + \tfrac{1}{2}MgBr_2$$
$$HgX_2 + RH \rightarrow RHgX + HX$$

Arylmercury compounds may be prepared from mercuric acetate:

$$C_6H_6 + (CH_3COO)_2Hg \rightarrow \underset{\text{phenylmercuric acetate}}{C_6H_5 \cdot HgOOC \cdot CH_3}$$

Organomercury compounds are important for the preparation of organometallic compounds of Groups I and II, Al, Ga, Sn, Pb, Sb, Bi, Se, Te, Zn and Cd.

$$R_2Hg + 2Na \rightarrow 2RNa + Hg$$

HgR_2 compounds are covalent liquids or low melting solids They are extremely poisonous. RHgX compounds are solids, and they are extremely toxic. Many RHgX compounds, particularly EtHgCl, PhHgCl and PhHgOOC > CH_3, and $Hg(Me)_2$ have been widely used for treating seeds, and as pesticides and fungicides.

Mercuric salts will add to double bonds in alkenes.

$$HgCl_2 + \rangle C{=}C\langle \rightarrow Cl{-}\rangle C{-}C\langle{-}HgCl$$

Mercuric salts also act as catalysts in the hydration of alkynes (acetylenes). The production of ethanal (acetaldehyde) from ethyne (acetylene) is commercially important.

$$CH{\equiv}CH \xrightarrow{H_2O} CH_2{=}CHOH \xrightarrow{H_2O} CH_3{-}CHO$$

BIOLOGICAL ROLE OF ZINC

Zn has an important biological role in the enzyme systems of animals and plants. Humans contain about 2 g of Zn. This is the second largest amount of a transition metal after Fe, 4 g. There are about 20 enzymes containing Zn, and some of the best known are as follows:

1. Carbonic anhydrase, which is present in red blood cells, is involved in respiration. It speeds up the absorption of CO_2 by red blood cells in muscles and other tissues, and the reverse reaction involving the release of CO_2 in the lungs. At the same time it regulates the pH.

$$CO_2 + OH^- \rightleftharpoons HCO_3^-$$

2. Carboxypeptidase, which is present in the pancreatic juice, is involved in the digestion of proteins by animals, and protein metabolism in plants and animals. The enzyme catalyses the hydrolysis of the terminal peptide (amide) link at the carboxyl end of the peptide chain. The enzyme is selective. It works only when the group R in the terminal amino acid is aromatic or a branched aliphatic chain, and has the *L* configuration.

$$-\underset{\displaystyle R''}{\overset{|}{CH}}-\overset{\displaystyle O}{\overset{\|}{C}}-NH-\underset{\displaystyle R'}{\overset{|}{CH}}-\overset{\displaystyle O}{\overset{\|}{C}}-NH-\underset{\displaystyle R}{\overset{|}{CH}}-\overset{\displaystyle O}{\overset{\|}{C}}-OH + H_2O \rightarrow$$

$$-\underset{\displaystyle R''}{\overset{|}{CH}}-\overset{\displaystyle O}{\overset{\|}{C}}-NH-\underset{\displaystyle R'}{\overset{|}{CH}}-\overset{\displaystyle O}{\overset{\|}{C}}-OH + NH_2-\underset{\displaystyle R}{\overset{|}{CH}}-\overset{\displaystyle O}{\overset{\|}{C}}-OH$$

3. Alkaline phosphatase (energy release).
4. Dehydrogenases and aldolases (sugar metabolism).
5. Alcohol dehydrogenase (metabolism of alcohol).

TOXICITY OF CADMIUM AND MERCURY

It is strange that Zn is an essential element for life, but that the other two elements in the group, Cd and Hg, are both extremely toxic. The main threat from Cd is in places near Zn smelters, as Cd may escape as dust with flue gases. The manufacture of Ni/Cd batteries has caused problems in Sweden and Japan. There is also concern at the amount of Cd in cigarette smoke. If Cd is ingested it accumulates in the kidneys. It causes malfunction of the kidneys and also replaces Zn in some enzymes, thus preventing them from working.

Mercury vapour is toxic, and if inhaled can cause giddiness, tremors, lung damage and brain damage. In the laboratory, mercury should be covered with oil or toluene, and spillages should be treated with flowers of sulphur, forming HgS.

Inorganic compounds such as $HgCl_2$, Hg_2Cl_2 and HgO are also poisonous if eaten. Mercury is a cumulative poison (like Cd, Sn, Pb and Sb). Because they have no biological function, there is no mechanism for excreting them from the body. Hg^{II} inhibits enzymes, particularly those containing thiol groups.

These inorganic mercury compounds have been used for controlling the growth of slime moulds in wood pulp and paper making factories. They are also used in antifouling paints for boats, as a fungicide for treating seeds and plants, and for treating a plant disease called clubroot in brassicas. More recently organomercury compounds such as dimethyl mercury $Hg(Me)_2$ and MeHgX have been used for treating seeds against attack by fungi. Alkyl mercury compounds such as dimethyl mercury are much more toxic than inorganic mercury compounds. Aryl mercury compounds are even more dangerous. They cause brain damage giving numbness, loss of vision, deafness, madness and death.

There have been several alarming incidents involving mercury poisoning.

1. In the early 1900s mercury salts were used in the production of felt hats. The dust affected the central nervous system of workers, causing tremor of muscles called 'hatter's shakes'. This led to the expression 'mad as a hatter'!
2. More recently 52 people died at Minamata (Japan) in 1952 through eating fish contaminated by mercury. The Hg came from losses from a factory using Hg^{II} salts to catalyse the production of ethanal (acetaldehyde) from ethyne (acetylene). The $HgCl_2$ was converted to an organomercurial compound MeHgSMe by anaerobic bacteria in the mud on the sea bed. This is concentrated in the food chain. First it is taken up by plankton, which is eaten by the fish, and other seafood, which in turn is eaten by man. There were outbreaks of mercury poisoning in Japan in 1960 and again in 1965 arising through eating contaminated shellfish.
3. Corn seeds which had been treated with an organomercurial compound to prevent fungal attack were eaten as food rather than being planted to grow a crop in Iraq in 1956 and again in 1960. This resulted in many deaths. There have also been problems through loss of mercury salts used as antifungicides in paper mills in Sweden.

The problems of mercury poisoning are now better understood. Though inorganic mercury salts are poisonous, they are nothing like as toxic as alkyl and aryl mercury compounds. However, the loss of inorganic mercury compounds must be prevented as Hg^{2+} and Hg_2^{2+} can be methylated by bacteria present in rivers, lakes and the sea. Algae, molluscs and fish may concentrate the small amounts present by a factor of 2–10 times, and they may be eaten by something else and concentrated again. $HgMe_2$ and $Hg—Me^+$ are both very stable and persist for a long time because they have a strong Hg—C bond. Considerable efforts are now being made to prevent the discharge of mercury compounds with industrial effluent. The main problems come from acetaldehyde and vinyl chloride monomer plants where mercury compounds are used as a catalyst, and from the electrolytic production of NaOH and Cl_2 where it is used as the cathode.

4. The most recent problem is from the widespread use of mercury to extract gold in Brazil. River water and silt are passed over mercury, when the gold dissolves in the mercury, forming an amalgam. Losses of mercury have poisoned considerable stretches of the Amazon river.

FURTHER READING

Bertini, I., Luchinat, C. and Monnanni R. (1985) Zinc enzymes, *J. Chem. Ed.*, **62**, 924–927.

Bryce-Smith, D. (1989) Zinc deficiency – the neglected factor, *Chemistry in Britain*, **25**, 783–786.

Coates, G.E. and Wade, K. (1967) *Organometallic Compounds*, 3rd ed., Vol. 1, Zinc, Cadmium and Mercury, Methuen, London.

Dirkse, T.P. (ed.) (1986) *Copper, Silver, Gold, Zinc, Cadmium, Mercury Oxides and Hydroxides*, Pergamon, Oxford.

Friberg, L.T. and Vostal, J.J. (1972) *Mercury in the Environment*, CRC Press, Cleveland.

Glocking, F. and Craig, P.J. (1988) *The Biological Alkylation of Heavy Elements*, Special Publication No. 66, Royal Society of Chemistry, London. (Mercury alkyls and aryls in the environment.)

McAuliffe, C.A. (ed.) (1977) *The Chemistry of Mercury*, Macmillan, New York.

Ochai, Ei-I. (1988) Uniqueness of zinc as a bioelement, *J. Chem. Ed.*, **65**, 943–946.

Prince, R.H. (1979) Some aspects of the biochemistry of zinc, *Adv. Inorg. Chem. Radiochem.*, **22**, 349–440.

Richards, A.W. (1969) Zinc extraction metallurgy in the UK, *Chemistry in Britain*, **5**, 203–206.

Roberts, H.L. (1968) The chemistry of mercury, *Adv. Inorg. Chem. Radiochem.*, **11**, 309.

Valentine, J.S. and de Freitas, D.M. (1985) Copper–zinc dismutase, *J. Chem. Ed.*, **62**, 990–996.

PROBLEMS (CHAPTERS 19–28)

1. Explain why La_2O_3 is not reduced by Al.

2. Explain why TiO_2 is white but $TiCl_3$ is violet.

3. Draw the structure of rutile. What is the coordination number of the ions and what is the radius ratio?

4. Describe how Ti metal is obtained, and explain why it has been called the wonder metal.

5. Describe the uses of titanium compounds in paint, in the polymerization of ethene and in the fixation of nitrogen.

6. Explain why the physical and chemical properties of Zr and Hf compounds are much more similar than the properties of Ti and Zr.

7. Describe what happens when the pH of a solution containing $[VO_4]^{3-}$ ions is gradually reduced.

8. When ammonium vanadate is heated with oxalic acid solution, a compound (Z) is formed. A sample of (Z) was titrated with $KNmO_4$ solution in hot acidic solution. The resulting liquid was reduced with SO_2, the excess SO_2 boiled off, and the liquid again titrated with $KMnO_4$.

   The ratio of the volumes of $KMnO_4$ used in the two titrations was 5:1. What conclusions can you make regarding the nature of compound (Z). ($KMnO_4$ oxidizes all oxidation states of vanadium to vanadium (+V). SO_2 reduces vanadium (+V) to vanadium (+IV).)

9. Give equations to show the reactions between CrO, Cr_2O_3 and CrO_3 with aqueous acids and alkalis.

10. Describe the structures of chromium(II) acetate and copper(II) acetate, and comment on any unusual features they have.

11. How may the last traces of oxygen be removed from nitrogen?

12. Manganese has been described as 'the most versatile element'. Explain this, and show the similarities and differences between the chemistry of manganese and rhenium.

13. Explain why $[Mn(OH)_6]^{2+}$ is pale pink, MnO_2 is black and MnO_4^- is intensely coloured purple.

14. Explain why a green solution of potassium manganate(VI) K_2MnO_4 turns purple and a brown solid is precipitated when carbon dioxide is bubbled into the solution.

15. Draw the crystal structure of ReO_3, and compare it with the perovskite structure.

16. Why do metals such as iron corrode, and how may this be prevented?

17. Describe how iron is extracted from its ores, and how it is converted into steel.

18. What are pig iron, wrought iron, mild steel, high carbon steel and stainless steel?

19. Predict the electronic arrangement on the metal from crystal field splitting and the nature of the ligands in the following complexes: (a) $[Fe(H_2O)_6]^{2+}$, (b) $[Fe(CN)_6]^{4-}$ and (c) $[Fe(CN)_6]^{3-}$. How many unpaired electrons are there in each complex, and what would you expect their magnetic moments to be?

20. Iron pentacarbonyl has a trigonal bipyramidal structure, yet the ^{13}C nmr spectrum of $Fe(^{13}CO)_5$ shows only a single peak even when the sample is held at low temperatures. Two peaks would be expected for a trigonal bipyramidal molecule. (The ^{13}C nucleus has a spin of $\frac{1}{2}$ and behaves in much the same way as a proton in nmr.) Explain the unusual spectrum.

21. Compare and contrast the structures of haem, vitamin B_{12} and chlorophyll.

22. Outline the essential features involved in the binding of O_2 to the myoglobin molecule.

23. Write a critical account of the various ways in which the nine elements in the iron, cobalt and nickel groups have been grouped together for comparison purposes.

24. Draw and name the possible isomers of the following complexes: (a) $[Co(en)_3]^{3+}$, (b) $[Co(en)_2(SCN)_2]^+$.

25. A monomeric complex of cobalt gave the following results on analysis:

	Co	NH_3	Cl^-	SO_4^{2-}
%	21.24	24.77	12.81	34.65

The compound is diamagnetic, and contains no other groups or elements, except water might be present.
   Calculate the empirical formula of the compound, give the structural formulae of all possible isomers, and suggest methods to distinguish between the isomers.
26. Give examples of metal clusters in transition metal ions.
27. Draw the structures of six different carbonyl complexes.
28. Discuss the origin of square planar complexes formed by Ni(II), and explain what other shapes may be formed.
29. Suggest reasons why the noble metals are relatively unreactive.
30. Give examples of compounds containing metal–metal bonds.
31. Compare and contrast the chemistry of the elements copper, silver and gold.
32. What are the electronic structures of Zn, Cd and Hg, and of their 2+ ions? Discuss their position in the periodic table. Would you expect them to behave as typical transition elements?
33. Draw the structure of zinc blende. What is the coordination number of the ions and what is the radius ratio?
34. Explain how Zn is sacrificed in the extraction of Cd.
35. Explain why a solution of zinc sulphate gives a white precipitate when added to an aqueous solution of ammonia, but not when added to an aqueous solution of ammonia containing ammonium chloride.
36. What is cadmium used for, how is it produced, and why are there problems of cadmium poisoning in the vicinity of zinc smelters?
37. List the main ways in which mercury compounds are discharged into the environment. Comment on the toxicity of inorganic and organo-mercury compounds.
38. Give an account of disproportionation. Why is the mercurous ion written Hg_2^{2+}, whilst the cuprous ion is written Cu^+?
39. What evidence is there that the mercurous ion is Hg_2^{2+} rather than Hg^+?
40. When mercury is oxidized with a limited amount of oxidizing agent (i.e. an excess of Hg) then Hg^I compounds are formed. If there is an excess of oxidizing agent then Hg^{II} compounds are formed. Explain this.

The *f*-Block Elements

Part Five

The lanthanide series

29

INTRODUCTION

These 14 elements are called the lanthanons, or lanthanides. They are characterized by the filling up of the antepenultimate $4f$ energy levels. They are extremely similar to each other in properties, and until 1907 they were thought to be a single element. In the past they were called the rare earths. This name is not appropriate because many of the elements are not particularly rare. Furthermore, the name is not precise since in addition to the 14 lanthanides it has been used to include La, or Sc, Y and La which are d-block elements, and sometimes Th (an actinide) and Zr (another d-block element).

ELECTRONIC STRUCTURE

The electronic structures of the lanthanide metals are shown in Table 29.1. Lanthanum (the d-block element preceding this series) has the electronic structure: xenon core $5d^1\ 6s^2$. It might be expected that the 14 elements from cerium to lutetium would be formed by adding 1, 2, 3, . . . 14 electrons into the $4f$ level. However, it is energetically favourable to move the single $5d$ electron into the $4f$ level in most of the elements, but not in the cases of Ce, Gd and Lu. The reason why Gd has a $5d^1$ arrangement is that this leaves a half filled $4f$ level, which gives increased stability. Lu has a $5d^1$ arrangement because the f shell is already full. The lanthanides are characterized by the uniform (+III) oxidation state shown by all the metals. They typically form compounds which are ionic and trivalent. The electronic structures of the ions are $Ce^{3+}\ f^1$, $Pr^{3+}\ f^2$, $Nd^{3+}\ f^3$, . . . $Lu^{3+}\ f^{14}$.

The $4f$ electrons in the antepenultimate shell are very effectively shielded from their chemical environment outside the atom by the $5s$ and $5p$ electrons. Consequently the $4f$ electrons do not take part in bonding. They are neither removed to produce ions nor do they take any significant part in crystal field stabilization of complexes. Crystal field stabilization is very important with the d-block elements. The octahedral splitting of f orbitals Δ_O is only about $1\ \text{kJ mol}^{-1}$. Whether the f orbitals are filled or empty has little effect on the normal chemical properties. However, it does affect their spectra and their magnetic properties.

Table 29.1 Electronic structures and oxidation states

Element		Electronic structure of atoms	Electronic structure of M^{3+}	Oxidation states*
Lanthanum	La	[Xe] $5d^1$ $6s^2$	[Xe]	**+III**
Cerium	Ce	[Xe] $4f^1$ $5d^1$ $6s^2$	[Xe] $4f^1$	**+III** +IV
Praseodymium	Pr	[Xe] $4f^3$ $6s^2$	[Xe] $4f^2$	**+III** (+IV)
Neodymium	Nd	[Xe] $4f^4$ $6s^2$	[Xe] $4f^3$	(+II) **+III**
Promethium	Pm	[Xe] $4f^5$ $6s^2$	[Xe] $4f^4$	(+II) **+III**
Samarium	Sm	[Xe] $4f^6$ $6s^2$	[Xe] $4f^5$	(+II) **+III**
Europium	Eu	[Xe] $4f^7$ $6s^2$	[Xe] $4f^6$	+II **+III**
Gadolinium	Gd	[Xe] $4f^7$ $5d^1$ $6s^2$	[Xe] $4f^7$	**+III**
Terbium	Tb	[Xe] $4f^9$ $6s^2$	[Xe] $4f^8$	**+III** (+IV)
Dysprosium	Dy	[Xe] $4f^{10}$ $6s^2$	[Xe] $4f^9$	**+III** (+IV)
Holmium	Ho	[Xe] $4f^{11}$ $6s^2$	[Xe] $4f^{10}$	**+III**
Erbium	Er	[Xe] $4f^{12}$ $6s^2$	[Xe] $4f^{11}$	**+III**
Thulium	Tm	[Xe] $4f^{13}$ $6s^2$	[Xe] $4f^{12}$	(+II) **+III**
Ytterbium	Yb	[Xe] $4f^{14}$ $6s^2$	[Xe] $4f^{13}$	+II **+III**
Lutetium	Lu	[Xe] $4f^{14}$ $5d^1$ $6s^2$	[Xe] $4f^{14}$	**+III**

* The most important oxidation states (generally the most abundant and stable) are shown in bold. Other well-characterized but less important states are shown in normal type. Oxidation states that are unstable, or in doubt, are given in parentheses.

OXIDATION STATES

The sum of the first three ionization energies for each element are given in Table 29.2. The values are low. Thus the oxidation state (+III) is ionic and Ln^{3+} dominates the chemistry of these elements. The Ln^{2+} and Ln^{4+} ions that do occur are always less stable than Ln^{3+}. (In this chapter the symbol Ln is used to denote any of the lanthanides.) In just the same way as for other elements, the higher oxidation states occur in the fluorides and oxides, and the lower oxidation states occur in the other halides, particularly bromides and iodides. Oxidation numbers (+II) and (+IV) do occur, particularly when they lead to:

1. a noble gas configuration, e.g. Ce^{4+} (f^0)
2. a half filled f shell, e.g. Eu^{2+} and Tb^{4+} (f^7)
3. a completely filled f level, e.g. Yb^{2+} (f^{14}).

In addition (+II) and (+IV) states exist for elements that are close to these states. Thus Sm^{2+} and Tm^{2+} occur with f^6 and f^{13} arrangements and Pr^{4+} and Nd^{4+} have f^1 and f^2 arrangements. The (+III) state is always the most common and the most stable. The only (+IV) and (+II) states which have any aqueous chemistry are Ce^{4+}, Sm^{2+}, Eu^{2+} and Yb^{2+}.

The lanthanide elements resemble each other much more closely than do a horizontal row of the transition elements. This is because the lanthanides effectively have only one stable oxidation state, (+III). Thus in this series

Table 29.2 Ionization energies and standard electrode potentials

		Sum of first three ionization energies ($kJ\,mol^{-1}$)	$E°$ $Ln^{3+} \mid Ln$ (volts)	Radius Ln^{3+} (Å)
Lanthanum	La	3493	−2.52	1.032
Cerium	Ce	3512	−2.48	1.020
Praseodymium	Pr	3623	−2.46	0.99
Neodymium	Nd	3705	−2.43	0.983
Promethium	Pm	–	−2.42	0.97
Samarium	Sm	3898	−2.41	0.958
Europium	Eu	4033	−2.41	0.947
Gadolinium	Gd	3744	−2.40	0.938
Terbium	Tb	3792	−2.39	0.923
Dysprosium	Dy	3898	−2.35	0.912
Holmium	Ho	3937	−2.32	0.901
Erbium	Er	3908	−2.30	0.890
Thulium	Tm	4038	−2.28	0.880
Ytterbium	Yb	4197	−2.27	0.868
Lutetium	Lu	3898	−2.26	0.861

Radii are for six-coordination.

it is possible to compare the effects of small changes in size and nuclear charge on the chemistry of these elements.

ABUNDANCE AND NUMBER OF ISOTOPES

The lanthanide elements are not particularly rare. Cerium is about as abundant as copper. Apart from promethium, which does not occur in nature, all the elements are more abundant than iodine.

The abundance of the elements and the number of naturally occurring isotopes vary regularly (Table 29.3). Elements with an even atomic number (i.e. an even number of protons in the nucleus) are more abundant than their neighbours with odd atomic numbers (*Harkins' rule*). Elements with even atomic numbers also have more stable isotopes. Elements with odd atomic numbers never have more than two stable isotopes. Throughout the periodic table the stability of a nucleus is related to both the number of neutrons and the number of protons in the nucleus (Table 29.4).

Element 61, promethium, does not occur naturally. It was first made and studied at Oak Ridge in Tennessee, USA, in 1946. Its absence may be explained by Mattauch's rule. This states that if two elements with consecutive atomic numbers each have an isotope of the same weight, one of the isotopes will be unstable. Since elements 60 and 62 have seven isotopes each, there are not many stable mass numbers available for promethium, element 61 (Table 29.5).

According to Mattauch's rule, if promethium is to have a stable isotope, it must have a mass number outside the range 142–150. The only isotopes of Pm which have been made so far are radioactive.

continued overleaf

Table 29.3 Abundance of the elements in the earth's crust, by weight and number of natural isotopes

Atomic number	Element	Abundance (ppm) in earth's crust	Relative abundance	Naturally occurring isotopes
58	Ce	66	26	4
59	Pr	9.1	37	1
60	Nd	40	27	7
61	Pm	0		0
62	Sm	7	40	7
63	Eu	2.1	49 =	2
64	Gd	6.1	41	7
65	Tb	1.2	56 =	1
66	Dy	4.5	42	7
67	Ho	1.3	55	1
68	Er	3.5	43	6
69	Tm	0.5	61	1
70	Yb	3.1	44	7
71	Lu	0.8	59	2

Table 29.4 Numbers of stable nuclei with odd and even numbers of neutrons and odd and even atomic number

Atomic number	Number of neutrons	Stable nuclei
Even	Even	164
Even	Odd	55
Odd	Even	50
Odd	Odd	4

Table 29.5

Element 60	142,	143,	144,	145,	146,		148,		150		
Element 62			144,			147,	148,	149,	150,	152,	154

EXTRACTION AND USES

World production of rare earth minerals was 72 100 tonnes in 1988, containing 41 800 tonnes of lanthanide oxides Ln_2O_3. The main sources of minerals are China 24%, the USA and Australia 17% each and India 5.5%.

1. Monazite sand is the most important and most widespread mineral. It accounts for 78% of the rare earths mined. Before 1960 monazite was the only source of lanthanides. It is a mixture containing mostly La phosphate and trivalent phosphates of the lighter lanthanide elements (Ce, Pr and Nd). In addition it contains smaller amounts of Y and the heavier lanthanides, and thorium phosphate. Th is weakly radioactive,

and traces of its daughter product Ra are also present, and are more radioactive.

2. Bastnaesite is a mixed fluorocarbonate $M^{III}CO_3F$ where M is La or the lanthanides. Large amounts are mined in the USA, and it provides 22% of the total supply of lanthanides. It is only found in the USA and Madagascar.
3. Very small amounts of another mineral, xenotime, are also mined.

Monazite is treated with hot concentrated H_2SO_4. Th, La and the lanthanides dissolve as sulphates, and are separated from insoluble material. Th is precipitated as ThO_2 by partial neutralization with NH_4OH. Na_2SO_4 is used to salt out La and the light lanthanides as sulphates, leaving the heavy lanthanides in solution. The light lanthanides are oxidized with bleaching powder $Ca(OCl)_2$. Ce^{3+} is oxidized to Ce^{4+} which is precipitated as $Ce(IO_3)_4$ and removed. La^{3+} may be removed by solvent extraction with tri-n-butylphosphate. The individual elements can be obtained by ion exchange if required. The treatment of bastnaesite is slightly simpler as it does not contain Th.

Once the different lanthanide elements have been separated completely or partially, the metal may be obtained as follows:

1. By electrolysis of the fused $LnCl_3$, with NaCl or $CaCl_2$ added to lower the melting point.
2. La and the lighter metals Ce to Eu are obtained by reducing anhydrous $LnCl_3$ with Ca at 1000–1100 °C in an argon-filled vessel. The heavier elements have higher melting points and so require a temperature of 1400 °C. At this temperature $CaCl_2$ boils, so LnF_3 are used instead, and in some cases Li is used instead of Ca.

About 5000 tonnes of La and 13 000 tonnes of the lanthanides are produced annually. The metals are of little use on their own. The main use is for an unseparated mixture of La and the lanthanides called Mischmetal (50% Ce, 40% La, 7% Fe, 3% other metals). This is added to steel to improve its strength and workability. It is also used in Mg alloys. Mischmetal is also used in small amounts as 'lighter flints'. La_2O_3 is used in Crooke's lenses, which give protection from UV light by absorbing it. CeO_2 is used to polish glass and as a coating in 'self-cleaning' ovens. $Ce^{IV}(SO_4)_2$ is used as an oxidizing agent in volumetric analysis. Gas mantles are treated with a mixture of 1% CeO_2 and 99% ThO_2 to increase the amount of light emitted by coal gas flames. Other lanthanide oxides are used as phosphors in colour TV tubes. 'Didymium oxide' (a mixture of praseodymium and neodymium oxides) is used with $CuCl_2$ as the catalyst in the new Deacon process to make Cl_2 from HCl. Nd_2O_3 is used dissolved in $SeOCl_2$ as a liquid laser. (Selenium oxochloride is used as the solvent because it contains no light atoms which would convert the input energy into heat.) Lanthanide elements are present in *warm superconductors* such as $La_{(2-x)}Ba_xCuO_{(4-y)}$ and $YBa_2Cu_3O_{7-x}$, and others (Sm, Eu, Nb, Dy and Yb) have been substituted. These are described at the end of Chapter 5.

SEPARATION OF THE LANTHANIDE ELEMENTS

The properties of metal ions are determined by their size and charge. The lanthanides are all typically trivalent and are almost identical in size, and so their chemical properties are almost identical. The separation of one lanthanide from another is an exceedingly difficult task, almost as difficult as the separation of isotopes of one element. The classical methods of separation exploit slight differences in basic properties, stability or solubility. These are outlined below. However, in recent years the only methods used are ion exchange and valency change.

Precipitation

With a limited amount of precipitating agent the substance with the lowest solubility is precipitated most rapidly and most completely. Suppose hydroxyl ions are added to a solution containing a mixture of $Ln(NO_3)_3$. The weakest base $Lu(OH)_3$ is precipitated first, and the strongest base $La(OH)_3$ is precipitated last. The precipitate contains more of the elements at the right of the series. Thus the solution contains more of the elements at the left of the series. The precipitate can be filtered off. Only partial separation is effected, but the precipitate can be redissolved in HNO_3 and the process repeated to obtain greater purity.

Thermal reaction

If a mixture of $Ln(NO_3)_3$ is fused, a temperature will be reached when the least basic nitrate changes to the oxide. The mixture is leached with water. The nitrates dissolve and can be filtered off, leaving the insoluble oxides. The oxides are dissolved in HNO_3 and the process repeated.

Fractional crystallization

This can be used to separate lanthanide salts. The solubility decreases from La to Lu. Thus salts at the Lu end of the series will crystallize out first. Nitrates, sulphates, bromates, perchlorates and oxalates have all been used as also have double salts such as $Ln(NO_3)_3 \cdot 3Mg(NO_3)_2 \cdot 24H_2O$ because they crystallize well. The process needs repeating many times to obtain good separations. Non-aqueous solvents such as diethyl ether have been used to separate $Nd(NO_3)_3$ and $Pr(NO_3)_3$.

Complex formation

A mixture of lanthanide ions is treated with a complexing agent such as EDTA (ethylenediaminetetraacetic acid). All the ions form complexes. Those ions at the right hand side of the lanthanide series such as Lu^{3+} form the strongest complexes as they have the smallest ions. Oxalates of the lanthanides are insoluble. However, addition of oxalate ions to this solution does not give a precipitate since the Ln^{3+} ions are all complexed with EDTA.

```
HOOC—CH2                              CH2—COOH
         \                           /
          N—CH2—CH2—N
         /                           \
HOOC—CH2                              CH2—COOH
```

If some acid is added to the solution, the least stable EDTA complexes are dissociated. This releases ions at the left hand side of the series Ce^{3+}, Pr^{3+}, Nd^{3+} which are immediately precipitated as the oxalates. These are filtered off. Separation is not complete, so the oxalates are redissolved and the process repeated many times.

Solvent extraction

The heavier Ln^{3+} ions are more soluble in tri-n-butylphosphate than are the lighter Ln^{3+} ions. Their solubilities in water and ionic solvents, however, are reversed. The ratios of the partition coefficients of $La(NO_3)_3$ and $Gd(NO_3)_3$ between a solution of the metal ions in strong HNO_3 and tri-n-butylphosphate is 1 : 1.06. This difference is quite small, but by using a continuous counter-current apparatus a very large number of partitions can be performed automatically. This is much less tedious than performing 10 000 or 20 000 crystallizations. Kilogram quantities of 95% pure Gd have been obtained by this method. The technique was originally developed in the early days of atomic energy to separate and identify the lanthanide elements produced by fission of uranium.

Valency change

A few lanthanides have oxidation states of (+IV) or (+II). The properties of Ln^{4+} or Ln^{2+} are so different from those of Ln^{3+} that separation is fairly easy.

Cerium can be separated from lanthanide mixtures quite easily as it is the only lanthanide which has Ln^{4+} ions stable in aqueous solution. Oxidizing a solution containing a mixture of Ln^{3+} ions with NaOCl under alkaline conditions produces Ce^{4+}. Because of the higher charge, Ce^{4+} is much smaller and less basic than Ce^{3+} or any other Ln^{3+}. The Ce^{4+} is separated by carefully controlled precipitation of CeO_2 or $Ce(IO_3)_4$, leaving the trivalent ions in solution.

Alternatively Ce^{4+} can readily be extracted from other Ln^{3+} lanthanides by solvent extraction in HNO_3 solution using tributyl phosphate. Ninety-nine per cent pure Ce can be obtained in one stage from a mixture containing 40% Ce.

In a similar way the properties of Eu^{2+} are very different from those of Ln^{3+}. Europium sulphate $Eu^{2+}SO_4^{2-}$ resembles the Group II sulphates and is insoluble in water. Ln^{3+} sulphates are soluble. If a solution of Ln^{3+} ions is reduced electrolytically using a mercury cathode, or by using zinc amalgam, then Eu^{2+} will be produced. If H_2SO_4 is present $EuSO_4$ will be precipitated. This can be filtered off. (Sm^{2+} and Yb^{2+} may also be produced in the same way, but these are oxidized slowly by water.)

Valency change is still a useful method for purifying Ce and Eu despite the advent in recent years of ion exchange.

Ion exchange

This is the most important, the most rapid and most effective general method for the separation and purification of the lanthanides. A solution of lanthanide ions is run down a column of synthetic ion-exchange resin such as Dowex-50. This is a sulphonated polystyrene and contains the functional groups —SO_3H. The Ln^{3+} ions are absorbed onto the resin and replace the hydrogen atom on —SO_3H.

$$Ln^{3+}_{(aq)} + 3H(resin)_{(s)} \rightleftharpoons Ln(resin)_{3(s)} + 3H^{+}_{(aq)}$$

The H^+ ions produced are washed through the column. Then the metal ions are eluted, that is are washed off the column in a selective manner. The eluting agent is a complexing agent, for example a buffered solution of citric acid/ammonium citrate, or a dilute solution of $(NH_4)_3H \cdot EDTA$ at pH 8. Consider the citrate case. An equilibrium is set up:

$$Ln(resin)_3 + 3H^+ + 3(citrate) \rightleftharpoons 3H(resin) + Ln(citrate)_3$$

As the citrate solution flows down the column, Ln^{3+} ions are removed from the resin and form the citrate complex. A little lower down the column the Ln^{3+} ions go back onto the resin. As the citrate solution runs down the column the metal ions form complexes alternately with the resin and the citrate solution many times. The metal ion gradually travels down the column, and eventually passes out of the bottom of the column as the citrate complex. The smaller lanthanide ions such as Lu^{3+} form stronger complexes with the citrate ions than do the larger ions like La^{3+}. Thus the smaller and heavier ions spend more time in solution, and less time on the column, and are thus eluted from the column first. The different metal ions present separate into bands which pass down the column. The progress of the bands may be followed spectroscopically by atomic fluorescence. The solution leaving the column is collected by means of an automatic fraction collector in separate containers. By this means the individual elements can be separated. The metals may be precipitated as insoluble oxalates, and then heated to give the oxides.

The chromatographic process is analogous to carrying out many separations or many crystallizations, but the separation is carried out on a single column. By using a long ion-exchange column the elements may be obtained 99.9% pure with one pass.

CHEMICAL PROPERTIES OF (+III) COMPOUNDS

The metals are all soft and silvery white. They are electropositive and therefore they are very reactive. The heavier metals are less reactive than the lighter ones because they form a layer of oxide on the surface. The chemical properties of the group are essentially the properties of trivalent ionic compounds.

The sum of the first three ionization energies varies with minima at La^{3+}, Gd^{3+} and Lu^{3+} which are associated with attaining an empty, half full or full *f* shell. Maxima occur at Eu^{3+} and Yb^{3+} associated with breaking a half full or full shell.

The standard reduction potentials (E:) are all high (Table 29.2). They vary in a regular way over a small range from −2.48 to −2.26 volts, depending on the size of the ions.

The lanthanides are all much more reactive than is Al ($E^\circ = -1.66$ volts) and are slightly more reactive than Mg ($E^\circ = -2.37$ volts). Thus they react slowly with cold water, but more rapidly on heating.

$$2Ln + 6H_2O \rightarrow 2Ln(OH)_3 + 3H_2$$

The hydroxides $Ln(OH)_3$ are precipitated as gelatinous precipitates by the addition of NH_4OH to aqueous solutions. These hydroxides are ionic and basic. They are less basic than $Ca(OH)_2$ but more basic than $Al(OH)_3$ which is amphoteric. The metals, oxides and hydroxides all dissolve in dilute acids, forming salts. $Ln(OH)_3$ are sufficiently basic to absorb CO_2 from the air and form carbonates. The basicity decreases as the ionic radius decreases from Ce to Lu. Thus $Ce(OH)_3$ is the most basic, and $Lu(OH)_3$, which is the least basic, is intermediate between scandium and yttrium in basic strength. The decrease in basic properties is illustrated by the hydroxides of the later elements dissolving in hot concentrated NaOH, forming complexes.

$$Yb(OH)_3 + 3NaOH \rightarrow 3Na^+ + [Yb(OH)_6]^{3-}$$
$$Lu(OH)_3 + 3NaOH \rightarrow 3Na^+ + [Lu(OH)_6]^{3-}$$

The metals tarnish readily in air, and on heating in O_2 they all give oxides Ln_2O_3. Yb and Lu form a protective oxide film, which prevents the bulk of the metal forming the oxide unless it is heated to 1000 °C. The one exception is Ce which forms $Ce^{IV}O_2$ rather than Ce_2O_3. The oxides are ionic and basic. Basic strength decreases as the ions get smaller.

The metals react with H_2, but often require heating up to 300–400 °C to start the reaction. The products are solids of formula LnH_2. Eu and Yb both have a tendency to form divalent compounds and EuH_2 and YbH_2 are salt-like hydrides and contain M^{2+} and two H^-. The others all form hydrides LnH_2 which are black, metallic and conduct electricity. These are better formulated as Ln^{3+}, $2H^-$ and an electron which occupies a conduction band. In addition Yb forms a nonstoichiometric compound approximating to $YbH_{2.5}$. The hydrides are remarkably stable to heat, often up to 900 °C. They are decomposed by water, and react with oxygen.

$$CeH_2 + 2H_2O \rightarrow CeO_2 + 2H_2$$

These 'dihydrides' take up H if heated under pressure, and all except Eu form salt-like hydrides LnH_3 made up of Ln^{3+} and three H^-. These do not have a delocalized electron, and do not show metallic conduction.

The anhydrous halides MX_3 can be made by heating the metal and halogen, or by heating the oxide with the appropriate ammonium halide.

$$Ln_2O_3 + 6NH_4Cl \xrightarrow{300\,°C} 2LnCl_3 + 6NH_3 + 3H_2O$$

The fluorides are very insoluble, and can be precipitated from solutions of Ln^{3+} by addition of Na^+F^- or HF. This is used as a test for the lanthanides in qualitative analysis. However, with excess F^-, the smaller lanthanide ions may form soluble complexes $[LnF(H_2O)_n]^{2+}$. The chlorides are deliquescent and soluble, and crystallize with six or seven molecules of water of crystallization. If the hydrated halides are heated, they form oxohalides instead of dehydrating to anhydrous halides.

$$LnCl_3 \cdot 6H_2O \xrightarrow{heat} LnOCl + 5H_2O + 2HCl$$

Heating $CeX_3 \cdot (H_2O)_n$ results in CeO_2. The bromides and iodides aresimilar to the chlorides.

At elevated temperatures, the lanthanides react with B, giving LnB_4 and LnB_6.

On arc-melting the metals with C in an inert atmosphere they form carbides of stoichiometry LnC_2 and $Ln_4(C_2)_3$. The carbides can also be made by reducing Ln_2O_3 with C in an electric furnace. LnC_2 are more reactive than CaC_2. They react with water, giving ethyne and also some hydrogen, C_2H_4 and C_2H_6. They also show metallic conductivity. They do not contain Ln(+II) and are best described as acetylides of Ln^{3+} and C_2^{2-} with the extra electron in a conduction band.

$$2LnC_2 + 6H_2O \rightarrow 2Ln(OH)_3 + 2C_2H_2 + H_2$$

$$C_2H_2 + H_2 \rightarrow C_2H_4 \xrightarrow{+H_2} C_2H_6$$

At elevated temperatures the metals also react with N, P, As, Sb and Bi, giving LnN etc. The latter is hydrolysed by water in a similar way to AlN:

$$LnN + 3H_2O \rightarrow Ln(OH)_3 + NH_3$$

A wide variety of oxosalts are known, including nitrates, carbonates, oxalates, sulphates, phosphates and also salts of strongly oxidizing ions such as perchlorates.

OXIDATION STATE (+IV)

The only (+IV) lanthanide which exists in solution and has any aqueous chemistry is Ce^{4+}. It is rare to find 4+ ions in solution. The high charge on the ion leads to it being heavily hydrated, and except in strongly acidic solutions the hydrated Ce^{4+} is hydrolysed, giving polymeric species and H^+. Ce(+IV) solutions are widely used as an oxidizing agent in volumetric analysis instead of $KMnO_4$ and $K_2Cr_2O_7$. In classical analysis burettes containing Ce^{4+} must be washed with acid, since washing with water gives the hydrated ion. Aqueous 'ceric' solutions can be prepared by oxidizing a Ce^{3+} solution with a very strong oxidizing agent such as ammonium peroxodisulphate $(NH_4)_2S_2O_8$. Ce(+IV) is also used in organic reactions, for example the oxidation of alcohols, aldehydes and ketones at the α-carbon atom. The common compounds are CeO_2 (white when pure) and $CeO_2 \cdot (H_2O)_n$ (a yellow gelatinous precipitate). CeO_2 can be obtained by heating the metal, or $Ce(OH)_3$ or $Ce_2^{III}(oxalate)_3$, in air. CeO_2 has a fluorite type of structure. It is insoluble in acids and alkalis, but dissolves if reduced, giving Ce^{3+} solutions.

Ceric sulphate $Ce(SO_4)_2$ is well known and is yellow like K_2CrO_4. CeF_4 is obtained from CeF_3 and F_2. It is white, and is rapidly hydrolysed by water. It has a three-dimensional crystal structure with the metal at the centre of a square antiprism. A number of complexes are stable, for example ceric ammonium nitrate $(NH_4)_2[Ce(NO_3)_6]$. The crystal structure is unusual and contains bidentate NO_3^- groups. The Ce atom has a co-ordination number of 12, and the shape is an icosahedron. This structure is stable even in solution. Two of the NO_3^- ions may be replaced by phosphine ligands Ph_3PO, giving a neutral 10-coordinate complex $[Ce^{IV}(NO_3)_4(Ph_3PO)_2]$.

The other (+IV) compounds are not stable in water and are known only as oxides, fluorides and a few fluoro complexes. Thus PrO_2, PrF_4, $Na_2[PrF_6]$, TbO_2, TbF_4, TbO_2, DyF_4 and $Cs_3[DyF_7]$ are all known.

$$Ce + O_2 \rightarrow CeO_2$$
$$2Ce(OH)_3 + \tfrac{1}{2}O_2 \rightarrow 2CeO_2 + 3H_2O$$
$$Ce_2(C_2O_4)_3 + 2O_2 \rightarrow 2CeO_2 + 6CO_2$$

The elements Pr, Nd, Tb and Dy also form (+IV) states. These are generally unstable, occur only as solids, and are found as fluorides or oxides which may be nonstoichiometric.

OXIDATION STATE (+II)

The only (+II) states which have any aqueous chemistry are Sm^{2+}, Eu^{2+} and Yb^{2+}.

The most stable divalent lanthanide is Eu^{2+}. This is stable in water, but the solution is strongly reducing. $Eu^{II}SO_4$ can be prepared by electrolysing $Eu_2^{III}(SO_4)_3$ solutions, when the divalent sulphate is precipitated. $Eu^{II}Cl_2$ can be made as a solid by reducing $Eu^{III}Cl_3$ with H_2.

$$2EuCl_3 + H_2 \rightarrow 2EuCl_2 + 2HCl$$

Aqueous Eu^{3+} solutions can be reduced by Mg, Zn, zinc amalgam or electrolytically to give Eu^{2+}. EuH_2 is ionic and similar to CaH_2. Eu(II) resembles Ca in several ways:

1. The insolubility of the sulphate and carbonate in water.
2. The insolubility of the dichloride in strong HCl.
3. The solubility of the metals in liquid NH_3.

One major difference between Eu and Ca is that the dihalides EuX_2 have a magnetic moment of 7.9 Bohr magnetons corresponding to seven unpaired electrons, whereas Ca compounds are diamagnetic.

The couple $Eu^{3+}|Eu^{2+}$ has a standard reduction potential of −0.41 volts. This is about the same as for $Cr^{3+}|Cr^{2+}$, and these are both about the strongest reducing agents that do not reduce water.

Yb^{2+} and Sm^{2+} can be prepared by electrolytic reduction of their trivalent ions in aqueous solution. However, the Ln^{2+} ions are readily oxidized by air. These two elements form hydroxides, carbonates, halides, sulphates and phosphates.

The states Nd(+II), Pm(+II), Sm(+II) and Gd(+II) are only found in solid dihalides $LnCl_2$ and LnI_2. These dihalides can be made by reducing the trihalide with hydrogen, with the metal, or with sodium amalgam. The dihalides such as LaI_2 and NdI_2 tend to be nonstoichiometric. They show metallic conduction, and are better represented as $La^{3+} + 2I^- +$ electron.

A detailed study of the third ionization energy shows the stability of a half filled and completely filled shell. The ionization energies also suggest that there may also be extra stability associated with a three quarters filled shell.

SOLUBILITY

Salts of the lanthanides usually contain water ofcrystallization. Solubility depends on the small difference betweenthe lattice energy and the solvation energy, and there is no obvioustrend in the group. The solubility of many of the salts follows thepattern of Group II elements. Thus the chlorides and nitrates aresoluble in water and the oxalates, carbonates and fluorides are almostinsoluble. Unlike Group II, however, the sulphates are soluble. Manyof the lanthanides form double salts with the corresponding Group I orammonium salts, e.g. $Na_2SO_4Ln_2(SO_4)_3 \cdot 8H_2O$, and as these double salts crystallize well, they have been used to separate the lanthanides from one another.

COLOUR AND SPECTRA

Many trivalent lanthanide ions are strikingly coloured both inthe solid state and in aqueous solution. The colour seems to depend on the number of unpaired *f* electrons. Elements with (n) *f* electrons often have a similar colour to those with $(14 - n)$ *f* electrons. (SeeTable 29.6.) However, the elements in other valency states do not all havecolours similar to their isoelectronic 3+ counterparts (Table 29.7).

Colour arises because light of a particular wavelength is absorbed in the visible region. The wavelength absorbed corresponds to the energy required to promote an electron to a higher energy level. In the lanthanides spin orbit coupling is more important than crystal field splitting. In the spectra of transition metals, crystal field splitting is of major importance. All but one of the lanthanide ions show absorptions in the visible or near-

Table 29.6 Colour of Ln^{3+} ions

	Number of 4*f* electrons	Colour		Number of 4*f* electrons	Colour
La^{3+}	0	Colourless	Lu^{3+}	14	Colourless
Ce^{3+}	1	Colourless	Yb^{3+}	13	Colourless
Pr^{3+}	2	Green	Tm^{3+}	12	Pale green
Nd^{3+}	3	Lilac	Er^{3+}	11	Pink
Pm^{3+}	4	Pink	Ho^{3+}	10	Pale yellow
Sm^{3+}	5	Yellow	Dy^{3+}	9	Yellow
Eu^{3+}	6	Pale pink	Tb^{3+}	8	Pale pink
Gd^{3+}	7	Colourless	Gd^{3+}	7	Colourless

Table 29.7 Colours of Ln^{4+}, Ln^{2+} and their isoelectronic Ln^{3+} counterparts

	Electronic configuration	Isoelectronic M^{3+}
Ce^{4+} Orange–red	$4f^0$	La^{3+} Colourless
Sm^{2+} Blood-red	$4f^6$	Eu^{3+} Pale pink
Eu^{2+} Pale greenish yellow	$4f^7$	Gd^{3+} Colourless
Yb^{2+} Yellow	$4f^{14}$	Lu^{3+} Colourless

UV regions of the spectrum. The exception is Lu^{3+} which has a full *f* shell. These colours arise from *f–f* transitions. Strictly these transitions are Laporte forbidden (since the change in the subsidiary quantum number is zero). Thus the colours are pale because they depend on relaxation of the rule. The *f* orbitals are deep inside the atom. Thus they are largely shielded from environmental factors such as the nature and number of ligands which form the complexes, and from vibration of the ligands. Thus the position of the absorption band (i.e. the colour) does not change with different ligands. Vibration of the ligands changes the external fields. However, this only splits the various spectroscopic states by about $100\,cm^{-1}$, so the absorption bands are unusually sharp. The lanthanides are used for wavelength calibration of instruments because of their sharp absorption bands. For an *f* electron the subsidiary quantum number $l = 3$, so m_l may have values 3, 2, 1, 0, −1, −2, −3. Thus a large number of transitions are usually possible. This is in marked contrast to the transition elements where *d–d* spectra give absorption bands whose position changes from ligand to ligand, and the width of the peak is greatly broadened because of the vibration of the ligands. It is also possible to get transitions from the 4*f* to the 5*d* level. Such transitions give broader peaks and their position is affected by the nature of the ligands.

Absorption spectra of lanthanide ions are useful both for thequalitative detection and the quantitative estimation of lanthanides. Lanthanide elements are sometimes used as biological tracers for drugsin humans and animals. This is because lanthanide elements can quite easily be followed in the body by spectroscopy, because their peaksare narrow and very characteristic.

Ce^{3+} and Yb^{3+} are colourless because they do not absorb in the visible region. However, they show exceptionally strong absorption in the UV region, because of transitions from $4f$ to $5d$. Absorption is very strong for two reasons. Since $\Delta l = 1$ this is an allowed transition and so gives stronger absorption than forbidden f–f transitions. Furthermore, promotion of electrons in these ions is easier than for other ions. The electronic configuration of Ce^{3+} is f^1 and Yb^{3+} is f^8. Loss of one electron gives the extra stability of an empty or half full shell. f–d peaks are broad, in contrast to the narrow f–f peaks.

Charge transfer spectra are possible due to the transfer of an electron from the ligand to the metal. This is more probable if the metal is in a high oxidation state or the ligand has reducing properties. Charge transfer usually produces intense colours. The strong yellow colour of Ce^{4+} solutions arises from charge transfer rather than f–f spectra. The blood red colour of Sm^{2+} is also due to charge transfer.

MAGNETIC PROPERTIES

La^{3+} and Ce^{4+} have an f^o configuration, and Lu^{3+} has an f^{14} configuration. These have no unpaired electrons, and are diamagnetic. All other f states contain unpaired electrons and are therefore paramagnetic.

The magnetic moment of transition elements may be calculated from the equation:

$$\mu_{(S+L)} = \sqrt{4S(S+1) + L(L+1)}$$

$\mu_{(S+L)}$ is the magnetic moment in Bohr magnetons calculated using both the spin and orbital momentum contributions. S is the resultant spin quantum number and L is the resultant orbital momentum quantum number. For the first row transition elements, the orbital contribution is usually quenched out by interaction with the electric fields of the ligands in its environment. Thus as a first approximation the magnetic moment can be calculated using the simple spin only formula. (μ_S is the spin only magnetic moment in Bohr magnetons. S is the resultant spin quantum number and n is the number of unpaired electrons.)

$$\mu_S = \sqrt{4S(S+1)}$$
$$\mu_S = \sqrt{n(n+2)}$$

This simple relationship works with La^{3+} (f^0), and two of the lanthanides Gd^{3+} (f^7) and Lu^{3+} (f^{14}).
La^{3+} and Lu^{3+} have no unpaired electrons, $n = 0$ and $\mu_S = \sqrt{0(0+2)} = 0$

Gd^{3+} has seven unpaired electrons, $n = 7$ and

$$\mu_S = \sqrt{7(7+2)} = \sqrt{63} = 7.9\,\text{BM}$$

The other lanthanide ions do not obey this simple relationship. The $4f$ electrons are well shielded from external fields by the overlying $5s$ and $5p$

electrons. Thus the magnetic effect of the motion of the electron in its orbital is not quenched out. Thus the magnetic moments must be calculated taking into account both the magnetic moment from the unpaired electron spins and that from the orbital motion. This also happens with the second and third row transition elements. However, the magnetic properties of the lanthanides are fundamentally different from those of the transition elements. In the lanthanides the spin contribution S and orbital contribution L couple together to give a new quantum number J.

$J = L - S$ when the shell is less than half full

and $J = L + S$ when the shell is more than half full

The magnetic moment μ is calculated in Bohr magnetons (BM) by:

$$\mu = g\sqrt{J(J+1)}$$

where

$$g = 1\tfrac{1}{2} + \frac{S(S+1) - L(L+1)}{2J(J+1)}$$

Figure 29.1 shows the calculated magnetic moments for the lanthanides using both the simple spin only formula, and the coupled spin plus orbital momentum formula. For most of the elements there is excellent agreement between the calculated values using the coupled spin + orbital momentum formula and experimental values measured at 300 K. The range of experimental values are shown as bars.

The agreement for Eu^{3+} is poor, and that for Sm^{3+} is not very good. The reason is that with Eu^{3+} the spin orbit coupling constant is only about 300 cm^{-1}. This means that the difference in energy between the ground

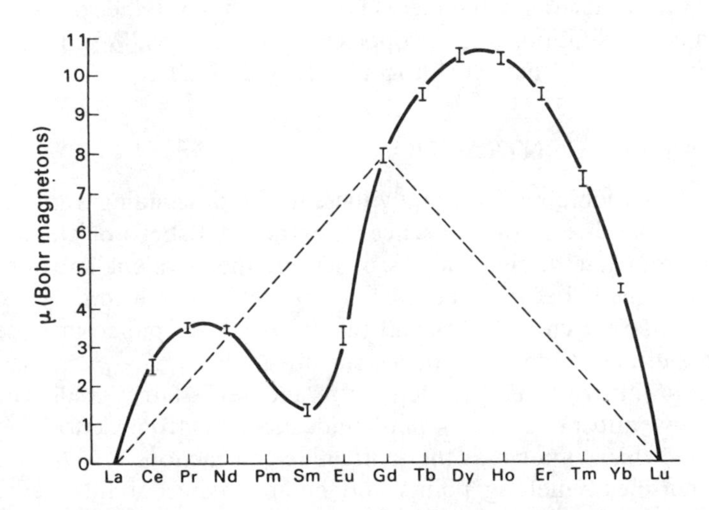

Figure 29.1 Paramagnetic moments of Ln^{3+} lanthanide ions at 300 K. Spin-only values are shown as a broken line, and the spin plus orbital motion as solid lines.

Table 29.8 Magnetic moments of La^{3+} and the lanthanide^{3+} ions

Element		Electronic structure of M^{3+}	Magnetic moment	
			Calculated (BM)	Observed (BM)
Lanthanum	La	[Xe] $4f^0$	0	0
Cerium	Ce	[Xe] $4f^1$	2.54	2.3–2.5
Praseodymium	Pr	[Xe] $4f^2$	3.58	3.4–3.6
Neodymium	Nd	[Xe] $4f^3$	3.62	3.5–3.6
Promethium	Pm	[Xe] $4f^4$	2.68	2.7
Samarium	Sm	[Xe] $4f^5$	0.84	1.5–1.6
Europium	Eu	[Xe] $4f^6$	0	3.4–3.6
Gadolinium	Gd	[Xe] $4f^7$	7.94	7.8–8.0
Terbium	Tb	[Xe] $4f^8$	9.72	9.4–9.6
Dysprosium	Dy	[Xe] $4f^9$	10.63	10.4–10.5
Holmium	Ho	[Xe] $4f^{10}$	10.60	10.3–10.5
Erbium	Er	[Xe] $4f^{11}$	9.57	9.4–9.6
Thulium	Tm	[Xe] $4f^{12}$	7.63	7.1–7.4
Ytterbium	Yb	[Xe] $4f^{13}$	4.50	4.4–4.9
Lutetium	Lu	[Xe] $4f^{14}$	0	0

state and the next state is small. Thus the energy of thermal motion is sufficient to promote some electrons and partially populate the higher state. Because of this the magnetic properties are not solely determined by the ground state configuration. Measuring the magnetic moment at a low temperature prevents the population of higher energy levels. The magnetic moment of Eu^{3+} at low temperature is close to zero as expected. (The measurement of magnetic moments is described in Chapter 17.

The unusual shape of the spin plus orbital motion curve arises because of Hund's third rule. When the *f* level is less than half full the spin and orbital momenta contributions work in opposition ($J = L - S$). When the *f* shell is more than half full they work together ($J = L + S$).

LANTHANIDE CONTRACTION

Covalent and ionic radii normally increase on descending a group in the periodic table due to the presence of extra filled shells of electrons. On moving from left to right across a period, the covalent and ionic radii decrease. This is because the extra orbital electrons incompletely shield the extra nuclear charge. Thus all the electrons are pulled in closer. The shielding effect of electrons decreases in the order $s > p > d > f$. The contraction in size from one element to another is fairly small. However, the additive effect over the 14 lanthanide elements from Ce to Lu is about 0.2 Å, and this is known as the lanthanide contraction.

The hardness, melting points and boiling points of the elements all increase from Ce to Lu. This is because the attraction between the atoms increases as the size decreases.

The properties of an ion depend on its size and its charge. The Ln^{3+} lanthanide ions change by only a small amount from one element to the next (Table 29.9), and their charge is the same, and so their chemical properties are very similar. Since Lu^{3+} is the smallest ion it is the most heavily hydrated. Though the lanthanides do not form complexes very extensively, since Lu^{3+} is the smallest ion the complexes formed by Lu^{3+} are the strongest. La^{3+} and Ce^{3+} are the largest ions so $La(OH)_3$ and $Ce(OH)_3$ are the strongest bases.

The lanthanide contraction reduces the radii of the last four elements in the series below that for Y in the preceding transition series. Since the size of the heavier lanthanide ions, particularly Dy^{3+} and Ho^{3+}, are similar to that of Y^{3+} it follows that their chemical properties are also very similar. As a result the separation of these elements is very difficult.

Table 29.9 Ionic radii of Sc^{3+}, Y^{3+} and La^{3+} and the Ln^{3+} ions (Å)

Sc														
0.745														
Y														
0.900														
La	Ce	Pr	Nd	Pm	Sm	Eu	Gd	Tb	Dy	Ho	Er	Tm	Yb	Lu
1.032	1.02	0.99	0.983	0.97	0.958	0.947	0.938	0.923	0.912	0.901	0.890	0.880	0.868	0.861

For simplicity covalent radii are compared in this table. Ionic radii depend on how many electrons are removed. However, a similar variation is observed if the sizes of ions of the same charge are compared.

Because of this contraction in size across the lanthanide series, the elements which follow in the third transition series are considerably smaller than would otherwise be expected. The normal size increase Sc → Y → La disappears after the lanthanides. Thus pairs of elements such as Zr/Hf, Nb/Ta and Mo/W are almost identical in size. The close similarity of properties in such a pair makes chemical separation very difficult. The sizes of the third row of transition elements are very similar to those of the second row of transition elements (see Table 29.10). Thus the second and third rows of transition elements resemble each other more closely than do the first and second rows.

Table 29.10 Covalent radii of the transition elements (Å)

Sc 1.44		Ti 1.32	V 1.22	Cr 1.17	Mn 1.17	Fe 1.17	Co 1.16	Ni 1.15
Y 1.62		Zr 1.45	Nb 1.34	Mo 1.29	Tc –	Ru 1.24	Rh 1.25	Pd 1.28
La 1.69	*	Hf 1.44	Ta 1.34	W 1.30	Re 1.28	Os 1.26	Ir 1.26	Pt 1.29

* 14 Lanthanide elements

COMPLEXES

The lanthanide ions Ln^{3+} have a high charge, which favours the formation of complexes. However, the ions are rather large (1.03–0.86 Å) compared with the transition elements (Cr^{3+} = 0.615 Å, Fe^{3+} = 0.55 Å (low spin)) and consequently they do not form complexes very readily. Complexes with amines are not formed in aqueous solution because water is a stronger ligand than the amine. However, amine complexes can be made in non-aqueous solvents. Very few stable complexes are formed with CO, CN^- and organometallic groups. This is in contrast to the transition metals. The difference arises because the $4f$ orbitals are well shielded and are 'inside the atom'. Thus they cannot take part in π back bonding, whereas in the transition elements the *d* orbitals are involved in π bonding. The most common and stable complexes are those with chelating oxygen ligands such as citric acid, oxalic acid, $EDTA^{4-}$ and acetylacetone. These complexes frequently have high and variable coordination numbers, and water or solvent molecules are often attached to the central metal. β-Diketone complexes of Eu^{3+} and Pr^{3+} dissolved in organic solvents are used as lanthanide shift reagents in nmr spectroscopy.

Coordination numbers below 6 are uncommon, and occur only with bulky ligands such as $(2,6\text{-dimethylphenyl})^-$ and $[N(SiMe_3)_2]^-$. In contrast to the transition elements, the coordination number 6 is not common. The most common coordination numbers are 7, 8 and 9 and these give a variety of stereochemistries. Coordination numbers 10 and 12 occur with the larger (lighter) lanthanides and small chelating ligands NO_3^- and SO_4^{2-} (Table 29.11).

Complexes with monodentate oxygen ligands are much less stable than the chelates, and tend to dissociate in aqueous solution. There are hardly any complexes with nitrogen donor ligands except ethylenediamine and NCS^-, and these are decomposed by water. Fluoride complexes $LnF^{2+}_{(aq)}$ are formed particularly by the smaller ions, but chloride complexes are not formed in aqueous media or concentrated HCl. This is an important distinction between the lanthanide and actinide groups.

Ce^{4+} is smaller and more highly charged, and $[Ce(NO_3)_6]^{2-}$ is formed in the non-aqueous solvent N_2O_4, and is 12-coordinate. Each NO_3^- uses two oxygen atoms to coordinate to the metal.

The lanthanides form no complexes with π bonding ligands, and the lack of π bonding is attributed to the unavailability of the *f* orbitals for bonding.

It is difficult to explain the bonding in complexes with high coordination numbers. If one *s* orbital, three *p* orbitals and all six *d* orbitals in the valency shell are used for bonding, this accounts for a maximum coordination number of 9. The higher coordination numbers of 10 and 12 present a problem. They imply either participation of *f* orbitals in bonding, or bond orders of less than one.

There are few organic compounds of the lanthanides. Alkyls and aryls can be made with lithium reagents in ether solution:

$$LnCl_3 + 3LiR \rightarrow LnR_3 + 3LiCl$$

$$LnR_3 + LiR \rightarrow Li[LnR_4] \qquad \text{and } [LnMe_6]^{3-}$$

Table 29.11 Some lanthanide complexes

Coordination number	Complex	Shape
4	$[Lu(2,6\text{-dimethylphenyl})_4]^-$	Tetrahedral
6	$[Ce^{IV}Cl_6]^{2-}$	Octahedral
6	$[Er(NCS)_6]$	Octahedral
7	$[Y(\text{acetylacetone})_3H_2O]$	Mono-capped trigonal prism
8	$[La(\text{acetylacetone})_3(H_2O)_2]$	Square antiprism
8	$[Ce^{IV}(\text{acetylacetone})_4]$	Square antiprism
8	$[Eu(\text{acetylacetone})_3(\text{phenanthroline})]$	Square antiprism
8	$[Ho(\text{tropolonate})_4]^-$	Dodecahedral
9	$[Nd(H_2O)_9]^{3+}$	Tri-capped trigonal prism
10	$[Ce^{IV}(NO_3)_4(Ph_3PO)_2]$	Complex (each NO_3^- is bidentate)
12	$[Ce^{IV}(NO_3)_6]^{2-}$	Icosahedral (each NO_3^- is bidentate)

Cyclopentadienyl compounds $[Ln(C_5H_5)_3]$, $[Ln(C_5H_5)_2Cl]$ and $[Ln(C_5H_5)Cl_2]$ are known but are sensitive to water and air.

FURTHER READING

Bagnall, K.W. (ed.) (1975) *MTP International Review of Science*, Inorganic Chemistry (Series 2), Vol. 7, Lanthanides and Actinides, Butterworths, London.

Bevan, D.J.M. (1973) *Comprehensive Inorganic Chemistry*, Vol. 3 (Chapter 49: Nonstoichiometric compounds), Pergamon Press, Oxford.

Brown, D. (1968) *Halides of the Transition Elements*, Vol. I (Halides of the lanthanides and actinides), Wiley, London and New York.

Bunzli, J.G. and Wessner, D. (1984) Rare earth complexes with neutral macrocyclic ligands, *Coordination Chem. Rev.*, **60**, 191.

Burgess, J. (1988) *Ions in Solution*, Ellis Horwood, Chichester.

Callow, R.J. (1967) *The Industrial Chemistry of the Lanthanons, Yttrium, Thorium and Uranium*, Pergamon Press, New York.

Cotton, S.A. and Hart, F.A. (1975) *The Heavy Transition Metals*, Macmillan, London. (See Chapter 10.)

Emeléus, H.J. and Sharpe, A.G. (1973) *Modern Aspects of Inorganic Chemistry*, 4th ed. (Chapter 22: Inner transition elements I, the lanthanides), Routledge and Kegan Paul, London.

Evans, W.J. (1985) Organolanthanide chemistry, *Adv. Organometallic Chem.*, **24**, 131–173.

Evans, W.J. (1987) Organolanthanide chemistry, *Polyhedron*, **6**, 803–835.

Greenwood, N.N. (1968) *Ionic Crystals, Lattice Defects and Non- Stoichiometry* (Chapter 6), Butterworths, London.

Gschneidner, K.A. Jr, and LeRoy, K.A. (eds) (1988) *Handbook on the Physics and Chemistry of the Rare Earths*, North Holland, Amsterdam.

Johnson D.A. (1977) *Adv. Inorg. Chem. Radiochem.*, **20**, 1. (An excellent review on recent advances in the chemistry of the less common oxidation states of the lanthanide elements.)

Johnson D.A. (1980) Principles of lanthanide chemistry, *J. Chem. Ed.*, **57**, 475–477.

Lanthanide and Actinide Chemistry and Spectroscopy (1980) ACS Symposium, Series 131, American Chemical Society, Washington.
Moller, M., Cerny, P. and Saupe, F. (eds) (1988) *Rare Earth Elements* (Special Publication of the Society for Geology Applied to Mineral Deposits, Vol. 7), Springer Verlag, Berlin.
Muetterties, E.L. and Wright, C.M. (1967) High coordination numbers, *Q. Rev. Chem. Soc.*, **21**, 109.
Subbarao, E.C. and Wallace, W.E. (eds) (1980) *Science and Technology of Rare Earth Materials*, Academic Press, New York.
Yatsimirskii, K.B. and Davidenko, N.K. (1979) Absorption spectra and structure of lanthanide coordination compounds in solution, *Coordination Chem. Rev.*, **27**, 223–273.

PROBLEMS

1. Name the lanthanide elements in the correct order, and give their chemical symbols and electronic structures.
2. In what way are the observed oxidation states of the lanthanides related to their electronic structures?
3. Why is it difficult to separate compounds of the lanthanide elements? What methods have been used, and which of these is still used?
4. What is the lanthanide contraction, and what are its consequences?
5. In what ways does the filling of the $4f$ energy level affect the rest of the periodic table?
6. Contrast the electronic spectra of the lanthanide and transition metal ions. Why do the lanthanide ions give rise to very sharp bands in their electronic spectra, and why are the magnetic properties of their complexes little affected by the nature of the ligands?
7. Compare the coordination numbers and stereochemistries commonly found in lanthanide complexes with those commonly found in transition metal complexes.
8. Work out the number of unpaired electrons in the ground state of the following ions:

$$La^{3+}, Ce^{4+}, Lu^{3+}, Yb^{2+}, Gd^{3+}, Eu^{2+}, Tb^{4+}.$$

The actinides

30

Table 30.1 The elements and their oxidation states

Atomic number	Element	Symbol	Outer electronic structure	Oxidation states*
89	Actinium	Ac	$6d^1\ 7s^2$	**III**
90	Thorium	Th	$6d^2\ 7s^2$	III **IV**
91	Protactinium	Pa	$5f^2\ 6d^1\ 7s^2$	III IV **V**
92	Uranium	U	$5f^3\ 6d^1\ 7s^2$	III IV V **VI**
93	Neptunium	Np	$5f^4\ 6d^1\ 7s^2$	III IV **V** VI VII
94	Plutonium	Pu	$5f^6\ 7s^2$	III **IV** V VI VII
95	Americium	Am	$5f^7\ 7s^2$	II **III** IV V VI
96	Curium	Cm	$5f^7\ 6d^1\ 7s^2$	**III** IV
97	Berkelium	Bk	$5f^9\ 7s^2$	**III** IV
98	Californium	Cf	$5f^{10}\ 7s^2$	II **III**
99	Einsteinium	Es	$5f^{11}\ 7s^2$	II **III**
100	Fermium	Fm	$5f^{12}\ 7s^2$	II **III**
101	Mendelevium	Md	$5f^{13}\ 7s^2$	II **III**
102	Nobelium	No	$5f^{14}\ 7s^2$	**II** III
103	Lawrencium	Lr	$5f^{14}\ 6d^1\ 7s^2$	**III**
104	Rutherfordium	Rf	$4f^{14}\ 6d^2\ 7s^2$	

* The most important oxidation states (generally the most abundant and stable) are shown in bold. Other well-characterized but less important states are shown in normal type.

ELECTRONIC STRUCTURE AND POSITION IN THE PERIODIC TABLE

Comparison with the previous row in the periodic table shows that francium Fr and radium Ra belong to Groups I and II and their outermost electrons must be in $7s$ orbitals. The next element actinium Ac begins to fill the penultimate d shell ($—6d^1 7s^2$). It has properties typical of the Sc, Y, La group. By analogy with what happened after La, it might be expected that in the following 14 elements electrons would enter the $5f$ shell and form a second inner transition series. The next 14 elements from atomic number

90 thorium to atomic number 103 lawrencium are called the actinide elements. However, the electronic structures of the actinides do not follow the simple pattern found in the lanthanides.

Immediately after La the 4*f* orbitals become appreciably lower in energy than the 5*d* orbitals. Thus in the lanthanides the electrons fill the 4*f* orbitals in a regular way (apart from minor differences where it is possible to attain a half filled shell). It might have been expected that after Ac the 5*f* orbitals would become lower in energy than the 6*d* orbitals. However, for the first four actinide elements Th, Pa, U and Np the difference in energy between 5*f* and 6*d* orbitals is small. Thus in these elements (and their ions) electrons may occupy the 5*f* or the 6*d* levels, or sometimes both. Later in the actinide series the 5*f* orbitals do become appreciably lower in energy. Thus from Pu onwards the 5*f* shell fills in a regular way, and the elements become very similar.

Before 1940 the only actinides known were Th, Pa and U. These elements were (wrongly) thought to be part of the *d* series. The reasons for this were some chemical similarity with groups of transition metals Ti, Zr, Hf. . . .Th? and Cr, Mo, W. . . .U? The increase in the number of oxidation states formed by the elements Ac, Th, Pa and U is reminiscent of the inverted pyramid of oxidation states obtained with the *d*-block elements (see Table 17.2). Also the increased stability of the higher states follows the same pattern as found in the *d*-block. This is in contrast to the almost uniform (+III) oxidation state of the lanthanides. U is the heaviest naturally occurring element. As a result of work on the atomic bomb during World War II, and of later work on atomic energy, at least 12 more elements have been made artificially. Since these man-made elements have atomic numbers higher than $_{92}$U they are sometimes called the trans-uranium elements.

However, as the transuranium elements were discovered and studied it became apparent that they were *f*-block elements from:

1. the sharpness of the lines in their UV–visible spectra
2. magnetic studies
3. the increasing importance of the (+III) oxidation state.

It is now generally accepted that the actinides are a second inner transition series, beginning with thorium and ending with lawrencium.

The lanthanide and actinide elements may be compared (Table 30.2).

Table 30.2 The lanthanide and actinide elements

Transition elements	Lanthanides													
La	Ce	Pr	Nd	Pm	Sm	Eu	Gd	Tb	Dy	Ho	Er	Tm	Yb	Lu
Ac	Th	Pa	U	Np	Pu	Am	Cm	Bk	Cf	Es	Fm	Md	No	Lr
	Actinides													

There are many similarities between the lanthanides and the later actinides. Cm closely resembles Gd, and both have the electronic configuration f^7, d^1, s^2. The elution of Am, Cm, Bk and Cf from an ion-exchange column exactly parallels that of the lanthanides Eu, Gd, Tb and Dy. The melting points and densities of the actinide elements do not fit with the values for the *d*-block (see Appendices II and III).

The elements Pa, U, Np, Pu and Cm have very sharp lines in their absorption spectra. This is a characteristic feature of *f–f* spectra. Spectral lines from the actinides are about ten times as intense as those from the lanthanides. If there is only one *f* electron present there will be only one peak in the spectrum, and therefore it will be easy to interpret. Usually the spectra are very complex, and are very difficult to interpret. The magnetic properties of the actinides are also difficult to interpret.

Whether the elements possess any *d* electrons in their ground state configuration is of little practical importance. In the most common oxidation state (+III) the two *s* electrons and the *d* electron (if present), will be removed. The energies of the 5*f* and 6*d* orbitals are very close. The bond energy is greater than the promotion energy $5f \rightarrow 6d$. The 7*s* and 7*p* orbitals are of comparable energy. Thus the levels occupied by electrons may change depending on the nature of the ligands, or between the solid state and a solution. It is often impossible to say which orbitals are being used.

The 5*f* orbitals extend into space beyond the 6*s* and 6*p* orbitals and participate in bonding. This is in direct contrast to the lanthanides where the 4*f* orbitals are buried deep inside the atom, totally shielded by outer orbitals and thus unable to take part in bonding. The participation of the 5*f* orbitals explains the higher oxidation states shown by the earlier actinide elements. The greater extension of the 5*f* orbitals compared with the 4*f* is shown by the difference in electron resonance spectra of Nd^{3+} and U^{3+} ions in CaF_2 or SrF_2. Both ions have an f^3 ground state (spectroscopic term symbol $^4I_{9/2}$ – see Chapter 29). The U^{3+} signal shows hyperfine structure caused by interaction with fluorine nuclei, whilst Nd^{3+} ions do not.

OXIDATION STATES

The known oxidation states of the elements are shown in Table 30.1.

The (+II) state is quite rare. Am^{2+} has an f^7 configuration. It is the analogue of Eu^{2+} in the lanthanides, but it only exists in the solid as the fluoride. In contrast Cf^{2+}, Es^{2+}, Fm^{2+}, Md^{2+} and No^{2+} exist as ions in solution. Their properties are like the Group II metals, particularly Ba^{2+}. It is the most stable oxidation state for No, and corresponds to an f^{14} arrangement.

The actinides all have an oxidation state of (+III), like the lanthanides. However, this is not always the most stable oxidation state in the actinides. (+III) is not the most stable oxidation state for the first four elements Th, Pa, U and Np. For example, U^{3+} is readily oxidized in air, and in solution. The (+III) state is the most stable state for the later elements $_{95}Am \rightarrow$

$_{103}$Lw (excluding $_{102}$No). Their properties are similar to those of the lanthanides.

The most stable oxidation states for the first four elements are Th (+IV), Pa (+V) and U (+VI). These high oxidation states involve using all the outer electrons (including *f* electrons) for bonding. Though Np(+VII) exists, it is oxidizing and the most stable state is (+V). Pu shows all the oxidation states from (+III) to (+VII), but the most stable is Pu(IV). Am has a range of oxidation states from (+II) to (+VI). However, for Am and almost all the remaining elements the (+III) state is the most stable.

The (+IV) state exists for all the elements from $_{90}$Th to $_{97}$Bk, and it is the most important state for Th and Pu. M^{4+} ions are known in acid solution, and are precipitated by F^-, PO_4^{3-} and IO_3^- ions. The elements all form solid dioxides MO_2 and fluorides MF_4.

The (+V) state occurs for the elements $_{91}$Pa → $_{95}$Am, and it is the most stable state for Pa and Np. A few solid compounds are known. M^{5+} ions do not occur in solution, but MO_2^+ ions exist between pH 2–4, and these oxo-ions are linear $[O—M—O]^-$. These ions disproportionate rapidly in solution, but are found in solid compounds.

$$\overset{(+V)}{2UO_2^+} + 4H^+ \rightarrow \overset{(+IV)}{U^{4+}} + \overset{(+VI)}{UO_2^{2+}} + 2H_2O$$

The (+VI) state exists as fluorides MF_6 for the elements U, Np, Pu and Am. The (+VI) state is more widely found as the dioxo ion MO_2^{2+}. This ion is linear $[O—M—O]^{2+}$, and is stable. It exists both in solution and in crystals. The crystal structure of uranyl nitrate $UO_2(NO_3)_2(H_2O)_2$ consists of the linear $(O—U—O)^{2+}$ ion surrounded by two NO_3^- groups and two H_2O molecules. The NO_3^- groups are bidentate, so two O atoms from each NO_3^- bond to the U. O atoms from the two H_2O atoms also bond to U, giving a coordination number of 8. Similarly in the crystal structure of sodium uranyl acetate $Na[UO_2(CH_3COO)_3]$ the acetate groups are bidentate using both O atoms, so U is eight-coordinate.

The lower oxidation states tend to be ionic, and the higher ones covalent. M^{2+}, M^{3+} and M^{4+} ions are all known. Hydrolysis of these ions occurs quite readily, but can be suppressed by using acid solutions. Perchloric acid is often the most suitable as it has little tendency to form complexes. Hydrolysis of compounds in the higher oxidation states give (+V) → MO_2^+ ions and (+VI) → MO_2^{2+} ions.

OCCURRENCE AND PREPARATION OF THE ELEMENTS

All the elements after $_{82}$Pb, that is from $_{83}$Bi onwards, have unstable nuclei and undergo radioactive decay. The elements up to and including $_{92}$U occur in nature and have been known for a long time. Even though they undergo radioactive decay, Th and U are by no means rare. They make up 8.1 ppm and 2.3 ppm of the earth's crust respectively. The fact that Th and U occur at all on the earth is because the isotopes $^{232}_{90}Th$, $^{235}_{92}U$ and $^{238}_{92}U$ have half lives sufficiently long for some to have remained since the earth was formed ($t_{1/2}$ for $^{238}_{92}U$ is 4.5×10^9 years, and $t_{1/2}$ for $^{235}_{92}U$ is 7.04×10^8 years). The elements following U have shorter half lives, and any present when the earth was formed has already decayed.

If the elements are significantly radioactive they must be handled with care. The later actinides have very short half lives (often a few minutes or less). Thus it is not possible to get high concentrations, or perform anything other than quick tracer experiments. Studying some of the elements is complicated because the radiation decomposes water into H and OH radicals. These radicals may reduce higher oxidation states such as Pu(+VI), Pu(+V), Am(+VI), Am(+V) and Am(+IV).

The radioactivity produces self-heating. Ten grams of ^{239}Pu generates 0.02 watts of heat. This cannot be used as a large scale power source since this isotope is fissile and thus undergoes nuclear fission. (The critical mass of ^{239}Pu is only about 1 kg.) The heat may decompose some compounds. It also prevents accurate structure determination by X-ray diffraction, since the atoms have an unusually high degree of thermal motion. The production of heat in this way by some of the actinides is used in light-weight power sources. For example, the heat is used to produce electricity with a thermopile in heart 'pace-makers'. They were used in the first moon probes, the Apollo space mission and in satellites. The isotopes $^{238}_{94}Pu$ and $^{242}_{96}Cm$ are used for this purpose. They are α emitters, and very little shielding is required as α particles are easily stopped by surrounding material. $^{241}_{95}Am$ has also been used, but it emits γ rays in addition to α rays, and thus requires extensive shielding.

Up to 10% Th is found in Monazite sand, mixed with the lanthanides as $(ThLn)PO_4$. It is also found as the ore thorite $ThSiO_4$. U is mined as the ore pitchblende UO_2. Very small amounts of Ac, Pa, Np and Pu have been detected in these ores. These four elements are only available by synthetic routes. Plutonium is formed in large amounts from uranium fuel in nuclear reactors. This is because plutonium is fissile, and can be used for military purposes (to make atomic weapons) and also as a fuel for nuclear generating stations to make electricity.

The chemistry of Th and U resembles that of the Ti and Cr groups of transition metals in several respects. The elements with higher atomic numbers than U are called the transuranium elements. These have all been produced artificially in the period since 1940. These elements were produced using a nuclear reactor to irradiate suitable elements with neutrons. They are also made using an accelerator to bombard a sample with α particles (He nuclei), or the nuclei of light atoms such as C, B, N, O or Ne. Most of the transuranium elements were discovered (first made) at the University of California.

PREPARATION OF THE ACTINIDES

The early members of the series are usually formed by (n, γ) reactions which are usually followed by β emission. They were first made in 1940 by bombardment of U in a cyclotron at Berkeley. They are now obtained from spent U fuel rods. Though the main reaction in a reactor is fission of $^{235}_{92}U$ into two smaller nuclei with the release of a lot of energy, several secondary reactions occur. The U fuel rod is irradiated with *slow neutrons* (of energy 1 MeV). A neutron may be captured by the nucleus in a (n, γ) reaction. The neutron increases the mass number of the nucleus by one, and some energy is released as γ radiation. Further neutrons may be added in a similar way. Addition of neutrons increases the neutron to proton ratio (the n/p ratio). This eventually makes the nucleus unstable because it contains too many neutrons. The nuclei decay by converting a neutron into a proton and a β particle (electron). This reduces the n/p ratio and also increases the atomic number by one. Thus a new element is formed, one place to the right in the periodic table of the original element (see Chapter 31, under Stability and the ratio of neutrons and protons). When the fuel rod is eventually removed from the reactor, it is processed, and the new elements can be recovered. There is not much use for Np so normally only Pu is recovered. (Pu is useful both as a nuclear fuel, and for weapons.)

$$^{235}_{92}U + ^{1}_{0}n \rightarrow ^{236}_{92}U + ^{1}_{0}n \rightarrow ^{237}_{92}U \xrightarrow[t_{1/2}\ 6.7\ \text{days}]{\beta} ^{237}_{93}Np$$

$$^{238}_{92}U + ^{1}_{0}n \rightarrow ^{239}_{92}U \xrightarrow[t_{1/2}\ 23.3\ \text{min}]{\beta} ^{239}_{93}Np \xrightarrow[t_{1/2}\ 2.3\ \text{days}]{\beta} ^{239}_{94}Pu$$

The yield of the heavier elements is controlled by two factors:

1. The half lives of the various isotopes.
2. By their ability to absorb neutrons, that is their neutron cross-section.

Isotopes of elements after Pu can be made by a succession of (n, γ) reactions starting with Pu in a nuclear reactor.

$$^{239}_{94}Pu \xrightarrow{(n,\gamma)} ^{240}_{94}Pu \xrightarrow{(n,\gamma)} ^{241}_{94}Pu \xrightarrow[t_{1/2}\ 13.2\ \text{years}]{\beta} ^{241}_{95}Am$$

The stepwise addition of slow neutrons is tedious. A quicker method is to subject the sample to a very high flux or density of fast neutrons, without allowing time for the intermediate products to decay. This happened during the hydrogen bomb explosions when elements $_{99}Es$ einsteinium and $_{100}Fm$ fermium were formed. As yet this does not provide a convenient and practicable synthetic route! In reactor fuel elements ^{238}U adds a fast neutron and then loses two neutrons.

$$^{238}_{92}U \xrightarrow{(n,2n)} ^{237}_{92}U \xrightarrow[t_{1/2}\ 6.7\ \text{days}]{\beta} ^{237}_{93}Np$$

An alternative method is to bombard the sample with small ions. These must have sufficient energy to overcome the coulombic repulsion between the ion and the heavy nucleus. These ions are given a high kinetic energy of motion by accelerating them to a great speed in a linear accelerator, or in a

Table 30.3 The main isotopes and their sources

Atomic number Z	Element	Main isotopes	Half life	Source
89	Actinium	^{227}Ac	21.7 years	Natural $^{226}_{88}Ra \xrightarrow{n\gamma} {}^{227}_{88}Ra \xrightarrow[41\ min]{\beta} {}^{227}_{89}Ac$
90	Thorium	^{232}Th	1.4×10^{10} years	Naturally occurring ores
91	Protactinium	^{231}Pa	3.3×10^{4} years	Natural (0.1 ppm in U ores) and from ^{235}U fuel elements
92	Uranium	^{235}U	7.1×10^{8} years	Natural (0.7% abundance in U ores)
		^{238}U	4.5×10^{9} years	Natural (99.3% abundance in U ores)
93	Neptunium	^{237}Np	2.2×10^{6} years	Formed from U fuel elements $^{235}_{92}U \xrightarrow{n\gamma} {}^{236}_{92}U \xrightarrow{n\gamma} {}^{237}_{92}U \xrightarrow[6.7\ d]{\beta} {}^{237}_{93}Np$; $^{238}_{92}U\ (n, 2n) \rightarrow {}^{237}_{92}U$
94	Plutonium	^{238}Pu	86.4 years	Several isotopes are formed in fuel elements $^{237}_{93}Np \xrightarrow{n\gamma} {}^{238}_{93}Np \xrightarrow[2.1\ d]{\beta} {}^{238}_{94}Pu$
		^{239}Pu	2.4×10^{4} years	$^{238}_{92}U \xrightarrow{n\gamma} {}^{239}_{92}U \xrightarrow[23\ min]{\beta} {}^{239}_{93}Np \xrightarrow[2.3\ d]{\beta} {}^{239}_{94}Pu$
		^{242}Pu	3.8×10^{5} years	$^{239}_{94}Pu \xrightarrow{three\ (n\gamma)} {}^{242}_{94}Pu$
		^{244}Pu	8.2×10^{7} years	$^{239}_{94}Pu \xrightarrow{five\ (n\gamma)} {}^{244}_{94}Pu$; $^{239}_{94}Pu \xrightarrow{two\ (n\gamma)} {}^{241}_{94}Pu$
95	Americium	^{241}Am	433 years	$^{238}_{92}U \xrightarrow{\alpha n} {}^{241}_{94}Pu \xrightarrow[13.2\ yr]{\beta} {}^{241}_{95}Am$
		^{243}Am	7.7×10^{3} years	$^{239}_{94}Pu \xrightarrow{four\ (n\gamma)} {}^{243}_{94}Pu \xrightarrow{\beta} {}^{243}_{95}Am$; $^{241}_{95}Am \xrightarrow{(n\gamma)} {}^{242}_{95}Am \xrightarrow[16.0\ h]{\beta} {}^{242}_{96}Cm$
96	Curium	^{242}Cm	162 days	$^{239}_{94}Pu \xrightarrow{\alpha n} {}^{242}_{96}Cm$
		^{244}Cm	17.6 years	$^{239}_{94}Pu \xrightarrow{four\ (n\gamma)} {}^{243}_{94}Pu \xrightarrow[5.0\ h]{\beta} {}^{243}_{95}Am \xrightarrow{n\gamma}$ $^{244}_{95}Am \xrightarrow[26\ min]{\beta} {}^{244}_{96}Cm$
97	Berkelium	^{249}Bk	314 days	Intense and prolonged neutron bombardment of ^{239}Pu in nuclear reactors
98	Californium	^{249}Cf	360 years	(as above)
		^{252}Cf	2.6 years	(as above)
99	Einsteinium	^{254}Es	250 days	(as above)
100	Fermium	^{253}Fm	4.5 days	(as above)
101	Mendelevium	^{256}Md	1.5 hours	Bombardment of ^{252}Cf with He^{2+} followed by β
102	Nobelium	^{254}No	3 seconds	Bombardment of ^{246}Cm with C^{6+}
103	Lawrencium	^{257}Lr	8 seconds	Bombardment of ^{252}Cf with B^{5+}
104	Rutherfordium	^{261}Rf	Approx. 70 seconds	

cyclotron. The simplest ion used is the α particle (that is a He nucleus). These increase the mass number by four and the atomic number by two.

$$^{244}_{94}Pu + ^{4}_{2}He \rightarrow ^{248}_{96}Cm$$

Often the addition of the helium nucleus upsets the ratio of neutrons to protons (see Chapter 31), and one or more neutrons are emitted. The equations for nuclear reactions may either be written showing all the particles in the equation, e.g.

$$^{239}_{94}Pu + ^{4}_{2}He \rightarrow ^{241}_{96}Cm + 2(^{1}_{0}n)$$

or in a shorthand way with the particles added and lost shown in brackets:

$$^{239}_{94}Pu \xrightarrow{(\alpha, 2n)} ^{241}_{96}Cm$$

$$^{243}_{95}Am \xrightarrow{(\alpha, n)} ^{246}_{97}Bk$$

$$^{242}_{96}Cm \xrightarrow{(\alpha, 2n)} ^{244}_{98}Cf$$

$$^{249}_{98}Cf \xrightarrow{(\alpha, 2n)} ^{251}_{100}Fm$$

$$^{253}_{99}Es \xrightarrow{(\alpha, n)} ^{256}_{101}Md$$

The heaviest elements were obtained by bombarding the sample with accelerated ions B^{5+}, C^{6+}, N^{7+} or O^{8+}.

$$^{238}_{92}U + ^{14}_{7}N \rightarrow ^{249}_{99}Es + 3(^{1}_{0}n)$$
$$^{238}_{92}U + ^{16}_{8}O \rightarrow ^{250}_{100}Fm + 4(^{1}_{0}n)$$
$$^{246}_{96}Cm + ^{12}_{6}C \rightarrow ^{254}_{102}No + 4(^{1}_{0}n)$$
$$^{252}_{98}Cf + ^{11}_{5}B \rightarrow ^{257}_{103}Lr + 6(^{1}_{0}n)$$

The sources, half lives and mass numbers of the most accessible isotopes are given in Table 30.3.

Other isotopes are known, and some have long half lives. ^{247}Bk is only prepared with difficulty by ion bombardment in an accelerator, but it has a half life of about 7000 years.

The quantities of these elements which are available are given in Table 30.4. The elements above atomic number 100 fermium exist only as short-lived species, and only minute quantities (a few atoms) have been prepared. The most stable isotopes are $^{258}_{101}Md$ = 53 days, $^{255}_{102}No$ = 185 seconds, $^{256}_{103}Lr$ = 45 seconds and $^{261}_{104}Rf$ = approx. 70 seconds.

GENERAL PROPERTIES

Some properties of the elements are given in Table 30.5. The elements are all silvery metals. Their melting points are moderately high, but are considerably lower than those for the transition elements. The size of the ions decreases regularly along the series, because the extra charge on the

Table 30.4 Availability of various isotopes

Tonnes	Kilograms	100 grams	Milligrams	Micrograms
^{232}Th	^{237}Np	^{231}Pa	^{244}Pu	^{257}Fm
^{238}U	^{239}Pu	^{238}Pu	^{249}Bk	
		^{242}Pu	^{242}Cm	
		^{241}Am	^{252}Cf	
		^{243}Am	^{253}Es	
		^{244}Cm	^{254}Es	

Table 30.5 Some properties of the actinides

	m.p. (°C)	b.p. (°C)	Density ($g\,cm^{-3}$)	Radius M^{3+} (Å)	Radius M^{4+} (Å)
Ac	817	2470	–	1.12	–
Th	1750	4850	11.8	(1.08)	0.94
Pa	1552	4227	15.4	1.04	0.90
U	1130	3930	19.1	1.025	0.89
Np	640	5235	20.5	1.01	0.87
Pu	640	(3230)	19.9	1.00	0.86
Am	1170	2600	13.7	0.975	0.85
Cm	1340		13.5	0.97	0.85
Bk	986		14.8	0.96	0.83
Cf	(900)			0.95	0.82
Es	(860				

nucleus is poorly screened by the *f* electrons. This results in an 'actinide contraction' similar to the lanthanide contraction. Comparison of the M^{3+} ionic radii with those for lanthanides (Table 29.6) shows that the actinide and lanthanide ions are very similar in size. Hence their chemical properties are alike. However, the actinides have much higher densities and a much greater tendency to form complexes.

The actinides are reactive metals like lanthanum and the lanthanides. They react with hot water, and tarnish in air, forming an oxide coating. In the case of Th this coating is protective, but this is not so with the others. The metals react readily with HCl, but reaction with other acids is slower than expected. Concentrated HNO_3 passivates Th, U and Pu. The metals are basic and do not react with NaOH. They react with oxygen, the halogens and with hydrogen. The hydrides are nonstoichiometric and have ideal formulae MH_2 or MH_3.

The metals are usually obtained by electrolysis of fused salts, or by reducing the halides with Ca at high temperatures.

THORIUM

Thorium is by no means rare. It comprises 8.1 ppm of the earth's crust and is the thirty-ninth most abundant element. The main source is monazite

Table 30.6 Some high coordination numbers

	Coordination number	Shape
$K_4[Th(oxalate_4)] \cdot 4H_2O$	8	Square antiprism
$(NH_4)_4[ThF_8]$	9	Tricapped trigonal prism (ThF_9) sharing two edges to give infinite chains
$Mg[Th(NO_3)_6]$	12	Icosahedral. The NO_3^- groups are bidentate

sand in which it occurs up to 10% as the phosphate, mixed with phosphates of the lanthanides. Thorium is also found as uranothorite (a mixed silicate of Th and U) in the Sudbury ores (Canada). Monazite is treated with NaOH. The insoluble hydroxides are filtered off and dissolved in HCl. The pH is adjusted to 6, when the hydroxides of Th(IV), U(IV) and Ce(IV) are precipitated. This separates them from the trivalent lanthanides. The hydroxide precipitates are dissolved in 6 M HCl and extracted with tributylphosphate and kerosene. If required the metal can be obtained by reducing ThO_2 with Ca, or $ThCl_4$ with Ca or Mg. These reactions must be carried out under an atmosphere of argon, as Th is very reactive when hot.

The only stable oxidation state is Th(+IV), and the Th^{4+} ion is known both in the solid and in solution. $Th(NO_3)_4 \cdot 5H_2O$ is the best known salt, and it is very soluble in water. In dilute solutions hydrated ions $[Th(H_2O)_n]^{4+}$ exist, and on adding NaOH, a precipitate of $Th(OH)_4$ is produced. The oxide ThO_2 is formed by heating the nitrate, or by heating the metal in air. The oxide is white and has the highest melting point of any oxide (3220 °C). It is unreactive except that it dissolves in a mixture of HNO_3 and HF. The other actinide oxides are also refractory.

Anhydrous halides ThX_4 are formed by direct reaction. They are also formed by strongly heating the oxide with the appropriate halogen acid or heating the oxide with CCl_4 at 600 °C. The halides are high melting and white. On strong heating ThI_4 decomposes to the elements. This has been used to purify the metal by the van Arkel method (see Chapter 6, under 'Thermal decomposition methods of extraction'). The halides hydrolyse in moist air, giving oxohalides $ThOX_2$. The white colour of Th(IV) compounds is associated with the absence of *d* or *f* electrons. The high charge of Th^{4+} favours the formation of complexes. These often have high coordination numbers and uncommon structures (Table 30.6).

Th(III) compounds are rare, and are only found as solids. ThI_3 and ThOF have been made, but they react with water, forming Th(IV) and liberating hydrogen. ThI_3 and ThI_2 can be made:

$$2ThI_4 + Th \xrightarrow{\text{heat}} 2ThI_3 + ThI_2$$

ThI_2 is a gold coloured solid, and is a good electrical conductor. These lower-valent compounds all have some metallic conduction. ThI_2 is probably Th^{4+}, $2I^-$ and 2 electrons in a conduction band. ThI_3 is probably Th^{4+}, $3I^-$ and an electron in a conduction band. ThS has also been made.

About 500 tonnes of thorium compounds are produced annually. The two industrial uses are as follows:

1. When thorium dioxide containing 1% cerium is heated in a gas flame it emits a brilliant white light. Because of this it was widely used for making incandescent gas mantles. At one time gas lighting provided the main source of artificial light. (Electric light bulbs and fluorescent tubes have largely replaced gas lighting except for mobile use, e.g. in caravans. However, making gas mantles still accounts for half the Th produced.)
2. Naturally occurring thorium is almost entirely ^{232}Th. This isotope is not fissionable, but if irradiated in the outer part of a nuclear reactor, ^{233}U is formed.

$$^{232}_{90}Th \xrightarrow{n\gamma} {}^{233}_{90}Th \xrightarrow[22\ min]{\beta} {}^{233}_{91}Pa \xrightarrow[27\ days]{\beta} {}^{233}_{92}U$$

This isotope of uranium does not occur in nature, and has a half life of 1.6×10^5 years. It is fissionable. Since more reactor fuel is produced than is consumed by the reactor it is important as the basis of 'breeder-reactors'.

PROTACTINIUM

Traces of Pa (about 0.1 ppm of ^{231}Pa) are found in the uranium ore pitchblende UO_2. The Pa is formed as a decay product of ^{235}U in the actinium decay series (see Chapter 31). It also occurs in the neptunium and uranium natural decay series. It is difficult to isolate Pa from minerals. It is usually prepared artificially either from Th by the reaction given above, or obtained as ^{231}Pa from processing spent uranium fuel elements removed from nuclear reactors.

The study of protactinium is a story of scientific cooperation which is rarely encountered. The chemistry of this element was almost unknown until 1960 when A.G. Maddock and a team at the UK Atomic Energy Authority extracted 130 g of the element from over 50 tonnes of waste from the extraction of U. They sent samples to major laboratories throughout the world. This rapidly produced most of the information on this element.

The chemistry of Pa is particularly difficult to study because the compounds hydrolyse readily. In addition the ions polymerize forming colloidal precipitates in water and most acids. These precipitates are adsorbed on to the vessel walls or onto other precipitates. Colloidal precipitates are not formed when Pa is handled in fluoride solutions, because it forms fluoride complexes.

Pa can be extracted from solutions in HCl or HNO_3 by tributyl phosphate. It is then precipitated as PaF_4, and reduced to the metal with

Ba at 1400 °C. Pa has been obtained in 100 g quantities. The metal is shiny, malleable, and tarnishes in air.

The most stable oxidation state is (+V). Pa_2O_5 is obtained as a white solid by igniting Pa compounds in air. The oxide is weakly acidic as it is attacked by fused alkali. Heating in a vacuum, or reduction with hydrogen at 1500 °C, gives a black nonstoichiometric phase $PaO_{2.3}$ and eventually PaO_2. PaF_5 can be made by the action of F_2 on PaF_4, or BrF_3 on Pa_2O_5. It is reactive, and can be sublimed. $PaCl_5$ and oxohalides $PaOX_3$ and PaO_2X are also known. Pa(V) complexes with oxalate, citrate, tartrate, sulphate and phosphate ions are known, and some unusual fluoride complexes have been studied. In $Rb[PaF_6]$ the Pa atom is eight-coordinate. The complex $K_2[PaF_7]$ contains nine-coordinate PaF_9 groups. These form two F bridges to neighbouring groups on either side, giving a chain. In $Na_3[PaF_8]$ the $[PaF_8]^{3-}$ ion is a slightly distorted cube. In the (+IV) state, PaO_2 and the halides PaX_4 are all known, and also oxohalides $PaOX_2$. Reduction of aqueous Pa(V) solutions with zinc amalgam or Cr^{2+} gives Pa(IV), but this is readily oxidized by air. Spectra of $PaCl_4$ in HCl or $HClO_4$ are similar to those for $Ce^{3+}(4f^1)$. This suggests that Pa(IV) has a $5f^1$ configuration, and is an actinide. Pa(III) has been detected polarographically.

URANIUM

Uranium ores were originally mined as a source of Ra, which was used for radiotherapy treatment of cancer. Small amounts of U were (and still are) used to produce pale yellow or green coloured glass. This glass fluoresces under UV light. Some uranium oxide is used for colouring ceramics.

The discovery of uranium fission by Otto Hahn in December 1938 stimulated a very detailed study of nuclear physics and of uranium chemistry. The liberation of energy by splitting a nucleus was of such importance that on 2 August 1939 Albert Einstein wrote about it to Franklin D. Roosevelt (the President of the USA). He said 'Some recent work by E. Fermi and L. Szilard leads me to expect that the element uranium may be turned to a new and important source of energy in the immediate future'. History shows how right he was.

Enrico Fermi (a refugee from Italy working at the University of Chicago) used the fission process to create the first man made nuclear chain reaction on 2 December 1942. His reactor consisted of a pile of alternate layers of fuel (U and UO_2) and moderator (graphite). Strips of Cd served to absorb neutrons and thus control the chain reaction. Fermi used 400 tonnes of graphite, 50 tonnes of UO_2 and 6 tonnes of U metal.

This led to the Manhattan Project to make atomic bombs, to the discovery of the transuranium elements (elements with higher atomic numbers than U) and the development of nuclear power. Two atomic bombs were used against Japan in 1945. Uranium is now of great commercial importance as a nuclear fuel. In 1989 there were over 120 nuclear power plants producing electricity in the USA, and over 400 in the rest of the world.

Occurrence

Uranium vanadates such as carnotite $K_2(UO_2)_2(VO_4)_2 \cdot 3H_2O$ constitute the chief ore of U. It occurs as a distinctive yellow or green–yellow crust, or in sandstones and soft aggregates. U is also mined as oxide ores, the most important being uraninite and pitchblende. These are black–brown coloured, nonstoichiometric and approximate to UO_2. They are often oxidized, and gummite is orange–yellow, brown or black and is a mixture of weathered U and Th ores. The stoichiometric oxides UO_2, U_3O_8 and UO_3 are black–brown, green–black and orange–yellow respectively. U is the forty-eighth most abundant element in the earth's crust (2.3 ppm). It is more abundant than some 'well known' elements such as Ag, Hg, Cd and I. World production of U (excluding the USSR) was 37 300 tonnes in 1988. This is based on the metal content, and is equivalent to 44 000 tonnes of U_3O_8. The main producers of U ores are Canada 35.5%, the USA 13%, South Africa 10%, Australia 9.5%, Namibia and France 9% each and Niger 8%. Half of the USA production comes from the Grants area of New Mexico.

Extraction

The extraction processes are complex. Ores are crushed and concentrated by physical and chemical means. Ore may contain 0.2% U, so 1 tonne of ore yields only 4 lbs or under 2 kg of U_3O_8. First the ore is concentrated using the very high density of U in flotation methods. Then it is roasted in air, and leached with H_2SO_4 (often with an oxidizing agent such as MnO_2 to ensure conversion to U(+VI)). This is precipitated as sodium diuranate $Na_2U_2O_7$, a bright yellow solid called 'yellowcake'. This dissolves in HNO_3, forming uranyl nitrate $UO_2(NO_3)_3(H_2O)_n$. Uranyl nitrate is purified either by adding ammonia to precipitate UO_3, or by solvent extraction of uranyl nitrate from aqueous solution into tributyl phosphate in an inert hydrocarbon diluent. The final steps are conversion to UF_4 followed by reduction of UF_4 with Ca or Mg to give the metal.

Nuclear fission

Naturally occurring uranium contains three isotopes: 99.3% ^{238}U, 0.7% ^{235}U and traces of ^{234}U. The isotope ^{235}U is fissile, and if it is irradiated with thermal (slow) neutrons the nucleus breaks up into two smaller nuclei. At least a million times more energy is liberated by this fission than from a chemical reaction. The nucleus may split giving several products (see 'Induced nuclear reactions', Chapter 31), one such reaction being:

$$^{235}_{92}U + ^{1}_{0}n \rightarrow ^{138}_{53}I + ^{95}_{39}Y + 3(^{1}_{0}n) + 2 \times 10^{10}\,\text{kJ mol}^{-1}\ \text{energy}$$

If one of the neutrons evolved splits another ^{235}U nucleus then a self-perpetuating nuclear chain reaction will be started. This liberates energy at a constant rate. Since three neutrons are evolved per fission, these could in principle split three further ^{235}U nuclei. This would liberate even more neutrons and start a branched chain reaction. This would run out of control, liberating energy at an ever increasing rate, resulting in an explosion.

It is very difficult to start and maintain a chain reaction. The neutrons produced by the fission are 'fast neutrons', and have energies of 2×10^8 kJ mol^{-1}. They tend to escape and are not very effective at causing further fission. They are much less effective than 'slow neutrons' with an energy of about 2 kJ mol^{-1}. (These are sometimes called 'thermal neutrons', because their energy is equivalent to thermal energy attainable at room temperature.) Two things can be done to increase the chance of a chain reaction.

1. The fast neutrons may be slowed down by collision with atoms of hydrogen, deuterium or carbon. These materials are called moderators.
2. A large enough mass of $^{235}_{92}U$ is needed to ensure that sufficient neutrons hit another fissile U nucleus rather than escape. Thus there is a critical mass. The size of this depends on the shape of the material and on its purity. A sphere has the minimum surface area, which minimizes the chance of neutrons escaping. Rods or sheets have a much larger area: hence neutrons can escape more easily. The chance of neutrons causing fission increases if the proportion of the fissile isotope is increased. Thus it is usual to 'enrich' the fuel.

Control rods are used to make sure that the reaction does not get out of control. The control rods absorb neutrons. The rods are lowered into the reactor to absorb neutrons and slow it down, or are raised out of the reactor to allow it to speed up. Control rods may be made of boron steel, cadmium or hafnium.

$$^{11}_{5}B + ^{1}_{0}n \rightarrow ^{12}_{5}B + \gamma$$
$$^{113}_{48}Cd + ^{1}_{0}n \rightarrow ^{114}_{48}Cd + \gamma$$

Paradoxically chain reactions can only be made to result in an explosion with difficulty. To get an explosion conditions must be such that the neutron propagation factor is greater than one. This means that more than one neutron from each fission is effective at causing another fission. This is called a branched chain reaction. To attain these conditions bomb grade fuel is very highly enriched. It may contain up to 80% ^{235}U. There are great difficulties in achieving this magnitude of enrichment. Normally the energy liberated melts the radioactive material, thus allowing it to spread out. Because of the increased surface area, more neutrons escape: hence the chain reaction ceases. Only if the fissile material is prevented from spreading out by some form of containment will a chain reaction result in an explosion. The temperatures reached are similar to that of the sun, and there is no casing that will withstand such temperatures. The first atomic bomb dropped on Hiroshima used ^{235}U. Two sub-critical masses of enriched ^{235}U at opposite ends of a gunbarrel were shot together by a small conventional explosion, which held them together long enough for a nuclear explosion to occur.

In a nuclear reactor, the neutron propagation factor is very close to one. This means that only one neutron from each fission is allowed to cause another fission. Thus the release of energy is controlled, and can be used for peaceful purposes such as generating electricity. The surplus neutrons may escape, or may be absorbed by the neutron absorbing control rods or by ^{238}U which is also present in the fuel rods.

The preparation of enriched U fuels containing large amounts of fissile ^{235}U is difficult. Fuel is usually enriched to between 2% and 4% of ^{235}U for civil nuclear reactors in power stations. A much higher enrichment is required for bombs and reactors for submarines (70% or 80%). There are four methods of separating the isotopes of U: thermal diffusion, gaseous diffusion, electromagnetic separation and using a gas centrifuge. Large scale separation is now carried out using the different rates of diffusion of gaseous $^{235}UF_6$ and $^{238}UF_6$ (see Chapter 15). The gas centrifuge method is increasing in importance.

$$UO_2 + HF \rightarrow UF_4 \xrightarrow{F_2} UF_{6\ (gas)}$$

When ^{238}U absorbs neutrons it forms the heavier element Pu which is itself fissile and can be used as a nuclear fuel. Since a larger quantity of Pu may be formed than the quantity of ^{235}U that is consumed, the reactor is called a fast breeder reactor.

Chemical properties

Uranium and the next three elements Np, Pu and Am show oxidation states of (+III), (+IV), (+V) and (+VI). These are similar to each other, except that the most stable state drops from U(VI) to Np(V) to Pu(IV) to Am(III). In the (+III) and (+IV) states the compounds are similar to lanthanides. The ions formed in the oxidation states (+III), (+IV), (+V) and (+VI) are M^{3+}, M^{4+}, MO_2^+ and MO_2^{2+} respectively. Oxidation–reduction reactions are rapid between M^{3+} and M^{4+}, or MO_2^+ and MO_2^{2+}, as these only involve the transfer of an electron. Oxidation of M^{4+} to MO_2^{2+} is slow because it involves transfer of oxygen.

Uranium is a reactive metal. Finely divided metal reacts with boiling water to give a mixture of UH_3 and UO_2. The metal dissolves in acids, and reacts with hydrogen, oxygen, the halogens and many elements.

Hydrides

Uranium reacts with hydrogen even at room temperature, though the reaction is faster at 250 °C, giving UH_3 as a black pyrophoric powder. The hydride is very reactive, and is often more suitable than the metal for making other compounds:

$$2UH_3 + 4H_2O \rightarrow 2UO_2 + 7H_2$$
$$2UH_3 + 4Cl_2 \rightarrow 2UCl_4 + 3H_2$$

$$2UH_3 + 8HF \rightarrow 2UF_4 + 7H_2$$
$$UH_3 + 3HCl \rightarrow UCl_3 + 3H_2$$

Oxides

The U–O system is complicated because there are several stable oxidation states, and the compounds are often nonstoichiometric. UO_2 is brown–black and occurs in pitchblende. U_3O_8 is greenish black, whilst UO_3 is orange–yellow. Some reactions are given:

$$UO_2(NO_3)_2 \cdot 2H_2O \xrightarrow{350\,°C} UO_3 \xrightarrow{CO\ 350\,°C} UO_2$$
$$UO_3 \xrightarrow{700\,°C} U_3O_8$$
$$U + O_2 \xrightarrow{heat} U_3O_8$$

All three oxides are basic and dissolve in acids. UO_3 dissolves in HNO_3, forming the yellow uranyl ion $[O{=}U{=}O]^{2+}$. Crystallizing this solution gives uranyl nitrate $UO_2(NO_3)_2(H_2O)_n$. The number of molecules of water of crystallization may be two, three or six depending on whether it is crystallized from fuming, concentrated or dilute HNO_3. Crystals of the dihydrate have an unusual eight-coordinate structure. This comprises a linear $[O{=}U{=}O]^{2+}$ group perpendicular to a hexagon of six oxygen atoms (four from two bidentate NO_3^- groups and two from water molecules) (Figure 30.1). The main uranium halides and their colours are listed in Table 30.7.

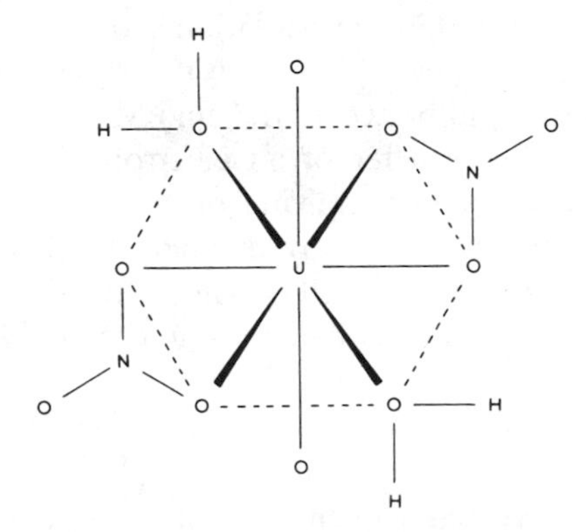

Figure 30.1 Uranyl nitrate dihydrate $UO_2(NO_3)_2 \cdot 2H_2O$.

Halides

Some of the reactions of the fluorides are given:

$$
\begin{array}{c}
UO_2 \xrightarrow{HF} UF_4 \xrightarrow{F_2\ 240\,^\circ C} UF_5 \xrightarrow[HBr\ 65\,^\circ C]{HF} [UF_6]^- \text{ and } [UF_8]^- \\
UF_4 \xrightarrow{Al\ 900\,^\circ C} UF_3 \qquad UF_4 \xrightarrow{F_2\ 400\,^\circ C} UF_6
\end{array}
$$

UF_6 and UCl_6 are octahedral but all the other halides are polymeric, and have high coordination numbers. The halides are all hydrolysed by water. The hexahalides dissolve in strong acid and give the UO_2^{2+} ion. At higher pH values this hydrolyses and polymerizes via hydroxyl bridges, giving:

$$
>UO_2{<}^{O(H)}_{O(H)}{>}UO_2{<}^{O(H)}_{O(H)}{>}UO_2<
$$

UF_6 is obtained as colourless crystals, m.p. 64 °C. It is a powerful fluorinating agent, and is rapidly hydrolysed by water. UF_5 tends to disproportionate to UF_4 and UF_6. The tetrahalides are the most stable.

Table 30.7 Halides of uranium

Oxidation state	Fluorides	Chlorides	Bromides	Iodides
+III	UF_3 green	UCl_3 red	UBr_3 red	UI_3 black
+IV	UF_4 green	UCl_4 green	UBr_4 brown	UI_4 black
+V	UF_5 white–blue	U_2Cl_{10} red–brown		
+VI	UF_6 white	UCl_6 black		
	U_2F_9 black			
	U_4F_{14} black			
	U_5F_{22} black			

NEPTUNIUM, PLUTONIUM AND AMERICIUM

Except for $^{239}_{94}Pu$, which is very important as a nuclear fuel and for bombs, the other transuranium elements have few uses outside research. $^{237}_{93}Np$ and $^{239}_{94}Pu$ can be extracted in kilogram quantities, and $^{241}_{95}Am$ and $^{243}_{95}Am$ in 100 g quantities from spent uranium fuel rods which have been used in a nuclear reactor. Their separation is extremely difficult and hazardous. Not only are they mixed with highly radioactive fission products, but the material is also fissile. The critical mass for a sphere of Pu metal is about 10 kg but in solution less than 1 kg may be critical. Furthermore, Pu is extremely toxic (a dose of 10^{-6} g may be fatal and smaller doses are carcinogenic). Am can be obtained from spent fuel rods. It is produced by intense neutron irradiation of pure Pu at the Oak Ridge National Laboratories in Tennessee.

The reprocessing of nuclear fuel rods is an important new technology. Plainly, reprocessing is necessary in breeder reactors to extract the new elements produced so that they may be used as fuel. Reprocessing is also essential in normal thermal reactors. This is because some of the fission products that are produced absorb neutrons. Thus they will stop the chain reaction before all the ^{235}U or ^{239}Pu has been used. There are over 30 different elements produced, including Sr, the second row transition elements, I, Xe, Cs, Ba, La and the lanthanides. (Many of the isotopes formed are radioactive: the best known are ^{90}Sr and ^{138}I.) There are large amounts of U and Pu, and small amounts of the other transuranium elements.

Fuel rods removed from a reactor are immersed in water for 100 days. This keeps them cool while the highly radioactive isotopes with short half lives such as $^{131}_{53}I$ ($t_{1/2}$ = 8 days) lose most of their activity.

In the Purex process the fuel rods are dissolved in 7 M HNO_3 and extracted with tributyl phosphate. UO_2^{2+} and Pu(+IV) are extracted in the same way. This leaves the other transuranium elements (mainly Np, Am and Cm) together with other fission products (mainly second row transition elements and lanthanides) in the aqueous solution. Careful reduction of the UO_2^{2+}/Pu(+IV) solution with SO_2, NH_2OH or iron(II)sulphamate $Fe(NH_2SO_3)_2$ gives Pu(+III). This is easily separated from UO_2^{2+} and U^{4+} by solvent extraction. U is precipitated as uranyl nitrate, and Pu as the oxalate or fluoride. Eventually UO_2 and PuO_2 are recovered.

Np, Pu and Am are reactive metals similar to U. They dissolve in acids, and react with hydrogen, oxygen, the halogens and many elements. The oxidation states (+III) → (+VI) are present in solution as M^{3+}, M^{4+}, MO_2^{+} and MO_2^{2+}. The (+VI) state becomes increasingly oxidizing in the order U → Np → Pu → Am. AmO_2^{2+} is as strongly oxidizing as $KMnO_4$. The most stable states are Np(+V), Pu(+IV) and Am(+III). Very strong oxidation of alkaline solutions of NpO_2^{2+} and PuO_2^{2+} with ozone or HIO_4 yields the (+VII) state which has been isolated as $Li_5[NpO_6]$ and $Li[PuO_6]$. The (+VII) state is strongly oxidizing and when acidified it is rapidly reduced to the (+VI) state and H_2O is oxidized to O_2.

The most important oxides are the dioxides MO_2, but the oxide systems may be nonstoichiometric, and contain various solid solutions. The halides are similar to those of uranium in structure, properties and preparation. A list of known compounds is given in Table 30.8.

All isotopes of plutonium are important as nuclear fuels. ^{237}Np is converted by neutron irradiation to ^{238}Pu for use as a power source in satellites. ^{241}Am is a valuable laboratory source of α particles.

THE LATER ACTINIDE ELEMENTS

Much less is known about the later elements curium, einsteinium, fermium, mendelevium, nobelium and lawrencium. This is due in part to their limited availability, and partly to their instability. Table 30.4 shows that only Cm is available in macroscopic quantities, and information on the others is largely from tracer studies.

Table 30.8 Halides of neptunium, plutonium and americium

+III	NpF_3 purple–black	PuF_3 purple	AmF_3 pink
	$NpCl_3$ white	$PuCl_3$ emerald green	$AmCl_3$ pink
	$NpBr_3$ green	$PuBr_3$ green	$AmBr_3$ white
	NpI_3 brown	PuI_3 brown	AmI_3 yellow
+IV	NpF_4 green	PuF_4 brown	AmF_4 tan
	$NpCl_4$ red–brown		
	$NpBr_4$ red–brown		
+VI	NpF_6 brown	PuF_6 red–brown	

Interest in these elements is largely concerned with showing that the second half of the actinides resemble the lanthanides quite closely. In spite of similarities, the actinides can be separated from the lanthanides quite easily, as the actinides form complexes more readily. For example, with concentrated HCl the actinides form chloro complexes. If both groups of ions are adsorbed on a cation exchange column, the actinides can be eluted with concentrated HCl. The actinide ions are separated from each other by ion exchange using citrate solutions to elute them. The order in which the actinides are eluted shows a close similarity to the order in which the lanthanides are eluted.

The (+III) oxidation state is the most stable for all but one of the elements with atomic numbers 96–103, i.e. Cm, Bk, Cf, Es, Fm, Md, (No) and Lw. The exception is No, which is most stable as No(+II). This is stable because it has a favourable f^{14} electronic configuration.

There is evidence of (+II) states for the elements 98–102, i.e. californium to nobelium. Except for nobelium this state is reducing or strongly reducing in nature. Higher oxidation states of Cm(+IV) are found in the solid but not in solution with compounds such as CmF_4 and $Rb[CmF_6]$. Bk(+IV) compounds are oxidizing. They exist in both the solid and in solution, and BkO_2 and $Cs_2[BkCl_6]$ have been isolated. Lawrencium exists only in the (+III) state and resists both oxidation and reduction, again illustrating the stability of an f^{14} electronic arrangement.

The first three elements curium, berkelium and californium have been obtained in milligram quantities. Their chemistry has been studied by normal small scale methods, and compounds have been isolated. The remaining elements have been studied by radioactive tracer methods, because they are only available in such minute amounts. In this technique the traces of the heavy elements are precipitated, or form complexes, in the presence of major quantities of a carrier element which has similar chemical properties. Thus mendelevium is studied using the lanthanide element europium, and the mendelevium is detected and followed by its radioactivity.

The elements up to 100 fermium undergo radioactive decay mainly by emitting α particles or β particles (see Chapter 31). The elements become increasingly unstable as the atomic number increases, and nobelium has a half life of only three seconds (Table 30.3). With these very heavy elements, spontaneous nuclear fission becomes the most important method of decay. ^{252}Cf could become a valuable neutron source.

FURTHER EXTENSION OF THE PERIODIC TABLE

The actinide series is complete at element 103, lawrencium. Elements 104–109 have been reported recently and are *d*-block elements. There are currently two major groups working on producing 'superheavy' elements, one in California, USA and the other at Dubna near Moscow in the USSR. By convention, the workers who discover a new element have the right to name it. Element 104 (Unq) was first reported by Russian workers and named kurchatovium Ku (after Igor Kurchatov). However, their work was repeated by American workers who obtained different results and named the element rutherfordium Rf (after Ernest Rutherford). It appears to resemble hafnium in the *d*-block. Tracer studies have been carried out, and $UnqCl_4$ seems to be similar to $HfCl_4$. Element 105 (Unp), provisionally named hahnium Ha (after Otto Hahn), seems to resemble tantalum. $UnpCl_5$ and $UnpBr_5$ have both been studied. Element 105 Unp has a half life of about 1.5 seconds and $UnpBr_5$ was observed from only 18 atoms. Both elements 104 and 105 have been made by bombarding actinides with the accelerated nuclei of light atoms. For example, element 105 was made from californium by ion bombardment with ^{15}N nuclei. The very short half lives of the isotopes and the increasing importance of spontaneous fission as the mode of decay of the elements with atomic numbers greater than 100 would suggest that extension of the periodic table to much higher atomic numbers is not very likely.

The IUPAC proposed a system for naming elements with $Z > 100$.

1. The names are derived by using roots for the three digits in the atomic number of the element and adding the ending -ium. The roots for the numbers are:

0	1	2	3	4	5	6	7	8	9
nil	un	bi	tri	quad	pent	hex	sept	oct	enn

2. In certain cases the names are shortened; for example, bi ium and tri ium are shortened to bium and trium, and enn nil is shortened to ennil.
3. The symbol for the element is made up from the first letters from the roots which make up the name. The strange mixture of Latin and Greek roots has been chosen to ensure that the symbols are all different.

Though the names are written as a complete word, in the examples below a hyphen has been inserted between each part of the name to make them more understandable. These hyphens should be omitted.

Isotopes of the superheavy elements which were known with reasonable certainty in 1989 are listed in Table 30.10 together with their half lives.

Table 30.9 IUPAC nomenclature for the superheavy elements

Atomic number	Name	Symbol	Atomic number	Name	Symbol
			110	un-un-nilium	Uun
101	un-nil-unium	Unu	111	un-un-unium	Uuu
102	un-nil-bium	Unb	112	un-un-bium	Uub
103	un-nil-trium	Unt	113	un-un-trium	Uut
104	un-nil-quadium	Unq	114	un-un-quadium	Uuq
105	un-nil-pentium	Unp	115	un-un-pentium	Uup
106	un-nil-hexium	Unh	116	un-un-hexium	Uuh
107	un-nil-septium	Uns	117	un-un-septium	Uus
108	un-nil-octium	Uno	118	un-un-octium	Uuo
109	un-nil-ennium	Une	119	un-un-ennium	Uue
			120	un-bi-nilium	Ubn
			130	un-tri-nilium	Utn
			140	un-quad-nilium	Uqn
			150	un-pent-nilium	Upn

Hyphens have been put in the names for clarity. They should be omitted.

Table 30.10 Superheavy elements and their half lives

Atomic number	Name	Symbol	Half life
104	un-nil-quadium	$^{259}_{104}Unq$	3 s or 255 257 s
		$^{261}_{104}Unq$	65 s
105	un-nil-pentium	$^{258}_{105}Unp$	4 s
		$^{260}_{105}Unp$	1.5 s
		$^{262}_{105}Unp$	34 s
106	un-nil-hexium	$^{260}_{106}Unh$	180 s
		$^{263}_{106}Unh$	0.9 s or 259 264 s
107	un-nil-septium	$^{262}_{107}Uns$	0.12 s
108	un-nil-octium	$^{265}_{108}Uno$	2×10^{-6} s or 263 264 s
109	un-nil-ennium	$^{266}_{109}Une$	5×10^{-6} s

Hyphens have been put in the names for clarity. They should be omitted.

Elements with an even number of protons in the nucleus (even atomic number) are usually more stable than their neighbours with odd atomic numbers (Harkins' rule). This means that they are less likely to decay, and are more abundant. Also nuclei with both an even number of protons and an even number of neutrons are more likely to be stable. The nucleus has a shell structure, with different energy levels, broadly similar to the energy levels of extra-nuclear electrons. A nucleus is more stable than average if the number of neutrons or protons is 2, 8, 20, 28, 50, 82 or 126. These are called 'magic numbers', and can be explained by the shell structure of the nucleus. This theory also requires the inclusion of numbers 114, 164 and 184 in the series of magic numbers. The stability is particularly high if both the number of protons and the number of neutrons are magic numbers. Thus $^{208}_{82}Pb$ is very stable with 82 protons, and 208 − 82 = 126 neutrons. This suggests that nuclides such as $^{278}_{114}Uuq$, $^{298}_{114}Uuq$ and $^{310}_{126}Ubh$ might be stable enough to exist.

It is just possible that stable isotopes of the very heavy elements could be made, but that the preparative techniques have so far only succeeded in producing unstable isotopes. Considerable efforts are being made to produce elements 114 and 126 which seem to be favourable nuclear arrangements. The elements up to atomic number 105 have been made by bombarding actinides with light nuclei such as He, B, C, N and O. Instead of continuing to build up the elements gradually in small steps, attempts are being made to make elements beyond $Z = 105$ by bombarding fairly heavy nuclei such as $_{82}Pb$ or $_{83}Bi$ with nuclei of medium sized atoms. The nucleus formed when they fuse is chosen to be close to a magic number. The newly formed nucleus will decay, emitting various particles, and the energy of the accelerated particle is kept as low as possible to lessen the chance of fission. There are enormous practical difficulties. In addition it is extremely expensive to build accelerators capable of imparting sufficient energy to the medium weight nuclei prior to bombardment. Element 107 has been made by bombarding $^{209}_{83}Bi$ with accelerated $^{54}_{24}Cr$.

$$^{209}_{83}Bi + ^{54}_{24}Cr \rightarrow ^{261}_{107}Uns + 2^{1}_{0}n$$

Element 109 has been reported from the USSR and was made by bombarding $^{209}_{83}Bi$ with accelerated $^{56}_{26}Fe$, forming $_{109}Une$. Only a few atoms have been prepared. There are islands of stability around $Z = 114$ which should be like lead, and around $Z = 126$. The latter is interesting, and could contain a new series with *g* electrons.

FURTHER READING

Abelson, P.H. (1989) Products of neutron irradiation, *J. Chem. Ed.*, **66**, 364–366. (Reflections on nuclear fission at its half century.)

Bagnall, K.W. (ed.) (1975) *MTP International Review of Science*, Inorganic Chemistry (Series 2), Vol. 7, Lanthanides and Actinides, Butterworths, London.

Brown, D. (1968) *Halides of the Transition Elements*, Vol. I (Halides of the lanthanides and actinides), Wiley, London and New York.

Cadman, P. (1986) Fuel processing and reprocessing, *Education in Chemistry*, **23**, 47–50.

Callow, R.J. (1967) *The Industrial Chemistry of the Lanthanons, Yttrium, Thorium and Uranium*, Pergamon Press, New York.

Cleveland, J.M. (1970) *The Chemistry of Plutonium*, Gordon and Breach, New York.

Cordfunke, E.H.P. (1969) *The Chemistry of Uranium*, Elsevier, Amsterdam.

Edelstein, N.M. (ed.) (1982) *Actinides in Perspective*, Pergamon Press, New York.

Edelstein, N.M., Navratil, J.D. and Schulz, W.W. (eds) (1985) *Americium and Curium Chemistry and Technology*, Dordrecht, Lancaster, Reidel. (Papers from a symposium in 1984 at Honolulu.)

Emeléus, H.J. and Sharpe, A.G. (1973) *Modern Aspects of Inorganic Chemistry*, 4th ed. (Chapter 23: Inner transition elements, the actinides), Routledge and Kegan Paul, London.
Erds, P. and Robinson, J.M. (1983) *The Physics of Actinide Compounds*, Plenum, New York and London.
Evans, C.H. (1989) The discovery of the rare earth elements, *Chemistry in Britain*, **25**, 880–882.
Handbook on the Physics and Chemistry of the Actinides, North Holland, Amsterdam and Oxford.
Herrmann, C. (1979) Superheavy element research, *Nature*, **280**, 543.
Katz, J.J., Seaborg, G.T. and Morss L.R. (eds) (1986) *The Chemistry of the Actinide Elements*, 2nd ed., Chapman and Hall, London and New York.
Kumar, K. (1989) *Superheavy Elements*, Hilger.
Lanthanide and Actinide Chemistry and Spectroscopy (1980) ACS Symposium, series 131, American Chemical Society, Washington.
Lodhi, M.A.K. (1978) *Superheavy Elements*, Pergamon Press, New York.
Organessian, Y.T. et al. (1984) Experimental studies on the formation of radioactive decay isotopes with $Z = 104–109$, *Radiochim. Acta*, **37**, 113–120.
Seaborg, G.T. (1963) *Man-Made Transuranium Elements*, Prentice Hall, New York.
Seaborg G.T. (ed.) (1978) *The Transuranium Elements: Products of Modern Alchemy*, Dowden, Hutchinson and Ross, Stroudsburg. (This gives 120 of the most important papers in the history of making these man-made elements.)
Seaborg, G.T. (1989) Nuclear fission and transuranium elements, *J. Chem. Ed.*, **66**, 379–384. (Reflections on nuclear fission at its half century.)
Sime, R.L. (1986) The discovery of protoactinium, *J. Chem. Ed.*, **63**, 653–657.
Spirlet, J.C., Peterson, J.R. and Asprey, L.B. (1987) Purification of actinide metals, *Adv. Inorg. Chem.*, **31**, 1–41.
Steinberg, E.P. (1989) Radiochemistry of the fission products, *J. Chem. Ed.*, **66**, 367–372. (Reflections on Nuclear Fission at its half century.)
Taube, M. (1974) *Plutonium: A General Survey*, Verlag Chemie, Weinheim.
Thompson, R. (ed.) (1986) *The Modern Inorganic Chemicals Industry*, Special Publication No. 31, The Chemical Society, London. (See the chapter 'The Inorganic Chemistry of Nuclear Fuels' by Findlay, J.R. et al.)
Tominaga, T. and Tachikawa, E. (1981) *Modern Hot-Atom Chemistry and Its Applications* (Inorganic Chemistry Concepts Series), Springer-Verlag, New York.
Wolke, R.L. (1988) Marie Curie's Doctoral Thesis: Prelude to a Nobel Prize, *J. Chem. Ed.*, **65**, 561–573.

PROBLEMS

1. Name the actinide elements in their correct order, and give their chemical symbols.

2. Which actinide isotopes are available (a) in tonnes, (b) in kilograms and (c) in gram quantities?

3. Compare and contrast the pyramid of oxidation states found in the first row of the transition elements with the oxidation states found in lanthanides and in the actinides.

4. What are the main sources of Th and U? How are the metals obtained?

5. Describe the methods which have been used to separate the isotopes of uranium. Explain the difficulties.

6. Give some typical reactions and comment on the structures of +3 and +4 ions of the actinides.
7. What elements would you expect rutherfordium (atomic number 104) and the elements of atomic numbers 105, 106, 107 and 112 to resemble?
8. The electronic configurations and position of the heavier elements in the periodic table are controversial. What are the possibilities, and what is the evidence?

Other Topics

Part Six

The atomic nucleus

31

STRUCTURE OF THE NUCLEUS

An atom consists of a positively charged nucleus, surrounded by a cloud of one or more negatively charged electrons. The charges balance exactly and the atom is electrically neutral. The nucleus is made up of positively charged *protons* and electrically neutral *neutrons*, which are bound together by very strong forces. These are short range forces, which fall off very rapidly as the distance between the particles increases. The particles which make up the nucleus are called collectively *nucleons*.

The radius of a nucleus is incredibly small, roughly 10^{-15} m. Nuclear distances are measured in femtometres (1 fm = 10^{-15} m). (Most atoms are 1–2 Å, i.e. 1–2 × 10^{-10} m in radius. The nucleus of oxygen, for example, has a radius of 2.5 fm.) To get some feel for how small the nucleus really is, imagine the nucleus measures 1 cm in diameter. On the same scale the diameter of the atom would be about 1000 m. Most of the mass of an atom is concentrated in the nucleus. As a result its density is very high, approximately 2.4 × 10^{14} g cm^3 or about 10^{13} times the density of the densest element iridium (22.61 g cm^{-3}).

The *atomic number Z* of an element is the number of protons in the nucleus. This is equal to the number of orbital electrons round the atom. The *mass number A* is the sum of the neutrons and protons in the nucleus.

Liquid drop model

The nucleus is sometimes described in terms of a 'liquid drop'. A small liquid drop is almost spherical. It is held together by short range forces – the attraction to near neighbours. Molecules at the edge of the drop feel the attractive force on one side only: this is the surface tension effect. A large liquid drop becomes elongated, and needs only a little disturbance to make it break into pieces.

In a similar way a nucleus is held together by short range forces (the exchange of π mesons). The 'surface tension effect' ensures that small nuclei are spherical. As the mass of the nucleus increases, the repulsion between protons increases more rapidly than the attractive forces. To minimize the repulsive force, the shape of the nucleus is deformed, just like the elongation of a large liquid drop.

This model was suggested by the Dutch physicist Niels Bohr to explain why heavy nuclei undergo nuclear fission. Uranium (element number 92) of mass number 235, $^{235}_{92}U$, is so deformed that the addition of a little extra energy, e.g. absorption of a neutron, causes the nucleus to break into two smaller nuclei. This is called nuclear fission, and when it occurs a large amount of energy is released. Nuclei with mass numbers larger than that of uranium are so deformed that they undergo spontaneous fission. This means that the nuclei disintegrate (break up on their own) without any external peturbation.

The density of all nuclei except the very lightest is almost constant. Thus the volume of the nucleus is directly proportional to the number of nucleons present. Nuclei with differing numbers of nucleons are regarded as different sizes of drop. The range of nuclear attractive forces is very small (2 fm–3 fm). The nucleons can move inside the nucleus rather like the particles in a liquid.

Shell model

In the electron cloud surrounding an atom we can distinguish different electronic energy levels. The electrons are arranged in different shells and orbitals which may be described by four quantum numbers. The nucleons are thought to be arranged in a definite way in the nucleus. Thus the nucleons are arranged in shells, corresponding to different energy levels. When the nucleons occupy the lowest energy levels this corresponds to the ground state. Under different conditions the nucleons may occupy different (higher or excited) energy levels. Usually the population of these higher nuclear energy levels is so short-lived that it cannot be observed. Thus the properties of the nucleus depend only on the number of neutrons and protons, and not on the energy levels they occupy.

In a few cases the excited nuclear states have a measurable life, and when this occurs nuclear isomers can exist. Nuclear isomers are simply nuclei with the same number of neutrons and protons, but whose energies differ. It follows that the masses of the isomers differ by a very small amount corresponding to the difference in energy.

Many nuclei are *unstable* even when they correspond to the ground state. Unstable nuclei decompose by emitting various particles and electromagnetic radiation, and this is called radioactive decay. Over 1500 unstable nuclei are known. If no radioactive decomposition can be detected the nucleus is said to be *stable*.

The shell structure is supported by a periodicity in nuclear properties. Certain combinations of neutrons and protons are particularly stable:

1. Elements of even atomic number are more stable and more abundant than neighbouring elements of odd atomic number. This is known as Harkin's rule. (The rule applies almost universally, but ^{1}H is a notable exception.)

2. Elements of even atomic number are richer in isotopes and never have less than three stable isotopes (average 5.7). Elements with odd atomic numbers often have only one stable isotope and never have more than two.
3. There is a tendency for the number of neutrons and the number of protons in the nucleus to be even (Table 31.1).

Table 31.1 The number of neutrons, protons and stable nuclei

Number of protons P (= Z)	Number of neutrons N	Number of stable isotopes
Even	Even	164
Eeven	Odd	55
Odd	Even	50
Odd	Odd	4

This suggests that nucleons may be paired in the nucleus in a similar way to the pairing of electrons in atomic and molecular orbitals. If two protons spin in opposite directions, the magnetic fields they produce will mutually cancel each other. The small amount of binding energy generated is sufficient to stabilize the nucleus. However, this is not the most important source of energy in the nucleus.

Certain nuclei are extra-stable, and this is attributable to a filled shell. Nuclei with 2, 8, 20, 28, 50, 82 or 126 neutrons or protons are particularly stable and have a large number of isotopes. These numbers are termed 'magic numbers'. The numbers 114, 164 and 184 should also be included in the series of magic numbers. When both the number of protons and the number of neutrons are magic numbers the nucleus is very stable. For example, $^{208}_{82}Pb$ is very stable and has 82 protons, and 208 − 82 = 126 neutrons. The emission of γ rays by the nucleus is readily explained by the shell model. If nucleons in an excited state fall to a lower nuclear energy level, they will emit energy as γ rays.

Thus some nuclear properties imply that the nucleons are free to move within the nucleus, and others suggest that nucleons exist in energy levels.

FORCES IN THE NUCLEUS

Protons have a positive charge. In any nucleus containing two or more protons there will be electrostatic repulsion between the like charges. In a stable nucleus, the attractive forces are greater than the repulsive forces. In an unstable nucleus the repulsive forces exceed the attractive forces and spontaneous fission occurs. The attractive forces in the nucleus cannot be electrostatic for two reasons:

1. There are no oppositely charged particles.
2. The forces only act over a very short distance of 2–3 fm.

If the nuclear particles are separated by a distance much greater than this, attraction ceases. Electrostatic forces diminish only slowly with distance.

The attractive force does not depend on the charge as the same attractive force binds protons to protons, protons to neutrons and neutrons to neutrons.

Two atoms may be held together by a sharing of electrons, because the exchange forces result in a covalent bond. By analogy, two nuclear particles may be held together by sharing a particle. The particle exchanged is called a π *meson*. Mesons may have a positive charge π^+, a negative charge π^- or no charge π^0. Exchange of π^- and π^+ mesons accounts for the binding energy between neutrons and protons. The transfer of a charge converts a neutron to a proton or vice versa. The resultant attractive forces are indicated by dotted lines in the examples below.

n, p $\pi-\rightarrow$ p, n   p, n $\pi+\rightarrow$ n, p

A π^0 meson is exchanged between two protons or between two neutrons.

p, p $\pi o\rightarrow$ p, p   n, n $\pi o\rightarrow$ n, n

The attractive forces between p–n, n–n, and p–p, are probably all similar in strength. The different types of mesons have similar masses. The mass of a π^0 meson is 264 times that of an electron and the masses of both π^+ and π^- mesons are 273 times that of an electron. All mesons are very unstable outside the nucleus. There are many other less common elementary particles in the nucleus. This topic is really within the field of particle physics and is beyond the scope of this book. The number of charged particles in a nucleus remains constant. However, the continual transfer of mesons means that the particles representing neutrons and protons are constantly changing. The transformation of neutrons into protons and vice versa are first order reactions. The rates of the reactions depend on the relative numbers of neutrons and protons present. In a stable nucleus, these two changes are in equilibrium.

STABILITY AND THE RATIO OF NEUTRONS TO PROTONS

The stability of a nucleus depends on the number of protons and neutrons present. For elements of low atomic number (up to $Z = 20$, i.e. Ca) the most stable nuclei exist when the nucleus contains an equal number of protons P and neutrons N. This means that the ratio $N/P = 1$. Elements with higher atomic numbers are more stable if they have a slight excess of neutrons as this increases the attractive force and also reduces repulsion between protons. Thus the ratio N/P increases progressively up to about

1.6 at $Z = 92$ (uranium). In elements with still higher atomic numbers, the nuclei have become so large they undergo spontaneous fission. These trends are shown in the graphs of neutron number N against proton number P, and N/P ratio against proton number for the stable nuclei (Figure 31.1).

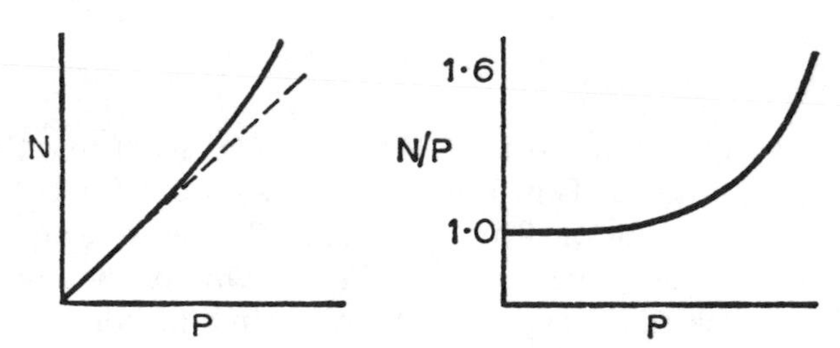

Figure 31.1 Neutron to proton ratio.

MODES OF DECAY

Stable nuclei lie near to these curves. Nuclei with N/P ratios appreciably higher or lower than the stable ratio are radioactive. When they decay they form nuclei closer to the line of maximum stability.

If the N/P ratio is high, the isotope lies above the curve. Such a nucleus will decay in such a way that it reduces the N/P ratio and forms a stable arrangement. The ratio can be reduced in two ways.

Beta emission

Electrons or β radiation may be emitted from the nucleus when a neutron is converted into a proton, an electron and a neutrino. This reduces the N/P ratio. The neutrino ν is a strange particle. It has zero mass and zero charge, and is postulated to balance the spins. A neutrino is emitted in almost all nuclear transformations. The change of a neutron into a proton may be written:

$$^{1}_{0}\text{n} \rightarrow {}^{1}_{1}\text{p} + {}^{0}_{-1}\text{e} + \nu$$

The mass numbers are shown at the top and must be balanced on both sides of this equation. The nuclear charges are shown at the bottom. These too must be balanced on both sides of the equation. The loss of an electron from the nucleus in this way decreases the N/P ratio. If an isotope is not far from the stable N/P line one β decay process may be sufficient:

$$^{14}_{6}\text{C} \rightarrow {}^{14}_{7}\text{N} + {}^{0}_{-1}\text{e} + \nu$$

$$^{29}_{13}\text{Al} \rightarrow {}^{29}_{14}\text{Si} + {}^{0}_{-1}\text{e} + \nu$$

Isotopes further from the stable line may undergo a series of β decays. The resultant nuclei become progressively more stable and have a longer half life period, until eventually a stable isotope is formed.

$$^{141}_{56}\mathrm{Ba} \xrightarrow[t_{1/2} = 18\,\mathrm{min}]{\beta} {}^{141}_{57}\mathrm{La} \xrightarrow[t_{1/2} = 3.7\,\mathrm{h}]{\beta} {}^{141}_{58}\mathrm{Ce} \xrightarrow[t_{1/2} = 28\text{ days}]{\beta} {}^{141}_{59}\mathrm{Pr}$$

In β decay the mass number remains unchanged, but the nuclear charge increases by one unit. Thus when β decay occurs the element moves one place to the right in the periodic table.

Neutron emission

An obvious way to decrease the N/P ratio would seem to be to emit a neutron from the nucleus. This form of decay is rare and only takes place with highly energetic nuclei. This is because the binding energy of the neutron in the nucleus is high (about 8 MeV). One of the few examples involves $^{87}_{37}\mathrm{Kr}$, which can decay either by neutron emission or β decay.

$$^{87}_{36}\mathrm{Kr} \longrightarrow {}^{86}_{36}\mathrm{Kr} + {}^{1}_{0}\mathrm{n}$$

$$^{87}_{36}\mathrm{Kr} \xrightarrow{\beta} {}^{87}_{37}\mathrm{Rb} \xrightarrow{\beta} {}^{87}_{38}\mathrm{Sr}$$

If the N/P ratio is too low, the isotope lies below the curve. There are three possible modes of decay.

Positron emission

Positrons or β+ radiation (positive electrons) result from the transformation of a proton to a neutron. The positron is ejected from the nucleus together with an anti-neutrino $\bar{\nu}$.

$$^{1}_{1}\mathrm{p} \rightarrow {}^{1}_{0}\mathrm{n} + {}^{0}_{1}\mathrm{e} + \bar{\nu}$$

The anti-neutrino is postulated to balance the spins. When the positron is ejected from the nucleus it very quickly collides with an electron in the surroundings. The two particles annihilate each other and their energy is released as two γ ray photons. These photons are released in opposite directions (180° apart), so there is no resultant linear or angular momentum. Thus each photon has exactly half of the annihilation energy of 1.022 Mev, i.e. each photon has an energy of 0.511 Mev. This energy came originally from the parent nucleus. It follows that the mass of the parent nucleus is greater than the mass of the daughter nucleus. This process increases the N/P ratio. Gamma radiation usually arises in another way, as described later. Some examples of positron emission are:

$$^{19}_{10}\mathrm{Ne} \rightarrow {}^{19}_{9}\mathrm{F} + {}^{0}_{1}\mathrm{e} + \bar{\nu}$$

$$^{11}_{6}\mathrm{C} \rightarrow {}^{11}_{5}\mathrm{B} + {}^{0}_{1}\mathrm{e} + \bar{\nu}$$

Orbital or K-electron capture

The nucleus may capture an orbital electron and thus convert a proton into a neutron with the emission of a neutrino:

$$^{1}_{1}p + ^{0}_{-1}e \rightarrow ^{1}_{0}n + \nu$$

This process increases the N/P ratio. Usually an electron from the shell closest to the nucleus is captured. This is called the K shell, so the process is called K-electron capture. An electron from a higher energy level drops back to fill the vacancy in the K shell and characteristic X-radiation is emitted. Electron capture is not common. It occurs in nuclei where the N/P ratio is low and the nucleus has insufficient energy for positron emission, that is $2 \times 0.51 = 1.02$ MeV. Some examples are:

$$^{7}_{4}Be + ^{0}_{-1}e \rightarrow ^{7}_{3}Li + \nu$$
$$^{40}_{19}K + ^{0}_{-1}e \rightarrow ^{40}_{18}Ar + \nu$$

Where the difference in mass of parent and daughter nuclei is equivalent to more than the required 1.02 MeV for positron emission, both positron emission and K capture occur.

$$^{48}_{23}V \xrightarrow{\beta^{+}\ (58\%)} ^{48}_{22}Ti$$
$$\text{K capture } (42\%)$$

Proton emission

Except for nuclei in a very high energy state, proton emission is unlikely as the energy needed to remove a proton is about 8 MeV.

GAMMA RADIATION

Immediately following any nuclear change, the neutrons and protons often have not arranged themselves in their most stable positions. The newly formed daughter nucleus is thus in an excited state, and has a higher energy than the ground state. Generally the nucleons rearrange themselves quite rapidly, thus lowering the energy of the daughter nucleus to the ground state. The corresponding amount of energy is emitted. This is in the range 0.1–1 MeV, and is emitted as electromagnetic radiation of very short wavelength, called γ rays.

HALF LIFE PERIOD

The time taken for half the radioactive nuclei in a sample to decay is called the half life period. This is a characteristic of a particular isotope.

For a single radioactive decay process, the number of nuclei which disintegrate in a short time period depends only on the relative number of radioactive atoms present. Thus the size of the sample does not affect the time taken to undergo radioactive decay. Nuclear decay is therefore a first order reaction. If n is the number of radioactive nuclei and t the time interval, the rate of decay (that is the change in the number of radioactive nuclei with time) is given by:

$$\frac{dn}{dt} = -\lambda n \qquad (31.1)$$

λ is the decay constant for the process, which indicates how rapidly the sample is decaying. (λ has the units time^{-1}.) It is usual to quote the time of half life rather than the decay constant. The two are related. Integrating equation (31.1) between limits from time = 0 to time = t gives:

$$\frac{n}{n_0} = e^{-\lambda t}$$

where n_0 is the original number of nuclei at time 0, and n the number remaining at time t. The half life $t_{1/2}$ is the time taken for the number of radioactive nuclei to fall to half the original number, that is to reach $n = \frac{1}{2}n_0$. Since there is no way of counting the number of radioactive nuclei present, we cannot calculate λ in this way. However, we can substitute the activity at time = 0 (A_0) and the activity at time = t (A) for the number of atoms. The activity can be measured and is the number of counts recorded in a fixed time on a Geiger counter or a scintillation counter. Thus we can evaluate λ:

$$\frac{A}{A_0} = \frac{n}{n_0} = e^{-\lambda t} = \frac{1}{2}$$

Taking natural logarithms of both sides of the equation

$$-\lambda t_{1/2} = \ln\left(\tfrac{1}{2}\right)$$

hence

$$t_{1/2} = -\frac{\ln(2)}{\lambda} = -\frac{0.693}{\lambda}$$

Nuclear energies are of the order of $10^9\,\mathrm{kJ\,mol^{-1}}$ of nucleons. The energies involved in chemical reactions are about $10\text{–}10^2\,\mathrm{kJ\,mol^{-1}}$. Clearly transmutation of one element to another by chemical means is impossible, because of the exceedingly high nuclear energy. For the same reason a change in temperature has no observable effect on the rate of decay.

The radioactivity of a sample is traditionally measured in curies (Ci). Originally a curie was taken as the amount of radioactive disintegration in 1 g of $^{226}_{88}Ra$. It is more precisely defined as the amount of a radioisotope which gives 3.7×10^{10} disintegrations per second. Thus one curie of different radioisotopes contains a widely different number of atoms. The SI unit of activity is the becquerel (Bq), and is defined as the amount of a radioisotope which gives one disintegration per second. Thus $1\,\mathrm{Bq} = 27 \times 10^{-12}\,\mathrm{Ci}$.

BINDING ENERGY AND NUCLEAR STABILITY

The mass of a hydrogen atom ^1_1H is equal to the sum of the mass of one proton and the mass of one electron. For all other atoms the mass of the atom is less than the sum of the constituent neutrons, protons and electrons. The difference is called the *mass defect*. The mass defect is related to the *binding energy* holding the neutrons and protons together in the nucleus. A stable nucleus must have less energy than its constituent particles or it would not form.

Energy and mass are related by the Einstein equation $\Delta E = \Delta mc^2$ were ΔE is the energy liberated, Δm the loss in mass (the mass defect) and c the velocity of light ($2.998 \times 10^8\,m\,s^{-1}$). The mass defect can be calculated, and converted to the binding energy in the nucleus. The larger the mass defect, the larger the binding energy, and therefore the more stable the nucleus.

Mass of ${}^1_1p$ = 1.007 277 a.m.u., Mass of ${}^1_0n$ = 1.008 665 a.m.u.
(931 is the conversion factor from atomic mass units a.m.u. to MeV).

Mass of ${}^4_2He$ nucleus = 4.0028 amu	Mass of ${}^6_3Li$ nucleus = 6.0170
Mass of 2n + 2p = 4.0319	Mass of 3n + 3p = 6.0478
Mass defect = 0.0291	Mass defect = 0.0308
Binding energy =	Binding energy =
$0.0291 \times 931 = 27.1$ MeV	$0.0327 \times 931 = 28.7$ MeV
or $2.6 \times 10^9\,kJ\,mol^{-1}$	or $2.8 \times 10^9\,kJ\,mol^{-1}$

It is quite easy to calculate whether a nucleus is stable against decay. Some possible decay processes are given below.

$${}^4_2He \rightarrow {}^1_0n + {}^3_2He \qquad {}^6_3Li \rightarrow {}^1_0n + {}^5_3Li$$
$${}^4_2He \rightarrow {}^1_0n + {}^1_0n + {}^2_2He \qquad {}^6_3Li \rightarrow {}^1_1p + {}^5_2He$$
$${}^4_2He \rightarrow {}^1_1p + {}^3_1H \qquad {}^6_3Li \rightarrow {}^4_2He + {}^2_1H$$

In each of the above reactions the mass of the parent is less than the combined mass of the suggested products. Thus none of the above decay processes occur.

The total binding energy of the nucleus increases with the number of nucleons present. To compare the stability of nuclei of different elements we calculate the average binding energy per nucleon:

$$\text{binding energy per nucleon} = \frac{\text{total binding energy}}{\text{number of nucleons}}$$

The graph of binding energy per nucleon against atomic numbeı, for the different elements,shows that nucleons are held together with increasing force up to a mass number of about 65. The binding energy for each additional nucleon decreases as the nuclei get larger (Figure 31.2). For most nuclei the average binding energy per nucleon is about 8 MeV. Consequently 8 MeV of energy is needed to remove either a proton or a neutron from the nucleus.

ALPHA DECAY

As nuclei get larger, the repulsive force between protons increases and the energy of the nucleus increases. A point is reached where the attractive

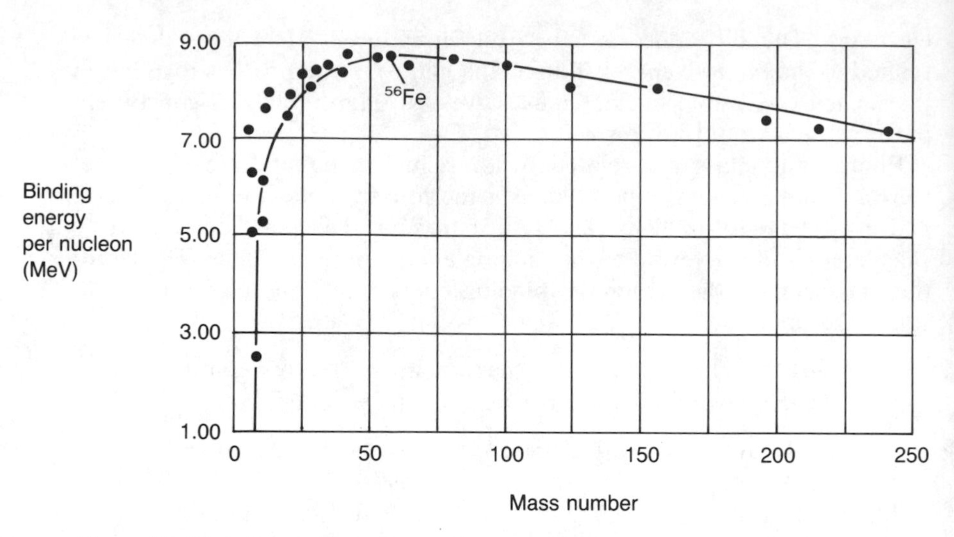

Figure 31.2 Binding energy per nucleon.

forces are unable to hold the nucleons together. Thus part of the nucleus breaks off. An α particle is emitted from the nucleus. An α particle is a helium nucleus $^{4}_{2}He$. This is a particularly stable nuclear fragment, as all the nucleons are in the lowest possible energy level, and both the number of neutrons and the number of protons are magic numbers. At the same time energy is liberated. In the decay of $^{238}_{92}U$, 4.2 MeV of energy is released because the mean binding energy per nucleon of the two daughter nuclei $^{234}_{90}Th$ and $^{4}_{2}He$ is greater than for the parent nucleus $^{238}_{92}U$. This is an enormous amount of energy ($1\,MeV = 96.48 \times 10^{6}\,kJ\,mol^{-1}$).

$$^{238}_{92}U \rightarrow {}^{234}_{90}Th + {}^{4}_{2}He + \text{energy}$$
$$^{210}_{84}Po \rightarrow {}^{206}_{82}Pb + {}^{4}_{2}He + \text{energy}$$

Alpha particles have no electron cloud, and have a charge of +2. Once emitted, an α particle quickly takes up two electrons from any atom in the vicinity, thus becoming a neutral He atom. The formation of He can be detected when α decay occurs.

The mass of the parent nucleus must provide both the mass of the daughter nucleus and that of the α particle, plus the small amount of mass which is converted into energy. It can be calculated from the mass of the nucleus whether α decay in any element is energetically possible. Natural α activity is only possible among elements with mass numbers greater than 209, as only these elements have the required energy. Conversely 209 is the largest number of nucleons which will fit into a stable nucleus.

If ejection of one α particle does not completely stabilize the nucleus then further α particles may be emitted. However, α decay raises the *N*/*P* ratio so it is often followed by β emission.

Nuclei of mass number above 230 may undergo spontaneous fission, forming two lighter nuclei. Thus two elements of lower atomic number are formed. Elements of low atomic number have a smaller N/P ratio, so some neutrons will be left over after fission. Most of these neutrons are emitted, but a few may change to protons with β emission. The neutrons emitted may be captured (absorbed) by another nucleus. This nucleus then becomes unstable and itself undergoes fission, liberating more neutrons. Thus a chain reaction has been started.

Alpha decay occurs only very rarely with nuclides with $Z < 83$ (atoms lighter than Bi). However, the great stability of the α particle causes a few very light, unstable atoms whose nuclear composition corresponds to, or nearly corresponds to, two or three α particles to undergo decay. When $^{8}_{4}Be$ undergoes this type of α decay the energy released is 0.09 MeV.

$$^{8}_{4}Be \rightarrow 2(^{4}_{2}He) + \text{energy}$$
$$^{8}_{3}Li \rightarrow 2(^{4}_{2}He) + {}^{0}_{-1}e + \text{energy}$$

RADIOACTIVE DISPLACEMENT LAWS

1. Emission of an α particle produces an element which is four mass units lighter and the atomic number decreases by two. The daughter element is therefore two places to the left of the parent in the periodic table.
2. When a β particle is emitted the mass number remains the same. However, the atomic number increases by one and the new element is one place to the right of the parent in the periodic system. These changes are shown in the following series:

$$^{223}_{88}Ra \xrightarrow{\alpha} {}^{219}_{86}Rn \xrightarrow{\alpha} {}^{215}_{84}Po \xrightarrow{\alpha} {}^{211}_{82}Pb \xrightarrow{\beta} {}^{211}_{83}Bi \xrightarrow{\alpha} {}^{207}_{81}Tl \xrightarrow{\beta} {}^{207}_{82}Pb$$

RADIOACTIVE DECAY SERIES

The heavy radioactive elements may be grouped into four decay series. The common radioactive elements thorium, uranium and actinium occur naturally and belong to three different series named after them. They are the parent members of their respective series and have the longest half life periods. They decay by a series of α and β emissions, and produce radioactive elements which are successively more stable until finally a stable isotope is reached. All three series terminate with lead ($^{206}_{82}Pb$, $^{207}_{82}Pb$ and $^{208}_{82}Pb$). Following the discovery of the artificial post-uranium elements, the neptunium series has been added, which ends with bismuth, $^{209}_{93}Bi$.

Thorium ($4n$) series
Neptunium ($4n + 1$) series
Uranium ($4n + 2$) series
Actinium ($4n + 3$) series

The numbers in brackets indicate that the parent and all the members of a particular series have mass numbers exactly divisible by four, or divisible by four with a remainder of one, two or three. There is no natural cross-linking between the four series, although this can be performed artificially.

Thorium (4n) series

$$^{232}_{90}\mathrm{Th} \xrightarrow{\alpha} {}^{228}_{88}\mathrm{Ra} \xrightarrow{\beta} {}^{228}_{89}\mathrm{Ac} \xrightarrow{\beta} {}^{228}_{90}\mathrm{Th} \xrightarrow{\alpha} {}^{224}_{88}\mathrm{Ra} \xrightarrow{\alpha} {}^{220}_{86}\mathrm{Rn} \xrightarrow{\alpha} {}^{216}_{84}\mathrm{Po}$$

$$^{216}_{84}\mathrm{Po} \xrightarrow{\beta} {}^{216}_{85}\mathrm{At} \xrightarrow{\alpha} {}^{212}_{83}\mathrm{Bi}$$
$$^{216}_{84}\mathrm{Po} \xrightarrow{\alpha} {}^{212}_{82}\mathrm{Pb} \xrightarrow{\beta} {}^{212}_{83}\mathrm{Bi}$$
$$^{212}_{83}\mathrm{Bi} \xrightarrow{\beta} {}^{212}_{84}\mathrm{Po} \xrightarrow{\alpha} {}^{208}_{82}\mathrm{Pb}$$
$$^{212}_{83}\mathrm{Bi} \xrightarrow{\alpha} {}^{208}_{81}\mathrm{Tl} \xrightarrow{\beta} {}^{208}_{82}\mathrm{Pb}$$

Neptunium (4n + 1) series

$$^{241}_{94}\mathrm{Pu} \xrightarrow{\beta} {}^{241}_{95}\mathrm{Am} \xrightarrow{\alpha} {}^{237}_{93}\mathrm{Np}$$
$$^{237}_{92}\mathrm{U} \xrightarrow{\beta} {}^{237}_{93}\mathrm{Np}$$
$$^{237}_{93}\mathrm{Np} \xrightarrow{\alpha} {}^{233}_{91}\mathrm{Pa} \xrightarrow{\beta} {}^{233}_{92}\mathrm{U} \xrightarrow{\alpha} {}^{229}_{90}\mathrm{Th} \xrightarrow{\alpha} {}^{225}_{88}\mathrm{Ra} \xrightarrow{\beta} {}^{225}_{89}\mathrm{Ac} \xrightarrow{\alpha} {}^{221}_{87}\mathrm{Fr} \xrightarrow{\alpha} {}^{217}_{85}\mathrm{At} \xrightarrow{\alpha} {}^{213}_{83}\mathrm{Bi}$$
$$^{213}_{83}\mathrm{Bi} \xrightarrow{\beta} {}^{213}_{84}\mathrm{Po} \xrightarrow{\alpha} {}^{209}_{82}\mathrm{Pb}$$
$$^{213}_{83}\mathrm{Bi} \xrightarrow{\alpha} {}^{209}_{81}\mathrm{Tl} \xrightarrow{\beta} {}^{209}_{82}\mathrm{Pb}$$
$$^{209}_{82}\mathrm{Pb} \xrightarrow{\beta} {}^{209}_{83}\mathrm{Bi}$$

Uranium (4n + 2) series

$$^{238}_{92}\mathrm{U} \xrightarrow{\alpha} {}^{234}_{90}\mathrm{Th} \xrightarrow{\beta} {}^{234}_{91}\mathrm{Pa} \xrightarrow{\beta} {}^{234}_{92}\mathrm{U}$$
$$^{238}_{93}\mathrm{Np} \xrightarrow{\beta} {}^{238}_{94}\mathrm{Pu} \xrightarrow{\alpha} {}^{234}_{92}\mathrm{U}$$
$$^{234}_{92}\mathrm{U} \xrightarrow{\alpha} {}^{230}_{90}\mathrm{Th} \xrightarrow{\alpha} {}^{226}_{88}\mathrm{Ra} \xrightarrow{\alpha} {}^{222}_{86}\mathrm{Rn} \xrightarrow{\alpha} {}^{218}_{84}\mathrm{Po}$$
$$^{218}_{84}\mathrm{Po} \xrightarrow{\beta} {}^{218}_{85}\mathrm{At} \xrightarrow{\alpha} {}^{214}_{83}\mathrm{Bi}$$
$$^{218}_{84}\mathrm{Po} \xrightarrow{\alpha} {}^{214}_{82}\mathrm{Pb} \xrightarrow{\beta} {}^{214}_{83}\mathrm{Bi}$$
$$^{214}_{83}\mathrm{Bi} \xrightarrow{\beta} {}^{214}_{84}\mathrm{Po} \xrightarrow{\alpha} {}^{210}_{82}\mathrm{Pb}$$
$$^{214}_{83}\mathrm{Bi} \xrightarrow{\alpha} {}^{210}_{81}\mathrm{Tl} \xrightarrow{\beta} {}^{210}_{82}\mathrm{Pb}$$
$$^{210}_{82}\mathrm{Pb} \xrightarrow{\beta} {}^{210}_{83}\mathrm{Bi} \xrightarrow{\beta} {}^{210}_{84}\mathrm{Po} \xrightarrow{\alpha} {}^{206}_{82}\mathrm{Pb}$$

Actinium (4n + 3) series

$$^{239}_{92}\mathrm{U} \xrightarrow{\beta} {}^{239}_{93}\mathrm{Np} \xrightarrow{\beta} {}^{239}_{94}\mathrm{Pu} \xrightarrow{\alpha} {}^{235}_{92}\mathrm{U} \xrightarrow{\alpha} {}^{231}_{90}\mathrm{Th} \xrightarrow{\beta} {}^{231}_{91}\mathrm{Pa} \xrightarrow{\alpha} {}^{227}_{89}\mathrm{Ac}$$
$$^{227}_{89}\mathrm{Ac} \xrightarrow{\beta} {}^{227}_{90}\mathrm{Th} \xrightarrow{\alpha} {}^{223}_{88}\mathrm{Ra}$$
$$^{227}_{89}\mathrm{Ac} \xrightarrow{\alpha} {}^{223}_{87}\mathrm{Fr} \xrightarrow{\beta} {}^{223}_{88}\mathrm{Ra}$$
$$^{223}_{88}\mathrm{Ra} \xrightarrow{\alpha} {}^{219}_{86}\mathrm{Rn} \xrightarrow{\alpha} {}^{215}_{84}\mathrm{Po} \xrightarrow{\alpha} {}^{211}_{82}\mathrm{Pb} \xrightarrow{\beta} {}^{211}_{83}\mathrm{Bi}$$
$$^{211}_{83}\mathrm{Bi} \xrightarrow{\beta} {}^{211}_{84}\mathrm{Po} \xrightarrow{\alpha} {}^{207}_{82}\mathrm{Pb}$$
$$^{211}_{83}\mathrm{Bi} \xrightarrow{\alpha} {}^{207}_{81}\mathrm{Tl} \xrightarrow{\beta} {}^{207}_{82}\mathrm{Pb}$$

Natural radioactivity also occurs in nine of the lighter elements. It is possible that as the sensitivity of detecting instruments increases, other radioactive elements will be found. Of these, the two most important are $^{14}_{6}\mathrm{C}$ and $^{40}_{19}\mathrm{K}$. The $^{40}_{19}\mathrm{K}$ isotope was probably formed when the earth was created. Its existence on earth is due to its long half life of 1.25×10^9 years. This isotope only constitutes 0.01% of naturally occurring potassium, but its presence makes living tissue appreciably radioactive. It may decay either by β emission or K capture.

$$^{40}_{19}K \xrightarrow{\beta} {}^{40}_{20}Ca$$
$$^{40}_{19}K \xrightarrow{\text{K capture}} {}^{40}_{18}Ar$$

$^{14}_{6}C$ has a half life of 5720 years and any originally present in the earth will have decayed before now. The fact that some exists shows that it must have been produced since the earth was formed. $^{14}_{6}C$ is produced continuously from the action of the neutrons in cosmic rays on nitrogen in the atmosphere. This involves a nuclear reaction:

$$^{14}_{7}N + {}^{1}_{0}n \rightarrow {}^{14}_{6}C + {}^{1}_{1}p$$

It is important in radiocarbon dating (see Chapter 12 under 'Carbon dating').

INDUCED NUCLEAR REACTIONS

Many nuclear reactions can be brought about by bombarding the nucleus with γ rays or various particles. The particles include electrons, neutrons, protons, α particles or the nuclei of other atoms such as carbon. The nuclei of C atoms which have had the orbital electrons removed are called 'stripped carbon'. The particle may be captured by the nucleus (fusion), or the nucleus may undergo fission, depending on the conditions of the bombardment.

Natural radiation may be used to induce nuclear reactions, but this limits the energy of the bombarding particle. More usually the charged particles are accelerated by a high voltage using a linear accelerator, or by alternate attraction and repulsion using a cyclotron or a synchrotron. (There are very large accelerators at CERN in France and Switzerland, at Berkeley USA and in the USSR.) In this way particles can be accelerated, giving them a very high kinetic energy which can be used to promote the nuclear reaction. This high energy is necessary to overcome the repulsion between a positively charged nucleus and a positively charged bombarding particle. Neutrons have no charge so they would not be repelled by the nucleus, but the absence of a charge means that they cannot be accelerated in this manner. Thus neutron bombardment can only be carried out using neutrons produced by nuclear reactions.

Some nuclear transformations are given below.

$$^{14}_{7}N + {}^{4}_{2}He \rightarrow {}^{17}_{8}O + {}^{1}_{1}H$$
$$^{27}_{13}Al + {}^{4}_{2}He \rightarrow {}^{30}_{15}P + {}^{1}_{0}n$$
$$^{23}_{11}Na + {}^{1}_{1}H \rightarrow {}^{23}_{12}Mg + {}^{1}_{0}n$$
$$^{113}_{48}Cd + {}^{1}_{0}n \rightarrow {}^{114}_{48}Cd + \text{energy}$$

In the first example the nitrogen is bombarded with α particles and a proton is formed. This is described as an (α, p) reaction. An alternative way of writing the reaction is:

$$^{14}_{7}\text{N} \xrightarrow{(\alpha,\text{p})} {}^{17}_{8}\text{O}$$

This reaction was first carried out by E. Rutherford in 1919, and this was the first induced nuclear transformation. (He was awarded the Nobel Prize for Chemistry in 1908.) The second example was carried out in 1932 by F. Joliot and Irène Joliot Curie and is an (α, n) reaction. (They were awarded the Nobel Prize for Chemistry in 1935.)

$$^{27}_{13}\text{Al} \xrightarrow{(\alpha,\text{n})} {}^{30}_{15}\text{P}$$

In a similar way the third example is a (p, n) reaction. In the last example the energy is emitted as γ rays, and so this is an (n, γ) reaction. The nuclei formed in this way may be stable, or may subsequently decay. The transuranic elements are all obtained by bombarding a heavy nucleus with α particles, stripped carbon, or the nuclei of other light atoms to produce an even heavier nucleus.

NUCLEAR FISSION

Very heavy nuclei have a lower binding energy per nucleon than nuclei with an intermediate mass. Thus nuclei of intermediate mass are more stable than heavy nuclei. When a slow neutron enters a nucleus of a fissionable atom such as uranium (which is already distorted) the extra energy may cause the nucleus to split into two fragments and spontaneously emit two or more neutrons. This is called fission. The fission process results in the release of large amounts of energy (about 8×10^9 kJ mol^{-1}). In the case of $^{235}_{92}\text{U}$, several different primary fission products are formed, depending on exactly how the nucleus splits up. Three of the more common reactions are:

$$^{235}_{92}\text{U} + {}^{1}_{0}\text{n} \rightarrow \begin{cases} ^{144}_{56}\text{Ba} + {}^{90}_{36}\text{Kr} + 2({}^{1}_{0}\text{n}) \\ ^{138}_{53}\text{I} + {}^{95}_{39}\text{Y} + 3({}^{1}_{0}\text{n}) \\ ^{140}_{55}\text{Cs} + {}^{92}_{37}\text{Rb} + 4({}^{1}_{0}\text{n}) \end{cases}$$

Note that the daughter nuclei formed fall into two classes. The heavier group have masses from 130 to 160, and the lighter group have masses from 80 to 110. It is rare for the two daughter nuclei of about the same mass to be formed (Figure 31.3).

The total mass of the fission products is some 0.22 mass units less than the mass of the uranium atom and neutron. This corresponds to an energy release of over 200 MeV. This is more than twelve times the energy liberated in a normal nuclear reaction. The complete fission of 1 lb (0.45 kg) of uranium releases as much energy as the explosion of 8000 tonnes of TNT.

The nuclei formed as primary fission products have a high neutron to proton ratio, and decay by β emission to lower this ratio. Usually several such steps are required before a stable nucleus is obtained. Thus each of the primary decay products is associated with a decay chain, for example:

$$^{138}_{53}I \xrightarrow{\beta} ^{138}_{54}Xe \xrightarrow{\beta} ^{138}_{55}Cs \xrightarrow{\beta} ^{138}_{56}Ba \quad \text{(stable)}$$

$$^{95}_{39}Y \xrightarrow{\beta} ^{95}_{40}Zr \xrightarrow{\beta} ^{95}_{41}Nb \xrightarrow{\beta} ^{95}_{42}Mo \quad \text{(stable)}$$

About 90 decay chains have been identified from the fission of $^{238}_{92}U$, giving a total of several hundred different nuclides.

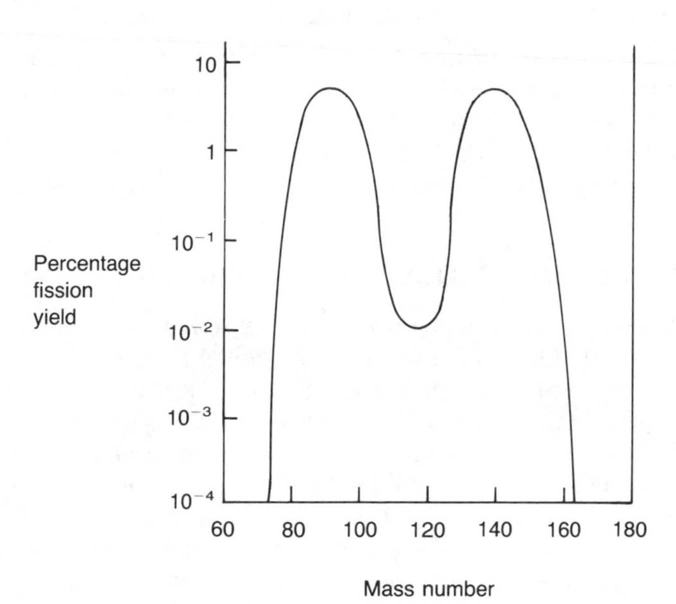

Figure 31.3 Percentage yields of elements from slow neutron fission of $^{235}_{92}U$.

Chain reaction

One fission reaction can cause another. The neutron produced by a fission will either:

1. Cause another nuclear fission reaction.
2. Be lost from the fissionable material
3. Be used in a non-fission reaction.

When neutrons produced in fission reactions initiate new fissions, this is called a chain reaction. There are two major applications of chain reactions – the atomic bomb (A-bomb) and nuclear powered electricity generating stations.

If more than one neutron per fission causes another fission we have a branched chain reaction, and a rapidly increasing release of energy takes place. This is what happens in the atomic bomb. In a nuclear reactor, control rods are used to absorb some of the neutrons, so that on average only one neutron per fission causes another fission. A chain reaction of this kind is self-perpetuating. The energy is released at a slow enough rate to be used.

Critical mass

There is a minimum amount of fissionable material that can produce a self-sustaining chain reaction. This is called the critical mass. The critical mass depends on several factors:

1. The purity of the sample.
2. The density of the material.
3. Its geometric shape, and its surroundings.

These all affect the neutron propagation ratio. If the sample is impure, many neutrons will be lost through colliding with non-fissile atoms. The more dense the material the greater the chance that neutrons will collide with another nucleus. The shape of the sample is also important since neutrons are more likely to escape from a long thin strip of material than from a sphere.

THE STORY BEHIND PRODUCTION OF THE ATOMIC BOMB

Nuclear fission of heavy nuclei was first achieved in Berlin in December 1939, and in 1944 Otto Hahn received the Nobel Prize for Chemistry for this work. The first German reactor (sub-critical Pile B-III) was surprisingly housed in the 'Virus House' at the Kaiser Wilhelm Institute for Biology at Dahlem. At that time nuclear reactors were called atomic piles, and the German reactor consisted of alternate layers of uranium metal and paraffin wax in an aluminium sphere which was immersed in water. Though the pile contained 551 kg of uranium, it was sub-critical. This means that less than one neutron produced from each fission caused another fission. Thus a neutron source was placed in a chimney at the centre of the sphere to keep the pile running.

Two years later Enrico Fermi made a pile go critical (self-supporting) at the University of Chicago, USA. (Fermi was an Italian, and was awarded the Nobel Prize for Physics in 1938, for producing new radioactive elements by neutron irradiation. He fled from World War II and Fascist Italy to the USA.) The Chicago pile consisted of a huge block of graphite surrounded by several feet of high density concrete. Horizontal channels in the graphite were used to insert enough slugs of uranium to produce the chain reaction. Air was blown through to cool the pile and to remove the energy produced. One thousand kilowatts of heat were extracted. The pile was housed in the squash courts! Larger piles were then built at Clinton, Tennessee and Hanford, Washington.

A secret military base was built at Oak Ridge in Tennessee as part of the 'Manhattan Project' to conduct research into making an atomic bomb. Naturally occurring U consists largely of two isotopes, 99.3% of ${}^{238}_{92}U$ which is non-fissile, and 0.7% of ${}^{235}_{92}U$ which is fissile.

Separation of isotopes

Enormous efforts were made to separate the isotopes to obtain sufficient fissile ${}^{235}U$ to make a bomb. Four methods were used:

Thermal diffusion

Two gases of different density are placed in a long vertical tube with an electrically heated wire down the centre of the tube. The lighter gas diffuses more readily towards the hot wire, where it is heated and rises in a convection current. Thus the lighter gas accumulates at the top of the tube, and the heavier gas streams downwards on the surface of the tube.

Electromagnetic separation of UCl_4 using a mass spectrometer

Ions must first be produced by bombarding the sample with electrons at low pressure. The ions are attracted by a high voltage across a vacuum chamber. A magnetic field deflects the particles by different amounts depending on their mass. This method can deliver pure isotopes, and though it is primarily used for small quantities, kilogram quantities have been produced.

Gaseous diffusion of UF_6

UF_6 is gaseous above its sublimation temperature of 56 °C. The gas is pumped through thousands of filter barriers (pinholes) in the enrichment process. The slightly smaller ^{235}U passed through the barriers a little more easily. The rate of diffusion of a gas is determined by Graham's law of diffusion:

$$\text{rate of diffusion} = K/\sqrt{D}$$

where K is a constant and D is the density of the gas. The vapour density is equal to the molecular weight/2. Thus the rate of diffusion of $^{235}UF_6$ is marginally faster than that of $^{238}UF_6$ by a factor $\sqrt{(352/349)} = 1.0043$. This operation is repeated many thousands of times by pumping the gas in a cascade process. After each stage the lighter fraction is passed forwards and the heavier fraction backwards. Gaseous diffusion plants are very large. The K25 building at Oak Ridge housed the original gaseous diffusion plant and occupied 10^6 square feet of floor space. A second plant was built at Hanford in Washington State. The method uses an enormous amount of electrical energy. (The Oak Ridge site had two advantages in the wartime period. Its remoteness and low population were advantages as the work had unknown hazards. In addition cheap electricity was available from numerous hydroelectric power stations built in the Tennessee valley in the 1930s to make work during the years of the depression.) Gaseous diffusion is still carried out commercially at two sites in the USA.

Using a gas centrifuge to separate $^{235}UF_6$ and $^{238}UF_6$

This has subsequently become the method used in the UK and the Netherlands. Centrifuges rotating at 1700 revolutions per second concentrate $^{235}UF_6$ towards the centre and $^{238}UF_6$ towards the walls of a cylindrical centrifuge.

There were immense difficulties in obtaining a sufficient quantity of ^{235}U to make a bomb. The critical mass depends on the purity, and the bomb had to be light enough to carry in an aeroplane.

Meanwhile a second fissile element plutonium Pu was discovered by Glenn Seaborg at the University of California (Berkeley) in 1940. Pu is produced by irradiating the relatively plentiful ^{238}U with neutrons in a nuclear reactor.

$$^{238}_{92}U + ^{1}_{0}n \rightarrow ^{239}_{92}U \xrightarrow[t_{1/2}\ 23.5\ \text{min}]{} ^{239}_{93}Np \xrightarrow[t_{1/2}\ 2.3\ \text{days}]{} ^{239}_{94}Pu$$

All isotopes of Pu are fissile, and Pu became important as a nuclear fuel. Work on plutonium was mainly carried out at Hanford.

Bomb making

A third site at Los Alamos (New Mexico) concentrated mainly on the technical problems of how to make bombs. The problems are how to transport sub-critical masses of Pu or U in an aeroplane, and combine them to give a critical mass when over the target. Then the material must be contained for a long enough time for the nuclear reaction to occur.

In the case of the Pu bomb an implosion device was used. Several sub-critical masses of ^{239}Pu were placed round the edges of a sphere and surrounded by high explosives. The conventional explosives were detonated. The implosion blew the sub-critical masses of Pu into the centre where they formed a critical mass. The force of the explosion held the Pu together for long enough for a nuclear explosion to occur. The first atomic bomb used this principle and was tested on a 100-foot-high tower at Trinity in New Mexico at 5.29 a.m. (just before sunrise) on 16 July 1945. It worked! Fermi estimated that the temperature produced was four times that at the centre of the sun, and a pressure of over 100 billion atmospheres was produced. The radioactivity emitted was a million times greater than from the world's total radium supply. Parts for another similar atomic bomb were shipped to the Far East to be assembled and dropped on Japan.

Meanwhile sufficient ^{235}U to make a bomb had laboriously been collected. The most practical way to make a bomb was to use a gun barrel, and shoot a sub-critical mass of ^{235}U down the barrel into another sub-critical mass at the end of the barrel. This had not been tested, since up till then there had not been enough ^{235}U for a test. Parts to assemble this bomb were shipped to the Far East.

The Pu bomb was called 'Fat Man', and was 10 feet 8 inches long and 60 inches in diameter, and weighed 10 800 lb. The bomb was so wide that the doors on the bomb bay of the aeroplane had to be altered to get the bomb in and to allow it to be dropped. The U bomb was called 'Little Boy' and was 10 feet long, but only 28 inches in diameter, and weighed 8900 lb. The smaller diameter allowed it to fit in the aeroplane, and made it possible to use the U bomb first.

The U bomb 'Little Boy' was dropped on Hiroshima on 6 August 1945.

The yield was equivalent to 13 kilotonnes of TNT. Four square miles were devastated, and there were 70 000 immediate deaths. On 9 August 1945 the Pu bomb 'Fat Man' was dropped on Nagasaki. The yield was equivalent to 23 kilotonnes of TNT. Two square miles were devastated, and there were 45 000 immediate fatalities.

NUCLEAR POWER STATIONS

A nuclear power station consists of a nuclear reactor in which a controlled chain reaction occurs using either U or Pu as fuel. The heat produced is extracted from the reactor and used to generate steam, which drives a turbine and produces electricity. The earliest nuclear reactors were built to irradiate U and produce plutonium for bombs, and for experimental purposes. The first commercial nuclear power station was commissioned in 1956 at Calder Hall (Cumberland, UK). In 1989 there were over 120 nuclear power plants in the USA and over 400 in the rest of the world. France produces two thirds of its electricity from nuclear plants.

Table 31.2 Countries with seven or more nuclear plants

USA	129	Spain	18
France	67	Czechoslovakia	13
USSR	61	Sweden	12
UK	42	India	10
Japan	42	Korea	9
West Germany	28	East Germany	7
Canada	22		

Moderators

Neutrons can only be obtained by nuclear reactions, and they are divided into three groups depending on their kinetic energy:

1. Slow neutrons, with an energy $< 0.1\,\text{eV}$
2. Intermediate neutrons, with an energy $0.1\,\text{eV}$ to $2\,\text{MeV}$
3. Fast neutrons, with an energy $> 2\,\text{MeV}$

Neutrons ejected from a nucleus usually have a very high energy and are called 'fast neutrons'. These travel so fast that they escape. 'Slow neutrons' are needed to cause fission and maintain a chain reaction.

In a *thermal reactor* a moderator is used to slow down some of the fast neutrons. The neutrons collide with the nuclei of the moderator, and thus lose some of their kinetic energy. The best moderators are light atoms which do not capture neutrons, for example $^{2}_{1}H$, $^{4}_{2}He$, $^{9}_{4}Be$ and $^{12}_{6}C$. Graphite is the most widely used. Heavy water D_2O which contains $^{2}_{1}H$ is also used. Sometimes ordinary H_2O is used. Be and He are not used because Be is poisonous and expensive, and He, being gaseous, is not sufficiently dense. *Fast breeder reactors* do not use a moderator.

Fuel

Early reactors used uranium metal. Most thermal reactors now use UO_2 as this has a higher melting point and is less chemically reactive. Natural U may be used as fuel (99.3% ^{238}U and 0.7% ^{235}U). However, it is usual to enrich the fuel to between 2% and 3% ^{235}U to allow for some neutrons being absorbed by the metal case cladding the fuel rods, or by the moderator. Enriching U reduces the critical mass, and hence the size of the reactor. However, it is very expensive. Enrichment beyond 3% is only carried out for military purposes – to make bombs or for special small high temperature reactors for submarines. The latter use UC_2 enriched to over 90% as fuel.

Fast (breeder) reactors use plutonium oxide as fuel. They have no moderator, so the fast neutrons produced can convert non-fissile ^{238}U into fissile Pu by the reactions:

$$^{238}_{92}U + ^{1}_{0}n \rightarrow ^{239}_{92}U + \gamma$$
$$^{239}_{92}U \rightarrow ^{239}_{93}Np + ^{0}_{-1}e$$
$$^{239}_{93}Np \rightarrow ^{239}_{94}Pu + ^{0}_{-1}e$$

More Pu is produced than is used: hence the name 'breeder reactor'. Sometimes thorium (which is non-fissile) is incorporated into the fuel. When this is irradiated with fast neutrons, the isotope $^{233}_{92}U$ is formed, and this is fissionable by slow neutrons.

$$^{232}_{90}U + ^{1}_{0}n \rightarrow ^{233}_{90}Th + \gamma$$
$$^{233}_{90}Th \rightarrow ^{233}_{91}Pa + ^{0}_{-1}e$$
$$^{233}_{91}Pa \rightarrow ^{233}_{92}U + ^{0}_{-1}e$$

TYPES OF REACTOR IN USE

Gas cooled thermal reactors (all use graphite as moderator)

Magnox reactors

These use U metal rods as fuel, enclosed in a Mg/Al (magnox) casing. The fuel is natural, i.e. not enriched, and CO_2 gas is used as the coolant. Most of the early reactors in the UK are of this type. These are now nearing the end of their life.

Advanced gas cooled reactors (AGR)

These use UO_2 pellets enriched to 2% as fuel, with CO_2 as the coolant.

High temperature reactor (HTR)

These are used for military purposes such as submarines. The fuel is UC_2, which is enriched to over 90%, thus allowing the reactor to be small. The coolant is He, and the control rods are made of Cd.

Water cooled thermal reactors (all use H_2O or D_2O as moderator)

Canadian deuterium uranium reactor (CANDU)

These are a Canadian design, and use natural (not enriched) UO_2 as fuel, and heavy water D_2O as both moderator and coolant.

Pressurized water reactor (PWR)

These are a US design, and use UO_2 pellets enriched to 3% as fuel. Water is used as both moderator and coolant. The casing must withstand the huge pressure from the steam produced, and is typically 10 inches (25 cm) of stainless steel surrounded by concrete.

Boiling water reactor (BWR)

This is similar to the PWR except that the UO_2 fuel is only enriched to 2.2%. It works at a much lower pressure, and so the reactor casing need not be so strong.

Steam generating heavy water reactor (SGHWR)

These use UO_2 pellets enriched to 2.3% as fuel, with D_2O as moderator and H_2O as coolant.

Fast breeder reactors (these do not use a moderator)

These are much less developed than are thermal reactors. They use PuO_2 as fuel. Enrichment is not necessary as all isotopes of Pu are fissile. No moderator is used, and so the neutrons in the reactor are 'fast neutrons'. Some reactors use liquid Na as coolant, whilst others use He gas under a high pressure. If depleted UO_2 (i.e. UO_2 which has had the fissile ^{235}U removed) is put in such a reactor, then the non-fissile ^{238}U is converted into fissile Pu. The name 'breeder reactor' arises because more fissile material is produced than is used in the process. Potentially this process could provide an unlimited source of energy.

In Great Britain there are three generations of nuclear power stations:

1. The ageing Magnox reactors (British design).
2. The second generation advanced gas cooled reactors (British design).
3. A limited number of American designed pressurized water reactors. These are claimed to be the design which will be used into the twenty-first century. However, the cost, the time taken to commission and the safety aspects (particularly of the pressure casing) of these PWRs have been called into question.

NUCLEAR FUSION

Nuclear fusion is the process of releasing energy from matter, which occurs in the sun, the stars and the hydrogen bomb. During fusion, atoms of light elements combine to form heavier elements. The binding energy per nucleon for light elements is less than that for elements of intermediate mass. Thus the fusion of two light nuclei results in a more stable nucleus, and a large amount of energy is liberated. Both fusion and fission are methods of releasing large amounts of energy. In fusion, light atoms are combined to give heavier elements, whilst in fission heavy radioactive atoms are split into atoms of intermediate mass.

The simplest nuclear fusion reaction involves the isotopes of hydrogen: deuterium ${}^{2}_{1}H$ and tritium ${}^{3}_{1}H$. A large amount of energy is required to overcome the repulsion between the positively charged nuclei to get them close enough (1–2 fm) to react. One way of producing high energy particles
is an accelerator. This is not appropriate in this case. The other way to produce high energy nuclei is to raise them to a very high temperature (roughly 10^8 K). If deuterium and tritium are heated to a temperature of over a million degrees a gas plasma is produced. (A plasma is a fourth state of matter, which is composed essentially of gaseous ions and a matrix of free electrons.) Since the atoms have been stripped of their electrons, collisions will be between nuclei. Some nuclear collisions will occur with enough energy for the nuclei to approach closely enough to experience each others' strong attraction, and a fusion reaction occurs. The mass lost in this reaction is converted into energy according to Einstein's equation $E = mc^2$.

$${}^{3}_{1}H + {}^{2}_{1}H \rightarrow {}^{4}_{2}He + {}^{1}_{0}n + \text{energy}$$

This fusion reaction has a relatively low ignition temperature, and produces a large amount of energy. Deuterium is available from natural sources, but tritium is difficult to obtain and is extremely expensive. Tritium could be generated in a fusion reactor by bombarding a blanket of lithium with neutrons:

$${}^{1}_{0}n + {}^{6}_{3}Li \rightarrow {}^{3}_{1}H + {}^{4}_{2}He$$
$${}^{1}_{0}n + {}^{7}_{3}Li \rightarrow {}^{3}_{1}H + {}^{4}_{2}He + {}^{1}_{0}n$$

Similar reactions are carried out using only deuterium, but these reactions require a temperature of several million degrees. Several reactions could occur, of which the simplest are:

$${}^{2}_{1}H + {}^{2}_{1}H \rightarrow {}^{3}_{1}H + {}^{1}_{1}H + 4.0\,\text{MeV energy}$$
$${}^{2}_{1}H + {}^{2}_{1}H \rightarrow {}^{3}_{2}He + {}^{1}_{1}n + 3.3\,\text{MeV energy}$$

Fusion is in principle a thermal reaction not inherently different in its kindling from an ordinary fire. Unlike fission, it does not require a critical mass. Once ignited its extent depends on the amount of fuel available. However, for fusion to occur, extreme physical conditions must be achieved:

1. A very high temperature must be attained.
2. Sufficient plasma density is required.

3. The plasma must be confined for an adequate time to allow fusion to occur.

These and other 'hydrogen burning' processes occur at the centre of the sun. This provides the enormous amount of solar energy which is radiated to earth and the rest of the solar system.

Thermonuclear weapons

The only fusion processes carried out successfully on earth have been the hydrogen bomb and other similar thermonuclear weapons. The very high temperatures and high densities required to bring about nuclear fusion reactions have been brought about by means of a small uranium fission bomb. This heats a jacket of lighter elements to a sufficiently high temperature to start the fusion reactions in the hydrogen and nitrogen bombs. These are called thermonuclear reactions. The first thermonuclear device was detonated at Eniwetoc in 1952. The energy yield is 100 times greater than that from an atomic bomb using U or Pu fission. The complete fusion of the nuclei in 1 lb (0.4 kg) of deuterium would result in the same energy release as 26 000 tonnes of TNT.

Controlled fusion reactions

Many attempts have been made to build apparatus in which controlled fusion reactions will occur. So far none have been successful. The problem is how to handle very hot gas plasmas. If the plasma touches a solid (e.g. a steel vessel), the solid is vaporized and the plasma cools down rapidly. The two main methods of confinement are magnetic and inertial. Since a plasma consists of charged particles moving at high speed, it can be deflected by a magnetic field. Plasma can be contained inside a doughnut shaped 'magnetic bottle'. (The extremely high magnetic fields required are obtained with electromagnets using a superconducting niobium/titanium alloy cooled in liquid helium to about 4 K.) Inertial confinement involves the rapid collapse of the fuel container to make the fuel so dense that the fusion reactions occur. Alternatively laser fusion can be used, when high powered, pulsed laser beams are used to heat and compress small 'pellets' of fuel.

It is just possible that fusion may be achieved by some totally different technique without using plasma to attain the high energy conditions. There was great excitement in March 1989 when Fleischmann and Pons claimed to have achieved 'cold fusion' in the laboratory at the University of Utah, USA. They electrolysed 99.5% enriched heavy water D_2O made conducting by dissolving in it some LiOD (D is ^2_1H). Heat appeared to be generated, and this was attributed to D–D fusion. In a similar experiment Jones at Brigham Young University claimed neutrons were released. Unfortunately the rest of the scientific world has so far failed to confirm these results.

Another interesting technique which may show promise is to replace an electron in a D_2^+ molecule by a negatively charged muon which weighs 207 times as much as an electron. This should reduce the D–D spacing by a factor of 200 which should make fusion easier.

Nuclear fusion holds the promise of being an important future source of energy. World energy consumption is high, and fuel resources are finite and limited. Oil and natural gas reserves may well be exhausted in 50 years. Coal may last rather longer, perhaps 200 years. Uranium resources are finite, and the use of nuclear powered electricity generating stations will only delay an eventual energy shortfall. All these fuels pose environmental problems. Fossil fuels (oil, gas and coal) contribute to the Greenhouse Effect and acid rain. A long term energy replacement needs to be found. There is concern over the safety of nuclear power stations, and even greater concern over the storage of nuclear waste products. If fusion can be fully developed:

1. The fuel for fusion (hydrogen) is almost infinitely available.
2. The nuclear processes in fusion are inherently safer than those of fission.
3. Fusion promises to have minimal pollution problems.
4. Difficulties with spent fuel rods and reaction by-products are far less than with fission.

Fusion is an advancing research programme, but many breakthroughs are required. The severe and demanding conditions for controlled fusion in the laboratory have yet to be achieved. If a controlled fusion reactor can be built it will supply almost unlimited power.

THE GENESIS OF THE ELEMENTS

It is an interesting philosophical point to consider how the universe was formed, how the various elements were formed, and why the different elements and their individual isotopes occur in the relative abundances we observe on earth.

The Doppler effect provides evidence that the universe is expanding. Light from the outermost galaxies has a longer wavelength than usual, that is it is towards the red end of the spectrum, because these galaxies are moving away from us. There are several theories for the origin of the universe.

The 'steady state theory' suggests that hydrogen is created continuously to fill the gaps in space created by the expanding universe. The other elements are formed from hydrogen by nuclear reactions.

The 'big bang theory' is currently the most favoured. It assumes that all the matter in the universe was packed as elementary particles into a 'nucleus' of immense density, temperature and pressure. This exploded, hence the name big bang, and dispersed the matter uniformly throughout space as neutrons. The neutrons then decayed, giving protons, electrons and anti-neutrinos.

$$^{1}_{0}n \rightarrow ^{1}_{1}p + ^{0}_{-1}e + \bar{\nu}$$

During the big bang and the fireball which followed, temperatures of 10^6-10^9 K occurred, and in the first hour or so a number of nuclear reactions occurred:

$$^{1}_{1}H + ^{1}_{0}n \rightarrow ^{2}_{1}H$$
$$^{2}_{1}H + ^{1}_{1}H \rightarrow ^{3}_{2}He$$
$$^{3}_{2}He + ^{1}_{0}n \rightarrow ^{4}_{2}He$$
$$^{4}_{2}He + ^{1}_{0}n \rightarrow ^{5}_{2}He$$

The isotope $^{5}_{2}He$ has a half life of only 2×10^{21} s, so building progressively heavier nuclei by the sequential addition of neutrons or protons stopped at this point. Once the temperature had fallen all of these reactions stopped. Thus most of the universe was in the form of H, with a small amount of He, and from this matter the galaxies of stars condensed.

Regardless of the origin of the universe, it is generally accepted that heavier elements may be produced by reactions in the stars. H still accounts for 88.6% and He for about 11.3% of all the atoms in the universe. Together these constitute 99.9% of the atoms, and over 99% of the mass of the universe.

The first process in the synthesis of heavier nuclei in stars is hydrogen burning. Stars are extremely dense (10^8 g cm^{-3}) and there is an enormously large gravitational force. Some of this force is converted into heat and the temperature rises to about 10^7 K. It has been mentioned previously (under 'Nuclear fusion') that this temperature is sufficient to overcome the repulsion between two positively charged H nuclei, and so these nuclei can undergo nuclear fusion, forming deuterium.

		Energy evolved (MeV)
	$^{1}_{1}H + ^{1}_{1}H \rightarrow ^{2}_{1}H + \beta^+ + \nu$	1.44
	$^{2}_{1}H + ^{1}_{1}H \rightarrow ^{3}_{2}He + \gamma$	5.49
	$^{3}_{2}He + ^{3}_{2}He \rightarrow ^{4}_{2}He + 2(^{1}_{1}H)$	12.86
overall	$4(^{1}_{1}H) \rightarrow ^{4}_{2}He + 2\beta^+ + 2\nu + 2\gamma$	

The process is exothermic. A small amount of mass is lost and energy is evolved. Thus more stable nuclei are formed. The process is also slow. The sun is estimated to be 5 billion years old, but still has about 90% of the hydrogen left.

As hydrogen is used up, helium accumulates in the core. The temperature at the core of the star drops, and the star expands to conserve heat. The star is cooler than before and is called a *red giant*. The core eventually collapses under intense pressure and the temperature rises to over 10^8 K. At this point the nuclei begin to fuse.

$$^{4}_{2}He + ^{4}_{2}He \rightarrow ^{8}_{4}Be + \gamma$$

The nuclei formed in this way fuse with more He:

$$^{8}_{4}Be + ^{4}_{2}He \rightarrow ^{12}_{6}C + \gamma$$
$$^{12}_{6}C + ^{4}_{2}He \rightarrow ^{16}_{8}O + \gamma$$
$$^{16}_{8}O + ^{4}_{2}He \rightarrow ^{20}_{10}Ne + \gamma$$

These reactions use up the He in the core and replace it with C, O and Ne. When most of the H and He have been used, small stars contract and become hotter and are called *white dwarfs*.

However, in larger stars (1.4 times the mass of the sun or greater) contraction gives even higher temperatures than before (6×10^8 K) and a carbon–nitrogen cycle occurs, involving the reaction of these elements with hydrogen, provided some $^{12}_{6}C$ is available as catalyst:

$$^{12}_{6}C + ^{1}_{1}H \rightarrow ^{13}_{7}N + \gamma \qquad ^{13}_{7}N \xrightarrow{\text{decays}} ^{13}_{6}C + \beta^+ + \nu$$
$$^{13}_{6}C + ^{1}_{1}H \rightarrow ^{14}_{7}N + \gamma$$
$$^{14}_{7}N + ^{1}_{1}H \rightarrow ^{15}_{8}O + \gamma \qquad ^{15}_{8}O \xrightarrow{\text{decays}} ^{15}_{7}N + \beta^+ + \nu$$
$$^{15}_{7}N + ^{1}_{1}H \rightarrow ^{4}_{2}He + ^{12}_{6}C$$

$$\text{overall} \quad 4(^{1}_{1}H) \rightarrow ^{4}_{2}He + 2\beta^+ + 2\nu + 2\gamma$$

In addition these nuclei may fuse with helium:

$$^{12}_{6}C + ^{4}_{2}He \rightarrow ^{16}_{8}O + \gamma$$
$$^{16}_{8}O + ^{4}_{2}He \rightarrow ^{20}_{10}Ne + \gamma$$
$$^{20}_{10}Ne + ^{4}_{2}He \rightarrow ^{24}_{12}Mg + \gamma$$

At these temperatures carbon and oxygen burning may occur:

$$^{12}_{6}C + ^{12}_{6}C \rightarrow ^{24}_{12}Mg$$
$$\text{or } ^{22}_{11}Na + ^{1}_{1}H$$
$$\text{or } ^{20}_{10}Ne + ^{4}_{2}He$$
$$^{16}_{8}O + ^{16}_{8}O \rightarrow ^{31}_{16}S + ^{1}_{0}n$$
$$^{16}_{8}O + ^{16}_{8}O \rightarrow ^{28}_{14}Si + ^{4}_{2}He$$
$$^{12}_{6}C + ^{4}_{2}He \rightarrow ^{16}_{8}O + \gamma$$
$$^{16}_{8}O + ^{4}_{2}He \rightarrow ^{20}_{10}Ne + \gamma$$

In some cases the star may contract further and the temperature rises to 10^9 K. At this point the γ radiation produced in many of the nuclear changes has sufficient energy to promote endothermic reactions such as:

$$^{20}_{10}Ne + \gamma \rightarrow ^{16}_{8}O + ^{4}_{2}He$$

The He so produced reacts further:

$$^{20}_{10}Ne + ^{4}_{2}He \rightarrow ^{24}_{12}Mg + \gamma$$
$$^{28}_{14}Si + ^{4}_{2}He \rightarrow ^{32}_{16}S + \gamma$$
$$^{32}_{16}S + ^{4}_{2}He \rightarrow ^{36}_{18}Ar + \gamma$$
$$^{40}_{20}Ca + ^{4}_{2}He \rightarrow ^{44}_{22}Ti + \gamma$$

These fusion reactions are exothermic up to $^{56}_{26}Fe$, and Figure 31.2 shows that the binding energy per nucleon increases from H up to Fe, and then decreases with the heavier elements.

The discussion so far explains why H and He make up so much of the universe. The most abundant and stable nuclei up to atomic number 20 have a 1 : 1 ratio of neutrons to protons, e.g. $^{4}_{2}He$, $^{14}_{7}N$, $^{16}_{8}O$, $^{24}_{12}Mg$ and $^{40}_{20}Ca$. Most of these (except N) have an even number of neutrons and an even number of protons (even atomic number) A few light isotopes are common and abundant but do not have a 1 : 1 *N*/*P* ratio, e.g. $^{19}_{9}F$, $^{23}_{11}Na$ and $^{27}_{13}Al$. These elements have an odd atomic number and hence an odd number of protons, but they have an even number of neutrons. These odd–even nuclei are more stable than the corresponding odd–odd nucleus which has a 1 : 1 *N*/*P* ratio. The elements Li, Be and B are of very low abundance compared with their neighbours. It is surprising that they occur at all since the small amounts produced by H and He burning are converted into heavier elements. The small amounts that are found are probably produced by *spallation reactions* where cosmic rays collide with C, N and O nuclei and cause them to break into lighter nuclei. This is sometimes called the *x-process*. A number of elements such as $^{12}_{6}C$, $^{16}_{8}O$ and $^{20}_{10}Ne$ are more abundant than their neighbours, and they differ by a $^{4}_{2}He$ nucleus, which reflects their mode of formation by fusing with helium. The nucleus $^{56}_{26}Fe$ is particularly abundant because it has the largest binding energy per nucleon. It is thus the most stable nucleus and is formed by fusion. The abundances of the elements preceding Fe (Sc, Ti, V and Cr) are also higher than expected, and this is probably due to spallation reactions where high speed cosmic rays collide with Fe, producing Sc, Ti, V or Cr as well as some Li, Be and B.

Nuclei heavier than $^{56}_{26}Fe$ are endothermic, and can only be produced by supplying energy. Thus it becomes increasingly difficult to make these elements by fusion. The heavier elements are synthesized by neutron capture reactions in stars. There are two main processes by which this may occur, called the *s-process* (slow neutron capture) and the r-process (rapid neutron capture).

In the slow neutron capture process neutrons are added one by one to the nucleus. The addition of a neutron increases the *N*/*P* ratio, and eventually the addition of a neutron makes the nucleus unstable. Because of the long time scale the nucleus has time to decay, and the unfavourable *N*/*P* ratio is corrected by β decay. Then the process is repeated and another neutron is added.

$$^{56}_{26}Fe + ^{1}_{0}n \rightarrow ^{57}_{26}Fe + ^{1}_{0}n \rightarrow ^{58}_{26}Fe + ^{1}_{0}n \rightarrow ^{59}_{26}Fe \rightarrow ^{59}_{27}Co + ^{0}_{-1}e$$

The neutrons are produced in red giants and in second generation stars by the normal processes in the star, such as:

$$^{13}_{6}C + ^{4}_{2}He \rightarrow ^{16}_{8}O + ^{1}_{0}n$$

or

$$^{16}_{8}O + ^{16}_{8}O \rightarrow ^{31}_{16}S + ^{1}_{0}n$$

The slow addition of neutrons can produce nuclei up to $^{209}_{83}Bi$. There are much higher than expected abundances of elements round about mass numbers 90, 138 and 208, and the isotopes $^{89}_{39}Y$ and $^{99}_{40}Zr$, $^{138}_{56}Ba$ and $^{140}_{58}Ce$, and $^{208}_{82}Pb$ and $^{209}_{83}Bi$ occur in relatively high abundance. In the discussion on the number of neutrons and protons in the nucleus at the beginning of this chapter it was noted that nuclei with 2, 8, 20, 28, 50, 82 or 126 neutrons or protons are particularly stable, and that these numbers are termed 'magic numbers'. The three clusters of elements with high abundances are close to the neutron magic numbers 50, 82 and 126. This means that they have unusually low neutron absorption cross-sections, so they do not capture neutrons very readily and hence their concentration builds up. The s-process produces the lighter or proton-rich isotopes of an element.

A different source of neutrons occurs just prior to and during a *supernova* period in a star. In this, big stars become extremely hot (8×10^9 K), and nuclei in the core break down into neutrons, protons and α particles which undergo a variety of reactions. The core contracts and eventually implodes, resulting in an explosion of the outer shell which scatters material out into space. In the r-process, many neutrons are added in a period of a few seconds, before β decay eventually takes place.

$$^{56}_{26}Fe + 13(^{1}_{0}n) \rightarrow ^{69}_{26}Fe$$

$$^{69}_{26}Fe \rightarrow ^{69}_{27}Co + ^{0}_{-1}e$$

The very heavy elements are produced in this way by adding many neutrons at once, and traces of $^{254}_{98}Cf$ are present in stars, and this isotope is also formed during nuclear explosions. In a similar way the elements $_{99}Es$ einsteinium and $_{100}Fm$ fermium were formed during the hydrogen bomb explosions. The r-process gives neutron-rich isotopes. It is also possible that neutron-rich isotopes of several lighter isotopes, e.g. $^{36}_{16}S$, $^{46}_{20}Ca$ and $^{48}_{20}Ca$, may be formed by the r-process. It is possible that proton capture may also occur in a supernova on a very short time scale. This is called the p-process, and it is possible that a range of isotopes from $^{74}_{34}Se$, $^{113}_{50}Sn$, $^{114}_{50}Sn$ and $^{115}_{50}Sn$, up to $^{196}_{80}Hg$ are formed in this way.

The short half life periods of all the isotopes of technetium and promethium explains why they are absent on earth.

SOME APPLICATIONS OF RADIOACTIVE ISOTOPES

The applications of radioisotopes are so numerous that only a small selection of them is covered here.

Measurements of radioactivity are used to estimate the age of various objects. $^{14}_{6}C$ occurs naturally and is used in radiocarbon dating of materials of plant and animal origin. This is described in Chapter 12. It depends on calculating when the sample was removed from free exchange with its surroundings. $^{14}_{6}C$ is produced continuously from the action of the neutrons in cosmic rays on nitrogen in the atmosphere. Whilst the plant or animal is alive it will have the same proportion of $^{14}_{6}C$ as the surroundings. When the plant or animal dies, the intake of radiocarbon stops and that already present gradually decays. In a similar way helium dating may be used to estimate the age of certain mineral deposits. Uranium minerals decay by emitting α particles, thus producing helium. One gram of U produces about 10^{-7} g of He in a year. The age of the mineral can be estimated if both the U and He contents are known. Corrections must be made for some He escaping, and for the production of He from other elements such as Th which may be present.

Isotope dilution analysis can be used to determine the solubility of sparingly soluble materials. For example, a solid sample containing radioactive $^{90}_{38}SrSO_4$ and normal $SrSO_4$ can be prepared and the activity per gram measured. A known volume of a saturated solution is evaporated and the activity of the residue measured. Comparison of the activities gives the solubility. In a similar way the isotope $^{32}_{15}P$ has been used to measure the solubility of $MgNH_4PO_4 \cdot 6H_2O$, and $^{131}_{53}I$ has been used to measure the solubility of PbI_2. A similar technique can be applied to vapour pressures of rather involatile materials.

Activation analysis is used to determine the amount of an element in a sample. The sample to be analysed and another sample containing a known amount of the element are placed in a nuclear reactor where they are bombarded (usually with neutrons). After irradiation for a suitable period (several times the half life of the expected radioisotope) the samples are removed from the reactor. The induced radioactivity in both samples is measured. The amount of the element present in the unknown sample is obtained from the ratio of the activities of unknown and standard samples. More than 50 elements (or strictly isotopes) can be determined in this way. The method is most used to determine trace quantities. It has the advantage that the sample is not destroyed, but the disadvantage that it requires access to a reactor. Not all elements can be determined in this way; for example, it is not feasible to determine C because $^{12}_{6}C$ has a very low absorption cross-section for neutrons, though activation analysis of C can be performed by bombarding with deuterium in a cyclotron.

Isotope exchange reactions provide information on the mechanisms of certain reactions. Thus exchange of $^{2}_{1}D$ in heavy water D_2O for $^{1}_{1}H$ in a compound occurs rapidly if H is bonded to N or O, but exchange is slow or hardly occurs in most cases where H is bonded to C. This is related to the mobility of protons and the higher polarity of N—H and O—H bonds.

$$D_2O + RO{-}H \rightleftharpoons HDO + RO{-}D$$

If labelled $^{15}_{7}NH_4Cl$ is dissolved in liquid NH_3, and the solvent evaporated, the activity from $^{15}_{7}N$ is evenly distributed between the NH_4Cl and the

NH_3. Since $^{15}_{7}NH_4Cl$ ionizes into $^{15}_{7}NH_4^+$ and Cl^-, this provides evidence for the ionization of liquid NH_3:

$$2NH_3 \rightleftharpoons NH_4^+ + NH_2^-$$

Similarly the lack of exchange between $^{14}_{6}C$ in a solution containing labelled CN^- and the complex $[Fe(CN)_6]^{3-}$ shows that the CN^- groups in the complex are not labile. Proof that the two S atoms in the thiosulphate ion $S_2O_3^{2-}$ are not identical can be obtained by heating labelled $^{35}_{16}S$ with an aqueous solution of sodium sulphite to form the thiosulphate.

$$SO_3^{2-} + {}^{35}_{16}S \rightarrow {}^{35}_{16}SSO_3^{2-}$$

If the thiosulphate so formed is decomposed by treatment with acid, all the radioactivity returns to the sulphur, and none to the SO_2 formed.

$$2H^+ + {}^{35}_{16}SSO_3^{2-} \rightarrow SO_2 + {}^{35}_{16}S + H_2O$$

The isotope $^{60}_{27}Co$ emits both β and γ radiation. It is used for γ-radiography to detect cracks and flaws in metal parts such as pipes, aircraft parts and welded joints, for example in the pressure vessels for nuclear reactors. It is also used to irradiate malignant tumours in the body as one method of cancer therapy.

The nuclide $^{131}_{53}I$ is used to locate brain tumours and also in the diagnosis and treatment of disorders of the thyroid gland. In the case of brain tumours, the isotope is incorporated into a dye and injected into the patient. It is preferentially absorbed by the cancerous cells, and the position of the tumour can be found by scanning the skull. For thyroid cases the patient consumes a low dose of this isotope as NaI. The radioactive I concentrates in the thyroid gland and can be measured by counting the γ activity.

Labelled NaCl solution which contains the isotope $^{24}_{11}Na$ is injected into the veins to discover the location and extent of blood clots and other circulation disorders.

SOME UNITS AND DEFINITIONS

amu = atomic mass unit = $\frac{1}{12}$ the mass of the ^{12}C carbon atom.

Mass number = number of neutrons + number of protons.

Mass of hydrogen atom $^{1}_{1}H$ = 1.007 825 amu.

Mass of proton $^{1}_{1}p$ or $^{1}_{1}H$ = 1.007 277 amu.

Mass of neutron $^{1}_{0}n$ = 1.008 665 amu.

Mass of electron $^{0}_{-1}e$ = 0.000 548 59 amu.

Mass of helium atom $^{4}_{2}He$ = 4.002 60 amu.

Mass of helium nucleus (α particle $^{4}_{2}He$) = 4.001 50 amu.

MeV = million electron volts ($1\,MeV = 9.648 \times 10^7\,kJ\,mol^{-1}$).

$1\,amu = 931.4812\,MeV = 8.982 \times 10^{10}\,kJ\,mol^{-1}$.

FURTHER READING

Abelson, P.H. (1989) Products of neutron irradiation, *J. Chem. Ed.*, **66**, 364–366.
Ahrens, L.H. (ed.) (1979) *Origin and Distribution of the Elements*, Pergamon Press,
Oxford. (Proceedings of second UNESCO symposium.)
Cadman, P. (1986) Energy from the nucleus, *Education in Chemistry*, **23**, 8–11.
Choppin, G.R. and Rydberg, J. (1980) *Nuclear Chemistry – Theory and Applications*, Pergamon, Oxford.
Cunninghame, J.G. (1972) *Chemical Aspects of the Atomic Nucleus*, Monographs for Teachers, No. 23, Royal Society for Chemistry, London.
Fergusson, J.E. (1982) *Inorganic Chemistry and the Earth* (Chapter 1, Origins: the Chemical Elements and the Earth), Pergamon, Oxford.
Grabowski, K.F.M. (1986) The early use of radioactive tracers in chemistry, *Education in Chemistry*, **23**, 174–176.
Nier, A.O. (1989) Some reminiscences of mass spectrometry and the Manhattan Project, *J. Chem. Ed.*, **66**, 385–388. (Reflections on nuclear fission at its half century.)
Peacocke, T.A.H. (1978) *Radiochemistry: Theory and Experiment*, Wykeham Publications, London. (A good elementary text.)
Rhodes, R. (1989) The complementarity of the Bomb, *J. Chem. Ed.*, **66**, 376–379. (Reflections on nuclear fission at its half century.)
Seaborg, G.T. (1989) Nuclear fission and transuranium elements, *J. Chem. Ed.*, **66**, 379–384. (Reflections on nuclear fission at its half century.)
Selbin, J. (1973) Stellar nucleosynthesis, *J. Chem. Ed.*, **50**, 306, 380.
Sime, R.L. (1989) Lise Meitner and the discovery of fission, *J. Chem. Ed.*, **66**, 373–376. (Reflections on nuclear fission at its half century.)
Steinberg, E.P. (1989) Radiochemistry of the fission products, *J. Chem. Ed.*, **66**, 367–372. (Reflections on nuclear fission at its half century.)
Taylor, R.J. (1972) *The Origin of the Chemical Elements*, Wykenham Publications, London. (A relatively straightforward account.)
Thompson, R. (ed.) (1986) *The Modern Inorganic Chemicals Industry*, Special Publication No. 31, The Chemical Society, London. (See the chapter 'The Inorganic Chemistry of Nuclear Fuels' by Findlay, J.R. et al.)
Vertes, A. and Kiss, I. (1987) *Nuclear Chemistry* (Topics in Inorganic and General Chemistry Series: No. 22), Elsevier.

PROBLEMS

1. What is the nature of the binding forces in atomic nuclei? How does the average binding energy per nucleon vary with the atomic number of the element?
2. Define the following terms: atomic number, mass number, isotope, α decay, β decay, γ radiation, nuclear fission, nuclear fusion.
3. In what way is the mode of decay of a particular nucleus related to (a) the ratio of neutrons and protons, and (b) the size?
4. Explain the terms mass number, isotopic mass and nuclear binding energy used in the description of an isotope.
5. Draw a diagram to illustrate how the binding energy per nucleon varies with the mass number. Comment on the shape of the curve.

6. How would you work out the binding energy per nucleon for a given atomic nucleus? What is the significance of this value?
7. Work out the binding energy per nucleon (in MeV per nucleon) for the isotope $^{56}_{26}Fe$ given the masses: ^{56}Fe 55.93494 amu, neutron 1.008665 amu, proton 1.00783 amu, electron 0.00054859 amu. (Answer: 8.79 MeV per nucleon.)
8. $^{24}_{11}Na$ is an unstable isotope of sodium with a half life of 15 hours. Calculate the value of the radioactive decay constant. Explain how you would attempt to predict the mode of decay.
9. What is the radioactive displacement law? Illustrate the radioactive displacement law by reference to the four radioactive decay series.
10. Write equations showing how the following nuclei undergo α decay: (a) $^{238}_{92}U$, (b) $^{232}_{90}Th$, (c) $^{193}_{83}Bi$, (d) $^{212}_{86}Rn$, (e) $^{215}_{84}Po$
11. Write equations showing how the following nuclei undergo β decay: (a) $^{27}_{12}Mg$, (b) $^{211}_{82}Pb$, (c) $^{24}_{10}Ne$, (d) $^{60}_{27}Co$, (e) $^{14}_{6}C$
12. Write equations showing how the following nuclei undergo electron capture:
(a) $^{40}_{19}K$, (b) $^{7}_{4}Be$, (c) $^{71}_{32}Ge$, (d) $^{119}_{51}Sb$, (e) $^{75}_{34}Se$
13. Write equations showing how the following nuclei undergo positron β^+ emission:
(a) $^{13}_{7}N$, (b) $^{18}_{9}F$, (c) $^{22}_{11}Na$, (d) $^{19}_{10}Ne$, (e) $^{60}_{30}Zn$
14. Write an equation to show what would happen if a nuclide $^{256}_{100}Fm$ underwent spontaneous fission producing two identical daughter nuclei and liberating four neutrons. Write another equation assuming that the daughter nuclei differ from each other by 40 mass numbers.
15. In the naturally occurring uranium decay series, the nuclide $^{238}_{92}U$ decays in succession by α, β, β, α, α, α, α, β decay. Write an equation showing the mass numbers, atomic numbers and symbols for the nuclides formed by these decays.
16. In the naturally occurring thorium decay series, the nuclide $^{232}_{90}Th$ decays in succession by α, β, β, α, α, α, β decay. Write an equation showing the mass numbers, atomic numbers and symbols for the nuclides formed by these decays.
17. The following reaction is one of the processes which occurs during fission:

$$^{235}_{92}U \rightarrow {}^{140}_{58}Ce + {}^{94}_{40}Zr + {}^{1}_{0}n + 6\,{}^{0}_{-1}e$$

Given that the masses are: U 235.0439 amu, Ce 139.9054 amu, Zr 93.9063 amu, n 1.008665 amu, e 0.00054859 amu, calculate how much energy is released in MeV per fission. (Answer: 205 MeV per fission.)

18. Explain the main differences between the two atomic bombs that were dropped on Japan.

19. (a) What is the difference between a chain reaction and a branched chain reaction.
    (b) What is a moderator? Give examples.
    (c) Explain what the critical mass of a fissile material is, and why does the critical mass vary?

20. Compare the processes of nuclear fission and fusion as sources of energy.

21. Calculate the energy released in MeV per fusion in the following process:

$$^{2}_{1}D + ^{1}_{1}H \rightarrow ^{3}_{1}H$$

given that the atomic masses are: ^{2}D 2.01410 amu, ^{1}H 1.007825 amu, ^{3}He 3.01603 amu. (Answer: 5.5 MeV per fusion.)

22. Explain how nuclei heavier than uranium can be prepared.

23. List uses for radioactive isotopes.

32 Spectra

Electronic spectra from transition metal ions and complexes are observed in the visible and UV regions. Absorption spectra show the particular wavelengths of light absorbed, that is the particular amount of energy required to promote an electron from one energy level to a higher level, whilst emission spectra show the energy emitted when the electron falls back from the excited level to a lower level. Transitions involving the outer shell of electrons are generally observed in the wavenumber region 100 000 cm^{-1} to 10 000 cm^{-1}, but most spectra are measured in the 50 000–10 000 cm^{-1} region (200–1000 nm). The interpretation of spectra provides a most useful tool for the description and understanding of the energy levels present.

ENERGY LEVELS IN AN ATOM

The energy levels in an atom are described in Chapter 1 in terms of four quantum numbers:

1. n the principal quantum number which may have values 1, 2, 3, 4... corresponding to the first, second, third or fourth shell of electrons around the nucleus.
2. l the subsidiary quantum number, which may have values 0, 1, 2... $(n - 1)$, and describes the orbital angular momentum or shape of the orbital. Thus

$n = 1$	$l = 0$	spherical s orbital
$n = 2$	$l = 0$	spherical s orbital
	$l = 1$	dumb-bell shaped p orbital
$n = 3$	$l = 0$	s orbital
	$l = 1$	p orbital
	$l = 2$	d orbital

3. m the magnetic quantum number may have values from $+l$, $(l - 1)$... 0... $-l$.
4. m_s the electron spin quantum number which has a value of either $+\frac{1}{2}$ or $-\frac{1}{2}$.

The build-up of electrons in the elements follows three simple rules:

1. Electrons normally occupy the orbitals of lowest energy.
2. Hund's rule: when several orbitals have the same energy, electrons are not paired if this can be avoided. Thus in the ground state, an atom will contain the maximum number of unpaired electron spins.
3. The Pauli exclusion principle: no two electrons in one atom can have all four quantum numbers the same.

This can be illustrated using boxes for orbitals, and arrows for electrons (Figure 32.1). The two electrons in the 1*s* orbital for He have opposite spins; thus ↑ denotes $m_s = +\frac{1}{2}$ and ↓ denotes $m_s = -\frac{1}{2}$. The filling of the 1*s* and 2*s* energy levels is straightforward.

At boron, an electron occupies a 2*p* orbital. The three 2*p* orbitals have identical values of $n = 2$, and $l = 1$, but have different values of m (+1, 0

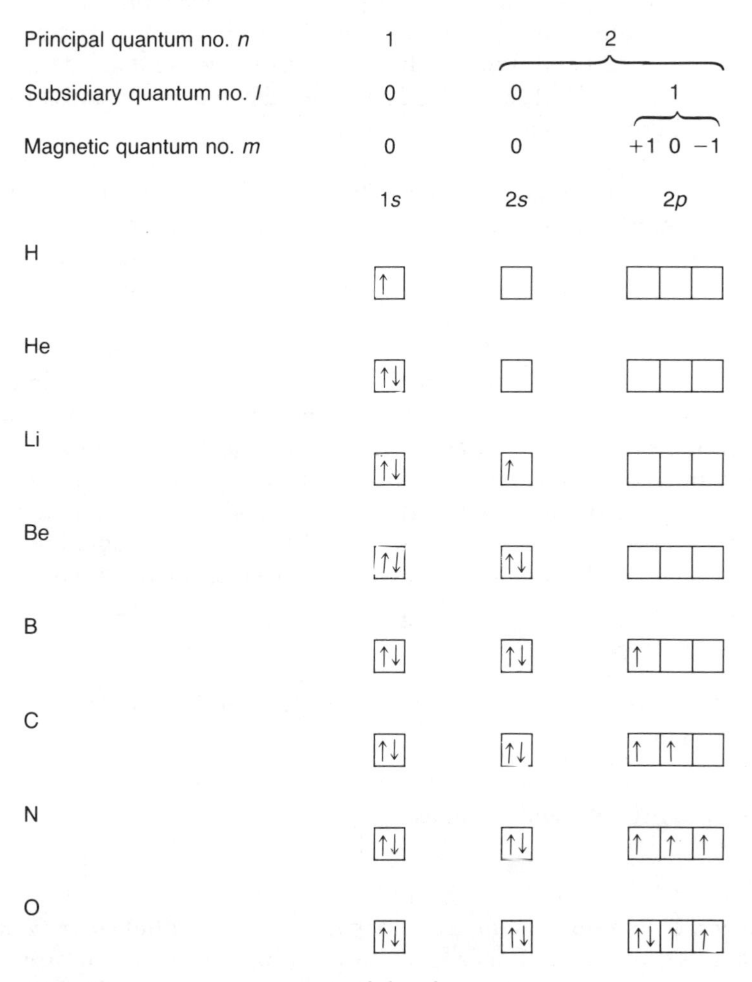

Figure 32.1 Electronic arrangements of the elements.

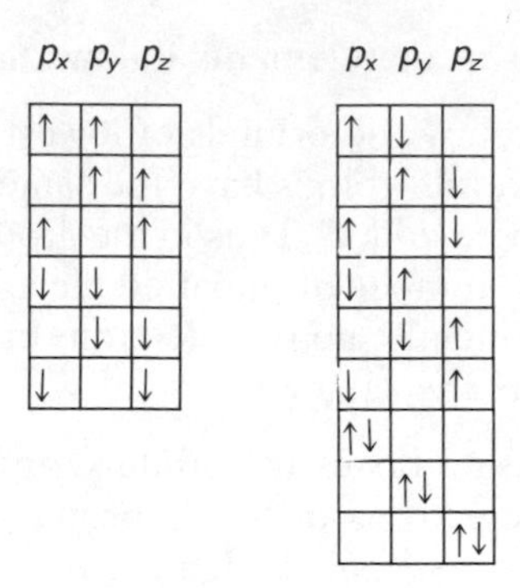

Figure 32.2 Electronic arrangements of microstates for p^2 configuration.

and -1), giving the p_x, p_y and p_z orbitals. Allowing for two values for m_s for each of these, there is a total of six possible arrangements for this single electron. The three p orbitals are degenerate, so it does not matter which arrangement is adopted. When only one electron is present in a degenerate energy level or subshell such as $2p$, $3p$ or $3d$, the energy depends on l, the orbital quantum number.

In the case of carbon, two electrons occupy the $2p$ level, and there are 15 possible electronic arrangements (Figure 32.2). These can be divided into three main groups of different energy – called three energy states. Thus even though the p orbitals are degenerate and have the same energy, the electrons present in them interact with each other and result in the formation of a ground state (lowest energy) and one or more excited states for the atom or ion. In addition to the electrostatic repulsion between electrons, they influence each other (1) by the interaction or coupling of the magnetic fields produced by their spins, and (2) by coupling of the fields produced by the orbital motion of the electrons (orbital angular momentum). When several electrons occupy a subshell, the energy states obtained depend on the result of the orbital angular quantum numbers of each of the electrons. This resultant of all the l values is denoted by a new quantum number L, which defines the energy state for the atom

$L =$	0	1	2	3	4	5	6		7	8	...
state	S	P	D	F	G	H	I	~~J~~	K	L	...

(The letter J is omitted since this is used for another quantum number described later.)

Coupling of orbital angular momenta

p^2 configuration

Angular momentum is quantized into 'packets' of magnitude $h/2\pi$ (where h is Planck's constant). For a p electron, the subsidiary or azimuthal quantum number $l = 1$, and the orbital angular moment $= 1(h/2\pi)$ and is

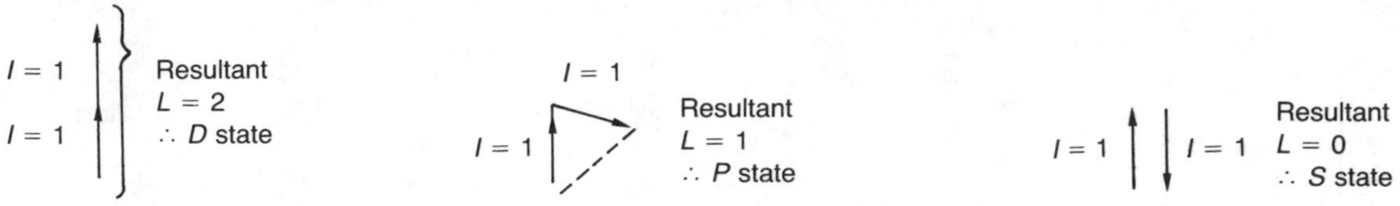

$l = 1$ / $l = 1$ — Resultant $L = 2$ ∴ D state

$l = 1$ / $l = 1$ — Resultant $L = 1$ ∴ P state

$l = 1$ / $l = 1$ — Resultant $L = 0$ ∴ S state

Figure 32.3 Resultant of l terms for p^2 configuration.

shown as an arrow of unit length. The ways in which the l values for two p electrons may interact with each other are shown diagrammatically using vector diagrams (Figure 32.3). Because angular momentum is quantized, the only permissible arrangements are those where the resultant is a whole number of quanta. Thus three possible states of spectroscopic terms D, P and S arise. For the P state, the vectors l must be at an angle to each other such that the resultant is a whole number of quanta and in this case $L = 1$.

p^3 configuration

The coupling of the l values of three p electrons by vectorial addition can be considered in an analogous way, and for simplicity this is considered as the interaction of a third p electron on the states obtained for the p^2 case (Figure 32.4). The result of coupling the orbital angular momenta is the production of one F state, two D states, three P states and one S state.

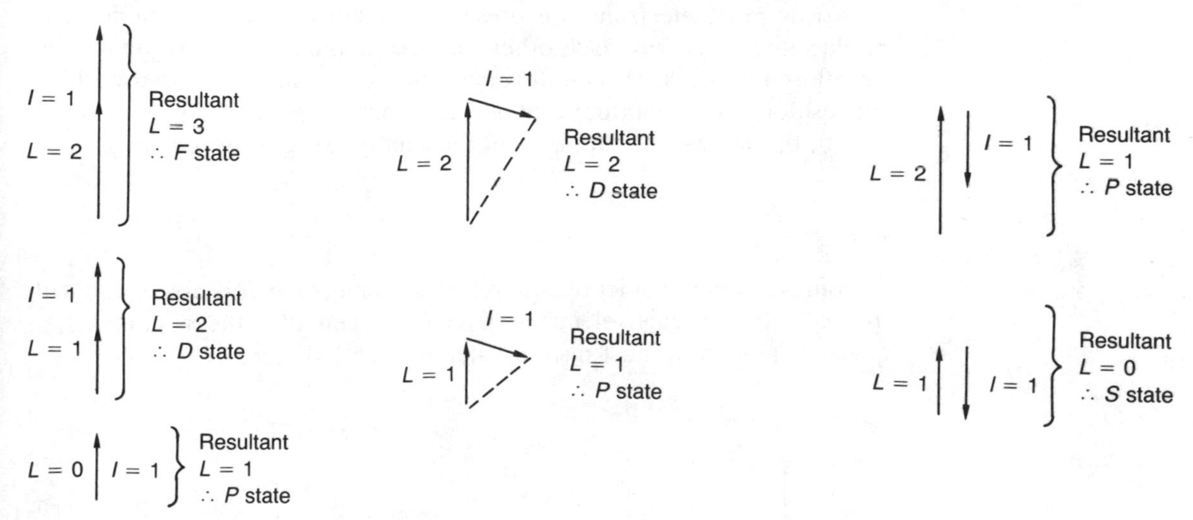

Figure 32.4 Resultant of l terms for p^3 configuration.

d^2 configuration

The coupling of l values for d electrons follows a similar pattern, except that for a d electron $l = 2$, so the arrows are of double length (Figure 32.5).

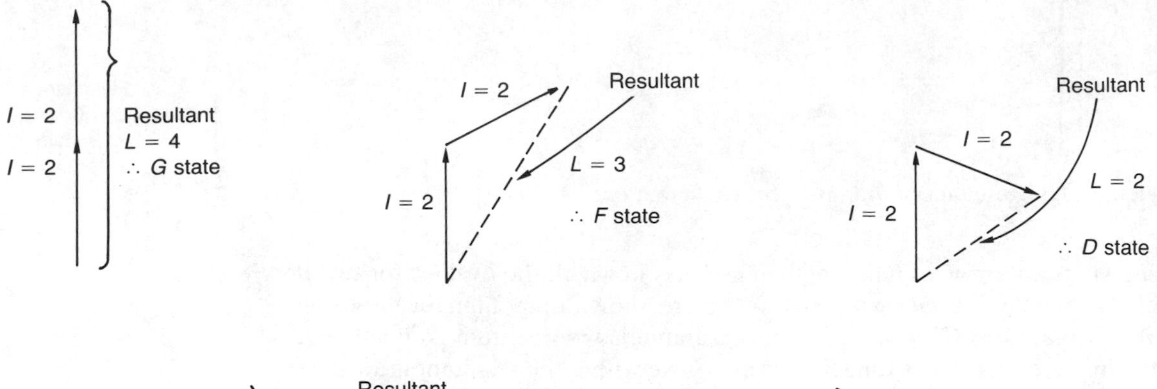

Figure 32.5 Resultant of l terms for d^2 configuration.

Coupling of spin angular momenta

For a single electron, the spin quantum number m_s has a value of $+\frac{1}{2}$ or $-\frac{1}{2}$. If two or more electrons are present in a subshell, the magnetic fields produced interact with each other, that is 'couple', giving a resultant spin quantum number S. (It is unfortunate that the symbol S is used both for the resultant spin quantum number, and for the spectroscopic state when $L = 0$, but in practice this does not normally cause confusion.)

p^2 or d^2 case

By using arrows to depict the quantized amounts of energy associated with the m_s value of each electron, it can be seen that the resultant spin quantum number S must have a value of 0 or 1 (Figure 32.6).

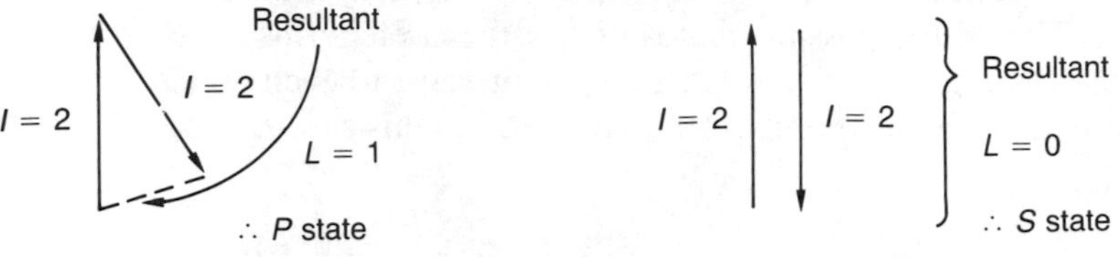

Figure 32.6 Resultant of m_s terms for p^2 or d^2 configuration.

p^3 or d^3 case

Here S has a value of $1\frac{1}{2}$ or $\frac{1}{2}$ (Figure 32.7).

Resultant $S = 1\frac{1}{2}$

Resultant $S = \frac{1}{2}$

Figure 32.7 Resultant of m_s terms for p^3 or d^3 configuration.

Spin orbit coupling

When several electrons are present in a subshell, the overall effect of the individual orbital angular momenta l is given by the resultant angular quantum number L, and the overall effect of the individual spins m_s is given by the resultant spin quantum number S. In an atom, the magnetic effects of L and S may interact or 'couple', giving a new quantum number J called the total angular momentum quantum number, which results from vectorial combination of L and S. This coupling of the resulting spin and orbital quantum numbers is called Russell–Saunders or LS coupling.

Spin orbit coupling p^2 case

It has been shown previously that with a p^2 arrangement, resultant orbital quantum number values of $L = 2$, 1 and 0, and also resultant spin quantum number values of $S = 1$ and 0, are obtained. These may be coupled to give the total angular quantum number J (Figure 32.8).

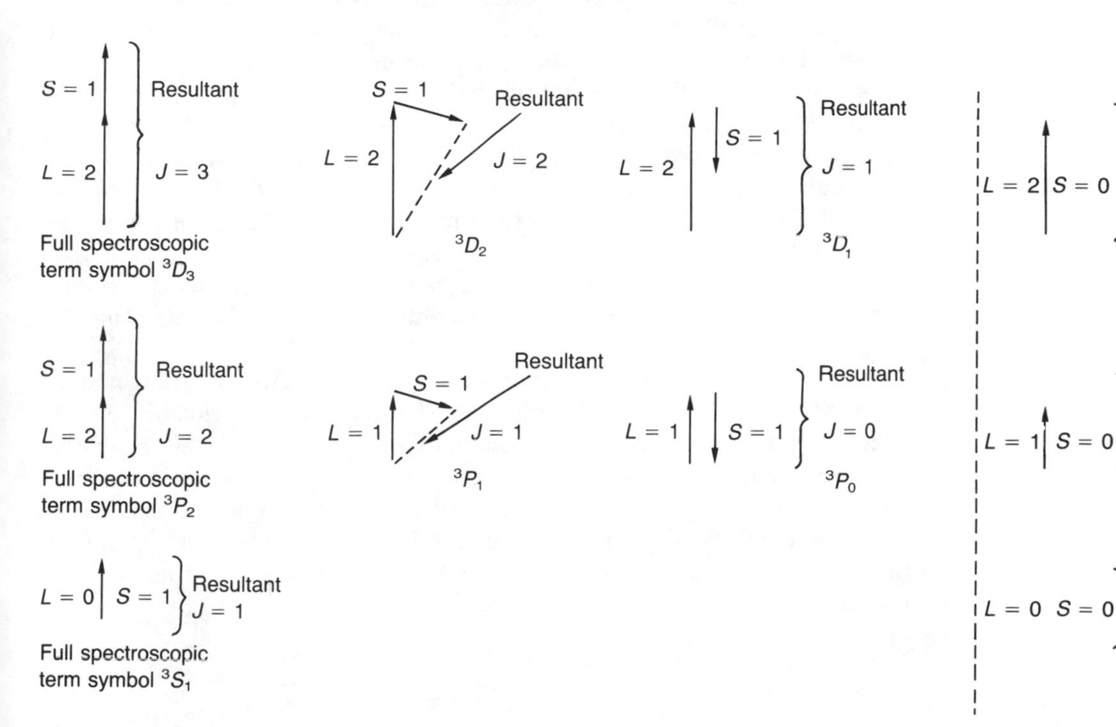

Figure 32.8 Obtaining spectroscopic term symbols by combining the resultant L and S terms.

Each of these arrangements corresponds to an electronic arrangement sometimes called a spectroscopic state, which is described by a full term symbol. The letter D indicates that the L quantum number has a value of two, P indicates that $L = 1$ and S denotes a value of $L = 0$ as described previously. The lower right hand subscript denotes the value of the total quantum number J, and the upper left superscript indicates the multiplicity, which has the value of $2S + 1$ (where S is the resultant spin quantum number). The relation between the number of unpaired electrons, the resultant spin quantum number S, and the multiplicity is given in Table 32.1.

Table 32.1

Unpaired electrons	S	Multiplicity	Name of state
0	0	1	Singlet
1	$\frac{1}{2}$	2	Doublet
2	1	3	Triplet
3	$1\frac{1}{2}$	4	Quartet
4	2	5	Quintet

Thus the symbol ${}^3D_2$ (pronounced triplet D two) indicates a D state, hence $L = 2$; the multiplicity is three, hence $S = 1$ and the number of unpaired electrons is 2; and the total quantum number $J = 2$.

All of the spectroscopic terms derived above for a p^2 configuration would occur for an excited state of carbon $1s^2$, $2s^2$, $2p^1$, $3p^1$. However, in the ground state of the atom $1s^2$, $2s^2$, $2p^2$ the number of states is limited by the Pauli exclusion principle since no two electrons in the same atom can have all four quantum numbers the same. In the ground state configuration, the two p electrons both have the same values of $n = 2$ and $l = 1$ so they must differ in at least one of the remaining quantum numbers m or m_s. This restriction reduces the number of terms from ${}^3D$, ${}^3P$, ${}^3S$, ${}^1D$, ${}^1P$ and ${}^1S$ to ${}^1D$, ${}^3P$ and ${}^1S$.

This can be shown by writing down only those electronic arrangements of m and m_s which do not violate the Pauli exclusion principle. For p electrons, the subsidiary quantum number $l = 1$, and the magnetic quantum number m may have values from $+l \rightarrow 0 \rightarrow -l$, giving in this case values of $m = +1$, 0 and -1. There are 15 possible combinations (Table 32.2). The values of M_s and M_L (the total spin and total orbital quantum numbers in the z direction) are obtained by adding the appropriate m_s and m values:

$$M_s = \Sigma m_s$$

$$M_L = \Sigma m$$

M_L has values from $+L \ldots 0 \ldots -L$ (a total of $2L + 1$ values) and M_s has values from $+S \ldots 0 \ldots -S$ (a total of $2S + 1$ values)

Table 32.2 Allowed values of m and m_s for the p^2 configuration

	$m = +1$	0	-1	M_S	M_L	Term symbol
1	↑↓			0	2	^1D
2			↑↓	0	−2	^1D
3		↑↓		0	0	} ^3P, ^1D, ^1S
4	↑		↓	0	0	
5	↓		↑	0	0	
6	↑	↓		0	1	} ^3P, ^1D
7	↓	↑		0	1	
8		↑	↓	0	−1	} ^3P, ^1D
9		↓	↑	0	−1	
10	↑	↑		1	1	^3P
11	↑		↑	1	0	^3P
12		↑	↑	1	−1	^3P
13	↓	↓		−1	1	^3P
14	↓		↓	−1	0	^3P
15		↓	↓	−1	−1	^3P

The L and S quantum numbers associated with each electronic configuration (and hence the spectroscopic term symbol) can be worked out from the M_L and M_S quantum numbers in Table 32.2. First choose the maximum M_S value, and select the maximum M_L associated with it. This gives $M_S = 1$ and $M_L = 1$ (number 10 in table), and corresponds to a group of terms where $L = 1$ and $S = 1$. Since $L = 1$, this must be a P state, and since $S = 1$, the multiplicity $(2S + 1) = 3$, so it is in fact a triplet P state ^3P. Using the equations above:

if $L = 1$, M_L may have the values $+1$, 0 and -1

and

if $S = 1$, M_S may have the values $+1$, 0 and -1

There are nine combinations of these two terms:

$$M_L = +1 \qquad M_S = +1, 0, -1$$
$$M_L = 0 \qquad M_S = +1, 0, -1$$
$$M_L = -1 \qquad M_S = +1, 0, -1$$

Examination of Table 32.2 shows that 13 of the allowed values could be assigned a ^3P term symbol.

From the unassigned combinations we next pick out the maximum M_S and M_L. In this case $M_S = 0$ and $M_L = 2$. From this it is deduced that $L = 2$ and $S = 0$. Since $L = 2$ this must be a D state. The value of $S = 0$ gives a multiplicity of $2S + 1 = 1$, so it is a singlet D state ^1D.

If $L = 2$, then M_L may have values $+2$, $+1$, 0, -1 and -2, and since $S = 0$, $M_S = 0$. This gives five combinations of M_L and M_S. Examination of

Table 32.2 shows that nine of the allowed values could be assigned a ^1D term symbol.

The ^3P and ^1D states account for 9 + 5 = 14 combinations, and the remaining one which corresponds to $M_L = 0$ and $M_S = 0$, must correspond to $L = 0$ and $S = 0$. This gives a singlet S state ^1S. Thus all 15 permissible electronic arrangements are accounted for by the ^1D, ^3P and ^1S states. Where two or more allowed electronic configurations have the same values for M_L and M_S (for example configurations 3, 4 and 5 in Table 32.2), more than one term symbol will describe the arrangement. In these cases a linear combination of the functions should be taken, and it is incorrect to attribute any one term to a particular arrangement.

DETERMINING THE GROUND STATE TERMS – HUND'S RULES

Once the terms are known, they can be arranged in order of energy, and the ground state term identified by using Hund's rules:

1. The terms are placed in order depending on their multiplicities and hence their S values. The most stable state has the largest S value, and stability decreases as S decreases. The ground state therefore possesses the most unpaired spins because this gives the minimum electrostatic repulsion.
2. For a given value of S, the state with the highest L value is the most stable.
3. If there is still ambiguity, for given values of S and L, the smallest J value is the most stable if the subshell is less than half filled, and the biggest J is most stable if the subshell is more than half filled.

(Hund's rules should not be used to predict the order of excited configurations such as C $1s^2$, $2s^2$, $2p^1$, $3p^1$.)

Applying the first rule to the terms arising from p^2 in the ground state of carbon the ^3P state must be the ground state since there is only one triplet state, ^1D and ^1S being singlets. Using the second rule, the ^1D state corresponds to a value of $L = 2$ and is more stable than the ^1S state where $L = 0$. Finally, the triplet P state has three terms $^3P_2$, $^3P_1$ and $^3P_0$, so from the third rule $^3P_0 < {}^3P_1 < {}^3P_2$. The experimentally measured energies for the terms arising from the ground state of carbon are shown in Figure 32.9. It can be seen that for a light atom like carbon, the splitting of the ^3P terms because of the J terms from spin orbit coupling is much smaller than the splitting into ^1S, ^1D and ^3P terms resulting from coupling of l quantum numbers. For the lighter elements below atomic number 30, the splitting of levels of different J is small compared with the splitting of levels of different L (see Figure 32.9); hence Russell–Saunders coupling gives the correct result for the sequence of energy levels or terms for the first row of transition elements. For the heavier elements, the J splitting is greater than the L splitting and Russell–Saunders coupling can no longer be used and an alternative form of j–j coupling is used instead.

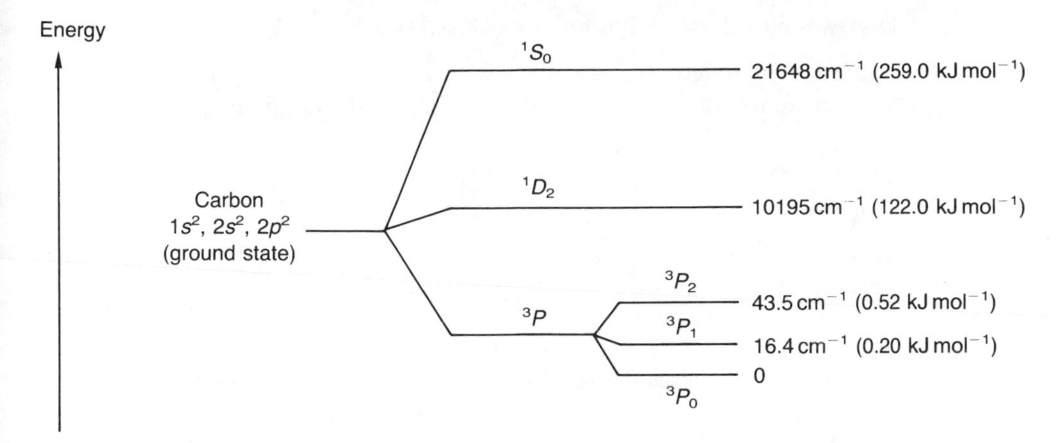

Figure 32.9 Splitting of terms in carbon ground state.

HOLE FORMULATION

When a subshell is more than half full, it is simpler and more convenient to work out the terms by considering the 'holes' – that is the vacancies in the various orbitals – rather than the larger number of electrons actually present. The terms derived in this way for the ground state of oxygen which has a p^4 configuration and hence two 'holes' are the same as for carbon with a p^2 configuration, that is ^1S, ^1D and ^3P. However, oxygen has a more than half filled subshell, and hence when applying Hund's third rule, the energy of the triplet P states for oxygen are $^3P_2 < {}^3P_1 < {}^3P_0$, making $^3P_2$ the ground state. In a similar way, by considering 'holes', the terms which arise for pairs of atoms with p^n and p^{6-n} arrangements, and also d^n and d^{10-n}, give rise to identical terms (Table 32.3).

Table 32.3 Terms arising for p and d configurations

Electronic configuration	Ground state term	Other terms
p^1, p^5	^2P	
p^2, p^4	^3P	^1S, ^1D
p^3	^4S	^2P, ^2D
p^6	^1S	
d^1, d^9	^2D	
d^2, d^8	^3F	^3P, ^1G, ^1D, ^1S
d^3, d^7	^4F	^4P, ^2H, ^2G, ^2F, ^2D, ^2P
d^4, d^6	^5D	^3H, ^3G, ^3F, ^3D, ^3P, ^1I, ^1G, ^1F, ^1D, ^1S
d^5	^6S	^4G, ^4F, ^4D, ^4P, ^2I, ^2H, ^2G, ^2F, ^2D, ^2P, ^2S
d^{10}	^1S	

Derivation of the term symbol for a closed subshell

If a subshell is completely filled with electrons, for example p^6 or d^{10} arrangements, the derivation of terms is greatly simplified.

	$m =$ +1 0 1	M_S	M_L
p^6	↑↓ | ↑↓ | ↑↓	0	0

	$m =$ +2 +1 0 −1 −2	M_S	M_L
d^{10}	↑↓ | ↑↓ | ↑↓ | ↑↓ | ↑↓	0	0

In both of these cases the total spin quantum number in the z direction M_S which equals the sum of all the individual m_s values is zero, hence $S = 0$, and the multiplicity $2S + 1 = 1$. Also in both p^6 and d^{10} cases the total orbital angular quantum number in the z direction $M_L = \Sigma m = 0$, hence $L = 0$ corresponding to an S state. Thus a closed shell of electrons always produces a singlet S state $^1S_0$.

Derivation of the terms for a d^2 configuration

For d electrons, the subsidiary quantum number l is 2, and the magnetic quantum number m has values from $+l \rightarrow 0 \rightarrow -l$, giving in this case $m = +2, +1, 0, -1$ and -2. There are 45 ways in which two d electrons may be arranged which do not violate the Pauli exclusion principle. These arrangements are shown in Table 32.40.

The terms are assigned in a similar way as for the p^2 case. The highest value of $M_L = 4$ can only arise if $L = 4$ corresponding to a G state, and since $M_S = O$, S must be 0, and it is a singlet G term ^1G. M_L can have values from $+L \ldots 0 \ldots -L$, in this case +4, +3, +2, +1, 0, −1, −2, −3 and −4, but M_S has only one value so there are nine configurations associated with this term.

The highest unassigned M_L value is +3, indicating an F state, and this occurs with $M_S = +1$, 0 and −1, suggesting a triplet F state ^3F. M_L may have values of +3, +2, +1, 0, −1, −2 and −3, and since M_S has three values there are 21 configurations associated with the ^3F term.

Thirty of the 45 configurations have been accounted for, and examination of the groups of configurations which have the same M_L and M_S values shows that the remaining 15 terms are the same as for the p^2 case, that is ^1D, ^3P and ^1S. The full list of terms for a d^2 configuration is therefore ^1G, ^3F, ^1D, ^3P and ^1S. Applying Hund's rules, the ground state is ^3F, and the energy of the various states is $^3F < {}^3P < {}^1G < {}^1D < {}^1S$.

The spectra of a number of d^2 ions have been measured, and the energy of the ^3P state is shown to be higher than the ^3F ground state in each case (Table 32.5). Quantitative data are available for the energy levels in free metal ions in the gas phase, and the next step is to find how these energy levels change when ligands approach to form a complex.

Table 32.4 Allowed values of m and m_s for the d^2 configuration

	$m = +2$	$+1$	0	-1	-2	$\Sigma m_S = M_S$	$\Sigma m = M_L$	Term symbol
1	↑↓					0	4	^1G
2	↑	↑				1	3	^3F
3	↑	↓				0	3	} ^1G, ^3F
4	↓	↑				0	3	
5	↓	↓				−1	3	^3F
6	↑		↑			1	2	^3F
7	↑		↓			0	2	} ^1G, ^3F, ^1D
8	↓		↑			0	2	
9		↑↓				0	2	
10	↓		↓			−1	2	^3F
11	↑			↑		1	1	} ^3F, ^3P
12		↑	↑			1	1	
13	↓			↓		−1	1	} ^3F, ^3P
14		↓	↓			−1	1	
15	↑			↓		0	1	} ^1G, ^3F, ^1D, ^3P
16	↓			↑		0	1	
17		↑	↓			0	1	
18		↓	↑			0	1	
19	↑				↑	1	0	} ^3F, ^3P
20		↑		↑		1	0	
21	↑				↓	0	0	} ^1G, ^3F, ^1D, ^3P, ^1S
22	↓				↑	0	0	
23		↑		↓		0	0	
24		↓		↑		0	0	
25			↑↓			0	0	
26	↓				↓	−1	0	} ^3F, ^3P
27		↓		↓		−1	0	
28		↑			↓	0	−1	} ^1G, ^3F, ^1D, ^3P
29		↓			↑	0	−1	
30			↑	↓		0	−1	
31			↓	↑		0	−1	
32		↑			↑	1	−1	} ^3F, ^3P
33			↑	↑		1	−1	
34		↓			↓	−1	−1	} ^3F, ^3P
35			↓	↓		−1	−1	
36			↑		↑	1	−2	^3F
37			↑		↓	0	−2	} ^1G, ^3F, ^1D
38			↓		↑	0	−2	
39				↑↓		0	−2	
40			↓		↓	−1	−2	^3F
41				↑	↑	1	−3	^3F
42				↑	↓	0	−3	} ^1G, ^3F
43				↓	↑	0	−3	
44				↓	↓	−1	−3	^3F
45					↑↓	0	−4	^1G

Table 32.5 Energy of ^3F and ^3P states for d^2 free ions

	Ti^{2+}	V^{3+}	Cr^{4+}
^3P	10 600 cm^{-1}	13 250 cm^{-1}	15 700 cm^{-1}
^3F	0	0	0

CALCULATION OF THE NUMBER OF MICROSTATES

Each different arrangement of electrons in a set of orbitals has a slightly different energy and is called a microstate, e.g. Table 32.2 for a p^2 arrangement or Table 32.4 for a d^2 arrangement. The number of microstates may be calculated from the number of orbitals and the number of electrons, using the formula:

$$\binom{n}{r} = \frac{n!}{r!(n-r)!}$$

where n is twice the number of orbitals and r is the number of electrons. (Note $n!$ is factorial n and $r!$ is factorial r.) Thus for a p^3 case there are three orbitals so $n = 6$, and three electrons so $r = 3$. Hence

$$\binom{6}{3} = \frac{6!}{3!(6-3)!} = \frac{6!}{3! \times 3!} = \frac{6 \times 5 \times 4 \times 3 \times 2 \times 1}{3 \times 2 \times 1 \times 3 \times 2 \times 1} = 20 \text{ microstates}$$

Similarly, for a d^2 case there are five orbitals so $n = 10$, and two electrons so $r = 2$. Hence

$$\binom{10}{2} = \frac{10!}{2!(10-2)!} = \frac{10!}{2! \times 8!}$$

$$= \frac{10 \times 9 \times 8 \times 7 \times 6 \times 5 \times 4 \times 3 \times 2 \times 1}{2 \times 1 \times 8 \times 7 \times 6 \times 5 \times 4 \times 3 \times 2 \times 1} = 45 \text{ microstates}$$

In a similar way the number of microstates can be worked out for all of the electronic arrangements p^1–p^6 and d^1–d^{10}.

Table 32.6 The number of microstates for various electron arrangements

Electronic arrangement	Number of microstates	Electronic arrangement	Number of microstates
p^1	6	d^1	10
p^2	15	d^2	45
p^3	20	d^3	120
p^4	15	d^4	210
p^5	6	d^5	252
p^6	1	d^6	210
		d^7	120
		d^8	45
		d^9	10
		d^{10}	1

ELECTRONIC SPECTRA OF TRANSITION METAL COMPLEXES

Spectra arise because electrons may be promoted from one energy level to another. Such electronic transitions are of high energy, and in addition much lower energy vibrational and rotational transitions always occur. The vibrational and rotational levels are too close in energy to be resolved into separate absorption bands, but they result in considerable broadening of the electronic absorption bands in *d–d* spectra. Band widths are commonly found to be of the order of 1000–3000 cm^{-1}.

The spectrum of a coloured solution may be measured quite easily using a spectrophotometer. A beam of monochromatic light obtained using a prism and a narrow slit is passed through the solution and on to a photoelectric cell. The amount of light absorbed at any particular frequency can be read off, or a whole frequency range can be scanned, and the absorbance A plotted as a graph on a paper chart recorder. The absorbance was formerly called the optical density. If I_0 is the intensity of the original beam of light, and I the intensity after passing through the solution, then

$$\log\left(\frac{I_0}{I}\right) = A$$

The molar absorption coefficient ε is usually calculated from the absorbance

$$\varepsilon = \frac{A}{cl}$$

where c is the concentration of the solution in $mol\,l^{-1}$, and l is the path length in centimetres. (Cells are commonly 1 cm long.)

Not all of the theoretically possible electronic transitions are actually observed. The position is formalized into a set of selection rules which distinguish between 'allowed' and 'forbidden' transitions. 'Allowed' transitions occur commonly. 'Forbidden' transitions do occur, but much less frequently, and they are consequently of much lower intensity.

Laporte 'orbital' selection rule

Transitions which involve a change in the subsidiary quantum number $\Delta l = \pm 1$ are 'Laporte allowed' and therefore have a high absorbance. Thus for Ca, $s^2 \rightarrow s^1p^1$, l changes by +1 and the molar absorption coefficient ε is 5000–10 000 litres per mol per centimetre. In contrast *d–d* transitions are 'Laporte forbidden', since the change in $l = 0$, but spectra of much lower absorbance are observed (ε = 5–10 $l\,mol^{-1}\,cm^{-1}$) because of

Table 32.7 Molar absorption coefficients for different types of transition

Laporte (orbital)	Spin	Type of spectra	ε	Example
Allowed	Allowed	Charge transfer	10 000	$[TiCl_6]^{2-}$
Partly allowed, some p–d mixing	Allowed	d–d	500	$[CoBr_4]^{2-}$, $[CoCl_4]^{2-}$
Forbidden	Allowed	d–d	8–10	$[Ti(H_2O)_6]^{3+}$, $[V(H_2O)_6]^{3+}$
Partly allowed, some p–d mixing	Forbidden	d–d	4	$[MnBr_4]^{2-}$
Forbidden	Forbidden	d–d	0.02	$[Mn(H_2O)_6]^{2+}$

slight relaxation in the Laporte rule. When the transition metal ion forms a complex it is surrounded by ligands, and some mixing of d and p orbitals may occur, in which case transitions are no longer pure d–d in nature. Mixing of this kind occurs in complexes which do not possess a centre of symmetry, for example tetrahedral complexes, or asymmetrically substituted octahedral complexes. Thus $[MnBr_4]^{2-}$ which is tetrahedral and $[Co(NH_3)_5Cl]^{2+}$ which is octahedral but non-centrosymmetric are both coloured. Mixing of p and d orbitals does not occur in octahedral complexes which have a centre of symmetry such as $[Co(NH_3)_6]^{3+}$ or $[Cu(H_2O)_6]^{2+}$. However, in these cases the metal–ligand bonds vibrate so that the ligands spend an appreciable amount of time out of their centrosymmetric equilibrium position. Thus a very small amount of mixing occurs, and low-intensity spectra are observed. Thus Laporte allowed transitions are very intense, whilst Laporte forbidden transitions vary from weak intensity if the complex is non-centrosymmetric to very weak if it is centrosymmetric (Table 32.7).

Spin selection rule

During transitions between energy levels, an electron does not change its spin, that is $\Delta S = 0$. There are fewer exceptions than for the Laporte selection rule. Thus in the case of Mn^{2+} in a weak octahedral field such as $[Mn(H_2O)_6]^{2+}$ the d–d transitions are 'spin forbidden' because each of the d orbitals is singly occupied. Many Mn^{2+} compounds are off white or pale flesh coloured, but the intensity is only about one hundredth of that for a 'spin allowed' transition (Table 32.7). Since the spin forbidden transitions are very weak, analysis of the spectra of transition metal complexes can be greatly simplified by ignoring all spin forbidden transitions and considering only those excited states which have the same multiplicity as the ground state. Thus for a d^2 configuration the only terms which need be considered are the ground state ^3F and the excited state ^3P.

SPLITTING OF ELECTRONIC ENERGY LEVELS AND SPECTROSCOPIC STATES

An *s* orbital is spherically symmetrical and is unaffected by an octahedral (or any other) field. *p* orbitals are directional, and *p* orbitals are affected by an octahedral field. However, since a set of three *p* orbitals are all affected equally, their energy levels remain equal, and no splitting occurs. A set of *d* orbitals is split by an octahedral field into two levels t_{2g} and e_g. The difference in energy between these may be written as Δ_o or $10D_q$. The t_{2g} level is triply degenerate and is $4D_q$ below the barycentre, and the e_g level is doubly degenerate and is $6D_q$ above the barycentre. For a d^1 configuration, the ground state is a ^2D state, and the t_{2g} and e_g electronic energy levels correspond with the T_{2g} and E_g spectroscopic states. A set of *f* orbitals is split by an octahedral field into three levels. For an f^1 arrangement the ground state is a ^3F state and is split into a triply degenerate T_{1g} state which is $6D_q$ below the barycentre, a triply degenerate T_{2g} level which is $2D_q$ above the barycentre and a single A_{2g} state which is $12D_q$ above the barycentre (Figure 32.10). We will see later that the same arrangement of states occurs with d^2 arrangement.

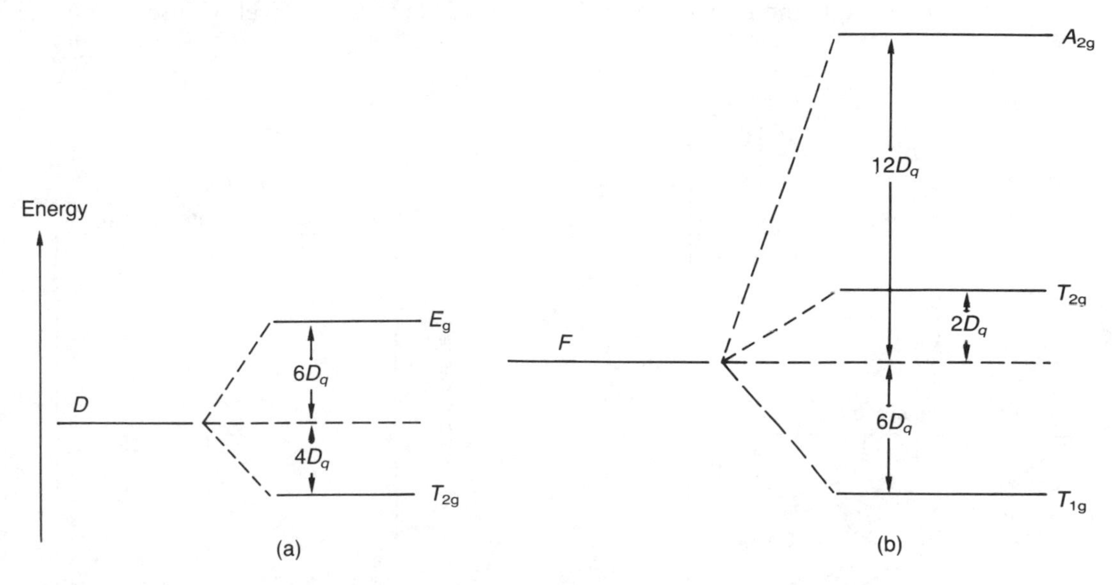

Figure 32.10 Splitting of spectroscopic terms arising from (a) d^1 electronic arrangement and (b) d^2 electronic arrangement.

In the 'one electron cases' s^1, p^1, d^1 and f^1 there is a direct correspondence between the splitting of electronic energy levels which occurs in a crystal field and the splitting of spectroscopic states. Thus in an octahedral field the *S* and *P* states are not split, *D* states are split into two states and *F* states are split into three states.

Table 32.8 Transforming spectroscopic terms into Mulliken symbols

Spectroscopic term	Mulliken symbols	
	Octahedral field	Tetrahedral field
S	A_{1g}	A_1
P	T_{1g}	T_1
D	$E_g + T_{2g}$	$E + T_2$
F	$A_{2g} + T_{1g} + T_{2g}$	$A_2 + T_1 + T_2$
G	$A_{1g} + E_g + T_{1g} + T_{2g}$	$A_1 + \mathrm{E} + \mathrm{T}_1 + T_2$

SPECTRA OF d^1 AND d^9 IONS

In a free gaseous metal ion the d orbitals are degenerate, and hence there will be no spectra from d–d transitions. When a complex is formed, the electrostatic field from the ligands splits the d orbitals into two groups t_{2g} and e_g. (This crystal field splitting is described in Chapter 28.) The simplest example of a d^1 complex is Ti(III) in octahedral complexes such as $[TiCl_6]^{3-}$, or $[Ti(H_2O)_6]^{3+}$. The splitting of the d orbitals is shown in Figure 32.11a. In the ground state the single electron occupies the lower t_{2g} level, and only one transition is possible to the e_g level. Consequently the

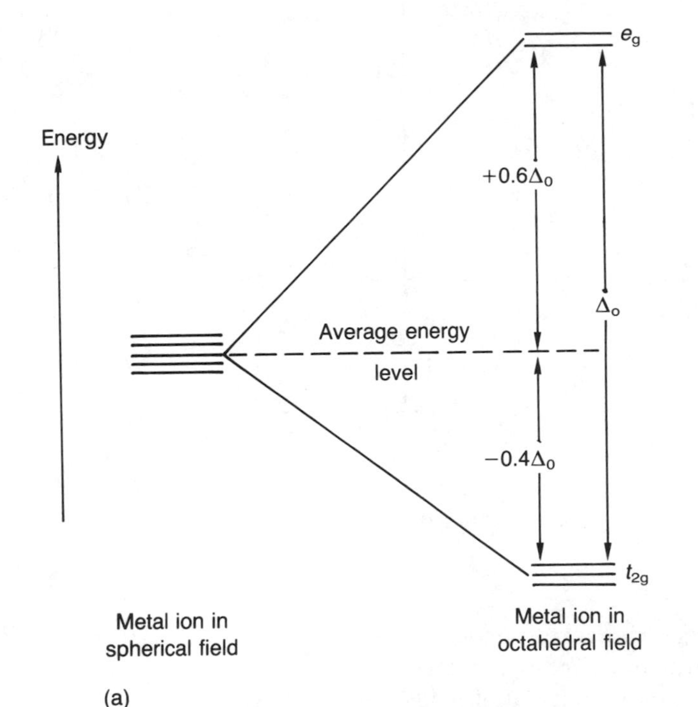

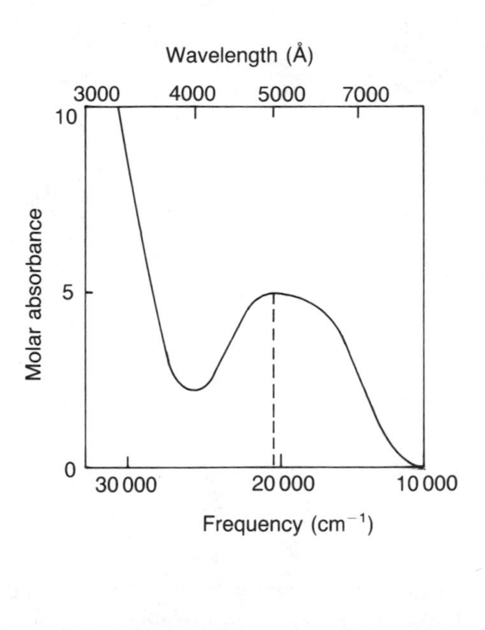

Figure 32.11 (a) Diagram of energy levels in octahedral field. (b) Ultraviolet and visible absorption spectrum of $[Ti(H_2O)_6]^{3+}$.

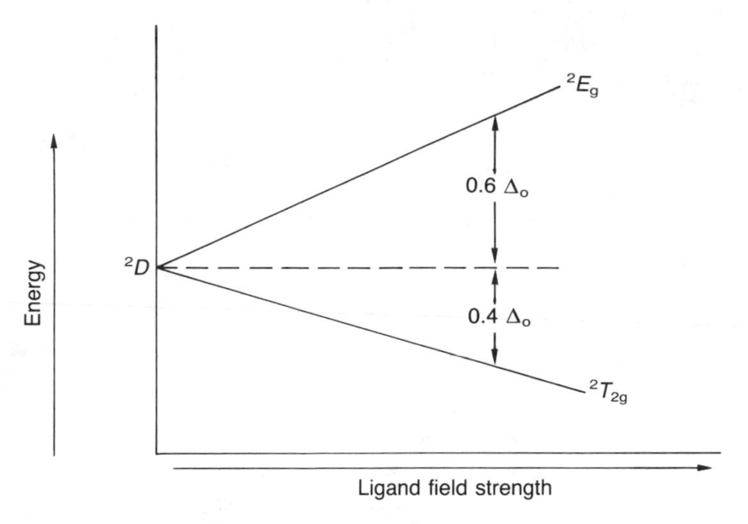

Figure 32.12 Splitting of energy levels for d^1 configuration in octahedral field.

absorption spectrum of $[Ti(H_2O)_6]^{3+}$ which is shown in Figure 32.11b, shows only one band with a peak at 20 300 cm^{-1}. The magnitude of the splitting Δ_0 depends on the nature of the ligands, and affects the energy of the transition, and hence the frequency of maximum absorption in the spectrum. Thus the peak occurs at 13 000 cm^{-1} in $[TiCl_6]^{3-}$, 18 900 cm^{-1} in $[TiF_6]^{3-}$, 20 300 cm^{-1} in $[Ti(H_2O)_6]^{3+}$ and 22 300 cm^{-1} in $[Ti(CN)_6]^{3-}$. The amount of splitting caused by various ligands is related to their position in the spectrochemical series (see Chapter 28). The effect of an octahedral ligand field on a d^1 ion is shown in Figure 32.12. The symbol ^2D at the left is the ground state term for a free ion with a d^1 configuration (see Table 32.3). Under the influence of a ligand field this splits into two states which are described by the Mulliken symbols ^2E_g and $^2T_{2g}$. (These symbols originate in group theory, and are used here without attempting to derive them. Useful references are given at the end of the chapter.) The lower T_{2g} state corresponds to the single d electron occupying one of the t_{2g} orbitals, and the ^2E_g state corresponds to the electron occupying one of the e_g orbitals. The two states are separated more widely as the strength of the ligand field increases.

Octahedral complexes of ions with a d^9 configuration such as $[Cu(H_2O)_6]^{2+}$ can be described in a similar way to the Ti^{3+} octahedral complexes with a d^1 arrangement. In the d^1 case there is a single electron in the lower t_{2g} level whilst in the d^9 case there is a single hole in the upper e_g level. Thus the transition in the d^1 case is promoting an electron from the t_{2g} level to thc e_g level, whilst in the d^9 case it is simpler to consider the promotion of an electron as the transfer of a 'hole' from e_g to t_{2g}. The energy diagram for d^9 is therefore the other way round, that is the inverse of that for a d^1 configuration (Figure 32.13).

If the effect of a tetrahedral ligand field is now considered, the degen-

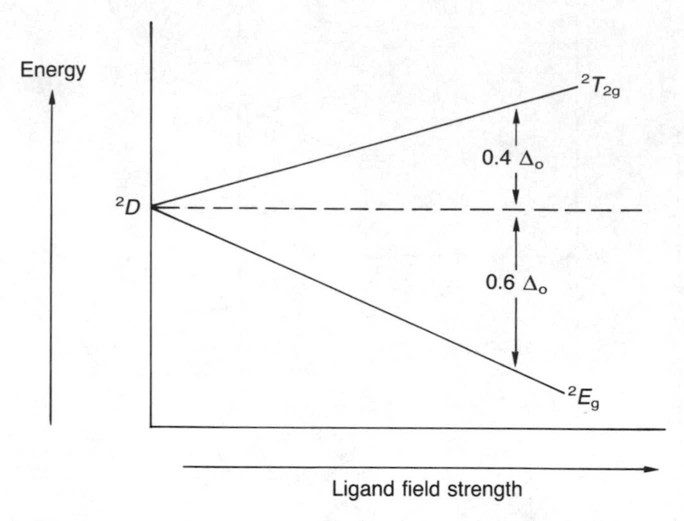

Figure 32.13 Splitting of energy levels for d^9 configuration in octahedral field.

erate d orbitals split into two e_g orbitals of lower energy and three t_{2g} orbitals of higher energy (see Chapter 28). The energy level diagram for d^1 complexes in a tetrahedral field is the inverse of that in an octahedral field, and is similar to the d^9 octahedral case (Figure 32.12), except that the amount of splitting in a tetrahedral field is only about $\frac{4}{9}$ of that in an octahedral field.

In a similar way the d^6 high-spin octahedral arrangement (Figure 32.14a) is related to the d^1 octahedral case. Since transitions which involve reversal of the electron spin are 'forbidden', and hence give extremely weak bands, the only 'permitted' transition is the paired electron in the t_{2g} level, which

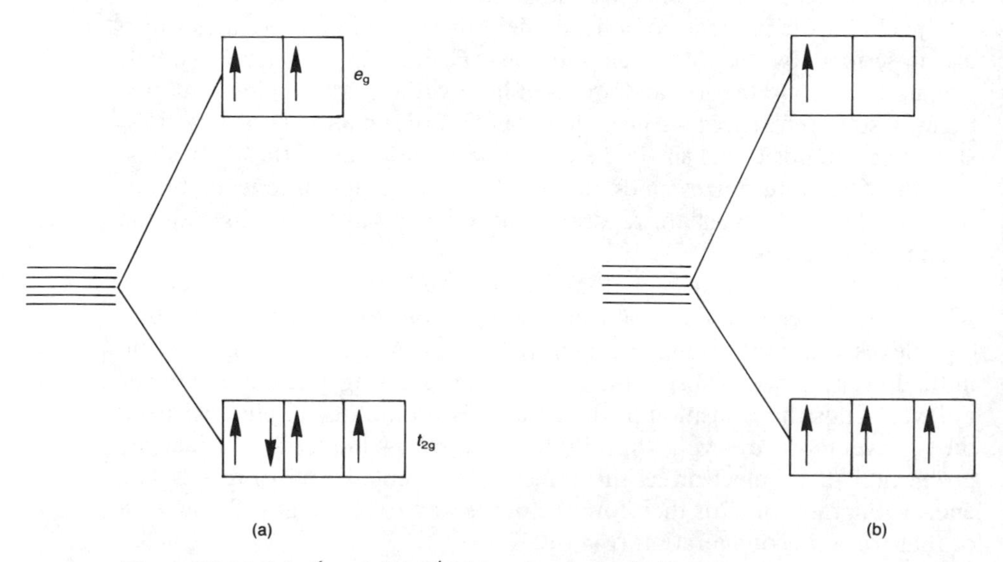

Figure 32.14 (a) d^6 and (b) d^4 high-spin octahedral arrangements.

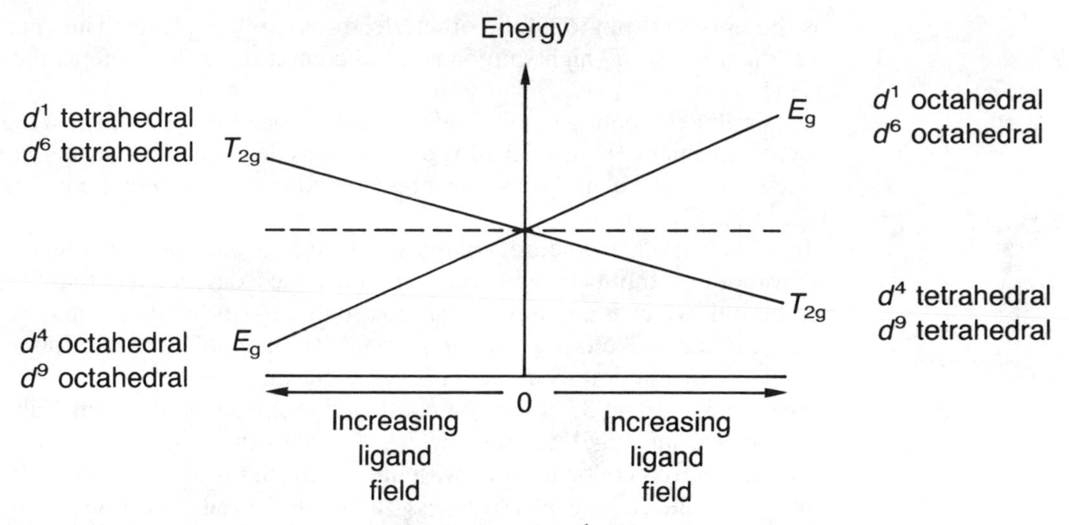

Figure 32.15 Orgel combined energy level diagram for d^1. (Note that g subscripts are dropped in tetrahedral cases.)

Table 32.9 Ground terms for d^1–d^{10} configurations

Configuration	Example	Ground term	m_l: 2 1 0 −1 −2	M_L	S
d^1	Ti^{3+}	^2D	↑	2	$\frac{1}{2}$
d^2	V^{3+}	^3F	↑ ↑	3	1
d^3	Cr^{3+}	^4F	↑ ↑ ↑	3	$1\frac{1}{2}$
d^4	Cr^{2+}	^5D	↑ ↑ ↑ ↑	2	2
d^5	Mn^{2+}	^6S	↑ ↑ ↑ ↑ ↑	0	$2\frac{1}{2}$
d^6	Fe^{2+}	^5D	↑↓ ↑ ↑ ↑ ↑	2	2
d^7	Co^{2+}	^4F	↑↓ ↑↓ ↑ ↑ ↑	3	$1\frac{1}{2}$
d^8	Ni^{2+}	^3F	↑↓ ↑↓ ↑↓ ↑ ↑	3	1
d^9	Cu^{2+}	^2D	↑↓ ↑↓ ↑↓ ↑↓ ↑	2	$\frac{1}{2}$

has the opposite spin to all the other electrons, to the e_g level. The energy level diagram for d^6 high-spin octahedral complexes is the same as the d^1 case (Figure 32.12).

By similar reasoning, octahedral complexes containing d^4 ions in a high-spin arrangement (Figure 32.14b) may be considered as having one 'hole' in the upper e_g level and thus they are analogous to the d^9 octahedral case (Figure 32.13).

In addition, d^6 tetrahedral complexes have only one electron which can be promoted without changing the spin, and have a diagram like that for d^1 tetrahedral, which is qualitatively similar to that for the d^9 octahedral case. Finally, d^4 and d^9 tetrahedral complexes with one 'hole' are qualitatively like the d^1 octahedral example with one electron.

Figures 32.12 and 32.13 can be combined into a single diagram (Figure 32.15) called an Orgel diagram, which describes in a qualitative way the effect of electron configurations with one electron, one electron more than a half filled level, one electron less than a full shell, and one electron less than a half filled shell.

SPECTRA OF d^2 AND d^8 IONS

The ground state for a d^2 electronic arrangement has two electrons in different orbitals. In an octahedral field the d orbitals are split into three t_{2g} orbitals of lower energy and two e_g orbitals of higher energy, and the electrons will occupy two of the t_{2g} orbitals. When an electron is promoted $(e_g)^2 \rightarrow (e_g)^1(t_{2g})^1$ there are two possibilities. It requires less energy to promote an electron from the d_{xz} or d_{yz} orbitals to the d_{z^2} orbital than it does to promote the electron to the $d_{x^2-y^2}$ orbital. This difference in energy arises because the former transition gives a $(d_{xy})^1(d_{z^2})^1$ arrangement where the electrons are spread around all three directions x, y and z which reduces electron–electron repulsion compared with the $(d_{xy})^1(d_{x^2-y^2})^1$ arrangement where the electrons are confined to the xy plane. If both electrons are promoted another high energy state will be formed. Thus from a consideration of the electrons we would expect four energy levels.

Table 32.3 shows that the terms arising for a d^2 configuration are the ground state ^3F and the excited states ^3P, ^1G, ^1D and ^1S. The ground state contains two electrons with parallel spins, but the ^1G, ^1D and ^1S states contain electrons with opposite spins. Thus transitions from the ground state to these three states are spin forbidden, will be very weak, and can be ignored. The two remaining states ^3F and ^3P can have spin permitted transitions. It will be remembered that p orbitals are not split by an octahedral field, but that f orbitals are split into three levels. Similarly in an octahedral field P states are not split (but are transformed into a T_{1g} state, and F states split into $A_{2g} + T_{1g} + T_{2g}$ (see Figure 32.10 and Table 32.8). The energy level diagram for these is shown in Figure 32.16a.

Three transitions are possible from the ground state $^3T_{1g}(F)$ to $^3T_{2g}$, $^3T_{1g}(P)$ and $^3A_{2g}$ respectively, and hence three peaks should appear in the spectrum. (This may be compared with only one transition in the d^1 case.)

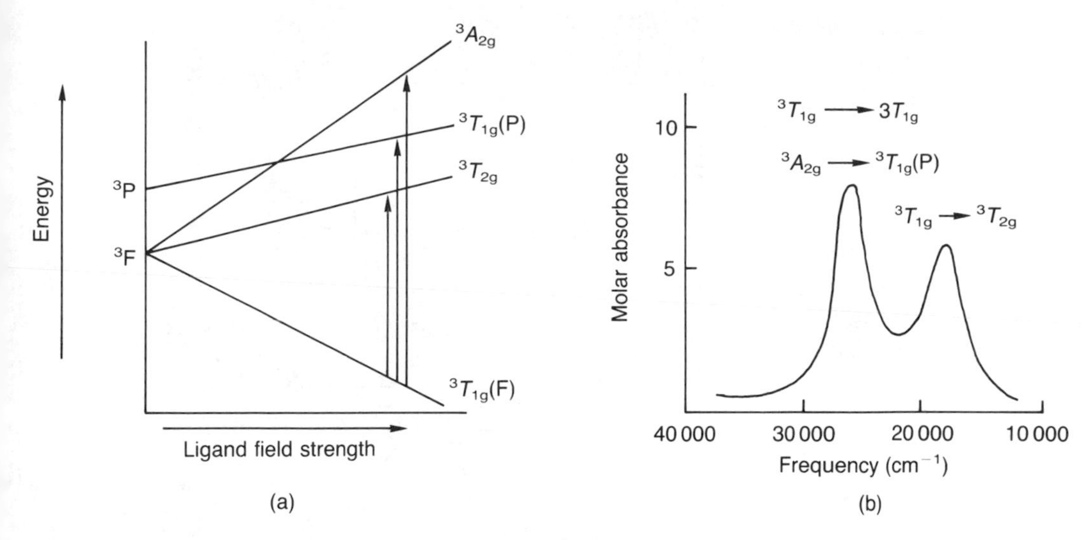

Figure 32.16 Orgel diagram and spectrum for a d^2 ion. (a) Energy diagram for d^2 configuration in octahedral field (simplified by including only the triplet states) showing three possible transitions. (b) Ultraviolet–visible absorption spectrum for the d^2 complex $[V(H_2O)_6]^{3+}$.

The spectrum of a d^2 complex ion $[V(H_2O)_6]^{3+}$ is shown in Figure 32.16. Only two peaks occur in this spectrum because the ligand field strength of water results in transitions occurring close to the cross-over point between the $^3T_{1g}(P)$ and $^3A_{2g}$ levels, and hence these two transitions are not resolved into two separate peaks. A V^{3+} ion complexed with a different ligand would show three peaks.

Complexes of metals such as Ni^{2+} with a d^8 configuration in an octahedral field can be treated in a similar way. There are two 'holes' in the e_g level, and hence promoting an electron is equivalent to transferring a 'hole' from e_g to t_{2g}. This is the inverse of the d^2 case. The ^3P state is not split, and is not inverted, but the ^3F state is split into three states and is inverted (Figure 32.17). Thus the ground state term for Ni^{2+} is $^3A_{2g}$. Note that in both the d^2 case and the d^8 case the ^3F state is the lowest in energy.

Three spin allowed transitions are observed in the spectra of $[Ni(H_2O)_6]^{2+}$, $[Ni(NH_3)_6]^{2+}$ and $[Ni(ethylenediamine)_3]^{2+}$.

Using the same arguments as applied to the d^1 case, the d^2 octahedral energy diagram is similar to the high-spin d^7 octahedral, and d^3 and d^8 tetrahedral cases. The inverse diagram applies to the d^3 and d^8 octahedral as well as d^2 and d^7 tetrahedral complexes. As before, the g subscript is omitted in tetrahedral complexes.

In a similar way Cr(III) has a d^3 configuration, and for octahedral complexes the inverted diagram at the left hand side of Figure 32.18 applies. This is similar to the d^2 diagram except that the energies of the three states derived from the ^3F state are inverted in order. The spectra of chromium complexes would be expected to show three absorption bands from the ground state $^3A_{2g}$ to $^3T_{2g}$, $^3T_{1g}(F)$ and $^3T_{1g}(P)$ respectively.

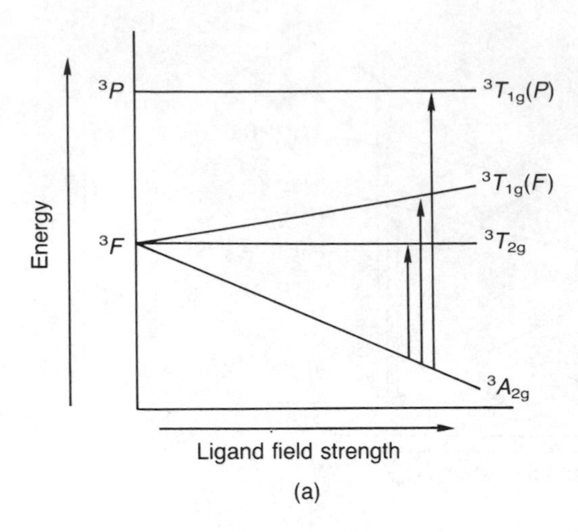

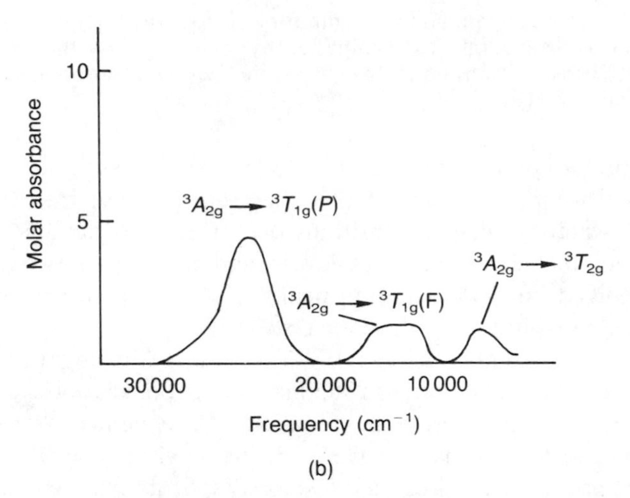

Figure 32.17 Orgel diagram and spectrum for a d^8 ion. (a) Energy diagram for d^8 configuration in an octahedral field (simplified by including only the triplet states) showing three spin allowed transitions. (b) Ultraviolet–visible absorption spectrum for the d^8 complex $[Ni(H_2O)_6]^{2+}$. Note the middle peak shows signs of splitting into two peaks because of Jahn–Teller distortion.

Chromium(III) complexes show at least two well defined absorption peaks in the visible region. In some cases the third band can be seen, though it is often hidden by a very intense charge transfer band.

The combined Orgel energy level diagram for two-electron and two-'hole' configurations is shown in Figure 32.18. Note that there are two T_{1g} states, one from the P state and the other from the F state. The two T_{1g} states are slightly curved lines, because they have the same symmetry, and they interact with one another. This interelectronic repulsion lowers the energy of the lower state and increases the energy of the higher state. The

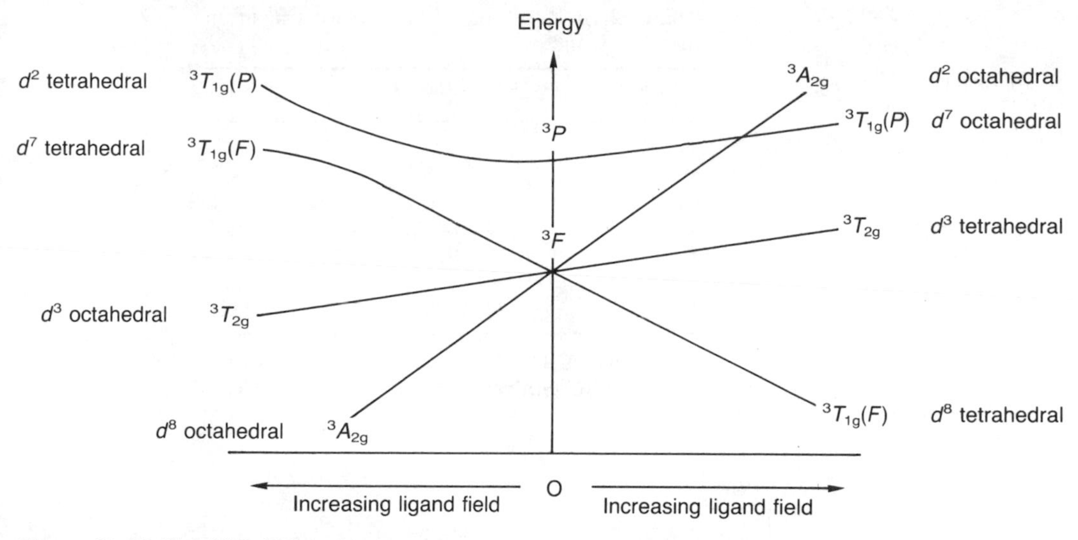

Figure 32.18 Combined Orgel energy level diagram for two-electron and two-'hole' configurations. (Note that the g subscript is omitted in tetrahedral cases. The same diagram also applies to the d^3 and d^8 arrangements except that the spectroscopic terms are ^4F and ^4P, and the Mulliken terms have a multiplicity of 4.)

effect is much more marked on the left of the diagram because the two levels are close in energy. If the lines had been straight, they would cross each other, implying at the cross-over point that two electrons in one atom may have the same symmetry and the same energy. This would be impossible, and is prohibited by the non-crossing rule, which says that states of the same symmetry cannot cross each other. The mixing or interelectronic repulsion which causes the bending of the lines is expressed by the Racah parameters B and C.

The Racah parameters can in principle be calculated from linear combinations of exchange integrals and coulomb integrals, but they are usually obtained empirically from the spectra of free ions. The difference in energy between states of the same multiplicity for example in a d^3 ion such as V^{2+} the separation between ^4F and ^4P is $15B$. Thus if all three bands are observed in the spectrum, B is readily obtained. In many cases it is sufficient to use only the parameter B to explain the positions of the bands in the spectrum. Both the B and C parameters are necessary for terms of different multiplicity. For example, in the d^3 V^{2+} ion the separation between ^4F and ^2G is $4B + 3C$. For most transition metal ions B is approximately 700–1000 cm^{-1} and C is approximately four times B.

Experimentally measured spectra may be compared with those expected from theory. Consider for example the spectra of Cr(III). Cr^{3+} is a d^3 ion. (See also Chapter 21.) In the ground state, the d_{xy}, d_{xz} and d_{yz} orbitals each contain one electron and the two e_g orbitals are empty. The d^3 arrangement gives rise to two states ^4F and ^4P. In an octahedral field the ^4F

Table 32.10 Racah parameters B for transition metal ions in cm^{-1}

Metal	M^{2+}	M^{3+}
Ti	695	–
V	755	861
Cr	810	918
Mn	860	965
Fe	917	1015
Co	971	1065
Ni	1030	1115

From Sutton, D., *Electronic Spectra of Transition Metal Complexes*, McGraw Hill, 1968.

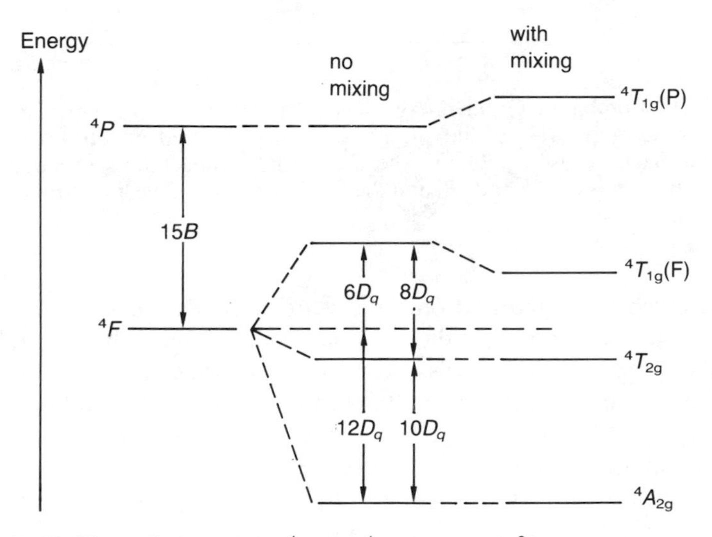

Figure 32.19 The splitting of the ^4F and ^4P states in Cr^{3+}.

state is split into $^4A_{2g}(F)$, $^4T_{2g}(F)$ and $^4T_{1g}(F)$ states, and the ^4P state is not split but transforms into a $^4T_{1g}(P)$ state (Figure 32.19).

Three transitions are possible $^4A_{2g} \rightarrow {}^4T_{2g}$, $^4A_{2g} \rightarrow {}^4T_{1g}(F)$ and $^4A_{2g} \rightarrow {}^4T_{1g}(P)$. The Racah parameters for the free Cr^{3+} ion are known exactly ($B = 918\,cm^{-1}$ and $C = 4133\,cm^{-1}$).

Table 32.11 Correlation of spectra for $[CrF_6]^{3-}$ (in cm^{-1})

		Observed spectra	Predicted	
$^4A_{2g} \rightarrow {}^4T_{1g}(P)$	ν_3	34400	30700	$(12D_q + 15B)$
$^4A_{2g} \rightarrow {}^4T_{1g}(F)$	ν_2	22700	26800	$(18D_q)$
$^4A_{2g} \rightarrow {}^4T_{2g}$	ν_1	14900	14900	$(10D_q)$

The lowest energy transition correlates perfectly, but agreement for the other two bands is not very good. Two corrections must be made to improve the agreement.

1. If some mixing of the P and F terms occurs (bending of lines on the Orgel diagram), then the energy of the ${}^4T_{1g}(P)$ state is increased by an amount x and the energy of the ${}^4T_{1g}(F)$ state is reduced by x.
2. The value for the Racah parameter B relates to a free ion. The apparent value B' in a complex is always less than the free ion value because electrons on the metal can be delocalized into molecular orbitals covering both the metal and the ligands. The use of adjusted B' values improves the agreement. This delocalization is called the nephelauxetic effect, and the nephelauxetic ratio β is defined:

$$\beta = \frac{B'}{B}$$

β decreases as delocalization increases, and is always less than one, and B' is usually $0.7B$ to $0.9B$). B' is easily obtained if all three transitions are observed since:

$$15B' = \nu_3 + \nu_2 - 3\nu_1$$

Using both of these corrections gives much better correlation between observed and improved theoretical results (Table 32.12). (The use of adjusted B and C terms is the basis of ligand field calculations.)

Table 32.12 Correlation of spectra for $[CrF_6]^{3-}$ (in cm^{-1})

		Observed spectra	Corrected theoretical	
${}^4A_{2g} \rightarrow {}^4T_{1g}(P)$	ν_3	34 400	34 800	$(12D_q + 15B' + x)$
${}^4A_{2g} \rightarrow {}^4T_{1g}(F)$	ν_2	22 700	22 400	$(18D_q - x)$
${}^4A_{2g} \rightarrow {}^4T_{2g}$	ν_1	14 900	14 900	$(10D_q)$

As a second example consider the spectrum of crystals of $KCoF_3$. There are three absorption bands at 7150 cm^{-1}, 15 200 cm^{-1} and 19 200 cm^{-1}. The compound contains Co^{2+} ions (d^7) surrounded octahedrally by six F^- ions. This case should be similar to the d^2 case and we would expect transitions ν_1 (${}^4T_{1g} \rightarrow {}^4T_{2g}$), ν_2 (${}^4T_{1g} \rightarrow {}^4A_{2g}$) ν_3 (${}^4T_{1g} \rightarrow {}^4T_{2g}(P)$). D_q may be calculated from ν_1 since:

$$\nu_1 = 8D_q$$

However, this makes no allowance for the configuration interaction between the ${}^4T_{1g}$ and ${}^4T_{2g}$ states (i.e. bending of lines on the Orgel diagram). It is therefore better to evaluate D_q from the equation:

$$\nu_2 - \nu_1 = 10D_q$$

since this is not affected by configuration interaction. Thus:

$$15\,200 - 7200 = 10D_q$$

hence $$D_q = 800\ \text{cm}^{-1}$$

The value of the configuration interaction term x is obtained either from the equation:

$$\nu_1 = 8D_q + x$$

or from $$\nu_2 = 18D_q + x$$

The Racah parameter B for a free Co^{2+} ion is 971 cm^{-1} (Table 32.10), but the corrected value B' may be calculated from the equation:

$$\nu_3 = 15B' + 6D_q + 2x$$

The pale pink colour of many octahedral complexes of Co(II) are of interest. The spectrum of $[Co(H_2O)_6]^{2+}$ is shown in Figure 32.20.

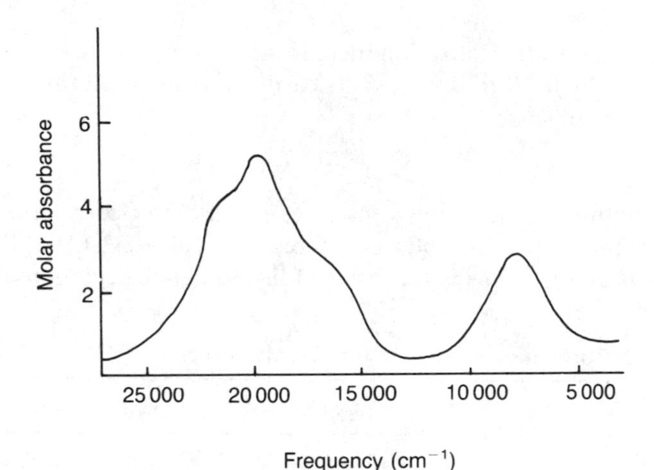

Figure 32.20 Electronic spectrum of $[Co(H_2O)_6]^{2+}$.

The spectrum of $[Co(H_2O)_6]^{2+}$ is less easy to interpret. It shows a weak but well resolved absorption band at about 8000 cm^{-1}, and a multiple absorption band comprising three overlapping peaks at about 20 000 cm^{-1}. The lowest energy band ν_1 at 8000 cm^{-1} is assigned to the ${}^4T_{1g} \rightarrow {}^4T_{2g}$ transition. The multiple band has three peaks at about 16 000, 19 400 and 21 600 cm^{-1}. Two of these are the ${}^4T_{1g} \rightarrow {}^4A_{2g}$ and ${}^4T_{1g} \rightarrow {}^4T_{2g}(P)$ transitions, and since the peaks are close together this indicates that this complex is close to the cross-over point between the ${}^4A_{2g}$ and ${}^4T_{1g}$ states on the energy diagram. This means that the assignments are only tentative, but the following assignments are commonly accepted:

$$\nu_2\ ({}^4T_{1g} \rightarrow {}^4A_{2g}) \qquad 16\,000\ \text{cm}^{-1}$$

and $$\nu_3\ ({}^4T_{1g} \rightarrow {}^4T_{2g}(P)) \qquad 19\,400\ \text{cm}^{-1}$$

The extra band is attributed either to spin orbit coupling effects or to transitions to doublet states.

Tetrahedral complexes of Co^{2+} such as $[CoCl_4]^{2-}$ are intensely blue in colour with an intensity ε of about 600 l mol^{-1} cm^{-1} compared with the pale pink colour of octahedral complexes with an intensity ε of only about 6 l mol^{-1} cm^{-1}. Co^{2+} has a d^7 electronic configuration, and in $[CoCl_4]^{2-}$ the electrons are arranged $(e_g)^4(t_{2g})^3$. This is similar to the Cr^{3+} (d^3) octahedral case since only two electrons can be promoted. There are three possible transitions: $^4A_2 \rightarrow {}^4T_2(F)$, $^4A_2(F) \rightarrow {}^4T_1(F)$ and $^4A_2(F) \rightarrow {}^4T_{1g}(P)$. Only two bands appear in the visible spectrum at 5800 cm^{-1} and 15 000 cm^{-1} (Figure 32.21). These two bands are assigned to ν_2 and ν_3. The lowest energy transition ν_1 is expected at 3300 cm^{-1} which is in the infrared region.

$^4A_2 \rightarrow {}^4T_1(P)$	ν_3	15 000 cm^{-1} in the visible region
$^4A_2 \rightarrow {}^4T_1(F)$	ν_2	5 800 cm^{-1} in the visible region
$^4A_2 \rightarrow {}^4T_2$	ν_1	(3300 cm^{-1} in the infrared region)

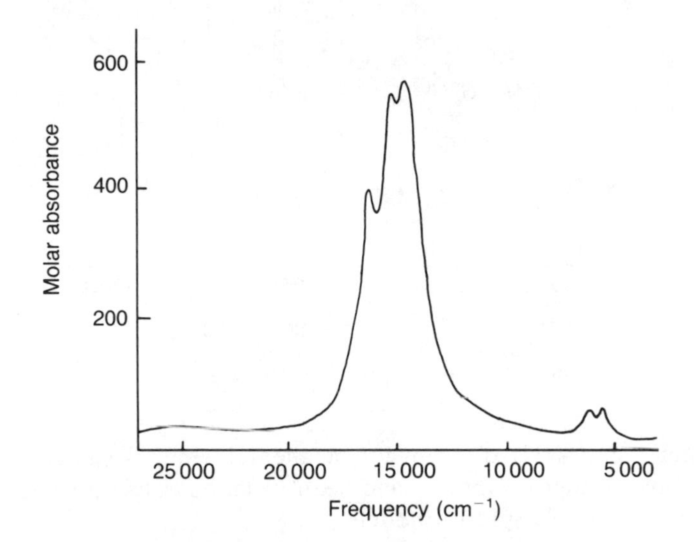

Figure 32.21 Electronic spectrum of $[CoCl_4]^{2-}$.

SPECTRA OF d^5 IONS

The d^5 configuration occurs with Mn(II) and Fe(III).In high-spin octahedral complexes formed with weak ligands, for example $[Mn^{II}F_6]^{4-}$, $[Mn^{II}(H_2O)_6]^{2+}$ and $[Fe^{III}F_6]^{3-}$, there are five unpaired electrons with parallel spins. Any electronic transition within the d level must involve a reversal of spins, and in common with all other 'spin forbidden' transitions any absorption bands will be extremely weak. This accounts for the very pale pink colour of most Mn(II) salts, and the pale violet colour of iron(III) alum. The ground state term is ^6S. None of the 11 excited states shown in Table 32.3 can be attained without reversing the spin of an electron, and hence the probability of such transitions is extremely low. Of the 11 excited states, the four quartets ^4G, ^4F, ^4D and ^4P involve the reversal of only one spin. The other seven states are doublets, are doubly spin forbidden, and are unlikely to be observed. In an octahedral field these four split into ten states, and hence up to ten extremely weak absorption bands may be observed. The spectrum of $[Mn(H_2O)_6]^{2+}$ is shown in Figure 32.22. Several features of this spectrum are unusual.

1. The bands are extremely weak. The molar absorption coefficient ε is about 0.02–0.03 l mol^{-1} cm^{-1}, compared with 5–10 l mol^{-1} cm^{-1} for spin allowed transitions.
2. Some of the bands are sharp and others are broad. Spin allowed bands are invariably broad.

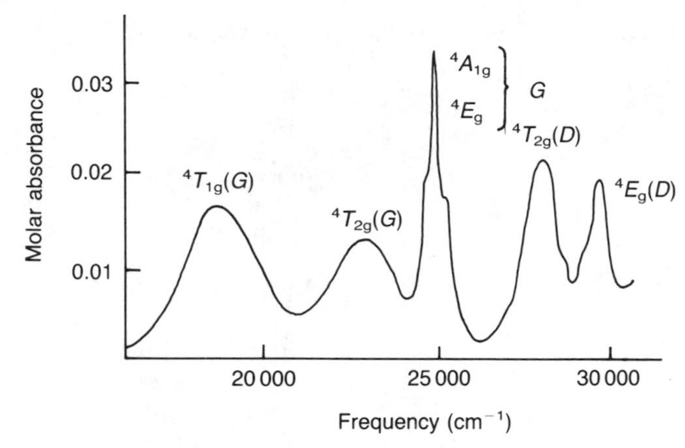

Figure 32.22 Electronic spectrum of $[Mn(H_2O)_6]^{2+}$.

The Orgel energy level diagram for octahedral Mn^{2+} is shown in Figure 32.23. Only the quartet terms have been included because transitions to the others are doubly spin forbidden.

Note that the ground state ^6S is not split, and transforms to the $^6A_{1g}$ state which is drawn along the horizontal axis. Note also that the $^4E_g(G)$, $^4A_{1g}$, $^4E_g(D)$, and $^4A_{2g}(F)$ terms are also horizontal lines on the diagram, so their energies are independent of the crystal field. The ligands in a complex vibrate about mean positions, so the crystal field strength $10D_q$ varies about a mean value. Thus the energy for a particular transition varies about a mean value, and hence the absorption peaks are broad. The degree of broadening of the peaks is related to the slope of the lines on the Orgel diagram. Since the slope of the ground state term $^6A_{1g}$ is zero, and the slopes of the $^4E_g(G)$, $^4A_{1g}$, $^4E_g(D)$, and $^4A_{2g}(F)$ terms are also zero, transitions from the ground state to these four states should give rise to sharp peaks. By the same reasoning transitions to states which slope appreciably such as $^4T_{1g}(G)$ and $^4T_{2g}(G)$ give broader bands.

The bands are assigned as follows:

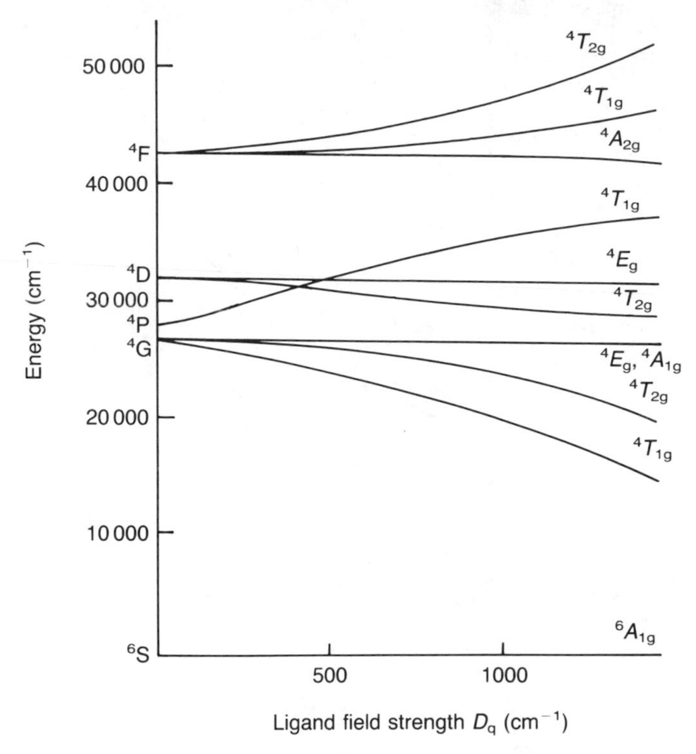

Figure 32.23 Orgel energy level diagram for $Mn^{2+}(d^5)$ octahedral.

$^6A_{1g} \rightarrow {}^4T_{1g}$	18 900 cm^{-1}
$^6A_{1g} \rightarrow {}^4T_{2g}(G)$	23 100 cm^{-1}
$^6A_{1g} \rightarrow {}^4E_g$ $^6A_{1g} \rightarrow {}^4A_{1g}$	} 24 970 and 25 300 cm^{-1}
$^6A_{1g} \rightarrow {}^4T_{2g}(D)$	28 000 cm^{-1}
$^6A_{1g} \rightarrow {}^4E_g(D)$	29 700 cm^{-1}

The same diagram applies to tetrahedral d^5 complexes if the g subscripts are omitted.

TANABE–SUGANO DIAGRAMS

The simple Orgel energy level diagrams are useful for interpreting spectra, but they suffer from two important limitations:

1. They treat only the high-spin (weak field) case.
2. They are only useful for spin allowed transitions when the number of observed peaks is greater than or equal to the number of empirical parameters crystal field splitting D_q, modified Racah parameter B' and bending constant x.

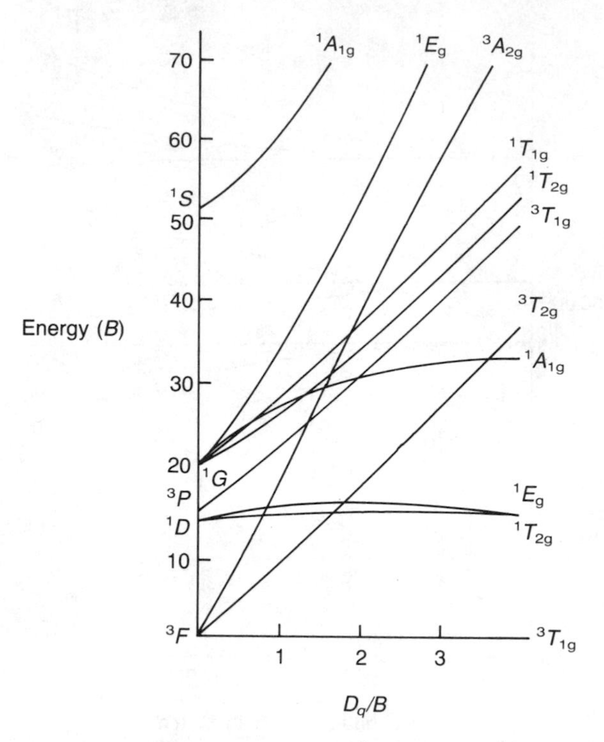

Figure 32.24 Tanabe–Sugano diagram for d^2 case, e.g. V^{3+}.

Though it is possible to add low-spin states to an Orgel diagram, Tanabe–Sugano diagrams are commonly used instead for the interpretation of spectra including both weak and strong fields. Tanabe–Sugano diagrams are similar to Orgel diagrams in that they show how the energy levels change with D_q, but they differ in several ways:

1. The ground state is always taken as the abscissa (horizontal axis) and provides a constant reference point. The other energy states are plotted relative to this.
2. Low-spin terms, i.e. states where the spin multiplicity is lower than the ground state, are included.
3. In order to make the diagrams general for different metal ions with the same electronic configuration, and to allow for different ligands, both of which affect D_q and B (or B'), the axes are plotted in units of energy/B and D_q/B.

A different diagram is required for each electronic arrangement. Only two examples are shown here. The T–S diagram for a d^2 case such as V^{3+} is shown in Figure 32.24. Note that in this case there is no fundamental difference between strong and weak fields.

The Tanabe–Sugano diagram for a d^6 ion such as Co^{3+} is shown in Figure 32.25. This is a simplified version and only the singlet and quintet

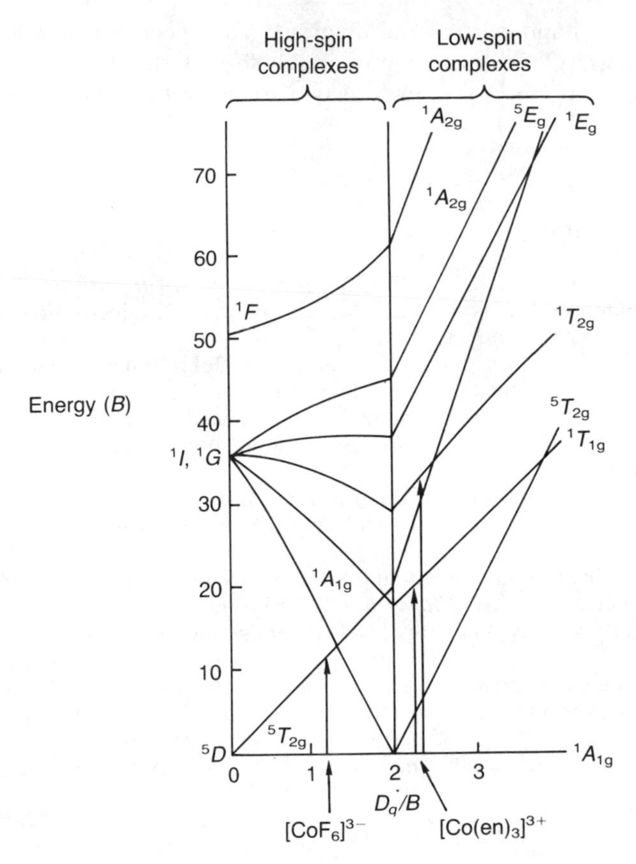

Figure 32.25 Tanabe–Sugano diagram for d^6 case, e.g. Co^{3+}.

terms are shown. There is a discontinuity at $10D_q/B = 20$, and this is shown by a vertical line. At this point spin pairing of electrons occurs. To the left of this line we have high-spin complexes (weak ligand field), and to the right we have low-spin complexes (strong ligand field). The free ion ground state is ^5D. This is split by an octahedral field into the $^5T_{2g}$ ground state and the ^5E_g excited state. The singlet ^1I state in the free ion is of high energy. This is split by the octahedral field into five different states, of which the $^1A_{1g}$ is important. This state is greatly stabilized by the ligand and drops rapidly in energy as the ligand field strength increases. At the point where $10D_q/B = 20$ the $^1A_{1g}$ line crosses the horizontal line for the $^5T_{2g}$ state (which is the ground state). At still higher field strengths the $^1A_{1g}$ state is the lowest in energy, and becomes the ground state. Since the ground state is taken as the horizontal axis, the right hand part of the diagram must be redrawn.

Since the fluoride ion is a weak field ligand, the complex $[CoF_6]^{3-}$ is high spin. The complex is blue in colour, and a single peak occurs at $13\,000\ cm^{-1}$. This is explained by the transition $^5T_{2g} \rightarrow {}^5E_g$ shown as an arrow in the left hand part of the diagram. The spectrum of a low-spin complex such as $[Co(ethylenediamine)_3]^{3+}$ should show the transitions $^1A_{1g} \rightarrow {}^1T_{1g}$ and $^1A_{1g} \rightarrow {}^1T_{2g}$ (shown as two arrows in the right hand part of the diagram).

FURTHER READING

Emeléus, H.J. and Sharpe, A.G. (1973) *Modern Aspects of Inorganic Chemistry*, 4th ed. (Complexes of Transition Metals: Chapter 17, Electronic Spectra), Routledge and Kegan Paul, London.

Figgis, B.N. (1966) *Introduction to Ligand Fields*, John Wiley, London. (UV spectra of hydrated transition metal ions. Chapter 7 gives an account of the splitting of spectroscopic terms. Suitable for advanced readers.)

Gerloch, M. (1986) *Orbitals, Terms and States*, John Wiley, Chichester. (A more basic approach.)

Hertzberg, G. (1944) *Molecular Spectra and Molecular Structure*, Vol III, Electronic spectra of polyatomic molecules, Van Nostrand Reinhold, New York.

Hyde, K.E. (1975) Methods for obtaining Russell–Saunders term symbols from electronic configurations, *J. Chem. Ed.*, **52**, 87–89.

Jotham, R.W. (1975) Why do energy levels repel one another? *J. Chem. Ed.*, **52**, 377–378.

Lever, A.B.P. (1964) *Inorganic Electronic Spectroscopy*, Elsevier, Amsterdam. (Up-to-date and comprehensive, and a good source of spectral data.)

McClure, D.S. and Stephens, P.J. (1971) *Electronic Spectra of Coordination Compounds in Coordination Chemistry* (ed. Martell, A.E.), Van Nostrand Reinhold.

Nicholls, D. (1974) *Complexes and First Row Transition Elements*, Macmillan, London. (Chapter 6 gives a good and readable explanation of electronic spectra of transition-metal complexes.)

Orgel, L.E. (1955) *J. Chem. Phys.*, **23**, 1004. (Original paper using Orgel diagrams.)

Richards, W.G. and Scott, P.R. (1976) *Structure and Spectra of Atoms*, John Wiley, London. (A more basic approach, particularly Chapters 3 and 4.)

Sutton, D. (1968) *Electronic Spectra of Transition Metal Complexes*, McGraw Hill, London.

Tanabe, Y. and Sugano, S. (1954) Semiquantitative energy-level diagrams for octahedral symmetry, *J. Phys. Soc. Japan*, **9**, 753–766. (Original paper on Tanabe–Sugano diagrams.)

Urch, D.S. (1970) *Orbitals and Symmetry*, Penguin, Harmondsworth.

Vicente, J. (1983) A simple method for obtaining Russell–Saunders term symbols, *J. Chem. Ed.*, **60**, 560–561.

An alternative approach to spectra is through Group Theory. This is a mathematical approach, which some may find particularly difficult:

Atkins, P.W., Child, M.S. and Phillips, C.S.G. (1970) *Tables for Group Theory*, Oxford University Press, London.

Ballhausen, C.J. (1962) *Introduction to Ligand Field Theory*, McGraw Hill.

Cotton, F.A. (1970) *Chemical Applications of Group Theory*, 2nd ed., Interscience, New York and Wiley, Chichester. (An excellent introductory text.)

Davidson, G. (1971) *Introductory Group Theory for Chemists*, Applied Science Publishers Ltd., Barking. (An excellent and readable introductory text.)

PROBLEMS

1. Electronic transitions of the *d–d* type displayed in spectra of octahedral transition metal complexes should be forbidden by the Laporte selection rule. Why are moderately strong spectra actually observed?

2. Why do tetrahedral complexes of an element give much more intense d–d spectra than its octahedral complexes?

Abundance of the elements in the earth's crust

APPENDIX

A

	Name	Symbol	Abundance*		Name	Symbol	Abundance
1	Oxygen	O	455 000	40	Samarium	Sm	7.0
2	Silicon	Si	272 000	41	Gadolinium	Gd	6.1
3	Aluminium	Al	83 000	42	Dysprosium	Dy	4.5
4	Iron	Fe	62 000	43	Erbium	Er	3.5
5	Calcium	Ca	46 600	44	Ytterbium	Yb	3.1
6	Magnesium	Mg	27 640	45	Hafnium	Hf	2.8
7	Sodium	Na	22 700	46	Caesium	Cs	2.6
8	Potassium	K	18 400	47	Bromine	Br	2.5
9	Titanium	Ti	6 320	48	Uranium	U	2.3
10	Hydrogen	H	1 520	49 =	Tin	Sn	2.1
11	Phosphorus	P	1 120	49 =	Europium	Eu	2.1
12	Manganese	Mn	1 060	51	Beryllium	Be	2.0
13	Fluorine	F	544	52	Arsenic	As	1.8
14	Barium	Ba	390	53	Tantalum	Ta	1.7
15	Strontium	Sr	384	54	Germanium	Ge	1.5
16	Sulphur	S	340	55	Holmium	Ho	1.3
17	Carbon	C	180	56 =	Molybdenum	Mo	1.2
18	Zirconium	Zr	162	56 =	Tungsten	W	1.2
19	Vanadium	V	136	56 =	Terbium	Tb	1.2
20	Chlorine	Cl	126	59	Lutetium	Lu	0.8
21	Chromium	Cr	122	60	Thallium	Tl	0.7
22	Nickel	Ni	99	61	Thulium	Tm	0.5
23	Rubidium	Rb	78	62	Iodine	I	0.46
24	Zinc	Zn	76	63	Indium	In	0.24
25	Copper	Cu	68	64	Antimony	Sb	0.20
26	Cerium	Ce	66	65	Cadmium	Cd	0.16
27	Neodymium	Nd	40	66 =	Silver	Ag	0.08
28	Lanthanum	La	35	66 =	Mercury	Hg	0.08
29	Yttrium	Y	31	68	Selenium	Se	0.05
30	Cobalt	Co	29	69	Palladium	Pd	0.015
31	Scandium	Sc	25	70	Platinum	Pt	0.01
32	Niobium	Nb	20	71	Bismuth	Bi	0.008
33 =	Nitrogen	N	19	72	Osmium	Os	0.005
33 =	Gallium	Ga	19	73	Gold	Au	0.004
35	Lithium	Li	18	74 =	Iridium	Ir	0.001
36	Lead	Pb	13	74 =	Tellurium	Te	0.001
37	Praseodymium	Pr	9.1	76	Rhenium	Re	0.0007
38	Boron	B	9.0	77 =	Ruthenium	Ru	0.0001
39	Thorium	Th	8.1	77 =	Rhodium	Rh	0.0001

* Units are ppm of the earth's crust, which is the same as g/tonne. Values mainly from *Geochemistry* by W.S. Fyfe, Oxford University Press, 1974, with some newer data.

APPENDIX

B

Melting points of the elements

Period \ Group	I	II											III	IV	V	VI	VII	0
1	H · −259																H · −259	He —
2	Li • 181	Be • 1287											B ● 2180	C ⬤ 4100	N · −210	O · −229	F · −219	Ne · −249
3	Na • 98	Mg • 649											Al • 660	Si • 1420	P • 44	S • 114	Cl · −101	Ar · −189
4	K • 63	Ca • 839	Sc ● 1539	Ti ● 1667	V ● 1915	Cr ● 1900	Mn • 1244	Fe ● 1535	Co • 1495	Ni • 1452	Cu • 1083	Zn • 420	Ga • 30	Ge • 945	As • 816	Se • 221	Br · −7	Kr · −157
5	Rb • 39	Sr • 768	Y ● 1530	Zr ● 1857	Nb ● 2468	Mo ● 2620	Tc ● 2200	Ru ● 2282	Rh ● 1960	Pd ● 1552	Ag • 961	Cd • 321	In • 157	Sn • 232	Sb • 631	Te • 452	I • 114	Xe · −112
6	Cs • 28.5	Ba • 727	La • 920	Hf ● 2222	Ta ● 2980	W ⬤ 3380	Re ⬤ 3180	Os ⬤ 3045	Ir ● 2443	Pt ● 1769	Au • 1064	Hg · −39	Tl • 303	Pb • 327	Bi • 271	Po • 254	At • 302[e]	Rn · −71[e]
7	Fr • 27[e]	Ra • (700)	Ac • 817															

Temperatures are in degrees Celsius. Large circles indicate high values and small circles low values; e indicates estimated value. Values given in parentheses are approximate.

<0 ·
1–1500 •
1501–3000 ●
>3000 ⬤

APPENDIX

C

Boiling points of the elements

Period \ Group	I	II											III	IV	V	VI	VII	0
1																	H −253	He −269
2	Li 1347	Be (2500)											B 3650	C	N −195.8	O −183.1	F −188	Ne −246
3	Na 881	Mg 1105											Al 2467	Si 3280	P 281	S 445	Cl −34	Ar −186
4	K 766	Ca 1494	Sc 2748	Ti 3285	V 3350	Cr 2960	Mn 2060	Fe 2750	Co 3100	Ni 2920	Cu 2570	Zn 907	Ga 2403	Ge 2850	As 615_s	Se 685	Br 60	Kr −153.6
5	Rb 688	Sr 1381	Y 3264	Zr 4200	Nb 4758	Mo 4650	Tc 4567	Ru (4050)	Rh 3760	Pd 2940	Ag 2155	Cd 765	In 2080	Sn 2623	Sb 1587	Te 1087	I 185	Xe −108.1
6	Cs 705	Ba (1850)	La 3420	Hf 4450	Ta 5534	W 5500	Re 5650	Os (5025)	Ir (4550)	Pt 4170	Au 2808	Hg 357	Tl 1457	Pb 1751	Bi 1564	Po 962	At	Rn −62
7	Fr	Ra (1700)	Ac 2470															

Temperatures are in degrees Celsius. Large circles indicate high values and small circles low values. Values given in parentheses are approximate. s indicates sublimation.

<0 •
1–1500 ●
1501–3000 ⬤
>3000 ⬤

APPENDIX

D

Densities of the solid and liquid elements

Group / Period	I	II											III	IV	V	VI	VII	0
1	H																H	He
2	Li 0.54	Be 1.85											B 2.35	C 2.26	N	O	F	Ne
3	Na 0.97	Mg 1.74											Al 2.70	Si 2.34	P 1.82	S 2.1	Cl	Ar
4	K 0.86	Ca 1.55	Sc 3.0	Ti 4.50	V 6.11	Cr 7.14	Mn 7.43	Fe 7.87	Co 8.90	Ni 8.91	Cu 8.95	Zn 7.14	Ga 5.90	Ge 5.32	As 5.77	Se 4.19	Br 3.19	Kr
5	Rb 1.53	Sr 2.63	Y 4.5	Zr 6.51	Nb 8.57	Mo 10.28	Tc 11.5	Ru 12.41	Rh 12.39	Pd 11.99	Ag 10.49	Cd 8.65	In 7.31	Sn 7.27	Sb 6.70	Te 6.25	I 4.94	Xe
6	Cs 1.90	Ba 3.62	La 6.17	Hf 13.28	Ta 16.65	W 19.3	Re 21.0	Os 22.57	Ir 22.61	Pt 21.41	Au 19.32	Hg 13.53	Tl 11.85	Pb 11.34	Bi 9.81	Po 9.14	At —	Rn
7	Fr —	Ra 5.5	Ac —															

Densities are given in $g\,cm^{-2}$ (which equals $10^{-3}\,kg\,m^{-3}$). The value for carbon is the value for graphite. Large circles indicate large values and small circles small values.

APPENDIX

E

Electronic structures of the elements

Z	Element	Symbol	Structure
1	Hydrogen	H	$1s^1$
2	Helium	He	$1s^2$
3	Lithium	Li	[He] $2s^1$
4	Beryllium	Be	[He] $2s^2$
5	Boron	B	[He] $2s^2$ $2p^1$
6	Carbon	C	[He] $2s^2$ $2p^2$
7	Nitrogen	N	[He] $2s^2$ $2p^3$
8	Oxygen	O	[He] $2s^2$ $2p^4$
9	Fluorine	F	[He] $2s^2$ $2p^5$
10	Neon	Ne	[He] $2s^2$ $2p^6$
11	Sodium	Na	[Ne] $3s^1$
12	Magnesium	Mg	[Ne] $3s^2$
13	Aluminium	Al	[Ne] $3s^2$ $3p^1$
14	Silicon	Si	[Ne] $3s^2$ $3p^2$
15	Phosphorus	P	[Ne] $3s^2$ $3p^3$
16	Sulphur	S	[Ne] $3s^2$ $3p^4$
17	Chlorine	Cl	[Ne] $3s^2$ $3p^5$
18	Argon	Ar	[Ne] $3s^2$ $3p^6$
19	Potassium	K	[Ar] $4s^1$
20	Calcium	Ca	[Ar] $4s^2$
21	Scandium	Sc	[Ar] $3d^1$ $4s^2$
22	Titanium	Ti	[Ar] $3d^2$ $4s^2$
23	Vanadium	V	[Ar] $3d^3$ $4s^2$
24	Chromium	Cr	[Ar] $3d^5$ $4s^1$
25	Manganese	Mn	[Ar] $3d^5$ $4s^2$
26	Iron	Fe	[Ar] $3d^6$ $4s^2$
27	Cobalt	Co	[Ar] $3d^7$ $4s^2$
28	Nickel	Ni	[Ar] $3d^8$ $4s^2$
29	Copper	Cu	[Ar] $3d^{10}$ $4s^1$
30	Zinc	Zn	[Ar] $3d^{10}$ $4s^2$
31	Gallium	Ga	[Ar] $3d^{10}$ $4s^2$ $4p^1$
32	Germanium	Ge	[Ar] $3d^{10}$ $4s^2$ $4p^2$

Z	Element	Symbol	Structure
33	Arsenic	As	[Ar] $3d^{10}$ $4s^2$ $4p^3$
34	Selenium	Se	[Ar] $3d^{10}$ $4s^2$ $4p^4$
35	Bromine	Br	[Ar] $3d^{10}$ $4s^2$ $4p^5$
36	Krypton	Kr	[Ar] $3d^{10}$ $4s^2$ $4p^6$
37	Rubidium	Rb	[Kr] $5s^1$
38	Strontium	Sr	[Kr] $5s^2$
39	Yttrium	Y	[Kr] $4d^1$ $5s^2$
40	Zirconium	Zr	[Kr] $4d^2$ $5s^2$
41	Niobium	Nb	[Kr] $4d^4$ $5s^1$
42	Molybdenum	Mo	[Kr] $4d^5$ $5s^1$
43	Technetium	Tc	[Kr] $4d^5$ $5s^2$
		Tc	[Kr] $4d^6$ $5s^1$
44	Ruthenium	Ru	[Kr] $4d^7$ $5s^1$
45	Rhodium	Rh	[Kr] $4d^8$ $5s^1$
46	Palladium	Pd	[Kr] $4d^{10}$ $5s^0$
47	Silver	Ag	[Kr] $4d^{10}$ $5s^1$
48	Cadmium	Cd	[Kr] $4d^{10}$ $5s^2$
49	Indium	In	[Kr] $4d^{10}$ $5s^2$ $5p^1$
50	Tin	Sn	[Kr] $4d^{10}$ $5s^2$ $5p^2$
51	Antimony	Sb	[Kr] $4d^{10}$ $5s^2$ $5p^3$
52	Tellurium	Te	[Kr] $4d^{10}$ $5s^2$ $5p^4$
53	Iodine	I	[Kr] $4d^{10}$ $5s^2$ $5p^5$
54	Xenon	Xe	[Kr] $4d^{10}$ $5s^2$ $5p^6$
55	Caesium	Cs	[Xe] $6s^1$
56	Barium	Ba	[Xe] $6s^2$
57	Lanthanum	La	[Xe] $5d^1$ $6s^2$
58	Cerium	Ce	[Xe] $4f^1$ $5d^1$ $6s^2$
59	Praseodymium	Pr	[Xe] $4f^3$ $5d^0$ $6s^2$
60	Neodymium	Nd	[Xe] $4f^4$ $5d^0$ $6s^2$
61	Promethium	Pm	[Xe] $4f^5$ $5d^0$ $6s^2$
62	Samarium	Sm	[Xe] $4f^6$ $5d^0$ $6s^2$
63	Europium	Eu	[Xe] $4f^7$ $5d^0$ $6s^2$
64	Gadolinium	Gd	[Xe] $4f^7$ $5d^1$ $6s^2$
65	Terbium	Tb	[Xe] $4f^9$ $5d^0$ $6s^2$
66	Dysprosium	Dy	[Xe] $4f^{10}$ $5d^0$ $6s^2$
67	Holmium	Ho	[Xe] $4f^{11}$ $5d^0$ $6s^2$
68	Erbium	Er	[Xe] $4f^{12}$ $5d^0$ $6s^2$
69	Thulium	Tm	[Xe] $4f^{13}$ $5d^0$ $6s^2$
70	Ytterbium	Yb	[Xe] $4f^{14}$ $5d^0$ $6s^2$
71	Lutetium	Lu	[Xe] $4f^{14}$ $5d^1$ $6s^2$
72	Hafnium	Hf	[Xe] $4f^{14}$ $5d^2$ $6s^2$
73	Tantalum	Ta	[Xe] $4f^{14}$ $5d^3$ $6s^2$
74	Tungsten	W	[Xe] $4f^{14}$ $5d^4$ $6s^2$
75	Rhenium	Re	[Xe] $4f^{14}$ $5d^5$ $6s^2$
76	Osmium	Os	[Xe] $4f^{14}$ $5d^6$ $6s^2$
77	Iridium	Ir	[Xe] $4f^{14}$ $5d^7$ $6s^2$
78	Platinum	Pt	[Xe] $4f^{14}$ $5d^9$ $6s^1$
79	Gold	Au	[Xe] $4f^{14}$ $5d^{10}$ $6s^1$
80	Mercury	Hg	[Xe] $4f^{14}$ $5d^{10}$ $6s^2$

Z	Element	Symbol	Structure
81	Thallium	Tl	[Xe] $4f^{14}$ $5d^{10}$ $6s^2$ $6p^1$
82	Lead	Pb	[Xe] $4f^{14}$ $5d^{10}$ $6s^2$ $6p^2$
83	Bismuth	Bi	[Xe] $4f^{14}$ $5d^{10}$ $6s^2$ $6p^3$
84	Polonium	Po	[Xe] $4f^{14}$ $5d^{10}$ $6s^2$ $6p^4$
85	Astatine	At	[Xe] $4f^{14}$ $5d^{10}$ $6s^2$ $6p^5$
86	Radon	Rn	[Xe] $4f^{14}$ $5d^{10}$ $6s^2$ $6p^6$
87	Francium	Fr	[Rn] $7s^1$
88	Radium	Ra	[Rn] $7s^2$
89	Actinium	Ac	[Rn] $6d^1$ $7s^2$
90	Thorium	Th	[Rn] $6d^2$ $7s^2$
91	Protactinium	Pa	[Rn] $5f^2$ $6d^1$ $7s^2$
92	Uranium	U	[Rn] $5f^3$ $6d^1$ $7s^2$
93	Neptunium	Np	[Rn] $5f^4$ $6d^1$ $7s^2$
94	Plutonium	Pu	[Rn] $5f^6$ $6d^0$ $7s^2$
95	Americium	Am	[Rn] $5f^7$ $6d^0$ $7s^2$
96	Curium	Cm	[Rn] $5f^7$ $6d^1$ $7s^2$
97	Berkelium	{ Bk	[Rn] $5f^9$ $6d^0$ $7s^2$
		{ Bk	[Rn] $5f^8$ $6d^1$ $7s^2$
98	Californium	Cf	[Rn] $5f^{10}$ $6d^0$ $7s^2$
99	Einsteinium	Es	[Rn] $5f^{11}$ $6d^0$ $7s^2$
100	Fermium	Fm	[Rn] $5f^{12}$ $6d^0$ $7s^2$
101	Mendelevium	Md	[Rn] $5f^{13}$ $6d^0$ $7s^2$
102	Nobelium	No	[Rn] $5f^{14}$ $6d^0$ $7s^2$
103	Lawrencium	Lr	[Rn] $5f^{14}$ $6d^1$ $7s^2$
104	Rutherfordium	Rf	[Rn] $5f^{14}$ $6d^2$ $7s^2$
105	Hahnium	Ha	[Rn] $5f^{14}$ $6d^3$ $7s^2$

APPENDIX F

Some average single bond energies ($kJ\ mol^{-1}$)

	I	Br	Cl	F	S	O	P	N	Si	C	H
H	297	368	431	565	339	464	318	389	293	414	435
C	238	276	330	439	259	351	263	293	289	347	
Si	213	289	360	539	226	368	213[e]	–	176		
N	–	243	201	272	–	201	209[e]	159			
P	213	272	330	489	230[e]	351[e]	213				
O	201	–	205	184	–	138					
S	–	213	251	284	213						
F	–	255	184	159							
Cl	209	217	243								
Br	180	193									
I	151										

[e] Indicates value estimated using electronegativity difference.

Some double and triple bond energies ($kJ\ mol^{-1}$)

C=C	611	C≡C	836
C=N	615	C≡N	891
C=O	740	C≡O	1071
N=N	418	N≡N	945

APPENDIX G

Solubilities of main group compounds in water

	F^-	Cl^-	Br^-	I^-	OH^-	NO_3^-	CO_3^{2-}	SO_4^{2-}	HCO_3^-
NH_4^+	vs	37	75	172		192	12	75	
Li^+	0.27[c]	83	177	165	12.8	70	1.33	35	
Na^+	4.0	36	91	179	109	87	21	19.4	9.6
K^+	95	34.7	67	144	112	31.6	112	11.1	22.4
Rb^+	131	91	110	152	177	53	450	48	
Cs^+	370	186	108	79	330	23	vs	179	
Be^{2+}	vs	vs	s	dec	ss	107		39	
Mg^{2+}	0.008	54.2	102	148	0.0009	70	ss	33	
Ca^{2+}	0.0016	74.5	142	209	0.156	129	ss	0.21	
Sr^{2+}	0.012	53.8	100	178	0.80	71	ss	0.013	
Ba^{2+}	0.12	36	104	205	3.9	8.7	ss	0.00024	
Al^{3+}	0.55	70 dec	dec	dec	ss	63		38	
Ga^{3+}	0.002	vs	s	dec	ss			vs	
In^{3+}	0.04	vs	vs	dec	ss				
Tl^+	78.6[b]	0.33	0.05	0.0006	25.9[a]	9.55	4.0[b]	s	
Tl^{3+}	dec	vs	s	s		s		4.87	
Ge^{II}	s	dec	dec	s					
Ge^{IV}	dec	dec	dec	dec					
Sn^{II}	s	270 dec[b]	s	0.98	ss	dec		33[d]	
Sn^{IV}	vs	dec	dec	dec					
Pb^{II}	0.064	0.99	0.844	0.063	0.016	55	0.00011	ss	
Pb^{IV}		dec							
As^{III}	dec	dec	dec	6.0[d]					
As^{V}									
Sb^{III}	dec	dec	dec	dec		dec		ss	
Sb^{V}		dec							
Bi^{III}	ss	dec	dec	ss	0.00014	dec		dec	

Solubilities in grams of solute per 100 grams of water at 20° C unless otherwise indicated. vs = very soluble; s = soluble; ss = slightly soluble; dec = decomposes. a = at 0° C; b = at 15° C; c = at 18° C; d = at 25° C.

Atomic weights based on $^{12}C = 12.000$

APPENDIX

H

Element	Atomic weight	Element	Atomic weight
Actinium	227	Holmium	164.93
Aluminium	26.982	Hydrogen	1.00797
Americium	(243)	Indium	114.82
Antimony	121.75	Iodine	126.904
Argon	39.948	Iridium	192.2
Arsenic	74.922	Iron	55.847
Astatine	(210)	Krypton	83.80
Barium	137.34	Lanthanum	138.91
Berkelium	(249)	Lead	207.19
Beryllium	9.102	Lithium	6.939
Bismuth	208.98	Lutetium	174.97
Boron	10.811	Magnesium	24.312
Bromine	79.909	Manganese	54.938
Cadmium	112.40	Mendelevium	(256)
Caesium	132.905	Mercury	200.59
Calcium	40.08	Molybdenum	95.94
Californium	(251)	Neodymium	144.24
Carbon	12.01115	Neon	20.183
Cerium	140.12	Neptunium	(237)
Chlorine	35.453	Nickel	58.71
Chromium	51.996	Niobium	92.91
Cobalt	58.933	Nitrogen	14.0067
Copper	63.54	Nobelium	(254)
Curium	(247)	Osmium	190.2
Dysprosium	162.50	Oxygen	15.9994
Einsteinium	(254)	Palladium	106.4
Erbium	167.26	Phosphorus	30.9738
Europium	151.96	Platinum	195.09
Fermium	(253)	Plutonium	(242)
Fluorine	18.993	Polonium	(210)
Francium	(223)	Potassium	39.102
Gadolinium	157.25	Praseodymium	140.91
Gallium	69.72	Promethium	(147)
Germanium	72.59	Protactinium	(231)
Gold	196.967	Radium	(226)
Hafnium	178.49	Radon	(222)
Helium	4.003	Rhenium	186.22

Element	Atomic weight	Element	Atomic weight
Rhodium	102.91	Terbium	153.92
Rubidium	85.47	Thallium	204.37
Ruthenium	101.07	Thorium	232.04
Samarium	150.35	Thulium	168.93
Scandium	44.96	Tin	118.69
Selenium	78.96	Titanium	47.90
Silicon	28.086	Tungsten	183.85
Silver	107.870	Uranium	238.03
Sodium	22.9898	Vanadium	50.942
Strontium	87.62	Xenon	131.30
Sulphur	32.064	Ytterbium	173.04
Tantalum	180.95	Yttrium	88.91
Technetium	(99)	Zinc	65.37
Tellurium	127.60	Zirconium	91.22

Values in parentheses are for the most stable isotope.

Values of some fundamental physical constants

APPENDIX I

Planck's constant h $= 6.6262 \times 10^{-34}$ J s

mass of electron $m = 9.1091 \times 10^{-31}$ kg

charge on electron e $= 1.60210 \times 10^{-19}$ C

permittivity of a vacuum $\varepsilon_0 = 8.854\,185 \times 10^{-12}\,\mathrm{kg^{-1}\,m^{-3}\,A^2}$

permeability of a vacuum $\mu_0 = 4\pi \times 10^{-7}\,\mathrm{kg\,m\,s^{-2}\,A^{-2}}$

velocity of light in a vacuum c $= 2.997\,925 \times 10^8\,\mathrm{m\,s^{-1}}$

Avogadro constant $N^o = 6.022\,045 \times 10^{23}\,\mathrm{mol^{-1}}$

Boltzmann constant k $= 1.3805 \times 10^{-23}\,\mathrm{J\,K^{-1}}$

Bohr magneton $\mu_B = 9.2732 \times 10^{-24}\,\mathrm{A\,m^2} = \mathrm{J\,T^{-1}}$

magnetic moment and dipole moment μ units are $\mathrm{A\,m^2} = \mathrm{J\,T^{-1}}$

APPENDIX J

Electrical resistivity of the elements at the stated temperature

Element	Symbol	Temp (°C)	Resistivity (μohm cm)	Element	Symbol	Temp (°C)	Resistivity (μohm cm)
Silver	Ag	20	1.59	Vanadium	V	20	25
Copper	Cu	20	1.673	Gallium*	Ga	20	27
Gold	Au	20	2.35	Ytterbium	Yb	25	29
Aluminium	Al	20	2.655	Uranium	U	20	30
Calcium	Ca	20	3.5	Arsenic α	As	20	33.3
Beryllium	Be	20	4.0	Hafnium	Hf	25	35.1
Sodium	Na	0	4.2	Zirconium	Zr	20	40
Magnesium	Mg	20	4.46	Antimony α	Sb	20	41.7
Rhodium	Rh	20	4.51	Titanium	Ti	20	42
Molybdenum	Mo	0	5.2	Barium	Ba	20	50
Iridium	Ir	20	5.3	Yttrium	Y	25	57
Tungsten	W	27	5.65	Dysprosium	Dy	25	57
Lanthanum	La	25	5.7	Scandium	Sc	22	61
Zinc	Zn	20	5.92	Neodymium	Nd	25	64
Potassium	K	0	6.15	Praseodymium	Pr	25	68
Cobalt	Co	20	6.24	Cerium	Ce	25	75
Cadmium	Cd	0	6.83	Thulium	Tm	25	79
Nickel	Ni	20	6.84	Lutetium	Lu	25	79
Ruthenium	Ru	0	7.6	Holmium	Ho	25	87
Indium	In	20	8.37	Samarium	Sm	25	88
Lithium	Li	20	8.55	Europium	Eu	25	90
Osmium	Os	20	9.5	Mercury	Hg	20	95.8
Iron	Fe	20	9.71	Erbium	Er	25	107
Platinum	Pt	20	10.6	Bismuth α	Bi	20	120
Palladium	Pd	20	10.8	Gadolinium	Gd	25	140.5
Tin β	Sn	0	11	Plutonium	Pu	107	141.4
Tantalum	Ta	25	12.45	Manganese α	Mn	23	185
Rubidium	Rb	20	12.5	Tellurium	Te	25	4.5×10^{5}
Niobium	Nb	0	12.5	Germanium	Ge	22	4.7×10^{7}
Chromium	Cr	0	12.9	Silicon	Si	20	4.8×10^{7}
Thorium	Th	0	13	Boron	B	20	6.7×10^{11}
Thallium	Tl	0	18	Iodine	I	20	1.3×10^{15}
Rhenium	Re	20	19.3	Selenium	Se	20	1.2×10^{16}
Caesium	Cs	20	20	Phosphorus #	P	11	1×10^{17}
Lead	Pb	20	20.648	Carbon	C	20	$\cong 1 \times 10^{20}$
Strontium	Sr	20	23	Sulphur	S	20	2×10^{23}

* Resistivity of Ga is 8.2, 17.5 or 55.3 μohm cm depending on direction.
# White phosphorus.

Top fifty chemicals in the USA, 1989

APPENDIX

K

Position	Chemical	Billion lb/year	Million tons/year
1	**Sulphuric acid**	**86.80**	**43.399**
2	**Nitrogen**	**53.77***	
3	**Oxygen**	**37.75**†	
4	Ethylene	34.95	
5	**Ammonia**	**33.76**	**16.873**
6	**Lime**	**32.99**	**16.495**
7	**Phosphoric acid**	**23.12**	**11.559**
8	**Chlorine**	**22.32**	**11.161**
9	**Sodium hydroxide**	**22.15**	**11.075**
10	Propylene	20.23	
11	**Sodium carbonate**	**19.76**	**9.896**
12	**Nitric acid**	**15.98**	**7.991**
13	**Urea**	**15.47**	**7.733**
14	**Ammonium nitrate**	**15.11**	**7.557**
15	Ethylene dichloride	13.68	
16	Benzene	11.67	
17	**Carbon dioxide**	**10.83**	**5.413**
18	Vinyl chloride	9.62	
19	Ethyl benzene	9.22	
20	Terephthalic acid	8.31	
21	Styrene	8.13	
22	Methanol	7.14	
23	Formaldehyde	6.37	
24	Toluene	5.84	
25	Xylene	5.80	
26	Ethylene glycol	5.50	
27	*p*-Xylene	5.49	
28	Ethylene oxide	5.32	
29	**Hydrochloric acid**	**5.26**	**2.628**
30	Methyl tert-butyl ether	4.98	
31	**Ammonium sulphate**	**4.71**	**2.354**
32	Cumene	4.54	
33	Phenol	3.89	
34	Acetic acid	3.83	
35	**Potash**	**3.35**	**1.521**
36	Propylene oxide	3.20	
37	Butadiene	3.09	

Position	Chemical	Billion lb/year	Million tons/year
38	**Carbon black**	**2.91**	
39	Acrylonitrile	2.61	
40	Acetone	2.50	
41	Vinyl acetate	2.47	
42	Cyclohexane	2.39	
43	**Aluminium sulphate**	**2.35**	
44	**Titanium dioxide**	**2.22**	**1.173**
45	**Calcium chloride**	**1.92**	**1.111**
46	**Sodium silicate**	**1.75**	**0.960**
47	Adipic acid	1.64	**0.877**
48	**Sodium sulphate**	**1.60**	
49	Isopropyl alcohol	1.43	**0.800**
50	Caprolactam	1.31	

Inorganic chemicals are shown in bold type. Data are reproduced from *Chemical & Engineering News*, page 37, 18 June, 1990 issue. These figures are updated annually and published in early June. * 742 billion cubic feet. † 456 billion cubic feet.

Inorganic chemicals manufactured in large tonnages (worldwide)

APPENDIX

L

Chemical	Million tonnes/year	Chemical	Million tonnes/year
Cement	1172	Soap	7.8
Cast iron and steel	776	Talc	7.6
Coke	347	Vermiculite	7.5
NaCl	180	Zn	7.1
H_2SO_4	163	(Polypropylene)	6.5
CaO	126	Bentonite	6.5
O_2	100	(Polystyrene)	6.3
NH_3	98	CaC_2	6.2
$CaSO_4$	86.5	$CO(NH_2)_2$	6
S	60.5	Pb	5.7
N_2	60	CaF_2	5.1
Phosphates	49	$BaSO_4$	5
(based on P_2O_5 content)		TiO_2	4.3
CO_2	33 USA	Na_2SO_4	4.2
NaOH	34	Carbon black	4
Potassium salts	32.1	Fullers' earth	3.7
(based on K_2O content)		$Al_2(SO_4)_3$	3.4
Glass	31.4	Cr	3.4
HNO_3	30	Ferrosilicon	3.2
Cl_2	23	Ferrochrome	3
Kaolin	23	Soluble silicates	3
Na_2CO_3	22		(in terms of SiO_2 content)
Al	21.6	Explosives	2.9
Chalk	18	Ferromanganese	2.6
(Polythene)	17.9	$Na_2B_4O_7$ and borates	2.4
O_2	14	Kieselguhr	2
Tin plate	11.7	CaNCN	1.3
$MgCO_3$	11.7		(max now reduced)
(Polyvinyl chloride)	11.5	P	1.2
HCl	9.0		(max now reduced)
Mn	9	Si	>1
Cu	8.7	Na_2SO_3	>1
Asbestos	8.1		

Organic compounds in parentheses.

Chemical	Tonnes/year	Chemical	Tonnes/year
Zircon/baddeleyte	918 000	As_2O_3	50 000
Ni	840 000	W	43 000
H_2O_2	824 400	Lanthanides	41 800
Ar	700 000		(based on Ln_2O_3 content)
PbO/Pb ('black oxide')	700 000	Organotin	40 000
Freons	700 000	U	37 300
	(max now reduced)	V	31 600
Active charcoal	658 000	Co	31 200
Natural graphite	605 000	$AlCl_3$	25 000 USA
CS_2	584 700	Cd	22 000
Sodium peroxoborate	550 000	PCl_5	20 000
$PbEt_4$	>500 000	N_2H_4	20 000
	(max now reduced)	Pb_3O_4	18 000
Br_2	398 500	Nb	17 900
Mg	393 000	I_2	15 300
Ferronickel	385 000	Ag	14 861
ZnO	352 300	Lithium salts	7 800
$Na_2Cr_2O_7$	304 700		(based on Li content)
SiC	300 000	Hg	5 700
Silicones	300 000	He	5 000
HCN	300 000	La	5 000
PbO (litharge)	250 000	Bi	4 300
PCl_3	250 000	BF_3	4 000
P_4S_{10}	250 000	Au	1 785
Mica	249 000	Se	1 478
Sn	205 000	Ca	1 000
Na	200 000 USA	*n*-Butyl lithium	1 000
$NaHCO_3$	200 000	Th compounds	500
NaOCl	180 000	Ta	400
H_3BO_3	170 000	Te	144
Strontium compounds	163 000	Pt group metals	267
NaCN	120 000	In	52
Garnet	100 975	Ge	49
Ti	100 000	Diamonds	18.48
Mo	92 000	Ga	10
$Ca(OCl)_2$	80 000	Tl	10
Sb	63 900	Re	2.4
Li_2CO_3	50 000		

Minerals used in large amounts

APPENDIX

M

Mineral	Million tonnes/year	Uses
Coal	4749	Energy, chemicals, coke
Crude oil	2944	Energy, petrochemicals
Iron ore	970	Iron, steel and ferrous alloys
Sodium chloride	180	Chlor-alkali industry, NaOH, Cl_2, Na_2CO_3 and HCl
Phosphate rock	159	Fertilizers, detergents, water treatment, phosphorus
Bauxite	100	Al, Al_2O_3, $Al(OH)_3$, $Al_2(SO_4)_3$
Gypsum	86.5	Used in plasterboard
Sulphur	60.5	Sulphuric acid manufacture
Limestone	60	$CaCO_3$, CaO, $Ca(OH)_2$, CaC_2, Ca
Potassium compounds: sylvine, sylvinite, carnallite	32.1	Fertilizers, KOH, KO_2, K
Kaolin	23	Filling paper, refractory ceramics
Pyrolusite	22.7	Mn alloys, MnO_2, paint driers
Magnesite	11.7	Mg and its compounds
Chromite	11.7	Chromates, dichromates, ceramics, pigments, chromium
Copper ores	8.7	Cu
Ilmenite and rutile	8.2	TiO_2 pigments, Ti
Sodium carbonate	8	Cleaning products
Talc	7.6	Ceramics, cosmetics
Vermiculite	7.5	Packing, artificial soil
Zinc ores	7.1	Zn alloys and rust-proofing
Bentonite	6.5	Drilling mud, thixotropic paint
Fluorite	5.1	AlF_3, $Na_3[AlF_6]$, HF, F_2, organic fluorides
Barytes	5.0	Slurry used for well drilling, Ba compounds, fillers, pigments
Asbestos	4.3	Heat resistant, filler
Fullers' earth	3.7	Absorbent
Galena	3.4	Pb batteries, Pb sheet, $PbEt_4$
Borates	2.4	Borax, boric acid, perborates, glazes
Sodium sulphate	1.5	Paper, glass, detergents
Sand and silica (chemical uses only)	1.0	Silicates, silicones, detergents

Mineral	Million tonnes/year	Uses
Zircon/baddeleyite	0.92	Zr
Graphite	0.61	Used for electrodes and as a lubricant in reactors
Sodium nitrate	0.55	Fertilizers, explosives
Vermiculite	0.53	Packing and artificial soil
Bromine compounds (sea water)	0.399	Bromides, agricultural chemicals
Mica	0.25	Electrical insulation, filler
Cassiterite	0.205	Sn
Strontianite	0.17	Sr compounds
Molybdenite	0.09	Alloys, MnO_2
Monazite/bastnaesite	0.072	Lanthanide and thorium compounds
Uranium ores	0.037	Nuclear power and weapons
Cobalt ores	0.031	Alloys
Iodine compounds	0.015	Iodates, iodides, iodine
Beryl	0.01	Be
Spodumene/lepidolite	0.008	Li stearate, Li_2CO_3, LiOH

Hardness of minerals – Mohs' scale

APPENDIX

N

Sometimes rocks are called soft because individual particles, which may be quite hard in themselves, are loosely held as an aggregate and fall apart fairly readily. The hardness of a mineral is something quite different and refers to the resistance of the whole surface to be scratched.

More than a century ago a mineralogist called Friedrich Mohs devised a scale of hardness called after him. He arbitrarily assigned talc (the softest known mineral) a hardness of 1, and diamond, (the hardest known mineral) a value of 10. The scale is arbitrary and does not indicate any exact hardness. Thus a mineral of hardness 6 can scratch those below it in the scale but it does not imply that it is twice as hard as a mineral of hardness 3. Two minerals of the same hardness will both scratch each other. There is a very large difference between 9 and 10 on the scale.

Table N.1 Mohs' scale of hardness

Mohs' scale	Mineral
10	Diamond
9	Corundum
8	Topaz
7	Quartz
6	Microcline
5	Apatite
4	Fluorite
3	Calcite
2	Gypsum
1	Talc

Common objects such as a fingernail, a copper coin, a knife blade or a metal file can also be used as test instruments. It may be convenient to use these to test samples for hardness quickly and easily when out in the field.

Table N.2 Tools for hardness testing

Mohs' scale	Tool
6.5	Steel nail file
5.5	Penknife blade or window glass
3.5	Copper coin
2.5	Fingernail

Minerals of hardness below 6.5 can be scratched by a nail-file, those under 5.5 are scratched by a penknife, those under 3.5 are scratched by a copper coin and those under 2.5 are scratched by a fingernail and will leave a mark on paper. Conversely minerals over 5.5 will scratch glass.

Standard textbooks

APPENDIX

O

FURTHER READING

In addition to the further reading lists given at the end of each chapter, further information or a different treatment may be found in the following.

Advanced and reference

Bailar, J.C., Eméleus, H.J., Nyholm, R.S. and Trotman-Dickinson, A.F. (eds) *Comprehensive Inorganic Chemistry*, (5 vols), Pergamon, 1973. (A collection of reviews systematically covering the elements with a few special topics. Slightly dated, it still provides much detailed information and many earlier references.)

Cotton, F.A. and Wilkinson, G., *Advanced Inorganic Chemistry*, 5th ed., John Wiley, 1988. (A comprehensive research level text by the well-known Nobel laureate. It contains extensive up-to-date references to the original literature.)

Eméleus, H.J., gen. ed., *MTP International Review of Science*, (10 vols), Butterworths, 1975. Individual volumes are edited by Lappert, M.F., Sowerby, D.B., Gutmann, V., Aylett, B.J., Sharp, D.W.A., Mays, M., Bagnall, K.W., Maddock, A.G., Tobe, M.L. and Roberts, L.E.J. (Like *Comprehensive Inorganic Chemistry* it collects reviews systematically covering the elements with a few special topics. It too is slightly dated but still provides much detailed information and many earlier references.)

Greenwood, N.N. and Earnshaw, A., *Chemistry of the Elements*, Pergamon, 1984. (A large and comprehensive advanced text for undergraduates and postgraduates. Best in its class, it provides an excellent treatment of systematic inorganic chemistry including both theoretical and practical aspects and many up-to-date references to the literature.)

Kirk, R.E. and Othmer, D.F. (eds), *Encyclopedia of Chemical Technology*, 3rd. ed., (26 vols), Wiley, 1984. (A comprehensive treatment of the chemicals used in the chemical industry, how they are made and what they are used for.)

Other general university-level texts

Burns, D.T., Carter, A.H. and Townshend, A., *Inorganic Reaction Chemistry: Reactions of the Elements and Their Compounds*, (Ellis Horwood Series in Analytical Chemistry), Halsted Press, 1981.

Cotton, F.A., Gaus, P.L. and Wilkinson, G., *Basic Inorganic Chemistry*, 2nd ed., John Wiley, 1986.

Douglas, B., McDaniel, D.H. and Alexander, J.J., *Concepts and Models of Inorganic Chemistry*, 2nd ed., John Wiley & Sons, 1983.

Jolly, W.L., *Principles of Inorganic Chemistry*, McGraw-Hill, 1985.
Mackay, K. and Mackay, A., *Introduction to Modern Inorganic Chemistry*, 4th ed., Prentice Hall, 1989.
Massey, A.G., *Main Group Chemistry*, Ellis Horwood, 1990.
Mortimer, C.E., *Chemistry*, 6th ed., Wadsworth, 1986.
Sharpe, A.G., *Inorganic Chemistry*, 2nd ed., John Wiley, 1987.
Shriver, D.F., Atkins, P.W. and Langford, C.H., *Inorganic Chemistry*, Oxford University Press, 1990.

Bioinorganic chemistry

Bioinorganic Chemistry, Journal of Chemical Education, 1986. (ISBN 0–910362–25–4).
Hay, R.W., *Bio-Inorganic Chemistry*, (Ellis Horwood Series in Chemical Science), Halsted Press, 1984.
Hughes, M.N., *The Inorganic Chemistry of Biological Processes*, 2nd ed., John Wiley, 1981.

Index

Page numbers in **bold** refer to the most important reference for the entry.